W9-BFN-233

SPI Plastics Engineering Handbook

of the Society of the Plastics Industry, Inc.

SPI Plastics Engineering Handbook

of the Society of the Plastics Industry, Inc.

Fifth Edition

Edited by

Michael L. Berins

VNR VAN NOSTRAND REINHOLD
New York

Library of Congress Catalog Card Number: 90-22784
ISBN: 0-442-31799-9

Manufactured in the United States of America

Published by Van Nostrand Reinhold
115 Fifth Avenue
New York, New York 10003

Chapman & Hall
2-6 Boundary Row
London SE1 8HN, England

Thomas Nelson Australia
102 Dodds Street
South Melbourne, Victoria 3205, Australia

Nelson Canada
1120 Birchmount Road
Scarborough, Ontario M1K 5G4, Canada

15 14 13 12 11 10 9 8 7 6 5 4 3

Library of Congress Cataloging in Publication Data

Society of the Plastics Industry.
Plastics Engineering Handbook of the Society of the Plastics Industry/[edited by] Michael L. Berins.—5th ed.
p. cm.
Includes index.
ISBN 0-442-31799-9
1. Plastics—Handbooks, manuals, etc. I. Berins, Michael L. II. Title.
TP1130.S58 1991
668.4—dc20 90-22784
CIP

Foreword

I am pleased to present the Fifth Edition of the *Plastics Engineering Handbook*. Last published in 1976, this version of the standard industry reference on plastics processing incorporates the numerous revisions and additions necessitated by 14 years of activity in a dynamic industry.

At that last printing, then-SPI President Ralph L. Harding, Jr. anticipated that plastics production would top 26 billion pounds in 1976 (up from 1.25 billion in 1947, when the First Edition of this book was issued). As I write, plastics production in the United States had reached almost 60 billion pounds annually.

Indeed, the story of the U.S. plastics industry always has been one of phenomenal growth and unparalleled innovation. While these factors make compilation of a book such as this difficult, they also make it necessary. Thus I acknowledge all those who worked to gather and relate the information included in this 1991 edition and thank them for the effort it took to make the *Plastics Engineering Handbook* a definitive source and invaluable tool for our industry.

Larry L. Thomas
President
The Society of the Plastics Industry, Inc.

Preface

In this fifth edition of the *SPI Plastics Engineering Handbook*, we continue a tradition of providing readers with a comprehensive manual for plastics processing. Since the last edition was published in 1976, there have been many changes in the types of materials available for manufacturing plastic products. Likewise, there have been new processes introduced and substantial improvements in those that already existed. All in all, the plastics processing industry has become much more sophisticated than it was in the 1970s, making use of higher-performance materials and computer-controlled equipment to widen the range of applications for both thermoplastics and thermosets.

Extensive revisions have resulted in an up-to-date handbook that reflects the plastics industry's state of the art. Some changes in chapter order have been made to guide the reader through the maze of processes for these versatile materials. For example, the Extrusion chapter now precedes the other chapters on thermoplastic processing because extrusion is generic to most of them.

Chapter 1 provides a glossary of words and current expressions in use in the plastics industry. Specialized glossaries with illustrations are placed at the end of the chapters on extrusion and injection molding.

Chapters 2 and 3 cover the range of polymer materials in use in today's industry. The first provides an understanding of the basic chemistry of polymers; the second supplies a description of the families of plastics, their physical and chemical properties, and a brief discussion of the applications for which they are suited. The information has been updated to include descriptions of the newest, high-temperature plastics, e.g., liquid crystal polymers.

Chapters 4 through 20 are devoted to the most widely used methods of plastics processing. The chapter on extrusion has been completely rewritten and provides a thorough understanding of the theory and practice of melting and conveying thermoplastics. For easier reference, the three chapters on injection molding and tooling have been placed in sequence in the new edition. Coverage of blow molding has been expanded considerably to include discussions of new technologies such as injection stretch blow molding. The chapter on thermoforming has also been extensively revised to reflect the many changes in this ever more versatile processing method.

For each process, the *Handbook* covers all aspects of the technology: machinery and equipment, tooling, materials, process variables, and, in some cases, troubleshooting techniques.

Chapters 18 and 19, instead of concentrating on a single processing method, furnish a comprehensive presentation of the materials and methods used in two major industry segments. Chapter 18, on reinforced plastics, covers the full range of composite materials and manufacturing processes that are widely used in making products for automotive, aircraft and aerospace, marine, and construction applications. Chapter 19, on cellular plastics, covers both thermoset and thermoplastic foams as well as such new techniques as structural reaction injection molding, resin transfer molding, and gas-assisted injection molding.

Chapter 20, on radiation processing, begins the *Handbook*'s coverage of secondary processing techniques. Chapter 21 is devoted to descriptions of equipment for handling incoming raw materials, their distribution to individual processing machines, and for handling finished parts and scrap. Chapter 22 discusses preparation of the raw materials for processing.

In Chapters 23 through 26, the *Handbook* moves to secondary operations, with coverage of machining, assembly, and decorating.

Finally, in Chapter 27, there is a presentation of accepted testing methods for plastic materials and manufactured products.

The last portion of the *Handbook* is probably the most used section. Special attention has been paid to the index for the Fifth Edition to help the reader easily locate the information he or she needs.

Acknowledgments

In preparing this Fifth Edition of the *SPI Plastics Engineering Handbook*, we have built upon a base of information that has grown along with the plastics industry over the past 45 years.

As a work whose contents are largely contributed, the *Handbook* owes a great deal to the authors and reviewers whose time and efforts went into its publication. They all have our sincere appreciation.

In compiling the new work, much text from the previous edition has been retained, with the necessary revisions and updates provided by current industry experts. We therefore thank all the contributors and reviewers who participated in the Fourth Edition as well.

Acknowledgment is due once again to the many suppliers of plastic materials and processing equipment who have generously offered literature and illustrations. We have provided credit for those contributions throughout the handbook.

Thanks are also in order to the Society of the Plastics Industry, which supported the publication of the Handbook and supplied us with association literature, data, standards, and other information contained in this volume.

I would also like to give special thanks to Joel Frados, editor of the Fourth Edition and my mentor in the plastics publishing business, to Mel Friedman and Ed Galli of Edgell Communications for their help in gathering information, to Stephanie Seber for her long hours of manuscript preparation, to Emil Davidson for his inspiration and assistance, and to Tina Berins for her constant encouragement.

MICHAEL L. BERINS, *Editor*

Contents

1 Glossary

Over the years, the plastics industry has built up a language and a terminology of its own. In this chapter, the most commonly used words and expressions are classified and defined. Definitions of other terms can also be found in the text, and can be located by means of the index at the back of the book.

In many instances, the words being defined are peculiar to the plastics industry and the way in which it manufactures its products. In other cases, the expression used by the industry may derive from words commonly used in other branches of manufacturing (e.g., the concept of forging plastics from metalworking terminology); as applied to plastics, however, these definitions may differ from common usage.

Readers also are referred to the special glossaries included in the chapters on extrusion and injection molding.

A-stage—An early stage in the reaction of certain thermosetting resins, in which the material is still soluble in certain liquids and fusible. Sometimes referred to as resol. See **B-stage** and **C-stage**.

ablative plastic—Description applied to a material that absorbs heat (while part of it is being consumed by heat) through a decomposition process known as pyrolysis, which takes place in the near surface layer exposed to heat. This mechanism essentially provides thermal protection of the subsurface materials and components by sacrificing the surface layer.

accelerator—A chemical additive that increases the speed of a chemical reaction.

accumulator—An auxiliary cylinder and piston (plunger) mounted on injection molding or blow molding machines to provide fast delivery of plasticated melt. The accumulator cylinder is filled during the time between "shots" with melted plastic coming from the main (primary) extruder.

activation—The process of inducing radioactivity in a specimen by bombardment with neutrons or other types of radiation.

additive—Substance compounded into a resin to modify its characteristics (i.e., antistats, stabilizers, plasticizers, flame retardants, etc.).

adiabatic—An adjective used to describe a process or transformation in which no heat is added to or allowed to escape from the system under consideration. It is used, somewhat incorrectly, to describe a mode of extrusion in which no external heat is added to the extruder although heat may be removed by cooling to keep the output temperature of the melt passing through the extruder constant. The heat input in such a process is developed by the screw as its mechanical energy is converted to thermal energy.

aging—The process of exposing plastics to an adverse environment (i.e., heat, chemicals, light) for an interval of time, to determine the effect on properties.

air ring—A circular manifold used to distribute an even flow of the cooling medium, air, onto a hollow tubular film (bubble) passing through the center of the ring. In extrusion blown films, the air cooling provides uniform thickness.

air-slip forming—A variation of snap-back forming in which the male mold is enclosed in a box in such a way that when the mold moves forward toward the hot plastic, air is trapped between the mold and the plastic sheet. As the mold advances, the plastic is kept away from it by the air cushion formed as described above, until the full travel of the mold is reached, at which point a vacuum is applied, destroying the cushion and forming the part against the plug.

air vent—See **vent**.

ambient temperature—Temperature of the medium surrounding an object.

amorphous—Having no crystalline structure.

anchorage—Part of the insert that is molded inside of the plastic and held fast by the shrinkage of the plastic.

anisotropy—The tendency of a material to react differently to stresses applied in different directions.

anneal—To heat an article to a predetermined temperature and slowly cool it to relieve stresses. Annealing is employed on parts produced from both metals and plastics. (Annealing of molded or machined parts may be done dry as in an oven or wet as in a heated tank of mineral oil.)

antiblocking agent—A substance added to plastic resin to prevent adhesion between touching layers of film caused by pressure, heat, and contact during fabrication and storage.

antistatic agent—A substance that can be applied to the surface of a plastic article, or incorporated in the plastic from which the article is to be made. Its function is to render the surface of the plastic article less susceptible to accumulation of electrostatic charges which attract and hold fine dirt or dust on the surface of the plastic article.

aramid fiber—Lightweight, high strength polymeric fibers used as ballistic armor and as reinforcements for plastics.

arc resistance—Time required for a given electrical current to render the surface of a material conductive because of carbonization by the arc flames. Ref.: Standard Method of Test for High-Voltage, Low-Current Arc Resistance of Solid Electrical Insulating Materials (ASTM Designation: D 495).

atactic—Description applied to a random arrangement of units along a polymer chain. See **isotactic**.

autoclave—A closed vessel for conducting a chemical reaction or other operation under pressure and heat.

autoclave molding—Procedure used in reinforced plastic molding, in which, after lay-up, the entire assembly is placed in a steam autoclave at 50 to 100 psi. Additional pressure achieves higher reinforcement loadings and improved removal of air. (Modification of pressure bag method.)

automatic mold—A mold for injection, compression, or transfer molding that repeatedly goes through the entire molding cycle, including ejection, without human assistance.

average molecular weight—Term used to indicate the chain length of the most typical chain in a given plastic. Molecular weight of polymers is determined by measuring the viscosity of the material in solution at a specific temperature. The value is independent of specific chain length and falls between weight average and number average molecular weight.

B-stage—An intermediate stage in the reaction of a thermosetting resin in which the material softens when heated and swells in contact with certain liquids but does not entirely fuse or dissolve. Resins in thermosetting molding compounds usually are in this stage. See also **A-stage** and **C-stage**.

back pressure—Resistance of a material, because of its viscosity, to continue flow when mold is closing.

back-pressure-relief port—An opening from an extrusion die for the escape of excess material.

back taper—Reverse draft used in a mold to prevent the molded article from drawing freely. See **undercut**.

backing plate—In injection molding equipment, a heavy steel plate used as a support for the cavity blocks, guide pins, busings, etc. In blow molding equipment, it is the steel plate on which the cavities (i.e., the bottle molds) are mounted.

baffle—A device used to restrict or divert the passage of fluid through a pipe line or channel. In hydraulic systems, the device, which often consists of a disc with a small central perforation, restricts the flow of hydraulic fluid in a high pressure line. A common location for the disc is in a joint in the line. When applied to molds, the term is indicative of a plug or similar device located in a stream or water channel in the mold and designed to divert and restrict the blow to a desired path.

bag molding—A method of applying pressure during bonding or reinforced plastics molding in which a flexible cover, usually in connection with a rigid die or mold, exerts pressure on the material being molded through the application of air pressure or the drawing of a vacuum.

Bakelite—A proprietary name for phenolic and other plastics materials, often used indiscriminately to describe any phenolic molding material or molding. The name is derived from that of Dr. Leo Hendrik Baekeland (1863–1944), a Belgian who developed phenolic resins in the early 1900s.

Banbury—An apparatus for compounding materials composed of a pair of contra-rotating rotors that masticate the materials to form a homogeneous blend. This is an internal-type mixer that produces excellent mixing.

band heater—Electrical heating units fitted to extruder barrels, adaptors, dies, nozzles, etc., utilized for heating the polymer to a desired temperature.

barrel—The tubular portion of an extruder in which the conveying screw rotates.

barrier plastics—A general term applied to a group of lightweight, transparent, and impact-resistant plastics, usually rigid copolymers of high acrylonitrile content. The barrier plastics generally are characterized by gas, aroma, and flavor barrier characteristics approaching those of metal and glass.

beta gauge—A thickness-measuring device used for sheeting or extruded parts. The device operates by beta radiation being emitted on one side of the part and a detector placed on the opposite side. When a part is passed through the beam, some of the beta radiation is absorbed, providing an indication of the part's thickness.

biaxial orientation—The process of stretching a hot plastic film or other article in two directions under conditions resulting in molecular orientation in two directions.

binder—A component of an adhesive composition that is primarily responsible for the adhesive forces that hold two bodies together. See **extender**, **filler**, and **matrix**.

biscuit—See **preform**.

blanking—The cutting of flat sheet stock to shape by striking it sharply with a punch while it is supported on a mating die. Punch presses are used. Also called **die cutting** (which see).

bleed—(1) To give up color when in contact with water or a solvent. (2) Undesired movement of certain materials in a plastic (e.g., plasticizers in vinyl) to the surface of the finished article or into an adjacent material; also called migration. (3) An escape passage at the parting line of a mold, like a vent but deeper, that allows material to escape or bleed out.

blind hole—Hole that is not drilled entirely through.

blister—Undesirable rounded elevation of the surface of a plastic, whose boundaries may be indefinitely outlined, somewhat resembling in shape a blister on the human skin. A blister may burst and become flattened.

blocking—An adhesion between touching layers of plastic, such as that which may develop under pressure during storage or use.

bloom—(1) A noncontinuous surface coating on plastic products that comes from ingredients such as plasticizers, lubricants, antistatic agents, etc., which are incorporated into the plastic resin. It is not always visible. Bloom is the result of ingredients coming out of solution in the plastic and migrating to its surface. (2) Also, term used to describe an increase in diameter of the parison as it comes from the extruder die(s) in the blow molding process.

blow molding—A method of fabrication in which a warm plastic parison (hollow tube) is placed between the two halves of a mold (cavity) and forced to assume the shape of that mold cavity by use of air pressure. Pressurized air is introduced through the inside of the parison and thereby forces the plastic against the surface of the mold that defines the shape of the product.

blow pin—Part of the tooling used to form hollow objects or containers by the blow molding process. It is a tubular tool through which air pressure is introduced into the parison to create the air pressure necessary to form the parison into the shape of the mold. In some blow molding systems, it is a part of, or an extension of, the core pin.

blow pressure—The air pressure required to form the parison into the shape of the mold cavity in a blow molding operation.

blow rate—The speed or rate at which the air enters or the time required for air to enter the parison during the blow molding cycle.

blowing agent—A substance that alone or in conjunction with other substances is capable of producing a cellular structure in a plastic mass.

blown film extrusion—Technique for making film by extruding the plastic through a circular die, followed by expansion (by the pressure of internal air admitted through the center of the mandrel), cooling, and collapsing of the bubble.

blown tubing—A thermoplastic film produced by extruding a tube, applying a slight internal pressure to the tube to expand it while still molten, and subsequent cooling to set the tube. The tube is then flattened through guides and wound up flat on rolls. The size of blown tubing is determined by its flat width in inches as wound, rather than by the diameter (which is used for rigid tubing).

blow-up ratio—In blow molding, the ratio of the diameter of the product (usually its greatest diameter) to the diameter of the parison from which the product is formed. In blown film extrusion, the ratio between the diameter of the final film tube and the diameter of the die orifice.

bolster—Space or filler in a mold.

boss—Projection on a plastic part designed to add strength, to facilitate alignment during assembly, to provide for fastenings, etc.

bottom blow—A specific type of blow molding technique in which hollow articles are formed by injecting the blowing air into the parison from the bottom of the mold (as opposed to introducing the blowing air at a container opening).

bottom plate—Part of the mold that contains the heel radius and the push-up.

branched—Chemistry term referring to a configuration having side chains attached to the original chain (in a direction different from that of the original chain) in the molecular structure of a polymer.

breakdown voltage—The voltage required, under specific conditions, to cause the failure of an insulating material. See **dielectric strength** and **arc resistance**.

breaker plate—A metal plate installed across the flow of the stock between the end of an extruder screw and the die, with openings through it such as holes or slots. It usually is used to support a screen pack.

breathing—The opening and closing of a mold to allow gases to escape early in the molding cycle. Also called degassing. When referring to plastic sheeting, breathing indicates permeability to air.

brinell hardness—Similar to **Rockwell hardness** (which see).

bubble—A spherical, internal void or globule of air or other gas trapped within a plastic. See **void**.

bubbler—A device inserted into a mold force, cavity, or core that allows water to flow deep inside the hole into which it is inserted and to discharge through the open end of the hole. Uniform cooling of molds and of isolated mold sections can be achieved in this manner.

bulk density—The density of a molding material in loose form (granular, nodular, etc.) expressed as a ratio of weight to volume (e.g., g/cm^3 cubed or lb/ft^3 cubed).

bulk factor—Ratio of the volume of loose molding powder to the volume of the same weight of resin after molding.

bulk molding compound (BMC)—Thermosetting resins mixed with stranded reinforcement, fillers, etc., into a viscous compound for injection or compression molding.

burned—Showing evidence of thermal decomposition through some discoloration, distortion, or localized destruction of the surface of the plastic.

burning rate—A term describing the tendency of plastics articles to burn at given temperatures.

burst strength—The internal pressure required to break a pipe or a fitting. This pressure will vary with the rate of buildup of the pressure and the time during which the pressure is held.

butt fusion—A method of joining pipe, sheet, or other similar forms of a thermoplastic resin wherein the ends of the two pieces to be joined are heated to the molten state and then rapidly pressed together.

C-stage—The final stage in the reaction of certain thermosetting resins, in which the material is relatively insoluble and unfusible. The resin in a fully cured thermosetting molding is in this stage. Sometimes referred to as resite. See **A-stage** and **B-stage**.

calcium carbonate ($CaCO_3$)—A filler and extender used in thermoplastics. It occurs naturally in the form of minerals such as calcite, chalk, limestone, marble, and whiting.

calender—To prepare sheets of material by pressure between two or more counter-rotating rolls. Also, the machine performing this operation.

carbon black—A black pigment produced by the incomplete burning of natural gas or oil. It is widely used as a filler, particularly in the rubber industry. Because it possesses useful ultraviolet protective properties, it is also much used in molding compounds intended for outside weathering applications.

carbon fiber—Fibers produced by pyrolysis of an organic precursor fiber in an inert atmosphere at temperatures higher than 1800°F. The material is used as reinforcement for lightweight, high strength, and high stiffness structures. The high stiffness and the high strength of fibers depend on the degree of preferred orientation.

cartridge heater—Cylindrical-bodied, electrical heater for providing heat for injection, compression, and transfer molds, injection nozzles, runnerless mold systems, hot stamping dies, sealing, etc.

case harden—To harden the surface of a piece of steel to a relatively shallow depth.

cast—(1) To form a ''plastic'' object by pouring a fluid monomer-polymer solution into an open mold where it finishes polymerizing. (2) Forming plastic film and sheet by depositing liquid resin, either molten or in solution or dispersion, onto a chilled surface.

cast film—A film made by depositing a layer of liquid plastic onto a surface and stabilizing this form by evaporation of solvent, by fusing after deposition, or by allowing a melt to cool. Cast films usually are made from solutions or dispersions.

casting—The finished product of a casting operation; it should not be used for molding.

catalyst—A substance that markedly speeds up the cure of a compound when added in minor quantity as compared to the amounts of primary reactants. See **hardener**, **inhibitor**, and **promoter**.

cavity—Portion of a mold that usually forms the outer surface of the molded part. Depending on the number of such depressions, molds are designated as a single-cavity or multi-cavity. See **core**.

cavity retainer plate—Plates in a mold that hold the cavities and forces. These plates are at the mold parting line and usually contain the guide pins and bushings. Also called force retainer plate.

cell—A single cavity formed by gaseous displacement in a plastic material. See **cellular plastic**.

cellular plastic—A plastic whose density is decreased substantially by the presence of numerous cells disposed throughout its mass. See **cell** and **foamed plastics**.

center gated mold—An injection or transfer mold wherein the cavity is filled with molding material directly into the center of the part.

centipoise—A unit of viscosity, conveniently and approximately defined as the viscosity of water at room temperature.

centrifugal casting—A method of forming thermoplastic resins in which the granular resin is placed in a rotatable container, heated to a molten condition by the transfer of heat through the walls of the container, and rotated so that the centrifugal force induced will force the molten resin to conform to the configuration of the interior surface of the container. It is used to fabricate large-diameter pipes and similar cylindrical items.

chalking—Dry, chalklike appearance or deposit on the surface of a plastic. See **haze** and **bloom**.

charge—The measurement or the weight of material used to load a mold at one time or during one cycle.

chill roll—A cored roll, usually temperature-controlled with circulating water, that cools the web before winding. For chill roll (cast) film, the surface of the roll is highly polished. In extrusion coating, either a polished or a matte surface may be used, depending on the surface desired on the finished coating.

chiller—A self-contained system comprised of a refrigeration unit and a coolant circulation mechanism consisting of a reservoir and a pump. Chillers maintain the optimum heat balance in thermoplastic processing by constantly recirculating chilled cooling fluids to injection molds, extruder chill rolls, etc.

chromium plating—An electrolytic process that deposits a hard film of chromium metal onto working surfaces of other metals where resistance to corrosion, abrasion, and/or erosion is needed.

C.I.L. (flow test)—A method of determining the rheology or flow properties of thermoplastic resins developed by Canadian Industries Ltd. In this test, the amount of the molten resin that is forced through a specified size orifice per unit of time when a specified, variable force is applied gives a relative indication of the flow properties of various resins.

clamping area—The largest rated molding area an injection or transfer press can hold closed under full molding pressure.

clamping force—In injection molding and in transfer molding, the pressure applied to the mold to keep it closed, in opposition to the fluid pressure of the compressed molding material, within the mold cavity (cavities) and the runner system. In blow molding, the pressure exerted on the two mold halves (by the locking mechanism of the blowing table) to keep the mold closed during formation of the container. Normally, this pressure or force is expressed in tons.

clamping plate—A plate fitted to a mold and used to fasten the mold to a molding machine.

clamshell molding—A variation of blow molding and thermoforming in which two preheated sheets of plastic are clamped between halves of a split mold. Each sheet is drawn into the individual mold cavity by vacuum in the cavity and air injection between the sheets.

closed cell foam—Cellular plastic in which individual cells are completely sealed off from adjacent cells.

coefficient of expansion—The fractional change in dimension (sometimes volume) specified, of a material for a unit change in temperature. Values for plastics range from 0.01 to 0.2 mil/inch, °C. Ref.: Standard Method of Test for Coefficient of Linear Thermal Expansion of Plastics (ASTM Designation: D 696).

coextrusion—Process of combining two or more layers of extrudate to produce a multiple-layer product in a single step.

coinjection—Technique of injecting two materials into a single mold from two plasticating cylinders, either simultaneously or in sequence.

cold-cure foams—See **high-resiliency flexible foams**.

cold drawing—Technique for using standard metalworking equipment and systems for forming thermoplastic sheet (e.g., ABS) at room temperature.

cold flow—See **creep**.

cold molding—Procedure in which a composition is shaped at room temperature and cured by subsequent baking.

cold parison blow molding—Technique in which parisons are extruded or injection molded separately and then stored for subsequent transportation to the blow molding machine for blowing.

cold runner molding—Process in which sprue and runner system (the manifold section) is insulated from the rest of the mold and temperature-controlled to keep the plastic in the manifold fluid. This mold design eliminates scrap loss from sprues and runners.

cold slug—The first material to enter an injection mold, so named because in passing through the sprue orifice it is cooled below the effective molding temperature.

cold slug well—Space provided directly opposite the sprue opening in an injection mold to trap the cold slug.

cold stretch—A pulling operation with little or no heat, usually on extruded filaments to increase tensile properties.

color concentrate—A measured amount of dye or pigment incorporated into a predetermined amount of plastic. This pigmented or colored plastic is then mixed into larger quantities of plastic material to be used for molding. The "concentrate" is added to the bulk of plastic in measured quantity in order to produce a precise, predetermined color in the finished articles to be molded.

combination mold—See **family mold**.

compound—The plastic material to be molded or blown into final form. This includes the resin itself, along with modifiers, pigments, antioxidants, lubricants, etc., needed to process the resin efficiently and to produce the desired properties in the finished article.

compression mold—A technique whereby molding compound is introduced into an open mold and formed under heat and pressure.

compression ratio—In an extruder screw, the ratio of volume available in the first flight at the hopper to that in the last flight at the end of the screw.

compressive strength—Crushing load at the failure of a specimen divided by the original sectional area of the specimen. Ref.: Tentative Method of Test for Compressive Properties of Rigid Plastics (ASTM Designation D 695).

concentrate—A measured amount of additive (e.g., dye, pigment, foaming agent, antistatic agent, flame retardant, glass reinforcement, etc.) that is incorporated into a predetermined small amount of plastic. This (the concentrate) then can be mixed into larger quantities of plastic to achieve a desired color or end-property.

condensation resin—A resin formed by a chemical reaction in which two or more molecules combine with the separation of water or some other simple substance (e.g., the alkyd, phenol-aldehyde, and urea formaldehyde resins).

conditioning—The subjection of a material to a stipulated treatment so that it will respond in a uniform way to subsequent testing or processing. The term frequently is used to refer to the treatment given to specimens before testing. (Standard ASTM test methods that include requirements for conditioning are indexed in the Index of ASTM Standards.)

cooling channels—Channels or passageways located within the body of a mold through which a cooling medium can be circulated to control temperature on the mold surface. It also is possible to heat a mold by circulating steam, hot oil, or other heated fluid through the channels, as in the molding of the thermosetting and some thermoplastic materials.

cooling fixture—Block of metal or wood holding the shape of a molded piece that is used to maintain the proper shape or dimensional accuracy of a molding after it is removed from the mold until it is cool enough to retain its shape without further appreciable distortion. Also called shrink fixture.

copolymer—See **polymer**.

core—(1) Male element in die, which produces a hole or a recess in the part. (2) The portion of a complex mold that molds undercut parts. These cores usually are withdrawn to one side before the main sections of the mold open (and usually are called side cores). (3) A channel in a mold for circulation of a heat-transfer medium; also called force. (4) Portion of a mold that forms the interior surface of the part. See **cavity**.

core pin—Pin used to mold a hole.

core pin plate—Plate holding core pins.

coring—The removal of excess material from the cross section of a molded part to attain a more uniform wall thickness.

corona discharge—A method of rendering the surfaces of inert plastics such as polyethylene more receptive to inks, adhesives, or coatings by subjecting the surfaces to an electrical discharge. A typical method is to pass a film over a grounded metal cylinder, above which a high voltage electrode is spaced to leave a small air gap. The corona discharge oxidizes the film, leading to the formation of polar groups.

cratering—Depressions of coated plastic surfaces caused by excessive lubricant. Cratering results when paint thins excessively and later ruptures, leaving pinholes and other voids. The use of less thinner in the coating can reduce or eliminate cratering, as can the use of less lubricant on the part.

crazing—Fine cracks that may extend in a network on or under the surface or through a layer of plastic material.

creep—The dimensional change with time of a material under load, following the initial instantaneous elastic deformation. Creep at room temperature is sometimes called cold flow. Ref.: Recommended Practices for Testing Long Time Creep and Stress-Relaxation of Plastics under Tension or Compression Loads at Various Temperatures (ASTM Designation: D 674).

crosslinking—The establishment of chemical bonds between the molecular chains in polymers. Crosslinking can be accomplished by chemical reaction, vulcanization, degradation, or radiation.

crystallinity—A state of molecular structure in some resins that denotes uniformity and compactness of the molecular chains forming the polymer. It normally can be attributed to the formation of solid crystals having a definite geometric form.

cure—To change the physical properties of a material by chemical reaction, which may be condensation, polymerization, or vulcanization; usually accomplished by the action of heat and catalysts, alone or in combination, with or without pressure.

curing temperature—Temperature at which a cast, molded, or extruded product, a resin-impregnated reinforcing material, an adhesive, or other material is subjected to curing.

curing time—In the molding of plastics, the interval of time between the instant of cessation of relative movement between the moving parts of a mold and the instant that pressure is released. Also called molding time.

curtain coating—A method of coating that may be employed with low viscosity resins or solutions, suspensions, or emulsions of resins in which the substrate to be coated is passed through and perpendicular to a freely falling liquid "curtain" (or "waterfall"). The flow rate of the falling liquid and the linear speed of the substrate passing through the curtain are coordinated with the desired coating thickness.

cycle—The complete, repeating sequence of operations in a process or part of a process. In molding, the cycle time is the period, or elapsed time, between a certain point in one cycle and the same point in the next.

dancer roll—A roller used as a tension maintenance device in the production of films and sheeting.

daylight—Distance between the stationary platen and the moving platen on a molding press when the actuating system is fully retracted without ejector box and/or spacers.

debossed—An indented or depressed design or lettering that is molded into an article so as to be below the main outside surface of that article.

deflashing—Any technique or method that removes excess, unwanted material from a molded article. Specifically, the excess material is removed from those places on the article where parting lines of the mold that formed the article may have caused the excess material to be formed.

deflection temperature—The temperature at which a specimen will deflect a given distance at a given load under prescribed conditions of test. See ASTM D 648. Formerly called heat distortion.

degassing—See **breathing**.

degradation—A deleterious change in the chemical structure or physical properties of a plastic, caused by exposure to heat, light, or other agent.

delamination—The splitting of a plastic material along the plane of its layers. Physical separation or loss of bond between laminate plies. See **laminated plastics**.

deliquescent—Capable of attracting moisture from the air.

density—Weight per unit volume of a substance, expressed in grams per cubic centimeter, pounds per cubic foot, etc.

desiccant—Substance that can be used for drying purposes because of its affinity for water.

destaticization—Treatment of plastic materials that minimizes the effects of static electricity on the surface of articles. This treatment can be accomplished either by treating the surface with specific materials or by incorporating materials into the molding compound. Minimizing the surface static electricity prevents dust and dirt from being attracted to and/or clinging to the article's surface.

diaphragm gate—Gate used in molding annular or tubular articles. The gate forms a solid web across the opening of the part.

die adaptor—The part of an extrusion die that holds the die block.

die cutting—(1) Blanking. (2) Cutting shapes from sheet stock by striking it sharply with a shaped knife edge known as a steel rule die. Clicking and dinking are other names for die cutting of this kind.

dielectric constant (permittivity or specific inductive capacity)—The ratio of the capacitance of an assembly of two electrodes separated solely by a plastics insulating material to its capacitance when the electrodes are separated by air. (ASTM Designation: D 150.)

dielectric heating (electronics heating or RF heating)—The process of heating poor conductors of electricity by means of high-frequency (20 to 80 MHz) currents. Dielectric loss in the material is the basis. The process is used for sealing vinyl films, preheating thermoset molding compounds, and drying hygroscopic resins before processing.

dielectric strength—The electric voltage gradient at which an insulating material is broken down or "arced through," in volts per mil of thickness. Ref.: Standard Methods of Test for Dielectric Breakdown Voltage and Dielectric Strength of Electrical Insulating Materials at Commercial Power Frequencies (ASTM Designation: D 149).

differential thermal analysis (DTA)—An analytical method where the specimen is heated simultaneously with an inert material as a control, with each having its own temperature sensing and recording apparatus. The curves generated show the weight losses of both materials under the same rates of heating.

diffusion—The movement of a material, such as a gas or liquid, in the body of a plastic. If the gas or liquid is absorbed on one side of a piece of plastic and given off on the other side, the phenomenon is called permeability. Diffusion and permeability are not due to holes or pores in the plastic but are caused and controlled by chemical mechanisms.

dimensional stability—Ability of a plastic part to retain the precise shape in which it was molded, fabricated, or cast.

dip coating—Applying a plastic coating by dipping the article to be coated into a tank of melted resin or plastisol, then chilling the adhering melt.

disc gate—Mold gate having the same cross section as the mold runner.

discoloration—Any change from the original color, often caused by overheating, light exposure, irradiation, or chemical attack.

dispersion—Finely divided particles of a material in suspension in another substance.

dissipation factor—A measure of electrical power lost in the form of heat when insulating materials are used in alternating current circuits.

doctor roll; doctor bar—A device for regulating the amount of liquid material on the rollers of a spreader.

double-shot molding—Means of producing two color parts and/or two different thermoplastic materials by successive molding operations.

dowel—Pin used to maintain alignment between two or more parts of a mold.

draft—The degree of taper of a side wall or the angle clearance designed to facilitate removal of parts from a mold.

drape forming—Method of forming thermoplastic sheet in which the sheet is clamped into a movable frame, heated, and draped over high points of a male mold. Vacuum then is pulled to complete the forming operation.

draw down ratio—The ratio of the thickness of the die opening to the final thickness of the product.

drawing—The process of stretching a thermoplastic to reduce its cross-sectional area, thus creating a more orderly orientation of polymer chains with respect to each other.

drive—The entire electrical and mechanical system used to supply mechanical energy to the input shaft of a gear reducer. This includes the motor, constant or variable speed belt system, flexible couplings, starting equipment, etc.

drooling—Leakage of resin from a nozzle or around the nozzle area during the injection step in injection molding or around the screen-pack during extrusion.

dry blend—Term applied to a molding compound, containing all necessary ingredients, mixed in a way that produces a dry-free-flowing, particulate material. This term commonly is used in connection with polyvinyl chloride molding compounds.

dry coloring—Method commonly used by fabricators for coloring plastic by tumble-blending uncolored particles of the plastic material with selected dyes and pigments.

durometer hardness—The hardness of a material as measured by the Shore Durometer. Ref.: Tentative Method of Test for Indentation Hardness of Rubber and Plastics by Means of a Durometer (ASTM Designation: D 2240).

dwell—A pause in the application of pressure to a mold, made just before the mold is completely closed, to allow the escape of gas from the molding material.

dyes—Synthetic or natural organic chemicals that are soluble in most common solvents; characterized by good transparency, high tinctorial strength, and low specific gravity.

EDM—See electric discharge machining.

E-glass—A low-alkali borosilicate glass widely used in reinforcing plastics.

ejector pin (or **ejector sleeve**)—A rod, pin, or sleeve that pushes a molding off a force or out of a cavity of a mold. It is attached to an ejector bar or plate that can be actuated by the ejector rod(s) of the press or by auxiliary hydraulic or air cylinders.

ejector return pins—Projections that push the ejector assembly back as the mold closes; also called safety pins and position pushbacks.

elastic deformation—The part of the deformation of an object under load that is recoverable when the load is removed.

elasticity—That property of a material by virtue of which it tends to recover its original size and shape after deformation. If the strain is proportional to the applied stress, the material is said to exhibit Hookean or ideal elasticity.

elastomer—A material that at room temperature stretches under low stress to at least twice its length and snaps back to the original length upon release of stress.

electric discharge machining (EDM)—A metalworking process applicable to mold construction in which controlled sparking is used to erode away the workpiece.

electroforming—Moldmaking method whereby a thin layer of metal is deposited onto a pattern. Molten metal then may be sprayed on the back of the mold to increase its strength.

electroplating—Deposition of metals on certain plastics and molds for finish.

elongation—The fractional increase in length of a material stressed in tension.

embossing—Techniques used to create depressions of a specific pattern in plastics film and sheeting. Such embossing in the form of surface patterns can be achieved on molded parts by the treatment of the mold surface by photoengraving or another process.

encapsulating—Enclosing an article (usually an electronic component or the like) in a closed envelope of plastic, by immersing the object in a casting resin and allowing the resin to polymerize or, if hot, to cool. See **potting**.

environmental stress cracking (ESC)—The susceptibility of a thermoplastic article to crack or craze when stressed, in the presence of surface-active agents or in other environments.

exotherm—(1) The temperature/time curve of a chemical reaction giving off heat, particularly the polymerization of casting resins. (2) The amount of heat given off. The term has not been standardized with respect to sample size, ambient temperature, degree of mixing, etc.

expandable plastic—A plastic compound that can be made cellular during processing by chemical or thermal means.

extender—A substance, generally having some adhesive action, added to a plastic composition to reduce the amount of the primary resin required per unit area. See **filler**.

extrudate—The product or material delivered by an extruder, such as film, pipe, the coating on wire, etc.

extruder—Basically, a machine that accepts solid particles (pellets or powder) or liquid (molten) feed, conveys it through a surrounding barrel by means of a rotating screw, and pumps it, under pressure, through an orifice. The nomenclature used encompasses the barrel, the screw, and other extruder elements. The metering section of the screw is a relatively shallow portion of the screw at the discharge end with a constant depth and lead, and having a length of one or more turns of the flight.

extrusion—Process of compacting and melting a plastic material and forcing it through an orifice in a continuous fashion. Material is conveyed through the heated machine barrel by a helical screw or screws, where it is heated and mixed to a homogeneous state and then forced through a die of the shape required for the finished product.

extrusion coating—Process in which a resin is coated on a substrate by extruding a thin film of molten resin and pressing it onto or into the substrates, or both, without the use of an adhesive.

fabricate—To work a material into a finished form by machining, forming, or other operation.

family mold—A multicavity mold wherein each of the cavities forms one of the component parts of the assembled finished object. The term often is applied to molds wherein parts from different customers are grouped together in one mold for economy of production. Sometimes called a combination mold.

fan gate—A shallow gate somewhat wider than the runner from which it extends.

feed section—First section or zone of an extruder screw, which is fed from the hopper and conveys solids to the melting zone.

fiber reinforcement—Thin fibers of glass, carbon, metal, or synthetic resin incorporated into resin to increase strength. Forms include continuous, chopped, knitted, and woven.

filament—Fiber of extreme length used in yarns and other compositions.

filament winding—Process in which roving or single strands of glass, metal, or other reinforcement are wound in a predetermined pattern onto a suitable mandrel. The pattern is so designed as to give maximum strength in the directions required. The strands can be run from a creel through a resin bath before winding, or preimpregnated materials can be used. When the right number of layers have been applied, the wound mandrel is cured at room temperatures or in an oven.

fill-and-wipe—Process whereby parts are molded with depressed designs, and after application of paint the surplus is wiped off, with paint remaining only in depressed areas. Sometimes called wipe-in.

filler—Inert substance added to a plastic to make it less costly. However, fillers may also improve physical properties, particularly hardness, stiffness, dimensional stability, and impact strength.

fillet—A rounded filling of the internal angle between two surfaces.

film—Sheeting having a nominal thickness not greater than 0.010 inch.

fines—Very small particles (usually under 200 mesh) accompanying larger grains, usually of molding powder.

fisheye—Small globular mass that has not blended completely into the surrounding material; particularly evident in a transparent or translucent material.

fixture—Means of holding a part during a machining or other operation.

flame retardant—A chemical compounded into a resin to make it fire-resistant.

flame spraying—Method of applying a plastic coating in which finely powdered fragments of the plastic, together with suitable fluxes, are projected through a cone of flame onto a surface.

flame treating—A method of rendering inert thermoplastic objects receptive to inks, lacquers, paints, adhesives, etc., in which the object is bathed in an open flame to promote oxidation of the surface of the article.

flash—Extra plastic attached to a molding along the parting line; under most conditions it would be objectionable and must be removed before the parts are acceptable.

flash gate—Usually a long gate extending from a runner that runs parallel to an edge of a molded part along the flash or parting line of the mold.

flash line—A raised line appearing on the surface of a molding and formed at the junction of mold faces. See **parting line**.

flash mold—A mold in which the mold faces are perpendicular to the clamping action of the press, so that the higher the clamping force, the tighter the mold seam.

flexible molds—Molds made of rubber or elastomeric plastics used for casting plastics. They can be stretched to remove cut pieces with undercuts.

flexural modulus, psi—The ratio of stress to strain for a given material within its proportional limit under bending load conditions. (ASTM test methods D 790.)

flexural strength—Ability of a material to flex without permanent distortion or breaking. Ref.: Standard Method of Test for Flexural Properties of Plastics (ASTM Designation: D 790).

flight—The outer surface of the helical ridge of metal on an extrusion or injection molding screw.

floating core—Mold member, free to move vertically, that fits over a lower plug or cavity, and into which an upper plug telescopes.

floating platen—Movable platen(s) between the stationary platen and the actuated platen on a vertically operating compression press.

flock—Short fibers of cotton, wood, or glass used as a filler for resins.

flour—An organic filler. Such organic fillers as wood flour and shell flours are used as extenders and reinforcements. Wood flour is a finely ground product commonly made from soft woods, whereas shell flours are derived from peanut and rice hulls.

flow—A qualitative description of the fluidity of a plastic material during the process of molding.

flow line—A mark on a molded piece made by the meeting of two flow fronts during molding. Also called the weld line.

fluidized bed coating—A method of applying a coating of a thermoplastic resin to an article in which the heated article is immersed in a dense-phase fluidized bed of powdered resin and thereafter heated in an oven to provide a smooth, pinhole-free coating.

foamed plastics—Plastics with internal voids or cells. The foam may be flexible or rigid, the cells closed or connected, and the density anything from slightly below that of the solid parent resin down to, in some cases, 1 lb/ft cubed, or less.

foaming agents—Chemicals added to plastics and rubbers that generate inert gases on heating, causing the resin to assume a cellular structure.

foam-in-place—A type of foam deposition that requires that the foaming machine be brought to the work, which is "in place," as opposed to bringing the work to the foaming machine.

foil decorating—Printing method where metallic or pigmented designs are transferred to a plastic surface using heat and pressure. See **hot-stamping**.

force—The portion of the mold that forms the inside of the molded part. Also called a core or a plunger.

forging—See **solid phase forming**.

forming—A process in which the shape of plastic pieces such as sheets, rods, or tubes is changed to a desired configuration. See also **thermoforming**. (*Note*: The use of the term "forming" in plastics technology does not include such operations as molding, casting, or extrusion, in which shapes or pieces are made from molding materials or liquids.)

friction welding—A method of welding thermoplastics materials whereby the heat necessary to soften the components is provided by friction. See **spin welding** and **vibration welding**.

frost line—In the extrusion of polyethylene lay-flat film, a ring-shaped zone located at the point where the film reaches its final diameter. In this zone, the film has a ''frosty'' appearance caused by the film's temperature falling below the softening range of the resin.

frothing—Technique for applying urethane foam in which blowing agents or tiny air bubbles are introduced under pressure into the liquid mixture of foam ingredients.

fusion bonding—Process for joining plastic parts where mating surfaces are heated to melting and held together under pressure. Also known as hot plate welding.

glass transition temperature—The temperature at which an amorphous polymer changes from a hard, brittle (glassy) condition to a viscous, elastomeric form. Also called second-order transition, gamma transition, rubber transition, and rubbery transition. The word transformation also has been used instead of transition.

gate—The short, usually restricted, section of the runner at the entrance to the cavity of an injection or transfer mold.

gate mark—A surface discontinuity on the part caused by the presence of the mold orifice through which material enters the cavity.

gauge—Thickness of plastic film or sheet.

gel—(1) A semisolid system consisting of a network of solid aggregates, in which liquid is held. (2) The initial jellylike solid phase that develops during the formation of a resin from a liquid. Both types of gel have very low strengths and do not flow like a liquid; they are soft and flexible, and will rupture under their own weight unless supported externally. (3) Small globular mass not completely blended into film or sheet, causing a defect.

gelatin—(1) Formation of a gel. (2) In vinyl dispersions, formation of a gel in the early stages of fusion.

glass finish—A material applied to the surface of a glass reinforcement to improve its effect upon the physical properties of the reinforced plastic.

gloss—The shine or luster of the surface of a material.

granular structure—Nonuniform appearance of finished plastic material due to retention of, or incomplete fusion of, particles of composition, either within the mass or on the surface.

granulator—Machine used for size reduction of plastic scrap for reuse. Also called grinder.

grid—Channel-shaped mold-supporting members.

grit blasting—A surface treatment of a mold in which steel grit or sand materials are blown on the walls of the cavity to produce a roughened surface. Air escape from mold is improved, and a special appearance of the molded article is often obtained by this method.

guide pins—Devices that maintain proper alignment of force plug and cavity as mold closes. Also called leader pins.

guide-pin bushing—A guiding bushing through which the leader pin moves.

gussets—Inward folds on sides of collapsed, blown film to reduce width and produce bags with rectangular form.

hardener—A substance or mixture of substances added to a resin or adhesive to promote or control the curing reaction by taking part in it. The term is also used to designate a substance added to control the degree of hardness of the cured film. See **catalyst**.

hardness—The resistance of a plastic material to compression and indentation. Among the most important methods of testing this property are Brinell hardness, Rockwell hardness, and Shore hardness.

haze—Cloudiness in plastic film.

heat-distortion point—The temperature at which a standard test bar deflects 0.010 inch under a stated load of either 66 or 264 psi.

Ref.: Standard Method of Test for Deflection Temperature of Plastics under Load (ASTM Designation: D 648).

heat forming—See **thermoforming**.

heat gun—Electrically heated gun for softening, curing, drying, preheating, and welding plastics, coatings, and compounds as well as shrinking of heat-shrinkable plastic tubing and plastic films.

heat-sealing—A method of joining plastic films by simultaneous application of heat and pressure to areas in contact. Heat may be supplied conductively or dielectrically.

heat-treating—Term used to cover annealing, hardening, tempering, etc.

helix—See **extruder**.

helix angle—The angle of a screw flight at its periphery relative to a plane perpendicular to the axis.

high-frequency heating—The heating of materials by dielectric loss in a high-frequency electrostatic field. The material is exposed between electrodes, and by absorption of energy from the electrical field is heated quickly and uniformly throughout.

high-pressure laminates—Laminates molded and cured at pressures not lower than 1000 psi and more commonly in the range of 1200 to 2000 psi.

high-resiliency flexible foams—Urethane foams that offer low hysteresis and modulus; they can be made more flame retardant than rigid foams, and process at lower oven temperatures and with shorter molding cycles. In trade parlance, these foams have a sag factor of 2.7 and above (i.e., better cushioning). They can be produced ''cold cure'' (with no additional heat needed over that supplied by the exothermic reaction of the foaming process) or with heated molds and heat cures.

hob—A master model used to sink the shape of a mold into a soft steel block.

hobbing—A process of forming a mold by forcing a hob of the shape desired into a soft steel blank.

homopolymer—The result of the polymerization of a single monomer, a polymer that consists of a single type or repeating unit.

honeycomb—Manufactured product consisting of sheet metal or a resin-impregnated sheet material (paper, fibrous glass, etc.) that has been formed into hexagonal-shaped cells. Used as core material for sandwich constructions.

hopper—Feed reservoir into which molding powder is loaded and from which it falls into a molding machine or extruder, sometimes through a metering device.

hopper dryer—A combination feeding and drying device for extrusion and injection molding of thermoplastics. Hot air flows upward through the hopper containing the feed pellets.

hopper loader—A curved pipe through which molding powders are pneumatically conveyed from shipping drums to machine hoppers.

hot gas welding—A technique of joining thermoplastic materials (usually sheet) whereby the materials are softened by a jet of hot air from a welding torch, and are joined together at the softened points. Generally a thin rod of the same material is used to fill and consolidate the gap.

hot plate welding—See **fusion bonding**.

hot-runner mold—A thermoplastic injection mold in which the runners are insulated from the chilled cavities and remain hot so that the center of the runner never cools in normal cycle operation. Runners are not, as is the case usually, ejected with the molded pieces. Sometimes called insulated runner mold.

hot-stamping—Operation for marking plastics in which roll leaf is stamped with heated metal dies onto the face of the plastics.

hydraulic clamp—Device used in variety of molding and forming machines that consists basically of a high-speed, variable hydraulic pump, valving, a fast-acting cylinder, and a high-pressure cylinder. Cylinders can be single or combination units. The clamp closes the mold halves to form the part.

hygroscopic—Readily absorbing and retaining environmental moisture.

impact strength—(1) The ability of a material to withstand shock loading. (2) The work done in fracturing, under shock loading, a specified test specimen in a specified manner.

impregnation—The process of thoroughly soaking a material such as wood, paper, or fabric with a synthetic resin so that the resin gets within the body of a material.

impulse sealing—A heat-sealing technique in which a pulse of intense thermal energy is applied to the sealing area for a very short time, followed immediately by cooling. It usually is accomplished by using an RF heated metal bar that is cored for water cooling or is of such a mass that it will cool rapidly at ambient temperatures.

inhibitor—A substance that prevents or retards a chemical reaction.

initiator—Peroxide used as source of free radicals. These substances are used in free radical polymerizations, in curing thermosetting resins, as crosslinking agents for elastomers and polyethylene, and for polymer modification.

injection mold—A mold into which a plasticated material is introduced from an exterior heating cylinder.

injection molding—A molding procedure whereby a heat-softened plastic material is forced from a cylinder into a cavity that gives the article the desired shape. It is used with both thermoplastic and thermosetting materials.

inorganic pigments—Natural or synthetic metallic oxides, sulfides, and other salts, calcined during processing at 1200 to 2100°F. They are outstanding in producing heat and light stability, weather resistance, and migration resistance.

insert—An integral part of a plastic molding consisting of metal or other material that may be molded into position or may be pressed into the part after the molding is completed.

insert molding—Process by which components such as fasteners, pins, studs, and terminals can be incorporated into a part as it is being molded.

integral-skin foams—As applied to urethane or structural foams, designation for molded foams that develop their own integral surface skins. The surface skin is generally "solid," as contrasted to the cellular construction in the interior of the part.

iridescence—Loss of brilliance in metallized plastics and development of multicolor reflectance. It is caused by cold flow of plastic or coating and from extra heat during vacuum metallizing.

irradiation (atomic)—As applied to plastics, term that refers to bombardment with a variety of subatomic particles, generally alpha-, beta-, or gamma-rays. Atomic irradiation has been used to initiate polymerization and copolymerization of plastics and in some cases to bring about changes in the physical properties of a plastic material.

isotactic—Description applied to a chain of unsymmetrical molecules combined head to tail, with their methyl groups occupying the same relative positions in space along the chain.

jet molding—Processing technique characterized by the fact that most of the heat is applied to the material as it passes through a nozzle or jet, rather than in a heating cylinder as is done in conventional processes.

jetting—Turbulent flow of resin from an undersized gate or thin section into a thicker mold section, as opposed to laminar flow of material progressing radially from a gate to the extremities of the cavity.

jig—Means of holding a part and guiding the tool during machining or assembly operation.

joint—The location at which two adherends are held together with a layer of adhesive.

kirksite—An alloy of aluminum and zinc used for the construction of blow molds; it imparts a high degree of heat conductivity to the mold.

kiss-roll coating—Roll arrangement that carries a metered film of coating to the web; at the line of web contact, it is split with part remaining on the roll and the remainder of the coating adhering to the web.

knife coating—A method of coating a substrate (usually paper or fabric) in which the substrate, in the form of a continuous moving web, is coated with a material whose thickness is controlled by an adjustable knife or bar set at a suitable angle to the substrate.

knit line—See **weld lines**.

knockout bar—A bar or plate in a knockout frame used to back up a row or rows of knockout pins.

knockout pin—See **ejector pin**.

***L/D* ratio**—The ratio of the length (*L*) to the diameter (*D*) of an extruder screw or barrel.

laminar flow—Flow of thermoplastic resins in a mold that is accompanied by solidification of the layer in contact with the mold surface, which acts as an insulating tube through which material flows to fill the remainder of the cavity. This type of flow is essential to duplication of the mold surface.

laminated plastic (synthetic resin-bonded laminate, laminate)—A plastics material consisting of superimposed layers of a synthetic resin-impregnated or -coated substrate (paper, glass mat, etc.) that have been bonded together, usually by means of heat and pressure, to form a single piece.

land—(1) The horizontal bearing surface of a semipositive or flash mold by which excess material escapes. (2) The bearing surface along the top of the flights of a screw in a screw extruder. (3) The surface of an extrusion die parallel to the direction of melt flow.

lay-up—(1) As used in reinforced plastics, the reinforcing material placed in position in the mold; also the resin-impregnated reinforcement. (2) The process of placing the reinforcing material in position in the mold.

leach—To extract a soluble component from a mixture by the process of percolation.

light-resistance—The ability of a plastics material to resist fading after exposure to sunlight or ultraviolet light. Ref.: Tentative Recommended Practice for Exposure of Plastics to Fluorescent Sunlamp (ASTM Designation: D 1501).

liquid injection molding (LIM)—(1) A process that involves an integrated system for proportioning, mixing, and dispensing two-component liquid resin formulations and directly injecting the resultant mix into a mold that is clamped under pressure. It generally is used for the encapsulation of electrical and electronic devices. (2) Variation on reaction injection molding that involves mechanical mixing rather than the high-pressure impingement mixing used with reaction injection molding. However, unlike mechanical mixing in other systems, the mixer here does not need to be flushed, as a special feed system automatically dilutes the residue in the mixer with part of the polyol needed for the next shot, thereby keeping the ingredients from reacting.

loading tray—A device in the form of a specially designed tray that is used to load the charge of material or metal inserts simultaneously into each cavity of a multicavity mold by the withdrawal of a sliding bottom from the tray. Also called charging tray.

locating ring—A ring that serves to align the nozzle of an injection cylinder with the entrance of the sprue bushing and the mold to the machine platen.

loss factor—The product of the power factor and the dielectric constant.

low-pressure laminates—In general, laminates molded and cured in the range of pressures from 400 psi down to and including pressures obtained by the mere contact of the plies.

low-profile resins—Designation applied to special polyester resin systems for reinforced plastics. These systems are combinations of thermoset resins and thermoplastic resins used to minimize surface waviness in molded parts.

lubricant—Additive to plastic resin to promote mixing and improve flow properties.

lug—(1) A type of thread configuration, usually thread segments disposed equidistantly around a bottle neck (finish). (2) A small indentation or raised portion on the surface of a product, provided as a means of indexing for operations such as multicolor decoration or labeling.

luminescent pigments—Pigments that produce striking effects in darkness or light. Forms include fluorescence and phosphorescence.

mandrel—(1) In blow molding, part of the tooling that forms the inside of the container neck and through which air is forced to form the hot parison to the shape of the molds. (2) In extrusion, the solid, cylindrical part of the die that forms tubing or pipe. (3) In filament winding of reinforced plastic, the form (usually cylindrical) around which the filaments are wound.

manifold—Mainly with blow molding and sometimes with injection molding equipment, the distribution or piping system that takes the single channel flow output of the extruder or injection cylinder and divides it to feed several blow molding heads or injection nozzles.

masterbatch—A plastics compound that includes a high concentration of an additive or additives. Masterbatches are designed for use in appropriate quantities with the basic resin or mix so that the correct end concentration is achieved. For example, color masterbatches for a variety of plastics are used extensively, as they provide a clean and convenient method of obtaining accurate color shades.

mat—A fabric or felt of glass or other reinforcing fiber used in manufacturing plastic composite parts.

material distribution—The variation in thickness of various parts of a product (i.e., body, wall, shoulder, heel, base, etc.).

matched metal molding—Method of molding reinforced plastics between two close-fitted metal molds mounted in a press.

material well—Space provided in a compression mold to care for the bulk factor of the material load.

matrix—The continuous phase of a composite material; the resin component in a reinforced plastics material.

matte finish—A type of dull, nonreflective finish.

mechanically foamed plastic—A cellular plastic whose structure is produced by physically incorporated gases.

melt fracture—An instability in the melt flow through a die, starting at the entry to the die. It leads to surface irregularities on the finished article such as a regular helix or irregularly spaced ripples.

melt index—The amount, in grams, of a thermoplastic resin that can be forced through a 0.0825 inch orifice when subjected to 2160 grams force in 10 minutes at 190°C. Ref.: Tentative Method of Measuring Flow Rates of Thermoplastics by Extrusion Plastometer (ASTM Designation: D 1238).

melt strength—The strength of a plastic while in the molten state.

melting point—The temperature at which a resin changes from a solid to a liquid.

memory—The tendency of a plastic article to return to a size and shape that existed during the manufacturing process.

metallizing—Application of a thin coating of metal to a nonmetallic surface. It may be done by chemical deposition or by exposing the surface to vaporized metal in a vacuum chamber.

metallic pigments—A class of pigments consisting of thin opaque aluminum flakes (made by a ball milling either a disintegrated aluminum foil or a rough metal powder and then polishing to obtain a flat, brilliant surface on each particle) or copper alloy flakes (known as bronze pigments). Incorporated into plastics, they produce unusual silvery and other metal-like effects.

metering screw—An extrusion screw that has a shallow constant depth, and a constant pitch section over, usually, the last three to four flights.

metering section—A relatively shallow portion of an extruder screw at the discharge end with a constant depth and lead, and having a length of at least one or more turns of the flight.

migration of plasticizer—Loss of plasticizer from an elastomeric plastic compound with subsequent absorption by an adjacent medium of lower plasticizer concentration.

modulus of elasticity—The ratio of stress to strain in a material that is elastically deformed. Ref.: Standard Method of Test for Flexural Properties of Plastics (ASTM Designation: D 790).

moisture vapor transmission—The rate at which water vapor permeates through a plastic film or wall at a specified temperature and relative humidity. Ref.: Standard Methods of Test for Water Vapor Transmission of Materials in Sheet Form (ASTM Designation: E 96).

mold—A hollow form or cavity into which

molten plastic material is introduced to give the shape of the required component. The term generally refers to the whole assembly of elements that make up the section of the molding equipment in which the parts are formed. Also called tool or die.

mold base—The assembly of all parts making up an injection mold, other than the cavity, core, and pins.

mold insert (removable)—Part of a mold cavity or force that forms undercut or raised portions of a molded article.

mold mark—Identifying symbol of the molder who produced the part; usually molded into an unobtrusive area.

mold release—A lubricant used to coat a mold cavity to prevent the molded piece from sticking to it, and thus to facilitate its removal from the mold. Also called release agent. See **parting agent**.

mold seam—A line formed by mold construction such as removable members in cavity, cam slides, etc. (not to be confused with mold parting line).

molding cycle—See **cycle**.

molding material—Plastic material in varying stages of granulation, often comprising resin, filler, pigments, plasticizers, and other ingredients, ready for use in the molding operation. Also called molding compound or powder.

molding pressure—The pressure applied directly or indirectly on the compound to allow the complete transformation to a solid dense part.

molding shrinkage—The difference in dimensions, expressed in inches per inch, between a molding and the mold cavity in which it was molded, both the mold and the molding being at normal room temperature when measured. Also called mold shrinkage, shrinkage, and contraction.

molecular weight—The sum of the atomic weights of all atoms forming a molecule.

molecular weight distribution—A measure of the relative amounts of polymers with different molecular weights within a batch of material. This measure may be indicated by the ratio of the weight-average molecular weight to the number-average molecular weight.

monomer—A relatively simple compound that can react to form a polymer (i.e., polymerize).

movable platen—The moving platen of an injection or compression molding machine to which half of the mold is secured during operation. This platen is moved by either a hydraulic ram or a toggle mechanism.

multicavity mold—A mold with two or more mold impressions; i.e., a mold that produces more than one molding per molding cycle.

multiple-flighted screw—A screw having more than one helical flight, such as double flighted, double lead, double thread, or two starts, triple flighted, etc.

multiple-screw extruders—As contrasted to conventional single-screw extruders, these machines involve the use of two or four screws (conical or constant depth). Types include machines with intermeshing counter-rotating screws and those with nonintermeshing counter-rotating screws.

neck-in—In extrusion coating, the difference between the width of the extruded web as it leaves the die and the width of the coating on the substrate.

needle blow—A specific blow molding technique where the blowing air is injected into a hollow article through a sharpened hollow needle that pierces the parison.

nip rolls—A pair of rolls on a blown film line that close the bubble and regulate the rate at which the film is pulled away from the extrusion die.

nitriding—A hardening process for ferrous alloys used on extruder screws.

notch sensitivity—The extent to which the sensitivity of a material to fracture is increased by the presence of a break in the homogeneity of the surface, such as a notch, a sudden change in section, a crack, or a scratch. Low notch sensitivity is usually associated with ductile materials, and high notch sensitivity with brittle materials.

nozzle—The hollow cored metal nose screwed into the extrusion end of (a) the heating cylinder of an injection machine or (b) a transfer chamber where this is a separate structure. A nozzle is designed to form under pressure a seal between the heating cylinder or the transfer chamber and the mold. The front end of a nozzle may be either flat or spherical in shape.

nucleating agent—A chemical substance that provides sites for crystal formation in polymer melts.

offset printing—A printing process in which the image to be printed first is applied to an intermediate carrier such as a roll or plate and then is transferred to a plastic film or molded article.

olefins—A group of unsaturated hydrocarbons of the general formula C_nH_{2n}, and named after the corresponding paraffins by the addition of "ene" to the stem. Examples are ethylene and propylene.

one-shot molding—In the urethane foam field, term applied to a system whereby the isocyanate, polyol, catalyst, and other additives are mixed together directly, and a foam is produced immediately (as distinguished from prepolymer).

opaque—Description of a material or substance that will not transmit light; opposite of transparent. Materials that are neither opaque nor transparent sometimes are described as semiopaque, but are more properly classified as translucent.

open-cell foam—A cellular plastic in which there is a predominance of interconnected cells.

orange peel—Uneven leveling of coating on plastic surfaces, usually because of high viscosity. Simple spray gun adjustments and/or addition of high boiling solvent to coating for a wetter spray is helpful.

organic pigments—Pigments characterized by good brightness and brilliance, which are divided into toners and lakes. Toners, in turn, are divided into insoluble organic toners and lake toners. The insoluble organic toners usually are free of salt-forming groups. Lake toners are practically pure, water-insoluble heavy metal salts of dyes without the fillers or substrates of ordinary lakes. Lakes, which are not as strong as lake toners, are water-insoluble heavy metal salts or other dye complexes precipitated upon or admixed with a base or filler.

organosol—A suspension of a finely divided resin in a volatile organic liquid. The resin does not dissolve appreciably in the organic liquid at room temperature, but does so at elevated temperatures. The liquid evaporates at the elevated temperature, and the residue on cooling is a homogeneous plastic mass. Plasticizers may be dissolved in the volatile liquid.

orientation—The alignment of the crystalline structure in polymeric materials so as to produce a highly uniform structure. It can be accomplished by cold drawing or stretching during fabrication.

orifice—The opening in the extruder die formed by the orifice bushing (ring) and mandrel.

orifice bushing—The outer part of the die in an extruder head.

outgassing—Devolatilization of plastics or applied coatings during exposure to vacuum in vacuum metallizing. Resulting parts show voids or thin spots in plating with reduced and spotty brilliance. Additional drying prior to metallizing is helpful, but outgassing is inherent in plastic materials and coatings ingredients, including plasticizer and volatile components.

parallels—The support spacers placed between the mold and the press platen or clamping plate. Also called risers.

parison—A precursor to a blow-molded part created by extrusion or injection molding.

parison programmer—A device that allows the extrusion of a parison that differs in thickness along its length in order to equalize the wall thickness of a blow-molded product. It can be done with a pneumatic or hydraulic device that activates the mandrel shaft and adjusts the mandrel position during parison extrusion (parison programmer, controller, or variator). It also can be done by varying the extrusion speed on accumulator-type blow molding machines or, in parison reheat systems, by varying the amount of heat applied.

parison swell—In extrusion blow molding, the ratio of the cross-sectional area of the parison to the cross-sectional area of the die opening.

part—In its proper literal meaning, a component of an assembly. However, the word is widely misused to designate any individual manufactured article, even when (like a cup, a comb, a doll) it is complete in itself, not part of anything.

parting agent—A lubricant, often wax, used to coat a mold cavity to prevent the molded piece from sticking to it, and thus to facilitate its removal from the mold. Also called release agent.

parting line—Mark on a mold or casting where halves of a mold met in closing.

pearlescent pigments—A class of pigments consisting of particles that are essentially transparent crystals of a high refractive index. The optical effect is one of partial reflection from the two sides of each flake. When reflections from parallel flakes reinforce each other, the result is a silvery luster. Possible effects range from brilliant highlighting to moderate enhancement of the normal surface gloss.

permanence—Resistance of a plastic to appreciable changes in characteristics with time and environment.

permeability—(1) The passage or diffusion of a gas, vapor, liquid, or solid through a barrier without being physically or chemically affected. (2) The rate of such passage.

pigment—A coloring agent mixed with plastic material prior to processing to provide a uniform color.

pill—See **preform**.

pinch-off—A raised edge around a cavity in the mold that seals off the part and separates the excess material as the mold closes around the parison in the blow molding operation.

pinhole—A very small hole in a plastic container, film, etc.

pinpoint gate—A restricted orifice of 0.030 inch or less in diameter through which molten resin flows into a mold cavity.

pipe—A hollow cylinder of a plastic material in which the wall thicknesses are usually small when compared to the diameter, and in which the inside and outside walls are essentially concentric. See **tubing**.

pipe train—Term used in extrusion of pipe to denote the entire equipment assembly used to fabricate the pipe (e.g., extruder, die, cooling bath, haul-off, and cutter).

pitch—The distance from any point on the flight of a screw line to the corresponding point on an adjacent flight, measured parallel to the axis of the screw line or threading.

plastic—(1) One of many high-polymeric substances, including both natural and synthetic products, but excluding the rubbers. At some stage in its manufacture, every plastic is capable of flowing, under heat and pressure if necessary, into the desired final shape. (2) Made of plastic; capable of flow under pressure or tensile stress.

plastic, rigid—A plastic with a stiffness or apparent modulus of elasticity greater than 100,000 psi at 23°C. (ASTM D 747.)

plastic, semirigid—A plastic with a stiffness or apparent modulus of elasticity between 10,000 and 100,000 psi at 23°C. (ASTM D 747.)

plastic deformation—A change in dimensions of an object under load that is not recovered when the load is removed; opposed to elastic deformation.

plasticate—To soften by heating or kneading. Synonyms are plastify, flux, and (imprecisely) plasticize.

plasticity—A property of a plastic that allows the material to be deformed continuously and permanently without rupture upon the application of a force that exceeds the yield value of the material.

plasticize—To soften a material and make it plastic or moldable, by means of either a plasticizer or the application of heat.

plasticizer—A material incorporated in a plastic to increase its workability and its flexibility or distensibility; normally used in thermoplastics. The addition of the plasticizer may lower the melt viscosity, the temperature of the glassy transition, or the elastic modulus of the plastic.

plastics tooling—Tools (e.g., dies, jugs, fixtures, etc.) for the metal forming trades constructed of plastics, generally laminates or casting materials.

plastify—See **plasticate**.

plastigel—A plastisol exhibiting gel-like flow properties; one having an effective yield value.

plastisols—Mixtures of vinyl resins and plasticizers that can be molded, cast, or converted to continuous films by the application of heat. If the mixtures contain volatile thinners also, they are known as organosols.

plate dispersion plug—See **breaker plate**.

platens—The mounting plates of an injection or compression molding press, to which the entire mold assembly is bolted.

plate-out—The undesirable deposition of additives onto machinery during processing.

plug forming—A thermoforming process in which a plug or male mold is used to partially preform the part before forming is completed using vacuum or pressure.

plunger—That part of a transfer or injection press that applies pressure on the unmelted plastic material to push it into the chamber, which in turn forces plastic melt at the front of the chamber out the nozzle. See **ram**.

plunger machines—Injection molding machine whose plasticating system consists of a piston-in-cylinder arrangement. The plunger, on each stroke, pushes unmelted plastic into the heating cylinder, forcing melt out through a nozzle on the opposite end.

polishing roll(s)—A roll or series of rolls, with a highly polished chrome-plated surface, that are utilized to produce a smooth surface on sheet as it is extruded.

polymer—A high-molecular-weight organic compound, natural or synthetic, whose structure can be represented by a repeated small unit, the monomer (e.g., polyethylene, rubber, cellulose). Synthetic polymers are formed by addition or condensation polymerization of monomers. If two or more different monomers are involved, a copolymer is obtained. Some polymers are elastomers, some plastics.

polymerization—A chemical reaction in which the molecules of a monomer are linked together to form large molecules whose molecular weight is a multiple of that of the original substance. When two or more different monomers are involved, the process is called copolymerization or heteropolymerization.

porosity—The existence in a plastic material of very small voids.

porous molds—Molds that are made up of bonded or fused aggregate (powdered metal, coarse pellets, etc.) in such a manner that the resulting mass contains numerous open interstices of regular or irregular size, by means of which either air or liquids may pass through the mass of the mold.

postcure—Operation whereby thermoset parts are subjected to elevated temperatures for a period of time after being removed from the mold to attain maximum property levels.

postforming—The forming, bending, or shaping of fully cured, C-stage thermoset laminates that have been heated to make them flexible. On cooling, the formed laminate retains the contours and shape of the mold over which it has been formed.

pot—Chamber to hold and heat molding material for a transfer mold.

pot life—See **working life**.

pot plunger—A plunger used to force softened molding material into the closed cavity of a transfer mold.

potting—A procedure similar to encapsulating except that here steps are taken to ensure complete penetration of all the voids in the object before the resin polymerizes.

powder molding—General term used to denote several techniques for producing objects of varying sizes and shapes by melting plastic powder, usually against the inside of a mold. The techniques vary according to whether the molds are stationary (e.g., as in variations on slush molding techniques) or rotating (e.g., as in variations on rotational molding).

preform—(1) A compressed tablet or biscuit of plastic composition used for efficiency in handling and accuracy in weighing materials. (2) To make plastic molding powder into pellets or tablets.

preheating—The heating of a compound prior to molding or casting in order to facilitate the operation, reduce the cycle, and improve the product.

preplastication—Technique of premelting injection molding powders in a separate chamber, then transferring the melt to the injection cylinder. The device used for preplastication commonly is known as a preplasticizer.

prepolymer—A chemical structure intermediate between that of the initial monomer or monomers and the final polymer or resin.

prepolymer molding—In the urethane foam field, term used for a system whereby a portion of the polyol is pre-reacted with the isocyanate to form a liquid prepolymer in a viscosity range suitable for pumping or metering. This component is supplied to end-users with a second premixed blend of additional polyol, catalyst, blowing agent, etc. When the two components are mixed together, foaming occurs. See **one-shot molding**.

prepreg—A term generally used in reinforced plastics to mean the reinforcing material containing or combined with the full complement of resin before molding.

preprinting—In sheet thermoforming, the distorted printing of sheets before they are formed. During forming the print assumes its proper proportions.

pressure forming—A thermoforming process wherein pressure is used to push the sheet to be formed against the mold surface, as opposed to using a vacuum to suck the sheet flat against the mold.

pressure pads—Reinforcements distributed around the dead areas in the faces of a mold to help the land absorb the final pressure of closing without collapsing.

primer—A coating applied to a surface, prior to the application of an adhesive or lacquer, enamel, or the like, to improve adhesion or finishing.

promoter—A chemical, itself a feeble catalyst, that greatly increases the activity of a given catalyst.

prototype mold—A simplified mold construction often made from a light metal casting alloy, an epoxy resin, or an RTV silicone rubber, in order to obtain information for the final mold and/or part design.

pultrusion—Automated method for producing continuous reinforced plastic shapes by pulling preimpregnated reinforcing fibers through a heated die where the resin is cured.

purging—Cleaning one color or type of material from the cylinder of an injection molding machine or extruder by forcing it out with a new color or material to be used in subsequent production. Purging materials also are available.

quench (thermoplastics)—A process of shock cooling thermoplastic materials from the molten state.

quench bath—The cooling medium used to quench molten thermoplastic materials to the solid state.

quench-tank extrusion—Process in which the extruded film is cooled in a quench-water bath.

radio frequency (RF) preheating—A method of preheating molding materials to facilitate the molding operation and/or reduce the molding cycle. The frequencies most commonly used are between 10 and 100 mc/sec.

radio frequency welding—A method of welding thermoplastics using a radio frequency field to apply the necessary heat. Also known as high frequency welding.

ram—Rod or plunger that forces melted plastic through the barrel and into a mold.

ram travel—Distance a ram moves when operating a complete molding cycle.

reaction injection molding (RIM)—Process that involves the high-pressure impingement mixing of two (or more) reactive liquid components; after mixing, the liquid stream is injected into a closed mold at low pressure. The finished parts can be cellular or solid elastomers, with a wide range of hardness and modulus values. It is used especially with urethanes. Variations include reinforced reaction injection molding or RRIM (where reinforcements are injected along with the reacting chemicals and structural reaction injection molding) and SRIM (where a reinforcing mat is placed in the mold before injection).

reaming—A method used to trim and size plastic bottle finishes. A special rotating cutting tool trims the sealing surface smooth and simultaneously reams (bores) the bottle opening to the desired size.

reciprocating screw injection molding—A combination injection and plasticizing unit in which an extrusion device with a reciprocating screw is used to plasticize the material. Injection of material into a mold can take place by direct extrusion into the mold or by reciprocating the screw as an injection plunger, or by a combination of the two methods, when the screw serves as an injection plunger, this unit acts as a holding, measuring, and injection chamber. See **injection molding**.

recycle—Material from flash, trimmings, scrap, rejects, etc., that can be ground up or repelletized and fed back into the processing machine.

reinforced molding compound—A material reinforced with special fillers or fibers to meet specific requirements (glass, synthetic fibers, minerals, etc.).

reinforced plastics—Plastics with some strength properties greatly superior to those of the base resin, resulting from the presence of high-strength fillers embedded in the composition. The reinforcing fillers usually are fibers, fabrics, or mats made of fibers.

reinforcement—A strong inert material bound into a plastic to improve its strength, stiffness, and impact resistance. Reinforcements usually are long fibers of glass, sisal, cotton, etc., in woven or nonwoven form.

release agent—See **mold release**.

resin—Any of a class of solid or semisolid organic products of natural or synthetic origin, generally of high molecular weight with no definite melting point. Most resins are polymers.

resin, liquid—An organic, polymeric liquid that, when converted to its final state for use, becomes a solid.

resin pocket—An apparent accumulation of excess resin in a small localized section visible on cut edges of molded surfaces. Also called resin segregation.

resin transfer molding—A process whereby catalyzed resin is transferred into an enclosed mold into which a reinforcing mat has been placed.

resistivity—The ability of a material to resist the passage of an electrical current, either through its bulk or on a surface. The unit of volume resistivity is the ohm-cm; that of surface resistivity is the ohm.

restricted gate—A very small orifice between the runner and the cavity in an injection or transfer mold. When the piece is ejected, this gate breaks cleanly, simplifying separation of the runner from the piece.

restrictor bar—An extension into the flow channel of an extrusion sheet die at its widest point to equalize pressure and produce a balanced flow.

retainer plate—The plate on which demountable pieces, such as mold cavities, ejector pins, guide pins, and bushings, are mounted during molding; usually drilled for steam or water.

retarder—See **inhibitor**.

reverse-roll coating—Coating that is premetered between rolls and then wiped off on the web. The amount of coating is controlled by the metering gap and also by the speed of rotation of the coating roll.

rheology—The study of material flow under varying conditions of heat and pressure.

rib—A reinforcing member of a fabricated or molded part.

ring gate—An annular opening for introducing melt into a mold, used to make cylindrical parts.

Rockwell hardness—A common method of testing material for resistance to indentation, in which a diamond or steel ball, under pressure, is used to pierce the test specimen. Ref.: Standard Method of Test for Rockwell Hardness of Plastics and Electrical Insulating Materials (ASTM Designation: D 785).

roller coating—Method used to apply paints to raised designs or letters.

roll mill—Two rolls placed in close relationship to one another, used to admix a plastic material with other substances. The rolls turn at different speeds to provide a shearing action to the materials being compounded.

rotating spreader—A type of injection torpedo consisting of a finned torpedo that is rotated by a shaft extending through a tubular-cross-section injection ram behind it.

rotational casting (or **molding**)—A method used to make hollow articles from thermoplastic materials. Material is charged into a hollow mold capable of being rotated in one or two planes. The hot mold fuses the material into a gel after the rotation has caused it to cover all surfaces. The mold then is chilled and the product stripped out.

roving—A form of fibrous glass in which spun strands are woven into a tubular rope. The number of strands is variable, but 60 is usual. Chopped roving is commonly used in preforming.

rubber—Any elastomer capable of rapid elastic recovery after being stretched to at least twice its length at temperatures from 0°F to 150°F at any humidity. Specifically, Hevea or natural rubber is the standard of comparison for elastomers. See also **thermoplastic elastomers**.

runner (refers to mold)—In an injection or transfer mold, the channel that connects the sprue with the gate to the cavity.

runner system (refers to plastic)—The term usually applied to all the material in the form of sprues, runners, and gates that lead material from the nozzle of an injection machine or the pot of a transfer mold to the mold cavity.

rupture strength, psi—A quantity whose true value is the stress of a material at failure, based on the ruptured cross-sectional area itself.

S-glass—A magnesia-alumina-silica glass used in high-strength reinforcements.

SRIM—See **reaction injection molding**.

sag—The extension locally (often near the die face) of the parison during extrusion by gravitational forces. This causes necking-down of the parison. The term also refers to the flow of a molten sheet in a thermoforming operation.

sag streaks—Uneven plastic surface due to heavy coating application and poor flow-out. It can be eliminated by thinning the coating, adjusting the spray gun, or changing the stroke.

sandwich constructions—Panels composed of a lightweight core material (honeycomb, foamed plastic, etc.) to which two relatively thin, dense, high-strength faces or skins are adhered.

sandwich heating—A method of heating a thermoplastic sheet prior to forming in which both sides of the sheet are heated simultaneously.

scrap—Any product of a molding operation that is not part of the primary product. In compression molding, this includes flash, culls, and runners, and the material is not reusable as a molding compound. Injection molding and extrusion scrap (runners, rejected parts, sprues, etc.) usually can be reground and remolded.

screen—A woven metal screen or equivalent device that is installed across the flow of stock between the tip of the screw and the die and supported by a breaker plate to strain out contaminants or to increase the back pressure, or both.

screen changer—A device for replacing filtering screens without interrupting the extrusion process.

screw plasticating injection molding—A technique in which the plastic is converted from pellets to a viscous melt by means of an extruder screw that is an integral part of the molding machine. Machines are either single stage (in which plastication and injection are done in the same cylinder) or double stage (in which the material is plasticated in one cylinder and then fed to a second for injection into a mold).

segregation—A separation of components in a molded article, usually denoted by wavy lines and color striations in thermoplastics. In thermosets, it usually means segregation of the resin and the filler on the surface.

self-extinguishing—A term indicating that a material will stop burning after the source of flame is removed.

self-reinforcing—A term describing the development of strength in liquid crystal polymers because of their internal structure.

shear strength, psi—The stress at which a material fails under a shear loading condition. (ASTM test method D 732.)

shear stress—The stress developing in a polymer melt when the layers in a cross section are gliding along each other or along the wall of the channel (in laminar flow).

$$\text{shear stress} = \frac{\text{force}}{\text{area sheared}} = \text{psi}$$

sheet (thermoplastic)—A flat section of a thermoplastic resin with the length considerably greater than the width and 10 mils or greater in thickness.

shelf life—The time that a molding compound can be stored without losing any of its physical or molding properties.

Shore hardness—A method of determining the hardness of a plastic material using a scelroscope. This device consists of a small conical hammer fitted with a diamond point and acting in a glass tube. The hammer is made to strike the material under test, and the degree of rebound is noted on a graduated scale. Generally, the harder the material is, the greater the rebound will be. A single indentor, without hammer, can be used to obtain Shore A or Shore D durometer measurements. Ref.: Tentative Method of Test for Indentation Hardness of Rubber and Plastics by Means of a Durometer (ASTM Designation: D 2240).

short or **short shot**—A molded part produced when the mold has not been filled completely.

shot—The yield from one complete molding cycle, including cull, runner, and flash.

shot capacity—The maximum weight of material that a machine can produce from one forward motion of the plunger or screw.

shrink fixture—See **cooling fixture**.

shrink wrapping—A technique of packaging in which the strains in a plastic film are released by raising the temperature of the film, thus causing it to shrink over the package. These shrink characteristics are built into the film during its manufacture by stretching it under controlled temperatures to produce orientation of the molecules. Upon cooling, the film retains its stretched condition, but it reverts toward its original dimensions when it is heated.

shrinkage—See **mold shrinkage**.

side coring or **side draw pins**—Projections used to core a hole in a direction other than the line of closing of a mold, which must be withdrawn before the part is ejected from the mold.

silicone—Chemical derived from silica; used in molding as a release agent and a general lubricant.

silk screen printing—Printing method that, in its basic form, involves laying a pattern of an insoluble material, in outline, on a finely woven fabric, so that when ink is drawn across it, the ink passes through the screen only in the desired areas. Also called screen process decorating.

single cavity mold (injection)—An injection mold having only one cavity in the body of the mold, as opposed to a multiple cavity mold or family mold, which has numerous cavities.

sink mark—A depression or dimple on the surface of an injection-molded part due to collapsing of the surface following local internal shrinkage after the gate seals. It also may be an incipient short shot.

sintering—In forming articles from fusible powders (e.g., nylon), the process of holding the pressed-powder article at a temperature just below its melting point for about half an hour. Particles are fused (sintered) together, but the mass, as a whole, does not melt.

sizing—The process of applying a material to a surface to fill pores and thus reduce the absorption of the subsequently applied adhesive or coating or to otherwise modify the surface. Also, the surface treatment applied to glass fibers used in reinforced plastics, for improving the bond between glass and plastic. The material used sometimes is called size.

slip additive—A modifier that acts as an internal lubricant which exudes to the surface of the plastic during and immediately after processing. In other words, a nonvisible coating blooms to the surface to provide the necessary lubricity to reduce the coefficient of friction and thereby improve slip characteristics.

slip forming—Sheet-forming technique in which some of the plastic sheet material is allowed to slip through the mechanically operated clamping rings during a stretch-forming operation.

slot extrusion—A method of extruding film sheet in which the molten thermoplastic compound is forced through a straight slot.

slurry preforming—Method of preparing reinforced plastics preforms by wet processing techniques similar to those used in the pulp molding industry.

slush molding—Method for casting thermoplastics, in which the resin in liquid form is poured into a hot mold where a viscous skin forms. The excess slush is drained off, the mold is cooled, and the molding is stripped out.

softening range—The range of temperature in which a plastic changes from a rigid to a soft state. Actual values will depend on the method of test. Sometimes erroneously referred to as softening point.

solid phase forming—Using metalworking techniques to form thermoplastics in a solid phase. Procedure begins with a plastic blank that is then heated and fabricated (i.e., forged) by bulk deformation of the materials in constraining dies by application of force.

solvent—Any substance, usually a liquid, that dissolves other substances.

solvent molding—Process for forming thermoplastic articles by dipping a male mold in a solution or dispersion of the resin and drawing off the solvent to leave a layer of plastic film adhering to the mold.

specific gravity—The density (mass per unit volume) of any material divided by that of water. Ref.: Standard Methods of Test for Specific Gravity and Density of Plastics by Displacement (ASTM Designation: D 792).

spider—(1) In a molding press, the part of an ejector mechanism that operates the ejector pins. (2) In extrusion, a term used to denote the membranes supporting a mandrel within the head/die assembly.

spider gate—Multi-gating of a part through a system of radial runners from the sprue.

spin welding—A process of fusing two objects together by forcing them together while one of the pair is spinning, until frictional heat melts the interface. Spinning is then stopped and the pressure held until they are frozen together.

spinneret—An extrusion die with tiny holes for producing fibers.

spiral flow test—A method for determining the flow properties of a thermoplastic resin in which the resin flows along the path of a spiral cavity. The length of the material that flows into the cavity and its weight give a relative indication of the flow properties of the resin.

spiral mold cooling—A method of cooling injection molds or similar molds wherein the cooling medium flows through a spiral cavity in the body of the mold. In injection molds, the cooling medium is introduced at the center of the spiral, near the sprue section, as more heat is localized in this section.

splay marks—Lines found in part after molding, usually due to flow of material in mold. Sometimes called silver streaking.

split cavity—Cavity made in sections.

split-ring mold—A mold in which a split cavity block is assembled in a chase to permit the forming of undercuts in a molded piece. These parts are ejected from the mold and then separated from the piece.

spray coating—Coating usually accomplished on continuous webs by a set of reciprocating spray nozzles traveling laterally across the web as it moves.

sprayed metal mold—Mold made by spraying molten metal onto a master until a shell of predetermined thickness is achieved. Shell is then removed and backed up with plaster, cement, casting resin, or other suitable material. It is used primarily as a mold in sheet-forming processes.

spray-up—Term covering a number of techniques in which a spray gun is used as the processing tool. In reinforced plastics, for example, fibrous glass and resin can be simultaneously deposited in a mold. In essence, roving is fed through a chopper and ejected into a resin stream, which is directed at the mold by either of two spray systems. In foamed plastics, very fast-reacting urethane foams or epoxy foams are fed in liquid streams to the gun and sprayed on the surface. On contact, the liquid starts to foam.

spreader—A streamlined metal block placed in the path of flow of the plastics material in the heating cylinder of extruders and injection molding machines to spread it into thin layers, thus forcing it into intimate contact with the heating areas.

sprue—Feed opening provided in the injection or transfer mold; also, the slug formed at this hole. Spur is a shop term for the sprue slug.

sprue-bushing—A hardened steel insert in an injection mold that contains the tapered sprue hold and has a suitable seat for the nozzle of the injection cylinder. Sometimes called an adapter.

sprue gate—A passageway through which molten resin flows from the nozzle to the mold cavity.

sprue lock or **puller**—In injection molding, a portion of the plastic composition that is held in the cold slug well by an undercut; used to pull the sprue out of the bushing as the mold is opened. The sprue lock itself is pushed out of the mold by an ejector pin. When the undercut occurs on the cavity block retainer plate, this pin is called the sprue ejector pin.

spunbonded sheet—A nonwoven material made by heat-sealing webs of randomly arranged thermoplastic fibers.

stabilizer—An ingredient used in the formulation of some plastics to assist in maintaining the physical and chemical properties of the compounded materials at their initial values throughout the processing and service life of the material.

stack mold—A multilevel injection mold in which two (stacked) sets of cavities are filled simultaneously.

static mixer—A device that causes multiple splitting and rejoining of a stream of material as it passes through the device.

stationary platen—The plate of an injection or compression molding machine to which the front plate of the mold is secured during operation. This platen does not move during normal operation.

steam molding—Process used to mold parts from pre-expanded beads of plastic, especially polystyrene, using steam as a source of heat to expand the blowing agent in the material. The steam in most cases makes intimate contact with the beads directly, or it may be used indirectly to heat mold surfaces that are in contact with the beads.

strain—Elastic deformation caused by stress, measured as change in length per unit of length.

stress-crack—External or internal crack in a plastic caused by tensile stresses less than that of its short-time mechanical strength. The development of such cracks frequently is accelerated by the environment to which the plastic is exposed. The stresses that cause cracking may be present internally or externally or may be combinations of these stresses. The appearance of a network of fine cracks is called crazing.

stretch blow molding—A blow molding process in which the parison is stretched and blown at controlled temperature to cause orientation of the polymer molecules.

stretch film—Polyethylene or PVC film used to wrap pelletized loads. The film is capable of extensions of up to 30% and maintains tension on the load.

striation—A separation of colors resulting in a linear effect of color variation.

stripper-plate—A plate that strips a molded piece from core pins or force plugs. It is set into operation by the opening of the mold.

structural foams—Expanded plastics materials having integral skins and outstanding rigidity. Structural foams involve a variety of thermoplastics resins as well as urethanes.

submarine gate—A type of edge gate where the opening from the runner into the mold is located below the parting line or mold surface, as opposed to conventional edge gating where the opening is machined into the surface of the mold. With submarine gates, the item is broken from the runner system on ejection from the mold.

surface finish—Finish of molded product.

surface resistivity—The electrical resistance between opposite edges of a unit square of insulating material, commonly expressed in ohms. Ref.: Standard Methods of Test for D-C Resistance or Conductance of Insulating Materials (ASTM Designation: D 257).

surface treating—Any method of treating a material so as to alter the surface and render it receptive to inks, paints, lacquers, and adhesives, such as chemical, flame, and electronic treating.

surfacing veil—A very thin mat of glass fibers used to produce a smooth surface on reinforced plastic parts.

surging—Unstable pressure build-up in an extruder leading to variable throughput and waviness of the parison.

sweating—Exudation of small drops of liquid, usually a plasticizer or softener, on the surface of a plastic part.

synergism—Term applied to the use of two or more stabilizers in an organic material where the combination of such stabilizers improves the stability to a greater extent than could be expected from the additive effect of each stabilizer.

syntactic foam—Composite manufactured by incorporation of hollow spheres of glass or plastic into liquid resin systems before curing.

T-die—A center-fed, slot extrusion die for film that, in combination with the die adapter, resembles an inverted T.

tab gated—A small removable tab of approximately the same thickness as the mold item, usually located perpendicular to the item. The tab is used as a site for edge gate location, usually on items with large flat areas.

take-off—The mechanism for drawing extruded material away from the die.

talc—A natural, hydrous magnesium silicate used in plastics as a reinforcing filler.

temper—To reheat after hardening to some temperature below the critical temperature, followed by air cooling to obtain desired mechanical properties and to relieve hardening strains.

tensile bar—A compression- or injection-molded specimen of specified dimensions that is used to determine the tensile properties of a material.

tensile modulus—The ratio of stress to strain for a given material within its proportional limit under tensile loading conditions. (ASTM test method D 638.)

tensile strength—The pulling stress, in psi, required to break a given specimen. Area used in computing strength is usually the original rather than the necked-down area. Ref.: Tentative Method of Test for Tensile Properties of Plastics (ASTM Designation: D 638).

thermal conductivity—Ability of a material to conduct heat; physical constant for quantity of heat that passes through unit cube of a substance in unit of time when difference in temperature of two faces is 1°. Ref.: Standard Method of Test for Thermal Conductivity of Materials by Means of the Guarded Hot Plate (ASTM Designation: C 177).

thermal expansion—See **coefficient of expansion**.

thermal stress cracking (TSC)—Crazing and cracking of some thermoplastic resins that results from overexposure to elevated temperatures.

thermocouple—A temperature-sensitive device consisting of a pair of wires of dissimilar metal welded together at one end. The electric current created at different temperatures is measured by a calibrated potentiometer.

thermoelasticity—Rubberlike elasticity exhibited by a rigid plastic, resulting from an increase of temperature.

thermoforming—Any process of forming thermoplastic sheet that consists of heating the sheet and pulling it down onto a mold surface.

thermoforms—The product of a thermoforming operation.

thermogravimetric analysis—The measurement of changes in weight of a specimen as it is heated, either in air or in a vacuum.

thermoplastic—(1) Capable of being repeatedly softened by heat and hardened by cooling. (2) A material that will repeatedly soften when heated and harden when cooled. Typical of the thermoplastic family are the styrene polymers and copolymers, acrylics, cellulosics, polyethylenes, polypropylene, vinyls, nylons, and various fluorocarbon materials.

thermoplastic elastomers (or **thermoplastic rubbers**)—The family of polymers that display the rubberlike property of returning to their original shape after repeated stretching and the ability to be melted and processed like other thermoplastics.

thermoset—A material that will undergo or has undergone a chemical reaction by the action of heat and pressure, catalysts, ultraviolet light, etc., leading to a relatively infusible state. Typical of the plastics in the thermosetting family are the aminos (melamine and urea), most polyesters, alkyds, epoxies, and phenolics.

thixotropic—Term describing materials that are gel-like at rest but fluid when agitated. Liquids containing suspended solids are apt to be thixotropic.

thread plug, **ring**, or **core**—A part of a mold that shapes a thread and must be unscrewed from the finished piece.

thrust bearing—The bearing used to absorb the thrust force exerted by the screw.

tie rods or **beams**—Those members of the clamping unit that join and align the stationary platen with the clamping force actuating mechanism, and that serve as the tension members of the clamp when it is holding the mold closed.

toggle action—A mechanism that exerts pressure developed by the application of force on a knee joint. It is used as a method of closing presses and also serves to apply pressure at the same time.

toggle clamp (hydraulic actuated, **mechanical actuated)**—A clamping unit with a toggle mechanism directly connected to the moving platen. A hydraulic cylinder, or some mechanical force device, is connected to the toggle system to exert the opening and closing force and hold the mold closed during injection.

tolerance—A specified allowance for deviations in weighing, measuring, etc., or for deviations from the standard dimensions or weight. Ref.: SPI Standards and Practices of Plastics Custom Molders.

tracking—See **arc resistance**.

transducer—A force-measuring device providing a measurable electrical output proportional to the load applied.

transfer molding—A method of molding thermosetting materials in which the plastic is first softened by heat and pressure in a transfer chamber and then is forced by high pressure through suitable sprues, runners, and gates into a closed mold for final curing.

transition section of screw (compression section)—The portion of a screw between the feed section and the metering section in which the flight depth decreases in the direction of discharge.

translucent—Descriptive term for a material or substance capable of transmitting some light but not clear enough to be seen through.

transparent—Descriptive term for a material or substance capable of a high degree of light transmission (e.g., glass).

tubing—A particular size of plastics pipe in which the outside diameter is essentially the same as that of copper tubing. See **pipe**.

tumbling—Finishing operation for small plastic articles by which gates, flash, and fins are removed and/or surfaces are polished by rotating them in a barrel together with wooden pegs, sawdust, and polishing compounds.

tunnel gate—See **submarine gate**.

twin-screw extrusion—See **multiple-screw extruders**.

twin-sheet thermoforming—Technique for thermoforming hollow objects by introducing high-pressure air in between two sheets and blowing the sheets in the mold halves (vacuum is also applied).

ultrasonic insertion—The inserting of a metal insert into a thermoplastic part by the application of vibratory mechanical pressure at ultrasonic frequencies.

ultrasonic sealing, bonding, or **welding**—A method by which plastics are joined through the application of vibratory mechanical pressure at ultrasonic frequencies (20 to 40 kHz). Electrical energy is converted to ultrasonic vibrations through the use of either a magnetostrictive or a piezoelectric transducer. The vibratory pressures at the interface in the sealing area developed localized heat, which melts the plastic surfaces, effecting the seal.

ultraviolet (UV) stabilizer—A chemical additive that selectively absorbs or filters out light waves at the ultraviolet end of the spectrum, protecting plastics from their harmful effects (embrittlement, discoloration, crazing, and disintegration).

undercut—A protuberance or indentation that impedes withdrawal from a two-piece, rigid mold. Flexible materials can be ejected intact even with slight undercuts.

unit mold—Mold designed for quick changing of interchangeable cavity parts.

vacuum forming—Method of sheet forming in which the plastic sheet is clamped in a stationary frame, heated, and drawn down by a vacuum into a mold. In a loose sense, it sometimes is used to refer to all sheet-forming techniques, including drape forming involving the use of vacuum and stationary molds.

vacuum metallizing—Process in which surfaces are thinly coated with metal by exposing them to the vapor of metal that has been evaporated under vacuum (one-millionth of normal atmospheric pressure).

vent, venting—In molds, shallow channels or minute holes cut in the cavity to allow air to escape as the material enters. Also called breathers.

vented extruder—A piece of extrusion equipment with provision for the removal of volatiles, either through a port in the barrel (vented barrel) or through the screw core (vented screw).

venturi dispersion plug—A plate having an orifice with a conical relief drilled therein that is fitted in the nozzle of an injection molding machine to aid in the dispersion of colorants in a resin.

vibration welding—Assembly technique in which frictional heat is generated by pressing the surfaces of parts together and vibrating the parts through a small relative displacement, which displacement can be either linear or angular. Vibration welding machines operate at relatively low frequencies, 90 to 120 Hz.

vibratory feeder—Device comprising a tray or tube vibrated electrically or mechanically to convey dry materials to a processing machine.

Vicat softening temperature—Measurement of the heat distortion temperature of a plastic material. Also called the heat deformation point. Ref.: Tentative Method of Test for Vicat Softening Point of Plastics (ASTM Designation: D 1525).

virgin material—Plastic resin or compound that has not been processed except to prepare it for its intended use.

viscoelastic—A term referring to plastics that store energy and dissipate it during mechanical deformation. The concept explains the "flow" of plastic materials under stress.

viscosity—Internal friction or resistance to flow of a liquid; the constant ratio of shearing stress to rate of a shear. In liquids for which this ratio is a function of stress, the "apparent viscosity" is defined as this ratio.

void—A space or bubble occurring in the center of a heavy thermoplastic part, usually caused by excessive shrinkage.

volume resistivity—The electrical resistance between opposite faces of a 1-cm cube of insulating material. It is measured under prescribed conditions using a direct current potential after a specified time of electrification, and commonly is expressed in ohm-cm. Also called the specific insulation resistance. Ref.: Standard Methods of Test for D-C Resistance or Conductance of Insulating Materials (ASTM Designation: D 257).

warpage—Dimensional distortion in a plastic object after molding.

water absorption—The ability of a thermoplastic material to absorb water from an environment. Ref.: Standard Method of Test for Water Absorption of Plastics (ASTM Designation: D 570).

weatherometer—An instrument utilized to subject articles to accelerated weathering conditions (e.g., rich UV source and water spray).

web—A thin sheet in process in a machine. The molten web is that which issues from the die. The substrate web is the substrate being coated.

weld lines (also **weld marks** or **flow lines**)—Marks on a molded plastic piece made by the meeting of two flow fronts during the molding operation.

welding—Joining thermoplastic pieces by one of several heat-softening processes. In hot-gas welding, the material is heated by a jet of hot air or inert gas directed from a welding "torch" onto the area of contact of the surfaces that are being welded. Welding operations to which this method is applied normally require the use of a filler rod. In spin-welding (which see), the heat is generated by friction. Welding also includes heat sealing, and the terms are synonymous in some foreign countries.

wet-out—Condition of a reinforcement or other porous material where all voids are completely filled with resin.

wet strength—The strength of paper when saturated with water, especially used in discussions of processes whereby the strength of paper is increased by the addition, in manufacture, of plastic resins. Also, the strength of an adhesive joint determined immediately after removal from a liquid in which it has been immersed under specific conditions of time, temperature, and pressure.

wetting agent—Compound that allows a liquid to penetrate into or spread over the surface of another material more easily.

wheelabrating—Deflashing molded parts by bombarding them with small particles at a high velocity.

window—A defect in a thermoplastic film, sheet, or molding, caused by the incomplete "plasticization" of a piece of material during processing. It appears as a globule in an otherwise blended mass. See also **fisheye**.

working life—The period of time during which a liquid resin or adhesive, after mixing with catalyst, solvent, or other compounding ingredients, remains usable.

yellowness index—A measure of the tendency of plastics to turn yellow upon exposure to heat or light.

yield value (yield strength)—The lowest stress at which a material undergoes plastic deformation. Below this stress, the material is elastic; above it, the material is viscous.

Young's modulus of elasticity—The modulus of elasticity in tension; the ratio of stress in a material subjected to deformation.

2
Polymer Chemistry

The chemical structure and the nature of plastics materials have a significant relationship not only to the properties of the plastic but to the ways in which it can be processed, designed, or otherwise translated into an end-product. Throughout this book the reader will find various references to polymer chemistry (e.g., the section entitled "Theory of Injection Molding," in Chapter 5 on "Injection Molding").

This chapter therefore will provide a basic review of polymer chemistry, with emphasis on the distinctions between various structures and their influence on the engineering of plastics products.

Forming Polymers

Basically, all polymers are formed by the creation of chemical linkages between relatively small molecules or monomers to form very large molecules or polymers; the same idea as connecting boxcars on a railroad to form a train, the boxcars being monomers and the train a polymer. Like boxcars, the molecules must have the ability to be coupled at either end.

Actually, in polymer formation, the process is more like forming many, many trains in a railroad yard simultaneously from the boxcars available in the yard in a competitive fashion, so that the switching engine that moved the fastest would form the longest train while the slowest switching engine would form the shortest train, due to depletion of the available rolling stock by the concurrent train-forming process. The train-forming process of polymerization comes to a stop when factors prevent any additional boxcars from being added to any of the trains being assembled. Thus, we ultimately end up with trains having a variety of lengths, yet all composed of the same kinds of boxcars.

The railroad analogy described above is basically what happens in an "addition" type of polymerization.

The above process is characterized by the simple combination of molecules without the generation of any by-products formed as a result of the combination. The molecules that combine do not decompose to produce fission products which then remain as part of the reaction debris or need to be removed from the reaction to either allow it to continue the molecule-building process or to ensure the formation of a pure polymer.

In reality the addition-type process can occur in several ways. One way simply involves the external chemical activation of molecules that causes them to start combining with each other in a chain-reaction-type fashion (by the bonding of atoms directly within the reacting molecule). Another way for an addition polymerization to occur is through a rearrangement of atoms within both reacting molecules, but still without the net loss of any atoms from the polymer molecule. And still a third way for addition polymerization to occur is for a molecule composed of a ring of atoms to open up and connect with other ring-type molecules being opened up under the influence of the proper catalytic activators, once again with no net loss of any atoms from the polymer structure.

In another type of polymerization reaction, which has been called "condensation" polymerization, the chemical union of two molecules can be achieved only by the splitting out of a molecule (usually small) formed by the atoms, which must be removed from the two molecules being joined to allow the coupling process or polymerization to continue. This is the type of polymerization involved in the formation of some nylons, phenolics, amino resins, and polyester pre-polymers.

Normally the reaction by-product in a condensation type of polymerization must be immediately removed from the reacting polymer because it may either inhibit further polymerization or appear as an undesirable impurity in the finished polymer.

There is yet another method by which polymers may be formed, but it is in reality simply a sequential combination of the previous two processes. Such a process is used in the formation of plastics such as the polyesters and the polyurethanes.

In such a polymerization, a condensation reaction usually is carried out first to form a relatively small polymer, which is then capable of undergoing further reaction by addition polymerization to form larger polymer molecules with a third ingredient. This is what is done when a polyester is formed first by a condensation reaction, and the then still-active polyester is reacted with styrene to form what is essentially a polyester–styrene copolymer.

The various types of polymerization reactions are shown in Figs. 2-1a, 2-1b, and 2-1c.

Catalyst activated bond opening (ethylene polymerization)

$$2CH_2{=}CH_2 + R{-}R \rightarrow 2R{-}\left[CH_2{-}CH_2\right]{-}$$

Initial reaction

$$R{-}\left[CH_2{-}CH_2\right]{-} + (n-1)CH_2{=}CH_2 \rightarrow$$

$$R{-}\left[CH_2{-}CH_2\right]_n{-}$$

Propagation reaction

$$R{-}\left[CH_2{-}CH_2\right]_n{-} + R{-} \rightarrow$$

$$R{-}\left[CH_2{-}CH_2\right]_n{-}R$$

Termination reaction (combination)

Rearrangement (polyurethane polymerization)

$$nO{=}C{=}N{-}R{-}N{=}C{=}O + nH{-}O{-}R{-}OH \rightarrow$$

$$O{=}C{=}N{-}\left[R{-}\overset{H}{\overset{|}{N}}{-}\overset{O}{\overset{\|}{C}}{-}O{-}R\right]_n{-}OH$$

Ring-opening reaction (nylon 6 from caprolactam)

$$nH{-}N{-}(CH_2)_5{-}C{=}O \rightarrow$$

$${-}\left[\overset{H}{\overset{|}{N}}{-}(CH_2)_5{-}\overset{O}{\overset{\|}{C}}\right]_n{-}$$

Fig. 2-1a. Typical addition polymerizations (no by-products).

Polymerization Techniques

In actual practice there are many different techniques used to carry out polymerization reactions. However, most involve one of four general methods of polymerization: the polymerization of the monomer or reactants in bulk, in solution, in suspension, and in emulsion forms. The bulk and solution methods are used for the formation of both addition and condensation type polymers, whereas suspension and emulsion techniques are used largely for addition polymerizations.

Bulk Polymerization. This type of polymerization involves the reaction of monomers or reactants among themselves without placing them in some form of extraneous media such as is done in the other types of polymerizations.

Two types of behavior are observed in bulk polymerizations. In one case, the polymer is soluble in the monomer during all stages of the polymerization, and a monomer-soluble initiator is used. As polymerization progresses, viscosity increases significantly, and chain growth takes place in the monomer or polymers dissolved in the monomer until all of the monomer is consumed.

In the second case, the polymer is insoluble in the monomer system. In such systems, the polymerization is believed to occur within the growing polymer chains because very high molecular weights are formed even though the polymer chain drops out of the monomer solution.

Phenol-aldehyde reaction

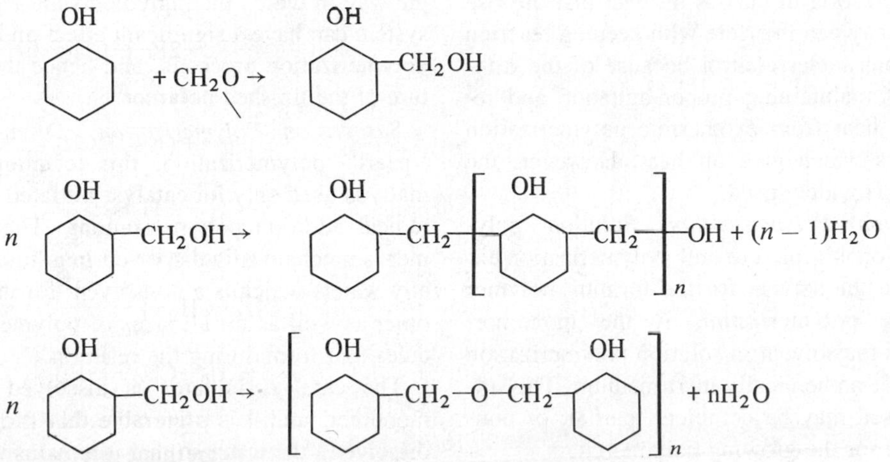

$$C_6H_5OH + CH_2O \rightarrow HOC_6H_4CH_2OH$$

$$n\ HOC_6H_4CH_2OH \rightarrow HOC_6H_4CH_2-[C_6H_3(OH)-CH_2]_n-OH + (n-1)H_2O$$

or

$$n\ HOC_6H_4CH_2OH \rightarrow [HOC_6H_4-CH_2-O-CH_2-C_6H_4OH]_n + nH_2O$$

Polyesterification (reaction between organic acids and alcohols)

$$n\ HO{-}R{-}COOH \rightarrow HO{-}[R{-}C(=O){-}O]_n{-}H + (n-1)H_2O$$

or

$$n\ HO{-}R{-}OH + n\ HOOC{-}R'{-}COOH \rightarrow HO{-}[R{-}O{-}C(=O){-}R'{-}C(=O){-}O]_n{-}H + (n-1)H_2O$$

Reaction when byproduct is other than water (polycarbonate)

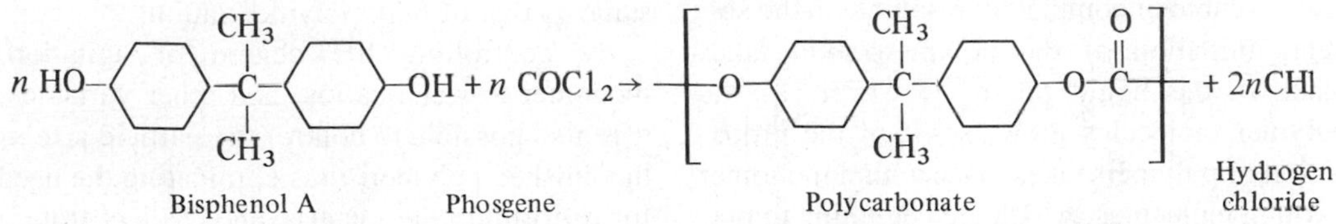

$$n\ HO{-}C_6H_4{-}C(CH_3)_2{-}C_6H_4{-}OH + n\ COCl_2 \rightarrow [{-}O{-}C_6H_4{-}C(CH_3)_2{-}C_6H_4{-}O{-}C(=O){-}]_n + 2nCHl$$

Bisphenol A — Phosgene — Polycarbonate — Hydrogen chloride

Fig. 2-1b. Typical condensation polymerizations (production of by-product).

Step 1: Condensation reaction

$$n\ HO{-}R{-}OH + n\ HOOC{-}CH{=}CH{-}COOH \rightarrow n\ HO{-}[R{-}O{-}C(=O){-}CH{=}CH{-}C(=O){-}O]{-}H + n\ H_2O$$

Step 2: Addition reaction

$$n\ HO{-}R{-}O{-}C(=O){-}CH{=}CH{-}C(=O){-}O{-}H + n\ CH_2{=}CHC_6H_6 \rightarrow [HO{-}R{-}O{-}C(=O){-}CH{-}CH(CH_2{-}CH{-}){-}C(=O){-}OH]_n$$

Fig. 2-1c. Combination polymerizations (curing of polyesters).

One of the disadvantages of carrying out a polymerization in bulk is the fact that the rise in viscosity can interfere with keeping reaction conditions under control because of the difficulty of maintaining proper agitation and removing heat from exothermic polymerization reactions which give off heat. However, the process is widely used.

Solution Polymerization. Solution polymerization is similar to bulk polymerization except that the solvent for the forming polymer in bulk polymerization is the monomer, whereas the solvent in solution polymerization is usually a chemically inert medium. The solvents used may be complete, partial, or nonsolvents for the growing polymer chain.

When the monomer and the polymer are both soluble in the solvent, initiation and propagation of the growing polymer chains take place in the oil or organic phase. Because of the mass-action law, rates of polymerization in solvents are slower than in bulk polymerizations, and the molecular weight of the polymers formed is decreased.

In another case, when the monomer is soluble in the solvent but the polymer is only partially soluble or completely insoluble in the solvent, initiation of the polymerization takes place in the liquid phase. However, as the polymer molecules grow, some of the propagation of polymers takes place within monomer swollen molecules, which are beginning to precipitate from the reaction. When this occurs, it again becomes possible to build up molecular weights because of the decreased dilution within the polymers. Thus, molecular weights as high as those possible with bulk polymerizations can also be achieved in solution polymerizations, provided the polymer precipitates out of solution as it is formed and creates a propagation site.

In the third case, in which the polymer is completely insoluble in the solvent and the monomer is only partially soluble in the solvent, rates of reaction are reduced, and lower molecular weights, below those possible in bulk polymerizations, are formed. However, the formation of relatively high molecular weight polymers is still possible in such a system.

In addition to the relative solubilities of monomer, polymer, and solvent in the system, the way in which the ingredients are fed to the system can have a significant effect on how the polymerization proceeds, and hence the structure of the finished polymer.

Suspension Polymerization. Often called "pearl" polymerization, this technique normally is used only for catalyst-initiated or free-radical addition polymerizations. The monomer is mechanically dispersed in a liquid, usually water, which is a nonsolvent for the monomer as well as for all sizes of polymer molecules that form during the reaction.

The catalyst initiator is dissolved in the monomer, and it is preferable that it does not dissolve in the water so that it remains with the monomer. The monomer and polymer being formed from it remain within the beads of organic material dispersed in the phase. Actually suspension polymerization is essentially a finely divided form of bulk polymerization. The advantage of suspension polymerization over the bulk type is that it allows the operator to effectively cool exothermic polymerization reactions and thus maintain closer control over the chain building process. Other behavior is the same as that of bulk polymerization.

By controlling the degree of agitation, monomer-to-water ratios, and other variables, it is also possible to control the particle size of the finished polymer, thus eliminating the need for re-forming the material into pellets from a melt, as is usually necessary with bulk polymerizations.

Emulsion Polymerization. This is a technique in which addition polymerizations are carried out in a water medium containing an emulsifier (a soap) and a water-soluble initiator. It is used because emulsion polymerization is much more rapid than bulk or solution polymerization at the same temperatures and produces polymers with molecular weights much greater than those obtained at the same rate in bulk polymerizations. The polymerization reaction in emulsion polymerization involves causing the reaction to take place within a small hollow sphere composed of a film of soap molecules, called a *micelle*. Monomer diffuses into these micelles, and control of the soap concentration, overall reaction-mass recipe, and reaction conditions provides additional controls over the reaction.

Polymerization techniques can have a significant effect on the number, size, and characteristics of the polymer molecules formed and thus will have a significant effect on the properties of the polymer. Thus, batches of a polymer such as polystyrene, which can be made by any of the four polymerization techniques described above, will differ, depending on which type of polymerization method was used to make the material.

Weight, Size, and Distribution of Molecules

Because there is such diversity among polymer molecules, a number of techniques for defining and quantifying these characteristics are in use by the industry—and are also of value to processors and end-users as a determinant of polymer properties.

One such parameter relates to the size of the molecules in the polymer and is known as *molecular weight* (MW). MW refers to the average weight of the molecules in the mixture of different size molecules that make up the polymer. It is expressed either as a number average, based on the sum of the number fractions of the weight of each species or size of molecule present, or as a weight average, based on the weight fractions of each species or size of molecule present in the polymer.

The molecular weight of a polymer has a significant effect on its properties. Thus, higher molecular weight polymers tend to be tougher and more chemically resistant, whereas low molecular weight polymers tend to be weaker and more brittle. In the polyethylene family, for example, low molecular weight polyethylenes are almost waxlike in characteristics, whereas ultra-high molecular weight polyethylenes offer outstanding chemical resistance and toughness (although, conversely, the higher the molecular weight is, the more energy in the form of temperature and pressure required to process the material.)

Another expression applicable to molecular weight is the *degree of polymerization* (DP). This refers to the number of monomer molecules that combine to form a single polymer molecule and is estimated by dividing the number-average molecular weight of the polymer by the molecular weight of the monomer. The relationship can be expressed as:

$$\text{MW of polymer} = \text{DP} \times \text{MW of monomer}$$

And, finally, it is important to know something about the molecular weight distribution within the polymer, that is, the relative proportions of molecules of different weight. Obviously, if one could create a mono-disperse polymer, all of the molecules would be of a single size. This has not been achieved commercially, however, and so another parameter of definition used to describe polymers is the distribution of the various sizes of molecules within a poly-disperse polymer, that is, the breadth of distribution or ratio of large, medium, and small molecular chains in the resin. If the resin is made up of chains close to the average length, it is called narrow; if it is made up of chains of a wider variety of lengths, it is called broad.

In general, narrow-range resins have better mechanical properties than broad-range resins, although, as with the case of the higher molecular weight materials, they are somewhat more difficult to process.

The molecular weights of polymers and the molecular weight distribution are determined indirectly by measuring the properties of the polymers themselves or of their solutions and correlating this information with the type of molecular weight believed to correspond to the type of property measured.

For example, chemical analysis of the end-groups present in polymer molecules, studies of the boiling points and freezing points of solutions of polymers, and osmotic pressure studies on polymer solutions yield data on the number-average molecular weight of the polymer. Light scattering in polymer solutions and sedimentation methods in the ultra-centrifuge yield data related to the weight-average molecular weights of polymers. Such methods yield the overall average molecular weight of the polymer samples. To get direct data on molecular weight distributions, the polymers first must be separated by fractionation methods into rather sharp fractions of samples of relatively mono-disperse molecular weight.

Polymer Structure

In addition to the size of molecules and the distribution of sizes in a polymer, the shapes or structures of individual polymer molecules also play a major role in determining the properties of a plastic.

Earlier, in the discussion of the analogy between the connecting of railroad boxcars to form a train and the formation of polymers, it was implied that polymers form by aligning themselves into long chains of molecules without any side protrusions or branches, or lateral connections, between molecules. Some polymers do largely this and nothing more; however, it is also possible for polymers to form more complex structures. Thus, polymer molecules may form in the shape of branched molecules, in the form of giant three-dimensional networks, in the form of linear molecules with regular lateral connections to form "ladder-type" polymers, in the form of two-dimensional networks or platelets, and so forth, depending on how many connections or bonds can exist between the mono-disperse monomeric molecules that were used to form the polymer and between sites on the forming or already formed polymer molecules. See Fig. 2-2.

Because of the geometry of such molecules, some can come closer together than others in which the structure prevents more intimate contact. The structural obstruction to close approach is called *steric hindrance*. Thus, polymers that can be packed closely or exhibit little steric hindrance ordinarily can more easily form crystalline structures in which the molecules are aligned in some regular pattern; others, such as polymers that are cross-linked prior to crystallization, are prohibited from aligning themselves in crystals because of the hindrance created by the multiple interconnections and hence tend to be amorphous, or noncrystalline.

Amorphous polymers do not have melting points, but rather softening ranges; are normally transparent; and undergo only small volume changes when solidifying from the melt, or when the solid softens and becomes fluid. Crystalline polymers, on the other hand, have considerable order to the molecules in the solid state, indicating that many of the other atoms are regularly spaced; have a true melting point with a latent heat of fusion associated with the melting and freezing process; and have a relatively large volume change during the transition from melt to solid.

Thus, many different structures are possible with plastics—and each will affect the basic properties of the polymer. For example, linear polymers, such as high-density polyethylene, are made of molecules that resemble spaghetti in a bowl and are relatively free to slide over one another or to pack more closely together (in the absence of steric hindrances due to large pendant side groups). Branched polymers, such as low-density polyethylene, on the other hand, have side appendages and interconnections that cause the molecules to resemble clumps of tree branches that cannot be easily compressed or compacted. Thus, branched polymers (with more voids) are more permeable to gases and solvents than linear polymers, lower in density (since the molecules are not compacted together), and more flexible. Linear polymers, on the other hand, are higher in tensile, stiffness, and softening temperatures.

Cross-linked structures, in which the individual chain segments are strongly bound together by chemical unions, also have special characteristics (as in the family of thermosetting plastics). They do not exhibit creep or relaxation unless such primary bonds actually are broken by continually applied stress or by elevated temperatures high enough to cause chemical decomposition of the polymer. Cross-linked polymers also are fairly resistant to solvent attack; solvents may swell such polymers, but seldom cause complete rupture or dissolution.

Ladder structures have unusual stability and have become important in terms of the new heat-resistant plastics. Aromatic compounds (such as benzene) and heterocyclic compounds (such as benzimidazole) that have semiladder, ladder, or spiro structures offer heat stabilities in excess of 900°F.

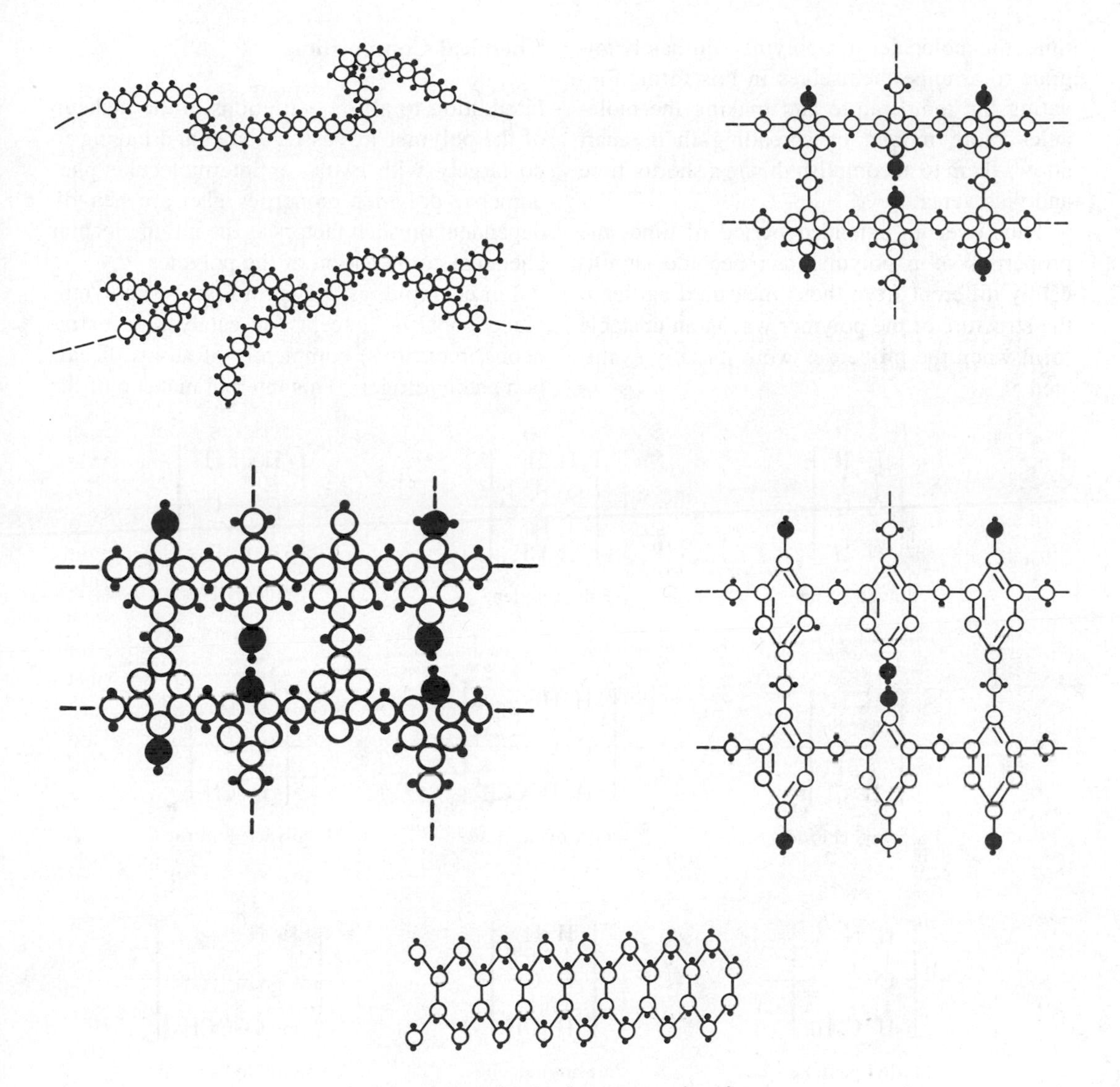

Fig. 2-2. Types of molecular structures in polymers.

Effect of Time and Temperature

It should also be noted that whereas the effect of molecular weight and molecular weight distribution on properties is relatively fixed and stable with temperature (barring decomposition of the polymer), the arrangement of the molecules within the structure of a polymer mass is in most cases relatively sensitive to temperature. Thus, the structure of any given polymer can be significantly changed by exposing it to different temperatures and thermal treatments.

For example, heating a crystalline-type polymer above its melting point and then quenching it can produce a polymer that is far more amorphous or noncrystalline in structure than the original polymeric sample. Such a quenched material can have properties that are significantly different from the properties of a sample that is cooled slowly and allowed to recrystallize.

The effects of time on a polymer structure are similar to those of temperature in the sense that any given polymer has a "most preferred" or equilibrium structure in which it would prefer to arrange itself, but it is prevented from doing so instantaneously on short notice by steric hindrances. However, given enough time, the molecules in a polymer ultimately migrate to arrange themselves in this form. Elevating the temperature and making the molecules more mobile or spreading them apart allows them to accomplish this in a shorter time and vice versa.

Thus over an extended period of time, the properties of a polymer can become significantly different from those measured earlier if the structure of the polymer was in an unstable form when the properties were initially evaluated.

Chemical Composition

In addition to all the variations in the makeup of the polymer discussed above and having to do largely with extra- or intermolecular phenomena, polymer properties also are heavily dependent on such factors as the intramolecular chemical composition of the polymer

For example, as shown in Fig. 2-3, polyethylene consists (except for catalyst or extraneous impurities) completely of atoms of carbon and hydrogen. This internal makeup of the

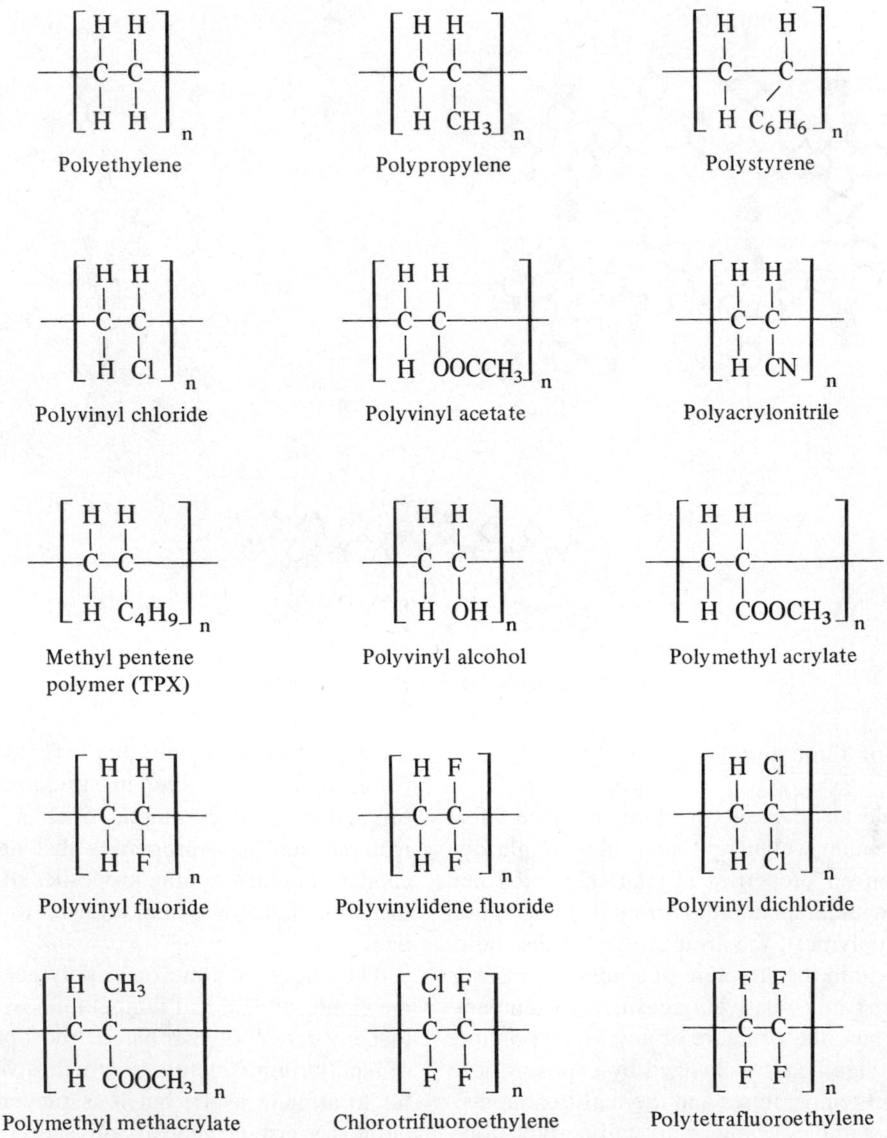

$$\left[-\overset{H}{\underset{H}{C}}-\overset{H}{\underset{H}{C}}-\right]_n$$

Polyethylene

$$\left[-\overset{H}{\underset{H}{C}}-\overset{H}{\underset{CH_3}{C}}-\right]_n$$

Polypropylene

$$\left[-\overset{H}{\underset{H}{C}}-\overset{H}{\underset{C_6H_6}{C}}-\right]_n$$

Polystyrene

$$\left[-\overset{H}{\underset{H}{C}}-\overset{H}{\underset{Cl}{C}}-\right]_n$$

Polyvinyl chloride

$$\left[-\overset{H}{\underset{H}{C}}-\overset{H}{\underset{OOCCH_3}{C}}-\right]_n$$

Polyvinyl acetate

$$\left[-\overset{H}{\underset{H}{C}}-\overset{H}{\underset{CN}{C}}-\right]_n$$

Polyacrylonitrile

$$\left[-\overset{H}{\underset{H}{C}}-\overset{H}{\underset{C_4H_9}{C}}-\right]_n$$

Methyl pentene polymer (TPX)

$$\left[-\overset{H}{\underset{H}{C}}-\overset{H}{\underset{OH}{C}}-\right]_n$$

Polyvinyl alcohol

$$\left[-\overset{H}{\underset{H}{C}}-\overset{H}{\underset{COOCH_3}{C}}-\right]_n$$

Polymethyl acrylate

$$\left[-\overset{H}{\underset{H}{C}}-\overset{H}{\underset{F}{C}}-\right]_n$$

Polyvinyl fluoride

$$\left[-\overset{H}{\underset{H}{C}}-\overset{F}{\underset{F}{C}}-\right]_n$$

Polyvinylidene fluoride

$$\left[-\overset{H}{\underset{H}{C}}-\overset{Cl}{\underset{Cl}{C}}-\right]_n$$

Polyvinyl dichloride

$$\left[-\overset{H}{\underset{H}{C}}-\overset{CH_3}{\underset{COOCH_3}{C}}-\right]_n$$

Polymethyl methacrylate

$$\left[-\overset{Cl}{\underset{F}{C}}-\overset{F}{\underset{F}{C}}-\right]_n$$

Chlorotrifluoroethylene

$$\left[-\overset{F}{\underset{F}{C}}-\overset{F}{\underset{F}{C}}-\right]_n$$

Polytetrafluoroethylene

Fig. 2-3. Polymers based on the ethylene chain.

polymer molecule affects in turn all the previously discussed variables and hence contributes its own basic characteristics to the overall properties of the polymer.

Going further, if one takes every fourth hydrogen atom occurring in polyethylene and replaces it with a methyl (CH_3) group in regular intervals along the length of the chainlike molecule as shown in Fig. 2-3, the polyethylene is transformed into polypropylene. In such a case, the degree of polymerization and the molecular weight distribution remain roughly the same; but the spacing of the molecules in the polymer matrix or the morphology of the polymer changes, and thus its macroscopic physical properties also change.

Similarly, if every fourth hydrogen atom of the polyethylene were substituted with a benzene ring structure (C_6H_6), or with a chlorine atom (Cl), as shown in Fig. 2-3, the ethylene would become either polystyrene or polyvinyl chloride, respectively.

The simple substitution of a single hydrogen atom with another atom or chemical group can cause a drastic change in the properties of a polymer. Whereas polyethylene is translucent, flexible, and crystalline, polystyrene is transparent, brittle, and amorphous.

Copolymers

In addition to making changes in the basic repeating unit by substitution, as illustrated above, it is possible to change the chemical composition and hence the morphology and properties of a polymer by mixing the types of structural groups or basic repeating units within the chain of a polymer. This is done by a process called *copolymerization*. In such a reaction, for example, monomers of styrene and acrylonitrile can react to form styrene-acrylonitrile copolymers; or styrene may react with butadiene to form styrene-butadiene copolymers; or acrylonitrile, butadiene, and styrene may all react simultaneously to form the ABS (acrylonitrile-butadiene-styrene) copolymers. Any of the polymerization techniques previously discussed can be used.

In addition to varying the types of starting ingredients used to form copolymers, the relative amounts of each monomer used in the reaction also may be varied to produce a literally unlimited number of possible permutations and combinations of types and amounts of monomers. The ability of the polymer chemist to react different monomers together in different amounts has given rise to the concept of "tailor-making" plastic materials.

Figure 2-4 illustrates some of the more common copolymers now available commercially. The subscripts of x, y, and z denote differing amounts of each of the structural units in the polymer chain. It should be noted that the structural formulas in Fig. 2-4 are somewhat fictional, in the sense that there is an implication that a long chain of one group is connected to a long chain of another group, and successively to a long chain of still a third type of polymer. Actually, the distribution of each of the species within the chain will depend both on the amount present and how the polymerization was carried out.

For example, an acrylonitrile-styrene copolymer might be made by first reacting a good portion of the acrylonitrile to form polyacrylonitrile and then adding the styrene and remaining acrylonitrile to complete the polymerization of the desired end product. Or, in another case, both ingredients might be added to a reactor simultaneously and allowed to react concurrently.

Other Polymers

From a study of Figs. 2-3 and 2-4, it will be noted that all of the polymers listed there are formed by combinations made possibly by reaction of the double bonds that exist between adjacent carbon atoms in the monomers used as starting materials. However, as mentioned earlier, this is not the only way in which polymers are formed. When polymers are created by rearrangements or condensation reactions, other types of polymeric chains can be formed. Figure 2-5 lists the structural formulas of several commercially important polymers in which oxygen (O) is an integral part of the bonds holding the polymer together. Similarly, Fig. 2-6 lists several types of polyamides and a polyurethane in which nitrogen (N) atoms form

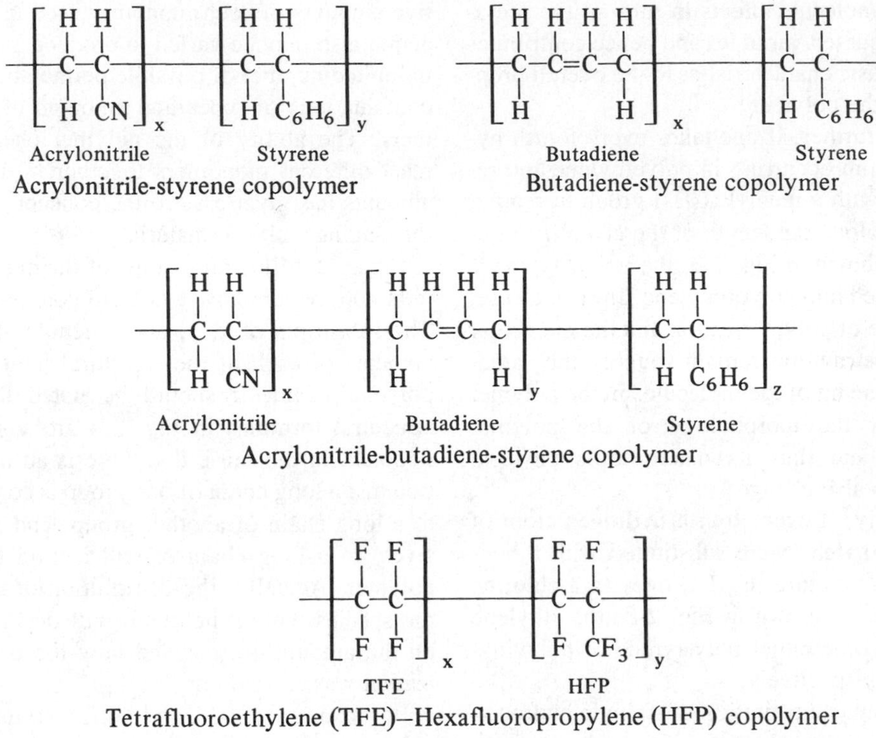

$$-\left[\begin{matrix}H & H\\ | & |\\ C- & C\\ | & |\\ H & CN\end{matrix}\right]_x-\left[\begin{matrix}H & H\\ | & |\\ C- & C\\ | & |\\ H & C_6H_6\end{matrix}\right]_y-$$

Acrylonitrile Styrene

Acrylonitrile-styrene copolymer

$$-\left[\begin{matrix}H & H & H & H\\ | & | & | & |\\ C- & C= & C- & C\\ | & & & |\\ H & & & H\end{matrix}\right]_x-\left[\begin{matrix}H & H\\ | & |\\ C- & C\\ | & |\\ H & C_6H_6\end{matrix}\right]_y-$$

Butadiene Styrene

Butadiene-styrene copolymer

$$-\left[\begin{matrix}H & H\\ | & |\\ C- & C\\ | & |\\ H & CN\end{matrix}\right]_x-\left[\begin{matrix}H & H & H & H\\ | & | & | & |\\ C- & C= & C- & C\\ | & & & |\\ H & & & H\end{matrix}\right]_y-\left[\begin{matrix}H & H\\ | & |\\ C- & C\\ | & |\\ H & C_6H_6\end{matrix}\right]_z-$$

Acrylonitrile Butadiene Styrene

Acrylonitrile-butadiene-styrene copolymer

$$-\left[\begin{matrix}F & F\\ | & |\\ C- & C\\ | & |\\ F & F\end{matrix}\right]_x-\left[\begin{matrix}F & F\\ | & |\\ C- & C\\ | & |\\ F & CF_3\end{matrix}\right]_y-$$

TFE HFP

Tetrafluoroethylene (TFE)–Hexafluoropropylene (HFP) copolymer

Fig. 2-4. Carbon "backbone" copoloymers.

the bonds between portions of the polymer molecule. In some cases the polymers are formed by molecular rearrangements (in the case of polyurethanes and epoxies) and in other cases by condensation reactions (some nylons).

Other condensation polymers are illustrated in Fig. 2-7, which shows some of the complex structures proposed for the more common types of thermosets. Note that the thermosets always have more than two linkages connecting the various structural units in the polymer, and they are commonly referred to as cross-linked materials involving a networklike molecular structure.

There are at least 30 to 40 different families of thermoplastics now offered and about 10 different basic families of thermosets. And this takes into account only the more common types of polymers. It should also be noted that copolymers, mixtures, and chemically modified versions of the polymers listed make the total number of polymeric materials that are theoretically possible entirely too great to list them in any one place.

In addition, the properties of a polymer may be radically altered by mixing it with nonpolymeric materials and chemicals. In fact, some polymers would be worthless to the molder or extruder if additives were not used to modify them before they were processed into finished products.

For example, phenolic, urea, and melamine resins, as they come out of the polymerization kettle, are largely brittle, frangible solids with low impact strength. To make them usable, such materials must be mixed with a filler of some kind to reinforce the strength of the polymer. In the case of the three thermosets mentioned, cellulose fillers made from wood flour or cotton are necessary to make the resins commercially moldable. Similarly, cellulosics and vinyl resins must also be mixed with semisolvents, called *plasticizers*, to soften them to some degree (so they may be molded or extruded) or to modify their properties (to make them soft and pliable, rather than hard and stiff).

In actual practice, the number of different

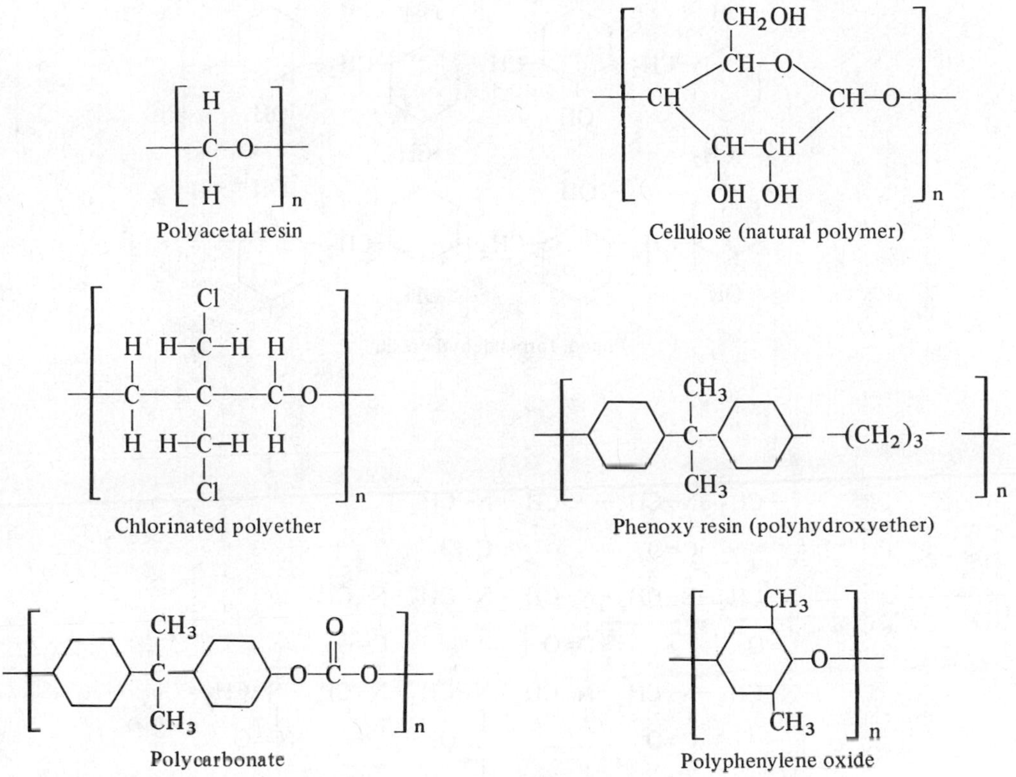

$$\left[-\overset{\displaystyle H}{\underset{\displaystyle H}{\overset{|}{\underset{|}{C}}}}-O- \right]_n$$

Polyacetal resin

Cellulose (natural polymer)

Chlorinated polyether

Phenoxy resin (polyhydroxyether)

Polycarbonate

Polyphenylene oxide

Fig. 2-5. Polymers with oxygen in the chain.

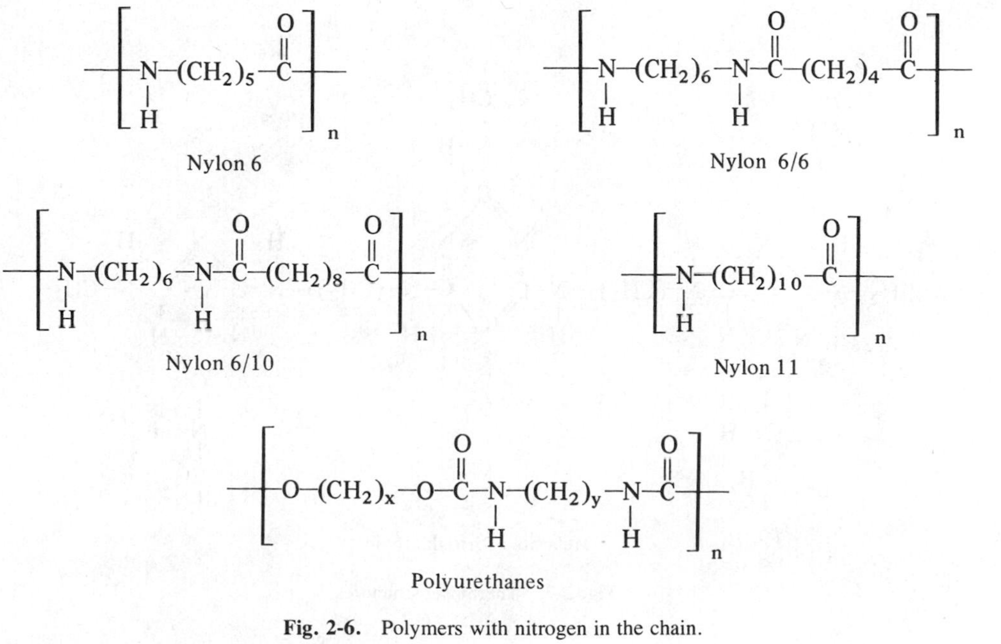

$$\left[-\underset{\displaystyle H}{\underset{|}{N}}-(CH_2)_5-\overset{\displaystyle O}{\overset{\|}{C}}- \right]_n$$

Nylon 6

$$\left[-\underset{\displaystyle H}{\underset{|}{N}}-(CH_2)_6-\underset{\displaystyle H}{\underset{|}{N}}-\overset{\displaystyle O}{\overset{\|}{C}}-(CH_2)_4-\overset{\displaystyle O}{\overset{\|}{C}}- \right]_n$$

Nylon 6/6

$$\left[-\underset{\displaystyle H}{\underset{|}{N}}-(CH_2)_6-\underset{\displaystyle H}{\underset{|}{N}}-\overset{\displaystyle O}{\overset{\|}{C}}-(CH_2)_8-\overset{\displaystyle O}{\overset{\|}{C}}- \right]_n$$

Nylon 6/10

$$\left[-\underset{\displaystyle H}{\underset{|}{N}}-(CH_2)_{10}-\overset{\displaystyle O}{\overset{\|}{C}}- \right]_n$$

Nylon 11

$$\left[-O-(CH_2)_x-O-\overset{\displaystyle O}{\overset{\|}{C}}-\underset{\displaystyle H}{\underset{|}{N}}-(CH_2)_y-\underset{\displaystyle H}{\underset{|}{N}}-\overset{\displaystyle O}{\overset{\|}{C}}- \right]_n$$

Polyurethanes

Fig. 2-6. Polymers with nitrogen in the chain.

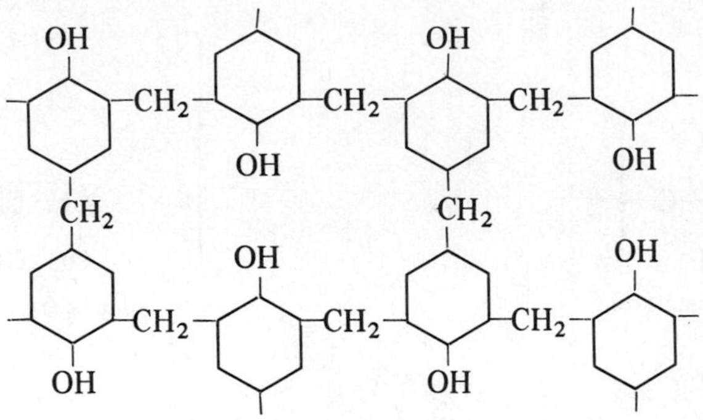

OH OH
$-CH_2-$ $-CH_2-$ $-CH_2-$
OH OH
CH_2 CH_2
OH OH
$-CH_2-$ $-CH_2-$ $-CH_2-$
OH OH

Phenol-formaldehyde resin

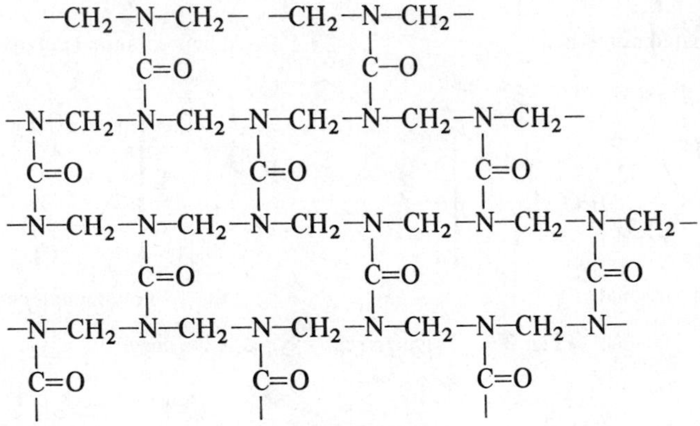

$-CH_2-N-CH_2-$ $-CH_2-N-CH_2-$
C=O C–O
$-N-CH_2-N-CH_2-N-CH_2-N-CH_2-N-CH_2-$
C=O C=O C=O
$-N-CH_2-N-CH_2-N-CH_2-N-CH_2-N-CH_2-N-CH_2-$
C=O C=O C=O
$-N-CH_2-N-CH_2-N-CH_2-N-CH_2-N-CH_2-N-$
C=O C=O C=O

Urea-formaldehyde

CH_2
N–H
C
N N
H N H N H N H
$-CH_2-N-C$ $C-N-(CH_2)_2-N-C$ $C-N-(CH_2)_2-N-C$ $C-N-CH_2-$
N N H N H N N
C C
N–H N–H
CH_2 CH_2

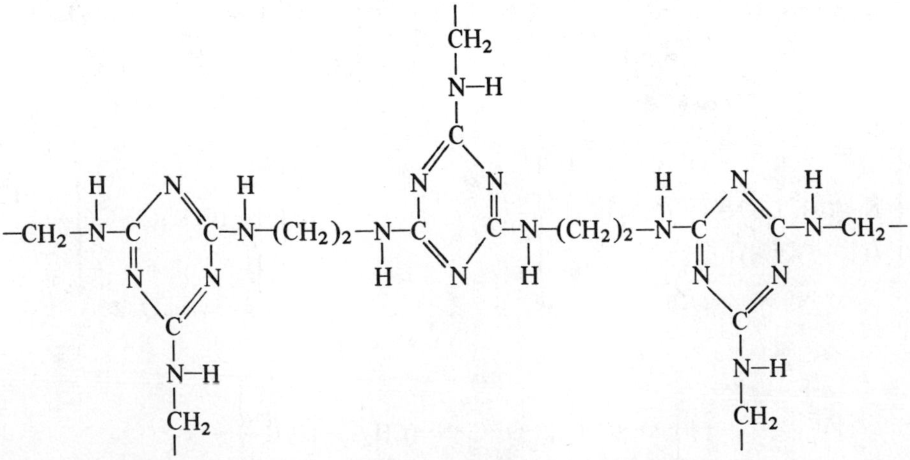

Melamine-formaldehyde

Fig. 2-7. Thermoset structures.

types of chemical additives and fillers that are mixed with the fundamental polymer is so large that the chemistry of plastic compounds becomes extremely complex.

It also is possible to vary existing polymers by any number of chemical, mechanical, or irradiation techniques to change the structure of the polymer. During processing, for example, it is possible to align or orient the molecules in a polymer to significantly increase the strength of the polymer in the direction of the orientation. Chemical or irradiation methods also can be used to cross-link a plastic (such as polyethylene) to improve its toughness and chemical resistance. Nucleation techniques and chemical modification (as in the case of chlorinated polyethylene) are further adaptable to changing polymer structures.

3

Plastic Materials/Properties and Applications

As indicated in Chapter 2, the family of plastics is extraordinarily varied and complex. There are, however, some fairly broad and basic guidelines for engineering and designing products made of plastic materials. This chapter provides a brief introduction to the range of materials that comprise the family of plastics. For each, a description of its important properties is provided, along with some specific processing requirements and examples of products in which the polymer is used. For a quick, broad comparison, see Table 3-1 on pages 48–50. To find information on a specific resin, the reader should consult the index to this book.

THE FAMILY OF PLASTICS

Plastics generally are organic high polymers (i.e., they consist of large chainlike molecules containing carbon) that are formed in a plastic state either during or after their transition from a small-molecule chemical to a solid material. Stated very simply, the large chainlike molecules are formed by hooking together short-chain molecules of chemicals (monomers) in a reaction known as polymerization. When units of a single monomer are hooked together, the resulting plastic is a homopolymer, such as polyethylene, which is made from the ethylene monomer. When more than one monomer is included in the process, for example, ethylene and propylene, the resulting plastic is a copolymer.

The two basic groups of plastic materials are the thermoplastics and the thermosets. Thermoplastic resins consist of long molecules, each of which may have side chains or groups that are not attached to other molecules (i.e., are not crosslinked). Thus, they can be repeatedly melted and solidified by heating and cooling so that any scrap generated in processing can be reused. No chemical change generally takes place during forming. Usually, thermoplastic polymers are supplied in the form of pellets, which often contain additives to enhance processing or to provide necessary characteristics in the finished product (e.g., color, conductivity, etc.). The temperature service range of thermoplastics is limited by their loss of physical strength and eventual melting at elevated temperatures.

Thermoset plastics, on the other hand, react during processing to form crosslinked structures that cannot be remelted and reprocessed. Thermoset scrap must be either discarded or used as a low-cost filler in other products. In some cases, it may be pyrolyzed to recover inorganic fillers such as glass reinforcements, which can be reused. Thermosets may be supplied in liquid form or as a partially polymerized solid molding powder. In their uncured

Table 3-1. Typical property ranges for plastics

Thermoplastics		Specific Gravity	Trans-parency	Tensile strength 10^3 psi	Tensile modulus, 10^5 psi	Impact strength Izod[c] ft. lb/inch	Dielectric constant at 60 cps	Dielectric strength, volts/mil	Max use temp, F (no load)	DTUL at 66 psi	DTUL at 264 psi	Weather resist	Chemical resistance[e] W ac	S ac	W al	S al	Solv
ABS	GP	1.05–1.07	No	5.9	3.1	6	2.8–3.2	385	160–200	210–225	190–206	R-E	R	A[k]	R	R	A[m]-R
	Hi. imp.	1.01–1.06	No	4.8	2.4	7.5	2.8–3.5	300–375	140–210	210–225	188–211	R-E	R	A[k]	R	R	A[m]-R
	Ht. res.	1.06–1.08	No	7.4	3.9	2.2	2.7–3.5	360–400	190–230	225–252	226–240	R-E	R	A[k]	R	R	A[m]-R
	Trans.	1.07	Yes	5.6	2.9	5.3	—	—	130	180	165	R-E	R	A[k]	R	R	A[m]-R
Acetals	Homo	1.42	No	10	5.2	1.4	3.7	320	195	338	255	R[j]	R	A	R	A-D	R
	Copoly	1.41	No	8.8	4.1	1.2–1.6	3.7	500	212	316	230	R[j]	R	A	R	R	R
Acrylics	GP	1.11–1.19	Yes	5.6–11.0	2.25–4.65	0.3–2.3	3.0–3.7	450–500	130–230	175–225	165–210	R	R	A[k]	R	A	A[m]-R
	Hi. imp.	1.12–1.16	No	5.8–8.0	2.3–3.3	0.8–2.3	3.5–3.9	400–500	140–195	180–205	165–190	R	R	A[k]	R	R	A[m]-R
		1.21–1.28	No	8.0–12.5	3.5–4.8	0.3–0.4	3.5–5.1	400–440	125–200	170–200	155–205	R	R	A[k]	R	A	A[m]-R
	Cast	1.18–1.28	Yes	9.0–12.5	3.7–5.0	0.4–1.5	3.5–5.1	400–530	140–200	165–235	160–215	R	R	A[k]	R	A	A[m]-R
	Multi poly-mer	1.09–1.14	Yes	6–8	3.1–4.3	1–3	3.3–3.5	495	165–175	—	185–195	E	R	A[k]	R	S	A[m]
Cellulosics	Acetate	1.23–1.34	Yes	3.0–8.0	1.05–2.55	1.1–6.8	3.5–7.5	250–600	140–220	120–209	111–195	S	S	D	S	D	D-S
	Butyrate	1.15–1.22	Yes	3.0–6.9	0.7–1.8	3.0–10.0	3.5–6.4	250–400	140–220	130–227	113–202	S	S	D	S	D	D-S
	Ethyl cellu-lose	1.10–1.17	Yes	3–8	0.5–3.5	1.7–7.0	3.2	350–500	115–185	—	115–190	S	S	D	R	S	D
	Nitrate	1.35–1.40	Yes	7–8	1.9–2.2	5–7	7.0–7.5	300–600	140	—	140–160	E	S	D	S	D	D
	Propionate	1.19–1.22	Yes	4.0–6.5	1.1–1.8	1.7–9.4	3.7–4.0	300–450	155–220	147–250	111–228	S	S	D	S	D	D-S
Eth. copolymers	EEA	0.93	Yes	2.0	0.05	NB	2.7	550	190	—	—	S	R	A[k]	R	R	A-D
	EVA	0.94	Yes	3.6	0.02–0.12	NB	2.50–3.16	525	—	140–147	93	S	R	A	R	R	A-D
Fluoropolymers	FEP	2.14–2.17	No	2.5–3.9	0.5–0.7	NB	2.1	500–600	400	158	—	R	R	R	R	R	R
	PTFE	2.1–2.3	No	1–4	0.38–0.65	2.5–4.0	2.1	400–500	550	250	—	R	R	R	R	R	R
	CTFE	2.10–2.15	Yes-No	4.6–5.7	1.8–2.0	3.5–3.6	2.6–2.7	530–600	350–390	258	—	R	R	R	R	R	S[n]
	PVF_2	1.77	No	7.2	1.7	3.8	10.0	260	300	300	195	S	R	A[l]	R	R	R
	ETFE & ECTFE	1.68–1.70	No	6.5–7.0	2–2.5	NB	2.4–2.6	400	300	220	160	R	R	R	R	R	R
Nylons	6/6	1.13–1.15	No	9–12	3.85	2.0	4.0	385	180–300	360–470	150–220	R	R	A	R	R	R-D[o]
	6	1.14	No	12.5	—	1.2	4.0–5.3	385	180–250	300–365	140–155	R	R	A	R	R	R-A[o]
	6/10	1.07	No	7.1	2.8	1.6	3.9	470	180	300	—	R	R	A	R	R	R-A[o]
	8	1.09	No	3.9		>16	9.3	340				R	R	A	R	R	R-A[o]
	12	1.01	No	6.5–8.5	1.7–2.1	1.2–4.2	3.6	840	175–260	—	120–130	R	R	A	R	R	R-A[o]
	Copolymers	1.08–1.14	No	7.5–11.0		1.5–19	3.2–4.5	440–450	180–250	—	130–350	R	R	A	R	R	R-A[o]
Polyarylate		1.2	Yes	9.5	2.9	4.2	2.6	400	—	—	345	S	R	A[k]	A	A	A
Polyarylsulfone		1.36	No	13	3.7	2	3.94	350	500	—	525	Darkens	R	R	R	R	R

Polybutylene		0.910	No	3.8	0.26	NB	2.25	—	225	215	130	E	R	A^k	R	R	—
Polycarbonate		1.2	Yes	9	3.45	12–16	3.17	380	250	270–290	265–285	R	R	A^k	A	A	A
PC/ABS alloy		1.14	No	8.2	3.7	10	2.74	500	220	235	220	R-E	R	A^k	R	S	A
Polyesters	PET	1.37	No	10.4	—	0.8	3.65	—	175	240	185	R	R	A^k	R	A	$R\text{-}A^o$
	PBT	1.31	No	8.0–8.2	3.6	1.2–1.3	3.3–3.7	590–700	280	310	130	R	R	R	R	A	R
	PTMT	1.31	No	8.2	—	1.0	3.16	420	270	302	122	R	R	R	R	A	R
	Copol.	1.2	Yes	7.3	—	1.0	—	—	—	—	154	—	—	—	—	—	—
Polyetheretherketone		1.3	No	13	—	1.6	—	—	480	—	320	E	R	A	R	R	R
Polyetherimide		1.27	Yes	15	4.3	1.1	3.1	710	338	405	390	R	R	R	S	A	$R\text{-}A^p$
Polyethersulfone		1.37	Yes	12	3.5	1.4–2.2	3.5	400	356	—	397	E	R	A	R	R	A^p
Polyethylenes	LD	0.91–0.93	No	0.9–2.5	0.20–0.27	NB	2.3	480	180–212	100–120	90–105	E	R	A^k	R	R	R
	HD	0.95–0.96	No	2.9–5.4		0.4–14	2.3	480	175–250	140–190	110–130	E	R	$R\text{-}A^k$	R	R	R
	HMW	0.945	No	2.5	1	NB	2.3	480	—	155–180	105–180	E	R	A^k	R	R	R
Ionomer		0.94–0.95	Yes	3.4–4.5	0.3–0.7	6-NB	2.4	1000	160–180	110	100–120	E	A	A^k	R	R	R
Polymethylpentene		0.83	Yes	3.3–3.6	1.3–1.9	0.95–3.8	—	700	275	—	—	E	R	A^k	R	R	A
Polyphenylene oxide based mtls.		1.06–1.10	No	7.8–9.6	3.5–3.8	5.0	2.65–2.69	400–500	175–220	230–280	212–265	R	R	R	R	R	R-A
Polyphenylene sulfide		1.34	No	10	4.8	0.3		595	500	—	278	R	R	A^k	R	R	R
Polyimide		1.43	No	5–7.5	5.4	5–7	4.12	310	500	—	680	—	R	R	A	A	R
Polypropylenes	GP	0.90–0.91	No	4.5–6.0	1.6–3.0	0.4–1.2	2.20–2.28	650	225–300	200–230	125–140	E	R	A^k	R	R	R
	Impact copolymer	0.88–0.91	No	3.5–5.0	1.3	2-NB	2.20–2.28	450–650	200–250	160–200	120–135	E	R	A^k	R	R	A
	Random copolymer	0.89–0.91	No	4.0–5.5	1.0–1.7	1.1–12	2.25–2.30	450–600	190–240	185–230	125–140	E	R	A^k	R	R	R
Polystyrenes	GP	1.04–1.07	Yes	6.0–7.3	4.5	0.3	2.45–2.65	400–600	150–170	—	180–220	S	R	A^k	R	R	D
	Hi. impact	1.04–1.07	No	2.8–4.6	2.9–4.0	0.7–1.0	2.45–4.75	400–500	140–175	—	175–210	S	R	A^k	R	R	D
Polysulfone		1.24	No	10.2	3.6	1.2	3.14	425	300	360	345	S	R	R	R	R	R-A
Polyurethanes		1.11–1.25	No	4.5–8.4	0.1–3.5	NB	5.4–7.6	460	190	—	—	R-S	S-D	S-D	S-D	S-D	R
PVC Rigid		1.3–1.5	No-Yes	5–8	3–5	0.5–20	3.2–3.6	425–1300	150–175	135–180	130–175	R	R	R-S	R	R	R-A
PVC Flexible		1.2–1.7	Yes & No	1.4	—	0.5–20	5–6	250–800	140–175	—	—	S	R	R-S	R	R	R-A
Rigid CPVC		1.49–1.58	Yes-No	7.5–9.0	3.6–4.7	1.0–5.6	2.8–3.6	—	230	215–245	200–235	R	R	R	R	R	R
PVC/acrylic		1.30–1.35	No	5.5–6.5	2.75–3.35	15	3.9–40	400	—	180	170	R	R	S	R	R	A
PVC/ABS		1.10–1.21	No	2.6–6.0	0.8–3.4	10–15	—	600	—	—	—	S	R	R-S	R	R	R-D
SAN		1.08	Yes	10–12	5.0–5.6	0.4–0.5	3.0–3.8	1775	140–200	—	190–220	S-E	R	A	R	R	A
SMA		1.05–1.15	Yes-No	5–9	2.7–4.4	0.5–12	3.5	420	200	—	205–260	S-E	R	S	S	A	A

[a]All values at room temperature unless otherwise listed. [b]Per ASTM. [c]Notched samples. [d]Deflection temperature under load. [e]Ac is acid and Al is alkali; R is resistant, A is attached, S is slight effects, E is embrittles and D is decomposes. [j]Chalks slightly. [k]By oxidixing acids. [l]By fuming sulfuric. [m]By ketones, esters, and chlorinated and aromatic hydrocarbons. [n]Halogenated solvents cause swelling. [o]Dissolved by phenols and formic acid. [p]Dissolved by halogenated hydrocarbons.

Table 3-1. (*Continued*)

Thermosets[a]	Specific gravity	Trans-parency	Tensile strength, 10^3 psi	Tensile modulus Ten mod, 10^5 psi	Impact strength Izod, ft-lb[c]/inch	Dielectric constant at 60 cps	Dielectric strength, volts/mil	Max use temp, F (nol load)	DTUL at 264 psi[d]	Weather res	Chemical resistance[e] Weak acid	Strong acid[f]	Weak alkali	Strong alkali	Solvents
Alkyds															
Glass filled	2.12–2.15	No	4–9.5	20–28	0.6–10	5.7	250–530	450	400–500	R	A	A	A	A	A
Mineral filled	1.60–2.30	No	3–9	5–30	0.3–0.5	5.1–7.5	350–450	300–450	350–500	R	R	A	A	D	A
Syn. fiber filled	1.24–2.10	No	4.5–7	20	0.5–4.5	3.8–5.0	365–500	300–430	245–430	R	R	S	R	S	A
Diallyl phthalates															
Glass filled	1.61–1.87	No	6–11	14–22	0.4–15	4.3–4.6	400–450	300–400	330–540	R	R	S	R-S	S	R
Mineral filled	1.65–1.80	No	5–9	12–22	0.3–0.5	5.2	400–420	300–400	320–540	R	R	S	R-S	S	R
Epoxies (bis A)															
No filler	1.06–1.40	Yes	4–13	2.15–5.2	0.2–1.0	3.2–5.0	400–650	250–500	115–500	R	R	A	R	S	R-S
Graphite fiber reinf.	1.37–1.38	No	185–200	118–120	—	—	—	—	—	S	R	R	R	R	R-S
Mineral filled	1.6–2.0	No	5–15	—	0.3–0.4	3.5–5.0	300–400	300–500	250–500	S	R	R	R	R	R-S
Glass filled	1.7–2.0	No	10–30	30	10–30	3.5–5.0	300–400	300–500	250–500	S	R	R-S	R	R	R-S
Epoxies (novolac)															
No filler	1.12–1.24	No	5–11	2.15–5.2	0.3–0.7	3.11–4.0	360–600	400–500	450–500	R	R	R	R	R	R
Epoxies (cycloaliphatic)															
No filler	1.12–1.18	Yes	10–17.5	5–7	—	3.6	—	480–550	500–550	R	R	R-A	R	R-A	R
Melamines															
Cellulose filled	1.45–1.52	No	5–9	11	0.2–0.4	6.2–7.6	350–400	250	270	S	R-S	D	R	D	R
Flock filled	1.50–1.55	No	7–9	—	0.4–0.5	—	300–330	250	270	S	R-S	D	R	D	R-S
Fabric filled	1.5	No	8–11	14–16	0.6–1.0	7.6–12.6	250–350	250	310	S	R	D	R	A	R-S
Glass filled	1.8–2.0	No	5–10	24	0.6–18	0.7–11.1	170–300	300–400	400	S	R	D	R	R-S	R
Phenolics															
Woodflour filled	1.34–1.45	No	5–9	8–17	0.2–0.6	5–13	260–400	300–350	300–370	S	R-S	S-D	S-D	A	R-S
Mica filled	1.65–1.92	No	5.5–7	25–50	0.3–0.4	4.7–6	350–400	250–300	300–350	S	R-S	S-D	S-D	A	R-S
Glass filled	1.69–1.95	No	5–18	19–33	0.3–18	5–7.1	140–400	350–550	300–600	S	R-S	S-D	S-D	A	R-S
Fabric filled	1.36–1.43	No	3–9	9–14	0.8–8	5.2–21	200–400	220–250	250–330	S	R-S	S-D	S-D	A	R-S
Polyesters															
Glass filled BMC	1.7–2.3	No	4–10	16–25	1.5–16	5.3–7.3	300–420	300–350	400–450	R-E	R-A	S-A	S-A	S-D	A-D
Glass filled SMC	1.7–2.1	No	8–20	16–25	8–22		320–400	300–350	400–450	R-E	R-A	S-A	S-A	S-D	A-D
Glass cloth reinf.	1.3–2.1	No	25–50	19–45	5–30	4.1–5.5	350–500	300–350	400–450	R-E	R-A	S-A	S-A	S-D	A-D
Silicones															
Glass filled	1.7–2.0	No	4–6.5	10–15	3–15	3.3–5.2	200–400	600	600	R-S	R-S	R-S	S	S-A	R-A
Mineral filled	1.8–2.8	No	4–6	13–18	0.3–0.4	3.5–3.6	200–400	600	600	R-S	R-S	R-S	S	S-A	R-A
Ureas															
Cellulose filled	1.47–1.52	No	5.5–13	10–15	0.2–0.4	7.0–7.5	300–400	170	260–290	S	R-S	A-D	S-A	D	R-S
Urethanes															
No filler	1.1–1.5	No & Yes	0.2–10	1–10	5-NB	4.0–7.5	400–500	190–250	—	R-S	S	A	S	S-A	R-S

condition, they can be formed to the finished product shape with or without pressure and polymerized by using chemicals or heat.

The distinction between thermoplastics and thermosets is not always clearly drawn. For example, thermoplastic polyethylene can be extruded as a coating for wire and subsequently crosslinked (either chemically or by irradiation) to form a thermoset material that no longer will melt when heated. Some plastic materials even have members in both families; there are, for instance, both thermoset and thermoplastic polyester resins.

THERMOPLASTIC RESINS

The first thermoplastics discussed in this section are the so-called commodity thermoplastics and their chemical relatives. These polymers have found their way into wide commercial use, with millions of tons produced annually. Their ranks include the polyolefins, the styrenics, and the vinyls. Also covered here are the acrylics and the cellulosics.

The remaining thermoplastic materials generally are categorized as engineering resins or engineering thermoplastics (ETPs). These resins are produced in substantially smaller quantities than the commodity thermoplastics, are priced higher, and tend to compete with metals rather than with glass, paper, and wood, as the commodities do.

Discussions of thermoplastic elastomers and of alloys and blends concludes the section, along with a list of some types of these materials that have been commercialized to date.

The Polyolefins

Polyethylenes. Polyethylenes are characterized by toughness, near-zero moisture absorption, excellent chemical resistance, excellent electrical insulating properties, low coefficient of friction, and ease of processing.

In general, they are not outstanding load-bearing materials, but high-density polyethylene can be used for some short-term light loads.

Few thermoplastics have the excellent chemical resistance and dielectric properties of polyethylenes. Soluble in some organic solvents above 140°F, polyethylenes resist bases and acids at room temperature. Their resistivity (both volume and surface) and dielectric strength are high.

Polyethylenes can be crosslinked to form infusible thermosetting materials with high heat resistance and crack resistance. Applications are in wire and cable coating, foams, and rotationally molded products.

Polyethylenes are classified by density as follows: (a) 0.880 to 0.915 g/cu cm (called ultra- or very low density), (b) 0.910 to 0.925 g/cu cm (low density), (c) 0.926 to 0.940 g/cu cm (medium density), and (d) 0.941 to 0.965 g/cu cm (high density). The primary differences among the types are in rigidity, heat resistance, chemical resistance, and ability to sustain loads. In general, as density increases, hardness, heat resistance, stiffness, and resistance to permeability increase. So-called conventional low-density polyethylenes (LDPE) are produced in high-pressure reactors. Such polymers have highly branched structures with moderate crystallinity (50–65%).

Low-density polyethylenes are quite flexible, with high impact strength and relatively low heat resistance (maximum recommended service temperature is 140–175°F), although grades are available with heat resistance up to 200°F.

Traditional markets for LDPE are in packaging films, extrusion coating of paper, wire and cable coating, injection molding, and pipe and tubing. Since the introduction of linear low-density polyethylene (see below), conventional LDPE has been gradually displaced in some of these areas.

The high-pressure route also is used to make copolymers of ethylene with polar monomers such as vinyl acetate or ethyl acrylate. The EVAs and EEAs that are produced have low-temperature flexibility and are used in tough films and as a component of multilayer constructions for low-temperature heat sealing. Acid copolymers (with acrylic or methacrylic acid) are used for their hot tack and adhesive properties. Neutralized with metallic ions, these materials become clear ionomers, used in coating applications.

Linear polyethylenes, as the name suggests,

have very little branching along the polymer chains. The polymerization of linear polyethylenes at low pressures has been used to manufacture high-density resins since the mid-1950s. Methods for making linear low-density polyethylenes (LLDPE) did not become commercial until the 1970s.

Commercial LLDPE resins are made in gas-phase reactors; the polymers typically contain up to 10% alpha olefin comonomers (butene, hexene, methyl pentene, or octene). The reactor output is in granular form, but normally is extruded and pelletized to incorporate additive systems.

LLDPE polymers, with little long-chain branching, have much greater elongation than LDPE. Their higher tear, tensile, and impact strength, along with improved resistance to environmental stress cracking, allow stronger products to be produced with less material. This has been particularly important in film markets, where considerable downgauging has been accomplished. (See Fig. 3-1.) Ultra- and very low-density polyethylenes (ULDPE and VLDPE) are essentially synonymous designations for linear polyethylenes with densities down to 0.880 g/cu in. Produced in gas-phase reactors, they are finding application as impact modifiers for other polyolefins and in film and sheet markets.

Fig. 3-1. Bag for ice cubes from linear low density polyethylene. (*Courtesy of Union Carbide Chemicals and Plastics*)

Linear high-density polyethylene (HDPE) can be produced via a slurry process or in gas-phase reactors identical to those used to make LLDPE. HDPE polymers are highly crystalline, tough materials that can be formed by most processing methods. Much HDPE is blow-molded into containers for household and industrial chemicals. It is injection-molded into items such as crates, housewares, pails, and dunnage containers; extruded into pipe, tubing, and wire insulation; blown into film for packaging; and rotationally molded into containers, toys, and sporting goods. (See Fig. 3-2.)

Within the density range of HDPEs, the stiffness, tensile strength, melting point, and chemical resistance all improve at the high end. However, materials with the highest densities have lower stress crack resistance and low-temperature impact strength.

High molecular weight HDPE polymers are a special class of linear resins with molecular weights in the 200,000 to 500,000 range. To obtain processability along with end-use properties, control of the molecular weight distribution is critical. Some materials are produced with a "bimodal" molecular weight distribution to obtain the necessary balance.

HMW-HDPE is made into blown film for packaging, extruded into pressure pipe, and blow-molded into large shipping containers. Extruded sheet is used to form truck bed liners and pond liners.

At the highest end of the spectrum is ultra-high molecular weight high-density polyethylene. Its weight average molecular weight is over 3 million. Because the resin does not flow when melted, it normally is compression-molded into thick sheets or ram-extruded.

High strength, chemical resistance, and lubricity make UHMW-HDPE ideal for gears, slides, rollers, and other industrial parts. It also is used to make artificial hip joints. In fiber form, the UHMW-HDPE's linear structure exhibits liquid crystal properties that are useful in

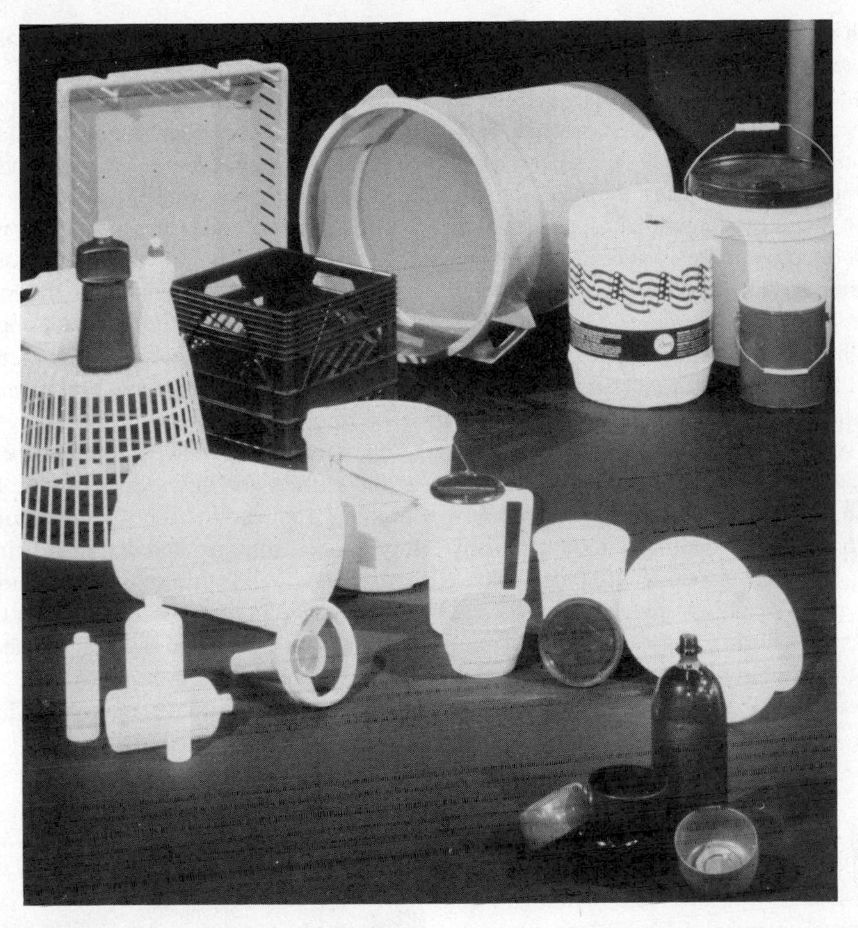

Fig. 3-2. Examples of high-density polyethylene products. (*Courtesy of Union Carbide Chemicals and Plastics*)

reinforcing composite structures. The fiber also can be woven into lightweight ultrastrong fabrics.

Polymethylpentene (PMP). Based on 4-methylpentene-1, this crystalline polyolefin is characterized by transparency (90% light transmission), low specific gravity (0.83), and a high melting point (464°F). Its drawbacks are brittleness and poor UV resistance. PMP has found application in making injection-molded and blow-molded laboratory ware and medical products, food processing equipment, and microwavable packaging. Glass fiber–reinforced grades, with heat distortion temperatures up to 440°F, can compete with more expensive engineering resins for electrical/electronic and automotive applications.

Polypropylene. This polyolefin has turned out to be the most versatile of the family, accounting for the continuing rapid increase in its use. First produced in the 1950s, early polypropylenes (PPs) suffered from low yields in polymerization, high percentage (about 10%) of atactic polymer, and poor control of molecular weight.

The development of high-activity catalysts in the 1970s improved the yields dramatically and almost eliminated production of the atactic form. The resulting isotactic polymers were highly stereoregular.

Polypropylenes have better resistance to heat (heat distortion temperature at 66 psi: 200–250°F) and resist more chemicals than do other thermoplastic materials of the same cost. Also, polypropylenes have negligible water absorption and excellent electrical properties, and they are easy to process.

In much the same way that density is important in determining the mechanical properties of polyethylenes, the stereoregularity (related to the repeated units in the stereoregular molecular chain) of a polypropylene very often determines the characteristics of the material. An increase in the stereoregularity of a polypropylene will sharply increase the yield strength of the material. The hardness, stiffness, and tensile strength also increase. On the other hand, as the stereoregularity decreases, elongation and impact strength increase.

The ability to carry light loads for long periods and over wide temperature ranges is one of the properties that make polypropylenes valuable in engineering applications. Polypropylenes do not have outstanding long-term creep resistance, but their fatigue endurance limit is excellent. In fact, polypropylene often is referred to as the ''living hinge'' thermoplastic.

One of the limitations most often mentioned for polypropylenes is their low-temperature brittleness (−4°F). However, polypropylene copolymers have been developed with brittleness points of about −20°F.

Like all other polyolefins, polypropylenes have excellent resistance to water and to water solutions, such as salt and acid solutions that are destructive to metals. They also are resistant to organic solvents and alkalis. Above 175°F, polypropylene is soluble in aromatic substances such as toluene and xylene, and in chlorinated hydrocarbons such as trichlorethylene.

Polypropylenes have excellent electrical resistivity (both volume and surface), and their dielectric strength is high.

The greatest commercial uses for homopolymer PP are in fibers and filaments. PP fibers are woven into fabrics and carpets, and they also are used to produce nonwoven fabrics for disposables. Slit tape-filaments are used as jute replacements in carpet backings and sacks. PP also is made into unoriented and oriented films for packaging, which have largely replaced cellophane and glassine. Homopolymer PP is injection-molded into caps and closures, appliance components, and auto parts.

Random copolymer PP (with up to 7% ethylene) has higher impact strength and better clarity than the homopolymer. Its heat distortion temperature is lower than that of the homopolymer, 150 to 205°F under 66 psi load. These materials are used in blow-molded containers (including oriented and multilayer bottles), injection-molded packaging, and flexible monolayer and coextruded films.

Impact copolymers are produced in a secondary reactor. The ethylene comonomer provides a flexible component in the otherwise rigid, crystalline structure. Impact copolymers have replaced impact PPs made by blending the homopolymer with ethylene–propylene rubber.

PP impact copolymers are tough, even at low temperatures, and yet retain a high percentage of the stiffness of the homopolymer. Injection molding applications include automobile battery cases, interior and exterior trim parts, housewares, and furniture. Coextruded with barrier polymers such as EVOH and PVDC, impact PP is made into multilayer sheet that is thermoformed into food packages that can withstand freezer storage and microwave cooking. (See Fig. 3-3.)

Fig. 3-3. Multilayer package from polypropylene with EVOH barrier. (*Courtesy of Rampart Packaging, Inc.*)

The rigidity of polypropylenes can be improved significantly by the addition of inexpensive mineral fillers. Filled PPs and even glass-reinforced PPs are made into chemically bonded compounds that are truly engineering thermoplastics.

Polybutylene. Polybutylene resins are flexible, linear polyolefins offering a unique combination of properties. They are produced from 1-butene as homopolymers and as copolymers with ethylene. Some of their important characteristics are: excellent resistance to creep at room and elevated temperatures—no failures after long-term loading at 90% of yield strength; good toughness (high yield strength, high impact strength, high tear strength, high puncture resistance); and exceptional resistance to environmental stress cracking.

Other important advantages include good moisture barrier properties, excellent electrical insulation characteristics, and resistance to most chemical environments.

Polybutylene can be processed on the same equipment used for low-density polyethylene, but it should be noted that after cooling the polymer undergoes a crystalline transformation. During this change, the tensile strength, hardness, specific gravity, and other crystallinity-dependent properties attain their ultimate values. This behavior presents opportunities for post-forming techniques such as the cold forming of sheeting or molded parts. The delay in aging does not pose any unusual processing problems.

The most important commercial application for polybutylene is in plumbing pipe for residential and commercial use. Large-diameter polybutylene pipe is used for transporting abrasive fluids. The material's abrasion resistance also is put to use in sheet form for vessel and chute linings.

Polybutylene is used to make hot melt adhesives and sealants and as a carrier resin for additive and color concentrates. High loadings (up to 80%) can be achieved. The resin also is used as a modifier for other polyolefins; in polypropylene it improves processability, impact, and weld line strength, and in polyethylene it imparts processability and stress crack resistance.

Ethylene-vinyl alcohol (EVOH). Copolymers of ethylene and vinyl alcohol provide excellent barriers to the permeation of oxygen and other gases, odors, and flavors. EVOH resins generally are used within packaging structures because of their strong tendency to absorb moisture, which leads to deterioration of the barrier properties.

Multilayer structures that contain EVOH are produced by coating, laminating, coextrusion, or coinjection molding. In such structures, the EVOH is encapsulated by other polymers, often polyolefins, which keep moisture out. In most cases, adhesive tie layers are required to bond the hydrophilic EVOH to hydrophobic polymers. A notable exception is with nylon, where bonding is achieved without adhesives.

The major use of EVOH barrier packaging is for food products although there are applications involving solvents and other chemicals. The most common combination is with polypropylene. Blow-molded bottles for ketchup, barbecue sauces, and salad dressings and thermoformed containers for jellies, sauces, and prepared foods all are sold commercially.

Styrenic Plastics

In general, styrenics are characterized by ease of processing, hardness, and excellent dielectric properties. They are relatively limited in heat resistance and are attacked by aliphatic and aromatic hydrocarbon solvents.

ABS Resins. ABS materials are composed of acrylonitrile, butadiene, and styrene in varying proportions, combined by a variety of methods including graft copolymerization and physical blending. Originally an outgrowth of polystyrene modification, the materials long have been in a category of their own.

ABS materials provide a balanced combination of mechanical toughness (Izod impacts can range from 2 ft-lb/in. of notch to 12 ft-lb/in. of notch), wide service temperature range (−40 to 240°F), good dimensional stability, chemical resistance, electrical insulating properties, and ease of fabrication.

ABS materials are produced in a wide range of grades including medium- and high-impact, heat-resistant, plateable fire-retardant, and both low- and high-gloss varieties.

ABS is used as a component in alloys with other plastics, notably polycarbonate, nylon, PVC, polysulfone, and SMA. Because of their stiff melt flow, most ABS polymers are difficult to color. ABS makers have installed sophisticated color matching and compounding equipment to meet customers' needs.

ABS materials have relatively good electrical insulating properties, which make them suitable for secondary insulating applications.

In general, ABS materials have very good resistance to a wide range of chemicals. They exhibit low water absorption, are very good in weak acids, and in both weak and strong alkalis; are generally good in strong acids; and are poor in solvents such as esters, ketones, aldehydes, and some chlorinated hydrocarbons. ABS is somewhat hygroscopic and has to be dried before processing.

ABS materials are available as compounds for injection molding, blow molding, extrusion, and calendaring, as sheet for thermoforming or cold forming, and in expandable grades for foam molding. Although most ABS materials are opaque, a transparent grade has been developed, using methyl methacrylate to match the refractive index to that of the rubber component.

The range of applications for ABS plastics is extremely broad. Falling in between the commodity plastics and the engineering resins, ABS fills the property requirements for many parts at a reasonable price. In the automotive market (Fig. 3-4), ABS is used to injection-mold interior panels and trim (low-gloss grades), grilles, wheelcovers, and mirror housings (electroplatable grades); flame-retardant grades are used to mold housings and keyboards for computers. General purpose ABS is the current material of choice for $3\frac{1}{2}$-inch computer disc housings, and it has long been used to mold telephones, calculators, and business machines and appliances. ABS sheet is thermoformed into refrigerator door liners and food storage compartments. Grades of ABS are made into DWV pipe for the building and construction market and are joined with injection-molded ABS fittings. Other ABS applications include

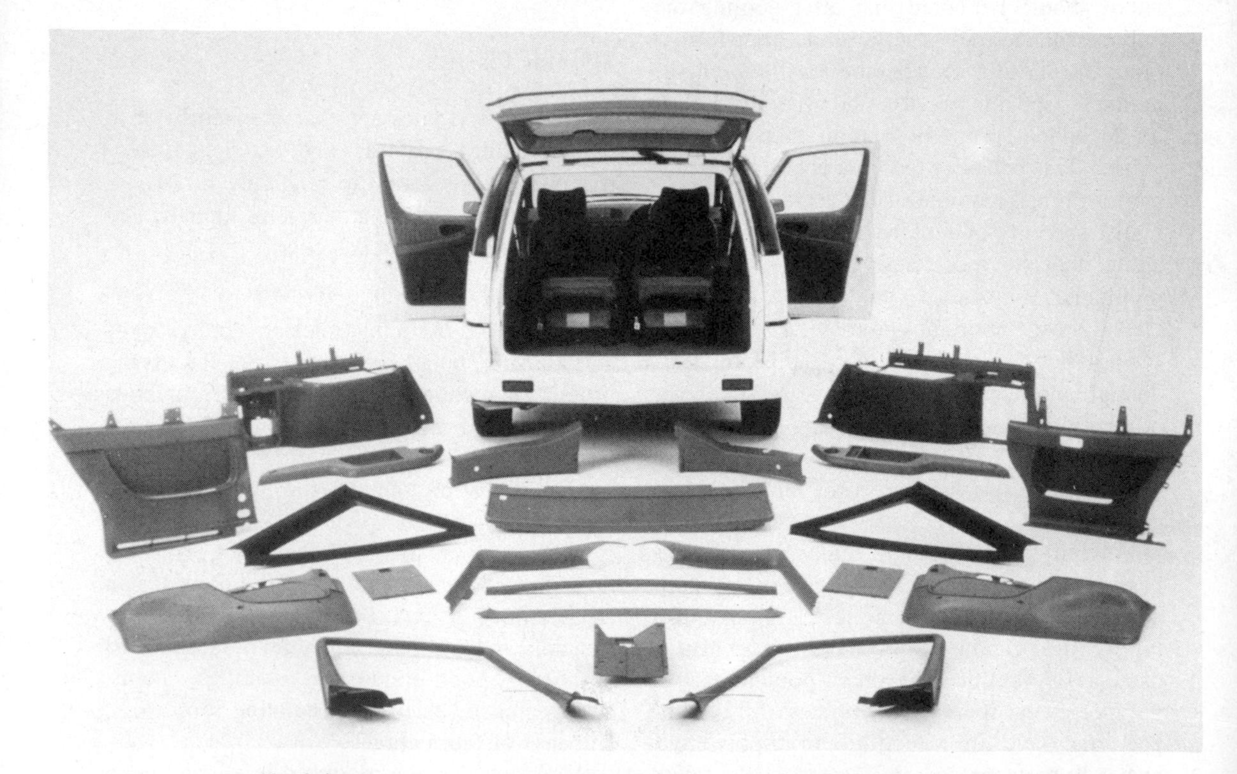

Fig. 3-4. Auto interior trim parts from ABS. (*Courtesy of the Dow Chemical Co.*)

housewares, luggage, toys, and sporting goods. ABS has been used to blow-mold bumper fairings for large trucks.

Polystyrene (PS). General-purpose polystyrene is the low-cost member of the family. Properties of the amorphous polymer include hardness, rigidity, optical clarity, dimensional stability, and excellent processibility. Most crystal polystyrene is injection-molded. Applications include packaging for food, cosmetics, and pharmaceuticals, audio cassette cases, dust covers, disposable drinking cups, and cutlery (see Fig. 3-5). Oriented crystal PS sheet is thermoformed into trays and blister packaging, and foamed crystal PS is used in a wide variety of insulating and packaging applications (see below).

Modified or impact polystyrene (with polybutadiene elastomers) extends the uses of polystyrenes into those areas where high impact strength and good elongation are required. Most products of this type have the rubber molecularly grafted to the polystyrenes; however, some are produced by mechanical mixing.

High-impact polystyrene is generally recognized as the best styrenic material for load-bearing applications. The other types of polystyrene offer various improvements in chemical, thermal, and optical performance, usually at the sacrifice of mechanical properties. High-impact polystyrene is a very rigid material, but it does not withstand long-term tensile loads as well as some other thermoplastics do.

Typical applications include food packaging (cups, lids, take-out containers), housewares, office products, and video and audio cassettes. Ignition-resistant grades are used in radio and TV housings.

Foamed crystal polystyrene has cushioning and insulation properties at very low densities. Extruded foam sheet is used as a protective wrapping material and is thermoformed into fast food packages, egg cartons, and meat and produce trays for supermarkets. Expandable polystyrene (EPS) in supplied in the form of small beads containing a chemical blowing agent. The beads are pre-expanded with steam to establish the density of the finished product and then steam-chest-molded into lightweight products such as coffee cups, cooler chests, and protective package inserts. (See Fig. 3-6.)

Fig. 3-5. Crystal polystyrene case for compact audio disk. (*Courtesy of Huntsman Chemical Corp.*)

Styrene–Acrylonitrile (SAN). These copolymers are transparent, amorphous materials with higher heat and chemical resistance than crystal polystyrene has. Because they are polar in nature, SAN resins are hygroscopic and require drying before processing. SAN is injection-molded into such products as dishwasher-safe housewares, refrigerator shelves, medical devices, ovencaps, connectors for PVC tubing, and lenses.

When olefin-modified, SAN becomes a tough, weatherable polymer that can be extruded into sheets and profiles as well as injection-molded. Its primary use, however, is as an extruded capstock for ABS and other less expensive substrates to provide weatherability for applications such as spas, boats, swimming pools, and recreational vehicles.

Styrene–Butadiene (SB) Copolymers. These block copolymers combine transparency and high impact strength. Unlike SAN, SB copolymers do not have to be dried before processing. Packaging is a major market in items such as cups, lids, trays, and clamshells for the fast-food industry.

Styrene–Maleic Anhydride (SMA) Copolymers. With heat distortion temperatures reaching almost 250°F, SMA copolymers can be used in applications outside the range of other styrenics. The maleic anhydride compo-

Fig. 3-6. Foamed polystyrene insulation panel. (*Courtesy of Huntsman Chemical Corp.*)

nent also imparts extra adhesion to glass fibers in reinforced grades. Impact grades are produced by copolymerization with butadiene or by alloying with ABS. SMA materials require thorough drying before processing.

Applications for SMA include automotive instrument panels, headliners, and interior trim, instrument panel supports (glass-reinforced), appliances, business machines, and pumps. Electroplatable grades are used to mold automotive and appliance parts.

Vinyl Plastics

Polyvinyl Chloride (PVC). Polyvinyl chloride is one of the world's most widely used plastics. Its acceptance comes from the polymer's versatility; it can be utilized in rigid compounds or blended with plasticizers to produce flexible grades.

Most PVC is made via suspension polymerization. A small amount is made by mass polymerization, and plastisols and organosols are produced by the emulsion process.

Rigid PVC compounds normally contain resin, a heat stabilizer, and an impact modifier such as ABS or chlorinated polyethylene. Powder compounds are made by combining the ingredients in a high-intensity mixer. Pelletized compounds are made on twin-screw extrusion lines.

Flexible PVC compounds contain plasticizers to soften the resin. These additives are high-voiling solvents for PVC such as dioctyl phthalate (DOP) and didecyl phthalate (DDP). The plasticizers also act as processing aids.

Because it degrades relatively easily and because the degradation products are corrosive, the processing of PVC requires special care. Machinery surfaces that come in contact with the polymer should be corrosion-resistant (e.g., chrome-plated). Flow paths should be as smooth as possible with no "dead areas" where material can collect and degrade. With proper precautions, rigid PVC can be processed as easily as other thermoplastics.

PVC resin is self-extinguishing. In a fire, however, it produces hydrochloric acid and other toxic and corrosive chemicals. It can be burned in a properly designed incinerator without releasing any of these chemicals into the atmosphere.

A chief limitation of PVC is its heat resistance. Its heat distortion temperature is only about 160°F at 264 psi. Alloying with ABS and other polymers can improve PVC's performance at higher temperatures.

Almost three-quarters of the rigid PVC produced goes into building and construction applications (see Fig. 3-7). Most is processed via extrusion into products such as pipe, siding, and window profiles. Packaging is another major market for PVC. Rigid grades are blown into bottles and made into sheet for thermoforming boxes and blister packs. Flexible compounds are made into food wrap. Other markets for PVC include wire and cable coating, flooring, garden hose, and toys.

PVC dispersion resins in the form of plastisols and organosols are processed by spray and dip coating, casting, and slush molding. The resin is fused by heat, and the solvents are removed. Applications include toys, wall coverings, and footwear.

Chlorinated Polyvinyl Chloride (CPVC). Based on PVC, the chlorinated resin has a higher distortion temperature under load (200–230°F at 265 psi) and more combustion resistance than PVC. CPVC is used to make pipe and fittings for potable water and industrial chemicals, dark-color window frames, and housings for appliances and business equipment.

Polyvinylidene Chloride (PVdC). Copolymers of vinylidene chloride with vinyl chloride or other monomers, these materials exhibit exceptional barrier resistance to oxygen, carbon dioxide, water, and many organic solvents. Care in processing PVdC includes keeping melt temperatures below 400°F and using corrosion-resistant metals wherever hot resin contacts the equipment surface. PVdC is made into monolayer films (Saran) for food wrap and medical packaging, and it also is used in coextruded film and sheet structures as a barrier layer. PVdC coatings are applied to containers to prevent gas transmission but have to be removed before the packages are recycled.

Other Thermoplastics

Acrylics. Acrylics have outstanding resistance to long-term exposure to sunlight and weathering. Polymethyl methacrylate (PMMA), a hard, rigid, and transparent material, is the most widely used member of the acrylic family. Cast PMMA sheet has excellent optical properties (it transmits about 92% total light) and is more resistant to impact than glass. It is not so resistant to surface scratching as glass, but surface coatings can partially overcome this limitation.

In addition to excellent optical properties,

Fig. 3-7. Vinyl R&D laboratory building with PVC siding. (*Courtesy of Borden Chemicals and Plastics*)

Fig. 3-8. Acrylic automotive lenses. (*Courtesy of CYRO Industries*)

acrylics have low water absorption, good electrical resistivity, and fair tensile strength. The heat resistance of acrylics is on the order of 200°F.

Recently, modified acrylics and acrylic multipolymers that offer high impact strength and toughness in addition to the standard acrylic properties have been made available. These grades incorporate elastomeric or alloying constituents that impart added strength (up to 10 to 20 times as much as that of general-purpose acrylic crystal). Acrylics are available as compounds for extrusion, injection molding, blow molding, and casting. Extruded or cast sheet and film also are marketed.

Typical applications include outdoor signs, glazing, aircraft canopies, skylights, auto taillights, dials, buttons, lighting applications, knobs, and machine covers (see Fig. 3-8). Some of the transparent acrylic multipolymers have found applications in the drug and food packaging industry.

Acrylonitrile-Based Resins. Sometimes called *barrier resins*, this family of plastics is intended primarily for use in packaging. Major characteristics of the materials are barrier properties relating to the transmission of gas, aroma, or flavor of the package's contents. The resins are similar in melt processibility to rigid PVC and some grades of ABS; they can be extruded, blow-molded, and thermoformed.

Cellulosics. The family of cellulosics includes cellulose acetate, cellulose acetate butyrate, cellulose propionate, and ethyl cellulose. There are other cellulosics as well, but these are the most widely used. Cellulosics are characterized by good strength, toughness, and transparency and a high surface gloss. In addition, they have good chemical resistance. Generally, these thermoplastics should not be used at temperatures much above 170 to 220°F.

Cellulose acetate, which is the lowest-cost cellulosic material, has good toughness and rigidity. This easily molded material is available in a variety of grades ranging from "soft" to "hard."

Cellulose acetate butyrate, although a little more expensive than the straight acetate, is somewhat tougher, with a hornlike quality, and has lower moisture absorption. It has relatively good weatherability and excellent transparency.

Cellulose propionate is similar to cellulose acetate butyrate in both cost and properties, but it has somewhat higher tensile strength, modulus, and impact strength than the latter material.

Ethyl cellulose is what might be called the impact grade of the cellulosics. The excellent toughness of this material (Izod impact strengths range from 1.7 to 8.5 ft-lb/in. of notch, depending on the specific resin) is maintained over a wide temperature range. Ethyl cellulose also has moderately low moisture absorption. In addition, this cellulosic is available in self-lubricating grades.

Nearly all cellulosics are noted for their toughness, but none of these materials is generally recommended for applications involving anything more than relatively low loads. The main feature of cellulosics is their excellent moldability, which results in a brilliant, high-gloss finish.

Cellulose acetate is subject to dimensional changes due to cold flow, extreme heat, and moisture absorption. Cellulose acetate butyrate is a slightly more stable material, but still not outstanding in comparison to some other thermoplastics. This material, however, is one of the few thermoplastics that resist weathering. This property, combined with its good optical properties, makes cellulose acetate butyrate an excellent material for outdoor signs. Cellulose propionate is similar in properties to the butyrate but has higher hardness and stiffness (see Fig. 3-9). Ethyl cellulose has several outstanding properties. It withstands heavy abuse and has good environmental resistance, and it retains its toughness at low temperatures. Like other cellulosics, however, it is not primarily a load-bearing material.

Cellulosic compounds are available for extrusion, injection molding, blow molding, and rotational molding. Cellulosics also are widely used in the form of film and sheet.

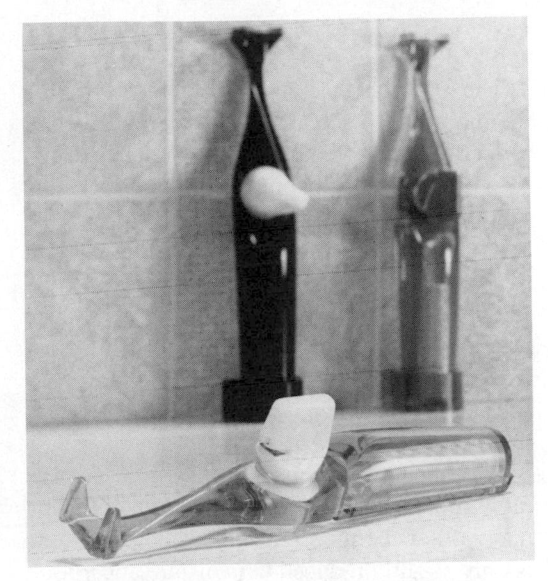

Fig. 3-9. Dental flossing instrument molded from cellulose propionate. (*Courtesy of Eastman Chemical Co.*)

Typical applications of the materials include: cellulose acetate—knobs, tool handles, and face shields; cellulose acetate butyrate—tool handles, knobs, tubular packaging, toys, and signs; cellulose propionate—automobile steering wheels, pen and pencil barrels, packaging sheet, and toys; ethyl cellulose—flashlight housings, tool handles, and roller wheels.

Engineering Thermoplastics

Acetals. Polymers of formaldehyde, correctly called polyoxymethylenes (POM), acetals are among the strongest (tensile strength: 10,000 psi) and stiffest (modulus in flexure: 410,000 psi) thermoplastics; acetals also are characterized by excellent fatigue life and dimensional stability. Other outstanding properties include low friction coefficients, exceptional solvent resistance, and high heat resistance for extended use up to 220°F.

At present there are two basic types of acetal: homopolymer and copolymer. The homopolymers are somewhat tougher and harder than the copolymers but suffer from instability in processing. Recent stabilization improvements have reduced odor and mold deposit problems and widened the processing temperature range.

The fact that acetals are highly crystalline thermoplastics accounts for their excellent properties and predictable long-range performance under load. In creep resistance, acetal is one of the best thermoplastics; however, the apparent modulus falls off consistently with long-term loading.

Acetals have an excellent fatigue endurance limit; at 100% RH it is 5000 psi at 77°F, and it is still 3000 psi at 150°F. Furthermore, lubricants and water have little effect on the fatigue life.

The impact strength of acetals does not fall off abruptly at subzero temperature like that of many other thermoplastics, and their hardness is only slightly reduced by moisture absorption or temperature below 215°F. Although less good than that of nylons, the abrasion resistance of acetals is better than that of many other thermoplastics. Like nylons, acetals have a slippery feel.

Acetals are especially notable among thermoplastics because of their resistance to organic solvents. However, when in contact with strong acids, acetals will craze.

In addition to good mechanical properties, acetals have good electrical properties. The dielectric constant and the dissipation factor are uniform over a wide frequency range and up to temperatures of 250°F. Aging has little effect on an acetal's electrical properties.

Acetals are available as compounds for injection molding, blow molding, and extrusion. Filled, toughened, lubricated, and UV-stabilized versions currently are offered. Grades reinforced with glass fibers (higher stiffness, lower creep, improved dimensional stability), fluoropolymer, and aramid fibers (improved frictional and wear properties) also are on the market.

Many applications involve replacement of metals where the higher strength of metals is not required and costly finishing and assembly operations can be eliminated (see Fig. 3-10). Typical parts include gears, rollers, and bearings, conveyor chains, auto window lift mechanisms and cranks, door handles, plumbing components, and pump parts.

Fluoroplastics. Outstanding properties of fluorocarbon polymers (or fluoroplastics, as they also are called) include inertness to most chemicals and resistance to high temperatures. Fluoroplastics have a rather waxy feel, extremely low coefficients of friction, and excellent dielectric properties that are relatively insensitive to temperature and power frequency. Their mechanical properties normally are low, but change dramatically when the fluorocarbons are reinforced with glass fibers or molybdenum disulfide fillers.

Numerous fluoroplastics are available, including polytetrafluoroethylene (PTFE), polychlorotrifluoroethylene (PCTFE), fluorinated ethylene propylene (FEP), ethylene chlortrifluoroethylene (ECTFE), ethylene tetrafluoroethylene (ETFE), polyvinylidene fluoride (PVDF), polyvinylfluoride (PVF), and perfluoroalkoxy (PFA) resin.

PTFE is extremely heat-resistant (up to 500°F) and has outstanding chemical resistance, being inert to most chemicals. Its coefficient of friction is lower than that of any other plastic, and it can be used unlubricated. TFE has a tensile strength on the order of 1500 to 5000 psi and an impact strength of 2.5 to 3.0 ft-lb/in. of notch. TFE also has outstanding low-temperature characteristics and remains usable even at cryogenic temperatures.

PTFE is extremely difficult to process via melt extrusion and molding. The material usually is supplied in powder form for compression and sintering or in water-based dispersions for coating and impregnating.

CTFE has a heat resistance of up to 390°F, and is chemically resistant to all inorganic corrosive liquids, including oxidizing acids. It also is resistant to most organic solvents except certain halogenated materials and oxygen-containing compounds (which cause a slight swelling).

In terms of processibility, CTFE can be molded and extruded by conventional thermoplastic processing techniques. It can be made into transparent film and sheet with extremely low water vapor transmission.

FEP, a copolymer of TFE and hexafluoropropylene, shows some of the same properties as TFE (i.e., toughness, excellent dimensional stability, and outstanding electrical insulating

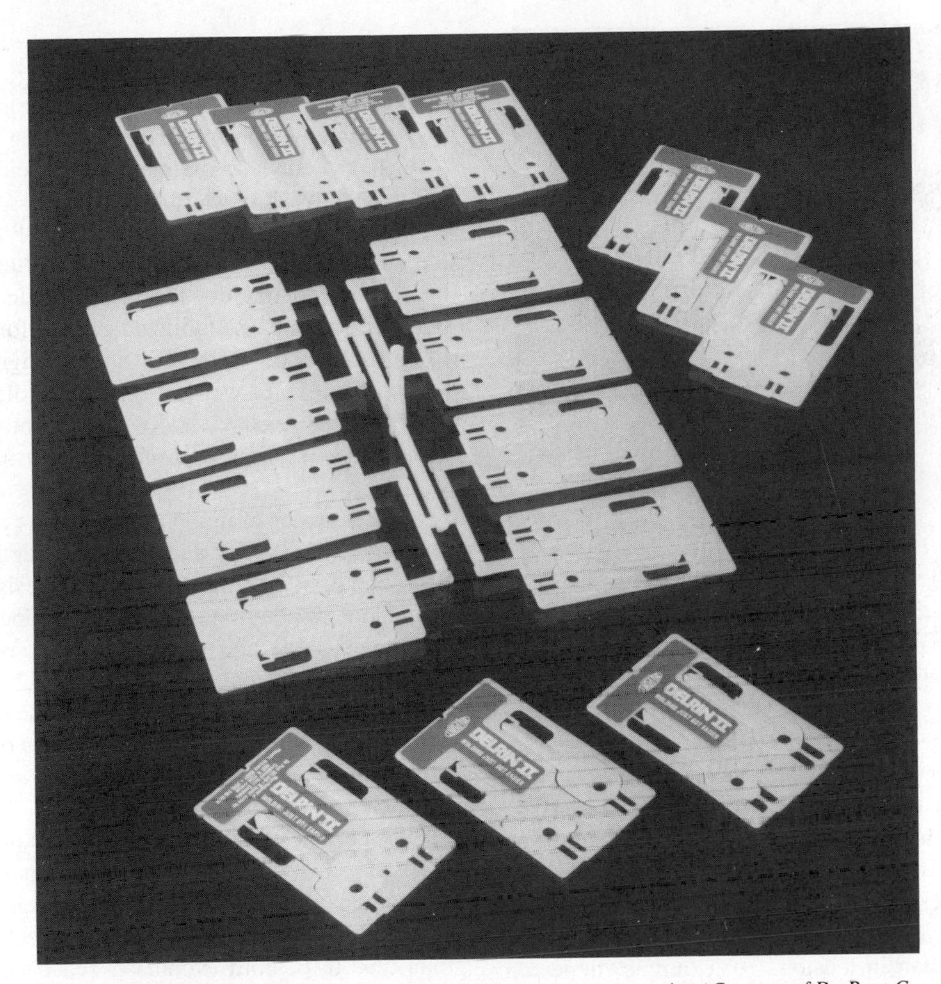

Fig. 3-10. "Credit card" spare car keys molded from acetal homopolymer resin. (*Courtesy of Du Pont Company*)

characteristics over a wide range of temperatures and humidities), but it exhibits a melt viscosity low enough to permit it to be molded by conventional thermoplastic processing techniques. FEP is used as a pipe lining for chemical process equipment, in wire and cable applications, and for glazing in solar collectors.

ECTFE, an alternating copolymer of ethylene with chlorotrifluoroethylene, is significantly stronger and more wear-resistant than PTFE and FEP. Its ignition and flame resistance makes it suitable for wire and cable applications, including plenum cable jacketing. Other uses include tank linings, tower packings, and pump and valve components for the chemical process industry and corrosion-resistant film and coatings.

ETFE is a related copolymer to ECTFE, consisting of ethylene and tetrafluoroethylene. Although its continuous-use temperature is somewhat higher (355°F) than that of ECTFE, it melts and decomposes when exposed to a flame. (See Fig. 3-11.)

PVDF also is a tough fluoroplastic with high chemical resistance and weatherability. It is self-extinguishing and has a high dielectric loss factor. A melt-processible material, it is used in seals and gaskets for chemical process equipment and in electrical insulation.

PVF, available only in film form, is used as a protective lamination on plywood, hardboard, and other panel constructions, as well as for interior truck linings.

Lastly, PFA is a newer family of melt-processible fluoropolymers, similar to FEP but with higher temperature resistance.

Fig. 3-11. Ball valve lined with glass fiber reinforced ETFE. (*Courtesy of ICI Advanced Materials*)

When reinforced with such materials as molybdenum disulfide, graphite, asbestos, or glass, the fluorocarbons show increased stiffness, hardness, and compressive strength, and reduced elongation and deformation under load.

Stress–strain data on fluorocarbons show these materials to be much better in compression than in tension. For example, depending on crystallinity, TFE has a tensile yield strength at 73°F of about 1800 psi (30% strain); in compression, however, this value can be as high as 3700 psi (25% strain).

Polyamides (Nylons). The nylon family members are identified by the number of carbon atoms in the monomers. Where two monomers are involved, the polymer will carry two numbers (e.g., nylon 6/6). Crystalline nylons have high tensile strength, flex modulus, impact strength, and abrasion resistance. Nylons resist nonpolar solvents, including aromatic hydrocarbons, esters, and essential oils. They are softened by and absorb polar materials such as alcohols, glycols, and water. Moisture pickup is a major limitation for nylons because it results in dimensional changes and reduced mechanical properties. However, overly dry material can cause processing problems, and particular attention needs to be paid to moisture control.

Several different types of nylon are on the market, the two most widely used being nylon 6/6 (hexamethylene diamine adipic acid) and nylon 6 (polycaprolactam).

Nylon 6/6 is the most common polyamide molding material. Its special grades include: (1) heat-stabilized grades, for molding electrical parts; (2) hydrolysis-stabilized grades, for parts to be used in contact with water; (3) light-stabilized grades, for weather-resistant moldings; and (4) higher-melt-viscosity grades, for molding of heavy sections and for better extrudability.

Special grades of nylon 6 include: (1) grades with higher flexibility and impact strength; (2) heat- and light-stabilized grades, for resistance to outdoor weathering; (3) grades for incorporating nucleating agents to promote consistent crystallinity throughout sections, thereby providing better load-bearing characteristics; and (4) higher viscosity grades, for extrusion of rod, film, pipe, large shapes, and blow-molded products.

Other types of nylon include: 6/9, 6/10, 6/12, 11, 12, 4/6, and 12/12. A transparent, amorphous nylon has been commercialized. Although type 6 absorbs moisture more rapidly than type 6/6, both eventually reach equilibrium at about 2.7% moisture content in 50% RH air, and about 9 to 10% in water.

Creep rates for nylons at various stress levels under both tension and compression show only a small deformation within the initial 24-hour period and increase from this point in a linear manner. This means that long-term deformation usually can be predicted accurately on the basis of short-term tests.

General-purpose nylon molding materials are available for extrusion, injection molding, blow molding, rotational molding, and (for the nylon 6 materials) casting or anionic polymerization. Nylon sheet and film also are marketed.

The properties of nylon resins can be improved by filling and reinforcing. Mineral-filled and glass-fiber-reinforced compounds are widely available. (See Fig. 3-12.) Several manufacturers offer specially toughened grades where extra impact strength is required.

Fig. 3-12. Housing for power blower is molded from reinforced nylon 6 resin. (*Courtesy of Akzo Engineering Plastics*)

For specific engineering applications, a number of specialty nylons have been developed, including; molybdenum disulfide-filled nylons (to improve wear and abrasion resistance, frictional characteristics, flexural strength, stiffness, and heat resistance); glass-fiber-filled nylons (to improve tensile strength, heat distortion temperatures, and, in some cases, impact strength); and sintered nylons. Sintered nylons are fabricated by processes similar to powder metallurgy (the same as those used for TFE fluorocarbons). The resulting materials have improved frictional and wear characteristics, as well as higher compressive strength.

Nylons have also found increasing use in alloys, notably with ABS and with PPE. The materials' aroma barrier properties have led to applications in multilayer packaging film constructions. In many electrical applications, the mechanical strength and resistance to oils and greases also are important properties.

Polyamide-imide (PAI). Among the highest-temperature amorphous thermoplastics, polyamide-imide resins have a useful service temperature range from cryogenic to almost 500°F. Their heat resistance approaches that of polyimides, but their mechanical properties are distinctly better than those of the polyimides.

PAI is inherently flame-retardant, with UL94 V-O ratings down to less than 0.010 inch thickness. The material burns with very low smoke and passes FAA standards for aircraft interior use.

The chemical resistance of PAI is excellent; it generally is unaffected by aliphatic and aromatic hydrocarbons, acids, bases, and halogenated solvents. At high temperatures, however, it is attacked by steam and strong acids and bases.

Polyamide-imide can be injection-molded, but some special modifications are needed because the material is reactive at processing temperatures. Special screws and accumulators are recommended. To develop the full physical properties of PAI, molded parts have to be post-conditioned by gradually raising their temperature.

Applications for polyamide-imide include engine and generator components, hydraulic bushings and seals, and mechanical parts for electronics and business machines.

Polyarylates. These relatively new materials are amorphous, aromatic polyesters. They fall between polycarbonate and polysulfone in terms of temperature resistance, with a deflection temperature under load (DTUI) of 300 to 350°F at 264 psi. Other features are toughness, dimensional stability, UV resistance, flame retardancy, and good electrical properties.

Polyarylate polymers are transparent but tend toward yellowness. One producer has an almost water-white material.

Markets for polyarylate are in automotive parts such as headlamp and mirror housings, brackets, and door handles, and in electrical/electronics components such as connectors, fuses, and covers. The material's temperature resistance allows it to withstand the soldering temperatures encountered in making circuit boards.

Polycarbonates. Polycarbonates are among the strongest, toughest, and most rigid thermoplastics. In addition, they have a ductility normally associated with the softer, lower-modulus thermoplastics.

These properties, together with excellent electrical insulating characteristics, are maintained over a wide range of temperatures (−60 to 270°F) and loading rates. Although there may be a loss of toughness with heat aging, the material still remains stronger than many thermoplastics.

Polycarbonates are transparent materials and resistant to a variety of chemicals, but they are attacked by a number of organic solvents including carbon tetrachloride solvents.

The creep resistance of these materials is among the best for thermoplastics. With polycarbonates, as with other thermoplastics, creep at a given stress level increases with increasing temperature; yet even at temperatures as high as 250°F, their creep resistance is good.

The characteristic ductility of polycarbonate provides it with very high impact strength. Typical values are about 14 ft-lb/in. of notch on a 1/8-inch-thick specimen, although grades are available as high as 18 ft-lb. Unnotched specimens show impact resistance greater than 60 ft-lb. Their fatigue resistance also is very good.

The moisture absorption for polycarbonates is low, with equilibrium reached rapidly. However, the materials are adversely affected by weathering (slight color change and slight embrittlement can occur on exposure to ultraviolet rays).

Polycarbonate molding compounds are available for extrusion, injection molding, blow molding, and rotational molding. Film and sheeting with excellent optical and electrical properties also are available. Among the specialty grades, glass-reinforced polycarbonates have proved especially popular by virtue of their improved ultimate tensile strength, flexural modulus, tensile modulus, and chemical resistance.

Typical applications include the following: safety shields, lenses, glazing, electrical relay covers, helmets, pump impellers, sight gauges, cams and gears, interior aircraft components, automotive instrument panels, headlights, lenses, bezels, telephone switchgear, snowmobile components, boat propellers, water bottles, housings for hand-held power tools and small appliances, and optical storage disks. (See Fig. 3-13.)

Polyesters, Thermoplastic. The two dominant materials in this family are polyethylene terephthalate (PET) and polybutylene terephthalate (PBT). The thermoplastic polyesters are similar in properties to types 6 and 6/6 nylons but have lower water absorption and higher dimensional stability than the nylons.

To develop the maximum properties of PET, the resin has to be processed to raise its level of crystallinity and/or to orient the molecules. Orientation increases the tensile strength by 300 to 500% and reduces permeability. PET is a water-white polymer and is made into fibers, films and sheets, and blow-molded and thermoformed containers for soft drinks and foods. Glass-reinforced PET compounds can be injection-molded into parts for automotive, electrical/electronic, and other industrial and consumer products.

PBT, on the other hand, crystallizes readily, even in chilled molds. Most PBT is sold in the form of filled and reinforced compounds for engineering applications. Its uses include appliances, automotive, electrical/electronic, materials handling, and consumer products.

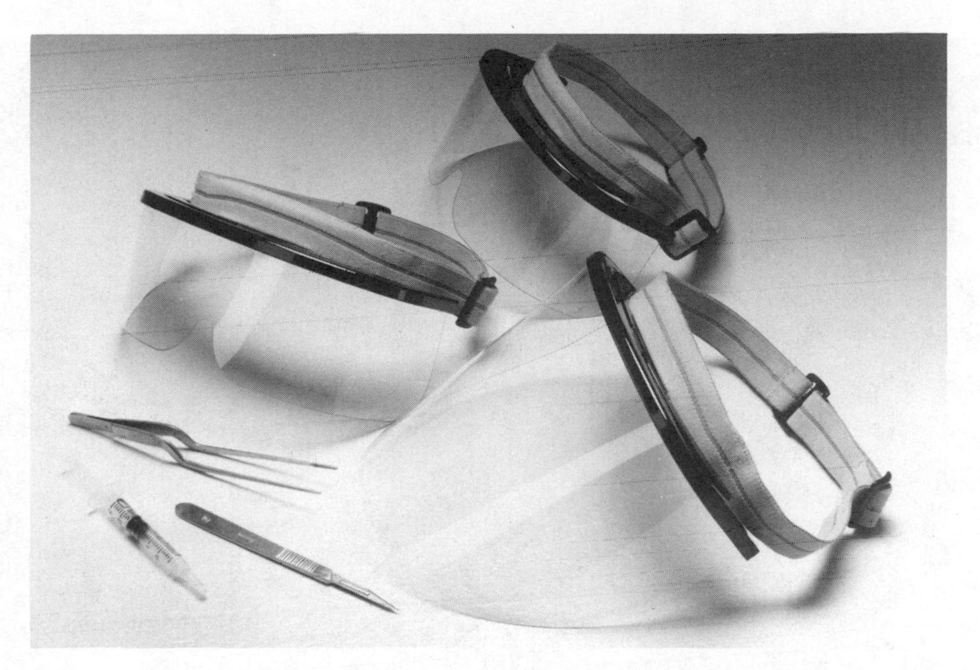

Fig. 3-13. Transparent polycarbonate protective eyewear. (*Courtesy of GE Plastics*)

Some newer members of the polyester family are starting to find application in specific niche markets. PCT (poly-1-4-cyclohexylene-dimethylene terephthalate) has higher heat resistance than either PET or PBT, and is used mainly as a component of blends. PEN (polyethylene naphthalate) has a combination of higher heat resistance, mechanical strength, chemical resistance, and dimensional stability, and can be processed into films, fibers, and containers. PEN has up to five times the oxygen barrier of PET.

Thermoplastic polyesters can be used as a component in alloys and blends. Widely used combinations include those with polycarbonate, polysulfone, and elastomers.

Also in the family of thermoplastic polyesters is the class of materials known as liquid crystal polymers (LCPs), aromatic copolyesters with a tightly ordered structure that is self-reinforcing. LCPs generally flow very well in processing, but they have to be thoroughly dried and molded at high temperatures. They also exhibit very high mechanical properties although molded structures tend to be quite anisotropic. The anisotropy can be reduced by proper gating and mold design and by incorporating mineral fillers and glass fiber reinforcements. LCPs are resistant to most organic solvents and acids. They are inherently flame-resistant (UL94 V-O) and meet federal standards for aircraft interior use. Some LCPs withstand temperatures over 1000°F before decomposing.

LCPs have found application in aviation, electronics (connectors, sockets, chip carriers), automotive underhood parts, and chemical processing, and are used to mold household cookware for conventional and microwave ovens. (See Fig. 3-14.)

Polyetherimide (PEI). An amorphous, transparent amber polymer, polyetherimide combines high temperature resistance, rigidity, impact strength, and creep resistance. PEI has a glass transition temperature of 420°F and a DTUL of 390°F at 264 psi. Thorough drying is required before processing, and typical melt temperatures for injection molding run from 650 to 800°F. PEI resins qualify for UL94 V-O ratings at thicknesses as low as 0.010 inch and meet FAA standards for aircraft interiors.

Polyetherimide is soluble in partially halogenated solvents but resistant to alcohols, acids,

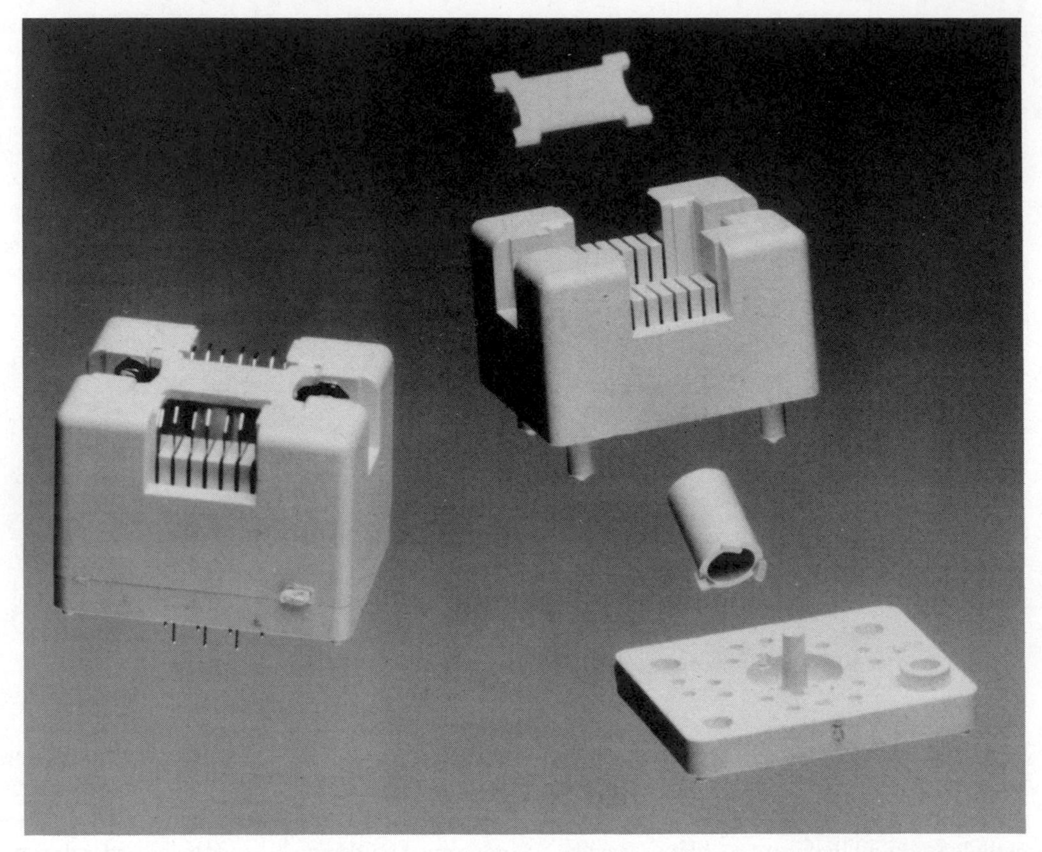

Fig. 3-14. Chip carrier test socket molded from aromatic polyester liquid crystal polymer. (*Courtesy of Amoco Performance Products*)

and hydrocarbon solvents. It performs well under humid conditions and withstands UV and gamma radiation.

Glass-fiber-reinforced PEI grades are available for general-purpose molding and extrusion; carbon-fiber-reinforced and other specialty grades also are produced for high-strength applications; and PET itself can be made into a high-performance thermoplastic fiber.

PEI has found use in medical applications because of its heat and radiation resistance, hydrolytic stability, and transparency; in the electronics field, it is used to make burn-in sockets, bobbins, and printed circuit substrates; automotive uses include lamp sockets and underhood temperature sensors; PEI sheeting is used in aircraft interiors; and extruded PET has been used as a metal replacement for furnace vent pipe. (See Fig. 3-15).

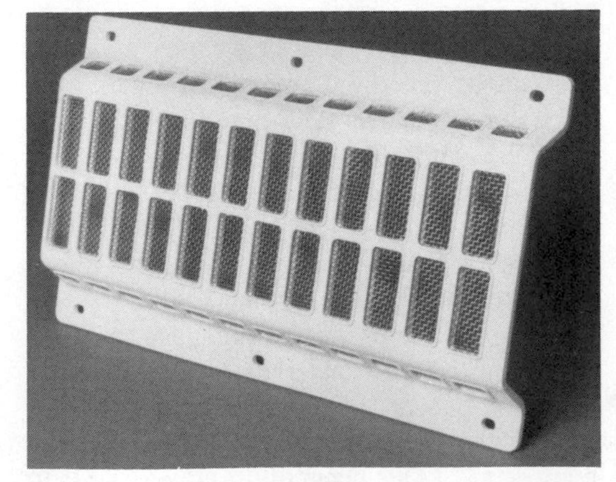

Fig. 3-15. Aircraft cargo vent formed from polyetherimide sheets. (*Courtesy of Transparent Products Corp.*)

Polyimide, Thermoplastic. These linear, aromatic polymers are prized for their high-temperature performance. While nominally thermoplastic, polyimides will degrade at temperatures below their softening point and thus have to be processed in precursor form. Material is supplied in powder form for compression molding and cold forming, and some injection molding grades are being offered.

The outstanding physical properties of the polyimides make them valuable in applications involving severe environments such as high temperature and radiation, heavy load, and high rubbing velocities. In terms of heat resistance, the continuous service of polyimide in air is on the order of 500°F. At elevated temperatures, polyimide parts retain an unusually high degree of strength and stiffness, but prolonged exposure at high temperatures can cause a gradual reduction in tensile strength.

Polyimides are resistant to most dilute or weak acids and organic solvents; they are not attacked by aliphatic or aromatic solvents, ethers, esters, alcohols, and most hydraulic fluids. However, they are attacked by bases such as strong alkali and aqueous ammonia solutions; and the resin is not suitable for prolonged exposure to hydrazine, nitrogen dioxide, or primary or secondary amine components. Parts fabricated from unfilled polyimide resin have unusual resistance to ionizing radiation and good electrical properties.

Polyimides are used to mold high-performance bearings for jet aircraft, compressors, and appliances. In film form, the material is used for electric motor insulation and in flexible wiring. Polyimides also are used in making printed circuit boards.

Polyketones. These materials are crystalline polymers of aromatic ketones that exhibit exceptional high-temperature performance.

Polyetheretherketone (PEEK) is tough and rigid, with a continuous-use temperature rating of 475°F. The polymer is self-extinguishing (UL94 V-O rated) and exhibits very low smoke emission on burning. Its crystallinity gives it resistance to most solvents and hydrolytic stability. PEEK also is resistant to radiation.

Polyetherketone (PEK) polymers are similar to PEEK but have slightly higher (about 10°F) heat resistance.

Polyketones can be processed by conventional methods, with predrying a requirement. Injection-molding melt temperatures range from 710 to 770°F, and mold temperatures have to be over 300°F. The materials also can be extruded for wire coating and for slot cast films. Powdered resin is supplied for rotational molding and for coating.

Glass- and carbon-fiber-reinforced polyketone compounds are standard, and PEEK also is combined with continuous carbon fiber to produce a thermoplastic prepreg for compression molding.

Because of their high price, uses for polyketones generally are in the highest-performance applications such as metal replacement parts for aircraft and aerospace structures, electrical power plants, printed circuits, oven parts, and industrial filters.

Polyphenylene Ether (PPE). Also referred to as polyphenylene oxide (PPO), these resins are combined with other polymers to make useful alloys. PPE is compatible with polystyrene and is blended with (usually high-impact) PS over a wide range of ratios, yielding products with DTUL ratings from 175 to 350°F.

Because both PPE and PS are hydrophobic, the alloys have very low water absorption rates and high dimensional stability. They exhibit excellent dielectric properties over a wide range of frequencies and temperatures.

The resin will soften or dissolve in many halogenated or aromatic hydrocarbons. If an application requires exposure to or immersion in a given environment, stressed samples should be tested under operating conditions.

PPE/PS alloys are supplied in flame-retardant, filled and reinforced, and structural foam molding grades. They are used to mold housings for appliances and business machines (see Fig. 3-16), automotive instrument panels and seat backs, and fluid handling equipment. Blow molding sometimes is used to make large structural parts.

PPE can also be alloyed with nylon to provide increased resistance to organic chemicals and better high-temperature performance. PPE/nylon blends have been used to injection-mold automobile fenders.

Fig. 3-16. Housing for portable marine air conditioner from foamed polyphenylene ether blend. (*Courtesy of GE Plastics*)

Polyphenylene Sulfide (PPS). This resin is characterized by excellent high-temperature performance, inherent flame retardance, and good chemical resistance. The crystalline structural of PPS is best developed at high mold temperatures (250–300°F).

Most PPS is sold in the form of filled and reinforced compounds for injection molding. Because of its heat and flame resistance, the material has been used to replace thermoset phenolics in electrical/electronics applications. PPS will withstand contact with metals at temperatures up to 500°F, and in the electrical/electronics market, its biggest use is in molding connectors and sockets. Because it can withstand vapor-phase soldering temperatures, the material is gaining wide use in molding surface-mounted components, circuit boards, and so on.

Another major application for PPS is in industrial pump housing, valves, and downhole parts for the oil drilling industry. PPS is used to mold appliance parts that must withstand high temperatures; applications include heaters, dryers, microwave ovens, and irons.

Its heat and chemical resistance have led to the use of PPS in automotive under-the-hood components, electronics, and lighting systems, as well as medical uses.

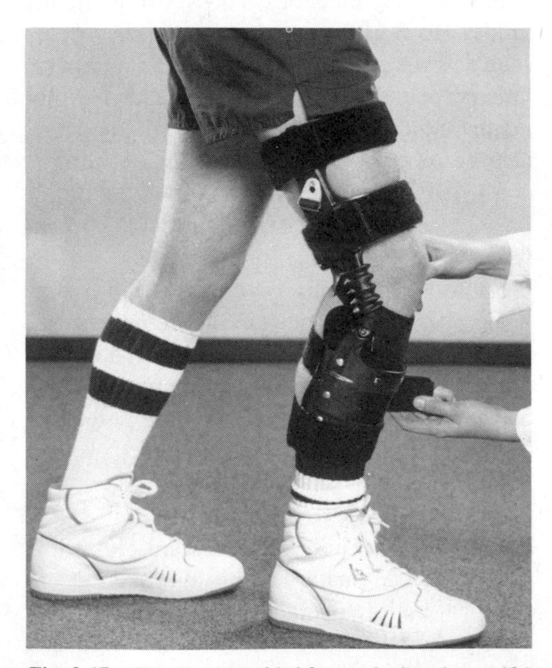

Fig. 3-17. Knee brace molded from polyphenylene sulfide composite. (*Courtesy of Phillips 66 Co.*)

Sulfone Polymers.

Polysulfone (PSO). Polysulfone is a transparent, amorphous thermoplastic that is stable, heat-resistant, and self-extinguishing, and can be molded, extruded, and thermoformed into a wide variety of shapes. Drying is required before processing. Characteristics of special significance to the design engineer are its heat distortion temperature of 345°F at 264 psi and its UL-rated continuous-service temperature of 320°F. Its glass-transition temperature, T_g, is 374°F.

Polysulfone is rigid, with a flex modulus of almost 400,000 psi, and strong, with a tensile strength of 10,200 psi at yield. Its electrical properties are good; it has a high dielectric strength and a low dissipation factor. The electrical and mechanical properties are retained at temperatures up to 350°F and after immersion in water or exposure to high humidity.

Polysulfone is highly resistant to mineral acid, alkali, and salt solutions. Its resistance to detergents, oils, and alcohols is good even at elevated temperatures under moderate levels of stress. PSO is attacked by polar organic solvents such as ketones and chlorinated and aromatic hydrocarbons.

Mineral-filled and glass-reinforced grades of polysulfone are offered where additional strength is required, and plating grades also are available.

PSO is used in medical instruments and in sterilizing equipment (see Fig. 3-18). Its high-temperature capabilities and transparency are useful in the food industry, where it is used for processing equipment, coffee carafes, beverage dispensers, and piping. Polysulfone also is used to mold microwave oven cookware. In the electronics field, PSO goes into circuit boards, capacitor films, switches, and connectors.

Fig. 3-18. Transparent polysulfone laboratory filter funnel. (*Courtesy of Amoco Performance Products*)

Polyarylsulfone. This amorphous material consists of phenyl and biphenyl groups linked by thermally stable ether and sulfone groups. It is distinguished from polysulfone polymers by the absence of aliphatic groups, which are liable to oxidative attack.

Polyarylsulfone is characterized by a high heat-deflection temperature, 400°F at 264 psi, and maintains its mechanical properties up almost to that temperature. At normal ambient temperatures, polyarylsulfone is a strong, stiff, tough material that in general offers properties comparable to those of other engineering thermoplastics.

Polyarylsulfone has good resistance to wide variety of chemicals, including acids, bases, and common solvents. It is unaffected by practically all fuels, lubricants, hydraulic fluids, and cleaning agents used on or around electrical components.

Polyarylsulfone may be injection-molded on conventional equipment with sufficient injection pressure and temperature capabilities. Generally, the cylinder and nozzle must be equipped to reach temperatures of 800°F, and the injection system should be capable of providing from 20,000 to 30,000 psi injection pressure. Polyarylsulfone also can be extruded on equipment of varying design. The material must be dried before processing.

Polyarylsulfone can be supplied in transparent or opaque colors and in filled and reinforced grades. Its high-temperature resistance and combustion characteristics have led to its use in many electrical/electronics applications, such as motor parts, lamp housings, connectors, and printed circuit boards.

Polyethersulfone (PES). This transparent, amorphous polymer is one of the highest-temperature engineering thermoplastics, carrying a UL rating for continuous service of 356°F. The polymer is self-extinguishing and, when burned, emits little smoke.

PES is rigid, tough, and dimensionally stable over a wide temperature range. It is resistant to most solvents, including gasoline and aliphatic hydrocarbons; however, polar, aromatic hydrocarbons will attack the resin, as will methylene chloride, ketones, and esters.

PES can be processed on conventional injection molding, blow molding, extrusion, and thermoforming equipment. Mold shrinkage is low.

The low flammability and low smoke generation of PES make it a good candidate to meet FAA regulations for aircraft interior parts. Transparent grades of PES are used in medical and food processing applications and in electrical components such as illuminated switches. Both neat and reinforced grades are used in appliances and power tools, pumps, sockets and connectors, and printed circuit boards.

Thermoplastic Elastomers (TPEs)

Though not a single family of materials, thermoplastic elastomers have a set of properties in common. They can be processed by conventional thermoplastic extrusion and molding techniques but function much like thermoset rubbers.

The earliest commercial TPEs, introduced in the 1950s, were the thermoplastic polyurethanes (TPUs). Since that time, other types of materials have been commercialized, among them styrenic block copolymers, copolyesters, olefin blends, and rubber olefin alloys.

TPUs are on the high end of the price and property range of thermoplastic elastomers. They can be processed easily at low melt temperatures (under 450°F) but must be thoroughly predried. Because of their natural adhesive properties, mold releases usually are required.

TPUs are extruded into hoses and tubing for automotive, medical, and electronics applications. Blown film is used for meat wrapping. Injection-molded TPU parts are used in auto exteriors, shoes and boots, drive wheels, and gears. TPUs can be blended with PVC styrenics, nylons, or polycarbonate to achieve a wide range of properties.

Styrenic block copolymers are the most commonly used TPEs. Their structure normally consists of a block of rigid styrene on each end with a rubbery phase in the center, such as styrene/butadiene/styrene (SBS), styrene/isoprene/styrene (SIS), styrene/ethylene-butylene/styrene (SEBS), and styrene/ethylene-propylene/styrene (SEPS).

Styrenic TPEs can be processed via conventional extrusion and molding techniques. With a hardness range from 28 to 95 Shore A, they can be made into a wide variety of products. Wire and cable jacketing, medical tubing and catheters, shoe soles, and flexible automotive parts are some examples. (See Fig. 3-19.)

Compatibility with many other polymers has led to the use of styrenic TPEs as impact modifiers and as compatibilizers for resin blends. They also are used as tie layers in coextruded products.

Olefinic TPEs, or TPOs as they are typically called, are blends of polypropylene with rubber (EPDM or EP) and polyethylene. They are characterized by high impact strength, low density, and good chemical resistance. TPOs can be easily processed by extrusion, injection, and blow molding.

Gas-phase polyolefin producers can make in-reactor TPOs with properties similar to those made by compounding. The in-reactor products offer cost advantages for large-volume applications. The largest single market for TPOs is in automotive exterior parts. Paintable for-

Fig. 3-19. Automotive shifter boot molded from styrenic block copolymer thermoplastic elastomer. (*Courtesy of Shell Chemical Co.*)

mulations are used to mold bumper fascia, air dams, and rub strips. Weatherable grades with molded-in accent colors are top-coated with a clear finish.

Olefin blends with vulcanized rubber (partially or fully crosslinked) form a special category of TPE. These materials can cover the broad property range from thermoplastic to thermoset rubber. They process easily in conventional extrusion and molding equipment and are replacing thermoset rubbers in applications such as hose and tubing, cable covering, gaskets and seals, and sheeting. Special grades are available for medical and food contact applications.

Recently, olefinic TPEs have been developed that use proprietary olefinic materials in place of vulcanized rubber as the soft phase. They are made by special blending and crosslinking methods.

Polyester TPEs are high-strength elastomeric materials. They are block copolymers with hard and soft segments providing strength and elasticity. The primary characteristic of polyester TPEs is their ability to withstand repeated flexing cycles. They retain their high impact strength at low temperatures and may have heat resistance up to 300°F. Polyester TPEs process well (they should be predried) by extrusion and molding. Applications include automotive parts (fascias, constant-velocity joint boots), sporting goods, and geomembrane sheeting.

Alloys and Blends

In addition to the variations on polymers created by copolymerization, there is also the possibility of combining finished thermoplastics to create new materials. When the combination of polymers has a single glass transition temperature and yields a synergistic effect (i.e., when the properties of the mix are significantly better than either of the individual components), the result is termed an alloy. When the resulting product has multiple glass transition temperatures and properties that are an average of the individual components, the product is referred to as a blend. To keep the phases of polymer blends from separating, especially where the components are chemically incompatible, compatibilizers are included in the mix.

Alloying and blending can be used to tailor-make materials with specific sets of properties for certain applications. The advantages of crystalline polymers (chemical resistance, easy flow) and amorphous polymers (low shrinkage, impact strength) can be combined in a single material. One of the earliest commercially successful blends was acrylonitrile butadiene styrene (ABS), which combines the chemical resistance, toughness, and rigidity of its components. Today the range of materials that can be produced by alloying and blending is almost limitless, and this has become the fastest growing segment of the engineering thermoplastics business.

The following is a partial list of alloys and blends available commercially:

ABS/nylon
Nylon/olefin elastomer
Polycarbonate/ABS
Polycarbonate/polyester
Polycarbonate/polypropylene
Polycarbonate/thermoplastic polyurethane

Polyphenylene ether/nylon
Polyphenylene ether/impact polystyrene
PVC/ABS
Polyester/elastomer
Polysulfone/ABS
Polysulfone/polyester
SMA/polycarbonate

THERMOSETTING RESINS

Among plastic materials, thermosetting materials generally provide one or more of the following advantages: (1) high thermal stability, (2) resistance to creep and deformation under load and high dimensional stability, and (3) high rigidity and hardness. These advantages are coupled with the light weight and excellent electrical insulating properties common to all plastics. The compression and transfer molding methods by which the materials are formed, together with the more recent evolution of thermoset injection molding techniques, offer low processing cost and mechanized production.

Thermosetting molding compounds consist of two major ingredients: (1) a resin system, which generally contains such components as curing agents, hardeners, inhibitors, and plasticizers; and (2) fillers and/or reinforcements, which may consist of mineral or organic particles, inorganic or organic fibers, and/or inorganic or organic chopped cloth or paper.

The resin system usually exerts the dominant effect, determining to a great extent the cost, dimensional stability, electrical qualities, heat resistance, chemical resistance, decorative possibilities, and flammability. Fillers and reinforcements affect all these properties to varying degrees, but their most dramatic effects are seen in strength and toughness and, sometimes, electrical qualities.

The following are summaries of the generic families of molding compounds. These descriptions cannot be all-inclusive in the space allowed; the attempt here is to characterize the materials in a brief and general manner.

Alkyds

Primarily electrical materials, alkyds combine good insulating properties with low cost. They are available in granular and putty form, permitting incorporation of delicate complex inserts. Their moldability is excellent, cure time is short, and pressures are low. In addition, their electrical properties in the RF and UHF ranges are relatively heat-stable up to a maximum use temperature of 250 to 300°F.

General-purpose grades normally are mineral-filled; compounds filled with glass or synthetic fibers provide substantial improvements in mechanical strength, particularly impact strength. Short fibers and mineral fillers give lower cost and good moldability; longer fibers give optimum strength.

Although the term ''alkyd'' in paint terminology refers to a fatty-acid-modified polyester, in molding compound parlance the term merely means a dry polyester molding compound, usually crosslinked with a diallyl phthalate (DAP) monomer.

Typical uses for alkyds include circuit breaker insulation coil forms, capacitor and resistor encapsulation, cases, housings, and switchgear components.

Allylics

Of the allylic family, diallyl phthalate (DAP) is the most commonly used molding material. A relatively high-priced, premium material, DAP has excellent dimensional stability and a high insulation resistance (5×10^6 megohms), which is retained to an extremely high degree after exposure to moisture (e.g., the same value is retained after 30 days at 100% RH and 80°F).

Both DAP and DAIP (diallyl isophthalate) compounds are available, the latter providing primarily improved heat resistance (maximum continuous-service temperatures of 350–450°F vs. 300–350°F for DAP). Compounds are available with a variety of reinforcements (e.g., glass, acrylic, and polyester fibers). The physical and mechanical properties are good, and the materials are resistant to acids, alkalis, and solvents.

DAP compounds are available as electrically conductive and magnetic molding grades, achieved by incorporating either carbon black or precious metal flake.

Most DAP compounds are used in electrical/electronic (they can withstand vapor-phase soldering temperatures), aircraft, and aerospace applications. Prepregs are used to manufacture composite parts.

Bismaleimides (BMI)

The bismaleimides are very-high-temperature-resistant condensation polymers made from maleic anhydride and a diamine such as methylene diamine (MDA). BMIs are used in applications where the heat resistance of epoxies is not sufficient, and have the capability of performing at 400 to 450°F continuous-use temperatures. They have similar processing characteristics to epoxy resins and can be handled on equipment used for epoxies. Composite structures made from BMI resins are used in military aircraft and aerospace applications. These resins also are used in the manufacture of printed circuit boards and as heat-resistant coatings.

Epoxies

These resins offer excellent electrical properties and dimensional stability, coupled with high strength and low moisture absorption.

The most important of the epoxy resins are based on bisphenol A and epichlorhydrin; they can be produced in either liquid or solid form. Epoxy novolacs are made by reacting a phenol formaldehyde with epichlorhydrin; the resulting resins have high heat resistance and low levels of ionic impurities. All epoxies have to be reacted with a hardening agent (e.g., an aliphatic amine or a polyamide) to crosslink into thermoset structures.

Epoxies are used by the plastics industry in several ways. One is in combination with glass fibers (i.e., impregnating fiber with epoxy resin) to produce high-strength composites that provide heightened strength, electrical and chemical properties, and heat resistance. Typical uses for epoxy-glass RP are in aircraft components, filament-wound rocket motor casings for missiles, pipes, tanks, pressure vessels, and tooling jigs and fixtures.

Epoxies also are used in the encapsulation or casting of various electrical and electronic components and in the coating of various metal substrates. (See Fig. 3-20.) Epoxy coatings are used on pipe, containers, appliances, and marine parts. Charged epoxy powders are electrostatically deposited on car bodies and other auto parts for corrosion protection.

Another major application area for epoxies is adhesives. The two-part systems cure with

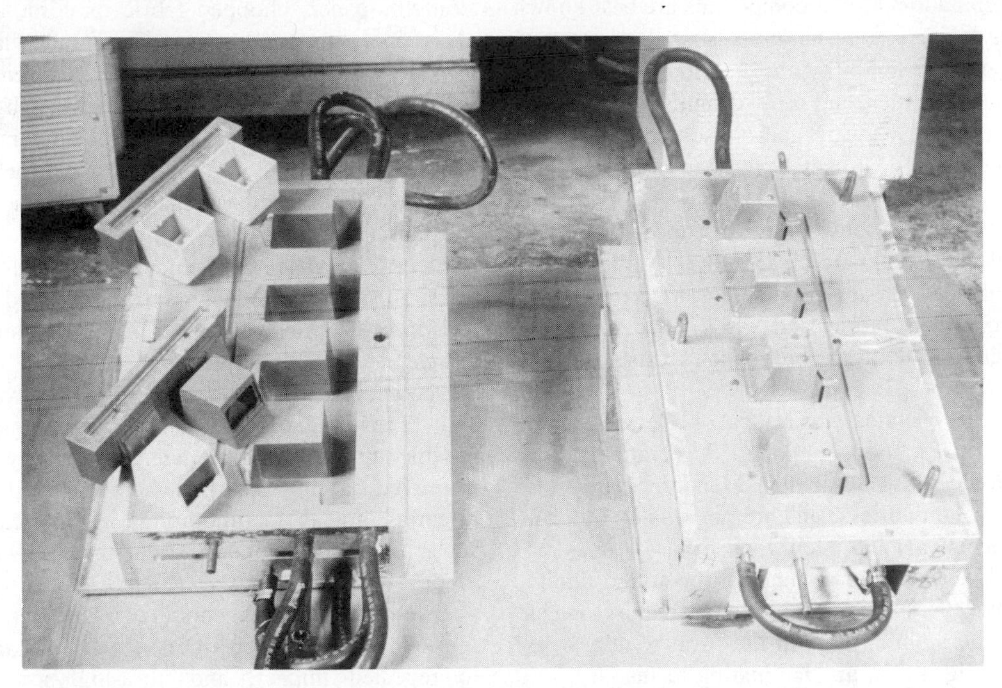

Fig. 3-20 Aluminum-filled epoxy tool for molding polyurethane parts. (*Courtesy of Ciba-Geigy Corp.*)

minimal shrinkage and without emitting volatiles. Formulations can be created to cure at different rates at room or elevated temperatures. The auto industry has started to use epoxy adhesives in place of welding and for assembling plastic body parts.

Epoxies can further be transfer- or injection-molded. Transfer presses, for example, are being used today to encapsulate electronic components in epoxy or to mold epoxy electrical insulators. Injection molding presses also are available for liquid epoxy molding.

Epoxy molding compounds are available in a wide range of mineral-filled and glass-, carbon-, and aramid-fiber-reinforced versions. The low-glass-content (up to about 12–50%) or mineral-reinforced compounds are primarily used for electrical applications and can be molded at relatively low pressures and short cycles. The high-glass-content compounds (over 60%) offer extremely high mechanical strengths, but are more difficult to process. Carbon fiber–epoxy composites are used to produce strong, lightweight parts for military and aircraft uses.

Melamines

Melamine molding compounds are best known for their extreme hardness, excellent and permanent colorability, arc-resistant nontracking characteristics, and self-extinguishing flame resistance. Dishware and household goods plus some electrical uses have been primary applications.

General-purpose grades for dishes or kitchenware usually are alpha-cellulose-filled. Mineral-filled grades provide improved electrical properties; fabric and glass fiber reinforcements provide higher shock resistance and strength.

The chemical resistance of the material is relatively good, although it is attacked by strong acids and alkalis. Melamines are tasteless and odorless, and are not stained by pharmaceuticals and many foodstuffs.

Like the other member of the amino family, urea, the melamines find extensive use outside of the molding area in the form of adhesives, coating resins, and laminating resins.

Phenolics

Phenolics were among the earliest commercial synthetic materials and have been the workhorse of the thermoset molding compound family. In general, they provide low cost, good electrical properties, excellent heat resistance, and fair mechanical properties (they suffer from low impact strength), coupled with excellent moldability. They are generally limited in color (usually black or dark brown) and color stability. Many phenolic applications have been converted to high-temperature thermoplastics, which have advantages in processing efficiencies.

General-purpose grades are usually wood-flour, fabric, and/or fiber-filled. They provide a good all-around combination of moderately good mechanical, electrical, and physical properties at low cost. They are generally suitable for use at temperatures up to 300°F.

The materials are severely attacked by strong acids and alkalis, whereas the effects of dilute acids and alkalis and organic solvents vary with the reagents and with the resin formulation.

Impact grades vary with the type and the level of reinforcement. Phenolic molding compounds can be reinforced or filled at levels higher than 70%. In order of increasing impact strength, paper, chopped fabric or cord, and glass fibers are used. Glass-fiber grades also provide substantial improvement in strength and rigidity. Glass-containing grades can be combined with heat-resistant resin binders to provide a combination of impact and heat resistance. Their dimensional stability also is substantially improved by glass.

Electrical grades are generally mineral- or flock-filled materials designed for improved retention of electrical properties at high temperatures and high humidities.

Many specialty-grade phenolics are available. These include chemical-resistant grades in which the resin is formulated particularly for improved stability to certain chemicals; self-lubricating grades incorporating dry lubricants such as graphite or molybdenum disulfide; grades with improved resistance to moisture and detergents; rubber-modified phenolics for improved toughness, resilience, and resistance to repeated impact; and ultra-high-strength

Fig. 3-21. Polimotor engine with reinforced phenolic block and head. (*Courtesy of Rogers Corp.*)

grades (glass content over 60% by weight), designed to provide mechanical strengths comparable to those of cloth-reinforced plastics laminates.

Phenolics are processed by thermoset injection molding, compression molding, and transfer molding, and have recently been formed by pultrusion and made into sheet molding compound. Their fire resistance and low smoke emission have found phenolics new applications in panels for buildings and transport facilities and in aircraft parts. (See Fig. 3-21.) For the same reasons, phenolic foams are gaining use as fire-resistant insulation materials.

Polyesters

The unsaturated polyesters are extremely versatile in terms of the forms in which they are used. Polyester resins can be formulated to be brittle and hard, tough and resilient, or soft and flexible. In combination with reinforcements such as glass fibers, they offer outstanding strength, a high strength-to-weight ratio, chemical resistance, and other excellent mechanical properties.

Among the members of the unsaturated polyester family are the orthophthalic (lowest cost) and isophthalic polyesters, vinyl esters, and blends of the various types. The crosslinking of unsaturated polyesters is initiated by a peroxide catalyst, selected by the requirements for curing temperature and rate.

Polyester resins generally are dissolved in a crosslinking monomer, with styrene the common solvent. Styrene has come under fire as a potential carcinogen, and its emissions are being strictly regulated at lower and lower levels. Resin suppliers are helping processors to reduce emissions by incorporating additives that keep the styrene from evaporating. It also is possible to replace styrene with paramethyl styrene, which has a much lower vapor pressure than styrene.

The prime outlet for unsaturated polyesters is in combination with glass fibers in high-strength reinforced plastics, or composites. Using various lay-up, spray-up, filament winding, pultrusion, injection, compression, and resin transfer molding techniques, polyester-glass is being used for such diverse products as boat hulls, automotive body parts, building panels, housings, bathroom components, tote boxes, pipes and pressure vessels, appliances, and electronic and electrical applications. (See Fig. 3-22.) For a detailed discussion of unsaturated polyester processing, the reader is referred to Chapter 18.

Where polyesters are used in building applications, they normally are filled with gypsum or alumina trihydrate (ATH) to make them flame-retardant. In sheet molding compounds (SMC) for automotive body panels, they are heavily filled with talc and reinforced with glass fibers; thermoplastic additives are used to provide a smooth Class A surface. Another major application for polyesters is in "cultured marble," highly filled materials that are made to resemble natural materials for sinks and vanities.

Polymer concretes based on polyesters are used to patch highways and bridges; their rapid cure minimizes the disruption of traffic. Polyester patching compounds also are used to repair auto body damage.

Polyimides

Thermoset polyimide resins are among the highest-temperature-performance polymers in

Fig. 3-22. Chevrolet Lumina van with major body panels of reinforced polyester. (*Courtesy of Owens-Corning Fiberglas*)

commercial use, with the ability to withstand 1000°F for short periods. The materials also perform extremely well under cryogenic conditions. They exhibit high impact and tensile strength and dimensional stability. Polyimides are inherently combustion-resistant. Applications for thermoset polyimides include films for electric motors, various aircraft/aerospace uses (engine components, insulation for wire and cable, bearings), chip carriers for integrated circuits, printed wiring boards, and coatings.

Silicones

These polymers are based on silicon rather than carbon and are partly inorganic. They have unparalleled long-term heat resistance, excellent electrical insulating properties (stable under both frequency and temperature changes), and good moisture and chemical resistance. Silicone polymers come in liquid, elastomeric, and rigid forms, depending on their molecular weight and degree of crosslinking. In this discussion, only the elastomeric and rigid forms are considered.

Elastomeric silicones usually are based on dimethylsiloxane and come in one- or two-part systems. Room temperature vulcanizing (RTV) silicone rubbers are used to make molds for casting various plastics and as adhesives and sealants for glass and metal structures. Two-component systems are used for potting electronic circuits and can be processed by liquid injection molding (LIM) into flexible parts such as O-rings and other seals. Ultraviolet light (UV)-curing silicones are used to coat printed circuit boards.

The much less familiar rigid silicones are polysiloxane resins with thermal stability well over 500°F. One growing application area is the hard-coating of softer plastics such as acrylics and polycarbonates to provide scratch resistance for lenses and windows.

Ureas

Excellent colorability, moderately good strength, and low cost are the primary attributes of urea molding materials. Their dimensional stability and impact strength are poor.

The materials are used for such products as decorative housings, jewelry casings, lighting fixtures, closures, wiring devices, and buttons.

Polyurethanes

For a full discussion of polyurethane and polyurea systems, the reader is referred to Chapter 19.

4 Extrusion Processes

The earliest extruders were used predominantly in rubber processing. The machines were short extruders because that process does not require melting per se as with plastics but merely softening and pumping through a shaping die. The traditional feed form of rubber is strips, which must be pulled into the screw by the screw's feeding flights and the feed section design features.

As plastics were developed, the extruder's design started to change in order to meet the new melting requirements. Barrel lengths increased, and heating of the barrel was more critical as the extruder needed to be heated in the 300–500°F range as opposed to the 100–200°F range for most rubber processing. The screw design for conveying rubber had to be noticeably altered to process the new polymers. Plasticating extruders as they are known today were first developed in the 1930s and 1940s as altered rubber extruders. Since that time understanding of the melting process has grown, as has processing experience with the various materials, leading to much more efficient designs of the extruders and the accompanying equipment.

Extrusion Concepts

Extrusion of thermoplastic materials can be accomplished through various means, depending upon the product being manufactured. Typically, extrusion with polymeric materials (plastics) involves a continuous operation as opposed to making a product with an intermittent process as done in injection molding. The various products made by extrusion include pipe, tubing, coating of wire, plastic bottles (blow molding), plastic films and sheets, various plastic bags (blown film), coatings for paper and foil, fibers, filaments, yarns, tapes, plastic plates and cups (thermoformed sheet), and a wide array of profiles.

Extrusion is accomplished by melting the material and forcing the melt through a forming die. The polymer material is fed to the extruder through a feed opening and can be introduced to the extruder in pellet (or cube) form or alternately as a powder, a granulate, or, in some processes, a melt. Extruders used in rubber extrusion and with some adhesives must accept a strip as the feed form. The extruders that are fed a melt are used for pumping to pressurize and to force the material through the die system or to aid in such parameters as cooling the melt from a melting extruder. The typical extruder is required to take a solid feed material and to melt, homogenize and pump the melt through the die system with acceptable output uniformity. The output consistency is measured by the uniformity of the dimensions of the finished product.

The extruded melt is continuously shaped and cooled by downstream equipment placed after the extruder. This sizing/cooling equipment can be comprised of cooling rolls, water tanks, vacuum sizing fixtures, air cooling tables, pulling devices, cutting equipment, coiling or winding equipment, and so on.

The extruders used to produce these products are overwhelmingly of the single screw vari-

By William A. Kramer and Edward L. Steward, Davis-Standard, Pawcatuck, CT.

ety, with several other types of machines used in some situations. These single screw extruders are simply comprised of a flighted screw that rotates within a heated cylinder (barrel). The screw is rotated by a drive motor through a gear reducer. Alternate extruders to the single screw include multiscrew machines (usually twin screw), rotary extruders (screwless), and ram extruders. The twin screw and also the less popular quad-screw extruders are comprised of multiple screws within a heated barrel and are most popular for making rigid PVC (polyvinyl chloride) powders into pipe and various profiles (window profiles, house siding profiles, etc.). The melting performance of this material lends itself to the low shear pumping seen with these types of extruders. Most polymers require more energy to thoroughly melt and homogenize them than a typical twin screw extruder can efficiently produce. When a twin screw extruder is designed to develop shear levels comparable to those of single screw extruders, the performance is not improved over the single screw machines, and the economics and operational advantages favor the single screw extruder. That is why twin screw extruders have not widely penetrated the single screw marketplace, except for the PVC powder extrusion applications. Another use of twin screw extruders is in compounding where additives must be dispersed into a polymer. Here special mixing twin screw machines have been developed, which do a good job and are very expensive. These compounding twin screw extruders typically can deliver high output levels (2000–10,000 pounds/hour and greater) and are not economically practical for the extrusion of everyday products as described above.

Rotary extruders have been in use for the last ten to fifteen years but have seen limited use due to their sealing problems against melt leakages and their low pressure-generating capabilities. These machines are made up of heated discs that rotate with polymer between the plates where shear is developed and melting takes place. Some pressure can be developed, but nowhere near the typical 2000 to 10,000 psi levels of single screw or twin screw extruders. The possibility of melt leakage leads to concern for contamination due to degrading polymer because the system is not totally self-cleaning. Some large rotary units are being used for pelletizing or compounding applications, and a few smaller units are being used for products such as polypropylene sheet; but the single screw extruder still is and will remain the workhorse in polymer extrusion for the foreseeable future.

There are some materials in the fluoropolymer area and some materials such as ultra-high molecular weight polyethylene that will not process acceptably on the screw extruders mentioned above. For these materials, a ram extruder is employed. This device is a non-steady-state machine that discharges its volume by using a ram or a plunger to extrude the melted material. The polymer is melted by conducted heat through the barrel in which the ram travels. This extruder is not a substantial influence on today's extrusion markets.

SINGLE SCREW FUNCTIONAL DESCRIPTION

Because the single screw extruder is by far the predominant machine used in polymer extrusion, its operation is mainly described, with comments regarding alternative extrusion means added as appropriate. The basic function of this type of extruder is to accept material in the feed section of the screw and convey this material along a flighted screw enclosed in a barrel. (See Fig. 4-1.) The conveying is forced by the rotation of the screw via a drive motor and gear reducer. The material usually must be melted (plasticated) along the path through the extruder screw although some processes introduce the feed material already in melt form so the extruder need only convey it. The melting of the polymer is aided by heaters that tightly encapsulate the barrel's outside diameter and are separated on the barrel into zones. These zones can be set at different temperatures as appropriate for the particular process involved. The screw must develop enough pumping efficiency to force the material through the die system. The pressure developed can be substantial for highly restrictive die systems and can reach 8000 to 12,000 psi. Typically the die pressure levels encountered are in the 1000 to 5000 psi range.

The design of the screw greatly impacts the

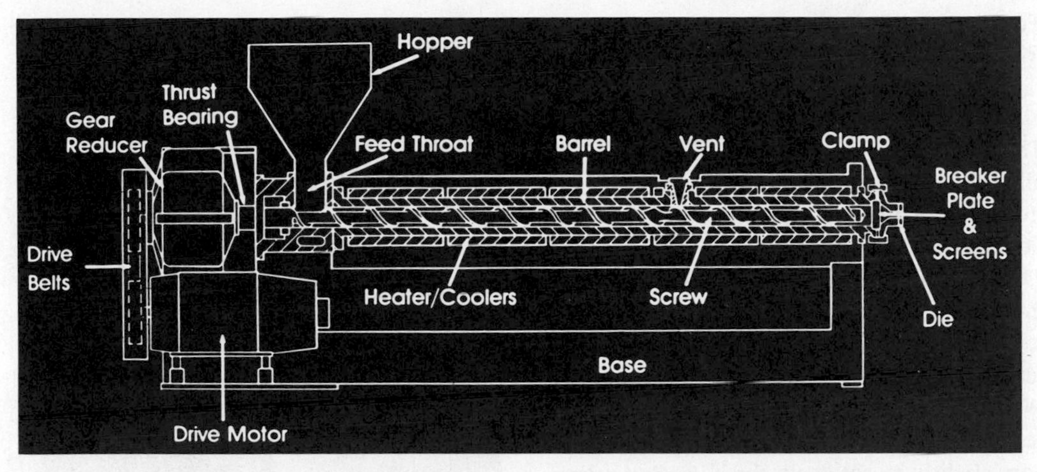

Fig. 4-1. Schematic figure of extruder. (*Courtesy Davis-Standard*)

extruder's performance and will be discussed in more detail in a later section. Much of the performance gains made by extruder manufacturers over the past 35 years have been due to improved screw designs. Numerous types of screws have been introduced over the years, some of them of great advantage while others are good for marketing but do little to improve the extruder's performance. Processors must recognize the benefits or shortcomings of particular screw design offerings to protect their interests and avoid the purchase of an ineffective design.

The extruder must produce its output (extrudate) within certain stipulations, as defined by the material, the end-product, and the process being accomplished. The extrusion goals and problem areas for various processes will be considered in a later section.

This section will cover the extruder's various functions and some ideas about how these functions can most effectively be accomplished. The basic function discussions will include feeding, melting, melt pumping, mixing, devolatilizing (venting) and the effects of some extruder add-on devices such as melt pumps, melt filters, and material preheaters.

Feeding and Solids Conveying

The extrusion process begins with the introduction of the material to the feed opening (feed throat) of the machine. The feed throat is located in a section of the extruder placed between the gear reducer and the barrel. This feed section typically is made from cast iron with cooling passages molded into the unit, and is designed to accept a material container (hopper) over the feed opening. This part of the extrusion process is of extreme importance because the failure of a consistent feed source will impact negatively on the stable performance of the extruder. Often in production situations, an unstable product gauge or dimension results from some poor feeding factor due to the material, screw design, or extruder feed area design. Most extrusion processes today are using polymers in solid form, either pellet, cube, powder, or granule. These require the full plasticating (melting) abilities of the extruder and will be the material forms of most importance to this discussion. Alternate forms of feed, including premelted material, rubber strips, reground material, and so on, will be touched upon as appropriate.

The conveying forward of the solid particles of material along the early portion of the screw is initiated by friction between the material and the feed section's bore. The conveying forces theoretically can become very high in a short distance and hence could lead to very high pressures. Actually, the pressures in the first portion of the barrel typically are 500 to 4000 psi. There are some materials that can exhibit very low pressures in the rear of the extruder due to poor solids-conveying efficiencies.

These materials include any highly lubricated formulations, members of the polypropylene family, and some of the high molecular weight high-density polyethylenes (HMW-HDPE). These low solids-conveying efficiencies usually lead to reduced output rates from the extruder, poor melting performance, and sometimes poor output consistency (output stability). The polypropylenes have enough conveying impetus to deliver moderate output levels and reasonable output stability as compared to easy-feeding materials such as most polyethylenes. When a material feeds very poorly, i.e., as screw speed is increased, the output increases very slightly or not at all, the extrusion process usually is unacceptable, and some feeding assistance is required.

Feed Section Design; Smooth Feed vs. Grooved. Typical commercially available polymers exhibit fair-to-excellent solids conveying; so the smooth bore of the feed section and barrel yield enough conveying force for stable extrusion at acceptable output rates. In these cases, melting and pumping performance can be maximized through screw design because no feeding inadequacies exist. When materials are encountered that do not have adequate solids conveying to allow the screw's melting or pumping functions to be reasonably maximized, feed assistance devices such as grooved feed sections are considered. (See Figs. 4-2 and 4-3.) These special feed sections contain grooves that start near the rear of the extruder's feed opening and continue past the feed opening area and usually beyond it by a distance of approximately three times the barrel inner diameter. The grooves usually are evenly spaced around the feed area's inner diameter and typically are deeper at their starting point at the rear of the feed section (gear case side). The grooves' depths then run out to the barrel inner diameter at their end. Typically, the grooves are cut axially along the feed section's bore; they also can be made in a helical configuration, which is believed to be more effective for the feeding of some materials but is more difficult to machine. Selection of the grooved feed section usually is justifiable only when there is a definite feeding problem. When the material feeds moderately effectively, as most polypropylenes do, the use of a grooved feed section is debatable. When the use of a slightly larger extruder with a smooth bore feed section can match the performance of a grooved feed section extruder, the smooth bore usually is favored. Whenever additional extrusion features can be avoided in favor of operational simplicity with no sacrifice of performance, that is the favored choice.

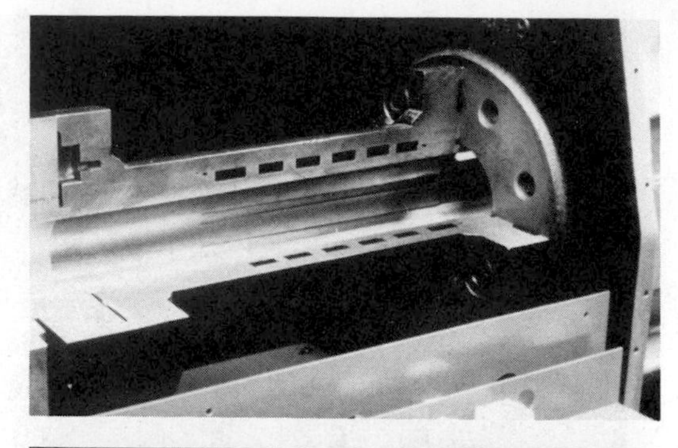

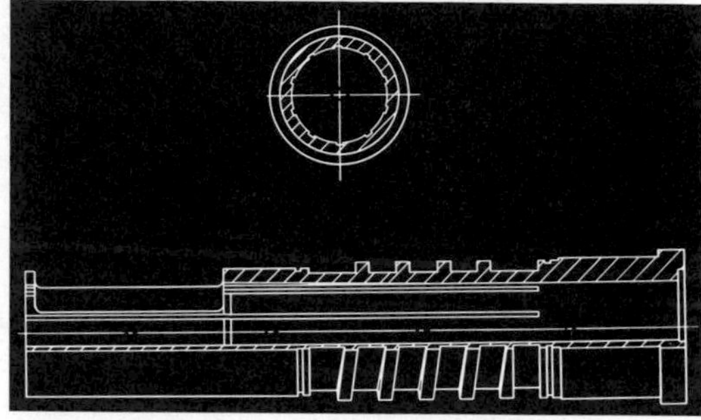

Figs. 4-2 and 4-3. Grooved feed section. (*Courtesy Davis-Standard*)

The grooved feed section will increase output under most conditions with all materials, but if the screw is already pumping at or near the melting limits with a smooth bore feed section, any added output will not be useful. Grooved feed sections require intensive cooling to avoid melting in the grooved area, which defeats the conveying efficiency. Smooth bore feed sections are cooled typically, but only modest cooling levels are desired and required to avoid material softening and "bridging" in the feed area of the screw or in the feed throat.

The strong positive conveying efficiency of the grooved feed sections, when efficiently designed, causes high pressure levels at the end of the groove section, up to 10,000 to 20,000 psi, as seen in added feed section and barrel wear as well as higher power usage from the drive motor. Most grooved feed sections thus are desirably produced from highly wear-resistant materials such as tungsten carbide lined cylinders.

Whether the feed section is smooth bore or grooved, the amount of wear definitely can affect feeding effectiveness and hence output rate and output stability. Worn smooth bore feed sections can be repaired with a sleeved section that will bring the bore back to original tolerances. Repair of grooved feed sections involves more effort because of the complexity of the machining.

Material Effects on Feed Section Design. The form of the material entering the feed hopper or feed section has an effect on the processing success of the extruder. Powders and fluffy regrinds, for instance, generally lead to more feed and processing difficulties than pellets, cubes, and heavier regrinds. The bulk density of the feed material determines how effectively the screw's feed flights are filled and how well the extrusion process can then commence. Most low-bulk-density regrinds and some powders (especially filled powders) will not readily flow down the hopper and through the feed throat to fill the feed flights adequately. When hopper flow problems are evident, special material forcing devices, such as compacting screws in the hopper/feed throat, sometimes are used to ensure a filled screw feed flight. Alternately, the materials that cause feeding difficulties can be pelletized or otherwise densified on other equipment to alleviate feed difficulties and hence processing inefficiencies on the production extruder.

Feeding melt to an extruder introduces the difficulty of obtaining a free-flowing situation through the feed throat area and may require a pressure-building source to push the material into the feed flights. Some processes drop a melted ribbon of material into the extruder's feed section, which makes a filled feed flight difficult to ensure. The screw feed flight design can help the feeding efficiency, but extrusion stability is not usually optimum.

Feed throat opening designs can vary, depending on the manufacturer and the process being performed. Today's typical, efficient throat design is a large rectangular opening directly above the screw. Through the years, feed openings have evolved from round shapes to oval to "obround" (lengthened oval-shaped) to rectangular. Today's rectangular throat design has an opening length of 1.5 to 2.5 times the barrel inner diameter dimension. The larger feed openings allow a free flow of material even with moderately high regrind percentages to ensure properly filled screw feed flights. The only uses of small feed openings in this era involve hoppers with force feeding screws (compactors) or force-fed melt conveying extruders.

Tangential feed throats enter the screw area from one side and have added clearance around part of the screw's diameter. They are used for feeding rubber strips to allow partial wrapping around the screw.

Most extrusion processes perform with best product uniformity when the screw is operated with full feed flights. Sometimes a metered feeder is used to run the process with starved feed flights for some processing reason; the extruder's stability must be acceptable or added processing devices must be used, such as melt pumps (see discussion of melt pumping later in this section). Twin screw extruders appear generally less sensitive than single screw machines to the starve feeding mode as far as output stability is concerned; but as the starving level is increased, even their output stability deteriorates.

Material Lubrication Effects. Some plastics formulations call for additives that create feeding and/or melting problems. The easiest place on the extruder to introduce additives is in the feed throat where the material is at atmospheric pressure. Should the material still feed and melt acceptably, the application of additives in the feed area is the most economical approach. When additives create feeding or melting difficulties, the possibility of injecting the additives somewhere along the barrel length is considered. Introducing liquid additives away from the feed area will help avoid solids conveying deterioration and, if far enough down the barrel, will avoid disrupting the melting performance. The concern for injecting late along the barrel is for mixing efficiency. Sometimes a mixer after the screw is helpful in blending the additives. One of the difficulties encountered when the additive is injected along the barrel is the need for a pump system to overcome the pressure in the barrel at the injection port. This pressure can be substantial, up to 5000 psi, so the pump system must be specified to meet the injection rate goal and to perform against whatever pressure is present. An additional advantage to injecting the additive later along the barrel is that the injected material does not remain in the barrel as long as the main polymer and is affected less by heat and shear. This is important where heat-sensitive materials, such as some liquid colors, are injected.

Melting Considerations

As solid material is conveyed from the feed throat and travels through the feed section of the screw, some compaction takes place. When it reaches the heated barrel, a melt film immediately forms on the barrel I.D. The melt film grows in thickness as the material moves down the barrel until it is thicker than the screw flight clearance with the barrel. Then the melt begins to collect at the rear of the screw channel (the pushing side of the flight). As the melt film goes through this thickness growth, transporting forces are developed by the shearing of the melt film. This conveying mechanism is termed *viscous drag*. (The material in the channel during the early melting process is sketched in Fig. 4-4.) The shearing of the melt film creates most of the energy for melting of the material at moderate to high screw speeds. The higher the viscosity of the melted material, the more heat is generated via melt film shearing. Stiffer materials such as rigid polyvinyl chloride (PVC) or high-density polyethylene (HDPE) generate much heat in the melt film, and the melt temperature reflects that fact as screw speed is pushed to moderate to high levels. Low viscosity materials exhibit much lower melt temperature rises as the screw speed is increased.

The barrel heat contribution at high or moderate screw speeds is often minimal, with melt film shear producing enough energy for the entire melting process. In fact, many extrusion situations require some of the barrel zones (especially those farthest from the feed area) to be cooled to remove excess heat created by the melt film. Barrel cooling has only a modest effect on lowering melt temperature levels in high-speed extrusion of high viscosity materials because of the short residence time in the barrel and poor conductivity of most polymers.

The screw designer's typical goal is to design a screw geometry to maximize output and control the melt temperature level required for the particular process. The discussion of ''Extruder Screws'' later in the chapter will further expound on melt temperature controlling factors.

Melt Pool Development. As the solid bed moves along the screw, the energy from the barrel heaters and the shearing of the melt film contribute to further melting. The melt being scraped off the barrel wall as the flight passes is trapped on the pushing side of the channel (rear) and forms a melt pool. The idealized

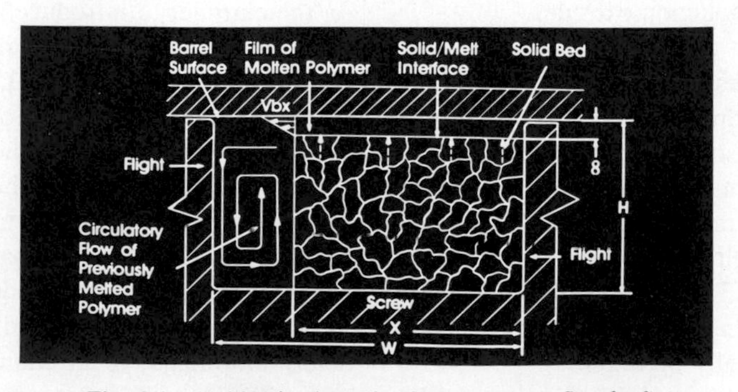

Fig. 4-4. Melting in channel. (*Courtesy Davis-Standard*)

nelting process suggests that this melting continues in a well-defined pattern with the melt pool increasing in size as the solid bed declines in size until all is molten. This full melt condition optimally would occur before the material reached the die system to ensure a good end product. The screw channel does not normally maintain an organized melt and solid bed arrangement (as shown in Fig. 4-4) but instead exhibits a random breakup of the solid bed.

The breakdown of the melt/solids relationship in the screw channel leads to several potential extrusion problems. First, stability is adversely affected because of pressure surges created when the solids break up. This pressure variation typically is seen to start in the early (near feed section) to mid-barrel locations through the use of pressure recording devices. Should the pressure variations be large enough, the upsets can be seen all along the barrel and into the die system. Pressure variations in the die system relate to output variations and hence product irregularities. A second concern of the breakdown is that a less-than-efficient melting situation created by solids breakup can cause unmelted material to find its way farther down the screw and sometimes into the die system. Screw design can control the melting situation and/or damp out the negative performance aspects to some extent. (Screw features will be discussed in the screw design section.) The effects of solid bed breakup are magnified as screw speed is increased on conventional screw designs, a fact that has led screw design to today's specialized screw geometries.

The melting capacity of the extrusion screw must be satisfied by an ample amount of feeding capacity. This is handled through screw design.

Material Effects On Melting Performance. The materials being extruded have a great effect on melting performance, just as they can affect feeding efficiency. The amount of heat developed in the melt film is a function of the viscosity of the material and the effects of any additive. Just as lubricating additives can disrupt the solids friction that drives the feeding, they can have an effect on melting performance.

The melting rate of the material has a determining effect on extrusion melting performance. This melting rate is determined by such properties as thermal conductivity and specific heat. Any fillers in the material usually will alter the melting and conveying performance of the material to some extent. Polymer blends are quite popular today because of the properties available from each of the components. The makeup of the blend will affect the melting performance because of the varied melting characteristics of the blend's basic materials. These blends can create sizable screw design challenges in the efficient handling of the melting performance.

Inconsistencies in the material also pose a threat to extruder output consistency. The use of reground extruder scrap and off-spec product can lead to inconsistency due to varying feed particle sizes and differing percentages of the regrind as time passes. When regrind is used from thin gauge sheet or film extrusion, the bulk density of the feed material has an effect on the material's hopper flow and how the screw's feed flights are filled, which can lead to output reduction and at some point a starved screw. In processes that are controlled by feeding limitations, such as the starved screw case in low-bulk-density reground films, it is generally difficult to maintain good output consistency and high output rates as screw speeds are increased. Inconsistency of the base material due to its manufacturing accuracies can yield a feed material that will process with variations as the lot of material is used. The extruder's feeding and melting performance can be noticeably altered by these variations in the feed materials.

The form of the material being introduced to the extruder also has an effect on solids conveying and melting efficiency. The material forms available include pellets of various sizes, granulate (like sugar consistency), powders of various particle size, regrind of various sizes, and strips (as in rubber extrusion or some melt extrusion). The shape of the feed material usually is determined by either economics (cost/pound) of material manufacture or extrusion performance features. For example, the most economical form of rigid PVC is powder, and many of the processors using large volumes of this material produce some or all of their extruded product from powder. The negative aspects of powder include the requirement of a vented extruder (see below) and extra plant maintenance caused by fine powder that travels through the air and can corrode electrical components such as drive systems and temperature control circuitry. Producers of PVC profiles who can dedicate an extruder to a small number of items can use a vented extruder and powder feed material. When part versatility is required, a vented extruder has operational problems due to varying restrictions on the extruder; it is recommended that producers choose PVC in pellet or cube form. Producers of thin, clear PVC sheet typically use pellets on nonvented single screw extruders; the melting mechanics and shear input of the single screw extruder better fit the higher temperature processing. Also, the high die pressures (5000–8000 psi) would preclude effective venting on this type of extruder. Economics provides the strongest driving force in the selection of the material's feed form as long as the processing considerations are not highly negative for the lowest-priced material.

Melt Pumping

Pumping of the material against the die resistance can begin back near the screw's feed section, especially when the die pressure levels are high. Melting starts early on the screw in most cases, and pressurization of the melt can begin there. In actuality the three basic functions of the extruder—solids conveying, melting, pumping—cannot be separated into three discrete regions along the extruder. The functions intermesh so strongly that they all must be studied together. As the material advances down the screw toward the die, more melt is present, and the predictions of the pumping theory developed many years ago can be understood. This theory is well known,* and includes the prediction of the pumping capacity of a simple metering section against no die resistance (drag flow) and the output-reducing tendencies of the die resistance (pressure flow). These equations reduce to fairly simple terms and give a form of rough output calculation for conventional metering screws with materials that feed well. The conclusions of the melt pumping analysis are stated below:

$$OUTPUT = Q_{drag} - Q_{pressure}$$

$$Q_d = \frac{F_d \pi^2 D^2 N h \left(1 - \frac{ne}{t}\right) \sin \phi \cos \phi}{2}$$

$$Q_p = \frac{F_p \pi D h^3 \left(1 - \frac{ne}{t}\right) \sin^2 \phi}{12 \mu L} \Delta P$$

where:

Q_a = Drag flow pumping term
Q_p = Pressure flow resisting pumping
$F_d = .140(h/w)^2 - .645(h/w) + 1$ (channel correction factor)
$F_p = .162(h/w)^2 - .742(h/w) + 1$ (channel correction factor)
D = screw diameter
N = screw speed (rpm)
h = screw's meter section channel depth
w = screw channel width (normal direction, not along axis)
n = number of flights on the screw
e = thickness of flight
t = flight lead (pitch)
ϕ = flight helix angle
μ = viscosity of melt (shear rate $= \pi ND/h$)
L = length of the metering section being investigated

This estimation of the output pumping of a screw is applied to the shallowest section of the screw because that is the region that limits the screw's output. Several simplifying assumptions were used to derive this flow estimation, including (1) a Newtonian material, (2) a fully developed melt flow situation, (3) no screw flight-to-barrel clearances, and (4) some other factors that help make the equation work rea-

*See E. C. Bernhardt, *Processing of Thermoplastic Materials*, Van Nostrand Reinhold, New York, 1959.

sonably well when some experience factors are employed.

Some examples of these alterations to the drag flow estimation include a 0.85 multiplier when mixing sections are employed and a 0.5 multiplier when cold water is circulated through the screw's core. When a barrier screw is used, the metering section may not control the output, so this equation could give poor results.

Wear Effects on Pumping

As screw flights and barrel I.D.'s wear, the pumping ability of the screw is diminished. Some materials and some additives will cause higher wear than others; for example, linear low-density polyethylene (LLDPE) will cause more wear than conventional LDPE or polypropylene. Many fillers, such as titanium dioxide (used for white colors) and reinforcing fibers, also create high wear situations. Under some conditions, screw/barrel wear can lead to instability of the extruder's output, but typically the main effect is output reduction. At some point, extruder wear will create an unacceptable situation that necessitates rebuilding or replacement of machinery parts such as the screws, barrel, and feed sections. Changes in the ability to increase the screw speed and still produce an acceptable melt contribute to the decision about when wear has passed acceptable limits.

Variations involved in production operations, including the materials run, screw speeds used, die system pressures, barrel set temperatures, screw design, screw flight hardening material, and barrel lining material, make it impossible to predict wear life accurately. The suggested way to understand the wear in a process is to set base conditions when the equipment is new and unworn; that is, run a commonly used material, and record all performance parameters, including output rate, screw speed, drive amperage, barrel temperature profile, product quality, and dimensional consistency. Whenever the opportunity to perform scheduled maintenance occurs, measure equipment clearances and rerun the process at the base conditions to compare performances to determine the extent of deterioration. The wear pattern then can be plotted to show the screw and barrel life for the given production case.

Mixing

Many extrusion processes require better mixing than a single screw delivers. For example, good dispersion of color masterbatches and proper mixing of polymer blends need higher shear mixing devices. This is particularly true where the masterbatch is based on a dissimilar material to the one being colored, and where polymers being blended have different melting temperatures and flow properties.

Two types of mixing usually are discussed in polymer processing, distributive and dispersive. In distributive mixing the material(s) are uniformly blended on a scale where any small particles or agglomerates are not broken down. These particles can be "gels" from various sources or small clusters of a material such as carbon black, which is not physically broken down without very high shear levels. Dispersive mixing includes very high shear mixing from extremely tight clearances (about 0.001 to 0.005 inch) where small particles or agglomerates are physically broken down to smaller pieces and distributed into the main mass of material. Such high shear, dispersive mixing can be seen with some of the kneading block geometries on compounding twin screw extruders. Mixing in single screw extruders is generally not on the scale of the very high shear levels characteristic of dispersive mixing. Typical mixing elements can yield very good distribution of a material's mass and its additives, but any small particles (gels) or agglomerates will still be present, although well spread throughout the melt. Should the presence of these undispersed particles create a problem in the final product, the origin of the particles must be determined, and their cause must be avoided in polymer blending methods or base material particle size selection. Examples of common mixing devices can be seen later in the screw design discussion. These mixing devices can be placed on the screw proper, can be attached to the screw tip, or can be placed after the screw in the die adaptor pieces.

The need for specialized and highly effective

mixing devices becomes more apparent as the mixing requirement becomes more difficult. For example, when lower amounts of additives are to be blended into a base material, the mixing requirements are tougher. Also, when the blended ingredients are more dissimilar, the mixing chore is harder to accomplish. A system that uses an injection technique along the barrel to add ingredients can present complications because the length of screw to accomplish the mixing is shortened, as compared to blending in the extruder feed hopper, and the additives (sometimes liquids) can disrupt the screw's melting and pumping functions, also adding to homogeneity problems.

Devolatilizing (Venting)

Some materials will produce a melt that becomes porous as it exits the conventional extruder, for several reasons. Usually this phenomenon is unacceptable except where a foamed product is desired. The porosity can be caused by moisture that has been absorbed by the material, or collected on its surface, which expands into steam as the hot melted material exits the extruder and drops to atmospheric pressure. The material enclosed in the extruder and die system is usually under sufficient pressure to keep the moisture in the melt and not expanded. Other causes for porosity in the melt may include trapped air (typical of powder materials) and certain volatiles, which escape from the material and expand at atmospheric pressures. Depending upon the cause of the porosity, the approach to eliminating the bubbles can be varied. Placing an opening (vent) through the extruder barrel wall is the typical method to remove volatiles before they can reach the die system. Use of a vented extruder requires that the material in the screw flights be at atmospheric pressure (partially empty flights) under the vent opening to keep material from flowing out of the vent opening and defeating the degassing function by blocking the escape path for the volatiles. The open vent hole is fitted with an insert (vent stack) that aids in streamlining the material's flow as it passes the barrel opening, to prevent material from getting caught on any blunt edges, which would lead to a blocked opening and cause venting problems. This requirement complicates screw efficiency and leads to long extruder barrel lengths in most cases (30 : 1, 34 : 1, up to 40 : 1 extruder L/D).

There are sometimes ways to avoid venting through the extruder barrel, depending upon the cause of the porosity. Trapping air with powder feed materials can be avoided by using pelletized materials instead of the powders, or by using a vacuum on an enclosed hopper system to remove the air before it becomes trapped in the early part of the screw. The pellet approach adds the expense of an additional extrusion operation but is sometimes the more efficient choice versus the venting approach. The vacuum hopper method includes some difficulties, such as the need to load the hopper—which is now a closed system—under vacuum. Also, there are seals on the screw shank at the rear of the feed section that must be maintained or air will be drawn into the feed throat (along the screw shank), and feeding will be disrupted as this air "bubbles" up through the feed throat. Air removal utilizing an extruder vent system usually can be accomplished by a simple open vent hole, but removal of moisture and most volatiles requires having a vent vacuum system to aid in boiling off the trapped gas. Moisture can be removed through the use of desiccant drying systems, which can operate in the extruder hopper or in a remote unit. Some materials that absorb moisture from the atmosphere (are hygroscopic) must be dried even if venting is chosen, to avoid foaming at the vent opening, which blocks the volatile's escape and defeats venting. The level of moisture in the material entering the extruder thus must be reduced to allow proper and continuous venting with some highly hygroscopic materials. The choice of the method to prevent porosity is driven by the processing difficulties of the available choices versus the economics of materials and/or production methods.

Effects of Extruder Accessories

Melt Pumps. Melt pumps attach to the exit end of an extruder and are made up of two intermeshing gears that are driven. (See Fig.

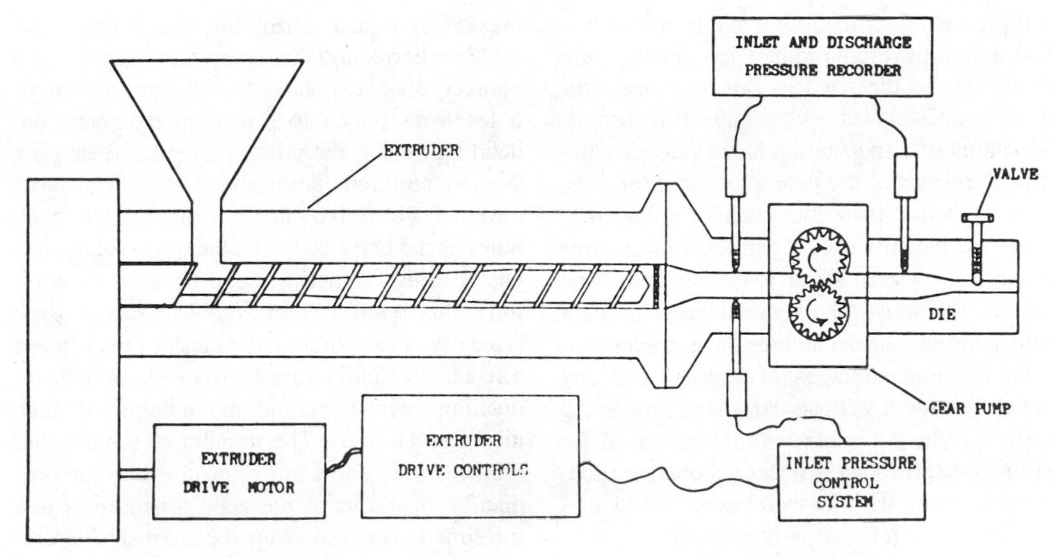

Fig. 4-5. Extrusion line with melt pump. (*Courtesy Davis-Standard*)

4-5.) The melt from the extruder enters the gear mesh and is pumped into the die system with very accurate flow consistency. Typically, the pressure on the extruder side of the gears (suction pressure) is lower than the exit pressure (die system pressure), with the gears developing the pressure rise. The gear pump thus can reduce the pressure required at the end of the extruder dramatically, a most important effect in vented extrusion. These devices have been on the extrusion scene for many years but have only recently found increased favor in general extrusion systems.

There are many published reports on the benefits of the use of melt pumps on the ends of extruders. Many of these claims are very dependent upon operational specifics (polymer material viscosities, die pressures, etc.), but the principal strength of a melt pump is its ability to ensure extrusion output consistency (pumping stability) where the screw design cannot.

Most single stage (nonvented) screw designs using today's technology can ensure excellent output stability and thus make economic justification of the melt pump very questionable. However, in some cases even a good screw design's ability to ensure stability is problematic, as with varying feed densities when there is random regrind addition to the main feed material. These cases sometimes can be shown to benefit from a melt pump. Some products demand ultimate precision in their final dimensions, which may also justify use of a melt pump if the screw cannot be proved to process the material with the required stability. An obvious application for melt pumps is with vented extruders, where the screws cannot be designed with the versatility or excellent output stability of nonvented designs; then the melt pump can restore stability to the extrusion system. The vented screw design also is limited by die pressure, which must be overcome to maintain an open vent; so the melt pump's ability to remove much of the pressure from the screw is a major advantage. There are limitations to the pressure differential across the gear mesh; very high die pressure applications (5000 psi and greater) should be carefully evaluated with the melt pump supplier.

The fiber industry has used melt pumps for years to split the melt stream of a large extruder into multiple paths that each have a pump. This setup allows consistent flow into the separate die systems, which each require only a small percentage of the extruder's output. The alternative to this system is to use many small extruders to fill each small die system individually, which would be much less efficient and prohibitive in cost, with added operational difficulties.

Coextrusion system users have shown added interest in and use of melt pumps. The use of multiple extruders feeding a single die to produce a multilayered product has added complexity due to the need to closely control the layers' consistencies and relative thicknesses. The output of a melt pump is not perfectly linear with respect to the gear's rotational speeds, but it is better than the extruder's linearity. There are usually a few extruders in a large coextrusion system (four to seven extruders) that are the most likely candidates for melt pump addition. These include the extruders utilizing large percentages of regrind and any vented extruders. Those extruders processing mostly virgin polymers without the need for venting typically can product acceptable product consistency without melt pumps, as long as the screw design is properly selected.

Melt Filters. There typically is a filtration of the melt after it exits the extruder barrel to trap any contaminants or impurities before they enter the die system. Large contaminants that enter the extruder with the material by mistake can be damaging to dies, melt pumps, and so on; so they are best caught at the end of the extruder. Many extruders are equipped with a tramp metal removal system in the feed throat to catch the most destructive contaminants before they reach the screw because the screw and barrel also can be damaged by metal pieces. These metal systems usually are magnets to catch the carbon steel pieces, which are most typical; but other systems are available that detect and remove all metals as the material drops through the feed throat.

The simplest melt filter is a screen pack held at the end of the extruder barrel by a perforated disc called a breaker plate. (See Fig. 4-6.) The breaker plate also forms the seal between the extruder barrel and the die system adaptor. The breaker plate can have a circular pattern of holes from $\frac{1}{8}$ inch to $\frac{3}{8}$ inch in diameter, depending on the extruder size and screen pack support required. Some materials exhibit visible streaks caused by the flow through the many holes as the extrudate exits the die system; slotted openings sometimes can be used to minimize this quality deterrent. The screen pack typically is specified by the mesh of the screens selected (which is a measure of the number of openings per inch) and the number of each mesh screen used. The number of screens and their restrictiveness are determined by the production operators as the need for more or less filtering is realized from the extrusion operation, usually based on melt quality requirements. A modest screen pack would be defined as having one 14 mesh, one 40 mesh and one 60 mesh screen. A fairly restrictive screen pack, as may be used with some low viscosity flexible vinyl applications, would include these screens: one 14 mesh (placed first against the breaker plate), one 40 mesh, one 60 mesh, one 120 mesh, two 200 mesh, and one 14 mesh (placed nearest the end of the screw). The coarse 14 mesh screen placed after the tight 200 mesh screens helps hold the fine screens in place as the breaker plate is being installed and the extruder is started up. The normal approach to screen pack selection is to choose the minimum screen pack to perform the job. The tighter the screen pack restriction, the sooner the pack will become plugged with contaminants.

Fig. 4-6. Breaker plate and screens. (*Courtesy Davis-Standard*)

When the pressure at the screw tip reaches a designated level above that of a clean and nonrestrictive screen pack, the extruder is stopped, and the breaker plate is removed for screen replacement. Should the extrusion line be difficult to tear down to gain access to the breaker plate, an automatic screen changer can be installed; this allows removal of the breaker plate via a sliding mechanism with a replacement moving into place as the plugged screen pack is being moved out, affording minimum down time. Some systems can remain running as the change is made. These screen changers can be supplied for manual or hydraulic powered operation. There are also systems that utilize a screen cartridge in a roll which can be continuously fed over a stationary breaker plate to avoid any stoppage difficulties. Seals at either end of the screen strip passing through the melt stream are formed by small cooling passages that freeze the melt. As the screen is moved, the sealing areas are allowed to heat up enough to allow movement of the screens. Materials that are very sensitive to heat and are easily degraded over short periods of time (such as some polyvinyl chlorides) cause concern with most screen changers due to the areas of material stagnation around the breaker plate and in the melt sealing regions around the slide plate.

When a very fine filtration is desired, as with some very low viscosity melts, a canister-type filter is utilized. This type of device can filter particles at least down to 100 to 200 microns in size. The need for this type of filtration is rare in the vast majority of extrusion applications.

Material Driers or Preheaters. Materials that are hygroscopic (e.g., acrylics, polycarbonates, ABS, nylons, polyesters, etc.) typically are predried before extrusion, even when barrel vents are used to remove volatiles. Driers are usually of the desiccant type; air is heated and circulated through desiccant beds where its moisture level is reduced before traveling through the plastic in the hopper. These driers can be mounted directly on the extruder as long as adequate volume is provided to allow the desired drying residence time. Some driers have moisture level indication because there are cases where the moisture level in the material must be held quite low, such as for polyester, where the material's properties are reduced if moisture is present in even small quantities. Some materials require moisture levels to be reduced to the 0.01% range in the drier, which is within the capacity of today's desiccant units. The air temperature must be kept below the softening point of the material to avoid premelting, bridging, or sticking.

This drying situation preheats the material and increases the output rate of the extruder. The energy imparted before extrusion reduces the amount of energy required for melting; output gains of 10 to 20% typically are seen with preheated material as compared to room temperature (or colder) feed material. Some or all of the energy required to operate the drier or for preheating can be recouped from the extrusion output gains. The use of driers is well known, but the use of preheaters for enhancing extrusion output is not typical. This method of performance improvement would only aid in extrusion cases where the system is operating under some extruder limitation such as output rate and/or melt temperature. The cost and maintenance of preheaters is hard to justify in the majority of cases.

TWIN SCREW VS. SINGLE SCREW OPERATION

This discussion is centered on single screw extruders, but a brief discussion of twin screw extruders may help the reader to achieve a fuller understanding of the equipment used in the industry to process polymers. Twin screw extruders (or more-than-one-screw extruders) have multiple screws within the same barrel that may have fully intermeshing, partially intermeshing, or totally nonintermeshing flights. The screws can rotate in the same direction (co-rotating) or opposite to each other (counter-rotating).

The use of twin screw extruders to form most thermoplastic materials into final products (film, profiles, wire coatings, etc.) typically is not economically justifiable compared with the simpler single screw extruder. Single screw machines continue to greatly outnumber the multiscrew machines in general extruder sales volume, but compounding applications usually use twin screws for their dispersive mixing ability; twin screws also are favored for rigid PVC powder applications because of their low shear processing and venting capabilities.

Co-rotating screws (intermeshing or not) are primarily used for compounding and mixing, because of their ability to disperse additives of very small particle size (e.g., carbon black agglomerates). Counter-rotating and fully intermeshing flight extruders are used primarily for the low-shear extrusion of rigid PVC into pipe and profiles, typically from powder feed. This setup creates the lowest-shear extrusion situation.

The most numerous multiscrew extruders presently in operation are, by far, the twin screw type. The screw O.D.'s may be constant, as in single screw extruders, or conical, where the feed end of the screw has a larger diameter than the exit end. The conical design allows for more feeding area for powder entry to the screw and reduces the thrust due to die pressure at the smaller screw tip. The conical twin screw is used mostly for rigid PVC powder extrusion at rates up to 700 to 1000 pounds/hour. Above those output rates, a parallel twin screw is used. The twins used in profile and pipe extrusion are almost exclusively for processing rigid PVC powder, because of the low-shear fusing characteristics of this unique material and the need for venting to remove trapped air. The die pressures encountered in profile extrusion (3000–7000 psi) are difficult to handle on single screw equipment, whereas the fully intermeshing twin screw creates enough pumping efficiency to allow an open vent situation at moderate to high output rates.

High output compounding applications, where various materials are mixed in the extruder and formed into pellets for subsequent product extrusion, often are performed with large twin screw machines. This application requires very high output rates (3000–20,000+ pounds/hour), probably has a venting requirement, and does not need to deliver a melt with high product quality in terms of surface smoothness, output consistency, final material properties, and so on. Twin screw extruders with special dispersive mixing sections are typical in these compounding operations, with very large costs involved. (See Fig. 4-7.)

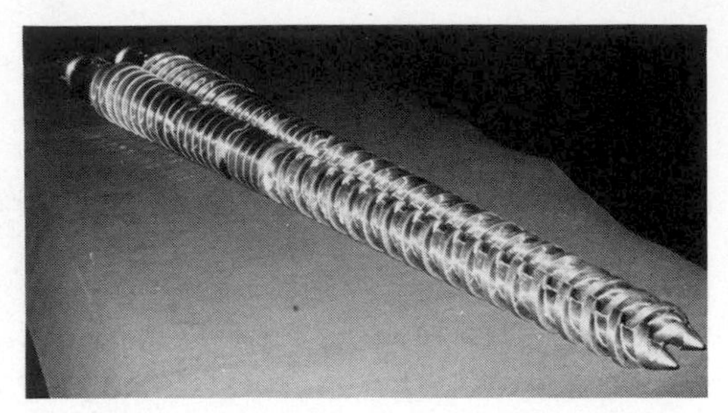

Fig. 4-7. Conical twin screws. *(Courtesy Davis-Standard)*

SINGLE SCREW EXTRUDER PHYSICAL DESCRIPTION

Barrels

The extruder barrel not only houses the screw, but serves as the primary heat transfer medium in the process. It must be designed to resist wear, contain the process pressures, and oppose screw torque. (See Fig. 4-8 for cutaway view of an extruder.)

Materials of Construction. Barrels normally are steel cylinders with a flange at the feed end for mounting to the machine and another at the discharge end for mounting the die system. The flanges are threaded on or welded. Alloy steel typically is used because of its strength, high-temperature performance, and economy.

Liner Materials for Wear Resistance. Nitrided barrels were once common in the United States and still are used frequently in Europe. They offer moderate wear life for a reasonable cost, but the hard surface is very thin, and once it is penetrated, wear progresses rapidly. Most extruders now have a cast-in bimetallic liner that provides superior wear resistance at a moderate cost and lasts two or three times as long as nitrided barrels. General-purpose liner alloys are iron-based and are adequate for most applications. Nickel-based alloys are available for highly corrosive materials, and alloy matrixes containing tungsten and other carbides can be supplied for extreme wear and abrasive conditions.

Barrel Strength. An extruder barrel must be designed to withstand the internal pressures generated in the process. The industry standard is a design limit of 10,000 psi (70 MPa), but one should always confirm an individual machine's limit before operating at high pressures. Under normal circumstances, an outside diameter of 1.5 to 2 times the inside barrel diameter is necessary.

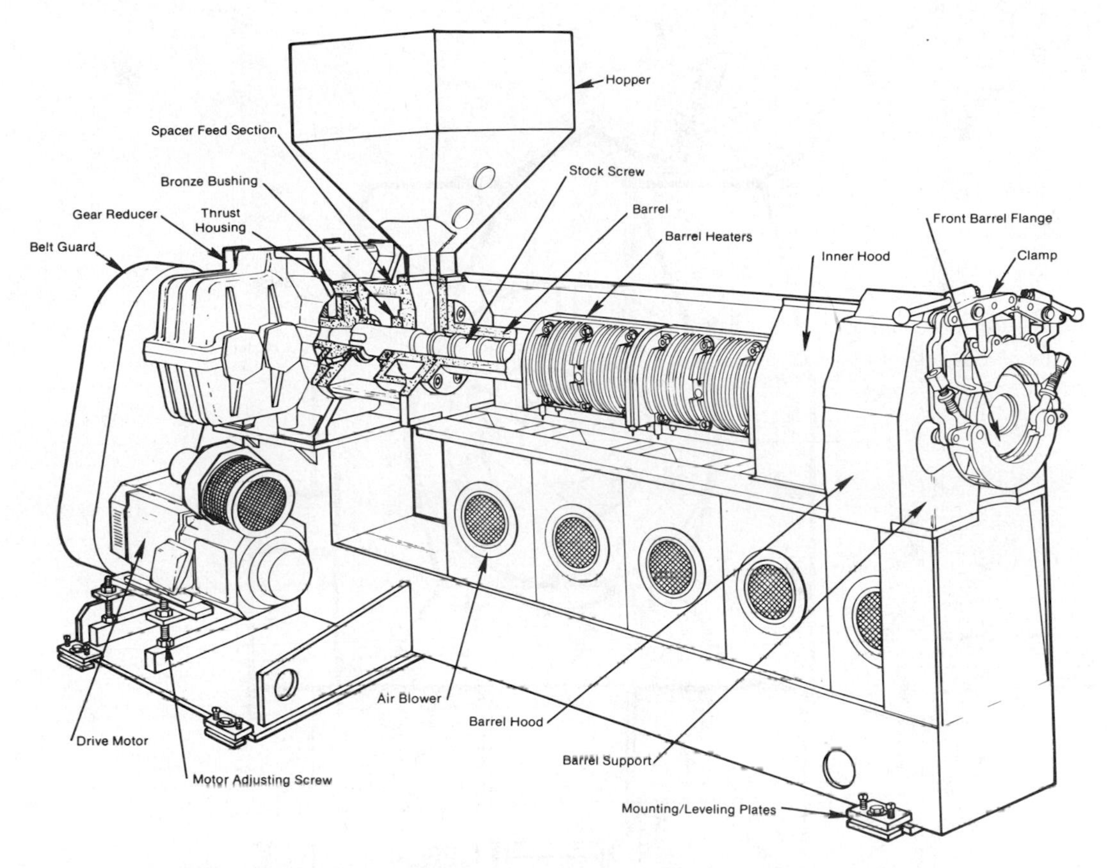

Fig. 4-8. Cutaway view of extruder. (*Courtesy Davis-Standard*)

Heating/Cooling Systems

Extruder barrel heating and cooling is necessary both for the addition and removal of heat and for the maintenance of the desired inner barrel surface temperature. The heating/cooling design must provide the required quantity of heat management as well as controllability. A typical extruder barrel is divided into control zones, each of which can be individually controlled to a desired setpoint.

Heating Designs. Many different heater designs have been used for barrel heating although, presently, electrical resistance heaters are used almost exclusively throughout the extrusion industry. Mica, ceramic, and other forms of strapped-on band heaters are common and economical, and are found on many low-cost and small-size extruders, but they suffer from poor life and performance. Cast aluminum heaters with electrical elements integrally cast inside now are accepted as the best design and are found on most modern extruders. These heaters are machine-bored in matched halves to fit closely to the barrel outside diameter when bolted or strapped on. The aluminum acts to eliminate temperature differentials and "hot spots" within the heater, gives excellent conduction to the barrel if properly mounted, and generally provides long service life.

Air Cooling. Air is an obvious choice for barrel cooling due to its simplicity, cleanliness, and economy. Air cooling can range from nothing more than convection for low-demand applications to forced air blower designs for more demanding processes. (See Fig. 4-9.) Inefficiencies in heating designs can be compensated for fairly easily by increased power levels, but air cooling designs, which must work with the available ambient air, are much

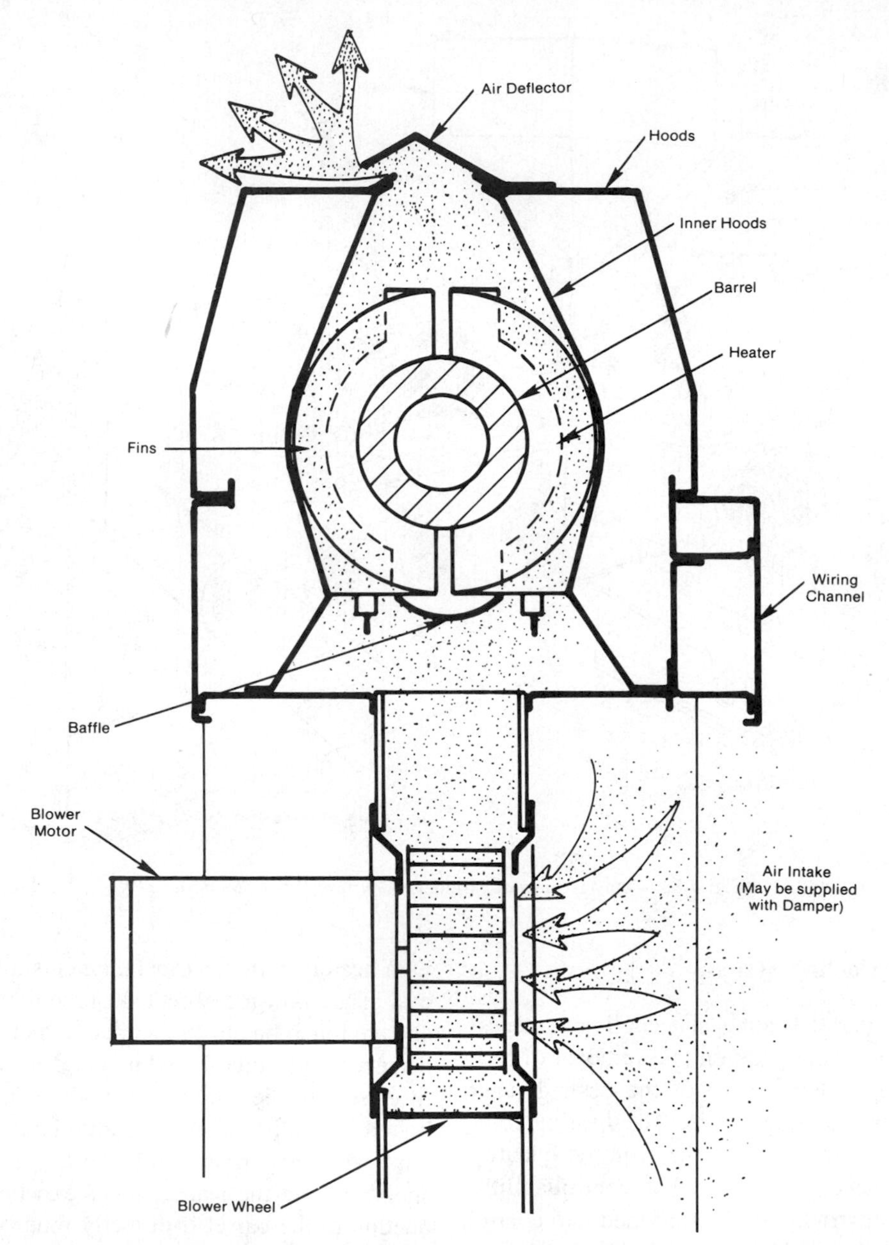

Fig. 4-9. Air flow diagram. (*Courtesy Davis-Standard*)

less tolerant than heating designs. Finned cast aluminum designs offer the best cooling efficiency, especially when properly ducted and supplied with high capacity blowers. Fins drastically increase the surface area exposed to the flow of air, the limiting factor for cooling efficiency. Sophisticated designs compensate for the cooler air at the entrance by increasing velocity at the exit. The better air-cooled designs can accommodate most extrusion processes.

Water Cooling. Water cooling provides for the sizable heat removal demand of high-load processes, offering as much as twice the efficiency of air cooling. (See Fig. 4-10.) Modern water-cooled extruders are usually equipped with cast aluminum heaters that include cast-in cooling tubes of stainless steel or Incoloy. Older designs utilized swaged-in tubing in a spiral groove on the barrel's outer diameter, over which band heaters were mounted, but these designs were plagued with poor heat

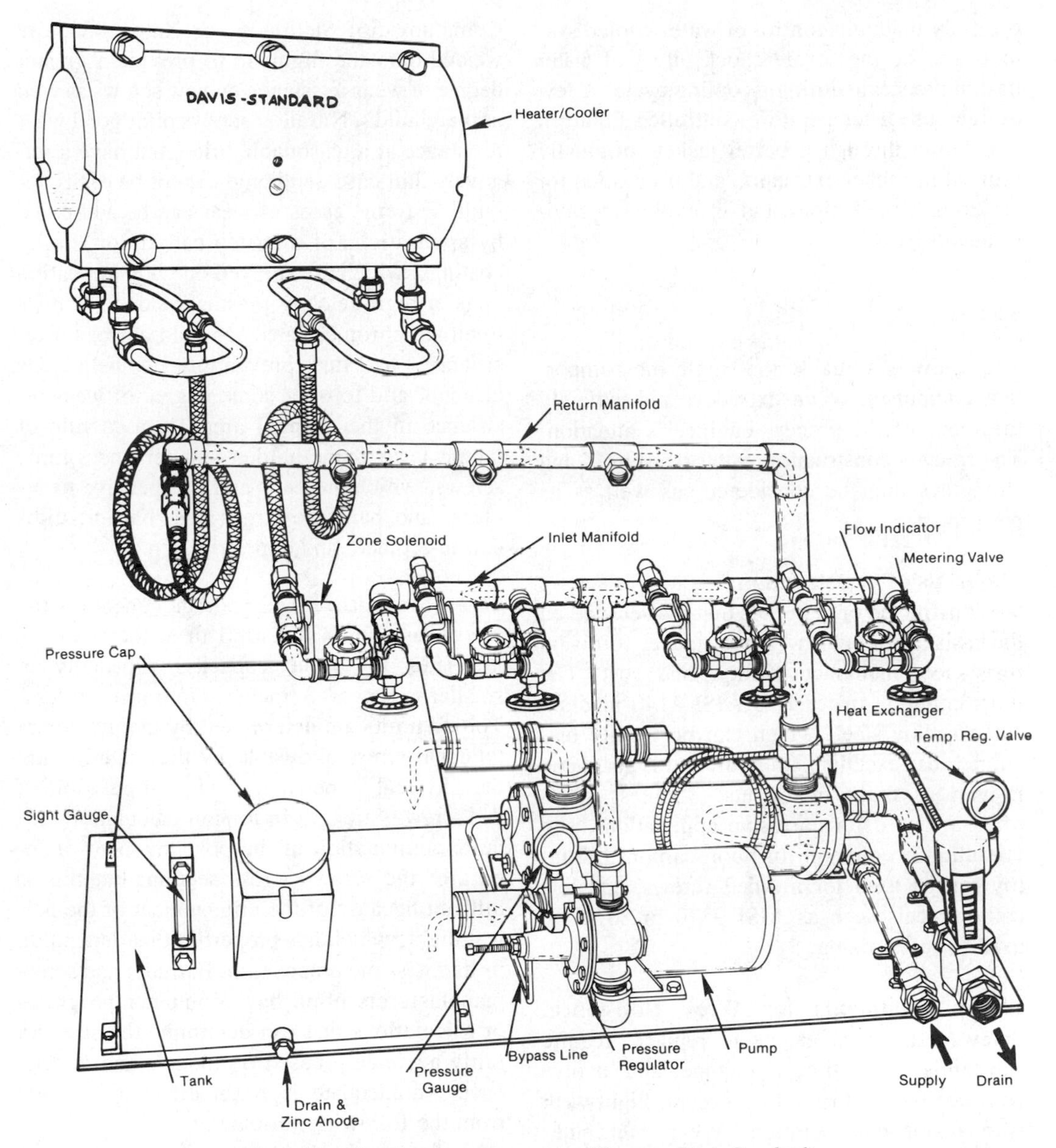

Fig. 4-10. Closed loop cooling system. (*Courtesy Davis-Standard*)

transfer and maintenance problems. Water cooling efficiency depends upon the total surface area or length of the cooling tube and on water velocity. High-temperature processes require sophisticated controls to accommodate the tremendous heat removal resulting from vaporization of the cooling water.

Water-cooled extruders are usually equipped with a self-contained closed-loop water supply system so that distilled or treated water can be circulated through the heaters to avoid scale buildup. The system consists of a tank, pump, heat exchanger, and distribution manifolds. Better systems include pressure regulation, temperature regulation, flow control, and flow indication. Water cooling is modulated by activating solenoids in response to temperature controller timed outputs.

Other Cooling Systems. Oil or other heat transfer fluids sometimes are used for barrel cooling. These fluids used to be more popular for stability reasons, prior to the advent of modern temperature controls, which can successfully maintain control of water-cooled systems despite the severe shock effect of water flashing to steam during a cooling cycle. A few designs use a temperature-controlled liquid in circulation through a barrel jacket, originally utilized in rubber extruders, and used often for controlling melt pipes and in explosive environments.

Screws

The screw is arguably the single most important component of an extruder, and certainly the focus of the process engineer's attention. The screw's construction, wear resistance, and strength should be considered, as well as its performance.

Materials of Construction. Extruder screws are constructed of materials that are selected on the basis of strength, wear resistance, corrosion resistance, manufacturability, and cost. The most common material is AISI 4140 medium carbon alloy steel, which can be readily machined, has excellent strength, and is adaptable to various flight treatments. Stainless steels often are used for corrosive applications, and Hastalloy is common for fluorocarbons. Nitraloy can be used for nitrided screws, and low carbon steels such as AISI 9310 or 8620 are used for carburizing.

Flight Treatments for Wear Resistance. Screws can be expensive to replace, require substantial down time to change, and involve considerable lead time to order; so flight wear is a major consideration. Wear contributes drastically to decreases in performance, quality, and productivity. Flights generally are manufactured for maximum wear by hardening, special coatings, or the welding of a hard layer on the surface. The most common configuration is flame-hardened 4140, which gives moderate life at reasonable cost, can be rebuilt when worn, and is suitable for most nonabrasive materials. Wear-resistant alloys, such as Colmonoy 56, Stellite 6, or Xaloy 830, are welded onto the flight tip to provide a greater degree of wear resistance on new screws as well as on rebuilds. Nitralloy screws offer good wear resistance at a reasonable price, but have a relatively thin case depth and cannot be easily rebuilt. Extreme cases of wear can be addressed by applications of tungsten carbide or similar coatings, which are sprayed on with detonation guns or high-velocity plasma welders. Screws often are chrome-plated for mild corrosion resistance, for rust prevention, to help ease cleanup, and to offer some degree of wear resistance in the channel area. A good rule of thumb is to flame-harden smaller and simple screws, which are easy and inexpensive to replace, and hardface larger-size, barrier, difficult-to-replace, and expensive screws.

Screw Strength. An important concern to the screw designer is the strength of the screw as it relates to torque capacity, especially in smaller sizes of 3 inches (75 mm) or less. Torque limits are determined by the maximum torsional stress allowable by the material and the physical dimensions and configuration of the screw. Strength-in-torsion calculations require computation of the polar moment of inertia of the screw's cross section; but this is difficult because of the arrangement of the helical flight, which has properties that depend on its hardness or composition. Extruder and screw manufacturers often have computer programs or calculations that can determine the strength fairly accurately based on models and experience. An adequate approximation can be made from the following formula:

$$\text{Torque} = \frac{\pi\left\{\left(\text{Dia.} - [\text{Channel depth} \times 2]\right)^4 - \left(\text{Core dia.}\right)^4\right\}}{16 \times \text{Dia.}} \times \text{Permissible stress}$$

This formula ignores any strength contribution of the flight, but then does not include any safety factor. If the screw is of normal proportions, the formula should prove adequate for estimations of strength.

Screw Heating and Cooling. Most screws benefit from a core at least as deep as the feed area so that bridging or compound sticking problems can be avoided by cooling. Screw temperature control of the first stage of a vented screw can help to balance the pumping characteristics of the two stages. Tip cooling can help prevent material from burning at the screw tip. Screw heating and cooling usually are accomplished through water or oil circulation. The fluid is supplied to a small-diameter coaxial tube inside the core hole and returns through a rotary union with inlet and outlet connections. Temperature control is especially important because the screw surface temperature can significantly affect the extruder flow rate and melt temperature.

Thrust Bearings

Extruder Screw Thrust. The extrusion process generates a tremendous amount of axial load or thrust directed toward the rear of the screw. The major component of thrust is caused by the head pressure acting upon the frontal area of the screw, which is like a hydraulic cylinder. A 6-inch (152 mm) screw at 5000 psi (34.5 MPa) generates thrust in excess of 140,000 pounds (620 kN). Another contributor to the total force is the pressure differential acting along the entire length of the flight, which can be as high as the head pressure component. Extruders must be designed to reliably accommodate this thrust.

Bearings. Modern extruders commonly accommodate thrust with a thrust bearing mounted integrally to the gear reducer's output shaft where the screw engages, usually of cylindrical or tapered roller design. Thrust bearings typically are long-lasting and trouble-free, but, in demanding applications of high head pressures or speeds, they require attention to their selection and, most important, to their lubrication. Adequate oil flow is necessary for bearing cooling as well as lubrication, the lack of which is the leading cause of problems and failure.

B_{10} Life Calculations and Comparisons. Extruder manufacturers normally list the thrust bearing rating in terms of a B_{10} or L_{10} life, which is the time in hours that 90% of a controlled test sample of bearings would exceed without failure at similar conditions of load and speed. This figure also implies that 10% of the bearings fail during this time. In order to compare the capacity of different machines, the B_{10} lives must be rated at the same speed and load. The following formula can be used to calculate the adjusted B_{10} life from speed S_1 to S_2 and from load L_1 to L_2:

$$(B_{10})_2 = (B_{10})_1 \times \left(\frac{S_1}{S_2}\right) \times \left(\frac{L_1}{L_2}\right)^{3.33}$$

The formula shows the linear relationship with speed and the exponential relationship with load. Most manufacturers list the B_{10} life in the literature and in manuals at 5000 psi (34.5 MPa) and 100 rpm, but occasionally the average life is listed—which is five times the B_{10} life.

Gear Case

Gear Case Designs. A single-screw extruder is a very simple machine with only one functional moving part, the screw. Normal screw speeds in the range of 100 rpm usually are achieved by reducing typical motor speeds of 1750 rpm through a gear case with an appropriate ratio. Worm gear reducers once were the norm and still are used extensively for small machines, but their low efficiency and today's increased power demands have led to the almost universal use of helical gearing. Helical gears are available in a variety of hardness and precision levels to accommodate the various speeds and loads found in extrusion. The gear case should always be supplied with a safety factor of 1.25 for smaller machines and 1.50 for large extruders.

Lubrication. The gear case lubrication system must provide proper lubrication for the gear meshes and, perhaps more important, the bearings. Oil supplies are used for cooling the bearings and gears, and on large machines and in demanding applications, a forced lubrication system should be employed to ensure an adequate supply to each bearing. All gear reducers have a thermal limit beyond which some sort of cooling is necessary. A heat exchanger in the oil circulation system of high load machines is the normal means for providing cooling and maintaining proper temperatures.

Power–Speed–Torque Relationships. Extruder timing involves careful consideration of several factors in order to accommodate a particular application. Timing refers to the speed at which the full motor horsepower is applied to the extruder screw. It involves the reduction ratio of the gearbox combined with the ratio of the belts and pulleys (when used) that connect the motor to the extruder. Motors are rated at full horsepower at a base speed, usually 1750 rpm. A DC motor's power output is proportional to speed; so it is rated at $\frac{1}{2}$ power at $\frac{1}{2}$ speed, and so on. An extruder timed at 75 rpm, for instance, has a total reduction ratio of 23.3 : 1 between a 1750 rpm motor and the screw. More important, the motor torque is multiplied 23.3 times to the screw. Torque is the quantity that best describes the force or power applied to turn the screw, and it is directly related to motor load or amps. Torque is best understood if expressed in terms of horsepower per rpm rather than as foot-pounds or inch-pounds.

Any particular extruder screw and material combination has a specific torque requirement. The drive initially is timed at a speed that provides the best compromise between torque available to the process and speed required. A processor who tries to run a new material on an extruder and experiences high motor loads actually is experiencing high torque demands. The torque (hp/rpm) available to the process can be increased either by increasing horsepower or by decreasing the timing (maximum screw speed). Belt drives are used on most extruders so that the timing can be changed easily by changing the pulley reduction ratio. Similarly, when higher speeds are required, and motor load is low, the extruder can be timed higher by changing pulleys—but one should be aware that this also reduces the available torque.

In changing timing, the gearbox capacity must always be considered. Gears have a torque limit that restricts the low end of the timing range and sometimes a thermal capacity that limits the upper end. One should always check the gear rating nameplate or consult with the manufacturer before retiming an extruder.

Drive Motor

The drive motor provides power for the conveying, melting, and pumping processes and establishes extruder output by maintaining a desired speed. DC electric motors are by far the most popular drive type because of their economy, serviceability, and accuracy. A DC drive is a constant torque device, which meets most processing requirements quite well and matches the capabilities of a typical helical gear case. AC variable frequency drives are gaining in popularity as their cost comes down but offer little advantage for most extrusion applications. Hydraulic drives are seen in some machines, but are much more complex and costly than the typical DC drive; and even though they allow for the elimination of the gear case, they offer no advantage in most applications.

Drive Coupling

The drive motor can be connected to the gear case directly by a coupling or by an arrangement of pulleys and belts. Direct coupling is necessary in large machines over 300 hp (225 kW); but it requires precision motor alignment and a specific gear ratio for each application. Belts and pulleys, the better system for smaller machines, are much more flexible as far as timing changes are concerned and give many motor mounting options. Modern designs operate with very little maintenance and high efficiency.

EXTRUSION CONTROLS AND INSTRUMENTATION

Basic Control Requirements

Extruders require some level of control of basic functions and can benefit from enhanced control of other functions. Minimum requirements include some form of barrel temperature control, drive speed control, and drive load indication. Head, or breaker plate, pressure indication also is typically included in the most basic systems. The demands of modern manufacturing methods, quality issues, and process control have dictated much more extensive control systems and instrumentation on many of today's extrusion lines. Extrusion controls have benefited from the overall rapid technology growth in the electronics, controls, computer, and instrumentation industries.

Sensors and Monitors

Barrel Temperature Measurement. Barrel temperature control is one of the most critical functions, so temperature sensing is extremely important. A sensor typically is located in the center of each barrel zone, traditionally at the bottom. Extruder barrels have substantially thick walls in order to withstand internal pressures, but the critical control surface is at the inside. The ideal sensor would be mounted at the inside surface, but this is generally impractical; so a deep hole normally is drilled to within $\frac{1}{8}$ inch or less of the inside, usually against the hard barrel lining. Thermocouples are the most commonly used sensor because of their durability, simplicity, and accuracy. Thermocouples also are the most accurate and best-suited sensor for the application because they sense the temperature at their tip, which is in contact with the bottom of the drilled hole and very close to the control surface. RTDs (Resistive Temperature Detectors), are often specified in sophisticated control systems, and as a sensor, are more accurate; but they are not well suited to this application because of the length of their sensing element, which measures a temperature some distance from the desired inner surface. The most advanced barrel control systems use a second thermocouple in the heater as part of their control scheme.

Melt Temperature Measurement. The polymer melt temperature often is a limiting parameter in the process, so its accurate measurement is extremely important to the serious processor. Measurements typically are made by utilizing a melt thermocouple installed in a hole that exposes the sensor directly to the polymer. Production lines often use flush designs, combination pressure transducer/thermocouples, or probes with only a small projection into the melt stream because these designs are fairly durable. Conduction effects from the die body and the relative slow velocity layer of melt near the wall may render these sensors ineffective, however.* Accurate melt temperature measurement requires a thermocouple, ideally extended into the center of the flow channel but at least $\frac{1}{2}$ inch into the stream; for maximum accuracy, an exposed-junction low-mass thermocouple is used. Fiber-optic infrared probes recently have been developed, but because of limited optical penetration into the melt they also must be mounted on a protruding probe. Adjustable-depth probes, either manual or automated, allow for optimum positioning but provide for retraction to protect the element during cleaning and start-up.

Pressure Measurement. Pressure measurement is of critical importance to the extrusion process as a window into the operation. Early designs utilized a high-temperature grease in a tube connecting a hole in the barrel to a Bourdon tube type mechanical gauge, but were plagued with maintenance and durability problems. Modern gauges all use a diaphragm that isolates the hot polymer from the sensing device and either directly actuates a mechanical gauge or utilizes a strain gauge transducer. Two basic concepts dominate the industry: push rod types, which mechanically connect the diaphragm to the gauge or transducer, and liquid (usually mercury)-filled capillary types, which transmit pressure hydraulically from the diaphragm. The push rod has the reputation of being more durable and less costly, whereas the capillary type is generally thought of as more accurate and much less temperature-sensitive. These reputations are well deserved, but recent designs have resulted in improvements in both types.

At a minimum, extruders should have one pressure transducer located before the breaker

*See E. L. Steward, "Making the Most of Melt Temperature Measurement," *Plastics Engineering*, Society of Plastics Engineers, July 1985.

plate to monitor pressure for safety and to indicate when filter screens are plugged, and ideally one in the head for process monitoring. Transducers along the barrel are vital in test labs for acquiring screw design data, and can be used in production for feed loss and vent port flooding alarms. The industry has standardized on a 0.312-inch-diameter hole in the barrel with a 45° seat and $\frac{1}{2}$–20 threads for gauge interchangability and easy replacement.

Drive Power Measurement. Drive power consumption, an important process indication, is readily available in DC or AC drives in the form of a motor ammeter showing either percent load or actual amps. The correlating parameter in hydraulic drives is pump pressure.

Screw Speed Sensors. Extruders, although not positive displacement pumps, normally exhibit a fairly linear relationship between screw speed and output; so screw speed indication is a critical control parameter. On approach is to mount rpm counters on the screw drive shaft or any of the other shafts on the gear case, but the simplest method is to utilize the output of the drive motor's tachometer scaled to screw speed.

Temperature Control

Barrel temperature control is one of the most important and difficult requirements of the extrusion process. Barrel temperature setpoints affect the final output rate, quality, temperature, and stability. The barrel heating and cooling system not only must maintain a set inner-surface temperature, but must do so while providing varying amounts of heating to compensate for the melting requirements or cooling because of excessive internal heat generation. The control problem is complicated by the physical arrangement of the barrel. Of necessity, the barrel has a substantial wall thickness; so the source of heating or cooling is at a significant thermal distance from the desired control surface, causing a sort of "thermal inertia." A simple on–off controller cycles wildly about the setpoint because by the time the sensor "sees" the required temperature, the heater/cooler is already much too hot or cold. The ideal extrusion temperature controller not only must compensate for this thermal inertia, but should anticipate the required heating or cooling during a process change or upset.

PID Controllers. Most commercial temperature controllers incorporate a PID control scheme with proportional, integral, and derivative functions that can be adjusted to provide a relatively stable temperature. The proportional band adjustment on heating and cooling is the most important feature. It simply reduces the amount of heating or cooling as the indicated temperature approaches the setpoint. For instance, the controller might call for 100% heating at a 50° error, 50% at 25°, and 10% at 5°. Other features help to reduce over- and undershoot, and "offset" allows the controller to stabilize at a fictitious setpoint that results in the desired setting. Modern controllers usually are offered with some type of auto-tuning, thus eliminating the extremely difficult chore of correctly adjusting a PID controller. Barrel controllers are typically heat/cool models, whereas die, head, and adaptor zones are normally heat only.

Dual Sensor Controllers. Even the most sophisticated PID controllers rely on schemes and "fixes" to compensate for the inertia caused by the thermal distance between the sensor and the heat or cooling source. A unique solution to the problem is a patented, two-thermocouple control system that incorporates a second sensor mounted directly in the heater/cooler and, through the use of a special computer algorithm, eliminates the problem of thermal inertia and temperature cycling. These controllers are microprocessor-based and often are incorporated into complete line computer control or supervisory systems.

Controller Output. The output of any temperature controller, even with a proportional heating band, is usually on–off. The required heating or cooling ratio is achieved by time-proportioning the output during a set duty cycle. The heat output is energized for 5 out of every 20 seconds, for example, for 25% output. With the stabilizing effect of aluminum heaters and the large thermal mass of the barrel, this approach is entirely satisfactory. Low voltage heat outputs in turn activate line voltage contactors, preferably mercury relays, in a heat control cabinet. Cooling solenoids are connected directly, and electric blower motors can be either direct-wired for small single-phase types or connected to three-phase motor starter relays for larger designs. Temperature controllers often are available with proportional control output connected to an SCR or other type of variable power supply for the barrel heaters. Although this is theoretically a superior system, it is of little value in extrusion barrel control.

Pressure Control

Ideally behaved fluid processes normally control pressure directly because the output flow rate is a direct function of pressure. In plastics extrusion, however, this direct relationship does not exist. Variations in melt temperature, shear rate, material deviation, and other uncontrollable parameters all can independently affect pressure and output. In extreme circumstances, an increase in screw speed, which increases shear rate and melt temperature, not only increases the output rate; because of the lower viscosity that results from the higher melt temperature, the head pressure actually can decrease! Controlling the extruder speed directly with a head pressure controller is rarely done because of these problems. Pressure indicators often are used for alarmingly high or low conditions. Full control is applied to extrusion melt pump systems, where the pump establishes the output rate, and the extruder is controlled to maintain a constant pump inlet pressure. The other common use of pressure control is in the operation of automatic screen changers, where quite often two pressure transducers, one before and one after the filter, are used to control a constant pressure differential.

Melt Temperature Control

Extrudate melt temperature control has long been a desirable goal in extrusion, but in practice it is extremely difficult to achieve. Practical approaches change the setpoint temperature of the last one or two barrel zones, but their use must be limited because the barrel temperature also can affect the output rate and upset the process; hence they are rarely used.

Size Controls

At the output of most extrusion heads or dies, whether the product is sheet, wire, film, tubing, or any of the wide variety of shapes encountered in industry, it is always desirable to make some sort of measurement of the extrudate. This is the first opportunity to "see" directly into the process and effect some sort of change. Extrusion, being a continuous process, is a difficult one for on-line inspection, as the product cannot be held and measured or inspected in traditional ways until the end of the line. Often, the line involves a large inventory of product, which all can be off-spec if adjustments are not made until the end. Size monitoring and control devices on-line allow immediate feedback and reduction of scrap. Whatever the method of measurement or gauging, corrective action usually is taken on the most easily controlled parameter. Single extruder lines with simple downstream equipment configurations should change line speed in response to size variations. More complex lines and tandem extrusion systems need to change extruder speed, but must do so gradually and with an awareness of possible related effects, such as melt temperature and pressure changes. Large-step extruder speed changes often have a two-stage response—initially an immediate reaction, followed by a gradual change due to temperature, viscosity, and pressure changes, barrel control setting, and screw inventory renewal. Coextrusion lines must discriminate measurement as well as control responses among the machines.

Diameter and Profile Gauges. Among the first and presently among the most successful on-line gauges are laser gauges, which project a beam across a critical dimension of the part causing a "shadow" on the receiver on the other side. Optical gauges differ only in their light source. These gauges have evolved into highly reliable units, with compensation for environment and product movement, and with sophisticated controls. They usually represent very good value and can be justified easily in most round-section, continuous extrusion lines such as wire coating, tubing, hose, and pipe. They also can be applied to profile extrusions that have a critical prominent dimension. Compound gauge heads can provide *X–Y* data, and rotary heads can give ovality information.

Thickness Gauges. Sheet and film thickness gauging usually is accomplished by a scanning head that either transmits to a receiver on the other side of the web or reflects back through it, utilizing radiation to measure density. The important aspect of web thickness gauging is that transverse as well as in-line profiles are detected. The transverse profile data can be used to automatically control adjustable die gap devices. Wall thickness gauging for pipe is available utilizing ultrasonic sensors and for wire using capacitance devices; both types can be combined with diameter gauging.

Rate Controls

Hopper Weight-Loss Feeders Extruder throughput is not directly measurable until the end of the process line, but output rate can be derived and controlled in several ways. The simplest method is to accurately measure the amount of material flowing through the feed hopper. This can be accomplished by timing and weighing small batches in an intermediate hopper and calculating the rate or by doing weight-loss calculations on a continuously monitored feed hopper. Both systems require accurate load cell sensing, isolation of outside effects, and a means of ensuring a constant-flood feed supply to the extruder. They work well with pellet or granule feedstocks and not as well with powders. Feed hopper rate determination has a few drawbacks, as it is blind to changes within the inventory time of the screw; however, it provides a very accurate long-term average rate and can be used to control extruder speed in a slow response manner.

Gravimetric and Volumetric Feeders. Gravimetric and volumetric feeders require starve feeding; and although long-term average rates can be accurate, starve-fed single screw extruders are notorious for surging and short-term instability.

Calibration Curves. The extrusion output rate can be calibrated and calculated quite accurately in a three-dimensional grid of screw speed, pressure, and output. Test data must be collected at several speed and pressure levels and input into a simple curve fit program to find the proper function and constants; once this is done, the output rate is quickly calculated from speed and pressure. The technique can be incorporated into a computerized control scheme and used to adjust the extruder to compensate for the slow plugging of filters or other gradual changes.

EXTRUSION PROCESSES

The extruder's function is complete when a good melt is pumped through the die system at an acceptable output rate, but success is not at hand until a usable product is formed in the downstream sizing and cooling equipment. Various products of countless shapes can be formed in a continuous manner with extrusion, making this process quite versatile. A brief description of some of the most popular products made with the extrusion process follows. The reader also is referred to Chapters 12 and 22, on blow molding and compounding.

Blown Film

In the blown film process, plastic melt passes through a die that forms it into an annular shape, usually directed upward. (See Figs. 4-11 and 4-12.) Air is introduced into the tube to inflate it and around the outside of the "bubble" to cool and solidify the melt. The cooled bubble is closed in a set of rolls (nip rolls) placed at some distance above the die. The amount that the bubble diameter is expanded is called the blow-up ratio (BUR), and is normally in the range of 2 : 1 to 4 : 1. Both stretching the bubble radially and pulling it away from the die axially impart orientation to the plastic, improving its strength and other properties.

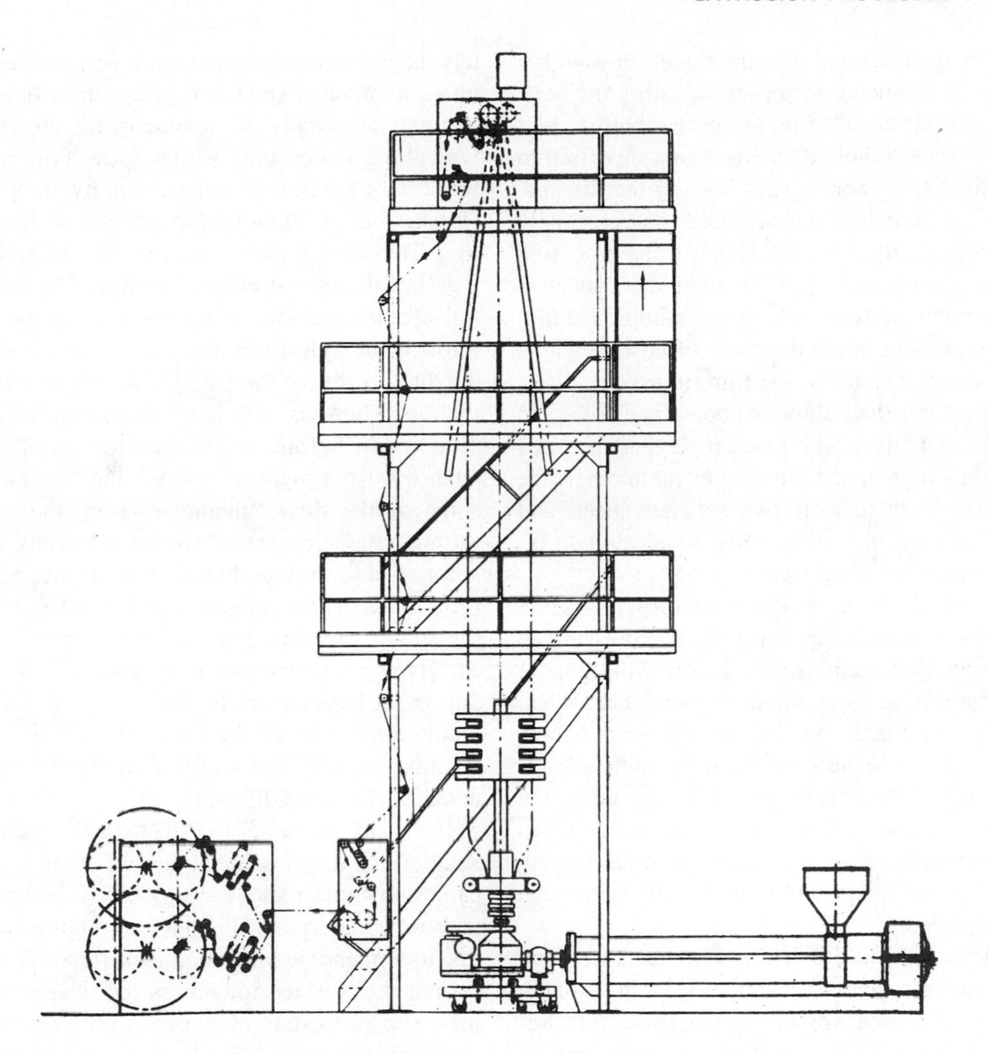

Figs. 4-11 and 4-12. Blown film process. (*Courtesy Davis-Standard*)

The distance that the tube travels upward before its diameter increases is called the *stalk height*. Most materials are blown with a short stalk, with bubble diameter being increased to its final size shortly after leaving the die lips. Some materials (such as HDPE and especially high molecular weight HDPE) are processed with a long stalk (six to eight die diameters above the die lips), which gives improved film strength both in the direction of flow (machine direction) and across the film (transverse direction). After the bubble is cooled and collapsed into a flat shape, the product (layflat) is rolled up on a high-speed winder. In some cases, the layflat is slit to form two separate sheets and wound onto individual rolls, requiring two independent winding shafts.

Cooling air is directed upward along the bubble's outer surface at high speed, using an air ring. Some sophisticated air ring designs are available that allow minimal turbulence. When high output rates are desired, additional bubble cooling can be accomplished by using a cooling device that sits on top of the die inside the bubble. Cooling fluid is passed through orifices inside the die body and into the internal cooling device, known generally as an IBC or internal bubble cooler.

After traveling a fairly short distance from the die exit, the melt becomes solid in a usually noticeable area known as the frost line. Because the solidified material is more opaque than the melt, the frost line is easily visible. If the extruder is pumping melt consistently with uniform temperature, the die is doing an efficient flow distribution job, and the cooling air is being uniformly applied, then the frost line will appear quite well defined and in a horizontal plane. If there is instability in the extruder/die/cooling system, there usually is a ragged frost line and poor film thickness (gauge) uniformity, due to uneven melt pumping and solidification.

The dies used to form the thin annular melt can be fed from either the bottom or the side, with the bottom-fed die being prevalent today. The extruder usually has a low centerline height (15 to 20 inches, instead of the typical 42 inches for other processes) to reduce the distance from the top of the die to the floor. The material enters the die through an adaptor pipe and is split into a number of smaller flow channels (ports) that extend radially in the base of the die. The typical die is designed with a spiral flow path for each of the flow channels, with overlapping spirals to aid in flow uniformity out of the die lips. This design eliminates the use of "spider legs" of the type used in pipe dies. The material spreads radially and upward as it moves through the spiral channel section of the die; the die lips set the thickness of the melt exiting the die. There usually is an increased volume section just before the die lip area, to offer a chance for the melt to "relax" and to remove some of the flow "memory" from the path through the die system. Some materials develop visible flow problems from the die ports (port lines) or from uneven radial exit flow from the spirals. The die lips are set by experience for given polymers and product sizes. The lip gap is set larger than the final gauge to allow ample drawdown of the melt. The die lip gap for a 0.001 inch thick LDPE film may be set at about 0.020 to 0.040 inch.

If the lip gap is set too tight (0.010 inch or smaller) the die pressure at the end of the screw can reach undesirable levels (6000–10,000 psi), and some materials will exhibit a rough surface due to a phenomenon called melt fracture. Reducing this surface roughness (because of the poor slip properties of the melt on restrictive metal passages) is difficult except by opening the lip clearance to reduce the shear experienced by the melt. This phenomenon is quite prevalent with materials such as LLDPE and HDPE.

Die designs that must process high viscosity materials such as LLDPE and HDPE are given special consideration, in order to maximize the internal clearances and reduce pressure drops while still ensuring good flow uniformity as the melt exits the die lips. It is easier to achieve uniform die flow with tight die lips and restrictive internal passages, but the negative effect of the resulting pressure drop on output rate and melt temperature of the screw is seldom acceptable to the blown film producer. Bubble formation in this case is aided by having the lowest possible melt viscosity and, hence, the lowest practical melt temperature. The experience and the knowledge gained in today's die design decisions allow a much lower pressure drop through the die while preserving good flow uniformity.

Blown film is typically thin, with 0.0001 inch film possible with some materials. Film thicknesses of 0.125 inch and thicker are possible under some circumstances, but the typical range of film gauges encountered with this process is 0.0001–.050 inch. Other than the gauge dimension, a layflat width usually is given, which is simply the width of the bubble after collapsing into a flat film of double width. The layflat width dimension is the bubble circumference (3.14 × final bubble diameter) divided by 2.

A growing trend in this process is coextrusion, just as is seen in the flat die sheet process. The die is designed with multiple flow distribution channels so the layers come together after they have been evenly distributed in the annular configuration, prior to the die lips. Many layer configurations are used to achieve a variety of film properties. Barrier materials (for oxygen, moisture, odor barriers, etc.) usually require adhesive layers for adherence to the outer layers of the film structure. The barrier layer typically must be placed inside the film structure because environmental effects (moisture, etc.) can noticeably reduce the barrier material's effectiveness. The number of layers can be from the simple two materials up to eight to ten layers with very complex die systems. Typically, four to seven layers are found to cover most situations.

There is no trimming necessary in this process to make the film edges uniform in dimension, because of the cylindrical die opening. There is some amount of regrind material created with the blown film process, due to startup scrap while the lines are being tuned to the final gauge and scrap formed when the layflat is slit to form sheets. In coextruded film, regrind can be hidden in an inside layer. The regrind usually is added to the virgin feed material in the extruder hopper at 10 to 30% levels to avoid disruption of extruder processing efficiency due to the bulk density reduction seen with fluffy regrinds of these films. An alternative approach to reusing the regrind is to repelletize the fluff in a reclaim extruder and add the pellets to the virgin material on the production line, with little or no feeding efficiency loss.

Some extrusion lines have printing equipment on-line as well as bag-making machines, which form the sealed edges and the opening configurations. The folds, or gussets, typical on the sides of some bag constructions can be made in-line with this process by using a folding apparatus in the collapsing area. Simpler extrusion lines form the film or sheet and perform the printing and bag-making functions off-line. Depending upon the line speeds, there are arguments that support both methods of performing post-extrusion functions. Films made by the blown film process usually are of lower clarity than their cast film counterparts.

Some of the products formed by the blown film process are garbage bags, can liners, agricultural films, grocery bags (T-shirt bags, etc.), and thin films for paper and tissue products.

Common polymer materials used in this process include polyethylenes, polypropylenes, EVA, and flexible PVC.

Sheet, Extrusion Coating, and Cast Film

Sheet: The extrusion process used to form sheets continuously is illustrated in Fig. 4-13. The melt is formed through a die having a narrow slit aperture that forms the sheet and directs the melt to cooling and polishing rolls. The die takes the melt from the extruder, typically in a continuous rod shape, and uniformly spreads the melt into a wide sheet of uniform thickness. The die that spreads the melt into a uniform sheet is shown in Fig. 4-14.

The manifold of the typical sheet or film die resembles the shape of a coathanger and is designed, along with the remainder of the die clearances, to uniformly distribute the melt from the extruder into a wide, thin shape. Because many sheet and film extrusion lines must produce a range of products with different thicknesses over a range of output rates, melt temperatures, and sometimes different polymers, the die must perform its uniform distribution function under various conditions. The versatility of one die manifold configuration has limitations that are sometimes eliminated

Fig. 4-13. Sheet extrusion line. (*Courtesy Davis-Standard*)

through the aid of a distributor or choker bar, especially with thicker sheet products. The die's uniformity can be roughly adjusted by lowering or raising the choker bar, and final gauge adjustments are made through the use of lip adjusting bolts. Some materials (such as vinyls) are more degradable than others and make the use of the choker bar impractical, because of the small potential hang-up areas on each side of the bar.

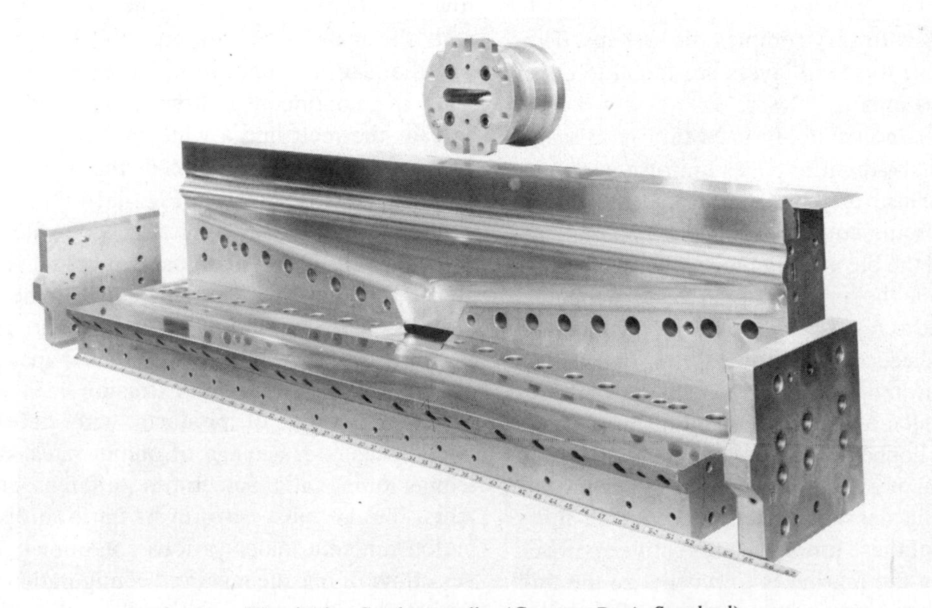

Fig. 4-14. Coathanger die. (*Courtesy Davis-Standard*)

The die lips are the final surfaces (lands) through which the material passes before exiting the die. The lengths of the lip lands are designed through experience and are usually 10 to 20 times the expected lip opening. The final lip opening usually is adjusted to approximately the thickness of the final sheet gauge. Based on experience, the amount of polymer swell or the desired drawdown of the material coming from the die lips will determine if the initial lip opening should be larger or smaller than the desired sheet gauge. Should a die be required to extrude products over a wide thickness range, the die must be designed so that the lower lip is removable for changing lip gaps over the broad range. The die's interior surfaces are highly polished and usually plated with a material such as chrome to ensure no flow interruption and to avoid any melt hang-up and resultant degradation or sheet uniformity problems.

The die has several heating sections (zones), which are controlled near the material's extrusion temperature. Adjusting these die heater zones independently of one another also will aid in altering material flow through the die and will give an additional parameter for rough sheet gauge adjustment, as the polymer flowing through the die is affected somewhat by the metal temperature of the die surfaces. Some sophisticated sheet extrusion systems incorporate an individual heater on each die lip adjusting bolt. Each of the lip bolt heaters is closely controlled to a temperature that will give a required expansion or contraction of the bolt and a resulting lip adjustment. These fine lip adjustments are determined by a microprocessor controller, which reads the sheet gauge from a scanning device placed after the cooling rolls, determines where the gauge needs correcting, and controls the proper die lip heaters to effect the change. The gauge must be set manually first, to a range close to the required uniform gauge, and final slight adjustments can be made by the lip bolt controller as the line is running.

The melted material flows out of the die lips and is wrapped around several cooling rolls, which impart the final gauge and surface finish characteristics. The most typical setup utilizes three rolls in a vertical arrangement with the melt entering between the top and middle rolls. The middle roll, which determines the surface finish, may be highly polished for a high-gloss sheet or embossed to produce a pattern on the molten material.

The rolls are driven either by a single DC drive connected to the rolls through a chain system, or by individual drives that all are closely coordinated for accurate speed control. The separate drive package usually is chosen for critical applications such as optical-quality sheet, where chain drives can impart "chatter marks" at a frequency related to the chain's pitch dimension.

The roll entry point (called the nip) is adjusted to a gap dimension 5 to 10% larger than the final sheet gauge to adjust for shrinkage of the material as it cools along the later roll surfaces. After passing through the nip, the material travels in an "S" wrap pattern, with one surface in contact with the center roll for 180°, and then does a reverse wrap, with the other surface contacting the lower roll for about 270°. The nip point between the middle and lower rolls usually is set close to the final sheet gauge because much of the shrinkage occurs as the material cools prior to entering the lower nip.

Sometimes the material is extruded "upstack," entering the rolls between the lower and middle rolls. Upstack extrusion allows the sheet to travel down the line with its embossed side up. An alternate arrangement is for the rolls to be set up on a horizontal plane with the material extruded downward. This allows very low viscosity materials to flow more uniformly to the rolls without drooping.

The cooling rolls are temperature-controlled by oil or water passing through their interior. Maintaining a minimal temperature differential across the roll surface is important, and the design of the cooling flow path must ensure high velocities and turbulent flow (a high Reynolds number) for the cooling fluid passing through the roll. The temperature of the cooling fluid for each roll is controlled by a separate system to allow different roll temperatures as required by the process. Usually the upper and middle rolls (downstack extrusion) are controlled at a higher temperature than the lower roll to temper the initial sheet cooling, control embossing, and control initial sheet surface cooling (hence surface appearance, etc.). The lower roll normally will be quite cool to maximize heat removal. Some atmospheric conditions may produce moisture condensation on cold roll surfaces, and increased roll temperatures may be required to avoid sheet surface marring by roll surface water droplets.

The size of the cooling rolls will depend upon the cooling time required, based on the extruder's output rate and melt temperature along with the sheet dimensions, which determine the line speeds and hence the time in contact with the rolls. The rolls are typically between 8 and 24 inches in diameter, with larger and smaller rolls possible under some conditions. The frame that holds the main rolls must be of sufficient mass and design to avoid vibrations and sheet quality problems (such as the chatter marks caused by the vibrations). The sheet line may require additional cooling after passing the three main rolls, which is accomplished through the use of auxiliary rolls placed downstream. The auxiliary rolls are often aluminum because of its good heat removal properties and the fact that the sheet's surface already is determined by the highly polished main steel rolls. The auxiliary rolls typically are controlled at as cool a temperature as is practical.

The die typically is a heavy block of steel with a wedged-shaped exit end. Because of the drawdown of the melted material after it exits the die and the trimming of the sheet's edges, the width of the die is greater than the finished sheet width by about 10 to 20%. When narrower sheet is required from a wide die, deckle bars may be used, which are restrictive bars that clamp onto and block off some distance from the die edges and across the lip surface. Instead of bars, rods may be used, which slide into grooves machined along or just prior to the die land. The pressures present in the die along with the sizable widths typically seen (up to 100 inches) require thick steel die body halves with many large connecting bolts, set on the extruder side of the flow manifold, to resist deflection and hence increased opening at the die lips, which would cause uncontrollable gauge. The die lips must extend between two of the main cooling rolls to a point fairly close to the nip to avoid excessive sagging of the melt. When very low viscosity materials are processed (such as polyesters), it becomes very critical to use sharp angles on the front of the die to allow close proximity of the lips to the nip point. Stiffer melts are not as great a concern in this respect because their melt strength will allow them to travel greater distances prior to the nip without drooping drastically and causing sheet forming problems, compared to the lower-viscosity materials.

After passing through the roll stack and any auxiliary rolls, the sheet usually is trimmed on each edge (with slitter knives) to produce uniform sheet dimensions, and is passed through a two-roll nip with rubber rolls that are speed-controlled to create some tension in the sheet. The sheet line tension can be set by using a torque controller or just through the speed setting versus the main roll stack speed. The material trimmed off the sheet edges normally is ground into fairly small pieces and re-extruded as a blend with the virgin material. Typically, a 10 to 40% blend of regrind is utilized in this manner. Downstream from the rubber rolls there is a winder or shear, depending upon the sheet's thickness and pliability. Thin sheet (0.020–0.030 inch thickness and thinner) or very soft sheet (flexible vinyl) can be easily rolled up on a winder; thicker sheet (up to 1 inch thick) or stiff sheet is typically cut to specified lengths and stacked for subsequent operations. The very thin sheet (0.010 inch and thinner) usually is called film. When the film thicknesses become very small (0.005 inch or thinner), the nip point of a three-roll stack will start to present as problems. Maintaining a uniform contact across the nip surfaces becomes difficult because of the tiny nip dimensions required. Sometimes the top roll can be lifted away from the middle roll, and having no nip may produce an acceptable film. If the material is not contacting the roll uniformly, and perhaps is creating air entrapment between the material and the roll surface, an ''air knife'' sometimes may be used to force the material to the roll through a uniform jet of air blowing on the melt toward the roll surface. Sometimes the very thin films must be produced through a cast film method (see below).

Some of the uses for sheet and film made by the extrusion technique discussed here include thermoforming in molds, packaging, window glazing, and laminating.

One of the fastest-growing extrusion procedures is coextrusion, where several material layers are combined to create unique product properties compared to monolayer materials. Sheet and film production are enjoying much growth in the area of coextrusion in applications such as food packaging, where the moisture and oxygen barrier requirements are satisfied with different materials (EVOH, PVDC, etc.) and, hence, multilayered sheet structures. The combining of the various layers for a multilayered coextrusion sometimes is done in the die system and at other times is done after the melt exits the die, in a lamination process. The most efficient coextrusion is accomplished with the melt streams combined in the die or die adaptor system. The melt streams usually are combined in a coextrusion adaptor block positioned prior to the die, and the die flow channels form the final sheet dimensions of the combined layers of melt. The melt streams meet in a square, rectangular, or round adaptor geometry and then flow in a laminar fashion through the die system. As the coextrusion geometries become more complex with increasing layers because of the necessary product end properties, the various adhesive layers required, and the typical need for an adhesive layer in the structure, the number of extruders and the complexity of the die coextrusion adaptor system increase. A coextrusion sheet line can consist of a simple two-extruder setup or may contain five or seven extruders in a system, making seven to ten layers in the product.

Some materials exhibit noticeable property improvements, such as tensile strength and tear resistance, when the sheet is oriented both in the machine direction and across the sheet. This is a termed biaxial orientation. The natural pulling of the solidifying melt in the machine direction as the cooling and polishing is being performed will give some orientation, but cross-sheet orientation and high-machine-direction orientation are achieved by using tenter frames (see Fig. 4-15). These devices stretch the film's length and width by a desired amount up to 10:1), depending on the properties sought. The temperature of the film is controlled in ovens on the sheet line during orientation to yield the proper physical properties in the final sheet.

Some of the many polymers used in sheet production are polypropylenes, polyethylenes, PVC, polycarbonate, ABS, styrenes, and acrylics.

Extrusion Coating. The extrusion of a melt through a flat sheet die also can be used to produce a thin film for coating on a paper, foils, fabric, or some other substrate material, which is unwound off a feeder roll on an unwind station and passed through the first nip point of the roll stack. (See Fig. 4-16.) Sometimes the substrate is treated prior to the coating for enhanced adhesion. The coating typically is 0.0002 to 0.015 inch thick, and the substrate materials typically are 0.0005 to 0.024 inch thick. Common coating materials involved in this process include polyethylenes, polypropylenes, urethanes, ionomers, EVAs, nylons, polyesters, and PVCs. Extrusion coating often requires an elevated temperature (as high as 650°F) to produce a very low viscosity melt to enhance the flow properties and adhesion of the material. Parameters determining adhesion of the coating material to the substrate include the chemical structure of the resin, oxidation of the melted material between leaving the die lips and meeting the substrate, melt viscosity, line speed, coating thickness, pretreatment used on the substrate (corona, chemical, flame, etc.), and cooling rates of the coated structure.

To provide elevated melt temperatures, coating extruders typically are long (30:1 or 34:1 L/D) and employ a valve in the die adaptor to increase the resistance that the screw has to overcome.

The melt covers the surface of the moving substrate, which then travels through nip rolls where the surface finish is determined and cooling is performed. The rolls are cooled, as

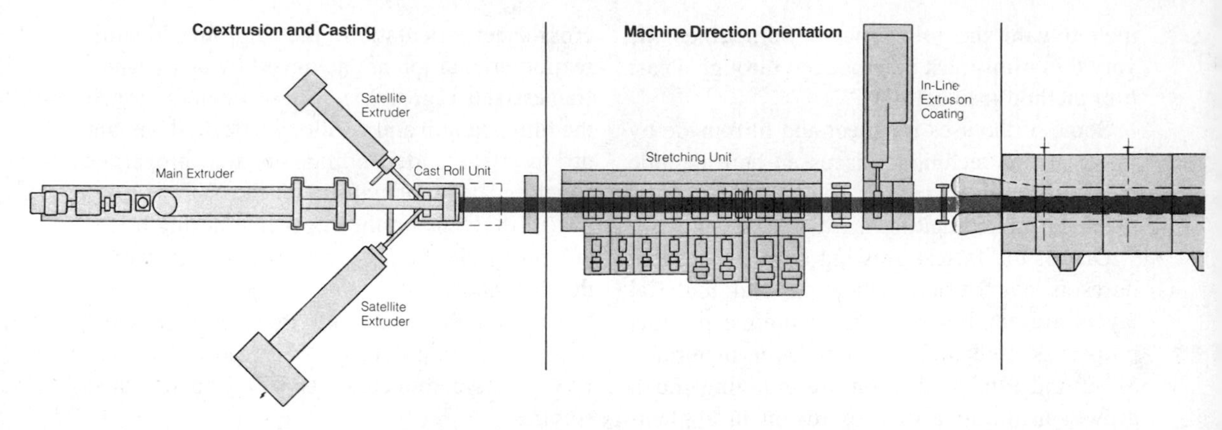

Fig. 4-15. Biaxially oriented film line using tenter frame stretcher. (*Courtesy Marshall & Williams Co. Providence, RI*)

with the sheet process, and the temperature of the cooling fluid usually is quite low to maximize heat removal. Some products are laminated with other layers, (e.g., scratch-resistant films), which are applied from rolls after the coating process.

In extrusion coating, the die normally is positioned so that the melt exits downward between a larger cooling roll and a smaller pressure roll. The material progresses around the large roll and exits to a secondary pressure roll about 270° from the initial contact. Sometimes, coating can be run with a vertical stack and a sharp die front angle to obtain close proximity of the lips and the nip point. Because of the low viscosity and weak melt strength of the thin extrudate, however, downward extrusion is most typical.

After cooling, the surface of the coated stock may be treated on-line for subsequent printing; untreated polyethylene, a common coating material, does not accept printing inks well. The coated material is trimmed to provide uniform edges and rolled up on a winder.

Cast Film. The cast film process is used to produce very thin films, down to gauges of 0.002 to 0.003 inch. The die, as with the extrusion coating process, is directed downward to lay a thin, low viscosity melt onto the primary cooling roll in a tangential orientation. The materials utilized for cast film are similarly low in viscosity, with melt temperatures of 350 to 550°F used to enhance thin film drawdown. Long extruders and die adaptor valves are used to elevate the melt temperature. The gap between the die lips is typically 0.010 to 0.020 inch, and the melt draws down from that dimension. (See Fig. 4-17.)

If the melt will not deposit uniformly on the

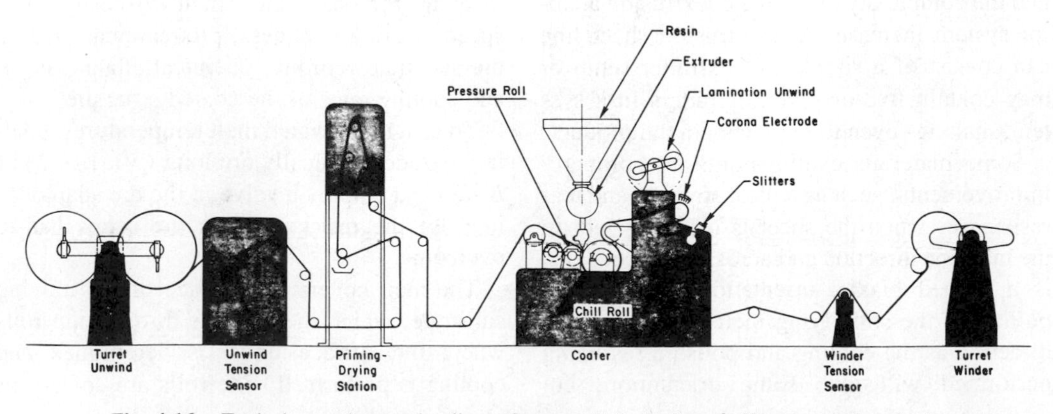

Fig. 4-16. Typical extrusion coating line. (*Courtesy E. I. du Pont de Nemours & Co., Inc.*)

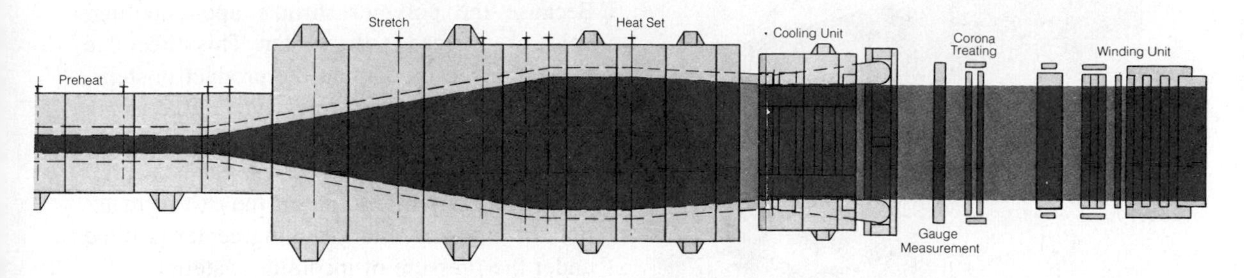

Fig. 4-15. (*Continued*)

cooling roll with no air entrapment, an air knife is required; it uses air jets to force the thin melt curtain onto the roll surface. The melt travels about 270° around the main roll, which is cooled with chilled water. The majority of the cooling and surface finish production occurs on the main roll, which usually is highly polished and chrome-plated to produce a smooth clear film.

The cooling process is completed as the film reverses its orientation and the opposite film surface contacts the secondary roll. The temperature of the cooling rolls is set on the basis of experience with the material being extruded and the degree of clarity required. Because the film is thin, it cools quickly, but line speeds are quite high and residence times thus are short. The rolls are driven at very precise speeds with DC drive systems to ensure consistent film dimensions. Gauge uniformity also depends on the extruder's pumping stability, die adjustment, cooling rate, alignment of die and cooling rolls (to avoid wrinkles), and melt uniformity from the screw.

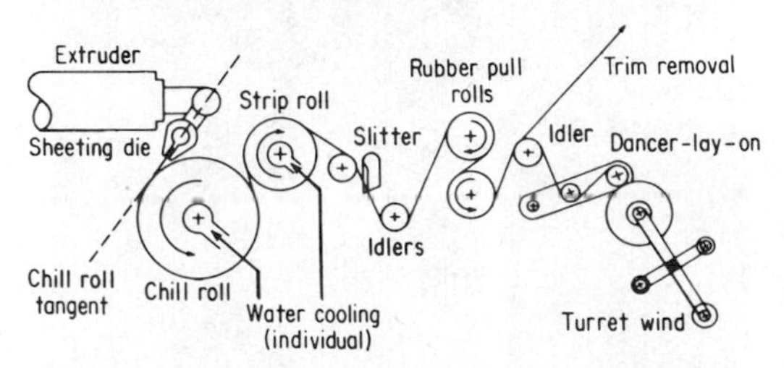

Fig. 4-17. Chill roll extrusion setup. (*Courtesy E. I. du Pont de Nemours & Co., Inc.*)

After cooling, the film is slit for edge trimming and perhaps for producing multiple film products at once. Edge trim is removed for regrinding and typically is re-extruded at low percentages (5–20%) with the virgin material. The low-bulk-density trim from these very thin products does not readily flow in a hopper and can cause feeding problems in the extruder at any high or even moderate percentage level. The possibility of introducing contamination in the regrind process, which would be very detrimental in thin film production, often precludes any reuse of trim or scrap.

The film winding tension is controlled by the speed of a set of rubber rolls and their relation to the cooling rolls' speed. The film is wound on a high-speed turret winder, which is capable of high-speed roll changes.

Profiles and Pipe

Profile Extrusion. Profile shapes can be produced by the extrusion process, from simple tubing or rods to complex custom shapes such as those used in window frame components (see Fig. 4-18). Methods of cooling and shaping the final products vary as widely as the number of shapes possible, from simple water trough cooling to complex vacuum sizing equipment that holds the outer surfaces fixed during solidification. The most common material formed by profile extrusion is rigid PVC, which has the needed processing and physical property balance. Other materials, such as polyolefins,

Fig. 4-18. Window profiles. (*Courtesy Davis-Standard*)

styrenics, and various elastomers, also are utilized in profile applications.

Extruders utilized in profile extrusion typically have barrel lengths of 24 : 1 L/D, with 20 : 1 and 30 : 1 occasionally used. The output of a profile line is limited by the downstream cooling efficiency and the related ability to hold tolerances on the finished product; as a result, the extruder is not so fully utilized as in a process such as sheet extrusion. The allowable melt temperatures in most profile applications are lower than the temperatures for making film or sheet, where lower viscosities aid in forming the end product. The custom profile extruder may have to process multiple materials over wide processing ranges. Enhanced operational versatility may be achieved by using a modern barrier screw design.

Any shape with a large, solid cross section tends to present problems due to voids that develop in the center after the outer surfaces have cooled. The cooling rate at the center is much slower than at the outer surfaces because the material is insulated from the cooling medium. Because the polymer shrinks upon cooling, voids are formed in the center. This effect can be minimized by cooling the product in steps, starting with air or warm water before going into the cold water bath. Voids in rod extrusion also may be eliminated by cooling the die itself and using the puller to retard the exit of material. In this way, the shrinking center is filled under the pressure of incoming material.

Difficulties of predicting polymer flow in a three-dimensional die, coupled with drawdown unknowns, cooling rates, and so on, make the design of complex profile die systems very difficult. Dimensional corrections to the die are usually necessary after it is tried for the first time. Persons proficient in profile die work lean heavily on experience, sometimes combined with computerized die flow analysis, when building a new die. (See Figs. 4-19 and 4-20).

Very narrow die openings can produce a rough surface on the extrudate due to a phenomenon called melt fracture, which is related to the ability of the material to slide smoothly across the die's surface. The die opening may be increased to reduce shear, forcing higher drawdowns to achieve the final dimensions.

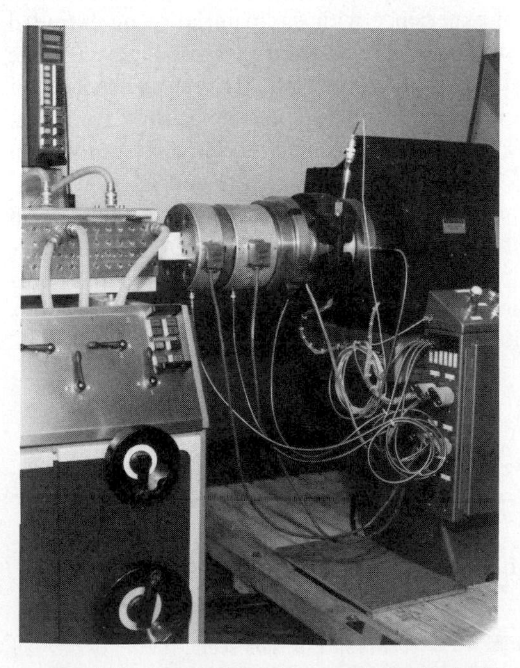

Fig. 4-19. Profile die. (*Courtesy Davis-Standard*)

Fig. 4-20. Profile extrusion line. (*Courtesy Davis-Standard*)

Alternatively, the die metal may be polished further and/or coated with a material to reduce friction. If mechanical adaptations are not successful, it may be necessary to substitute a less viscous polymer.

Downstream from the profile extrusion line die, the material is pulled through a cooling area to a cutoff or coiling station. Cooling methods are designed to produce an accurate shape for the profile, minimize cooling voids and other distortions, ensure a good surface finish, and maximize line speeds. Simple shapes made from high viscosity materials such as rigid PVC can be successfully sized and cooled by air blowers, using jigs that hold the part as it passes over a table. Thicker areas of the part will cool more slowly than thinner areas and must be favored with extra cooling medium if possible. When water cooling baths are used, straightness sometimes is maintained by using radiant heaters positioned to alter the cooling process.

When hollow shapes are made, vacuum usually is employed to hold the outer walls of the part in place during cooling. Vacuum cooling of intricate parts such as window profiles is done in one or more cooling stations built with the part's outer shape on their internal openings. Vacuum passages and cooling fluid passages are bored in these stations (vacuum calibrators) so that heat is transferred from the part as it is held in position. (See Fig. 4-21.) Allowances for part shrinkage are designed into the calibrator(s). As the output rate increases (along with the extruded part's linear speed), the cooling time in each calibrator decreases, requiring more units to ensure part tolerances. Some systems use water sprays between and directly on the calibrators' outer surfaces to add cooling capacity. These calibrators usually are made of aluminum with wear-resistant coatings applied to the inner contact surfaces. The calibrators are from about 6 inches to 36 inches long. As the number of calibrators increases, the additive drag over the units increases, requiring a fairly large-capacity puller.

Profile pullers have long contact lengths (up to about 40–80 inches) and high squeezing force capacity. There are some limitations to the amount of force that the puller can exert in the holding function, due to possible part deformation. These forces usually are quite high

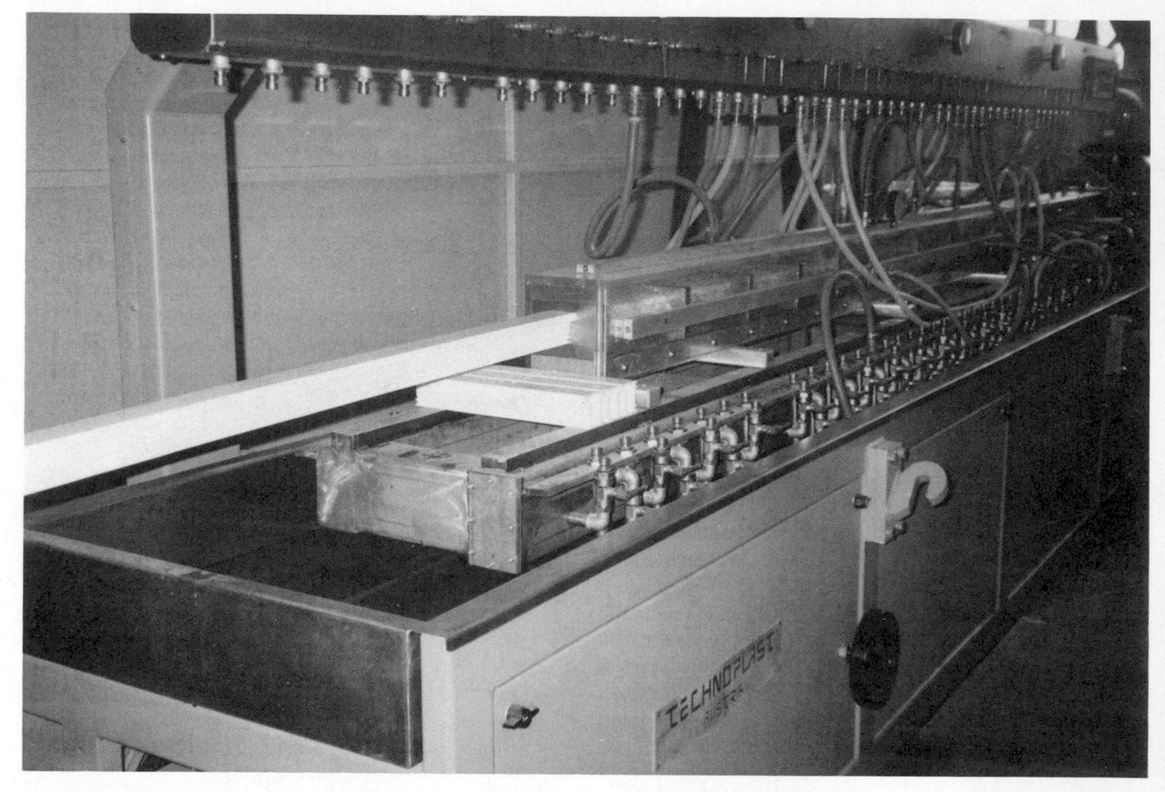

Fig. 4-21. Vacuum calibrator. (*Courtesy Davis-Standard*)

with rigid PVC window profiles, and an intricate part with small protrusions can be run on a puller with formed belts or cleats for the particular profile. This way, the part's main body will absorb the forces, and the protrusions are protected by reliefs cut into the belts or cleats. The part then is cut to length with a traveling saw and stacked on a collection table.

Less intricate parts are extruded and pulled through a water bath, and their shape can be held without external fixtures. Between the simple and complex shapes are a host of profiles that are cooled and sized with differing levels of shaping fixture complexity. The difficulty in cooling and sizing a particular part determines the output rate possible from an extruder. Melt temperature and corresponding viscosity also play a large part in determining attainable line speeds.

Coextrusion is seen in the profile processes with such products as window profiles, where there may be a rigid PVC profile with flexible PVC sealing ribs for the window panes, which are slid into some of the window frame profiles. Some pipe or tubing also is produced with coextruded wall geometries to achieve some unique properties, such as foam inner layers for weight and material cost reductions or coextruded weather-resistant material over a cheaper substrate for parts that must be used outside.

When a part is to be made of foam for weight reduction, sizing usually is initiated in a sizing block at the beginning of the cooling tank. Once the outside of the part is set, the sizing block is not needed and thus usually is quite short (2–12 inches long). Often, vacuum is applied to the sizing box through small orifices, forcing the material's outer surfaces to be in positive contact with the sizing block. There are natural forces from the foaming material that cause it to swell and contact the block's surfaces; nonetheless, some materials yield better part shape uniformity and surface finish when vacuum is used. Most profile foam shapes are produced by using chemical blowing agents (CBAs), which are added to the polymer prior to extrusion. This method allows density reductions of up to about 50%. If lower density is required, a physical blowing agent such as nitrogen, pentane, or a chlorofluorocarbon must be employed. Coextrusion may be selected to form a profile with low density and a smooth surface.

Pipe and Tubing Extrusion. Pipe and tubing are produced from similar die designs, except for very large pipe (10 to 60 inches in diameter), where the dies more closely resemble those used in making blown film. To form the rod-shaped extrudate into a hollow shape, pipe dies contain a spider, a contoured inner section held in position by several ribs. After the pipe exits the die, it is pulled through a cooling and sizing apparatus. (See Fig. 4-22.)

As the pipe comes out of the die, the melt normally swells slightly; the extrudate first is drawn down into a slightly oversized (compared to the final pipe O.D.) sleeve at the end of a water-filled tank. Air is introduced inside the pipe through a hole in the die pin, helping to maintain the internal dimension. Most commonly used pipe materials (rigid PVC, ABS, HDPE) will easily slip through the sizing sleeve; for materials that tend to stick, a precooling stream of water, perhaps containing a lubricant, is dripped onto the melt as it exits the die. The cooling tank contains enough water to cover the pipe completely, and vacuum is drawn in the air space between the water and the tank cover.

Fig. 4-22. Extruding vinyl pipe. (*Courtesy Goodyear*)

The length of the initial cooling tank may be as short as 6 feet, depending upon cooling requirements of the pipe and line speeds being run. Additional cooling is performed in secondary water tanks with no vacuum. After the final tank, the pipe passes through a belt puller and is sawn into lengths.

Some pipe processes do not use vacuum sizing but instead have a plug attached to the die pin. The plug is placed at a distance of 1 to 12 inches from the die and sets the I.D. rather than the O.D. of the product.

The extra buoyancy of large-diameter pipes in water baths makes it difficult to maintain straightness; so spray cooling often is chosen. High-velocity water sprays provide improved heat removal for thick-walled large pipes. The takeoff equipment for large pipe also is different; in place of belt pullers, special devices are used that have driven rubber tires located around the periphery. Cutting saws rotate around the pipe's circumference while moving along the line with the pipe.

Smaller-diameter pipe and tubing (under $\frac{1}{2}$ inch O.D.) usually can be sized without the use of vacuum. The dimensions of the product are determined by the tooling dimensions and by fine control of the air pressure inside the tube as it goes into the cooling tank. Some small tubing (e.g., with thin walls or made from soft melts) can benefit from vacuum sizing.

Pipe and tubing also can be produced by using a crosshead die, which directs the melt perpendicular to the extruder axis. This type of die, which is widely used in wire and cable coating, is described in the following paragraphs. Its advantages in making tubing are the ability to coat a core (another tube, a metal pipe, etc.) and to eliminate the die spider, which sometimes gives "knit" lines where the melt reconnects. The crosshead die has one knit line on the side away from the extruder, whose effect must be minimized by a well-streamlined flow path.

Coextrusion is used in the pipe and tubing industry to create products with improved properties over monolayer constructions. Foam inner layers, for example, are used to reduce weight and cost in ABS and PVC pipe. Coextruded tubing can be made in a tandem process where a finished tube is passed through a crosshead die that deposits the second layer on its outer surface. Today's die technology, however, allows the different layers of the tube to be produced in a single extrusion operation.

Wire and Cable

Wire coating is one of the oldest extrusion processes for thermoplastics. The die used on a wire coating extruder is unique, as a wire must pass through the die for coating. (See Fig. 4-23.) There were some early attempts at passing the wire along the screw's core, exiting out of a clearance hole in the screw tip and through a straight die. Sealing problems involved with this method led to the development of today's crosshead wire coating dies. A simplified sketch is shown in Fig. 4-24.

Two tooling designs are typical in wire coating with crossheads. One is pressure tooling (shown in Fig. 4-24), where the melt is applied to the wire inside the die under whatever pressure exists after the wire exits the guider tip. The space between the guider tip and the tapered shape of the inside of the die is termed the gum space; it can be adjusted by using bolts on the back of the crosshead (where the wire is entering). With the second design, called tubing tooling, the melt is applied as the wire exits the die. Here there is no pressure forcing the melt onto the wire; in some instances, a low level of vacuum is drawn in the die core to aid in pulling the melt onto the wire and avoiding trapped air. (See Fig. 4-25.)

Fig. 4-23. Wire line and crosshead die. (*Courtesy Davis-Standard*)

In wire coating, the bare wire is unwound from a reel and fed through the extrusion die. Line speed is determined by a puller called a belt wrap capstan. (See Fig. 4-26.) The coated wire then runs through a cooling trough for solidification of the polymer. A winding device at the end of the line pulls the wire from the trough and produces reels of finished product. The trough lengths are determined by the size of the wire and the coating thickness. Some materials require cooling in stages, with warmer water used closer to the die.

Preheating the wire prior to coating enhances adhesion of the insulation being applied. Both flame and induction heating are commonly used. Wire straightening devices also are popular for ensuring uniform coating. Another optional auxiliary device is a spark tester, which is placed after the cooling trough to detect wire breaks or insulation defects. Printing equipment and footage counters are used on the majority of wire lines.

Because the melt is supported by the wire, high line speeds are possible, and melt temperatures can be relatively high compared with profile processing. Some small-diameter wire coating (e.g., under 0.030 inch O.D.) can be run at 7000 to 10,000 ft/min. Winding of the finished wire requires a sophisticated takeup with automatic reel changeovers. Small wire that is being coated at high line speeds may utilize cooling troughs with the ability to handle several loops of the wire (multi-pass) so that the cooling residence time is multiplied, as opposed to using single-pass troughs of very long lengths. Also with the high-speed wire lines, a series of sheaves on a sliding frame is used to accumulate the wire as a winder reel is being

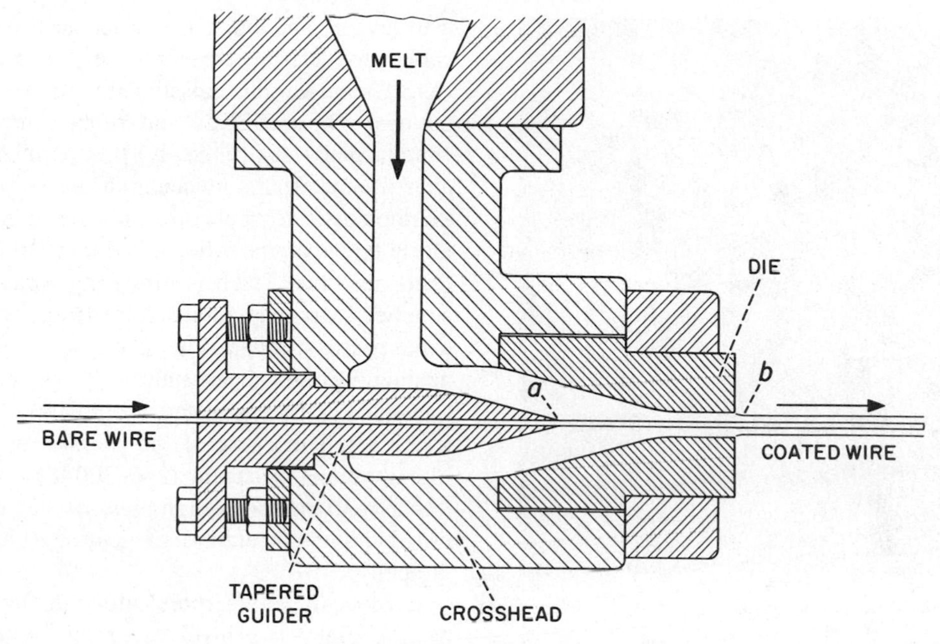

Fig. 4-24. A crosshead holds the wire-coating die and the tapered guider. (*Courtesy Quantum Chemical Corp.*)

changed while the line continues to produce wire at high speeds. (See Fig. 4-27.)

When the coated wire O.D. is measured for quality control, which is quite normal in today's wire coating plants, a gauge is placed after the cooling trough to "look" at the O.D. in both *X* and *Y* axes. Microprocessors allow data gathering for statistical analysis and control and allow accurate "real time" product knowledge for control of the line. The extruder speed and the line speed can be altered automatically by a controller that maintains the wire coating O.D.

Some wire production does not immediately cool the coated wire but involves further heating to a point that causes the special plastic coating to achieve crosslinking of molecules. This crosslinking changes the properties so that reheating to temperatures that normally would melt the thermoplastic wire coating does not soften the material. The main usage of these crosslinked products is in power cables, where high voltage electrical transmission creates high heat levels that the wire must withstand. Some wiring used in automobile engine compartments must be crosslinked to handle the high

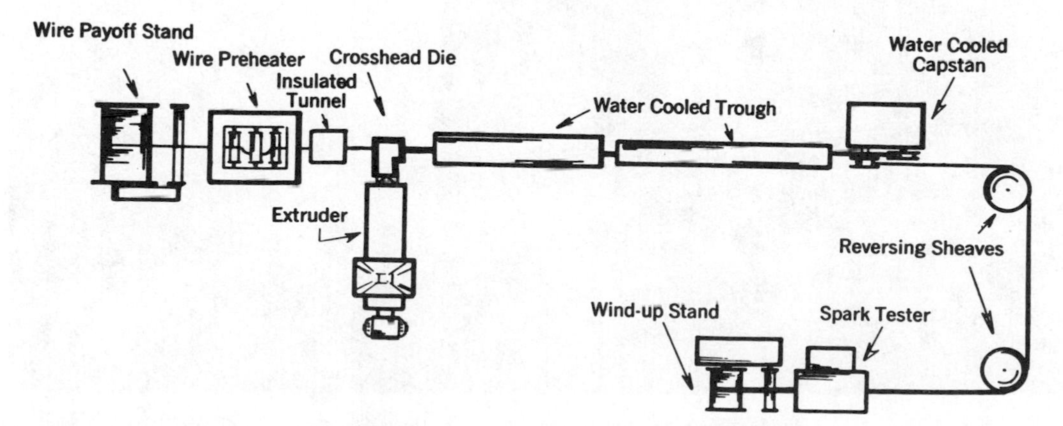

Fig. 4-25. Schematic of flow line for wire coating, showing preheater between payoff and die. (*Courtesy Solvay America*)

Fig. 4-26. Belt wrap capstan. (*Courtesy Davis-Standard*)

heat levels. The major materials used in these cable products are crosslinkable polyethylene (XLPE) and various crosslinkable rubber compounds. These XPLEs and rubber materials contain peroxides (typically) that permanently alter the material's molecular structure from a thermoplastic (or noncured rubber) to a thermoset type material, after sufficient heating and residence time. The heat for curing typically is supplied by a steam tube that the freshly coated cable passes through. The tube is split into the heating portion (steam) and the cooling portion (water). The tube is sealed at both ends for the steam setup, because of the pressurization of high-temperature steam (250–300°F). An alternative to the use of high pressure steam is to utilize high temperature gas curing (nitrogen typically).

A process getting more attention over the past several years involves producing crosslinked cable without the peroxide and heat

Fig. 4-27. Wire accumulator. (*Courtesy Davis-Standard*)

method. A special additive is introduced to the polyethylene in the extruder hopper (a silane material and a catalyst developed by the Union Carbide Corp.), and the cable insulation is extruded and cooled as normally done for thermoplastic wire coating. The finished cable is placed in a high-humidity atmosphere for a time, which reacts the silane and crosslinks the coating material.

Coextrusion is popular in wire and cable production to allow the efficient production of many wire constructions. (See Fig. 4-28.) Previously, many multilayered coatings were made by passing the cable through an independent crosshead for each layer. The development of coextrusion crossheads allowed the extrusion of several insulation layers at one time and cooling of the "assembly" on one extrusion line. The production of many different colors on a single extrusion setup often uses special coextrusion dies to allow switching of the layers, to bring a given layer to the outside of the insulation and hide the previous color inside the wire coating. The alternative for a color change is to change the material in the coating extruder, which takes longer and wastes wire and coating material.

Some of the many materials seen in the coating of wire and cables include rubber, polyethylene, polypropylene, nylon, fluoropolymers, and PVC.

Fiber

Fibers have been produced from thermoplastics for some time. The materials used in normal fiber spinning operations are of lower viscosity than their blow molding, profile, or blown film counterparts. The polymers are extruded at fairly high melt temperatures into a piping manifold with a series of melt pumps. Each melt pump feeds a spinnerette, which is a die with many small circular orifices. The fine melt strands produced are pulled from the die and cool rapidly. As the fibers are pulled farther downstream, they are oriented in the machine direction to impart good tensile strength. The orientation is performed by a series of rolls that are wrapped by the fibers and have successively faster speeds. The increase in speed pulls the fiber and elongates the strands, imparting orientation in the machine direction. The many strands then are spooled on a rack of many rotating cores. Were the same higher-viscosity materials as seen in blown film and blow molding used in the traditional fiber operations, the pressure resulting from the small-diameter die openings would be prohibitive, and the drawdown ability of the fibers would be reduced.

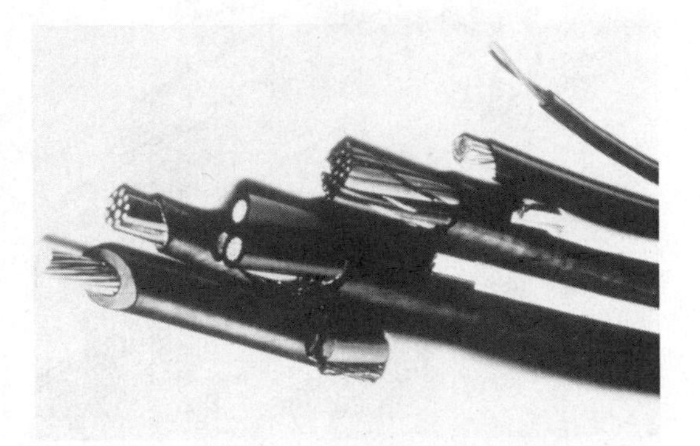

Fig. 4-28. Wire and cable configurations. (*Courtesy Davis-Standard*)

Some fiber operations use a wide slit die (15 to 30 inches) and cut the strip into numerous narrow ribbons, which subsequently are drawn down and oriented into a coarse fiber (tape yarn). This method uses somewhat higher viscosity materials than the spinnerette process and operates at lower temperatures—more like a sheet operation than a traditional fiber process.

Materials commonly used for making fibers include polypropylene, nylon, and polyesters.

EXTRUDER SCREWS

Extruder Screws (General)

This section will present some ideas about selecting screws for specific situations and the design considerations that lead to a screw's selection. (See Fig. 4-29.) Screw design is a blend of some basic engineering, some theory, and a good amount of experience with the material and process in question. Many factors are involved in design selection to define a screw's overall performance, and most of these factors

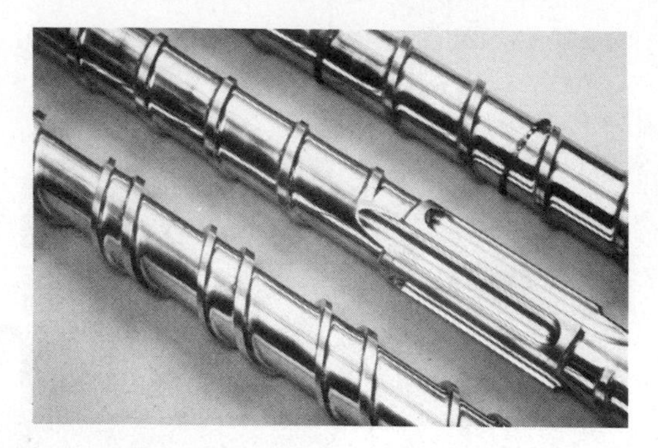

Fig. 4-29. Three extrusion screws. (*Courtesy Davis-Standard*)

must be defined to set the goals of the screw design selection process.

The main performance goals are listed below:

1. *Output rate requirements:* The amount of material a screw must deliver usually is defined as the maximum that the process can handle when sizing and cooling the melt into the final product. This is defined in well-known terms of lbs/hr, ft/min of product, lbs/hr/rpm, etc. The use of lbs/hr/rpm is the least precise because of the nonlinearity of this number as screw speed changes or as die pressure levels change. The other two measurements of output are well defined by common terms.
2. *Output stability (screw pumping consistency):* This parameter often is assumed by the purchaser of a screw to be ensured by the screw designer. The product dimensional consistency required leads to the levels of pumping consistency necessary. This parameter is easily measured through the use of a pressure transducer at the end of the barrel or in the die adaptor system. A chart recorder or data acquisition system can be utilized to allow long-term stability to be recorded. A good measure of pressure variation at the screw tip or in the die system for most processes would be 0–2.0% (±1% or less). A pressure variation level in the range of 2.5–3.5% would be judged as fair, with some measurable product variation expected, and a variation of 4% or higher is above the acceptable limits for most product variation allowances. The output variation can be up to three times as large as the pressure variation, depending on the degree that the material acts like a Newtonian fluid in the melt phase, but typically the above ranges are accepted, based on much accumulated extrusion experience. An exception to low output variations would include some pelletizing operations, where a 20–40% variation of pellet dimension may be totally acceptable.
3. *Melt temperature level:* Melt temperature is often the limiting factor to an extruder's output performance. The limit usually is due to a melt-handling limitation, as the viscosity is reduced by increasing the melt temperature or when a temperature is reached that will start to degrade the extruded material. The method by which this parameter is measured is critical to its usefulness in defining a screw's limitations or comparing one screw to another. (More information on the measurement of melt temperature is given below.)
4. *Melt temperature consistency:* This parameter is best measured on a chart recorder or with a data acquisition system so that long-term variation can be accurately defined. The variation measured is a good indicator of the homogeneity of the screw's melt and the degree of distributive mixing that has occurred. A good melt temperature variation level for most situations would be 2°F or lower (±1°F). This would be measured with a probe protruding an adequate distance into the melt stream.
5. *Melt quality level:* The appearance of the final product usually is defined for the screw designer. This can be based on a product smoothness specification, the surface finish (gloss vs. dull), or some other aesthetic concern.
6. *Mixing efficiency:* Sometimes colors or the blending of other additives is necessary; so some minimum mixing level

must be reached. If the mixing specification is visible, the evaluation can be simple. If a nonvisible ingredient must be blended, product properties may need to be evaluated to determine if the screw has performed sufficient mixing.

The selection of the final screw design usually must fulfill specifications defined in terms of most or all of the above parameters. A good understanding of their measurement and evaluation is necessary to good screw definition. One of the most elusive quantities to define is the melt temperature. The use of a thermocouple placed in the die adaptor system is a reasonable method, but if the tip of the thermocouple is not protruding at least one-third of the way into the melt stream, an inconsistent temperature is recorded. Flush thermocouples, as seen with pressure transducer/melt thermocouple combination devices, thus are not true indicators of melt temperature. Just like melt temperature, the other screw design parameters must be well understood to ensure that any values used to compare one screw's performance to another's are measured consistently.

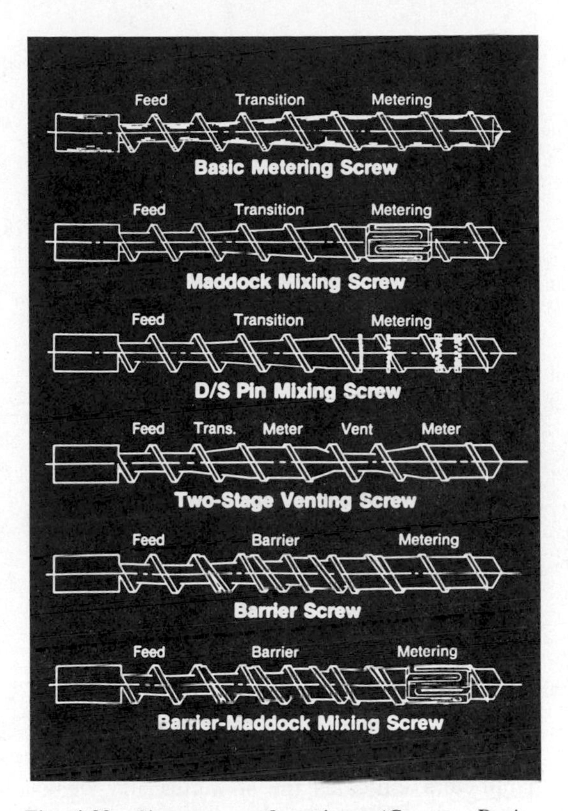

Fig. 4-30. Six screw configurations. (*Courtesy Davis-Standard Co.*)

Screw Types

Many screw designs have been utilized over the years for the various extrusion processes and materials. A few of these designs are seen in Fig. 4-30. The best-known screws can be loosely separated into two categories—conventional screws and barrier screws. Conventional screws usually are single flighted designs with the flight lead (or flight pitch) equal to the screw's nominal diameter. This type is termed a square-pitched screw, and when no mixing devices are present, it usually is referred to as a metering screw. Barrier screws are seen in many design configurations used in the extrusion industry, depending upon the designer. These screws have a second flight, which acts as a melt/solid separator designed to enhance output stability and allow higher usable output performance levels. Like conventional screws, barrier screws must be well designed to ensure that optimized performance is reached and that screw design parameters are successfully met. A poorly designed barrier screw can yield worse performance than a well-designed conventional screw. Either screw can give improved performance with the addition of a mixing section of minimized shear levels for the particular material and processing conditions being utilized. (Such mixing devices will be discussed in a later part of this section.) The type of screw selected will depend on the performance required and the extruder size selected. When the extruder is large for a required output level, a simple screw usually is adequate (conventional metering or mixing design). Most extruders are not large for the process involved, so maximum output rates and/or minimum melt temperatures are required from the screw. This means maximizing screw depths to generate a high output rate at low screw speeds and low melt temperatures. Maximizing screw depths while maintaining good output stability and melt quality is typically best done on barrier screws because of their better melting performance. One exception to the general practice of minimizing melt temperatures at moderate or high screw speeds occurs when extrusion coating or cast film is being produced. These cases require high melt temperatures to allow adherence of the melt to some substrate being passed through the cooling roll nip point or just to produce a low viscosity melt to allow very thin sheet gauges. In general, in the majority of cases the extruder should minimize the melt temperature.

Conventional Screws (Metering and Mixing). These screws usually include a constant-depth feed section, a constant-depth metering section at the output end of the screw, and a varying-depth transition section connecting the feed and metering sections (see Fig. 4-31). The lengths and depths of the sections are varied to alter the performance of the screw to meet given performance goals. There are numerous schools of thought about the optimum screw for a given processing situation, but measuring the different screws' performance parameters indicates the strengths and weaknesses of each design. There has been much extrusion experience with metering and mixing screws with most materials, due to the long history of their usage; so a good design starting point usually is determined when conventional screw designs are desired. Low or moderate output performance usually is certain with these designs; but as the extruder is asked to yield higher and higher output rates, work is needed to select a design that will ensure the output rate, the output stability, an acceptable melt temperature, and good melt quality (and possibly other important design parameters). As screw speeds are increased, the melting performance along the screw becomes quite chaotic compared to that defined in the earlier section on melting, and the conventional screw typically presents problems in output stability and perhaps melt quality parameters. The years of optimizations performed with these designs have led to an array of screw designs for various situations, with the higher output screws usually containing mixing devices and fairly long metering sections to control as much as possible, the melting situations that cause quality and instability problems. This long metering section design can only be taken to the point where the shortened feed section and transition will adequately convey the solids and early melt and ensure completely filled metering flights. Some materials (polyethylenes, flexible PVCs, etc.) will process successfully with a 3–5-diameter feed section length and a 3–5-diameter transition section length, thus allowing a long metering section (12–17 diameters). The longer metering section gives higher shear in the channels, but that factor is offset by deepening the channel for maximized output rates. Screw channel depths also vary widely, depending on the material extruded, the length of screw section, and the output performance required from a given extruder size. Usually, metering section depth is based on the screw's output requirement. This can be roughly calculated by the melt pumping equations (drag and pressure flow equations) given previously. The feed section depth then is chosen, based on a factor called the compression ratio, typically in the range 2 : 1 to 4 : 1. The specific value has been determined for most materials from the expe-

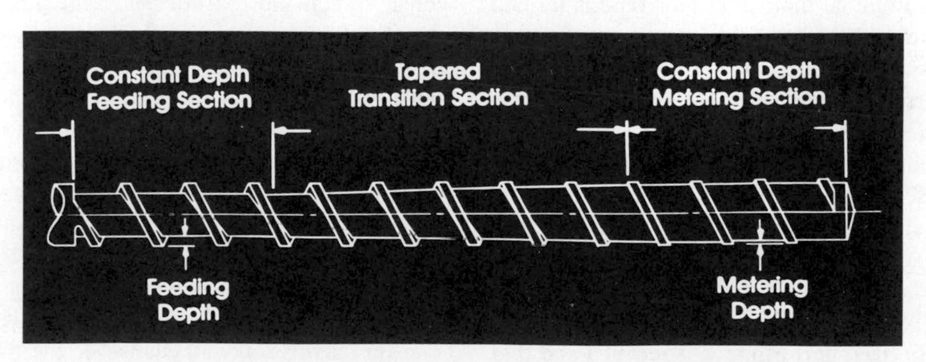

Fig. 4-31. Metering screw design. (*Courtesy Davis-Standard Co.*)

rience of polymer manufacturers and/or screw manufacturers.

True metering screws (no mixing elements) usually are used where low or moderate output rates are required from an extruder; mixing screws are used in high output situations, or where added mixing is desired at lower output levels. A fairly shallow design (shallow channel depths) can be made to perform well up to the screw speed where a melt temperature limit is reached. Deepening this design to improve output rates usually leads to deterioration in output stability (and usually melt quality) as screw speeds are increased for increased output requirements. There are some rare exceptions to this phenomenon in processing areas where the material is unusually easy to fuse into extrudable form, such as most true rubber compounds and some highly filled polymers.

Offering specific screw designs for a given material is always of questionable value because several screws will process a material successfully over a range of conditions. Optimum screw design depends upon the conditions on a given extrusion line and the performance goals (extruder size and length, extruder cooling/heating available, die pressures reached, materials extruded, melt temperature levels allowable, etc.).

Table 4-1 shows some designs that may be obtained from various screw suppliers for a given material. These are four screws that might be specified by various suppliers for processing LDPE, depending on the conditions required on the 2.5″ 24:1 extruder. The first screw is a simple metering screw with reasonable results at low to moderate screw speeds (20–50 rpm). The second screw will operate at higher screw speeds (75–150 rpm) with acceptable quality and usually reasonable output stability, up to the point where the melt temperature limit is reached. The third screw, which may or may not be specified with a mixer, would also be a modest output performer and would have a slightly higher but less stable output than screw A. The 8*D*-8*D*-8*D* length setup is a typical design offered by numerous material suppliers and screw suppliers as a "generic" screw of modest output performance for small extruders (2.5″ and smaller) but is not an exceptional performer compared to the other three screws under most conditions. As die pressures increase (processes with restrictive dies, such as blown film and some small wire coating applications), the simpler screws usually exhibit improved output stability. If die pressures are quite high (6000–8000 psi), melt temperature limitations often are reached at low screw speeds (10–40 rpm), depending on the materials utilized, and a simple metering screw may do an acceptable job.

Barrier Screws. In applications where extruders must improve output rate, maximize output stability, and minimize melt temperature levels while maintaining acceptable melt quality, barrier screw designs are enjoying increased usage (see Fig. 4-32). Barrier screws are more difficult to design and are more expensive than conventional screws, but potential performance benefits in production under most conditions usually justify their use. When properly understood and properly designed, these screws make it possible to maintain good output stability over wide screw speed ranges and thus allow deeper channels and better out-

Table 4-1. Four possible screw designs for a material (extruder: 2.5″ (63.5 mm) 24:1 L/D; material: LDPE, 2.0 melt index).

	Screw A	*Screw B*	*Screw C*	*Screw D*
Feed section length (dias.):	4*D*	5*D*	8*D*	5*D*
Feed depth:	.330″	.450″	.390″	.480″
Transition:	8*D*	3*D*	8*D*	11*D* (Barrier sect.)
Meter section:	12*D*	16*D*	8*D*	8*D*
Meter depth:	.110″	.150″	.130″	.190″
Mixing sect.:	No	Yes	Yes	Yes

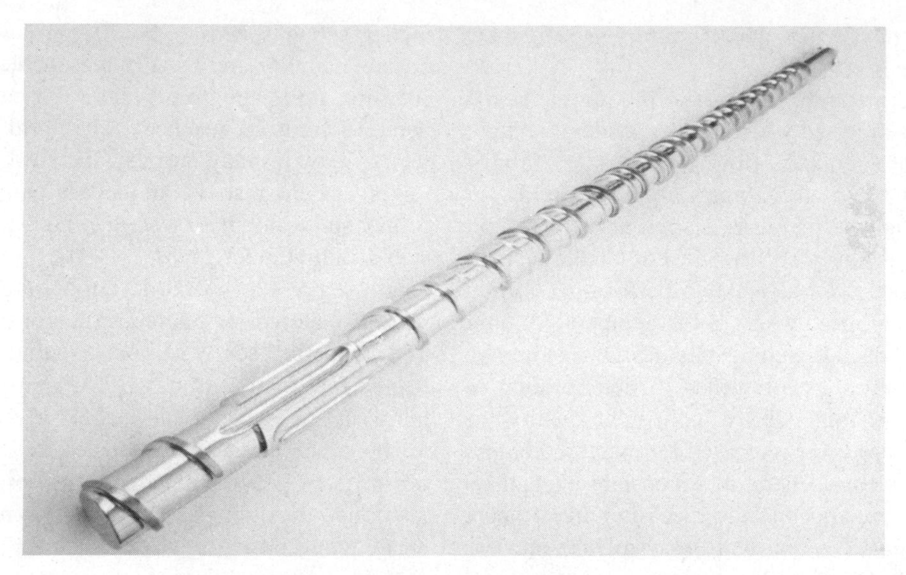

Fig. 4-32. Example of barrier screw. (*Courtesy Davis-Standard*)

put rates at a given screw speed. When the output rate for a given screw speed is maximized, the screw speed necessary to deliver the output rate requirement is reduced, thus reducing melt temperature levels. Alternately, the higher output rate at a given speed can be achieved at similar melt temperatures to those of the lower-output design. Therefore the barrier screw typically offers lower melt temperatures or higher output rates or a combination of the two parameters. Barrier screws have a feed section and a metering section, as with conventional screws, but there is also a barrier section made up of a second flight designed to separate the solid bed from the melt pool and maintain this configuration until all (or most) of the material has melted. The placement of the barrier flight, barrier section design (width of solids and melt channels), depth and length of feed and metering sections, solids channel depth, melt channel depth, and clearance of barrier flight all affect the screw's performance. Whether a mixer will be present and how to design the mixer's clearances (shear levels) are also important design considerations.

There are many barrier screws on the market that are available to the extrusion plants. The barrier screw is a more complex design to manufacture, and it requires a better understanding of polymer melting performance to achieve optimum results. The extrusion operator needs to be fully assured of successful operation when contemplating the purchase of a relatively expensive barrier screw in an effort to improve performance. Unless the screw is reasonably well optimized prior to delivery, there is some chance that a rework for improved performance will not be possible because of the barrier design; for example, if the barrier flight is not properly located on the screw. Also, a poor design cannot be easily analyzed on most production lines because of a lack of proper instrumentation. A random attempt at extrusion with a ''standard'' barrier screw can lead to disappointment and wasted production time.

Two Stage (Venting) Screws. When it is determined that a vented extruder is required, the screw design requires some special considerations, whether a conventional screw or a barrier screw design is selected. A vented screw usually is defined by the number of stages on the screw. A two-stage screw (see Fig. 4-30) has initial feed, transition (or barrier section), and meter sections that define the screw's first stage. There then is a rapid deepening of the screw flights to form a secondary feed section, also called the vent section. This is the area where the barrel has an opening to the atmosphere or to a vacuum source for removing moisture, air, and/or other volatiles. The screw then has a second transition and a final metering section. The vent section, second transition, and second metering section make up the second stage. The screw operates like two single-stage screws connected in tandem, which operate at the same screw speed. The first stage must operate like a typical single-stage screw, although full melting to an extrudable melt quality usually need not be reached. The first stage typically must perform 50–90% of the full melting, depending on the material involved and the extent of the volatile removal. For example, most of the powder fed materials, which only require air removal at the vent, do not need much more than 50–70% melting at the vent for the air removal to be successfully performed. This air removal often can be done with no need to use a vacuum system on the vent. On the other hand, when moisture or some internal volatiles must be removed, a fuller melt usually is necessary to ensure that the majority of the volatiles will be removed by the vent vacuum system. These volatiles will boil out of the melted material more easily at lower pressures (under vacuum), but if a substantial amount of the material passing by the vent is not adequately melted, the volatiles will not be removed and may show in the final product as voids.

A goal of increasing output rates and decreasing melt temperature would lead to deepening the first stage as much as possible and allowing as much partial melt to pass the vent opening as the venting efficiency (and final melt quality) of the process permitted. Should the screw be designed for full melting at the vent section, the shear contributed by the second stage would generate a sizable melt temperature increase. The particular material being processed will determine how complete the melting must be at the end of the first stage for a successful extrusion (i.e., void-free and of sufficient melt quality). A screw with two barrel vent openings would be a three-stage screw. The single-vent setup is the most popular choice by a wide margin, but the second vent of the three-stage design sometimes is believed desirable for more complete volatiles removal in some compounding operations. Single-screw processing would be limited for practical reasons to a double-vent setup, but some specialized twin-screw compounding extruders can be designed to have four or five barrel openings, used for venting or additive introduction to the screw.

One of the most difficult screw designs, used to provide good output stability, is the vented design. The "two screws" placed together in tandem to make up a two-stage screw design must operate at a reasonable output balance to ensure output stability. On the other hand, one could see vent flow from a poor screw "balance." The first stage is doing the majority of the melting and is setting the output rate of the process. The second stage must take away the material from the first stage and produce enough pressure pumping capability to push the material through the die system and avoid the occurrence of vent flow. Should the second stage have too much pressure pumping ability under an attempted extrusion condition, the second stage would operate with a low channel fill percentage, which usually results in a surging condition (product dimension variation). Two-stage screws typically can be designed to operate with 1.5 to 3% pressure variation at the end of the screw under a specific condition, that is, a specific material, a given screw speed, and a given die system resistance (pressure). Many two-stage screws must operate with some variation of conditions caused by product ranges (different die pressures) and alternative materials, and must operate over a screw speed range. These variations cause the screw design to be difficult because, inherently, there is limited versatility in two-stage screws. The use of an adjustable valve in the die adaptor system can offer some die pressure correction if the second stage is designed with too much pressure pumping ability; the valve can increase the pressure to improve the screw's balance. This is limited to cases where the die pressure is low or modest and where non-degrading materials are being extruded. When die pressures are high, due to thin product dimensions or restrictive die systems for some other reasons, the vent must be located as far from the screw tip as possible to allow a long second stage for pumping. High-pressure die systems (3000–

5000 psi) require a 12–15-diameter second stage length, whereas low-pressure dies (500–2000 psi) may require only an 8–10-diameter second stage. If there is an easily justifiable place for a melt pump in extrusion to enhance stability and reduce extrusion pressure from the screw, it is with vented screws. Most single-stage screws (especially barrier screws) can be designed to produce good extrusion pumping stability, but two-stage screws are not easily guaranteed to yield good extrusion consistency. Three-stage screws are even more difficult to design for good extrusion stability and to avoid vent flow over any modest range of processing conditions.

The need for adequate pressure pumping capacity against moderate or high die resistance causes the typical vented screw to be longer than 24:1 L/D. The typical length of a vented extruder of reasonable output capacity is 30:1 minimum, with 34:1 or longer also typically utilized.

Screw Cooling

Most screws have a core hole drilled from the rear (gearcase shank) end for some distance. Some cores are full, that is, they extend to within about 1 inch of the screw's tip, whereas other core holes are drilled to some shorter length. Today's screws are typically designed to operate "neutral," that is, with no cooling circulation, because that situation ensures maximum output rate with most polymer materials. Cooling reduces the screw's ability to convey material, and the colder the cooling fluid is, the lower the resultant output rate. Some older metering screw designs require screw cooling to allow acceptable melt quality on a screw that is too deep or one that has too short a metering section. Most conventional screw designs now contain some mixing section and/or a long enough metering section to provide adequate melting without retarding the output rate by screw cooling. Properly designed barrier screws also ensure adequate melting and so can avoid the full screw cooling mode. Most screws today are cored for a short distance, which reaches the exit side of the feed throat plus perhaps one to two diameters farther. This drilling allows cooling of the feed area, which is necessary under some conditions where the material tends to stick to the root of the screw (e.g., adhesives or nylon materials).

When a screw is fully cored today, that is done either because the screw will process a degradable material (such as some PVCs) where stagnation at the screw tip is avoided by cooling. The tip cooling setup typically utilizes oil rather than water as the medium to avoid freezing melted material to the end of the screw, where the slowest-moving material resides. A double cooling pipe is popular, where the outer pipe threads into the rear of a removable screw tip (which has a smaller core than the main screw with a pipe tap), and the feed tube runs inside the outer pipe. The return flow does not directly touch the screw core and therefore will not adversely affect the output rate. Oil cooling usually is performed at temperatures in the 200 to 300°F range.

Mixing Considerations and Mixing Devices

Mixing devices are used for several reasons on single-screw extruders and come in numerous designs. (See Figs. 4-33 and 4-34.) The mixing contributed by these devices can ensure improved melt quality and melt temperature uniformity and/or can perform some distributive mixing for color blending, polymer blending, and additive blending. The typical goal is to achieve the desired mixing quality level while inducing the minimum amount of shear and thus minimizing the rise in melt temperature. These devices can be placed on the screw, attached to the screw tip, or placed after the screw (in the die adaptor). The most common mixers utilized on the screw include fluted mixers (credited to Maddock of the Union Carbide Corp.; see Fig. 4-33) and mixing pins. Mixing pins are arranged in rings around the screw and can be of various sizes and shapes to break down and hold back any solids that have reached the ring position. This action reduces the possibility of unmelted polymer reaching the die system. The rings typically are placed along the metering section of the screw to offer successive blocks to solids moving along the screw channel. Placing multiple rings in close

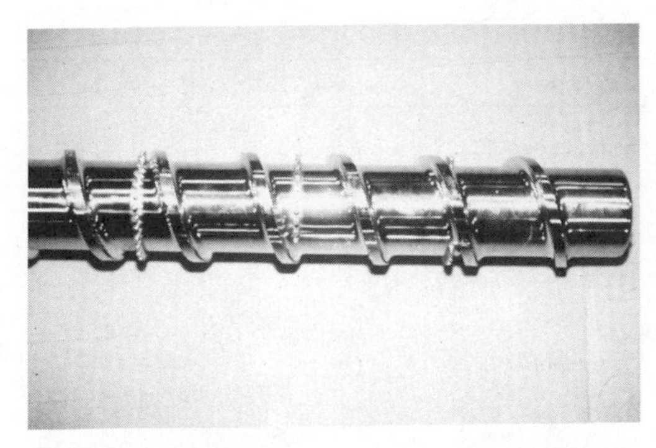

Fig. 4-33. Maddock mixing device. (*Courtesy Davis-Standard*)

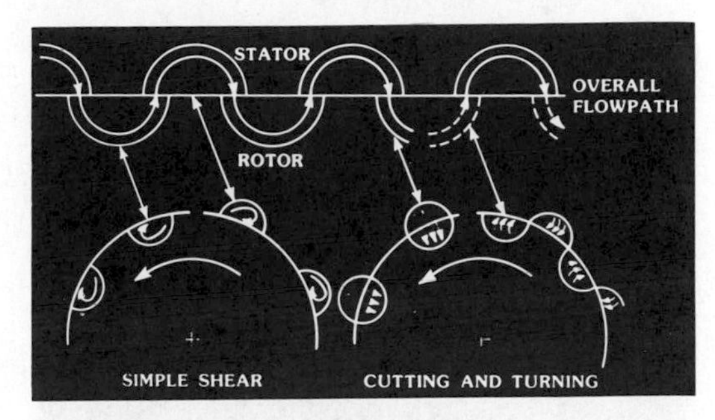

Fig. 4-35. Cutaway of cavity transfer mixer. (*Courtesy Davis-Standard*)

approximation to one another creates a more restrictive situation. Multiple rings are used occasionally in low-viscosity-material applications, where blocking and breaking down the soft solids is difficult unless fairly tight restrictions exist. The fluted (Maddock) mixing section creates several dam sections as the material enters the flutes on the feed end of the device, and the material then is forced over the dam, enters the exit flutes, and is forced into the following screw section. The dams of this mixer create a blockage to the remaining solids, breaking them down to ensure improved melt uniformity. Altering the shear input of this design is accomplished by changing the dam clearance or altering the dam widths. There are many other variations of these mixing devices, but the performances are similar to the basic designs discussed.

When simple mixers on the screw cannot give a high enough degree of distributive mixing (as with low amounts of a color additive, 0.25–1.5%), there are some advanced dynamic mixer designs that attach to the screw tip and noticeably improve mixing. An example of this type of mixer is the cavity transfer mixer or CTM (see Figs. 4-35 and 4-36). The spherical "pockets" on the stator and rotor are misaligned by half a pocket length to allow material to pass from the stator to the rotor and back while moving toward the die system. The melt is split as it moves from stator to rotor, and with the rotation of the screw, the number of "splits" grows rapidly (exponentially) with increasing rows of pockets. Even a CTM with four or five rows does a noticeably better distributive mixing job, compared with mixers on the screw.

Mixers installed in the die adaptor system are of the static variety. Several manufacturers produce these mixers and the amount of mixing depends on the length of the mixing elements selected. These devices split the stream, like dynamic mixers, but without rotation they require longer devices to do a comparable job. The pressure drop across any of these mixers obviously increases with element length and must be taken into account in the design of the device because pressure affects melt temperature levels. The biggest volume usage of the static mixers has occurred where long adaptor pipes are necessary between the extruder and the final die. This condition leads to tempera-

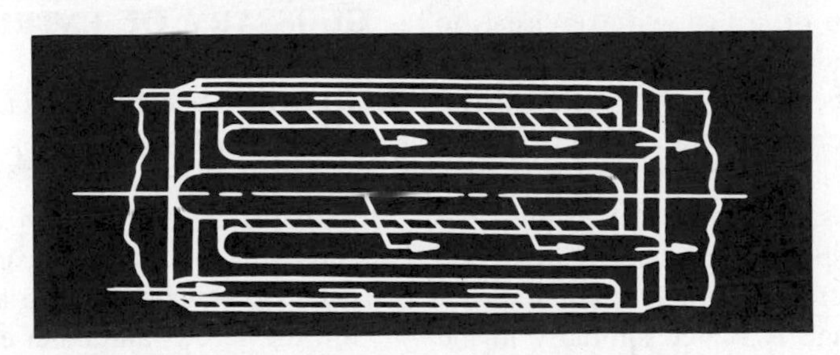

Fig. 4-34. Pin mixing device. (*Courtesy Davis-Standard*)

Fig. 4-36. Cavity transfer mixer—anticipated mixing. (*Courtesy Davis-Standard*)

Table 4-2.

Parameter	*Symbol*	*Scaling Law*
Metering depth	h	$\frac{h_1}{h_2} = \left(\frac{D_1}{D_2}\right)^{0.75}$
Screw speed	N	$\frac{N_1}{N_2} = \left(\frac{D_1}{D_2}\right)^{-0.75}$
Output	Q	$\frac{Q_1}{Q_2} = \left(\frac{D_1}{D_2}\right)^{2}$
Power	P	$\frac{P_1}{P_2} = \left(\frac{D_1}{D_2}\right)^{2}$
Torque	Md	$\frac{Md_1}{Md_2} = \left(\frac{D_1}{D_2}\right)^{2.75}$

ture gradients forming along the long pipe system, which are corrected with static mixers at the exit end of the piping, just before the die. The melt entering the die thus is assured of having a uniform melt temperature throughout. Systems with short adaptor pipes (2 or 3 feet or less) are not typically fitted with static mixers because the screw-type mixers can ensure a uniform melt into the die system.

Scaling Screw Design and Extruder Performance

When the performance is known for a given extruder size, it is sometimes necessary to predict the performance on a different-size machine using the same material. Some mathematical relationships have been developed to perform this function, as summarized in Table 4-2. The output rate is scaled by using the square of the diameter ratio, which relates to the barrel surface area and, hence, the melting capacity. The goal of a scaleup or scaledown exercise is to predict the equivalent extrusion performance at similar melt temperature conditions. The shear created on different-diameter extruders at similar screw speeds is different; thus for equal melt temperatures, a larger extruder must run at slower screw speeds. The scaled output rate is therefore delivered at a different screw speed. Drive power usage is scaled similarly to the output rate because it is assumed that the scaled condition uses equal lb/hr/hp. In scaling from a smaller to a larger extruder, it is difficult to achieve accurate results because of the melt temperature difference usually seen. A good practice when using the scaling equations is to keep the distance of scale as small as possible. The scale from a 3.5-inch-diameter extruder to a 4.5-inch machine is reasonably accurate, while a 1.5-inch scaled to a 4.5-inch extruder is quite risky and will not give good confidence that the scaled numbers will closely match reality. The use of a very small lab extruder ($\frac{3}{4}$ or 1 inch) thus will not allow good data to be used to predict production-sized extruder results. As long as the scaling laws are used with this limitation in mind, a reasonable result can be achieved.

Simplified extruder scaling laws are illustrated in Table 4-2 for comparing the basic parameters of extruders of diameters D_1 and D_2.

GLOSSARY OF EXTRUSION TERMS

extruder—Basically, a machine that accepts solid particles (pellets or powder) or liquid (molten) feed, conveys it through a surrounding barrel by means of a rotating screw, and pumps it, under pressure, through an orifice. Nomenclature used applies to the barrel, the screw, and other extruder elements.

Nomenclature applicable to the barrel includes:

barrel—A cylindrical housing in which the screw rotates, including a replaceable liner, if used, or an integrally formed special surface material. Nomenclature covers:

feed openings—A hole through the feed section for introduction of the feed material into the barrel. Variations (see Fig. 4-37) include vertical, tangential, side feed openings.

feed section—A separate section, located at the upstream end of the barrel, that contains the feed opening into the barrel.

grooved liner (or barrel)—A liner whose bore is provided with longitudinal grooves.

barrel heaters—The electrical resistance or induction heaters mounted on or around the barrel.

barrel jacket—A jacket surrounding the outside of the barrel for circulation of a heat transfer medium.

barrel heating zone—A portion of barrel length having independent temperature control.

barrel vent—An opening through the barrel wall, intermediate in the extrusion process, to permit the removal of air and volatile matter from the material being processed.

decreasing lead screw—A screw in which the flight lead, or pitch, decreases over the full flighted length (usually of constant depth).

metering type screw (Fig. 4-38)—A simple screw design with a feed section, a transition section, and a metering section.

multiple-flighted screw—A screw having more

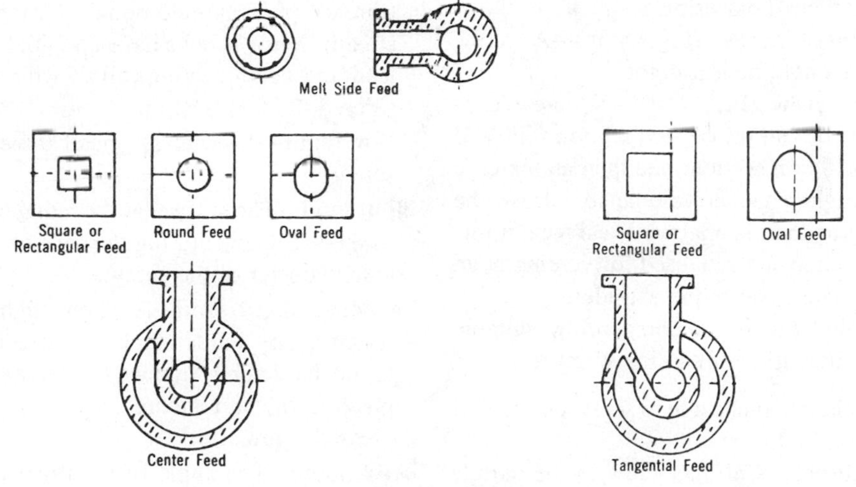

Fig. 4-37. Feed variations.

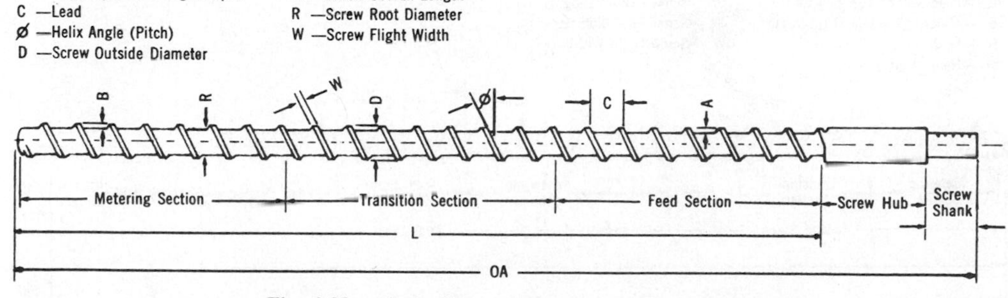

Fig. 4-38. Single flight, single-stage extrusion screw.

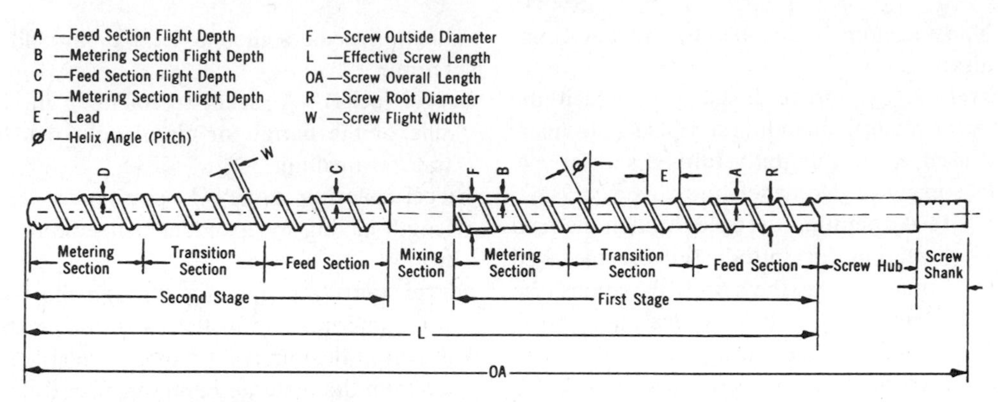

Fig. 4-39. Single flight, two-stage extrusion screw with mixing section.

than one helical flight such as: double flighted, double lead, double thread, or two starts, and triple flighted, etc.

multiple-stage screw (Fig. 4-39)—A two- or more-stage screw with introduction of special mixing sections, choke rings, or torpedoes for vented extrusion.

single-flighted screw (Fig. 4-38)—A screw having a single helical flight.

two-stage screw (Fig. 4-40)—A screw constructed with an initial feed section followed by a restriction section, and then an increase in the flight channel volume to release the pressure on the material while carrying it forward, such as a screw used for venting at an intermediate point in the extruder.

water-cooled screw—A cored screw suitable for the circulation of cooling water.

Nomenclature applicable to screw design and construction includes:

screw channel—With the screw in the barrel, the space bounded by the surfaces of flights, the root of the screw, and the bore of the barrel. This is the space through which the material is conveyed and pumped.

feed section of screw—The portion of a screw that picks up the material at the feed opening (throat) plus an additional portion downstream. Many screws have an initial constant lead and depth section, all of which is considered the feed section.

screw flight—The helical metal thread of the screw.

flight land—The surface at the radial extremity of the flight constituting the periphery or outside diameter of the screw.

hardened flight land—A screw flight having its periphery harder than the base metal by flame hardening, induction hardening, depositing of hard facing metal, etc., for increased screw life.

helix angle—The angle of the flight at its pe-

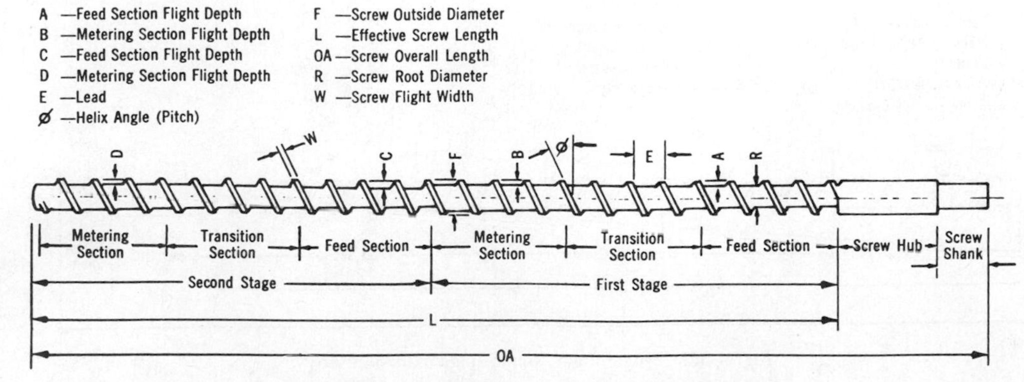

Fig. 4-40. Single flight, two-stage extrusion screw.

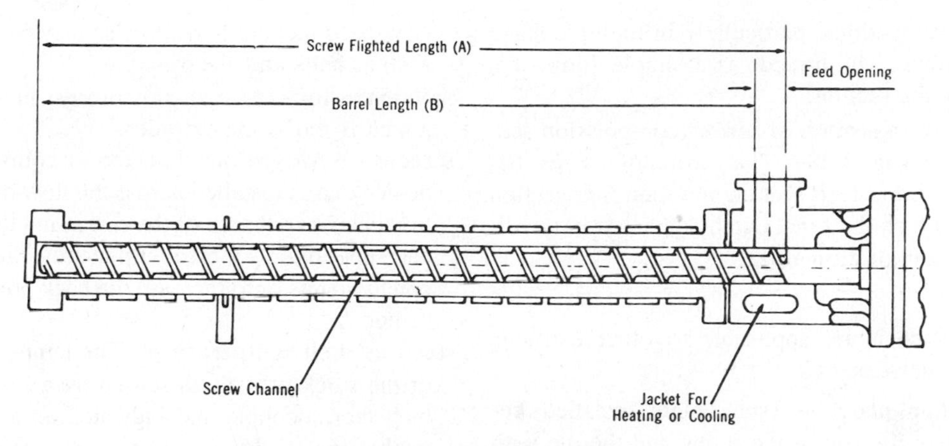

Fig. 4-41. Barrel cross-section.

riphery relative to a plane perpendicular to the screw axis.

screw hub—The portion immediately behind the flight which prevents the escape of the material.

length-to-diameter ratio (L/D ratio) (Fig. 4-41)—The length to diameter ratio (L/D) can be expressed in two ways: (a) L/D ratio based on the barrel-length—the distance from the forward edge of the feed opening to the forward end of the barrel bore divided by the bore diameter and expressed as a ratio wherein the diameter is reduced to 1 such as 20:1 or 24:1; and (b) L/D ratio based on the screw flighted length—the distance from the rear edge of the feed opening to the forward end of the barrel bore divided by the bore diameter and expressed as a ratio wherein the diameter is reduced to 1, such as 20:1 or 24:1. Note: Either definition of the L/D ratio can be considered correct, but machinery manufacturers should state which applies to a particular extruder.

metering section of screw—A relatively shallow portion of the screw at the discharge end with a constant depth and lead, and having a length of a least one or more turns of the flight.

restriction section or choke ring—An intermediate portion of a screw offering a resistance to the forward flow of material, with the intent to improve extrudate quality and/or output stability.

torpedo (Fig. 4-42)—An unflighted cylindrical portion of the screw usually located at the discharge end, but which can be located in

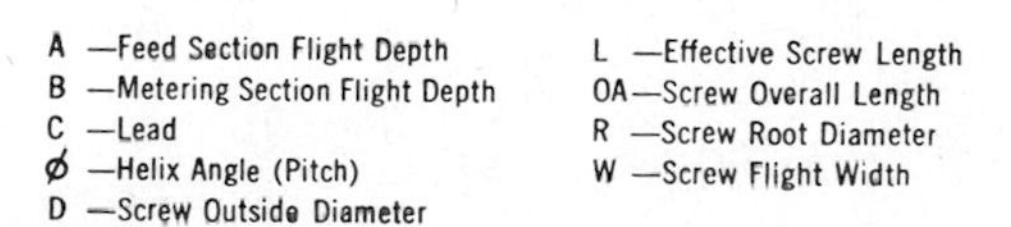

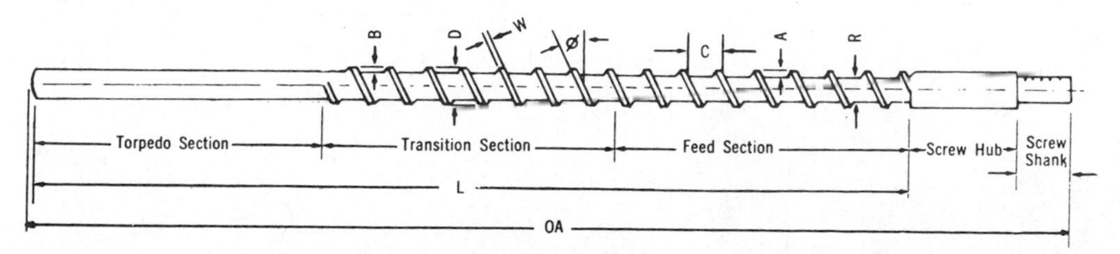

Fig. 4-42. Single flight, single-stage, with torpedo, extrusion screw.

other sections, particularly in multiple-stage screws. The torpedo is a simple form of a mixing section.

transition section of screw (compression section) (Fig. 4-38)—The portion of a screw between the feed section and metering section in which the flight depth decreases in the direction of discharge.

Nomenclature applicable to other extruder parts includes:

breaker plate—A metal plate installed between the end of the screw and the die with openings through it such as holes or slots. Usually used to support a screen pack.

drive—The entire electrical and mechanical system used to supply mechanical energy to the input shaft of gear reducer. This includes the motor, constant or variable speed belt system, flexible couplings, starting equipment, etc.

reduction gear (gear reducer)—The gear device used to reduce speed between the drive motor and the extruder screw. Supplementary speed reduction means also may be used, such as belts and sheaves, etc.

melt extrusion—An extrusion process in which a melt is fed to the extruder.

screens—A woven metal screen or equivalent device that is installed across the flow of material between the tip of the screw and the die and supported by a breaker plate to strain out contaminants or to increase the back pressure or both.

stock or melt temperature—The temperature of the stock or melt as sensed by a stock or melt thermocouple and indicated on a compatible meter. The location of the measurement must be specified for it to be meaningful.

surging—A pronounced fluctuation in output over a short period of time without deliberate change in operating conditions.

thrust bearing—The bearing used to absorb the thrust force exerted by the screw.

thrust—The total axial force exerted by the screw on the thrust bearing (for practical purposes equal to the extrusion pressure times the cross-section area of the bore).

5

Injection Molding of Thermoplastics

OVERVIEW

Injection molding is a major processing technique for converting thermoplastic materials, accounting for almost 20% of U.S. resin production.

The process was patented by John and Isaiah Hyatt in 1872 to mold camphor-plasticized cellulose nitrate (celluloid). The first multicavity mold was introduced by John Hyatt in 1878. Modern technology began to develop in the late 1930s and was accelerated by the demands of World War II. A similar surge in the technology of materials and equipment took place in the late sixties and early seventies.

This section will discuss the machinery and the practice of injection molding.

The basic concept of injection molding is the ability of a thermoplastic material to be softened by heating, formed under pressure, and hardened by cooling. In a reciprocating screw injection molding machine (Fig 5-1), granular material (the plastic resin) is fed from the hopper (a feeding device) into one end of the cylinder (the melting device). It is heated and melted (plasticized or plasticated), and it is forced out the other end of the cylinder (while still melted) through a nozzle (injection) into a relatively cool mold (cooling), held closed by the clamping mechanism. The melt cools and hardens (cures) until it is fully set up. Then the mold opens, ejecting the molded part (ejection).

Thus, the elements of injection molding are:

1. The way in which the melt is plasticized and forced into the mold (the injection unit).
2. The system for opening and closing the mold and holding it closed with a force (the clamping unit).
3. The type of machine controls.
4. The injection mold, which forms the part and acts as a heat exchanger.

Injection Ends

The four types of injection ends are:

1. The straight plunger machine (Fig. 5-2).
2. The two-stage plunger machine where one plunger plasticizes and the other shoots. (Fig. 5-3).
3. The reciprocating screw (Fig. 5-4).
4. The two-stage screw-plunger machine where the screw plasticizes material, which is forced into the shooting cylinder (Fig. 5-5).

Plunger-Type Injection Cylinder. The original injection molding machine was an adaptation of one used for molding rubber. Its method of plasticizing was a plunger (Fig.

By Irvin I. Rubin. The author gratefully acknowledges the permission of John Wiley & Sons, Inc., New York, to use material from his book *Injection Molding, Theory and Practice* (1973). Mr. Rubin is president of Robinson Plastics Corp.

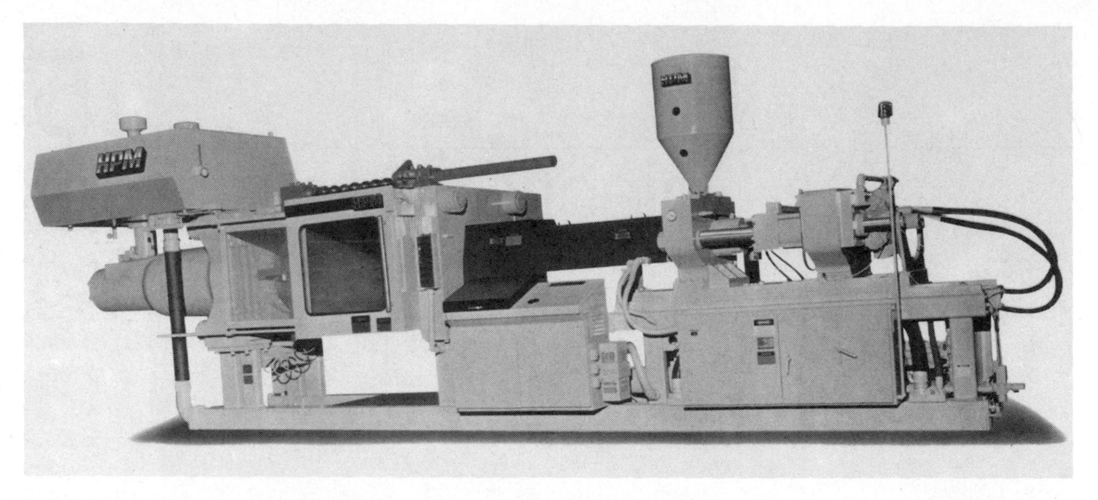

Fig. 5-1. Reciprocating screw injection molding machine with a 300-ton hydraulic clamp and a 28-oz injection shot capacity. (*Courtesy HPM Corp.*)

5-2), now obsolete except for molding variegated patterns such as "tortoise shell" cutlery parts and cosmetic containers.

Material from the hopper is measured volumetrically and fed into a chute to fall in front of the plunger. The injection plunger moves forward and forces the cold granules into the plasticizing chamber over the torpedo and out through the nozzle tip into the mold. The heat is supplied by resistance heaters on the outside of the barrel, and the material is melted by conduction of the heat from the heating bands through the cylinder wall into the plastic. Because there is relatively little convection in a plunger machine and the plastic has good insulating (or poor thermal conduction) proper-

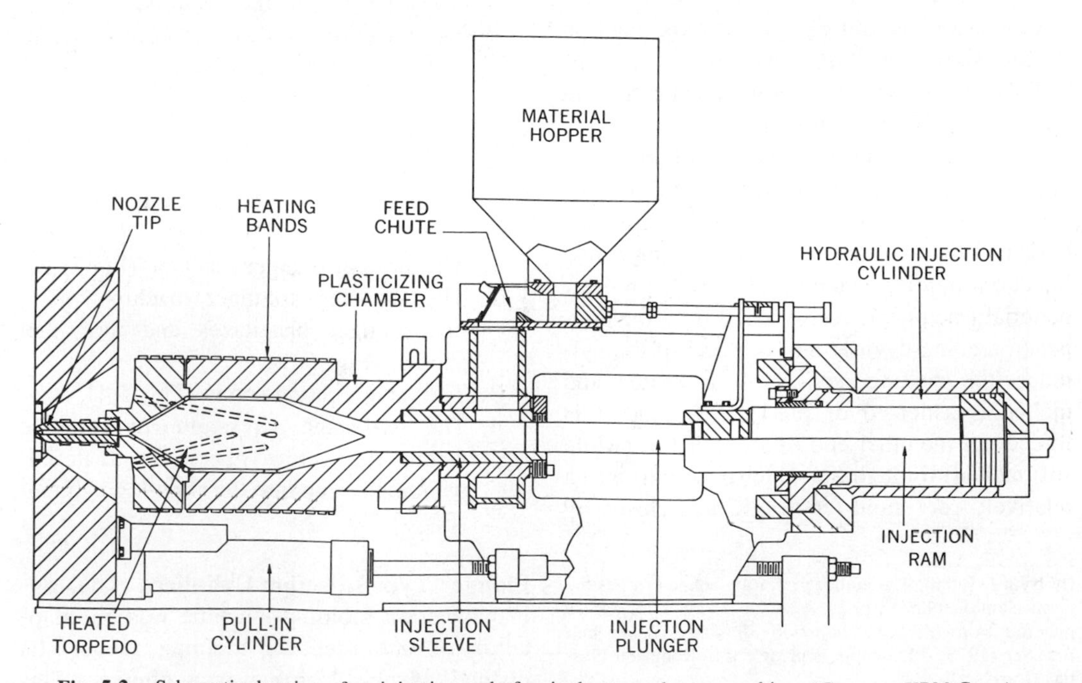

Fig. 5-2. Schematic drawing of an injection end of a single-stage plunger machine. (*Courtesy HPM Corp.*)

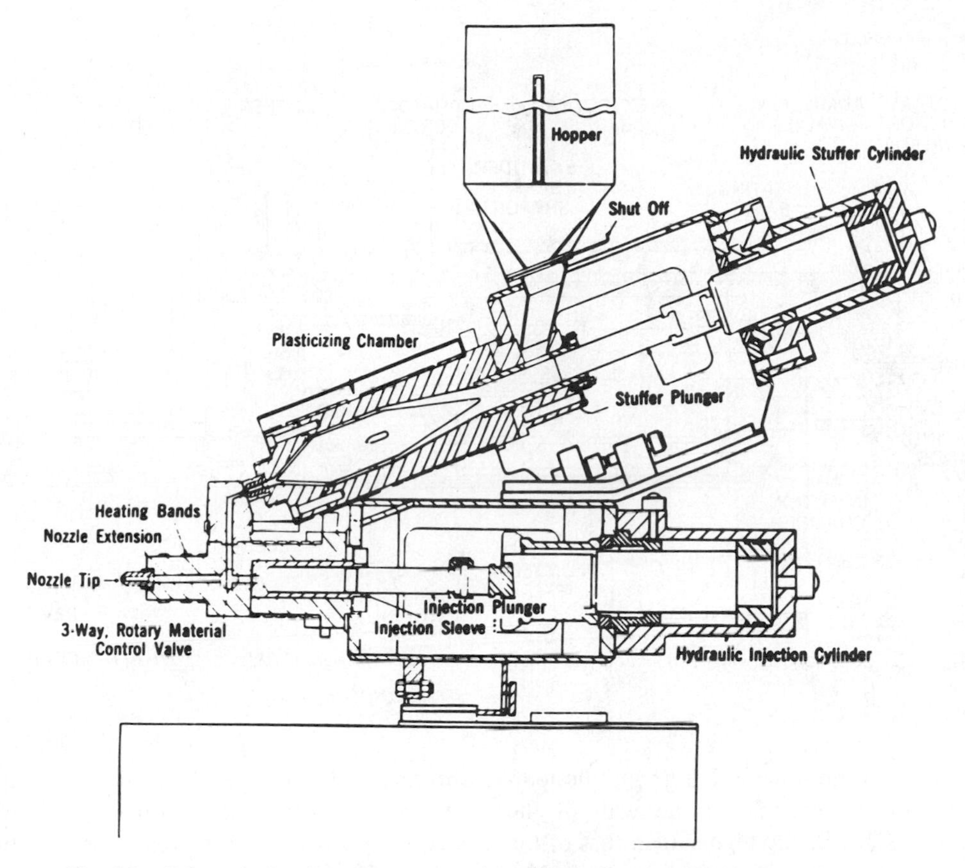

Fig. 5-3. Schematic drawing of a two-stage plunger unit. (*Courtesy SPI Machinery Division*)

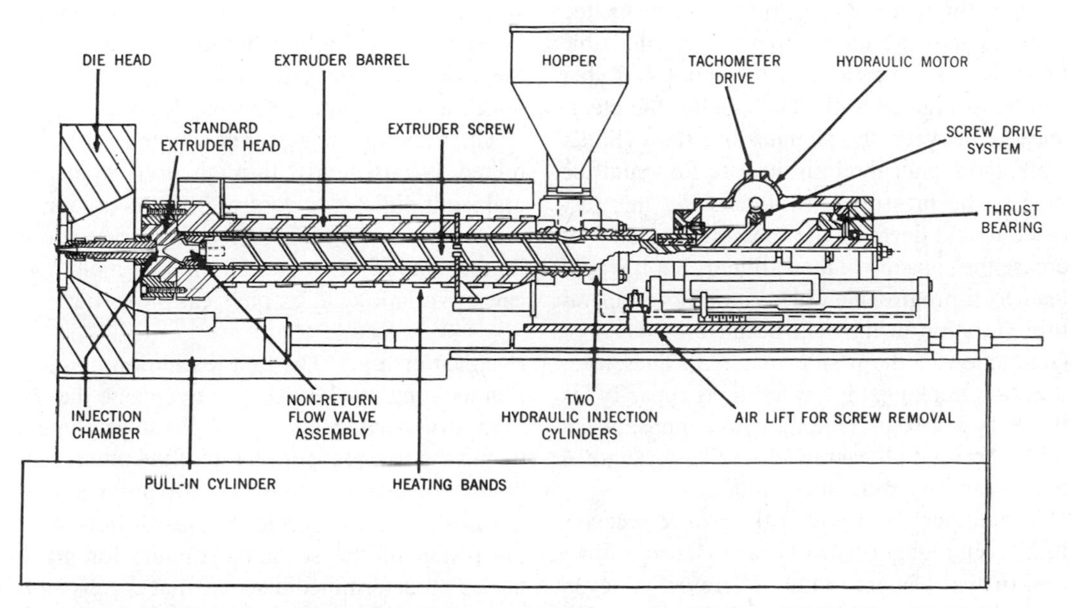

Fig. 5-4. Schematic drawing of an in-line reciprocating screw unit. (*Courtesy SPI Machinery Division*)

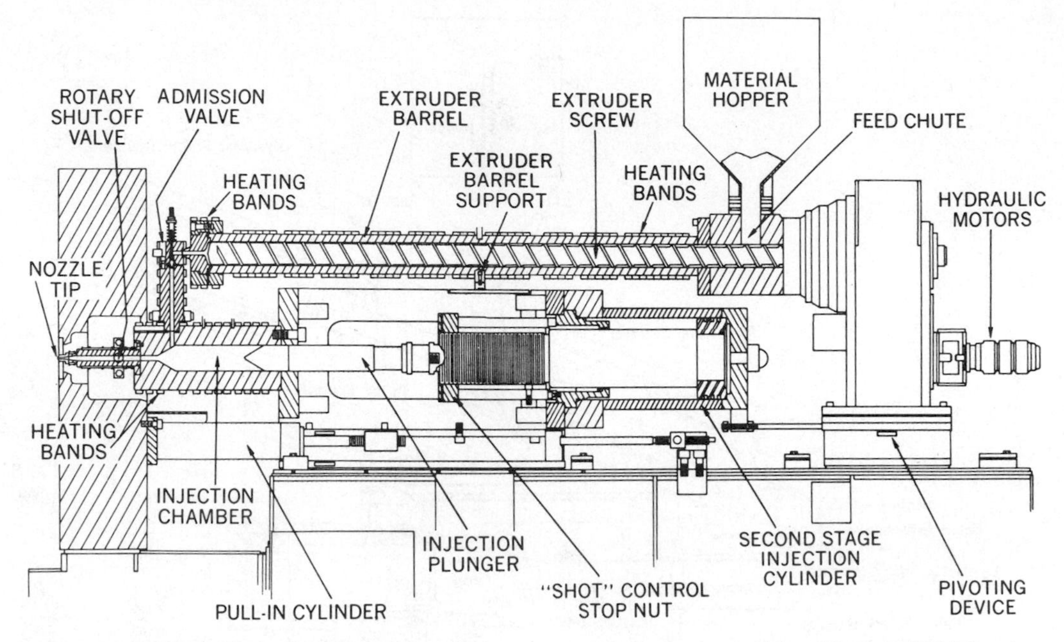

Fig. 5-5. Schematic drawing of injection end of a two-stage screw-plunger machine. (*Courtesy HPM Corp.*)

ties, there is little if any mixing, and the temperature of the material at the wall of the torpedo is significantly higher than that of the material in the center. This condition makes for poor temperature homogeneity, which will adversely affect the molded part.

Pressure is transmitted from the injection plunger through the cold granules and melted material into the mold. There is considerable pressure loss, which averages about 40% but can be as high as 80%. The smaller the clearance is between the plunger and the cylinder wall, the higher the resistance to flow and the greater the pressure loss. Increasing this distance would decrease the pressure loss and increase the capacity of the cylinder, but it would lead to a progressively deteriorating temperature variation in the melt because of the poor conduction of the plastic material. Therefore, the straight plunger is limited in its capacity, as these two design requirements oppose each other. Straight plunger machines larger than 16 oz. in capacity were rarely found.

The amount of heat that the plastic receives in a given plunger-type cylinder is also a function of the residence time. Therefore, uneven cycles change the temperature of the melt and prevent good moldings. When the machine is off cycle, even for a short period of time, it is necessary to purge out the overheated material and continue with the molding. One of the principal advantages of the screw plasticizer is that its heating is done by shearing action caused by the screw rotation. This mechanical working puts the heat directly into the material. When the screw is not turning, the only heat the material gets is from the barrel heaters, which is a comparatively small amount.

The pressure in a plunger system is transmitted inconsistently through cold granules, making it difficult to control the injection pressure in the mold. This increases the variations in the amount of material in the mold and causes variations in its physical properties.

Preplasticizing. The first revolution in injection molding machine design overcame the obvious disadvantages of the single-stage plunger by putting one plunger on top of the other, separating the two functions. The top plunger plasticizes the material, which is pushed in front of the piston of the second cylinder, forcing it back a predetermined distance that corresponds to the proper amount of hot melted material for the next shot. The second cylinder then shoots the material into the mold (Figs. 5-3 and 5-5). This approach is called preplasticizing. The second type of preplasticizing is accomplished in the reciprocating screw itself, where the shooting pot is the chamber that develops in front of the retracting screw. Preplasticizing offers a number of significant advantages:

1. The melt is more homogeneous because it mixes as it passes through the small opening connecting the two chambers.
2. There is direct pressure on the material by the injection plunger.
3. Faster injection is possible.
4. Higher injection pressures and better injection control are possible.
5. Better shot weight control is possible.
6. The weight of a single shot can easily be increased by making the stroke of the shooting cylinder longer, or its diameter larger. This is not true in the reciprocating screw type of preplasticizing.
7. All the material passes over the full length of the screw.

The two-stage plunger-plunger machine (Fig. 5-3) is obsolete, having been replaced by the two-stage screw-pot (Fig. 5-5) because of the superior plasticizing ability of the screw.

Reciprocating Screw Injection End. Figure 5-4 shows a schematic drawing of the injection end of a reciprocating screw machine. The extruder screw, which is contained in the barrel, is turned most often by a hydraulic motor (as contrasted to an electric motor attached to a gear system).

As the screw turns, it picks up material from the hopper. As it progresses down the screw, the resin is compacted, degassed, melted, and pumped past the nonreturn flow valve assembly at the injection side of the screw. This, in essence, is a check valve, which allows material to flow only from the back of the screw forward. As the material is pumped in front of the screw, it forces back the screw, hydraulic motor, and screw drive system. In so doing, it also moves the piston and rod of the hydraulic cylinder(s) used for injection. Oil from behind the piston(s) goes into a tank through a variable resistance valve, called the back pressure valve. Increasing this resistance requires higher pressures from the pumping section of the screw, and results in better mixing, a slower cycle, and greater energy consumption.

The screw will continue to turn, forcing the carriage back until a predetermined location is reached. Then the rotation is stopped, and an exact amount of melted material is in front of the screw and will be injected into the mold at the appropriate time in the cycle. This is accomplished by using the hydraulic injection cylinder(s). It is evident that the first particles of material from the hopper will pass over the full flight of the screw. As the screw moves toward the right, succeeding particles will travel through a diminishing length of the screw. This action of the reciprocating screw limits the amount of material that can be plasticized by any given screw. It also results in a less homogeneous melt compared to a two-stage screw-pot machine, though this is not important for most applications.

These limitations and others may be overcome by using a two-stage screw-plunger machine, shown in Fig. 5-5. Here, the material goes over the full length of the screw through a rotary valve into the injection or shooting chamber. Here, too, the injection plunger is forced back until it reaches a predetermined point, at which time the screw stops. The rotary shutoff valve is rotated so that when the injection plunger advances, the material is injected into the mold.

The main advantages of a two-stage screw are that the material passes over the whole length of the screw, and it receives additional mixing time as it goes through the rotary valve; also, pushing directly on the material with a plunger, rather than with a check valve at the end of the screw, gives significantly better shot control.

Cycle of a Hydraulic Clamp/Reciprocating Screw Machine. The cycle of a reciprocating screw molding machine may be summarized as follows:

1. Oil is sent behind the clamp ram, closing the mold. Pressure builds up to develop

enough force to keep the mold closed while the injection process occurs.

2. Previously plasticized material in front of the reciprocating screw is forced into the mold by the hydraulic injection cylinder(s).
3. Pressure is maintained on the material to mold a part free of sink marks, flow marks, welds, and other defects.
4. At the end of this period, the screw starts to turn, plasticizing material for the next shot. After melting, the material is decompressed to prevent drooling from the nozzle.
5. While this is occurring, the plastic is cooling in the mold and solidifying to a point where it can be successfully ejected. This cooling is accomplished by circulating a cooling medium, usually water, through drilled holes or channels in cavities, cores, and the mold base.
6. Oil is sent to the return port of the clamping ram separating the mold halves.
7. As the moving platen returns, the knockout or ejection mechanism is activated, removing the pieces from the mold.

Packing. Once the mold is filled initially, additional material is added to the mold by the injection pressure to compensate for thermal shrinkage as the material cools. This process is called packing. Too much packing will result in highly stressed parts and may cause ejection problems. Insufficient packing causes short shots, poor surface, sink marks, welds, and other defects. The proper amount of packing is determined by trial and error or with the assistance of computerized process simulation. The material will continue to flow into the mold as long as there is injection pressure, provided that the gate is not sealed. When no more material enters the mold, contraction of the cooling material results in a rapid decrease in the pressure in the mold. The residual pressure caused by the original deformation of the steel of the mold and the adhesion of the plastic to the steel must be overcome by the knockout system to eject the parts.

Injection Screw. Extrusion screws are used to plasticize material in almost all new equipment, so it is essential to understand their characteristics and how they work. Figure 5-6

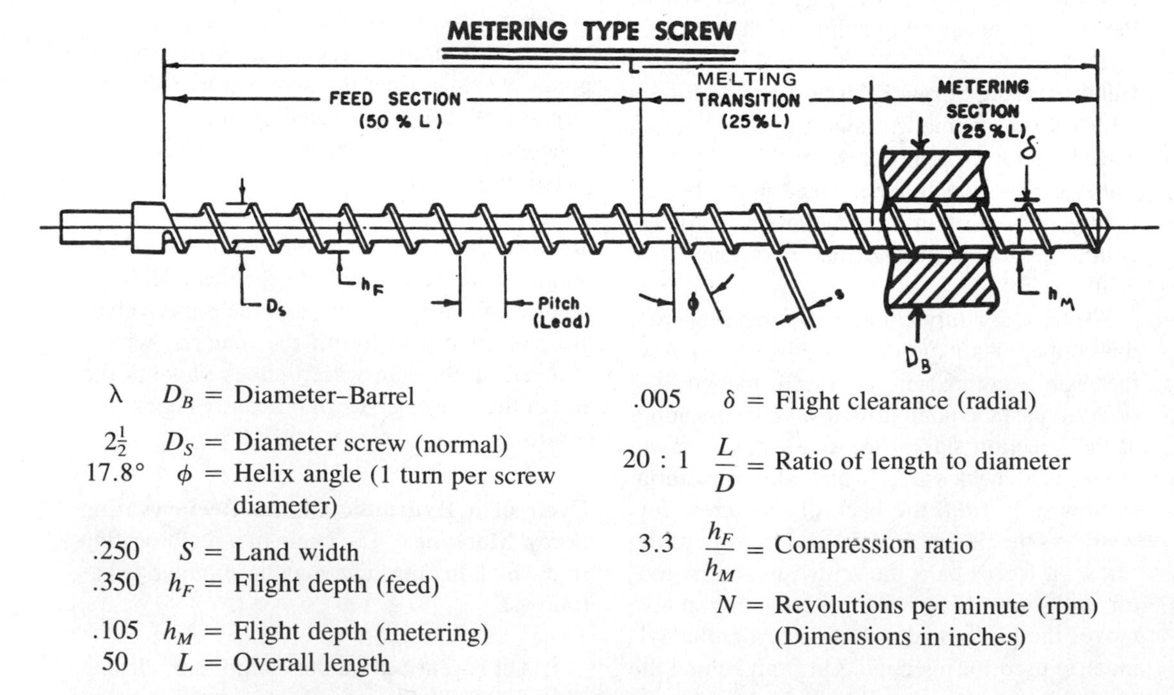

Fig. 5-6. Typical metering screw used for reciprocating screw type injection molding machines. (*Courtesy Robinson Plastics Corp.*)

shows a typical screw used in injection molding equipment. Its task is to convey the cold pellets at the hopper end, compact and degas the material in the feed section, plasticize it in the transition section, pump the screw back in the metering section, and act as a plunger forcing the material into the mold.

The feed section is approximately half the length of the screw, the melting or transition section one-quarter the length, and the metering or pumping section one-quarter the length. The outside diameter of the barrel is determined by the pressure requirements of the cylinder. An extruder barrel requires a pressure that generally does not exceed 8000 psi, whereas the barrel of a reciprocating screw must contain at least the 20,000 psi of injection pressure that is developed. The diameter of the screw is the nominal diameter of the hole in the barrel. Experience has shown that a pitch of one turn per screw diameter works well for injection molding. This helix angle (θ) is 17.8°, which is standard for injection screws. The land width is approximately 10% of the diameter. The radial flight clearance is specified by considering the effects of leakage flow over the flights, the temperature increase in the plastics caused by this clearance, the scraping ability of the flights in cleaning the barrel, the eccentricity of the screw and barrel, and the manufacturing costs. The depth of the flight is constant but different in the feed and metering sections; it is reduced from the deeper feed section to the shallower metering section in the transition section wherein the material is melted. The compression ratio of a screw is the ratio of the flight depth in the feed section to that in the metering section. The length of the screw is the axial length of the flighted section. The ratio of the length to the diameter (L/D) is a very important specification in a screw. Higher L/D ratios give a more uniform melt, better mixing, and higher output pressures. An L/D of 20 : 1 should be the minimum for molding machines, with higher ratios such as 24 : 1 more desirable. For more information on extruder design in general, refer to Chapter 4.

Melting a Plastic in a Screw

To be conveyed forward, the plastic must adhere more to the barrel than to the screw. If the coefficients of friction between the plastic and the screw and between the plastic and the barrel were identical, there would be no flow of material—it would just rotate as a plug within the flight of the screw. Because the polymer molecule is anchored more to the side of the barrel than to the screw, the polymer molecules slide over one another. This behavior is called a shearing action. In effect, molecular forces are being broken and turned into heat; this is what is meant when one refers to converting the mechanical action of the screw into heat.

The solid pellets are picked up from the hopper and moved only in the forward direction in a helical path (Fig. 5-8, part A). They touch the barrel, forming a thin film of melted plastic on the barrel surface (Fig. 5-7). This melting is due in part to conduction from the heater bands. The relative motion of the barrel and the screw drags this melt, which is picked up by the leading edge of the advancing flight of the screw. It flushes the polymer down in front of it forming a pool, which circulates. Heat enters the plastic by a shearing action (rubbing of molecules of plastic over each other) whose energy is derived from the turning of the screw. Unmelted material is drawn into the melted area in increasing amounts down the flights of the barrel. The melting is complete at the point where the solid bed of unmelted plastic is completely melted, which is the end of the compression or melting section.

Figure 5-8, part B shows a simplified view of the flight of a particle in a screw. If there is

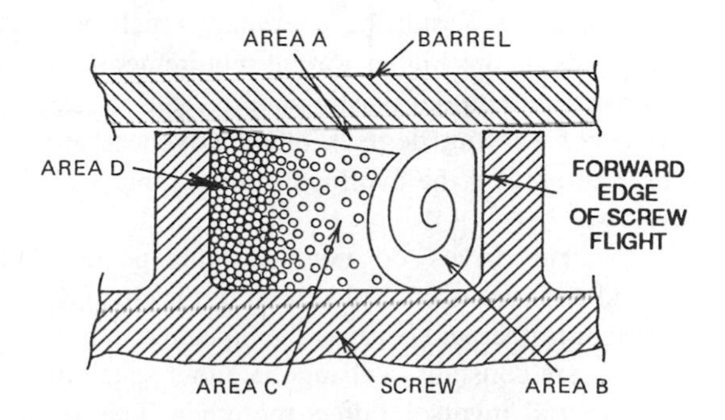

Fig. 5-7. Schematic drawing of melting of plastic in a screw. Area A—material melted by conduction; Area B—material melted by shearing; Area C—partially melted material; Area D—unmelted material (solid bed).

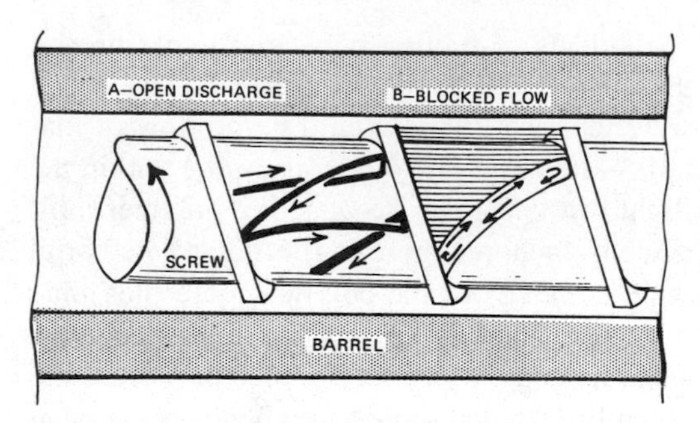

Fig. 5-8. Flow of plastic in a screw. A—open discharge. B—blocked flow. (*Courtesy Robinson Plastics Corp.*)

little or no resistance at all, as happens when the screw is retracted, a particle of material will move as shown in part A. It moves with a circulatory motion from flight to flight. Additionally, it goes forward along the axial direction of the barrel toward the nozzle. At no time does a particle ever flow rearward. If the screw could not retract as it rotated, the particles would circulate within the flight. Obviously, by regulating the amount of resistance to the return of the screw (by the back pressure control valve), the amount of mixing in the melt can be controlled. This is an important feature in reciprocating screw molding. The higher the back pressure is, the more homogeneous the melt and the better the color and additive dispersion. Higher back pressure often eliminates warpage and shrinkage problems.

Screw Drives. One method of applying the driving force to turn the screw is to attach it to an electric motor, through a speed-reducing gear train with different speed ranges. The available torque of an electric motor has a pattern of a 100% starting overload, which corresponds to the higher torque requirements of starting with colder plastic material. Safety devices of varying degrees of effectiveness are used to prevent the overload from snapping the screw.

Electric drives do not have independent speed and torque control. The speed is changed by gear trains, and because the input and output power are constant, a change in either speed or torque will inversely affect the other. This relationship makes it difficult to find optimum molding conditions. The electrical system has a high efficiency of approximately 95%, compared to the 60 to 75% of the hydraulic system.

A second method of turning the screw is to connect it to a nonvariable speed-reducing coupling driven by a hydraulic motor. A third method, which is used on most machines today, is to attach a special hydraulic motor directly to the screw. This can be done externally or internally in the injection housing. This type of drive has much to recommend it, as it has no gear speed reducers and develops minimum noise and vibration. No lubrication is required, and its maintenance is minimal.

A hydraulic motor can never develop more than its rated torque. Consequently, the hydraulic motor must be larger than an electric motor would be. The safety protection for a hydraulically operated screw is a simple relief valve, as the oil pressure on the hydraulic motor controls the torque, thus preventing screw breakage. Fixed-displacement hydraulic motors are used for the screw drive and have a constant torque output at a given pressure. Therefore, their horsepower varies with the speed. The maximum torque depends upon the pressure setting of the controlling relief valve, and the speed of the hydraulic motor is controlled by a hydraulic flow control valve. The outstanding characteristic of the hydraulic drive is its stepless control of torque and speed, which permits selection of optimum molding conditions. Additionally, the ease of making these changes encourages production people to make them. The hydraulic motor is smaller and lighter in weight than the equivalent electric system, thus permitting a lower initial back pressure. This gives better shot weight control.

To summarize, the hydraulic drive is less efficient, has a higher margin of screw safety, it smaller and lighter in weight, provides stepless torque and speed control, and gives better melt quality than an electrical drive; but the electrical drive is more efficient than the hydraulic drive and provides acceptable melt quality.

Figure 5-9 shows a cutaway view of a 300-tone toggle machine with a 2.5-inch reciprocating screw. The screw is driven by an hydraulic motor (labeled No. 27 in the diagram), which is internally mounted in the injection housing.

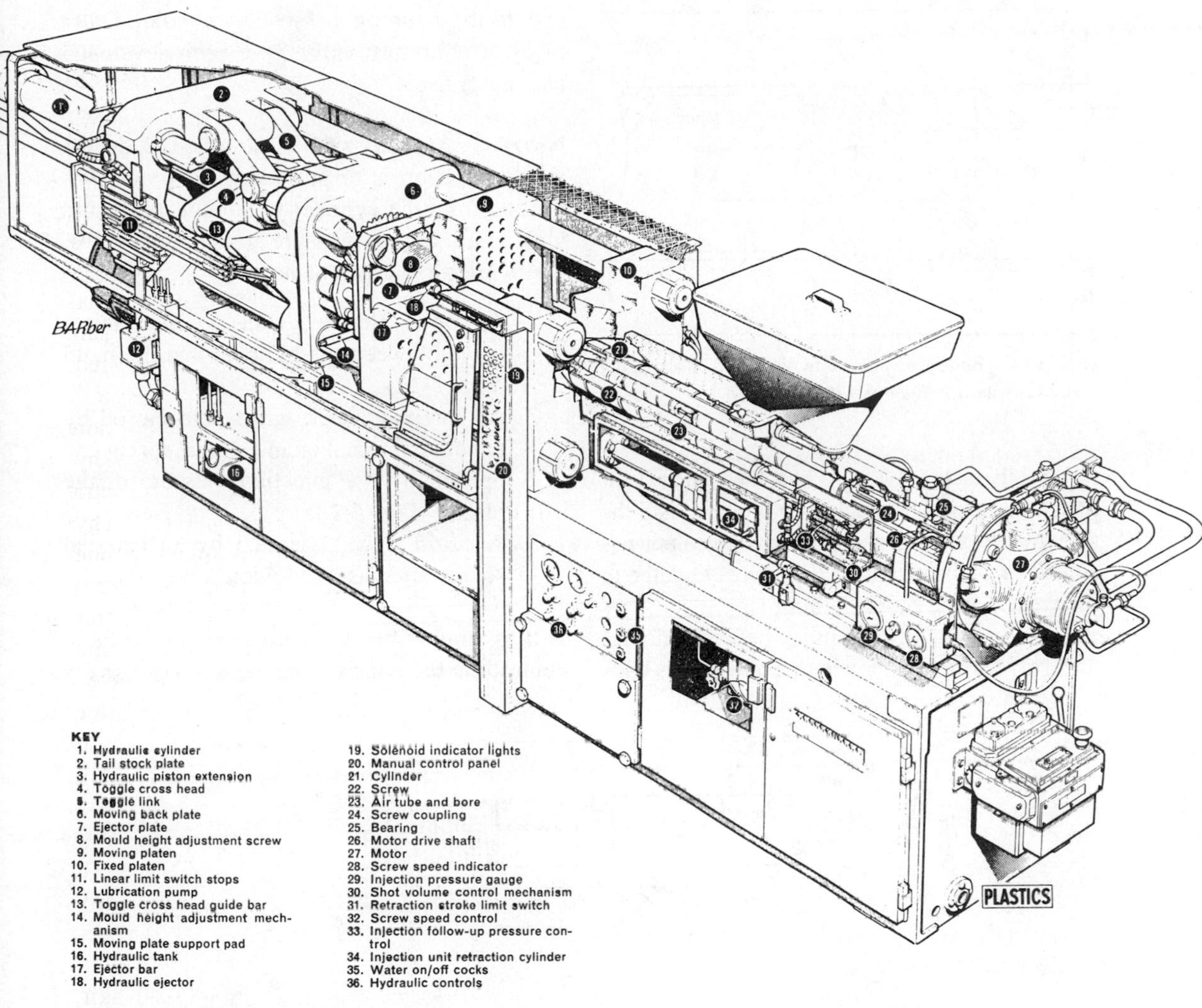

Fig. 5-9. Cutaway view of a 2.5-inch reciprocating screw, 300-ton toggle clamp machine (a PECO injection machine, as originally published in *British Plastics* magazine).

Reciprocating Screw Tips. The reciprocating screw machine uses the screw as a plunger. As the plunger comes forward, the material can flow back into the flights of the screw. For heat-sensitive materials such as PVC, a plain screw tip is used. For other materials this is not adequate, and a number of different check valves have been designed. Figure 5-10 shows a sliding ring-type nonreturn valve. The nozzle (not shown) is on the left and is plugged with the hardened sprue from the last shot. The ring is in the forward position while plasticizing, so that the material can flow past the seat and through its hollow portion in front of the screw,

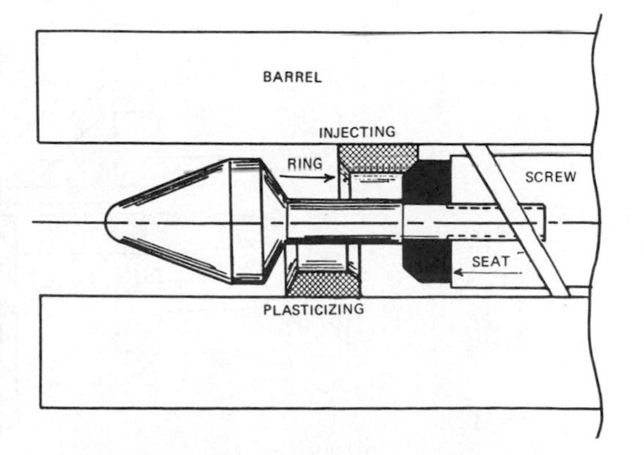

Fig. 5-10. Ring-type nonreturn valve for tip of screw. (*Courtesy Robinson Plastics Corp.*)

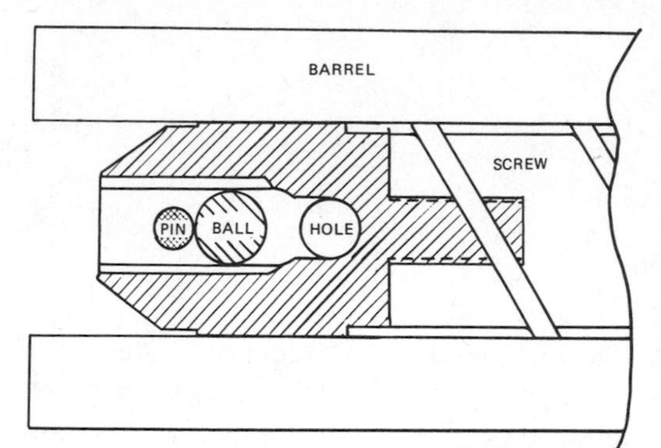

Fig. 5-11. Ball-type nonreturn valve in plasticizing position. (*Courtesy Robinson Plastics Corp.*)

forcing the screw back. When the material injects, the ring slides back onto the seat, and the screw acts as a plunger. This arrangement has a number of disadvantages, primarily related to the wear caused by the sliding ring.

A better method is to use a ball-type nonreturn valve (Fig. 5-11), where a ball moves back and forth, opening or sealing a hole. Other types of nonreturn valve have been developed and are in use.

Nozzles. The nozzle is a tubelike device whose function is to provide a leakproof connection from the cylinder to the injection mold with a minimum pressure and thermal loss. There are three types of nozzles:

1. An open channel with no mechanical valve between the cylinder and the mold (Fig. 5-12).
2. An internal check valve held closed by either an internal or an external spring and opened by the injection pressure of the plastic (Fig. 5-13).
3. A cutoff valve operated by an external source such as a cylinder.

It is impossible to mold correctly without controlling the nozzle temperature. With a very

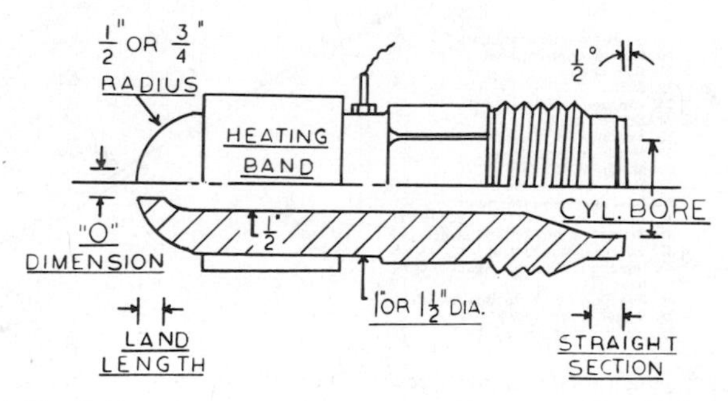

Fig. 5-12. Standard nozzle. (*Courtesy Robinson Plastics Corp.*)

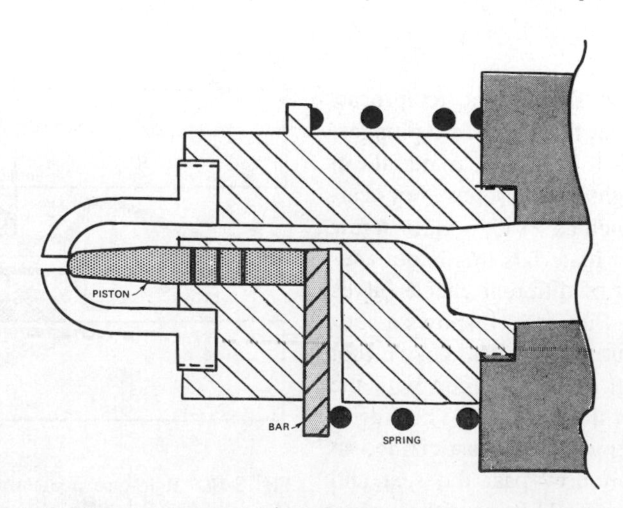

Fig. 5-13. Needle-type shutoff nozzle. (*Courtesy Robinson Plastics Corp.*)

short nozzle, heat conduction from the cylinder usually will be adequate to maintain it at the proper temperature. Usually nozzles are long enough to require external heat, which must be independently controlled and never attached to the front heating band of the cylinder. The cylinder requirements are completely different from those of the nozzle, and burned and degraded plastic may result as well as a cold nozzle plugging up the cylinder. Nozzles should be controlled by a thermocouple and a pyrometer. Heating bands can be used, although tubular heaters are preferable because they do not burn out when contacted by molten plastic. Nozzles are available with heating cartridges inserted therein, which produce up to four times as much heat as other nozzles. The internal configuration of nozzles is streamlined to prevent pressure loss and hangup of the material with subsequent degradation. The land length is kept to a minimum consistent with the strength requirement. The nozzle need not be manufactured in one piece. Very often, the tip is replaceable, as it is screwed into the nozzle body. This arrangement makes replacement and repair considerably less expensive.

The nozzle in Fig. 5-13 is sealed off internally by a spring acting through the bar onto the piston. When there is enough injection pressure to overcome the force of the spring, the piston is pushed back, and the material flows. When the injection pressure drops to the predetermined point, the piston closes again, sealing off the injection cylinder from the mold. Nozzles are available with thermocouples that extend into the melt stream to monitor the temperature of the plastic rather than that of the nozzle.

Temperature Control. Most machines use a bimetallic (iron–constantan) thermocouple as the sensing unit for the temperature of the heating cylinder. It feeds into a pyrometer, an instrument which uses that information to provide the proper amount of heat to the cylinder. Because there is a large thermal override from the heating band to the material, pyrometers have compensating circuits. The pyrometers use relays to control contractors that turn the heating bands on or off.

Newer equipment today uses solid state controls in the pyrometer circuit. Thermistors can be used for temperature sensing. By means of SCR units, the amount of heat is regulated by the requirements of the heating cylinder; the heating bands will give a percentage of their heat output as required rather than the on–off action of older controls. This provides a much smoother temperature control system with lower variations in material temperature. However, it should be noted that both of these systems measure the temperature of the heating barrel; it is more desirable to measure the temperature of the plastic melt itself.

Advantages of Screw Plasticizing. From the preceding discussion, we can see that there are major benefits to using the screw plasticizing method, where the melting is a result of the shearing action of the screw. As the polymer molecules slide over each other, the mechanical energy of the screw drive is converted into heat energy, and the heat is applied directly to the material. This action, plus the mixing action of the screw, gives this plasticizing method several important advantages:

1. This high shearing rate lowers the viscosity, making the material flow more easily.
2. Good mixing results in a homogeneous melt.
3. The flow is nonlaminar.
4. The residence time in the cylinder is approximately three shots, compared to the eight to ten shots of a plunger machine.
5. Most of the heat is supplied directly to the material.
6. Because little heat is supplied from the heating bands, the cycle can be delayed by a longer period before purging.
7. The method can be used with heat-sensitive materials, such as PVC.
8. The action of the screw reduces the chances of material holdup and subsequent degradation.
9. The preplasticizing chamber is in front of the screw.
10. The screw is easier to purge and clean than a plunger machine.

Injection End Specifications. The following items must be specified:

1. Type—reciprocating screw or screw-pot.
2. Diameter of the screw.
3. L/D ratio.
4. Maximum weight in ounces of polystyrene that can be injected in one shot; or, alternatively, the volume of material per shot.
5. The plasticizing capacity, which is in effect the amount of material that can be melted per unit time with the screw running continuously. In injection molding the screw runs about one-half of the time.
6. Maximum injection pressure on the screw, usually 20,000 psi.
7. Other specifications that will be provided by the manufacturer and are dictated by the above.

Clamping Ends of Injection Molding Machines

The clamping and injection ends of a molding machine are described and rated separately. All clamp ends use hydraulic force. Machines can be completely hydraulic, a combination of a hydraulic cylinder and a toggle mechanism, or completely electric. Clamp ends are rated by the maximum number of tons of locking force exerted. In a fully hydraulic machine the relationship is:

$$F = \frac{P \times A}{2000} \qquad (1)$$

where:

F = force (tons)
P = pressure (lb/in.^2)
A = area of clamp ram (in.^2)

As a rule of thumb, $2\frac{1}{2}$ tons of force is required for each square inch of projected area of whatever is molded. The projected area is the maximum area parallel to the clamping force (i.e., the platens). A part behind another part does not require extra clamping force. For example, a center-gated polystyrene box 10×14 inches or 140 sq. in. would require a 350-ton press ($140 \text{ in.}^2 \times 2.5 \text{ tons/in.}^2 = 350$ tons). The depth of the box would not be relevant in determining the clamp force requirements because the sides are not perpendicular to the clamping force.

Hydraulic Clamping System. Figure 5-14 shows a schematic drawing of a hydraulic clamping system. The stationary platen is attached to the molding machine and has four tie rods that go through it and support the moving platen and then go through the hydraulic mounting plate. At each end of the tie rods are tie rod nuts. The hydraulic clamp ram moves the moving platen, to which is attached the moving or knockout side of the mold. The stationary side of the mold is attached to the stationary platen. The ejection mechanism of the molding machine is not shown.

As can be seen from Fig. 5-14, when oil is put in the forward port, the clamp ram will move the moving platen until the mold parts make contact. As the pressure builds up, the force behind the clamp ram is transmitted through it, the moving platen, the mold, and the stationary platen to the tie rod nuts. This force stretches the tie bar, which provides the clamping action. When the mold is to be opened, oil is sent to the return port, and the forward port is vented to the tank. This retracts the clamp ram, the moving platen, and the moving part of the mold. The plastic part normally remains on the moving part of the mold and is ejected or knocked out of the mold by the ejection mechanism (see below). The ejection mechanism can be operated mechanically or hydraulically, with the latter means preferable because it allows for control of the timing, direction, and force of the stroke. In a mechanical system the knockout plate is stopped by the knockout bars of the machine.

Toggle Systems. Another type of clamp is the toggle, which is a mechanical device used to amplify force. In a molding machine, it consists of two bars joined together end to end with a pivot. The end of one bar is attached to a stationary platen, and the other end of a second

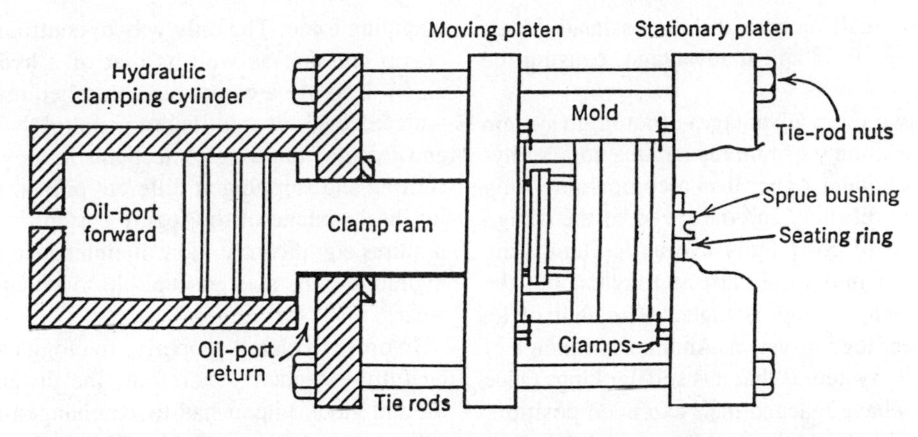

Fig. 5-14. Schematic drawing of an all-hydraulic clamp. (*Courtesy Robinson Plastics Corp.*)

bar is attached to the movable platen. When the mold is open, the toggle is in the shape of a V (Fig. 5-9, No. 5). When pressure is applied to the pivot, the two bars form a straight line. The mechanical advantage can be as high as 50:1. The force needed to straighten the toggle is applied by a hydraulic cylinder. A 350-ton machine with a 50:1 toggle would require only 7 tons of force from the hydraulic cylinder.

Figure 5-15 shows a standard double toggle clamping mechanism. In the mold open position, the hydraulic cylinder has retracted, pulling the crosshead close to the tail stock platen. This pulls the moving platen away from the stationary platen and opens the mold. It is difficult to stop the moving platen before completion of the full stroke. If it is important to achieve this, nylon buffers can be used as a mechanical stop. To close the mold, the hydraulic locking cylinder is extended. The moving plate moves rapidly at first and automatically decelerates as the crosshead extends and straightens out the

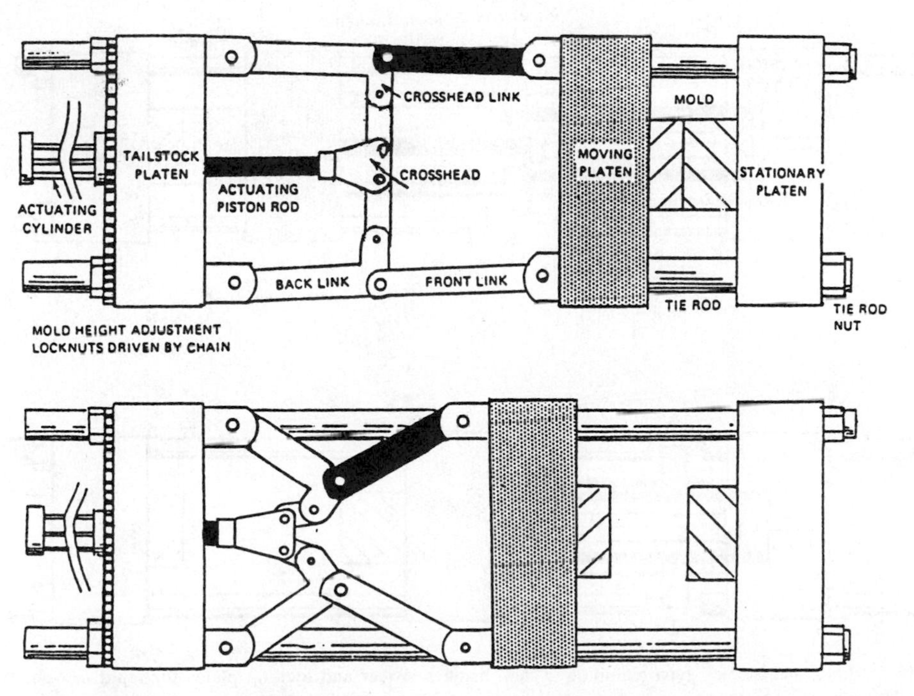

Fig. 5-15. Double toggle clamping system. (*Courtesy Robinson Plastics Corp.*)

links. A small motion of the crosshead develops a large mechanical advantage, causing the locking.

The two main advantages of a toggle system are the economy of running a much smaller hydraulic cylinder rather than a comparable fully hydraulic machine and the speed of the design per dollar of cost. Fully hydraulic clamps are capable of moving as fast as toggles, but the cost to achieve this is higher than that of an equivalent toggle system. Another advantage of the toggle system is that it is self-locking. Once the links have reached their extended position, they will remain there until retracted. The hydraulic system requires continuous maintenance of line pressure.

The toggle systems on molding machines have several disadvantages. A primary one is that there is no indication of the clamping force, so that it is difficult to adjust and monitor. The clamping force in a hydraulic system is read immediately by a pressure gauge and can be controlled in stepless increments, but in a toggle press it must be read with a strain gauge on a tie bar. As the room temperature changes, so does the length of the tie bar, thus changing the clamping force. The only way to control toggle clamp pressure as well as that of a hydraulic clamp is to use a computer-controlled machine with feedback. It is difficult to control the speed and force of the toggle mechanism, as well as starting and stopping at different points. A major disadvantage of the toggle system is that it requires significantly more maintenance than a hydraulic one; it is susceptible to much more wear.

In order to clamp properly, the toggles must be fully extended. Therefore, the distance of the tail stock platen has to be changed to accommodate different molds. This is done with a chain or gears that simultaneously move the four locking nuts on the tail stock platen. It can be turned mechanically, electrically, or hydraulically.

Clamping for Large Machines. As machines became large, larger clamp capacities and longer strokes were required, a design that became unwieldy and costly for all-hydraulic clamps. Many systems have been designed to overcome this problem, such as the one shown in Fig. 5-16, which is commonly called a lock

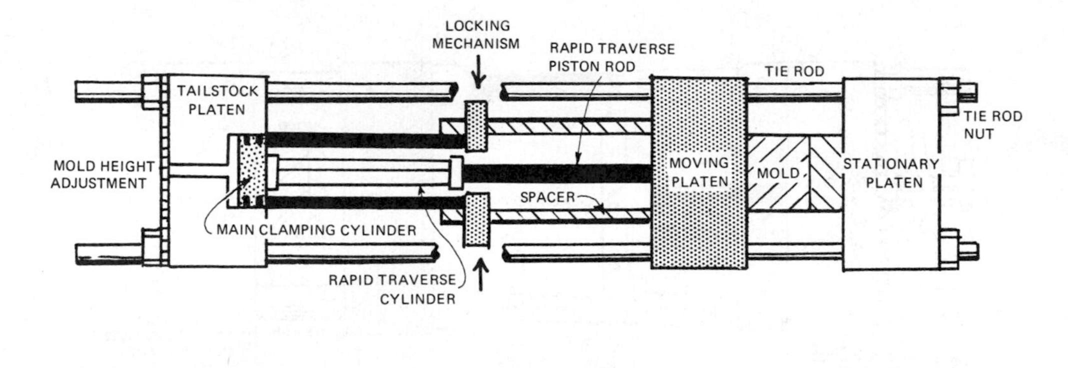

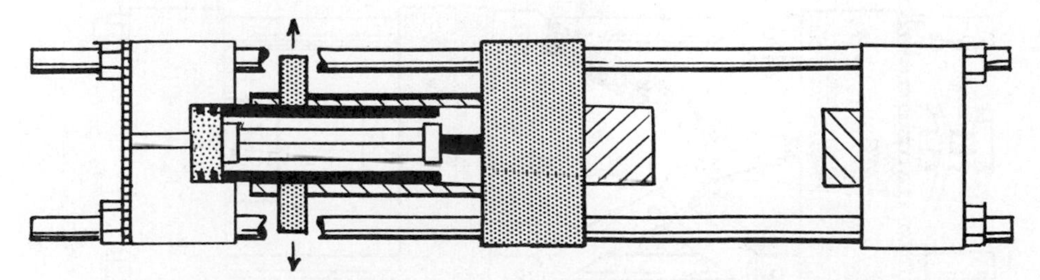

Fig. 5-16. "Lock and block" type clamping system using a spacer and locking plates for rapid motion. (*Courtesy Robinson Plastics Corp.*)

and block. A small-diameter high-speed rapid traverse cylinder is used to move the movable platen. Spacers, which may be hollow tubes, are attached to the movable platen and main clamping cylinder. At the end of the stroke of the rapid traverse cylinder, a locking mechanism, hydraulically operated, is inserted between the spacers. The large-diameter short-stroke hydraulic cylinder moves forward approximately an inch to provide the full locking force. This type of lock and block mechanism requires a mold height adjustment. Systems of this design, while losing a slight speed advantage because of the three motions, gain in economy because they have a smaller hydraulic cylinder size, lower power requirements, and no need for huge toggle links.

A disadvantage of this system is its long, unadjustable stroke, with a corresponding waste of floor space and long stressed tie bars. This is overcome in the design shown in Fig. 5-17. The rapid traverse cylinder moves the moving tail stock platen and clamping plate until the mold closes. The half nuts are closed over the tie bars, anchoring the moving tail stock platen. The large-diameter, limited-stroke main cylinder extends, providing the clamping force. To open the mold, the clamping cylinder is retracted, the half nuts are extended, and the rapid traverse cylinder is retracted. The tie bars are stressed only between the point where the half nuts are locked and the tie bar nuts behind the stationary platen. The advantages of this design are the adjustable stroke, minimum stress on the tie bars, increased rigidity, and smaller floor space.

Other Specifications. Aside from the type of clamp and the clamping force, there are a number of other important specifications. The clamp stroke is the maximum distance the moving platen will move. The maximum daylight is the maximum distance between the stationary platen and the fully retracted moving platen. These two specifications will determine the height of the mold that can fit in the machine, and, with the size of the part, will determine if there is enough opening to eject the part and remove it from the machine.

The clearance between the tie rods is the de-

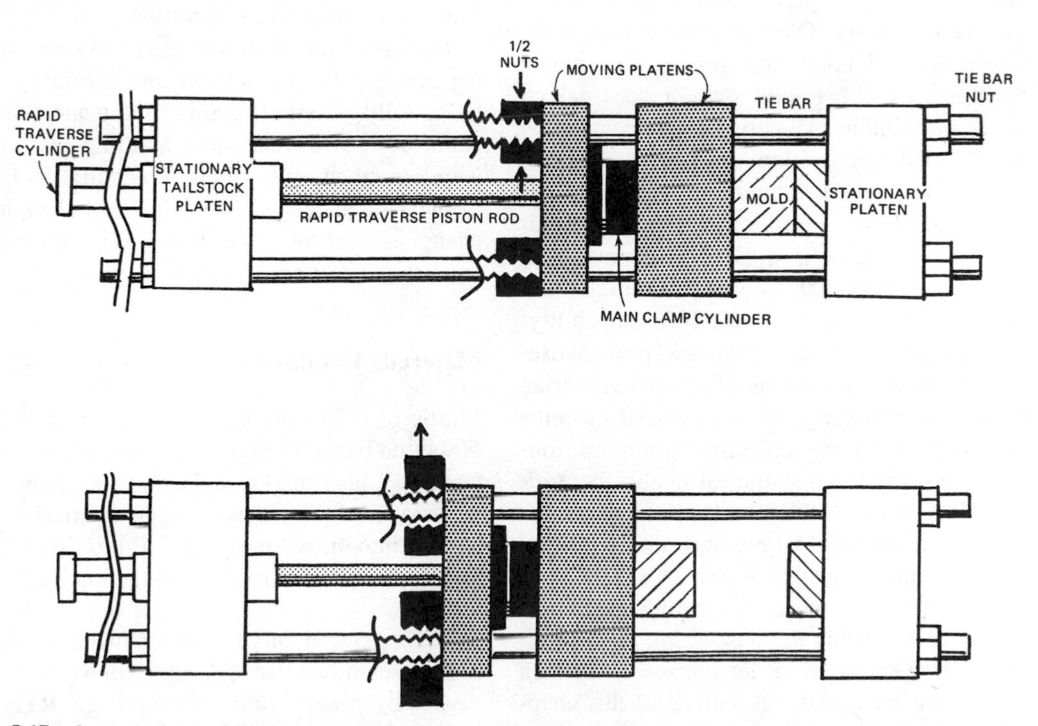

Fig. 5-17. Long-stroke, rapid-acting, clamping system, using half nuts that anchor on the tie bars to lock the moving platens. (*Courtesy Robinson Plastics Corp.*)

termining factor for whether a mold of a given length or width will fit into the press. The length and width dimensions of the molds are often determined by the side that is parallel to the knockout plate.

The clamp speed is an important specification. Losing a half second per shot on a machine producing 100 shots per hour will waste 110 productive hours a year.

The knockout stroke determines the maximum knockout movement available. There are two types of knockout system—a mechanical one that stops the knockout plate before the mold stops and a hydraulic one that is independent of the machine action. The hydraulic knockout system is preferable.

Machine knockout patterns all follow the SPI standard so that molds are interchangeable among machines. The other specifications are determined by the manufacturer to achieve the desired results.

Circuits

Space does not permit discussing electrical and hydraulic circuits. Older machines used electro-mechanical relays and timers. The newer machines are using solid state devices, which are more reliable. Monitoring of the injection speed, melt temperature, pressure, and pressure of the material in the mold forms the basis for the automatic control of the injection molding process. This is done in computer-controlled machines *with feedback*. There are machines with computers that lack this capability; but they, like those with feedback, provide useful information for the molder, such as setting molding conditions, keeping track of machine performance, giving real-time warning of troubles, and providing statistical quality control. Machines with feedback have increased in efficiency to the point where they are rapidly replacing nonfeedback machines. This is the third "revolution" in injection molding machines, the other two being preplasticizing and the reciprocating screw. (For details, see the section on "Process Control" at the end of this chapter.)

Automation

The expression "automatic machines" often is used in referring to automatic molding. This is a misnomer, as all machines today are automatic. The basis for automatic molding is the mold, and there are a number of requirements for the process. The machine must be capable of constant repetitive action, the mold must clear itself automatically, and low-pressure closing must be available so that any stuck parts will not damage the mold.

Automatic molding does not necessarily eliminate the operator. Many times an operator is present to package the parts and perform secondary operations. Automatic molding gives a better-quality piece at a more rapid cycle. Usually when automatic molding machines are used, an experienced person attends several machines. Unless the powder, feed, and part removal are automated, he or she will take care of them.

Automation is expensive to obtain. It requires excellent machinery, molds, trained employees, and managerial skills. When the quantity of parts permits, it is a satisfactory and economical method of operation.

The use of robots in both part and sprue/runner removal from machines and assembly has led to fully automatic plants. In the late 1980s, about 80% of the Japanese molding machines were equipped with robots, compared to 12% in the United States. This situation is rapidly changing, and the concept of total automation has begun to be evident in U.S. plants.

Materials Handling

Plastic materials are shipped in 25-pound tins, 50-pound bags, 300-pound drums, 1000-pound Gaylords, tank trucks, and tank cars. In the latter two instances, they are pneumatically removed into silos for storage. From there they are conveyed automatically to the molding machine.

The selection of the proper materials handling technique can greatly reduce wastage caused by contamination and spillage. Regardless of the size of the plant, the materials handling procedure should be constantly reviewed and updated. The subject is covered in Chapter 21.

Sprues, runners, and rejected parts can be reground and reused. A typical grinder consists of an electric motor that turns a shaft on which there are blades that cut the plastic. The reground material usually is blended with virgin material, a step that can be done automatically.

Certain materials such as nylon, polycarbonate, and cellulose acetate are hygroscopic, or moisture-absorbing. They must be dried by heating, as moisture causes surface defects and seriously degrades the material. Drying can be done in ovens that are thermostatically controlled and circulate heated air. The material is spread out in trays for drying. Alternatively, hopper driers fit on top of the molding machine and send filtered, heated, dried air through the material to dry it.

A third method is to use a vented extruder barrel. In this method the material is compressed, melted, and then decompressed under a vent. The material is at about 400°F, well above the boiling point of water. Thc steam usually escapes by itself; if not, a vacuum is used. The material then is recompressed and pumped in front of the screw.

Secondary Operations

Injection-molded parts are fabricated and decorated with the same techniques used on plastics manufactured by other methods. Milling off gates, drilling, and routing are routinely done on molded parts. Frequently the cycle time allows these operations to be done at the press. Cementing, ultrasonic and vibratory welding, riveting, eyeletting, tapping, and adding metallic threaded inserts and nuts and bolts are common methods of joining plastic to itself or to other materials.

Injection-molded parts can be decorated by all commercial processes, such as painting, silk screening, hot stamping, electrolytic plating, and vacuum metallizing. Figure 5-18 shows some vacuum metallized lamp parts that have completely eliminated brass for this application.

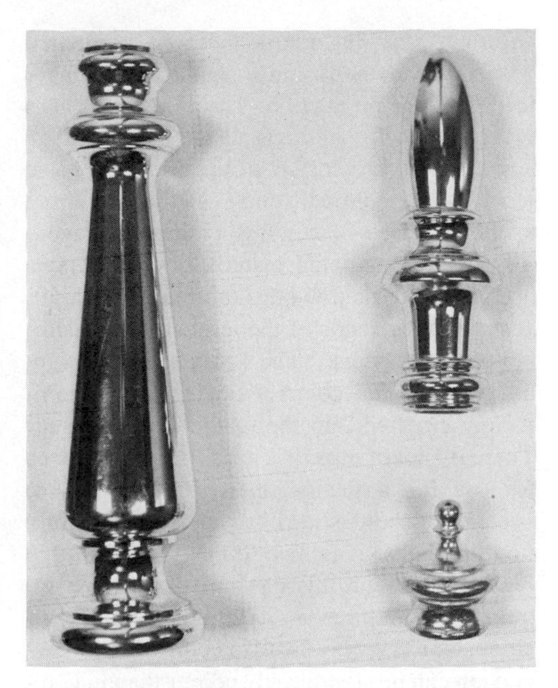

Fig. 5-18. Vacuum metallized plastics lamp parts. (*Courtesy Robinson Plastics Corp.*)

The full scope of these operations is covered in Chapters 23, 24, and 26.

Safety

The Occupational Safety Health Act of 1970 (OSHA) declared the congressional purpose to "assure as far as possible every working man and woman in the nation safe and healthful working conditions and to preserve our human resources." It is the employer's duty to furnish his or her employees a place of employment free from recognized hazards causing or likely to cause death or serious physical harm. The SPI has many resources to help injection molders comply with this law.

Coloring

There are three methods of obtaining colored material—compounded-in color, dry coloring, and liquid color concentrates.

Compounded-in color has the advantage of providing the best dispersion and the ability to match colors. It is the most costly method of coloring, requiring a large inventory compared with other methods and a significantly longer lead time for production. The base material is reextruded with colorants.

Dry coloring consists of tumbling the basic resin with a colorant, and sometimes wetting agents are used. It requires a minimum inventory of the uncolored material. Dry colorants usually are packed in bags for mixing with 50, 100, or 200 pounds of the resin. It has a number of disadvantages: The wrong colorant might be used, or the colorant might not be completely emptied into the tumbling equipment. There is a lot of materials handling and a chance for loss. It is basically a messy procedure, and, because of the dusting, may cause contamination problems. It is more difficult to clean the feed system with this method. Colors that require several pigments, particularly in small amounts, are very difficult to color evenly. Dispersion can be significantly poorer than in compounded material, and in many instances is unacceptable.

A compromise between compounded-in material and dry coloring is to use color concentrates or masterbatches. High concentrations of colorant are extruded into the same resin from which the parts will be molded. The concentrate usually is prepared so that 5 pounds of concentrate will color 100 pounds of base resin. Although this method is more expensive than dry coloring, it overcomes a number of that method's disadvantages. Colors difficult to match in dry coloring usually will work with color concentrates. Compared to compounded-in materials, these colorants usually are less expensive, require lower material inventory, and occupy less space.

Also, liquid color concentrates can be used with automatic material handling systems, which meter in liquid colorant during feeding. This is an economical means of coloring.

Coloring methods are covered in more detail in Chapter 22.

Multicylinder Machines

Machines are built with two or more cylinders to mold two different colors of the same material or two different materials. (Buttons on telephones, calculators, and computers are good examples of the applications of the process.) The mold design is different from the single-cylinder type in that two-shot molding requires a tool with two identical cores and two different cavities. The core side of the mold is able to rotate. To mold a two-shot button, for example, one core/cavity would mold the base in the primary color or material from the first cylinder, leaving an air space for the letter or symbol desired. The mold would rotate 180°, and the second cavity would seal off the first molding, which has an air space for the letter or symbol. The second color or material would be molded into it from the second cylinder. While this is happening, the first set would be molding the first color.

Two cylinders also can be used for sandwich molding, to combine functional materials (in the core) and "appearance" materials for the surface.

THEORY OF INJECTION MOLDING

This section will develop a qualititative theory of injection molding based on rheological data, concepts of energy levels, molecular structure and forces, and theories of heat transfer. This theory can be of great value in preventing problems, solving difficulties, and providing an understanding of the literature. Of necessity, coverage of this kind will contain certain generalizations, which, while not mathematically exact, adequately serve the purpose.

The reasons why a quantitative statement of injection molding is impossible become clear when one analyzes the machinery, materials, and process. The factors that affect the process—material temperature, temperature profile, pressure, material velocities in the cylinder and mold, mold temperature, viscosity, and flow patterns—are not measured at all or are measured intermittently at isolated points. Until recent years there was no feedback of the processing variables to the molding machine to compensate automatically for changing conditions.

The material used is not identical each time, having different heat histories, molecular weights, molecular weight distributions, degrees of polymerization, and impurities. It is exposed to moist air and compressed with it in the heating cylinder, with varying amounts of oxidation. It is heated by convection, conduction, and shearing. The pressure changes, from zero to possibly 30,000 psi.

The material is not linear in most of its functions. It is compressible, stretchable, elastic, and subject to changing properties after removal from the mold. It may differ according to the direction of the material flow in the mold. It has time-dependent properties that are strongly altered by its environment. In some materials, there are varying degrees of crystallinity, which are not reproducible or predictable. For these reasons, quantitative solutions are not yet possible. This is equally true of computer-generated information.

Material exists in three forms—solid, liquid, and gas. In plastics, one is concerned primarily with the liquid and solid phases. In crystalline materials, the change from solid to liquid is abrupt and easily discernible. In an amorphous (noncrystalline) polymer, the change is not abrupt or readily apparent; the material softens over a wide temperature range, and there is no dramatic visible change in its properties at any given point, such as would be found in the conversion of ice to water or water to ice. If certain properties of an amorphous plastic such as the specific volume or heat capacity are plotted against the temperature, at a certain point there is an abrupt change of the slope of the line. In the thermodynamic sense, this is called a second-order transition; in polymer science, it is called the glass transition point (T_g). Below the glass transition point the polymer is stiff and dimensionally stable, behaving like a solid. It has brittle characteristics with little elasticity, and its properties are relatively time-independent. Above T_g the polymer will behave as a viscous liquid, being elastomeric and having highly time-dependent properties.

The difference between the three forms of matter can be explained in terms of molecular attraction. In a solid, the closeness of the molecules to each other permits the strong cohesive force of molecular attraction to limit their motion relative to each other. Although solids can be deformed, it takes a comparatively large amount of energy to do so. If the solid is stressed below its elastic limit, it will be deformed. An ideal solid obeys Hooke's law, which states that the amount of strain (movement) is directly proportional to the stress (force). The constant of proportionality, E (stress/strain), is called Young's modulus or the modulus of elasticity. When the stress is removed, the molecular bonds that have been stretched contract, bringing the solid back to its original position. Plastics are not ideal solids; they exhibit, in common with all other materials, elastic properties that are combined with viscous or flow properties. Hence, they are called viscoelastic materials.

When a stress is applied to a polymer above its T_g, the initial movement is elastic, of a Hookean type. It represents stretching the chemical bonds. As more stress is applied, the plastic molecules slide over each other. This is viscous flow, which is not recoverable. There is a third type of flow, which is a retarded type of elasticity. When the stress is removed, the initial elastic deformation is instantaneously recovered, a response that is Hookean in nature. The retarded elasticity eventually will recover, but not instantaneously; the length of time will depend on the nature of the material.

There are two energy systems within a material. One is the potential energy of Newtonian gravitational energy, and is a measure of the forces between the molecules. The other is kinetic energy, which is the energy of motion and is related to the thermal or heat energy of the system; it is the random motion of the molecules, called Brownian movement. The higher the internal (heat) energy of the system is, the greater the random motion.

In a solid, the potential energy (forces of attraction between the molecules) is larger than the kinetic energy (energy of movement tending to separate the molecules). Hence, a solid has an ordered structure, with a molecular attraction strong enough to limit the molecules' motion relative to each other.

As more energy is put into the system, it turns into a liquid where the potential and the kinetic energy are equal. The molecules can move relative to each other, but the cohesive forces are large enough to maintain a molecularly contiguous medium.

It is appropriate here to briefly review the nature of the bonding forces in the polymer. The two major bonding forces in plastics are the covalent bond and the secondary forces, called van der Waals forces. The atom consists of a relatively small nucleus that contains most of the atom's mass and the positive charges, and negative electrons that spin around the nucleus. The orbits of the electrons form outer concentric shells, each of which has certain stable configurations. A covalent bond exists when the electron in the outer shells of two atoms is shared between them. This is typical of carbon—carbon bonds and is the primary bond found in polymers used commercially. It has a disassociation energy of 83 kcal/mole.

The van der Waals forces are electrostatic in nature and have a disassociation energy of 2 to 5 kcal/mole. These forces exist between molecules and molecular segments, and they vary as the sixth power of the distance. Obviously much weaker than the primary covalent bond, they are part of the resistance to flow. The energy that attracts molecules, as contrasted to the bonds holding atoms together, is sometimes called cohesive energy, and it is that energy required to move a molecule a large distance from its neighbor.

If one takes a cubic inch of plastic and raises its temperature, its volume will increase. As no molecules are being added to the cube, it is reasonable to believe that the distance between the molecules increases. Because the van der Waals forces decrease with the sixth power of the distance, the molecules and their segments become much more mobile. As these forces are decreasing in an exponential manner, there is a relatively narrow range in which the polymer properties change from ''solid'' to a ''liquid''—the glass transition point (T_g). These cohesive forces form a major portion of the strength of the polymer; so polymer properties are quite temperature-dependent.

Having a physical concept of a plastic molecule makes it easier to understand its flow properties and characteristics. For example, let us consider the polyethylene polymer, which is made by linking ethylene molecules. The double bond between the carbons in ethylene is less stable than the single bond in polyethylene. It is possible, with appropriate temperatures, pressures, and catalysts, to cause the ethylene molecules to react with each to form polyethylene:

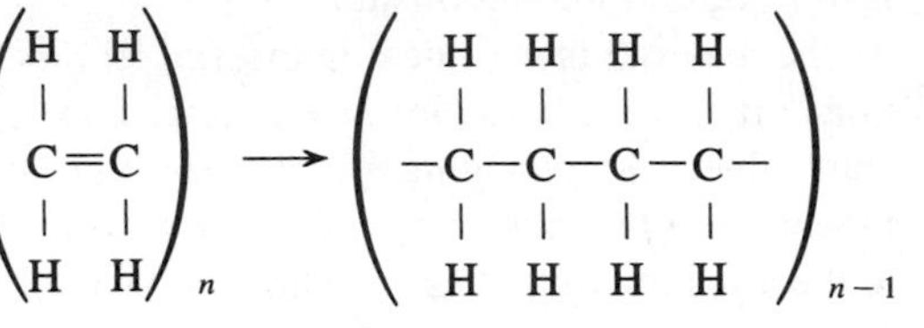

$$\left(\begin{array}{cc} H & H \\ | & | \\ C & = C \\ | & | \\ H & H \end{array}\right)_n \longrightarrow \left(\begin{array}{c} \begin{array}{cccc} H & H & H & H \\ | & | & | & | \\ -C- & C- & C- & C- \\ | & | & | & | \\ H & H & H & H \end{array} \end{array}\right)_{n-1}$$

Ethylene Polyethylene

The molecular weight of an ethylene molecule is 32. A typical Ziegler-type polyethylene polymer might have 7000 ethylene molecules with a molecular weight of approximately 200,000. The polymerization does not proceed as simply as indicated above. The polymer molecules will not all be the same length, nor will they polymerize in a straight line of carbon linkages. Also, references to the molecular weight of polymers mean the average molecular weight. The molecular weight distribution is an important characteristic of the polymer. If it is spread over a wide range, its properties will differ from those of a polymer with a narrow distribution. For example, a wide-spectrum material shows more elastic effects and extreme pressure sensitivity. The viscosity, solubility, and stress crack resistance are among the other properties affected.

To get some idea of the size of a polyethylene molecule, consider that if the methyl group, CH_2, were 0.25 inch in diameter, a typical polyethylene molecule would be one city block long. A molecule of water would be about the size of the methyl group. In view of the possibilities of entanglement, kinking, and partial crystallization of the huge polyethylene molecule, compared to the small size and simplicity of the water molecule., it would not be unexpected to find considerable differences in flow properties between these molecules. The flow of water is relatively simple (Newtonian), whereas the viscoelastic flow of polymers is much more complicated and has not yet been mathematically defined in a quantitative manner.

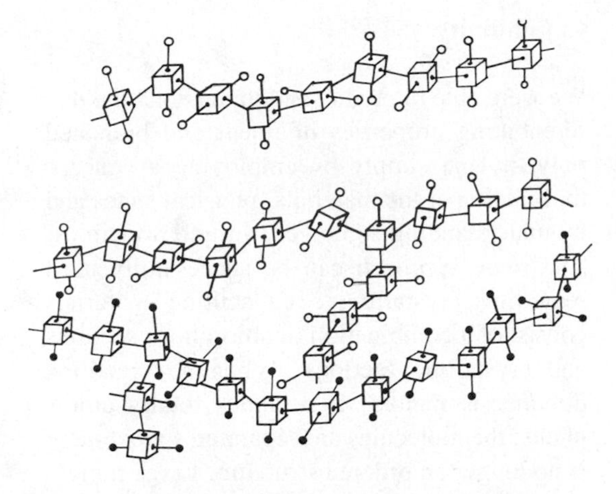

Fig. 5-19. Schematic representation of molecular structures of linear polyethylene (top) and branched polyethylene (bottom). Branching keeps the chains farther apart, reducing density, rigidity, and tensile strength. (*Courtesy Robinson Plastics Corp.*)

When ethylene molecules polymerize, theoretically they could do so to produce a straight line of carbon linkages, as shown in the top portion of Fig. 5-19. This would be a linear type of material. What happens when polymerization takes place is that the carbon atoms attach to each other in nonlinear fashion, branching out to form chains, as shown in the bottom section of Fig. 5-19. The amount of branching depends on the method of manufacture. High-pressure processes produce more branching than the low-pressure systems. To further understand the nature of the polymer molecule, it should be noted that the carbon atoms are free to rotate around their bonds and can bend at angles less than 180°. The effect of this swiveling and twisting is that the molecules and segments of the molecules become entangled with each other. The cohesive forces between the molecules consist of van der Waals type attraction. The other type of force in the polymer is in the carbon–carbon (C–C) linkages.

Properties of Linear and Branched Materials

With these simple concepts of molecule structure, it should be possible to predict the different properties of linear and branched materials:

Density. The linear structure of the polymer should permit the polymer molecules to come closer to each other, allowing more molecules to pack into a given volume (i.e., to be more dense). Obviously, the branching of molecules prevents this, so that they are less dense.

Yield. The higher the density is, the fewer the pieces of molded parts per pound of polyethylene that can be produced. This is not an unimportant consideration in material selection.

Permeability to Gas and Solvents. The branched materials physically create larger voids in the polymer, so that it is more permeable to gases and solvents than the linear or high-density material.

Tensile Strength. The linear materials, having molecules that are closer together, should have stronger intermolecular (van der Waals) forces than the nonlinear materials. Tensile strength is a measure of the strength of these molecular forces, and it is higher in linear materials than in branched. For example, a .96 resin has a typical tensile strength of 4300 psi and a .915 resin a strength of 1400 psi.

Percent Elongation to Failure. Because the linear molecules can entwine and kink more than the branched molecules, one would expect it to be more difficult to separate the linear molecules, so that applying a strong tensile force would rupture the molecule rather than cause it to flow and elongate. Branched material, having lower intermolecular strength because the molecules are farther apart and the van der Waals forces are weaker, would be expected to slide considerably more before rupturing. This is the case.

Stiffness. Linear polyethylene molecules, being closer together than the branched ones, have less room for segmental motion of the chains and bending of the backbone. Therefore, the linear materials are stiffer.

Heat Distortion. The heat deflection temperature under load is that temperature at which a specimen bar, under given conditions of loading to produce an outer fiber stress of 66 or 264 psi, will deflect 0.010. At a given temperature the molecular forces of attraction of a high-density material are greater than those of a low-density material because the segments are closer together. Therefore, it will take a given amount of heat energy to separate the linear molecules to such a distance that their attractive strength will be equal to that of the branched. The extra heat required means, in effect, that the linear material can absorb more heat for a given stiffness, so that its heat distortion temperature is higher. Typical heat distortion temperatures are: for a low-density polyethylene, 100°F; for a medium-density polyethene, 130°F; and for a high-density polyethylene, 160°F.

Softening Temperature. Similarly, the softening temperature of the high-density material is higher than that of the low-density material.

Hardness. Because linear molecules are closer together, the linear material should be harder. This is the case.

Resistance to Creep. Creep is the amount of flow (strain) caused by a given force (stress). As one would expect, the higher intermolecular force of the linear material make it more resistant to strain.

Flowability. Because of the stronger molecular attraction of the linear material, it should be more resistant to flow than the branched materials.

Compressibility. Because there is more open space in branched material, it should compress more easily than linear material.

Impact Strength. Because linear material has greater molecular attractive forces than the branched material, one normally would expect the linear material to have a higher impact strength, but this is not the case. Polyethylene crystallizes, and it is well known that impact forces travel along the interfaces of a crystalline structure, propagating breaks rapidly. Because the molecules in the linear material are closer together, they will crystallize more readily than those in the branched material. The higher crystallinity of the linear material is the reason for its lowered impact strength.

Thus, increasing the density of polyethylene increases its tensile strength, percent elongation to failure, stiffness, heat distortion temperature, softening temperature, and hardness; but increasing the density decreases the material's yield, permeability to gases and solvents, creep, flowability, compressibility, and impact strength.

Crystallinity

We were able to predict and understand the differentiating properties of linear and branched polyethylene simply by employing a conceptual catalog of the materials' physical states and by understanding some very simple principles. The same approach can be successfully used regarding crystallinity. Crystalline materials consist of a combination of amorphous sections and crystalline sections. When a crystalline polymer is melted; it becomes totally amorphous; the molecules are separated so that there is no longer an ordered structure. Large molecular segments vibrate and rotate to give a totally disordered structure. When the plastic cools, a point is reached where the forces of attraction are strong enough to prevent this free movement and lock part of the polymer into an ordered position. The segments now can rotate and oscillate only in a small fixed location. In an amorphous polymer the molecular configuration is the same throughout; the intermolecular distances are about the same and are controlled by the temperature. In a crystalline material the molecules are in an ordered structure, which takes up much less space than the amorphous state. Crystallization is indicated by a sharp decrease in volume; thus crystalline polymers show greater shrinkage than amorphous ones. Because the amount of crystallization varies with the material and molding conditions, it is much more difficult to hold tolerances in crystalline materials than in amorphous ones.

The molecular segments are much closer together in ordered crystalline lattices than in amorphous materials. To achieve a change in state, energy is required. For example, if ice and water are heated, the temperature remains the same, 32°F, until all the ice melts—the heat energy is being used to break up the crystalline structure. This also happens when polymer crystals break up into an amorphous condition. The fact that a crystalline structure has the molecules closer together with a corresponding increase in intermolecular forces, compared to the amorphous state, explains the properties of crystalline material.

As crystallinity results in a more compact

structure, the density increases with crystallinity. The flexibility of a plastic depends on the ability of its segments to rotate; thus, crystalline structures, which inhibit rotation, are stiffer than amorphous structures. Because a crystalline structure has its molecules closer together, the tensile strength increases with crystallinity. However, the impact strength decreases with crystallinity, primarily because of the propagation of faults along the crystalline structure. Shrinkage will increase with crystallinity because crystals take up less space. The heat properties will be improved with crystallinity because the crystalline material must absorb a significant amount of heat energy before the structure is analogous to an amorphous material. Increasing the crystallinity brings the molecules closer together and increases the resistance to permeability of gases and vapors. Increasing crystallinity lowers the resistance to stress cracking, probably following the same mechanism as the lowering of impact strength. Crystalline materials warp more than amorphous ones, probably because the different densities within the same material set up internal stresses.

Flow Properties

We have been considering the static properties of polymers. An understanding of flow properties is essential in polymer processing, and they too are amenable to simple analysis.

Two investigators, Hagan and Poiseuille, independently derived the volumetric flow rate for a Newtonian liquid through a tube:

$$Q = \frac{\pi \times R^4 \times \Delta P}{8L \times \mu} \tag{2}$$

where:

Q = Volumetric flow rate
R = Radius of tube
L = Length of tube
ΔP = Pressure drop
μ = Viscosity

Inspection shows that the volumetric flow rate depends on three things. The first is the physical constants of the tube, $\pi R^4/8L$. A Newtonian liquid is extremely sensitive to the radius of the tube, but this effect is less important with plastic, as we shall shortly see, where viscosity varies with the shear rate (velocity). Second, the greater the pressure is, the higher the flow rate. Finally, the more viscous the material is, the lower the flow rate.

To maintain the same volume of material in the cavity shot after shot after shot, it is necessary to maintain the same pressure and viscosity conditions; this is the basis for automation. In plastics (viscoelastic materials) the viscosity is both temperature- and speed-dependent; so the speed of the plunger also must be controlled. This need for constant conditions is one of the main reasons why the use of a computer-controlled machine with feedback is extremely productive.

In rheology (the study of flow), the word stress is not used in the sense of a force acting on a body; instead it is a measure of the internal resistance of a body to an applied force. This resistance is the result of the attraction of molecular bonds and forces. When we say that we increase the shear stress to increase the shear rate, we really mean that we have to overcome increasing molecular resistance to achieve a faster flow rate. Shear stress is a measure of the resistance to flow of molecules sliding over each other; it is reported in pounds per square inch (psi). Force, which is measured in the same unit, is different from shear stress in two respects. Force acts perpendicular to the body, whereas the shear stress acts parallel to the containing surface. Pressure is force per unit area, while shear stress is resistance to force. Newton developed this concept of viscosity, using concentric cylinders. It can be explained more easily by imagining a stationary plate over which, at a distance X, there is a movable plate with an area A, moving at a velocity U, pushed by a force F. Neglecting the slip of the molecules on the stationary plate, we assume the velocity of the liquid at the stationary plate to be zero and the maximum velocity U to occur at the moving plate. The rate of change of velocity is the slope of the line connecting the velocity vectors, or du/dx. The force is therefore proportional to the area and velocity:

$$F = f \times A\,(du/dx) \qquad (3)$$

The proportionality constant (f) is called the viscosity and designated μ (mu) for Newtonian liquids or η (eta) for non-Newtonian liquids. Shear force or stress is represented by the Greek letter τ (tau) and shear rate by γ (gamma). Rearranging the terms, we have the following classical definition of viscosity:

$$\text{Viscosity} = \frac{F/A\ \text{(shear stress)}}{du/dx\ \text{(shear rate)}} \quad \text{or}$$

$$\mu = \frac{\tau}{\gamma} \qquad (4)$$

In a Newtonian liquid the shear force is directly proportional to the shear rate; doubling the unit force doubles the unit rate. In thermoplastic materials this is not the case; in the processing range a unit increase in the shear force varies and may even quadruple the shear rate. The viscosity is dependent on the shear rate and drops exponentially with increasing shear rate.

This is shown in Fig. 5-20, which has arithmetic plots of the viscosity and shear rate and of the shear stress and shear rate. It can be noticed that in the latter graph, there is a Newtonian portion at the beginning and at the end of the curve. It is more practical to plot such data on log–log plots, which will characterize the flow properties of a material. The information is obtained by using rheometers. The most common one extrudes the polymer through a capillary tube while measuring the force and speed of the plunger; viscosity then is a simple calculation. Such an instrument is called a capillary rheometer. In looking at the viscosity/shear rate curves of plastics, it is obvious that to maintain the same viscosity one must maintain the same speed as well as the same temperature.

These flow properties can be easily understood with a conceptual illustration of how the molecules move. Figure 5-21 shows a representation of a number of different polymer molecules of the same kind. They are in a random pattern, their vibrations or movement being determined by their heat energy. This Brownian movement, named after its postulator, tends to locate the polymer segments in random positions, this being the lowest energy level. The plastic molecule is too large to move as a unit. Brownian motion occurs in segmental units of the polymer.

If a force is applied in one direction to a polymer above its glass transition point, it will begin to move in a direction away from the force. As it starts to move, the carbon–carbon chains of the molecule will tend to orient themselves in the direction of flow (Fig. 5-22). If the force is applied very slowly, so that the Brownian motion overcomes the orientation caused by flow, the mass of the polymer will move with a rate proportional to the applied stress. This is Newtonian flow and is the corresponding straight section at the beginning of the left curve in Fig. 5-22.

As the flow rate increases, two things happen. The chains move so rapidly that there is

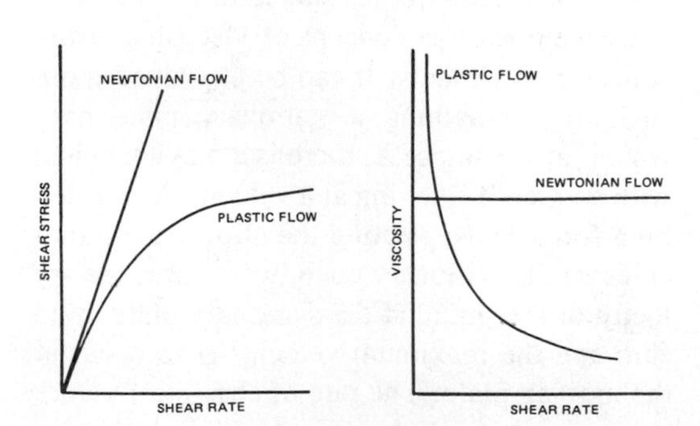

Fig. 5-20. Arithmetic plots of shear stress vs. shear rate and viscosity vs. shear rate for Newtonian and plastics materials. (*Courtesy Robinson Plastics Corp.*)

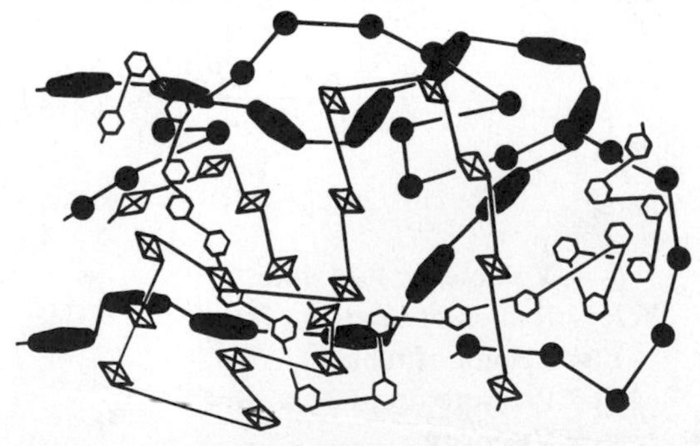

Fig. 5-21. Schematic representation of segments of polymer chains in their random position. This is a result of local vibration, thermally controlled, called Brownian movement. (*Courtesy Robinson Plastics Corp.*)

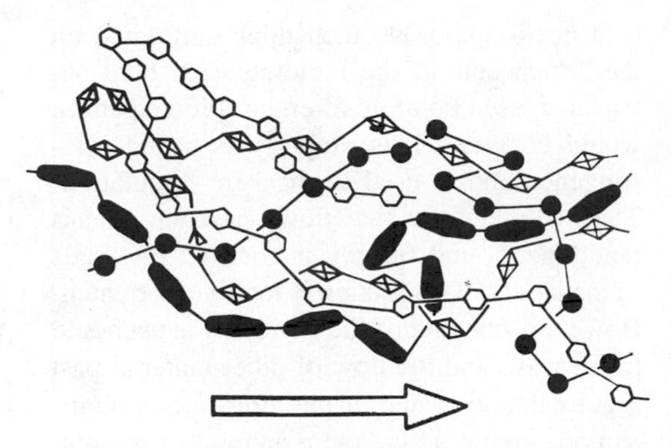

Fig. 5-22. Schematic representation showing the effect of a stress on the random structure of Fig. 5-21. Molecular segments tend to orient in the direction of flow. (*Courtesy Robinson Plastics Corp.*)

not sufficient time for the Brownian motion to have an appreciable effect; also, the molecules that are being oriented in one direction separate because their side chains untangle. This separation reduces the intermolecular forces exponentially (because they are van der Waals type attractions), permitting them to slide over each other much more easily. In other words, the increased shear rate (speed) is no longer proportional to the shear stress (force). A unit increase in shear stress will give a much larger increase in shear rate than would happen with a Newtonian liquid, where a unit increase in stress would give a unit increase in shear. The unit force, then, does two things: it accelerates the mass and separates the molecular segments. The proportion of each changes exponentially—this is the central portion of the shear stress/shear rate curve in Fig. 5-20, and is a characteristic of plastic polymer flow, where shear rate is no longer linearly proportional to shear stress. It is this central portion of the curve that is met in injection molding. As the flow rate increases, it reaches a final stage where all the polymer molecules have become oriented to their maximum level; there is no further untangling. Therefore, any increase in the shearing stress in this range will give a proportional increase in the shearing rate, and the material acts as a Newtonian fluid, which is indicated in the top portion of the curve in Fig. 5-20.

The basic difference between a Newtonian and a non-Newtonian fluid is in the length of the molecule. Newtonian liquids such as water, toluol, and glycerine all have very short molecular lengths. As is evident from the previous discussion, the rheological or flow properties of a polymer depend in some measure on its molecular structure. It is also evident that the flow properties are highly temperature-dependent, as the temperature is an indication of molecular motion and intermolecular distances. The relationship is exponential, and a plot of the log of the viscosity versus the temperature at a given shear rate is a straight line over narrow temperature ranges, describing viscous flow fairly accurately (Fig. 5-23). This type of information also has practical value. For example, if cavities were not filling out during the molding of cellulose acetate, increasing the temperature would not have a great effect on the viscosity or hence the filling. It would be necessary either to increase the pressure or to open the gate. On the contrary, acrylic material is very temperature-sensitive, and raising the temperature even a small amount will result in a considerable decrease in viscosity. This also means that temperature control of the material is more important for a viscosity/temperature-sensitive material such as acrylic than for cellulose acetate or polystyrene.

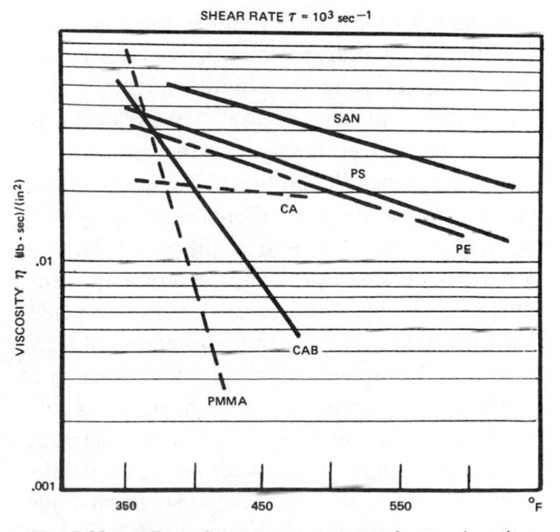

Fig. 5-23. Effect of temperature upon polymer viscosity. SAN—styrene-acrylonitrile; PS—polystyrene; PE—polyethylene; CA—cellulose acetate; CAB—cellulose acetate butyrate; PMMA—polymethyl-methacrylate (acrylic). (*Courtesy Robinson Plastics Corp.*)

From this discussion, the controlling parameters for consistent injection molding become evident. Equation (2) states that the viscosity and the pressure control the output in a given geometric system. We also have seen that the viscosity in thermoplastic materials depends on two conditions—temperature and rate of flow. The temperature of the material is controlled by the conditions of the heating cylinder, nozzle, and mold. Therefore, in order to have consistent molding, the temperature, the pressure, and the rate of flow must be controlled. It is not possible to have successful automatic molding or consistent molding without controlling the flow rate. When the machine is instrumented to make measurements, compare the measurements to a preset standard, and change the conditions during the molding to meet the standard, it is possible to have consistent automatic molding. That is what a computer-controlled machine with feedback provides.

Orientation

Orientation effects are very important. The term orientation means the alignment of the molecule and molecular segments in the direction of flow. The strength in that direction is that of the carbon–carbon linkage, whose disassociation energy of 83 kcal/mole is much greater than the 2 to 5 kcal/mole of the van der Waals type forces holding the polymer together perpendicular to the line of flow. Thus plastic that is oriented will be stronger in the direction of flow than perpendicular to it. The ratio will not be 83/5, as no material orients completely; but the greater the orientation is, the closer the material gets to this ratio. The second major implication of this concept is that the oriented plastic will shrink more in the direction of flow than it will perpendicular to it. Shrinkage is a result of two factors—a normal decrease in volume due to temperature change and relaxation of the stretching caused by carbon–carbon linkages. As there are more carbon–carbon linkages in the direction of the oriented flow than perpendicular to it, this phenomenon occurs.

Plastics do not all exhibit orientation to the same degree. Consider molding a rectangular plaque of clear polystyrene 2 inches wide and 6 inches long, 0.090 inch thick and gated on the 2-inch end. If the molding were held between crossed Polaroid filters, a colored pattern would be seen. This property is called birefringence and is used to measure orientation. The material front that flows past the gate is randomized, and freezes as such on the walls of the cavity. This section is totally unoriented. However, one end of the molecule is anchored to the wall, and the flow of other material past it pulls the other end of the molecule in its direction, giving a maximum amount of orientation. As the part cools, the orientation is frozen at the walls. The center of the section remains warm for the longest time, allowing Brownian motion to disorient many of its molecules. Therefore, the center section is the least oriented. This is shown by birefringence patterns.

This behavior can be easily demonstrated by milling off one-third (0.030 inch) of the thickness. In effect, one section is highly oriented, and the center section, which has been exposed by the milling, is less oriented. If the milled piece is heated, the stretched carbon–carbon linkages should return to their normal position. Because the oriented section has the carbon–carbon linkages lined up more in one direction than they are in the less oriented sections, that part should shrink more. In effect, then, it would be acting as a bimetallic unit, one side shrinking more than the other, and the piece should bend over. This is what happens.

As the amount of orientation depends on the flow and on the forces that aid or prevent the motion of the molecular segments, it is easy to see what conditions can affect orientation. Anything that increases the mobility of the segments decreases orientation. Therefore, higher material temperatures, higher mold temperatures, and slower cooling would decrease orientation. Pressure on the material would limit mobility. Thus, low injection pressures and a short ram forward time decrease orientation. The use of a thicker part would decrease orientation because a longer time is needed for the center portion to cool with increasing thickness. We shall now examine some practical situations involving orientation.

Practical Applications. Consider molding a lid or cover 6 inches in diameter in a polyolefin

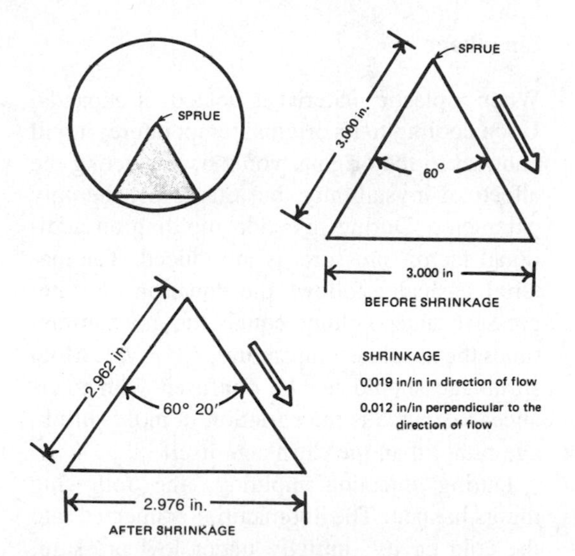

Fig. 5-24. Warping of center-gated polypropylene cover caused by the different shrinkages perpendicular and parallel to the direction of flow. (*Courtesy Robinson Plastics Corp.*)

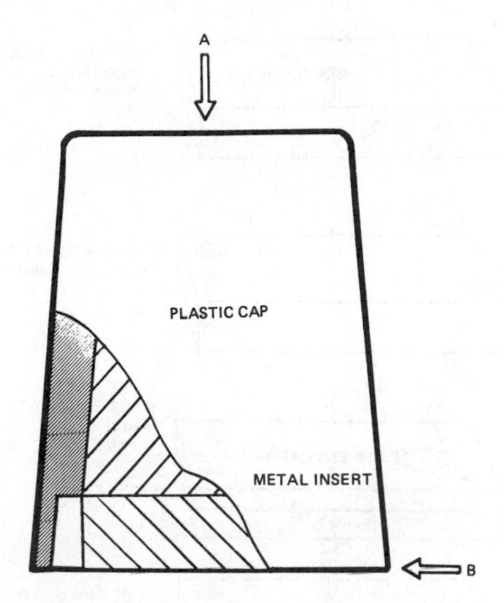

Fig. 5-25. Effect of orientation on a plastic cap molded with a metal insert. Gating at point A will give a cap strength along the walls. Gating at point B will give a cap strength in the hoop direction. (*Courtesy Robinson Plastics Corp.*)

(Fig. 5-24). The shrinkage in the direction of flow is 0.019 in./in., while the shrinkage perpendicular to flow is 0.012 in./in. The difference is caused by the different numbers of carbon–carbon linkages in the direction of and perpendicular to the flow.

Consider a 60° segment of the cover immediately upon molding. Each side will be 3.000 inches long. Upon cooling, the two sides in the direction of flow will have shrunk to 2.962 inches, and the segments perpendicular to flow will now be 2.976 inches. A simple trigonometric calculation shows the central angle is now 60°28′. The full 360° circle is now 362°48′. Obviously the extra material has to go somewhere. If it cannot lie in a flat plane, it will warp. If the thickness of the material and the ribbing provided enough strength, the part might not visibly warp, but it would be highly stressed. The way to minimize such warp or stress is to mold under those conditions that give the least orientation. Multiple gating also is effective, as is redesigning the cover.

Gate location affects the amount and the direction of orientation. Figure 5-25 shows a cap with a metal insert that was used as a protective guard over the fuse mechanism of a shell. The dimensions were controlled by a brass cap, which it replaced. The plastic was molded over a threaded metal insert originally gated at point *A*. After some time in the field, cracking developed around the metal insert. The main strength was in the direction of flow rather than in the hoop or circumferential direction. Because the thickness of the material could not be increased, the effects of orientation were used by changing the mold and regating at point *B*. The material flowed in the hoop direction and gave the maximum strength there. This slight difference was enough to prevent failure in the field.

Consider gating a deep polyolefin box (Fig. 5-26), using the thinnest possible wall section. Gating the box in the center (A) would give severe radial distortion for the same reasons illustrated in Fig. 5-24. It would be further complicated by the difference in flow length from the gate to point *X* and from the gate to point *Y*. The wall would have to be heavy enough to overcome this stress. Gating it diagonally with two gates (B) would give a radial twist, for the same reasons. It would be much less distorted than the center gate design and would require thinner walls for a stable part. It also would require a three-plate mold for the gating.

It would seem logical to gate on the edge of the *Y* portion, as shown in (C). This would be

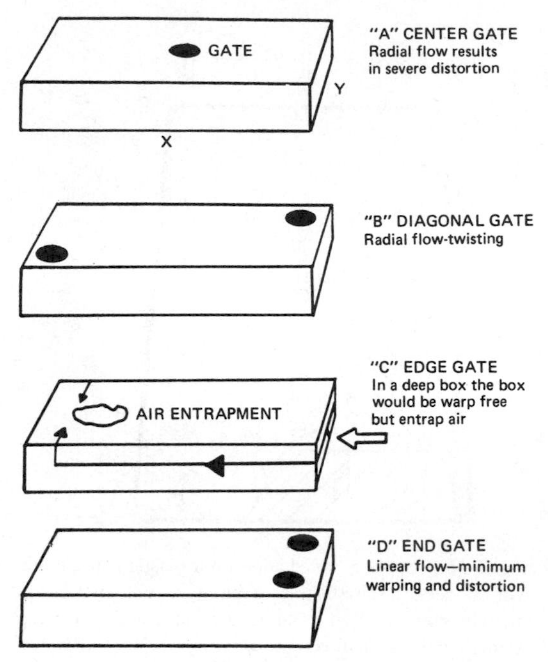

Fig. 5-26. Effect of gate location on a deep molded polyethylene box. (*Courtesy Robinson Plastics Corp.*)

true for a relatively shallow box. With a deep box, however, the material flows around the sides faster than over the top, and air is entrapped somewhere on the top, which virtually cannot be eliminated by venting. This still is not the best method of gating. The preferred method is shown in (D), where there are two gates on the top end of the box. This arrangement gives maximum linear flow without air entrapment and produces a part with the least amount of warp. In most instances, a satisfactory part could be molded with one gate located on the top end. Another possibility is to place two submarine gates near the top. For large parts, it is sometimes necessary to multiple-gate to ensure relatively even orientation patterns and flow lengths. The main problems that can be encountered are air entrapment and weld lines.

Warpage is the result of unequal stress in the molded part when the stress is strong enough to strain or distort the piece. Warping can be caused by the nature of the material, poor part design, poor mold design, and incorrect molding conditions. (See Chapter 11 for additional data.)

Shrinkage

When a plastic material is heated, it expands. Upon cooling to its original temperature, it will contract to the original volume, neglecting the effects of crystallinity, but this is not the only parameter. During injection molding an additional factor, pressure, is introduced. The material basically follows the equation of state: pressure times volume equals the gas constant times the absolute temperature, $PV = RT$. Mold shrinkage should not be confused with tolerance; tolerance is the variation in mold shrinkage rather than the shrinkage itself.

During injection molding, the following things happen: The hot material is injected into the cold cavity, initially under low pressure. Cooling starts immediately, as the parts in contact with the wall solidify. Because the specific volume (the volume of a unit weight of plastic) decreases with the temperature, the solid will occupy less room than the molten polymer. The material fills the cavity, and the pressure builds up rapidly. The pressure does two things: it adds more material to the cavity to make up for the decrease in volume of the material that is already solidified, and it adds more material to compensate for the decrease in volume that will occur when the rest of the material solidifies. If not enough material is put in, there will be excess shrinkage. If too much material is put in, there will be a highly stressed area at the gate. This process of material addition is called packing. The correct amount of material is found by trial and error. The effect of the machine pressure ceases when the injection pressure is stopped or the gate seals. A second, lower injection pressure is used until the machine is opened.

Overstressing of the gate section of an opaque part is impossible to detect during molding, yet highly stressed parts are much more likely to fail than less stressed parts. Thus it is one of the drawbacks of molding that quality cannot be immediately determined. This constraint is also a good reason why economic conditions should not force improper molding.

To decrease shrinkage, the molder should reduce wall thickness, increase injection pressure and injection forward time, increase injection speed, increase the overall cycle, raise the material temperature slightly, lower the mold temperature, decrease the molecular weight distribution, and usually increase the gate size.

Tolerance in molding is beyond the scope of this discussion. Suggested tolerances may be found on pages 821 through 844 or in the SPI Publication, ''Standards and Practices of Plastics Molders.'' Controlling molding tolerance requires a good mold, a good machine, good management, and proper calculation of the price of the item. It is hardly necessary to point out that unneeded tolerances are costly. Good engineering specifies the minimum tolerances required for any application.

CORRECTING MOLDING FAULTS

Injection molding faults may appear when the molder is starting a new mold, after mold changes or repairs, while changing to a new material, or during the regular operation of a mold. The causes of these faults can be as follows: the machine, the mold, operating conditions (time, temperature, pressure, injection speed), the material, parts design, and management, the last of these being the most important.

In order to correct a fault, it first must be found. The purpose of quality control is to find the fault during molding, rather than to discover the error some hours, shifts, or days later when it has become history rather than a force influencing productivity and profit. Quality control should be a continuing ongoing process, starting with the ordering of the raw materials and molds and ending with the shipping of the finished part.

It is obvious that conditions of the machines, molds, auxiliary equipment, and work area contribute to preventing and correcting molding faults.

Poor communications and record keeping are other obvious sources of molding problems.

Before an attempt can be made to correct faults, the machine must be operating consistently, and the temperature control and the mold operation must be consistent.

A single-cavity mold should fill evenly, as should a multiple-cavity mold. If one cavity fills first, the gate may seal off so that it will not be fully packed. The material destined for this cavity will be forced into other cavities, overpacking them. Shrinkages, sticking, and other problems result.

The machine should have a large enough clamping and plasticizing capacity. The plasticizing system should be appropriate for the job. The ejection system should be effective.

Most time functions are controlled by timers. If the mold is not run automatically, the major variable is the operator. The amount of time required to open the gate, remove the part, inspect the mold, and close the gate is variable. This factor often is a major cause of molding problems.

The temperature normally is controlled automatically. Poor design and malfunction of the temperature control system are not uncommon. In some locations, the voltage is not constant because of inadequacies in electrical power systems. Such inconsistency can be disastrous in the molding operation. The best way to recognize it is to use a recording voltmeter. Mold temperature should be controlled automatically.

The pressure also is controlled automatically. Sometimes, overheating will change the viscosity of the oil in the machine, thereby changing the operating conditions. This problem is difficult to detect, but an overheated machine is always suspect. A malfunctioning or worn screw tip will affect the injection pressure on the material.

A major variable in the molding process is the material, but this problem is inherent in the manufacturing process. Normally, there is little the molder can do other than recognize it. Changing materials will usually pinpoint this problem area.

Specific molding problems can be arbitrarily divided into several categories:

- *Short shots* are usually caused by the material solidifying before it completely fills the cavity. The problem usually is due to insufficient effective pressure on the material in the cavity. It can also be caused by a lack of material feeding to the machine. It may require increasing the temperature and pressure on the material, increasing the nozzle-screw-runner-gate system, improving the mold design, and redesigning the part.

• *Parts flashing* usually is caused by a mold deficiency. Other causes are an injection force that is greater than the clamping force, overheated material, insufficient venting, excess feed, and foreign matter left on the mold.

• *Sink marks* usually are caused by insufficient pressure on the parts, insufficient material packed into the cavity, and piece part design. They occur when there are heavy sections, particularly adjacent to thinner ones. These defects are predictable, and the end user should be cautioned about them before the mold is built. At times the solution lies in altering the finished part to make the sink mark acceptable. This might include putting a design over the sink mark or hiding it from view. (See Chapter 11.)

• *Voids* are caused by insufficient plastic in the cavity. The external walls are frozen solid, and the shrinkage occurs internally, creating a vacuum. Inadequate drying of hygroscopic materials and residual monomer and other chemicals in the original powder may cause the trouble. When voids appear immediately upon opening of the mold, they probably indicate a material problem. When the voids form after cooling, usually there is a mold or molding problem.

• *Poor welds (flow marks)* are caused by insufficient temperature and pressure at the weld location. Venting at the weld if possible and adding runoffs from the weld section may help. Increasing the thickness of the part at the weld is also useful. The material may be lubricated, or the mold locally heated at the weld mark. Poor welds have the dismaying proclivity of showing up as broken parts in the field. Flow marks are the result of welding a cooler material around a projection such as a pin. Their visibility depends on the material, color, and surface. They rarely present mechanical problems. Flow marks are inherent in the design of some parts and cannot be eliminated. This possibility should be explored thoroughly before the part design is finalized.

• *Brittleness* is caused by degradation of the material during molding or contamination, and it may be accentuated by a part that is designed at the low limits of its mechanical strength. Material *discoloration* results from burning and degradation; it is caused by excessive temperatures, and by material hanging up somewhere in the system (usually in the cylinder) and flowing over sharp projections such as a nick in the nozzle. The other major cause for discoloration is contamination, which can come from the material itself, poor housekeeping, poor handling, and colorants floating in the air.

• *Splays, mica, flow marks, and surface disturbances at the gate* are probably the most difficult molding faults to overcome. If molding conditions do not help, it usually is necessary to change the gating system and the mold temperature control. Sometimes localized heating at the gate will solve the problem. Splay marks or blushing at the gate usually is caused by melt fracture as the material enters the mold. Usually it is corrected by changing the gate design and by localized gate heating and reduced flow rates.

• *Warpage and excessive shrinkage* usually are caused by the design of the part, the gate location, and the molding conditions. Orientation and high stress levels are also factors.

• *Dimensional variations* are caused by inconsistent machine controls, incorrect molding conditions, poor part design, and variations in materials. Once the part has been molded and the machine conditions are set, dimensional variations should maintain themselves within a small given limit. The problem usually lies in detecting the dimensional variances during the molding operation.

• *Parts sticking in the mold* results primarily from mold defects, molding defects, insufficient knockout, packing of material in the mold, and incorrect mold design. If parts stick in the mold, it is impossible to mold correctly. Difficult ejection problems usually are the result of insufficient consideration of ejection problems during the design of the part and the mold.

• *Sprue sticking* is caused by an improperly fitted sprue–nozzle interface, pitted surfaces, inadequate pull-back, and packing. Occasionally the sprue diameter will be so large that it will not solidify enough for ejection at the same time as the molded part.

• *Nozzle drooling* is caused by overheated material. For a material with a sharp viscosity change at the molding temperature, such as nylon, the use of a reverse-taper nozzle or a positive-seal type of nozzle is recommended.

• *Excessive cycles* usually are caused by poor management. Proper records are not kept, standards are not established, and constant monitoring of output is not done. Other causes of excessive cycles are insufficient plasticizing capacity of the machine, inadequate cooling channels in the mold, insufficient cooling fluid, and erratic cycles.

PROCESS CONTROL

Process controls for injection molding machines can range from unsophisticated to extremely sophisticated devices. As this section will review, they can (1) have closed loop control of temperature and/or pressure; (2) maintain preset parameters for the screw ram speed, ram position, and/or hydraulic position; (3) monitor and/or correct the machine operation; (4) constantly fine-tune the machine, and (5) provide consistency and repeatability in the machine operation.

On-Machine Monitoring

There are different means available for monitoring injection molding machines. First, for clarity, let us separate *monitoring* from *controlling*. Monitoring means watching/observing—in our case, the performance of a molding machine. Traditionally, this is done in a variety of ways: by time and temperature indicators, screw speed tachometers, hour-meters, mechanical cycle counters, and the like. Controlling, on the other hand, means commanding the process variables to achieve the desired levels. Often, a control function is combined with monitoring in a single instrument. These devices may be called "indicating controllers."

This section will focus on monitoring as opposed to controlling: specifically, monitoring parameters such as cycle time, down time, rate, and totals, as opposed to temperature, pressure, and other process parameters.

Sophisticated "electronic stopwatches," or monitors, are available that take advantage of the fact that molding machines have numerous signals that are specifically indicative of the cycle. These signals can be utilized to trigger the electronic watch by direct electrical connection to molding machine contacts. With these direct connections, accurate cycle times are assured. For example, measuring from the injection forward relay (a frequent choice) can provide an accurate, continuous display of overall machine cycle times.

There are two proven benefits from monitoring the cycle time on a continuous basis:

1. Production can be maintained at the established optimum cycle time. Display resolution to .01 second quickly shows changes. For example, if a mechanical or hydraulic problem is developing, it can be detected before it progresses to a breakdown. If unauthorized people are meddling with machine settings, they can be observed. When changes are easily seen, unauthorized people are deterred from making them.
2. Product quality can be kept high because cycle variations due to the previously suggested potential causes are minimized. Further, material changes that contribute a small cycle effect but have a significant product effect can be picked up with continuous, accurate monitoring.

Implicit in maximizing these benefits is having the cycle time displayed on the machine. Many users post the standard cycle time in large numerals next to the digital display. This enables engineers, operators, mechanics, supervisors, foremen—anyone walking by the machine—to see and compare the current cycle with the desired one and to respond appropriately to deviation.

This section was excerpted from Chapter 11 of the *Injection Molding Handbook*, by D. V. Rosato, published by the Van Nostrand Reinhold Company. Reprinted with permission of the publisher.

In addition to monitoring overall cycle time, elapsed time displays can yield precise information about the individual elements that comprise the overall cycle. For instance, with a single-signal input cycle time display, the time a specific relay, switch, valve, etc., is energized can be measured and displayed. Other digital electronic stopwatches are available that accept input signals from two independent sources and can measure a variety of times between them.

An electronic stopwatch that accepts two input signals adds analytical capability beyond that available with one-signal input. For example, an engineer wants to set the optimum time for every element of a cycle. First, he or she must accurately determine where they are now. Then by ''tweaking'' the times down—while monitoring for verification—and checking product quality, the engineer can ''set'' each segment as fast as possible while maintaining good quality. If all active segments are optimized, and there is no ''dead time'' between segments, the cycle will, by definition, be as fast as it can be and still produce the desired product.

Note that ''dead time'' between active portions of the cycle must be eliminated. A dual-input digital display enables this to be done by switching between various signal sources in the machine. Once eliminated, dead time must also be kept out of the cycle in order to keep production up. By continuous monitoring of the most likely areas for dead time, it can be minimized. For instance, improper material additives have affected screw retraction adversely, to the extent of extending cycles because of screw slippage. With continuous monitoring of the retraction time, as measured between two limit switches or their equivalent, this problem could be detected quickly so the material could be changed as soon as it occurred.

The most sophisticated level of monitoring takes advantage of the evolution that has occurred in electronics. With the microprocessor, it is possible to add economical memory and multifunction capability to a machine display.

''Multifunction'' means that in addition to the important ''cycle-time measuring'' component, additional data can be acquired, stored, and displayed. Unless it is separately available on the machine, all monitors of this type for injection molders should include a cycle measurement function. This may be either cycle time directly (in seconds or minutes) or production rate (in shots per hour, cycles per minute, etc.). The availability of a production rate display is important because in many companies the ''shop floor language'' is shots per hour, and a digital display of these numbers directly is more meaningful than a time readout; for example, the change in rate from 120 to 119 shots per hour compared to the cycle time changing from 30.0 seconds to 30.25 seconds. The successful use of monitors hinges on operating personnel understanding them as an aid to production. Therefore, the display should be scaled and read out in the user's particular terminology.

Additional data that may be compiled with these powerful monitors include totals, run time, down time, etc. ''Down time'' may be defined in several ways. It can be as simple as a machine set on the manual switch setting (for setup) instead of automatic or semiautomatic; or, as complicated as the monitor ''learning'' a good cycle and comparing every subsequent cycle to it, then accumulating down time for any cycle that is not at least 90% of the ''good'' cycle. The ''learning'' of the good cycle may be via a user-set switch identifying a desired cycle, or by the monitor calculating an average cycle. The ability to specifically accumulate and record down time on the machine changes a notoriously inaccurate data source—down time is usually guessed—to a precise record that is used to improve performance of machines and people.

Monitors may also be obtained with outputs to drive typical machine audio/visual alarms. These outputs can be energized when down time occurs, when a slow cycle occurs, or when a rate is below a standard the user inputs. (The latter type is only available with the most sophisticated type of monitor—one that communicates bidirectionally to a keyboard/computer.)

These more sophisticated, powerful monitors can provide multiple functions displayed on the machine; in addition, they can communicate directly with a centrally located computer. The central computer eliminates the manual collection of production data; it summarizes data, prints reports, calculates efficiencies and utilization, etc., automatically and immediately, not hours or days later.

Temperature Control of Barrel and Melt

The viscosity of the melt and the speed and pressure of injection determine whether an acceptable molded part is produced. Viscosity is a function of the temperature of plastics, and temperature is a result of the forces of screw rpm, back pressure, and externally applied heat. Injection machine control specialists are generally agreed that one-third of the melt temperature is derived from external heat. Closed loop temperature control thus deserves in-depth attention.

Many excellent instruments are available today as a result of reliable and cost-effective solid state and digital technologies. The temperature control result is, of course, no better than the quality of other components and installation practices employed on the machine. Too many times we find the advantages of a sophisticated temperature control (TC) instrument completely negated by poor installation techniques. Before deciding prematurely that the instrument is at fault, you should make the following checks:

1. Is the thermowell too big for the TC protection tube? Air is an excellent insulator.
2. Is there contamination inside the thermowell? Rust, scale, and residue prevent proper contact of the protection tube with the thermowell.
3. Is the TC junction partially open?
4. Are there oxidation and corrosion inside the protection tube?
5. Is the proper extension wire being used? Copper wire allows another thermocouple junction.
6. Is extension wire polarity observed? A single reversal will give a downscale reading; a double reversal will result in an erratic input to the controller.
7. Are wire terminations properly isolated? False cold junctions are a common problem.
8. Is the cold junction compensation at the extension wire termination on the controller working properly? A poorly positioned or poorly connected compensation component will allow the input to vary.
9. In the panel, are the thermocouple leads isolated from the ac wiring as required? Are the TC wiring and the ac wiring run in separate conduits from the control cabinet to the machine as required?
10. Is the control cabinet thermal environment within the specification of the controller? Excessive cabinet temperatures can cause a controller to drift.
11. Examine the power contactor. If it is a mechanical contactor, deterioration of the contacts can result in reduced power delivered to the heaters.
12. Are the heaters sized correctly? Modern temperature controllers can compensate for limited mis-sizing, but cannot substitute for proper design.
13. Heater bands must be secured tightly to the barrel; again, air is an excellent insulator.
14. Check the voltage being supplied to the heaters. High voltage leads to premature heater failure.
15. Inspect wiring terminations at the heater band; connections must be secure.

If the integrity of the heating system has been verified, your attention can now be turned to the advantages of modern temperature control instrumentation. To demonstrate the improvements made available in recent years, a comparison of the three basic instrument designs is helpful. Millivoltmeter designs can hold setpoint within 20 to 30 degrees; solid state designs can hold within 10 to 20 degrees; microprocessor-based designs typically hold setpoint within 2 to 5 degrees.

Microprocessor-based designs provide several distinct advantages. Already mentioned is the inherent ability to control the temperature at setpoint. Microprocessors do not drift; they either work perfectly, or they experience a catastrophic failure. They are absolutely repeatable, allowing the operator to duplicate a log of setpoint temperatures perfectly the next time that particular job is run. Microprocessors allow a natural avenue to provide digital displays of process information. Values are not subject to inaccurate interpolations and misreadings.

Microprocessors allow the implementation of PID (proportional, integral, derivative) control at little or no cost. PID has been shown to reduce process variations by as much as three or four degrees. Discussions of PID advantages are available from all major temperature control suppliers. (See Fig. 5-27, which shows a PID temperature controller.)

Microprocessor technology is relatively trouble-free—about six times more reliable than analog solid state designs, and about twelve times more reliable than millivoltmeter designs.

Another significant cost reduction effort, being implemented recently with excellent results, focuses on the controller output and power handling. Although an analog controller output accepted by a phase angle or zero angle SCR power controller is ideal in terms of power factor and heater life, it is a relatively costly arrangement. A more acceptable method, in line with cost restraints and providing very nearly the same advantages, is to use a controller with a solid state time-proportioned pilot duty output along with inexpensive mercury contactors or solid state relays. The controller output cycle time can then be reduced to ten seconds or less, thus approaching the same constant temperature and heater life advantages available with the more costly design.

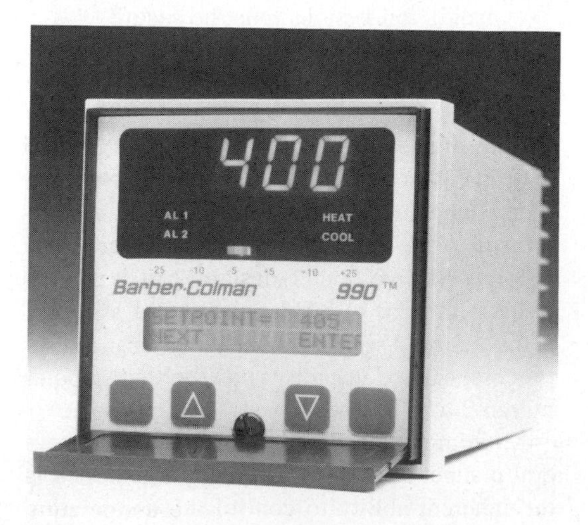

Fig. 5-27. PID temperature controller. (*Courtesy Barber Colman Co., Loves Park, IL)*

Many more advantages are available when the microprocessor is used as the core component for temperature control. Automatic tuning, introduced recently, has already established an enviable track record. Its benefits fall into three major areas:

1. The unit will identify varying thermal behavior and adjust its PID values accordingly. Variables affecting viscosity include screw rpm, back pressure, variations in heater supply voltage, resin melt index, resin contamination, room ambient temperature, percent colorant, screw wear, barrel lining wear, heater and thermocouple degradation, percent regrind, hygroscopic characteristics, and feed zone instabilities.
2. Savings in management and maintenance activity will result from auto-tuned temperature control. Documentation of PID values for various jobs and machines can be eliminated. Individual operator preference for PID values that vary from the norm is precluded. Maintenance personnel are not required to dedicate a particular unit to a specific zone; instruments can be interchanged at will, and spares can be installed with no attention other than selection of the appropriate setpoint. A payback through reduction of overhead costs alone can generally be expected in six to eight months.
3. Energy saving is another major benefit. One customer study showed a 50% reduction in power consumed by the heaters, solely because the automatic tuning feature eliminates the cycling around setpoint normally associated with ineffectively tuned instruments.

Microprocessors also provide a means to communicate digital data to information collection stations. Although the economic feasibility of including the function with an individual temperature control instrument has not been demonstrated in the plastics industry, the feature is beginning to enjoy significant exposure on multiple zone injection machine controllers because of the low cost of adding another digital card to an existing rack. More commonly found on discrete controllers is an analogy communications output that provides a signal to remote recorders.

The ultimate implementation of the microprocessor has been its design in systems installations. Available systems include multizone temperature control and multipoint, multiloop control of sequence. Systems that depend on a single central processing unit (CPU) are available from many suppliers to control temperature, sequence, position, velocity, or pressure. Even more cost-effective are the total machine controllers, which control all machine parameters from a single keyboard. (See Fig. 5-28.) As compared to individual instruments, these systems typically reduce the per-zone cost of control, and provide unlimited future control flexibility as needs change. As production pro-

Fig. 5-28. Injection molding machine control panel. (*Courtesy Cincinnati Milacron, Batavia, OH*)

fessionals discover the need to manage the process at the least possible cost, machine control systems that can communicate with a central management computer are of increasing importance. Central control systems are available that can simultaneously receive information from the injection machine and transmit required parameter changes or complete job setups at the same time. Many injection machines can thus be interfaced with a single control location. If central on-line control is not justified, but one-way machine reporting is required, a choice of several management information systems is available.

PID Injection Pressure Control

A trend to faster-acting, more precise, and more energy-efficient hydraulic systems and components is one response by injection molders and machine builders to a business climate that demands higher productivity and more consistent product quality. Examples of this trend are found in the growing popularity of accumulators, which can deliver a large amount of oil at high pressure, making possible very high injection speeds without the need for an extremely large, energy-consuming pump; of variable-volume pumps, either single or multiple, which provide just the amount of flow that is needed at any point in the cycle, for energy-efficient molding; of servovalves, whose fast response is necessary to control the high injection speeds that the more efficient hydraulic systems can provide; and of multistep injection speed and pressure profiling, providing more sensitive control of the process so as to improve part quality.

One thing that all the above have in common is the tendency for changes in hydraulic pressure during a machine cycle to occur faster than ever before, and this in turn necessitates application of pressure controls that are responsive enough to keep pace. Fortunately, meeting this need does not require inventing new control technology, but rather, more thorough application of what we already have.

Hydraulic pressure-control logic is, in fact, the same as that used for temperature control; its more sophisticated form uses three modes of control, known as PID, for proportional, integral, and derivative (also called gain, reset, and rate, respectively). Each of these mutually interrelated modes of control has an adjustable "tuning constant" that permits the operator to adjust the sensitivity of the pressure controls to the dynamics of the particular machine's hydraulic system.

Some molders may not realize that these tuning adjustments are variables that are just as important to good process control as the set-points for the actual pressure values that the controller is asked to achieve.

Most commercial process-control systems for injection molding to date have not provided full PID pressure control—usually only proportional, or perhaps proportional-plus-reset (integral), control. Furthermore, these systems have commonly offered at most a gain adjustment, or else no tuning adjustment at all. Consequently, the concept of PID pressure control is probably unfamiliar to most molders, as is the role of tuning in obtaining the maximum benefit from three-mode controls.

Yet it is our feeling that, in order to get the kind of cycle-to-cycle repeatability that today's market demands and that today's microprocessor-based control systems are designed to provide, molders should understand the value of PID control logic and must know how to keep such controls properly tuned. Fortunately, current microprocessor know-how can make full PID control available at little or no extra cost, and makes tuning an easy task for the average setup person or technician.

PID Tuning: What It Means. The following is a brief explanation of the three control modes and their tuning constants. It is important to remember that the three terms are not independent, but mutually interactive, and that both the order and the magnitude of adjustments made to the tuning constants can affect the settings of the others.

• *Proportional control (gain):* With this type of control, the magnitude of the control output is proportional to the difference between the actual pressure and the desired pressure—

in other words, the magnitude of the error signal. The "proportional band" is the range of error above and below setpoint, within which the control output is proportioned between zero and 100%.

Usually the proportional band is expressed in terms of its inverse, the gain. If the proportional band is set too wide (low gain), the controller will probably not be able to achieve the setpoint within the time frame of that segment of the cycle. On the other hand, if the proportional band is too narrow (high gain), it will cause violent oscillation of pressure around the setpoint, leading to intense machine vibration, shaking of hoses, and rapid movement of valve spools back and forth, all of which are hard on a machine's hydraulic system and can shorten the life of its components. In either case, inconsistent cycles will result.

The proportional band, or gain, setting is the most fundamental part of the tuning process, which strongly influences everything else. For that reason, it is usually performed first, although subsequent adjustment of the other tuning constants may require some readjustment of the gain.

• *Integral (or reset) control:* Unfortunately, it is a characteristic of purely proportional control that, in response to changing load conditions, it tends not to stabilize the process at setpoint, but rather, some distance away from it. Integral or reset control responds to this steady-state error, or "proportional droop," by shifting the proportional band up or down the pressure scale (without changing the band's width) so as to stabilize the process at setpoint. The amount of reset action to use, expressed in repeats per minute, is the second tuning constant.

• *Derivative (rate) control:* This type of control action responds to changes in error, or the rate at which the actual pressure approaches the setpoint. The faster the change in the magnitude of the error, the greater the rate control signal, and vice versa. It serves to intensify the effect of the proportional corrective action, causing the process to stabilize faster. Rate control's main effect is to prevent the undershoot/overshoot oscillation that may never be completely eliminated with proportional-plus-reset control alone. The amount of rate action, expressed in percent, is the third tuning constant, usually the last to be set.

Rate Control Necessary on High-Speed Machines. Until recently, it was not always necessary for an injection process controller to have rate or derivative control in addition to proportional and reset. Rate control has, however, become essential on newer, faster cycling machines with updated hydraulics.

For example, the high injection speeds of accumulator-assisted machines can create extremely fast changes in the conditions governing hydraulic pressure. In order to smooth out the resulting pressure fluctuations, rate control responds only to fast changes in hydraulic pressure, such as when the ram begins to feel resistance of the melt pushing through the runners and gates of the mold. Changing from one pressure setpoint to another, as in multistep injection profiling, can require the same fast stabilizing action; so the derivative control will help to bring about a faster setpoint change, with minimal overshoot.

A multiple-pump machine will experience a momentary drop in hydraulic pressure when the high-volume pump "drops out" and the smaller holding pump continues injection. This drop in pressure is sometimes so large that the injection ram will actually back up. Derivative control will help to lessen this short dip in pressure and smooth out the injection pressure curve.

Relating Process Control to Product Performance

Monitoring of the molding system can show the effects of mechanical and thermal strains. Strains are imposed upon the material as it is conveyed through the machine and mold. Instrumentation to sense, measure, and display changes in molding parameters helps the molder to determine process consistency.

Monitoring helps relate the process to the product. The sense molding parameters can show the relationship between pressure, temperatures, and position (movement) during the process.

Monitoring can also establish whether additional machine control is needed. The forgiving nature of the molding process and liberal product dimensions allow most parts to be produced with conventional ''open loop'' machine control systems. As product demands become more stringent, both dimensionally and physically, ''closed loop'' machine control may become advantageous.

Sensor Requirements. Any sensor used requires a power supply and an amplifier. A sensor is driven by an input voltage, usually called an ''excitation'' voltage. A resultant output signal is generated as the sensor responds to the monitored parameter. An amplifier is used to boost the output signal's strength. Increased signal strength or amplitude is needed for recording capabilities.

Sensors and electrical systems should be tested and calibrated before actual use. Variances do occur between sensors of the same type. Sensors should be maintained at a ''zero'' reference if precise monitoring or measuring is to be done. Electrical ''drifting'' destroys the accuracy of the information being obtained.

Molding Parameters.

Pressure.

• *Machine hydraulic pressure transducer:* A hydraulic pressure transducer is used to generate a signal. Monitoring the hydraulic pressure profile can help diagnose many machine problems. The hydraulic pressure transducer should be placed as close to the injection ram as possible; this location gives the most accurate pressure profile. Hydraulic pressure profiles can determine the following:

1. Hydraulic pressure relief valve setpoint consistency.
2. Timer accuracy for switching cutoff pressures.
3. Hydraulic back pressure setting during screw return.
4. Screw return time consistency.
5. Hydraulic pressure changes during injection, reflecting material viscosity changes.

• *Machine material pressure transducer:* Monitoring the material pressure can be done with a transducer in the machine nozzle. The material pressure profile will be similar to the machine hydraulic pressure profile. The pressure of the material and the hydraulics in the machine barrel become similar as the mold is filled. Sensing of material pressure at the machine nozzle can be done, but its usefulness is questionable.

• *Mold material pressure:* Material pressure transducers can be installed in the mold's runner system and in the cavity. Indirect and direct material sensors are available. The type of transducer selected depends upon the product configuration in the mold, mold construction, and type of runner system.

Pin-loaded-type material pressure transducers must be designed and installed with care. The use of pins to transmit material pressure can cause errors; the pins can stick, bend, and induce thermal effects during cure time. Location and pin diameter must be considered for monitoring. Because of the ''select point'' pressure sensing, the transducer output may be poor.

Direct material pressure transducers are now available. The accuracy of pressure sensing is much better, but there is a problem in selecting the location to sense and monitor the material pressure. Monitoring at a point located halfway into the cavity is a good general rule. Maintenance of built-in transducers should be considered when designing a mold.

The mold material pressure profile can determine the following:

1. Material filling time.
2. Material peak pressure consistency.
3. Machine nozzle contamination or freeze-off.

Temperature.

• *Machine barrel temperature:* Barrel temperatures are sensed and controlled with thermocouples (T.C.). One T.C. is needed for each zone that is being controlled. Usually three zones (front, middle, and rear) are sensed and controlled. The nozzle usually has its own control. For accurate temperature control and temperature setpoint, current-proportioning controllers should be used, not the time on–off type of temperature setpoint controllers.

Monitoring barrel temperatures can determine:

1. Temperature controller performance.
2. Barrel heater failure.

• *Mold temperature:* The control of mold temperature is usually done with an independent heater/chiller unit(s). The controller has temperature setpoints, and the mold usually balances out at some temperature around the setpoint. If the controller supply lines, mold water lines, and pressure losses are minimized, the control is acceptable.

Monitoring of the mold temperature is usually done with T.C.'s. Their accuracy depends on the T.C. placement. The T.C. location must be tried to determine the optimum location. This area of monitoring temperature in the mold is difficult because of the high thermal inertia in the heater/chiller/mold system.

• *Material temperature:* Material temperature can be measured in the machine nozzle. Commercial T.C. sensors are available to measure the material melt temperature. The T.C. devices are the simplest and most stable to install; infrared and ultrasonic systems are also available, but are much more complex.

Material temperature variances can exist in the melt because of screw mixing, barrel heating, and a varying shot-to-barrel ratio. Sensing the nozzle melt can show:

1. Material melt consistency.
2. A change in machine plasticating.
3. Heater failure on the barrel.

Position.

• *Machine ram position:* The ram position is monitored from a potentiometer mounted on the machine, either linear or rotary. The sensor indicates the ram during the molding process and can show the following:

1. Injection rate of material into the mold.
2. Consistency of ram profile during "open loop" or "closed loop" machine control.
3. Screw position during return to back position.
4. Screw return time consistency.

• *Machine tie bars:* Machine tie bars "stretch" when the mold is clamped. This mechanical strain or elongation can be measured with strain gauges, dial indicators, and linear variable displacement transducers (LVDTs). The LVDTs eliminate the need to drill holes in the tie bars or clamping on small indicating devices. Monitoring tie bar strain can show:

1. Balance of tie bar strain during clamp.
2. Mold clamp tonnage.
3. Machine/mold clamp tonnage changes occurring because of thermal effects of machine cycling and mold heating or cooling.

• *Mold part line:* Mold part line separation can be measured with indicator gauges and LVDTs. As material is packed into the mold, the part line can open. There is a direct relationship between machine clamp on the mold, material viscosity, and material injection rate. Monitoring for a mold's part line separation can show the following:

1. Dimensional changes in the product.
2. Mold flashing.

Display of Monitored Molding Parameters.

Analog Display. Analog devices include:

1. Chart recorder.
2. Voltmeter with a sweep needle.
3. Oscilloscope.

Analog signals are useful for seeing a continuous profile of the parameter being sensed. This profile is useful because it is time-related. Chart recordings show a continuous profile but are limited in the type of information that may be interpreted. Total span and peak changes are shown, but comparisons of one cycle to another are difficult.

Digital Display. Digital devices include:

1. Controllers with numerical setpoints.
2. Sensing devices with numerical readout display.

Digital monitoring devices give a numerical readout. The sensor's output signal is conditioned to give a discrete numerical readout(s). Data loggers are used to monitor multiple parameters digitally. Digitizing (displaying discrete numerical values at a certain rate) of analog signals can be a useful technique, but the rate at which information can be digitized must be considered. If any rapidly occurring events are being considered, this system can give erroneous or insufficient information.

CRT Display. Cathode ray tube (CRT) displays include:

1. Oscilloscopes (scope).
2. Storage scope.
3. Analog/digital scope.
4. Television.

Storage scopes can be utilized to monitor repeating cycles. A selected starting point is used to "trigger" the scopes. The storage scope display shows the excursion of a parameter over a period of time. A multichannel storage scope is very useful to relate more than one molding parameter on a single display.

Machine Control.

"Open Loop" Machine Sequence Control. In a conventional "open loop" machine sequence control system, input commands are set, and there is an unknown machine output response.

Monitoring of machine hydraulic pressure and ram position relates:

1. Screw return profile consistency.
2. Hydraulic pressure profile consistency.
3. Ram injection rate consistency.

An open loop machine control system *cannot* compensate for changes in material viscosity. Material viscosity changes result in:

1. Increased viscosity (increased stiffness)
   (a) Higher initial hydraulic pressure profile.
   (b) Slower ram injection rate.
   (c) Lower final in-mold material pressures.
2. Lower viscosity (more fluid)
   (a) Lower initial hydraulic pressure profile.
   (b) Faster ram injection rate.
   (c) Higher final in-mold material pressure.

The ram injection rate is controlled by the metering of oil into the hydraulic injection ram cylinder. Material viscosity establishes the hydraulic pressure profile during mold filling and packing. The hydraulic pressure profile is a valuable parameter to monitor for establishing mold/machine consistency.

"Closed Loop" Machine Sequence Control. In a "closed loop" machine sequence control system, input commands are set, and corrections are made to the machine output response. The correction can be either of the following:

1. *Real time:* A sensed deviation is corrected in cycle, as quickly as the machine electrohydraulic valve and fluid system can respond.
2. *Adaptive:* A sensed deviation is adjusted for on the next cycle. The system's ability to adjust depends upon how sensitive the molding process is and controller capability to correct the deviation.

A closed loop machine control system *can* compensate for changes in material viscosity. This capability improves the consistency of initial mold filling but does not fully address final packing pressure in the mold.

The ram position is programmed to establish a material filling rate into the mold. The hydraulic pressure compensates for material viscosity changes during the controlled filling of the mold's sprue, runner, and cavity.

The final packing pressure is controlled by switching from the ram position (velocity) profile to a hydraulic packing pressure.

Control of the molding process is better, but actual improvement in the product is not always realized. Monitoring the molding system can help the molder to:

1. Improve mold setup consistency.
2. Resolve molding problems.
3. Determine the effectiveness of the equipment.
4. See the process working.

Adaptive Ram Programmer

The injection molding process has a number of variables in material and machine conditions that tend to change during production. All these variables affect the critical properties of the molded part. When material properties change or the machine drifts outside the ideally preset operating parameters, the operator must reestablish the conditions best suited for making the part. He or she is faced with a complex situation, as the interdependency of machine functions and material conditions requires a thorough understanding of the process, and a series of complex adjustments on the machine must be made to maintain part quality. Often the variables are not controllable to the necessary degree, and the operator has to contend with imperfect production and a high rejection rate.

The Spencer and Gilmore equation, developed a number of years ago, is now widely utilized to predict the relationships that must be maintained to keep the critical functions that affect part quality constant. This equation indicates that plastic pressure and volume are inversely related if temperature (or material viscosity) is constant. During molding, filling, and packing, the plastic temperature drops only slightly because of the short time interval involved. The material viscosity tends to change, however, as a function of composition or long-term temperature conditions of the machine.

The shrinkage of the plastic during mold cooling is primarily determined by the number of molecules in a given cavity under a given pressure. For this reason, cavity pressure controls have been utilized in an effort to control the shrinkage parameters of the part. As viscosity changes, however, it is important to adjust the plastic volume so that the number of molecules packed in a mold cavity will remain constant. In order to accomplish this, the precompressed shot size must be adjusted so that when the desired pressure in the cavity is reached, the total volume under pressure that exists between the tip of the ram and the cavity will be held constant. As the two parameters, pressure and volume, are highly interdependent, continual adjustments must be made (on each shot) following the trends in material parameters.

Another critical condition to be maintained is plastic flow rate. The Poiseuille equation for fluid flow (see above) shows the significance of pressure on flow rates. Plastic viscosity deviates considerably from constant during flow. The effect is to make flow behavior dependent upon pressure. As the operator desires to maintain the flow surface velocity for the plastic constant, or to adjust the flow in accordance with the requirements of the mold, the injection velocity together with material volume and pressure form the most important parameters that have to be controlled to maintain part quality.

Heretofore, individual parameters such as cavity pressure, ram oil pressure, and ram velocity have been measured and even controlled. The interdependence of these three functions, however, demands that a control system be utilized that can control all three parameters simultaneously, and is capable of automatic adjustments and decision making to maintain the equations in balance during the molding process.

Microprocessor Advantages

Microprocessor-based process controllers have been achieving more widespread acceptance as their cost has come down. Whereas a few years ago these controls were used only for applications that required their precise control, now we find advantages in their application on almost any job.

• *Setup time reduction:* Time for setup can be greatly reduced by the ability to record and store timer settings, limit switch positions, and pressure levels. The data can then be fed to the controller in seconds to preadjust the machine to the new setup.

• *Easier operator "tuning":* Since the microprocessor inputs can be located at the operator station, adjustments can be made without crawling around the machine.

• *Smoother operation:* This is achieved through ramping of the control signals. We can now eliminate many of the readjustments necessary as the machine temperature changes, simply by setting these ramps such that the time is longer than the response under conditions of start-up. Since the signal is now slower than valve response, the signal is always in control, yielding a more uniform cycle.

• *Less down time:* The constant monitoring of machine performance made possible with these systems can allow lower pressures and eliminate shock peaks, thereby extending component life. A properly applied system will also have fewer components to troubleshoot when a problem does occur, and diagnostic programs can be included.

• *Input energy reduction:* By programming the hydraulic system to respond to the varying demands of the circuit, we have the potential to reduce input power requirements.

GLOSSARY OF TERMS RELATED TO INJECTION MOLDING EQUIPMENT

There are three essential parts to an injection molding machine: the mold clamping device, the injection unit, and the mold. Accepted terminology is as follows:

Clamping system terminology—

clamping unit—That portion of an injection molding machine in which the mold is mounted, and which provides the motion and force to open and close the mold and to hold the mold closed during injection. When the mold is closed in a horizontal direction, the clamp is referred to as a **horizontal clamp**. When closed in a vertical direction, the clamp is referred to as a **vertical clamp**. This unit can also provide other features necessary for the effective functioning of the molding operation.

daylight, open (Fig. 5-29)—The maximum distance that can be obtained between the stationary platen and the moving platen when the actuating mechanism is fully retracted without ejector box and/or spacers.

daylight, closed or minimum mold thickness (Fig. 5-29)—The distance between the stationary platen and the moving platen when the actuating mechanism is fully extended, with or without ejector box and/or spacers. Minimum mold thickness will vary, depending upon the size and kind of ejector boxes and/or spacers used.

daylight, maximum closed (Fig. 5-29)—That distance between the stationary platen and the moving platen when the actuating mechanism is fully extended without ejector box and/or spacers.

daylight, minimum closed (Fig. 5-29)—That distance between the stationary platen and the moving platen when the actuating mechanism is fully extended with standard ejector box and/or spacers.

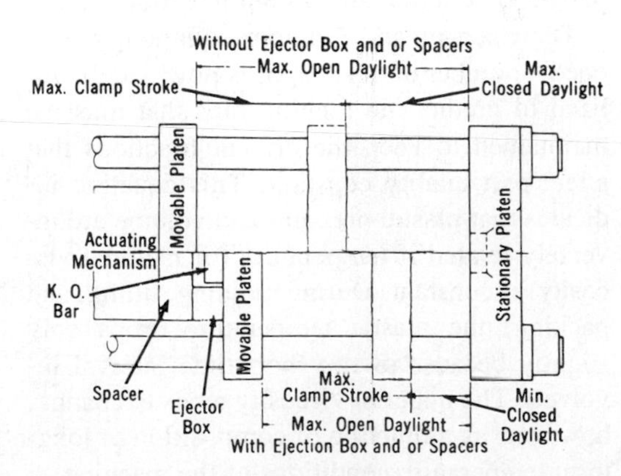

Fig. 5-29. Clamp die space nomenclature. *(All illustrations on injection molding courtesy SPI Machinery Division)*

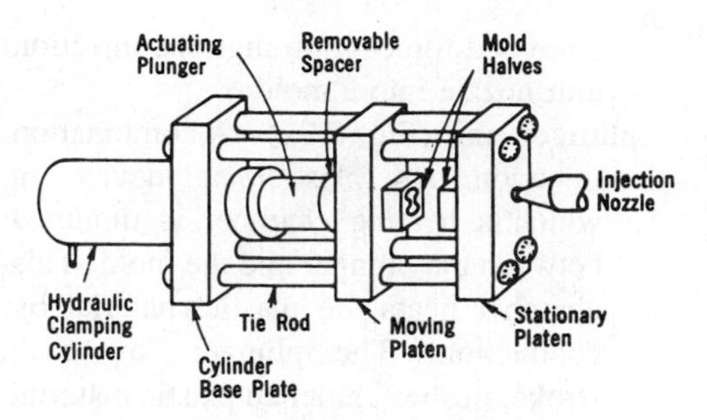

Fig. 5-30. Hydraulic clamp.

ejector (knockout)—A provision in the clamping unit that actuates a mechanism within the mold to eject the molded part(s) from the mold. The ejection actuating force may be applied hydraulically or pneumatically by a cylinder(s) attached to the moving platen or mechanically by the opening stroke of the moving platen.

full hydraulic clamp (Fig 5-30)—A clamping unit actuated by a hydraulic cylinder which is directly connected to the moving platen. Direct fluid pressure is used to open and close the mold, and to provide the clamping force to hold the mold closed during injection.

moving platen (Figs. 5-30 and 5.31)—That member of the clamping unit which is moved toward a stationary member. The moving section of the mold is bolted to this moving platen. This member usually includes the ejector (knockout) holes and mold mounting pattern of bolt holes or "T" slots. A standard pattern was recommended by SPI Standards Testing Method (Injection Machinery Division Standards, September 11, 1958).

stationary platen (Figs. 5-30 and 5-31)—The fixed member of the clamping unit on which the stationary section of the mold is bolted. This member usually includes a mold mounting pattern of bolt holes or "T" slots. A standard pattern was recommended by SPI Standards Testing Method (Injection Machinery Division Standards, September 11, 1958). In addition, the stationary platen

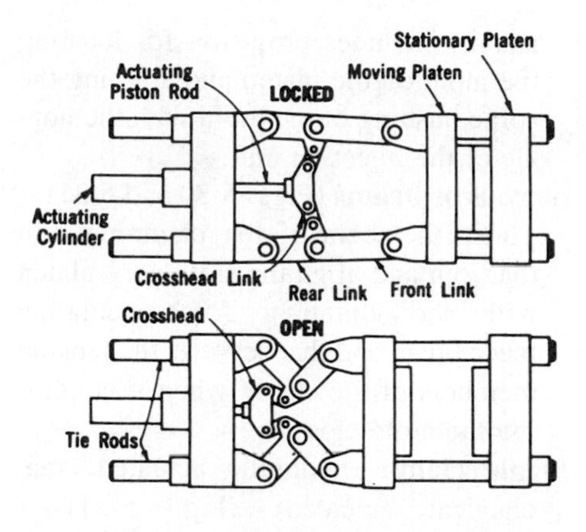

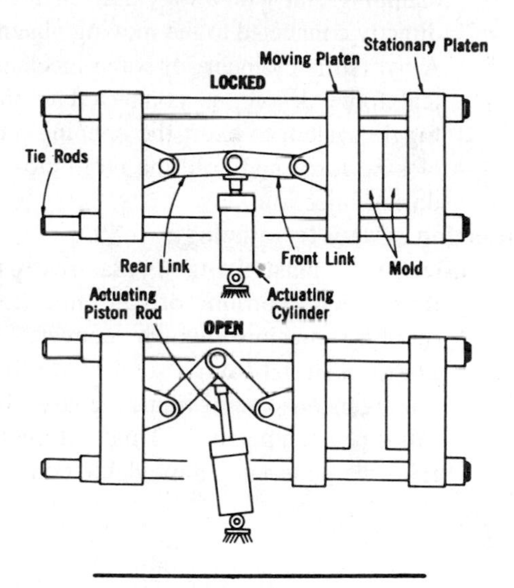

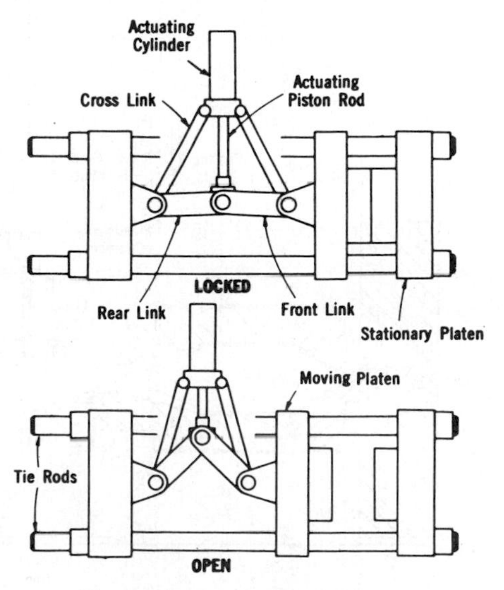

Fig. 5-31. Types of toggle clamps.

usually includes provision for locating the mold on the platen and aligning the sprue bushing of the mold with the nozzle of the injection unit.

tie rods or beams (Figs. 5-30 and 5-31)—Those members of the clamping unit that join and align the stationary platen with the clamping force actuating mechanism and that serve as the tension members of the clamp when it is holding the mold closed.

toggle clamp (hydraulic actuated, mechanical actuated) (Fig. 5-31)—A clamping unit with a toggle mechanism directly connected to the moving platen. A hydraulic cylinder, or some mechanical force device, is connected to the toggle system to exert the opening and closing force and hold the mold closed during injection.

Injection system terminology—

injection plasticizing (plasticating) unit—That portion of an injection molding machine which converts a plastic material from a solid phase to a homogeneous semi-liquid phase by raising its temperature. This unit maintains the material at a moldable temperature and forces it through the injection unit nozzle into a mold.

plunger unit (Fig. 5-32)—A combination injection and plasticizing device in which a heating chamber is mounted between the plunger and the mold. This chamber heats the plastic material by conduction. The plunger, on each stroke, pushes unmelted plastic material into the chamber, which in turn forces plastic melt at the front of the chamber out through the nozzle.

prepacking—Also called "stuffing," a method that can be used to increase the volumetric output per shot of the injector plunger unit by prepacking or stuffing additional material into the heating cylinder by means of multiple strokes of the injector plunger. (Applies only to plunger unit type injection machines.)

reciprocating screw (Fig. 5-33)—A combination injection and plasticizing unit in which an extrusion device with a reciprocating screw is used to plasticize the material. Injection of material into a mold can take place by direct extrusion into the mold, or by reciprocating the screw as an injection plunger, or by a

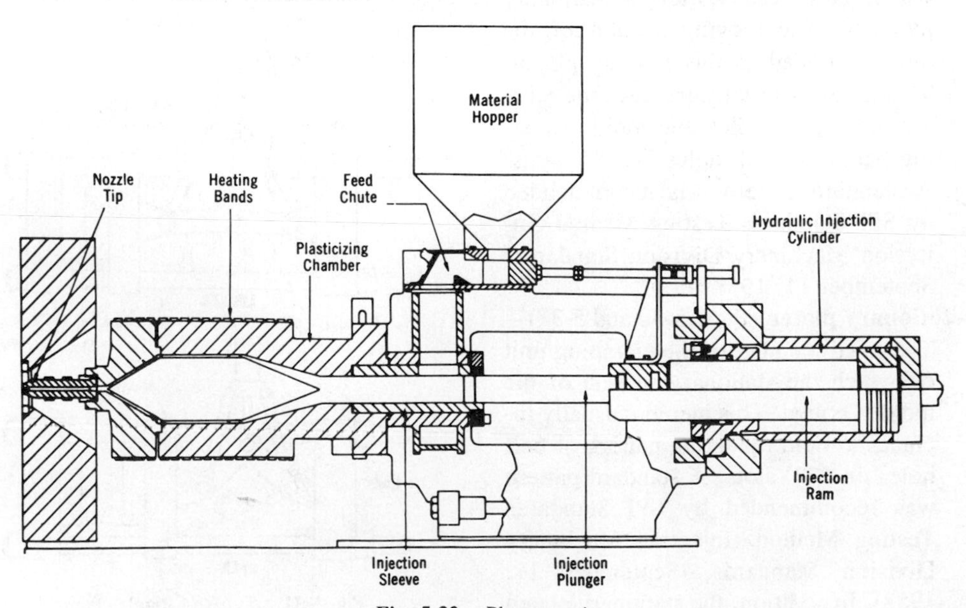

Fig. 5-32. Plunger unit.

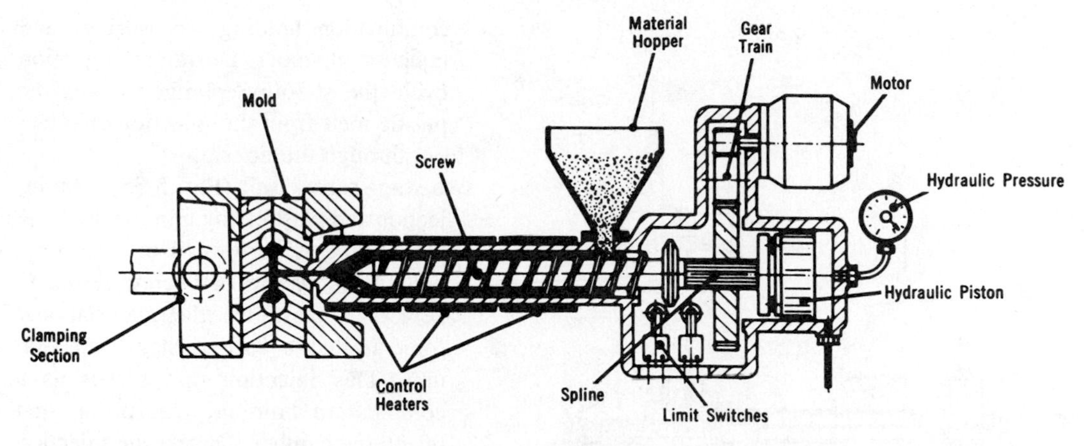

Fig. 5-33. Reciprocating screw unit.

combination of the two. When the screw serves as an injection plunger, this unit acts as a holding, measuring, and injection chamber.

two-stage plunger unit (Fig. 5-34)—An injection and plasticizing unit in which the plasticizing is performed in a separate unit. The latter consists of a chamber to heat the plastic material by conduction and a plunger to push unmelted plastic material into the chamber, which in turn forces plastic melt at the front of the chamber into a second stage injection unit. This injection unit serves as a

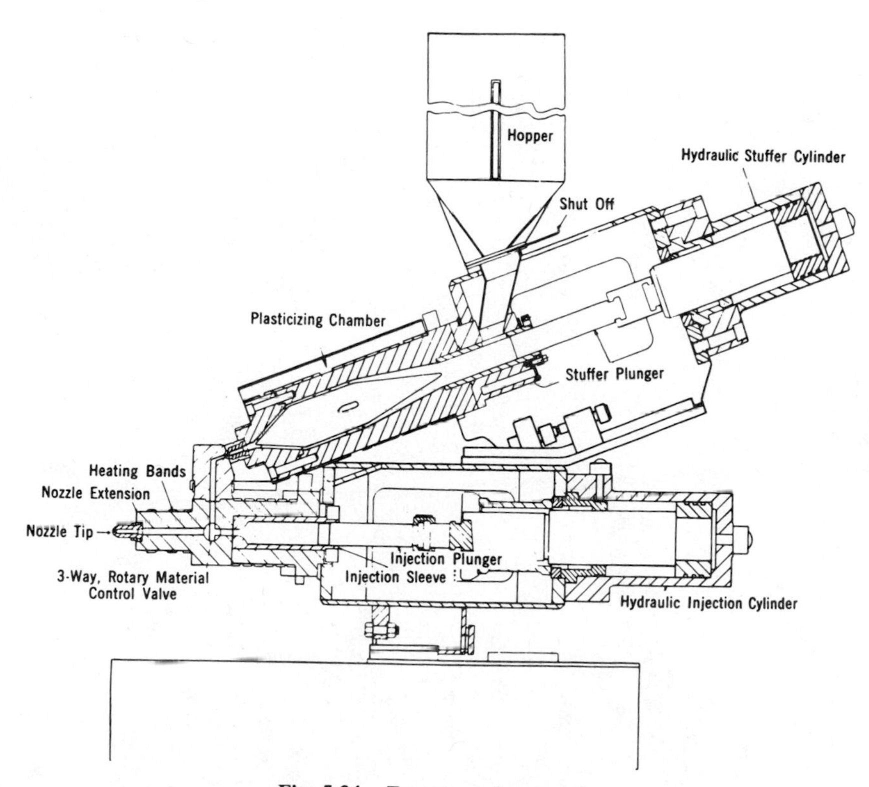

Fig. 5-34. Two-stage plunger unit.

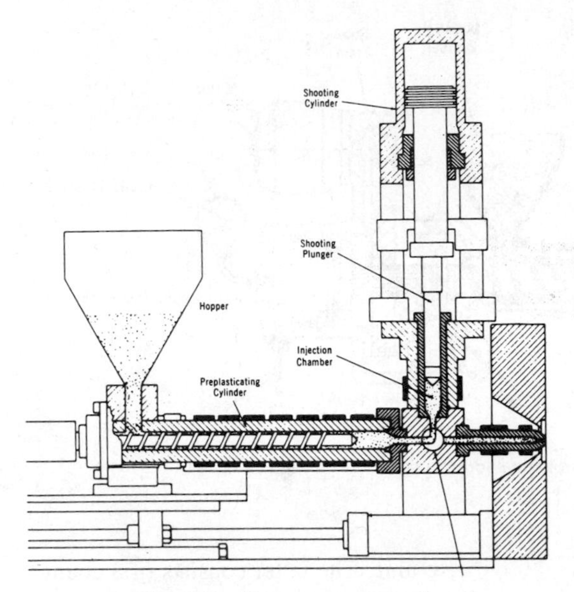

Fig. 5-35. Two-stage screw unit.

combination holding, measuring, and injection chamber. During the injection cycle the shooting plunger forces the plastic melt from the injection chamber out through the nozzle.

two-stage screw unit (Fig. 5-35)—An injection and plasticizing unit in which the plasticizing is performed in a separate unit which consists of a screw extrusion device to plasticize the material and force it into a second stage injection unit. This injection unit serves as a combination holding, measuring, and injection chamber. During the injection cycle a plunger forces the plastic melt from the injection chamber out through the nozzle.

See also molding definitions in Chapter 1.

6

Design Considerations for Injection Molds

The injection mold has two functions. It forms the air space into which the hot plasticized material is injected under pressure, and it acts as a heat exchanger, removing heat from the material in the mold until it is rigid enough to be ejected so that the final part will conform to all its specifications. The quality of the part and its cost of manufacture are largely dependent on mold design, construction, and excellence of workmanship.

Over the years, as the size of the molding machine increased, so did the cost of molds. Molds costing upward of $300,000 are no longer unusual.

The two critical areas in ensuring the right mold are the design of the piece and the design of the mold. An acceptable product is not possible with an incorrect part design. An acceptable part may be possible with an incorrect mold design, but always under extremely difficult and uneconomical conditions. Although the design of the part is not within the scope of this discussion, it nevertheless should be thoroughly reviewed by those people directly involved in the design of the mold and in the molding operation. The last opportunity to change the design of the part occurs when the mold is being designed.

It is imperative that the user, moldmaker, molder, part designer, quality control department, and packaging people be consulted before designing the mold. The design of the mold is usually done by the molder and the moldmaker. The selection of the method of gating, location of the gate, venting, parting line, and method of ejection affect the appearance and function of the part.

Moldmakers vary from an individual owner with several helpers to plants with hundreds of employees. The quality of the mold bears no relation to the size of the shop. In selecting a moldmaker, it is important to use one who has the best type of equipment for the job and who has had experience in building similar molds. Surveys have shown that moldmakers are leaving the industry about three times faster than they are being replaced. This situation has greatly increased the use of computer aided design and manufacturing (CAD/CAM) systems in moldmaking.

The overriding consideration in mold design is to be completely confident that the mold will work. Unfortunately, moldmaker selection often is done on the basis of initial price with no regard for the money that may have to be spent to get the mold to work correctly after it is delivered. Because of daily lost productivity of a bad mold versus a high-quality one, the lowest-priced mold has too often turned out to be the most expensive. If there are questions, they should be resolved before the mold is built, not afterward, as it is much easier to change a design on a piece of paper than to change a

---

By Irvin I. Rubin, President, Robinson Plastics Corp., Hoboken, NJ. The author gratefully acknowledges the permission of John Wiley & Sons, Inc., New York, to use material from his book *Injection Molding, Theory and Practice* (1973).

completed die. If, during the course of mold construction, questions of this kind arise, the work should be stopped and the problem resolved.

The injection mold normally is described by a variety of different criteria, including the following:

- *Number of cavities*
- *Material*
  Steel-hardened
  Stainless steel
  Prehardened steel
  Hardened steel
  Beryllium copper
  Aluminum
  Epoxy-steel
- *Surface finish*
  Polish (using SPI nomenclature)
  Chrome plate
  Electroless nickel
  EDM
  Sandblast
  Photoetch
  Impingement
- *Parting line*
  Regular
  Irregular
  Two-plate mold
  Three-plate mold
  Stack mold
- *Method of manufacture*
  Machined
  Hobbed
  Gravity cast
  Pressure cast
  Electroplated
  EDM (spark erosion)
- *Runner system*
  Hot runner
  Insulated runner
- *Gating*
  Edge
  Restricted (pin point)
  Submarine
  Sprue
  Ring
  Diaphragm
  Tab
  Flash
  Fan
  Multiple
- *Ejection*
  Knockout pins
  Stripper ring
  Stripper plate
  Unscrewing
  Cam
  Jiggler pins
  Removable insert
  Hydraulic core pull
  Pneumatic core pull

Following these criteria, a typical mold might be described as follows: a four-cavity, machined, hardened steel, chrome-plated, hot-runner, stripper-plate, tumbler mold.

MOLD DESIGN FEATURES

Two-Plate Mold

Figure 6-1 shows a two-plate mold that certainly has more than two plates. This is the common name for a mold with a single parting line. The parting line of a mold can best be defined as that surface where a mold separates to permit the ejection of plastic (either the molded parts or gates and runners). The part being made in the mold shown in Fig. 6-1 is a shallow dish, edge-gated. The temperature control channels are in both the cavity and the core. This arrangement is preferable to cooling the plate, as the boundary between the cavity or core and the plate has high thermal resistance and greatly lowers the rate of heat transfer. Note also the support pillars, which prevent the support plate from buckling under the pressure of the injecting material and the knockout bar, which is attached to the machine and is the actuator for the knockout plates. A sprue puller with a reverse taper is used.

Three-Plate Mold

Suppose, for example, that one had to mold several cups or similar-shaped items in one mold. It would not be possible to gate them except at the top. Otherwise, venting problems would make it impossible to mold. One way to

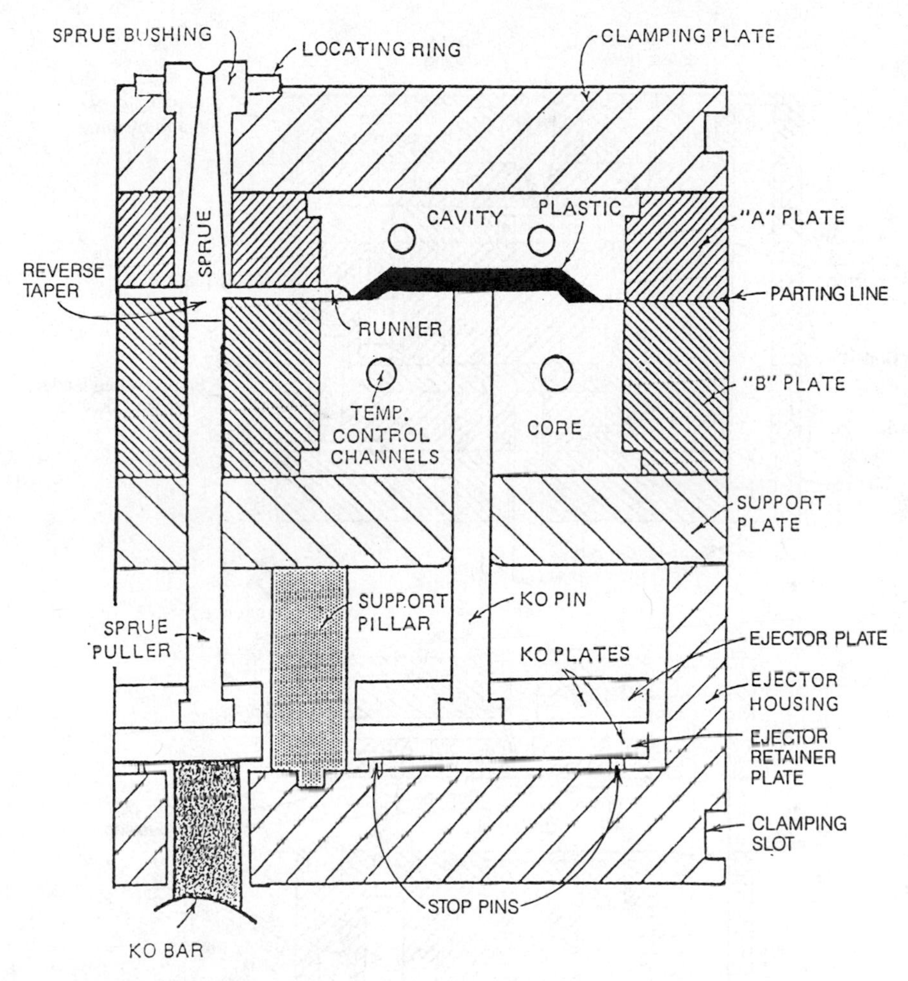

Fig. 6-1. Schematic drawing of a "two"-plate mold. (*Courtesy Robinson Plastics Corp.*)

produce such parts is to use a three-plate mold (Fig. 6-2).

To do this, a runner system is developed that received its material from the nozzle and distributes it through secondary nozzles to each individual part. In order to remove this runner, an extra plate is added, which, when separated, allows the runner to be removed. The runner is trapezoidal with a 7° per side taper and is cut the runner plate and molds between it and the pin plate. The pin plate moves on its own leader pins and bushings to prevent binding.

After molding, the mold opens up at parting line 1, breaking the gate while both the runner and the part are contained in steel. The mold then opens at parting line 2, a step that results in the ejection of the part; and if chains or latches attach the movable side to the A plate, the stationary side opens at parting line 3. The runner had been held flush against the pin plate by small undercut pins. When the pin plate moves forward, these undercut pins remain stationary, and the runner moves forward, separating itself from them. The runner is mechanically ejected. Another way to move the A plate, without using chains or latches, is to attach air cylinders to both sides, with the rods attached to the platens.

Hot Runner Molds

In a three-plate mold, the runner system must be reground and the material reused. However, it is possible to eliminate the runner system en-

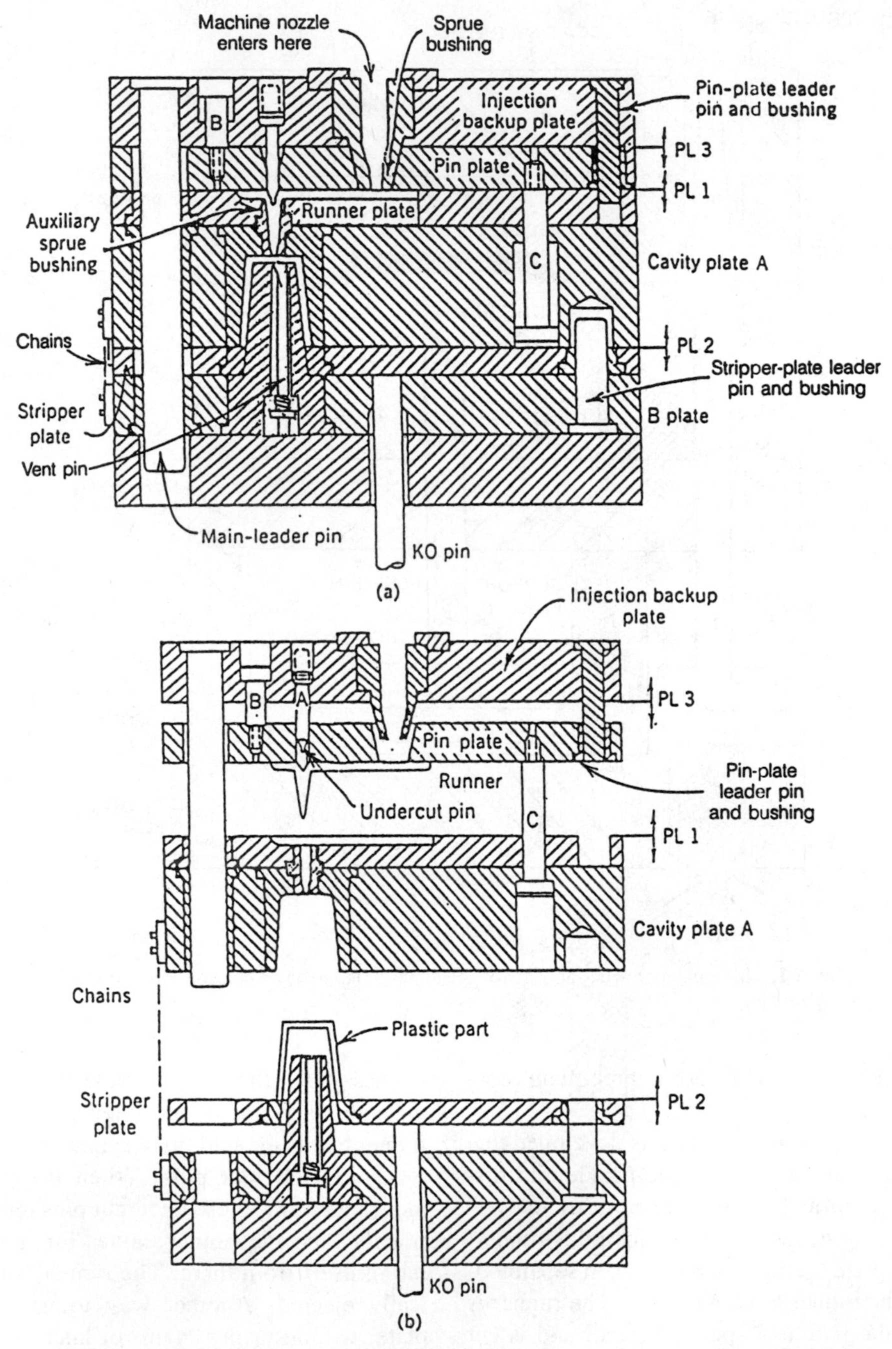

Fig. 6-2. Schematic drawing of a three-plate mold. (*Courtesy Robinson Plastics Corp.*)

tirely by keeping the material fluid. This is done by using a hot runner mold (Fig. 6-3). The material is kept plasticized by the hot runner manifold, which is heated with electric cartridges. The block is thermostatically controlled, and heated probes, controlled by thermocouples and pyrometers, are used. The material is kept fluid at the proper temperature, and the injection pressure is transmitted through the hot runner manifold and gate into the part. A probe is inserted from the runner to the gate area. There are two kinds of probes: one has no moving

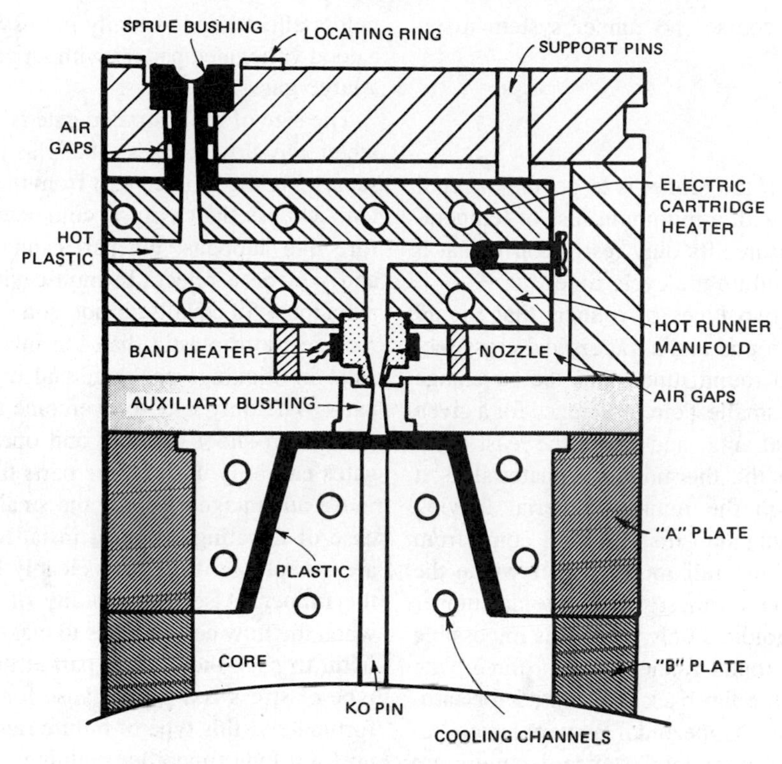

Fig. 6-3. Hot runner mold. (*Courtesy Mold-Masters Systems*)

parts and essentially depends on the insulating properties of the plastic at the gate area to prevent it from drooling, whereas the other has a reciprocating device or other method to block the flow between each shot.

A hot runner mold is more difficult to start up than a three-plate mold. It takes considerably longer to become operational. In multicavity molds, balancing the gate and the flow and preventing drooling can be difficult. The hot runner mold is highly susceptible to tramp metal, wood, paper, and other contaminants, which quickly clog up the nozzle. Various screen systems have been devised in attempts to prevent this. Cleaning out a plugged hot runner mold is a long process, and the molds themselves are relatively expensive; but they offer the advantage that in a long-running job, they provide the most economical way of molding. There is no regrinding, with its cost of handling and loss of material. The mold runs automatically, eliminating the variations caused by operators. The loss of temperature and pressure experienced with a cold runner in a three-plate mold is eliminated.

Insulated Runner Molds

An insulated runner mold is a cross between a hot runner mold and a three-plate mold. If a large-diameter runner, 2 inches or over, were used in a three-plate mold, the outside would freeze, and the inside would remain molten, acting as a runner. Half the runner is cut on each of the two plates so that during start-up, or if the runner freezes, the two plates are separated, and the runner system is quickly removed. A heated probe may be used. If it is, it is controlled in the same way as those used in a hot runner mold. The insulated runner system, like the hot runner system, was in use during the 1930s.

Insulated runner molds are more difficult to start and operate than three-plate molds, but are considerably easier than a hot runner mold.

There is, of course, no runner system to regrind.

Runners

The purpose of the runner is to get the material to the cavity with a minimum loss of temperature and pressure. Its only restriction is that it should not hold up the cycle time.

There are two types of runners that should be used: full round and a trapezoidal cross section. The full round runner has the advantage of having the smallest circumference for a given cross-sectional area, and hence the least chilling effect on the thermoplastic material as it passes through the runner. Material flowing from the runner into the gate will come from the center of the full round runner where the material is the warmest. A trapezoidal runner should be employed only when it is impossible to use a full round runner, or in a three-plate mold where the flat back is essential for automatic ejection. Trapezoidal runners have a taper of 7° per side, and standard cutters are available for them. Under no condition should a half round runner be considered, as a trapezoidal shape is much superior to a half round.

Gates

The gate is the connection between the runner system and the molded part. It must permit enough material to flow into the mold to fill out the part plus such additional material as is required to overcome the normal thermal shrinkage of the part. The location of the gate, its type, and its size strongly affect the molding process and the physical properties of the molded part.

There are two types of gates, (a) large and (b) restricted or pin pointed. Restricted gates usually are circular in cross section and for most materials do not exceed 0.060 inch in diameter. The restricted gate is successful because the apparent viscosity of the plastic is a function of the shear rate. The faster it moves, the less viscous it becomes. As the material is forced through the small opening, its velocity increases, decreasing its viscosity. A second characteristic is high mixing, due to the Reynold's effect. It is virtually impossible to mold a good variegated pattern without going through a large gate.

The size of the restricted gate is so small that when the flow ceases, the plastic solidifies, separating the molded part from the runner system. Unless the cavity is completely filled before this happens, the part cannot be packed and will be a reject. In multicavity molds all parts must fill equally; if not, some cavities will receive more plastic than the others. This results in sticking, packing, and different-sized parts. The only way to overcome this problem is to short-shoot the shot and open individual gates carefully until all the parts fill equally.

An advantage of using the small gate is the ease of degating. In most instances, the parts are acceptable if they are cleanly broken from the runner. Also, the rapidity of gate sealing when the flow ceases tends to make it more difficult to pack the molded part at the gate. This type of stress is a prime cause for failure. Unfortunately, this type of failure might not show up for a long time after molding.

Figure 6-4 shows various types of gating. A sprue gate feeds directly from the nozzle of the machine into the molded part. It has the advantage of direct flow with minimum pressure loss. Disadvantages include the lack of a cold slug, the possibility of sinking around the gate, the high stress concentration around the gate area, and the need for gate removal. Most single-cavity molds are gated this way.

A diaphragm gate has a sprue attached to a circular area that leads to the piece. In gating hollow tubes, flow considerations suggest the use of this type of gate. The part could be gated by a single, double, or quadruple runner coming from the sprue, but four runners, 90° apart, would give four flow or weld lines on the molded part, which is often objectionable. Additionally, the diaphragm gate gets rid of stress concentration around the gate because the whole area is removed. The cleaning of a diaphragm gate is more difficult and time-consuming than cleaning of a sprue gate. In some instances, gating internally in a hollow tube is not practical. Ring gates accomplish the same purpose from the outside.

The most common gate is the edge gate,

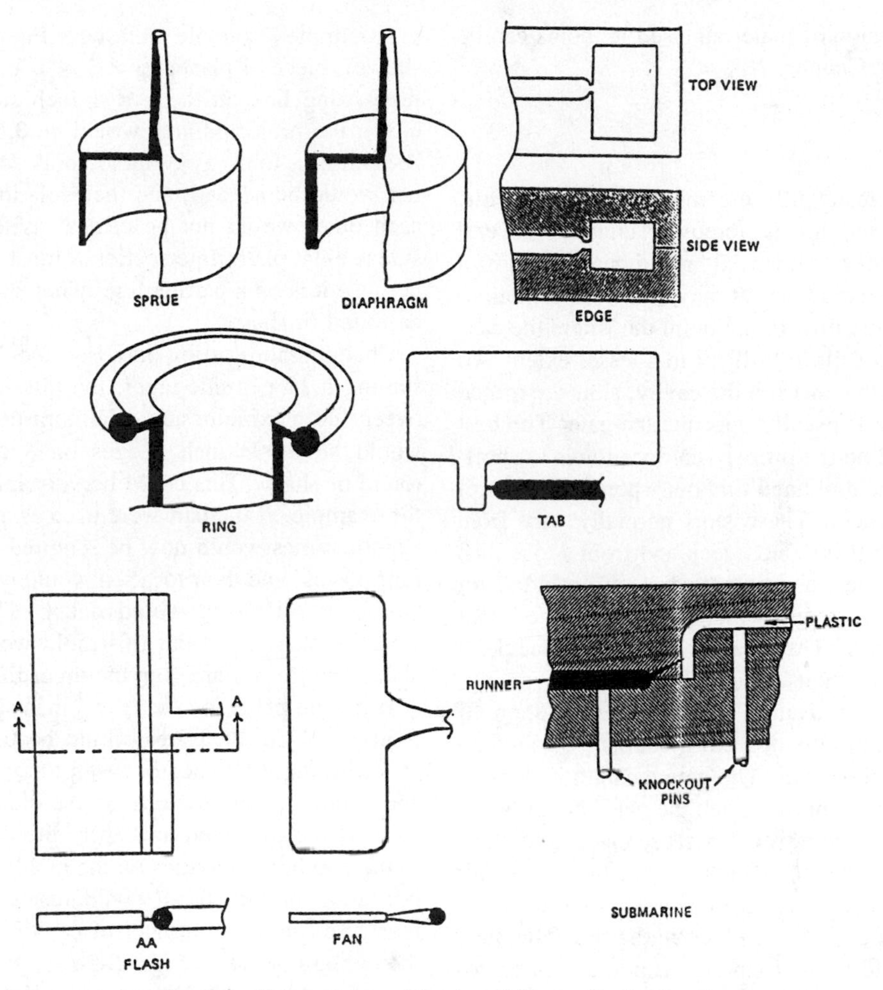

Fig. 6-4. Gating designs. (*Courtesy Robinson Plastics Corp.*)

where the part is gated by either a restricted or a larger gate on some point the on the edge. This is an easy method from the point of view of mold construction. For many parts it is the only practical way to gate. When parts are large, or there is a special technical reason, the edge gate can be fanned out, making what is called a fan gate. If the flow pattern is required to orient in one direction, a flash gate may be used. Here the fan gate is extended the full length of the piece but kept very thin, comparable to a piece of flash.

With some materials, gating directly the part can cause jetting or some other surface imperfection. To overcome this, a tab is extended from the part into which the gate is cut. The tab will have to be removed in a secondary operation. This approach is called tab gating.

A submarine gate goes through the steel of the cavity and forms an undercut (steel is in the way of direct plastic removal). When the mold opens, the gate breaks, and the part sticks to the core. The submarine gate works because of the flexibility of the runner and judicious placement of the knockout pin. This is a highly desirable means of gating and is very often used in automatic molds.

Moldmaking

Various techniques for manufacturing molds, as well as a basic guide to ordering molds, and a discussion of materials used in molds can be found in Chapter 7.

Venting

When plastic fills the mold, it displaces air. Unless the air is removed quickly, several things may happen. It may ignite, causing a characteristic burn. It may compress enough to restrict the flow to the point the where the cavity cannot fill or will fill to a lesser extent. To remove the air from the cavity, slots are milled or ground, usually opposite the gate. The best way to find the proper vent location is to short-shoot the mold and find out where the fronts of plastic meet. These slots normally vary from 0.001 to 0.002 inch deep and from $\frac{3}{8}$ to 1 inch wide. After the vents have been extended for $\frac{1}{2}$ inch, the depth is increased, usually to about 0.025 inch. The clearances between knockout pins and their holes also provide venting, though they usually cannot be relied upon. If more venting is needed, a flat can be found on the knockout pin. The gate location is directly related to venting. Often the gate cannot be located in certain places because that would make it impossible to vent the mold or prevent air entrapment. A typical example is the impossibility of gating a tumbler on the rim. The plastic would flow around the rim and then up toward the top, forcing the air into the upper part of the tumbler. No amount of vent pins or venting devices would successfully remove the air from that location because it is not on the parting line.

Parting Line

When a mold closes, the core and the cavity meet, producing an air space into which the plastic is injected. The mating junction appears as a line on the molded piece, called the parting line. A piece may have several parting lines. Selection of the parting line is largely influenced by the shape of the piece, method of fabrication, tapers, method of ejection, type of mold, aesthetic considerations, post-molding operations, inserts, venting, wall thickness, number of cavities, and location and type of gating.

A simple example follows: Figure 6-5a shows a piece of plastic $\frac{3}{16} \times \frac{1}{2} \times 1$ inch. With the parting line on the $\frac{1}{2} \times 1$ inch side (Fig. 6-5b), the projected area would be 0.5 sq. in. Assuming a four-cavity mold, only four cavities would be needed, and their relationship to each other would not be critical as the other side is a flat plate. Intersection *R* must be sharp because it is on a parting line, whereas *S* could be round or sharp.

Changing the parting line (Fig. 6-5c) and assuming a 1° per side taper, the difference between the maximum and minimum dimension would be 0.003 inch. Edges on *S* could be round or sharp. This could be very important, for example, if the part were used as a handle. Eight cavities would now be required to make four pieces, and their location would be important or a mismatch would occur. The parts would look similar, but this mold would cost about one-third more than the preceding one.

If it were parted on the $\frac{3}{16} \times \frac{1}{4}$ inch side (Fig. 6-5d), the projected area would be 0.093 sq. in., with the difference between the maximum and minimum dimensions of the sides 0.035 inch. If the projected area were the limitation to the number of cavities on the mold, the configuration in Fig. 6-5d would result in more than five times the number of cavities used in the configurations in Figs. 6-5b and 6-5c. The cavity could be EDM'd but not polished; so it would have to be either made in two pieces or electroformed. If the part were gated at the parting line, the air would be trapped on the top; so a three-plate, hot, or insulated runner would be required. This might double the original mold cost. Additionally, the part in Fig. 6-5d might not even be recognized as the original part.

It thus can be easily seen that selection of the parting line, like all other mold design items, will affect the piece, its cost, its physical properties, and its appearance.

Ejection

After a part has been molded, it must be removed from mold. It is mechanically ejected by KO (knockout) pins, KO sleeves, stripper plates, stripper rings, or air, either singly or in

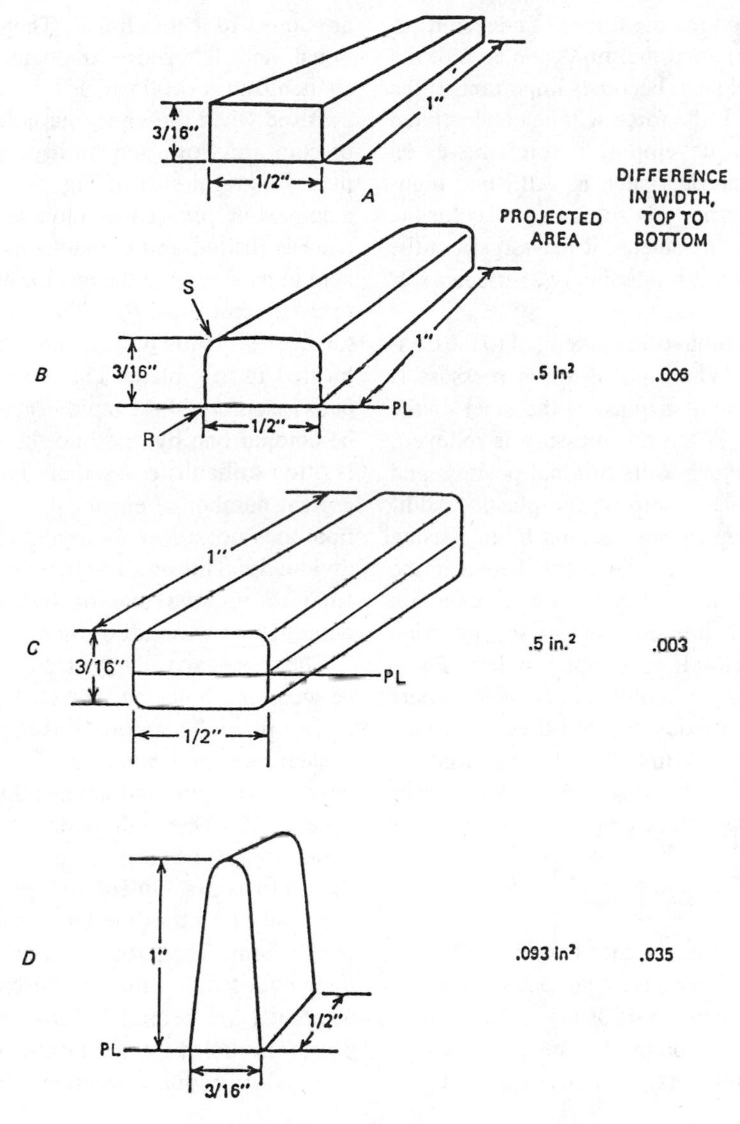

Fig. 6-5. Effect of parting line on the shape of a molded part.

combination. The selection of the type and location of the ejection mechanisms affects the molded part in ways similar to the selection of the parting line. The most frequent problem in new molds is with the ejection of the part, as there is no mathematical way to determine the amount of ejecting force needed. It is entirely a matter of experience. The larger the area of the ejection system, the lower the pressure on the part, thus permitting lower ejection temperatures and less stress from the knockout system. It is poor practice to build a mold unless the part will eject properly. Unless a mold ejects consistently, it is impossible to produce parts economically. Mold release never should be used except in an emergency, or unless it is more economical to finish the run by using it instead of stopping and repairing the mold.

The geometry of the part, the location of the gate(s), and the plastic material are the most important parameters in ejection. It is desirable to have a minimum draft of 1° per side, though many parts will work with less draft. The quality of polishing is important. Polishing and stoning only in the direction of ejection often will solve a difficult problem. Because ejection involves overcoming the forces of adhesion between the plastic and the mold, the area of the knockout mechanism becomes important. If the area is too small, the force will be concentrated (high pressure), developing severe stresses on the part. In materials such as ABS and high-impact polystyrene, this force can discolor the plastic. In other instances, it may so stress the part that it will fail immediately or in later service.

Sticking in a mold often is related to the elasticity of steel. When the injection pressure is applied to the molten plastic, the steel of the mold deforms. When the pressure is relieved, the steel will return to its original position and then will act as a clamp on the plastic. Additionally, packing causes sticking by increasing even more the adhesive forces between the plastic and the mold. Very often, a reduction of the injection pressure and/or the injection forward time will eliminate the problem. Packing is also common in multicavity molds where the individual cavities do not fill equally. One cavity will seal off first, and the material intended for that cavity will be forced into other cavities, causing overfilling.

Ejector Mechanisms

Ejector pins are made either from an H-11 or a nitriding steel. They have a surface hardness of 70 to 80 R_c, to a depth of 0.004 to 0.007 inch. The inside core is tough. The heads are forged and annealed for maximum strength, and they are honed to a fine finish. They come in fractional and letter-size diameters, each being available in a 0.005 inch oversized pin. They are used when the knockout holes in the cavity or core are worn and flash occurs around the pins. The right side of Fig. 6-6 shows the way a knockout pin (L) is mounted. The ejector plate is drilled and countersunk. The pins are held in by screwing the ejector retainer plate (J) to the ejector plate (G). The ejector pins, ejector sleeves, sprue puller, and return pins are all located in this plate. This construction facilitates assembly of the mold because the pins can be entered one by one into the cavity plate. It is often difficult to assemble large molds with a great number of ejector pins if the construction does not allow the pins to be inserted individually. This construction also makes it possible to remove one or two knockout pins without removing all of them.

There is nearly always a slight misalignment between the holes and the cavity plate and the ejector plate. Therefore, it is important to leave a clearance of from $\frac{1}{64}$ to $\frac{1}{32}$ inch around the heads of the pins and at least 0.002 inch clearance at K. This will permit the pins on the counterbore to find their proper location when the mold is assembled. Return pins should be used only to return the knockout plates, not to guide them. The knockout plates should be on their own leader pins and bushings. Chamfers at point I are helpful for easy insertion of the pins. The holes for the ejector pins should be relieved to within a fraction of an inch of the

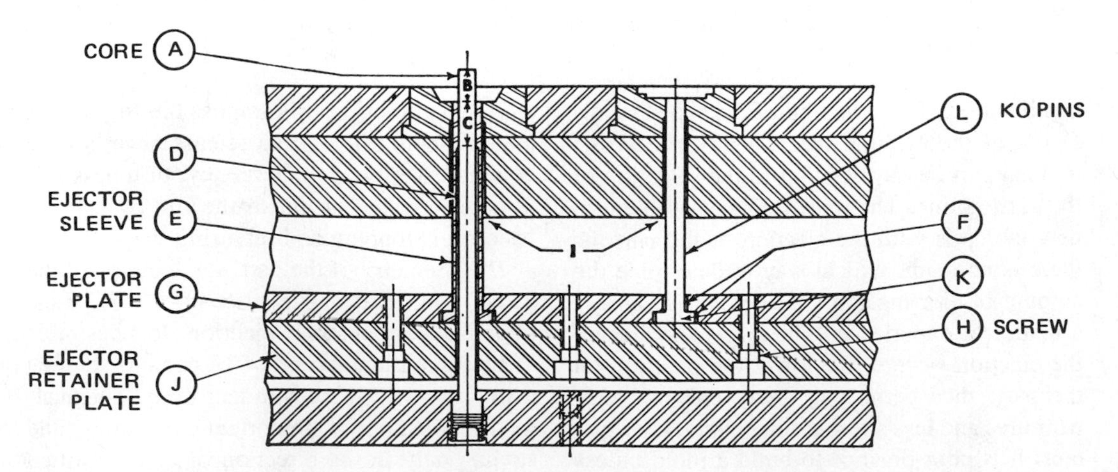

Fig. 6-6. Ejector (KO) pins, ejector plates, and stripper pin and sleeve. (*Courtesy D-M-E Co.*)

face of the cavity or core to facilitate the alignment and operation of the pins. The top of the knockout pins will leave a circle on the molded parts.

Ejector Sleeve

Ejector sleeves labeled E in Fig. 6-6 are preferred when molded parts have to be stripped off round cores. They are subjected to severe stress and wear, so their inside and outside surfaces must be hard and finely polished. If they are not sufficiently hard and are of different hardnesses, scoring of both the cavity and the core may take place. Additionally, both the cavity and the core must be of a different hardness compared to the ejector sleeve. Two parts of equal hardness, regardless of how hard they are, will scour. The lower portion of the sleeve should be drawn to obtain maximum toughness, while the upper part should be left hard for the full length of the ejector movement.

The outside diameter of the sleeve should be about 0.001 to 0.002 inch smaller than the hole in the cavity. An equal clearance between the core and the sleeve should be maintained for a distance *C*. The inside diameter of the sleeve should be about $\frac{1}{64}$ to $\frac{1}{32}$ inch larger than the core, leaving a clearance as indicated by D. The core (A) should be dimensioned so that the portion that extends into the molded article (distance *B*) is at least $\frac{1}{64}$ inch smaller in diameter than the lower part. If this is not done, the reciprocating movement of the sleeve will damage the fine finish of the core. Distance *C* should be at least $\frac{3}{8}$ inch longer than the entire movement of the ejector plate. If the clearance extends too far, the shoulder and the end of the core may be damaged when the sleeve is retracted. It is also important to leave a clearance of $\frac{1}{64}$ inch around the outside of the sleeve. This clearance, however, should not extend too far because it is necessary to have a bearing at least $\frac{1}{2}$ inch long at the cavity.

Stripper Ring

The ejector sleeve requires that a core be attached beyond the ejector plates, usually to the backup plate. In many instances a stripper ring

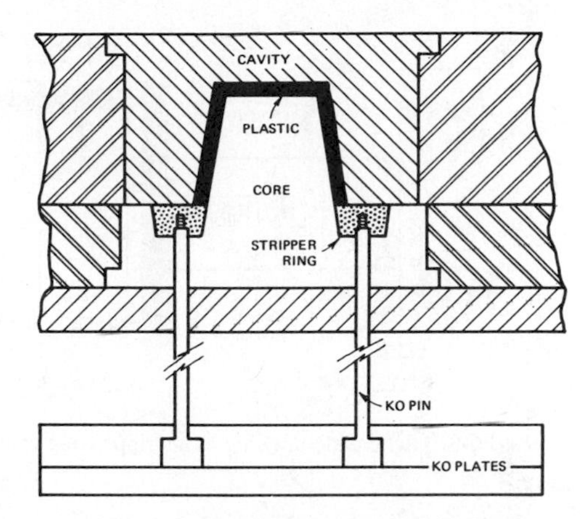

Fig. 6-7. Stripper ring type of ejection.

can be used (Fig. 6-7). The core then can be attached to the core-retaining plate. A ring is inserted, which is directly attached by stripper bolts to the knockout plates. An advantage of the stripper ring is that other knockout devices, such as knockout pins, can be attached to the same plate. This would be useful in molding a container and its cover at the same time.

Stripper Plates

If, instead of mounting the stripper rings directly on the knockout plate, the stripper rings were mounted on their own plate and that plate were mounted on the knockout plate, a stripper plate mold (Fig. 6-8) would result. Hardened inserts should be put around the core and inserted into the stripper plate to prevent wear and for ease of maintenance. Stripper plate molds give a larger surface area for ejection than knockout pins. Heretofore, they primarily were used for round cores. With the use of EDM equipment, irregularly shaped stripper plate molds are economically feasible.

Vent Pins

Consider the case of molding a deep container, as shown in Fig. 6-9. After the cavity separated from the core, the atmospheric air pressure would make it difficult, if not impossible, to remove the part. To overcome this difficulty a vent pin is used, which is held in its normally

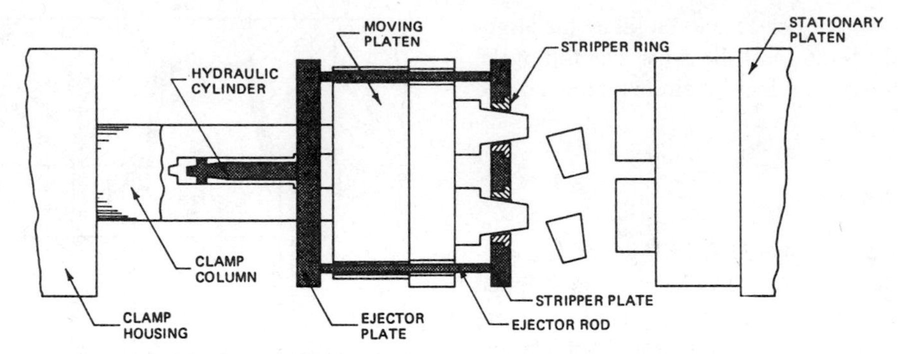

Fig. 6-8. Schematic drawing of a stripper ring ejection. (*Courtesy Husky Injection Molding Systems Ltd.*)

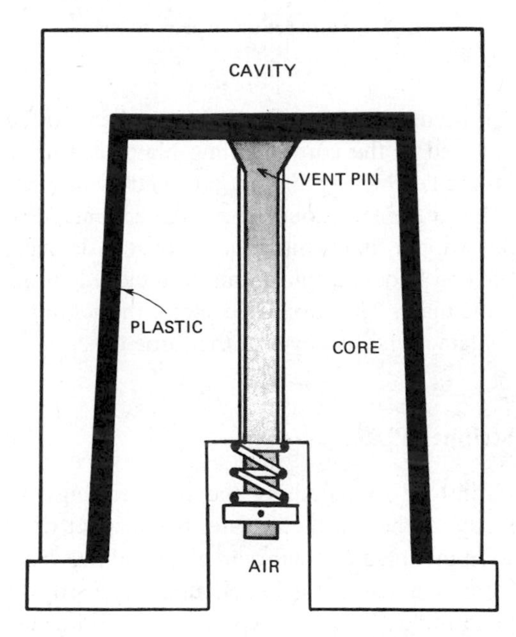

Fig. 6-9. Use of a venting pin to break the vacuum on the core. The pin is held closed by a spring; the pressure of the material on the head of pin forces it tightly closed. (*Courtesy Robinson Plastics Corp.*)

closed position by the spring. When the material is injected, the pressure of the material on the head of the pin forces it tightly closed. When the part is to be ejected, the pin will move up when the knockout system is activated, venting the interface between the core and the plastic. Additionally, air can be used to help blast the part off. In that instance, the plastic part acts like the piston of an air cylinder. Often this provides enough force for ejection.

Cam Actions

A cam-acting mold is used to overcome the effect of an undercut, which is an interference by the mold that will prevent mechanical ejection of the part. There are two types of cam action—one moves the cams independently of the machine action, and the other moves the cam by the molding machine action.

Figure 6-10 shows a schematic section of a cam used to make a hole in the side of a plastic box. The cam pin is attached to the cam slide, which moves on wear plates. The slide is activated by either an air or a hydraulic cylinder. The locking device is not shown. This cylinder can be activated at any time in the cycle. For the plastic part to stick on the core, the cam pin must be retracted before the mold opens. This would be impossible in a mold where the cam action is worked by the machine opening.

Figure 6-11 shows a standard cam action. The cam slide is attached to Plate A, which is the moving side of the mold. The cam pin and the cam lock are attached to the stationary side of the mold. In the mold closed position the cam pin is held in place by the cam lock. As the mold opens, the cam slides up on the cam pin, moving toward the right. The amount of the cam slide movement will depend on the cam angle and how far the slide moves on the cam pins. By the time that the mold is opened, the cam slide has moved enough that the cam pin has been fully withdrawn from the plastic part. Following that, the part is ready for normal ejection. There are other ways of camming mechanically, but they all work on the same prin-

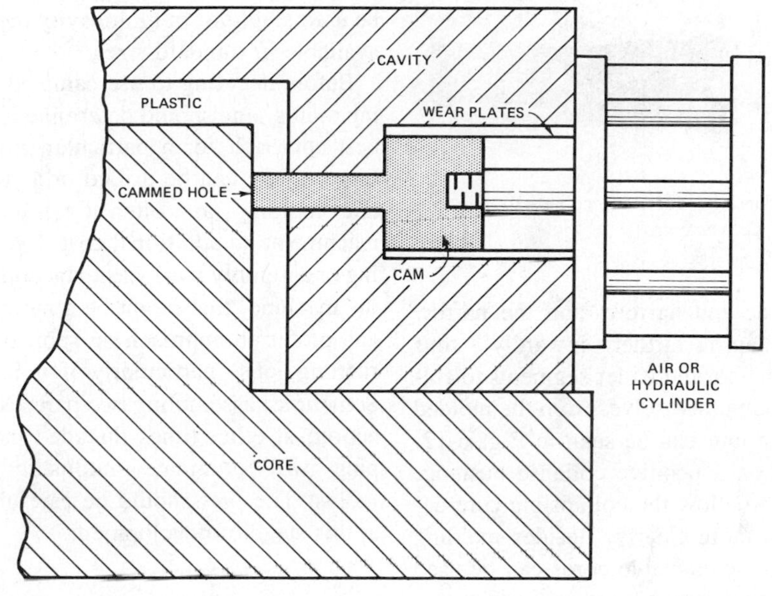

Fig. 6-10. Schematic drawing of an externally operated cam for molding a hole in plastic. (Locking device is not shown.)

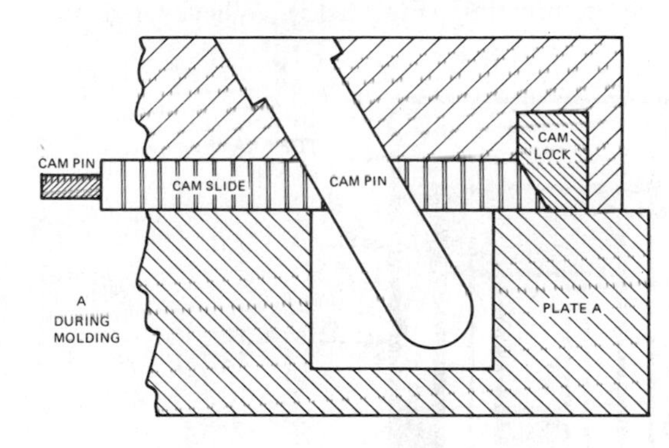

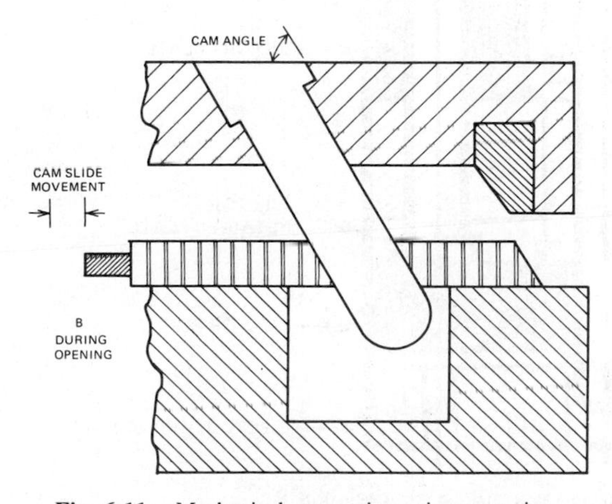

Fig. 6-11. Mechanical cam action using cam pins.

ciple as that of an angle pin. If the cam pin should come out of the cam slide, a method of holding the cam slide in place is required.

External threads can be molded readily by using cam pins and slides to separate the cavities themselves. Such movement of the cavity block is a common form of cammed mold.

Molding internal threads or external threads when cams are not used requires unscrewing devices. The usual methods include a rack and pinion, where the rack is moved by the machine action, a hydraulic cylinder, a pneumatic cylinder, or an electric motor. Unscrewing can also be obtained by the use of an individual hydraulic motor for each core, or by a gear and chain mechanism, motor-driven.

Another way to mold internal threads and undercuts is to use a collapsible core, which consists of a circular, segmented core or sleeve into which the internal threads or configuration is ground. When a center core pin, which backs up and holds the segments in the molding position, is retracted by normal ejector plate travel, the segments collapse inwardly, because of their inherent flexibility, pulling themselves away from the molded undercuts and allowing the part to be ejected. The segments are

Fig. 6-12. Collapsible core in collapsed position. (*Courtesy D-M-E Co.* [*patented*])

alternately wide and narrow, and the narrow segments collapse farther inwardly, thus ''making room'' for the wider segments to collapse also, freeing themselves from the molded undercuts. The unit can be seen in Fig. 6-12. A third member, a positive collapse bushing, is not shown, to allow the collapsible core detail to be seen more clearly. Besides molding internal threads, collapsible cores can be used to mold parts with other types of undercuts—such as internal protrusions, dimples, grooves for cap liners, interrupted threads, and window cutouts—completely around the part. They can be used singly or in multicavity molds, and are available in standard sizes.

Before deciding to use cammed or unscrewing molds, one should determine if they are the best approach for a particular project. Sometimes a hole can be drilled or a slot machined after molding, and often it can be done at the machine at no additional cost. Even if machining or assembly work has to be done away from the machine, that might be a more economical and preferred approach on short or even long-running jobs, particularly if it is automated. Sometimes assembling two parts can avoid cam action; at other times threaded inserts or tapping will eliminate automatic unscrewing molds. The part should be carefully reviewed at this stage of development.

Early Ejector Return Unit

A typical mold base accessory is an early ejector return unit (Fig. 6-13). Whenever a me-

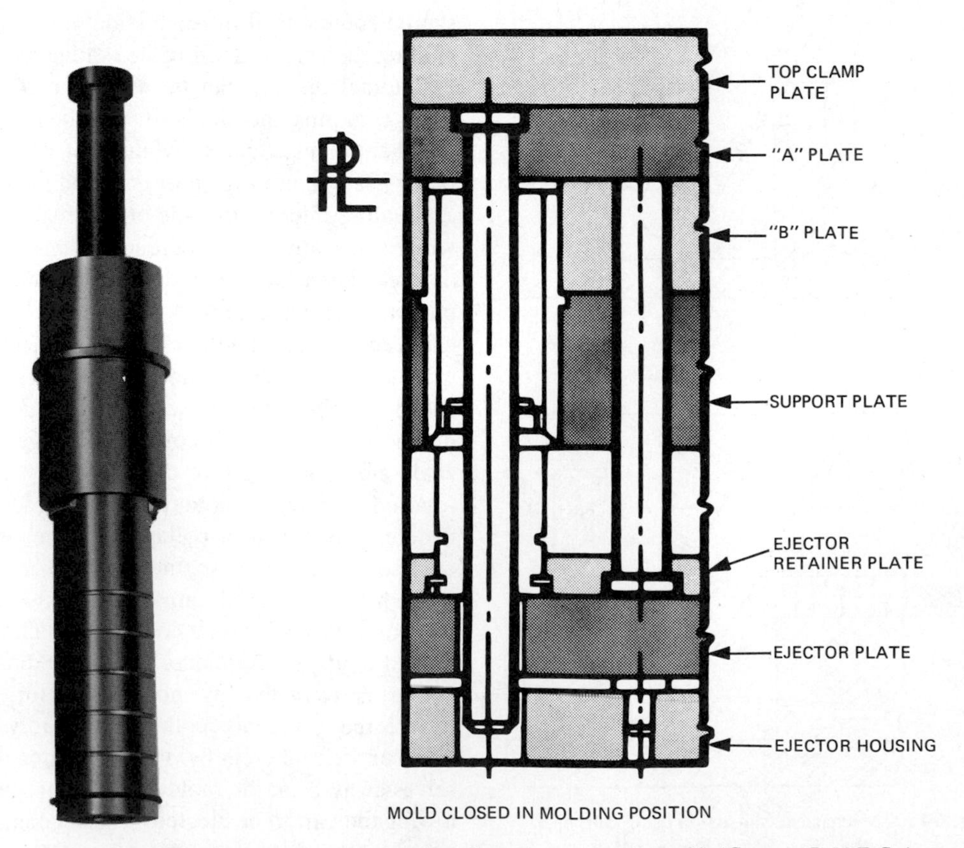

Fig. 6-13. Early ejector return units showing schematic assembly in mold. (*Courtesy D-M-E Co.*)

chanically operated cam slide passes over an ejector pin, the ejector plate must be returned "early," or before the mold is closed; otherwise the returning cam will slide and hit the ejector pins with mold damage. This is not the case if the machine or mold has air or hydraulically operated knockout plates. To prevent this problem in other types of molds, an early ejector return unit can be used. The unit consists of a bushing, a post with slidable cam fingers, and a cam actuating pin. The bushing is installed in the B plate, the post is attached to the ejector plate, and the cam actuating pin is installed in the A plate.

In operation, the early return of the ejector plate is accomplished while the press is closing, by the cam actuating pin pushing against the projecting cams on the post (and thus returning the ejector plate) until the cams are released into a matching countersink, which happens when the ejector plate is fully back. The cam pin then passes on through as the mold continues to close. Timing is regulated by adjusting the length of the pin.

MOLD COOLING

The mold, in effect, is the containment for the molten plastic and the vehicle through which heat is removed from the plastic to solidify it to the point of ejection. It is obvious that proper selection of mold cooling channels is essential for economic operation. Mold temperature control consists of regulating the temperature and the flow rate of the cooling fluid. The functioning of the mold and the quality of the part depend in large measure on the location of the cooling channel. It is difficult to have "too much" cooling because the amount is always regulated by the temperature and velocity of the circulating cooling fluid. If there are "too many" cooling channels, they can always be blocked off and not used.

The second factor in mold temperature control is the selection of the material itself. Beryllium copper molds have a high thermal conductivity, approximately three times that of steel and six times that of stainless steel. Because the time required for cooling plastic in the mold is a function of heat removal, a beryllium copper cavity will remove heat more rapidly than a steel one, although not in the three-to-one ratio; one still has to wait for the heat to come out of the plastic, and the steel cooling channel can be 50% closer to the plastic because of steel's greater strength compared to beryllium copper.

Cooling channels should be in the cavities and cores themselves, and in the retaining plate only if needed. The rate of heat transfer is lowered tremendously by the interface of two pieces of metal, no matter how close the contact. The cooling channels should be as close as possible to the surface of the plastic (taking into account the strength of the mold material), and should be evenly spaced to prevent an uneven temperature on the surface of the mold. For practical purposes, it can be assumed that the heat is removed in a linear fashion; that is, equal distances from a cooling channel will be at the same temperature.

The simplest means of cooling is to drill channels through the cavity or core. The minimum size should be $\frac{1}{4}$ inch pipe, with $\frac{3}{8}$ inch pipe preferable; $\frac{1}{8}$ inch pipe never should be used unless its use is dictated by the size limitations of the mold. A center distance of three to four diameters should work well.

Figure 6-14 shows one of the simplest types of water cooling, commonly used in injection molds for rectangular shapes such as boxes and radio cabinets. It consists of a number of channels drilled near the top of the core (the number depending on the size of the core), joined by several risers to circulate the cooling medium. The horizontal channels are drilled from one end of the core, and are drilled and counterbored to accommodate a soft brass plug, which should be a light drive fit, machined flush with the core surface and then polished. Usually a continuous flow is used through the core, and should it be necessary to cross over channels already drilled, the flow of the fluid can be directed by using blank plugs, which are drive fit in the channels.

Figure 6-15 shows a cooling system using channeled inserts within the cavity block and core block. The cavity block and core block are machined to hold the inserts (circular, as shown, or rectangular) before the cavity and

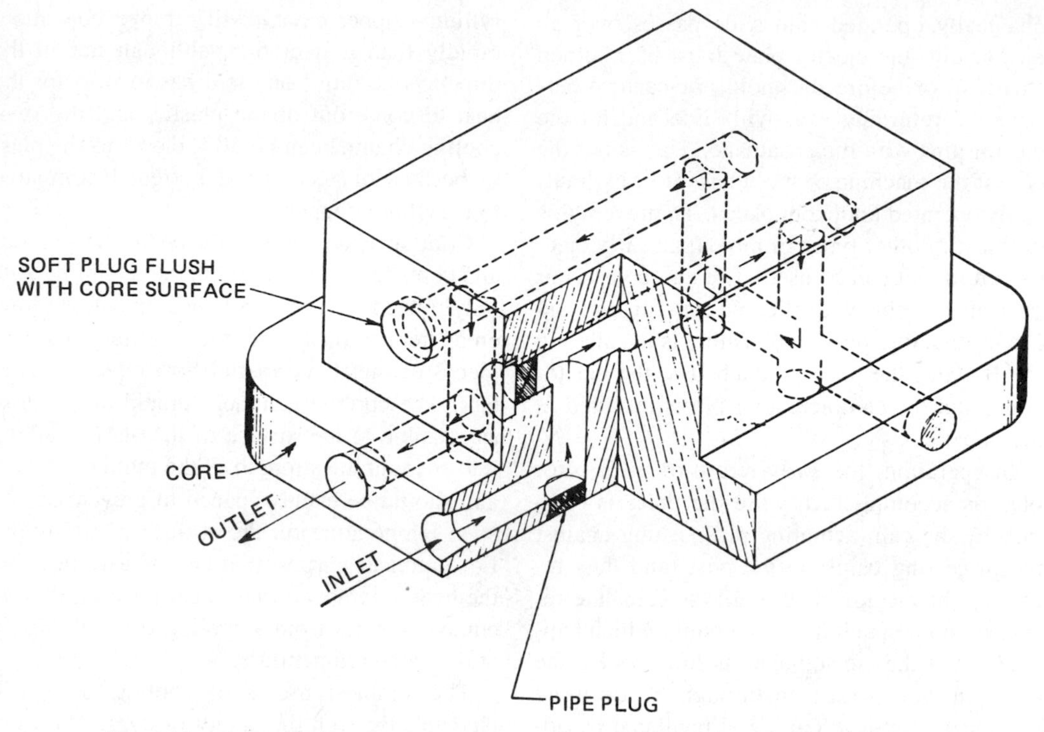

Fig. 6-14. Simple type of cooling channels. (*Courtesy Consolidated Molded Products Corp.*)

core are finished, hardened, and polished. The inserts are made to fit the spaces thus provided. Grooves for the channels are machined in the face of the inserts and then are interconnected and baffled to provide the desired route of flow. The inserts, not hardened, are fastened into place within the cavity and core blocks and sealed by O-rings. In designing the grooves, it is important to leave a sufficient area unmachined to support the cavity or core against distortion by molding pressure.

Figure 6-16 shows a bubbler in the core and O-rings to seal off the cooling medium or confine it to each individual cavity or core in the chase. The underside of the cavity or core may have a pattern of grooves similar to that of the channeled insert in Fig. 6-15, or it may be as shown.

The two main means of directing cooling fluid in the mold, aside from cooling channels, are bubblers (Fig. 6-17) and baffles (Fig. 6-18). A baffle is a flat strip of brass or stainless steel (to prevent rusting) inserted in the core to divide the channel in the core into two separate channels so that the cooling fluid is forced to flow up one side of the baffle to the end of the core, and down the opposite side and out. The thickness of these baffles ranges from $\frac{1}{32}$ to $\frac{1}{8}$ inch, depending on the size of the channel in the core. The baffle is never extended to the extreme end of the core because the fluid must pass over the end of the baffle to flow down to the opposite side. The baffle may be attached by brazing or machined with a mechanical force fit. The water seals of Figs. 6-17 and 6-18 can be achieved by threading or by the use of O-rings. It should be noted, in Fig. 6-17, that if the water connection were reversed and the pressure low, the pipe would be filled, and the water might dribble over the side without ever contacting the core itself. This would result in almost no cooling. On the contrary, when the water is connected correctly, no matter how low the pressure or flow, the core has to fill with water until it reaches the top of the tube, from which it is drained.

Cooling with baffles is an excellent method if there are not too many hooked up in series.

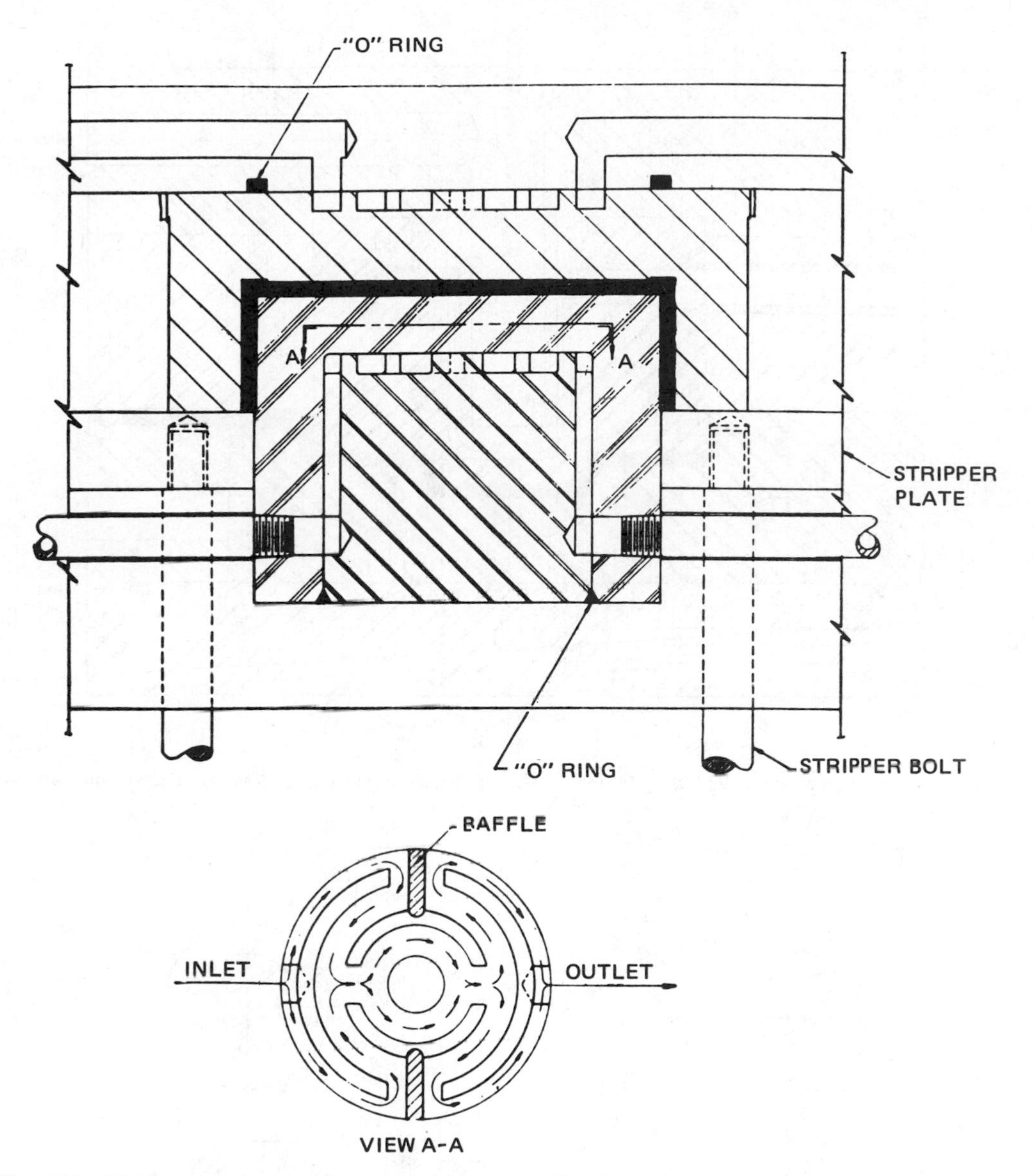

Fig. 6-15. Mold construction showing a more elaborate cooling system. (*Courtesy Consolidated Molded Products Corp.*)

Consider ten pins, 4 inches high and $\frac{1}{2}$ inch in diameter, with baffles hooked up in series (Fig. 6-19). The first pin will be cooler than the last pin. The total resistance to flow will drastically reduce the amount and velocity of the cooling fluid flowing through the system, and thus will markedly decrease the cooling capacity. Additionally, if there is any blockage because of dirt, or for any other reason in one pin, the whole unit is inoperative. In practice this design would not work, and the pins would be too hot for molding.

Figure 6-20 shows the correct way to cool the same pins, using parallel cooling. This can be done with pins or with any other series of bubbler action in a core or cavity. This type of cooling will give significantly faster molding compared to that shown in Fig. 6-19.

The excellence of the cooling system depends on the ingenuity and the ability of the

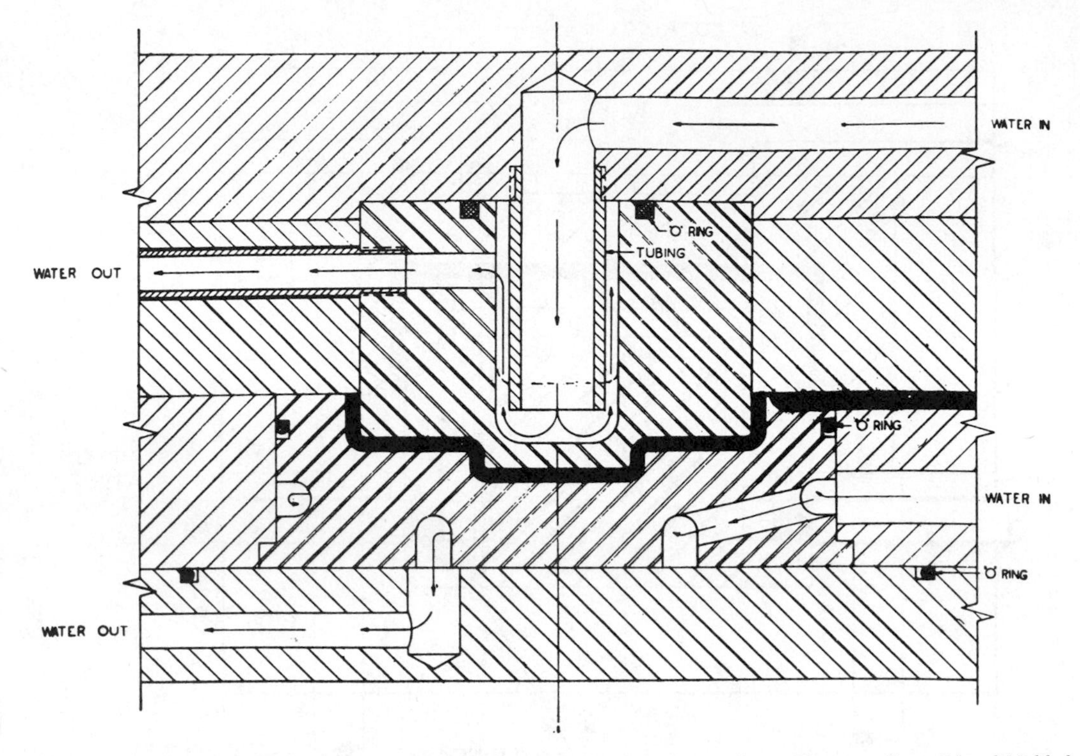

Fig. 6-16. Cooling with a bubbler and using O-rings to seal off the cooling medium. (*Courtesy Consolidated Molded Products Corp.*)

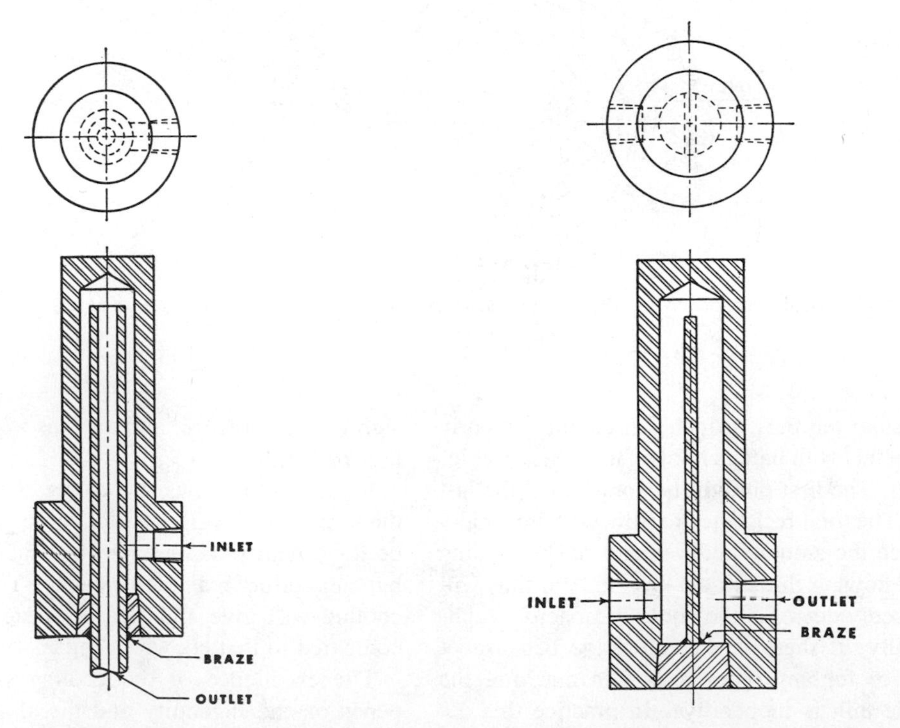

Fig. 6-17. Bubbler-type cooling. (*Courtesy D-M-E Co.*)

Fig. 6-18. Baffler-type cooling. (*Courtesy D-M-E Co.*)

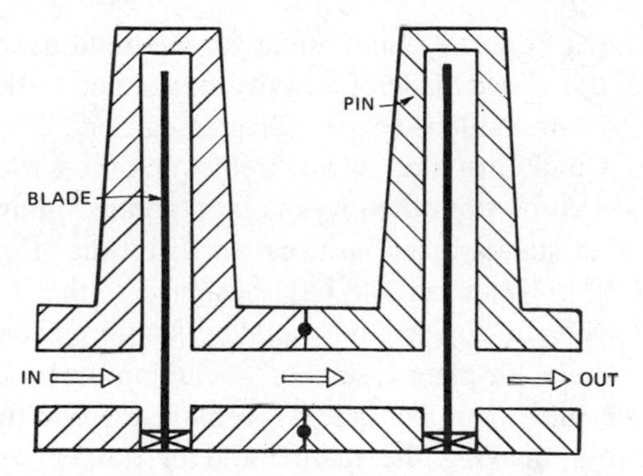

Fig. 6-19. Series cooling of pins, poor design. (*Courtesy Robinson Plastics Corp.*)

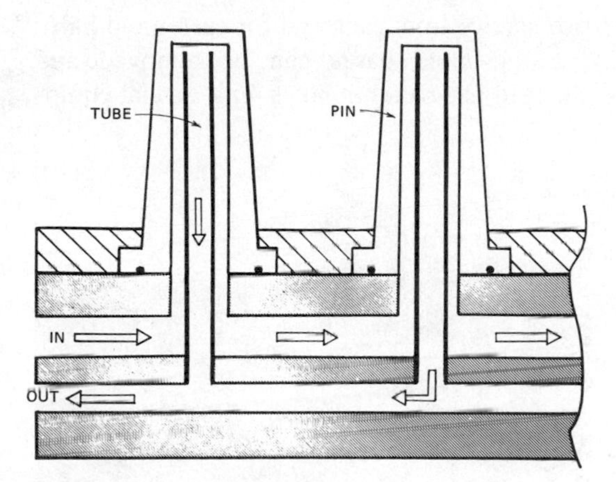

Fig. 6-20. Parallel cooling of pins, good design. (*Courtesy Robinson Plastics Corp.*)

mold designer and the moldmaker. No aspect of moldmaking affects the cycle and the quality of the molded part more than the selection and the location of the cooling channel.

STANDARD EQUIPMENT

Standard Mold Bases

In the early years of injection molding, moldmakers had to build their own mold bases, a costly and inefficient process, as most moldmakers do not have the large equipment for such machining. The manufacture of standard mold bases, an important factor in the development of more efficient moldmaking, was started in 1942 by I. T. Quarnstrom (a moldmaker is now known as the D-M-E Co.). Others have followed suit.

By standardizing on the mold assembly and the component parts, especially those that come in contact with the press platens such as the locating ring, sprue bushing, and ejector plates and housing, the mold became interchangeable among all makes of injection molding machines. All of these parts, together with other parts such as leader pins and bushings, return pins, sprue puller pins, and ejector pins, were made available as stock items for replacement. Standard mold base assemblies for injection molds also could be rearranged for use in compression or transfer molds.

All mold components incorporated in the mold base assemblies are also available as separate items so that moldmakers can make up their own special bases. The maximum-size mold plates (A and B) are $45\frac{3}{4} \times 66 \times 5\frac{7}{8}$ inches. The plates are prefinished; that is, they are finish ground on all sides to extremely close tolerances. The plates are available in medium carbon steel, SAE 1030, known as DME #1 mold steel; type AISI 4130 heat treated to 300 Bhn, known as DME #2 steel; and DME-#3, a P-20 AISI 4130 heat treated to 300 Bhn, that also can be used for cavities and cores.

A standard injection mold base (Fig. 6-21) consists of a stationary part that is attached to the stationary platen, through which the material comes, and an ejector or movable part attached to the movable platen, which contains the knockout mechanism. The top or stationary

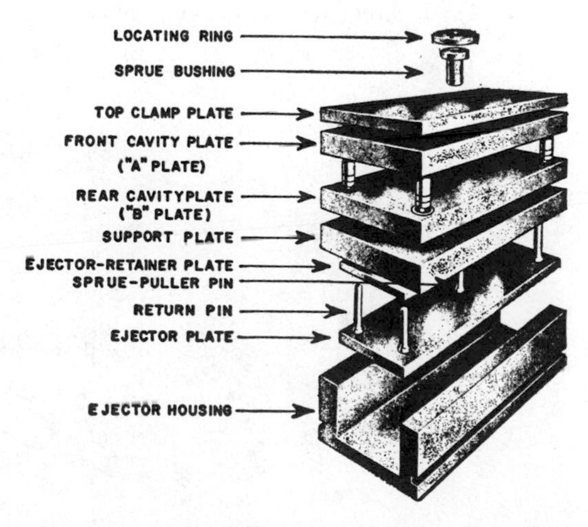

Fig. 6-21. Exploded view of a standard mold base, showing the component parts. (*Courtesy D-M-E Co.*)

side has a top clamp plate and a top cavity or A plate. This side contains a locating ring for positioning to the press platen so that the sprue bushing, which has an orifice through which the plastic material enters the mold under heat and pressure, is directly in front of the nozzle opening of the cylinder. Leader pins are also contained in the upper cavity plate and enter into leader pin bushings on the other side to align the mold halves.

The ejector, movable, or lower half of the mold has the lower cavity or B plate, which rests on its support plate. An ejector housing is mounted on the support plate. The ejector housing contains the lower clamp slots; it also incorporates within it the ejector plate and the ejector retainer plate, between which are held the ejector pins or other devices that eject the molded part from the mold. The return pins and the sprue puller pin also are held between these two plates. The return pins return the ejector plate as the mold closes and pull or "break" the sprue at mold opening. The sprue puller pin is located opposite the sprue. The ejector plates should ride on their own leader pins and bushings.

Cavities and cores can be mounted in any one of four ways: (1) surface-mounted on the A and B plates (or cavity retainer plates as they are often called); (2) inserted about 0.015 inch above the surface in blind pockets milled partway through the A and B plates; (3) inserted in holes cut completely through the A and B plates; (4) cut directly into the cavity plates. In the last case, it is important for the moldmaker to use plates made of a cavity steel, such as the P-20 prehardened type indicated earlier.

Quick installation and removal of cavities and cores from the press is incorporated in another standardized unit, called a unit die (Fig. 6-22). This is a mold base designed so that the upper and lower cavity plates, support plates and ejector plate assembly, including the ejector pins, can be quickly removed from the press, leaving the main mold or holder still clamped in position. A unit assembly for molding different parts then can be mounted back into the holder, being fastened down with only two screws from each end for each mold half.

Large mold bases can be removed and changed in several minutes with special equip-

Fig. 6-23. Standard cavity insert rounds, showing bored holes in upper and lower cavity plates to receive inserts.

Fig. 6-22. Standard two-cavity unit die, showing a standard replacement unit in a semiremoved position.

ment external to and on the machine. These quick mold changers are used for "just-in-time" deliveries and are expensive.

Standard Cavity Inserts

Standard, rectangular cavity insert blocks also are available in sizes of from 3 × 3 inches to 6 × 8 inches, and in thicknesses of from $\frac{7}{8}$ to $4\frac{7}{8}$ inches. In addition, round cavity inserts are also available from stock in 1- to 4-inch diameters in $\frac{1}{2}$-inch increments, and in thicknesses of from $\frac{7}{8}$ to $3\frac{7}{8}$ inches in $\frac{1}{2}$-inch increments. All of the cavity inserts have a ground finish on all sides and surfaces, ready for immediate cavity cutting. The rectangular inserts come in P-20 type steel, H-13 type steel, and T-420 type stainless steel. The rounds come in 20 and H-13 type steels. Round inserts are shown in Fig. 6-23.

7

Injection Mold Manufacturing

THE MOLDMAKER

The moldmaker, to a very large extent, must be credited with the many technical advancements achieved in the applications of plastics. The many talents and abilities of the moldmaker in turning raw steel into highly crafted, sophisticated process molds have been responsible for many of the plastic products used today. These molds are best utilized in combination with advanced microprocessor-controlled plastics processing equipment along with the latest in hydraulic technology. A judicious combination of processing, molds, and equipment has made possible the economical production of top-quality products in large quantities. These developments have enhanced the application and advancement of both commodity and engineering plastic resins, converted to products used in industry today.

Although the talents of these moldmakers are utilized in the production of a wide variety of molds including blow molding, compression, and transfer molding, this chapter will mainly concentrate on the injection mold phase of tooling.

Like most phases of manufacturing, the construction of any mold starts with a need to produce a desired plastic item of some useful shape and configuration. The idea for a product may come from the company's own design-engineering group or may be initiated by a customer. The common element is that an estimate or quote must be developed for the cost of the mold in order for a tooling appropriation or purchase order to be generated. This estimate is often the most difficult phase of the project, as the total cost is extremely dependent on the type and quality of the final mold. Therefore, it is imperative that specifications for the mold be established prior to the development of the estimate.

By Robert Dealey, D-M-E Company, Madison Heights, MI.

DESIGN AND CONSTRUCTION

Number of Cavities

The number of cavities that will be included in the mold has to be determined during the preestimation stage. Several factors influence the ultimate number of cavities that should be selected. The first consideration has to be given to the quantity of product required. Obviously, the mold must have the ability to meet production requirements. A general rule of thumb is that the mold must have the capability of producing twice the requirements as peak production times to allow for uninterrupted flow of product, while also allowing for maintenance downtime of the mold. The cycle time for injection molded articles varies, depending on the part thickness, size, plastic material, configuration, and operation of the mold. Although cycle times have varied from as short as a few seconds to as long as five minutes for extremely thick parts, the average cycle generally will fall between 20 and 60 seconds for a well-designed injection mold with the proper temperature control system. (See Fig. 7-1.)

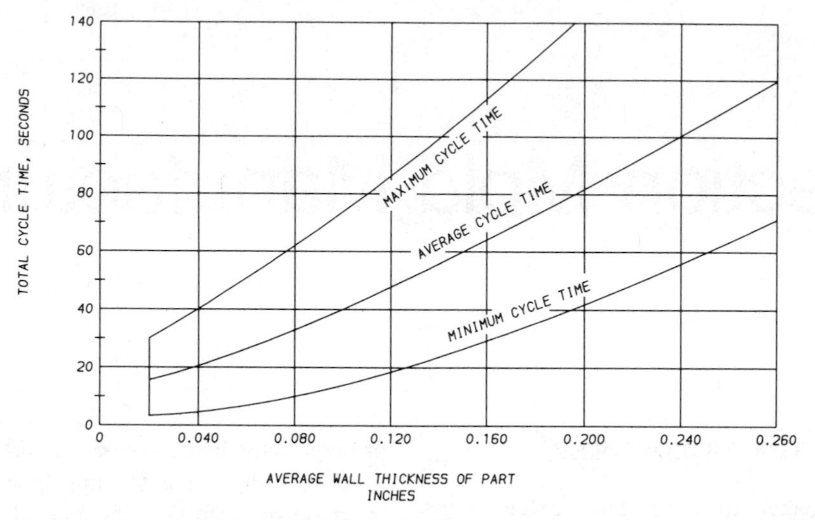

Fig. 7-1. Chart showing approximate cycle time as a function of part wall thickness. (*Courtesy D-M-E Co.*)

Machine Size

The second consideration for the optimum number of cavities is the size of the molding machine in which the mold will run. Obviously, the mold must physically fit between the tie bars of the machine. Mold bases are designed to be installed in standard SPI specified machinery with .125-inch total clearance between opposing tie bars. A third consideration is that the molding machine must have adequate clamping pressure to hold the mold closed at the parting line under injection pressure. This clamp force is easy to calculate, as the force generated by the material trying to overcome the clamp pressure is equal to the square inches of projected area (piece part area times number of cavities plus runners) times maximum injection pressure. (If maximum injection pressure is not known, most material manufacturers publish a figure of tons per square inch for their materials, which may be used as a reference.) Four to six tons per square inch is most common and normally can be used for calculations.

Plasticizing Capacity

The maximum amount of plastic required to fill the mold, including sprue, runner system, and cavities, must be less than 60 to 70% of the shot capacity of the molding machine to ensure proper plasticizing of the raw plastic material. Then, a judgement has to be made about whether the dimensional tolerances of the product will allow multicavity molding within the quality limits imposed by the customer. Many customers now are demanding statistical quality control methods, which call for product to be dimensionally produced with three sigma limits falling within 70% of the tolerance zone.

Economics

The last factor in selecting the optimum number of cavities must be an economic decision. For extremely low volumes, anything more than a single-cavity mold is not economically justifiable. At the other extreme, very high volumes of product call for large numbers of cavities. Molds are normally built in multiples that result in a balanced cavity layout, with 1, 2, 4, 8, 16, 32, 64, and 128 cavities normally constructed. A general rule is that each time the number of cavities is doubled, the additional cavity cost is approximately 65% of the preceding cavity cost. Typically, the additional costs for a greater number of cavities must be justified against manufacturing costs over a given length of time, with three years being the average.

Specifications

Other specifications for the mold also must be established prior to quotation. These specifications must include the type of steel for the required mold base, the steel hardness, and the finish of the completed cavities. The style of mold must be determined—whether it will be a conventional runner; three-plate; a runnerless mold, either internally or externally heated, or combination of the two, or even valve-gated. The moldmaker must decide on the placement and style of ejection. Part ejection could be accomplished with a stripper-plate, an ejector sleeve, a two-stage ejection system, or conventional pin ejectors.

Features

Other features of the mold must be specified to ensure that the customer receives the type of mold required from the supplier. Items that not only affect the performance of the mold but also influence the cost include the type, placement, and number of waterlines. The inclusion of guided ejector systems, accelerated knockouts, mold date inserts, wear plates for side actions, early ejector returns, vents to allow volatiles from the plastic to escape, and support pillars, and whether plating of cavities and cores is required, are all factors that will influence the final cost of the mold.

Mold Quotations

SPI has published a specification sheet that would be extremely helpful for the tool engineer to follow when establishing specifications for mold quotation and build, if the tool engineer did not have such a sheet. Where the quoted cost of molds varies, experience has shown that the greatest variables are quality of the finished mold and any extras that a mold supplier has included in the quotation. The advent of computer quoting methods, as they become available to industry, will be of tremendous assistance to mold suppliers in the development of more accurate quotes. Of great benefit to everyone concerned is the practice of dealing with the same group of mold vendors on a repeat basis. The rapport that is established over time provides the opportunity for both the customer and vendor to know what is expected from the tooling that is to be built. This business relationship has proved invaluable for both parties in the delivery of quality tooling within the quoted cost. The customer must remember that a reasonable profit must be made by the mold supplier to allow the continuation of any sound business.

Delivery Time

Delivery time for a mold can vary, depending upon the complexity and the amount of work required to build the mold. This lead time can vary from 6 to 20 weeks, and even longer for extremely large and complicated molds. With today's business climate of compressed lead times, the moldmaker is taking more advantage of standardized mold bases, cavity inserts, components, designer options, and off-the-shelf accessories and is having the basic mold base supplier do more of the cavity pocket, waterline, and mold base machining. With more complicated tooling, including runnerless molding systems, the need to go with production-proven, off-the-shelf standardized components is increasing. These components will help the initial delivery and provide convenient and immediate availability of those items, should maintenance be required during the useful life of the mold.

Placement

Once a mold is placed, the moldmaker has considerable and immediate expenses that must be paid. Generally, the mold is placed as design and build. The detail design of an injection mold varies from 10 to 25% of the total mold cost. In addition, the moldmaker must purchase the mold base and components, the mold steel, and expendable tooling for the construction of the mold. With these costs to the moldmaker, it has become standard practice to pay one-third down with mold placement, one-third with delivery of the mold, and one-third with approval of the completed mold.

Responsibilities

It also is a general practice for the customer to review and approve the mold design prior to the commencement of construction. The responsibilities for plastic material shrinkage and warpage, type and placement of the gate, and parting line location are normally with the customer. Recently, computer programs have been developed to assist in the accurate prediction of these variables. The moldmaker typically is responsible for delivering a mold built to the mold design with dimensional integrity of the cavities and cores consistent with the part drawing.

The mutual trust of both parties is developed by both a long and an ethical business practice. This relationship can be established only by their dealing in good faith with each other, which is important, as the placement into production of complicated mold tooling can be a lengthy and trying experience. It is not uncommon for a mold to require several adjustments to the gates, vents, cavities, and cores to achieve proper dimensional and cosmetic acceptance by the end-user. This phase of mold debugging can be exasperating unless both parties are cognizant of who is responsible for specific phases of development. Again, nothing is more valuable at these times than the ethical business relationships that the parties have developed by working together on previous mold programs.

Standard Mold Bases

Once the formalities of quoting, placement, and design approval have been completed, the actual construction of the mold will begin. The mold base becomes of extreme importance, as it not only will retain and position the cavities and cores, but contains the plastic feed system and the temperature control circuits. The mold base is the means by which the cavities and cores are coupled to the molding machine. Standardized mold bases are available in almost any form desired. (See Fig. 7-2.) These mold bases are designed and available with widths $\frac{1}{8}$ inch less than the standard tie bar spacing offered by machine manufacturers. The standardizing of mold bases was developed by I. T. Quarnstrom of Detroit Mold Engineering (later shortened to D-M-E) during World War II, to better meet the need to supply completed molds quickly for the war effort.

These standard mold bases are available in a variety of widths and lengths, which meet the majority of mold requirements. Standardized mold bases offer many advantages to the moldmaker, including extremely quick delivery, cost savings over one-of-a-kind construction, and a very high-quality product that will complement the quality of the cavities and cores, on which the moldmaker now can concentrate. Additionally, the use of standardized mold bases quickens design time, as placement of the basic components, such as leader pins and bushings, return pins, locating ring and sprue, has been predetermined.

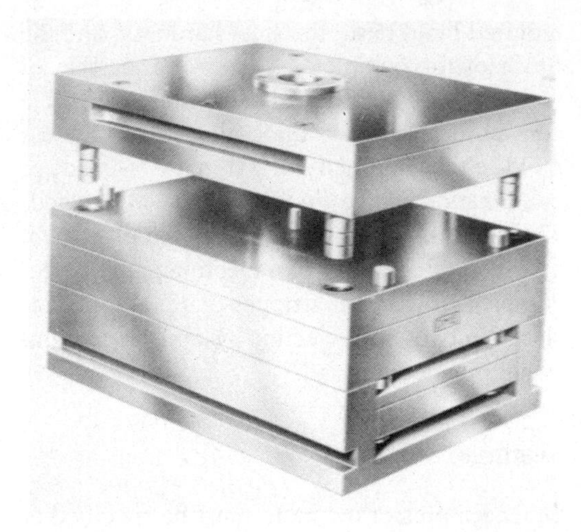

Fig. 7-2. Standard "A" Series Mold Base. (*Courtesy D-M-E Co.*)

Mold Plates

Mold plates also have been standardized to thicknesses of $\frac{7}{8}$, $1\frac{3}{8}$, $1\frac{7}{8}$, $2\frac{3}{8}$, $2\frac{7}{8}$, $3\frac{3}{8}$, $3\frac{7}{8}$, $4\frac{7}{8}$, and $5\frac{7}{8}$. This allows for the removal of any scale that is present on the steel as received from the mill, plus removal of sufficient stock that form a flat and parallel plate from the material. These thickness stackups are beneficial, as designer options are supplied that will match the buildup dimensions of the mold halves.

Mold Base Nomenclature

The "A" Series Mold Base consists of the following items: (1) top clamping plate, (2) A

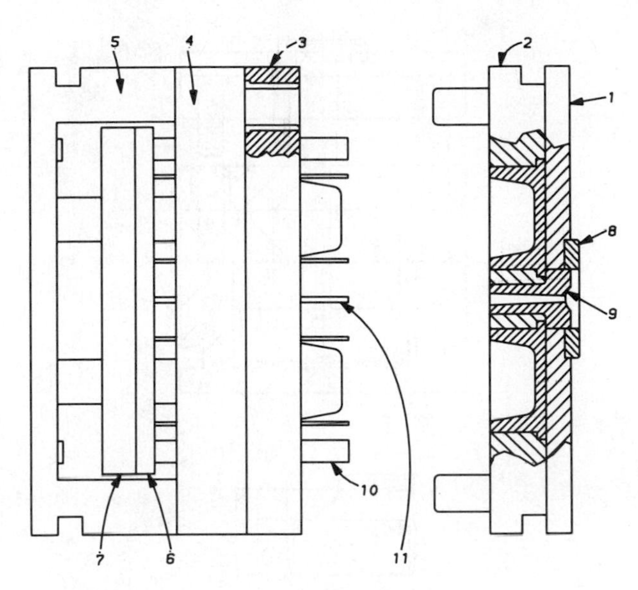

Fig. 7-3. Mold base nomenclature. (*Courtesy D-M-E Co.*)

plate, (3) B plate, (4) support plate, (5) ejector housing, (6) ejector retainer plate, (7) ejector plate, (8) location ring, (9) sprue bushing, (10) return pins, and (11) sprue puller pin. (See Fig. 7-3.)

"A" Series Mold Bases

These standard "A" Series Mold Bases are catalog items and available in widths and lengths from $7\frac{7}{8}$ by $7\frac{7}{8}$ to $23\frac{3}{4}$ by $35\frac{1}{2}$. Custom-designed, standard small mold bases are available for many of the popular small injection molding machines. All of the mold bases are described in detail in the D-M-E catalog, along with pertinent design data, specifications, weight, and price, for the convenience of the mold designer and the moldmaker. (See Fig. 7-4.)

Mold Base Number 1 Steel

Mold bases are available in No. 1, 2, or 3 Steel as standards and stainless steel as a "special" standard. No. 1 Steel (Fig. 7-5) is a medium carbon SAE 1030 silicon kiln forging quality steel, which has approximately 25% greater tensile strength than typical low-carbon warehouse steels. No. 1 Steel machines readily, allowing for both fast and smooth machining. This steel is commonly selected for prototype molds or short-run jobs, as the lower hardness of the material will more quickly succumb to peening or hobbing than the higher grades of mold base steels.

Mold Base Number 2 Steel

The No. 2 Steel (Fig. 7-6) is an AISI 4130 type steel, supplied preheat-treated to 300 Bhn, plus or minus 21 points. This high-strength but free-machining steel is ideal for A and B plates that will retain the cavities and cores, will stand up to constant clamping and unclamping when setting the mold in the machine, and will act as support plates. No. 2 Steel is considered the standard for medium- to long-run quality mold bases where the cavities and cores are inserted into the A and B plates.

Mold Base Number 3 Steel

For ultimate service life or in instances where the cavities or cores are cut directly into the A or B plates, No. 3 Steel normally is specified. No. 3 Steel (Fig. 7-7) is a P-20 (AISI modified

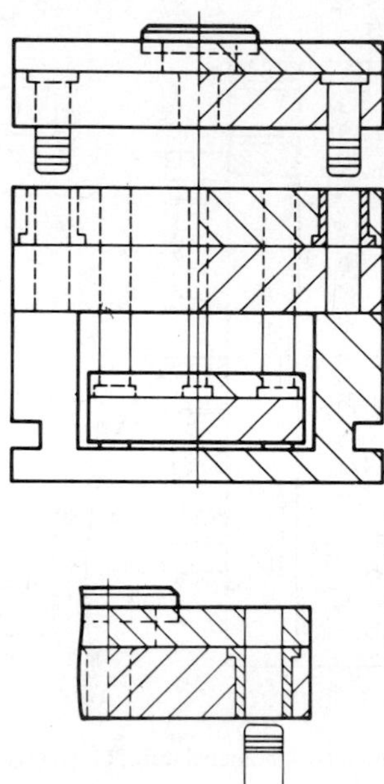

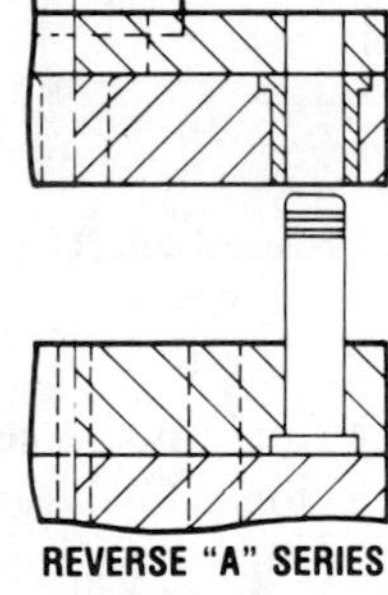

Fig. 7-4. "A" Series Mold Base configuration with leader pins extending from A side, and reverse "A" Series with leader pins from B side. (*Courtesy D-M-E Co.*)

4130) type cavity steel. This steel is electric-furnace-melted, carbon deoxidized, and then electroslag-remelted to provide an exceptionally clean, inclusion-free cavity quality steel. This steel, preheated to 300 Bhn, plus or minus 21 points, provides a high-hardness, good-machining steel with good polishability for the mold base and cavity plates.

Mold Base Number 5 Steel

H-13, No. 5 Steel (Fig. 7-8), is used where extremely high demands are placed on mold bases. Generally, this steel is used when great precision is required on smaller mold plates, or where the cavities and cores are integral with the mold plates. The electric connector field frequently utilizes a steel that will achieve a higher hardness. H-13 can be heat-treated to the low 50's R_c and still be readily machined with carbide to achieve very precise pocket sizes and locations.

Mold Base Stainless Steel

Mold bases that will be subjected to extremely corrosive conditions or will be used in any environment in which rusting is a problem are frequently being built from stainless steel (AISI 420 or 420 F), supplied at 290 HB plus or minus 20 points. Advantages of stainless steel are obvious. The mold base will not show signs of corrosion or rust. Also, cavities and cores will be easier to remove from the A and B plates, and any waterlines in the plates will remain free from mineral buildup. The penalties include an approximately 20 to 40% premium for the mold base because of the additional cost of raw material and a more difficult machining task. (See Fig. 7-9.)

Window Pockets

Once the proper mold base has been selected, pockets, either through or blind, must be cut

C	0.30%
Mn	0.75%
P	0.04% Max.
S	0.05% Max.
Si	0.25%

Fig. 7-5. Typical analysis of No. 1 mold base steel. (*Courtesy D-M-E Co.*)

C	0.30%
Mn	1.20%
P	0.03%
S	0.04%
Si	0.28%
Mo	0.20%
Cr	0.75%

Fig. 7-6. Typical analysis of No. 2 mold base steel. (*Courtesy D-M-E Co.*)

C	0.30%
Mn	0.85%
P	0.015%
S	0.003%
Si	0.30%
Cr	1.10%
V	0.08%
Mo	0.55%

Fig. 7-7. Typical analysis of No. 3 mold base steel. (*Courtesy D-M-E Co.*)

C	0.40%
Mn	0.40%
Si	1.10%
Mo	1.00%
Cr	5.00%
V	1.10%

Fig. 7-8. Typical analysis of No. 5 mold base steel. (*Courtesy D-M-E Co.*)

C	0.30% to 0.40%
Mn	1.00% Max.
Si	1.00% Max.
Cr	13.00%

Fig. 7-9. Typical analysis of stainless mold base steel. (*Courtesy D-M-E Co.*)

into the plates for the inserted cavities and cores. Through or window frame pockets are typically the least expensive to install. Cavity and core inserts usually are retained in through pockets with a heel existing around the perimeter of the insert. This heel, which fits into a corresponding counterbore pocket, ensures that the insert will not move forward on either mold opening or ejection. Conversely, the top clamp plate or support plate keeps the insert from receding under injection pressure. Pockets normally are square, rectangular, or round, as these are the shapes most conducive to precision machining of both pockets and inserts. If the pockets are round, a method of keying or registering the insert is required to ensure proper registration and mating with the corresponding insert.

Blind Pockets

Blind pockets normally are machined into "B" Series Mold Bases. "B" Series Mold Bases are similar to the "A" Series with the exception that the "B" Series utilizes a thicker cavity retainer plate, which becomes both the top clamping plate and the A plate. The thicker B plate eliminates the requirement of the support plate. This series mold base is often referred to as a two-plate design, whereas the "A" Series would be referred to as a four-plate design. Blind pockets afford a stronger support for the cavities, and this is the preferred method for deep cavities or when high or lateral injection pressures will be encountered.

Cavity Pockets

Pockets generally are machined in the plates on horizontal or vertical milling machines. Where extreme accuracy of location or size of the pocket is required, jig boring or grinding is a fabrication method utilized by the mold base supplier or moldmaker. When plates require large amounts of stock removal, it is prudent to have the mold base supplier either rough or finish machine the pockets, as the plates can be stress-relieved prior to final sizing and installation of the leader pins and bushings. This fabrication method is preferred for high-precision mold base requirements. The typical tolerance for pockets for length and width is plus or minus .001. When more precision fits are required, the cavity or core inserts normally are built oversize and then fit to the pocket by the moldmaker.

Wear Plates

When side cores, also referred to as slides, are installed in the mold base, wear plates of lamina-bronze or hardened steel are recommended and should be installed along with gibs of similar materials to provide a surface conducive to the moving member to ride. The mold base should not be utilized as a wear surface, as galling of the mold base or slides could result, partially due to the lack of a hardness differential between the components.

Mold Base Machining

Other machining on the mold base generally is in the form of round holes. These holes may be utilized to install designer-option components or ultimately will be tapped for installation of cap screws. Conventional and radial drill presses, NC machines, boring equipment, and jig boring and grinding equipment are all employed to perform the necessary machining.

Special Mold Bases

Special mold bases, whether they are due to nonstandard requirements of width and length dimensions or a result of special plate thickness, are available from the major mold base suppliers. These mold bases are built to the same specifications as standard mold bases, while incorporating the features required by the mold designer. All special mold bases are custom-constructed to the design supplied. Obviously, the lead time is longer and the cost is greater than for a comparable standard mold base; but the requirement that the mold base must fulfill its intended use is justification enough to "go special."

Quality of Mold Bases

The highest-quality mold bases manufactured today feature plates that display integrity of dimensionally located features within plus or minus .0005 while maintaining bore locations within plus or minus .0002. Plates that feature thickness tolerances of plus or minus .001, including flatness and parallelism, are standard in quality mold bases. These mold bases are square-machined, either ground and milled with special cutter heads to provide a perfectly square mold base with all edges in the same plane for the most exacting moldmaker. Through bores and pockets with a tolerance of plus .001 are standard, while blind pockets carry tolerances of plus or minus .001.

Mold Base Numbers

Catalog numbers for standard mold bases combine the nominal size (width and length) of the base and the thickness of the A and B plates. As all plate thicknesses are a combination of a whole number and either $\frac{3}{8}$ or $\frac{7}{8}$ inch, the designation 13 indicates $1\frac{3}{8}$ inches and 17 indicates $1\frac{7}{8}$ inches, and so on. A very convenient mold base number system has been developed, which will cover the many combinations of mold base widths, lengths, and plate thicknesses that are available. As an example, an ''A'' Series Mold Base that is $11\frac{7}{8}$ inches wide and 20 inches long with an A plate of $2\frac{7}{8}$ inches and a B plate that is $1\frac{3}{8}$ inches thick would be listed as a 1220A-27-13.

Specific-Purpose Mold Bases

In addition to standard and special mold bases, standard mold bases for a specific purpose are available as catalog items. These include stripper plate, ''AX,'' and ''T'' Series Mold Bases. The stripper plate mold base, commonly referred to as the ''X'' series, features five- or six-plate construction. The leader pins and bushings and screw and dowel locations are identical to those of the ''A'' Series Mold Bases. However, the leader pins are located on the B side to guide the X plate when actuated. Stripper plate mold bases are utilized when ejection is accomplished by engaging the edge of a part to remove it from a core, a practice most commonly followed in manufacturing round items such as cups and tumblers.

''AX'' Mold Bases

The ''AX'' Series Mold Bases are essentially ''A'' Series Mold Bases with the addition of a floating plate on the A side of the mold (Fig. 7-10). This feature is utilized, when it is desirable because of part requirements for core detail of the cavity side, to make it possible to remove part of the cavity detail, allowing the part to remain on the ejector side.

''T'' Series Mold Bases

Three-plate mold bases carry the designation ''T'' Series and are utilized when top runner molds are required (Fig. 7-11). These mold bases incorporate two floating plates. The ''T'' Series Mold Base can be equipped with a standard sprue bushing as supplied, or refit with a

D-M-E Standard "AX" Series Mold Bases

A D-M-E Standard ''AX'' Series Mold Base is basically an ''A'' Series type mold base with a floating plate (''X-1'' plate) added between the cavity plates. This type assembly is used when it is desirable to have the floating plate remain with the upper half of the assembly.

"AX" SERIES

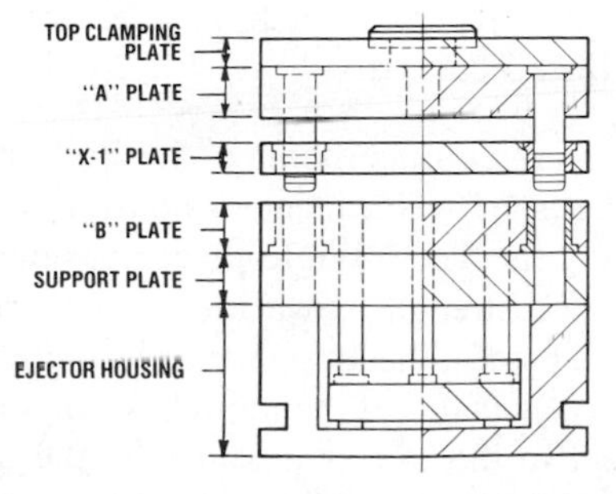

Fig. 7-10. ''AX'' Series Mold Base with floating X-1 plate. (*Courtesy D-M-E Co.*)

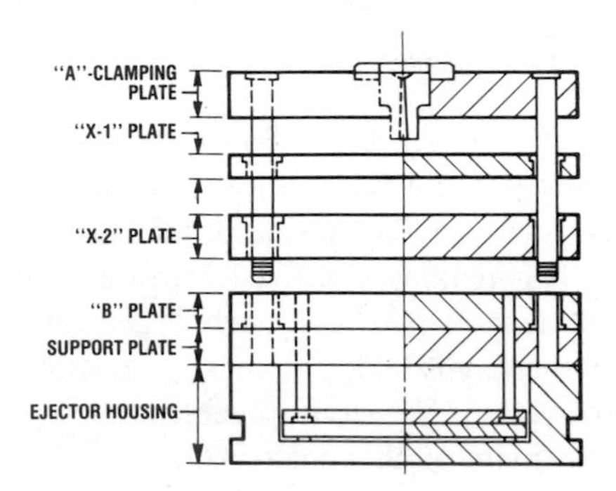

Fig. 7-11. "T" Series Mold Base for three-plate applications. (*Courtesy D-M-E Co.*)

Fig. 7-12. Single-piece ejector housing for the most demanding mold applications. (*Courtesy D-M-E Co.*)

Fig. 7-13. Plates for special mold bases are available in both standard mold bases and special sizes. (*Courtesy D-M-E Co.*)

hot sprue bushing to eliminate the sprue if that is desired. The mold base is designed with the intent that the moldmaker will install pulling devices to first break the gate by opening the parting line between the X-1 and X-2 plates, thereby separating the gate from the part prior to opening the main parting line. After the main parting line has opened the required distance, the X-1 plate is actuated, stripping the runner system from the sucker pins, which are installed in the top clamping plate, freeing the runner for ejection. All mold bases with floating plates can have optional lineal or preloaded type ball bushings installed to allow for free-floating plates.

Mold Base Features

Other features incorporated in high-quality mold bases are welded ejector housings and tubular dowels. These single-piece housings provide added strength, ensuring the integrity of the mold base in the most demanding situations (Fig. 7-12). Plates of the mold base are located and held in the proper position by the use of tubular dowels placed around retainer screws. These tubular dowels do not detract from the usable mold space, as they encircle the retaining cap screws.

Mold Base Plates

To assist the moldmaker, who for a variety of reasons has to construct a non-standard mold base, cavity retainer sets, plates, and single-piece ejector housings are offered as standard catalog items (Fig. 7-13). This availability allows the moldmaker to supply the customer with the highest-quality mold possible when a standard mold base cannot be utilized. The use of catalog items should be encouraged, as the availability of these items is essential to put a mold back into production in the shortest amount of time, should replacement be required due to some unfortunate incident during production.

TYPES OF MOLDS

Conventional Molds

The most commonly used injection mold is a conventional runner system, standard mold de-

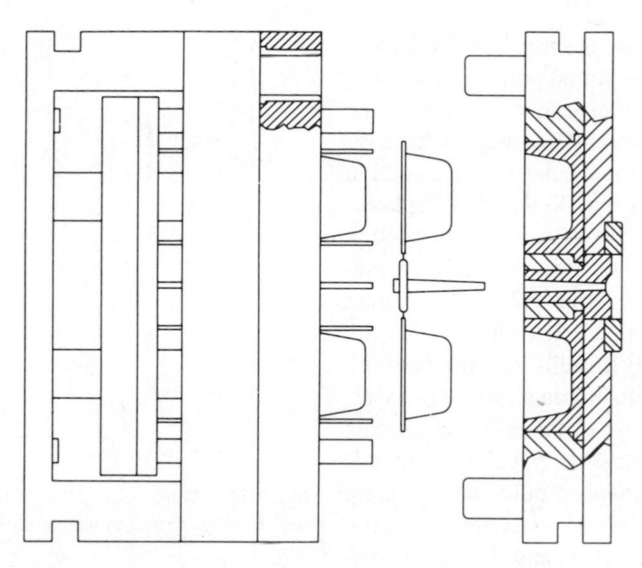

Fig. 7-14. Conventional mold with standard parting line opening to allow removal of parts with runner and gates attached. (*Courtesy D-M-E Co.*)

sign (Fig. 7-14). This design incorporates a traditional sprue bushing to accept the melt delivered from the molding machine nozzle. Standard sprue bushings are available in a variety of styles and lengths to meet every application. These bushings, with a spherical radius selected to mate with the nozzle of the molding machine, are available with orifice sizes of $\frac{5}{32}$, $\frac{7}{32}$, $\frac{9}{32}$, and $\frac{11}{32}$ inch. All feature tapers of $\frac{1}{2}$ inch per foot for ease of sprue release.

The runner system then is cut at the parting line to route plastic to the cavities. Runner systems can be round, half-round, square, trapezoidal, or a variety of other configurations. The full round runner is the most efficient and popular system employed. Runner diameters of $\frac{1}{8}$, $\frac{3}{16}$, $\frac{1}{4}$, $\frac{5}{16}$, $\frac{3}{8}$, and $\frac{1}{2}$ inch generally are selected, based on the type of plastic being molded, length of flow to reach the cavity, and volume of material required by that cavity. Occasionally, when using round runner systems, the moldmaker installs the runner with an end mill featuring a full radius and cuts two-thirds of the runner in the ejector side and one-third in the A side, creating a natural tendency for the runner to remain on the ejector side. The conventional runner system offers an economical approach to mold construction and is most frequently employed for short-run applications. A disadvantage of this style of mold is the requirement for degating of parts and regrinding of the runners and sprue, which are labor-intensive operations and do not lend themselves well to automatic molding.

The conventional runner system is complemented with a full array of gate configurations to meet the requirements of the part to be molded. These gates, which may be either edge, tab, fan, chisel, submarine, flash, disc, or other design configurations, form a transition from the runner feed system and the part. Gate design is in itself an art. The gate must be large enough to allow the part to be filled in the shortest length of time and must be small enough to freeze off quickly to seal the plastic in the cavity, while allowing for economical degating and leaving a minimal gate scar. The normal practice is to start with small gates and then progressively open them until the desired effect is achieved. If this is done, it should be made clear, prior to mold construction, whether the practice is the moldmaker's or the molder's responsibility, to avoid any misunderstanding.

Three-Plate Molds

Three-plate molds are used to automatically degate parts in the molding cycle and also allow the part to be gated on the top, usually a more desirable position for round parts. In the operation of the mold, plastic enters through the sprue bushing, flows into a runner system, which normally is trapezoidal in cross section and located between the X-1 and X-2 plates, through feeding gates which are located through the X-2 plate, and then into the cavity. Upon mold opening, the X-1 and X-2 plates separate first, breaking the gate from the part, as the moldmaker normally installs springs between the plates. The main parting line then is allowed to open, and when the mold has almost reached the fully opened position, a moldmaker-installed mechanism pulls the X-1 plate forward, stripping the runner system from the sucker pins. With the parts and the runner system cleared from the mold, the next cycle can begin with mold closing. (See Fig. 7-15.)

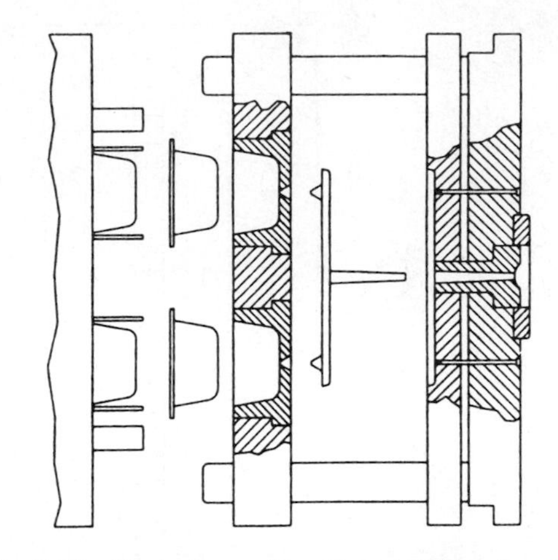

Fig. 7-15. Three-plate molds with two additional parting line openings to accommodate separation of part and runner.

Insulated Runner Molds

Very few insulated runner molds are built today, as other runnerless molding technologies perform much better than this type. This mold design relied on a very thick runner system to form insulation of the outside of the runner, while allowing the new melt to flow through the center of the runner with the residual melt heat maintaining an open flow path. This style of mold, while easy to build, was extremely difficult to run with when cycle interruptions occurred. If new material was not frequently introduced into the system, the insulated runner would freeze, and the cull would have to be physically removed from the mold. As this was a frequent occurrence, the runner plates were latched together, and the machine clamping pressure was relied on to keep the plates from separating under injection pressure. This worked fine when everything went well; but when the projected area of the runner was greater than the cavity surface, or the clamp pressure on the machine backed off, flashing of the runner plates was experienced. Insulated runner molding has some success with certain materials, including the polyolefins, for making large shots on fast cycles where cycle interruptions were infrequent. (See Fig. 7-16.)

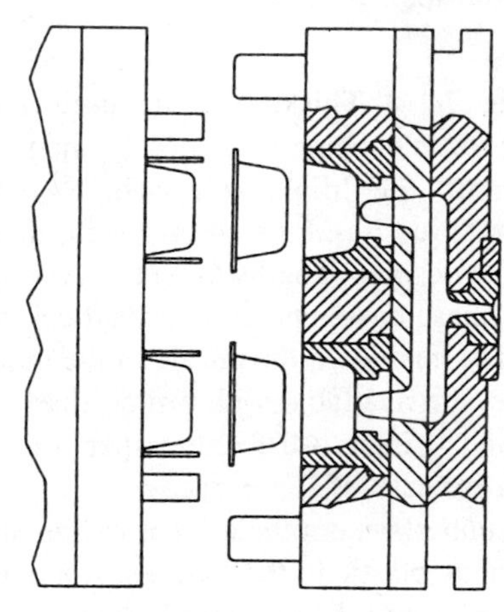

Fig. 7-16. Insulated runner mold with runner removed for easy start-up.

Internally Heated Runnerless Molding Systems

Typically the least expensive in terms of initial cost and continued maintenance is the internally heated runnerless molding system. In this system, material from the machine nozzle enters through a heated nozzle locator into a flow channel where heat is supplied to the plastic melt by a thermocouple cartridge heater located inside a distributor tube, held in position by end caps. Melt is distributed to probes or to secondary distributor channels through either

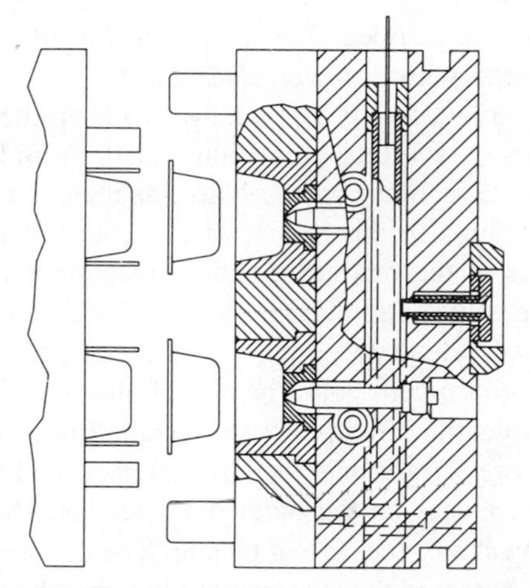

Fig. 7-17. Internally heated runnerless molding systems with heated nozzle locator, distributor tubes, and probes. (*Courtesy D-M-E Co.*)

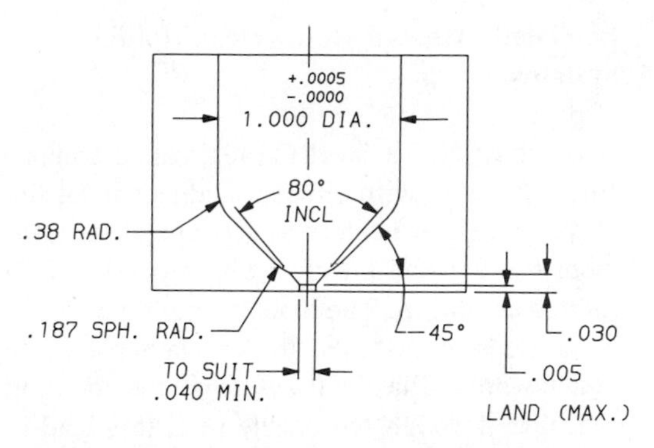

Fig. 7-18. Illustration of gate machining configuration for best results to keep gate area from freezing, while maintaining the best control of gate vestige.

round bores in solid plates or trapezoidal channels in split-plate designs. The molten plastic material flows along the gate probe to the gate and into the cavity. The gate probe is heated by a removable heater or an integrally installed thermocouple heater. (See Fig. 7-17.)

This system is the least susceptible to material leakage of all runnerless molding systems. On solid-block construction, the distributor channels are gun-drilled in a block of No. 3 Steel (4130), or trapezoidal channels are milled in the distributor plates for split-plate designs. The distributor plates are retained with allen head cap screws of sufficient strength to withstand the molding injection pressures. This system is comprised of pre-engineered components, and application information is supplied on varied mini-prints for the moldmaker to design and build this style of mold. Additionally, this type of runnerless molding system is the most energy-efficient design. As the plastic is heated from the inside out, it requires less wattage than externally heated systems. The probes, with either conventional or euro-style tips, provide excellent control of the gate, allowing drool-free molding without gate freeze-up. Cycle interruptions of up to five minutes are possible without freezing the gate in these systems. (See Fig. 7-18.)

Gate probes as small as .312 inch in diameter are available as off-the-shelf items in lengths from 2.720 inches in $\frac{1}{2}$-inch increments up to 4.220 inches long to encompass a wide range of applications while allowing for cavity spacing as close as .620 inch. Other probes, in diameters of .394, .500, .625, and .750 inch, are available in a variety of lengths to suit most applications. These probes, built from D-2 steel, provide extremely long life even in the most demanding applications, where both abrasion and corrosion from the plastic are encountered. (See Fig. 7-19.)

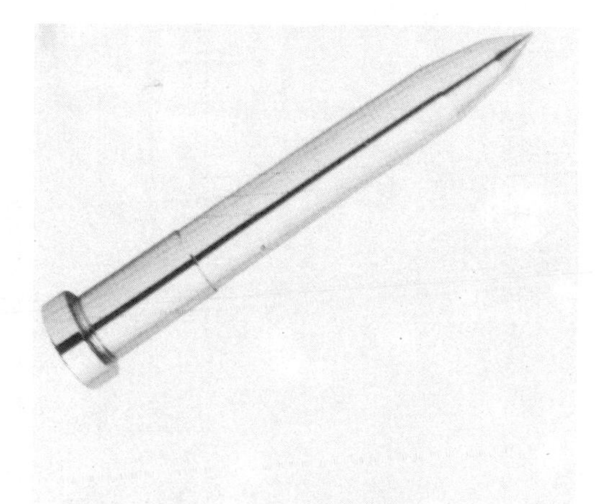

Fig. 7-19. Probe constructed from D-2 steel for resistance to abrasive materials and long life. Probe is heated by a replaceable, thermocoupled, flat-bottomed heater.

Externally Heated Runnerless Molding Systems

A block of No. 3 Steel (4140), called a manifold, is heated with either round cartridge, flat strip, or cast-in heaters to a temperature sufficient to keep the plastic being processed in a molten condition. The machine nozzle mates to a nozzle seat installed to form a replaceable contact area. Plastic flows from the molding machine, through the nozzle seat, to a lead-in channel, and then into the primary flow channel. This primary channel will deliver the melt to the bushing drops. The flow channels are bored into the manifold to form channels for the plastic to flow. Additional flow channels may connect, forming secondary and even tertiary flow paths before ending at the bushing drop locations. (See Fig. 7-20.)

Charts published by the component suppliers recommend a specific bushing size with corresponding flow channels, based on the volume of material and the viscosity of the plastic. The selection of the bushings will dictate the diameter of the flow channel in the manifold. Manifold channels must have a finely polished surface with no dead spots where the plastic could be trapped. End plugs are installed and secured with a dowel, and then locked with a set screw of sufficient strength to keep the end plugs in position, preventing plastic from leaking. Care must be taken to adequately retain any plugs in manifolds, as the injection pressures of the modern molding machine can exceed 30,000 pounds. If the plugs were not retained adequately, the operational personnel and equipment could be hurt if struck by such a projectile. End plugs are installed into counterbores larger than the channel itself and must form a seal on the outer diameter and face to prevent any seepage of plastic. The diameter of this bore and the subsequent plug must be large enough to prevent any sharp edges, which could break down and create a dead spot for the plastic.

Once the end plugs have been installed, the channels leading to the bushing drops are bored into the manifold. The end drops to the bushing where the end plug is located and the drop at the opposite end of the manifold are finished with a full radius cutting tool to provide smooth flow of the redirected plastic.

Thermocouple holes are installed in the manifold to measure the operating temperature of

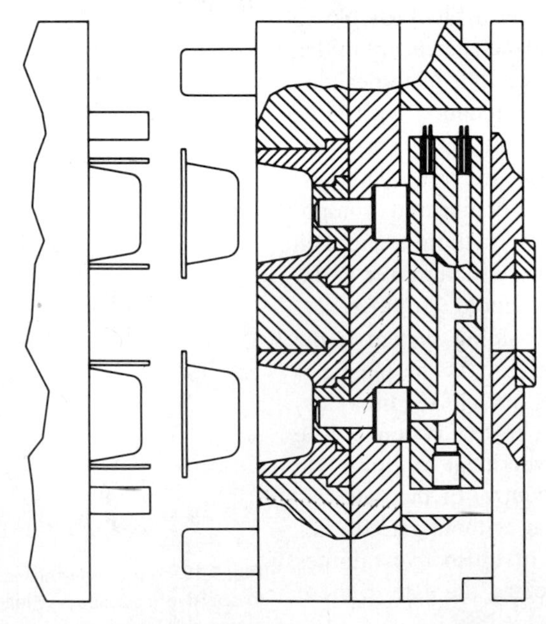

Fig. 7-20. Externally heated runnerless molding system with hot manifold for high-temperature engineering material applications. (*Courtesy D-M-E Co.*)

the steel and feed this information to modern microprocessor temperature controllers for accurate control of the manifold. The thermocouple location is critical to the proper operation of any manifold. Large manifolds, including X, Y, H, T, and variations, often utilize more than one thermocouple for better control of the manifold thermal profile. Typically, the location of the thermocouple is halfway between the heaters and the flow channel and centered between bushings.

Thermal expansion on the length of the manifold must be calculated to ensure that the drops feeding the bushings will align perfectly once the manifold has reached the desired operating temperature. The thermal expansion for the manifold steel is normally stated as six millionths of an inch-per-inch per degree Fahrenheit temperature rise over ambient. Manifolds have pressure pads installed above each drop and under the nozzle seat. Cap screws are installed to secure the manifold to the support plate, with spacers placed between the two for the purpose of squaring the manifold.

While it is debatable whether thermal expansion should be calculated and press added to the stack height of the manifold and pressure pads, experience has shown that zero preload provides the best performance, as growth of the assembly normally is sufficient to prevent leakage. This style of runnerless molding system technology is not new. The success of these systems has been particularly aided by advances in microprocessor temperature controls and consequent increased thermal control.

Runnerless Molding Nozzles

Nozzles are available in a variety of configurations. Flow channel sizes are designed to allow sufficient flow without creating large pressure drops for the plastic being molded. The nozzle and the manifold must form a sliding seal to prevent plastic leakage and most often utilizes an aluminum V-seal or hollow stainless steel O-ring. These seals, along with good surface finishes of the mating components, have proved to be very reliable. Seals should be replaced each time that the mold is disassembled or at the earliest sign of any leakage. Material

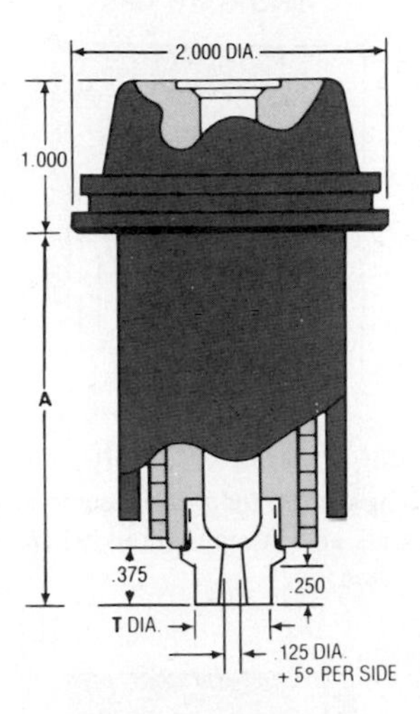

Fig. 7-21. Sprue gate nozzle for externally heated runnerless molding systems for maximum material flow. Sprue will remain on part or runner system.

selected for bushings is frequently 4140, 6530, or even H-13 and stainless steel for applications where abrasion or corrosion are of concern.

Gate tip selection is crucial for successful molding. Sprue gate nozzles (Fig. 7-21) are the most popular, as they result in the lowest pressure loss and will successfully handle most plastics. This type of tip is used when a small sprue is not objectionable. Unrestricted flow is recommended for filled materials as well as larger parts, and sprue gates frequently are used for feeding runners when plastic is being distributed to multiple conventional gates. Threaded-in tips are replaceable on this style of nozzle, and tips are available with extra stock to allow for machining either runners or the part configuration directly into the bushing tip.

Ring gate bushings are utilized for unfilled plastics in smaller-shot-size applications when a sprue vestige is either not desired or not allowed. The tip will leave a small circular witness mark and a slightly raised gate vestige. These tips, which are constructed from nickel-plated beryllium copper, provide excellent control of the gate area and eliminate problems with freeze-off and drooling. Ring gate tips are available in several standard gate sizes. Orifices may be enlarged by the moldmaker if flow requirements are more demanding. However, it must be understood that the larger the orifice

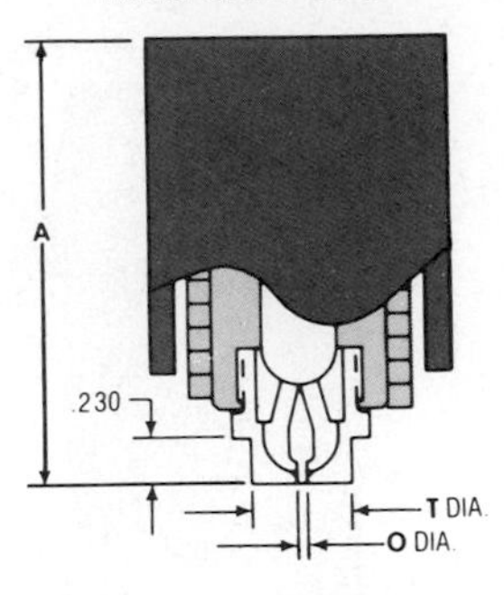

Fig. 7-22. Ring gate tip for nozzle assemblies for use with unfilled materials and in applications where gate vestige height is a concern.

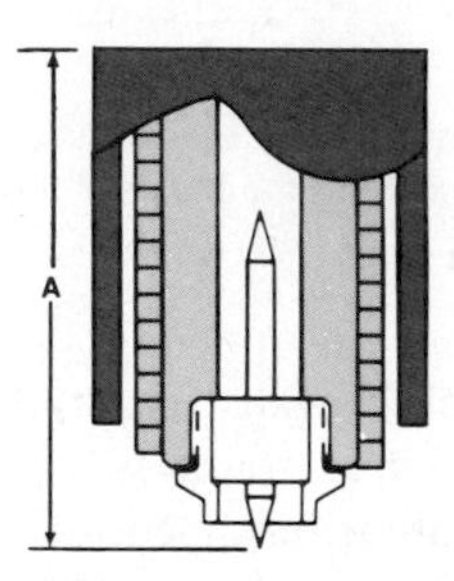

Fig. 7-23. Point gate tip for nozzle assemblies for installations where witness lines of the nozzle would be undesirable on the molded part.

is, the higher the remaining gate vestige. (See Fig. 7-22.)

Another style of tip is the point gate. This version is installed in the cavity without showing a witness line on the molded part. The point gate needle is constructed from beryllium copper and nickel-plated for improved wear resistance. This type of tip provides good flow to the gate while controlling gate freeze-off and drooling. The point gate needle is replaceable in the event of damage and works well with unfilled materials. (See Fig. 7-23.)

Valve-Gate Runnerless Molding Systems

As its name implies, the valve gate system opens and closes the gate with assistance from hydraulic or pneumatic cylinders. (See Fig. 7-24.) Normally, the cylinders are placed in line with the corresponding bushing, and a reciprocating pin is retracted or even advanced to open or close the gate. Most valve gate systems built today start with the hot manifold runnerless molding technology, while utilizing bushings that are modified to guide the valve pins. (See Figs. 7-25, 7-26, and 7-27.)

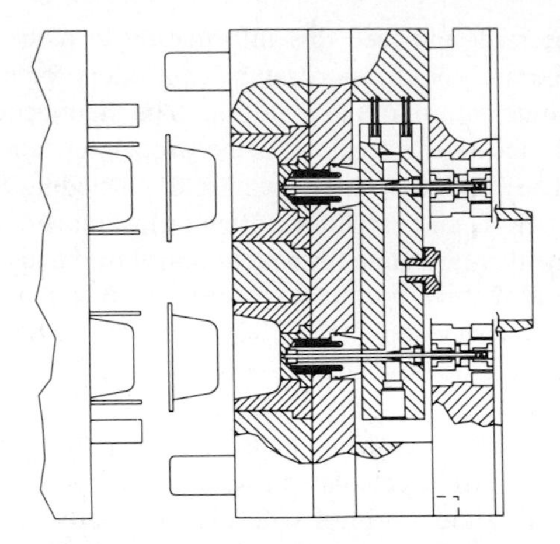

Fig. 7-24. Valve-gate nozzles, valve pins, and hydraulic cylinders for operating and controlling the opening and closing of gates. (*Courtesy D-M-E Co.*)

In the gate closed position, the valve pins form a seal against the gate inserts and are an effective way to control not only gate drool and gate vestige, but also the opening and closing of gates that control the flow of plastic in large and complicated molds. The entrance of the valve pin to the manifold requires a special guide that also acts as a seal, preventing the molten plastic from seeping out of the primary melt channel. Control panels allow for the independent movement of each valve gate, having both delay timers to initiate gate opening and timers to control the duration that the gate is open. Valve gates by their nature are more expensive to build and to maintain than conventional runnerless molding systems, but operate well and fill their niche well in the right applications.

FABRICATION OF COMPONENTS

Cavity Machining

Mold cavity and core inserts are fabricated by a wide variety of methods. However, conventional machining of the cavity and core inserts accounts for the most widely employed method of fabrication. The term machining is used here to denote milling, duplicating, drilling, boring, turning, grinding, and cutting. The first step

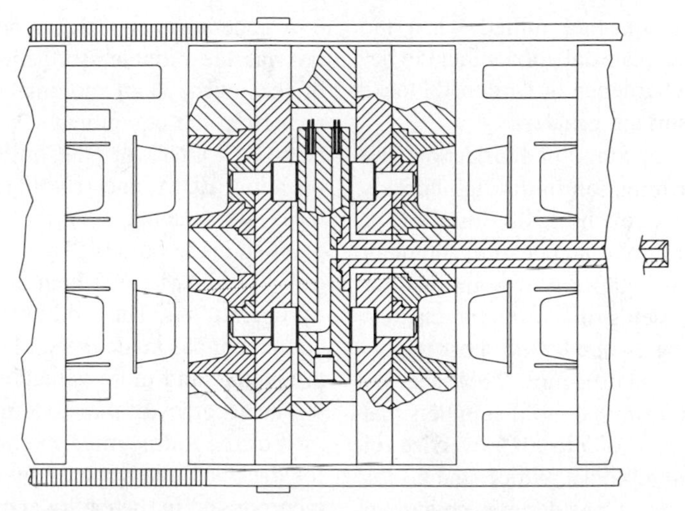

Fig. 7-25. Stack mold with externally heated manifold with nozzles feeding parts to both sides of a center section. This arrangement allows for twice the number of cavities to be molded without increasing the clamp tonnage requirements. (*Courtesy D-M-E Co.*)

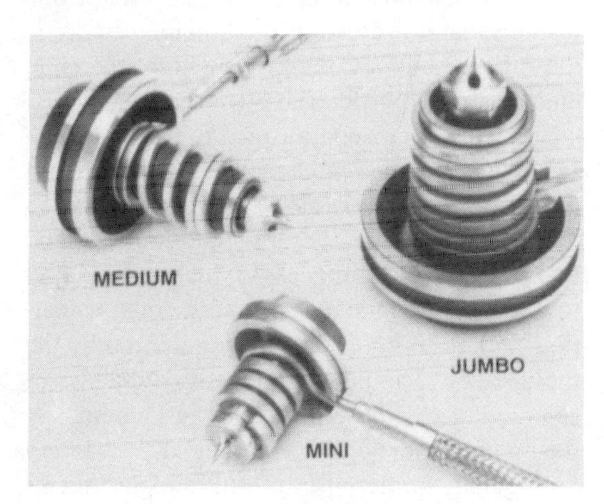

Fig. 7-26. Gate Mate bushings for use in single-point, direct part gating. With Gate Mate 4, these bushings can be used in conjunction with externally heated manifold for multiple-cavity molding. (*Courtesy D-M-E Co.*)

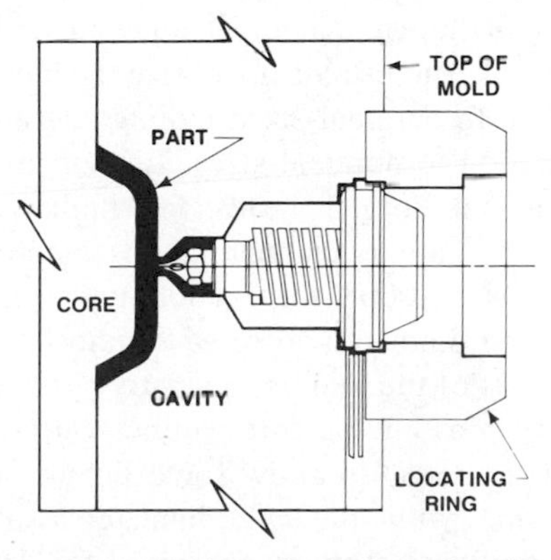

Fig. 7-27. Gate Mate shown in single-cavity, direct part gating application. (*Courtesy D-M-E Co.*)

after selection of the cavity and core material is to cut the raw material to the approximate size. This is normally accomplished with either vertical or horizontal power sawing equipment. The next operation is to square, true, and size the inserts. Normally, this stage of fabrication is carried out in the material's soft or annealed stage if hardened steels are being utilized for the final product. Round inserts generally are trued and sized on turning equipment, which includes lathes and cylindrical grinders. Square

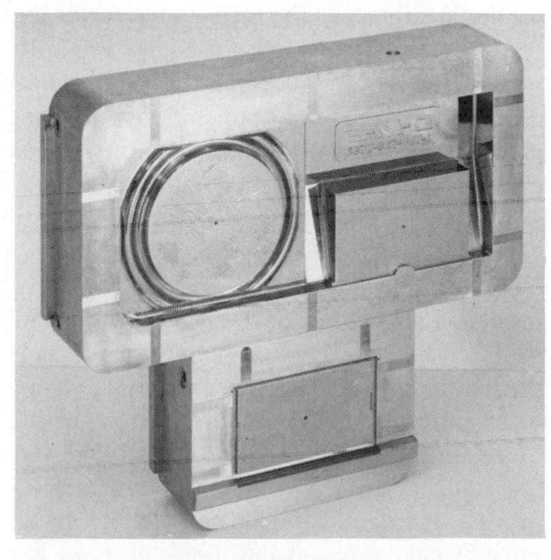

Fig. 7-28. Illustration of two-cavity family mold for making a card/cup holder. Cavities were manufactured from P-20 steel, polished to SPI No. 2 finish. (*Courtesy D-M-E Co.*)

and rectangular blocks are milled when the hardness of the raw material allows them to be. Final sizing of prehardened or hardened blocks is completed on surface grinders.

In this preliminary stage of fabrication, the heels required for retention in through pockets are established, or screw holes are installed for inserts that will fit into blind pockets. Although in these preliminary stages it appears that not much progress is being made on the cavities and cores, squaring is one of the most important steps in mold manufacture. Tool steel inserts are available from the mold suppliers that have been sized from .005 to .015 oversize for square or rectangular blocks, with round inserts furnished with heels. Considerable savings of in-house labor are possible by utilizing these inserts, which are generally available in P-20, H-13, and 420 stainless steel.

One of the next operations to be carried out in the inserts is installation of the mold temperature control circuits. Square and rectangular blocks normally have water, steam, or oil channels drilled directly into them, with other lines connecting to form internal loops. Frequently, internal water channels are blocked with threadless brass pressure plugs to direct the flow of the temperature control fluid. Generally, only one inlet and one outlet are used per insert with the remaining channels blocked, using brass or steel pipe plugs. Conventional drilling equipment is normally utilized for this operation, with gun drilling used for deep holes where accuracy of location is required. Great care must be taken in this operation, as nothing dampens the enthusiasm of the moldmaker more than hitting a waterline with a screw hole.

As modern molds require greater sophistication in temperature control to meet the challenges presented by today's design engineers, programs have been developed and are commercially available from suppliers to accurately predict the amount and placement of temperature control channels. Generally, the core will be required to remove approximately 67% of the Btu's generated in the injection molding process. This requirement presents one of the greatest challenges to the moldmaker, as less space usually is available in the core than in the cavity.

A generation ago, the standard of the industry was the reliable Bridgeport, with the skill of the journeyman moldmaker coaxing accuracy from the equipment. Today, the standard is the NC, CNC, or DNC milling, drilling, duplicating, EDM, and grinding machines. Often, tapes are generated by the mold designer, a practice made possible by the wide acceptance of CAD/CAM equipment in even the smallest of operations. The addition of CAD/CAM equipment has made the skilled craftsman more productive and more valuable.

The accuracy of the modern equipment or the skill of the journeyman moldmaker is required for the machining operations that now will be incorporated in the cavity and core inserts. Depending on subsequent machining operations, the knockout of ejector pins holes may be located and established at this time. Perhaps one of the greatest values in moldmaking today is the off-the-shelf availability of high precision ejector pins. Nitrided hotwork ejector pins, constructed from superior-quality thermal stock H-13 steel, are available in shoulder or straight type with outside diameter hardness in the 65 to 74 R_c range. Ejector pins, both imperial and metric, starting from $\frac{3}{64}$ (.046) inch and ranging to one inch are available in every popular fraction, letter, and millimeter size required by the moldmaker. Should requirements dictate nonstandard sizes, those pins, too, can be built by the component supplier as a special. It is worth noting that, for purposes of future replacement, the use of standard, off-the-shelf components is highly recommended and frequently demanded by most tool engineers. (See Fig. 7-29.)

Ejector pins are designed to fit into nominal-diameter holes, making it convenient for the moldmaker to establish and size the ejector pin holes with nominal-sized tooling and allowing inspection to nominal sizes. Ejector pins, featuring hot forged heads for higher tensile strengths, are manufactured to tolerances as close as −.0003 to −.0006 under nominal specified diameters to meet the most stringent demands of the molding industry. The nitrided ejector pins have a core hardness in the range of 30 to 38 R_c to allow some flexure without fatiguing, while the outer diameters exhibit the high hardness required for long life. Through-

Fig. 7-29. Hotwork nitrided ejector pins, manufactured from high-quality H-13 steel. (*Courtesy D-M-E Co.*)

hardened, close-tolerance ejector pins are built from high-speed steel and maintain a through hardness of 60 to 65 R_c. Ejector sleeves are designed to accept standard ejector pins and provide an excellent method of removing the plastic part from cored holes. All other dimensional data for designing and machining information are available in parts catalogs.

Machining of ejector and core pin holes is accomplished by many methods. Locations of these holes frequently are established by jig boring equipment, whereas locations that are less critical have been made on conventional milling and drilling machines. With more computer-controlled equipment available, very accurate locations are possible. If the core block of steel is to be heat-treated, the holes generally are located and established undersized with final sizing accomplished with reaming, jig grinding, or honing for ultimate accuracy. (See Fig. 7-30.)

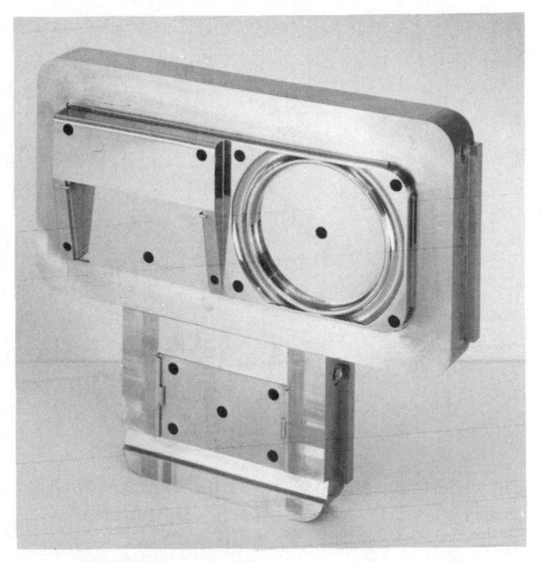

Fig. 7-30. Cores for the two-cavity family mold with ejector pin holes installed. (*Courtesy D-M-E Co.*)

Core and Cavity

The shape of the core or cavity will entirely depend on the configuration of the product that will be molded. Additionally, the mold designer will incorporate features that will result in the most economical manufacturing of the mold, while meeting the dimensional and functional requirements of the product as designed. Whereas the cavity and core blocks normally start out solid, core pins or core inserts are utilized to provide the necessary part definition. Wherever possible, round core pins are used, as they provide the most economical method for machining and installation. Standard core pins, constructed from H-13 steel, available in a hardness of 30 to 35 R_c and a high hardness or "CX" of 50 to 55 R_c, are obtainable and are an extremely cost-effective means of providing coring for holes or for material reduction (see Fig. 7-31). The core pins are available in lengths of 3, 6, and 10 inches, supplied with a plus tolerance of .001 to ensure proper fit in the blocks. These core pins, ranging in size from $\frac{3}{32}$ (.093) to $\frac{3}{4}$ (.750) inch, can cost as little as $2.50.

Other core inserts will be custom-fabricated to the specific requirement of the mold. These

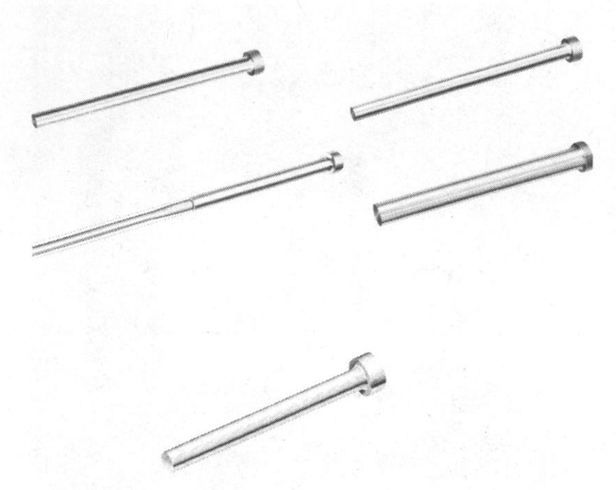

Fig. 7-31. Standard core pins are cost-effective method used in mold construction. (*Courtesy D-M-E Co.*)

inserts most frequently are installed in through pockets in the insert block and retained in position with heels. Round cores, utilizing standard components or custom-manufactured, frequently must be keyed to provide proper orientation for registration or to ensure that part specifications are adhered to. The most commonly utilized method is to machine a flat on the heel along with a corresponding groove in the mating part, and then to retain proper orientation with a dowel or drill rod section. As with all mold components, it is imperative that the moldmaker mark the individual item and place a corresponding mark on the block in which the item will be installed to provide a reference for subsequent assembly of the mold.

Stock removal of the cavity and core materials should be accomplished by the most economical method consistent with the machinery available. Cutting remains the most universally accepted method of roughing the cavity and core inserts. The evolution of cutting tools, including high-speed steel, carbide, and ceramics, as well as the development of coatings such as titanium nitriding, has enhanced the abilities of machining. Since these advances have been combined with computer-controlled equipment along with higher spindle horsepower, the cost of accurately removing a cubic inch of stock has reached an all-time low.

Electric-Discharge Machining

Electric-discharge machining (EDM) is another widely utilized method of producing cavity and core stock removal. Electrodes fabricated from materials that are electrically conductive are turned, milled, ground, and developed in a large variety of shapes, which duplicate the configuration of the stock to be removed. The electrode materials, selected for their ability to be economically fabricated while producing the desired wear characteristics, include graphite, copper, tungsten, copper-tungsten, and other electrically conductive materials. Wire EDM equipment is rapidly gaining popularity in the mold shop and is used for removing through sections of steel for later installation of inserts.

Slide and Unscrewing Devices

External thread features normally are formed with slides, actuated mechanically with cam pins or with hydraulic or pneumatic cylinders. Internal threads present a greater problem. Certain soft and flexible materials have been stripped when shallow threads or undercuts were employed in noncritical dimensional applications. When plastic part material requirements will not allow stripping, threads are formed on cores that are rotated out of the part or with collapsible cores. Full threads can be unscrewed, and molds with as many as 96 cavities have been built and successfully run in production. Unscrewing molds incorporate pinions as an integral part of the core. Upon mold opening, a hydraulic cylinder actuates a rack that rotates the pinions while in unison advancing a stripper plate. This action unscrews the part from the core. When the correct number of rotations of the core has been achieved, the stripper plate continues to advance, ejecting the part from the core. The rack then is returned to its previous position, the mold closes, and the cycle is repeated. As the threads are accurately ground, including shrinkage on the hardened cores, extremely precise threads result.

Collapsible Cores

When undercuts, threads, or other internal part configurations that cannot be ejected are required, standardized, off-the-shelf collapsible cores are available. These cores, built from 0-1 steel with D-6 center pins, are available in sizes that will mold threads or undercut configurations as small in diameter as .910 inches to a maximum outer diameter thread configuration of 3.535 inches. Smaller-size cores presently are being introduced that will result in a part with a thread of internal part configuration of .425 inches. The smaller collapsible cores will mold an interrupted thread. The interruptions are necessary to physically allow the core segments to collapse, freeing the undercut and allowing the part to be stripped clear of the core. (See Figs. 7-32 and 7-33.)

Collapsible Core Sequence

Technical manuals for the design and construction of molds to accommodate the cores are available from the supplier. As with unscrewing molds, the use of a center top gate is the most desired arrangement, as typically a round part is being produced. Also, stripper plate molds are utilized for proper removal of the part from the core. The operation of a collapsible core mold is relatively simple and follows an exact sequence. After the mold opens at the parting line, if runnerless molding systems are used, the ejector system is actuated by the molding machine. As the center pin remains stationary in the ejector housing and the core body is installed between the ejector retainer and the ejector plate, the body moves off the center pin, and the segments are allowed to collapse inward, freeing the undercut in the plastic part. The inward collapse is accomplished by the memory of the segments, which has been instilled in heat treatment. To ensure the full and final collapse, a positive collapse sleeve makes contact with bumps on the outer diameter of the core body at the end of the stroke.

Once the core has reached its full travel with all segments collapsed inward, a limit switch actuates hydraulic or pneumatic cylinders. The cylinders, which are mounted between the ejector bar and the stripper plate, move the stripper plate forward, ejecting the part from the mold. The mold is then closed in the reverse sequence to allow for the next cycle.

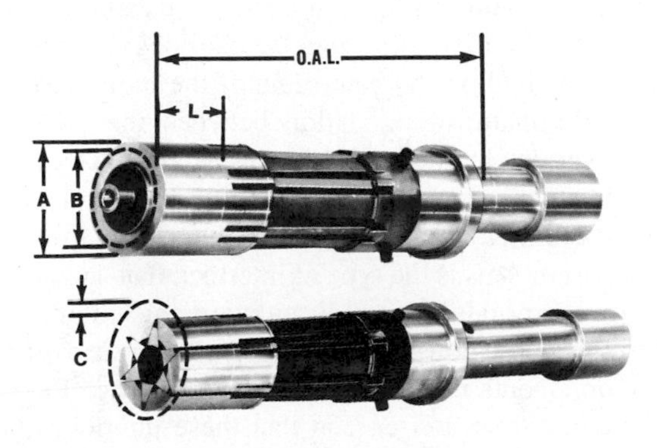

Fig. 7-32. Standardized Collapsible Core for molding round parts with internal threads or undercuts. A. With maximum O.D. of thread configuration. B. The minimum I.D. of thread part configuration; *L* is the maximum molding length including shutoff. C. With maximum collapse per side at the tip of the core. (*Courtesy D-M-E Co.*)

Mold Components

Many components are incorporated into molds to achieve specific objectives. Leader pins and bushings are, by design, utilized to guide the mold halves together. Leader pins normally are constructed from high alloy steel and carburized to establish a very hard wear surface while maintaining an extremely tough core. Bushings are built from steel or bronze laminate over alloy steel. Extensive testing has proved that grooveless leader pins outwear grooved leader pins when used with steel bushings. However, for maximum protection against galling, grooveless leader pins with bronze-plated bushings should be used. (See Figs. 7-34 and 7-35.)

Interlocks

As leader pins and bushings only act as guides, they are not intended to register and lock the mold halves together. Tapered interlocks,

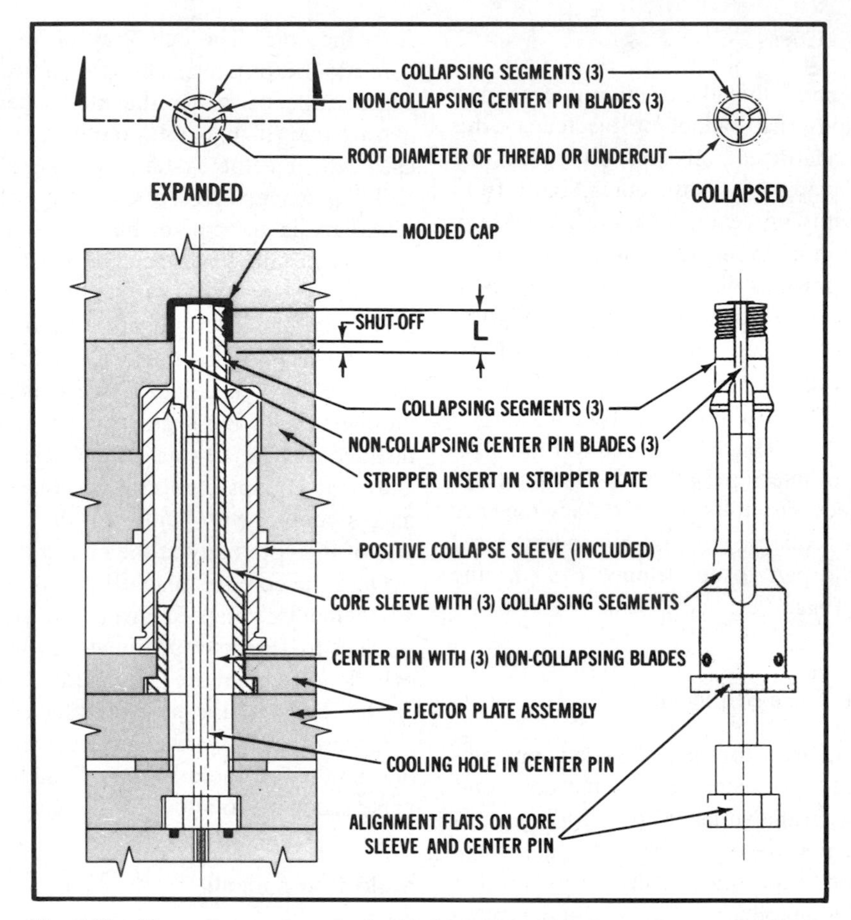

Fig. 7-33. Nomenclature of standardized D-M-E Collapsible Core. (*Courtesy D-M-E Co.*)

Fig. 7-34. Grooveless leader pins for alignment of mold halves. (*Courtesy D-M-E Co.*)

Fig. 7-35. Leader pin bushings to accept leader pins. (*Courtesy D-M-E Co.*)

round or rectangular, are designed to establish and maintain registration once the parting line has been closed. Rectangular interlocks should be installed on the centerline of the mold base to eliminate any variation between the mold halves due to uneven thermal expansion. To establish alignment between mold halves prior to the mold closing, straight side interlocks must be used. This is the type of interlock that should be incorporated in molds with vertical shutoffs, or in applications where floating plates ride over components that are subjected to galling. Remember that it is crucial that these interlocks, installed in accurately machined pockets, should be installed on the centerline of the mold base to eliminate alignment problems, which could result from uneven thermal expansion between mold halves. (See Figs. 7-36 through 7-40.)

Fig. 7-36. Round tapered interlocks for smaller molds. (*Courtesy D-M-E Co.*)

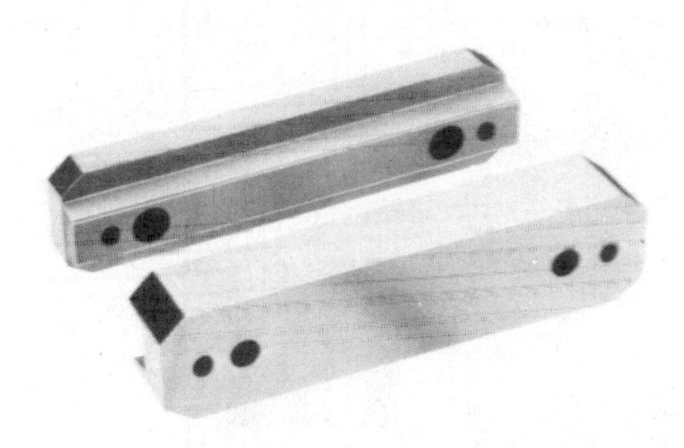

Fig. 7-37. Rectangular tapered interlocks for use with larger molds. (*Courtesy D-M-E Co.*)

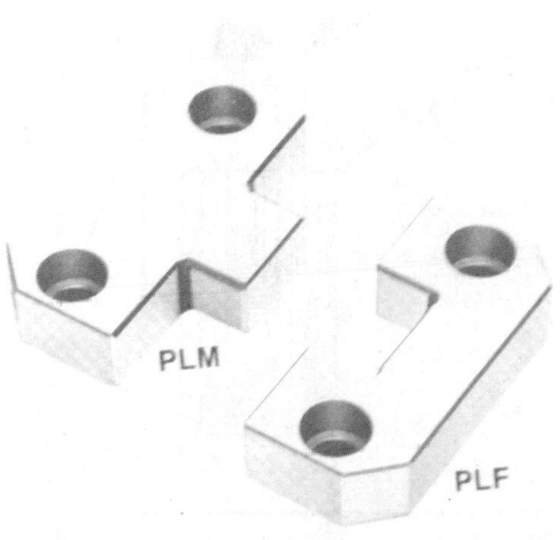

Fig. 7-38. Straight side interlocks. (*Courtesy D-M-E Co.*)

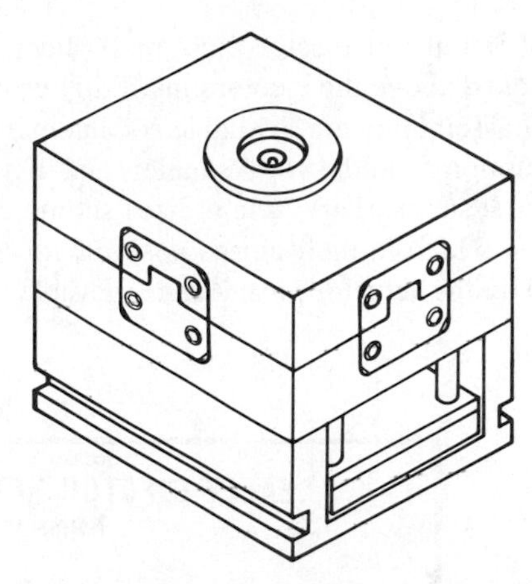

Fig. 7-39. Straight side interlocks showing proper installation on centerline of mold. (*Courtesy D-M-E Co.*)

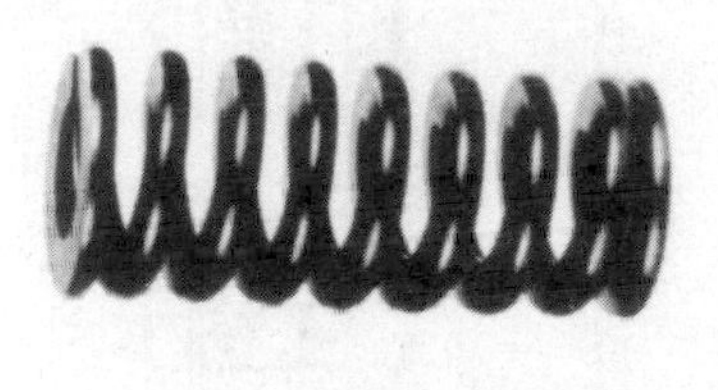

Fig. 7-40. Chromium vanadium springs utilized in assisting the return of the ejector system. (*Courtesy D-M-E Co.*)

Return Pins and Springs

Return pins are utilized to return the ejector plates to their molding position upon mold closing. This system is both a positive and a cost-effective method of accomplishing this phase of the molding cycle if the mold design allows it. Springs, installed around the return pins in the ejector housing, are incorporated in many molds and assist in freeing the molded part from the ejector pins. (See Fig. 7-40.)

Early Ejector Return

An early return mechanism must be employed when it is imperative that the ejector system be returned prior to mold closing because of the use of thumbnail ejector pins, or if slides are positioned above the ejector pins. Early ejector return assemblies are available for internal installation in molds with smaller and lighter ejector systems. This style utilizes sliding cam fingers, which in their closed position are contacted by the actuator pins. As the movable side of the mold, where the actuator pins are mounted, continues its closing movement, the ejector system is positioned to its home position. At this point, the cam fingers are allowed to open, and the actuator then is free to advance past the cam fingers until the mold is fully closed. (See Fig. 7-41.)

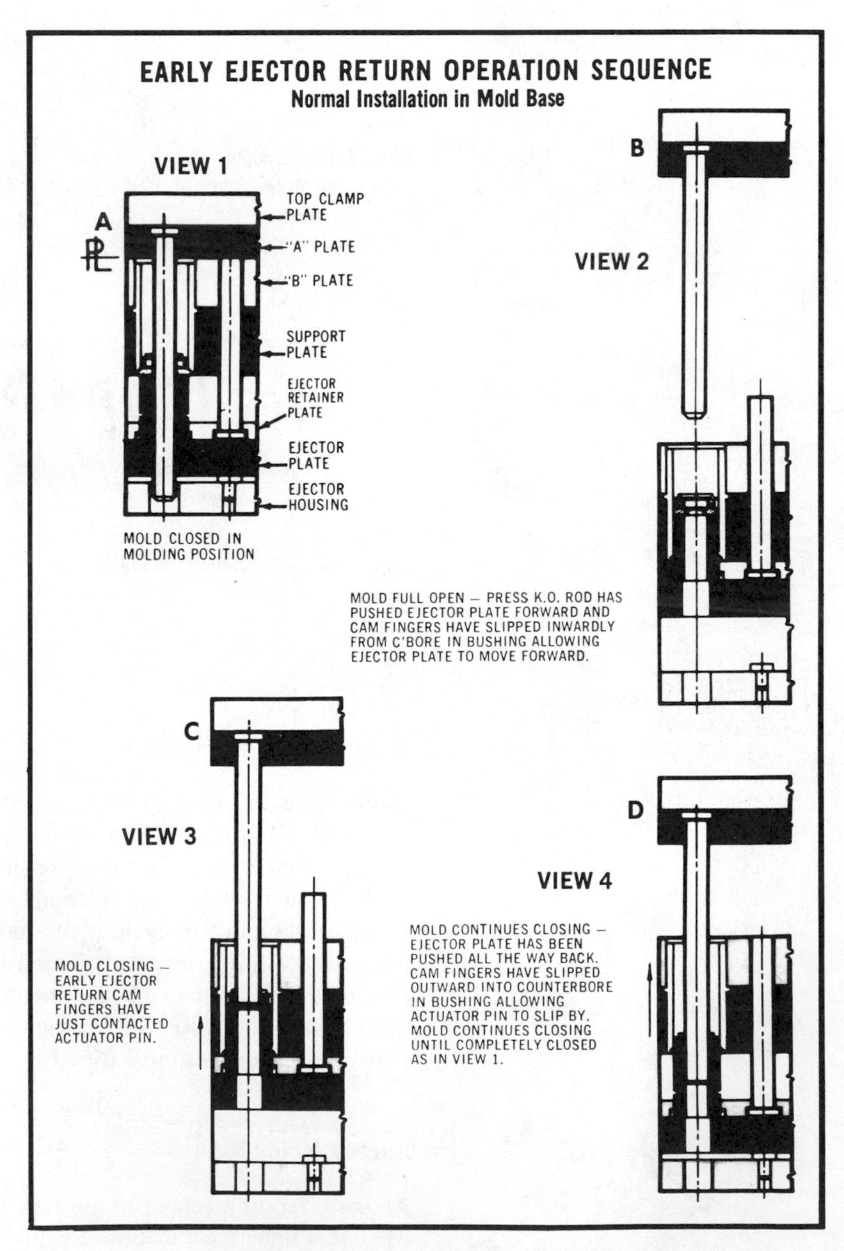

Fig. 7-41. Positive early ejector return system and operating sequence for lighter ejector system. (*Courtesy D-M-E Co.*)

External Positive Return System

With larger mold bases, which typically have both bigger and heavier ejector systems, Toggle-Loks are used to ensure early positive ejector return. These assemblies are externally mounted on the side of the mold. The lever is mounted to the stationary side of the mold with arms and joints located on the movable side connecting to the ejector plates. The arms and joints are positioned so that when the ejector system is actuated forward, the arms pivot inward, occupying the space vacated by the retracted lever. Before the mold reaches the fully closed position, the angle on the lever contacts the radius on the pivoted arm, applies pressure through the joint, and subsequently returns the ejector system to its fully retracted position. (See Fig. 7-42.)

Hobbing

Cavities also are formed by methods other than machining. After machining, hobbing ranks as the next most common method employed to produce cavities. Hobbing consists of forcing a hardened negative pattern into the selected cavity stock under extremely high pressure. This process has a limiting factor of obtainable cavity depth to a certain ratio of hob diameter. Hobs frequently are constructed from steels such as S-1, S-4, A-2, A-6, and D-2. The master hob must contain the polish and precise detail desired in the finished cavity, as the process faithfully reproduces the master hob's characteristics to the cavities. The next operation after the hobbing itself is to cut the outside of the cavity blank to the desired size, install any required water passages and gate details, harden, and then final polish. Molds that contain high cavitation are the most viable candidates. The closure industry takes great advantage of this process with the high-cavitation molds utilized.

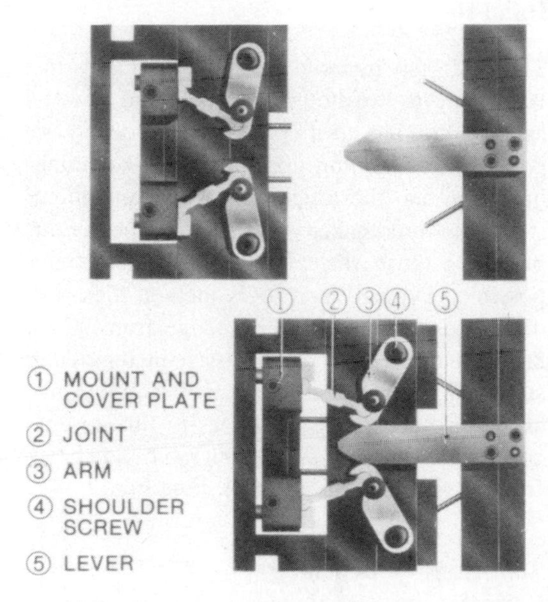

Fig. 7-42. Externally mounted positive ejector system return with mold open and ejector system advanced, and mold closing with ejector system fully returned. *(Courtesy D-M-E Co.) (Courtesy D-M-E Co.)*

Cast Cavities

Cast cavities have certain advantages not obtainable with either machining or hobbing. The advantage of cast cavities is that large amounts of excess stock do not have to be removed by machining. A variety of cavity materials are suitable for casting, including steel, beryllium copper, kirksite, aluminum, and others. In the casting process, a pattern must be constructed that not only incorporates the features of the desired cavity configuration, but also the shrinkage of the casting material as well as that of the plastic which is to be molded. As it is almost always less expensive to machine a pattern than the final cavity, fabrication economics favor this process in larger cavities. The prototyping of large automotive parts is frequently done in cast kirksite molds. Because the melting point of kirksite is relatively low, patterns can be built from wood or plaster. Another advantage is that in some materials, water lines can be cast internally. As with many specialty processes, the casting of cavities is best performed by companies specializing in this fabrication process. Steels that are cast include AISI 1020, 1040, 4130, 4340, and 8630, as well as S-7 and stainless steel.

Electroforming

Electroforming is a process in which metal is deposited on a master in a plating bath. Many companies engaged in electroforming use proprietary processes, which are closely guarded. In one method of forming cavities, the master is constructed of plastic and coated with silver to provide a conductive coating. The coated master then is placed in a plating tank and nickel or nickel-cobalt is deposited to the desired thickness, which could approach $\frac{1}{4}$ inch. With this method, a hardness of up to 46 R_c is obtainable. The nickel shell then is backed up with copper to a thickness sufficient to allow for machining a flat surface, to enable the cavity to be mounted into a cavity pocket. The electroforming process is used for production of single or low numbers of cavities, as opposed to hobbing. Some deep cavities are formed by this process instead of swaging, including long slender components such as ball-point pen barrels.

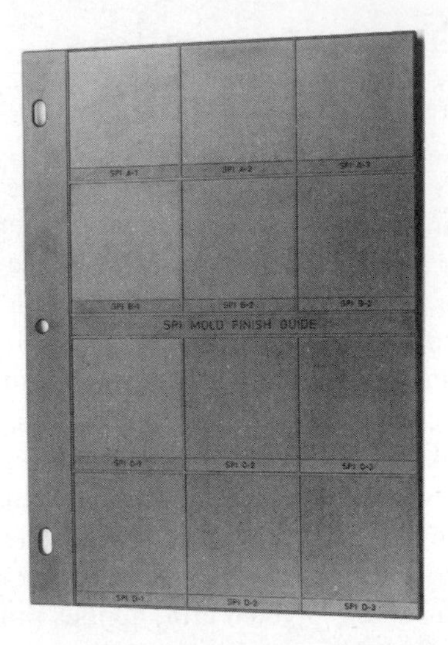

Fig. 7.43 SPI Mold Finish Standards

Cavity Finishing (SPI Specifications)

When machining of the cavity and core inserts has been completed, the components normally will require benching and final finishing or polishing. If the steel inserts are to be heat-treated, then only the benching phase will be completed, with polishing performed after the inserts have been returned from the heat treaters. As the plastic part will duplicate the surface of the mold cavity, it is extremely important to specify the correct finish for the mold. Mold finish is specified as an SPI number to properly communicate cavity and core finish requirements between the product and mold designer, moldmaker, and molder. This SPI standard uses 12 alphanumeric designations to define the distinct finishes. These finishes, with A-1 representing the highest polish and D-3 the roughest finish (See Fig. 7-43), correspond to the final finishing process utilized. The SPI standard finishes are listed from A-1 through D-3 and are categorized as follows:

A-1	Grit #3 Diamond Buff	C-1	600 Stone
A-2	Grit #6 Diamond Buff	C-2	400 Stone
A-3	Grit #15 Diamond Buff	C-3	320 Stone
B-1	600 Grit Paper	D-1	Dry Glass Bead #11
B-2	400 Grit Paper	D-2	Dry Blast #240 Oxide
B-3	320 Grit Paper	D-3	Dry Blast #24 Oxide

Hand Benching

The first step in mold finishing normally involves the use of both hand tools and power-assisted grinders to prepare the surfaces to be polished. Depending upon the last machining operation and the roughness of the remaining stock, the moldmaker will select the method of preparing the surface in the quickest manner possible. Typical hand tools include files and die maker rifflers. The files range from 20 to 80 teeth per inch and will assist in removing stock quickly and accurately. Rifflers are available in a wide range of shapes to fit into any conceivable contour found on a mold, and have from 20 to 220 teeth per inch. (See Fig. 7-44.)

Tool-Asslsted Benching

Power-assisted tools include electric, pneumatic, and ultrasonic machines, which can be

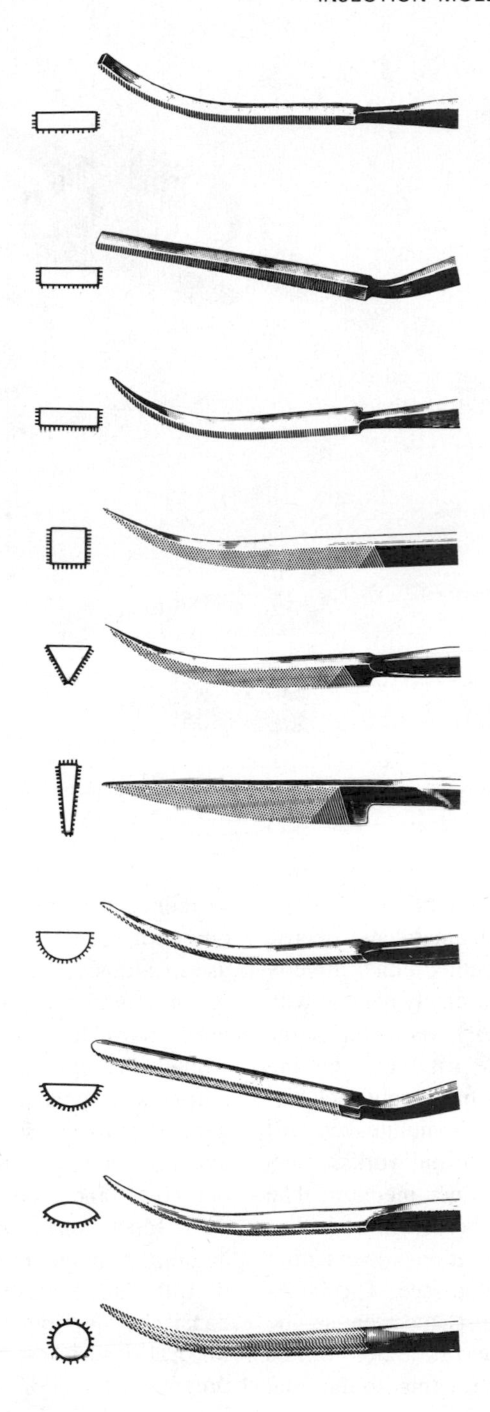

Fig. 7-44. Moldmaker's rifflers in a variety of shapes. (*Courtesy D-M-E Co.*)

Fig.7-45. Two variations of electrical power-assisted benching aids. (*Courtesy D-M-E Co.*)

equipped with a wide variety of tool holders. (See Fig. 7-45.) Rotary pneumatic and electric grinders, with the tool mounted either directly or in a flexible shaft, are extremely popular with the moldmaker for rapidly removing large amounts of stock. With the grinding or cutting medium mounted in a 90° or straight tool, depending on the surface, the moldmaker will start with the coarser medium and work the surface progressively to the finer medium. The finishing process can best be described as one where each preceding operation imparts finer and finer scratches on the surface. The initial stages are perhaps the most important in the finishing state. Spending too little time in this phase normally will be detrimental to the final surface finish.

Benching Material

A variety of cutting tools, from carbide rotary cutters, abrasive drum, band, cartridge, tapered cone, and disc to flap wheels, are used to reduce the roughness of the surface in the quest for the desired mold surface finish. The use of abrasive stones, either worked by hand or in conjunction with reciprocating power tools, normally is the next phase in mold finishing. These stones are a combination of grit particles suspended in a bonding agent. For general-purpose stoning, silicon carbide is available in type A stones for steel hardness under 40 R_c and type B stones for higher hardness. Most stones are available in square, rectangular, triangular, and round shapes in grits of 150, 240, 320, 400, and 600. Other stones consist of aluminum oxide in type E for working EDM surfaces, type M which has oil impregnated for additional lubrication, and type F for added flexibility for reduced breakage of thin stones. (See Fig. 7-46.)

Direction of Benching

The next step normally consists of using abrasive sheets or discs to continue smoothing of

Fig. 7-46. Abrasive stones are available in a variety of sizes, shapes, and grades to accommodate the moldmaker. (*Courtesy D-M-E Co.*)

the mold surface. All of the finish work is carried out in the direction of the scratches installed in the line of draw or ejection. Machine, file, stone, or abrasive marks installed perpendicular to the direction of draw are detrimental to removal of the part from either the cavity or the core, and must be avoided. At this point, the desired mold surface finish may have been achieved, or the piece parts grit blasted, and finishing is considered complete. These finishes may be acceptable for nonfunctional core surfaces where ejection does not pose a problem.

Mold Polishing

To achieve the highest polished surfaces, SPI No. 1 or 2, an extremely fine abrasive material, is used. Diamond particles, suspended in a carrier, are applied, and the mold surface worked with a variety of polishing aids. The application tools include fiber and brass brushes, round felt wheels for rotary hand pieces, square and rectangular long felt bobs for reciprocating and hand work, and polishing sticks of cast iron and wood. As with files and stones, the coarser diamond grades are used first to establish the mold surface for each succeeding step. Diamond micron sizes range from 0, the finest, to 60, the coarsest. Considerable time is saved by starting with a coarser grade material and, after proper preparation, moving to the next grade, rather than rushing to use the finest grade. Because of the expense of achieving an SPI No. 1 finish, it normally is reserved for optical lenses. As a rule of thumb in polishing, if the operation that you are performing does not fully remove the scratches from the previous operation, do not continue forward, but rather backtrack at this point to a preceding step. Continuing would result in a flaw that never would be removed but always would be apparent even if good gloss were achieved.

Ultrasonic Tools

The latest aid for mold finishing has been the introduction of ultrasonic finishing and polishing systems (Fig. 7-47). The ultrasonic controls deliver strokes that can be adjusted to between 10 and 30 microns at speeds up to 22,000 cycles per minute. This action eliminates costly hand finishing and is extremely effective in polishing deep narrow ribs that have been EDM-installed. These devices, along with diamond files, ruby or abrasive stones, and wood laps, have made the work of the mold polisher much more pleasant.

Textured Cavities

Texturing can add another dimension to a molded part. For example, the term "camera case finish" has been used to describe a textured surface finish that is desirable in certain situations. A surface that is to be textured should be worked up to a 320 grit (SPI No. 3) finish. A technician then applies the selected pattern to the mold surface. In this process, acid will be applied to the area where mold material will be dissolved; therefore, wax or a resistant coating is applied to the surface that will remain on the cavity. Selective texturing on a cavity will require that acid-resistant material be applied to areas of no texturing to guard against attack by the acid. The depth of the texturing will be a function of the mold material, its chemical resistance, the type and concentration of the acid, and time. As with many other

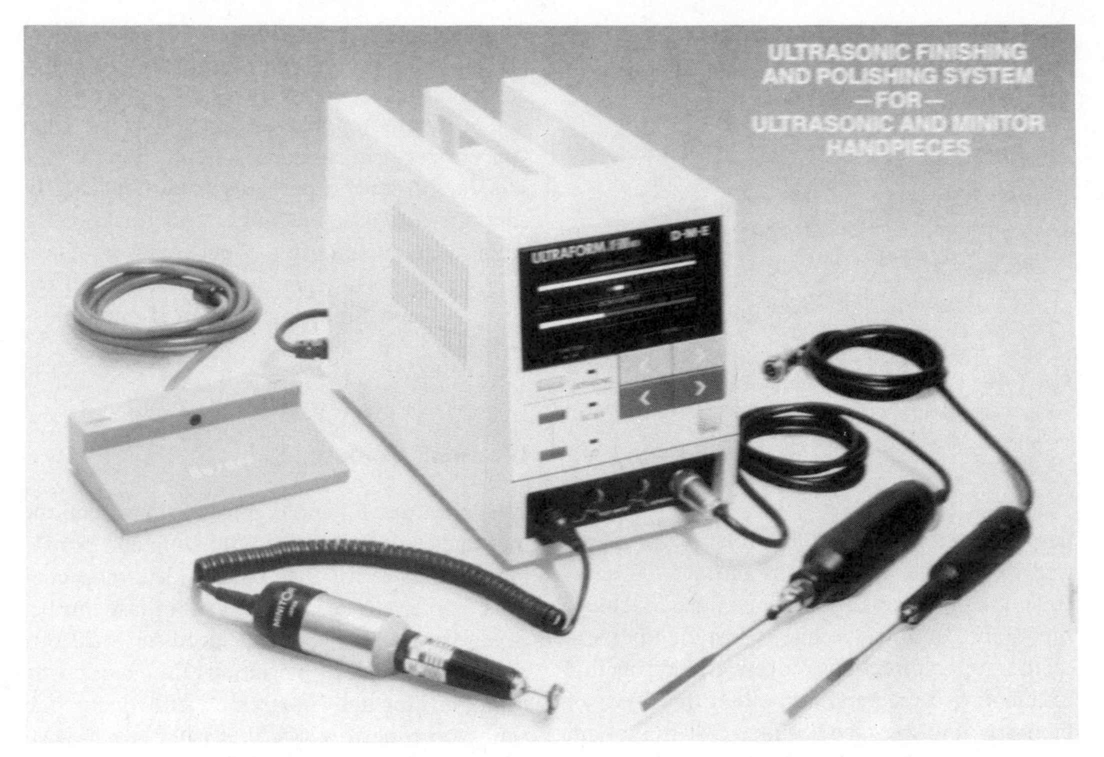

Fig. 7-47. Ultrasonic finishing devices enable the moldmaker to drastically reduce finishing time. (*Courtesy D-M-E Co.*)

processes, the quality of the finished part depends on the person performing the task, who must frequently check the depth of the texture to ensure that the desired effect is achieved. When the operator is satisfied with the texture depth, the acid is neutralized, and the wax resist is removed.

Texture Patterns

Almost any pattern imaginable is possible. Most texture sources have a wide range of standard patterns, and molded plaques are available to show the final results. The designer must careful to allow sufficient draft on the cavity walls for proper release of the plastic from the textured surfaces. First, the draft angles must be at least the minimum acceptable for the plastic that will be molded. Then, from 1 to $1\frac{1}{2}°$ of draft per .001 inch of texture depth must be allowed, to enable proper release of the textured plastic part. When special directional patterns, which may run perpendicular to the parting line, or low-shrinkage, less-flexible materials are to be molded, greater draft angles will be required for release.

Mold Steels

Most steels used in plastic molds texture well. Many textured P-20, 0-1, A-6, S-7, H-13, and 420 stainless steel cavities have been completed successfully. It is imperative to use the same cavity material and have the texturing completed at the same time for textured parts that will be joined later in assembly. The texturing of cavities is useful in concealing flow lines, sink marks, and gate blushes on molded parts, in addition to being aesthetically pleasing.

Compression Molds

Compression molds, normally semipositive or positive types, must be sturdily built to withstand the extremely demanding conditions inherent in the process. Compression molding requires loading the cavities with loose powder or preheated preforms and then subjecting the plastic material to extremely high pressures, compressing the thermosetting material to the part shape. Compression mold cavities, normally constructed of A-2, A-6, H-13, S-7, or even M-Series tool steels, must be designed to withstand tremendously high compression forces while requiring low maintenance for the life of the cavities and forces. Wall sections of cavities must be of sufficient thickness to resist the flexural stresses inherent in this process. The mold base steel must assist in holding the cavity side walls stable without breathing. Ejector pins must be as large in diameter as possible to allow ejection without extreme flexing of the pins, ensuring a long and useful life. Mold support plates must be of sufficient thickness and properly supported with parallels to withstand the compressive loads imposed on the mold by the molding machine. Compression mold bases normally are built from No. 2 or No. 3 steel.

Ratio of core pin length to diameters.

Diameter	*Compression mold*	*Transfer mold*
Less than $\frac{1}{8}''$	1.5 ×	2.5 ×
$\frac{1}{8}$ to $\frac{1}{4}''$	2.0 ×	3.0 ×
$\frac{1}{4}$ to $\frac{1}{2}''$	3.0 ×	4.0 ×
$\frac{1}{2}''$ and larger	4.0 ×	6.0 ×

Transfer Molds

Transfer molds, either top or bottom loading, must have cavities and cores constructed from steels that are highly resistant to abrasion. Typically, gates are inserted in transfer tools and built from superior, wear-resistant alloy steels including carbides. Shot sleeves constructed from 4140 steel are nitrided to extend the life and provide a good wear surface for the plunger. As in compression molds, A-2, A-6, D-2, and S-7 steels, along with carbide inserts in high-wear areas, are utilized to overcome the abrasion of the material and failures. Care must be taken when designing compression and transfer molds to maintain the ratios of core lengths to their diameters. (See tabulation.)

Reaction Injection Molds

Reaction injection molds, most of which are being built for automotive applications, have cavities constructed from steel, aluminum, kirksite, or epoxy. As many of the parts produced by the RIM process contain many contours, casting is a very popular method of cavity finishing. Extremely long-run tools normally are constructed of 4140, which is then chrome- or nickel-plated. Aluminum also is an excellent choice for the RIM process, and cavities are either cast or machined. This lighter-weight material is advantageous in larger molds with the tilting presses used in RIM molding. Epoxy and kirksite cavities generally are used for extremely low-run or prototype molds although a two-cavity epoxy mold was used to produce the first production run for an automobile bumper. The demolding of RIM parts usually is accomplished with the generous use of lifters and bar ejectors. Because of the RIM material's inherent difficulty in releasing from the core, ejector pins seldom are used.

Expandable Polystyrene Molds

In expandable polystyrene molding, a universal steam chest usually is adapted for the process and will be utilized with a variety of molds in an effort to reduce reoccurring mold costs. Steam chests typically are constructed from aluminum, as pressures rarely exceed 50 psi. To eliminate any possibility of thermal expansion problems, aluminum is the most frequent first choice for the cavities. As the process alternates between constant heating and cooling, the mold must offer exceptional heat transfer qualities while resisting the environmental effects of the ever-present moisture. With the low molding pressures, cavity wall thicknesses generally are between .375 and .500 inch. As always, care must be taken in the selection of dissimilar materials in contact with the aluminum to prevent galvanic corrosion in the presence of moisture. If a steel insert will be utilized, it should be fabricated from stainless steel.

Structural Foam Molds

Because of the relatively low molding pressures involved in structural foam molding, these molds can be constructed of soft materials. Most cavities are built from prehardened steel or aluminum with beryllium copper, or kirksite is used on occasion. Steel cavities have higher initial costs, but they normally pay back handsomely on long runs because of the lower maintenance involved. Generally, the mold steels selected are 4130 or 4140, with aluminum grades of 6061 or 7075 sometimes utilized.

Chrome Plating

Cavity and core steels may be heat-treated or plated by a wide variety of methods, either to obtain long mold life or to provide for a specific function. Molds have been hard-chrome-plated for many years, and this practice is extremely useful in protecting thermosetting cavities. Chrome in thicknesses as low as .0002 inch is plated over the molding surface with an approximate hardness of 65 to 69 R_c. The presence of chrome, or, better, the absence of chrome plating, can be checked by applying a solution of copper sulfate to the area in question. If the copper does not plate out on the surface, chrome is present. If the copper does plate out, the chrome protection is missing.

Electroless Nickel Plating

Electroless nickel plating is popular not only for protecting the molding surface but also for protecting the waterline and mold base surfaces. Whereas the hardness of nickel (between 45 and 50 R_c) does not approach that of chrome, its wear-resistant properties and surface protection are good, and this type of plating is applied evenly with no buildup. Nickel should not be used with materials containing large amounts of sulfur, and, conversely, chrome should not be in contact with plastics containing chloride or fluoride.

Nitriding and Carburizing

Steels also are nitrided and carburized to improve the surface hardness, thus making the surface more wear-resistant. Nitriding will penetrate the surface from .003 to .020, depending on the steel, and can result in hardnesses of 65 to 70 R_c. Another process for imparting a surface hardness is carburizing. In this process, carbon is introduced into the surface of the cavity or core steel, and the inserts are heated to above the steel's transformation temperature range while in contact with a carbonaceous material. This process frequently is followed by a quenching operation to impart the hardness case. Hardness as high as 64 R_c and a depth of .030 inch are possible with this process.

Proprietary Wear Processes

Several other proprietary processes are available to reduce wear between components and/or ease the force required to eject parts. These processes include Armoloy®, Drilube®, Microseal®, and Tuf-Triding®, to name a few. As requirements vary greatly, it is recommended that you contact the suppliers of these processes for their advice to ensure that the process chosen is the best suited to your application.

Mold Steel

The following information has been adapted from Irvin Rubin's discussion in Chapter 22 of the fourth edition of this handbook, as the author could not improve upon Mr. Rubin's information.

Steel Quality. Mold-quality steel is highly dependent upon the process used in its manufacture. Obviously, the chemical composition of the steel will affect steel properties. However, the method of melting the steel is the most critical factor in the quality of the final product. Other factors affecting quality include the cooling rate of the molten steel to solidification and the rate of cooling from solidification to room temperature. The entire cooling process is lengthy, and may encompass several days.

The next factor affecting the quality of the steel is the reduction of the ingot to the finished bar stock. Also affecting quality is the relationship of the size and shape of the ingot to the size and shape of the final product during the reduction process. Additionally, mold quality steel is a result of the temperature involved in both the beginning and the ending of the reduction process, as well as in the annealing and heat treatment procedure.

As ingots of molten steel solidify first on the bottom, then the sides, then the top, and finally the center, voids unfortunately will most likely occur in the middle of the block of steel. An extremely large block of mold steel is more likely to have internal voids than a smaller block of the same steel. On large P-20 blocks of mold steel, special forging practices are employed to ensure the soundness of the center of the block. The degree of soundness of the steel, the relative porosity, and the uniformity are dependent on the methods used in melting and reducing the steel ingot to the finished bar stock.

To ensure the integrity of a block of mold steel, inspection should be carried out prior to any machining. One such method is ultrasonic inspection. It is good practice to have large blocks of steel inspected and certified as to their soundness initially, to avoid expending large amounts of machining time only to discover internal porosity or voids in the final product.

Mold-quality steels frequently are referred to by their chemical composition, which denotes the quantity of the various elements particular to that grade of steel. The chemical composition, however, does not reflect the quality of the steel but rather its physical properties.

Steel Hardness. The hardness of the mold steel is extremely important in both machining and finishing the mold, as well as for the longevity of the final product. All rough machining of cavities and cores should be carried out with the steel in the soft or annealed stage. Once the mold steel has been heat-treated to the desired hardness, the only machining that typically is performed involves grinding or EDM'ing. Selecting the hardness of the mold steel also depends on the desired characteristics of the final product. Typically, a softer cavity will be tougher and more resistant to cracking than a hard cavity. Harder cavities will outwear softer cavities. The application will dictate the hardness required for the mold.

Polishability. Generally, the higher the hardness is that the steel is heat treated to, the higher the luster will be, if everything else remains the same. Steels with hardnesses of less than 30 R_c typically will be difficult to obtain and to maintain a good polish on. When a good finish is required, care must be taken to select the correct alloy and proper hardness to will yield the desired luster. The final plastic part appearance will be directly affected by the finish on the cavity. In polishing, it is important to remember that in addition to the luster produced, the surface must be smooth and free from all machine marks and scratches.

Wear Resistance. The mold cavity and core must perform well and provide maintenance-free production for many thousands of parts. Depending on the type of plastics and the filler used, the mold steel must resist the corrosive and abrasive effects of the plastics being molded. Plastics that are corrosive in nature, including polyvinyl chloride, typically need to have cavities and cores constructed from stainless steel. Other steels that are resistant to corrosive materials include those high in chrome content.

Fillers in thermoplastic materials, including glass, carbon, and mineral, can be extremely abrasive. Therefore, abrasion resistance is extremely critical to the cavities and cores when high volumes of dimensionally demanding parts are being molded from these plastics. The high-hardness tool steels, 54 R_c and above, perform best in these conditions. Although S-7 has been

Table 7-1. Chemical compositions of steels used in molds (percentage).

AISI	*C*	*Mn*	*Si*	*Cr*	*Ni*	*Mo*	*W*	*V*
				Air-Hardening Medium-Alloy Cold Work Steels				
A2	0.95–1.05	1.00 max	0.50 max	4.75–5.50	0.30 max	0.90–1.40	—	0.15–0.50
A6	0.65–0.75	1.80–2.50	0.50 max	0.90–1.20	0.30 max	0.90–1.40	—	—
A9	0.45–0.55	0.50 max	0.95–1.15	4.75–5.50	1.25–1.75	1.30–1.80	—	0.80–1.40
A10	1.25–1.50	1.60–2.10	1.00–1.50	—	1.55–2.05	1.25–1.75	—	—
				High-Carbon, High-Chromium Cold Work Steels				
D2	1.40–1.60	0.60 max	0.60 max	11.00–13.00	0.30 max	0.70–1.20	—	1.10 max
D7	2.15–2.50	0.60 max	0.60 max	11.50–13.50	0.30 max	0.70–1.20	—	3.80–4.40
				Chromium Hot Work Steels				
H11	0.33–0.43	0.20–0.50	0.80–1.30	4.75–5.50	0.30 max	1.10–1.60	—	0.30–0.60
H12	0.30–0.40	0.20–0.50	0.80–1.20	4.75–5.50	0.30 max	1.25–1.75	1.00–1.70	0.50 max
H13	0.32–0.45	0.20–0.50	0.80–1.20	4.75–5.50	0.30 max	1.10–1.75	—	0.80–1.20
				Tungsten Hot Work Steels				
H21	0.26–0.36	0.15–0.40	0.15–0.50	3.00–3.75	0.30 max	—	8.50–10.00	0.30–0.60
H23	0.25–0.35	0.15–0.40	0.15–0.60	11.00–12.75	0.30 max	—	11.00–12.75	0.75–1.25
				Low-Alloy Special-Purpose Tool Steels				
L2	0.45–1.00	0.10–0.90	0.50 max	0.70–1.20	—	0.25 max	—	0.10–0.30
L6	0.65–0.75	0.25–0.80	0.50 max	0.60–1.20	1.25–2.00	0.50 max	—	0.20–0.30
				Molybdenum High Speed Steels				
M1	0.78–0.88	0.15–0.40	0.20–0.50	3.50–4.00	0.30 max	8.20–9.20	1.40–2.10	1.00–1.35
M4	1.25–1.40	0.15–0.40	0.20–0.45	3.75–4.75	0.30 max	4.25–5.50	5.25–6.50	3.75–4.50
				Oil-Hardening Cold Work Steels				
O1	0.85–1.00	1.00–1.40	0.50 max	0.40–0.60	0.30 max	—	0.40–0.60	0.30 max
O2	0.85–0.95	1.40–1.80	0.50 max	0.35 max	0.30 max	0.30 max	—	0.30 max
O6	1.25–1.55	0.30–1.10	0.55–1.50	0.30 max	0.30 max	0.20–0.30	—	—
				Low-Carbon Mold Steels				
P2	0.10 max	0.10–0.40	0.10–0.40	0.75–1.25	0.10–0.50	0.15–0.40	—	—
P3	0.10 max	0.20–0.60	0.40 max	0.40–0.75	1.00–1.50	—	—	—
P4	0.12 max	0.20–0.60	0.10–0.40	4.00–5.25	—	0.40–1.00	—	—
P5	0.10 max	0.20–0.60	0.40 max	2.00–2.50	0.35 max	—	—	—
P6	0.05–0.15	0.35–0.70	0.10–0.40	1.25–1.75	3.25–3.75	—	—	—
P20	0.28–0.40	0.60–1.00	0.20–0.80	1.40–2.00	—	0.30–0.55	—	—
				Shock-Resisting Steels				
S1	0.40–0.55	0.10–0.40	0.15–1.20	1.00–1.80	0.30 max	0.50 max	1.50–3.00	0.15–0.30
S7	0.45–0.55	0.20–0.80	0.20–1.00	3.00–3.50	—	1.30–1.80	—	0.20–0.30
				Tungsten High-Speed Steels				
T1	0.65–0.80	0.10–0.40	0.20–0.40	3.75–4.00	0.30 max	—	17.25–18.75	0.90–1.30
T2	0.80–0.90	0.20–0.40	0.20–0.40	3.75–4.50	0.30 max	1.00 max	17.50–19.00	1.80–2.40
T4	0.70–0.80	0.10–0.40	0.20–0.40	3.75–4.50	0.30 max	0.40–1.00	17.50–19.00	0.80–1.20
T5	0.75–0.85	0.20–0.40	0.20–0.40	3.75–5.00	0.30 max	0.50–1.25	17.50–19.00	1.80–2.40
T6	0.75–0.85	0.20–0.40	0.20–0.40	4.00–4.75	0.30 max	0.40–1.00	18.50–21.00	1.50–2.10
T8	0.75–0.85	0.20–0.40	0.20–0.40	3.75–4.50	0.30 max	0.40–1.00	13.25–14.75	1.80–2.40
T15	1.50–1.60	0.15–0.40	0.15–0.40	3.75–5.00	0.30 max	1.00 max	11.75–13.00	4.50–5.25
				Stainless Steels				
420	0.15 min	1.00 max	1.00 max	12.00–14.00	—	0.75 max	—	—
440	0.75–0.95	1.00 max	1.00 max	16.00–18.00	—	—	—	—

All steels contain 0.25 max Cu, 0.03 max P, and 0.03 max S; D-2 has 1.00 max Co.

Table 7-2. Heat treating of mold steels (all temperatures in °F).

AISI-SAE Steel:	*A-2*	*A-6*	*A-9*	*A-10*	*D-2*
1. Hardening					
Preheat temp, °F	1450	1200	1450	1200	1200
Austentize temp, 0°F	1700–1800	1525–1600	1800–1875	1450–1500	1800–1875
Holding time, minutes	20–45	20–45	20–45	30–60	15–45
Quench medium	A	A	A	A	A
Quenched hardness, R_c	62–65	59–63	56–58	62–64	64
2. Annealing					
Temperature, 0°F	1550–1600	1350–1375	1550–1600	1410–1460	1600–1650
Cooling rate, 0°F/hr	40	25	25	15	40
Annealed hardness, HB	200–230	215–250	210–250	235–270	215–255
3. Normalizing temperature	Not recommended	Not recommended	Not recommended	1450	Not recommended
4. Depth of hardening	Deep	Deep	Deep	Deep	Deep
5. Nondeforming properties	Best	Best	Best	Best	Very good
6. Tempering range	350–1000	300–900	See manufacturer	See manufacturer	400–1000

AISI-SAE Steel:	*D7*	*H11*	*H12*	*H13*	*H21*
1. Hardening					
Preheat temp, °F	1200	1500	1500	1500	1500
Austentize temp, 0°F	1850–1950	1825–1875	1825–1875	1825–1900	2000–2200
Holding time, minutes	15–45	15–40	15–40	15–40	2–5
Quench medium	A	A	A	A	A-O
Quenched hardness, R_c	65	53–55	52–55	48–53	43–52
2. Annealing					
Temperature, 0°F	1600–1650	1550–1650	1550–1650	1550–1650	1600–1650
Cooling rate, 0°F/hr	40	40	40	40	40
Annealed hardness, HB	215–255	190–230	190–230	190–230	205–235
3. Normalizing temperature	Not recommended	Not recommended	Not recommended	Not recommended	Not recommended
4. Depth of hardening	Deep	Deep	Deep	Deep	Deep
5. Nondeforming properties	Best	Good	Good	Good	Fair
6. Tempering range	See manufacturer	1000–1200	1000–1200	1000–1200	900–1150

AISI-SAE Steel:	*H23*	*L2*	*L6*	*M1*	*M4*
1. Hardening					
Preheat temp, °F	1500	See manufacturer	See manufacturer	1350–1550	1350–1550
Austentize temp, 0°F	2200–2300	1450–1550	1450–1550	2175–2250	2200–2250
Holding time, minutes	2–5	10–30	10–30	2–5	2–5
Quench medium	A-O	W	O	A-O-S	A-O-S
Quenched hardness, R_c	33–35	50–63	62	64–66	64–66
2. Annealing					
Temperature, 0°F	1600–1650	1400–1450	1400–1450	1600–1650	1600–1650
Cooling rate, 0°F/hr	40	40	40	40	40
Annealed hardness, HB	205–235	160–200	185–215	210–240	225–255
3. Normalizing temperature	Not recommended	1600–1650	1600	Not recommended	Not recommended
4. Depth of hardening	Deep	Medium	Medium	Deep	Deep
5. Nondeforming properties	Fair	Fair	Fair	Fair	Fair
6. Tempering range	1200–1300	400–750	300–650	800–1200	800–1200

Table 7-2. (*Continued*)

AISI-SAE Steel:	*O1*	*O2*	*O6*	*P2*	*P3*
1. Hardening					
Preheat temp, °F	1200	1200	See manufacturer	Not required	Not required
Austentize temp, 0°F	1450–1500	1400–1475	1450–1500	1525–1550	1475–1525
Holding time, minutes	10–30	5–20	2–5	15	15
Quench medium	O	O	O	O	O
Quenched hardness, R_c	63–65	63–65	63–65	62–65*	62–64*
Carburizing temp, °F				1650	1650
2. Annealing					
Temperature, 0°F	1400–1450	1375–1425	1410–1450	1350–1500	1350–1500
Cooling rate, 0°F/hr	40	40	20	50	Slow
Annealed hardness, HB	185–210	185–210	185–215	103–123	109–137
3. Normalizing temperature	1600	1550	1600	Not required	Not required
4. Depth of hardening	Medium	Medium	Medium	Carburized	Carburized
5. Nondeforming properties	Good	Good	Good	Good	Good
6. Tempering range	350–400	350–400	350–400	300–500	300–500

AISI-SAE Steel:	*P4*	*P5*	*P6*	*P20*	*S1*
1. Hardening					
Preheat temp, °F	Not recommended	Not required	Not required	1650	
Austentize temp, 0°F	1775–1825	1550–1600	1450–1800	1500–1600	1650–1750
Holding time, minutes	15	15	15	15	15–45
Quench medium	A	O-W	A-O	O	O
Quenched hardness, R_c	62–65*	62–65*	60–62*	58–64*	57–59
Carburizing temp, °F	1650	1650	1650	1650	
2. Annealing					
Temperature, 0°F	1600–1650	1550–1600	1500–1550	1400–1500	1450–1500
Cooling rate, 0°F/hr	25	40	30	50	40
Annealed hardness, HB	116–128	105–116	183–217	150–180	185–230
3. Normalizing temperature	Not recommended	Not required	Not required	1650	Not recommended
4. Depth of hardening	Carburized	Carburized	Carburized	Prehard	Medium
5. Nondeforming properties	Very good	Good	Good	Good	Fair
6. Tempering range	300–500	300–500	300–450	300–500	400–1200

AISI-SAE Steel:	*S7*	*420*	*440*
1. Hardening			
Preheat temp, °F	1200–1300	See manufacturer	See manufacturer
Austentize temp, 0°F	1700–1750	1800–1950	1850–1950
Holding time, minutes	15–45	See manufacturer	See manufacturer
Quench medium	A-O	A-O	A-O
Quench hardness, R_c	60–61	48–56	53–60
2. Annealing			
Temperature, 0°F	1500–1550	1550–1650	1550–1650
Cooling rate, 0°F/hr	25	See manufacturer	See manufacturer
Annealed hardness, HB	185–225	86–95	94–98
3. Normalizing temperature	Not recommended	See manufacturer	See manufacturer
4. Depth of hardening	Best	Good	Very good
5. Nondeforming properties	Good	Good	Good
6. Tempering range	400–600	400–700	300–700

*Hardness after carburizing

found to be a good mold steel, D-2 and some of the "M" Series steels have proved to be more effective than it is.

Hobbing or Peening. To resist deformation, especially at the parting line, mold steels must have a hardness greater than 30 R_c. When the tools are used in high-volume, high-speed molding, hardness in the 50 R_c range is required to withstand frequent and repetitive compressive forces resulting from mold clamping.

Rigidity. To keep a mold from flashing, the cavity, retainer, and support plates must have enough rigidity to prevent flexing. Obviously, flex will result if the injection force is great enough to cause a separation of the mold at the parting line. Although harder steels will prevent hobbing due to flexing of the cavity, the steel composition or hardness does not affect the flexural strength of the steel. To achieve greater stiffness, an increase in the cross-sectional area or thickness of the steel is the only solution. Obviously, the greater the span is between supports of the plate, the greater the amount of flexing that can take place. This flexing can more readily be overcome with the use of support pillars than by using a thicker plate of steel.

Table 7-3. Identification of steels by AISI-SAE designations.

Steels	*AISI-SAE designations*	*Amount alloy*
Carbon	10XX	Plain with 1.00% max Mn
	11XX	Resulfurized
	12XX	Resulfurized and rephosphorized
	15XX	Plan with 1.00 to 1.65% Mn
Manganese	13XX	1.65% Mn
Nickel	23XX	3.50% Ni
	25XX	5.00% Ni
Nickel-chromium	31XX	1.25% Ni; 0.65 to 0.80% Cr
	32XX	1.75 Ni; 1.07% Cr
	33XX	3.50% Ni; 1.50 to 1.57% Cr
	34XX	3.00% Ni; 0.77% Cr
Molybdenum	40XX	0.20 or 0.25% Mo
	44XX	0.40 or 0.52% Mo
Chromium-molybdenum	41XX	0.50, 0.80, or 0.95% Cr; 0.12, 0.20, 0.25, or 0.30% Mo
Nickel-chromium-molybdenum	43XX	1.82% Ni; 0.50 or 0.80% Cr; 0.25% Mo
	47XX	1.05% Ni; 0.45% Cr; 0.20 or 0.35% Mo
	81XX	0.30% Ni; 0.40% Cr; 0.12% Mo
	86XX	0.55% Ni; 0.50% Cr; 0.20% Mo
	87XX	0.55% Ni; 0.50% Cr; 0.25% Mo
	88XX	0.55% Ni; 0.50% Cr; 0.35% Mo
	93XX	3.25% Ni; 1.20% Cr; 0.12% Mo
	94XX	0.45% Ni; 0.40% Cr; 0.12% Mo
	97XX	0.55% Ni; 0.20% Cr; 0.20% Mo
	98XX	1.00% Ni; 0.80% Cr; 0.25% Mo
Nickel-molybdenum	46XX	0.85 or 1.82% Ni; 0.20 or 0.25% Mo
	48XX	3.50% Ni; 0.25% Mo
Chromium	50XX	0.27, 0.40, 0.50, or 0.65% Cr
	51XX	0.80, 0.87, 0.92, 0.95, 1.00, or 1.05% Cr
Chromium with 1.00% C min	50XXX	0.50% Cr
	51XXX	1.02% Cr
	52XXX	1.45% Cr
Chromium-vanadium	61XX	0.60, 0.80, 0r 0.95% Cr; 0.10 or 0.15% V min
Tungsten-chromium	72XX	0.75% Cr; 1.75% W
Silicon-manganese	92XX	0.00 or 0.65% Cr; 0.65, 0.82, or 0.85% Mn; 1.40 or 2.00% Si

The first two digits denote the type of steel, and the second two digits indicate the carbon percentage.

Steel Tables: Types of Steels for Molds. Table 7-1 classifies steel by the AISI designation and groups of steel according to either the type of hardening or the major alloy. This chart is intended to be a guide and reference to the approximate chemical compositions (percentage) of the various steels used in mold making. The author wishes to acknowledge the contributions of Irvin Rubin, for supplying information for the charts in Chapter 22 of the fourth edition of this handbook, and gives special thanks to ASM International for allowing use of data from their *Metals Handbook Ninth Edition*.

Table 7-2 contains heat-treating data for mold steels, including information regarding hardening and annealing temperatures and times. The chart is intended as a general guide. It is recommended that steel suppliers be contacted for specific information regarding their steel types prior to the heat treating of any mold steel.

Table 7-3 is a guide to help identify steels by their AISI-SAE designations. In this table, the first two digits identify the family of steel, and the second two digits indicate the percent of carbon contained in the steel.

8

Injection Molding of Thermosets

OVERVIEW

In the mid-1960s, about two years after the introduction of automatic transfer molding machines and screw-preplasticizing/transfer molding machines, the concept of in-line screw injection molding of thermosets (sometimes referred to as direct screw transfer or DST) was developed. It had long been thought that this process, already highly successful with thermoplastic materials, would be impractical for thermosetting plastics because the critical time–temperature relationship would prove uncontrollable, and material would set up in the barrel.

Development progressed not only on screw and barrel designs but also on stability of thermoset materials at elevated temperatures, finally leading to successful processing by screw injection molding. (See Fig. 8-1.)

This technique has had a significant influence on the thermoset molding business by virtue of reduced molding cycle time and the potential it offers for low-cost, high-volume production of molded thermoset parts.

Today, thermoset injection molding machines are available in all clamp tonnages up to 1200 tons and shot sizes up to 20 lb. (see Figs. 8-2, 8-3, and 8-4).

Thermoset in-line screw assemblies are fitted to horizontal or vertical clamp machines. Most horizontal clamp thermoplastic injection machines can be converted to injection of thermoset by changing the screw, barrel, and nozzle. Because most thermoset barrels are shorter than their thermoplastic counterpart, it may be necessary to reposition the injection assembly closer to the clamp fixed platen.

The thermoset molding materials developed specifically for injection molding have a long shelf life in the barrel at moderate temperatures (approximately 200°F), and react very rapidly when the temperature is brought up to 350 to 400°F as the material is forced through the sprue, runners, and gates, and fills the cavities. This unique development in materials helped to gain acceptance of injection molding as a reliable production process.

How the Process Operates

A typical arrangement for in-line screw injection molding of thermosets is shown in Fig. 8-5.

The machine consists of two sections mounted on a common base. One section clamps and holds the mold halves together under pressure during the injection of material into the mold. The other section—the plasticizing and injection unit—includes the feed hopper, the hydraulic cylinder which forces the screw forward to inject the material into the mold, a motor to rotate the screw, and the heated barrel which encloses the screw.

Basically, the injection-molding press, whether for thermosets or for thermoplastics, is the same, utilizing the reciprocating screw to

Reviewed and revised by John L. Hull, Vice Chairman, Hull Corporation, Hatboro, PA.

Fig. 8-1. Reciprocating screw injection molding, 600-ton hydraulic clamp. (*Courtesy Hull Corporation*)

Fig. 8-2. A 300-ton clamp, reciprocating screw injection molding machine. (*Courtesy HPM Corp.*)

Fig. 8-3. A 150-ton toggle clamp BMC injection molding machine, rear side showing stuffing cylinder feeding into injection barrel. (*Courtesy HPM Corp.*)

Fig. 8-4. Close-up of BMC stuffer cylinder, at left, feeding through elbow into injection barrel of plunger-injection molding machine. (*Courtesy Hull Corp.*)

plasticate the material charge. In most injection molding of thermosets, the material, in granular or pellet form, is fed from the hopper by gravity into the feed throat of the barrel. It then is moved forward by the action of the flights of the screw. As it passes through the barrel, the plastic picks up conductive heat from the heating element on the barrel and frictional heat from the rotation of the screw.

For thermoset materials, the depth of flights

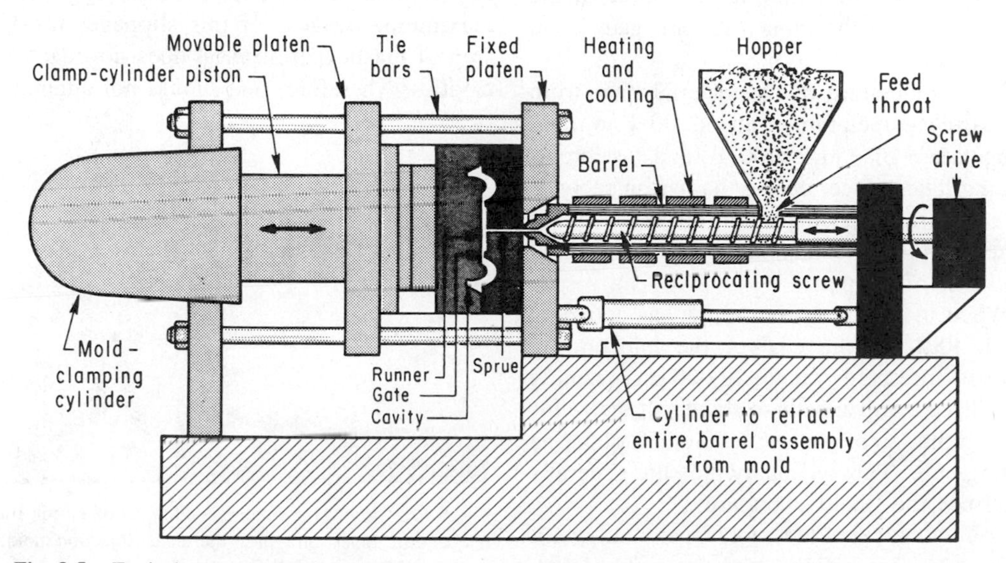

Fig. 8-5. Typical arrangement of direct screw transfer molding machine for thermosets. (*Courtesy S. Bodner*)

of the screw at the feed-zone end is normally the same as the depth of flight at the nozzle end; this is a 1 : 1 compression ratio screw. This compression ratio is the major difference between thermoset and thermoplastic molding machines—the latter having compression ratios such that the depth of flight at the feed end is $1\frac{1}{2}$ to 5 times that at the nozzle end, giving compression ratios of 3 : 1 to 5 : 1.

As the material moves forward in the barrel, it changes from a granular to a semiviscous consistency, and it forces the screw backward in the barrel against a preset hydraulic pressure. This back pressure is an important processing variable. The screw stops turning when the proper amount of material has reached the nozzle end of the barrel, as sensed by a vernier-set limit switch. This material at the nozzle end—the charge—is the exact volume of material required to fill the sprue, runners, and cavities of the mold.

The heated and plasticated slug of material at the front or nozzle end of the barrel is moderately stable for perhaps a minute or more—enough retention time for the mold to complete closing and the machine to be in the high-pressure clamp mode.

The screw is then moved forward at rapid speeds (up to 2000 inches per minute) by hydraulic pressure (up to 20,000 psi) on the plastic. The hot plastic melt is forced through the nozzle of the barrel, through the sprue of the mold, and into the runner system, gates, and mold cavities.

The temperature of the material rises from the barrel temperature of about 200°F to mold temperature of 350 to 400°F, and fast cross-linking takes place, curing the part in seconds or a few minutes, depending on the mass of material and the maximum cross-sectional thickness of the part.

When the reciprocating screw has delivered about 95% of the charge, the injection or "boost" pressure of up to 20,000 psi generally is reduced automatically to about half that value for completion of the cavity fill and for holding during cure. This secondary pressure generally is termed the "hold" pressure.

Following cure, the mold opens automatically, parts are ejected, and the mold closes. At a carefully selected time during the cycle, the plasticating process in the barrel is initiated and timed to be complete by the time that the mold is fully closed again. The molding cycle repeats automatically.

In the process, the granular plastic is fed directly into the barrel of the press, and no external auxiliary preforming or preheating equipment is required. Despite this, the temperature of the material entering the mold is higher and more uniform than that in other thermoset molding techniques because of the homogenizing effect of the screw upon the plastic.

Most barrels are covered with a metal jacket designed to permit heated water to flow across and around the barrel (Fig. 8-6). The water or temperature-controlled fluid not only assists in heating the thermoset material as it moves along the flight of the screw but helps prevent overheating, as the fluid will withdraw excessive internal heat should it occur because of excessive screw rpm or back pressure.

Screw check valves or sleeves, which are standard for thermoplastics (except for heat-sensitive materials such as PVC), are not used with thermosets. They provide restrictions to even flow of high viscosity materials, and eventually cause the material to set up in the valves or sleeves. Some material slippage or back flow along the flights of the screw does occur when the injection pressure approaches maximum. Because of this slippage, the full slug of plasticated material does not reach the cavities. Therefore, one should not attempt to

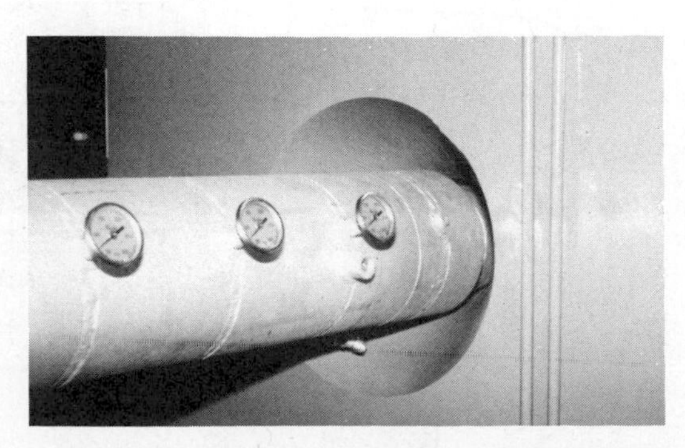

Fig. 8-6. Close-up of water-heated injection barrel, three temperature zones, on typical thermoset injection molding machine. (*Courtesy Hull Corp.*)

use the maximum rated shot capacity of the machine. A rule of thumb for most materials and machines is to use up to 80% of rated capacity.

Nozzles at the end of the barrel usually are water-cooled or temperature-controlled to maintain a proper balance between a hot mold (350–400°F), and a relatively cool barrel (150–200°F).

PROCESS AND DESIGN CONSIDERATIONS

Thermoset molding resins require curing by a chemical reaction, or polymerization. Thus, the closer to the curing temperature the material is when it fills the cavity, the shorter will be the in-cavity cycle time. The direct screw injection process substantially reduces cure time from that of other thermoset processes, particularly for parts having $\frac{1}{8}$ inch wall section or greater (Fig. 8-7). Cycle-time reductions on the order of 20 to 30% are common; even greater reductions have been reported.

Molding a thermoset material on an injection molding machine involves approximately the same cost factors as molding a thermoplastic material. Machinery costs, plant space requirements, and labor costs are about the same. Therefore, the costs that makes the difference are the basic material price and the cycle time required.

Materials

Most materials are slightly modified from conventional thermoset compounds. These modifications are required to provide the working

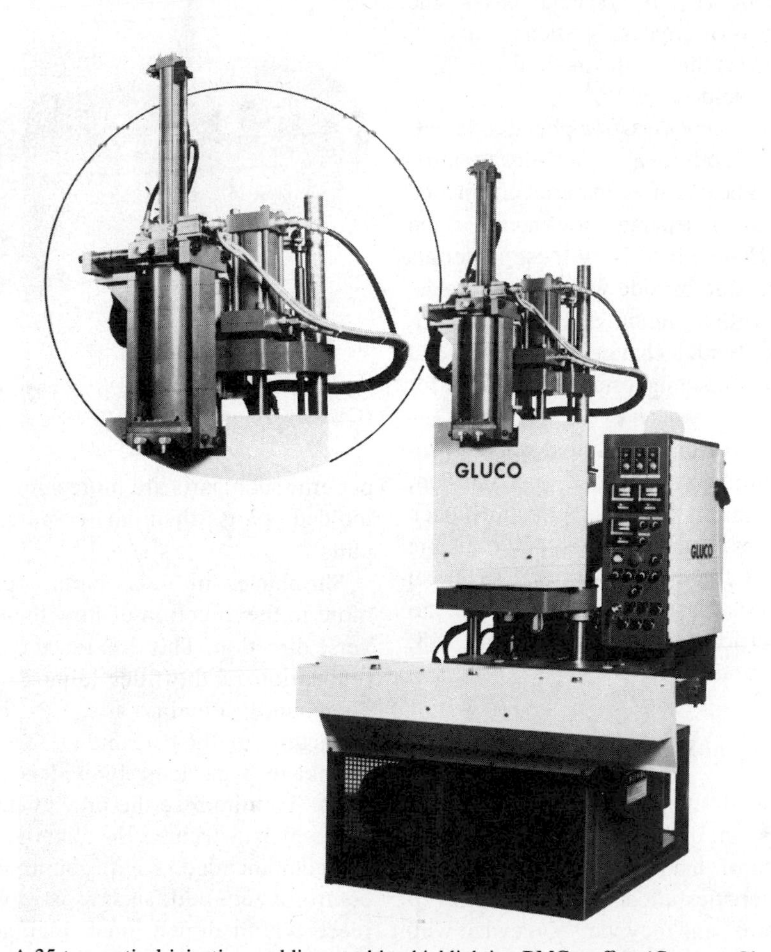

Fig. 8-7. A 35-ton vertical injection molding machine highlighting BMC stuffer. (*Courtesy Gluco, Inc.*)

time/temperature relationship needed for screw plasticating. Additionally, material formulations may be altered according to the geometry of the part being molded. The new compounds are priced approximately the same as conventional materials.

Although all thermosetting materials have been molded by this technique, the most commonly used materials are the phenolics. These materials were the first to be injection-molded, and today it is estimated that approximately 400 million pounds are being processed by injection molding annually. Other thermoset materials being molded by the process include melamine, urea, polyester, alkyd, diallyl phthalate (DAP), polyurethane, and alloys such as phenolic melamine and phenolic epoxy.

A wide variety of fillers can be used to achieve the properties required in the finished product. Except in a few special cases, the choice of filler is of greater significance in end-product properties than in the method by which the product is molded.

The introduction of dust-free phenolic in pellet form brought additional advantages to injection molding. The dust-free material eliminated the former need for separate molding areas and equipment for phenolics. Now these materials can be molded side by side with a thermoplastic operation, with minimum danger of contamination. In fact, with a change of the barrel and screw, the same machine often can be used for both materials.

Applications particularly suited for the process are automotive power brake, transmission, and electrical parts; wet/dry applications such as in steam irons, washer pumps, and moisture vaporizers; and communications and electrical distribution parts that require dimensional stability and electrical insulation. (See Figs. 8-8, 8-9, and 8-10.)

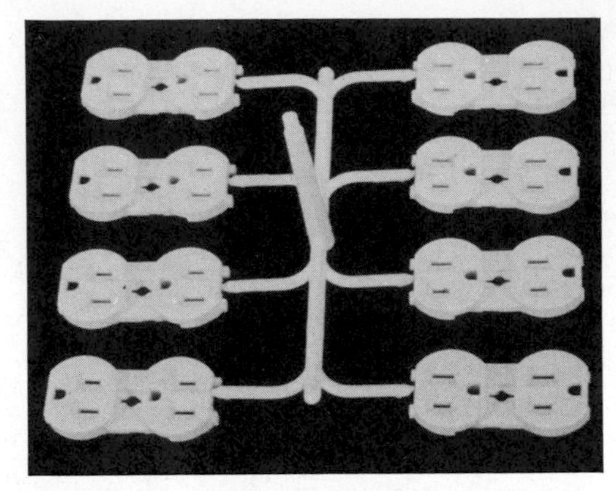

Fig. 8-8. Shot from eight-cavity mold for double outlet wiring device, injection-molded with urea formaldehyde. (*Courtesy Hull Corp.*)

Fig. 8-9. Steam iron handle screw molded from phenolic. (*Courtesy Westinghouse Electric*)

Process Effects on Design

Proper design of injection-molded thermoset parts requires an understanding of the flow characteristics of material within the mold. These characteristics affect some properties of the finished parts, and they vary somewhat with the different materials. In that they have a flow pattern, such parts are more similar to transfer-molded parts than to compression-molded parts.

Shrinkage, in most parts, occurs slightly more in the direction of flow than in the transverse direction. This difference is caused by the orientation of the filler (almost all thermoset compounds contain a filler) as the compound flows through the gate and into the cavity. Such shrinkage is particularly evident in edge-gated parts. To minimize the flow effect, center gating, which distributes the material more evenly, is recommended where optimum shrinkage control is required, such as in molding a round piece within dimensional tolerance. Technology is now well advanced in the use of warm-

Fig. 8-10. Two-cavity shot from conventional injection mold, large electrical switchgear housing, BMC polyester material. (*Courtesy GE Plastics*)

runner molds (equivalent to hot-runner molds as used with thermoplastics), which, when used with center gating, provide a substantial reduction in material consumed in the sprue and runner—a cost saving over conventional thermoset molding.

The flatness required in the finished molded part is another design parameter affected by process capabilities. Long, narrow parts that have to be flat may be difficult to mold because of the variation in shrinkage. Similarly, a nonuniform wall section may warp as a result of nonuniform shrinkage. The choice of gate location sometimes can compensate for such conditions; close cooperation with the mold designer is recommended at this stage.

The surface finish on molded parts also differs from that of traditionally molded thermosets. The greater the distance of material flow required, the more likely it is that flow marks will occur. Avoiding large, flat, polished areas helps reduce rejects and keeps molding costs down. Here, too, gate location is a factor. If large, flat areas are necessary, such areas can be textured or patterned to mask slight irregularities. A rough surface texture generally is not desirable, however, because it may cause sticking in the mold cavity.

Gates used for thermoset injection molding usually are smaller than those needed for transfer molding. This difference provides several advantages. A smaller gate gives a cleaner gate break that requires little or no hand finishing. A gate can be placed in an area where previously the appearance of a larger gate mark would have been objectionable. A small gate also contributes to faster cure cycles because it increases frictional heat within the molding compound during filling. However, the impact properties of compounds containing long-fiber reinforcement degrade when the material passes through a small gate. When such materials are specified, close cooperation between the material supplier and the molder is necessary. Other physical properties such as tensile and flexural strength can also be controlled by gate geometry.

Injection-Compression Molding

To minimize fiber orientation, and thereby improve the impact strength of injection-molded fibrous compounds, the conventional injection molding process is often modified to allow the mold to remain slightly open—perhaps $\frac{1}{4}$ inch—during cavity fill. When the cavity is essentially full, the mold is closed and held under full clamp pressure until cure is completed. In this process, the final mold closing acts like compression molding, permitting the fibers to shift from a longitudinal flow orientation to a more random orientation. The design and construction of the mold must be such that it prevents escape of the injected material to mold parting surfaces during the cavity filling step. Frontal flanges are incorporated to effectively close off the cavity when the mold is partially closed. The injection-compression process often reduces the clamping force required, as compared to conventional injection molding.

Part-Design Considerations

As in most other molding processes, wall-section uniformity is important. Molding cycles, and therefore costs, depend upon the cure time of the thickest section. Therefore, cross sections should be as uniform as design parameters allow, with a minimum wall thickness of $\frac{1}{16}$ inch. A good working average for wall thickness is $\frac{1}{8}$ to $\frac{3}{16}$ inch. Nevertheless, as part requirements dictate, heavy walls favor the use of thermosets because the cure rate of thermosetting materials is considerably faster than the cooling rate of thermoplastics. A rule of thumb for estimating cycle times for a $\frac{1}{4}$ inch wall section is 45 seconds for thermoplastics and 30 seconds for thermosets, when injection molding is used.

Generous radii and fillets are recommended, as in other plastic processes, for maximum strength. Most plastic materials are somewhat notch-sensitive, so avoiding sharp corners is important. Generous fillets are less likely to cause sink marks in the low-shrinkage thermoset materials than they are in some of the thermoplastics, so more freedom is available in the use of reinforcing fillets.

Draft to allow the release of the molded part from the cavity and force plug (core) should be at least $\frac{1}{4}°$ per side; greater draft is preferred if possible. Because the injection method is automatic, ease of mold release is essential for rapid ejection with minimum distortion of the part. Provision should be made in the part for the placement of ejector pins for this purpose. The use of knockout pins as large as possible (at least $\frac{1}{8}$ inch in diameter) consistent with part design and aesthetic considerations will provide trouble-free processing, which ultimately is reflected in lower costs.

Venting of gases from the cavity must be done in the short time available during the filling of the cavity. The most effective vent is the parting line of the mold; so try to visualize the material flow from the gate to the most distant point of the part where the gases will collect. If this point is on a mold parting line, then the part is well designed for venting. Avoid long, dead-end corridors of flow and trapped wall sections that prevent venting. Vacuum venting, as described in Chapter 9, is readily adapted to injection molding when complex shapes cannot be adequately vented from the parting line.

Molded-in inserts commonly are used with thermosetting materials. However, because the injection process is automatic, the use of post-assembled inserts rather than molded-in inserts is recommended. Molded-in inserts require holding the mold open each cycle to place the inserts. A delay in manual placement destroys the advantage of uniformity and consistency of the automatic cycle, affecting both production and quality.

Tolerances of parts molded by the injection method are comparable to those produced by compression and transfer methods. Tolerances have been held as low as ± 0.001 in./in., but ordinarily, tolerances of ± 0.003 to 0.005 in./in. are economically practical for production. See Chapter 28 for recommendations.

All mold techniques and variations such as cams and movable die sections used in other forms of molding are adaptable to thermoset injection molding. Therefore, parts can include such features as molded-in side holes, threads, and undercuts. These refinements increase mold cost, of course, so the decision of how many refinements to build into a mold is usually an economics problem rather than a technical one.

COLD RUNNER MOLDING

Standard injection molds for thermoset materials are very similar to standard mold designs for thermoplastics. Because one cannot grind and reuse the thermoset runner system, there is always interest in reducing the amount of material in a runner system required to produce parts on a production basis.

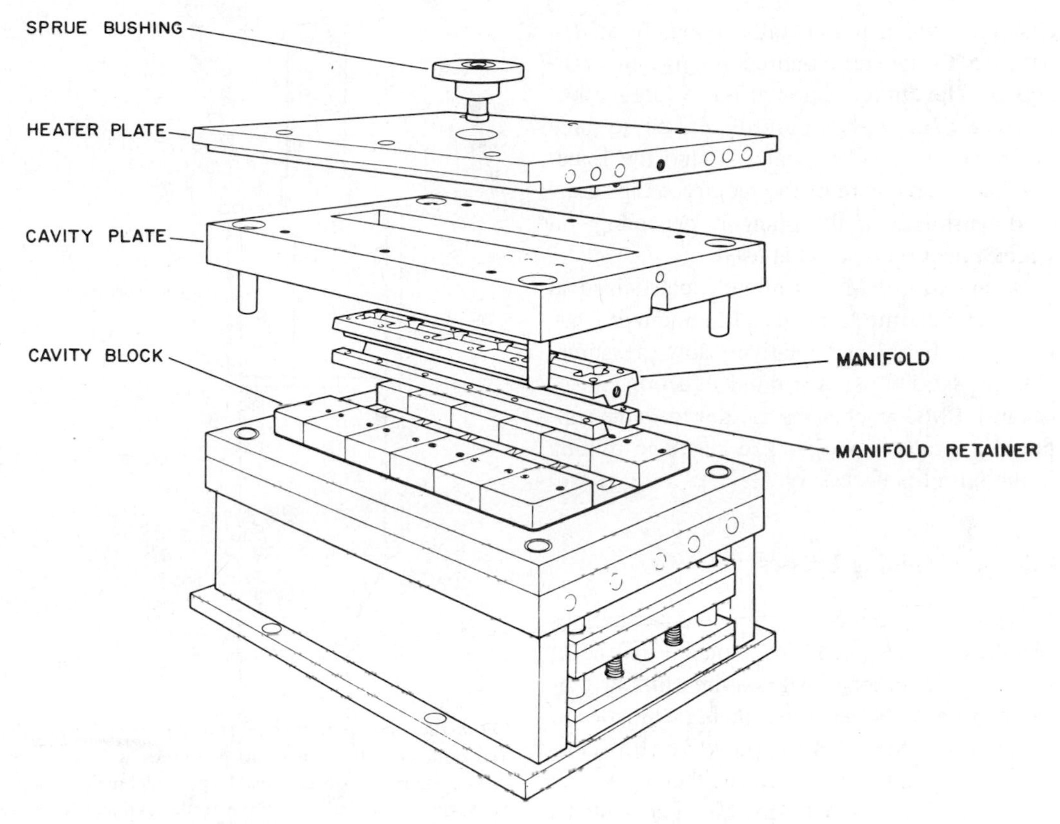

Fig. 8-11. Exploded view of a cold runner manifold. (*Courtesy Stokes Div., Pennwalt Corp.*)

Toward this end, molds may be designed to maintain thermoset material in a plastic state in the runner without ejecting it from the mold. This cold runner technique is not too different from "hot runner" thermoplastic mold designs, except that heated water is used to maintain a "runner" temperature of between 150 and 210°F, and cartridge heaters are used to maintain proper cavity and force temperatures for curing. This type of mold is known as a "cold runner" thermoset mold, or sometimes a "warm runner" mold (see Fig. 8-11). Parts may be separated from the runners right at the part surface; or short subrunners may be ejected with the parts, leaving most of the runner material in the warm manifold, to be used in the next shot.

A multicavity standard injection mold for small thermoset parts may have 50 to 150% runner or waste material. Using a cold runner mold, this waste can be reduced to as little as 10% of the shot.

INJECTION MOLDING REINFORCED POLYESTERS

A popular material increasingly used in thermoset injection molding is polyester reinforced with glass or other fibers. This material generally is referred to as bulk molding compound (BMC) in the United States and as dough molding compound (DMC) in Europe. Because it is doughlike in consistency at room temperature, it does not feed from a hopper but requires a special stuffing mechanism on the injection machine for automatic molding.

Although BMC often is molded by using a conventional reciprocating screw, many BMC injection machines use a simple plunger to force the material from the barrel to the mold. The plunger is used to lessen shear degradation of the fibers as well as to prevent back slippage of material during the injection and hold part of the cycle.

The stuffer mechanism includes a loading

chamber, which is generally manually loaded with up to several hundred pounds of compound. The stuffer chamber has a large-diameter piston, air- or hydraulically driven, to force the room-temperature material into the injection barrel upstream of the reciprocating screw or downstream of the plunger, depending on which injection method is used.

Because the BMC is already plasticized at room temperature, it needs a minimum of heating in the barrel and relatively low pressures, 6000 to 10,000 psi, for rapid injection. Also, because BMC reaches the molder from the supplier in a fully homogenized state, no mixing in the barrel is necessary.

Injection Molding Presses

There are a number of injection molding presses specifically designed to handle thermoset polyester operations. All require stuffing cylinders because of the physical characteristics of the material. Most FRP is puttylike (BMC) or a fiberlike coated material, neither of which flows freely through normal hopper systems (Fig. 8-12).

On some machines the material or compound is forced from the stuffer cylinder from the top or side into the rear of a conventional screw injection cylinder. The screw acts only as a conveyor that moves the material to the front of the cylinder instead of providing a plasticating function. Then, the screw acts as a plunger. It does not turn as it pushes the material into the mold.

On other machines a plunger instead of a screw pushes the material into the mold. This type densifies material as it pushes it into the mold cavities.

Screws and plungers can be interchanged within the same machine frame, requiring only changes in electricals.

Another type is the coaxial plunger machine. Material is dropped into a stuffer cylinder inline with a smaller cylinder that pushes the material into the mold. Advantages claimed for this structure include the short distance that the material has to flow. Angles and corners around which the material must move have been eliminated.

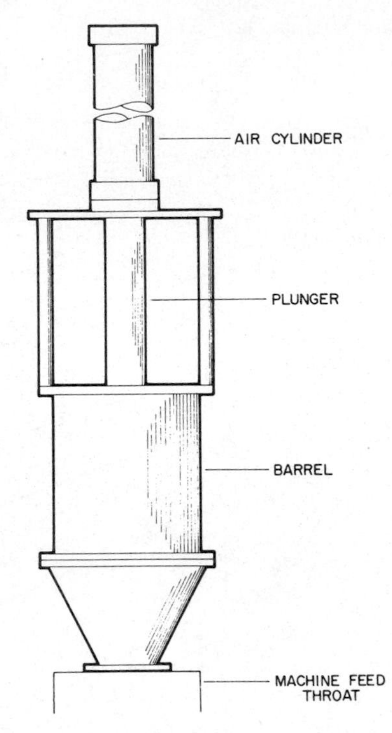

Fig. 8-12. Sketch of material stuffer for BMC materials. The stuffer usually consists of a cylinder with a plunger that maintains pressure against the polyester material while the screw is rotating. When the stuffer cylinder is nearly empty, the plunger is withdrawn and additional material is deposited into the stuffer. (*Courtesy Stokes Div., Pennwalt Corp.*)

Regardless of their basic design, machines should be capable of accurately controlling temperatures and pressures throughout the molding cycling with as little glass degradation as possible.

Temperature Control

The object of controlling temperatures is to warm the material in the barrel enough to permit free flow to preheat it as it goes through the nozzle and runners, and to cure it as fast as possible after it has stopped flowing in the cavity. Because of the temperature sensitivity of thermoset materials, control is very critical. Typical settings are: rear zone of barrel, 125°F, front zone, 150°F; cold runner, 175°F; and both halves of the mold, 335°F. As a practical limit, material coming out of the nozzle should not be more than 200°F. The mold tempera-

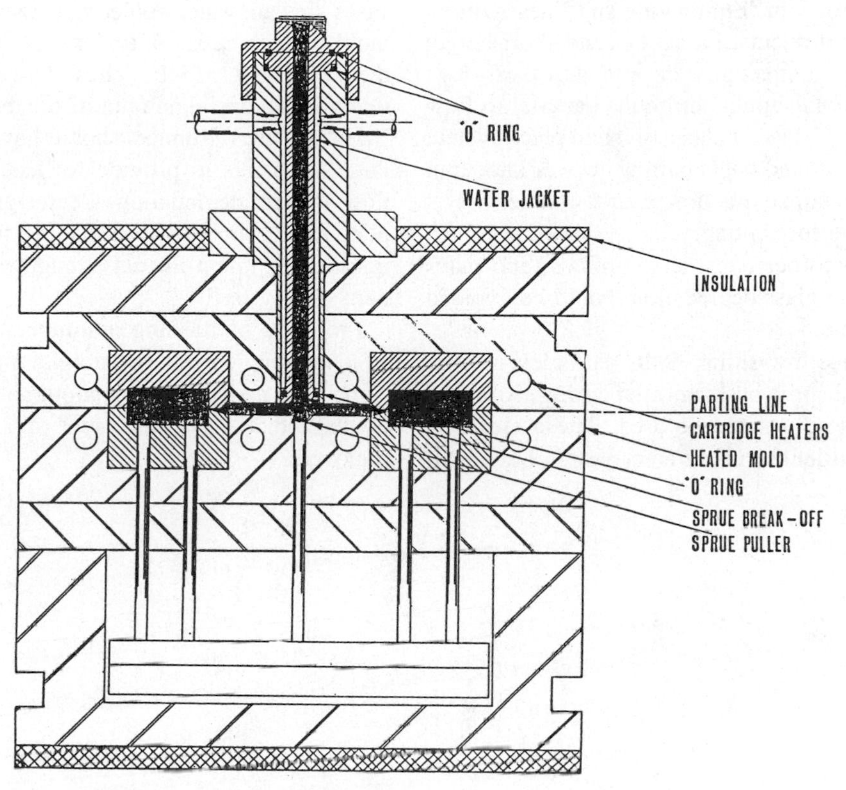

Fig. 8-13. Schematic of sprueless mold for BMC polyester injection molding. (*Courtesy Hull Corp.*)

tures for most materials are between 300 and 400°F.

Experience shows that the cure time of a $\frac{1}{8}$-inch section should be within 15 to 45 seconds, depending on part size.

Other typical press conditions are screw rpm, 20 to 75; back pressure up to 100 psi; injection time 1 to 5 seconds; injection pressure, 5000 to 20,000 psi during injection, and as little as half that during cure; and clamp pressure, 3 tons/sq. in. max (based on cavity projected area).

Besides presses specifically designed for thermoset molding, conversion kits also are available to change over machines originally designed for thermoplastic operations.

Tooling requirements are similar to those for thermoplastic injection molding. The major difference is that adequate provision must be made for heating the molds. Molds should be fully hardened, polished, and chrome-plated. Tool steels such as AISI H13 are recommended.

Another significant difference between thermoplastic and thermoset injection molds is the need for harder runner, gate, and cavity surfaces because of the extreme abrasiveness of most thermoset materials. Typical hardnesses are 56 to 58 Rockwell C, and such surfaces are often hard-chrome-plated for further resistance to wear.

The low-shrinkage characteristic of BMC makes ejection of pieces more difficult than with thermoplastics. Liberal draft, where possible, and provisions for positive ejection on both sides of the mold generally are required.

Properly hardened molds should wear well, but gate areas are susceptible to wear. Designing them as replaceable inserts makes their repair and replacement less costly.

Gates should be located so as to minimize effects of knit lines that form after material has flowed around an obstruction in the mold. Also, multiple gates should be avoided to reduce the number of knit lines that form where materials from gates join. Eliminating knit lines reduces chances of rejects. Large-area and short-length gates and runners give the strongest parts.

The mold should permit the material to flow so that it pushes air ahead of it into places where it can be vented out at parting lines or knockout pins. Vacuum extraction of air is another useful technique for venting.

Sharp corners, restricted orifices, and gates that cause glass degradation should be avoided if possible.

Because low-shrink BMC can stick in conventional sprue bushings after curing, cold runners and sprues can be used. These elements can be oriented in different ways, but in all cases they are water-cooled to somewhat below mold temperatures. A typical temperature for them is about 175°F. Thus the material actually gets a useful amount of preheating in the "cold runner." Runners should have relatively large diameters to provide for easier material flow and less degradation. Center-gated, three-plate molds should be considered for parts requiring maximum impact strength if cold runners are not used.

Problems of flashing common to compression molding to not occur with properly designed and maintained injection molds. The little flash that occurs is paper-thin and easily removed.

9 Compression and Transfer Molding

COMPRESSION MOLDING

Historians cannot establish definitely the date of origin of the art of molding. It might be said that the art of molding originated with prehistoric humans, when they learned how to make pottery from clay, using the pressure of their hands to form the shape and the heat of the sun to harden the clay.

The earliest application of compression molding as a manufacturing process occurred early in the nineteenth century, when Thomas Hancock perfected a process for molding rubber. The first patent on a process of molding in the United States was issued in 1870 to John Wesley Hyatt, Jr. and Isaiah S. Hyatt.

Dr. Leo H. Baekeland's development of phenol-formaldehyde resins in 1908 gave the industry its first synthetic molding material, which even today is one of the principal materials used in the compression molding process.

Technique and Materials

The process of compression molding may be simply described by reference to Fig. 9-1. A two-piece mold provides a cavity in the shape of the desired molded article. The mold is heated, and an appropriate amount of molding material is loaded into the lower half of the mold. The two parts of the mold are brought together under pressure. The compound, softened by heat, is thereby welded into a continuous mass having the shape of the cavity. The mass then must be hardened, so that it can be removed without distortion when the mold is opened.

If the plastic is a thermosetting one, the hardening is effected by further heating, under pressure, in the mold. If it is a thermoplastic, the hardening is effected by chilling, under pressure, in the mold. (The procedure is described in greater detail later in this chapter.)

Compression molding is used principally for thermosetting plastics, and much less commonly for thermoplastics (for which injection is the preferred method of molding).

Thermosetting Materials. Thermosetting materials are chemical compounds made by processing a mixture of heat-reactive resin with fillers, pigments, dyestuffs, lubricants, and so on, in preparation for the final molding operation. These materials or molding compounds are, in most cases, in powder, granulated, or nodular form, having bulk factors ranging from 1.2 to 10. Some are used in the form of rope, putty, or slabs.

The materials of lower bulk factor are usually those having wood flour or mineral compounds as fillers, whereas those of higher bulk factor have as fillers cotton or nylon flock, rag fibers, pieces of macerated rag, tire cord, sisal, and, for very high impact strengths, glass rov-

Reviewed and revised by John L. Hull, Vice Chairman, Hull Corporation, Hatboro, PA.

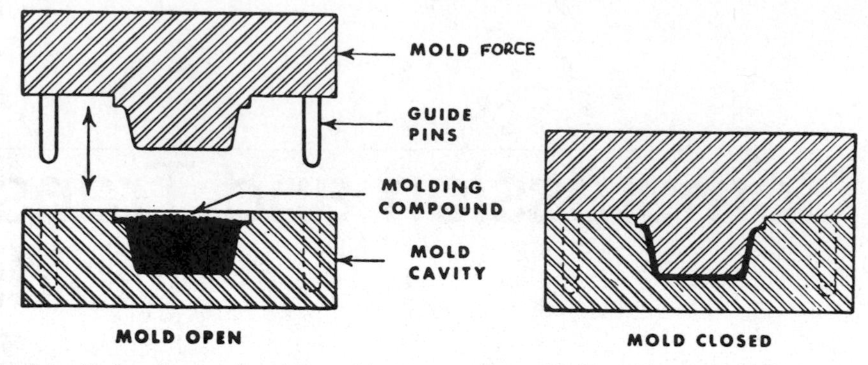

Fig. 9-1. Basics of a two-piece fully positive compression mold. (From *Plastics Mold Engineering*)

ings, carbon fibers, boron fibers and Kevlar fabric.

Phenol-formaldehyde is the single most common resin and catalyst combination, generally called phenolic molding compound. If the filler is mineral, such as mica, the molded part will have good electrical properties. If the filler is glass fibers, say one-quarter inch long, the molded part will have good impact strength. Small hollow glass micro-balloons have been used as fillers to make low-density parts. Such micro-balloon-filled compounds, when molded, often are termed "syntactic foam," achieving densities as low as 0.78 g/cc.

Other resin systems include melamine-formaldehyde (often used in plastic dinnerware), urea-formaldehyde (common in white or pastel heat-resistant handles for kitchenware, or outlet sockets for household use), alkyds and polyesters (often used in high voltage insulators in TV sets, or for arc resistance and insulation in circuit breakers and switch gear), diallyl phthalate (electrical connectors in computers), epoxy (housings for electronic components), and silicone (high-temperature requirements to 600°F or more). Common fillers include silica, glass, wood flour, natural or synthetic fibers, and combinations of these.

Although most thermosetting formulations are dry and granular at room temperature, some are puttylike, some in the form of dry or moist matted fibers, and some a fine powder.

When subjected to heat, thermosetting formulations first become liquid and then undergo an irreversible chemical reaction called cure or polymerization. If polymerization occurs under mechanical pressure, as in a closed mold, the resulting material is a dense solid. Polymerization generally is a time-temperature relationship, with shorter cure times when higher temperatures are used. Typical pressure, temperature, and time values for a phenolic wall socket in semiautomatic compression molding might be 3000 psi, 300°F, and $1\frac{1}{2}$ minutes.

Thermoplastics. Practically all compression molding uses thermosetting plastics. But in certain specialized applications, thermoplastic materials may be processed by compression molding.

These materials, when compression-molded, become plastic under pressure and heat in a heated mold and flow out to the contour of the cavity. Molds must be arranged for rapid heating and cooling because the molded articles cannot be removed from the mold until the material has been sufficiently cooled to harden. This process of softening the plastic by heating and hardening it by chilling can be repeated indefinitely.

Large plastic optical lenses, for example, may be compression-molded from methyl methacrylate (acrylic). In this particular instance, using compression molding rather than the normal injection molding helps eliminate flow marks, warpage, and shrink marks.

For molding articles of heavy cross section from thermoplastics (e.g., toilet seats), a combination of injection and compression is sometimes advantageous. The mold is filled with hot softened material by injection, and then is subjected to pressure by means of a compression force plug. This positive application of pressure during cooling minimizes the development of voids and shrink marks.

Molding of Phenol-Formaldehyde Compounds

The details of the procedure of molding thermosetting materials can be conveniently covered by a description of the molding of phenol-formaldehyde compounds. The minor differences in procedure required by other thermosetting materials will be discussed in later sections.

A typical mold is made in two parts, which, when brought together, enclose a cavity representing the article to be molded. The two parts are mounted in register, in a hydraulic, pneumatic, or mechanical press, which serves to open and close the mold and to apply pressure to its contents.

The most common method of heating molds for compression molding today uses electrical heating cartridges or strips. Electrical heating gives good results with molds that have a uniform distribution of heating elements with adequate capacity. It has the advantages of high efficiency, is clean and simple to hook up, is easily adjustable to 400°F, and, with modern temperature controllers, is accurate and reliable.

Steam molding, popular in the past, has declined in recent years, partly because of the difficulty of producing the higher pressures needed to reach the not uncommon 350°F and higher mold temperatures. With steam, the mold must be cored or channeled for circulation of steam under pressure. However, steam is useful in large, complicated molds because it automatically replaces itself as it condenses and holds the mold at an even temperature.

Molds occasionally are heated by other media such as oil, hot water, and gas flame. However, such media are limited in their usefulness because they lack the latitude and adaptability of steam and electricity.

Phenolic compression molding is carried out by inserting a predetermined amount of material into the lower half of the mold (Fig. 9-1). The mold is closed under pressure, and with heat and pressure, the material softens, filling out the contours of the cavity created by the two halves of the mold.

The molding powder may be volumetrically fed, weighed out, or shaped into preforms. The charge may be fed as cold powder, or preheated. It can be preheated by radio frequency, infrared, an oven, or other methods of heating. Preheating shortens cure time, reduces the molding pressure required, minimizes erosion and abrasion on mold cavity surfaces, and improves electrical properties.

Compression molding conditions generally fall in the following ranges:

Temperature: 300–400°F (340°F typical)
Pressure: 2000–10,000 psi on part (3,000 psi typical)
Cure time: 30–300 seconds (90 seconds typical)

For most molded articles, the plastic material must be confined during molding by telescoping one half of the mold into the other half. A slight clearance is allowed between the halves in order to permit a slight excess of molding compound to escape; usually a clearance of 0.001 to 0.005 inch is sufficient. The material filling the cavity is held under heat and pressure to harden it, and then the mold is opened and the molded article removed. Thermosetting materials, having been hardened by a chemical change caused by the heat, can be ejected after the proper curing cycle with no cooling of the mold.

Usually, articles requiring special dimensional control can be removed hot from the heated mold and placed on a suitable fixture that holds them during cooling and prevents distortion.

The polymerization reaction of phenolics releases water vapor and traces of other gases such as ammonia. For parts with section thicknesses of $\frac{1}{8}$ inch or more, gassing or breathing the mold may be necessary in compression molding of phenolic. The procedure followed is to release the pressure on the mold either just before or after it has closed on the material charge. The mold should be opened just far enough to allow entrapped air and gas to escape from each cavity. Certain molding compounds and/or molded articles require a timed "dwell" in this open position before the mold is closed again. And, on rare occasions, two complete breathe cycles are required for optimum results from a given mold. Gassing or breathing of the mold will result in denser moldings, reduce the chances of internal voids or blisters, and shorten the molding cycle.

The proper molding procedure produces cured articles of sound, uniform structure. There is usually some excess cured material at the parting line, still adhering to the molded part. This "flash" must be removed by hand operation, tumbling, or blasting with a mild abrasive, or finished as described in Chapter 23.

General-purpose phenolic molding compounds are supplied in granular form, and can be loaded into the mold in this form in weighed or measured charges. But in commercial semiautomatic operations there is economy in preforming such material into tablets of the correct size and weight by means of automatic preforming equipment. This is less expensive than weighing out the individual charges; the tablets are easily handled and can be loaded into the cavity conveniently, either manually or by means of a loading board. Frequently, through the use of preforms the cavity loading chamber of the lower half of the mold can be made less deep than it would have to be in order to hold the charge in the bulkier form of looser granules. (See Fig. 9-2.)

A few high-impact phenolic molding materials, which contain fabric or fibers as reinforcement for the molded article, can be preformed automatically only in specially constructed preformers with stuffing mechanisms. In many facilities they are loaded as weighed charges or preformed by hand.

All plastics require heat in order to be molded. Because of their inherently poor thermal conductivity, the penetration of heat from the hot mold into the cold material is slow, and it may not be uniform. Time is saved, and uniformity is promoted, by preheating the plastic before it is put into the mold. This converts the plastic into a uniformly softened mass, ready to flow cleanly and evenly as the mold closes.

Fig. 9-2. An automatic preformer, 20 tons pressing capacity, with hydraulic power pack, for compacting cold thermosetting granular compounds to produce preset uniform weight preforms for compression or transfer molding. (*Courtesy Hull Corp.*)

The practical benefits include better surface finish, freedom from flow marks, better uniformity of cure, and less difficulty in production of articles of thick section, as well as the economy of shorter cure cycles, and the prolonged life of the cavity because of reduced surface wear.

Metal parts, or inserts, placed in the mold and held firmly in position, can be molded into the article. (See Chapter 25.)

Molding of Urea-Formaldehyde and Melamine-Formaldehyde Compounds

The techniques employed in handling and molding urea-formaldehyde and melamine-formaldehyde are, in general, similar to those used for phenol-formaldehyde, but some differences in practice frequently are required.

As these are usually light-colored materials, in contrast to the typical browns and blacks of phenolics, attention must be directed to preventing any contamination that will show up in the molded article. Dust from adjacent presses, the soiling of preforms, and incomplete removal of flash from the mold are frequent sources of contamination. By making the necessary provisions to prevent contamination, urea and melamine can be run with a low percentage of rejection for dirt.

In molding urea and melamine, the design of the article and of the mold is particularly important because the translucency and light color of these plastics fail to conceal flow marks and gas pockets such as may be present but undetected in the dark, opaque phenolic materials, Hence it is desirable, whenever possible, to design both the article and the mold to minimize such defects in appearance. Ribs, variations in thickness, louvers, and molded holes can be molded satisfactorily in urea when the article and the mold are properly designed.

In molding these materials, it frequently is necessary to open the mold slightly and briefly, after it has been initially closed, to allow the escape of gas formed in the reaction of curing. Melamine and urea materials generate moisture under the heat of molding. Also, they entrap and effectively seal air during their compression from an initially high bulk down to one-half or less volume in molded form.

Both urea and melamine sometimes may be preheated advantageously before being put into the mold. Melamine-formaldehyde may be electronically preheated successfully, but urea does not react so well to electronic methods. For urea, heating by conduction, such as in a rotary-canister preheater or, for automatic presses, in an oil-bath heater, has gained more favor. Infrared lamps also have been used with some success. The mold temperature for urea materials is usually 325°F or less.

Molding of General-Purpose Polyester Molding Compounds

A line of general-purpose polyester molding compounds is available in pastel shades having excellent high stability. These materials have an outstanding characteristic in that there is little or no "after-mold shrinkage." This makes them desirable for electrical household appliance applications, especially when used in connection with metal inserts.

These polyester materials can be readily molded in either transfer or compression molds. The design of the molded article is facilitated by the fact that variations in wall thickness cause no trouble. Materials can be preheated to 250°F and molded at regular temperatures of 315°F with no danger of discoloration. The duration of cure is equal to or less than that of phenolic compounds.

Low pressure, glass-filled polyester molding compounds can be molded satisfactorily in regular compression equipment if the molds have adequate restriction to ensure filling out. Also, adequate vents are needed because the polyesters effectively seal air in blind holes in molds, with resulting porosity in the molded article.

In most cases, the actual molding cycle involves simply loading the mold, closing the press, degassing (although that is less often required for polyesters than for conventional formaldehydes), and completing the cure. However, there are applications that require more elaborate techniques, such as a slow close or a dwell period, to give the best results. In such cases, no overall rule may be prescribed; the cycle must be worked out for each job.

Figures 9-3 and 9-4 show two types of presses used for compression molding.

Molding of Alkyd Materials

The techniques of molding alkyd thermosetting materials are similar to the above, except that the chemical reaction is usually much faster for alkyds, so that faster-closing and -opening presses are required. In molding these materials, it frequently is necessary to breathe or degas the mold slightly and briefly, after it has been initially closed, to allow the escape of gas formed in the reaction of curing.

Molding of Thermoplastic Materials

The general procedure for thermoplastic materials is the same as for thermosetting materials, but the molded article is hardened under pressure by cooling in the mold (i.e., by shutting off the steam and circulating cold water through the coring of the mold). Some experimentation will be necessary to establish at what point in the cycle the cooling should begin and how long

Fig. 9-3. A down-acting compression press, 400 tons clamping capacity, with self-contained power pack mounted on press head. Control system includes microprocessor and keyboard with CRT to provide and store information for an Allen Bradley PLC2 controller. All press positions are sensed by an electronic feedback device. The control includes both velocity and pressure profiling of the clamp movement to ensure optimum part quality, especially with sheet molding compounds. (*Courtesy C. A. Lawton Company*)

Fig. 9-4. An upward-pressing combination transfer or compression press, 140-ton clamping capacity, console-style, with microprocessor cycle controls for semiautomatic operation. (*Courtesy Hull Corp.*)

it must be continued in order to harden the article sufficiently.

Thermoplastic molding compounds usually cannot be preformed.

Molding of Fluorocarbons

These materials are classed as thermoplastics, but are not fully amenable to the conventional technique of compression molding.

Chlorotrifluoroethylene (CTFE-fluorocarbon resin) requires mold temperatures in the range of 450 to 600°F. The necessity of alternately heating to so high a temperature and cooling, makes for long cycles and large power consumption in conventional integrally heated compression molds. Hand- or bench-type molds usually are preferable. Separate platen presses are used, one electrically heated and one water-cooled. The mold is heated, and the shaping of the article is accomplished in the hot press; then the mold is quickly transferred to the cold press, to chill the article, under pressure, so that it can be removed.

Tetrafluoroethylene resin (TFE-fluorocarbon resin) does not soften and flow in the manner of conventional thermoplastics, and it is molded by a special technique, as follows:

1. The granular material is preformed at room temperature by a pressure of 2000 to 10,000 psi.
2. The preform is sintered into a continuous gel by exposure to a temperature of about 700 to 740°F in an oven or a fluid bath.
3. The piece then is given its final shape by a pressure of 1000 to 20,000 psi in a confining die cavity, in which it is cooled. This final step, of which there are several variants (cold-coining, hobbing, hot-

coining), depending on the shape of the article and the dimensional accuracy required, may be omitted with articles of thin section for which accuracy is not important.

Fluorinated ethylene-propylene resin (FEP-fluorocarbon resin) can be molded into simple shapes by an essentially conventional compression-molding technique, with gradual application of pressure up to about 1500 psi. For rapid molding, a temperature of 650 to 700°F is required, but a 3-mil sheet of aluminum must be interposed to prevent sticking to the surface of the mold. When this is undesirable, the molding can be done at 550 to 600°F, but more time will be required.

Cold Molding

Some materials are formed in presses with unheated molds, giving a very fragile molding. When sintered or baked in ovens at appropriate temperatures, the material hardens or fuses. After cooling, the part is relatively strong and ready for use. The process is called cold molding.

This technique is used to make certain types of ceramic insulators. The molding compound may be ceramic powder with perhaps 15% phenolic resin by weight, homogeneously mixed. The cold pressing compacts the mixture. Subsequent heating cures the phenolic, which binds the ceramic particles together.

Advantages of Compression Molding

Thermosetting and thermoplastic materials can be compression-molded. Quantity production is possible through the use of multiple-cavity molds, and, when production volume warrants, in fully-automatic presses, which feed metered charges of granular material into each cavity and remove finished molded parts and initiate the next cycle, all automatically. Large housings (e.g., for computers, switch bases, furniture drawers, etc.) are commonly compression-molded. The size of the article that can be molded is limited only by the tonnage and size of available press equipment.

Compression molding of thermosetting materials has certain advantages over transfer or injection molding, as follows:

1. Waste of material in the form of sprue, runners, and transfer-culls is avoided, and there is no problem of gate erosion.
2. Internal stress in the molded article is minimized by the shorter and multidirectional flow of the material under pressure in the mold cavity. In the case of high-impact types with reinforcing fibers, maximum impact strength is gained. This results because reinforcing fibers are not broken up as they are when forced through runners and gates in transfer and injection molding, and because fibers are more randomly positioned, as compared to the more oriented fibers resulting from flow into transfer or injection molds.
3. A maximum number of cavities can be used in a given mold base without regard to demands of a sprue and runner system.
4. Compression molding is readily adaptable to automatic loading of material and automatic removal of molded articles. Automatic molding is widely used for small items such as wiring device parts and closures.
5. This technique is useful for thin wall parts that must not warp and must retain dimensions. Parts with wall thicknesses as thin as 0.025 inch are molded; however, a minimum wall thickness of 0.060 inch usually is recommended because thermosetting materials are brittle as compared to more resilient thermoplastics.
6. For parts weighing more than 3 pounds, compression molding is recommended because transfer or screw injection equipment would be more expensive for larger parts.
7. For high-impact, fluffy materials, compression molding normally is recommended because of the difficulty in feeding the molding compound from a hopper to the press or preformer.
8. In general, compression molds usually are less expensive to build than transfer or injection types.

Limitations of Compression Molding

In the case of very intricately designed articles containing undercuts, side draws, and small holes, the compression method may not be practicable, because of the need for complicated molds and the possibility of distorting or breaking mold pins during the flow of the material under high pressure. Articles of 0.35 inch or more thickness may be more advantageously made by transfer molding, particularly a thick article of small area, in which there is little flow. Thus, for a heavy handle, compression molding would be slower than transfer or injection because in transfer the plastic is thoroughly heated and is precompressed almost to its final density prior to entering the mold.

Frequently, insufficient consideration is given to the physical condition of thermosetting plastics at various stages of molding. The complete filling of a compression mold cavity is spoken of as resulting from the flow of the plastic, but because of its extremely high viscosity, the plastic must be mechanically forced to fill all parts of the cavity. To ensure complete filling, most articles to be molded require that two parts of the mold fit telescopically into each other to prevent escape of plastic prior to the final closing of the mold. Such molds are designated as fully positive (see Fig. 9-1). Also, in order to ensure complete filling out of the mold, it may be necessary to place the charge of plastic into an optimum position in the mold, and in some cases to use preforms of special shape. This is particularly important if the mold does not provide a means of confining the charge. Polyester and alkyd compounds are particularly troublesome, and require positive means of confinement in order to fill the cavity completely.

All thermosets, during their period of flow in a mold, have an apparent surface viscosity that is so low that clearances between mold parts, even when held to less than one thousandth of an inch, become filled with plastic. This often results in damage to the mold if adequate escape is not provided for this leakage. Overflow vents thus are provided for a restricted escape, making the mold a semipositive type. It also is necessary that mating surfaces, or land areas, for the molds be cleaned between successive shots. Also slight fins or flash must be expected on molded articles where the mold sections meet.

Another important consideration is the plastics' degree of rigidity at that point of final cure when ejection is to take place. Melamines are very hard and rigid, phenolics more flexible, and unreinforced polyesters quite weak. Thus, a compression mold for phenolic may work with undrafted or even moderately undercut cores. With melamines, the same mold would require enormous pressure to open, and would probably crack the molded articles at the undercut. Polyester articles require very careful adherence to all rules for draft; they also require generous ejector pin areas to avoid fracture where the ejector pins push against the molded part.

In some cases, compression molding of thermosetting material may be unsatisfactory for production of articles having extremely close dimensional tolerances, especially in multiple-cavity molds and particularly in relation to nonuniformity of thickness at the parting line of the molded article. In such cases, transfer or injection molding is recommended. For a further discussion of this method of molding see below, section on "Transfer Molding," as well as Chapter 8.

Procedure for Compression Molding

The sequence of operations constituting the molding cycle is as follows:

1. Open the mold.
2. Eject the molded article(s).
3. Place article in shrink or cooling fixtures when necessary to maintain close dimensional tolerances (if necessary).
4. Remove all foreign matter and flash from the mold, usually by air blast.
5. Place inserts or other loose mold parts, if any.
6. Load molding compound (powder or preforms, cold or preheated).
7. Close the heated mold (breathe if necessary).

8a. For thermosetting materials, hold under heat and pressure until cure is completed. Certain materials require cooling under pressure for best control of dimensions.
8b. For thermoplastic materials, hold under pressure, while cooling to harden the article.

The temperature of the mold and the pressure applied are extremely important, and it is advisable to follow the recommendations of the manufacturer for each grade of material used.

Thermosetting materials used in compression molding can be classified as conventional and low-pressure materials (the latter should not be confused with materials used in low-pressure molding of impregnated laminates).

There are five very important variables in the compression molding of thermosetting materials, which determine the pressure required to produce the best molding in the shortest length of time. They are as follows:

1. Design of the article to be produced:
   (a) Projected area and depth.
   (b) Wall thickness.
   (c) Obstruction to vertical flow (such as pins, louvers, and sharp corners).
2. Speed of press in closing:
   (a) Use of slow- or fast-acting self-contained press.
   (b) Use of fast-acting press served by hydraulic line accumulator system.
   (c) Capacity of accumulator to maintain constant follow-up of pressure on material.
3. Plasticity of material:
   (a) Degree and type of preheating.
   (b) Density of charge (preform or powder).
   (c) Position of charge in cavity.
   (d) Mobility of resin under pressure.
   (e) Type and concentration (usually expressed as percentage by weight) of filler (wood flour, cotton flock, macerated fabric, asbestos, glass or mica).
4. Overall temperature of mold:
   (a) Temperature variations within cavity and force of mold.
5. Surface condition of mold cavity and force:
   (a) Highly polished chrome-plated surface.
   (b) Polished steel.
   (c) Poor polish (chromium plating worn; pits, gouges, and nicks).

Molding pressures required for most thermosetting materials follow the pattern established for phenolic materials.

Conventional phenolic materials loaded at room temperature (i.e., without preheating) require a minimum pressure of 3000 psi on the projected land area for the first inch of depth of the molded article, plus 700 psi for each additional inch of depth. Efficient high-frequency preheating, however, may reduce the required pressure to as low at 1000 psi on the projected land area, plus 250 psi for each additional inch of depth. The pressure required on high-impact materials may reach 10,000 to 12,000 psi. These recommendations of pressure are predicated on minimum press-closing speeds of 1 in./sec. The flow characteristics of thermosetting molding materials are changing continually during the molding, and the effect of this change is particularly noticeable in slow-closing presses.

Low-pressure phenolic molding materials, efficiently preheated by high frequency, require a minimum of 350 psi on the projected mold area, plus about 100 psi for each additional inch of depth.

Table 9-1 may be used as a guide for the pressure required for the depth of a given molded article. The wide range of pressures given in the second column is necessary to cover the variety of molded pieces and the types of press equipment used. For example, an article of comparatively small area and great depth requires pressures at the lower end of the range given; an article of large area and great depth may require pressure at the upper end of the range. Also, the wall thickness of the article influences the pressure required; thin sections require more pressure than heavier sections. For articles involving a deep draw, fast-closing presses with speeds of over 20 in./min.

Table 9-1. Pressure table; pressure, psi, of projected land area.

Depth of Molding (in.)	Conventional Phenolic		Low-Pressure Phenolic	
	Preheated by high frequency	Not preheated	Preheated by high frequency	Not preheated
0-¾	1000-2000	3000	350	1000
¾-1½	1250-2500	3700	450	1250
2	1500-3000	4400	550	1500
3	1750-3500	5100	650	1750
4	2000-4000	5800	750	2000
5	2250-4500	*	850	**
6	2500-5000	*	950	**
7	2750-5500	*	1050	**
8	3000-6000	*	1150	**
9	3250-6500	*	1250	**
10	3500-7000	*	1350	**
12	4000-8000	*	1450	**
14	4500-9000	*	1550	**
16	5000-10000	*	1650	**

* Add 700 psi for each additional inch of depth; but beyond 4 in. in depth it is desirable (and beyond 12 in. essential) to preheat.
** Add 250 psi for each additional inch of depth; but beyond 4 in. in depth it is desirable (and beyond 12 in. essential) to preheat

for full depth of draw will make it possible to use lower pressures.

The time required to harden thermosetting materials commonly is referred to as the cure time. Depending upon the type of material, preheat temperature, and thickness of the molded articles, the time may range from seconds to several minutes.

Types of Compression Molds

For most economical production, mold cavities are made from high-grade tool steels so that they can be hardened and polished. Through-hardening, to about Rockwell C56 to C58 is preferred to case-hardened steels, for a long mold life.

A hand mold is so constructed that it must be removed from the press manually, following each cycle, taken apart to remove the molded article, and assembled again with the molding material charge placed in the cavity (cavities), for the next molding cycle. These molds are used primarily for experimental or for small production runs, or for molding articles that, by reason of their complexity, require dismantling of mold sections in order to release them. These molds usually are small and light and contain no more than a few mold cavities. They usually are heated by means of electrically heated or steam-heated platens attached to the press. (See Fig. 9-3.)

Semiautomatic molds are self-contained units that are firmly mounted on the top and bottom platens of the press. The operation of the press opens and closes the mold and also operates the ejector mechanism provided for the removal of the molded piece from the cavity or cavities. This type is employed particularly for multiple-cavity work and for relatively long production runs, or for articles too large or too deep in draw for hand molding. The use of ejector pins calls for careful consideration in the design of the heating provision in the mold, to ensure that there is no interference between ejector pins and the electric heating cartridges of the steam channels.

Fully automatic molds are of a special design and adapted to a completely automatic press. The complete cycle of operation, including the loading and unloading of the mold, is carried out automatically. A multiple-cavity mold may be used, and usually the molded article contains no insert or metal part.

Machinery and Equipment

Presses. Presses for hand molds range from small laboratory presses to production equipment of clamping capacities from 5 to 100 tons. The heating plates are fastened directly to the top and bottom press platens. The mold is placed between these plates for the transfer of heat and pressure to it during the molding operation. Presses for semiautomatic molds range in size from 10 to 4000 tons and up. These presses have top and bottom platens with grids and parallels so that the molds can be readily mounted. The presses are provided with either mechanical or hydraulic ejecting apparatus to remove molded articles from the molds.

In compression molding articles with high vertical walls, it is important that the pressure be applied sufficiently fast to maintain maximum pressure on the material as it becomes soft and flows within the mold cavity. Straight ram hydraulic presses, as opposed to hydraulic toggle-type presses, are considered best for articles with high vertical walls.

Self-contained hydraulic press units must have a pump capacity large enough and must be fast enough to take full advantage of the short period of maximum plasticity.

Hydraulic System. Hydraulic presses may be provided power, or a group of presses may be operated from a single line. A hydraulic pump provides the necessary volume of oil or (rarely) water at the required pressure, and an accumulator maintains a reserve sufficient to serve all presses on the line without fluctuation of pressure.

In some cases, a high- and low-pressure accumulator system is used for economy. The low-pressure system is especially advantageous for operating presses with a long stroke; the low-pressure system can advance the moving part of the mold up to the point where the material takes pressure, and then the high pressure is valved in to complete the molding cycle.

Heating and Cooling of Molds. A hand mold is heated and cooled by contact with cored platens mounted in the press, and its temperature is often held reasonably constant by keeping the mold halves on heated surfaces during the time when they are outside the press heating platens. Heat for automatic and semiautomatic molds may be supplied by steam flowing through channeled mold sections or by electric heaters installed in the platens of the mold or press.

For chilling thermoplastics, a supply of cold water is required, which flows through cooling channels in the mold, or the hot oil circulating system is arranged for chilling as well as heating.

Automatic Compression Molding

Description. Fully automatic compression molding involves the automatic sequencing of necessary functions performed in making finished molded articles from granulated or nodular thermosetting molding compounds. In essence, it involves a compression-molding press equipped with a mold and additional equipment designed (1) to store a quantity of molding compound; (2) to meter, volumetrically or gravimetrically, exact charges of compound; (3) to deposit these charges into appropriate mold cavities; (4) to remove the finished article from the mold area following each cycle; and (5) to remove any flash or granular molding material not ejected with the finished article, usually by means of an air blast from appropriately positioned nozzles. With reference to (2) and (3) above, many automatic compression presses also preheat the charge by infrared or high frequency heating just prior to filling the cavities. Once an automatic press has been put into operation, it can run unattended except for periodic recharging of the storage container (hopper) and removal of finished moldings, plus an occasional adjustment in the control system to accommodate minor variations in the compound, ambient temperature, and humidity.

Automatic compression molding has been done for many years. Many basic improvements in some of the earlier types have led to faster operations, better-molded products, and more economical production. Molders generally believe that automatic molding will expand many times over in years to come. Although screw-injection molding of thermosetting compounds (see Chapter 8) is gaining in usage and popularity, automatic compression molding continues to prove ideal for many applications, and to play a major role in thermoset molding.

Applicability. Probably the main reason for selecting automatic compression molding is the need to produce a large quantity of an article. An automatic press involves heavier financial investment than a manual or semiautomatic unit, and the full setup, including metering adjustment and alignment of the comb that removes articles from the mold, may take longer for automatic molding than for semiautomatic molding. These extra costs may be offset in a long production run by a high rate of production, a small number of rejects, and a sharp reduction in labor costs. Contrary to general belief, a mold for an automatic press often is no more expensive than a good semiautomatic mold because both must be precisely machined and have essentially the same arrangement of knockout pin, hold-down pin, and so on. In fact, an automatic mold usually requires fewer cavities than a semiautomatic mold to achieve a given production rate, and may therefore be less expensive.

For preliminary analysis of a particular molding requirement, the difference in cost between semiautomatic and fully automatic molding may be calculated on the basis of labor. For example, semiautomatic operation may cost several dollars more per hour than fully automatic molding, because one operator is used for one or possibly two presses in semiautomatic molding, whereas one person only is needed to monitor as many as eight fully automatic machines. Other factors may include a higher production rate, less wastage of raw material, fewer rejects, and less downtime. Not to be overlooked is the fact that production rates in automatic molding often can be geared to the demand requirements of the molded article; inventories of the molded article thus may be restricted, with additional savings.

Intelligent analysis of the various cost factors given above will enable the molder to determine when fully automatic molding is feasible.

On the other hand, automatic molding may not be feasible if the article to be molded is subject to one or more design changes early in the life of the mold. Even this consideration is altered if the fully automatic mold has, say, four cavities, and the semiautomatic mold has eight cavities.

Automatic molding generally is not feasible if the molded article must be made from molding compounds with a bulk factor in excess of 3 (except in the case of certain nodular molding compounds) or with poor dry flowing properties in feeding devices, or in parts requiring molded-in inserts.

Basic Requirements for Automatic Compression Molding

Design of Molded Article. For automatic compression molding, considerable attention must be given to the design of the molded article. Articles 3 inches or more in height may require daylight, platen stroke, or ejector pin stroke that exceeds the specifications of automatic presses now available. The need to keep such deep cavities free from flash or impurities also may make trouble-free operation difficult.

Articles of particularly fragile sections, or requiring fragile mold pins or mold sections, generally are not practicable for trouble-free automatic molding. The molded article should be free of sharp changes in section, and should have adequate draft (1° or more) to facilitate ready removal from cavity or force. Such a draft simplifies the manufacture of molds and helps to minimize their wear.

Articles normally requiring molded-in inserts should be studied to determine the possibility of molding them without inserts and putting the inserts in as a secondary operation. The shorter cycles possible with this method generally result in an overall cost saving. Some articles that normally require side cores or a split mold should be evaluated to determine whether a redesign is possible, or whether two plastic parts, both molded automatically, might be substituted for the one article and still prove more economical in the long run. The best results are obtained when the designer is thoroughly familiar with the techniques possible in automatic molding, the techniques used by the moldmaker, and the functional requirements of the final application of the article.

Mold Design. The design of a mold for automatic molding incorporates most of the features used in manual or semiautomatic molding, but two major features require special attention. First, the molded article must be made to withdraw with the same half of the mold, either top half or bottom half, on every opening stroke. If, for example, bottom ejection is required, every precaution must be taken that the molded piece will not stick to the top half of the mold on the opening stroke. Undercuts, or ridges to grip the article, are frequently required. Top hold-down pins are often used. These are essentially the same as knockout pins, but are designed to push the molded piece away from the top half of the mold as it is opened, to ensure that the piece remains in the bottom half of the mold. Similarly, bottom hold-up pins are used where top ejection is required. The design of the article indicates whether top ejection or bottom ejection is preferable. In automatic operation, the molded article must be under positive control at all times. Automatic takeoff devices and scanning or checking devices are set to conform with a situation that must remain essentially unchanged from cycle to cycle.

The second essential design feature is to ensure that any flash will cling to the molded piece, and not to the mold face or cavity or force. Automatic presses generally utilize an air blast as a cleaning device to remove flash or excess powder from a mold just prior to loading the new charge of material. But such an air-blast mechanism is never so thorough as an operator in ensuring that the mold is clean. Thus, every effort must be made to keep flash to a minimum, and to be sure that any flash that does develop is automatically removed with the molded article. In the mold, well-polished surfaces are necessary, chrome plating of the actual cavities and forces and land areas generally is recommended, and all sharp edges or corners should be rounded wherever possible, to minimize accumulation of flash.

Today, with miniaturization, small molded articles frequently are necessary. In automatic molds for small articles, subcavity design frequently is the most satisfactory solution. A number of the small cavities are arranged in a subcavity, which receives one charge of molding compound in each cycle. The complete cluster of molded pieces remains intact during ejection, being held together by flash of controlled thickness. Articles smaller than about $\frac{1}{2}$ inch in diameter usually should be molded in this way.

Selection of Molding Compound. In addition to the above requisites, consideration must be given to the automatic metering of molding compound. Generally speaking, automatic molding requires materials that flow easily in the dry state and do not tend to cake or bridge when in storage or when required to flow through feeder tubes. Bulk factors generally must be under 3.0, and sometimes the finer powders and coarser granules must be removed. The limitations on molding compounds develop not so much because of the molding operation itself, but because of the problems involved in automatic feeding. Manufacturers of molding compounds have recognized this problem and have produced standard materials that provide excellent pourability for volumetric feeding in automatic compression presses. They also are able to give specific advice on materials that can be used and will produce the best quality in the finished molded article.

Some automatic molding presses utilize vibrators or agitators in the hopper to minimize caking or bridging.

Equipment Available. Several different types of automatic molding presses are available, some of which have been in use for as long as 40 years or more. The conventional equipment is of the vertical type, with the platen moving either up or down during the compression stroke. As the press opens following each cycle, the molded pieces are stripped from the knockout pins by means of a comb or an air blast; following this operation, the loading board moves over the mold to drop charges into the cavities. The loading board then leaves the mold area, and the press closes for its next cycle. Operation generally is of a sequential nature, wherein each individual operation, when completed, initiates the next step. Controls generally are electrically actuated, and use pneumatic, hydraulic, or mechanical means to position the takeoff comb and loading board and other moving parts. Some smaller presses depart from sequential-type operation and remove the molded articles and load the new charge of powder simultaneously.

Another type of automatic press is the rotary press. As the name implies, these presses have a rotary movement of the main moving parts. In one type, a number of single-cavity molds are arranged in a large circle, the lower halves affixed to one platen and the upper halves to another platen. Generally speaking, these platens do not move in relation to one another. A traveling mechanism moves continuously around these two platens at a controlled speed. This mechanism performs, in the appropriate sequence, the functions of causing two individual mold halves to come together and to remain together while the mechanism completes essentially 300° of travel around the platen. The moving mechanism causes the mold to open, the molded pieces to be physically ejected and the force and cavities to be cleaned by an air blast, a new metered charge to be dropped into the cavity, and the mold halves to close again. Each mold thus is actuated individually and at a uniform sequence after each preceding mold. There may be 10 stations, or even 30 stations, on a rotary press of this type. The cavities may all be alike, or they may differ, but the cure times of each must be compatible with the others, as the overall cycle is dependent on the time required for one complete rotation of the moving mechanism.

Also, rotary presses are made in which the "mechanism" remains stationary and the round platens with the molds rotate about a central axis. In operation, however, the principle is the same as that described above, except that all molded pieces must be approximately the same size.

Some of the advantages of rotary presses are listed below:

1. Individual cavities are easy to change, thus reducing loss of production time.
2. Some presses can incorporate molds of different volumes, as the material feed can be varied for individual cavities.
3. High productivity can be gained with a minimum amount of labor.
4. Rotaries are especially adaptable to high-volume, small parts such as wiring devices and closures.
5. When different-size cavities are used in one press, molding cycles are determined by the part with the heaviest section.
6. The variable speed of rotation controls the length of cure.
7. Unloading mechanisms can be used (e.g., wheel to unscrew molded threaded caps, air jet, stripper plates, knockout pins, twist ram).
8. Cavities do not have to be precisely matched for applied pressure.

Several limitations to the rotary press are as follows:

1. Inserts cannot be loaded easily.
2. Part size usually is limited to small parts and is governed by the pressure available to each station. The mold load is usually in the range of 2 to 5 tons, with some machines as high as 15 tons.
3. Molds must be of the flash or semipositive type. Complicated molds with features such as split cavities, cores, or side draws cannot be readily adapted.

Types of Machines

Several commercially available automatic molding presses are shown in Figs. 9-5, 9-6, and 9-7. In selecting the equipment for a specific job, the different characteristics of the various presses should be studied to ensure that the job in hand can be done safely, effectively, and economically. Various safety features are

Fig. 9-5. Operator positions hand transfer mold on lower heating platen prior to molding cycle. The work loading station mounted on front of the press features temperature-controlled heated surfaces as well as a fixture for removal of molded parts from the mold. (*Courtesy Hull Corp.*)

available on such presses to ensure protection from double shots, malfunction of accessories, and so on.

Fig. 9-6. A 75-ton, hydraulically actuated, fully automatic compression molding press. Many such presses are still in use but they are no longer manufactured. (*Courtesy Stokes Div., Pennwalt Corp.*)

Preheating in Automatic Compression Molding

To obtain the fastest cycles in automatic compression molding, particularly when part thickness is 0.060 inch or more, the molding compound should be preheated prior to placing it in the mold cavities. There are three common methods used for such preheating.

Automatic Screw Preplasticating. This approach to preheating the molding compound prior to depositing it into the cavity utilizes a rotating screw inside a heated barrel, similar to the screw and barrel mechanism used in injection molding machines. (See Fig. 9-8.)

Machines are arranged with one, two, or three separate screws, each with its own speed control, water-jacket, and shot measuring device (Fig. 9-9). Cups at the open end of the barrel nozzle accept and compact the material, controlling the weight of the preheated charge.

At the proper sequence in the machine cycle, the cups are withdrawn from the nozzles, leav-

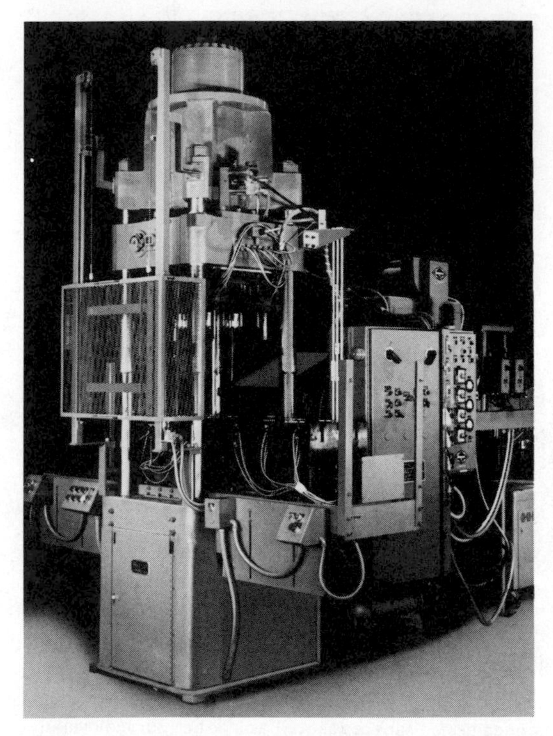

Fig. 9-7. A 200-ton shuttle press for compression molding with inserts, using automatic feeder for electronic preheating and loading molding compound. Two bottom halves of molds, arranged on the shuttle table, and the simple top half of the mold enable essentially continuous molding while the operator loads inserts in the bottom half for the next cycle. (*Courtesy Hull Corp.*)

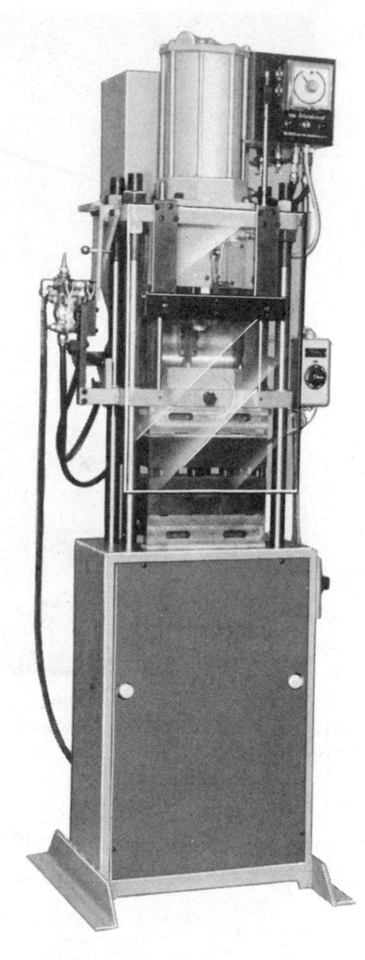

Fig. 9-8. An automatic compression press, downward acting, available with clamping capacities from 18 to 70 tons, utilizing double toggle action, material metering, cavity loading, and molded part ejection are mechanically interlocked and motivated by the moving press platen. (*Courtesy Gluco, Inc.*)

ing the preheated preform exposed (Fig. 9-10). A shearing device is then actuated to cut the extruded material from the nozzle. A chute assembly or a feed board located beneath the nozzle moves forward to direct the preforms into their respective cavities.

Preheating Process. The temperature of the preheated charge can be as high as 280°F, depending upon the material. Heat input is accomplished through convection from the water jackets (110 to 130°F), and frictional heat generated by forcing the material through the clearance between the screw tip and the nozzle wall.

Work input at the nozzle results in a minimum lapsed time between obtaining a high material temperature and depositing the charge into the mold.

The preform cups are designed with minimum and maximum material weight charges. Each is adjustable, to make preforms to any weight increment within its capacity. When the cup has been fully charged to its preset measurement, a limit switch stops the screw rotation.

Cure time comparisons, shown in Fig. 9-11, reflect the extremes of part thickness. A part with a cross-section thickness of 0.070 inch cures in an average of 24 seconds when using cold powder of an "eight" plasticity, and in almost 45 seconds using an "eighteen" flow. The same part using screw compression cures in 12 and 15 seconds, respectively, for the same materials.

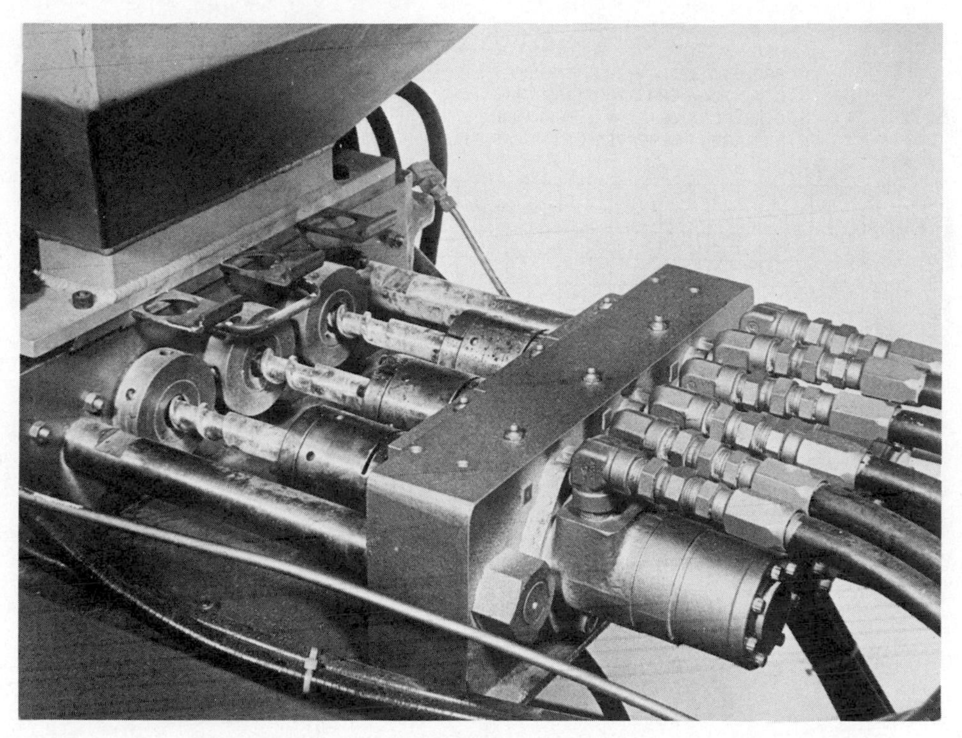

Fig. 9-9. A tri-screw unit after preheating a charge of thermoset material. (*Courtesy Stokes Div., Pennwalt Corp.*)

Fig. 9-10. A tri-screw assembly with measuring cups withdrawn. (*Courtesy Stokes Div., Pennwalt Corp.*)

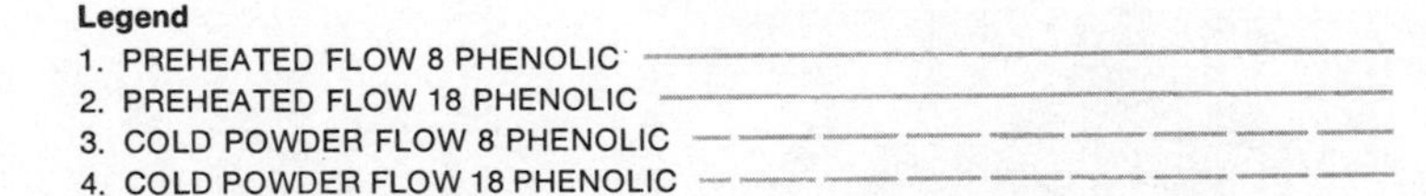

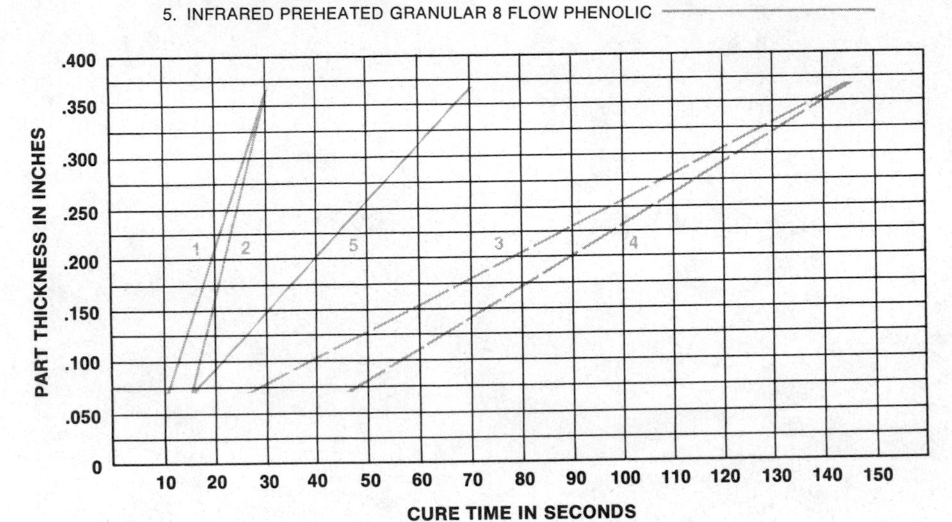

Fig. 9-11. Cure time comparison. (*Courtesy Stokes Div., Pennwalt Corp.*)

A part with a cross-section thickness of .370 inch cures in an average of 150 seconds, as compared with 30 seconds with screw compression. With cold powder, even a 150-second cure does not ensure that the part will be free from internal voids.

Infrared Preheating. This approach to preheating features an infrared heat source positioned between the hopper and the press, focused downward to the level of the powder loading board. When the individual cavity charges have been deposited from the hopper to the cups in the loading board, the loading board moves toward the mold, but it pauses directly under the infrared heat source for a time interval of approximately the same length as the cure time set for the molding. While parts are undergoing cure in the mold, the material in the loading cups is heated to temperatures approaching 200°F. As soon as the molding cycle has been completed, and the molded parts have been ejected and removed from the mold area and the mold blow-off has been accomplished, the loading board travels over the bottom mold half and deposits its charge of heated material. Then it returns to the hopper for the next charge while the press closes for cure.

High-Frequency Preheating. High-frequency (generally 70 MHz or 100 MHz) preheating is accomplished by bringing the material charge of molding compound between two flat plates, or electrodes, which reverse their electrical charge at very high frequencies. Because the molecules of thermosetting molding compounds are polar (i.e., have north and south poles), the molecules reverse their position with each change of frequency. Such fast mechanical activity generates heat fairly uniformly throughout the mass of compound and raises its temperature from ambient to about 200°F in 10 to 20 seconds, depending on the power of the preheater and the mass of molding compound per molding cycle.

Automatic compression presses using high-frequency preheating generally are equipped with an integrated loading unit incorporating the hopper, the powder metering mechanism, the loading board with cups, and the high-frequency generator and electrodes, with the loading board traveling over the bottom half of the mold to deposit the preheated charges at the appropriate moment in the press cycle.

TRANSFER MOLDING

The term "transfer molding" is now generally applied to the process of forming articles, in a closed mold, from a thermosetting material that is conveyed under pressure, in a hot, plastic state, from an auxiliary chamber, called the transfer pot, through runners and gates into the closed cavity or cavities.

History

Prior to the development of transfer molding, thermosetting materials were handled only in compression molds, by methods adapted from the prior art used in forming articles from rubber, shellac, and cold-molded compositions. The compression method is still widely used, and is entirely suitable for the production of a great variety of small and large articles of relatively simple outline and plain cross-section.

However, compression molding does not readily permit the forming of articles having intricate sections, thin walls, and fragile inserts, and those on which close tolerances on "build-up" dimensions are desired. When compression molding is applied to such articles, the molds and maintenance usually are very costly, because of excessive wear and breakage.

To overcome these difficulties, transfer molding was introduced by Shaw Insulator Company, Irvington, New Jersey, in 1926. At that time, the process was carried out in conventional single-ram compression presses, with three-piece molds, as described later. To promote flow, it frequently was necessary to employ special, long-flow, premium molding compounds, and to preheat them in an oven. This type of transfer molding usually is referred to as the *pot type*.

Later, presses equipped with two or more rams were developed for transfer molding. In these, the main ram is attached to the lower or upper press platen in the usual manner, and holds the mold closed. An auxiliary ram is used to force the material into the closed mold. This form of transfer molding frequently is called *plunger molding*.

In the most commonly used U.S. molding presses, the main ram is attached to the lower press platen and operates upward. When these presses are equipped for transfer molding, the auxiliary rams may be mounted on the side columns, operating at right angles to the main columns, operating at right angles to the main ram, or on the press head, operating downward in a direction opposite to that of the main ram, or on the press bed or within the main ram, operating upward and in the same direction as the main ram.

The introduction of high-frequency dielectric preheating during World War II greatly accelerated the growth of transfer molding, particularly with auxiliary-ram presses. Speeds of transfer and of cure were increased, and lower pressures could be used for transfer and for clamping the mold.

In 1956, two automatic transfer-molding machines were developed. One is a small horizontal machine, using loose powder and comparatively inexpensive molds. These and various other working parts of the machine are readily accessible and easily maintained. The other machine is much larger, with a 400-ton clamping capacity, and is extremely fast. It is designed to use preheated preforms.

In the mid-1960s, a high speed automatic transfer press, with integral preforming and high-frequency preheating, entered the market, and yielded fast cycles, as short as 25 seconds, for multi-pin connectors and similar parts.

The progress in molding technology and in equipment has been matched by the development of improved materials, particularly the phenolics, which are better suited to the large multiple-cavity transfer molds now in use than were the special transfer materials previously employed. These materials have more rapid rates of flow and shorter cure times than those previously available for transfer molding. With the aid of dielectric preheating, they exhibit the same total flow as did the special transfer materials formerly required, but accomplish this flow in a much shorter time. Rigidity on discharge from the mold has also been greatly improved. This all has made for faster production and lower molding costs, and has further enhanced the usefulness of transfer molding.

Utility of Process

As a molding technique, transfer molding offers the molder certain advantages. Some typical applications where the process is useful are as follows:

1. For parts requiring side draw core pins which must be readily withdrawn before discharging a part from the mold, and which may have to be extended into the cavity after the mold is closed and before material is introduced into the cavity.
2. For applications where inserts must be molded into intricate parts.
3. For complicated parts where tolerances are very close in three dimensions.
4. For production of various small molded parts that are assembled together. A "family" transfer or plunger mold is economical, and it also permits substitution of different cavities in one mold, depending upon production requirements.
5. In the reduction of finishing costs, because of less flash on the transfer-molded parts.

Description of Process

Essentially, this type of molding requires the transfer of material under pressure from a "pot" or "well" through runners and gates into cavities retained in a closed heated mold (Figs. 9-12 and 9-13). Usually, the charge has

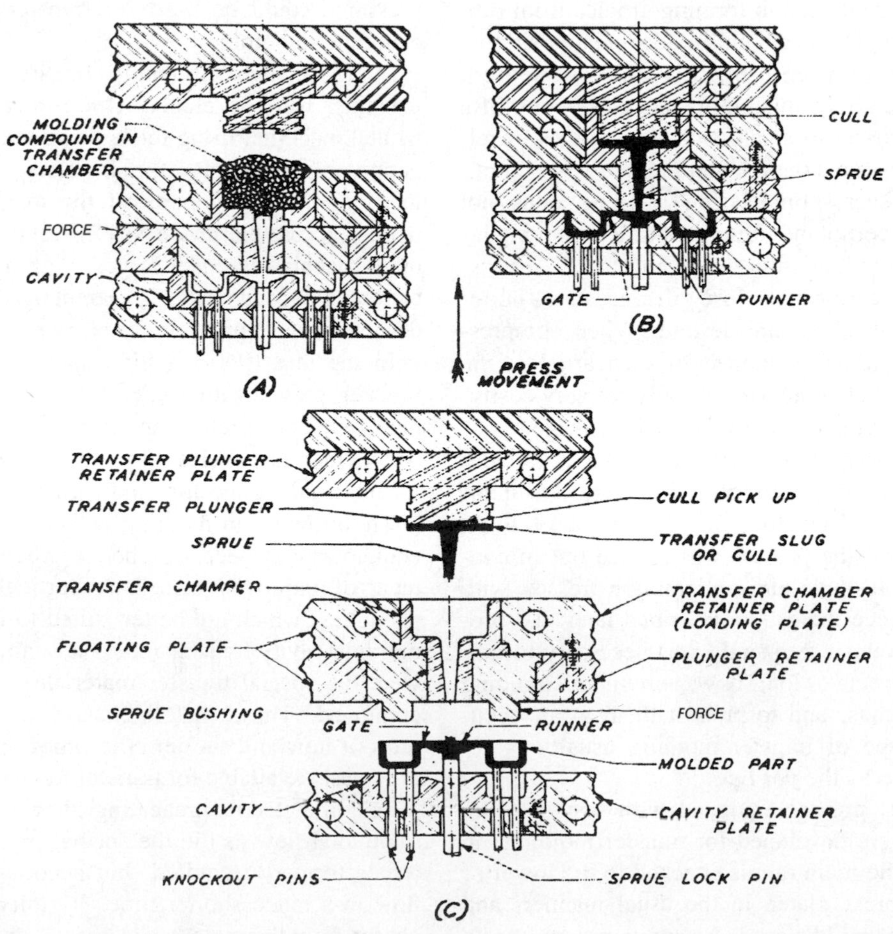

Fig. 9-12. True transfer or pot-type transfer molding, the forerunner of today's common "plunger" or conventional transfer molding. (From *Plastics Mold Engineering*)

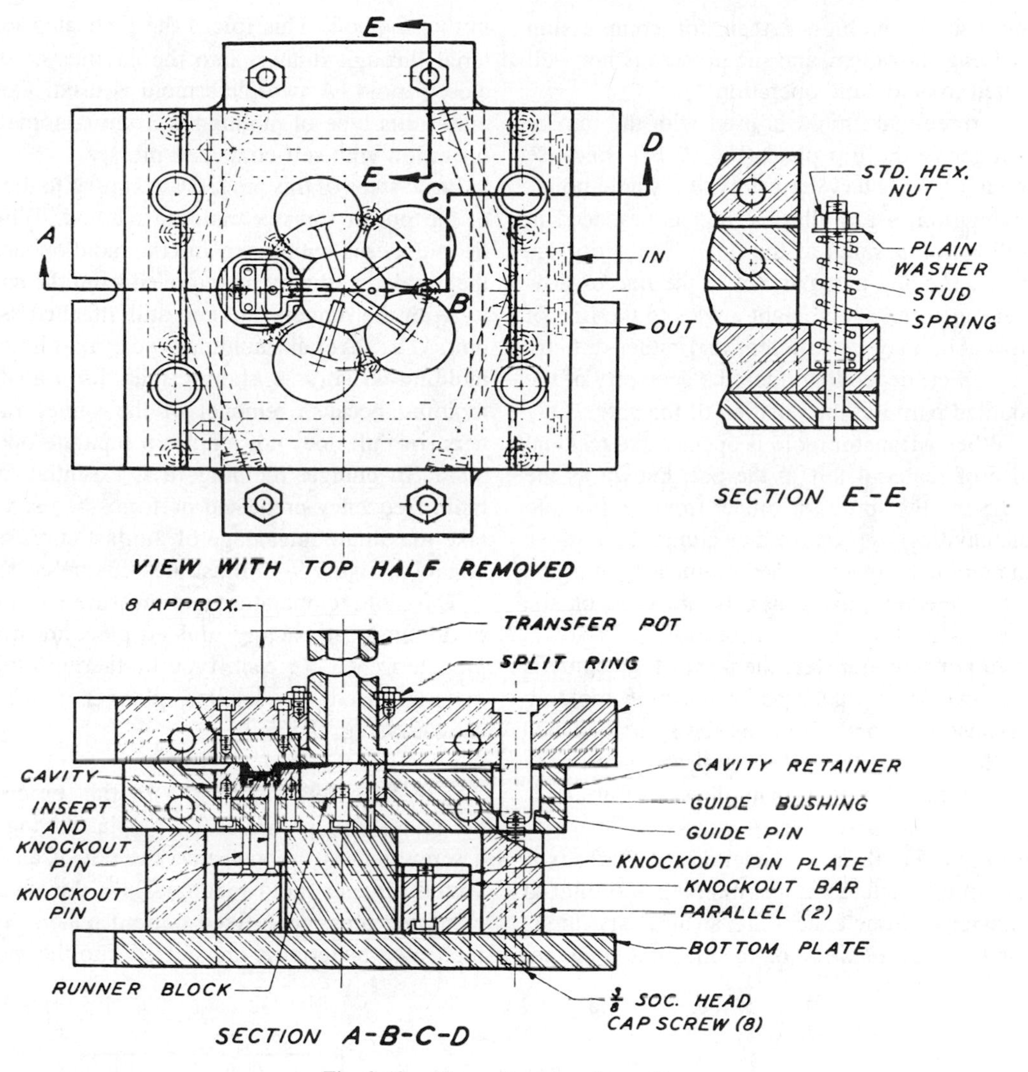

Fig. 9-13. Plunger-type transfer mold.

been preheated before being placed in the pot. With preheating, less pressure is required for transfer, and the mold cycle time and mold wear are reduced. Basically, there are three variations of this technique, as described in the following paragraphs.

Transfer Mold in Compression Press. With this type, a single, hydraulic ram is used. The plunger for the pot is clamped to the upper platen of the press. Pressure is developed in the pot by the action of the main hydraulic ram of the press. The area of the pot should exceed the area of the cavities by a minimum of 10%. Thus, the wedging action of the material will not force the mold cavities to open and flash. Pot-type molding often is a manual operation.

This method is usually faster than compression molding, as the cure time, particularly for thick sections, often is shorter because the preheated material is introduced into the cavity at high speed through a restricted gate, imparting considerable mechanical shear to the flowing compound with resultant frictional heat. The sharp increase in temperature throughout the material charge enables rapid curing of molded parts with good dimensional control and uniform density. The mold cost for transfer molding usually is higher than for compression molding, however, and the process is not well suited to automatic operation.

A three-plate mold is used with the ram or plunger in the top plate (Fig. 9-14). Because the material enters the cavity at a single point, orientation of any fibrous filler is produced in a direction parallel to the flow. The shrinkage of the molded part parallel to the line of flow and the shrinkage at right angles to the line of flow thus may be different and rather difficult to predict, depending upon the geometry of the molded part and the position of the gate.

When a transfer mold is opened, the residual disc of material left in the pot, known as the cull, and the sprue (or runner from the pot into the cavities) are removed as a unit. To remove the molded part from the bottom section of the mold, ejector pins generally are used on the parts as well as on the runners.

In pot-type transfer, the taper of the sprue is the reverse of that used in injection molding because the goal is to keep the sprue attached to the cull so that it will pull away from the part. A detailed pot type is shown in Fig. 9-12.

Plunger Molding. Sometimes called auxiliary ram transfer, plunger molding is similar to transfer molding except that an auxiliary ram is used to exert pressure on the material in the pot in this method. This forces the preheated material through runners into the cavities of the closed mold. A two-plate mold is used. Generally this type of molding is a semiautomatic operation with self-contained presses.

Basic steps of this process are similar to those of the pot or transfer molding method. When the plunger is withdrawn and the mold opened, the molded part may be removed from the cavity with the runners and cull still attached as a unit. The overall molding cycle in plunger molding usually is shorter than for transfer molding because removal of the sprue, runners, or cull does not require a separate operation. In plunger molding, it is essential that radio frequency preheated preforms be used to take maximum advantage of the fast cures obtainable. Figure 9-15 illustrates this process.

The mold temperature and pressure required to obtain a satisfactory molded piece must be predetermined for each type of thermosetting material used. A detailed plunger mold is shown in Fig. 9-13.

Screw-Transfer Molding. In this process, the material is preheated by preplasticizing in a screw and is dropped into the pot of an inverted plunger mold mounted in a downward clamping press with fixed bottom platen. The preheated material is transferred into the mold

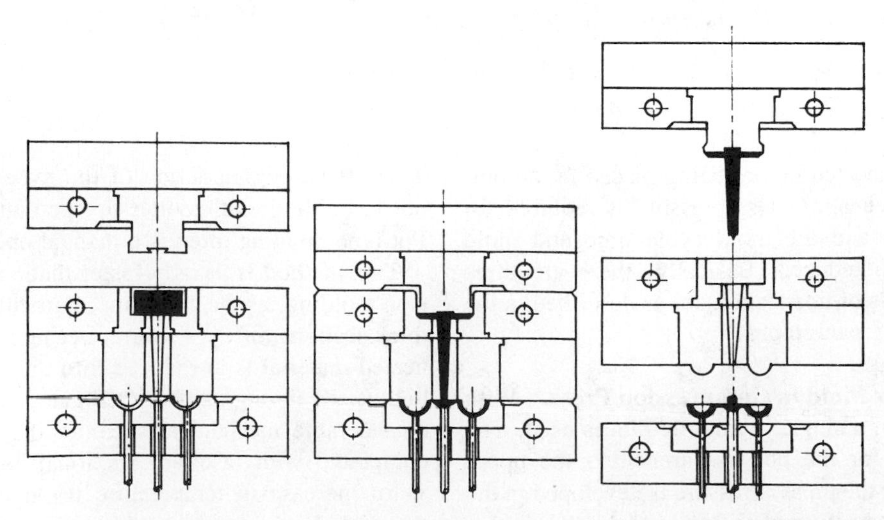

Fig. 9-14. Molding cycle of a transfer mold. Material is placed in the transfer pot (left), then forced through an orifice into the closed mold (center). When the mold opens (right), the cull and sprue are removed an a unit, and the part is lifted out of the cavity by ejector pins. (*Courtesy Chemicals and Plastics Company, Inc.*)

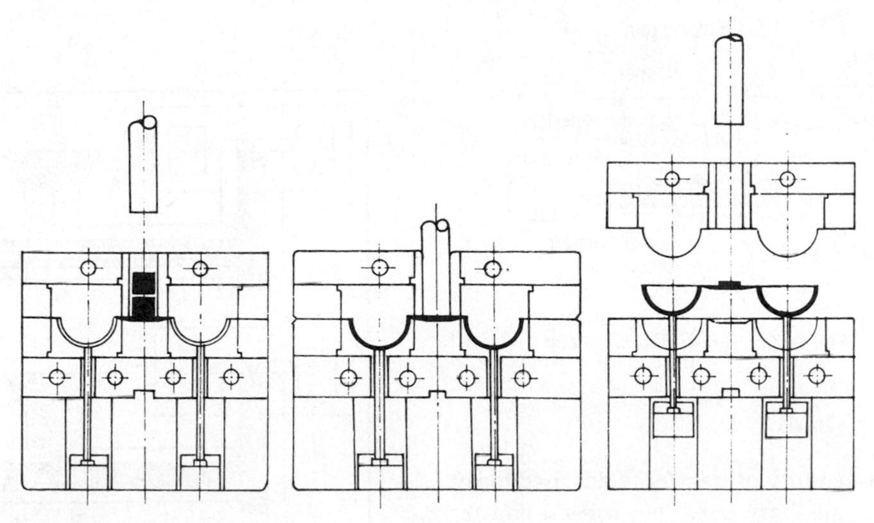

Fig. 9-15. Plunger molding. An auxiliary ram exerts pressure on the material in the pot (left) and forces it into the mold (center). When the plunger mold is opened (right), the cull and sprue remain with the molded piece. (*Courtesy Chemicals and Plastics Company, Inc.*)

by the same method as shown in Fig. 9-15. The screw transfer process and the sequence of operation are shown in Figs. 9-16 and 9-16a. This molding technique lends itself particularly well to fully automatic operation. The optimum temperature of a phenolic mold charge is 240° ±20°F, the same as that in RF preheating for transfer and plunger molding.

Screw-transfer presses are equipped with screw units capable of preparing from 3-ounce to 5-pound shots. To provide adequate control of preheating of the material in the screw units, certain requirements are necessary. The compression ratio (ratio of shallowest to deepest depth of flight) for a thermoset screw transfer is much less than the ratio for screws available for thermoplastic materials. The length-to-diameter screw ratio, in most cases, is considerably shorter. Because of the exothermic reaction of most thermoset resins, a means to carry away heat generated in the barrel also is provided. Aluminum water jackets surround the barrel, permitting the circulation of water. The temperature of this water is regulated by separate control units. Normally, one water control unit is provided for each temperature zone of the barrel. Some machines even are provided with plasticizing screws that are channeled to permit the circulation of controlled-temperature water. The object of the plasticizing units is to prepare material as hot as possible without precuring the resin. Variables such as the water temperature of the barrel jackets, screw rotational speed, and back pressure applied to the screw as it prepares the material serve to control the amount of heat input to the material.

In the screw-transfer process (Fig. 9-16a), thermosetting material is gravity-fed from a larger storage hopper through a hole in the barrel to the reciprocating screw preplasticizer. As the screw rotates, material travels forward along the flights and is thoroughly preheated by mechanical shearing action. The material flows off the end of the screw and begins to accumulate. This buildup of material pushes the screw back along its axis (away from the transfer pot) to a predetermined point, which can be set by a limit switch. The amount of reverse travel of the screw establishes the volume of the charge. While the shot is being preplasticized, the transfer ram is in the raised position, blocking the opening into the transfer pot.

After the shot is formed, the transfer ram returns to its lower position, leaving the opening to the transfer pot completely clear. At this point, the screw moves forward, pushing the preheated material into the transfer pot. With the press closed, the transfer ram advances, delivering the preplasticized material to the runners and cavities.

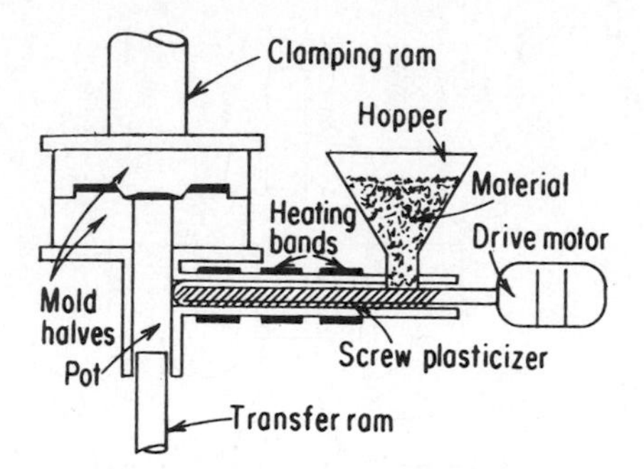

Fig. 9-16. Screw-transfer process.

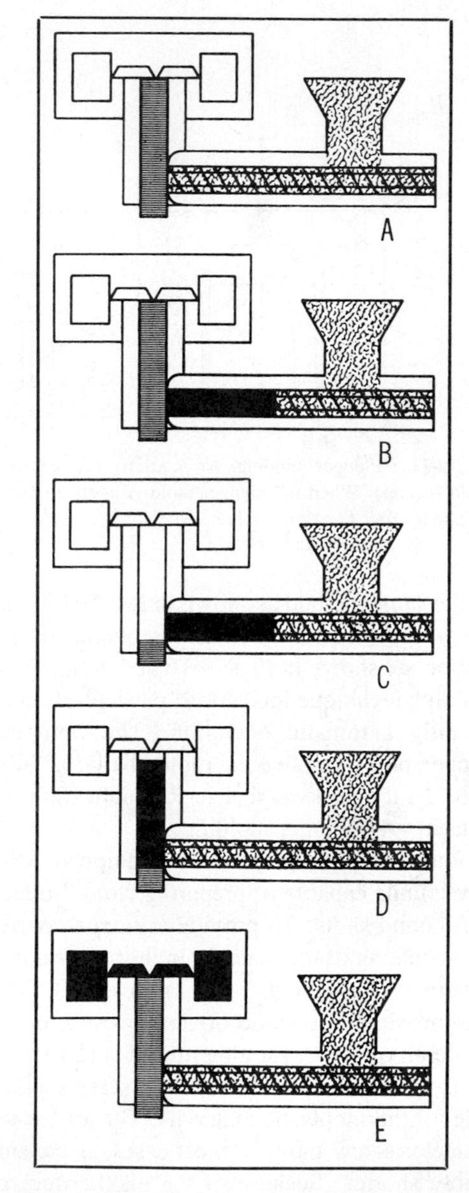

Fig. 9-16a. Screw-transfer sequence. In A, the material is preplasticized as it travels along the flights of the screw. In B, material builds up to the end of the screw and forces it to move backward a predetermined distance. In C, after the shot is formed, the transfer ram lowers to open the transfer pot. In D, the screw moves forward to push the material into the transfer pot. In E, the transfer ram then advances through a standard bottom-transfer molding operation to inject the material into the mold. (*Courtesy Stokes Div., Pennwalt Corp.*)

Transfer Molds

The vast majority of transfer molds used today are of the auxiliary ram type, using a transfer plunger actuated by a hydraulic ram that is part of the press and operates independently of the press clamping ram. For historical purposes, and also because some of the older-type transfer molds may still be in use, the following sections on loose-plate molds and integral molds have been included.

Loose-Plate Molds. This classification may be subdivided into manual and semiautomatic types, depending on the method of mounting and the operation.

One of the earliest and simplest hand-transfer molds is illustrated in Fig. 9-17. This mold is especially useful where the molded piece contains a group of fragile inserts extending completely through it. The mold consists of a plunger, a loose plate with orifices around its perimeter, and the cavity. The space in the cavity above the loose plate serves as the pot or transfer chamber.

In operation, the inserts are loaded into the loose plate, which is then inserted into the cavity so that the lower ends of the inserts enter the proper holes. The compound is loaded into the mold above the loose plate and is transferred by the plunger through the orifices into the closed mold. After-curing, the molded piece, loose plate, and cull are ejected by knockout pins, and disassembled at the bench. Two plates may be used alternately to speed up production by keeping the press continually in operation.

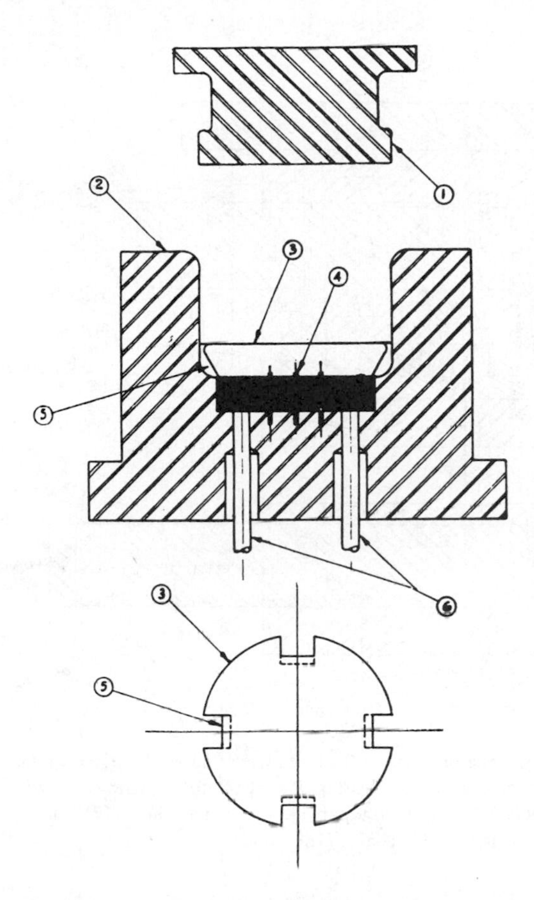

Fig. 9-17. Loose-plate or hand-transfer mold. (1) Plunger; (2) cavity; (3) loose plate; (4) inserts; (5) gate; (6) knockout pin.

These molds are most useful when the cost of the mold must be as low as possible, and when the volume of production is small.

Figure 9-18 shows a semiautomatic mold assembly of the loose- or floating-plate type. In this case, the floating plate is carried more or less permanently in the press and has a central opening that will accommodate stock-transfer pots and plungers of various sizes, as required by different molds. In this way the design of the mold itself is simplified. Presses equipped with these floating plates have been called transfer presses, but are not to be confused with auxiliary-ram transfer presses. Movement of these loose plates may be accomplished either by latches or bolts built into the mold or by auxiliary pneumatic or hydraulic cylinders attached to the press.

Integral Molds. As the name indicates, these molds are self-contained; each one has its own pot and plunger. This arrangement frequently increases the efficiency of the mold because the transfer pot can be designed for best results with a specific cavity. A simple type is illustrated in Fig. 9-19.

Molds of this type can be designed for either manual or semiautomatic operation. The transfer chamber may be located above or below the mold cavity, and the material may flow through a sprue, runner, and gate to the cavity, or the sprue may enter the cavity directly, as in the illustration.

Auxiliary-Ram Molds. Molds of this type are illustrated in Figs. 9-20 and 9-21. These are integral molds, in which the transfer plunger is operated by a separate double-acting cylinder mounted on the press head. Figure 9-22 shows a molded shot from the mold illustrated in Fig. 9-21.

As mentioned previously, this transfer cylinder and plunger may be mounted on the bottom press platen, within the main ram, or even on the tie rods or side columns of the press. These various methods of mounting are usually described as top-ram, bottom-ram, or side-ram mounting, and the molds are designated accordingly.

In comparing the advantages and disadvantages of the various designs of auxiliary-ram methods, it is difficult to generalize because of limiting factors in each design and the method of operation. The statements that follow are therefore made in the realization that special conditions in individual plants may alter considerably the conclusions given.

Top-ram molds are most prevalent today despite some awkwardness in loading the charge due to the restricted opening above the transfer pot and between the upper bolsters. Aside from this inconvenience, top-ram molds normally will permit a faster cycle, after loading, because the mold is already closed at the start of the cycle. There are certain exceptions to this statement, however, as will be noted later.

Another factor to be considered in any appraisal of top-ram molds is the difficulty of

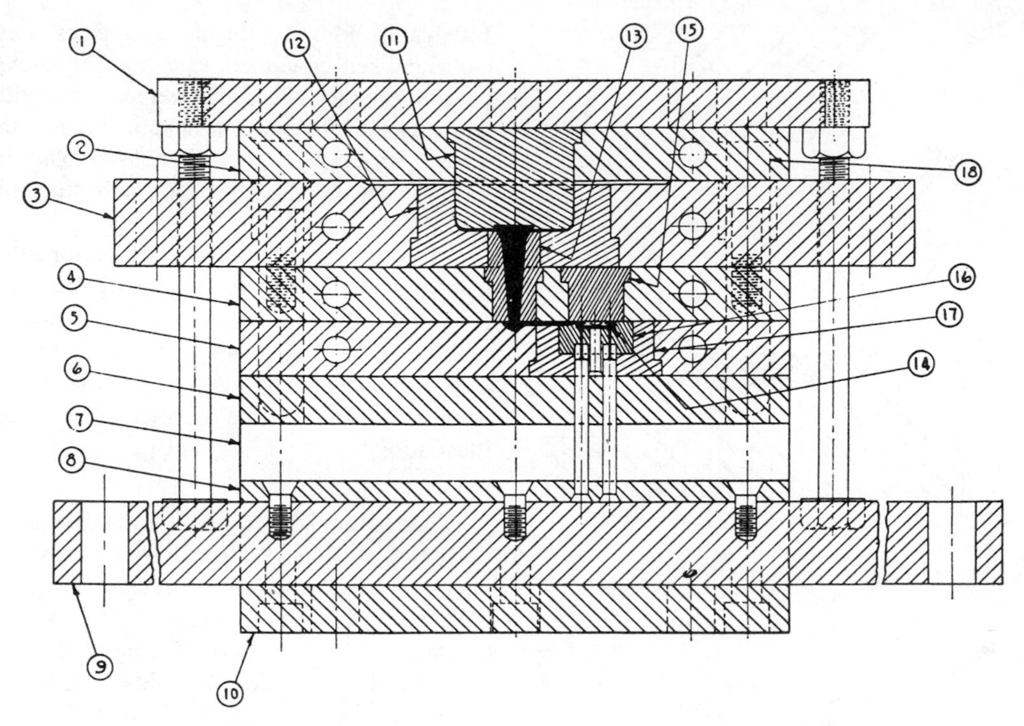

Fig. 9-18. Design of transfer mold for use in a transfer press. The press has a floating platen that will receive several standard sizes of transfer chamber. (1) Top plate; (2) upper force plate; (3) floating platen; (4) force plate; (5) cavity plate; (6) backing-up plate; (7) parallel; (8) pin plate; (9) knockout bar; (10) bottom plate; (11) upper force; (12) loading chamber; (13) sprue plug; (14) molded article; (15) force; (16) cavity; (17) chase; (18) guide pin.

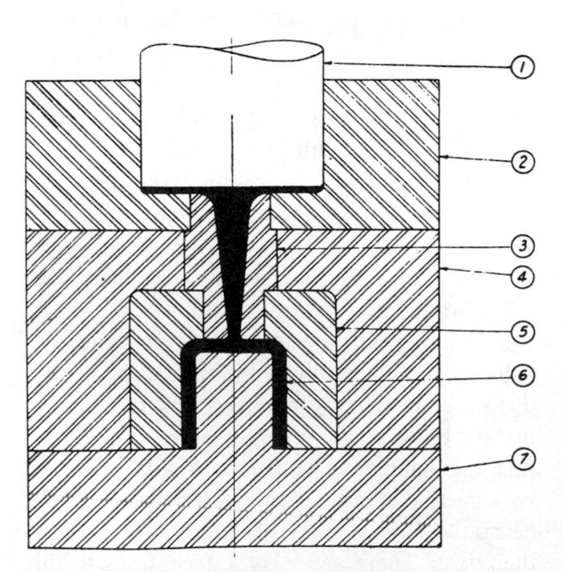

Fig. 9-19. The "integral" or "accordion" type of transfer mold. (1) Plunger; (2) loading chamber or transfer pot; (3) sprue; (4) cavity plate; (5) cavity; (6) molded article; (7) force.

using top ejection mechanisms with this design, because of the space required for the transfer cylinder and ram.

Bottom-ram molds are easier to load than the top-ram type because they are open, and the transfer well is readily accessible, at the beginning of the cycle. However, the molding cycle may be longer, because the mold must be closed, after loading, before the transfer ram is actuated.

When a bottom-ram mold is mounted in a conventional upstroke compression press, some of the stroke and opening must be sacrificed to provide mounting space on the lower press platen for the transfer cylinder. In some cases this can be offset by the use of longer side columns or strain rods.

In bottom-ram transfer and screw-transfer presses, the transfer cylinder and ram are contained within the main ram. This saves opening or daylight in the press.

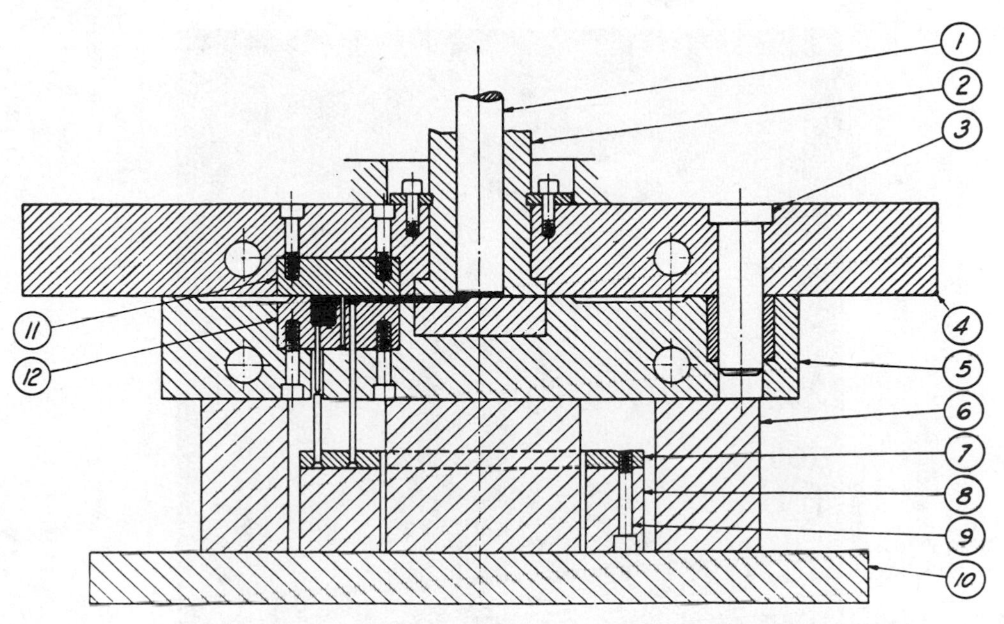

Fig. 9-20. Design of a pressure-type of transfer mold that makes use of the conventional press for clamping pressure only. Transfer is effected by means of an auxiliary ram. (1) Plunger; (2) loading chamber; (3) guide pin; (4) force plate; (5) cavity plate; (6) parallel; (7) pin plate; (8) knockout bar; (9) pin-plate screw; (10) bottom plate; (11) force; (12) cavity.

Side-ram molds are less common, and are used primarily where the design of the molded article requires injection of material at the parting line. Theoretically, this design permits use of the full clamping area of the mold, but in actual practice, unless the molding material is extremely soft and fluid and the transfer pressure low, there is a distinct tendency for the mold to spring and flash as a result of localized, unbalanced transfer pressure near the gate.

Whatever method of mounting is used, it is now generally conceded that the auxiliary-ram method of transfer permits the highest production with the lowest mold cost and minimum loss of material in sprues, runners, and culls. Other advantages, common to this and other types of transfer molding, are discussed below.

Mold design is covered in more detail in Chapter 10.

Advantages of Transfer Molding

Molding Cycles. Loading time usually is shorter in transfer molding than in compression molding, as fewer and larger preforms are used. They can be preheated most rapidly and effectively in dielectric equipment.

In entering the mold, the material flows in thin streams through small runners and gates. This promotes heat transfer, and may also momentarily add some heat to the material through friction and mechanical work. Most phenolic materials also are homogenized thereby, and volatile matter, which might otherwise remain in the piece and necessitate a longer cure to avoid blistering, is reduced in passing through these small channels and by escape through vents and clearance spaces around movable pins. All of these factors contribute to substantially shorter molding cycles than are possible in compression molds, without preheating.

Tool and Maintenance Costs. Deep loading wells are not necessary with transfer molds, and mold sections can be thinner than in compression molds because they are not required to withstand the higher stresses involved during closing of the latter. This difference obviously permits an initial savings in tool costs.

This is less wear on the mold in transfer molding and much less tendency toward break-

Fig. 9-21. Four-cavity auxiliary-ram transfer mold. (*Courtesy Eaton Corp.*)

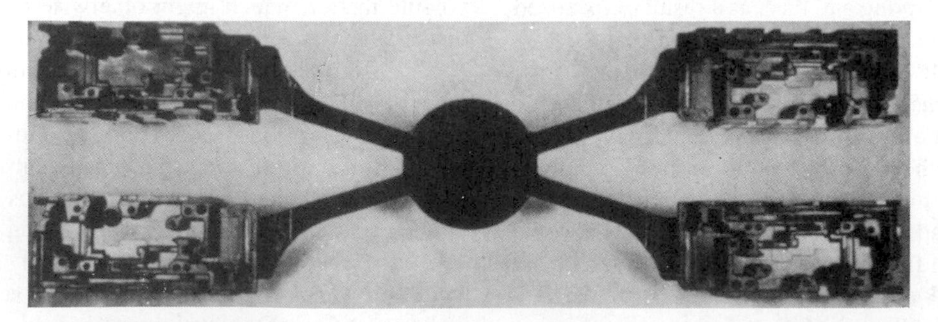

Fig. 9-22. Shot from plunger transfer mold illustrated in Fig. 9-21. (*Courtesy Eaton Corp.*)

age of pins. In transfer molds, core pins can readily be piloted in the opposite half of the mold. When so designed, core pins having a length-to-diameter ratio of 8 : 1 can safely be used; in compression molds the largest ratio that ordinarily can be used is 4 : 1 using dielectrically preheated preforms and about 2.5 : 1 using cold powder. Transfer molds retain their original accuracy and finish considerably longer than do compression molds.

Molding Tolerances. Because the articles are produced in closed molds which are subjected to less mechanical wear and erosion by the molding material than are compression molds, closer tolerances on all molded dimensions should be possible in transfer molding. This is particularly true of dimensions perpendicular to parting lines, because of the very small amount of flash in properly designed and operated molds of this type.

Also, it should be possible to hold closer tolerances on diameters of holes and dimensions between holes because forces in the mold that tend to distort or displace pins or inserts are much less in transfer molding.

This is the theoretical ideal that could be attained if no other factors than those mentioned above were at work in the process. Actually, warpage and dimensional variations have occurred in numerous instances in plunger-molded articles, and it is not always possible in these cases to achieve the close dimensional tolerances hoped for.

Careful study has indicated that these difficulties are caused, in most cases, by abnormal shrinkage and internal strains set up in the molded articles by improper gating and excessive transfer pressure. In some instances, the article, although complicated in outline and section, may not lend itself to transfer molding because of peculiarities of its design.

Those who have studied the auxiliary-ram transfer process most thoroughly are convinced that in most instances close tolerances can be held, and optimum dimensional stability realized if, and only if, the preheating, the plasticity of the material, and the transfer pressure are accurately controlled, and if the runners and gates are so located, and of such size and shape, as to permit rapid, free flow of the material into the closed mold. See Chapter 28 for recommendations.

Finishing Costs. Transfer molding reduces the costs of finishing all thermosetting materials, and especially those having cotton flock and chopped fabric as filler. In an article molded with the latter materials in a compression mold, the flash frequently is heavy, tough, and expensive to remove; this condition becomes even worse as the mold becomes worn. In a transfer-molded article, assuming a properly designed mold with adequate clamping pressure, the flash is quite thin or altogether absent.

Gates, except in certain cases with fabric-filled phenolics, usually can be made sufficiently thin and can be so located that their removal is easy and inexpensive. It may be possible to gate into a hole in the molded piece and to remove the gate by drilling.

Limitations of the Process

As might be expected, there are certain inherent limitations in transfer molding, are discussed below.

Mold Costs. The statements concerning lower mold costs apply particularly to auxiliary-ram transfer molds. Pot-type transfer molds may, in some cases, be more expensive than equivalent compression molds, without offsetting advantages.

Loss of Material. The material left in the pot or well (the cull), and also in the sprue and runners, is completely polymerized and must be discarded. This loss of material is unavoidable, and for small articles it can represent a sizable percentage of the weight of the pieces molded. In the auxiliary-ram molds, the cull is reduced to a minimum, and the main sprue is eliminated.

Some years ago, transfer molding was not well suited to the production of small molded items because of this loss of material. However, to meet the increasing trend toward miniaturization and increased demands for close tolerances and thin-wall sections in molded electronics parts, methods were developed to use small, fully automatic transfer molding presses for the economical production of such small items. Although loss of material in cull and runners is still a factor in molding costs for these small parts, the savings elsewhere, including mold costs and finishing costs, usually offset the loss of material.

On the other hand, where large-volume production (in excess of 500,000 pieces) is required, and where cross sections are simple and uniform, as in buttons and bottle caps, production by compression molding—in multicavity semiautomatic or automatic molds, with electronic preheating—may be cheaper.

Effect on Mechanical Strength. With wood-flour-filled phenolic materials, no significant loss of mechanical strength attributable to the transfer-molding process has been observed in laboratory tests on standard ASTM specimens. This conclusion has correlated reasonably well with general production molding experience.

Some instances of cracking around inserts and at weld lines have been noted in commercially molded articles, especially with mineral-filled phenolics and melamine materials, which have inherently lower mechanical strength. Changing the design and location of gates, altering the cycle of preheating, or reducing the volatiles content of the molding material, when tried separately or in combination, have overcome these difficulties in many cases; in some severe cases, however, no practicable remedy could be found within the framework of the transfer molding process, and it was necessary to mold the articles by compression in order to overcome the trouble.

With transfer molding of the improved-impact materials containing fibrous fillers, a definite and sometimes marked decrease in strength from values obtained in compression molding has been noted by a number of investigators. This effect on strength can be minimized by the use of lower transfer pressures and larger gates.

Comparison between Compression and Transfer Molding

A comparison of the two molding techniques is provided in Table 9-2, as many parts may be molded by either of these methods. Molding technique selection frequently depends upon the molder's own economics and capabilities.

Materials Used

Phenol-formaldehyde and melamine-formaldehyde molding compounds are the materials most widely used in transfer molding, generally giving excellent results.

Urea-formaldehyde materials have been transfer-molded quite successfully in some cases, principally for small articles, but because of their reactivity and critical behavior in dielectric preheating, the process has not been found as generally satisfactory with them as with the phenolics and melamines.

The alkyd or polyester molding materials have been tried in transfer molds. In certain instances, they have proved suitable for smaller articles, but their extremely fast rate of reaction has prevented their use in larger molds and in articles where considerable plastic flow is required.

Phenol-Formaldehyde. Phenolic materials of all of the standard types, containing any of the usual fillers, can be successfully molded by transfer. The principal types used are listed below:

Type	*Filler*
General-purpose	Wood flour
Heat-resistant	Asbestos
Medium-impact	Cotton flock or fabric
High-impact	Cotton fabric
Highest-impact	Chopped cotton cord
Low-electrical-loss	Mica

For satisfactory use in transfer molding, a phenolic material must be extremely fluid at molding temperature in order to flow readily through runners and gates and form dense homogeneous molded articles. For most economical production, this flow should be accomplished in the shortest possible time; that is, the material should have a rapid rate of flow. Phenolic materials possess these properties and, with the aid of dielectric preheating, give excellent results in all types of transfer molds.

Melamine-Formaldehyde. Mineral-filled, cellulose-filled, and fabric-filled melamine materials have been employed in transfer molding with good results. Certain melamine-urea formulations and melamine-phenolic formulations also have shown good characteristics in transfer molding.

Melamine materials may require some modification of basic properties in order to produce the best flow in transfer. These compounds as

Table 9-2. Compression vs. transfer molding.

Characteristic	*Compression*	*Transfer*
Loading the mold	1. Powder or preforms. 2. Mold open at time of loading. 3. Material positioned for optimum flow.	1. Mold closed at time of loading (assuming top transfer, bottom clamp). 2. RF heated preforms placed in transfer pot.
Material temperature before molding	1. Cold powder or preforms. 2. RF heated preforms to 220–280°F.	RF heated preforms to 220–280°F.
Molding temperature	1. One step closures—350–450°F. 2. Others—290–390°F.	290–360°F.
Pressures	1. 2000–10,000 psi (3000 optimum on part). 2. Add 700 psi for each inch of part depth.	1. Plunger head—2000–6000 psi on material. 2. Clamping ram—minimum tonnage should be 75% of load applied by plunger ram on mold.
Breathing the mold	Frequently used to eliminate gas and reduce cure time.	1. Neither practical nor necessary. 2. Accomplished by proper venting.
Cure time (time pressure is being applied on mold)	30–300 sec—will vary with mass of material, thickness of part, and preheating.	45–90 sec—will vary with part geometry.
Size of pieces moldable	Limited only by press capacity.	About 1 lb maximum.
Use of inserts	Limited—inserts apt to be lifted out of position or deformed by closing.	Unlimited—complicated. Inserts readily accommodated.
Tolerances on finished products	1. Fair to good—depends on mold construction and direction of local flow of material during final closing. 2. Flash—poorest. Positive—best. Semipositive—intermediate.	Good—close tolerances easier to hold.
Shrinkage	Least.	1. Greater than compression. 2. Shrinkage across line of flow is less than with line of flow.

a class show greater after-shrinkage than do phenolics. This tendency exists more or less independently of the method of molding; it can be considerably reduced by proper control of the resin and its compounding, and by suitable preheating before molding.

Silicones. Glass-filled silicone materials are molded by transfer methods for special high-impact heat-resistant applications. Dielectric preheating is recommended to improve the flow. As with other fiber-filled types, precautions must be taken to use gates of adequate cross section to maintain proper strength. After-baking at temperatures up to 200°C usually is practiced to obtain optimum strength and heat resistance.

Soft flowing epoxy and silicone molding compounds, molding at pressures of several hundred psi, are extensively used in transfer molding to "encapsulate" or mold plastic around electrical or electronic components,

Fig. 9-23. Typical mold and loading frame, after molding, for transfer encapsulation of electronic integrated circuits, with a soft-flowing epoxy or silicone molding compound. (*Courtesy Hull Corp.*)

such as diodes, transistors, integrated circuits, resistors, capacitors, solenoid coils, and modules containing several such components. In such applications, the materials provide electrical insulation, moisture protection, and mechanical protection (see Fig. 9-23).

Theoretical and Design Considerations

There are many different opinions within the industry about the design of transfer molds, as well as molding pressures and molding techniques. Because of the many factors involved, no hard-and-fast rules can be given to serve as infallible guides in the design and operation of these molds. The principal factors are listed below, and recommendations based on experience and successful current practice are given. (See also Chapter 10.)

Mold Design. It is considered good practice in multiple-cavity plunger-type transfer molds to mount the mold cavities, forces, and runners as separate hardened-steel inserts in the mold chases or retainer plates. This permits removal of individual mold cavities and runner blocks for repair or replacement when wear or breakage occurs. Where not precluded by other features of design, such as location of electric heating cartridges and ejector pins, a circular layout of mold cavities with short radial runners from the central well is preferred. However, there are numerous examples of successful transfer molds having various ladder and multiple-"T" arrangements.

The center pad beneath the transfer pot should be hardened and well supported to prevent deflection and consequent flashing of the mold. Similarly, all mating mold surfaces should be ground smooth and perfectly flat to provide uniform contact. It is not imperative that cavities and runners be chrome-plated, but this is usually advisable to improve release from the mold and to reduce wear, particularly in the runners.

Runners, Gates, and Vents. The design of runners and the size and location of gates can be controversial subjects. The recommendations given below are in accord with current practice, but are not intended to be inflexible rules.

Main runners are usually $\frac{1}{4}$ to $\frac{5}{16}$ inch wide and $\frac{1}{8}$ inch deep, and semicircular in section; branch runners, where used, should be about $\frac{1}{8}$ inch wide and $\frac{3}{32}$ inch deep. These runners are located in the half of the mold containing the ejector pins; this is usually the plate opposite the transfer pot. The runners should be kept to minimum length and should extend into the pad at the bottom of the pressure well at least $\frac{1}{2}$ inch.

The location, size, and shape of gates are quite important to proper operation of the mold. There are many conflicting opinions concerning location, and striking individual exceptions may be found on particular jobs, but it is considered good practice to place the gate at the thickest section of the molded article, and, whenever possible, at a readily accessible point on the article, so that it can be removed by simple sanding or filing.

The size of the gate will vary with the type of material molded, the size of the piece, and the molding pressure available. For small pieces, and where general-purpose wood-flour-filled phenolics are used, it is recommended that the gate be 0.80 to 0.100 inch wide and 0.015 to 0.020 inch deep. With mineral-filled materials, a gate 0.125 inch wide and 0.030 inch deep usually is necessary.

The shape of gates for transfer molds has received considerable attention. Depending on the size and shape of the molded part, gates of circular or rectangular cross section may be used. The circular gate generally permits most rapid filling with smallest gate area, but it may leave an unsightly scar on the finished part. Often fan gating or edge gating is used, with gates from 0.010 to 0.020 inch thick, and as wide as several inches (perhaps as wide as the part). Such gates, if shaped such that the runner approaching the gate is always of thicker and larger cross section than the gate itself, will permit clean break-off of the part from the runner with minimum scar and often no further finishing costs.

Because of the coarseness of the fillers used in high-impact materials, larger gates usually are necessary, both to lessen the transfer pressure required and to prevent impairment of the mechanical strength of the molded articles. In these cases, a width of 0.500 inch and a depth of 0.125 inch may be needed. A gate that is too small may become partially blocked with fiber fillers such that only the low viscosity resin flows into the cavity, without adequate filler, yielding a part of unacceptable strength.

These are general recommendations; it may well be necessary, in individual cases, to enlarge gates beyond these dimensions, depending on circumstances. It is always easier, in any case, to remove metal from a mold than to add it.

Vents are important to permit the escape of air, moisture, and other volatiles as the materials fill the mold cavity. It frequently is found that without proper vents the mold cavities will not fill properly even under high transfer pressures. Vents should be located on the same half of the mold as the runners, and opposite the gate. They usually are 0.003 to 0.005 inch deep, and about $\frac{1}{8}$ inch wide, and are extended to the outer surface of the mold block or land area. With epoxies and silicones, vent depths are only 0.001 to 0.002 inch.

Vacuum Venting. To overcome the trapping of air or gas in a cavity, in locations that are difficult to vent effectively, molds may be designed such that all cavity vents feed into a space that is sealed from the outside of the mold (when closed) by an O-ring seal, and is connected to a vacuum reservoir through a vacuum line containing a solenoid-operated valve. In operation, as soon as the mold is closed, and the transfer plunger enters the pot, the aforementioned solenoid valve is automatically opened, causing the cavities to vent rapidly into the vacuum reservoir before the molding compound has entered or filled the cavities. Two benefits result: First, the material, finding it unnecessary to ''push'' the air from the cavity through the vents, enters with a minimum of back pressure and thus fills the cavity more rapidly, leading to faster cures. Second, not only is there essentially no entrapped air, and there-

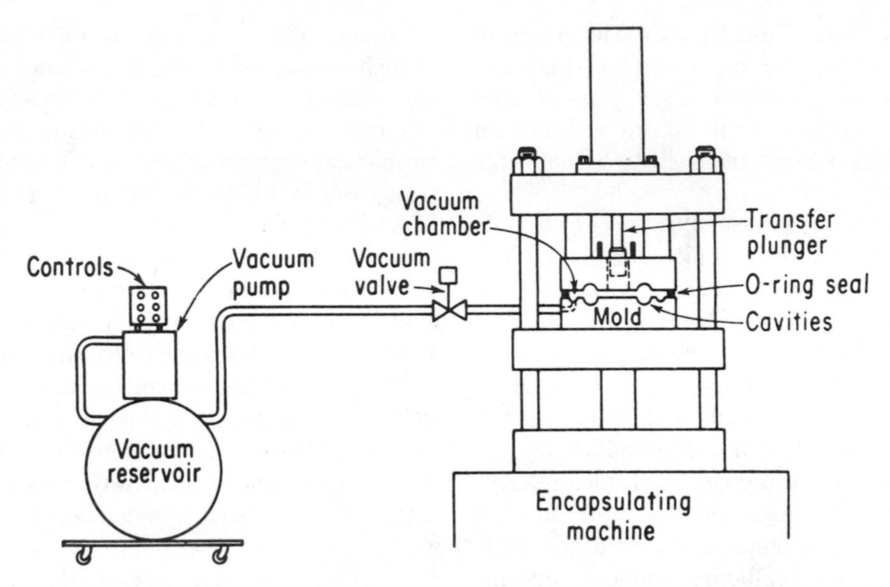

Fig. 9-24. Schematic diagram showing principle of vacuum venting with transfer molding. The principle also is used with automatic transfer and screw injection molding where the cavity configuration precludes adequate parting line venting. (*Courtesy Hull Corp.*)

Fig. 9-25. Multicavity transfer mold arranged for vacuum venting. Note rectangular O-ring seal surrounding the six bottom chases. (*Courtesy Hull Corp.*)

fore no voids in the part, but such minute quantities of air as may be present are readily absorbed "into solution" in the molding compound because of the molding pressure.

Vacuum venting has been used successfully in both semiautomatic and fully automatic transfer molding, as well as in automatic inline screw injection molding. (See Figs. 9-24 and 9-25.)

Selection of Material. The problem here is three-fold—to choose the type of material that will yield the desired physical properties in the molded article (e.g., impact strength, heat resistance, or low electrical loss); to choose from among a number of formulations of this type the one having the proper preheating and molding characteristics for the mold in question; and, finally, to choose the proper plasticity, which will provide the fastest transfer and cure. The last cannot be decided in advance; it must be determined by experiment.

Choice of the proper plasticity is very important. A material that is unduly stiff will be extremely critical in behavior during preheating, and may not fill the mold; a material that is too soft may flash the mold or require excessive preheating to attain a satisfactorily rapid cure.

A maximum transfer time of 10 to 15 seconds at a mold temperature of 320 to 350°F, with a preform temperature of 220 to 260°F, usually is found satisfactory. Individual conditions may necessitate some departure from these values, but these represent the average of general experience.

Molding Pressure. It would be extremely difficult, if not impossible, to measure accurately, or to calculate theoretically, the magnitude and distribution of pressure within thermosetting material flowing in a transfer mold. If all of the factors involved were in perfect dynamic balance, there would be no unbalanced fluid pressure in the mold cavity at the moment of complete filling.

Initially, with either pot or plunger molds, fluid pressure in the hydraulic line to the press is converted to mechanical pressure on the molding material as the latter is compressed and forced into the closed mold. As the mold fills, and polymerization of the material continues, this pressure builds up to a maximum. Ideally, the material continues to be fluid until the cavity is completely filled and "packed." At that point, and not before, the cross-linking reaction should proceed to harden the part sufficiently for ejection.

It is recognized that at the instant of cavity fill and packing, there is a hydrostatic pressure exerted against top and bottom mold halves, tending to open the mold. Calculation of the actual clamping force required to counterbalance this pressure is difficult because pressure throughout the cavity and runner system is not uniform; but if the force calculated by multiplying the projected area of cavities and runners and cull times the calculated hydraulic pressure at the plunger head (where the plunger meets the plastic) is greater than the clamping force, the mold will open and flash. Good mold design requires a minimum of a 10 to 15% margin of extra clamping force over the projected area "hydrostatic force" to preclude mold opening before cure.

Because of the many variables involved in the design of an article and of the mold and in the molding conditions, it is practically impossible to calculate the minimum transfer pressure required to mold a given article. Only empirical assumptions can be made, based on limited experimental data and on previous production-molding experience.

For soft-plasticity, general-purpose phenolic materials, electronically preheated, minimum transfer pressures of 8000 psi for plunger-type and of 12,000 psi for pot-type transfer molding generally are recommended. This assumes a total runner area of about 0.05 sq. in., and is valid primarily for quantities of material of the order of 100 grams. For quantities of material greater than 100 grams, enlarge the runner and gate areas by 50 to 100%.

For fabric-filled phenolic materials also, a 50% increase in transfer pressure usually will be required to keep transfer time to an economical minimum. Here, too, larger gates, as recommended previously, should be used.

Design Calculations. The various factors included in the design of plunger transfer molds, insofar as pressure is concerned, can be defined and illustrated as follows:

1. *Line pressure:* The pressure in psi in the hydraulic line to the press, supplied either by a self-contained pump or from a central pump and accumulator system.
2. *Clamping-ram force:* The total force delivered by the main ram, determined by multiplying the ram area by the line pressure.
3. *Transfer-ram force:* The total force applied by the transfer ram, expressed in pounds, determined by multiplying the hydraulic ram area by the line pressure.
4. *Plunger pressure:* The pressure exerted on the material in the transfer chamber by the plunger. It is equal to the transfer ram force divided by the area of the plunger.
5. *Mold-clamping pressure:* The effective pressure that holds the mold closed against the pressure exerted by the material within the mold cavities and runners. It is expressed in psi and is deter-

mined by dividing the clamping-ram force by the total projected area of cavities, lands, runners, and cull in square inches.

These factors can be shown symbolically as follows:

L_1 = line pressure, psi, clamp
L_2 = line pressure, psi, transfer
A_m = projected area of each mold cavity, land, and runner, sq. in.
A_r = area of clamping ram, sq. in.
A_i = area of transfer hydraulic piston, sq. in.
A_p = area of plunger, sq. in. (also equal to cull area)
CRF = clamping-ram force ($A_r \times L_1$), lb
TRF = transfer-ram force ($A_i \times L_2$), lb
PP = plunger pressure, psi $\dfrac{(A_i \times L_2)}{A_p}$
MCP = mold-clamping pressure $\dfrac{(A_r \times L_1)}{A_m + A_p}$, psi

By definition, to ensure safe operation and avoid flashing, mold-clamping pressure = plunger pressure + 15%, or:

$$MCP = 1.15\ PP$$

Practical application of this simple equation is demonstrated in the following example.

Example. Determine how many cavities can be placed in a mold that is to operate under the following conditions:

Projected area of molded article including land area = 3 sq. in.
Estimated runner area per cavity = 1 sq. in.
Total area per cavity = 4 sq. in.
Diameter of main ram = 14 in.
Diameter of transfer hydraulic piston = 6 in.
Plunger diameter = 3 in.
Line pressure = 2000 psi, both clamp and transfer

Let:

x = the number of cavities
L = 2000 psi (L_1 and L_2)

$$A_m = 4x$$

$$A_r = \frac{3.1416 \times 14^2}{4} = 154 \text{ sq. in.}$$

$$A_i = \frac{3.1416 \times 6^2}{4} = 28.3 \text{ sq. in.}$$

$$A_p = \frac{3.1416 \times 3^2}{4} = 7.1 \text{ sq. in.}$$

$$CRF = A_r \times L = 154 \times 2{,}000 = 308{,}000 \text{ lb}$$

$$TRF = A_i \times L = 28.3 \times 2{,}000 = 56{,}600 \text{ lb}$$

$$PP = \frac{A_i \times L}{A_p} = \frac{56{,}600}{7.1} = 8085 \text{ psi}$$

$$MCP = \frac{A_r \times L}{A_m + A_p} = \frac{308{,}000}{4x + 7.1} \text{ psi}$$

Since $MCP = 1.15\ PP$

$$\frac{308{,}000}{4x + 7.1} = 1.15 \times 8085 = 9298 \text{ psi}$$

$$4x + 7.1 = \frac{308{,}000}{9298} = 33.1$$

$$4x = 26$$

$$x = 6.5, \text{ or 6 cavities}$$

$$MCP = \frac{A_r \times L}{A_m + A_p} = \frac{308{,}000}{4x + 7.1} \text{ psi}$$

In designing pot-type transfer molds, where the only source of external pressure is the main ram, the pot area is equated to the total mold area, and, as mentioned previously, the latter is increased by 15% to prevent flashing.

Let:

L = line pressure, psi
A_r = area of main ram, sq. in.
A_p = projected area of pot or material chamber, sq. in.
A_m = projected area of mold cavities, lands and runners, sq. in.

CF = clamping force ($A_r \times L$), lb
MCP = mold-clamping pressure

$$\left\{\frac{A_r \times L}{A_m}\right\}, \text{psi}$$

Then for equilibrium, with the factor of safety:

$$(1)\ A_m = 1.15\ Ap$$

$$(2)\ A_m = \frac{A_r \times L}{MCP}$$

$$(3)\ \frac{A_r \times L}{MCP} = 1.15\ A_p$$

Other Considerations. It has been assumed in the foregoing discussion of transfer-molding pressure that flashing will be prevented if sufficient clamping pressure or area is provided. This is generally true if the mold is properly supported, particularly under the transfer chamber. However, if insufficient bolstering is used at this point, or if the mold is not properly aligned or hardened, deflection can occur and permit flashing in spite of the excess clamping pressure provided in the design.

Transfer pressure should be sufficient to fill the mold, under ordinary circumstances, in 10 to 15 seconds. Excessive transfer pressure should be avoided—it may cause flashing, it will probably cause undue wear of gates and runners, and it may decrease normal molding-shrinkage so much that molded articles will be outside normal tolerances.

The rate at which a phenolic material will be preheated in a dielectric preheater depends, among other factors, on the high-frequency output of the machine and on the electrical-loss characteristics of the material involved. Wood-flour-filled compounds having a higher electrical-loss factor, will preheat more rapidly than mica-filled materials. Modern dielectric preheaters of 2.5 kW output and with frequency of 100 MHz are designed to heat approximately 1 pound of wood-flour-filled material from room temperature to 300°F in 30 seconds.

Where a number of preforms are being preheated at one time, the uniformity of preheating of the material is quite important. If hardness and density of preforms vary, then the temperature will vary considerably throughout the preforms, and their behavior during transfer and cure will be erratic. Once established, optimum preheating conditions should be maintained as nearly constant as possible for a given batch of material. To facilitate uniform heating of cylindrical preforms, preheaters with automatic roller electrodes cause the preform to rotate under the top electrode during the actual heating.

Numerous other factors affect transfer molding, but they are already familiar to molders employing this method and need not be described in detail. It has been the purpose of this section, instead, to point out some of the important features of this method, and the advantages in increased production and lowered production costs that may be achieved with it, if certain fundamental considerations are known and observed.

Although automatic transfer molding was practiced extensively in the 1960s and 1970s, automatic thermoset injection molding generally has replaced it as the preferred process. (See Chapter 8.)

Liquid Resin Molding

Liquid resin molding (LRM) is a combination of liquid resin mixing and dispensing and transfer molding. The process equipment includes a machine to proportion, mix, and dispense a resin and catalyst directly into a transfer mold, generally through a parting line sprue feeding into the runner system. Another version, suitable for long pot-life liquid resin systems, dispenses the resin mix from a storage chamber, with the catalyst already added. (See Fig. 9-26.)

Because the pressures of injection are approximately 25 to 50 psi, very fragile inserts can be molded, and mold wear is at a minimum. Some formulations for LRM also may be molded at temperatures as low as 200°F, which permit encapsulation of some heat-sensitive electronic components that do not lend themselves to encapsulation at conventional transfer molding temperatures of 300°F or higher.

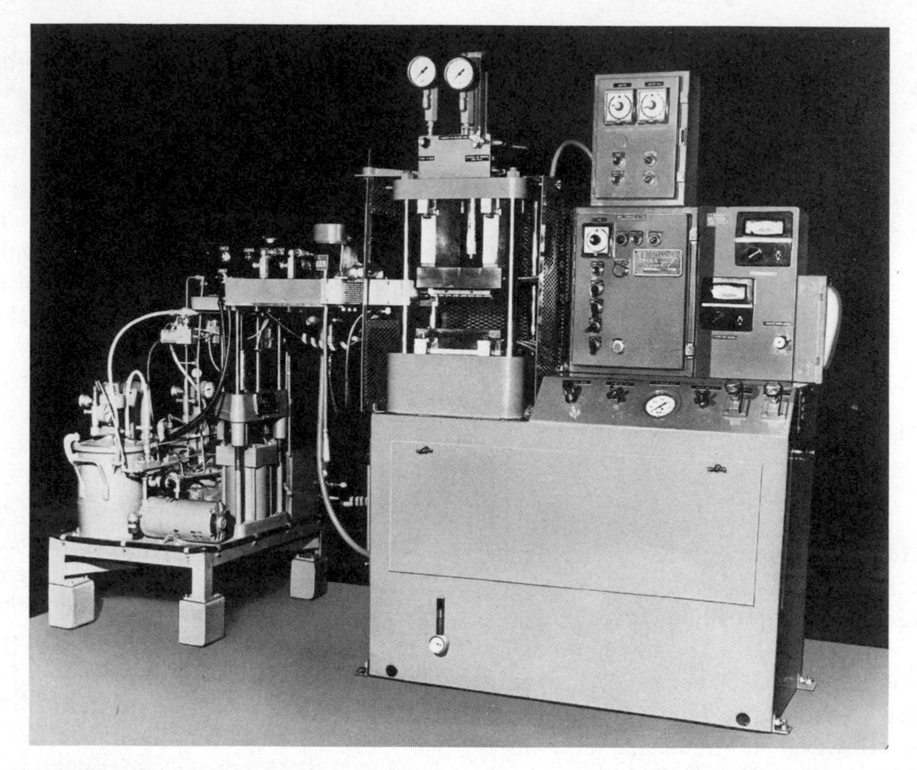

Fig. 9-26. Complete installation for liquid resin molding process. Apparatus at left proportions, mixes, and dispenses liquid resin and liquid curing agent, which is then injected into closed molds at parting line, at pressures as low as 25 psi. (*Courtesy Hull Corp.*)

Liquid resin systems for LRM may be filled or unfilled. They generally have a pot life of at least two hours at room temperature, but a "snap cure" of often less than one minute at mold temperature. Unlike liquid coating resins, such resin systems must include a mold release agent to enable easy removal of parts from cavities.

This process of transferring unfilled liquid resins into heated mold cavities also is being used increasingly in molding highly sophisticated aerospace parts reinforced with oriented reinforcing woven fabric. In molding these composites, the fiber mat or fabric layers are placed in the open mold, following which the mold is closed. Then the liquid unfilled low-viscosity catalyzed resin (often epoxy or polyester) is transferred into the cavity through one or more gates. The resin saturates the fabric "insert" and cures under continued heat and pressure, producing a highly engineered composite component offering maximum strength and rigidity with minimum weight. The reinforcing fabric used in such parts is sometimes glass, but often graphite or boron fibers or Kevlar synthetic fibers.

10

Designing Molds for Thermoset Processing

A successful and profitable thermoset molding operation is dependent on the selection of the proper type of molding equipment, the plant layout, the method of molding, and molds that have been engineered properly and are of superior construction. A poor mold design or an inferior mold will produce a part that is costly in that it may require subsequent finishing operations before it can be shipped. Processors of thermoset raw materials must constantly keep abreast of new molding methods and new and improved mold steels or nonferrous alloys. They also must make sure their mold suppliers are aware of the newer, more advanced techniques for making molds (computer-controlled machining equipment, spark or chemical erosion processes, etc.). This chapter will cover the methods of molding and mold design for thermosets. It does not discuss mold machining or mold manufacturing, which are covered in Chapter 7.

Method of Molding

The trend is toward automation wherever possible. Therefore, the processor has an opportunity to select one of four methods of molding (taking into account the geometric design of the part, the finished part tolerances, the operating conditions, the type of molding compound (long fiber filled, bulk handling characteristics, etc.), and environmental surroundings): (1) compression, (2) transfer, (3) injection, and (4) injection-compression.

Compression Molds

Compression molds may be designed for hand molds or semiautomatic or automatic operations. In the early stages of molding, hand molds were quite popular, but in recent years they have been used principally for prototype parts or parts with limited production. The custom molder does a large amount of molding using the semiautomatic method. The compression molds may be hand-loaded with loose powder, bulk molding compound (BMC), or preforms. The preforms generally are preheated to 200 to 260°F in high-frequency preheaters. The trend being to automate, both custom and captive molders are automatically compression molding a wide range of thermoset parts. This is done on a variety of press equipment. Conventional single-ram hydraulic or toggle presses may range from 15- to 300-ton capacity.

Rotary presses designed for compression molding are used to mold closures, wiring device parts, timer cams, outlet boxes, and so on. The size of the part is limited to the clamp capacity of each individual mold station, often 10 tons and rarely exceeding 30 tons.

The automatic molding presses for compression molding are so designed that various methods of loading the cavities may be used.

Reviewed and revised by John L. Hull, Vice Chairman, Hull Corporation, Hatboro, PA.

Material may be loaded with cold granular, high-frequency preheated or infrared preheated granular, preheated preforms, or a charge that is preheated by a heated barrel and reciprocating screw.

A compression mold consists of heated platens onto which are mounted the cavity and force block or blocks. A knockout system, either top, bottom, or both, is incorporated to remove the parts from the mold. A typical cycle is: mold open after cure period, remove parts, clean mold, load material, close press. On closing of the mold, the excess material is forced out through the flash escapement area.

In general, when parts are molded by compression, the part density is the same in all directions. For this reason it is recommended that parts that have very close tolerances over the face of the part be molded by compression; there is little chance of uneven density causing shrinkage problems or out-of-round conditions or differences in shrinkage from width to length.

As explained previously, in a compression mold the material is loaded into an open mold, and the excess material required to fill the part is forced out over the flash line. The least expensive compression mold is a flash mold, as illustrated by Fig. 10-1. The greatest disadvantage of a flash mold is that there is minimum restriction to the flow of material, and very little back pressure is built up in the molded part. The excess material is allowed to flow out horizontally with little restriction until the mold is essentially "metal to metal." The parts may lack uniform density; so they may be inferior with regard to their strength and molded finish. It is most difficult to maintain dimensional tolerances when density variations exist. Although the mold cost may be less than for other types of compression molds, the fact that there is no loading space except for the cavity of the part itself usually makes the material charge excessive. The percentage of flash loss may be extremely high, thus costly on a piece part basis. It is extremely difficult to maintain the centerline relationship between the force and the cavity, and if guide pins wear, the force and the cavity become mismatched. Very few semiautomatic or automatic molds, therefore, are flash molds.

The second type of compression mold, the semipositive horizontal flash, is illustrated by Fig. 10-2. The majority of the semiautomatic

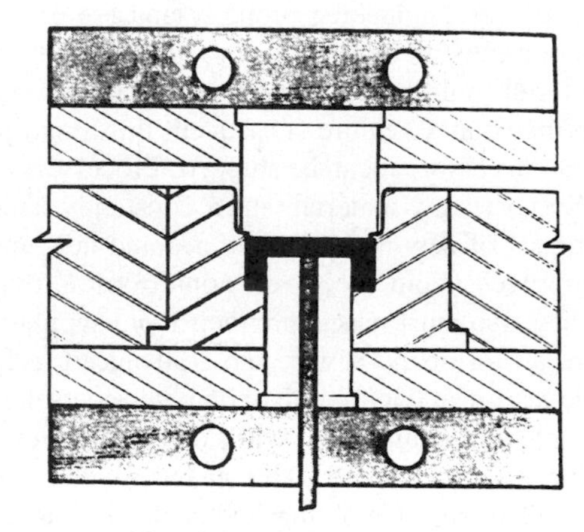

Fig. 10-2. Semipositive mold.

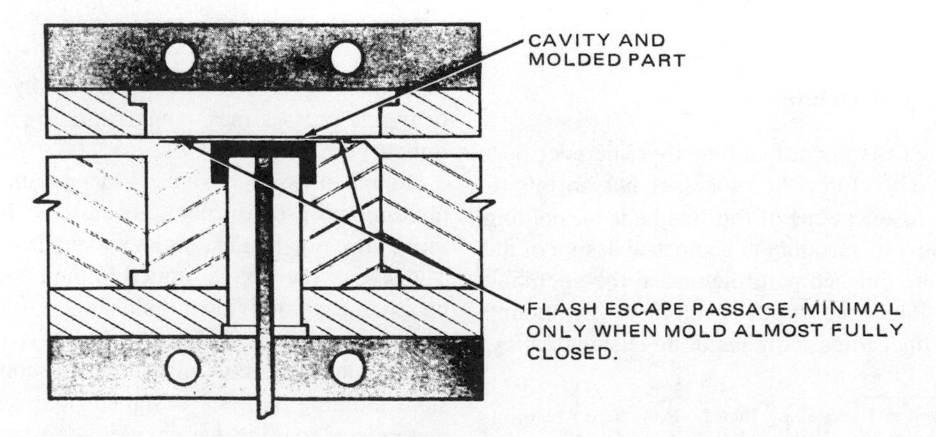

Fig. 10-1. Flash mold.

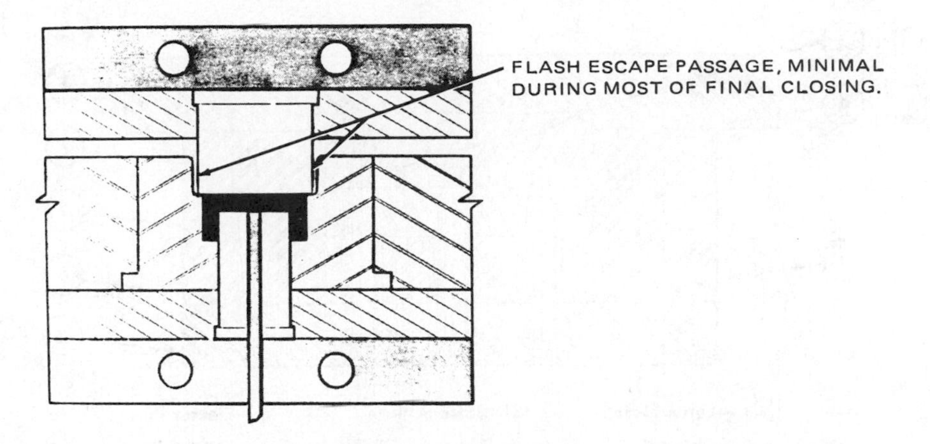

Fig. 10-3. Semipositive mold.

and automatic compression molds are of this design. Ample loading space may be provided so that the percentage of flash loss may be held to a minimum. If the clearance between the force and cavity is held to 0.003 to 0.004 inch, all the material is trapped in the cavity so that the part is of maximum density. Because there may be slight differences in the volume contained within the force and the cavity in a multicavity mold, it is well to provide flash escapements in the form of flats or grooves 0.015 to 0.020 inch in depth. These should always be put on the force because they then can be cleaned off each shot more easily than when put into the cavity. This type of mold produces parts having maximum physical strength. If molds are "breathed" just before final close, porosity may be eliminated by releasing the trapped volatiles. Dimensional tolerance may be closely held, especially over the face of the part.

The third type of compression mold, the semipositive vertical flash, is illustrated by Fig. 10-3. It is quite similar to the semipositive horizontal flash except that the force enters directly into the geometric design of the part itself for a distance of $\frac{1}{16}$ to $\frac{1}{8}$ inch. The clearance between the force and cavity at this point should be 0.005 to 0.007 inch to prevent scoring of the cavity side wall. If this is not the case, and scoring takes place, the molded article will have a scratched appearance. Again, flash escapements should be provided on the force. A part molded in such a mold will be of good density. The flash is vertical, which may be preferable, and can be removed easily. The molded article, because it has good density, should have a good finish and maximum physical strength. This type of mold is quite expensive because there are two areas between the force and the cavity that must fit perfectly. There is absolutely no chance for mismatch between the force and the cavity, so concentricity of the article from cavity to force is as close as possible. Again, good dimensional tolerances are achieved over the face of the part.

The fourth type of compression mold, fully positive, is illustrated by Fig. 10-4. This type of mold generally is used if the materials are of high bulk, or if the filler fibers are quite long, as in glass roving. It is difficult to pinch off such fibers to a thin flash. The fully positive molds give a vertical flash that may be easier to remove than that of other designs. It is dif-

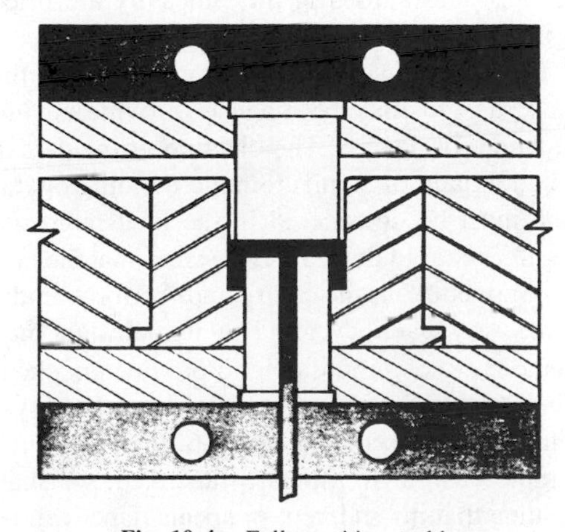

Fig. 10-4. Fully positive mold.

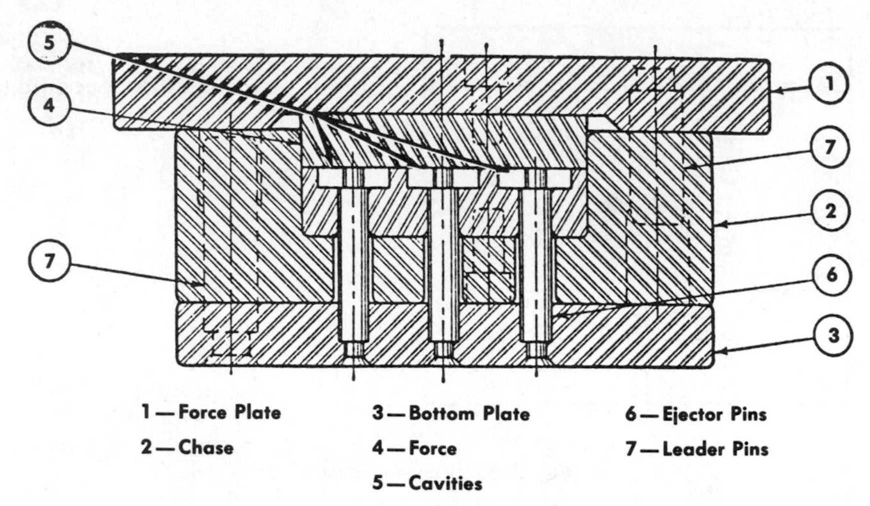

Fig. 10-5. Subcavity mold.

ficult to produce parts of equal density if a multicavity mold is used. The clearance between force and cavity must be excessive, or generous flash escapements should be provided. It may be desirable also to provide a slight taper on the force to help in flash escapement. Pressure pads, away from the loading area, must be provided to ensure proper control of the thickness of the part by making the mold close to the same stopping point each time. Material or flash must be kept free from the surface of such pads.

If parts are extremely small so that it becomes difficult to charge each individual cavity, a subcavity mold can be used, as illustrated in Fig. 10-5. The mold shown is a positive mold; however, a semipositive design may be used. Only one charge is needed in a subcavity because all cavities in the subcavity are filled by the single charge.

In the design of the mold, variations of the basic design may be made to provide an improved tool design. The ejection of the part in an automatic or semiautomatic molding operation must be considered in the final mold design. The mold design must *ensure* that the part is removed from the cavity and the force in the same way each cycle. The part design may make it almost impossible to provide knockout pins. In this case a stripper plate knockout system must be used. Figure 10-6 illustrates such a mold design. In the part illustrated, we shall assume that no undercut or special taper can be allowed, to ensure that the molded part will stay either on the force or in the cavity when the mold is opened. In this case, both top and bottom ejection are provided, the top being the stripper plate and the bottom a normal knockout system. In using the stripper plate design, care must be taken to provide for positive automatic flash removal with each cycle of the press. Stripper plate molds are used to mold parts such as pin connector boards containing a large number of holes.

One should observe certain details of the mold design here illustrated:

A. Clearance under the dowels or leader pins to keep flash from building up in this area, preventing full entry of pins.
B. The use of a hardened dowel or leader-pin bushing in a soft plate.
C. Slight positive entry action of the force into the cavity, to ensure alignment of the mold and full density of the molded articles.
D. Flash grooves in the force.
E. Clearance between underside of stripper plate and top of cavity to allow escape of material from flash grooves.
F. Clearance holes under inserts to allow any flash to fall completely through the mold.
G. Pressure pads or blocks to control the thickness.
H. Retainer plate for top ejection pins.
I. Clearance space above stripper plate so that flash working up between the top

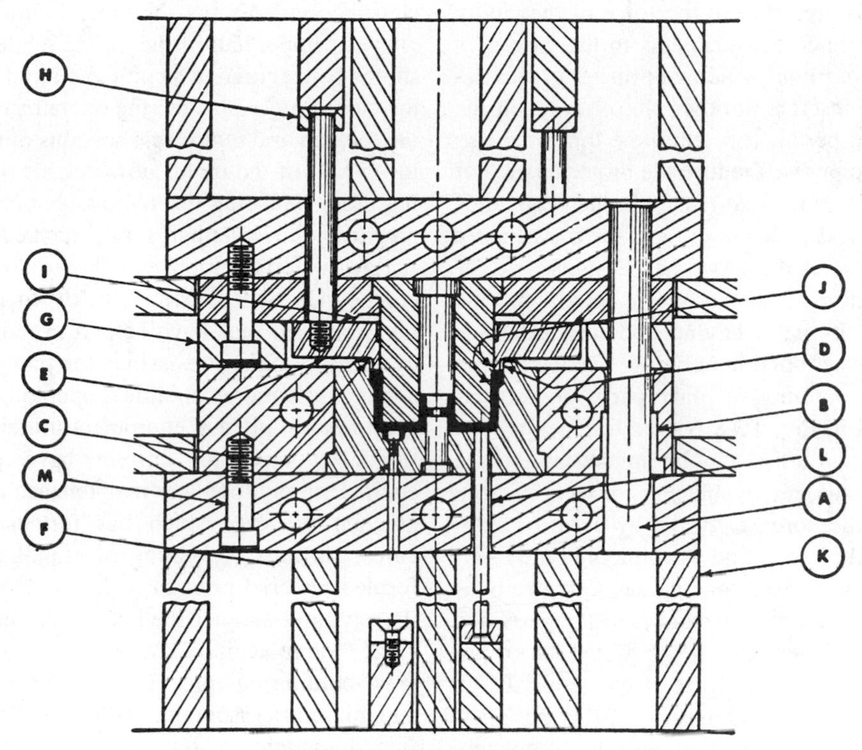

Fig. 10-6. Stripper plate knockout system.

force and the stripper plate will not be confined and can easily be blown out.

J. Inside diameter of stripper plate made slightly greater than outside diameter of the force that shapes the inside of the molded article, in order to avoid scoring of the force, which would mar the inside surface of the molded article.

K. Placement of lower supporting parallels to avoid blocking off the clearance holes under the dowel bushings.

L. Counterbore in underside of ejection pin holes to reduce friction and to allow flash escape.

M. Screws for holding mold parts together put in from undersides of plates, remote from the surfaces where molding material might get into the counterbored recesses for screw heads and cause trouble in cleaning and in screw removal for mold maintenance.

Pressure pads, illustrated as Item G in Fig. 10-6, are desirable on all compression molds. This is a safety measure that assures that the land area of the mold will not be damaged in mounting the mold in the press. The landing pads (or pressure pads) should be designed to be in contact when there is approximately 0.002 inch clearance between the force and cavity at the land area or other surfaces such as pins, blades, and so on.

In the design of compression molds, especially flash and/or semipositive types, it is important that the land area at cutoff in the outer landed area, or in the cavity itself, be carefully designed in relation to the molding pressure. If, for instance, a meter housing or an instrument door should have one or more rather large openings in the face surface, it is imperative that the plastic thickness at these areas be held to 0.005 to 0.010 inch for ease of deflashing. If the area of these openings exceeds 1 to 2 sq. in., the thermosetting material will advance in cure to the point where the equipment is endeavoring to flow a set-up material across this land area, which will prevent the mold from closing unless excessive pressure is used. To eliminate this condition, it is advisable to provide a relief area in the center portion, which will inhibit the cure effect. Provide a land that is $\frac{1}{8}$ to $\frac{3}{16}$ inch in width inside of the opening, and in this area relieve the force or the cavity to $\frac{1}{16}$ to $\frac{3}{64}$ inch in thickness. In the case of a rather large round window or opening, it is desirable to insert a through pin, or core, in the cavity and permit it to enter the top force. Inserting the pin will reduce the projected area of the piece by the area of the pin diameter, thus requiring less pressure to mold the part. Figure 10-7 illustrates these two part designs.

Semiautomatic or automatic molds may be designed for parts having undercut side surfaces such as found in bobbins or spools, pump housings, automotive brake parts, and so on. The mold in Fig. 10-8 is used for such a part. The undercut areas may be molded by the use of split molds, removable pins, or sections from molds. Side cores activated by cams or side-acting hydraulic cylinders commonly are used. In the latter case, locking devices (such as wedges) of sufficient strength must be provided to keep the cores from being forced out of position by the internal force of the material.

One of the critical aspects of this design is the flash removal; the flash must be completely removed from around the movable side cores. It is for this reason that most such molds are operated semiautomatically. In order to provide economical operations, shuttle presses or reciprocating table presses are used. This design incorporates two or more bottom halves with one upper half of the mold. While the one shot is being cured in a complete mold between the closed press platens, the operator is removing the part and removable sections of the other lower half of the mold, now outside the pressing area, replacing the removable section and preparing the lower half for the next cure cycle.

Raw-material suppliers have formulated compounds that respond quickly to heat and pressure, and thus have helped to reduce cure cycles and make it possible for processors to adopt automatic, economical operations. Manufacturers of preheat equipment utilizing high-frequency electrical current have provided equipment requiring less maintenance and more efficient preheat, which has further reduced cures. Infrared preheating of granular or preformed material prior to its being deposited in the cavity is another method of shortening the cycle. Manufacturers specializing in compression molding equipment now offer solid state and microprocessor controls to provide trouble-free automatic or semiautomatic operations. The most sophisticated machines often use closed-loop computer controls for optimum part quality and productivity. Reduced time for dry cycles has permitted shorter overall cycles. Also, reciprocating screws used in connection

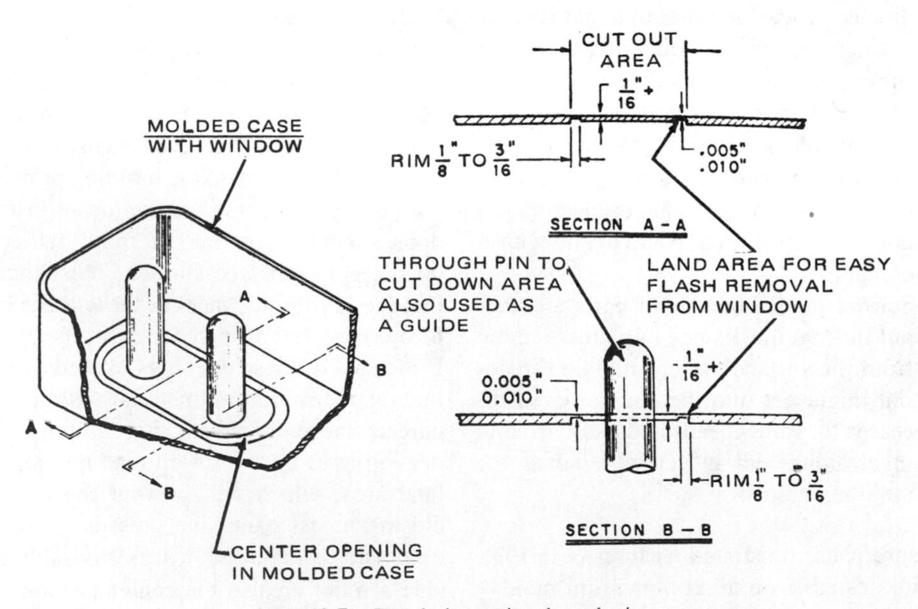

Fig. 10-7. Part designs using through pins.

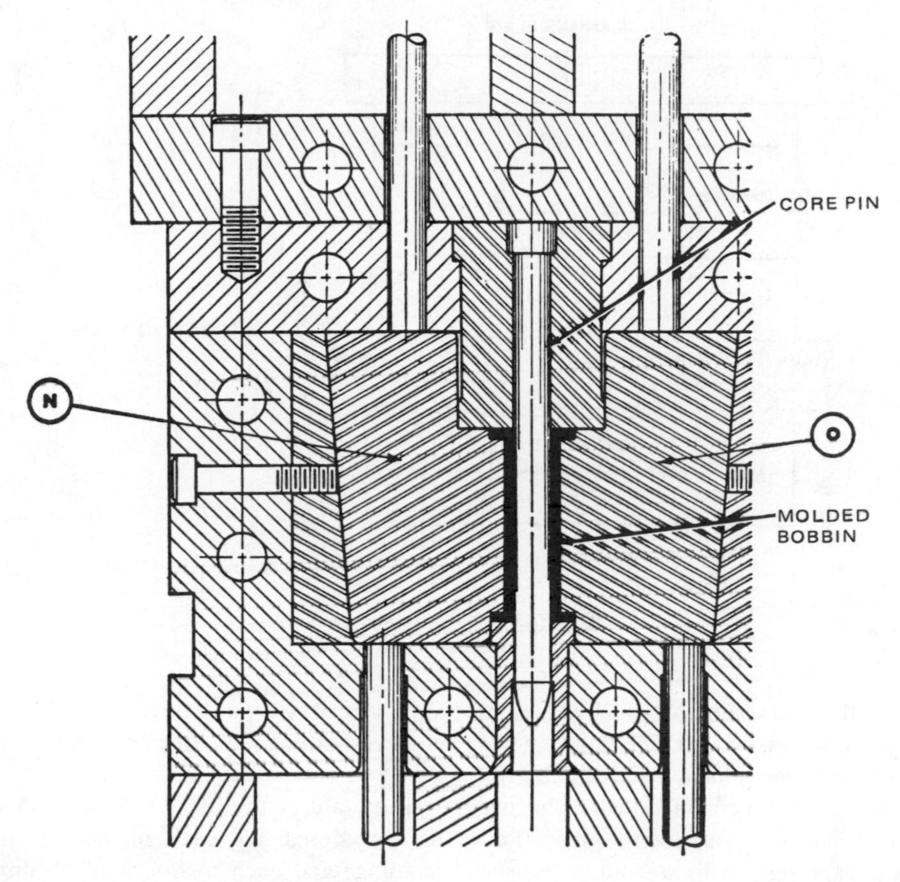

Fig. 10-8. Semiautomatic mold for part with undercuts in side walls. Mold segments N and O are automatically raised from the cavity plate when the press and mold open, and then are moved to the left and the right, respectively, to allow the molded part to be stripped from the core pin.

with heated barrels mounted on the molding press can provide an extruded charge of preheated material that will further reduce cure cycles. The charge can be deposited directly into the cavity or stored in a loading tray with several cavities fed at one time.

All of the above factors have contributed to the continuing use of compression molding for making a wide range of plastic parts.

Transfer Molds

Transfer molds will vary in design, depending on the method of transfer molding. There are three basic methods: (1) The original method—pot transfer—utilizes a three-plate mold with the plunger-retaining plate mounted to the head of the press, a floating plate containing the pot and sprue bushing and top cavity or cavities, and the third retainer plate (mounted on grids) containing the bottom cavity. Figure 10-9 shows such a mold, often operated as a hand mold, the bottom plates of which are removed from the press each cycle for manual separation of the plates and removal of parts. Figure 10-10 shows how the same mold can remain in the press each cycle, having the middle plate raised by pull- or push-rods when the press opens. (2) The second method is known as the plunger transfer method. In this case, the plunger is activated by an independent hydraulic cylinder. The transfer cylinder may be mounted at the top of the press or at the bottom of the press, or it may incorporate a ram within a ram if the main clamp cylinder and the transfer plunger are both on the top, or both on the

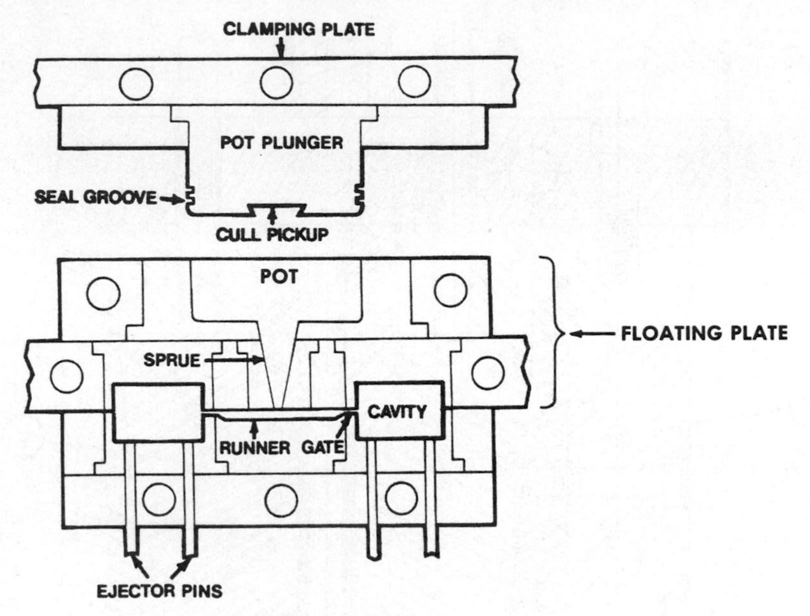

Fig. 10-9. Pot transfer mold.

bottom, of the press. Figure 10-11 illustrates the top plunger transfer. (3) The third method is the two-stage, reciprocating screw transfer method. In this case a conventionally designed hydraulic or toggle, top clamp press (i.e., downward pressing), with a bottom transfer cylinder and plunger, is used. The reciprocating screw and barrel is mounted horizontally, and material is discharged into the transfer chamber through an opening in the transfer sleeve or pot.

There are certain design features that must be considered for each method of transfer molding, and each method will be discussed separately.

The *pot method* of transfer molding gener-

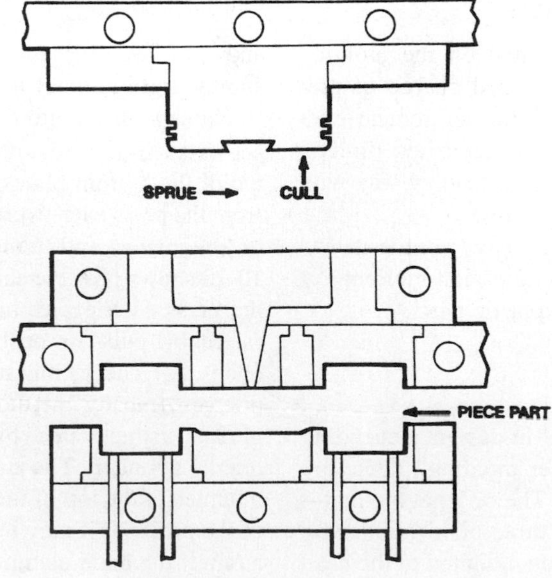

Fig. 10-10. Pot transfer mold.

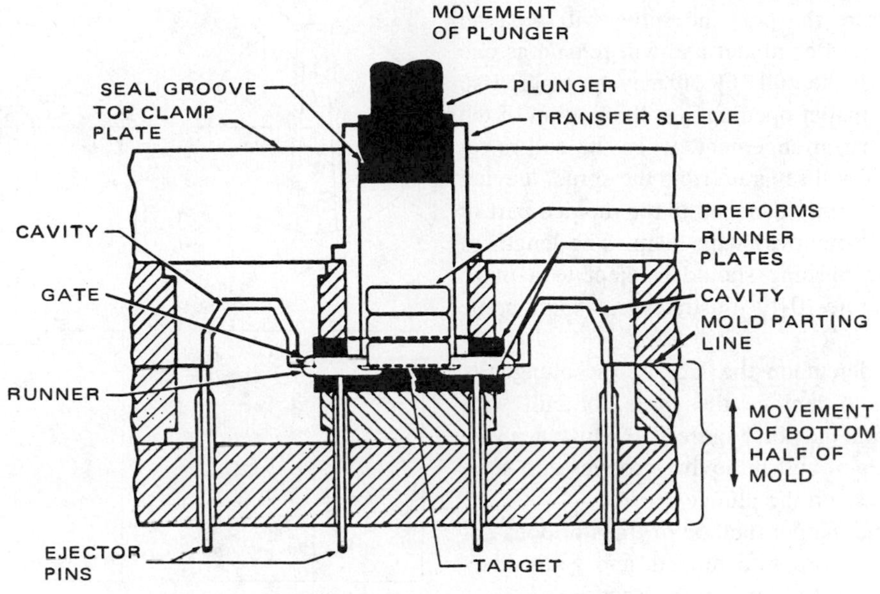

Fig. 10-11. Top plunger transfer mold.

ally is done in a conventional bottom ram clamp molding press. The molds usually are three-plate molds with the center section floating, as in Fig. 10-10.

Generally, the material is fed directly into the mold cavity or cavities from the pot through the sprue bushing. If multiple cavities are used, the sprues may come from a common pot, or multiple pots may be used. The principle may also employ a sprue into a runner system feeding one or more cavities.

After the mold layout of the individual cavity or cavities has been determined, the next step is to accurately calculate the projected area of the part or parts. If a runner system is used, its projected area must be added. The area, in square inches, of the pot or pots required to clamp the mold is subject to discussion.

It was considered good practice to make the area equal to or slightly greater than the total projected area of all articles to be molded, including the sprue and runners. However, because the material is advancing, it need not be treated as a hydrostatic liquid. Standard practice is to reduce the area in the pot to less than the above total area.

The next step is to determine the volume of charge to be transferred and to make the pot of sufficient depth to accept the preheated charge. It may be granules, preforms, or extruded material.

The mold design should provide for the pot or pots to be centrally located so that pressure is exerted uniformly in the center of the press.

The design of the steel retainer or mounting plate and section of steel containing the pot should be sufficiently strong to withstand the higher transfer pressures used to transfer material from the pot to the mold cavity. In order to prevent escapement of material around the plunger, the clearance between the plunger and the pot should be held to 0.002 to 0.003 inch on smaller-diameter plungers and 0.004 to 0.005 inch on larger-diameter or rectangular plungers. To ensure that there is no escapement of material, it is well to incorporate one or two sealing grooves, which may be $\frac{5}{32}$ to $\frac{3}{16}$ inch in width and $\frac{1}{16}$ to $\frac{3}{32}$ inch in depth and are generally half-round in cross section. A small radius is desirable at the bottom of the pot to provide extra strength to the pot and to assure that the cull remains on the force.

In pot transfer molding, the pot may have a sprue cut into the solid piece of steel; however, in most cases it is desirable to insert sprue bushings. The sprue bushing may be tapered so that the smaller opening is at the cavity or runner surface, which means that as the cull is pulled from the pot, the sprue will separate from the part or runner and will remain as one piece with the cull. Or, it may be tapered so that the smaller opening is at the bottom of the pot; in that arrangement, when the cull is removed, it will separate from the sprue, leaving the sprue firmly attached to the molded part or runner for removal with them. The length of the sprue bushing should be kept to a minimum. Figure 10-12 illustrates a standard sprue bushing.

An undercut on the face of the plunger is provided to remove the sprue, or cull with sprue, from the pot. Figure 10-3 illustrates various types of cull-removing grooves and sealing grooves on the plunger.

Because the pot method of transfer does not normally incorporate runners and gates, they will be discussed later in this section.

The second and more popular method of transfer molding is the *plunger transfer method.* In this method the press may be a toggle clamp or a full hydraulic. The clamping pressure may be applied from either the top or the bottom of the molding press, and it may vary from 25 to 500 tons. Individual or multiple hydraulic cylinders are mounted in opposite direction to the clamping pressure to provide the transfer pressure. Limited presses are available that have the transfer ram within the clamp ram. The clamp force in relation to transfer force is generally three or four to one (100 ton clamp/25–35-ton transfer). The majority of presses are constructed with a horizontal parting line; however, horizontally moving presses are available with vertical parting lines. The advantage of the latter is that no unloading mechanism is necessary to remove parts from the mold, the parts falling from the ejector pins by gravity.

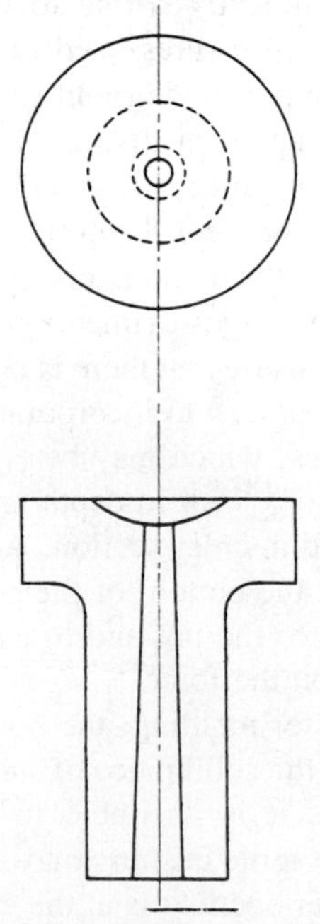

Fig. 10-12. Standard sprue bushing.

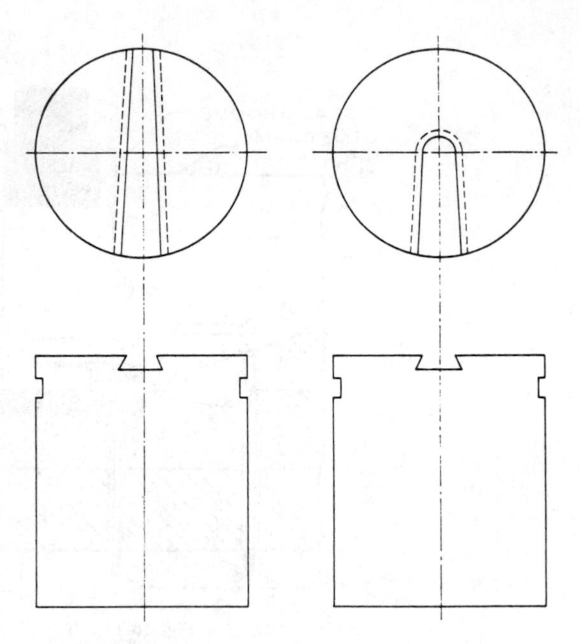

Fig. 10-13. Types of cull-removing grooves and sealing grooves on the plunger.

The plunger transfer method of molding can easily be designed for automatic molding, especially if the transfer ram is in the lower half of the mold. High-frequency preheated pills or extruded stock may be charged automatically (by the use of a loading tray) into the transfer pot. In case of a horizontal transfer cylinder, the preheated material is dropped into an opening on the upper side of the transfer sleeve or pot.

The clearance between the plunger and the transfer pot should be held to a minimum, and sealing grooves provided on the end of the plunger. Some transfer plungers are designed with removable tips to simplify maintenance.

Other designs incorporate a bronze sleeve on the plunger with no sealing grooves, permitting clearances between the plunger and the cylinder to be held very close. As the bronze sleeve wears, it can be replaced with a new sleeve at limited cost and minimum downtime for changeover. The cull developed between the face of the mold and the face of the transfer ram should be held to a minimum. The face of the mold and the plunger should be designed so as not to cut off the flow of material into the runner system.

In the design of plunger transfer molds, it is important that ample clamping pressure be provided to keep from flashing the mold. The surface of the mold should be relieved in those areas where a sealoff of part, runner, gate, and plunger is not necessary. This surface should be relieved $\frac{1}{32}$ to $\frac{1}{16}$ inch in depth. A sealoff surface, often termed a "land area," of $\frac{3}{16}$ inch width is sufficient, provided that landing pads are located in the area of the guide pins. Each mold half should be well supported with pillar supports to keep the center section of the mold from bowing or opening when under pressure of material. Any flashing of the mold will result in uneven pressure in the mold cavities and may prevent full "packing" pressure on the material if the plunger "bottoms out" before the cavities are filled.

The third method of transfer molding is known as the *two-stage, reciprocal screw method.* The method of transfer and the mold layout are similar to the plunger method of bottom ram transfer. Only one press manufacturer has produced this type of equipment in sizes of 150 and 300 tons, and no presses of this type were being produced in 1990. Several hundred such presses have been produced, however, and many are still in use. Such presses have a vertical toggle clamp, and the reciprocating screw mechanism is mounted horizontally, below the mold. The reciprocating screw is used as a means of preheating the charge of material, thus eliminating need for preform equipment and preheater. The material is preheated to 220 to 260°F and fed into the transfer pot or sleeve during the curing cycle. The operation is fully automatic, with parts being removed from the knockout pins by means of a comb. The parts may be degated from the cull and runner system, and then separated in one operation. If inserts are involved, a press can be provided with a rotating table to permit loading and unloading of one section of the mold while the other section is curing.

Injection Molding of Thermosets

Injection molding of thermosets has become popular during the last few years, and the process is rapidly displacing many compression and transfer molding installations.

The mold design for an injection mold is similar to that of a plunger transfer or two-stage reciprocal screw transfer method. The transfer pot or sleeve is replaced by a sprue bushing.

As most injection presses have a horizontal action and vertical cutoff area, knockout systems are somewhat different in this process, in that the parts and runner system fall by gravity from the mold area. It is not uncommon to have combs installed to assure that all molded material is removed from the mold area before the next cycle.

Faster molding cycles are obtained and higher mold temperatures are common, making injection molding more economical than the above-mentioned processes for a high rate of production. It is essential that more heating elements be provided to keep molds at the desired mold temperature. This is especially true with alkyd materials because their high specific gravity and mineral filler system tend to pull considerable heat from the mold.

Venting is extremely important because injection cavity fill times are faster than those in transfer molding.

Injection-Compression Molding of Thermosets

The process of injection-compression molding of thermosets combines features of both injection and compression molding. The mold contains a sprue bushing, which accepts heated molding compound from an injection barrel and nozzle. The mold is designed somewhat like a fully positive compression mold with a force plate (also containing the sprue bushing with wide opening toward the cavity plate) that enters partly into the cavities. In operation, the mold closes partway, within about $\frac{1}{4}$ inch of full close, and stops. The injection screw then forces material from the barrel through the nozzle and through the sprue bushing into the runner area and cavities. As soon as the full shot has flowed into the space between the mold halves, the clamp system is activated to close the mold the final $\frac{1}{4}$ inch of travel. Molding compound in the cavity or cavities thus is further densified, and, with the additional displacement of material during the final compression stroke, fiber orientation in the cavity or cavities will be less pronounced because of random movement, resulting in greater strength of the molded part. The process is used principally with fiber-reinforced molding compounds where optimum strength in the molded part is desired.

Runner and Gate Design

Molds for all types of transfer and injection molding should be designed for the ultimate in economical operations; use of reduced transfer and injection pressure; minimum waste material in cull, runners, and gates; maximum heat buildup to obtain short cure cycles; and ease of removing parts from the cavities, and gates and runners from the parts.

Ideally, the runners should be circular in cross section because such runners offer minimum resistance to flow and the most effective insulation of the center of the stock against the heat of the mold block, which tends to advance the cross-linking or polymerization of the material. Advanced material can cause uneven density, warpage, and poor surface appearance. If a full round runner cannot be incorporated, a trapezoidal cross section in one mold half only may be used. Avoid a half-round runner because it offers less effective insulation of the center of the flowing compound, and precuring may develop. Runners should be as short as possible to prevent waste and precure. One should design to avoid sudden changes in direction and with generous radii when changing direction of flow. The type of material, number of cavities, and part size and design all influence the size of the runners. In the design of the mold, it is recommended that the runner section be a separate section of the mold, hardened to 60 to 65 Rockwell C scale, and well polished. It is desirable to hard-chrome-plate this section of the mold to facilitate better flow and release and to prolong the life of the mold. Undercut wells may be provided with knockout pins below the wells to ensure holding the runner system on one side of the mold until ejection.

Gate design is extremely important when consideration is given to part density, warpage, shrinkage, weld lines, and overall physical strength. Gate size, shape, and location are responsible for material orientation, the pattern of flow, time and pressure to fill the cavity, and temperature of the material as it enters the cavity. In Figure 10-14, test specimens are being fed from constant cross-section runners through gates of various designs. Temperature of the molding compound rises from 270°F at the nozzle end of the barrel to as high as 351°F in the mold. The increase in temperature is caused by frictional heat developed in flow through the runner and the gate. Seven gate designs are illustrated—center, small fan, medium fan, "Y" with two gates, large fan, side on end, and edge. The large fan and edge gate on part of this design give the most uniform flow. The "Y" gate design is not recommended, as it produces parts filled with "ball joints" and weld lines.

Figure 10-15 shows a ring gate used to avoid weld lines and uneven density within the molded part, which causes warpage and internal stresses.

Figure 10-16 illustrates submarine or pin-point gates, which are becoming quite popular in transfer and injection molds. The main advantage of the pin-point gate is automatic degating. When pin-point gate design is used, the gate area must be a removable section of the mold. The pin-point gate will erode quite rapidly, and hardened gate sections may need to be replaced frequently. Erosion of the gate will depend on the type of filler used in the molding compound as well as cavity fill rates.

It is a good practice to gate into the heaviest

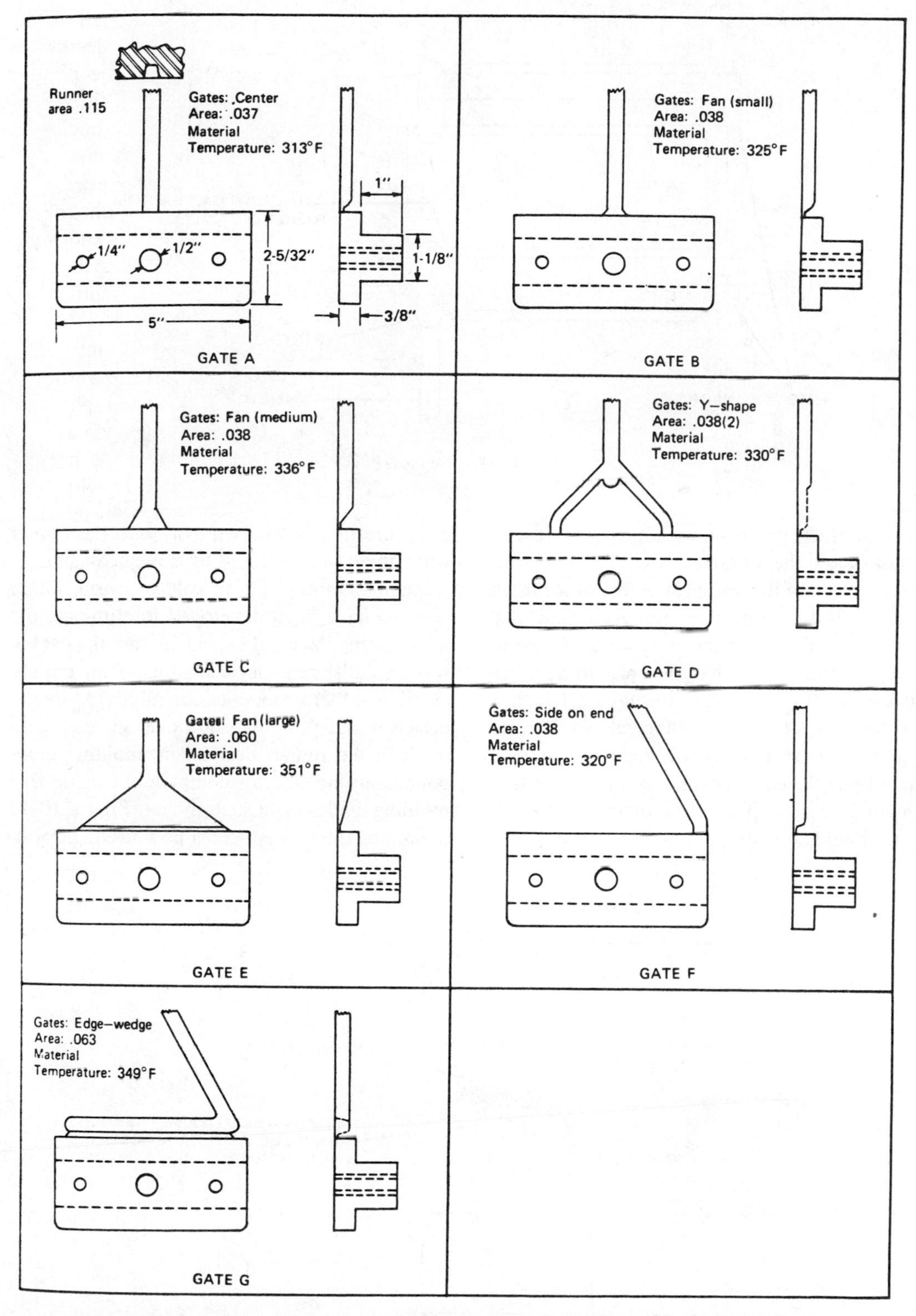

Fig. 10-14. Seven gate designs: (A) center; (B) small fan; (C) medium fan; (D) "Y" with two gates; (E) large fan; (F) side on end; and (G) edge. Areas given are in square inches and refer to the trapezoidal runner cross-section areas. The material temperature is measured as soon as the cavity is full.

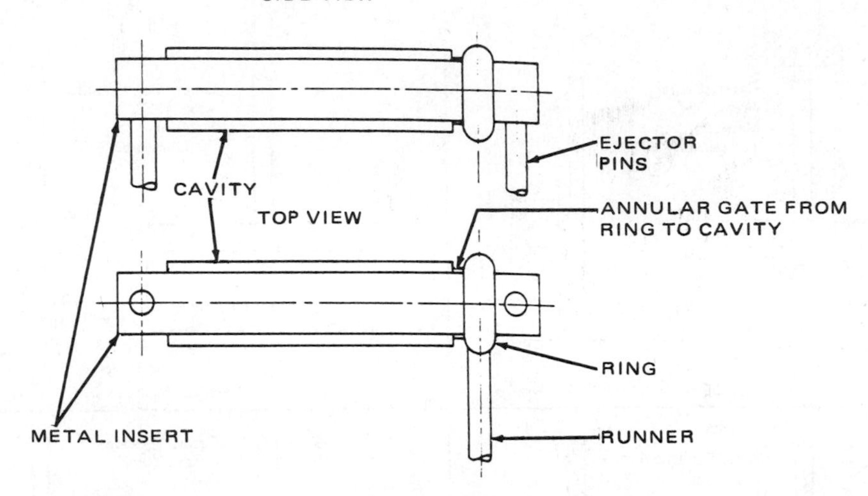

Fig. 10-15. Ring gate.

section of the molded part, if possible. Placing gates below the surface of the part will eliminate erosion of the top surface area of the mold. If the part design permits, removable gate sections should be incorporated in gate layout. They should be very hard in order to minimize gate wear. Frequently, a special steel such as tungsten carbide is used in the gate section. Removable gate sections provided in the mold may be replaced as needed to maintain a minimum gate size. The angle of the axis of the gate should be in excess of 45° from horizontal to ensure its positive removal with the runner when the runner is lifted by the ejector pins.

Runnerless molding or cold or warm runner systems have been developed to eliminate the large percentage of scrap material lost in sprues, culls, runners, and gates. They are applicable to thermoset injection molds. More experience and possibly additional development work in the runner design and molding compounds are needed for completely trouble-free molding cycles with such molds. Figure 10-17 illustrates the two concepts now used. In prin-

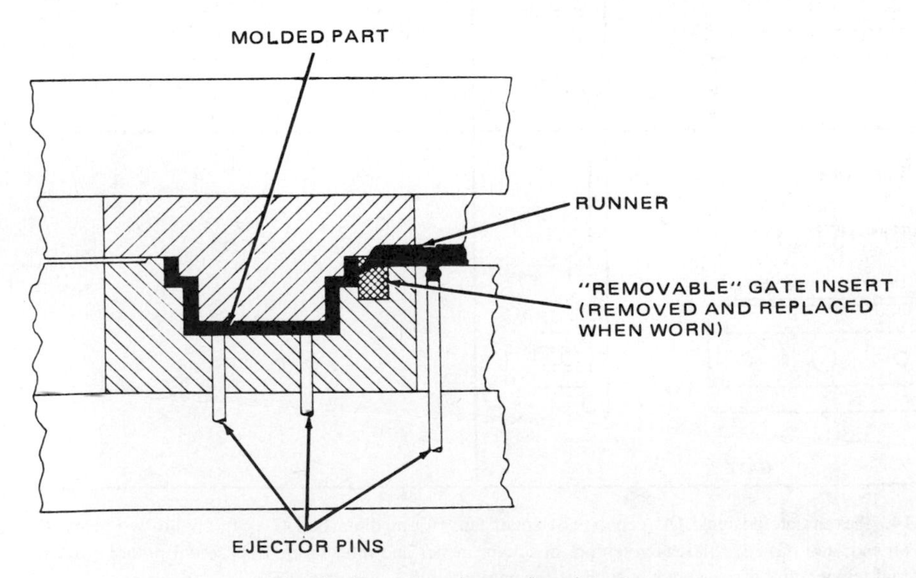

Fig. 10-16. Submarine or pin-point gate.

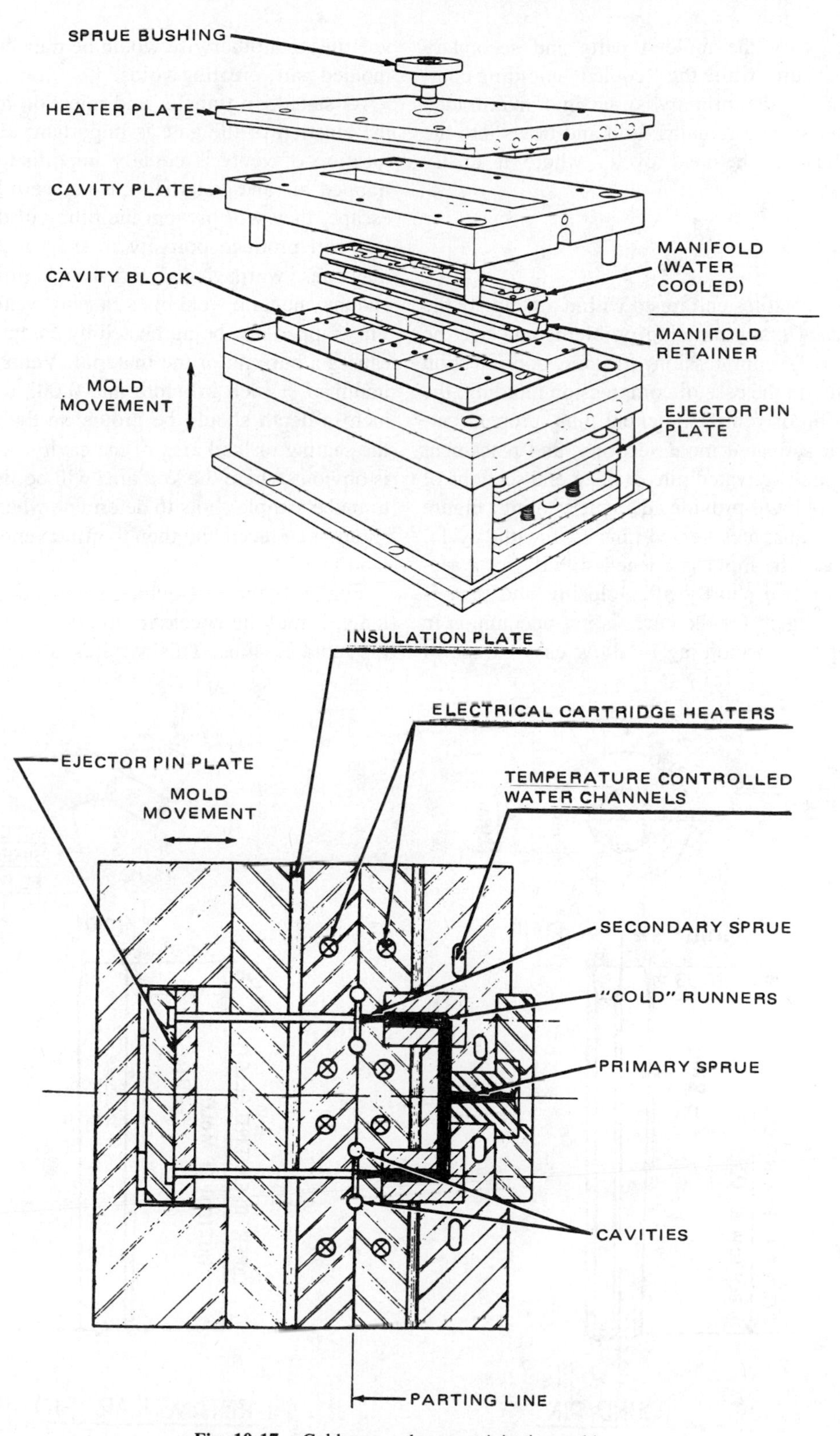

Fig. 10-17. Cold runner thermoset injection molds.

ciple, only the molded parts and secondary sprues cure, while the "cooled" molding compound in the primary sprue and manifolded runner system remains fluid and flows into the cavities on the next cycle, where it finally cures.

Vents

Air or volatiles entrapped within a mold are the cause of many molding problems and may be present in compression, transfer, and injection molds. In the case of compression molding, the location of vented knockout pins, proper venting at activated mold sections, and placement of vented activated pins at dead end sections of the mold will provide adequate venting. Figure 10-18 illustrates vented pins. A breathe cycle, wherein the mold is opened slightly for a second or two shortly after closing and then is closed again for the cure, is not uncommon in compression molding to allow easy escape of volatiles that otherwise would be trapped in the molded part, creating voids.

As stated, in transfer and injection molding the location of the gate is important; also, the location of vents is equally important. If entrapped air and volatiles are not permitted to escape, they will prevent the filling of the cavity, will produce porosity in the part, which may cause warpage and dimensional problems, and may contain weld lines causing weak parts. The trapped air, being heated by compression, causes a burning of the material. Vents a minimum of $\frac{3}{16}$ inch in width and 0.003 to 0.005 inch in depth should be ground in the face of the sealing or land area of the cavity. Unless it is obvious where the knit area will be, it is best to make sample shots to determine where vents should be placed and then to grind vents on the land area.

For particularly troublesome venting problems, it may be necessary to apply vacuum to the mold cavities. This requires all vents from

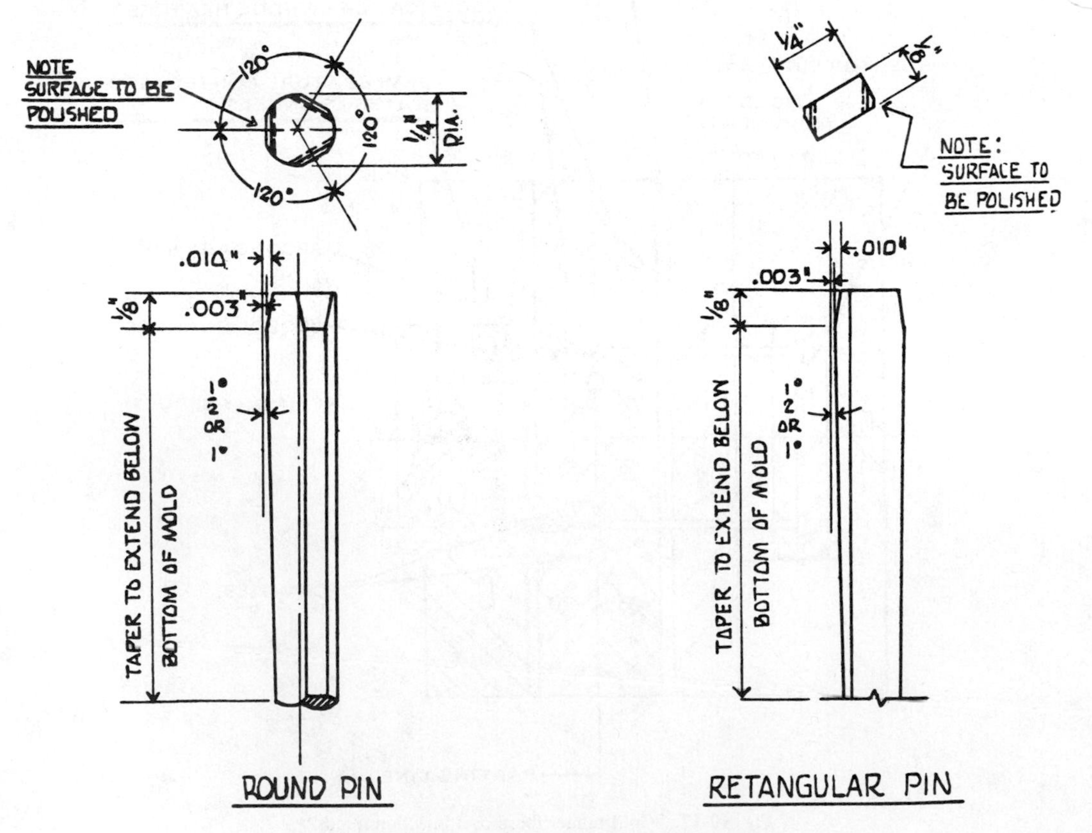

Fig. 10-18. Method for venting knockout pins.

the cavities to discharge into an air groove or passage leading to a vacuum pump and reservoir, and to have the entire cavity area enclosed within the confines of a Teflon or silicone rubber gasket or O-ring located between parting surfaces of the mold. An ample vacuum reservoir tank must be provided to pull 15 to 25 inches of vacuum immediately as the material enters and fills the cavities.

General Mold Design Considerations

All mold components contacted by the molding compound—runners, gates, cavities, land areas—should be made of through-hardened tool steels, hardened to 65 to 68 on the Rockwell C scale, highly polished, and hard-chrome-plated.

Because most thermoset compounds are slightly soft at the time of ejection from cavities, ejector pins should have an adequate cross-sectional area to minimize the possibility of distorting or puncturing the molded plastic at this point in the cycle.

In automatic molds, it is vital to ensure, with part design, undercuts, or hold-down pins, that the molded parts, during mold opening, consistently remain in the desired half of the mold (usually the moving half), so that when the parts and the runner system are ejected, the comb or extractor will always "find" them and effect positive removal.

Flash removal from the mold, each cycle, is critical for successful automatic molding (the "flash-free molds" are myths). Every effort must be made to have the flash ejected with the molded part. An air blast, directed appropriately to cavities and land areas each molding cycle when the mold is open, frequently is incorporated into the mold or press to further ensure the absence of flash each cycle. Should flash begin to accumulate in or on the mold, the mold should be cleaned, polished, and/or repaired.

Molds should be of uniform temperature in the cavities, and should have adequate heating capacity to ensure maintenance of the desired temperatures despite continual heat extraction by the relatively cooler molding compound each cycle. Temperature sensors and heating cartridges must be judiciously placed to provide this uniformity of temperature. Insulating blankets may prove helpful to minimize mold heat losses and variations due to local air currents around molds and presses. To minimize local temperature variations in large molds, heating cartridges often are grouped in zones, with each zone having its own temperature controller and sensor. Sensors should be positioned within $\frac{1}{4}$ inch of heating cartridges to prevent significant temperature overswings.

It also is prudent to provide for a "mold overtemp" sensor, which will cut off power to the heating cartridges whenever it senses a mold temperature more than a few degrees over the desired mold temperature. Excessive mold temperatures not only will result in reject parts, but also may anneal the mold steels and warp critical mold components.

11
Molded Part Design

Considering the enormous diversity of plastics materials and the processes by which they are made into products, it becomes a formidable task to try to summarize in a single chapter any basic rules of product design. Although this chapter concentrates on products made by the injection or compression molding process, readers should note that different parameters apply when the part is blow-molded or thermoformed or made by any of the reinforced plastics molding processes. Wherever possible, this type of information has been included in the specific chapter pertaining to an individual process (readers are especially referred to the excellent design charts in Chapter 18).

However, this guide to the fundamental principles of molded product design is useful as a starting point for one's understanding of how plastics differ from other materials and how they can best be approached in achieving an economical and serviceable design. Again, the reader should be aware of the physical differences in the many plastics materials and the influences these differences will have on design criteria. Obviously, for example, a flexible plastic such as polyethylene represents less of a problem in the design of a piece with undercuts (because it can be snapped out of the mold) than a rigid plastic such as polystyrene. Similarly, the molding of structural foam plastics can call different criteria into play from those applicable to the molding of solid plastic.

Bearing these variations in mind, in this chapter we shall attempt to formulate the basic ground rules for designing a molded product. The first section will deal with the mechanical properties of plastics and how they determine part performance. The main body of the chapter will concentrate on the details of design criteria (walls, ribs, tapers, bosses, threads, etc.). The last section of the chapter is a discussion of designing plastic parts to avoid warpage, which is included so that readers may better understand the relationship between part design, mold design, and molding conditions, in terms of their influence on the finished product.

The reader is also urged to contact material suppliers for available literature on the subject of design. Considerable data are available from these sources to help designers working with specific plastics. Recommendations for molding tolerances can be found in Chapter 28.

STRUCTURAL DESIGN CHARACTERISTICS OF PLASTICS

The purpose here is to describe some of the basic behavior characteristics of plastics materials as they affect the structural design of plastics parts.*

Mechanical Properties—The Stress–Strain Curve

In the tensile testing of plastics, three distinct types of stress–strain curves may be observed, illustrated in Fig. 11-1. Stress–strain curves Types A and B illustrate materials that have gradual and abrupt yielding. The Type C diagram shows a material that fails before yielding occurs.

*Portions of this section courtesy of the Du Pont Company.

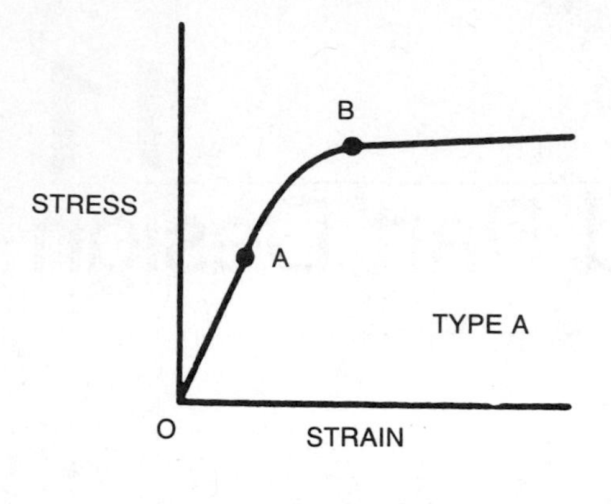

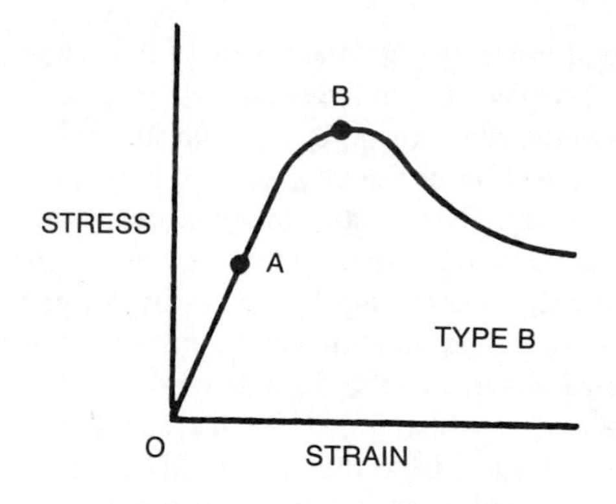

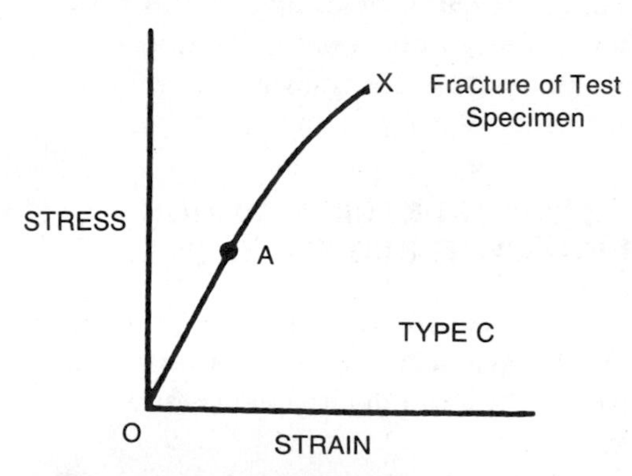

Fig. 11-1. Stress–strain curves. (A) Gradual yielding; (B) abrupt yielding; (C) fracture occurring at low strains before yielding. (*All sketches in "Structural Design Characteristics," courtesy of The DuPont Company*)

The designer usually is concerned with two distinct regions of the stress–strain curve. In the first region, designated by the line *OA*, elastic design principles may be applied. The second region, around Point *B*, which will be referred to as the yield point, can be used when fracture or failure due to large deformation is of prime concern. In the following discussion, both of these regions are considered in detail.

Proportional limit and elastic limit are two terms used to describe the point of termination of the linear portion of a stress–strain curve, the limit of Hooke's law. This is about Point *A*. However, a careful study of stress–strain curves for thermoplastics has shown that there is no linear section; rather, there is deviation from linearity from the origin. The deviation is very slight below 0.5% strain, and most published figures of the elastic modulus, E_o, are the slope of a line tangent to the low-strain portion of the stress–strain curve.

The secant modulus, E_s, is the slope of a line drawn from the origin to a point on the curve; the secant modulus decreases with higher strains. A plot of E_s/E_o emphasizes that application of the initial modulus becomes less accurate with increasing strain. This is shown in Fig. 11-2 for three different plastics. Thus a part in service may deflect more than calculations indicate because linearity is assumed in deriving equations for deformation.

With thermoplastics there is no "limit"; rather, we are concerned with the degree of deviation with increasing strain, and a concept of modulus accuracy is needed. Hence, we have chosen the term "modulus accuracy limit." As an arbitrary choice, an accuracy of 15% was selected as an allowable error in most design calculations.

On a specific design problem, after the deflection is calculated, the strain or the stress should be calculated. If the modulus accuracy limit is not exceeded, the calculation is sufficiently accurate. Otherwise, the geometry of the part must be changed to reduce the strain (or stress), or a correction must be applied to the calculation. Each designer and each application will have specific needs for accuracy, which are related to the concept and the use of a safety factor. It is through an understanding and application of a modulus accuracy limit that these specific needs are met.

It is convenient that strain at the 15% modulus accuracy limit turns out to be essentially a material constant regardless of temperature or loading time.

It is possible that beyond a limiting strain value, marked inaccuracies in calculations for creep deflection and stress relaxation may oc-

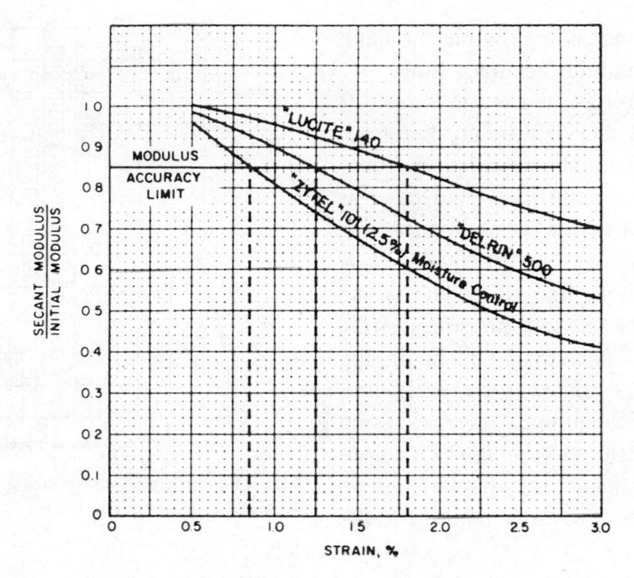

Fig. 11-2. Decrease of ratio of secant modulus to initial modulus with increasing strain (73°F, 23°C) for three different plastics (top to bottom): acrylic, acetal, and nylon 6/6.

cur. This possibility emphasizes the importance of using the modulus accuracy limit concept as a checkpoint until, or unless, future data show the limit is not necessary.

In order to define the yield point in plastics of Type A, as shown in Fig. 11-1, the true stress-logarithmic strain curve first must be calculated from the conventional stress–strain curve. In the conventional stress–strain curve, strain is calculated in terms of the original length of the test specimen, and stress is calculated on the basis of the original cross-sectional area. Because the cross-sectional area of the test specimen changes during a tensile test, the true stress-logarithmic strain curve, which is based on instantaneous dimensional changes, is more meaningful. The equations that convert conventional strain, ϵ, to logarithmic strain, $\bar{\epsilon}$, and conventional stress, S, to true stress, $\bar{S}$, are:

$$\bar{S} = S(1 + \epsilon) \tag{1}$$

and:

$$\bar{\epsilon} = \ln(1 + \epsilon) \tag{2}$$

The following examples illustrate how these equations may be used.

Example 1. A specimen is stressed 5000 psi at a strain of 1%. What are the corresponding true stress and logarithmic strain?

$$\text{True stress} = \bar{S} = 5000\,(1 + 0.01) = 5050 \text{ psi}$$

Logarithmic strain

$$= \bar{\epsilon} = 2.303 \log_{10}(1 + 0.01) = 0.0099 \text{ or } 0.99\%$$

Example 2. Based on the original dimensions, a specimen is stressed to 10,000 psi at a strain of 10%. What are the corresponding true stress and logarithmic strain?

True stress

$$= \bar{S} = 10{,}000\,(1 + 0.1) = 11{,}000 \text{ psi}$$

Logarithmic strain

$$= \bar{\epsilon} = 2.303 \log_{10}(1 + 0.1) = 0.095 \text{ or } 9.5\%$$

By applying Equations (1) and (2), the true stress-logarithmic strain curves are constructed from the conventional stress–strain curve. Examples 1 and 2 show that the differences between the conventional and the true stress-logarithmic strain curves are substantial for relatively large strains and negligible for small strains within the modulus accuracy limit.

In design terminology, there are many different definitions of the yield point; therefore, the danger of confusion always exists. For materials that yield gradually, the yield point is determined from the true stress-logarithmic strain curve as is shown in Figs. 11-3, 11-4, and 11-5. The straight line *BC* is the work-hardening region of the tensile stress–strain curve, and point *B*, which marks the beginning of work hardening, is defined as the yield point. For materials that yield abruptly, the yield point is defined as the maximum or hump of the conventional stress–strain curve, as shown in graph B of Fig. 11-1. The stress corresponding to the

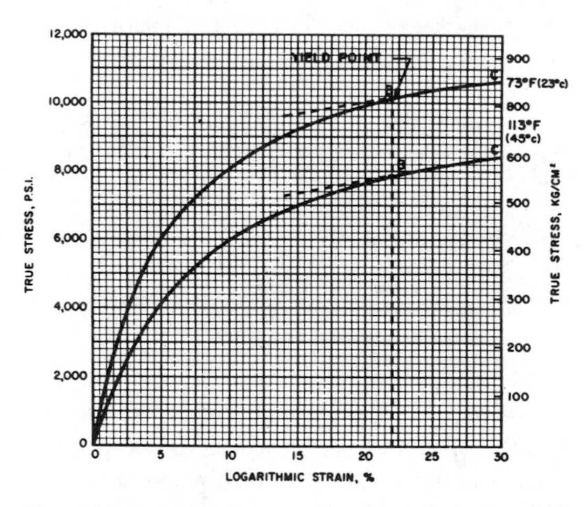

Fig. 11-3. Tensile stress–strain curves for nylon 6/6 (2.5% moisture content).

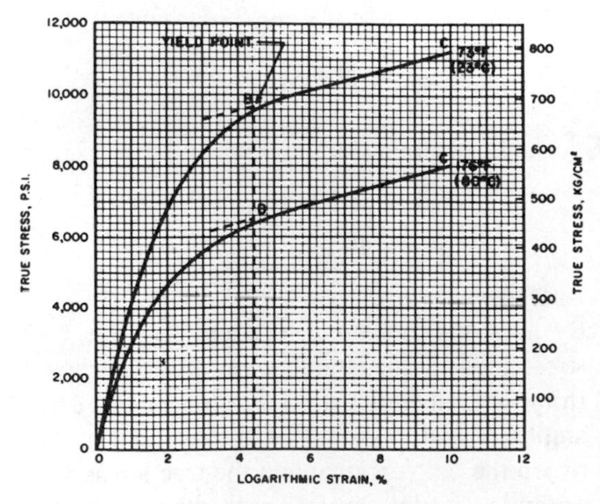

Fig. 11-4. Tensile stress–strain curves for acetal.

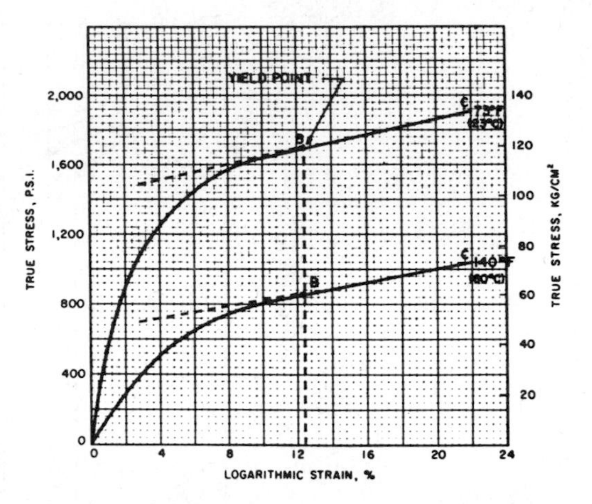

Fig. 11-5. Tensile stress–strain curves for low-density polyethylene.

yield point is called the yield-stress. Although this definition for materials that yield gradually differs from the ASTM definition, it is used in most texts on the theory of plasticity.

Figures 11-3, 11-4, and 11-5 are the true stress-logarithmic strain curves for nylon 6/6 (2.5% moisture content), acetal, and low-density polyethylene, measured at standard ASTM loading rates. These curves illustrate how the yield point may be determined. The yield strains are shown to be 22, 4.5, and 12.5% logarithmic strain, respectively. If a test specimen of a ductile plastic material is extended to relatively large strains, the test specimen will neck down, and the cross-sectional area of the specimen becomes uncertain. The stress–strain curve beyond the yield point is of little value to the design engineer except that it adds confidence that stress concentrations can be distributed without failure by fracture.

Effect of Temperature, Loading Time, and Environment

Tensile testing of materials at ordinary temperatures has shown that most metals are characterized by one type of stress–strain curve. This is not the case for plastics, which are, in general, more sensitive to temperature, rate of testing, and environmental conditions.

Nylon 6/6 is a good example. The testing rate and the moisture content can change the type of stress–strain diagram, thereby changing the basic design properties such as yield stress. It is essential that the designer of plastic parts know under what conditions changes in material behavior may occur and adapt the design procedure accordingly.

It is an established fact that the yield stress increases with the rate of testing. In many practical applications the load is applied at speeds that are considerably faster than the ASTM recommended testing rates; therefore, the material properties that are obtained from standard tests sometimes are misleading.

In some calculations, it is important for the designer to know the stress–strain curves in tension and compression. For example, in flexural design problems both types of stress–strain curves are needed because tensile and compressive stresses are present in the structure. Furthermore, various materials are considerably stronger in compression than in tension (e.g., concrete). It has been shown that for large strains the compressive strength properties of various plastics are substantially higher than the corresponding tensile properties.

The tensile and compressive stress–strain curves for several plastics are given in Figs. 11-6, 11-7, and 11-8. From these curves, which were obtained at ASTM strain rates, the following conclusions can be drawn:

1. The tensile and compressive stress–strain curves are, for all practical purposes, identical at small strains. Therefore, the modulus in compression is equal to the modulus in tension.
2. It follows that the flexural modulus is equal to the tensile modulus.
3. For relatively large strains, the compressive stress is higher than the corresponding tensile stress. This means that the yield stress in compression is greater than the yield stress in tension.

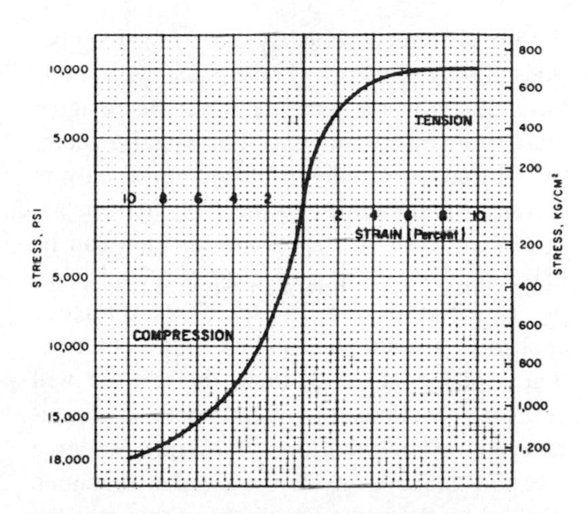

Fig. 11-7. Conventional stress–strain curves in tension and compression of acetal (73°F, 23°C).

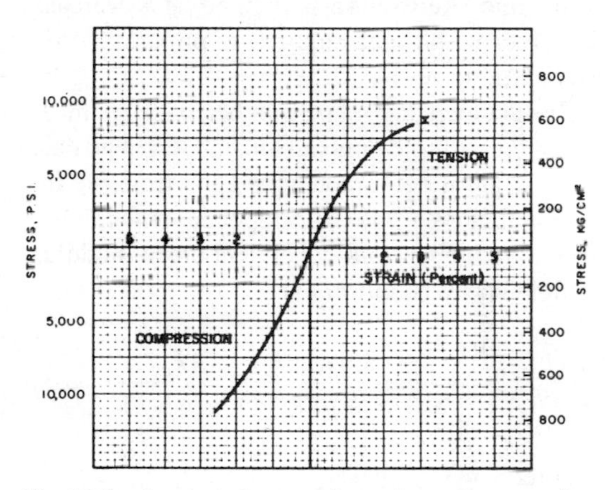

Fig. 11-8. Stress–strain curves in tension and compression of acrylic (73°F).

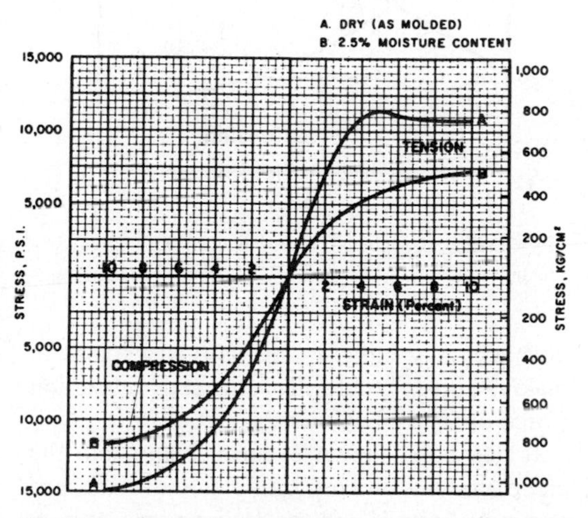

Fig. 11-6. Conventinal stress–strain curves in tension and compression of nylon 6/6 (73°F, 23°C).

Creep and Relaxation

The phenomena of creep and relaxation in plastic materials are of prime concern to the designer of plastic parts that, in use, can be only slightly deformed yet carry a load for long periods of time. Creep will take place even at low loads and low temperature, and the designer must estimate its amount. For metals, analogous behavior usually is encountered only at elevated temperatures; so these phenomena are of less concern. Creep and relaxation can be illustrated by considering a test bar that is simply loaded in tension. In Fig. 11-9a, a force is applied to a specimen, and an initial deformation is observed. With time, the test bar will permanently elongate. Creep is defined as nonrecoverable deformation, with time, under a constant load. Consider a similar specimen stressed by the same force, but in this case the specimen is confined to the initial deformation. As is shown in Fig. 11-9b, the force decreases with time. Relaxation is defined as a decrease of force, with time, required to produce a constant strain.

Basic creep and relaxation behavior are measured by a simple uniaxial tension test. The data usually are given graphically, and from the individual curves, the time-dependent moduli are calculated. The time-dependent creep modulus is defined by equation (3):

$$E_{t\,\text{creep}} = \frac{\text{Stress}}{\text{Strain at time } (t)} = \frac{F_o/A_o}{\Delta L_t/L_o} \qquad (3)$$

where:

F_o = Constant force, lb
A_o = Original cross-sectional area, in.
L_o = Original length, in.
ΔL_t = Increase in length at time (t), in.

The time-dependent relaxation modulus is similarly defined:

$$E_{t\,\text{relax}} = \frac{\text{Stress at time } (t)}{\text{Strain}} = \frac{F_t/A_o}{\Delta L_o/L_o} \qquad (4)$$

where:
A_o and L_o are as before
ΔL_o = Initial deformation, in.
F_t = Force at time (t), lb

CREEP

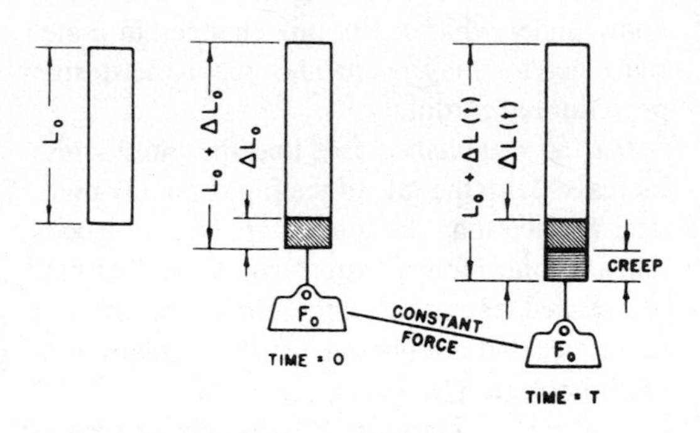

RELAXATION

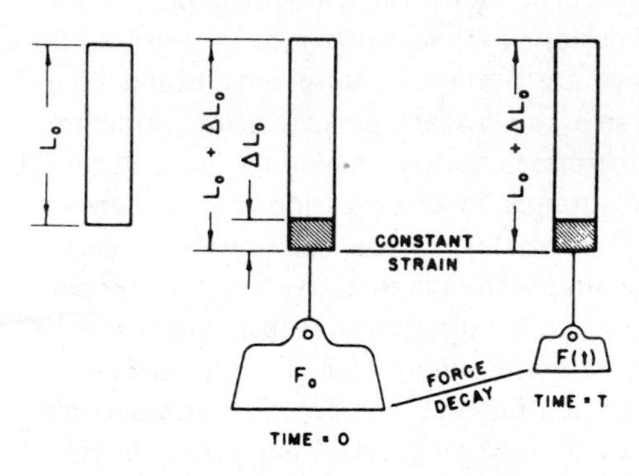

Fig. 11-9a. Creep and relaxation in tension.

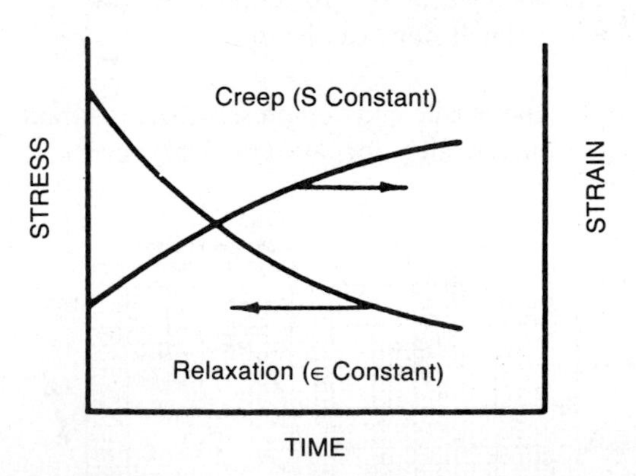

Fig. 11-9b. Graphical representation of creep and relaxation data.

Tests so far have confirmed that equations (3) and (4) are applicable when the initial strain is within the modulus accuracy limit. The modulus accuracy limit is used as a checkpoint in making creep (deformation) and relaxation (force decay) calculations.

It has been shown experimentally that the creep modulus and relaxation modulus are similar in magnitude and for design purposes may be assumed to be the same; that is, $E_{t\,\text{creep}} = E_{t\,\text{relax}}$. Therefore, only one time-dependent modulus, called the apparent modulus is necessary. The following examples illustrate the similarity between creep and relaxation moduli.

Example 1. Determination of the apparent modulus from a creep experiment would be accomplished by placing a test bar in tension under a load of 95 lb. The bar has an original gauge length of 5.0 in. and a cross-sectional area of 0.063 sq. in. After 1 year, the increase in length is 0.025 in.

$$\text{Stress (constant)} = \frac{95}{0.063} = 1500 \text{ psi}$$

$$\text{Strain after 1 year} = \frac{0.025}{5.0} = 0.005 \text{ or } 0.5\%$$

$$E(1 \text{ year})_{\text{creep}} = \frac{\text{Stress}}{\text{Strain after 1 year}} = \frac{1500}{0.005} = 300{,}000 \text{ psi}$$

Example 2. Similar information would be obtained by elongating a specimen of the same dimensions to 0.5% strain, maintaining a constant strain, and recording the force decay. After 1 year the force decreases to 95 lb.

$$\text{Strain after 1 year} = \frac{95}{0.063} = 1500 \text{ psi}$$

$$\text{Strain (constant)} = 0.005$$

$$E(1 \text{ year})_{\text{relax}} = \frac{\text{Stress after 1 year}}{\text{Strain}} = \frac{1500}{0.005} = 300{,}000 \text{ psi}$$

In using apparent modulus in design procedure, in general, the standard strength of materials design formulas are used. However, the modulus chosen is an apparent modulus selected on the basis of the time that the part is under load.

Design Methods

The major difference between metals and plastics design lies in the choice of structural properties to be used in standard design theory and formulas. For metals, these properties are relatively constant over wide ranges of temperature, time, and other measures of the environment. For plastics, structural properties are more sensitive to comparable changes in environment.

The well-known stress distribution of a simple beam in flexure is the background for the recommended methods for using the concepts discussed above in plastics design. When a beam is flexed, two distinct regions of stresses are obtained, which are separated by a neutral plane. On one side of the neutral plane the fibers are in tension, while on the other side the fibers are in compression. Since an idealized stress–strain curve in tension and compression is symmetrical and linear at small deformations, the stress distribution in the beam in flexure also is symmetrical about the neutral plane, as is shown in Fig. 11-10a. This type of linear distribution, which exists only at relatively small deflections, is assumed in the derivation of the standard elastic design equations.

When a beam is flexed to carry maximum load, the stress distribution in the beam approaches the rectangular shape shown in Fig. 11-10b. This can occur only if the material is ductile and does not fracture before this distribution is reached. It is important to note that the neutral plane is no longer in the center of the beam because the yield stress in compression is greater than the yield stress in tension. Yield design equations, derived on the basis of this type of behavior for metals, can be modified for plastics. The basic assumption is that a structure will fail when all the fibers have

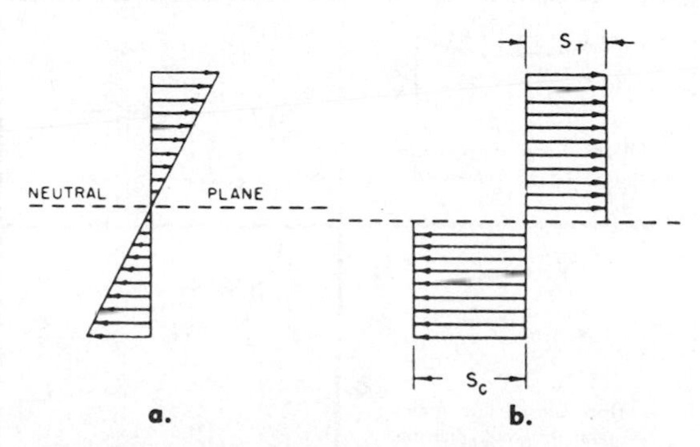

Fig. 11-10. Stress distribution in flexure: (a) linear (elastic) stress distribution; (b) stress distribution maximum load.

yielded. In both those sections, illustrations with simple design calculations are intended to familiarize the reader with the general procedure.

Designing within the Modulus Accuracy Limit

Elastic design equations that were developed for metals can be applied to plastics. The deflection equations are expressed in terms of two material variables, the elastic (Young's) modulus and Poisson's ratio, whereas the stress equations are dependent only on load and geometry. These formulas can be converted to the appropriate time-dependent equations by replacing the elastic modulus with the apparent time-dependent modulus and assuming Poisson's ratio to be a constant.

For practical purposes, Poisson's ratio is constant with time. The other major assumption is that the stress distribution is initially linear and remains linear with time.

The following example illustrates this method of designing within the modulus accuracy limit; terms in the equations are defined in Fig. 11-11.

Example (radial displacement of a pipe under internal pressure). In the construction of plastic piping networks, it sometimes is important to know the radial displacement of the wall of a pipe that is under internal pressure for a period of time. The radial displacement and the mean hoop stress of a thin closed-end pipe under internal pressure may be calculated by using the following equations:

$$\text{Radial displacement} = Y = \frac{R}{E_t}\left(1 - \frac{\mu}{2}\right)\frac{P_j R}{t} \quad (5)$$

$$\text{Mean hoop stress} = S_b = \frac{P_j R}{t} \quad (6)$$

Consider, for example, a pipe made of nylon 6/6 conditioned to 2.5% moisture content. The

Geometry	Deflection Equation	Stress Equation	Nomenclature
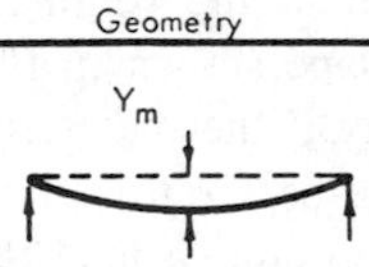 Y_m Centrally Loaded Beam	$Y_m = \frac{PL^3}{48 E_t I}$	$S = \frac{3PL}{2bd^2}$	P = Load L = Span d = Depth b = Width I = Moment of Inertia E_t = Apparent Modulus S = Outer Fiber Stress Y_m = Maximum Deflection
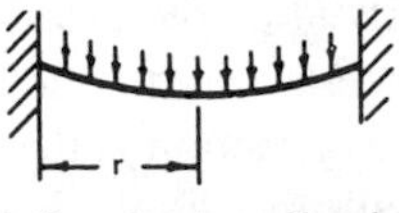 r Uniform Load on Circular Plate (Edges Fixed)	$Y_m = \frac{3p r^4 (1-\mu^2)}{16 E_t t^3}$	$S = \frac{3p r^2 (1+\mu)}{8 t^2}$	p = Load per Unit Area t = Thickness r = Radius μ = Poisson's Ratio S = Maximum Stress
 Uniform Load on Circular Plate (Edges Supported)	$Y_m = \frac{3p r^4 (5 - 4\mu - \mu^2)}{16 E_t t^3}$	$S = \frac{3p r^2 (3+\mu)}{8 t^2}$	p = Load per Unit Area t = Thickness r = Radius μ = Poisson's Ratio S = Maximum Stress
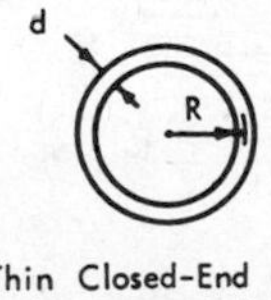 d R Thin Closed-End Pipe Under Internal Pressure	$Y = \frac{R}{E_t}\left(1 - \frac{\mu}{2}\right)\frac{P_i R}{t}$	$S_h = \frac{P_i R}{t}$	t = Wall Thickness R = Mean Radius P_i = Internal Pressure Y = Radial Displacement S_h = Hoop Stress

Fig. 11-11. Typical formulas for designing within the modulus accuracy limit.

pipe contains compressed air at 73°F and 50% relative humidity and is under a pressure of 150 psi for 5 years.
Geometry:

Mean radius of pipe $= R = 0.05$ in.

Wall thickness $= t = 0.10$ in.

Physical constants at 73°F after 5 years:

Apparent modulus

$$= E_t(5 \text{ years}) = 100{,}000 \text{ psi}$$

Poisson's ratio $= 0.40$

$$Y = \frac{0.50}{10 \times 10^4}\left[(1 - 0.20)\frac{150 \times 0.50}{0.10}\right]$$

$$= 0.0029 \text{ in.}$$

$$S_h = \frac{150 \times 0.50}{0.10} = 750 \text{ psi}$$

$$\epsilon = \frac{S_h}{E_o} = \frac{750}{175{,}000} = 0.0043 \text{ in./in. or } 0.43\%$$

The calculation shows that the radius of the pipe would increase by 2.9 mils in 5 years. The radial displacement is a realistic estimate because both the initial stress and the strain did not exceed the modulus accuracy limit for this specific nylon (1400 psi, 0.85%).

The preceding example applies only to calculating deformation. In biaxial stress systems, fracture due to excessive stress is another factor that must be considered in the overall design. Examples that illustrate the calculation of burst stress are given in the following discussion on yield design.

Yield Design

The previous discussion considered the deformation of structures under load. In ultimate load determinations, the designer is concerned primarily with the stress required for failure. Accurate design can be accomplished only if the effect of time and temperature on the mechanical properties is known.

Standard formulas used in limit design are found in various textbooks dealing with the theory of plasticity. The equations can be used in designing structures made of plastics if the yield stress at time of failure is known. Experimental evidence justifying this procedure has been obtained in a number of cases. The simplest example is the determination of the maximum load that can be sustained by a centrally loaded beam with supported ends. Failure occurs when all the fibers in tension reach the yield stress in tension, and all the fibers in compression reach the yield stress in compression (Fig. 11-12). The following equation is used to calculate the maximum load:

$$P_{\max} = K\frac{bh^2}{L}$$

where:

$$K = \frac{2S_t - S_c}{S_t + S_c} \tag{7}$$

and b, h, and L are the width, depth, and span, respectively, S_t is the yield stress in tension, and S_c is the yield stress in compression.

At present, the greatest use of this method is made in predicting the failure of pipe and pressurized containers. In order to obtain the solutions to these types of multiaxial stress problems, the von Mises-Hencky yield condition is applied to plastics:

$$(S_1 - S_2)^2 + (S_2 - S_3)^2 + (S_3 - S_1)^2 = 2S_t^2(t)$$

where S_1, S_2, and S_3 are the three principal stresses and $S_t(t)$ is the yield stress in tension at the time of failure.

DESIGNING THE PART

To ensure proper design, close cooperation is required between the industrial designer, the engineer, the draftsman, the tool-builder, the molder, and the raw material supplier. Prefer-

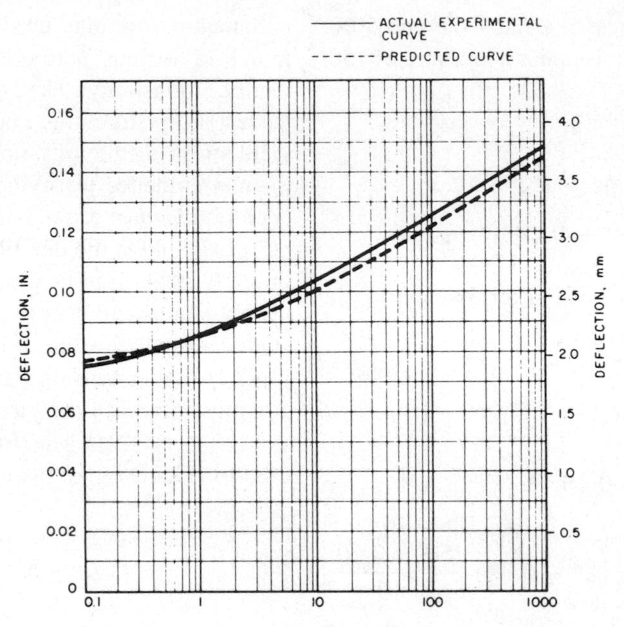

Fig. 11-12. Deflection of a centrally loaded beam with time (acetal, 73°C, end supports).

ably, each must become involved to some extent at the very beginning of the design process. A general stepwise procedure for the development of a commercial part would be about as follows:

• *Defining the end-use requirements:* As an initial step, the designer should list the anticipated conditions of use and the performance requirements of the article to be designed. Then the designer may determine the limiting design factors and, by doing so realistically, avoid various pitfalls that can cause loss of time and expense at a later stage of development. Use of the checklist in Table 11-1 will be helpful in defining the various design factors.

• *Drafting the preliminary design:* With the end-use requirements meaningfully defined, the designer is ready to start developing designs using the properties of the plastic being considered.

• *Prototyping the design:* This gives the designer an opportunity to see the product as a three-dimensional object, and the first practical check of the engineering design. Perhaps the most widely used technique for making prototypes is to machine them from rod or bar stock. In some cases, it may be preferable to build a single-cavity sample mold. The sample cavity technique not only allows economical production of a quantity of prototype parts but also serves to develop mold design data for building the production mold.

More recently, a prototype-making process called stereolithography has been developed, related to computer-aided design (CAD) technology. Using data for part dimensions, the design is "sliced" into sections that are traced by a laser on the surface of a liquid polymer. The laster light cures and solidifies the polymer, and the prototype is thus built up, layer by layer, in a matter of a few hours.

• *Testing the design:* Every design should be subjected to some form of testing while in the prototype stage to check the accuracy of calculations and basic assumptions. There are several types of testing:

1. Actual end-use testing of a part in service is the most meaningful kind of prototype testing. Here, all of the performance requirements are encountered, and a complete assessment of the design can be made.
2. Simulated service tests often are conducted with prototype parts. The value of

Table 11-1. Checklist for product design.

General Information
- What is the function of the part?
- How does the assembly operate?
- Can the assembly be simplified by using plastics?
- Could it be made and assembled more economically?
- How will it be made and assembled?
- What tolerances are necessary?
- What space limitations exist?
- What service life is required?
- Is wear resistance required?
- Is light weight desirable?
- Are there acceptance codes and specifications such as SAE, UL?
- Do analogous applications exist?

Structural Considerations
- How is it loaded in service?
- Magnitude of loads?
- For how long will it remain in service?
- How much deflection can be tolerated in service?

Environment
- Operating temperature?
- Chemicals, solvents?
- Humidity?
- Service life in the environment?

Appearance
- Style?
- Shape?
- Color?
- Surface finish?
- Decoration?

Economic Factors
- Cost of present part?
- Cost estimate of part in plastic?
- Will redesign of the part simplify the assembled product and thus give rise to savings in installed cost?
- Are faster assemblies and elimination of finishing operations possible?

this type of testing depends on how closely the end-use conditions are duplicated. An automobile engine part might be given temperature, vibration, and hydrocarbon resistance tests; a luggage fixture could be subjected to impact and abrasion tests; and a business machine component might undergo tests for electrical and thermal insulation.

3. Accelerated tests of a mechanical or chemical nature often are used as a basis for prototype evaluation. Such procedures, when used by an experienced and qualified person, can be very meaningful.
4. Standard test procedures such as those developed by the ASTM generally are useful as a design guide, but they normally cannot be drawn upon to predict accurately the performance of a part in service. Again, representative field testing may be indispensable.

• *Taking a second look:* A second look at the design helps to answer this basic question: "Will the product do the right job at the right price?" Even at this point, most products can be improved by redesigning for production economies or for important functional and aesthetic changes; weak sections can be strengthened, new features added, and colors changed. Substantial and vital changes in design may necessitate complete evaluation of the new design. If the design has held up under this close scrutiny, specifications and details of production can be established.

• *Writing meaningful specifications:* The purpose of a specification is to eliminate any variations in the product that would prevent it from satisfying functional, aesthetic, or economic requirements. The specification is a complete set of written requirements that the part must meet. It should include such things as: method of fabrication, dimensions, tolerances, surface finish, parting line location, flash, gating, locations where voids are intolerable, warpage, color, decorating, and performance specifications.

• *Setting up production:* Once the specifications have been carefully and realistically written, molds can be designed and built to fit the processing equipment. Tool design for injection molding should be left to a specialist or an able consultant in the field because inefficient and unnecessarily expensive production can result from improper design of tools or selection of manufacturing equipment.

• *Controlling the quality:* It is good inspection practice to schedule regular checking of production parts against a given standard. An inspection checklist should include all the items that are pertinent to satisfactory performance of the part in actual service. The end-user and the molder should jointly establish the quality control procedures that will facilitate production of parts within specifications.

BASIC DESIGN THEORY

In the sections that follow, individual design criteria relating to the placement of ribs, bosses, radii, and so on, are discussed in detail. (For a quick review of some design dos and don'ts, readers are referred to Fig. 11-13.)

Designers should be aware, however, of the uniqueness of plastic materials as used in molding processes. Chemically, the molecules of thermoplastics consist of long chains of repeating units. When melted and injected under high pressures into a closed mold, the polymer withstands forces and undergoes changes. Injection molding has been compared by some to stuffing coil springs into a cavity. If the cavity has generally rounded and uniform contours, it is relatively easy to fill. If, however, it has sharp corners and thick-and-thin areas, it not only will be more difficult to fill, but, when filling is completed, the springs will be more compressed, stretched, and distorted. When the analogue of this happens in molding plastics, a part is said to contain molding strains; though undesirable, such strains are present to some degree in every plastic part. When a molding is heated, it warps because it is pulled in different directions by the strains within it and is less able, because of softening (when the temperature is raised sufficiently), to maintain its shape.

When the polymer is injected into every portion of the mold cavity with about equal force and uniformly cooled, the distribution of internal stresses will tend to reach a balance and yield a part with less tendency to warp. High levels of mold strain, on the other hand, are especially detrimental when the part is subjected to further external strain or stress, whether it be physical force, heat, or a stress crack agent (i.e., certain materials that disturb the intermolecular bonds of strained molecules and provide an ''opening wedge'' for a crack to develop).

Wall Thickness

Under favorable conditions, the design of wall thickness normally depends upon the selection of the material. Occasionally, however, limitation of space precludes this, and the selection of material becomes predicated partly upon the wall thickness available. Whichever the path of approach, the determination of wall thickness should be the result of an analysis of the following requirements.

Requirements of use	*Manufacturing requirements*
1. Structure	1. Molding
2. Weight	(a) Flow
3. Strength	(b) Setting
4. Insulation	(c) Ejection
5. Dimensional stability	2. Assembly
	(a) Strength
	(b) Precision

The foregoing requirements are intimately related. From a purely economic standpoint, a wall thickness that is too great or too small can affect the design adversely.

Plastic parts should be designed with the minimum wall thickness that will provide the specific structural requirement. This thickness results in an economy of material and high production levels due to the rapid transfer of heat from the molten polymer to the cooler mold surfaces. Occasionally, the use of the article may involve little need of strength, but still adequate strength must be provided to withstand ejection from the mold and to facilitate assembly operations.

Tables 11-2 and 11-3 give preferred minimum, average, and maximum wall thicknesses for the various thermoplastics and thermosetting plastics.

It should also be remembered that molding phenomena, such as flow and cure, can influence the choice of wall thickness. Basically, wall thickness should be made as uniform as possible to eliminate part distortion, internal stresses, and cracking. Figure 11-14 illustrates the incorporation of uniform section thickness

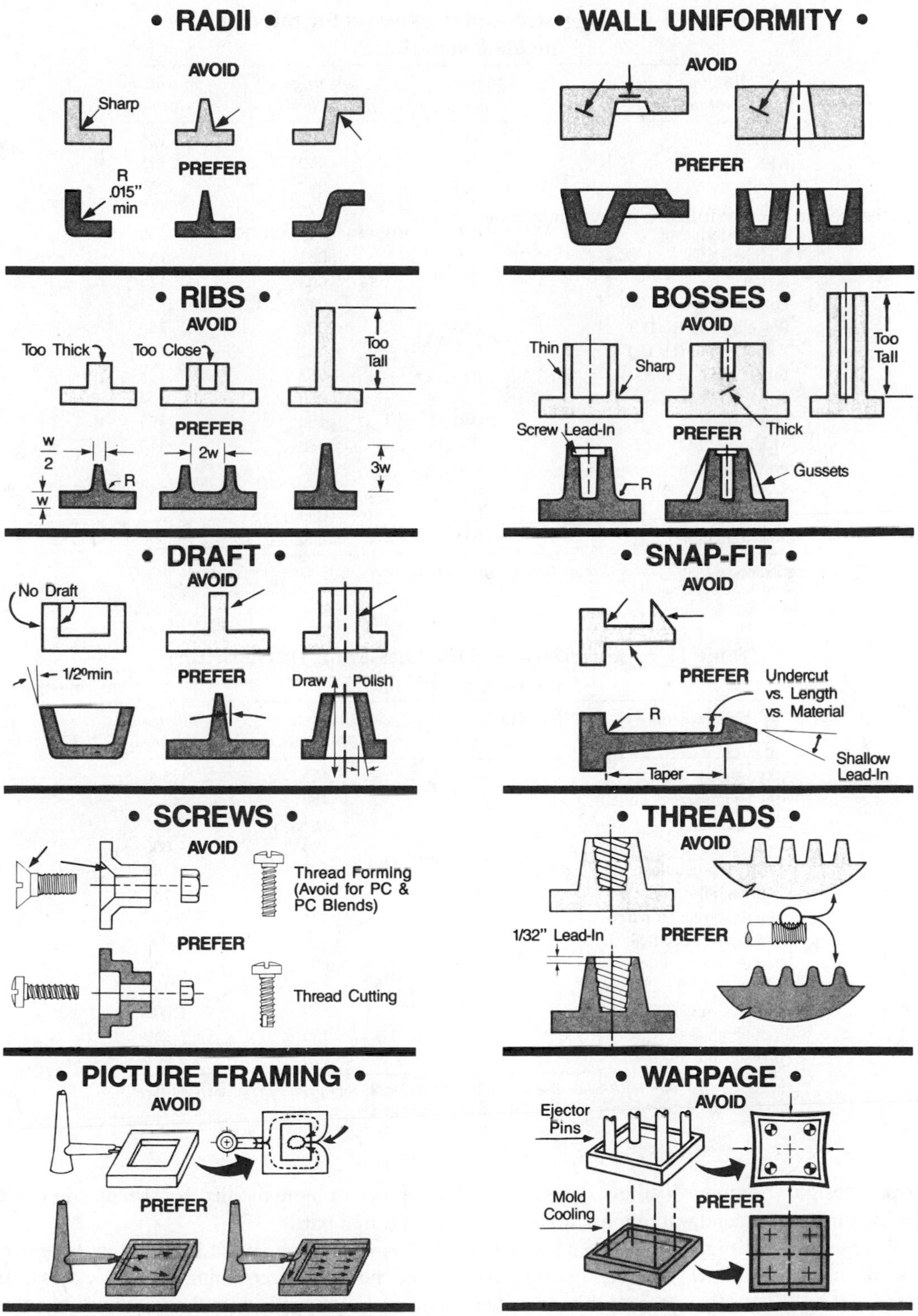

• RADII •

AVOID

Sharp

PREFER

R
.015"
min

• WALL UNIFORMITY •

AVOID

PREFER

• RIBS •

AVOID

Too Thick

Too Close

Too
Tall

PREFER

w
2

2w

R

w

3w

• BOSSES •

AVOID

Thin

Sharp

Too
Tall

Screw Lead-In

PREFER

Thick

R

Gussets

• DRAFT •

AVOID

No Draft

1/2°min

PREFER

Draw

Polish

• SNAP-FIT •

AVOID

PREFER

Undercut
vs. Length
vs. Material

R

Taper

Shallow
Lead-In

• SCREWS •

AVOID

Thread Forming
(Avoid for PC &
PC Blends)

PREFER

Thread Cutting

• THREADS •

AVOID

1/32" Lead-In

PREFER

• PICTURE FRAMING •

AVOID

PREFER

• WARPAGE •

Ejector
Pins

AVOID

Mold
Cooling

PREFER

Fig. 11-13. Some design dos and don'ts. (*Courtesy Mobay Corporation, Pittsburgh, PA*)

Table 11-2. Suggested wall thicknesses for thermoplastic molding materials.*

Thermoplastic Materials	*Minimum Inches*	*Average Inches*	*Maximum Inches*
Acetal	.015	.062	.125
ABS	.030	.090	.125
Acrylic	.025	.093	.250
Cellulosics	.025	.075	.187
FEP fluoroplastic	.010	.035	.500
Ionomer	.025	.062	.750
Nylon	.015	.062	.125
Polycarbonate	.040	.093	.375
Polyethylene (L. D.)	.020	.062	.250
Polyethylene (H. D.)	.035	.062	.250
Ethylene vinyl acetate	.020	.062	.125
Polypropylene	.025	.080	.300
Polysulfone	.040	.100	.375
Modified PPO	.030	.080	.375
Polystyrene	.030	.062	.250
SAN	.030	.062	.250
PVC—rigid	.040	.093	.375
Polyurethane	.025	.500	1.500

*Reprinted from *Plastics Product Design*, by Ronald D. Beck, with permission of the publishers, Van Nostrand Reinhold Co.

Table 11-3. Suggested wall thicknesses for thermosetting molding materials.*

Thermosetting Materials	*Minimum Thickness*	*Average Thickness*	*Maximum Thickness*
Alkyd—glass filled	.040	.125	.500
Alkyd—mineral filled	.040	.187	.375
Diallyl phthalate	.040	.187	.375
Epoxy/glass	.030	.125	1.000
Melamine—cellulose filled	.035	.100	.187
Urea—cellulose filled	.035	.100	.187
Phenolic—general purpose	.050	.125	1.000
Phenolic—flock filled	.050	.125	1.000
Phenolic—glass filled	.030	.093	.750
Phenolic—fabric filled	.062	.187	.375
Phenolic—mineral filled	.125	.187	1.000
Silicone glass	.050	.125	.250
Polyester premix	.040	.070	1.000

*Reprinted from *Plastics Product Design*, by Ronald D. Beck, with permission of the publishers, Van Nostrand Reinhold Co.

in part design. If different wall thicknesses must be used in a part, blend wall intersections gradually, as shown in Fig. 11-15. When possible, the material should flow from thick to thin sections (filling thick sections from thin ones can result in poorly molded parts). Also, under such circumstances, consideration should be given to the use of assembly techniques for assembling two or more molded components to make the desired part.

As indicated earlier, it is important to understand flow in determining wall thickness. If consideration is given to flow within the mold, one can see that it is possible for material to flow past a pocket or depression, leaving this area to be filled later as pressure builds up. It

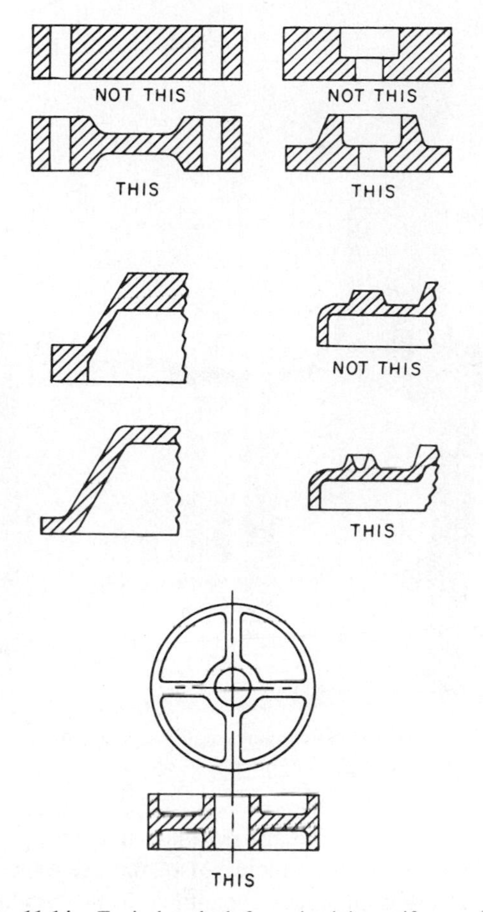

Fig. 11-14. Typical methods for maintaining uniform wall thickness.

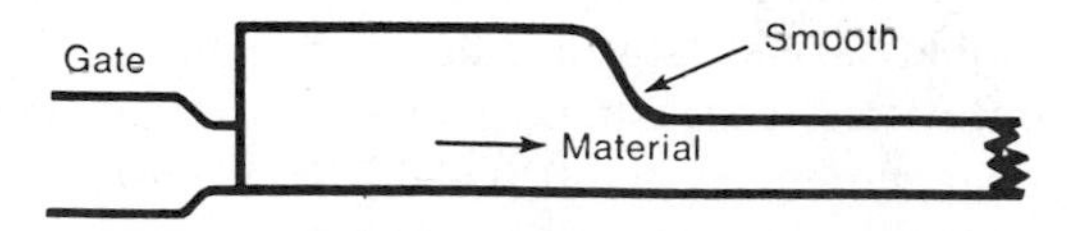

Fig. 11-15. Gradual blending between different wall thicknesses. (*Courtesy GE Plastics*)

is also possible, because of restrictions of flow, for material momentarily to hesitate until sufficient pressure has been developed to overcome resistance to flow, and then to spurt forward, causing part of the material to be chilled. Remixing with the oncoming melt can result in surface defects in both cases.

The formation of welds relates directly to wall thickness and is an important controlling factor. In molding terminology, a weld is created when two fronts of molten material meet, such as when the flow is interrupted by a pin. If the interruption is long or if the wall is thin, poor flow may cause a weak weld.

Flow within the mold could be severely restricted in walls thinner than the minimums recommended in Tables 11-2 and 11-3, and the mold might not fill properly. At the other extreme, voids or sink marks could develop in very thick cross sections.

Another factor to be taken into consideration in determining wall thickness is the process of curing—a function of heat transfer, from or to the mold. Obviously, thin sections set more rapidly than thick ones, and, because contraction or shrinkage occurs simultaneously with setting, irregularity in thickness causes irregularity in contraction and creates internal stresses. These stresses will tend to relieve themselves, either by forming concave depressions, known as sink marks, on the thick sections, or by causing warping. This difficulty frequently can be eliminated by coring a thick section so as to divide it into two thin sections, when that is feasible.

Core removal should be designed parallel to the motion of the mold platen. Cores at right angles to the platen movement require cam or hydraulic actions, which increase mold cost. Movable and loose-piece cores are expensive, but they can be used to mold internal undercuts and threads.

Core size generally is determined by part design. As an example, the following are recommendations on core design in polycarbonate materials: The recommended core length for blind cores larger than 0.187 inch in diameter is 2.5 times the diameter. For blind cores smaller than 0.187 inch in diameter, the suggested core length is twice the diameter. Coring recommendations are summarized in Fig. 11-16. The length of cored-through holes should not exceed six times the diameter for core diameters greater than 0.187 inch and four times the diameter for core diameters smaller than 0.187 inch. Draft should be added to all cores. If the polycarbonate parts require deep holes of small diameter, it may be necessary to drill the holes rather than mold them. Drilling usually proves more satisfactory than trying to maintain the alignment of fragile core pins. Also, if threads are required, the holes can be

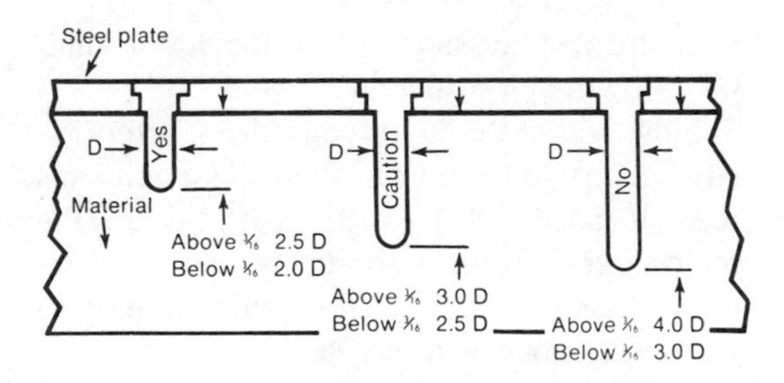

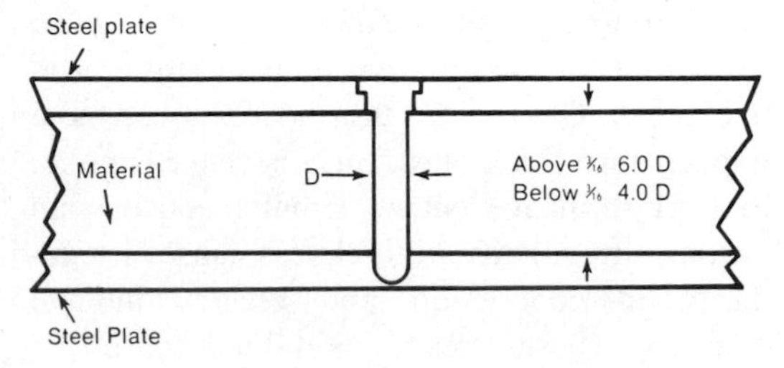

Fig. 11-16. Coring recommendations for polycarbonate. (*Courtesy GE Plastics*)

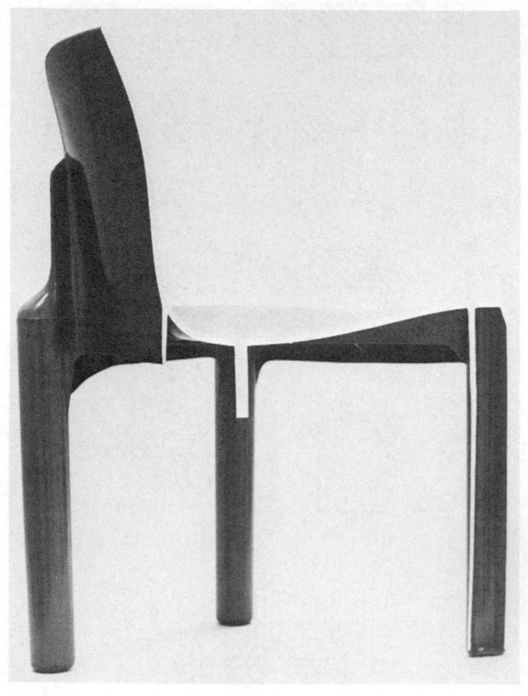

Fig. 11-18. Cross-section of wall design in molded ABS chair. (*Courtesy GE Plastics*)

tapped during the drilling setup. (See Fig. 11-17.)

Similar recommended practices for other plastics are available from appropriate suppliers. Figures 11-18 and 11-19 are included to indicate how these wall thickness selection criteria are put to successful commercial use in the design of ABS chairs and grilles.

Fillets and Radii

Sharp corners in plastic parts are perhaps the greatest contributors to part failure. Elimination of sharp corners reduces the stress concentration at these points and produces a molding with greater structural strength. Fillets provide streamlined flow paths for the molten polymer in filling the mold and permit easier ejection of the part from the mold. Also, the use of radii and fillets produces economies in the production of the mold because they are easier to machine and less subject to damage than sharp corners. All inside and outside rounded corners should have as large a radius as possible to reduce stress concentrations. The recommended minimum radius of 0.020 to 0.030 inch usually is permissible even where a sharp edge is required, and the radius ensures more economical, longer-life molds. A larger radius should be specified wherever possible.

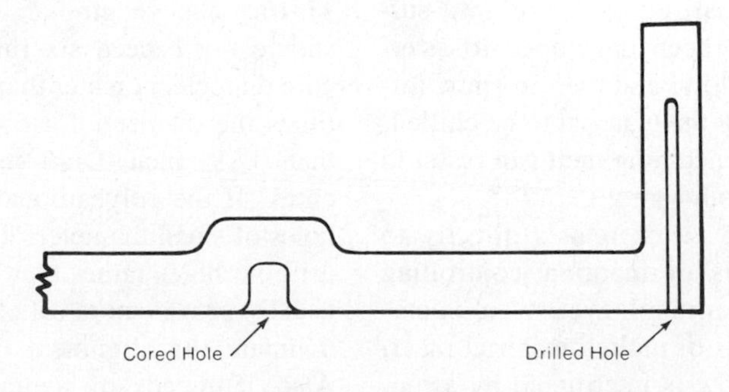

Fig. 11-17. Coring-out thick wall sections. (*Courtesy GE Plastics*)

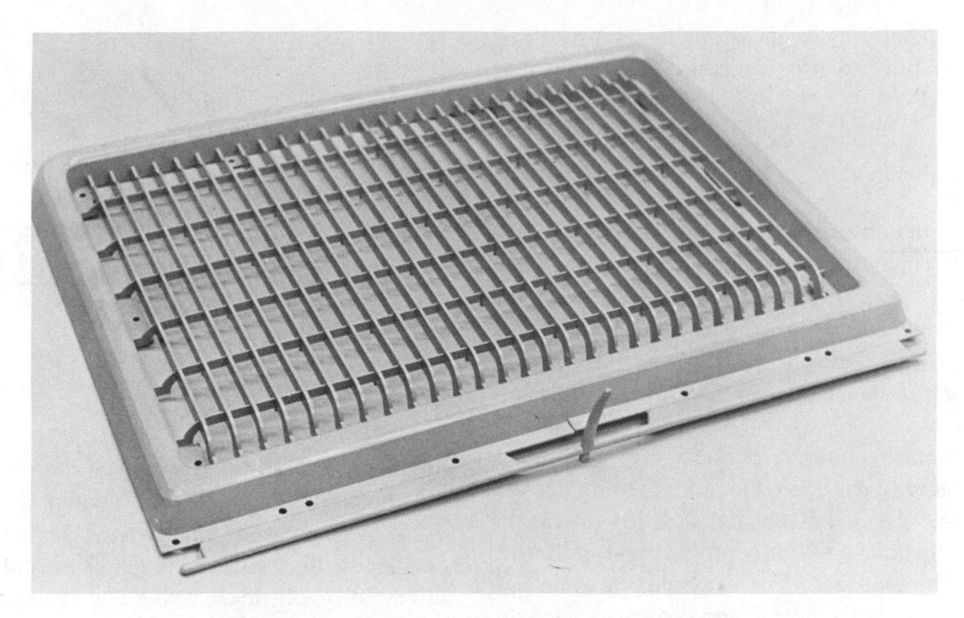

Fig. 11-19. Different wall thicknesses in a molded ABS grille. (*Courtesy GE Plastics*)

Figure 11-20 illustrates the effect of a fillet radius on stress concentration. Assume that a force P is exerted on the cantilever section shown. As the radius R is increased, with all other dimensions remaining constant, R/T increases proportionally, and the stress-concentration factor decreases, as shown by the curve. The stress-concentration factor has been reduced by 50% (3.0 to 1.5) by increasing the ratio of fillet radius to thickness sixfold (from 0.1 to 0.6). The figure illustrates how readily the stress-concentration factor can be reduced by using a larger fillet radius. A fillet of optimum design is obtained with an R/T of 0.6, and a further increase in radius reduces the stress concentration only a marginal amount. This is true in general for most shapes; however, other ratios may have to be used on specific parts because of other functional needs.

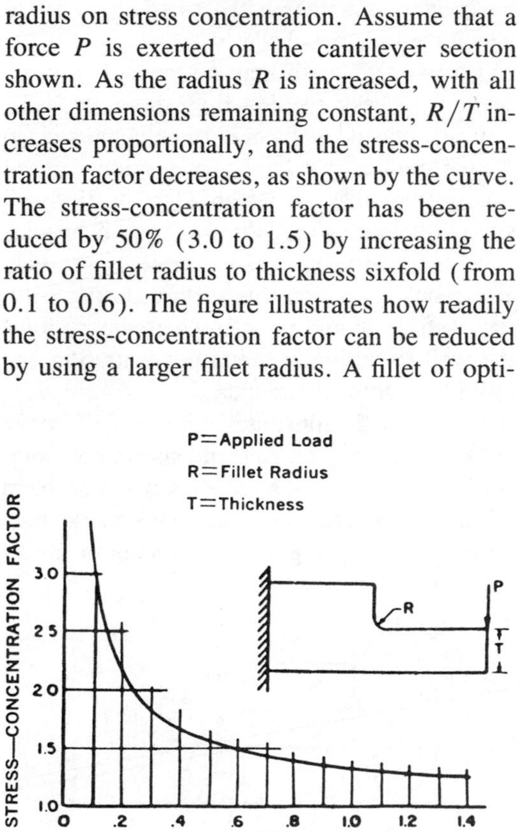

P=Applied Load
R=Fillet Radius
T=Thickness
STRESS—CONCENTRATION FACTOR
3.0
2.5
2.0
1.5
1.0
0 .2 .4 .6 .8 1.0 1.2 1.4
R/T
P
R
T

Fig. 11-20. Effect of a fillet radius on stress concentration. (*Courtesy The DuPont Company*)

The overall advantages of radii and fillets in the design of a molded article are several:

1. The molded article is stronger and more nearly free from stress.
2. Elimination of the sharp corner automatically reduces the hazard of cracking due to notch-sensitivity and increases the overall resistance to sudden shock or impact. Elimination of all internal sharp corners by using a radius of 0.015 or $\frac{1}{64}$ inch will greatly improve the strength.
3. The flow behavior of the plastic will be greatly improved. Rounded corners permit uniform, unretarded, and less stressed flow of plastic into all sections of the mold, and will improve the uniformity of density of the molded sections.
4. Mold members, force plugs, and cavities will be stronger because there will be less tendency for these parts of the mold to develop internal stresses due to notch-sensitivity created by concentration of stress at sharp corners. Mold parts fre-

quently crack in hardening as a result of failure to provide appropriate radii and fillets.

Ribs

The function of ribs is to increase the rigidity and strength of a molded piece (such as the ABS postal tray shown in Fig. 11-21) without increasing wall thickness. The proper use of ribs usually will prevent warpage during cooling, and, in some cases, they facilitate flow during molding.

Several features in the design of a rib must be carefully considered in order to minimize the internal stresses associated with irregularity in wall thickness. Width, length, and so forth, must be analyzed. For example, in some applications, thick heavy ribs can cause vacuum bubbles or sink marks at the intersection of mating surfaces and will result in appearance problems, structural discontinuity, high thermal stresses, and stress concentration. To eliminate these problems, long, thin ribs should be used. It also is possible to core ribs from the underside.

In general, the width of the base of a rib will be less than the thickness of the wall to which it is attached. This can be demonstrated by examination of Fig. 11-22. Ribs of the proportions shown frequently are used in molded articles. However, when the circle R_1 is placed at the junction of the rib and the wall, it will be seen that the thickness is increased by about 50%, and this may produce sink marks on the surface under normal molding conditions. By

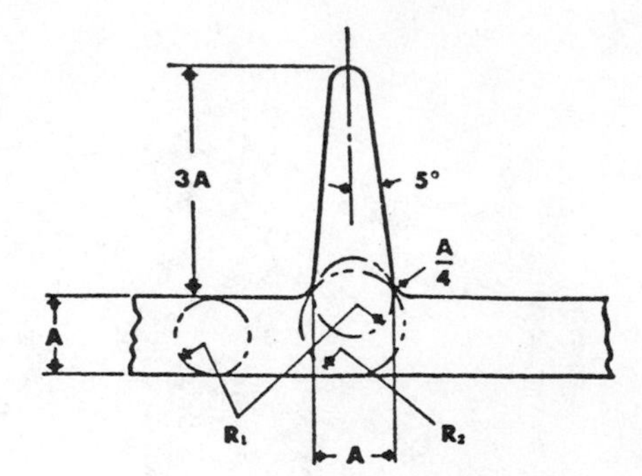

Fig. 11-22. Wide rib base results in sink marks.

reducing the width of the base of the rib one-half the wall thickness, as shown in Fig. 11-23, the increase in thickness at the junction is made less than 20%. Sink marks are unlikely with these proportions. Using two or more ribs is better than increasing the height of a single rib more than is shown. When two or more ribs are required, the distance between them should be greater than the thickness of the wall to which they are attached.

The base of a rib may be tapered in cross section for easy ejection from the mold. Ribs or beads on side walls may be perpendicular to the parting line to ensure easy ejection. Spaced ribs or a groove should be added to the wall surface behind a bead of any size, if appearance is important, because large beads may cause sink marks on this surface during cooling. A fillet should be used where the rib joins the wall to minimize stress concentration and provide additional strength.

Figure 11-24 illustrates the effect of heavy beads on the part surface and shows one solution (the bottom cross section) to the problem of surface imperfections (i.e., sink marks may be disguised by engraving emblems or corru-

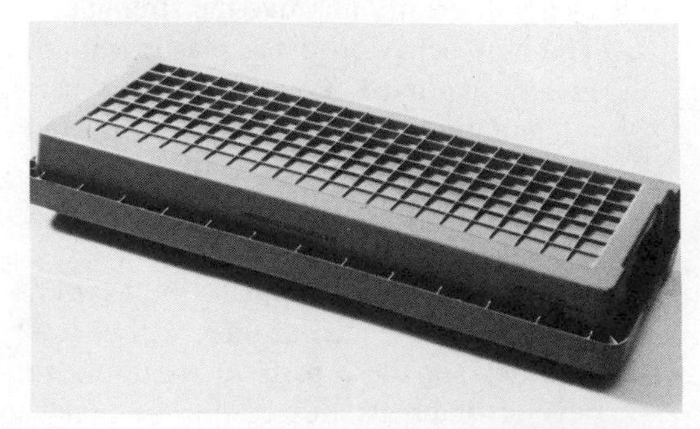

Fig. 11-21. Rib design in a molded ABS postal tray. (*Courtesy GE Plastics*)

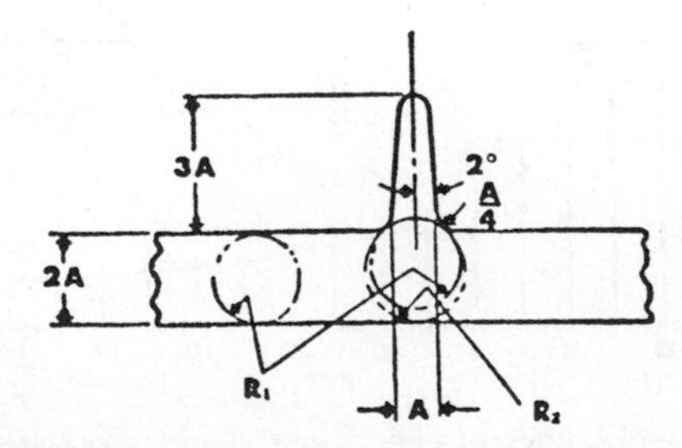

Fig. 11-23. Proper rib design to avoid sink marks.

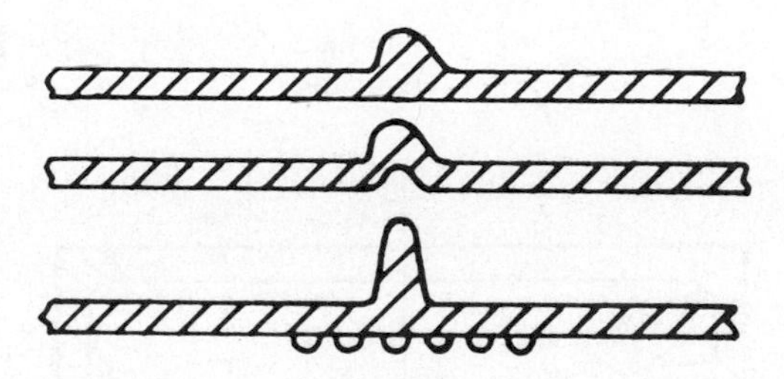

Fig. 11-24. Disguise for sinks: section design (top), sink on molded part (center), disguise (bottom). Disguise can consist of engraving emblems or corrugated patterns into the cavity. (Courtesy E. I. du Pont de Nemours & Co., Inc.)

gated patterns into the cavity so that they will be reproduced in the molded part).

Where wall thicknesses are increased to provide added strength, as for bosses surrounding holes, or where machining is to be done, varying thicknesses should be blended gradually into each other.

In some cases, it may be desirable to corrugate a surface instead of using ribs to increase stiffness and support. Another way to stiffen a panel is to give it a slight dome shape. A fairly flat conical shape may serve the same purpose more economically if the area is circular.

Undercuts

Modern molding practice dictates certain principles of design, which should be observed if molded articles are to be produced successfully. Most elementary is the fact that the piece must be easily removed from the mold after it is formed. This point frequently is overlooked, and many products are designed with undercuts that make it impossible to eject them directly from the mold cavity. If undercuts are essential, then split molds or removable mold sections are required, which increase the cost of molds and of the molded articles. Typical devices are the side pull core used for external undercut molding (see Fig. 11-25) and split-pin molding for internal undercuts (as shown in Fig. 11-26, in which a part of the pin is located in each mold platen).

Basically, undercuts can be classified as external or internal. As the names imply, undercuts located in the outside contours of the piece

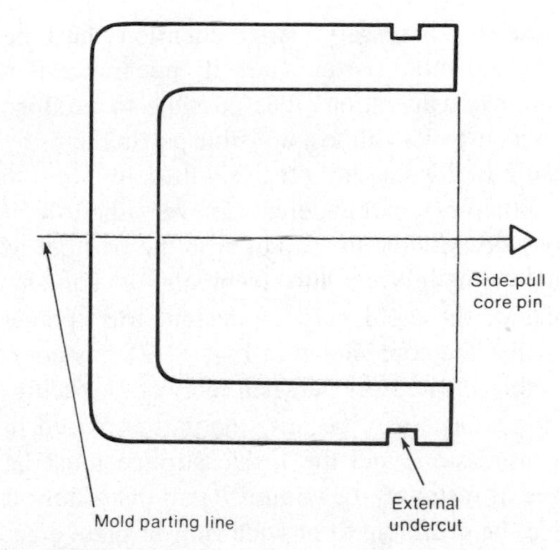

Fig. 11-25. Side pull core for external undercut molding. (*Courtesy GE Plastics*)

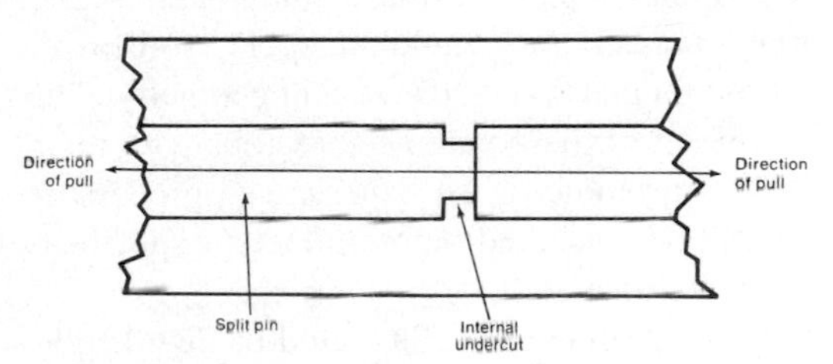

Fig. 11-26. Split-pin molding for internal undercuts. (*Courtesy GE Plastics*)

are called external undercuts; located on the inside contours, they are called internal undercuts.

An example of external undercuts is shown in Fig. 11-27. This relatively simple shape must be produced in a split-cavity mold to permit ejection of the part. When undercuts such as

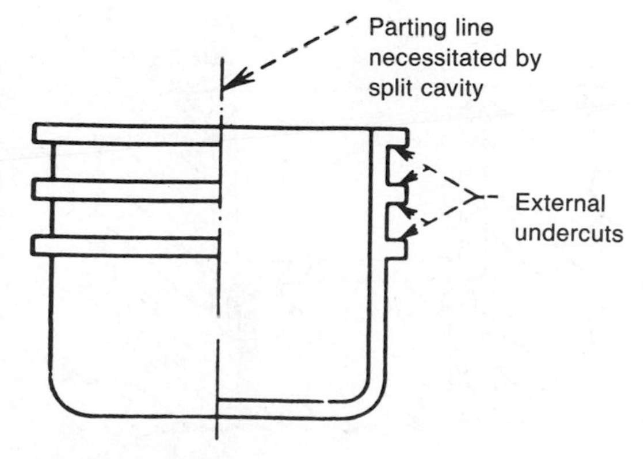

Fig. 11-27. External undercuts. (*Courtesy The DuPont Company, Inc.*)

these are necessary, strict attention must be given to mold parting lines if appearance is a major consideration. It is possible to produce such parts with almost invisible parting lines by using molds that are precisely fitted.

When external undercuts are very shallow, it may be possible to snap or strip the part out of the cavity. When this technique is contemplated, the mold must be designed to operate so that the core shown in Fig. 11-28 has completely cleared the part before ejection occurs. This is necessary because the part is placed in compression, and the inside surface must be free of restraint. To minimize part distortion, it may be desirable to provide ring or plate ejection, rather than ejector pins.

Among the important variables to consider in stripping a part are the allowable part strain, mold temperature, molding cycle, forces required for ejection, and undercut geometry. The definition of these parameters varies with resin and part geometry; so experienced processors should be consulted before undercut specifications are finalized.

Internal undercuts, illustrated in Fig. 11-29, can be provided in injection molding. However, as in the case of external undercuts, tool

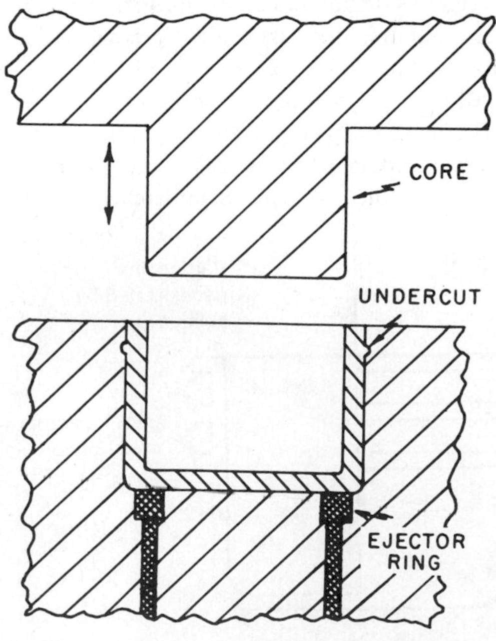

Fig. 11-28. Mold for external undercuts. (*Courtesy The DuPont Company, Inc.*)

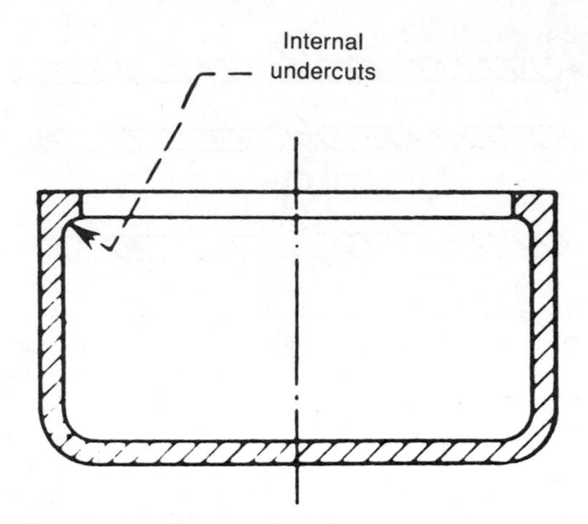

Fig. 11-29. Internal undercuts. (*Courtesy The DuPont Company, Inc.*)

and processing costs may be increased when internal undercuts are specified, particularly if collapsible cores are required.

As with external undercuts, it may be possible to snap or strip internal undercuts from the mold. With internal undercuts the part must be free to stretch around core projections during ejection. The basic concepts presented for stripping external undercuts also apply to internal undercuts. Again, suggestions and recommendations from experienced processors and suppliers are highly desirable.

Taper or Draft

In the design of articles produced from moldable rigid and elastomeric plastics, it is important that consideration be given to the easy removal of the piece from the mold cavity. Draft or taper should be provided, both inside and outside. Also, it is important that the surface of the molded piece, and hence that of the mold, particularly on the vertical or tapered walls, be polished by a high-luster finish. If the molded piece is straight-sided or vertical, it is necessary to exert a strong pull in order to open the mold; but by the use of taper and a highly polished surface, the article is easily released.

In deep-drawn articles, converging tapers assist by creating a wedging or compressing action as the mold is closed, and, in compression molding of thermosetting materials, tapers of this type increase the density of the plastic in the upper sections.

There are no precise calculations or formulas for taper. The amount of taper required will vary with the depth of draw, and common sense will dictate how much. When the draw is relatively deep and the part shape complex, for example, one would generally allow more draft on internal walls. Draft also will vary according to the molding process, the wall thickness, and the particular plastic being molded.

As designers should be careful to avoid details in design that would obstruct free ejection from the mold, they should provide the most liberal taper that the design will tolerate.

Figure 11-30 relates the degree of taper per side to the dimension, in inches per inch. It illustrates the effect of taper for various depths of piece (as shown in Fig. 11-31). A piece 4 inches in depth can carry a 4-degree taper, or 0.2796 inch per side. A 4-degree taper may be more than can be used in a deep article. For example, in a 10-inch piece, it will amount to 0.6990 inch per side.

Bosses

Bosses are protruding studs or pads that are used in design for the reinforcement of holes or for mounting an assembly. The same general precautions to be considered in the design and use of ribs may apply to the utilization and design of bosses (Fig. 11-32). Wherever practicable, they should be located at the apex of angles where the surface contours of the part changes abruptly (e.g., corners) to get a balanced flow of material. If possible, use beads as material runners.

It is considered good design practice to limit the height of the boss to twice its diameter to obtain the required structural strength. But higher bosses can be provided in an injection-molded part at the discretion of the designer.

In compression molding, high bosses tend to trap gas, which decreases both the density and the strength of the molded section. This is particularly true in straight compression molding, and where the boss is formed by the upper half or plunger of the mold.

Ribs may be employed on the sides of the boss to assist the flow of the material. In many cases, bosses are interlocked with other bosses or to side walls to provide structural stability. When the boss is in contact with external surfaces, proper rib design is essential to avoid sink marks on appearance surfaces. Fillets at the junction of the boss to the wall section are important. In some instances, where possible, a bubbler should be used on the cavity side of the mold opposite the boss to minimize sink on the outside wall.

Data are available from various suppliers (as in Fig. 11-33 for polycarbonate) providing specific information on boss design for individual plastics.

Holes

For a variety of reasons, holes often are required in a molded piece. They should be designed and located so as to introduce a minimum of weakness and to avoid complication in production. Several factors require careful consideration.

The distance between successive holes or between a hole and the adjacent side wall should be at least equal to the diameter of the hole, as shown in Fig. 11-34. It always is wise to provide as thick a wall section as is practicable, as cracking around holes in assembly generally is traceable to a disregard of this fundamental consideration.

The problem of design always is particularly complicated when a threaded hole must be used because of the concentration of stress, which causes notch-sensitivity in the region immediately surrounding the hole. Laboratory determinations provide a wealth of evidence indicating that the linear distance between the edge of the hole and the edge of the piece should be three times the diameter of the hole if the stress at the edge of the hole is to be reduced to a safe working figure. As shown in the discussion of bosses, increased boss areas often are used to provide additional strength.

Through Holes. Through holes usually are more useful for assembly than blind holes, and they also are easier to produce. Through holes should be preferred whenever possible because

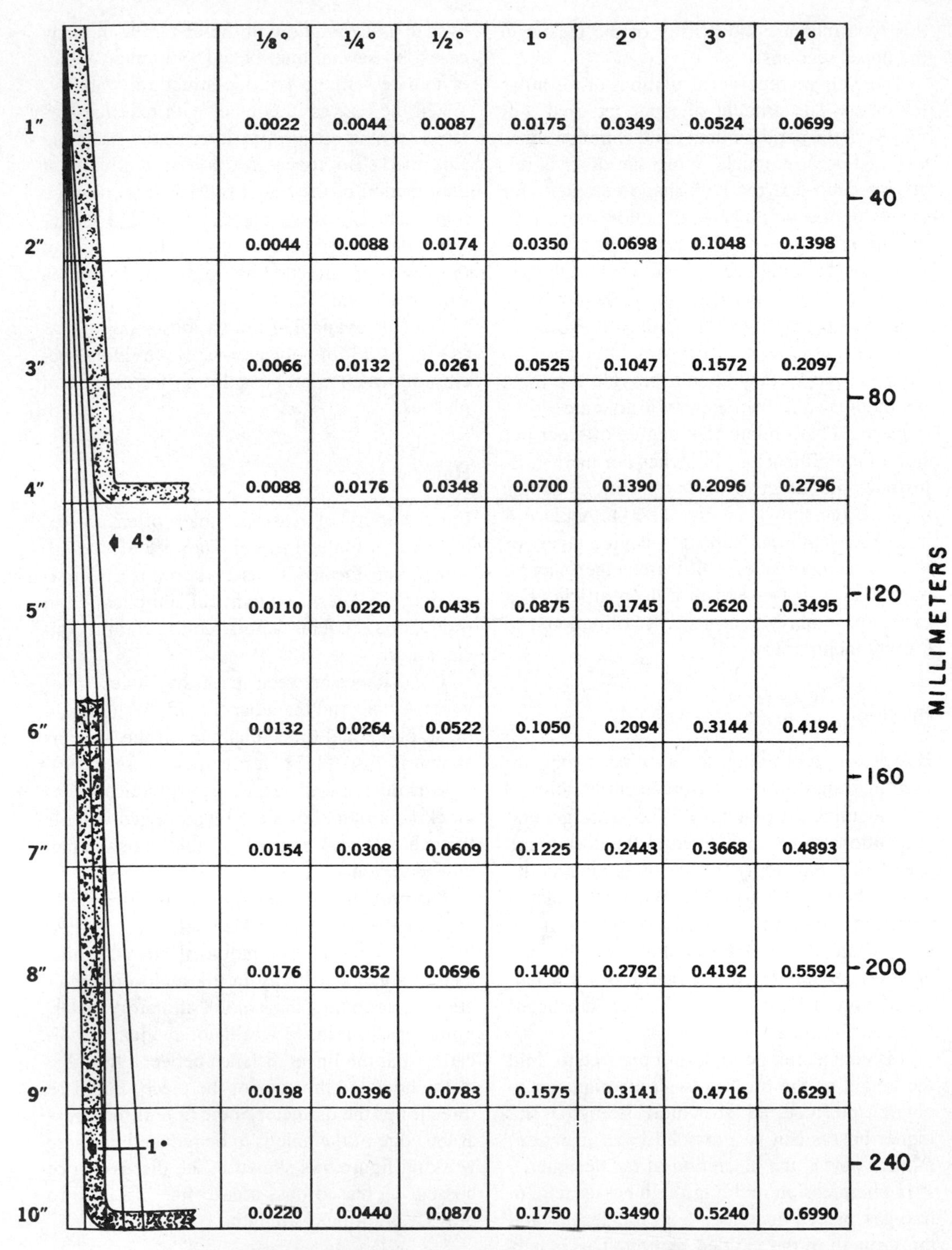

	⅛°	¼°	½°	1°	2°	3°	4°
1″	0.0022	0.0044	0.0087	0.0175	0.0349	0.0524	0.0699
2″	0.0044	0.0088	0.0174	0.0350	0.0698	0.1048	0.1398
3″	0.0066	0.0132	0.0261	0.0525	0.1047	0.1572	0.2097
4″	0.0088	0.0176	0.0348	0.0700	0.1390	0.2096	0.2796
5″	0.0110	0.0220	0.0435	0.0875	0.1745	0.2620	0.3495
6″	0.0132	0.0264	0.0522	0.1050	0.2094	0.3144	0.4194
7″	0.0154	0.0308	0.0609	0.1225	0.2443	0.3668	0.4893
8″	0.0176	0.0352	0.0696	0.1400	0.2792	0.4192	0.5592
9″	0.0198	0.0396	0.0783	0.1575	0.3141	0.4716	0.6291
10″	0.0220	0.0440	0.0870	0.1750	0.3490	0.5240	0.6990

Fig. 11-30. Relation of degree of tapere per side to the dimension in inch/inch.

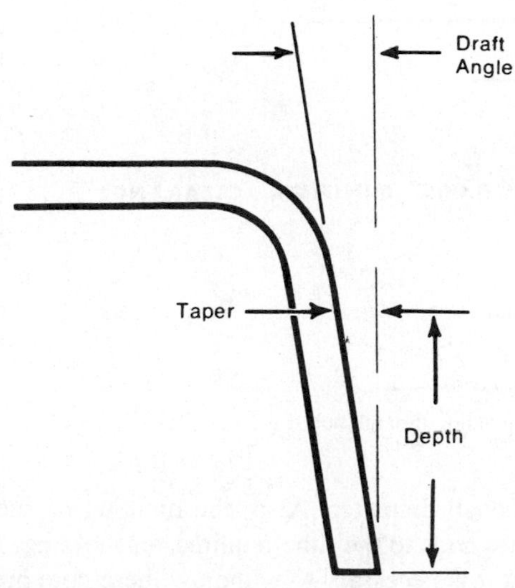

Fig. 11-31. Relationship of draft angle, depth, and taper. (*Courtesy GE Plastics*)

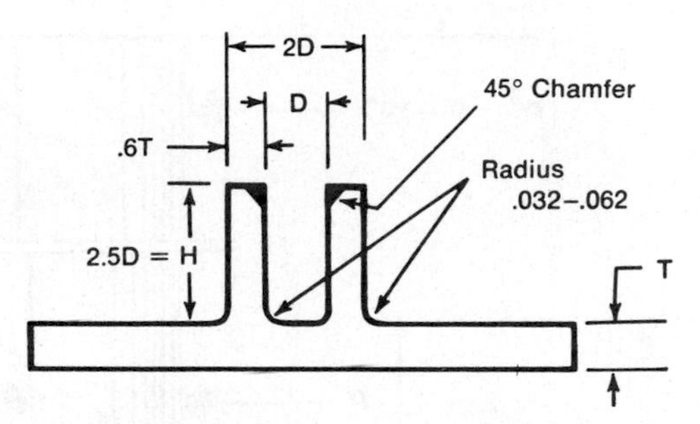

Fig. 11-33. Designing bosses for a polycarbonate part. (*Courtesy GE Plastics*)

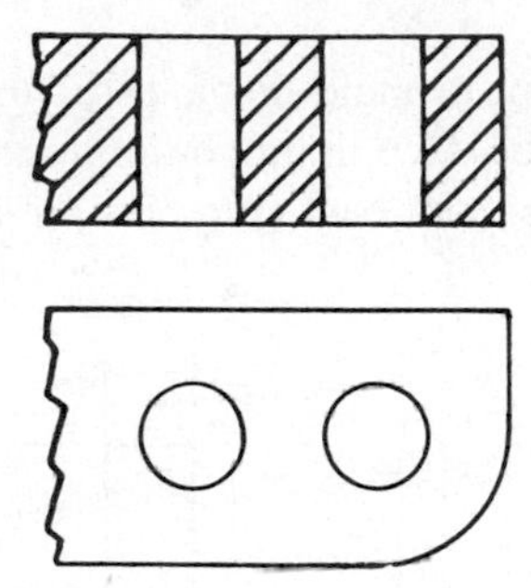

Fig. 11-34. Spacing recommendations for holes. (*Courtesy The DuPont Company*)

the mold pins that form the holes then can be supported in both parts of the mold. Through holes may be produced by either a single pin supported at each end or by two pins butted together. The first method generally is considered the better. Where the two-pin method is used, one pin should be slightly larger in diameter than the other to compensate for any misalignment (see Fig. 11-35). The butting ends of the pins should be ground flat, and provisions must be made in the mold assembly for sufficient clearance to avoid upsetting the ends of the pins. Although opinion differs regarding the amount required, a clearance of 0.005 inch appears to be good practice.

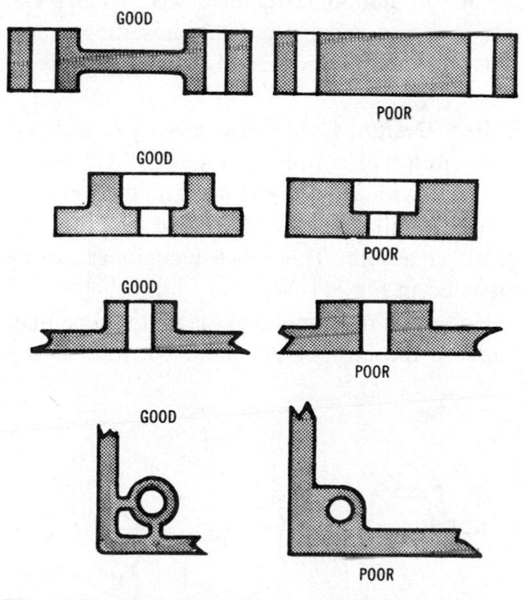

Fig. 11-32. Boss design. If possible, bosses should be located at the junction of two or more surfaces to favor balanced flow of material into the mold cavity. (*Courtesy The DuPont Company*)

Blind Holes. The designer must remember that a blind hole is formed by a core pin, which of necessity can be supported only at one end. Thus this pin can be distorted, bent, or sheared off during the molding operation as a result of the unbalanced pressure exerted by the flow of the plastic. The depth of blind holes (i.e., the length of the core pin) should be limited to twice the diameter of the hole, unless the diameter is $\frac{1}{16}$ inch or less, in which case the length should not exceed its diameter (see Fig. 11-36).

Drilled Holes. Often it will be found less expensive to drill holes after the molding, rather

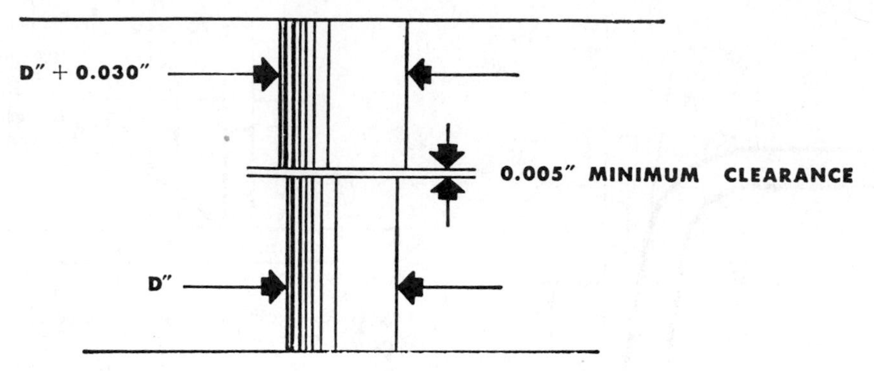

Fig. 11-35. Two-pin method for molding through holes.

than attempt to mold them, particularly when they must be deep in proportion to their diameter. Broken and bent core pins are an expensive item.

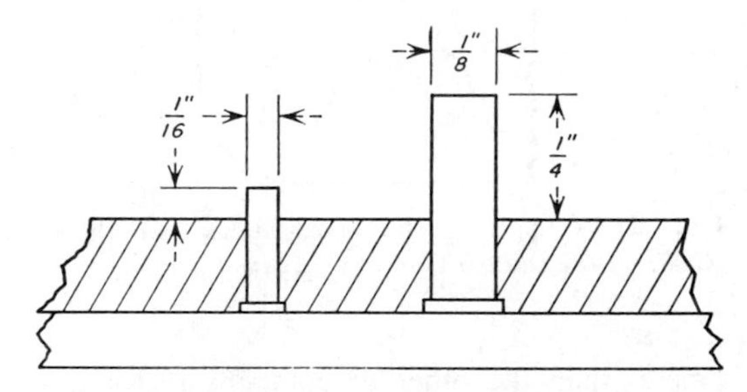

Fig. 11-36. Pin configurations for molding blind holes.

Good manufacturing practice calls for the provision of drill jigs for the accurate drilling of molded pieces. Drill jigs will do much to reduce the cost of broken drills and too-frequent resharpening. However, in some instances it is possible to provide spot points on the molded part, which can be used to locate the holes so that jigs are not needed.

Threaded Holes. See discussion of threads, below.

Side Holes. Side holes are difficult to produce, and present problems that are not easily solved because they create undercuts in molded pieces. Holes that must be molded at right angles to each other necessitate split molds or core pins and therefore are more costly, particularly in compression molding. The core pins, which are not parallel to the applied pressure, may be distorted or sheared off, particularly if they are of small diameter. Also, the molding of such holes adds to the time required for molding, in that it is necessary to withdraw these core pins from the molded piece before it can be removed from the cavity. These problems are less serious when transfer or injection methods are used, but there remains the necessity of withdrawing the core pins from the molded piece prior to its removal.

Another problem brought about by the presence of core pins is imperfect welding of the plastic, in the area back of or adjacent to the pin and on the side opposite to the direction of flow. This condition is encountered more frequently in thermoplastic materials, but also exists in pieces made from thermosetting materials in a compression mold.

Other Design Considerations. A stepped hole often will permit more depth than can be obtained with a single-diameter hole. Notching of a side wall also may be utilized to reduce the depth of a hole. These two techniques are illustrated in Fig. 11-37.

Holes that run parallel to the parting line may increase the cost of the mold because of the

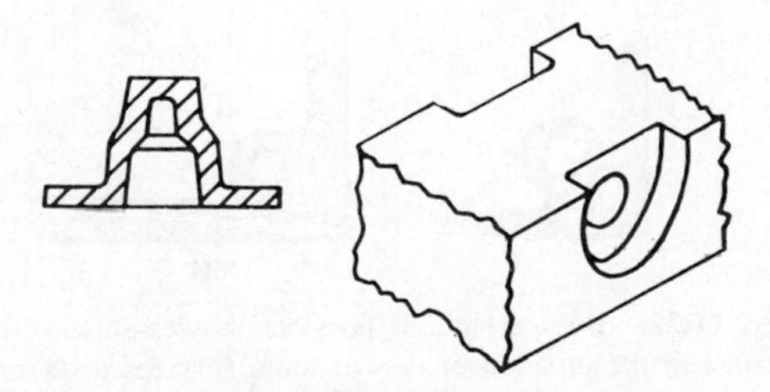

Fig. 11-37. Stepped hole (left); counterboring (right). (*Courtesy The DuPont Company*)

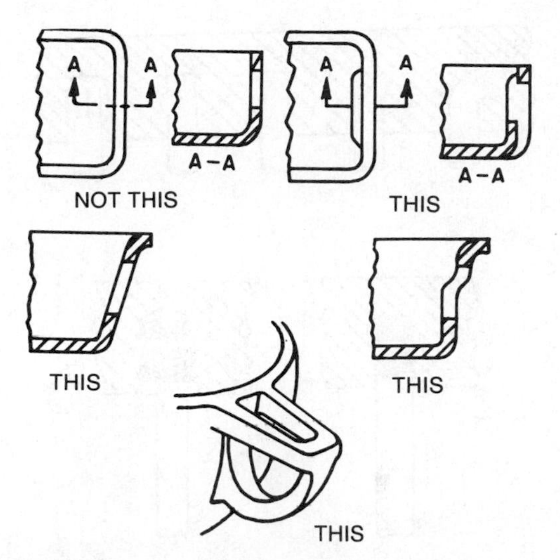

Fig. 11-38. Recommendations for placing holes in walls. Bottom sketch shows hole perpendicular to parting line. (*Courtesy The DuPont Company*)

necessity of retractable core pins or split dies. Frequently, this problem is overcome by placing holes in walls that are perpendicular to the parting line, using steps or extreme taper on the wall. Several of these possibilities are shown in Fig. 11-38.

Allowance should be made for a minimum $\frac{1}{64}$ inch (0.4 mm) vertical step at the open end of holes. A perfect chamfer or radius at the open end of a hole demands precision in the mold that may be unattainable from a practical or an economic standpoint. This is illustrated in Fig. 11-39.

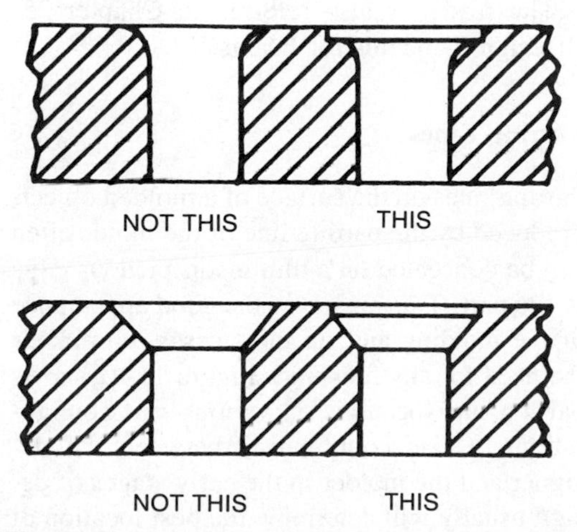

Fig. 11-39. Chamfer on open holes. (*Courtesy The DuPont Company, Inc.*)

External and Internal Threads

External and internal threads can be produced economically in plastic parts with modern molding techniques (or by machining or self-tapping). Screw threads, produced by the mold itself (using rotating core pins or loose-piece inserts) eliminate expensive post-molding threading operations. Although almost any thread profile can be produced, the Unified Thread Standard is the most suitable for molded parts. This form eliminates the feather edge at both the root and the tip of the thread. The Unified Thread Standards are divided into the following classes:

- Class 1A, 1B*—Most threaded nuts and bolts fall in this class and are used for most assembly work.
- Class 2A, 2B*—This is a tighter fit than Class 1 with no looseness and is widely used.
- Class 3A, 3B*—This is used in precision work and is obtainable under exacting molding conditions with constant dimensional checking.

Coarse threads can be molded more easily than fine threads, and threads less than 32 pitch should be avoided. Generally, threads of Class 1 or 2 are adequate for most applications. In many cases it is desirable to provide an interference fit between two threaded parts. This is easily accomplished with plastic materials and helps prevent a nut, bolt, or screw from loosening under mechanical vibration.

Parts with external threads can be removed from the mold either by unscrewing the parts from the mold cavity or by locating the mold parting line along the axial centerline of the screw profile. Internal threads can be produced by a threaded core that is unscrewed from the part; or if a part has internal threads of short engagement (few threads), it may be possible to strip the part over the threaded core, thus eliminating the unscrewing mechanism. An example of a threaded part that utilizes this economical principle is a cap for bottles and jars.

*The letter A refers to the external thread and B to the internal thread.

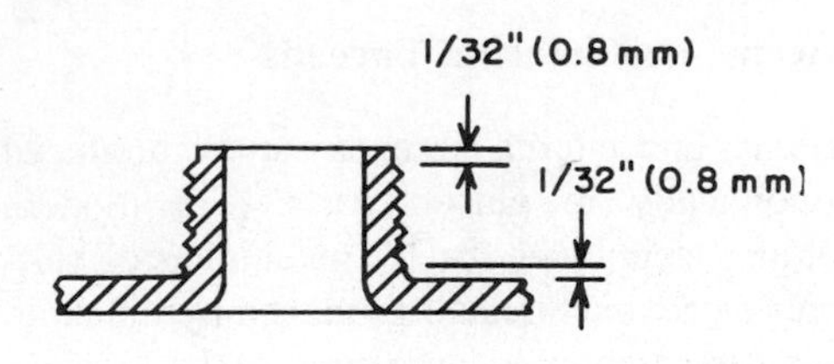

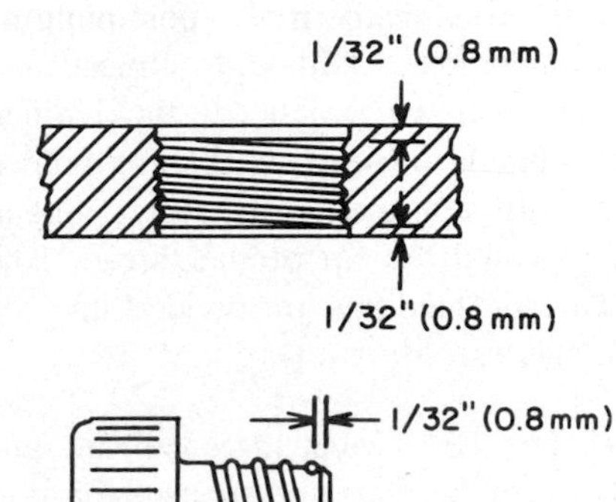

1/32" (0.8 mm)

Fig. 11-40. Thread design. (*Courtesy The DuPont Company, Inc.*)

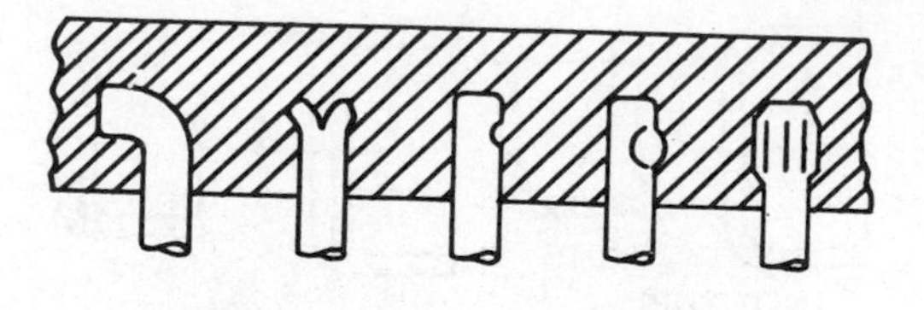

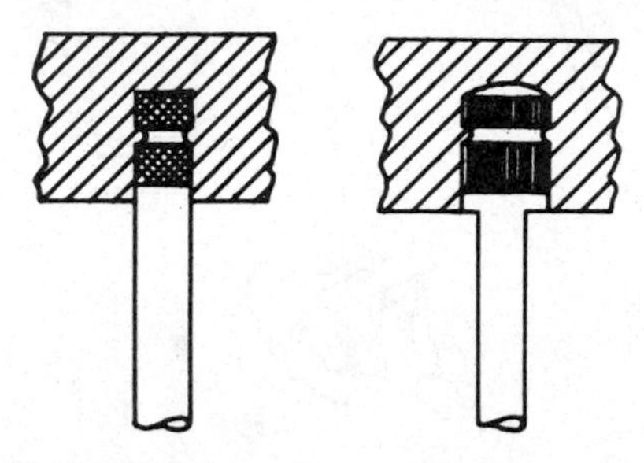

Fig. 11-41. Techniques for securing inserts. (*Courtesy The DuPont Company, Inc.*)

Figure 11-40 shows that parts should not be designed with threads that run to the very end of the part or completely up to a shoulder. A clearance of at least $\frac{1}{32}$ inch (0.8 mm) should be provided at each end of the thread. The wall thickness supporting either an external or an internal thread should be at least equal to the depth of the thread.

Inserts

Inserts in plastic parts can act as fasteners or load supports, or may simplify handling or facilitate assembly. Inserts may be functional or purely decorative, but they should be used sparingly because they increase costs.

Inserts derive a good deal of their holding power from the fact that plastic materials, when cooling, shrink around a metal insert.

The design of inserts must assure a secure anchorage to the plastic part to avoid rotation as well as pull-out. Sharp corners on inserts may result in areas of stress concentration and should be avoided. For techniques for securing inserts, see Fig. 11-41.

Knurling should be deep enough to permit material flow into the depressions and should be provided with a smooth surface where an insert protrudes from the molded part.

For ease of molding, inserts should be located perpendicular to the parting line. Also, inserts should be located so that they are not displaced during the injection portion of the molding cycle. Pin-supported inserts require a minimum hole diameter of $\frac{1}{8}$ inch (3.17 mm) to avoid damage to pins under molding pressures. For applications where the insert O.D. is less than $\frac{1}{4}$ inch (6.35 cm) the boss outside diameter should be at least twice that of the insert O.D. Radial ribs may be used to strengthen the boss and insert.

The reader is also referred to Chapter 25, ''Design Standards for Inserts.''

Parting Lines

Parting lines on the surface of a molded object, produced by the parting line of the mold, often can be concealed on a thin inconspicuous edge of the part. This preserves the good appearance of the molding and, in most cases, eliminates the need for any finishing. Figure 11-42 shows parting line locations on various part configurations. Close coordination between the designer and the molder in the early stages of design usually will determine the best location of the parting line.

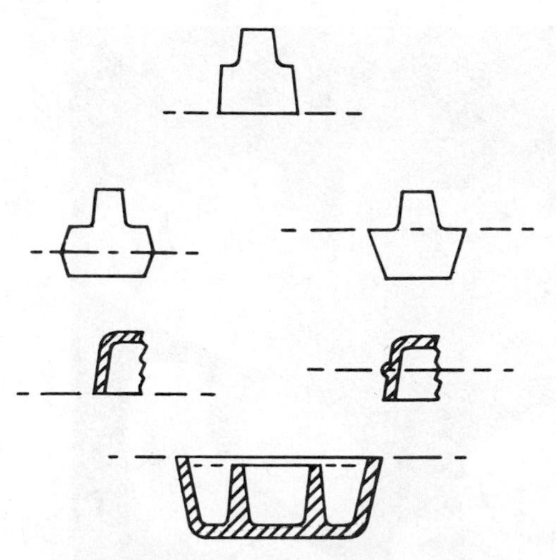

Fig. 11-42. Parting line locations. (*Courtesy The DuPont Company, Inc.*)

In compression molds, the closing or telescoping of the two parts of the mold results in a flow of material into the clearance between these parts. This material is known as flash and occurs at the parting line of the molds. The removal of flash from the article leaves a flash line, which is unavoidable and generally unsightly. The problem of flash lines requires careful consideration by the designer, as the attractiveness of the product may depend to a large degree on a careful location of the flash line where it will not be seen. When a piece is molded in a transfer or an injection mold, the problem of flash is greatly simplified because the mold is already locked in closed position before the plastic enters it.

Surface Treatment

The designer should make full use of the many different types of surface treatment to which plastics readily lend themselves. The possibilities are limited only by the ingenuity of the designer plus the fundamental requirement that the piece must be removed from the mold. They range from mirrorlike surfaces to a dull finish (created by sandblasting the molded surface, as one technique). Surface treatments such as fluting, reeding, stippling, fine straight lines, diamond-knurl cut, leather grain—to mention only a few—may be used to advantage. They tend to conceal surface blemishes that might prove objectionable on a highly lustrous surface.

Various designs can be created in the mold and transferred to the part by either debossing or embossing the mold surface. It usually is less expensive to use raised letters on a part, as they may be readily engraved in the mold. On the other hand, the converse means removing all metal but the letters, a more costly procedure. Either may be done, however, depending on the effect desired.

Knockout sleeves or pins generally are used to remove a part from the mold. Some consideration should be given to the placement of such knockout mechanisms because they do leave a visible mark on the part. Gate location should be considered carefully for the same reason. Consultation with the mold designer is recommended when it is thought that functional and aesthetic considerations may clash.

The appearance of a product is greatly influenced by the surface treatment employed. The designer should be acquainted with the problems encountered by the molder. This is particularly true when large flat surfaces with a high lustrous finish are required, as such surfaces are difficult to produce. The molder invariably must resort to special molding techniques that are both time-consuming and costly. The designer, understanding these problems, can do much to simplify the molder's difficulties through modification of the design.

Wide, sweeping curves and domed rather than flat surfaces should be employed. Improved flow and distribution of material, during the molding operation, will result, and the tendency to warp will be greatly reduced. As a result, the appearance of the molded piece improves.

Unfortunately there are no simple rules or formulas for solving the problems of surface treatment. What may prove to be a sound solution in one case may not work in another. Each new piece presents its own problems. Designers must draw freely upon their ingenuity, always remembering that the molded article must be easily removed from the mold after it has been formed.

The telephones shown in Fig. 11-43 are typical of the pleasing appearances that can be

Fig. 11-43. ABS telephones typify high quality of surface finish obtainable with proper mold surfacing. (*Courtesy General Electric Plastics*)

gained by combining wide sweeping curves, domed surfaces, and a lustrous finish.

Molded Lettering

Names, monograms, dial numbers, instructional information, and the like, are frequently required on molded articles. The lettering must be applied in such a manner as not to complicate the removal of the article from the mold. This is accomplished by locating it perpendicular to the parting line and providing adequate draft.

While both raised and depressed letters are possible, the method to be used in constructing the mold will dictate which is the more economical.

When the mold is to be made by machining, raised letters on the molded piece will be less costly. A raised letter on the molded piece is formed by a corresponding depression in the mold, and it is far less costly to engrave or machine the letters into the face of the mold than it is to form a raised letter on the surface of the mold by cutting away the surrounding metal. On the other hand, if the mold is to be formed by hobbing, then the letters on the molded piece should be depressed, as it is the hob that must be machined. In making the hob, the letters are engraved into its surface. As the hob is sunk into the steel blank to produce the mold cavity, the letters are raised on the surface of the latter. These raised letters on the mold in turn produce depressed letters on the molded piece.

To improve legibility, depressed lettering, filled in with paint, sometimes is required. When hobbing of the whole mold is not practicable, the desired result generally can be accomplished by setting in a hobbed block carrying the lettering. When this insert is treated as a panel and the fin line is concealed by fluting, the appearance is not unpleasant.

The application shown in Fig. 11-44 is indicative of the type of precision lettering that can be accomplished in plastics products.

AVOIDING WARPAGE PROBLEMS*

Warpage is defined as "dimensional distortion in a plastic object after molding." It is directly related to material shrinkage; that is, as shrink-

*By Nelson C. Baldwin, Technical Service Engineer, Hoechst Celanese Corp.

Fig. 11-44. Molded-in lettering in ABS typewriter head. (*Courtesy GE Plastics*)

age increases, the tendency for warpage to occur increases.

Anisotropic properties, such as are found in some filled and reinforced materials and in liquid crystal polymers, also can contribute to warpage.

The purpose of this discussion is to offer guidelines to the designer and the processor on the factors that can cause warpage through uneven shrinkage; namely, part design, mold design, and molding conditions.

Part Design

Warpage caused by part design is the worst type, being nearly impossible to correct by molding conditions. For this reason, it is imperative that the part be designed to *prevent* objectionable warpage. As shrinkage is directly proportional to wall thickness, wall thickness is directly related to warpage. Hence wall thickness must be uniform to provide uniform shrinkage. Different wall thicknesses in the same part *must* result in either warpage, through stress-relief, or molded-in stress.

Examples 1 and 2 illustrate warpage due to nonuniform wall thicknesses.

EXAMPLE 1:

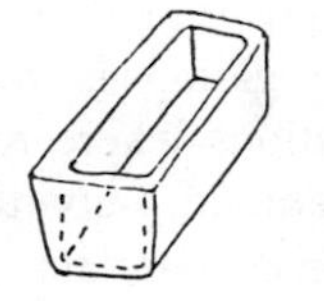

Results in

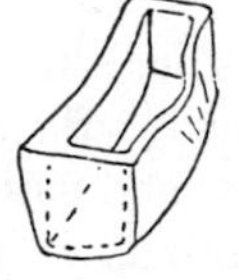

EXAMPLE 2: Results in

This varying wall thickness condition is probably the single largest cause of warpage.

Another type of part design warpage concerns ribs and bosses. Indiscriminate location of ribs and improper selection of rib thickness can result in shrinkage patterns that will alter the shape of the entire molding.

Ribs should be no more than 50% of the adjacent wall thickness of the part to avoid sinks and possible distortion. However, ribs that are very thin compared to the main body can cause distortion due to different degrees of shrinkage.

Bosses can affect the shape of the molded part if they are of a different wall thickness than the base to which they are attached, or if they are connected to a side wall of different thickness. Initial wall thickness for bosses should be the same as for ribs.

Example 3 depicts distortion from a rib tied to a side wall.

EXAMPLE 3:

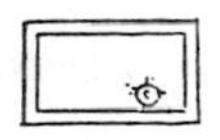

Results in

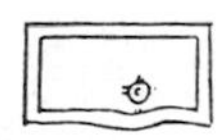

Ribs and bosses also can affect part geometry through changes in heat transfer from the mold. This subject will be covered later.

Mold Design

One of the most critical aspects of mold design, in unfilled and glass-filled crystalline polymers, is gate location. Its importance is due to many factors, including the inherent high shrinkage of the material and the anisotropic behavior it may exhibit.

Anisotropy refers to a shrinkage differential between the flow direction and the direction perpendicular (transverse) to flow. With unfilled materials, the greater shrinkage usually is encountered in the flow direction. Shrinkage in the transverse direction, conversely, usually ranges from 70 to 98% of the longitudinal shrinkage (direction of flow), depending on gate size and part thickness. Thinner parts do not exhibit the degree of anisotropy shown in thicker parts.

On the other hand, polymers reinforced with glass fibers show the opposite condition. Shrinkage in the flow direction is *less* than transverse shrinkage, because of orientation of the fibers to the flow direction. The percentage difference between the shrinkage in each direction is dependent on the wall thickness, gate size, fiber length, and so on, and thus is difficult to pinpoint. However, the average difference is approximately 50%, with the shrinkage in the flow direction being the lesser. It is recommended that material molding manuals or preferably a technical service representative be consulted prior to any actual mold fabrication to assist in determining shrinkage values.

It can be seen from the above discussion on anisotropy that orientation by gate location can have a significant effect on warpage, even in an ideally designed part.

Examples 4 and 5 illustrate warpage of this type with unfilled material. Glass reinforcement would have an opposite effect.

EXAMPLE 4:

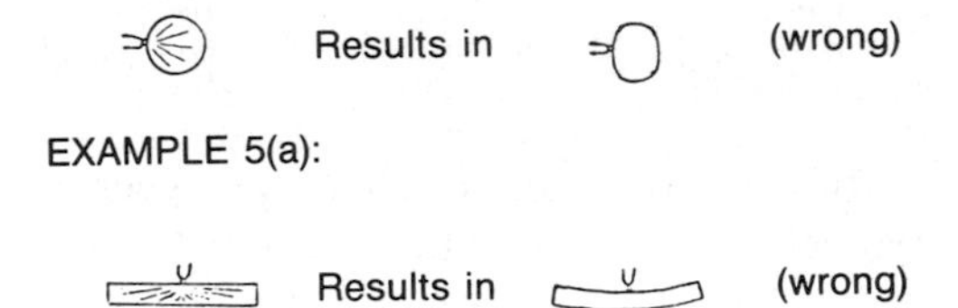

whereas

(b): or Results in (right)

The correct way to minimize warpage from orientation is to provide a longitudinal flow path for rectangular parts such as shown in Example 5, or a radial flow path for circular items as shown in Example 6.

EXAMPLE 6:

For circular parts that are cored in the center, other types of gating can be used to achieve uniform flow. Examples 7 and 8 depict gating systems of this type.

EXAMPLE 7:

GATE (3) Three gates 120° apart (3 plate mold)

EXAMPLE 8:

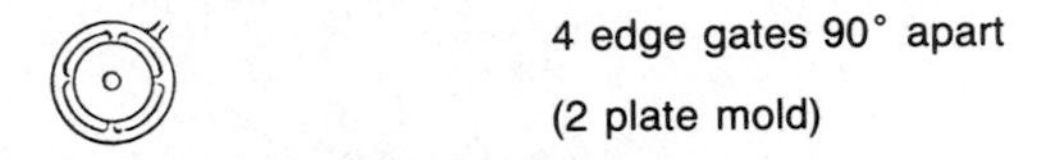

Because of the high degree of anisotropy in glass-filled material, gating systems such as shown in Example 8 will sometimes still give enough orientation to cause slight out-of-roundness or warpage. For this reason, it is recommended that wherever possible, full ring gating (or if on the I.D., full disc gating) be used for these materials.

Example 9 illustrates the orientation patterns developed by four-point gating systems with glass-reinforced material. For purposes of illustration, the resultant shapes are exaggerated.

EXAMPLE 9(a): (b):

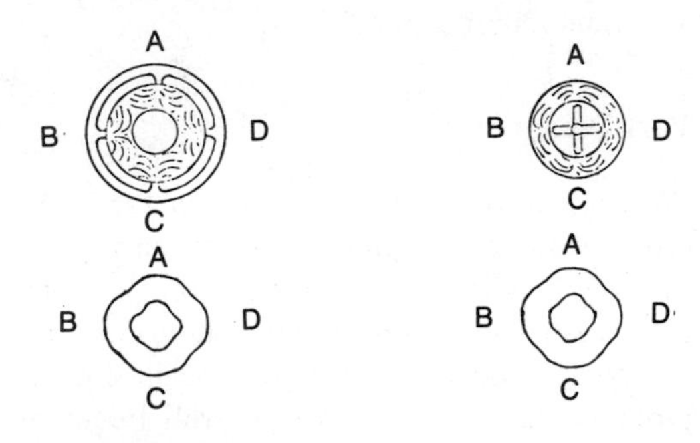

Example 10 also shows flow orientation for a cylindrical shape, utilizing four-point gating in a two-plate mold.

EXAMPLE 10:

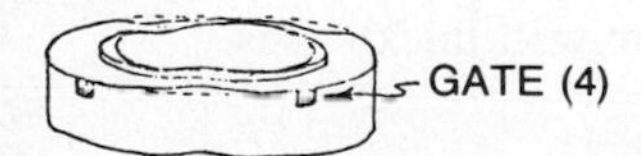

Thus, full ring or disc gating is recommended for critical applications where roundness and flatness are of paramount importance, especially for glass-reinforced materials.

Another form of warpage associated with gating involves pressure distribution within the cavity. Refer to Example 6, and assume that the disc has a uniform wall thickness.

Bear in mind the following three things: (1) warpage is the result of shrinkage differences in the same part, (2) shrinkage is affected by pressure (high pressure → low shrinkage), and (3) pressure decreases as flow increases.

Using these principles, it can be said that the pressure at the outside diameter of the part is less than the pressure at the gate. Therefore, shrinkage is greater at the O.D. than at the gate area even though the wall is constant. On this basis, warpage *must* result if stress relaxation is allowed to take place. The only way this part can be held flat through molding conditions is by using techniques that give the lowest shrinkage (i.e., decreased mold and melt temperature, fast fill, increased injection pressure, plunger forward time, and overall cycle time, large gate, etc.). This procedure actually locks in the stresses so that the part retains its shape.

Using the three principles mentioned above, a suggested remedy for a part of this shape is to taper the thickness from the center to the O.D., the thicker section being in the center. A gradual taper would reduce the pressure drop, allowing more effective pressure at the outer periphery. This would reduce shrinkage in this area, and at the same time increase shrinkage at the center sections because of the thicker wall.

This leads to the second consideration for minimizing pressure differences, to locate the gate at the thickest section of the part. Again referring to the factors governing shrinkage, thicker sections shrink more than thin ones, and high-pressure areas shrink less than low-pressure areas. Therefore, if a gate is located in a thin section that feeds a thicker one, the highest pressure (developed near the gate) is generated in the section that will shrink the least because of its wall. This is a direct invitation to warpage. The gate should be located in the thicker wall so that the highest pressures are developed in this area to help minimize shrinkage from wall thickness and thus minimize shrinkage differentials in the part.

The last comment regarding gating involves gate size. The size of the gate regulates not only the volume of material allowed to enter the cavity, but also the effective pressure transmitted for packing out this material.

If the gate is too small, there is the possibility that it will prematurely freeze off before the part is adequately filled. This would cause low effective pressure and inadequate packing in the mold cavity with resultant increased shrinkage. In addition to causing undersize dimensions, the high shrinkage would magnify any trends toward warpage that might be present.

Product bulletins and preferably a technical service representative should be consulted prior to actual mold construction for gating recommendations.

Another condition encountered in mold design that can lead to warpage is the use of dissimilar metals in cavity construction.

Thermal conductivity varies with different metals. If two metals are used in fabricating the cavity and the core, the one with the lower thermal conductivity will retain heat longer and thus create a differential mold temperature. Because shrinkage increases as mold temperature increases, the plastic part, in this case, will bow toward the hotter side, with consequent warpage.

Example 11 shows this condition when tool steel and beryllium copper are used for cavity construction.

EXAMPLE 11:

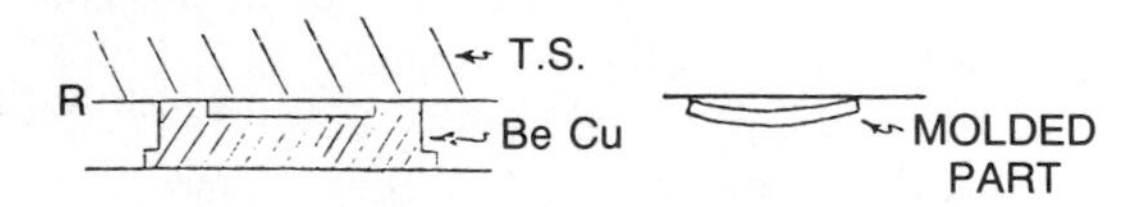

As beryllium copper dissipates heat much more rapidly than tool steel, it runs cooler, even though the same external heat is supplied to both mold halves. The solution for overcoming this condition is to use separate mold controllers for each half of the mold and make adjustments until both halves are at the same temperature. It also may be necessary to include additional coolant channels in the tool steel half to achieve balance.

A final comment on mold design is concerned with nonuniform heat dissipation from the mold. This condition, which is perhaps the second largest contributor of warped parts, usually is not taken into account during the design stages of mold fabrication.

Nonuniform heat dissipation occurs when one side of the mold must dissipate more heat than the other, because of the shape of the part and the area of the steel in contact with the hot resin. Ribs and bosses are prime causes of nonuniform temperatures in the mold, as shown in Examples 12 and 13.

EXAMPLE 12:

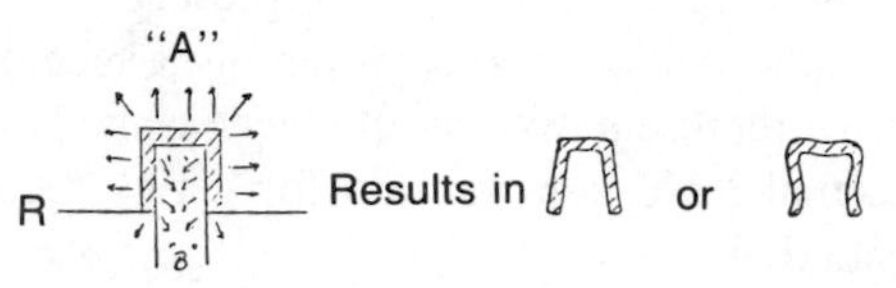

EXAMPLE 13:

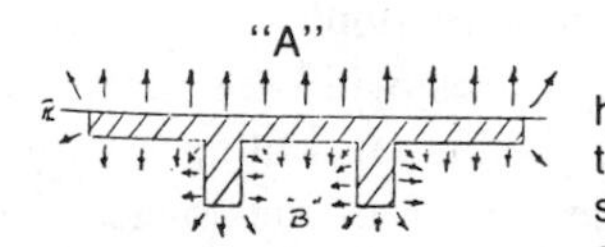

Note that the "A" plate has greater heat dissipation area than "B". Result – B gets hotter and causes more shrinkage

results in

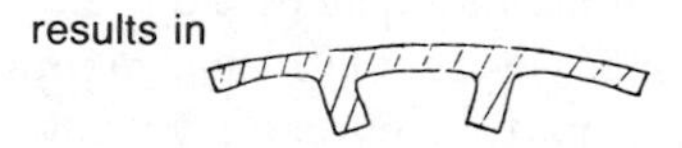

Nonuniform dissipation causes one side to run hotter than the other, and because thermoplastic resin shrinkage increases with increased mold temperature, more shrinkage occurs at the hotter face of the mold, with warpage inevitable.

Several things can be done to offset this condition, including the following:

1. Provide additional cooling channels, or bubblers, in the areas that must dissipate the most heat.
2. Use differential mold temperatures to provide extra heat in the areas that do not have much initial heat to remove. (This method also can be used to purposely warp parts the opposite way so that post-molding shrinkage will result in a flat part.)
3. Take advantage of thermal conductivity differences in dissimilar metals by using metals with high conductivity in areas requiring greater heat dissipation. An illustration would be to make the core pin in Example 12 of beryllium copper.

Evidence has suggested that in the case of glass-filled resins there may be a random "bunching" of fibers at sharp corners. These fibers are disoriented from the normal flow and may, because of anisotropic shrinkage, cause a pulling in of adjacent walls.

Example 14 shows this possibility, and Examples 15 and 16 show possible solutions to offset the condition.

EXAMPLE 14:

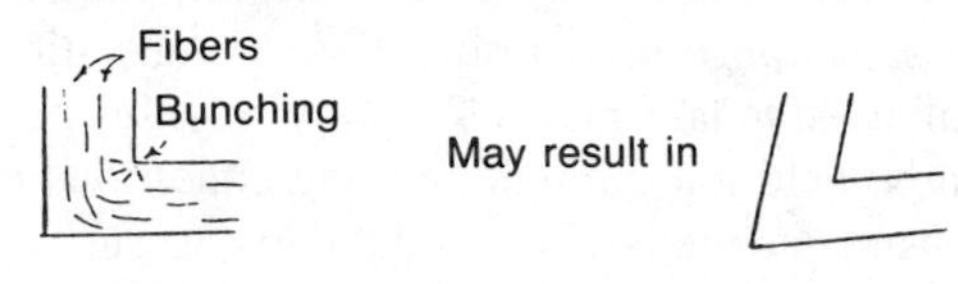

EXAMPLE 15:

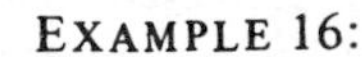

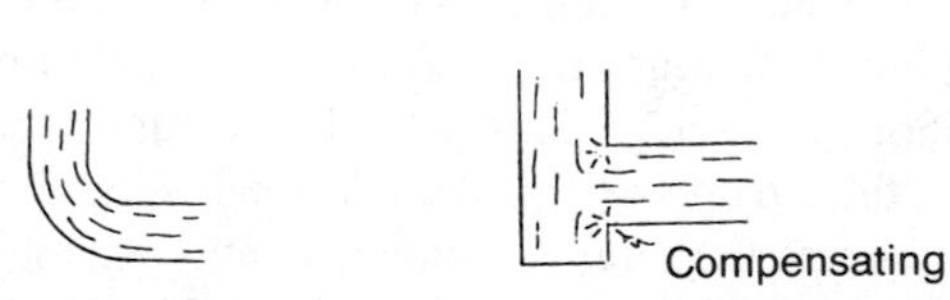

In any case, however, all sharp corners should be radiused 25 to 75% of the wall thickness to reduce stress concentrations that may lead to premature failure.

The improper ejection of molded parts may cause what appears to be warpage although it actually is deformation. This problem can be caused by undercuts in the cavity or an inadequate ejection system and should be checked along with the other design considerations.

Molding Conditions

Molding conditions can either contribute to warpage or aid in minimizing it. Thus, it is important to know how the various machine functions affect the molding material and its resultant post-molding behavior. The following paragraphs review the pertinent functions one by one, and offer comments from a shrinkage, or warpage, point of view.

Filling Speed. Filling speed is based to a large extent on gate size. However, assuming that the gate size is correct, it is recommended that the cavity be filled as rapidly as possible. A fast fill ensures that the resin temperature will be constant throughout the cavity with a resultant uniform shrinkage.

Mold Temperature. A high mold temperature is preferable for optimum properties and surface finish. It causes the molded parts to cool slowly, thus relieving molded-in stress. However, this stress-relieving action will cause a tendency to warp if the configuration of the part is such that uneven shrinkages will take place. If this is the case, then the mold temperatures must be lowered to retard shrinkage.

Mold temperatures can be as low as necessary, provided that the parts do not stress-relieve to a warped condition during post-molding shrinkage or in end-use operation.

If parts are to be used in a high-temperature environment, it is suggested that the following test be performed on parts molded with a cold mold:

Condition a small quantity of molded parts for 10 to 12 hours at 180 to 190°F. Follow this step with a slow cool to room temperature. If the parts are warped after cooling, the mold temperature is too low. Resample with a warmer mold, and repeat the test until the parts remain flat after testing.

Note that this test should only be used to complement actual end-use testing. The values obtained must be correlated with conditions known to give acceptable parts for the test to be of benefit.

Material Temperature. Temperature settings must be high enough to give a homogeneous melt. If settings are too low, incomplete melting will occur, and varying degrees of shrinkage will take place with resultant warpage. The shrinkage differential is caused by varying pressure.

This same condition can occur if there are localized hot spots in the machine barrel or if too much of the machine capacity is being used with each shot. The melt, in these cases, will have a varying temperature and therefore a varying shrinkage rate. Once again, warpage may result. The maximum shot size should be no more than 75% of the rated machine capacity.

Cycle Time. Too short a cycle will result in ejection of parts that have not cooled sufficiently to maintain structural stability. Post-molding shrinkage, higher because of the decreased cycle, will in this case follow a stress-relieving pattern. As stresses relieve, any uneven shrinkage will result in warpage. Also, the parts are subject to deformation during ejection from the short cycle.

Too short an injection forward time also can lead to warpage problems. If the ram is retracted prior to gate seal, material will backflow from the cavity. This causes low and nonuniform pressure in the cavity, resulting in increased and nonuniform shrinkage.

Injection Pressure. A low injection pressure allows increased material shrinkage. In addition to dimensions being undersize and erratic, the increased shrinkage promotes warpage if the configuration of the part is prone to differential shrinkage.

Too high an injection pressure, on the other hand, can lead to localized overpacking in the cavity. This causes differential shrinkage and possible warpage.

Lost Contact with Cavity Surface. This condition usually is due to part configuration and/or inadequate venting, rate of fill, injection pressure, and so on. The condition appears as a pulling away of the plastic from the cavity, leaving an air gap. The air gap acts as an insulator, causing a slower cooling rate. The result is the same as if the area were hotter than the rest of the mold. A higher shrinkage takes place, and warpage is the result.

A change in mold temperature, or the use of differential mold temperature, usually corrects this condition, assuming that other corrections such as venting, faster fill speed, or increased pressure have been tried first.

Troubleshooting Guide

It is evident that warpage is the result of nothing more than distortion from shrinkage differ-

Table 11-4. Trouble shooting guide—warpage.

Cause	*Solution*
1. *Part Design*	
A. Wall thickness variation*	Core heavy sections.
B. Ribs and bosses	Follow recommendations for wall thickness, especially when connecting to main body walls.
2. *Mold Design*	
A. Gate location	
1. Flow orientation*	Locate gate, or gates, for uniform orientation.
2. Pressure distribution	Gate into heavy section.
B. Gate size	Enlarge or reduce gate. Too small a gate causes premature freeze-off. Too large permits localized over-packing. in general, however, a large gate is preferable to a small one.
C. Materials of construction	Use materials to provide uniform dissipation. See also ''D'' below.
D. Non-uniform dissipation*	Provide extra coolant passages where required. Also use differential mold temperature or dissimilar metals.
E. Ejection	Provide adequate, uniform ejection. Check for undercuts.
3. *Molding Conditions*	
A. Fill speed	Use fast fill whenever possible.
B. Mold temperature	Alter to suit. Hot mold increases shrinkage, stress relieving and warpage. Cold mold results in molded-in stress.
C. Melt temperature	Ensure homogenous melt for particular molding conditions. Non-uniform melt gives non-uniform shrinkage. Melt too cold gives inadequate pressure and/or non-uniform shrinkage.
D. Cycle	Keep injection forward time long enough for gate seal. Overall cycle should be long enough for part to maintain structural stability after ejection.
E. Pressure	Use minimum pressure consistent with good fill. Too low a pressure results in excessive shrinkage. Too high a pressure results in localized over-packing.

*Key items in design of part and mold.

ences in a molded part. By recognizing this, and understanding the various conditions that affect shrinkage, it is possible to properly design and mold parts that are essentially warp-free.

The key, of course, is the initial design of the part. If this is not done properly, it is almost impossible to make corrections through changes in molding conditions.

Warpage caused by mold design or molding conditions sometimes can be corrected by annealing or fixturing. Warpage caused by part design, however, usually is made worse by annealing. Therefore, it should be emphasized that warpage must be *prevented* through proper part design so that it need not be corrected in molding.

Table 11-4 is a troubleshooting guide for warpage. It is meant to assist those who have already reached the point where warpage is a problem, as well as those who are just beginning to design a part.

12

Blow Molding of Thermoplastics

Historically, the blow molding of thermoplastic materials began during World War II. Polystyrene was the first material used with the newly developed blow molding machines, and low-density polyethylene was used in the first large-volume commercial application, a squeeze bottle for deodorant. In the beginning, the plastic bottle business was dominated by companies such as Owens-Illinois, Continental Can, American Can, Plax, Imco, and Wheaton Industries, using proprietary technology and equipment. The introduction of high-density polyethylene and the commercial availability of blow molding machines, mostly from such German companies as Fischer, Bekum, and Kautex, led to phenomenal industrial growth and diversity in the 1960s.

Basically, blow molding is intended for use in manufacturing hollow plastic products; a principal advantage is its ability to produce hollow shapes without having to join two or more separately molded parts. Although there are considerable differences in the available processes, as described below, all have in common production of a parison (precursor), enclosing of the parison in a closed female mold, and inflation with air to expand the molten plastic against the surface of the mold, where it sets up into the finished product.

Differences exist in the way that the parison is made (i.e., by extrusion or by injection molding); in whether it is to be used hot as it comes from the extruder or injection molding machine (as in conventional blow molding), or stored cold and then reheated (as in cold preform molding); and in the manner in which the parison is transferred to the blow mold or the blow mold is moved to the parison.

The basic process steps remain the same, however:

1. Melt the material.
2. Form the molten resin into a tube or parison.
3. Enclose the hollow parison in the blow mold.
4. Inflate the parison inside the mold.
5. Cool the blow-molded part.
6. Remove the part from the mold.
7. Trim flash, as needed.

In many cases, all these steps can be carried out automatically, with the finished products conveyed to downstream stations for secondary operations and packaging.

Although there are many variations, the two basic processes are extrusion blow molding and injection blow molding. Extrusion processes are by far the more widely used, but injection blow molding and injection stretch blow molding have captured significant market segments. While reviewing these methods, the reader is urged to refer to Chapters 4 and 5 for additional background material.

Reviewed and updated by Samuel L. Belcher, Sabel Plastechs Inc., Cincinnati, OH.

Fig. 12-1. Reciprocating screw, intermittent extrusion system for blow molding milk and juice containers. (*Courtesy Johnson Controls, Inc.*)

EXTRUSION BLOW MOLDING

The true secret of good extrusion blow molding lies in using a good plastifier, good die head design, and good tooling. The goal is to produce a uniform melt, form it into a tube with the desired cross section, and blow it into the exact shape of the product. Figure 12-3 shows a section through a typical extrusion die head, and Figure 12-4 shows three typical extrusion die head designs for blow molding. For information on extruder design and selection, refer to Chapter 4. Mold design is covered later in this chapter.

Fig. 12-2. Single-stage injection stretch blow molding machine for PET bottles. (*Courtesy Cincinnati-Milacron*)

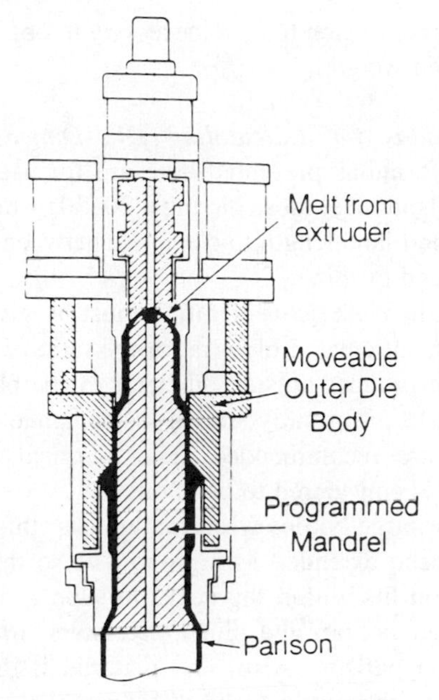

Fig. 12-3. Section through a typical extrusion die head.

Process Variables*

Obviously, the process parameters to be considered in blow molding will be conditioned by the type of resin used (e.g., making an acetal product would involve higher blow pressures than would be required for polyethylene), the type of blow molding unit used, and the product being made.

The discussion below deals primarily with the extrusion blow molding of high-density polyethylene bottles—the technique, material, and application in most common use today. The process variables discussed cover the extruder die (for making the parison) and the blowing air.

Die. In a sense, the parison die has become the key element in blow molding because it controls material distribution in the finished item and, in turn, the economics of the final product. Therefore, increasing attention has been devoted to making the programming die work to improve economics as well as properties. The main control factor in parison programming is the core pin. This pin can be given greater latitude by providing a taper at the die face and providing for movement of the pin so the opening at the face of the die can be made larger or smaller as required to deliver parisons with thicker or thinner walls. Such a movable

*This section courtesy of Soltex Polymer Corporation.

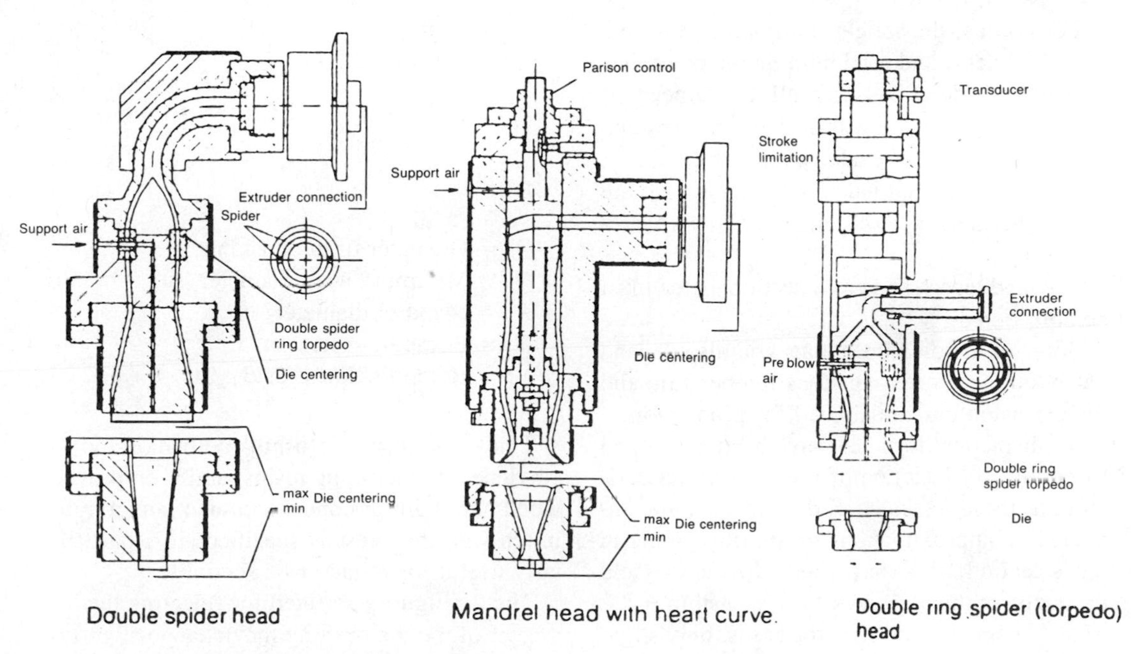

Fig. 12-4. Three basic parison extrusion die heads. (*Courtesy Battenfeld-Fischer*)

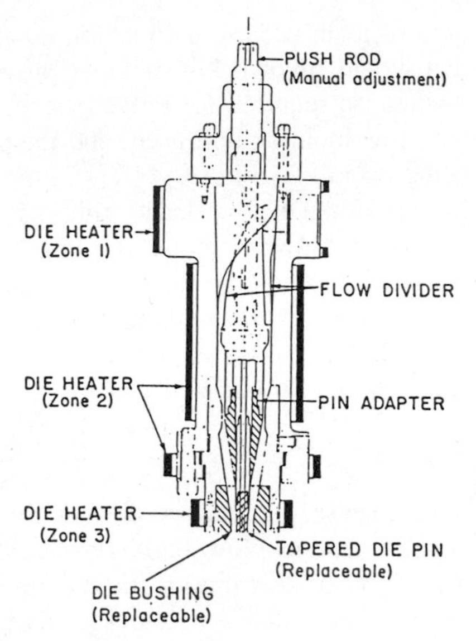

Fig. 12-5. Manually variable die. (All illustrations on Processing Variables, *Courtesy Soltex Polymer Corp.*)

core pin is schematically diagrammed in Fig. 12-5.

Die Dimension Calculations. In selecting the die bushing and mandrel dimensions to be used for the production of a blow-molded polyethylene product, several features must be considered.

For bottles, the weight, minimum allowable wall thickness, and minimum diameter are important considerations, as well as the need, if any, to use a parison within the neck area and whether there may be adjacent pinch-offs.

The type and melt index of the resin used are factors because of swell and elasticity characteristics.

Die land length and cross-sectional area must be considered.

Some of the die dimensions will also depend partly on the processing stock temperature and the extrusion rate anticipated for production.

Mathematical formulas have been developed to permit the selection of die dimensions. Although these calculated dimensions are intended as approximations or starting points in die selection, they have been found to yield products, in the majority of cases, within ±5% of the design weight. In some cases, only slight changes in mandrel size or stock temperature and/or extrusion rate are necessary to obtain the desired weight.

Formulas for Calculating Die Dimensions. The formulas presented here are for use with long land dies, those having a 20–30 : 1 ratio of mandrel land length to clearance between mandrel and bushing.

In their use, consideration must be given as to the anticipated blow ratio, the ratio of maximum product outside diameter to the parison diameter. Normally, ratios in the range of 2–3 : 1 are recommended. The practical upper limit is considered to be about 4 : 1.

For large bottles with small necks, this ratio has been extended as high as 7 : 1 so that the parison fits within the neck. In such a case, a heavier bottom and pinch-off results from the thicker parison. Also, less material is distributed in the bottle walls 90° from the parting line than in similar bottles with lower blow ratios.

When the neck size of a bottle or the smallest diameter of the item is the controlling feature (as when the parison must be contained within the smallest diameter), the following approximations may be used to calculate die dimensions:

For a free falling parison:

$$D_d \cong 0.5N_d$$

$$P_d \cong D_d^2 - 2B_d t + 2t^2$$

where:

D_d = Diameter of die bushing, in.
N_d = Minimum neck diameter, in.
P_d = Mandrel diameter, in.
B_d = Bottle diameter, in.
t = Bottle thickness at B_d, in.

This relationship is useful with most polyethylene blow molding resins, and is employed when bottle dimensions are known, and a minimum wall thickness is specified. It is particularly useful for round cross sections.

The 0.5 figure presented for selecting the diameter of the die bushing may change slightly, depending on processing conditions employed (stock temperature, extrusion rate, etc.), resin melt index, and die cross-sectional areas available for flow. It may be slightly lower for a very thin die opening (small cross section) and higher for large openings.

If product weight is specified rather than wall thickness for a process employing "inside-the-neck" blowing, the following approximation may be employed:

$$P_d = D_d^2 - 2W/T^2\ Ld$$

where:

W = Weight of object, g
L = Length of object, in.
d = Density of the resin, g/cc
T = Wall thickness, in.

This system is applicable to most shapes and is of particular advantage for irregularly shaped objects.

A controlled parison is one in which the dimensions are partially controlled through tension (i.e., the rotary wheel, the falling neck ring, etc.).

Because of this, the following relationships are employed:

$$D_d \cong 0.9N_d$$

$$P_d \cong \sqrt{D_d^2 - 3.6B_dt + 3.6t^2}$$

$$P_d \cong \sqrt{D_d^2 - 3.6W/T^2\ Ld}$$

Derivation of Formulas (core pin blow system). When a polymer is forced through a die, the molecules tend to orient in the direction of the flow. As the extrudate leaves the die, the molecules tend to relax to their original random order. Parison drawdown, the stress exerted by the parison's own weight, tends to prevent complete relaxation. This results in longitudinal shrinkage and some swelling in diameter and wall thickness.

Through laboratory and field experience it has been found for most high-density polyethylene blow molding resins that:

$$D_d \cong 0.5N_d$$

$$A_d \cong 0.5A_b$$

where:

D_d = Die diameter
N_d = Minimum neck diameter
A_d = Cross-sectional area of the die
A_b = Cross-sectional area of the bottle

and that:

$$A_D = \frac{\pi}{4}(D_2^2 - P_d^2)$$

$$A_B = \frac{\pi}{4}\left[B_d^2 - (B_d - 2t)^2\right]$$

where:

P_d = Mandrel diameter, in.
B_d = Product diameter, in.
t = Product thickness, at B_d, in.

$$\therefore A_D = 0.5A_B$$

$$= 0.5\frac{\pi}{4}(B_d^2 - B_d^2 + 4B_dt - 4t^2)$$

$$\frac{\pi}{4}(D_d^2 - P_d^2) = 0.5\frac{\pi}{4}(-4t^2 + 4B_dt^2)$$

Dividing through by $\pi/4$ and rearranging terms:

$$P_d^2 = D_d^2 - 2B_dt + 2t^2$$

or:

$$P_d = \sqrt{D_d^2 - 2B_dt + 2t^2}$$

Also:

$$A_B = \frac{W}{Ld}$$

where:

W = Object weight, g
L = Object length, in.
d = Resin density, g/cc

$\therefore$ Since $A_D = 0.5A_B$:

$$\frac{\pi}{4}(D_d^2 - P_d^2)\,4 = 0.5\,\frac{W}{Ld}$$

$$D_d^2 - P_d^2 = \frac{4}{\pi}\,0.5\,\frac{W}{Ld}$$

$$P_d^2 = D_d^2 - 2\,\frac{W}{\pi Ld}$$

$$P_d = \sqrt{D_d^2 - 2W/\pi Ld}$$

The same derivation is employed for controlled parisons except that:

$$D_d = 0.9N_d$$

$$A_D = 0.9A_B$$

Other Considerations. As shown, the sizes selected for the die bushing and mandrel depend on wall thickness of the finished blow molded part, the blow ratio, and certain resin qualities included in the above formulas for various polyethylene blow molding resins. These qualities are parison swell (increase in wall thickness as the parison exits the die) and parison flare (ballooning or puffing out of the parison as it exits the die). Both depend on processing conditions. It has been shown that calculations can be made for the general die dimensions. The other dimensions of the die—approach angles and lengths—vary widely with machinery capabilities and manufacturer's experience. Calculations for these dimensions thus will not be given here. Instead, a few rules of thumb suffice. For example, the land length of the die (see Fig. 12-6) generally is eight times the gap distance between the pin and the die. In simple tabular form, this works out to be:

Gap size (in.)	*Land length (in.)*
Above 0.100	1–2
0.030–0.100	$\frac{3}{4}$–1
Below 0.030	$\frac{1}{4}$–$\frac{3}{8}$

Notice that the land length is at least $\frac{1}{4}$ inch, regardless of gap size. This land length is necessary to get the desired parison flare.

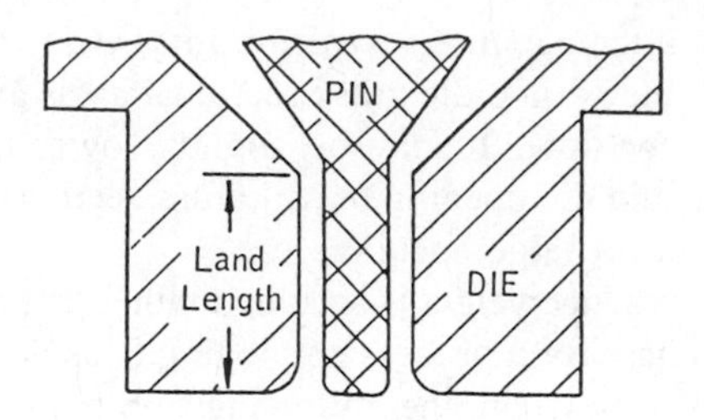

Fig. 12-6. Die and pin.

The die should be streamlined to avoid abrupt changes in flow, which could cause polymer melt fracture. When no further changes are expected in die dimensions, the die mandrel and bushing should be highly polished and chrome-plated. This helps to keep the surface clean and eliminates possible areas of resin hangup. Finally, the edges of the pin (mandrel) and die should have slight radii to minimize hangup within or at the exit of the die area. The face of the mandrel should extend 0.010 to 0.020 inch below the face of the die to avoid having a doughnut at the parison exit.

Air Entrance. In blow molding, air is forced into the parison, expanding it against the walls of the mold with such pressure that the expanded parison picks up the surface detail of the mold. Air is a fluid, just as is molten polyethylene, and as such it is limited in its ability to flow through an orifice. If the air entrance channel is too small, the required blow time will be excessively long, or the pressure exerted on the parison will not be adequate to reproduce the surface details of the mold. General rules of thumb to be used in determining the optimum air entrance orifice size when blowing via a needle are summarized below:

Orifice diameter (in.)	*Part size (vol.)*
$\frac{1}{16}$	Up to one quart
$\frac{1}{4}$	1 quart–1 gallon
$\frac{1}{2}$	1 gallon–55 gallons

Normally, the gauge pressure of the air used to inflate parisons is between 40 and 150 psig. Often, too high a blow pressure will "blow out" the parison. Too little, on the other hand, will yield end products lacking adequate surface detail. As high a blowing air pressure as possible is desirable to give both minimum blow time (resulting in higher production rates) and finished parts that faithfully reproduce the mold surface. The optimum blowing pressure generally is found by experimentation on the machinery with the part being produced. The blow pin should not be so long that the air is blown against the hot plastic. Air blowing against the hot plastic can result in freeze-off and stresses in the bottle at that point.

Moisture in the blowing air can cause pock marks on the inside product surface. This defective appearance is particularly objectionable in thin-walled items such as milk bottles. Use of a dryer is recommended to prevent this problem.

Parison Variations. To obtain even wall distribution in blow-molded products, the parison can be modified from its normal concentric tubular shape. Die bushings can be "notched" or "ovalized" to provide a nonuniform cross section to accommodate nonround product designs (see Fig. 12-7). Parison thickness can be varied in the lengthwise direction as well, by using a process called parison programming (see Figs. 12-8 and 12-9). Credit for developing the first parison programmer is given to Denes Hunkar of Cincinnati, Ohio. His system moved a tapered die mandrel in relation to a fixed die bushing during extrusion to increase or decrease the wall thickness. Others operate in one of the following ways:

1. By varying the extrusion rate.
2. By varying the extrusion pressure.
3. By moving a tapered die bushing in relation to a fixed mandrel.
4. By varying the take-off rate in a continuous parison operation.

Early programmers had the capability to set eight points along the parison length; today, parison programmers are available that can change the thickness up to 128 times. The additional control over wall thickness allowed the blow molding industry to expand rapidly into

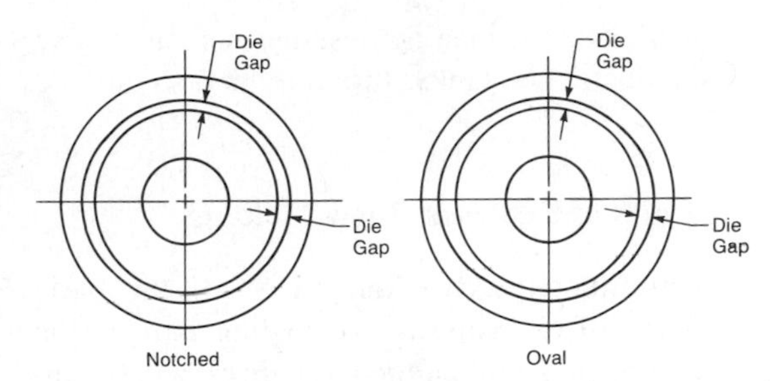

Fig. 12-7. Notched and ovalized die bushings for making non-round products.

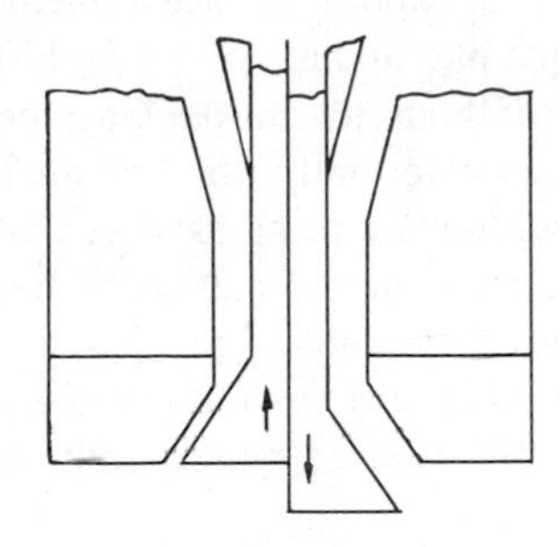

Fig. 12-8. Effect of moving core on thickness of parison wall.

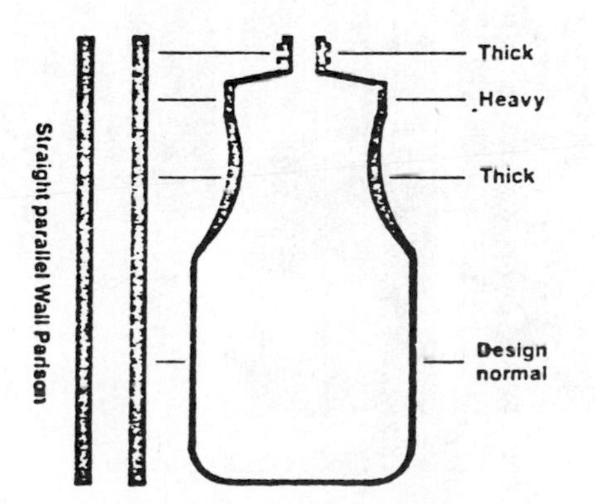

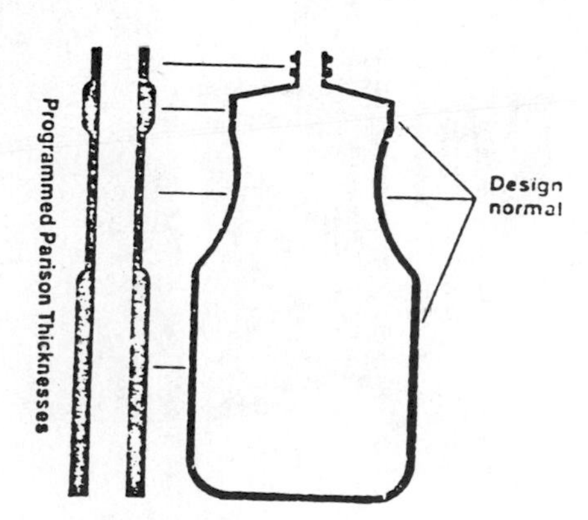

Fig. 12-9. Bottle wall distribution effect of parison programming.

markets other than bottles, such as automotive air ducts, fuel tanks, furniture, and so on.

Types of Extrusion Blow Molding

Continuous Extrusion. One of the basic forms of extrusion blow molding is based on producing a molten tubular parison without interruption. The many variations on continuous extrusion blow molding come about because of the need to move the blow molds in and out of the die area to capture the needed length of tubing for each part and remove it for blowing and cooling. Methods for introducing the blowing air also vary. Normally, the size and design of the product (handle or no handle, center or offset finish, etc.) and the number to be produced will govern the choice of process.

Shuttle mold systems remove the parison to a position below or to one or both sides of the extrusion die for blowing. When the tube reaches the proper length, the blow mold is moved under the die head, where it closes around the parison, pinching one end closed; the tube is severed by a knife or a hot wire, and the mold moves to the blow station to clear the way for the next parison. For higher productivity, more than one parison can be extruded from the die head at a time (see Fig. 12-10). In the common rising mold type of machine, the blow mold rises from below to close around the tube; the blow pin enters from the bottom (see Fig. 12-11). Other adaptations of the shuttle mold process move the blow mold on an incline or use alternating molds moving in from left and right. In these cases, the blow pin normally enters the precut parison from the top (see Figs. 12-12 and 12-13). Effects of moving heavy molds at high speeds limit the shuttle mold process to products of about 2 gallons (8 liters) in capacity.

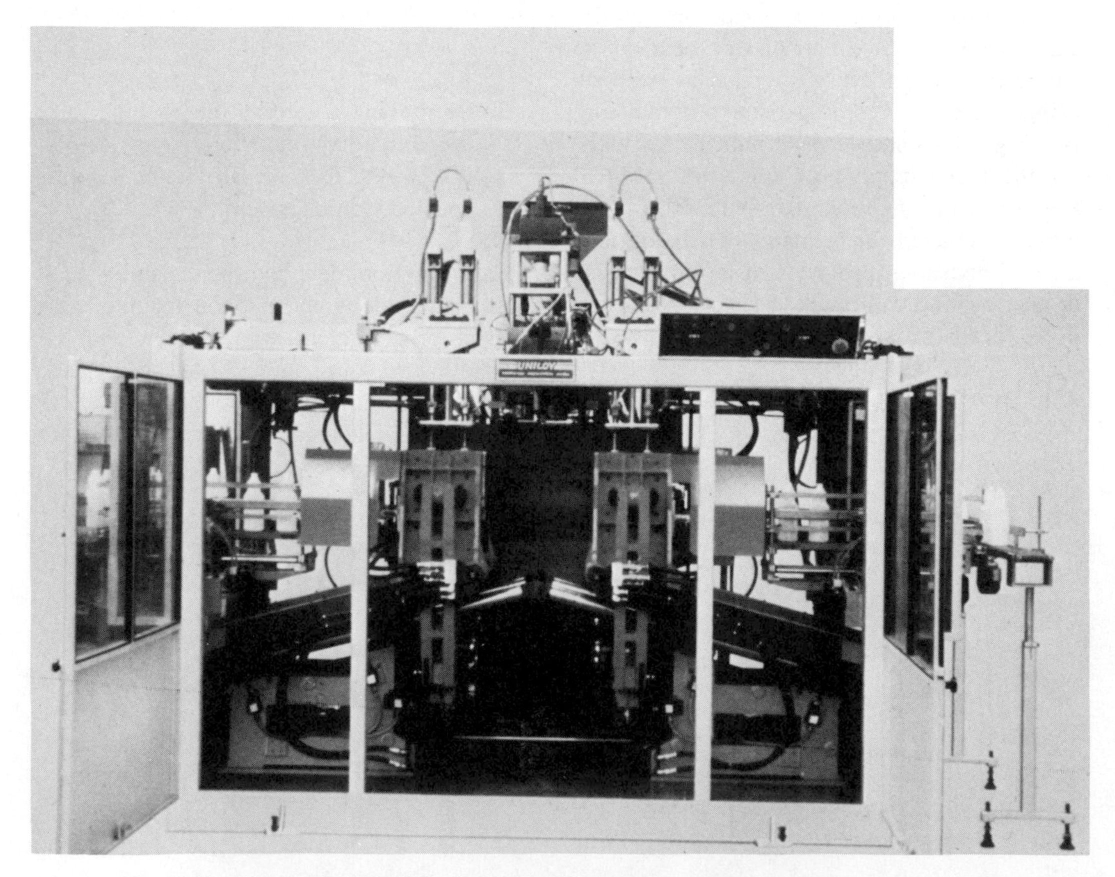

Fig. 12-10. Twin parison, dual shuttle blow molding machine. (*Courtesy Johnson Controls, Inc.*)

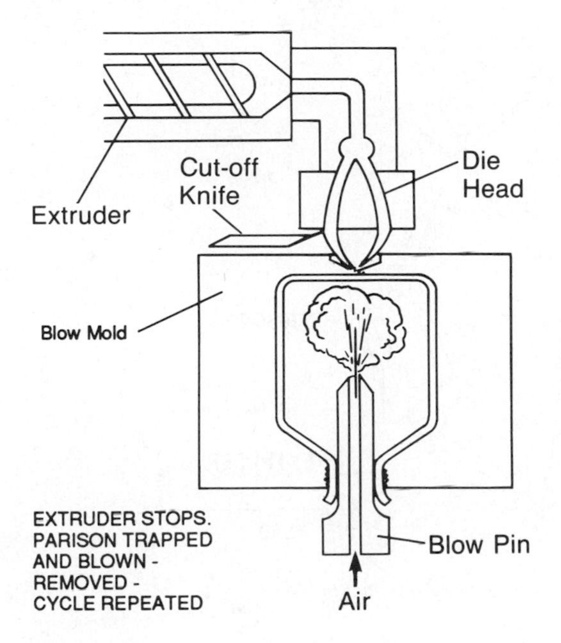

Fig. 12-11. Bottle blown neck down in rising mold machine.

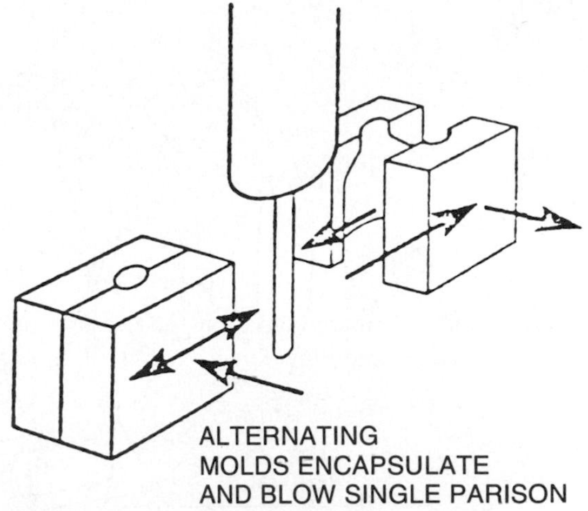

Fig. 12-13. Alternating molds encapsulate and blow single parison.

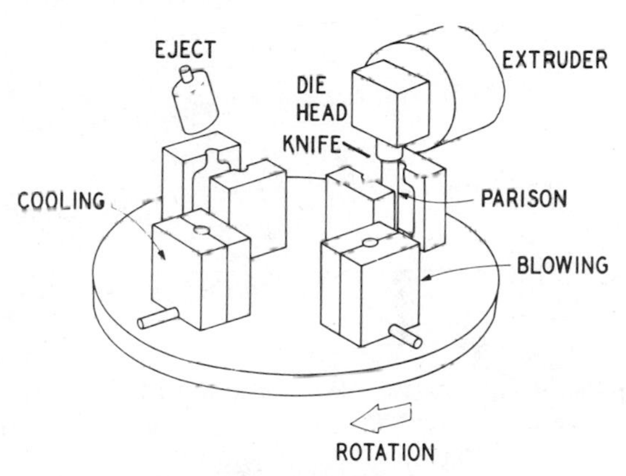

Fig. 12-14. Rotating horizontal table machine. (*Courtesy Phillips 66*)

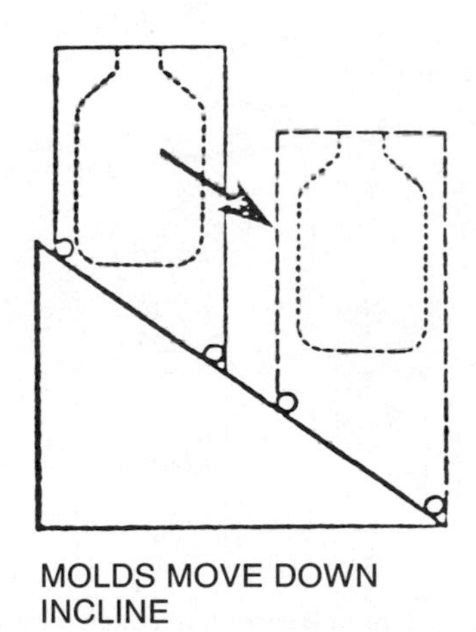

Fig. 12-12. Molds move down incline.

Rotary or wheel type blow molding machines (see Figs. 12-14 and 12-15) were originally developed by Continental Can Company and referred to as ''Mills'' machines. As the wheel rotates, molds close around the continuously extruded parison and move away for blowing and cooling. Blow air is introduced through a hollow needle inserted into the parison from the side. Sometimes, two containers are blown from a single section of parison. In this case, the needle enters in the middle of the parison for blowing both items (see Fig. 12-16). After blowing, excess material is removed in a downstream deflashing operation. Rotary equipment may use as many as 20 mold sets and provides very high output; the high tooling costs and complexity of setup limit this type of machine to long production runs, and they are generally used to mold containers for products such as bleach, detergents, motor oil, and foods. Companies utilizing rotary blow molding machines find that the complexity of the arrangement is more than offset by the productivity per labor dollar.

Another continuous extrusion method uses a transfer mechanism to remove the parison from the die and place it between the blow mold

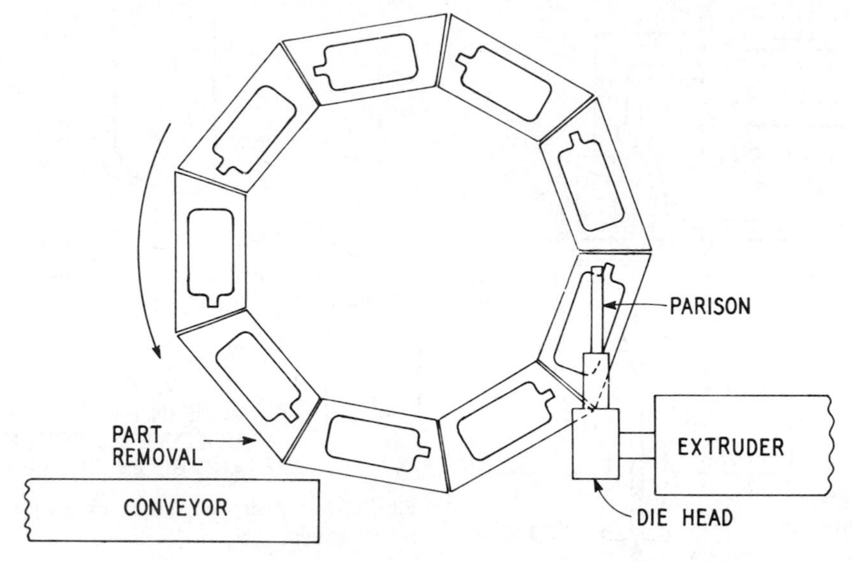

Fig. 12-15. Ferris Wheel type machine. (*Courtesy Phillips 66*)

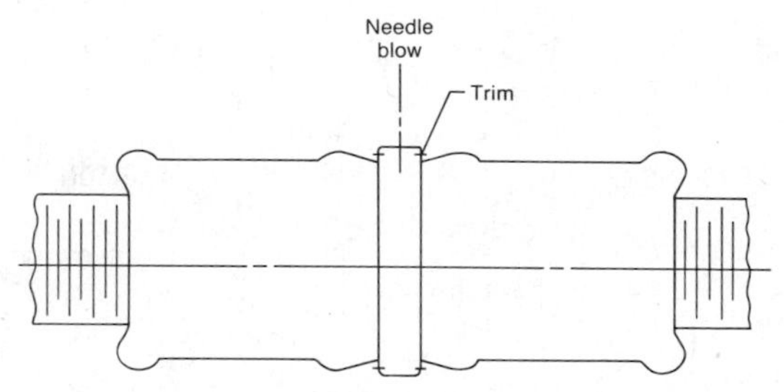

Fig. 12-16. Tandem blow one-quart HDPE oil containers.

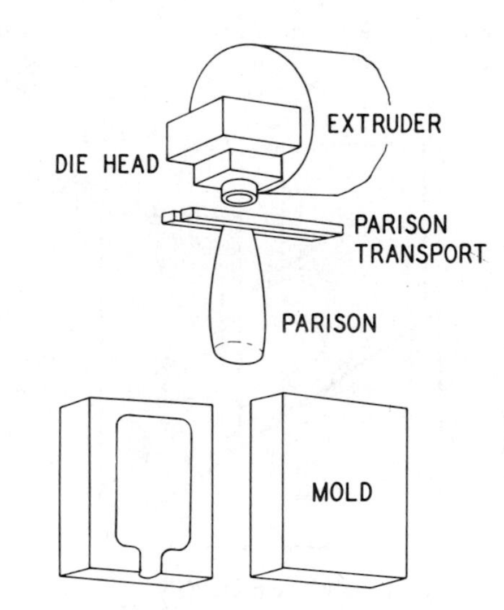

Fig. 12-17. Parison transfer method. (*Courtesy Phillips 66*)

halves. In most machines of this kind, the parison is placed over the blow pin, and the bottles are blown upside down (see Fig. 12-17).

Intermittent Extrusion. Instead of producing parisons on a continuous basis, this class of equipment stores the melted material generated by the extruder while the molded part is being blown and cooled. When the mold is opened, the accumulated melt is forced through the die to make the parison(s). Advantages of intermittent extrusion are that the molds do not have to be shuttled and that the stored melt can be used to produce large parisons rapidly. This is particularly important in using materials that are lacking in hot melt strength and for making very large items such as 55-gallon drums, fuel tanks, or industrial containers. Intermittent extrusion can be accomplished by using a reciprocating screw extruder, a ram accumulator, or an accumulator die head.

A reciprocating screw machine is shown in Fig. 12-18. As the screw rotates, melted plastic is pushed forward to the end of the extruder barrel, and the pressure pushes the screw in the opposite direction. When the necessary amount of melt has collected, a hydraulic cylinder moves the screw forward, forcing plastic out of the die head to form the parison(s). Typical die head and mold positions for making containers

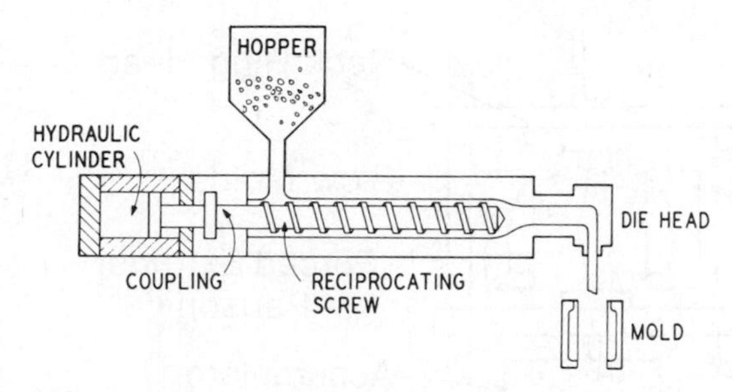

Fig. 12-18. Reciprocating screw blow molder. (*Courtesy Phillips 66*)

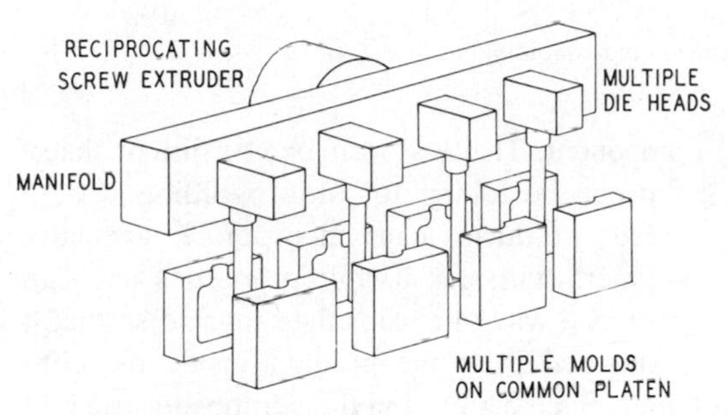

Fig. 12-19. Multiple molds mounted on a common platen. (*Courtesy Phillips 66*)

are shown in Fig. 12-19. With all the molds mounted on a single set of platens, removal of the blown products for secondary operations is greatly facilitated. Reciprocating screw machines generally are used where total shot weight does not exceed 5 pounds.

For larger parts, the ram accumulator is the machine of choice. This type of machine (see Fig 12-20) is used to make parts weighing up to 50 pounds. Typical applications include industrial parts, shipping containers, and toys. The size of the extruder is independent of accumulator size, and in some cases more than one extruder is used. The main disadvantage of the system is that the first material to enter the accumulator is the last to leave, making the process unsuitable for heat-sensitive materials.

A variation on ram accumulators that provides first-in, first-out material flow is the accumulator head machine. Melt from the extruder enters the accumulator head from the side and flows around a mandrel; a tubular plunger pushes the entire shot through the die. Parts as large as 300 pounds have been produced by this method.

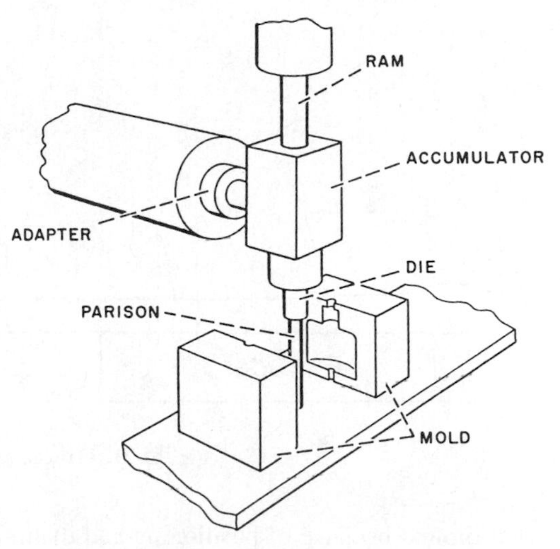

Fig. 12-20. Ram accumulator blow molding.

Forced Extrusion. Developed by Owens-Illinois around 1950, the method was utilized on the company's workhorse BC-3 bottle making machines. In this process, a split "neck ring" mold is moved down to rest against the annular extrusion die (see Fig. 12-21). The extruded melt, stored in an accumulator, is injected to form the threaded finish of the bottle. With the neck rings full, the head rises, and the attached parison is extruded upward to the proper length. The blow molds close and pinch off the lower end of the tube; blow air is introduced through the core pin in the neck ring mold. When the bottle is cool, the blow mold and the neck ring mold open, and the container is removed by gripping the tail flash.

Advantages of the process are:

1. The bottle neck is molded accurately and requires no further finishing.
2. Less scrap is generated because of the molded neck.
3. The parison can be pre-expanded with air pressure as it is being formed to enhance final wall distribution, especially in producing handleware.

Its major disadvantages are that it is relatively slow compared with wheel machines, and that offset finish and handleware designs are some-

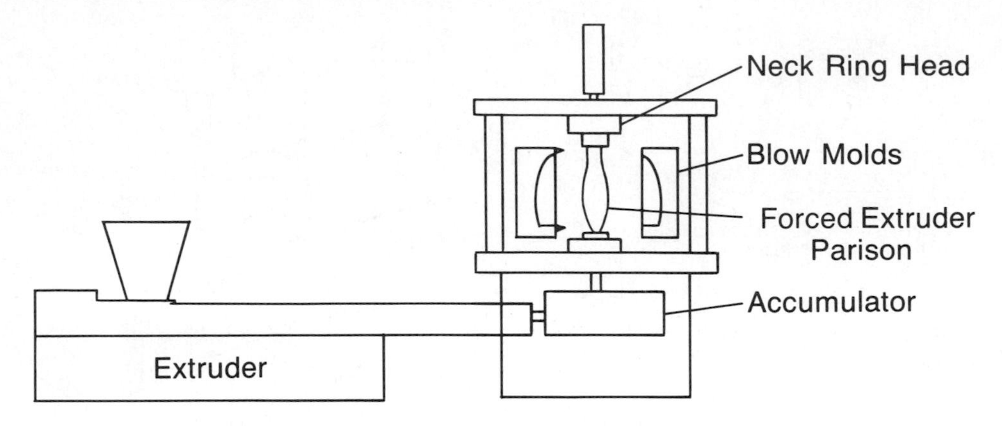

Fig. 12-21. Forced extrusion blow molding.

what limited because of positioning and diameter requirements for the parison.

Dip Blow Molding. A process being utilized in Europe but not commercialized in the United States is Schloemann-Siemag's dip blow molding (see Fig. 12-22). In steps 1 and 2, the core rod, with diameter equal to the inside of the container neck opening, is dipped into the melt accumulator, and the piston retracts. When the neck rings at the top of the core make contact, the piston forces plastic into the finish, in step 3. In steps 4 and 5, the core rod is withdrawn, and the piston moves upward at the same speed. When the rod is completely out of the accumulator, the parison is severed, and the accumulator begins to refill for the next cycle, as shown in steps 6 and 7. Steps 8 through 11 show the parison transferred to the blow mold and blown into the final article.

Form/Fill Seal. A variation on extrusion blow molding is being used on a limited scale to produce sterile, filled, sealed containers for medical and pharmaceutical products. After the package is formed by pulling a vacuum through the mold (see Fig. 12-23), product is filled through the neck opening, which is held open by the vacuum and heated. After the filling tube is removed, the opening then is pinched and sealed off, and the filled package is removed for trimming.

Coextrusion Blow Molding. The coextrusion technique is used to make multilayer materials that combine the advantages of each component. The final form can be film or sheet, or it can be tubing for blow molding. Coextruded products can incorporate specialty polymers with gas barrier properties and can serve as a way to encapsulate regrind so that it is shielded from the product inside the container. Each material in the composite structure requires its own extruder; because adhesive or tie layers may be needed to hold the dissimilar materials together, a typical coextrusion blow molding machine will be equipped with a main extruder and as many as four satellite extruders (see Fig. 12-24). As in single-layer extrusion blow molding, success depends on good tooling and, most important, a good die head design. Figure 12-25 shows a typical coextrusion wall in a bottle plus the coextrusion die head.

Recycling trim and scrap from some multilayer package production can be a problem. If it is put back into the structure as an inner or an outer layer, it can cause delamination and impact failure.

Coextrusion is also used to produce containers with a clear stripe to make the contents level visible. In this instance, the main extruder produces pigmented material for the body of the container, and a satellite extruder produces translucent material for the stripe. In this type of coextrusion, recycling of trim is not a problem.

Part Trimming and Handling

Neck Calibration. To produce a higher-quality threaded neck finish on extrusion blow-molded bottles, techniques have been devel-

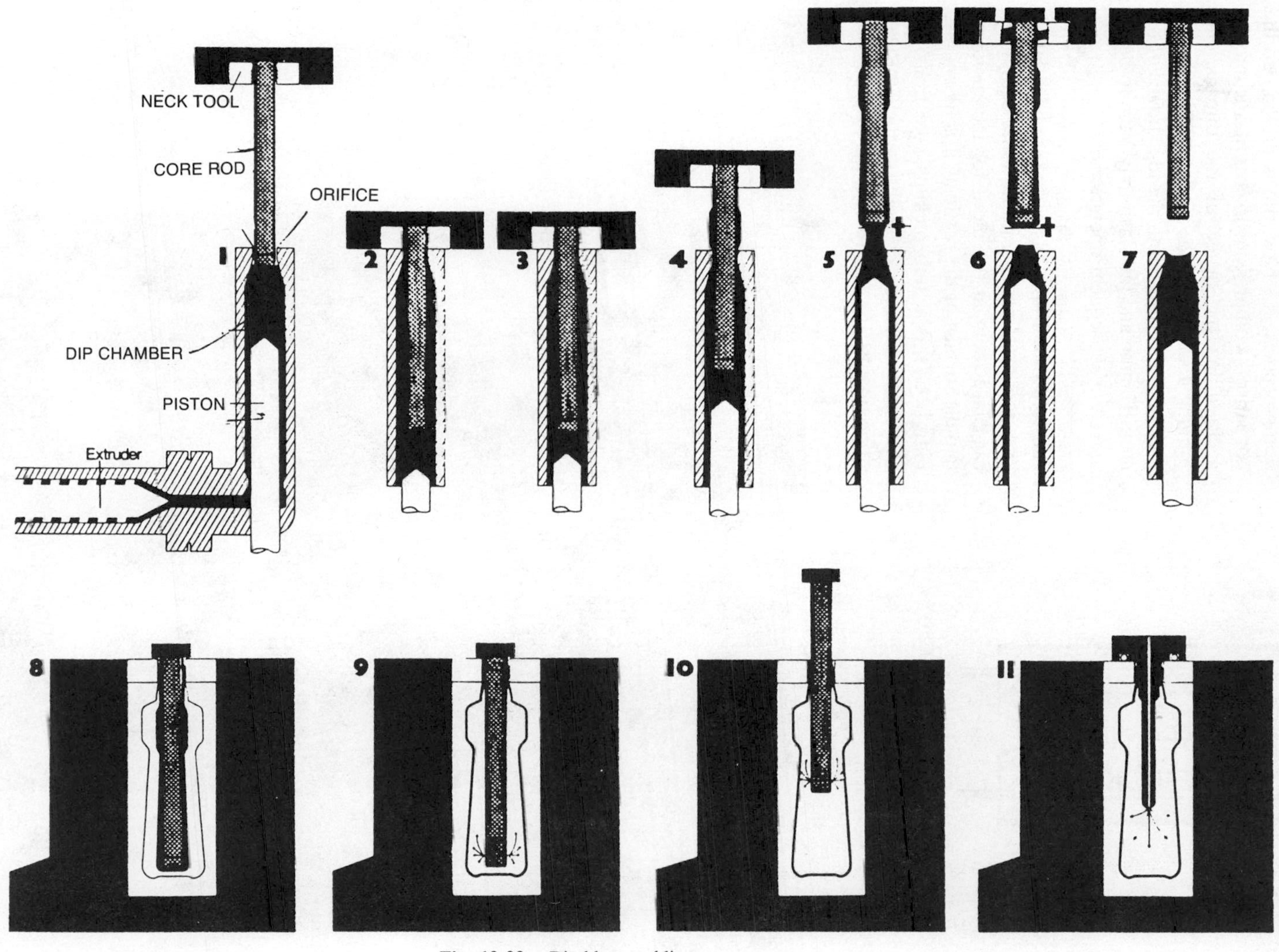

Fig. 12-22. Dip blow molding process.

Fig. 12-23. Form/fill/seal process.

oped to "compression-mold" that portion of the container (see Fig. 12-26). In such calibration processes, the blow pin is driven into the hot parison, compressing excess plastic into the neck, filling the inside of the threads, and forming a smooth inside surface. The pin also may be rotated to smooth the top surface of the neck. Handle and tail flash may have to be removed in subsequent operations.

In-Mold Labeling. Instead of applying labels in a post-molding operation, some equipment can place them in the mold for incorporation into the surface of the bottle as it is blown (see Fig. 12-27). Vacuum pickups take front and back labels from stacks and bring them into position against the open mold halves; the vac-

Fig. 12-24. Coextrusion blow molding machine in operation. (*Courtesy Battenfeld-Fischer*)

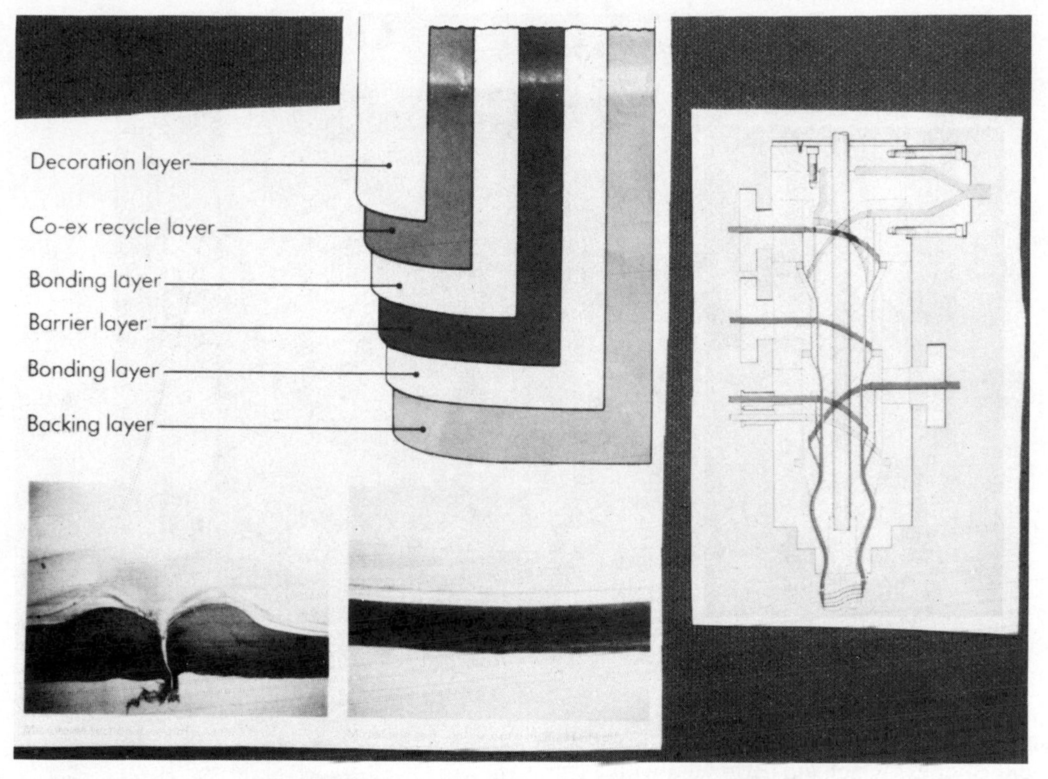

Fig. 12-25. Coextrusion die head and bottle wall section.

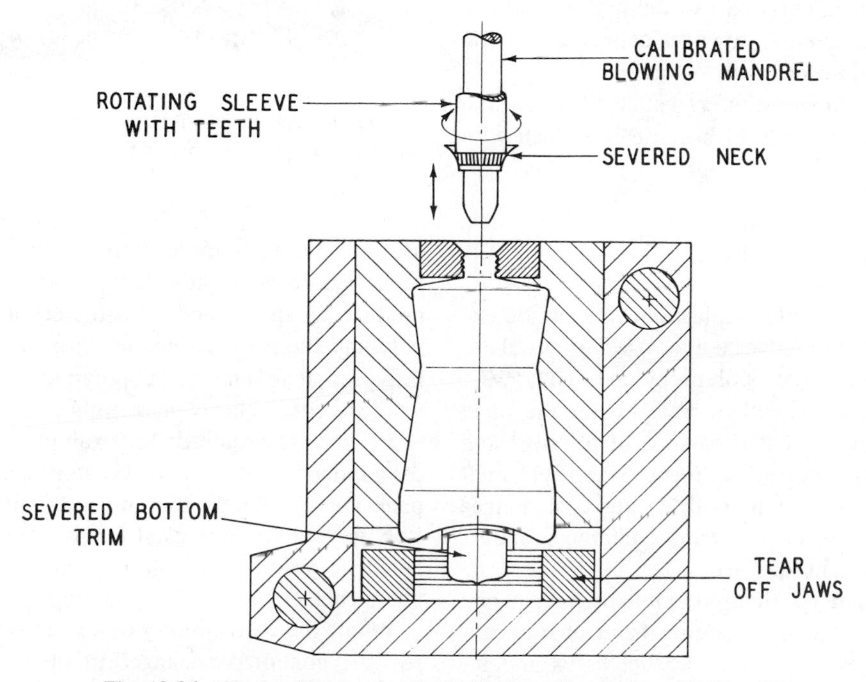

Fig. 12-26. Neck calibration and detabbing in mold. (*Courtesy Phillips 66*)

Fig. 12-27. In-mold labelling system. (*Courtesy Johnson Controls, Inc.*)

uum then is switched from the pickups to the mold cavities, which are perforated with a pattern of small holes. The vacuum holds the labels against the mold surface as the hot parison is expanded. The labels adhere to the molten plastic and, in the case of a plastic label, can become an integral part of the bottle. The use of in-mold labeling slows the overall cycle time by $1\frac{1}{2}$ to $3\frac{1}{3}$ seconds.

In-mold labeling offers better adhesion and protection of the label because it is flush with the surface of the package.

Fig. 12-28. Automobile fuel tank of high molecular weight HDPE. (*Courtesy Soltex Corp.*)

Materials

Although most thermoplastic resins can be extrusion-blow-molded, the most widely used are the polyethylenes, polypropylene, and PVC. High-density polyethylene is by far the most popular. Its excellent balance of physical and processing properties (heat stability, melt strength, etc.), along with its low cost, makes it well suited for producing containers, industrial parts, and toys.

Polyethylenes are used to make all types and sizes of containers, ranging from pharmaceutical bottles to automotive fuel tanks and industrial drums. Low-density polyethylene is used to blow squeeze bottles, high-density polyethylene goes into all types of household product packages, and high molecular weight polyethylene is the choice for large containers. Polypropylene and its copolymers are used where contact clarity and higher temperature resistance are required; for example, in medical products that need to be sterilized and in packages for foods that must be filled hot. Polyvinyl chloride is used for packaging some cosmetic products, edible oils, and household chemicals.

Where the permeability of a resin is too high for it to contain a packaged product, a barrier layer can be encapsulated between layers of the

Fig. 12-29. Returnable polycarbonate milk bottle. (*Courtesy General Electric Plastics*)

structural material by coextrusion. The most commonly used barrier plastics are ethylene vinyl alcohol (EVOH), polyvinylidene chloride (Saran), and nylon. Normally, tie layers made from an adhesive polymer have to be placed between the incompatible barrier and resin materials to hold the structure together.

Engineering thermoplastics, such as polycarbonate, polysulfone, amorphous nylon, and polyphenylene ether blends, are finding increased application in blow-molded products for packaging and structural uses (see Figs. 12-29 and 12-30).

INJECTION BLOW MOLDING

In the injection blow molding process, the parison is injection-molded instead of extruded. The still-molten preform is transferred on the core rod to the blow station, where it is expanded into its final form. Credit is given to Piotrowsky for developing the first true injection blow molding process, which was performed on a modified injection molding machine. Other methods appeared that shuttled parisons from the injection station to blowing stations, but industry has generally accepted the Gussoni method for production.

The Gussoni method utilizes a horizontal rotary table (see Fig 12-31) on which the parison cores are mounted. At the first station, parison molds close around the cores, and the preforms are injected by a reciprocating screw extruder. The table indexes, and at the second station, blow molds close around the hot parisons, and they are blown through openings in the core pins. After cooling, the table indexes again, and the blown items are stripped off the core rods. Figure 12-32 depicts the total process.

Figure 12-33 shows the core rods, open parison mold, secondary injection nozzles, and die set. Note the hoses running from the rotating head to each internally cooled core rod. Figure 12-34 shows a typical blow mold die set for a three-station machine. Figure 12-35 shows a four-cavity setup in the stripper position, with the blown bottles ready to be blown off onto a conveyor belt for packing.

Fig. 12-30. Blow molded Xenoy bumper for Hyundai. (*Courtesy General Electric Plastics*)

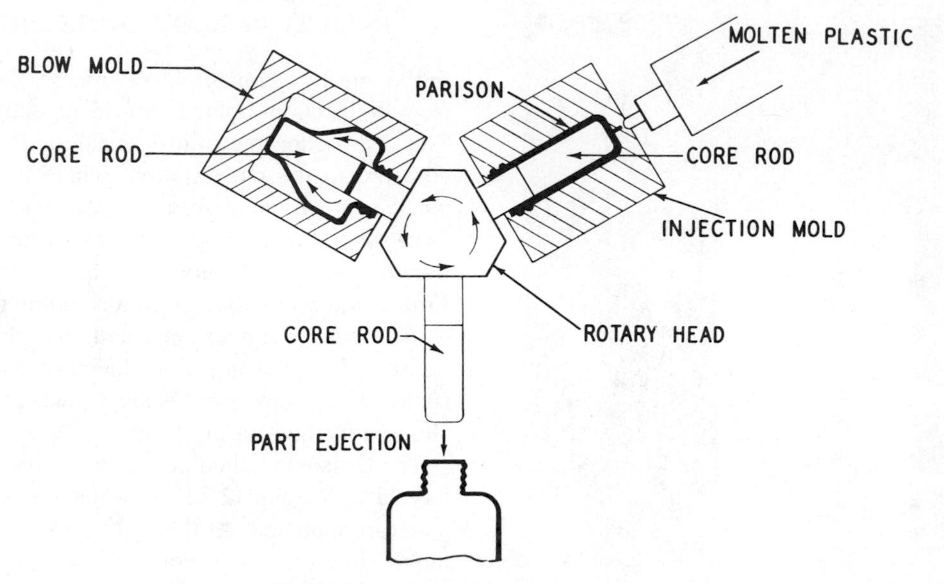

Fig. 12-31. Injection blow molding system.

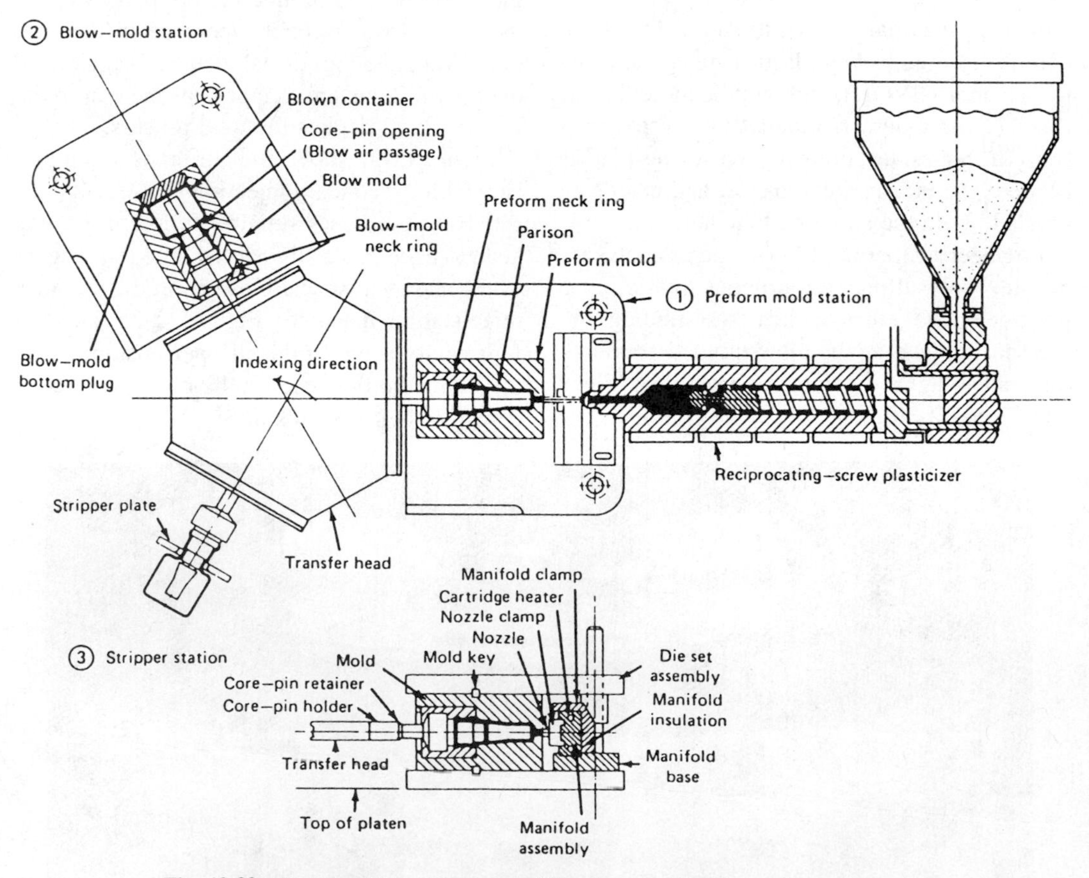

Fig. 12-32. Details of injection blow molding process. (*Courtesy Johnson Controls*)

Fig. 12-33. Core rods and open parison mold. (*Courtesy Bekum, W. Germany*)

Some injection blow molding machines have a fourth station that can be used for different purposes. Wheaton Industries pioneered the four-station design, primarily for safety. The extra station can be used for treating or labeling the bottle after blowing, or to preblow the parison before it goes to the blow station, or to condition the core rods in a temperature-controlled air chamber. (See Fig. 12-36.)

Most machines use a horizontal reciprocating screw extruder; Jomar uses a vertical extruder to reduce the amount of floor space needed for the machine.

Advantages of injection blow molding are:

1. The process produces completely finished products with no trim scrap.
2. High-quality neck molding provides dimensional accuracy for making snap-fit and child-resistant finishes and for accepting sprayers and other inserts.
3. Injection-molded parisons allow accurate control over bottle weight and wall thickness.
4. There is no pinch-off scar on the bottle.

On the negative side, injection blow molding cannot be used to produce bottles with handles, and its tooling costs are much higher than those of extrusion blow molding, as the production of a single container requires an injection mold, three or four core rods, and a blow mold. The higher cost can be offset by high productivity for small containers, where the number of cavities is high. Injection blow molding normally is not used for short runs or for bottles over $\frac{1}{2}$ liter in size.

Three main factors are involved in evaluating an injection blow molding machine. First is the capacity of the reciprocating extruder to plastify enough material for the maximum number of cavities within the desired cycle time. Second is injection station clamp tonnage needed to withstand filling pressure. Clamp tonnage is calculated by using the projected area of the preform, multiplied by the number of cavities, times the injection pressure required for filling. Normal injection pressure for polyolefins is in the 3000 psi range, whereas for PVC, polycarbonate, and PET it can be as high as 8000 psi. The third factor is the mini-

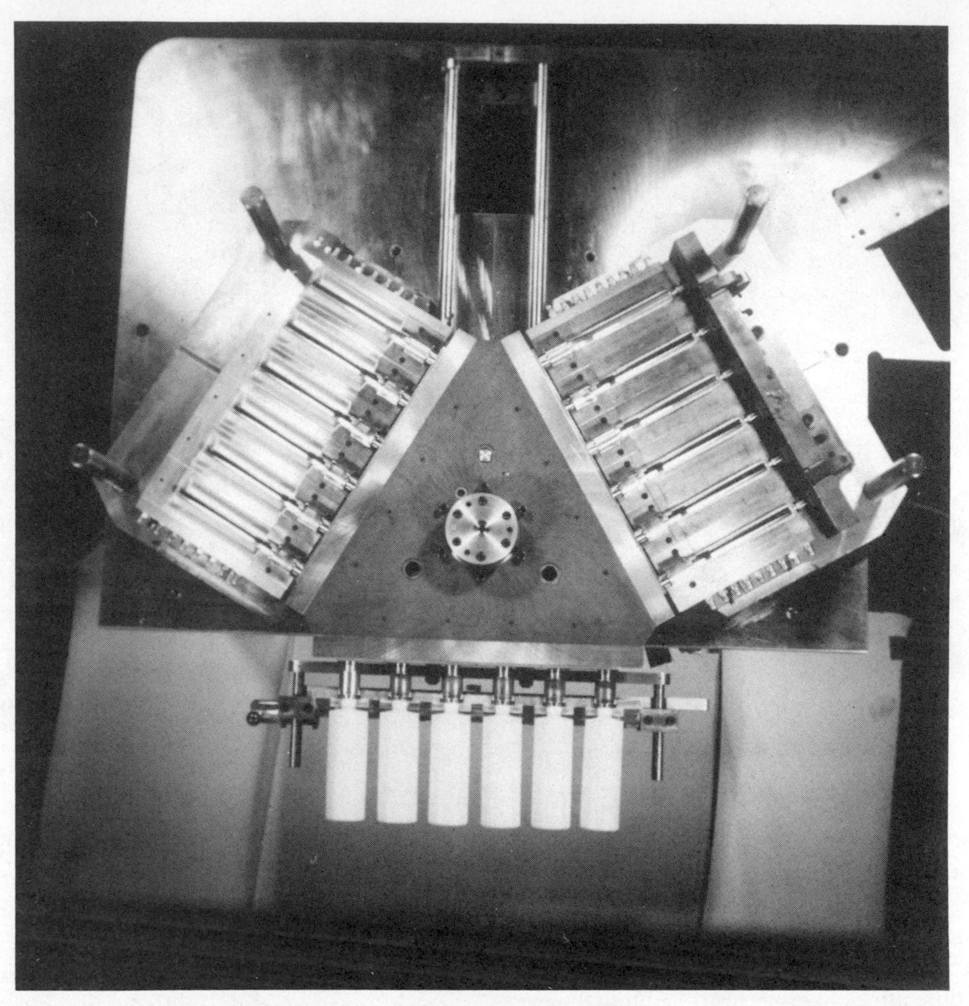

Fig. 12-34. Typical die set for three-station machine. (*Courtesy Jomar Corporation*)

Fig. 12-35. Stripper station on four-station machine. (*Courtesy Bekum, W. Germany*)

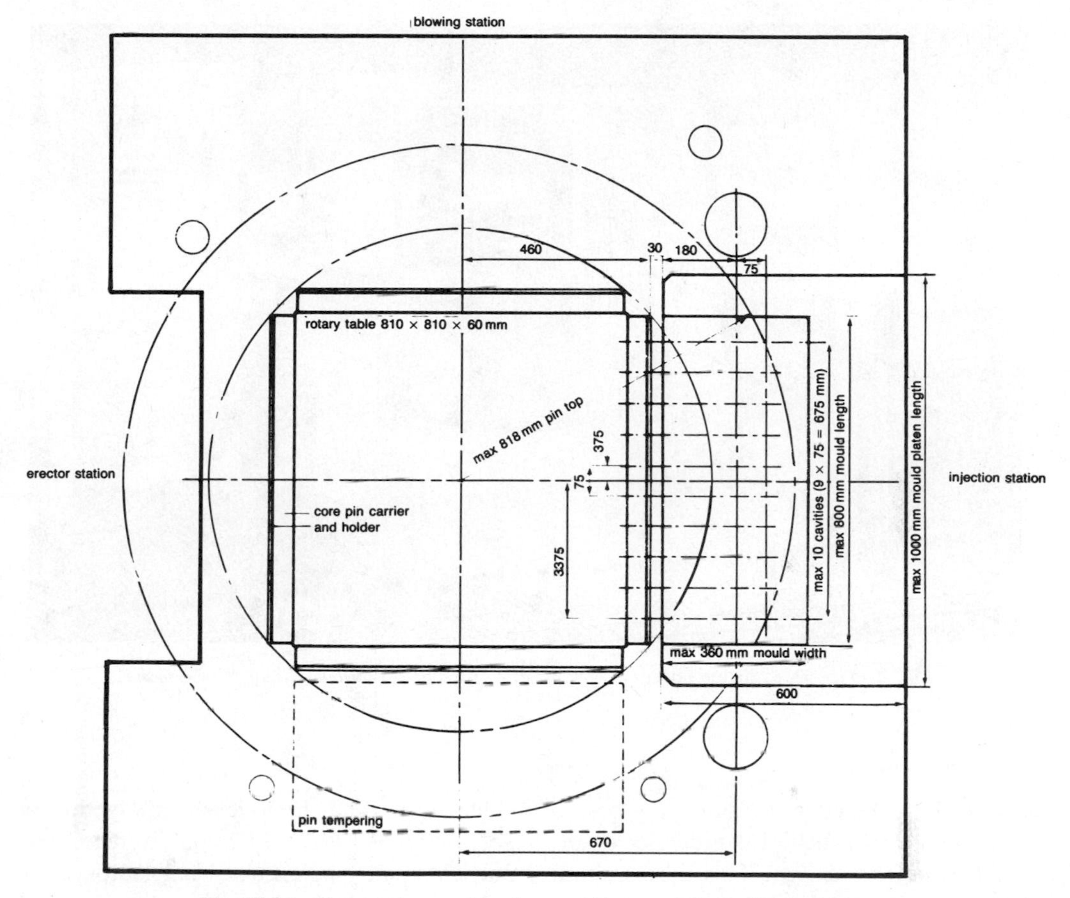

Fig. 12-36. Four-station machine layout. (*Courtesy Bekum, W. Germany*)

mum time needed to open the molds, index, and close the molds again. This can range from $1\frac{1}{2}$ to almost 3 seconds. Buyers should check out all three of these factors before selecting a production machine.

Design Considerations

The design of the preform in injection blow molding is critical. It should be designed to have a wall thickness in the body of the preform anywhere from approximately 0.035 inch (about 1 mm) to approximately 0.200 inch (5 mm). The preform length is designed to clear the inside length of the bottle in the blow mold by approximately .040 inch (1 mm). Thus there is minimum stretch in the axial direction of the preform when the bottle is blown. The diameter of the core rod is in all practicality determined by the maximum inside dimension (I-dimension) of the finish of the desired container. In determining the wall thickness of the preform in the main body, it is necessary to know what wall thickness is desired in the final blown article plus the maximum inside diameter of the desired blown article. The ratio of the inside diameter of the preform (D_1) to the inside diameter of the blown container (D_2) is known as the hoop ratio.

If the wall thickness in the blown article is to be 0.022 inch (approximately 0.56 mm) and the hoop ratio is 3, then the preform should have approximately 0.066 inch (1.67 mm) wall thickness in the main body. The tip of the preform should be designed for good plastic flow, yet not be bullet-shaped, to permit easy deflec-

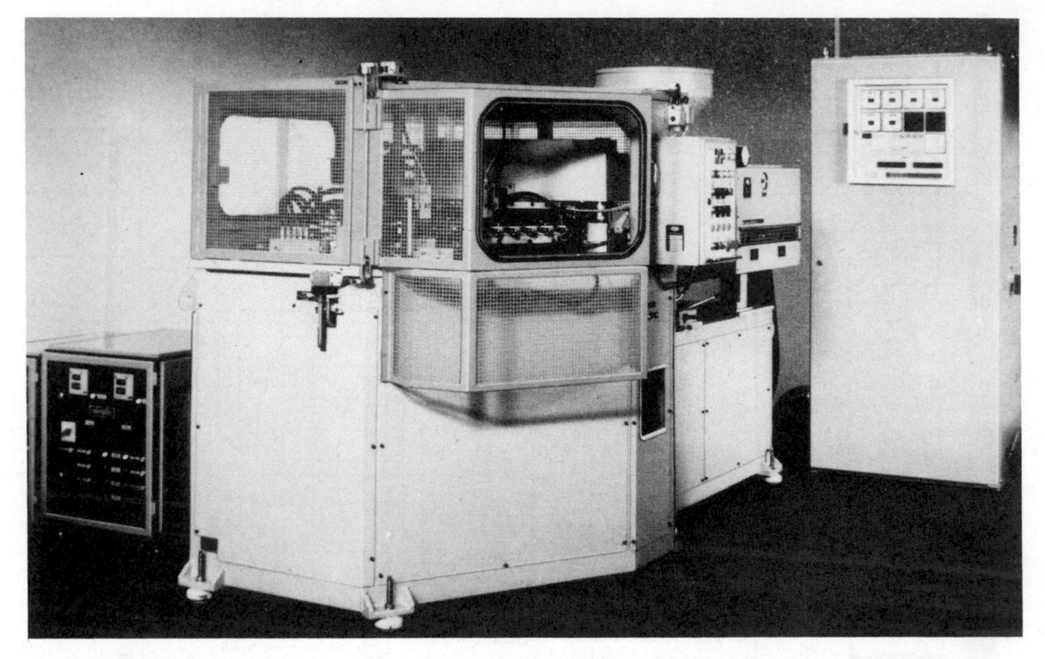

Fig. 12-37. Three station injection blow molding machine. (*Courtesy Bekum W. Germany*)

tion due to the injection pressure. It is best to have at least 0.045 inch (1.1 mm) flat on the tip of the preform design. This measurement can be greater, depending on the gate diameter or the orifice size of the gate. The outside radius of the core rod should be designed to approximately 20% extra material at this area because this is the material that expands the greatest amount and the farthest, as it makes the heel area of the blown container. Injection blow mold tooling is a subject in itself, as is most injection molding, blow molding, or any other plastic process tooling. Injection blow molding is both an art and a science, but with the new process controls it is becoming more scientific.

A detailed discussion of injection blow mold design is presented later in this chapter.

Many thermoplastics can be processed by injection blow molding, some of them being specially formulated for use on IBM machines. Easy flow, metal release, heat stability, and melt strength are among the properties that are specifically designed into compounds for IBM processing. Resins currently being used include low- and high-density polyethylene, polypropylene, PVC, PET, polystyrene, ABS, polycarbonate, polyacetal, polysulfone, and polyacrylonitrile.

Coinjection Blow Molding

Just as multilayer bottles can be produced from a coextruded parison, they can be made from a coinjection-molded parison. Equipment to produce coinjection blow-molded bottles has been run on a commercial basis.

STRETCH BLOW MOLDING

The soft drink industry's adoption of oriented polyethylene terephthalate (PET) bottles for its two-liter package catapulted injection stretch blow molding into the spotlight. In the late 1960s and early 1970s, several companies were working to produce a thermoplastic soft drink container. In cooperation with Coca-Cola, Monsanto marketed its Lopac polyacrylonitrile bottle in one-liter and 16-ounce sizes. In 1976, the Lopac bottle was removed from the market

Fig. 12-38. Injection blow molded container.

after coming under the scrutiny of the Food and Drug Administration.

The oriented PET bottle was invented at Du Pont by Nathaniel C. Wyeth, brother of painter Andrew Wyeth, in collaboration with Ronald N. Roseveare and Frank P. Gay. Du Pont was granted basic patents in 1973 and 1974 and has since issued licenses to more than 70 PET bottle makers around the world. In the United States alone, almost 10 billion PET soft drink bottles are being produced annually.

Both the strength and the barrier properties of plastics—specifically, tensile, drop impact, top load, and burst strength, as well as clarity,

Fig. 12-39. Stretch blow molded containers. (*Courtesy Nissei ASB*)

Table 12-1. Typical barrier properties of plastics.

	CO_2^a	O_2^a	Water vapor transmission[b]
Polyester (PET)	12–20	5–10	2–4
Polyacrylonitrile	3	1	5
PVC	20–40	8–15	2–3
High-density polyethyl ene	300	110	0.5
Polypropylene	450	150	0.5
Polycarbonate	550	225	75

[a]cm^3-mil / 100 in.^{2}-day-1 atm at 73°F, 0% RH.
[b]g-mil / 100 in.^{2}-day at 100°F, 100% RH.

oxygen, carbon dioxide, and moisture barrier (see Table 12-1)—can be improved by orienting the polymer molecules. Stretch blow molding normally allows a minimum 10% weight savings per container, and up to 30% in sizes above one liter.

Orientation is achieved by stretching the material at a temperature below its crystalline melting point but above its glass transition temperature. In stretch blow molding, the parison is brought to orientation temperature, and is stretched in the axial direction by mechanical means or by air pressure and in the hoop direction by blowing at high pressure into the mold. In this manner, orientation in both directions, or biaxial orientation, is achieved. Figure 12-40 shows the injected preform used to produce the two-liter beverage bottle (base cup design).

Although other thermoplastics, such as polypropylene, PVC, polyacrylonitrile, acetal, nylon, and ionomer, can be injection stretch blow-molded, it is PET that has gained widest commercial application because of its crystal clarity and its acceptance for use in contact with food products.

In order to understand stretch blow molding, it is necessary to understand the terms orientation temperature, hoop ratio, axial ratio, and blow-up ratio. *Orientation temperature* is defined as the range within which the preform must be maintained for the polymer molecules to become aligned during stretching. The orientation temperature range needed to achieve maximum properties from various materials is shown in Table 12-2. *Hoop ratio* is defined as

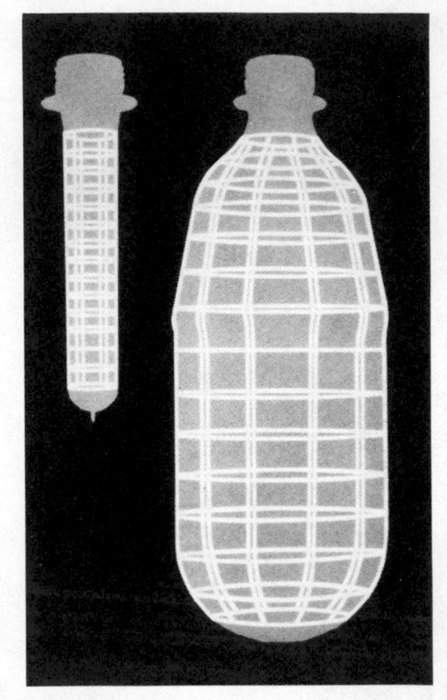

Fig. 12-40. Preform design for two-liter PET bottle.

Table 12-2. Orientation temperatures and stretch ratios for stretch blow molding materials.

Material	*Orientation temperature*	*Stretch ratio*
Polypropylene	260–280°F	6/1
Polystyrene	290–320°F	12/1
Polyethylene terephthalate	195–240°F	16/1
Polyacrylonitrile	240–260°F	12/1
Polyvinyl chloride	210–230°F	7/1

the ratio of the largest inside diameter (D_1) of the blown article to the inside diameter (D_2) of the preform, or D_1/D_2. *Axial ratio* is defined as the ratio of the length (A_1) from where stretching starts to the inside bottom of the blown article to the same measurement (A_2) on the preform, or A_1/A_2. The total *blow-up ratio* (BUR) is equal to the hoop ratio times the axial ratio (see Fig. 12-41). Orientation and BUR can be different throughout the blown article, as shown for a two-liter bottle in Fig. 12-42.

Each type of material also has a "natural" stretch ratio, above which it will burst or will not expand farther. Those ratios also are displayed in Table 12-2.

In designing a stretch-blown container for holding pressure, the hoop ratio is the most im-

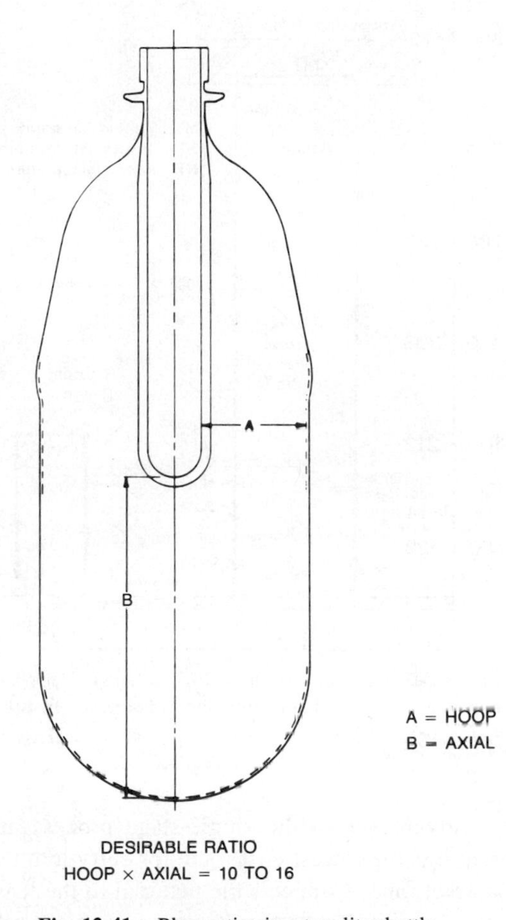

Fig. 12-41. Blow ratios in a two-liter bottle.

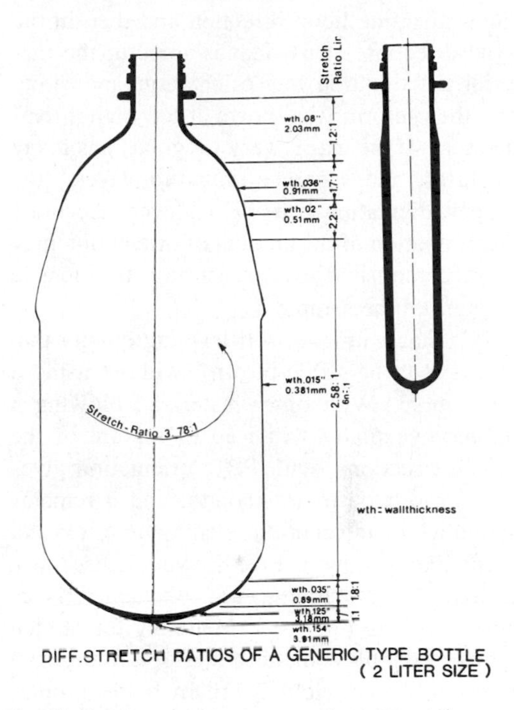

Fig. 12-42. Stretch ratios in a two-liter bottle. (*Courtesy Hoechst Celanese Corp.*)

portant factor determining creep resistance, burst resistance, and barrier performance. The BUR should be at least 10, with the hoop ratio at least 4.8. The original two-liter PET soda bottle with base cup had a BUR of 10.48. If pressure retention is not a factor, the BUR should be selected to ensure that the blown bottle has the necessary wall thickness to prevent collapsing and to offer sufficient top load strength for closing, stacking, filling, and so on. Loss of oxygen or moisture through the walls will create a partial vacuum inside a sealed bottle, causing paneling or distortion. The compatibility of the product to be packaged with the plastic being used should be carefully checked in all cases.

Another use for the BUR is in calculating the minimum wall thickness for the main body of the preform. For example, if the molder wishes to have a wall thickness of 0.020 inch in the bottle and a BUR of 8, the preform will have a minimum wall thickness of 0.020 × 8 or 0.160 inch.

Biaxial stretching of the preform not only produces orientation but induces crystallization in the material. Both orientation and crystallinity increase the barrier effectiveness of a container, but it must be understood that orientation and crystallinity are not directly related. Orientation will take place only if the material is stretched at temperatures within a certain range. Crystallinity, on the other hand, can be produced by orientation and/or heating.

Orientation can be checked in the laboratory by comparing the tensile strength of specimens cut from blown bottles with that of a standard molded plaque of unoriented PET. By cutting horizontal and vertical specimen strips, the degree of axial and hoop orientation can be determined.

A quick, easy check that can be used on the production floor to determine if orientation is being achieved in both directions is to cut the blown bottle in each direction and to tear the sidewall in the hoop direction and then in the axial direction. If the tear is smooth, the material is exhibiting low orientation, indicating that the preform was blown at too high a temperature. If the tear is very irregular, has many fractures, and separates into thin layers, the proper orientation is being achieved. Accurate determination of the amount of orientation may be accomplished by measuring the tensile strength of the sample.

PET has a unique, self-leveling quality that allows it to be "free-blown" without using a blow mold. With other materials, blowing a bubble eventually will lead to rupture of the thinnest sections; with PET, orientation gives the thin section extra strength, and it remains stable while adjacent areas continue to expand much like a balloon. Free blowing is used as a technique for studying stretch characteristics and designing preforms. Matching the stretch ratios of the blow mold to that of a free-blown preform should yield optimum bottle properties.

There are basically two ways to make injection stretch blow-molded bottles. One is to injection-mold preforms in a conventional manner and later to reheat them to orientation temperature for stretching and blowing. This two-stage system is known as the reheat/stretch/blow process. The first injection-molded preform was made by Broadway Companies of Dayton, Ohio; until then, PET preforms were made by cutting PET pipe into lengths and heat-forming the ends, one to create threads and the other to close the bottom.

The second basic method is a continuous process in which the molded parisons are immediately "conditioned" to orientation temperature and then stretched and blown. Crystallization in the single-stage (Method A) and two-stage (Method B) processes is depicted graphically in Fig. 12-43.

The single-stage process consists of the following: (a) plastifier, (b) injection mold, (c) conditioning station, (d) stretch blow molding station, and (e) eject station. Neck rings carry the preforms and blown bottles during the process. The core rods are removed from the preforms as they exit the injection molds (see Fig. 12-44).

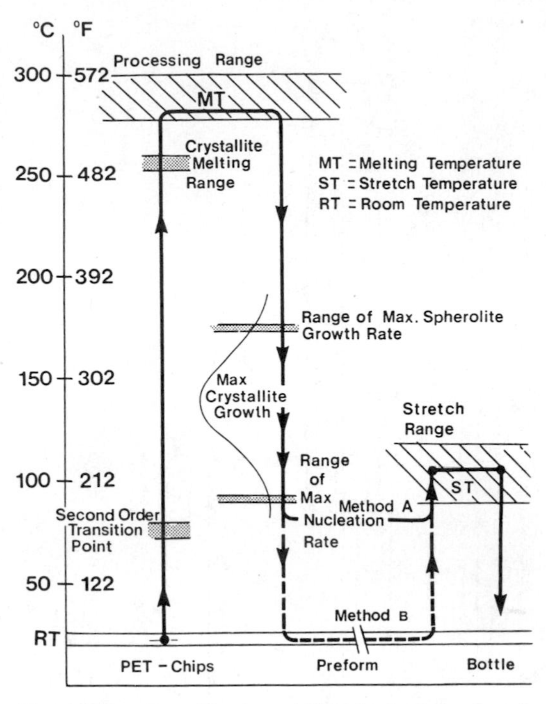

Fig. 12-43. Crystallization of PET in single-stage and two-stage stretch blow molding (*Courtesy Hoechst Celanese Corp.*)

Advantages of the single-stage process are that it is the lowest-cost route for entry into the market, and it subjects the material to the least heat history. Its disadvantages are that it is relatively slow because the injection cycle controls the machine's output, and it cannot achieve the highest levels of orientation because blowing takes place at a higher temperature.

The two-stage stretch blow molding system usually is referred to as the reheat blow molding system. The trademark RHB was registered by Cincinnati Milacron, the first company to offer a machine utilizing injection-molded preforms. In the Cincinnati Milacron process, the preforms are passed through an oven, rotating at approximately 60 rpm for even heating. Upon exiting the oven, the preforms are "equilibrated" to allow the inside temperature to rise to almost the same temperature as that of the outside skin. The preforms are transported on a shuttle pallet to the stretch blow clamp, blown in two stages (blow air is first introduced at approximately 200 psi, followed by 450 psi or

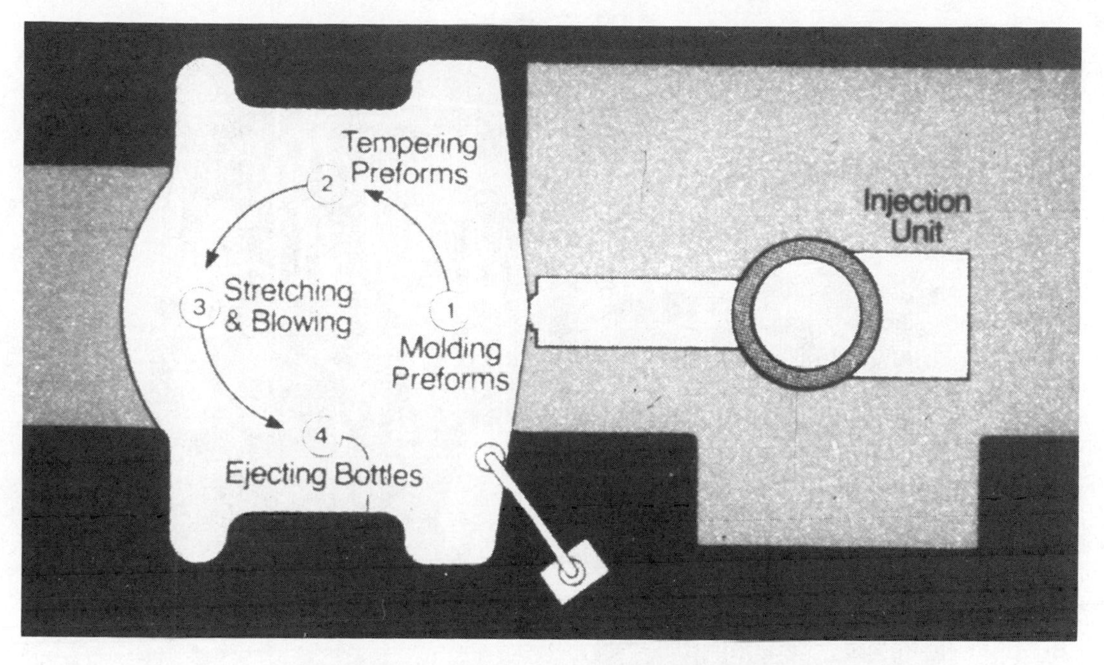

Fig. 12-44. Simplified schematic of single-stage process.

more), removed to the ejection station, and sent downstream for base cupping, labeling, and so on. (See Figs. 12-45 and 12-46.)

Injection molding systems for preforms are sold by Cincinnati Milacron, Husky, Netstal, Nissei, Meiki, Battenfeld, and others. These systems generally consist of an injection molding machine with multicavity tooling, high-temperature resin dryer, mold chiller, take-out robot with "soft drop" of the preforms, and post-conditioning station.

Reheat stretch blow molding machines are of-

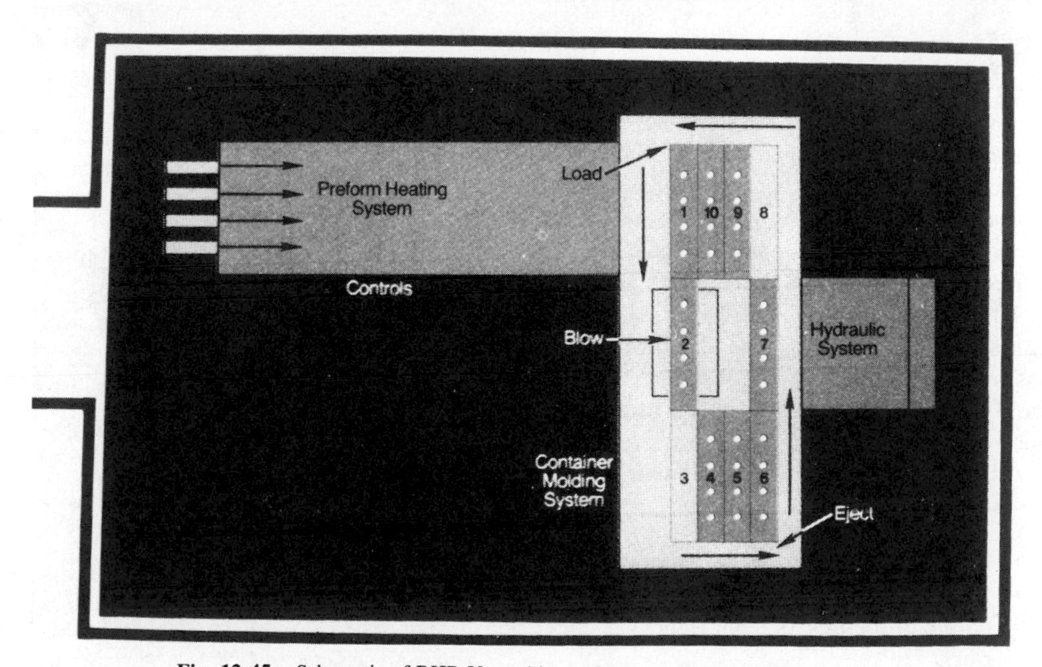

Fig. 12-45. Schematic of RHB-V machine. (*Courtesy Cincinnati Milacron*)

Fig. 12-46. Original RHB-V model machine. (*Courtesy Cincinnati Milacron*)

fered by Krupp Corpoplast, Sidel, Mag-Plastic, Bekum, and others. There are several differences between the major types: Cincinnati Milacron offers a rugged and versatile indexing type of machine; the ovens can be moved to allow the equilibration time to be adjusted for different preforms. The rotary machines built by Krupp Corpoplast and Sidel (Figs. 12-47 and 12-48) have the advantage that the blown bottle has more in-mold time, and if center rods and mold bottom are used, there is no sacrifice in the overall production cycle.

Advantages of the two-stage system are:

1. It offers the lowest total cost.
2. It produces the lowest-weight bottle.
3. Preforms can be designed for optimum bottle properties (strength and gas barrier).

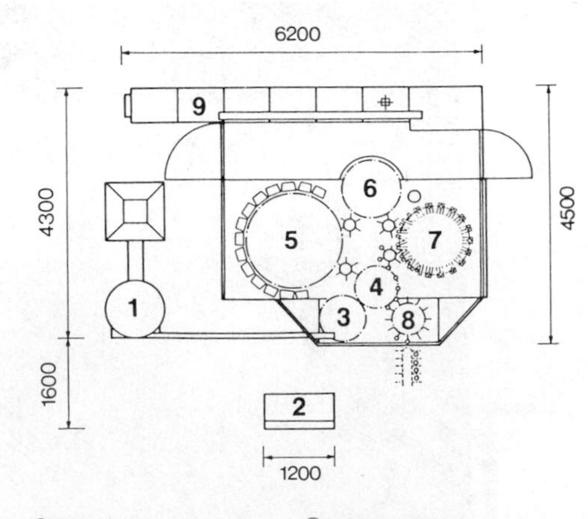

Fig. 12-47. Schematic of Corpopolast Model B40. (*Courtesy Krupp Corpoplast, W. Germany*)

Fig. 12-48. Sidel Model SB010. (*Courtesy Sidel, SMTP, France*)

4. Preforms can be produced in a single location, inventoried, and shipped for blow molding at other sites to meet peaks and valleys of demand.
5. It allows the highest efficiency of production of both preforms and bottles.

The main disadvantage of the two-stage process is the cost of investment, which can range from about $1 million to over $3 million, depending on the annual volume of bottles required.

Injection-molded preforms for PET bottles originally were molded in 8-cavity hot runner molds. With the rapid success of the two-liter carbonated beverage bottle, demand for high production led to increased mold size, and today's preform mold is likely to contain 24, 48, 64, or even 72 cavities (see Fig. 12-49). Today, the typical preform for a two-liter bottle weighs less than 48 grams, compared with 67 grams for the original preform. Molding cycles are in the range of 18 seconds, with postcooling used to prevent distortion of the hot preforms. The PET must be dried thoroughly before molding.

In the two-stage process, preform design and manufacture is critical. The saying in the industry is, "If you make a good preform, the blow molding machine will make a good bottle." The molded preforms are heated (via Cal-

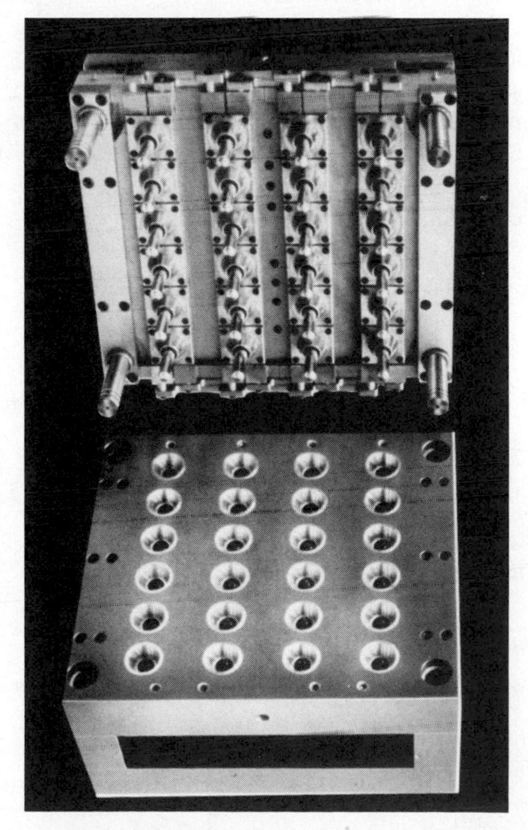

Fig. 12-49. 24-cavity PET preform mold. (*Courtesy Electra Form, Inc.*)

rod, quartz, or radio frequency heaters) to orientation temperature. The heat intensity must be regulated to prevent the exterior wall from crystallizing before the inside gets to the required temperature.

On all machines, two-stage and single-stage, there is a conditioning station for equilibration of the preform temperature before blowing. See Fig. 12-50, where sketch AA depicts the wall of a preform as it exits a reheat oven, sketch BB depicts the wall of a preform as it leaves the second station of a single-stage machine, and sketch CC depicts the ideal that is sought on both processes. The two-stage preform will produce better barrier and wall distribution; the single-stage preform will produce the more optically clear container.

The quality of stretch blow-molded bottles is affected by the pressure of the blowing air. PET should be blown with a minimum of 300 psi, and some designs require up to 600 psi. The air must be oil-free and dry.

Preforms for reheat stretch blow molding also can be formed from extruded tubing. Cut sections are heated at the ends, and the molten material is formed into the threaded neck and a closed bottom.

Another variation on stretch blow molding is extrusion stretch blow molding, an in-line process in which the preforms are extrusion blow-molded, cooled to orientation temperature, and then stretched and blown in a second mold set.

Injection Orientation Blow Molding. A process that Aoki of Japan introduced to the market in 1986 is the method referred to as injection orientation blow molding. The process was originally developed by Owens-Illinois in the 1960s, based on a patent by K. Allen, and known as the ''Flair'' process. Owens-Illinois was using materials such as polypropylene and impact polystyrene. Aoki has taken advantage of new resins such as PET and improved types of polypropylene and polycarbonate. The basic process begins with the injection molding of a saucer-like disc or, in other cases, a preform shaped like a small cup. The preform is transferred to a second station where a plug assist and/or center rod stretches it to the bottom of the blow cavity. Blow air is introduced to form the finished container.

Coinjection Stretch Blow Molding. Single-stage blow molding machines are used predominantly to produce PET bottles. Some installations, however, are using coinjected materials such as PET/nylon/PET, PET/EVOH/PET, and polycarbonate/PET/polycarbonate. The coinjection blow-molded bottles are produced on single-stage machines utilizing two plasti-

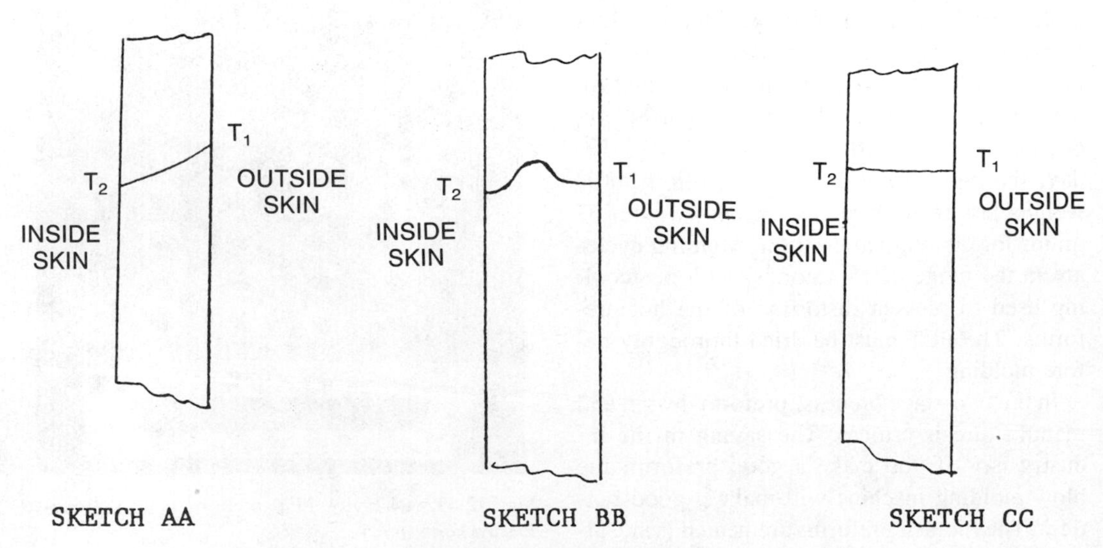

Fig. 12-50. Temperature profiles.

fiers. PET/EVOH/PET bottles for ketchup are being commercialized to replace coextrusion-blown barrier bottles made from polypropylene structures that require adhesive layers. The new bottles can be recycled in existing plants for PET soft drink bottles; when the bottles are granulated, the thin EVOH layer is removed in the process, along with paper labels and other fines.

AUXILIARY EQUIPMENT

All blow molding machines require certain auxiliary equipment. A complete blow molding system could include: (1) a resin storage silo with pneumatic transfer line; (2) a hopper loader that proportions virgin resin, regrind, and colorants; (3) a resin dryer; (4) a scrap granulator; (5) a chiller or a heater for molds; (6) an air compressor; (7) a flash trimmer; (8) a parts conveyor; (9) a parison programmer; (10) a crane or a mold cart; (11) a flame treater; and (12) decorating equipment.

In blow molding, air is a critical factor. The compressor must be properly sized and must produce dry, oil-free air to protect valves and molds. Also critical is the mold temperature control equipment, which should be sized to ensure that proper flow can be maintained and that the temperature differential between the inlet and the outlet is no more than 3°F.

MOLDS FOR EXTRUSION BLOW MOLDING

Most molds for extrusion blow molding in the United States are made with either aluminum, beryllium-copper, or a zinc alloy. All of these metals are good thermal conductors of heat and can be cast. Nearly all of the larger molds are cast with aluminum to minimize weight. With many small molds, there is a choice between a cast mold or a machined mold. Well-constructed molds made by either casting or machining will give good performance. Although zinc alloys are always cast, steel or beryllium-copper inserts are added where the mold must pinch the parison.

Aluminum is by far the most popular material for blow molds because of the high cost of beryllium-copper and the short life of the Kirksite (soft zinc alloy) molds. Aluminum is light in weight, a relatively good conductor of heat, very easy to machine (but also easy to damage—if the mold is abused), and low in cost.

Aluminum is used for single molds, molds for prototypes, and large numbers of identical molds, as might be used on wheel-type blowing equipment or equipment with multiple die arrangements. Aluminum may tend to wear somewhat after prolonged use. Thin areas, as at pinch-off regions, can wear and become damaged relatively easily. However, if carefully handled and used, these regions can be as durable as the rest of the mold. Steel pinch-off inserts, of course, could be replaced when damaged or worn.

Steel can be used for extremely long production runs where utmost durability is necessary. Cooling difficulties and the poor heat transfer characteristics of steel discourage its use in any great volume. It is not possible to cast steel molds, but multiple duplicates can be hobbed.

Table 12-3 shows approximate properties of materials used to produce blow molds.

Table 12-3. Properties of some mold materials.

Metal	*Cast/cut*	*Lb per cu. in.*	*Durability*	*Resistance to PVC*	*Heat conductivity*
Zinc	Cast	.259	Fair	None	.0017
Aluminum 70/75T6	Cut	.0975	Good	Fair	.002
Aluminum	Cast	*	Fair−	Fair−	*
Brass	Cut	.304	Good	Fair+	.0015
Beryllium-copper	Cut	.320	Very good	Very good	.004
Beryllium-copper	Cast	.320	Very good	Very good	.004
Stainless steel 300	Cut	.290	Fair	Very good	.0006
Stainless steel 400	H.T. cut	.290	Good	Fair	.0003

*Depends on density of casting.

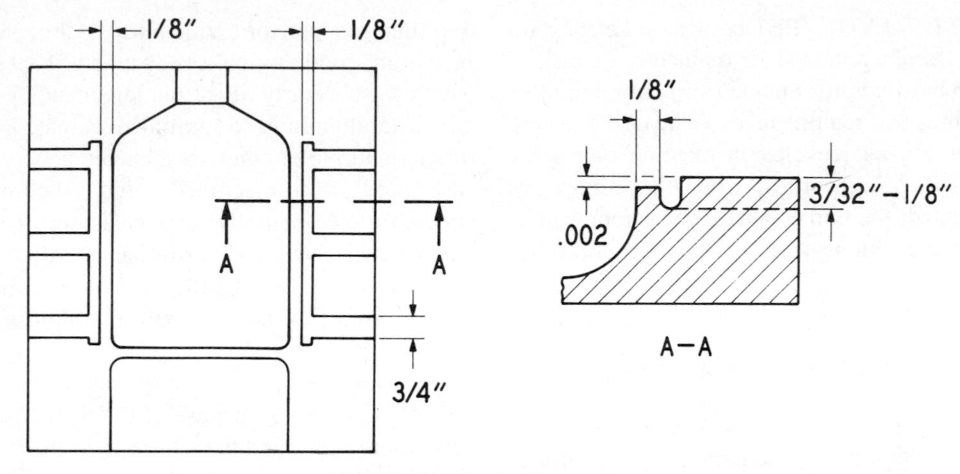

Fig. 12-51. Parting line venting.

Well-designed molds are vented, as entrapped air in the mold prevents good contact between the parison and the mold cavity surface. When air entrapment occurs, the surface of the blown part is rough and pitted in appearance. A rough surface on a shampoo bottle, for example, is undesirable because it can interfere with the quality of decoration and can detract from the overall appearance. Molds are easily vented by means of their parting line, with core vents and with small holes. A typical mold parting line venting system is shown in Fig. 12-51. The venting is incorporated only in one mold half. This type of venting can be used on all sizes of molds. When certain areas of the mold cavity are prone to trap air, core vents as shown in Fig. 12-52 can be used. Venting in the mold cavity should be anticipated in the

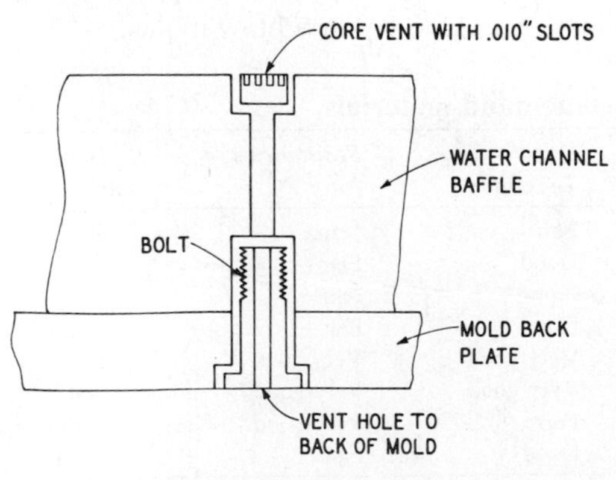

Fig. 12-52. Core venting.

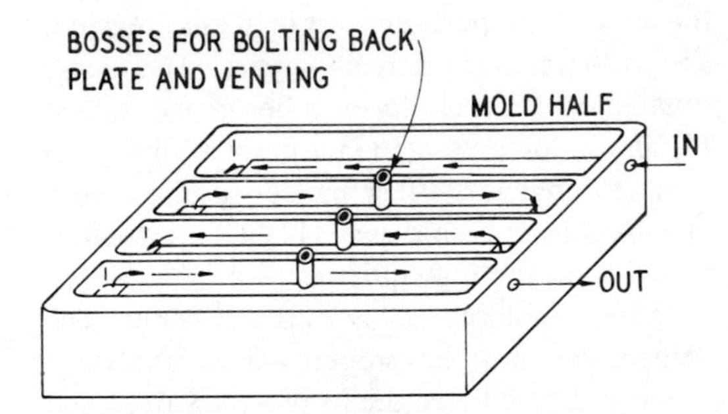

Fig. 12-53. Location of vents in baffles.

mold design and layout of the cooling channels so that provisions can be made for their locations. For cast molds the cooling channel baffles can be located over areas to be vented, as shown in Fig. 12-53. The vent opening will pass through a boss in the baffle to the back or outside of the mold. In machined molds, care must be taken so that vents miss the drilled cooling channels. When core vents cannot be used because the slots mark off on the blown part and show, small drilled holes can be used. The effect of the size of hole on the surface of the part is shown in Fig. 12-54. If the hole is too large, a protrusion will be formed; if it is too small, a dimple will be formed on the part. Venting also can be incorporated in molds that are made in sections. A 3- to 10-mil gap between the two sections with venting to the outside of the mold is a very effective vent. For small containers a 2- to 3-mil opening is used, and up to a 10-mil opening has been used on

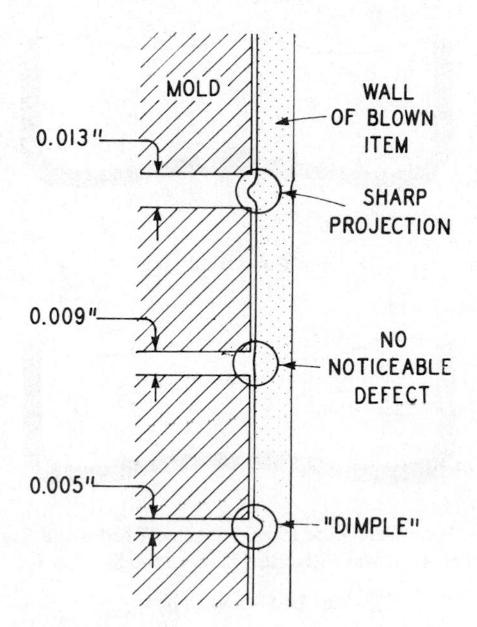

Fig. 12-54. Effect of vent hole size on part surface.

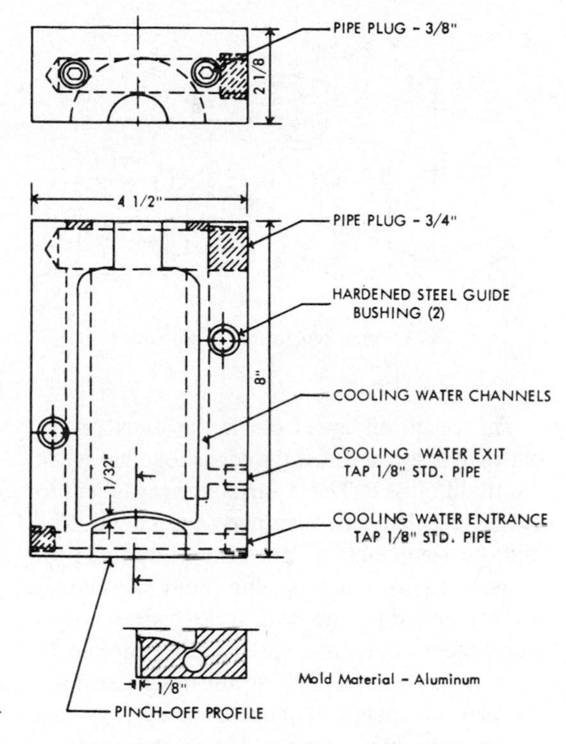

Fig. 12-55. Typical 12-oz container mold (half).

large parts such as a 20-gallon garbage container.

The mold cavity surface has an important bearing on mold venting and on the surface of the molded part. With polyethylenes and polypropylenes a roughened mold cavity surface is necessary for the smoothest surface. Grit blasting with 60 to 80 mesh grit for bottle molds and 30 to 40 mesh grit for larger molds is a common practice. The clear plastics such as PVC and styrene require a polished mold cavity for the best surface. A grit-blasted surface will reproduce on some clear plastics, an effect that is not normally desirable.

Compared with injection molds, molds used for extrusion blow molding can be considerably less rugged in construction. Clamping pressures need only overcome blowing pressures from 5 to 200 psi. This contrasts with pressures that reach tens of thousands of psi tons in injection molding applications. A typical bottle mold is schematically drawn in Fig. 12-55.

On many blow molding machines, the mold closes in two steps in conjunction with expansion of the parison by the blowing air. On these systems, the mold closes rapidly until $\frac{1}{3}$ to $\frac{1}{2}$ inch of daylight remains. The final clamping is done at reduced speed, but with increased clamping force. This two-speed operation protects the mold surface from being marred by objects that may fall between the mold halves, and it reduces mold surface wear.

The mold does not have to be positioned vertically between the platen area. It also does not have to be positioned so that there is equal resin all around the blow pin. ''Offsetting'' of the molds commonly is done when containers with large handles on one side (such as jugs for milk, bleach, or detergent) are blow-molded from polyethylene. This practice makes it especially easy to ''catch'' the hollow handles of such containers.

There are three main ways of making molds: machining, casting, and hobbing. Cast molds are most common because they require little machining or tooling work. Hobbed molds exactly duplicate the hob in the same way that cast molds duplicate the original pattern. These two types of construction are the most useful where several identical molds are required. The main concerns for a mold designer are the pinch-off areas at the top and bottom of the mold and the provisions for cooling.

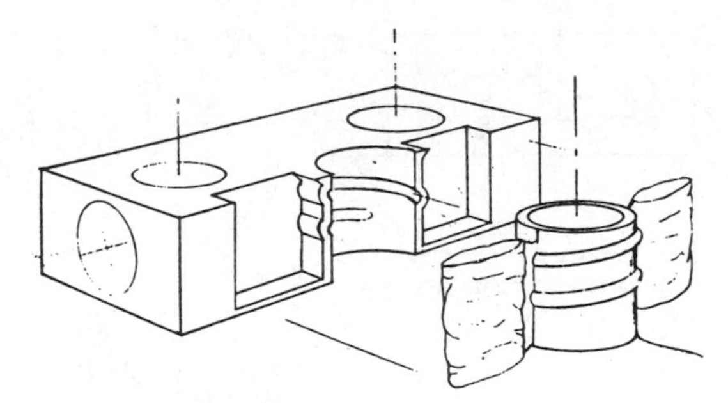

Fig. 12-56. Replaceable neck insert.

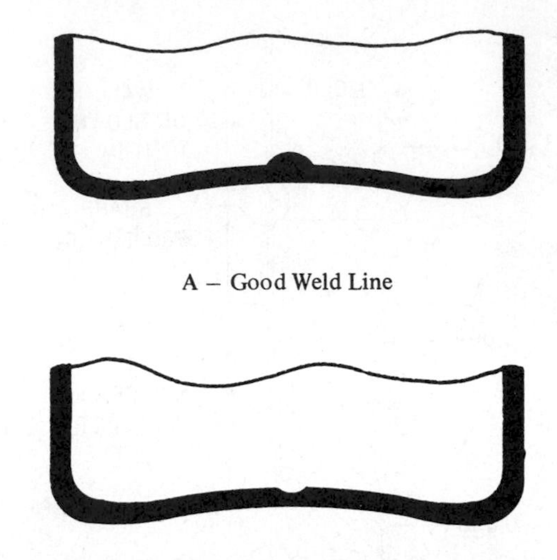

A – Good Weld Line

B – Poor Weld Line (Weak) Pinch-off had Knife Edge; Relief Angle was Either too Large or too Small.

Fig. 12-57. Weld lines.

The pinch-off areas pinch the ends of the plastic parison and seal the edges together when the mold closes. These surfaces are subject to more wear than any other part of the mold. The high-heat-conductive metals preferred for blowing molds, such as aluminum and copper alloys, generally are less wear-resistant than steel. Steel inserts often are used for the pinch-off areas of molds of the softer metals. An additional advantage of pinch-off inserts is that they can be made replaceable in the event of wear or damage. A neck pinch-off insert is sketched in Fig. 12-56.

Also important is the insert constructed at the "tail" section of the untrimmed plastic part. If improperly constructed, the end products may break along the bottom pinch-off weld line on drop impacting. Generally, the pinch-off insert is made of hard steel with the rest of the mold made of a nonferrous metal. This pinch-off insert should not have knife edges, as they will tend to yield a grooved weld as shown in Fig. 12-57b. A good pinch-off insert, as shown in Fig. 12-57a, has a 0.012 to 0.125 inch land before it flares outward in a relief angle of approximately 15°. Figure 12-58 illustrates the way a blowing mold with these features looks. The weld line should be smooth on the outside and form an elevated bead on the inside.

Other uses for inserts include replaceable neck finishes to allow making bottles with different neck sizes from a single mold, as well as to incorporate in products textured areas or lettered information or trademarks that can be changed, making some molds more versatile.

Molds must have pins that perfectly align the mold when closed. These guide pins and bushings can be arranged to suit the opening and closing operation required for a given set of equipment.

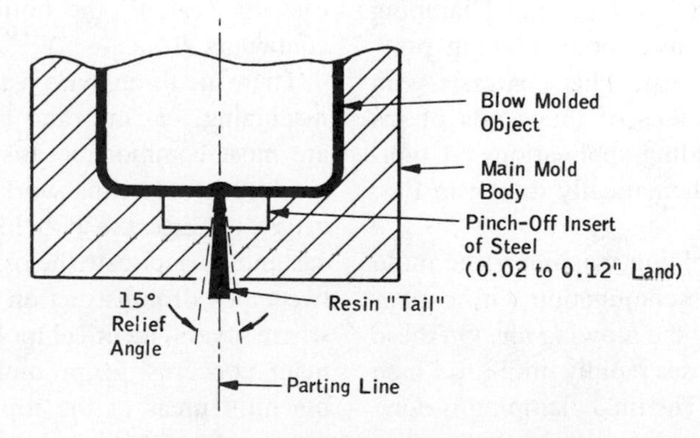

Fig. 12-58. Recommended pinch-off and insert.

Cooling. Cooling is particularly important in mold design because it consumes much of the cycle time and thus bears on product economics. Cooling can take two-thirds or more of the entire mold-closed time in a cycle. Best economics require cooling as quickly as possible. Mold cooling is a function of three factors: coolant circulating rate, coolant temperature, and efficiency of the overall heat transfer system.

Coolants can be circulated through the mold in several ways. One of the simplest is to "hog" out the back of each mold half, leaving an open area, and enclose this cavity with a gasketed back plate. Then, with simple entrance and exit openings, the coolant can be circulated in fairly large volume. Another method is to drill holes through the mold block from top to bottom and from side to side so that they intersect. Excess openings are plugged and hose connections inserted for the exit and entrance lines. This type of cooling arrangement can provide extra cooling where needed in the mold, such as at the neck finish and bottom pinch-off areas where the plastic material is thicker and requires more cooling.

Results are best when uniform temperatures are maintained throughout the mold.

The biggest problem, which is frequently encountered, is inadequate mold cooling. It can be improved by creating turbulent coolant flow through the mold. The flow of coolant can be improved by increasing the size of the cooling channels (including the adjacent piping) and by increasing the coolant flow rate. Heat transfer can also be enhanced by lowering the temperature of the coolant or by using a mold material with better heat transfer properties.

The most common mold coolant is tap water. Hard water should be softened before use. The overall efficiency of a mold cooling system would be greatly reduced if water scale were allowed to plate out on, or partially block, cooling channels. Generally, cooling water circulates first through a cooling system that lowers its temperature to between 40 and 70°F. If the temperature of the molding shop is high, cooling water can cause moisture to condense on mold surfaces. This will create defects in finished parts, small blotches or pock marks on the surface. Condensation problems can be eliminated by raising the mold coolant temperature, or by air conditioning the molding shop. This solution merits careful consideration, as overall production increases resulting from shorter cycles can pay for the initial capital investment in air conditioning.

It is often useful to cool the neck rings and the blow pin independently; here, as with mold cooling, basic rules of heat transfer optimization must be applied.

In addition to external systems that circulate liquid coolants through the molds to cool both the mold and the part simultaneously, there has been interest in internal cooling systems. These systems use circulating air, a mixture of air and water, or carbon dioxide injections to cool the inside of the parts while they are in the mold. The following commercial methods are typical of those now in use: (1) liquid carbon dioxide is injected into the blown part, followed by vaporization, superheating, and exhausting of the coolant as hot gas through the blow pin exhaust; (2) highly pressurized moist air is injected into the part, where it expands to normal blow pressures, producing a cooling effect by lowering the air temperature and freezing the moisture present into ice crystals (crystals and cold air strike the hot walls of the part where they melt and vaporize); (3) air is passed through a refrigeration system and into the hot parison; and (4) normal plant air is cycled into and out of the blown parts by a series of timers and valves.

Injection Blow Molding Design and Construction*

The design of tooling for injection blow molding as discussed herein is intended for equipment utilizing the horizontal rotary index method of core rod transfer. This principle has been the most successful to date and is used by over 90% of the machines now in service. Figure 12-59 shows the mold and machine.

---

*This section adapted from *Plastics Mold Engineering Handbook*, 4th edition, DuBois and Pribble, Van Nostrand Reinhold Co., by Robert D. DeLong.

Fig. 12-59. Operational principle of the injection blow molding process. (*Courtesy Johnson Controls, Inc.*) (*from Plastics Mold Engineering Handbook*, 4 ed., DuBois and Pribble.)

Principle of Operations. On typical three-station machines, two sets of molds are required, along with three sets of core rods. To ensure alignment and speed mold changeovers, each set of molds is mounted on its own mold base or die shoe. In the first or parison mold, melted resin is injected over the core rods, filling the cavities and fixing the weight of the container. At the same time, the neck finish is molded completely. This set of core rods carrying the preformed parisons is rotated to the second or blow station, where they are enclosed within the blow molds and blown to the desired shape. The blown containers then are rotated to the stripping station for removal, oriented if desired. Because there are three sets of core rods, all three operations are conducted simultaneously by sequential rotation of the core rods.

Mold Design. The constraints of container length, diameter, and number of cavities are imposed by the physical dimensions of the machine at hand. Parison molds normally are mounted so as to place their centroids of projected area directly under the centerline of clamp pressure. The distance, then, from the face of the rotating head to the center of pressure will determine the maximum container length. Machines are available to cover the range of $\frac{1}{2}$ to 12 inches in length. The maximum container diameter is about $\frac{1}{4}$ inch less than the mold opening, to allow a minimum of $\frac{1}{8}$ inch top and bottom clearance for the blown container to rotate from the blow station to the stripper station. Present machines allow from 4 to 6 inches for the mold opening, depending on the size.

The maximum number of cavities normally is limited by the clamp tonnage of the parison mold and the material being molded. The following clamping requirements are suggested as *minimums*:

LDPE—2800 psi/in.^{2} of projected area
HDPE—3200 psi/in.^{2} of projected area

SAN and PS—3800 psi / in.^{2} of projected area
PP—4100 psi / in.^{2} of projected area
PC—4400 psi / in.^{2} of projected area
PET—6000 psi/in^2 of projected area

The nature of the injection blow process places some limitations on the shapes of the containers that can be made. Injection blow molding is well suited to wider-mouth bottles and conventional shapes. Because the core rod is a cantilever beam, its L/D ratio should be 12/1 maximum to prevent deflection during injection. A 32 oz. cylinder round-style container requires a core rod about 9 inches long. Specified with a 24 mm finish, the L/D ratio is an unworkable 14/1. Moving up to a 28 mm or 33 mm finish would reduce the L/D to a manageable ratio.

Swing weight becomes a consideration with containers of over 32 oz. capacity that have wide necks. The core rod for a 48 oz. capacity container with a 110 mm finish, shown in Fig. 12-60, weighs about 9 pounds. Mounted at the extremes of the swing radius, the weight impose severe inertial loadings on the transfer mechanism, as indexing takes place in about 1 second.

One advantage of injection blow is the diametrical and longitudinal programming of the parison by shaping the parison mold, the core rod, or both. This two-dimensional programming becomes especially advantageous in the production of oval containers. Up to an ovality (container width/container thickness) ratio of 1.5/1 quite satisfactory containers can be blown from circular-cross-section parisons. Up to ovality ratios of 2.2/1 can be handled with oval-cross-section parisons. Generally the ovalization is done to the parison mold; the core rod remains round. Extensive parison ovalizing, perhaps above 35%, can lead to selective fill during injection and result in visible knit lines in the finished bottle. Under these circumstances it will be necessary to ovalize *both* the core rod and the parison cavity to obtain the desired distribution in the finished container. Some provisions to prevent the core rod from

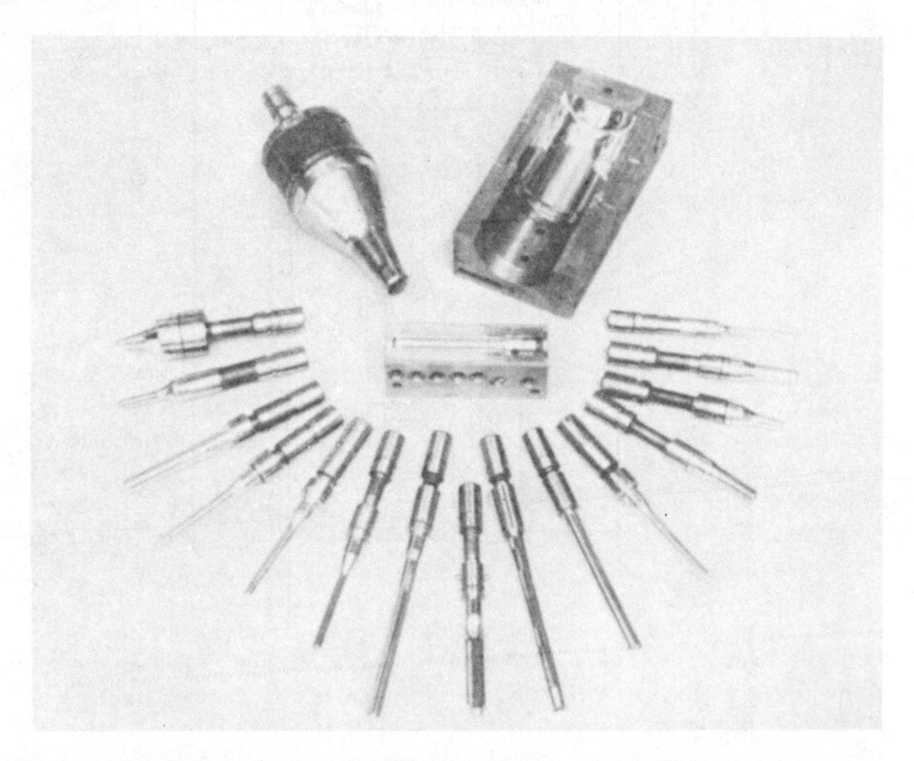

Fig. 12-60. Core rods, parison mold, and blow mold. Upper left shows 48 oz.—110 mm core rod. Lower center blade shape core rods illustrate ovalization and means of preventing rotation. (*Courtesy Johnson Controls.*) (from *Plastics Mold Engineering Handbook*, 4 ed., DuBois and Pribble.)

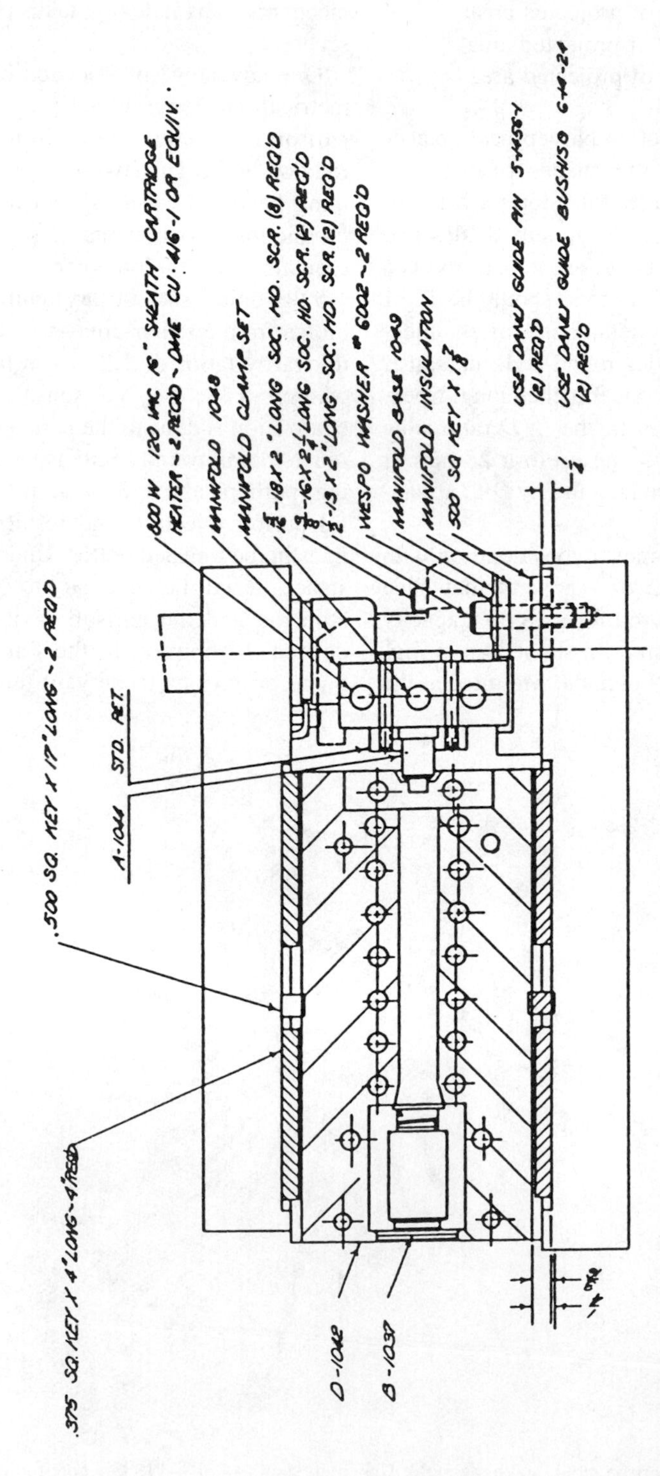

Fig. 12-61. Parison mold assembly-side view. (*Courtesy Johnson Controls, Inc.*) (from *Plastics Mold Engineering Handbook*, 4 ed., Dubois and Pribble.)

rotating then must be incorporated. Ovality ratios above 3/1 are not suggested. The center core rod in Fig. 12-60 shows an ovalized core rod. The preferred blow ratio is between 2 and 3/1, measured as the average parison O.D. against the bottle O.D. This normally will yield an ideal parison thickness of from .120 to .180 inch with predictable expansion characteristics. Increasing the blow ratio by reducing the core rod diameter reduces the projected area, often enough to allow another cavity to be fitted. Unavoidably, the parison thickness must be increased, to maintain a constant weight, and experience has shown parison thicknesses above .225 inch to be unpredictable in their blowing characteristics.

Alignment of Components. Examination of Fig. 12-59 suggests some of the alignment problems encountered in assuring that the core rods are concentric with their respective cavities. It is usual practice to utilize individual cavity blocks and locate them via keyways on the die shoes. Setup then is simplified, as it involves changing only the die shoe with the mounted cavities as a unit. The die shoe is located from a single center keyway mating with the machine; radial spacing from the center of rotation is adjusted as required by shims. Figure 12-61 illustrates a parison mold assembly.

Initial core rod/mold alignment is achieved by the index mechanism of the machine. Upon mold closing, the index head is guided by guide pins to further refine the mold/core alignment, with final alignment by the core rod shank nesting into its seat in the mold. This shank is a precise (.0000–.0005 in.) fit to its nest. Figure 12-62 shows 28-mm and 48-mm core rods nested in their respective parison mold halves. The core rod mounting ends are clearanced .004 to .006 inch to "float" in the rotating head to compensate for thermal expansion differences between the hot (200–300°F) parison mold and the cold (40–60°F) blow molds.

Core Rod Design and Construction. Core rods typically are of hardened steel, polished and hard-chromed. Location of the air entrance

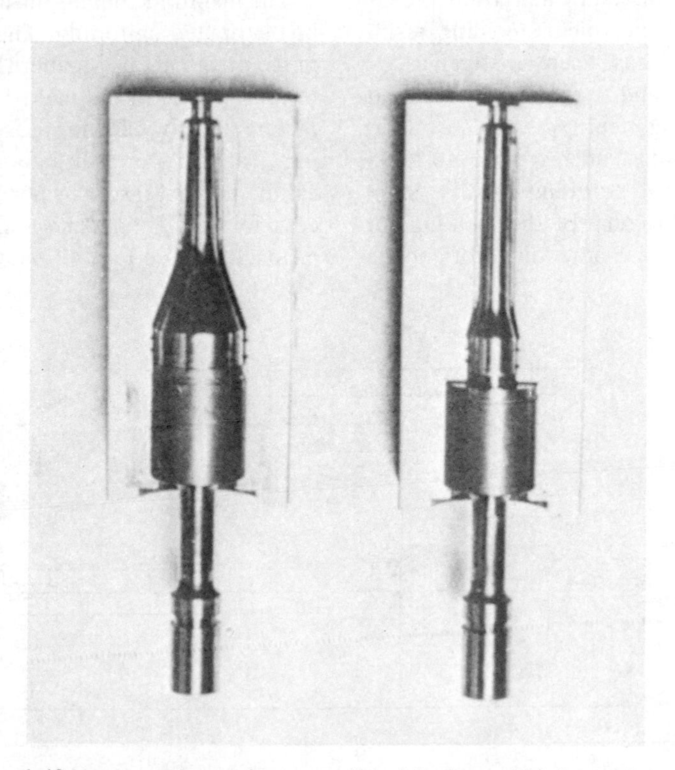

Fig. 12-62. 28mm and 48mm core rods are shown nested in injection mold halves. (*Courtesy Captive Plastics, Piscataway, NJ*). (from *Plastics Mold Engineering Handbook*, 4 ed., DuBois and Pribble.)

valve at the shoulder is preferable, but it often is moved to the tip when core rod L/D exceeds 6/1, because of mechanical considerations. Figure 12-63 details a typical core rod as shown in Fig. 12-60.

Normal construction of the core rod is solid, although coring-out to reduce mass may be used in larger sizes. Where space permits, some means to enhance heat transfer may be employed, such as aluminum or beryllium-copper inserts. Heat pipes are being installed in the center of some core rods to obtain faster cycles and close temperature control. Examination of Fig. 12-60 will reveal one or two annular grooves on the core rod near the seating shank. These grooves (.004–.010 in. deep) perform a dual function: They stabilize the parison against elastic retraction during transfer from the parison mold to the blow mold, which would result in thread mismatch. They also seal the bottle against excessive air loss during blowing; hence the name, blow-by grooves.

Parison Molds. Individual parison mold blocks normally are fabricated from prehard steel for use with polyolefin molding resins. Molds for rigid resins, such as styrenics, nitriles, carbonates, and so on, usually are made of an oil-hardening steel (A-2 or equivalent), rough-machined and hardened to R_c 40 to 45. In both cases, final polishing usually is followed by hard-chroming of the molding surfaces, although this is not required for polyolefin molds. Molds for PVC parisons usually are made from 400 series stainless steel and hardened.

Cooling lines for control of cavity temperature are placed close to the surface. The use of O-rings between adjacent cavities is standard practice.

No venting is used on the parison mold parting line. If required, the venting is over the core rod and out the back of the mold.

Neck rings are separate inserts; construction is normally of oil-hardening steel with hardening to R_c 54 to 56 for all resin types.

Nozzles seat directly into the parison bottom mold, with the top half of the parison mold clamping and unclamping with every cycle. Mold flash, if it occurs, is most probably at the nozzle, with potential nozzle or nozzle seat hobbing. Practice varies widely on nozzle hardness, but replaceable nozzle seats, as shown in Fig. 12-61, are in wide use. This eliminates the need for enlarging the nozzle seat with time and stocking nozzles of various diameters.

The manifold contour must cause the melt to be distributed uniformly. The critical elements in the design are flow pattern, temperature, and residence time of the melt. For ease of manufacture and for ease in disassembly for cleaning, manifolds are split into sections. The split can be either vertical or horizontal to give access for the cutting and shaping of the flow channel(s). (See Fig. 12-64.)

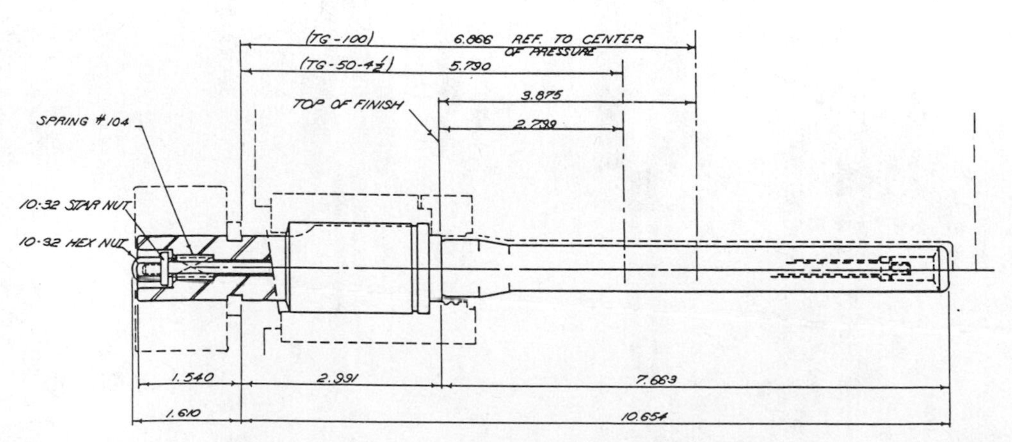

Fig. 12-63. Core rod for assembly in injection mold for a 16-oz alcohol bottle. (*Courtesy Johnson Controls, Inc.*, Manchester, MI). (from *Plastics Mold Engineering Handbook*, 4 ed., Dubois and Pribble.)

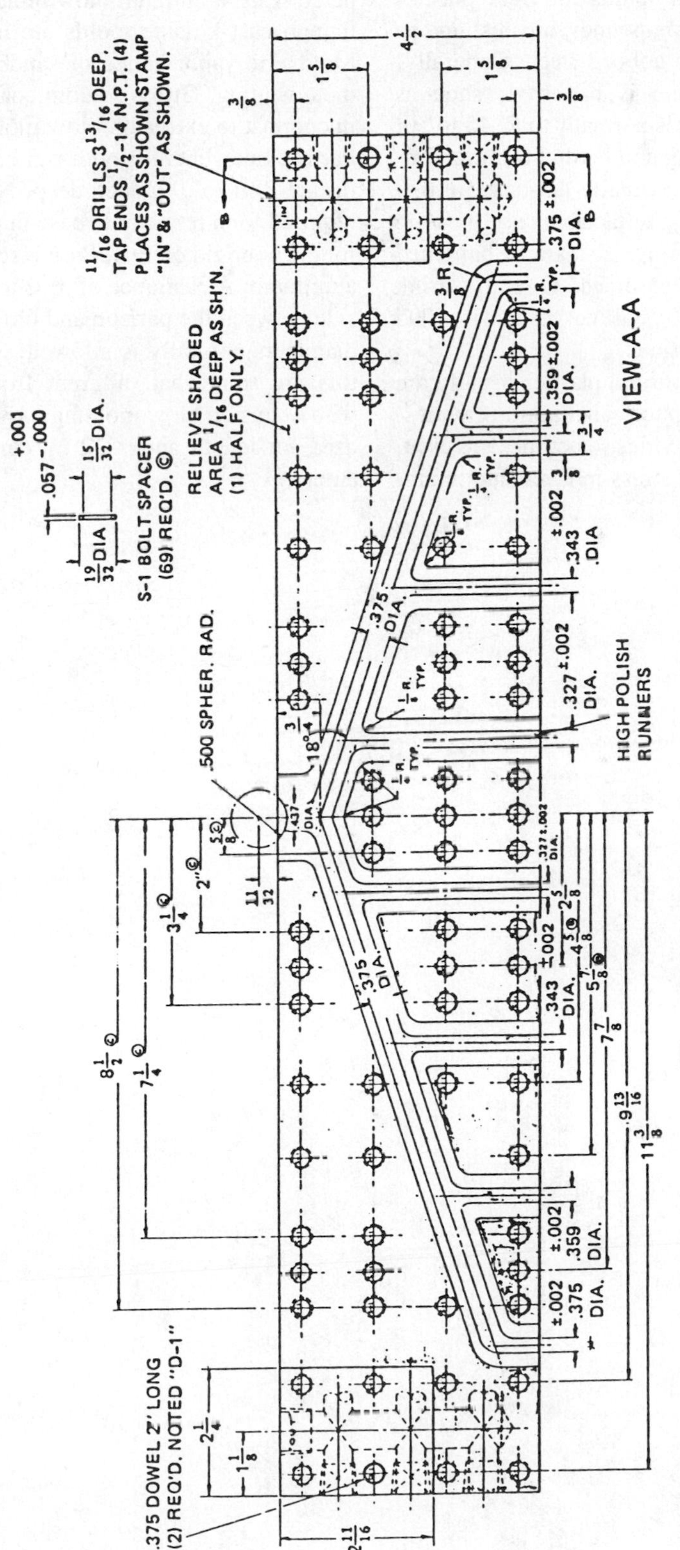

Fig. 12-64. Layout of 8-cavity, 2.625 in. centerline manifold for .375 in. diameter rod. (*Courtesy Captive Plastics, Inc., Piscataway, NJ*) (from *Plastics Mold Engineering Handbook*, 4 ed., Dubois and Pribble.)

Blow Molds. Blow molds for rigid plastics such as the styrene compounds, nitriles, and so on, are of turned or hobbed steel. Generally, an air-hardening steel is preferred, such as A-2 type. Hardening is normally to R_c 45 to 50, followed by polishing and hard-chroming. Unless the container is virtually flat-based (.030 in. push-up or less), a retractable base push-up must be used. Neck rings also are of hardened steel, polished and chromed. Overall shrink factor for the noncrystalline resins is .005 in./in. in both directions.

Blow molds for nonrigid plastics usually are turned or pantographed aluminum cavities. Hobbed aluminum cavities sometimes are used, but must be carefully stress-relieved and retempered. Cast aluminum blow molds are rare, although cast kirksite molds are in limited use. Molds are vapor-honed or sandblasted to promote venting. Grit is seldom coarser than 180, in contrast to extrusion blow molding practice. Parting line vents for both rigid and olefin resins are .001 to .002 inch deep. No provision is required for a retractable base in the blow mold for the nonrigid olefins. Base inserts usually are aluminum. A clearance of .003 to .004 inch per side between the parison and blow mold thread diameters normally is allowed. Shrinkage factors are somewhat different from those used with extrusion blow molding, with .012 in./in. used for length and .022 in./in. used for diameter.

13 Thermoforming of Plastic Film and Sheet

OVERVIEW

Thermoforming is the process of heating a plastic material in sheet form to its particular processing temperature and forming the hot and flexible material against the contours of a mold by mechanical means (e.g., tools, plugs, solid molds, etc.) or pneumatic means (e.g., differentials in air pressure created by pulling a vacuum or using the pressures of compressed air). When held to the shape of the mold and allowed to cool, the plastic retains the shape and detail of the mold. Because softening by heat and curing by the removal of heat are involved, the technique is applicable only to thermoplastic materials and not to thermosets.

Examples of thermoformed products are plastic or foam dinnerware, cups, meat and produce trays, egg cartons, refrigerator liners, computer housings, interior and exterior automotive parts, blisters for packaging, and countless others. These common and often taken-for-granted products are not usually thought of as the result of detailed tooling design, precise controlled heating and forming, expert material technology, and trimming/finishing operations. Clearly, the thermoforming process is an important link in the plastics industry.

The basic process as it is known today has been developing for over 40 years, but historians have found that the Pharoahs of Egypt developed a type of thermoforming by softening tortoise shells with hot oils to form food utensils. During World War II, with the development of thermoplastics, thermoforming was used to produce aircraft canopies and domes, as well as relief maps.

During the 1950s, thermoforming began to expand into new areas—acrylic windshields, outdoor signs, blister packaging, and refrigerator door liners.

Exceptional growth occurred during the 1960s, as thermoforming became one of the major plastics fabricating techniques. High-volume markets for individual-portion food packaging, blister packaging, and appliance housings were developed.

The 1970s brought an emphasis on high-volume, high-speed processing machinery—and viable competition with the pulp and paper industry resulted.

Advantages of thermoforming over most other methods of processing plastics include lower tooling and machinery costs, high output rates, the ability to use predecorated plastic sheet, and good-quality physical properties in finished parts.

Its disadvantages include the need to begin with sheet or film rather than less costly basic resins, trimming material used to clamp sheet for forming, and the problem of trim scrap reclamation.

METHODS OF THERMOFORMING

Because there are so many different techniques for thermoforming plastic sheet, the basic forming methods will be reviewed in this sec-

Reviewed and updated by Judith A. Penix, Marketing Coordinator, Lyle Industries, Inc., Beaverton, MI.

tion. Many of the illustrations (Figs. 13-1 through 13-14 inclusive) are based on equipment incorporating parallel-acting frames, heaters above and below the sheet stock (called sandwich heaters and capable of heating both sides of the sheet), and top and bottom movable platens with vacuum or compressed air, enabling the use of male or female molds and plug assists. With this flexibility, any of the known techniques can be used. There are, of course, many variations in thermoforming techniques that go beyond this simplified approach, such as folding book-action frames, single heaters above and below, alternating molds, and so on.

It should also be noted that, according to general terminology, the term "male molds" connotes a primarily convex mold; "female mold" is a primarily concave mold; the "ring" is the initial plane from which the sheet is drawn to shape it; a "plug" (which may or may not be the male mold) is the component of the mold that moves through the ring and carries the sheet of plastic with it.

Vacuum Forming into a Female Mold

To vacuum form a thermoplastic sheet into a female mold without prestretching, the ratio of depth to minor dimension of a given section should not be greater than 1 : 1, and no sharp inside radii are required. The sheet stock is locked in a frame around its periphery only, is heated to a predetermined temperature, and then is brought into contact with the edge of the mold. This contact should create a seal so that it is possible to remove the air between the hot plastic and the mold, allowing atmospheric pressure (about 14 psi) to force the hot plastic against the mold.

Most thermoplastics sheets (with the exception of cast acrylic) can be easily formed with vacuum. The reasons for using a female mold are: greater details can be achieved on the outer surface of the part (the side against the mold); multiple cavities can be placed closer together; and this type is easier to work with when close tolerances are needed on the outside of the part. Female molds are more expensive to make than male molds, however.

THERMOPLASTIC SHEETS

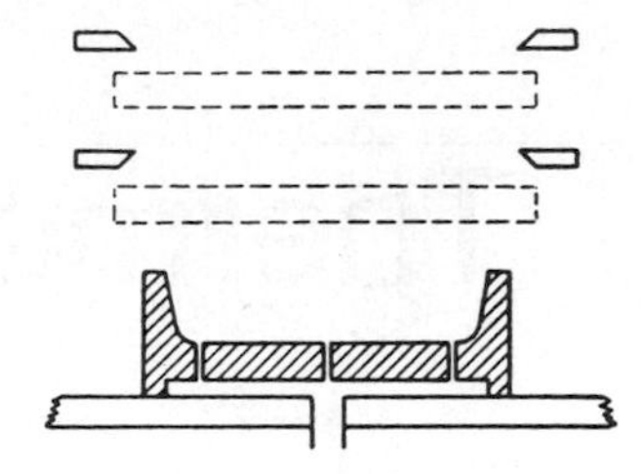

Female mold on platen—frames open—heaters idle.

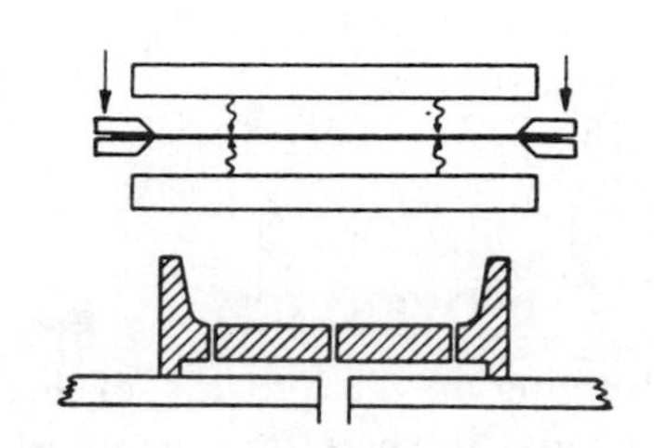

Stock in place—frames closed—heaters active.

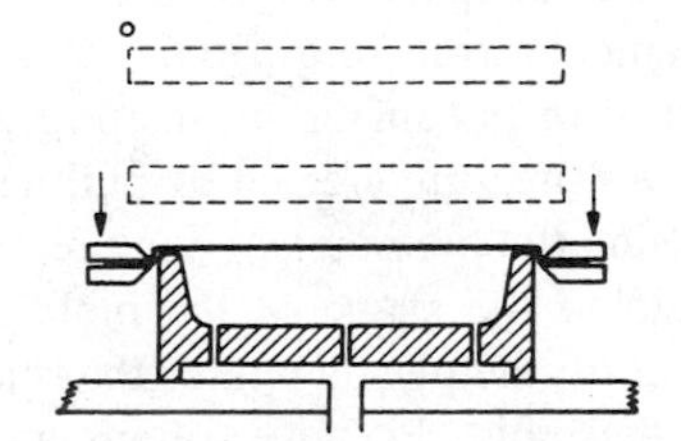

Heaters idle—frames lowered, drawing stock into contact with mold.

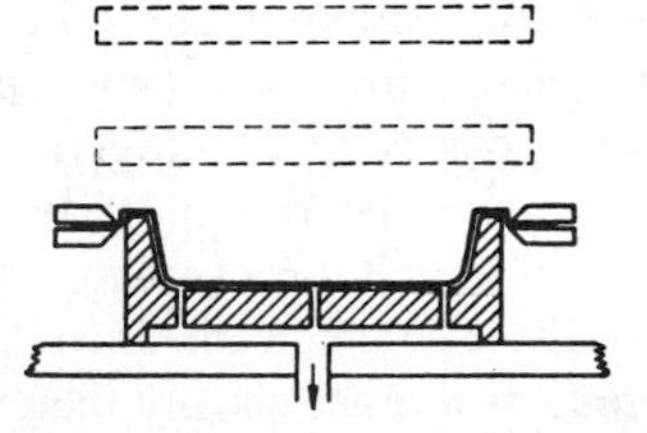

Vacuum applied—stock cooling.

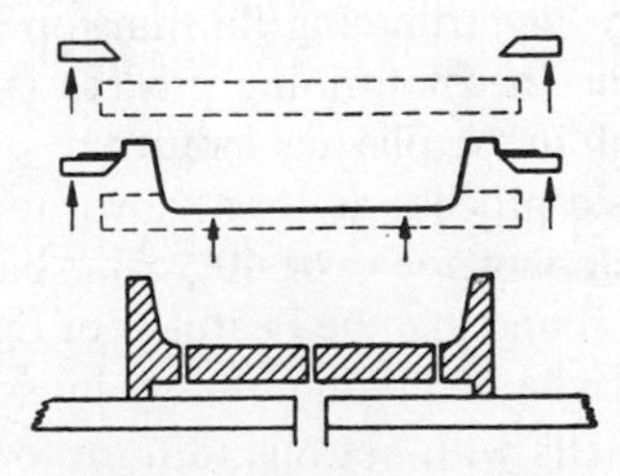

Cycle completed—equipment idle.

Fig. 13-1. Vacuum forming.

Vacuum forming is illustrated in Fig. 13-1.

When this method is used to make an article having an irregular ring line or periphery, the stock is not held by a frame, but is draped manually to the required contour in such a way as to make a seal possible. After that, the procedure described above is followed.

Pressure Forming into Female Cavity

Instead of relying on atmospheric pressure against a vacuum (as in vacuum forming, above), forming by compressed air at up to 500 psi is now also common. Positive air pressure is applied against the top of the sheet to force it into a female mold, and, at the same time, *full* vacuum also is applied. As contrasted to vacuum forming, pressure forming offers a faster production cycle (the sheet can be formed at a slightly lower sheet temperature), greater part definition, and greater dimensional control. (See Figs. 13-2 and 13-3.)

Fig. 13-3. Computer housing is an example of pressure forming, providing critical part definition and textured effect. (*Courtesy CAM/Central Automated Machinery, Inc.*)

Free Forming

This variation, also called *free blowing* has been used with acrylic sheeting to produce parts that require superior optical quality. The periphery is defined mechanically by clamping, but no mold is used, and the depth of draw is governed only by the vacuum or compressed air applied. Visual control or an "electric eye" is used to cut off the pressure when the required depth or height is reached.

Plug-Assist Forming

Straight cavity forming is not well adapted to forming a cup or box shape. The sheet, drawn down by vacuum, touches first along the side walls and then at the center of the bottom of the box-shaped mold and starts to cool there, with its position and its thickness becoming fixed. As the sheet continues to fill out the mold, solidification continues in such a way as to use up most of the stock before it reaches the periphery of the base; hence this part of the article will be relatively thin and weak. This

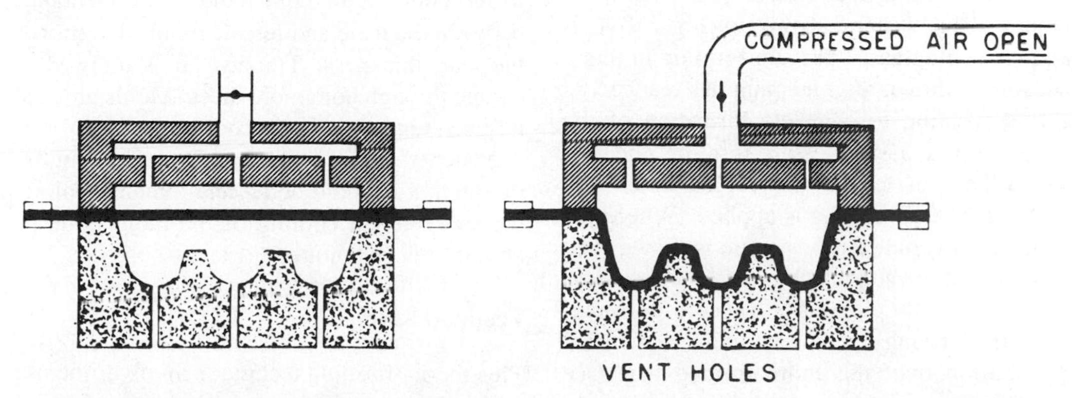

Fig. 13-2. Pressure forming into female cavity. Heated sheet is clamped over cavity, and compressed air pressure forces the sheet into the mold. (*Courtesy Dow*)

characteristic is undesirable in a cup shape, and usually unacceptable in a box shape, in which the thinness will be most marked at the corners of the base.

To promote uniformity of distribution in such shapes, designers use the plug assist—any type of mechanical helper that carries extra stock toward an area that otherwise would be too thin. Usually the plug is made of metal, and heated to a temperature slightly below that of the hot plastic, so as not to cool the stock before it can reach its final shape. Instead of metal, a smooth-grained hard wood can also be used, or a felt-covered phenolic or epoxy; these materials are poor conductors of heat and hence do not withdraw much heat from the sheet stock. Syntactic foam has been very popular for this application.

The plug-assist technique is illustrated in Fig. 13-4. Note that it is also possible to reverse mold and plug assist from the bottom platen to give better material utilization around the edges.

Both cavity and plug-assist forming make possible the production of shapes having protuberances on their inner surfaces, formed in contact with plugs projecting from the interior of the female mold. In the sides of an article, such protuberances constitute undercuts, so that ordinarily the sections that form them must be withdrawn in order to release the article from the mold. But some undercut articles made from tough flexible sheeting can be sprung out of the mold.

Plug-assist techniques are adaptable to both vacuum forming and pressure forming techniques. The system shown in Fig. 13-4 is known as plug-assist vacuum forming in that a vacuum is drawn after the plug has reached its closed position to complete formation of the sheet. In plug-assist pressure forming, the process differs in that after the plug enters the sheet, a partial vacuum is applied. Where the plug bottoms out, a full vacuum is drawn, and compressed air is applied to the opposite side. As opposed to plug-assist vacuum forming, pressure forming offers more uniform material distribution over the entire formed part (see Fig. 13-5).

Drape Forming

This method also is adaptable to either machine or manual operation. After framing and heating, the stock is mechanically stretched over a male mold to allow the framed edge to make a seal with the periphery of the mold. This stretching serves to redistribute or preform the sheet prior to application of the vacuum. It has the disadvantage of allowing the stock to touch prominent projections, freeze there, and perhaps rob other areas of sufficient mass to make acceptable articles. Careful control of the temperature of the mold, plus selective heating of the sheets, can alleviate some difficulties. Many articles are well adapted to this technique, which has the advantages of low cost for both mold and machine, and rapidity of operation. This technique is illustrated in Fig. 13-6.

The term ''drape forming'' sometimes is applied to forming a hot sheet by laying it over a mold (either male or female) and allowing it to conform by gravity or pressure.

Matched Mold Forming

In recent years, a number of mechanical techniques that use neither air pressure nor vacuum have evolved. Typical of these is matched mold forming (Fig. 13-7). In this operation, the plastic sheet is locked into the clamping frame and heated to the proper forming temperature. A male mold is positioned on either the top or the bottom platen with a matched female mold mounted on the other one.

The mold then is closed, forcing the plastic to the contours of both molds. The clearance between the male and female molds determines the wall thickness. Trapped air is allowed to escape through both mold faces. Molds are held in place until the plastic cools and cures.

Matched mold forming offers excellent reproduction of mold detail and dimensional accuracy. Internal cooling of the mold is desirable in this technique.

Trapped Sheet

This thermoforming technique involves the use of both contact heat and air pressure (Fig.

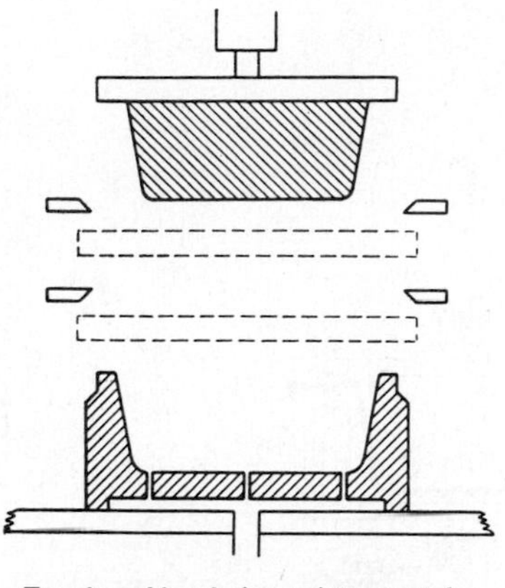

Female mold and plug assist mounted—frames open—heaters idle.

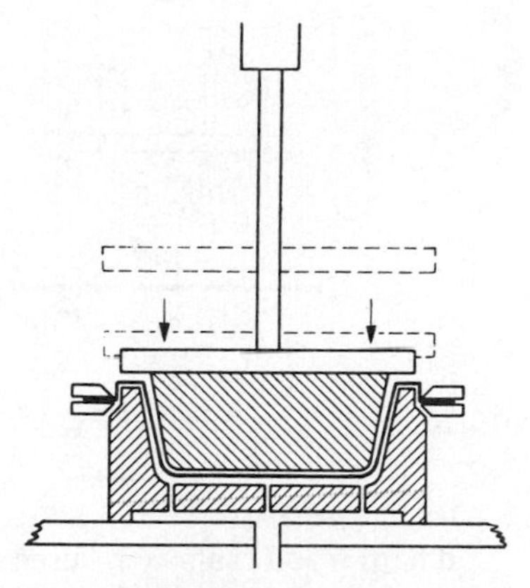

Plug assist lowered—prestretching the stock.

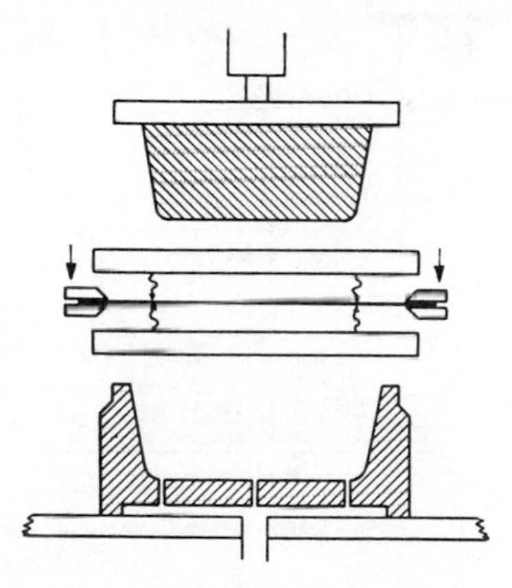

Stock in place—frames closed—heaters active.

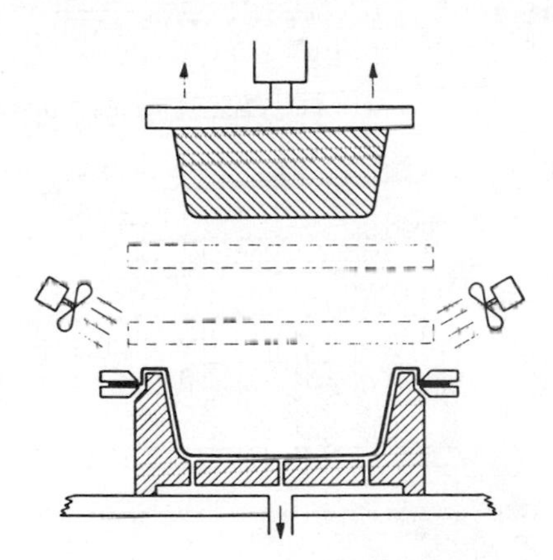

Vacuum applied—plug assist retracted—fans operating.

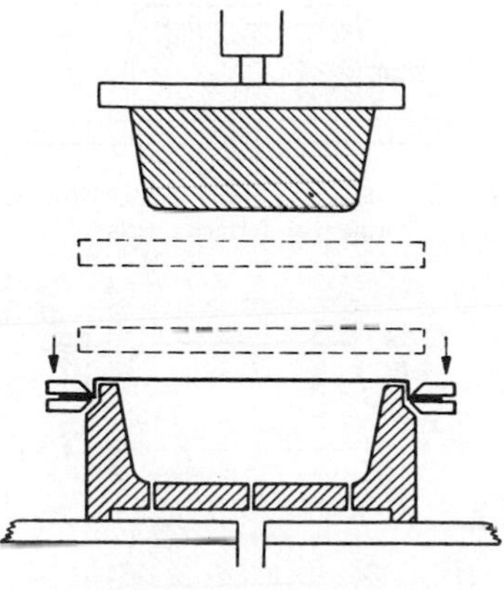

Heaters idle—frames lowered, drawing stock into contact with mold.

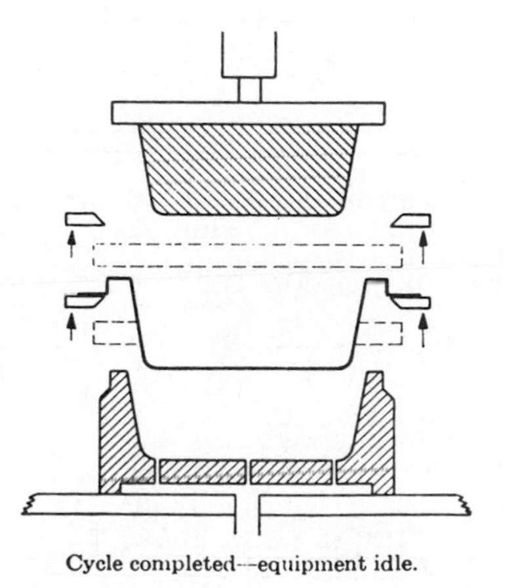

Cycle completed—equipment idle.

Fig. 13-4. Plug-assist forming, using vacuum.

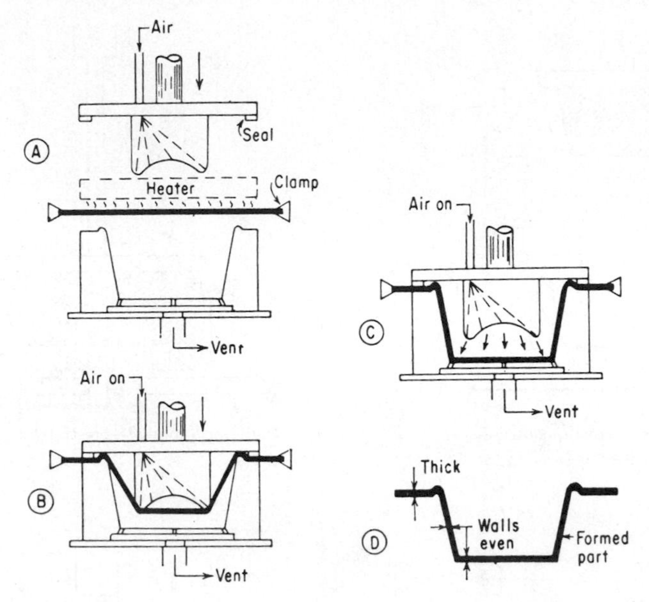

Fig. 13-5. Plug-assist pressure forming. (*Courtesy McGraw-Hill*)

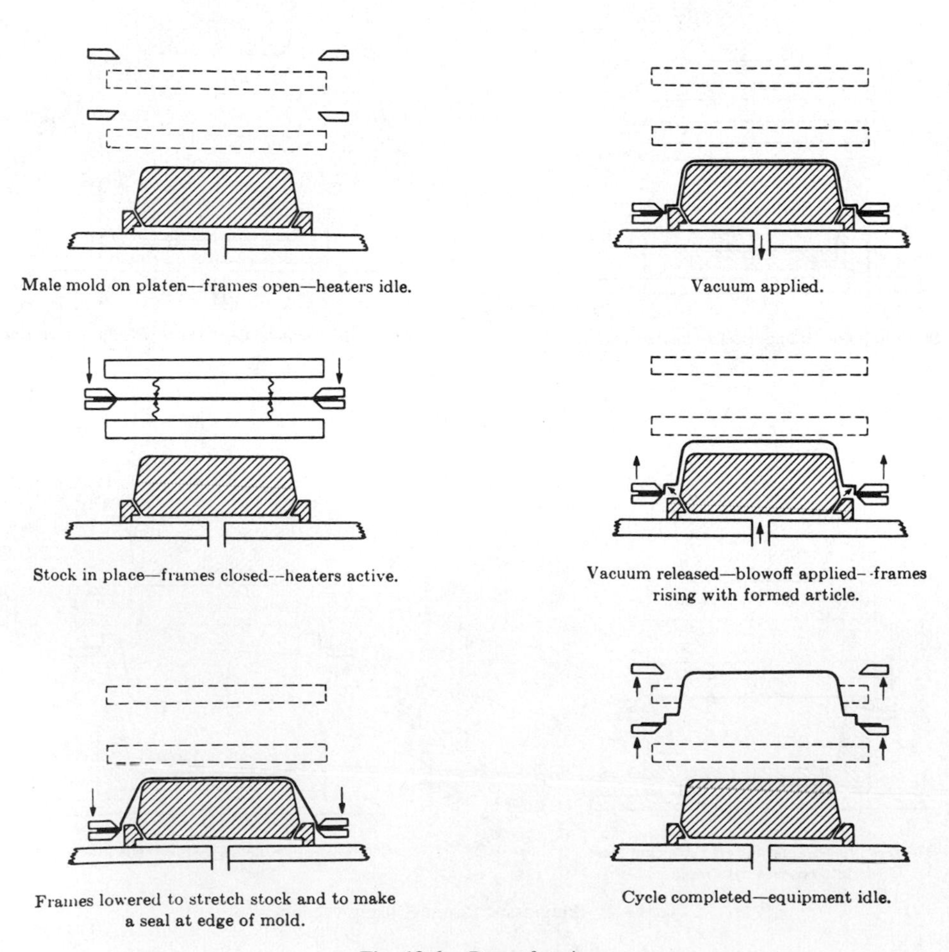

Fig. 13-6. Drape forming.

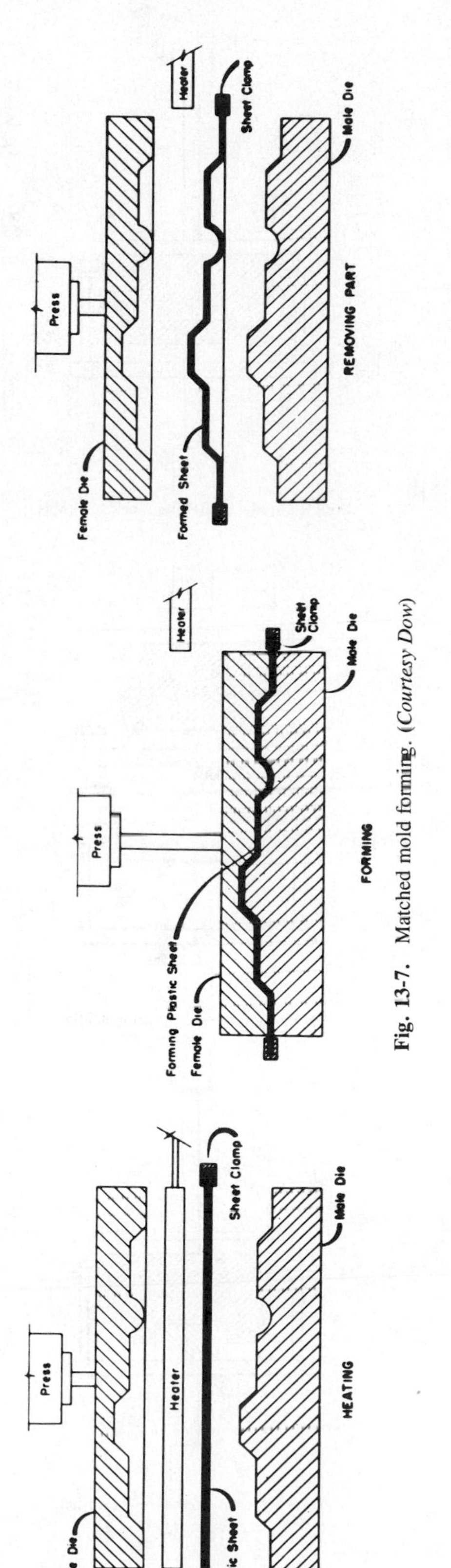

Fig. 13-7. Matched mold forming. (*Courtesy Dow*)

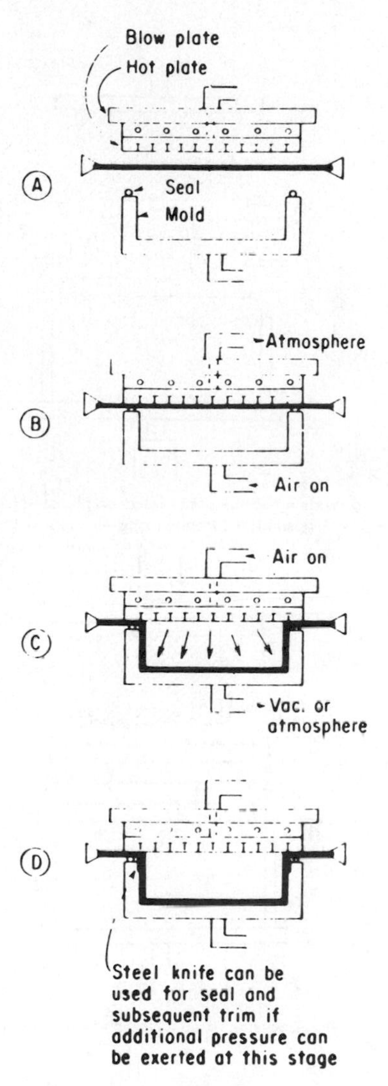

Fig. 13-8. Trapped sheet forming. (*Courtesy McGraw-Hill*)

13-8). The plastic sheet is inserted between the mold cavity and a hot blow plate (which is flat and porous to allow air to be blown through its face). The mold cavity seals the sheet against the hot plate. Air pressure is applied from the female mold beneath the sheet and blows the sheet against the contact hot plate. A vacuum also can be drawn on the hot plate. After heating, the plastic sheet is ready for forming.

Air pressure applied through the hot plate forms the sheet into the female mold. Venting or vacuum can be used on the opposite side, and steel knives can be inserted into the molds

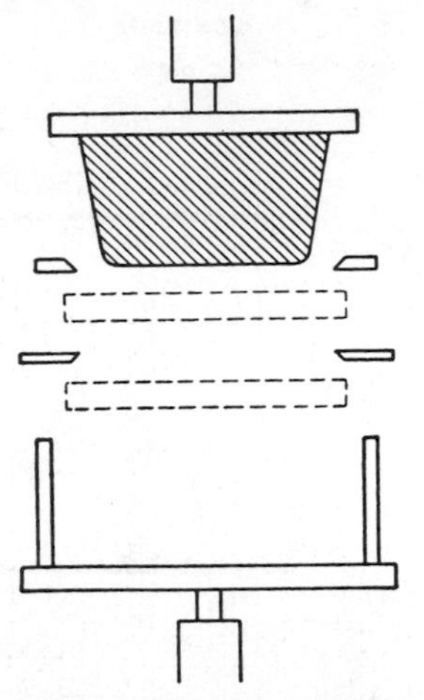

Male mold mounted—risers on platen—frame open above forming ring—heaters idle.

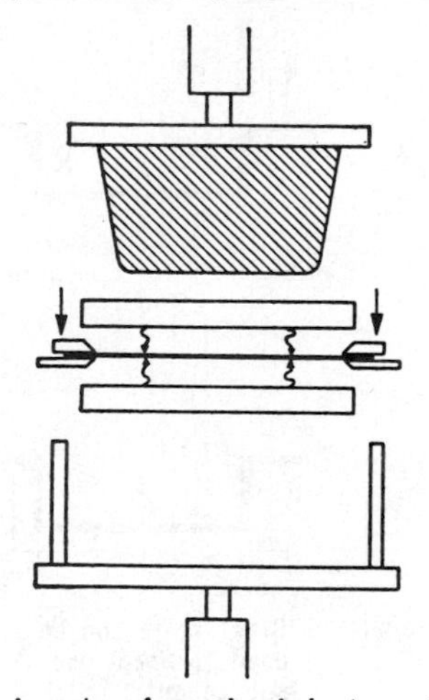

Stock on ring—frame closed—heaters active.

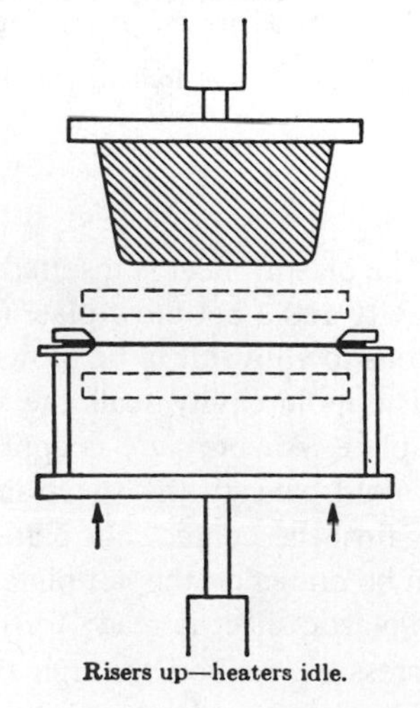

Risers up—heaters idle.

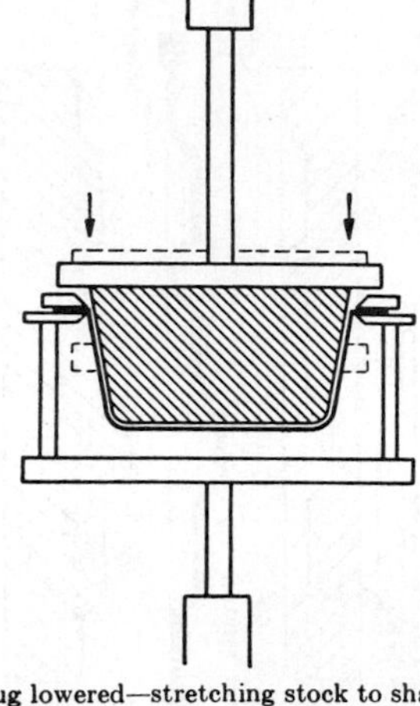

Plug lowered—stretching stock to shape.

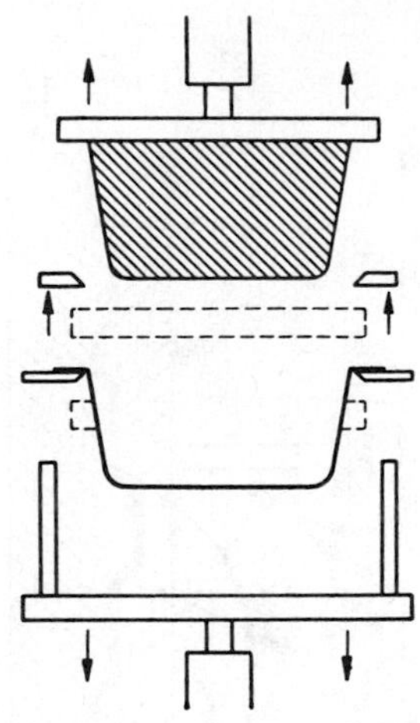

Cycle completed—equipment idle.

Variation: with female plug added.

Fig. 13-9. Plug-and-ring forming.

for sealing. After forming, additional closing pressure can be exerted for trimming.

Plug-and-Ring Forming

This is the simplest type of mechanical forming that involves more than a fold into two planes. The equipment consists of a plug, which is the male mold, and a ring matching the outside contour of the finished article. Stock may be heated away from the ring or on the ring, after which the plug is forced through the ring, drawing the plastic with it in such a way as to redistribute the stock over the shape of the plug. In order to prevent excessive chilling of the plastic, either the plug is made of a material of slow thermal conductivity, or, if it is made of metal, its temperature is controlled. This method, illustrated in Fig. 13-9, has been used for the manufacture of single- and multicavity trays, with excellent yields in large volumes.

Ridge Forming

This is a variation of plug-and-ring forming in which the plug is reduced to a skeleton frame that determines the shape of the article. Because the sheet comes into contact only with the ridges of this frame, the intervening flat areas are free from mold surface defects or "mark-off" and have better surface quality than if formed against a solid mold. (See Fig. 13-10.)

If a skeleton frame that surrounds a plane is used, the areas of the formed piece are plane surfaces. In other shapes with ridges that do not fall in a plane, the intervening surfaces tend to be concave.

Slip Forming

This method is adapted from the technique of drawing metal from a preshaped blank having approximately the same area as that of the drawn article; thus an article 4 × 4 inches and 2 inches deep is made from an irregularly shaped blank about 8 × 8 inches, held in pressure pads corresponding to the length and width of the article, and allowed to slip from them as drawing proceeds. As applied to the forming of plastics, this method is restricted by the hot strength of the plastic, and by the likelihood of its being scored in slipping out of the restraining pressure pads.

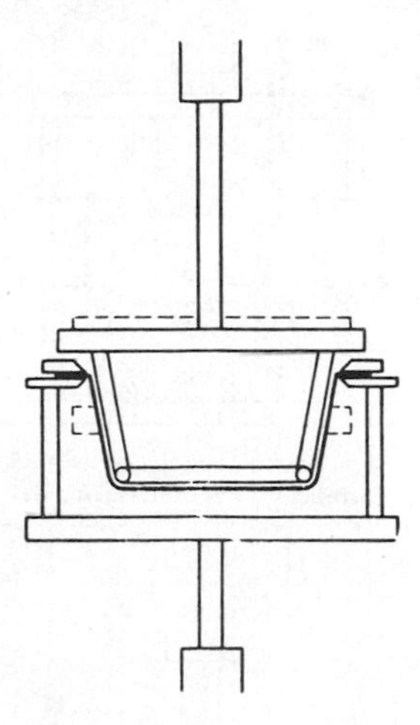

Fig. 13-10. Ridge forming.

This technique is illustrated in Fig. 13-11a; a variant of it is shown in Fig. 13-11b. The springs shown in these drawings can be replaced by air cylinders with relief valves, to improve the control. Automotive carpeting is formed in this way, as are woven fabric reinforced materials.

Snap-Back Forming

The vacuum snap-back techniques preceded so-called vacuum forming machines by several years. In the 1940s most deep draw point-of-sale plastic displays were made by this process.

The ring matches the desired periphery of the article. In mechanical operation, the cold sheet of plastic is clamped to the ring close to its forming edge, and is heated; the vacuum box beneath the ring is moved into contact with the ring to make a seal. In manual operation, the sheet is preheated and clamped to the ring, and the ring is placed over the vacuum box. A small bubble of the plastic is drawn into the box, by vacuum, just ahead of the male mold. As soon as the male mold reaches its final position, the vacuum in the prestretch box is released, and

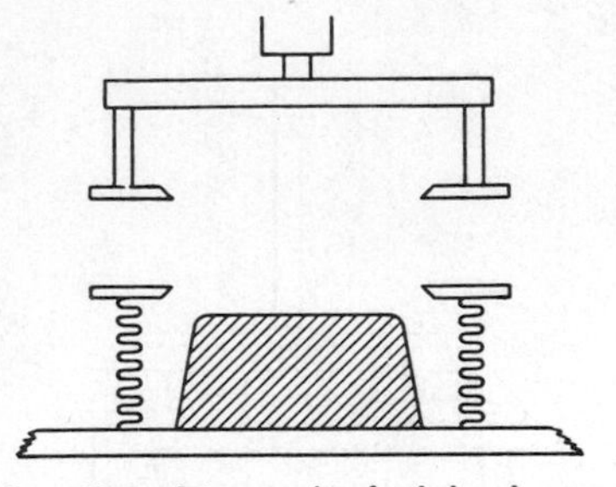

Male mold on platen—spring-loaded pads open—forming ring with risers above.

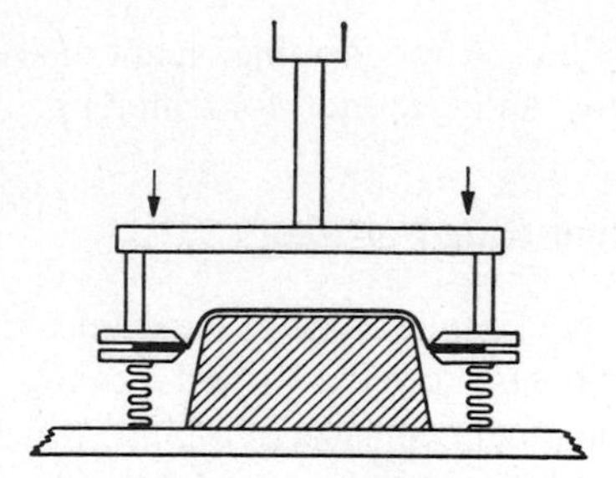

Stretching in progress—stock slipping inward.

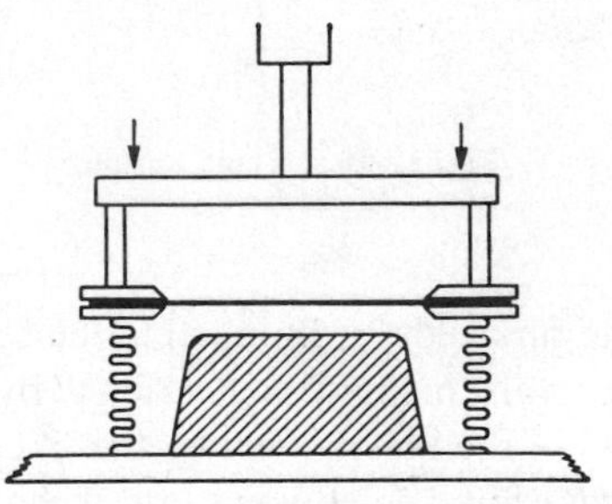

Stock in place—ring moving downward.

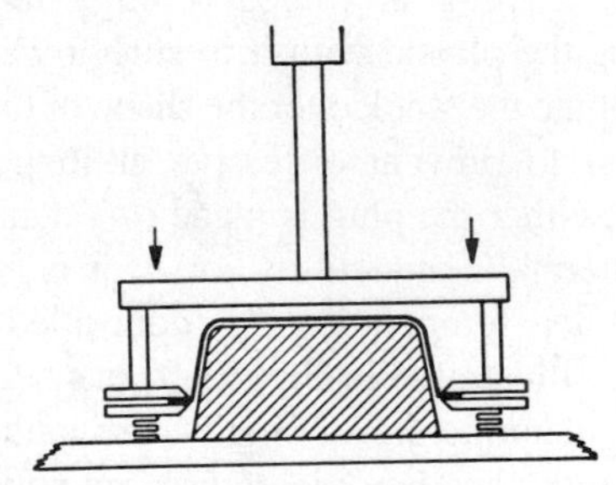

Forming completed—stock slipped to optimum.

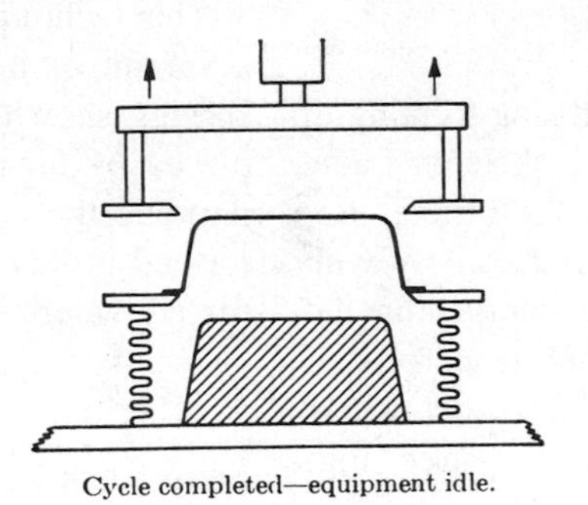

Cycle completed—equipment idle.

Fig. 13-11a. Slip forming.

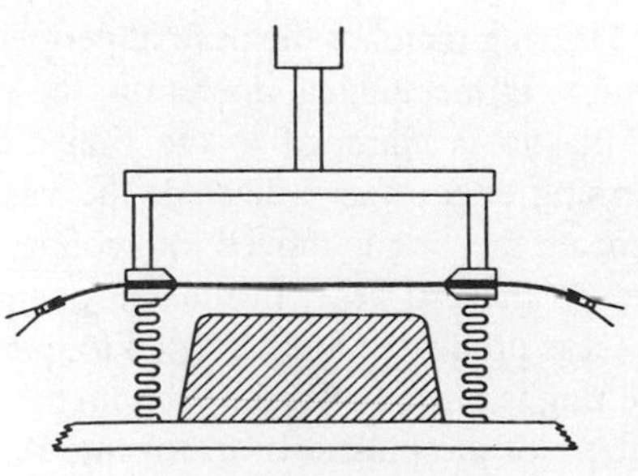

Variation: narrow ring--clips to restrict slippage.

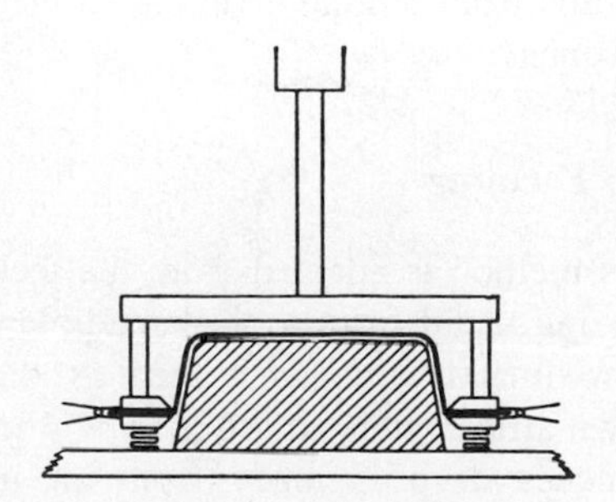

Forming completed--clips against ring.

Fig. 13-11b. Slip forming.

the hot plastic snaps back around the mold with the vacuum now applied to the mold. This method produces good uniform wall thickness. It is entirely possible to control all phases of this procedure with microprocessors or timers and limit switches, so that repetitive performance is guaranteed.

Many cases and luggage shells and computer housings made from grained or textured sheet are formed by this technique. Also, materials such as acrylic cast sheet, that have high hot strength and do not conform to small radii under vacuum alone, frequently are formed over a male mold by such a process. (See Fig 13-12.)

A variation of this process is known as the air-slip method. The male mold is confined in a box in such a manner that, as it moves toward the hot plastic, air is trapped. As pressure builds up, the plastic is pushed ahead of the mold just as is done with vacuum snap-back. When the mold is in final position, the air is released, and the vacuum is applied to the mold to define the details.

Billow-Up Prestretch Techniques

There are at least three variations of this process worthy of study, but all employ the principle of pushing the plug into the outside of a bubble of hot plastic, thus accomplishing a folding operation that permits deeper draws than are possible by any other common practice. The final shape may be determined by a male mold or a female mold. The latter is recommended except when a grained stock is used and the grain is on the outside of the article, or when the original surface gloss of the sheet might be impaired by contact with a mold.

The material is thinned at the center of the bubble before contact with the plug. A free-blown hemisphere has at its center a thickness about one-third that of the original sheet. The thinned area, however, remains in contact with the plug (or male mold), and is carried by it. Further stretching occurs in the thicker side area, so that the finished article has approximately uniform thickness.

Billow-Up Plug Assist (or Reverse Draw with Plug Assist)

The stock is placed in the frame and heated. Then the female mold is moved into the stock close to the edge of the frame, deep enough to maintain a seal while a bubble is blown. Air is introduced into the female mold to a pressure of about 1 to 3 psi, or more, sufficient to create a bubble toward the plug. The plug, in this case only a sheet-carrying assist, is heated to a temperature only slightly less than that of the hot stock itself. As it is moved, it pushes the redistributed stock into the cavity in a folding or rolling action. The plug is designed slightly undersize, about 85%, but should move the stock to a shape quite close to its final position in the female mold. The final shaping then is accomplished with vacuum and/or compressed air being applied to the female mold. (See Fig. 13-13.)

By regulating the temperature of the stock, the height of the bubble, and the temperature and speed of travel of the plug assist, as well as the release of air pressure and application of vacuum, it is possible to closely control the thickness of any given section of the article. One exception is that corners never will be thin because the folding or rolling action as the plug moves downward gathers extra stock around the corners. Articles formed into a cavity shrink away from the mold; so little draft is required in the mold.

By this process articles may be drawn with uniform wall thicknesses less than 25% of the original stock thickness, with repetitive quality, and usually without apparent impairment of the physical properties of the original material. But in some cases a uniaxial orientation of stress may develop, so that the finished article is crack-sensitive parallel to the direction of stretch. This becomes more noticeable if the flanges are trimmed off.

Billow-Up Vacuum Snap-Back

The third version of this method is used with grained or polished stock on a male mold, to preserve the finish. The blank must be oversize

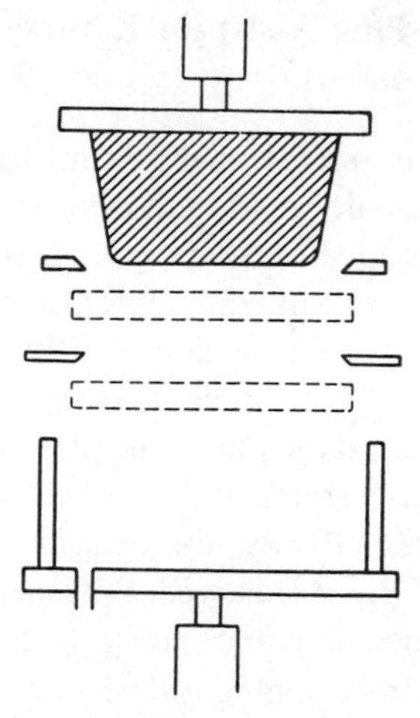

Male mold above—vacuum chamber below—frame open—forming ring and heaters idle.

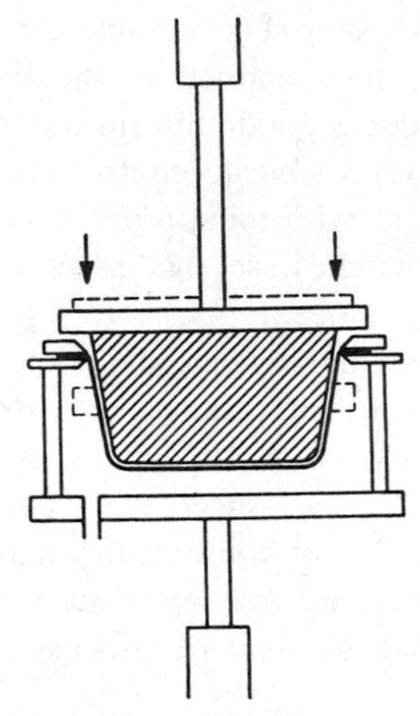

Forming completed—vacuum released for snap-back. (Vacuum should not be released until male mold is in place.)

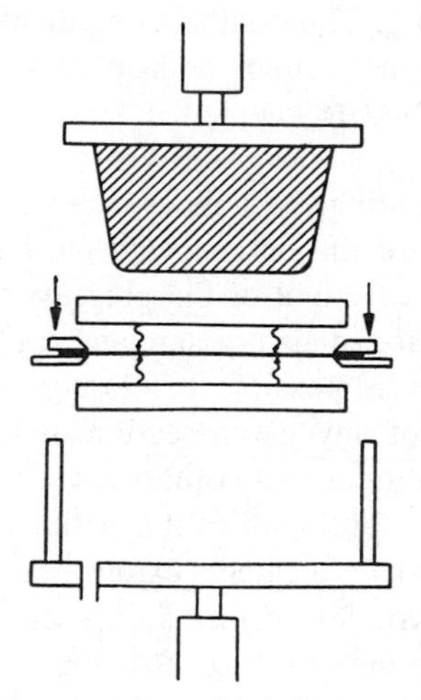

Stock on ring—frame closed—heaters active.

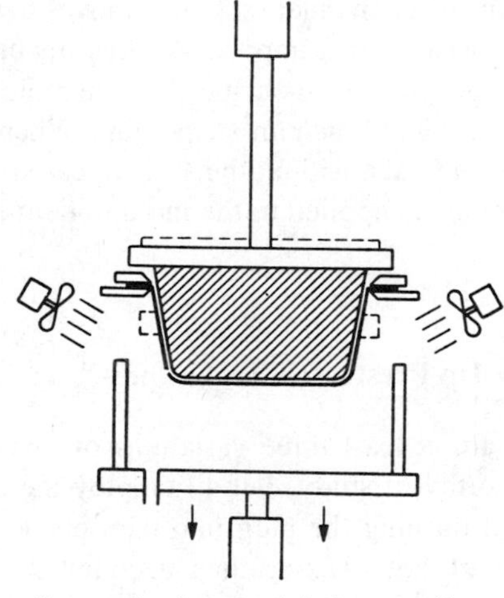

Vacuum chamber retracted—fans operating.

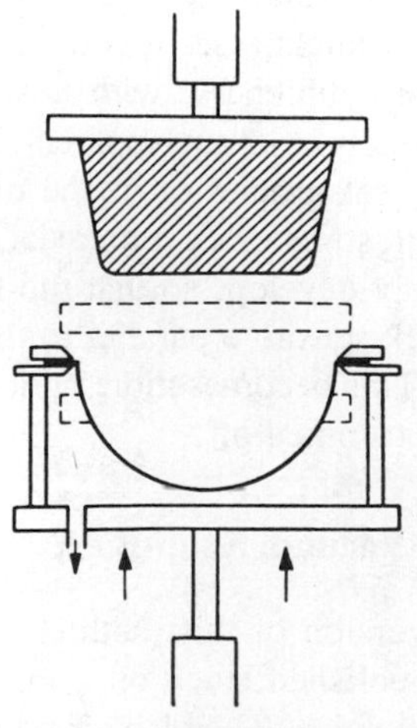

Vacuum chamber up—heaters idle—vacuum applied.

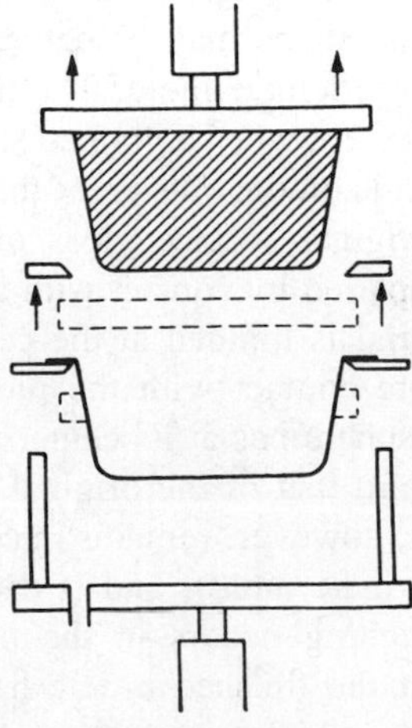

Cycle completed—equipment idle.

Fig. 13-12. Snap-back forming.

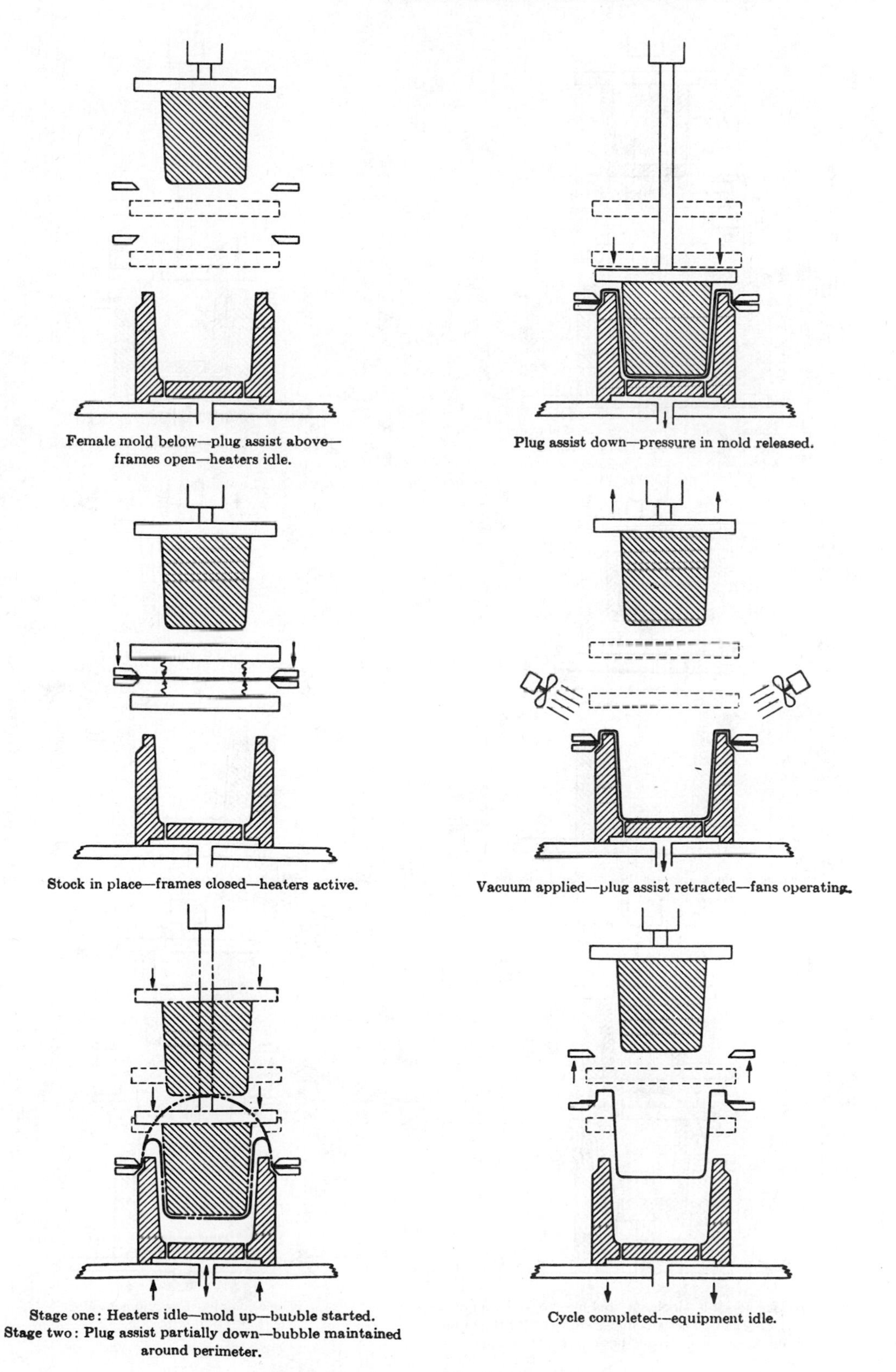

Fig. 13-13. Reverse draw with plug assist.

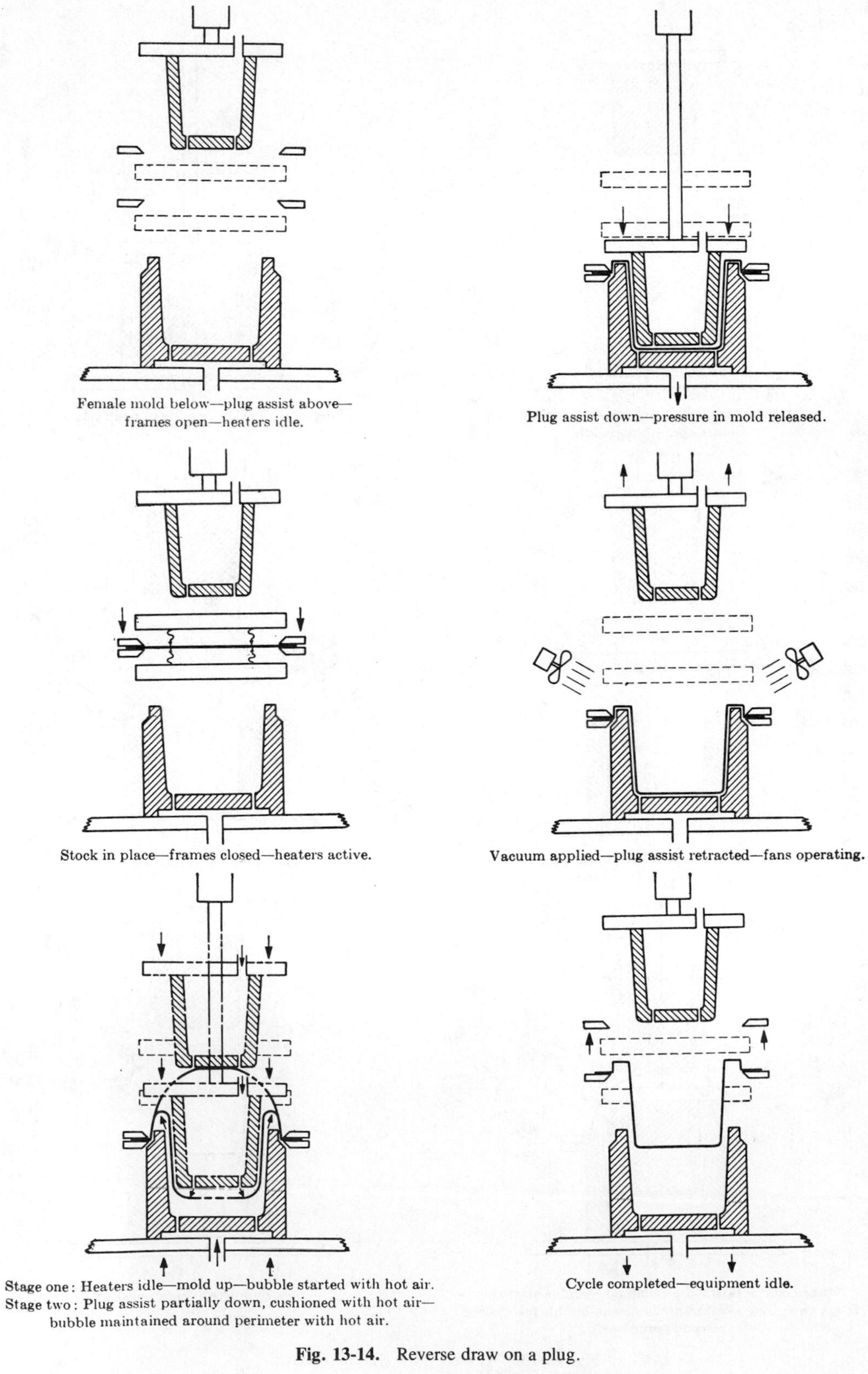

Female mold below—plug assist above—frames open—heaters idle.

Plug assist down—pressure in mold released.

Stock in place—frames closed—heaters active.

Vacuum applied—plug assist retracted—fans operating.

Stage one: Heaters idle—mold up—bubble started with hot air.
Stage two: Plug assist partially down, cushioned with hot air—bubble maintained around perimeter with hot air.

Cycle completed—equipment idle.

Fig. 13-14. Reverse draw on a plug.

to provide a bubble of sufficient size and shape, so that when the mold comes into contact with the outside of the bubble, the stock that touches the bottom of the mold freezes there and need not be reduced further in thickness or change its shape. Corners should be "cut" to avoid "webbing." The forming ring matches the finished article and assists in "wiping" the stock around the mold as the mold moves into final position. (See Fig. 13-14.)

The mold should be run as warm as is compatible with efficient cooling of the article. It should be a temperature-controlled aluminum mold.

Dual Sheet Forming

A number of techniques have been made available for the production of hollow products by thermoforming. Typical is the concept of dual sheet thermoforming (also known as twin-sheet forming). It operates as follows: Two rolls of plastic sheet are automatically fed, one above the other, with a predetermined space in between, through the heating stations and into the forming station. Here a blow pin enters at the central point of the hollow object (i.e., in between the two sheets), and the upper and lower halves of the tool close onto the sheets and pinch off around the entire perimeter. High-pressure air then is introduced between the two sheets from the blow pin, and a vacuum is applied to each of the two mold halves. The hollow object then indexes forward, and the next two segments of sheet move into place for forming.

In one variation of the process, urethane foam instead of air pressure is introduced between the two sheets. The urethane bonds to the two skins, forming a rugged sandwich construction. This technique is applicable to the manufacture of urethane foam-filled boat hulls.

Another technique for making hollow products is known as clam-sheel forming. This is a sheet-fed technique involving the use of a rotary-type machine. The individual sheets are placed in separate clamp frames, and then indexed through the heating station and into the forming station. Here, a vacuum is applied to both halves of the mold to draw the upper sheet into the upper mold half and the lower sheet into the lower mold half. (See Fig. 13-15.)

Still another method, known as twin-shell forming, involves the use of a series of continuously moving molds (traveling on belts) that clamp onto the sheets (feeding off rolls) and travel with them as the vacuum is pulled and the sheet halves are formed. (See Fig. 13-16.)

Fig. 13-15. Dual-sheet thermoforming is accomplished on this four-station rotary thermoformer by two sheets of thermoplastic material that are heated independently and are permanently bonded in the forming station. (*Courtesy CAM/ Central Automated Machinery, Inc.*)

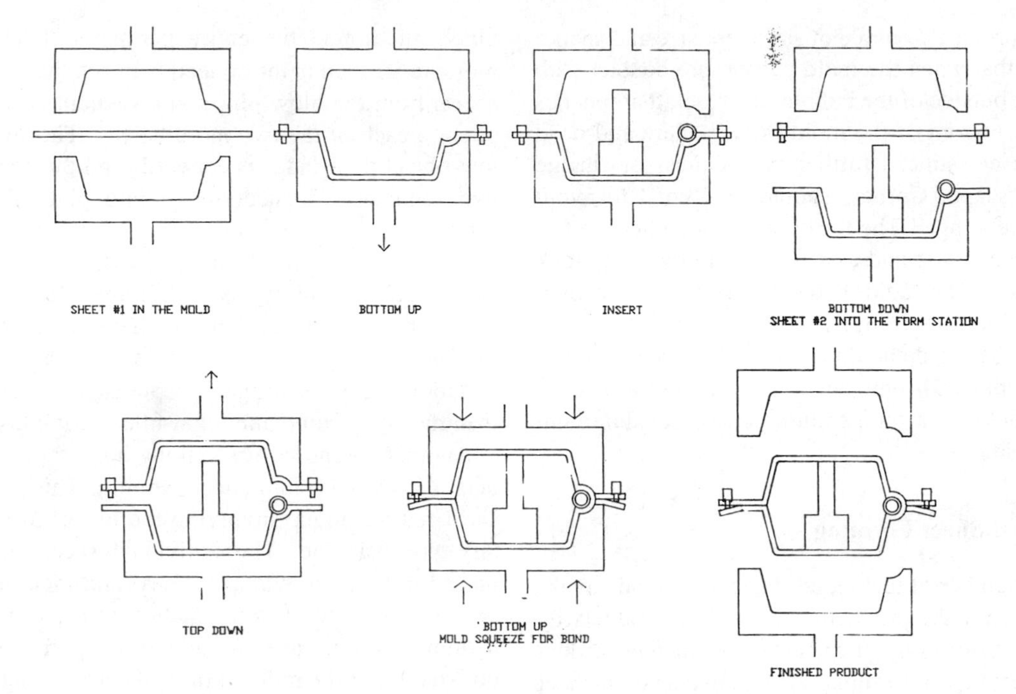

Fig. 13-16. Dual-sheet forming. (*Courtesy CAM/Central Automated Machinery, Inc.*)

THERMOFORMING MACHINES

Industrial thermoforming machines are as varied as the products they were designed to produce. They range from relatively simple, shallow-drawn, manually controlled equipment to complex, microcomputer-controlled rotary and in-line machines designed for production and efficiency, and to the recent "totally electric" thermoformers.

Any thermoforming machine must provide the following: (1) a method for heating sheet to the pliable plastic state called the "forming temperature; (2) a clamping arrangement to hold the plastic sheet for heating and positioning for forming; (3) a device to raise or lower the mold into the plastic sheet or to move the clamped sheet over the mold; (4) a vacuum system; (5) an air-pressure system; (6) controls for the various operations; and (7) safety devices.

Machines often are classified according to the number of operations they perform, as discussed below.

This section presents two types of thermoforming machinery—cut-sheet and roll-fed.

Cut-sheet or sheet-fed machinery utilizes material typically .060 inch up to .500 inch thick, being ideally suited to forming large durable goods such as industrial pallets, boat hulls, computer housings, automotive components, and refrigerator door liners.

Roll-fed or continuous-type thermoformers are fed from a continuous roll of plastic sheet, usually less than .035 inch thick, or fed directly from an extruder. Typical applications are high-production items such as plates, cups, trays, and blister packaging. (See Fig. 13-17.)

Single-Stage, Sheet-Fed Machines

A single-stage machine can perform only one operation at a time, and its total cycle will be the sum of the times required for loading, heating, forming, cooling, and unloading. In a typical operation, the sheet is clamped in a frame, and the frame is moved between the heaters (or under a single heater) and back to the forming station for thermoforming. (See Figs. 13-18 and 13-19.)

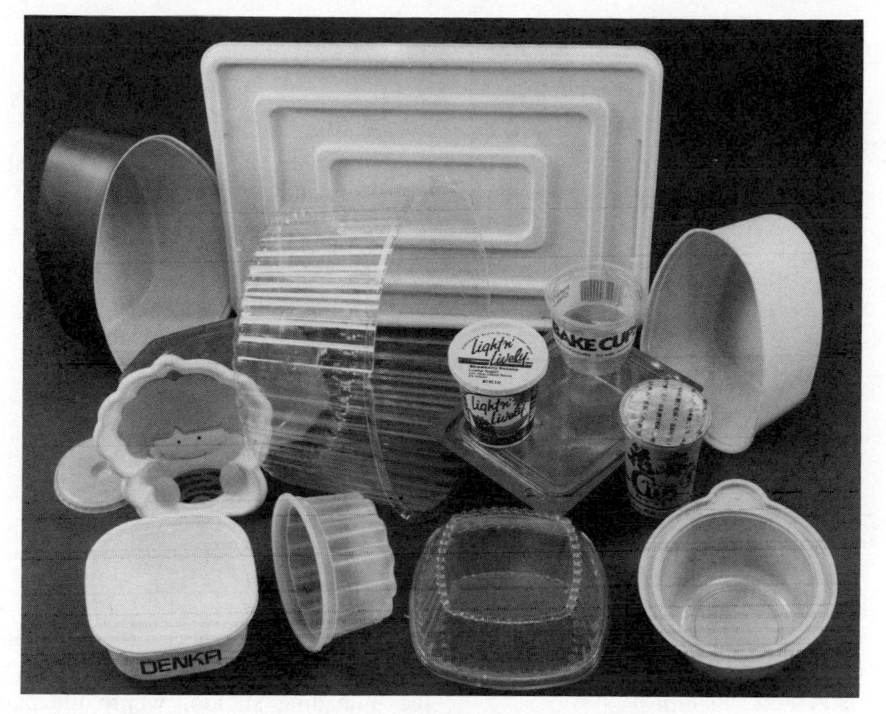

Fig. 13-17. Typical products on roll-fed continuous thermoformers. (*Courtesy Lyle Industries, Inc.*)

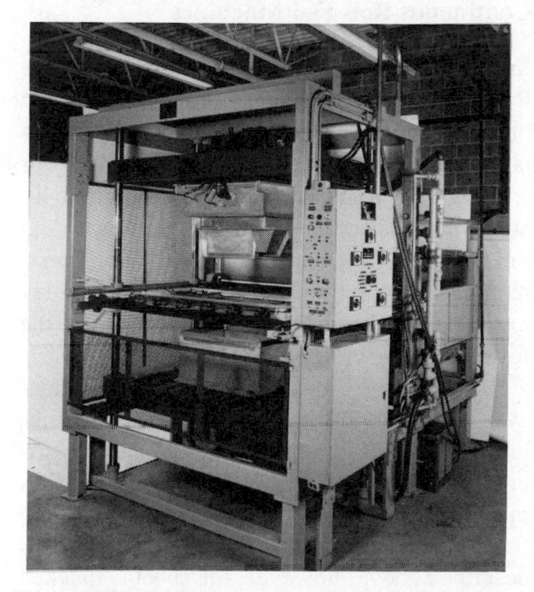

Fig. 13-18. Single-stage shuttle thermoformer for cut-sheet applications. Material is loaded into clamp frame, moved into oven area, shuttled back into form station, and then unloaded. (*Courtesy CAM/Central Automated Machinery, Inc.*)

Multiple-Stage, Sheet-Fed Machines

A two-stage machine can perform two operations simultaneously. It usually consists of two forming stations and a bank of heaters that move from one station to the other. Machines with three stages or more usually are built on a horizontal circular frame and are called rotaries. (See Fig. 13-20.) The rotary thermoformer operates like a merry-go-round, indexing through the various stations. A three-stage rotary machine would have a loading and unloading station, a heating station, and a forming station, where cooling also takes place, and would index 90° after each operation. Because there is always a sheet in each of the stations, it provides considerably higher output than a single-station machine. (See Fig. 13-21.)

Four-stage rotaries can incorporate various station configurations (i.e., load/unload, preheat, heat, form; or load, heat, form, and unload).

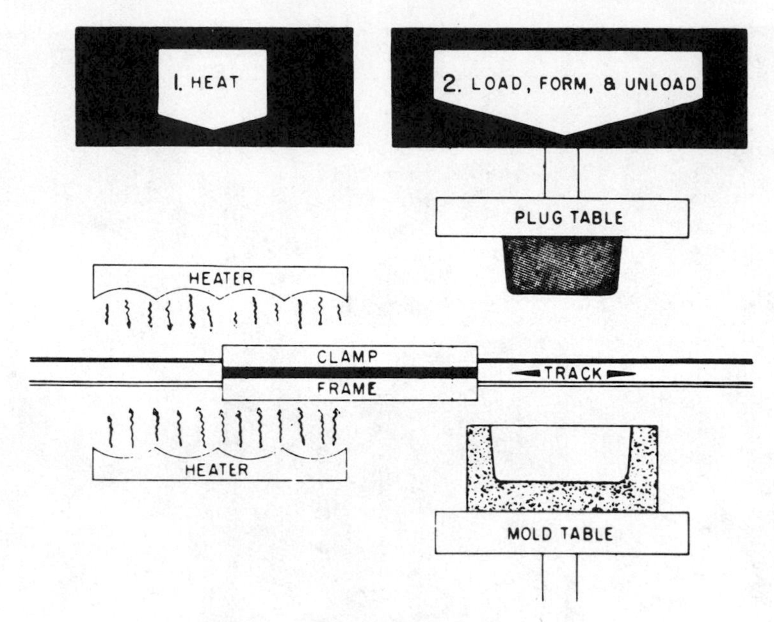

Fig. 13-19. Single-stage forming unit. (*Courtesy Dow*)

In-Line, Sheet-Fed Machines

Here, the sheet follows the same pattern as a caterpillar track. The sheet is clamped in a frame that travels into the heating station, then indexes through the forming station and on to the unloading station, where the part is removed by the operator. This type of machine has a total of five clamp frames in use at all times.

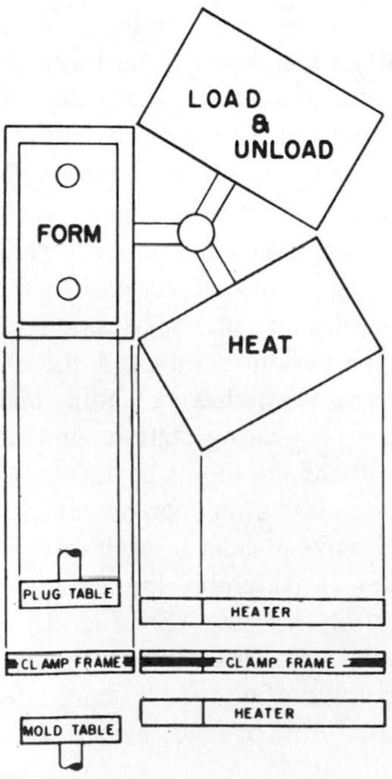

Fig. 13-20. Three-stage rotary unit. (*Courtesy Dow*)

Continuous Roll-Fed Machines

In the early days of thermoforming, continuous forming machines that fed off a roll of plastic or directly from an extruder used either a rotating cylinder as the mold (drum former) or conventional molds that traveled horizontally at the same speed as the sheet (the upper and lower platens moved with the sheet but had a reciprocating motion so that they could index to the next unformed section of sheet, once the forming cycle was complete).

The workhorse of the industry today, however, is the intermittently fed continuous forming machine (used for thin wall containers, disposable cups, lids, etc.). (See Fig. 13-22.) This machine is fed either from a roll or directly from an extruder. As opposed to the systems described above, however, the sheet is indexed through the machine intermittently. It is considerably faster than those techniques where the sheet is continuously moving through the machine, and the molds have to move back and forth.

Fig. 13-21. Three-state rotary thermoformer includes load/unloaded station, heat station, and form station. (*Courtesy CAM/Central Automated Machinery, Inc.*)

Fig. 13-22. Roll-fed continuous thermoformer in-line with trim press for high production of thin-wall products. (*Courtesy Lyle Industries, Inc.*)

In a typical operation, the sheet feeds off a roll at the rear of the thermoformer into a set of conveying chains that indexes the sheet intermittently forward through heating, forming, and trimming. Once the roll of material is threaded through the system, it functions completely automatically and can cycle as fast as two seconds.

Packaging Machines

Although most of the equipment already discussed can be used to form plastic packages, several machines have been adapted to combine the forming operation with other packaging functions. Such machines are used chiefly for skin or contour packaging and blister packaging. Both processes involve the sealing of paperboard or plastic to the formed plastic sheet, with the product to be packaged enclosed. Most conventional machines can be equipped to handle skin packaging, as the loading operation is usually manual. (See Fig. 13-23.) These machines may operate from roll or sheet stock, and some are equipped to trim the packages as they leave the machine.

Rather elaborate machines now are available for complete blister packaging. These machines will form the blisters, provide an area for manual or automatic loading, seal paperboard to the back side, and die-cut the finished packages. Generally, they operate with intermittent or continuous feed of roll stock, and the

Fig. 13-23. Typical thermoformed blister packaging and box inserts, unusually formed to the shape of the product to be packaged. (*Courtesy Lyle Industries, Inc.*)

various functions are performed simultaneously at a sequence of stations in-line.

Form/fill/seal equipment is widely used, in which a package is formed, immediately filled with the product to be packaged (in some cases, a liquid product also can be used as a pressure medium to assist in forming), and then sealed. Because aseptic features are required for machines of this type, all operations generally are enclosed in pressurized sterilized air chambers.

MATERIALS

The thermoplastic material used for thermoforming is one of the most important factors in the process. The properties and characteristics of a particular material will determine its thermoforming qualities. As the thermoforming process has become increasingly important, the emphasis of extruders has turned to producing sheet specifically to meet the demands of thermoforming markets (i.e., multilayer barrier materials for shelf-stable food packaging or larger sheet for the production of products such as boats, pallets, signs, siding for houses, etc.). Current maximum dimensions for thick sheet are limited by the presently available manufacturing facilities (sheet up to 10 feet wide × 14 feet long). But some manufacturers will purchase a thicker sheet than required and "stretch" it with a thermoformer to obtain a larger mold area (i.e., swimming spas). Future mold sizes may be almost unlimited in this case. Thinner sheeting (produced by extrusion, calendering, or continuous casting) is available in rolls of several hundred feet in length.

Various modifications to the body or surface of the plastic sheet can be incorporated during the extrusion process to produce specialized effects. The modified material may be:

- *Coextruded*—in an increasingly popular process that affects the surface quality for a desired application, such as weather resistance, or impact/mar resistance; multilayers provide various "barrier" qualities for food packaging and "shelf-stable" items.
- *Flocked*—usually providing a velvetlike layer for decorative box inserts.

- *Textured*—so that it can provide a leather- or woodlike look and feel, or an engraved look.
- *Tinted*—providing translucent color.
- *Pigmented*—giving solid coloring.
- *Metallized*—shiny and mirrorlike; here the material is difficult to thermoform because of the sheet's reflectiveness to radiant heat and thinning upon stretching.
- *Filled*—enabling cost reduction, as well as enhancement of physical qualities (i.e., improving rigidity; increasing impact resistance).
- *Foamed*—a type popular in fast-food packaging because of heat-insulating qualities; this modification is a resin-extender, making it cost-effective. (See Fig. 13-24.)

Two of the fastest-growing areas for thermoforming today utilize composites/engineered materials and polypropylene.

Engineered materials and composites are very popular in the automotive body panel and aerospace component markets because of high structural strength and flame resistance. New developments in thermoforming equipment are making thermoforming a viable and cost-effective alternative to blow molding and injection molding.

More competitive pricing of polypropylene in the last few years has made it the most cost-effective resin for the booming microwavable-container market. Resin suppliers are taking giant steps to produce more thermoformable polypropylene for processors with conventional melt-phase forming equipment. (See below, section on thermoforming of coextruded structures and other special materials.)

Let us now note some of the unique characteristics and properties of thermoplastic sheets that must be taken into account when considering thermoforming.

Plastic Memory

Most thermoplastics possess enough elasticity when they are hot that when stretched, by mechanical or pneumatic means, they tend to draw tight against the force that stretches them and to stretch uniformly. These characteristics permit forming against a single mold, which is a common practice in many thermoforming methods (as distinguished from matching male and female molds, used in most other plastic processing techniques). Reheating a formed part to the original forming temperature will activate the so-called plastic memory and cause it to relax back toward its original shape (i.e., a flat blank). Hence, errors in the contours of formed parts often can be corrected by re-forming. It also is possible to decorate a formed part, then relax it back to the original flat, and use the ''distorted'' pattern to print subsequent sheets ''in distortion.'' The printed sheets can then be formed into products in which the decorative elements will be in perfect register on the three-dimensional surface.

Hot Elongation or Hot Strength

All thermoplastic sheet materials can be stretched when hot, but this property varies greatly with different materials, and under different conditions, and it is measurable. It is intimately related to temperature and to speed of elongation, and, in many of the methods described earlier, it is of critical importance. Because of its dependence on the correct temperature, methods of heating, methods of stretching or forming, choice of material for molds, and related methods of cooling on the mold, hot strength is referred to frequently.

Some commercial sheet stocks can be stretched as much as 500 or 600% over their original area; others stretch as little as 15 to 10%. Naturally this characteristic has a great influence on what shapes can be produced and the quality of what is produced.

Some materials at forming temperatures become almost puttylike and respond to a minimum of pressure, either pneumatic or mechanical, in such a way as to pick up every detail of the mold. Others exhibit strong resistance, and thus require heavier equipment and tools. The limited differential of pressure available in vacuum methods may not suffice to provide small details in some formed articles. In such cases, compressed air can be added.

This property is somewhat related to the

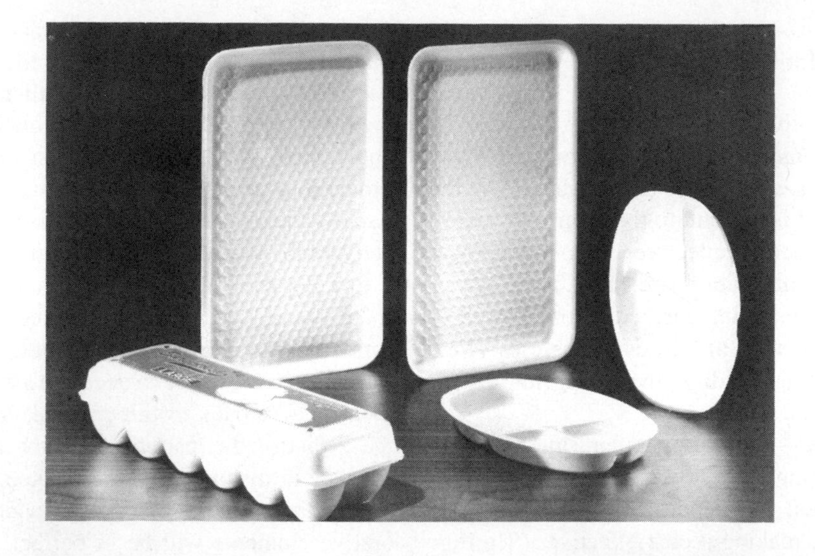

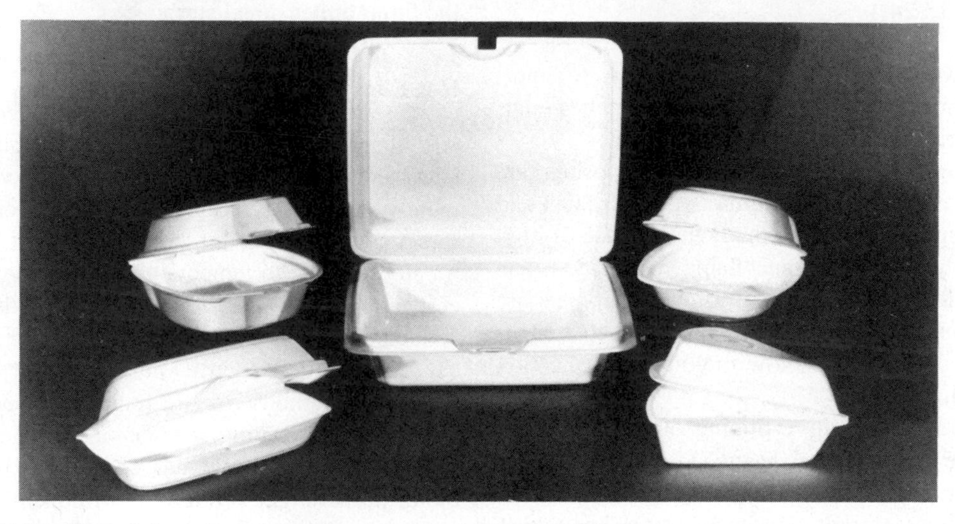

Fig. 13-24. Typical thermoformed foam products (fast-food packs, meat trays, egg cartons, and dinnerware) provide protection to the product as well as cost-effective packaging and heat-sealing qualities. (*Courtesy Brown Machine, Div. John Brown*)

ability to be stretched while hot, but does not run parallel to it. The hot strength of thermoplastics varies dramatically with temperature changes, but very little with gauge variations.

Temperature Range for Forming*

Amorphous thermoplastics (ABS, acrylic, styrene, polycarbonate, and vinyl, for example) do not have melting points. Their softening with increase of temperature is gradual, and each material has its own range of specific processing temperatures. Selection of the forming temperature comes by knowing the degradation temperature and then determining the highest temperature *under* that where the sheet has enough ''hot strength'' to handle and still form properly.

Crystalline thermoplastics, such as polyethylene, polypropylene, and nylon, have sharp melting points. Unfortunately, most of the forming temperatures are the same as the melting temperatures. Polypropylene should be

*By William K. McConnell, Jr., McConnell Co., Inc., Fort Worth, TX.

heated to 330°F for the proper thermoforming temperature; however, regular grades *melt* at 330°F. Special grades and modifiers have been developed recently to give good hot strength at these temperatures.

Table 13-1 shows the processing temperature ranges for the thermoforming of some popular thermoplastic materials.

Thermoforming Processing Temperatures (see Table 13-1).

1. *Mold and Set Temperature*. The set temperature is the temperature at which the thermoplastic sheet hardens and can be safely taken from the mold. This is generally defined as the heat distortion temperature at 66 psi (455 kPa). The closer the mold temperature is to the set temperature without exceeding it, the less one will encounter internal stress problems in the part. For a more rapid cycle time, if post-shrinkage is encountered, post-cooling fixtures can be used so that parts may be pulled early.

2. *Lower Processing Limit*. This column shows the lowest possible temperature for the sheet before it is completely formed. Material formed at or below this limit will have severely increased internal stress that later can cause warpage, lower impact strength, and other poorer physical properties—another reason for rapid vacuum or forming pressure.

The least amount of internal stress is obtained by a hot mold, hot sheet, and very rapid vacuum and/or compressed air.

3. *Orienting Temperatures*. Biaxially orienting the molecular structure of thermoplastic

Table 13-1. Thermoforming processing temperature ranges.

Material	*1. Mold and set temperature*		*2. Lower processing limit*		*3. Orienting temperature*		*4. Normal forming (core) temperature*		*5. Upper limit*	
	°F	°C	°F	°C	°F	°C	°F	°C	°F	°C
ABS	185	85	260	127	280	138	300	149	360	182
Acetate	160	71	260	127	280	138	300	149	360	182
Acrylic	185	85	300	149	325	163	350	177	380	193
Acrylic/PVC (DKE-450[2])	175	79	290	143	310	154	340	171	360	182
Butyrate	175	79	260	127	275	135	295	146	360	182
Polycarbonate	280	138	335	168	350	177	375	191	400	204
Polyester, thermoplastic (PETG[3])	170	77	250	121	275	135	300	149	330	166
Polyethersulfone	400	204	525	274	560	293	600	316	700	371
Polyethersulfone, glass-filled	410	210	535	279	560	293	650	343	720	382
Polyethylene, high-density	180	82	260	127	270	132	295	146	360	182
Propionate	190	88	260	127	270	132	295	146	360	182
Polypropylene	190	88	265	129	280	138	310–330	154–166	331	166
Polypropylene, glass-filled	195	91	265	129	280	138	400+	204+	450	232
Polysulfone	325	163	374	190	415	213	475	246	575	302
Styrene	185	85	260	127	275	135	300	149	360	182
Teflon (FEP[1])	300	149	450	232	490	254	550	288	620	327
Vinyl, rigid	150	66	220	104	245	118	280–285	138–141	310	154
Vinyl, rigid foam	162	72	240	116	260	127	300	149	350	177

Registered trademarks: [1]DuPont, [2]Polycast Corp., [3]Eastman
Courtesy of William K. McConnell, Jr.; McConnell Co., Inc., Fort Worth, TX.

sheet approximately 275 to 300% at these temperatures and then cooling greatly enhances properties, such as impact and tensile strength. Careful matching of heating, rate of stretch, mechanical stresses, and so on, is required to achieve maximum results.

In thermoforming oriented material, good clamping of the sheet must be used. The sheet is heated as usual to its proper forming temperature and thermoformed. The hot forming temperatures do not realign the molecular structure; therefore, the better properties of the oriented sheet are carried into the finished part.

4. *Normal Forming Temperature*. This is the temperature that the sheet should reach for proper forming conditions under normal circumstances. *The core (interior) of the sheet must be at this temperature*! The normal forming temperature is determined by heating the sheet to the highest temperature at which it still has enough hot strength or elasticity to be handled, yet below the degrading temperature.

5. *Upper Limit*. This is the temperature at which the thermoplastic sheet begins to degrade or decompose. It is crucial to ensure that the sheet temperature stays below this value. When radiant heat is used, the sheet surface temperature should be carefully monitored to avoid degradation while waiting for the "core" of the material to reach forming temperature. These limits can be exceeded, if for a short time only, with minimum impairment of the sheet properties.

HEATING

Provided that the thermoplastic sheet used meets specifications, the next most important factor in thermoforming is proper heating. The sheet must be heated precisely and uniformly from surface to core, except when profile or area heating is required, to obtain proper material distribution.

Heat demands energy; in fact, in the thermoforming process, the heating phase can consume up to 80% of the total energy demand. It is important to determine the most cost-effective method of heating, depending on the material, product specifications, and desired cycle speeds. There are three methods of heating: (1) convection heating, which circulates hot air; (2) contact heat (heated plates come in contact with the sheet); and (3) radiant heat (radiant energy is transmitted between the heater and the sheet). Far infrared gas-fired and electric heaters are made with accurate controls to produce even overall heat. Most heating now is done by far infrared radiant heat rather than air recirculating ovens, which were used more in the past.

Types of radiant heater elements include tubular rods (often called Calrod heaters), flat strip heaters, quartz heaters, and ceramic heaters. Tubular rods are the most widely used today. However, ceramic heating elements are becoming very popular, despite their higher initial cost compared to tubular rods, because of their longer wear and profitability. (See Table 13-2 and Fig. 13-25.)

Temperature Control

The thermoforming process demands precise controlled heating to ensure quality and repeatability from cycle to cycle. Two considerations are the temperature output rate of the heater elements and the time exposure of the sheet to the heat. Lower heat levels slow down cycle rates, but can improve forming conditions and final product outcome.

All heating elements must have a means of control, whether it be a simple on/off switch or a thermostat. A device offering more control is the percentage timer, which relays predetermined on/off settings to the heating elements.

For more precise temperature controls, thermocouples are used, which actually sense the temperature and activate the on/off control or a more complicated electronic relay system. Proportional controls provide variable power inputs—increasing up to setpoint temperatures, then decreasing power input proportionately. Solid-state relays offer even more control than this.

The ultimate controller today is the microprocessor. Precise and accurate temperature readings are made automatically by the system, calculated in accordance with the acceleration and deceleration deviations, and adjustments are made accordingly to bring the sheet to the setpoint temperature. (See Fig. 13-26.)

Table 13-2. Descriptions of most popular heating elements.

Type of heating element	*Efficient when new*	*Efficiency after 6 mo.*	*Average life*	*Comments*
Small-diameter coiled nichrome wire	16–18%	8–10%	1500 hr	Cheapest initially; very inefficient, heat nonuniformly with use.
Tubular rods and metal panels	42%	21%	3000 hr	Less expensive than ceramic or quartz panels, but heat not as uniform and profiling more difficult*
Ceramic panels	62%	55%	12,000–15,000 hr	Uniform heat; efficient—ideal for profile heating. More expensive than tubular rods.*
Quartz panels	55%	48%	8000–10,000 hr	
Gas-fired infrared	40–45%	25%	1000–6000 hr	Cheapest to operate but *many disadvantages*; very difficult to maintain uniform heat.

*Product application determines the most efficient heating element choice in order to accomplish required forming specs.

Fig. 13-25. Roll-fed thermoformer oven configuration shows quartz panel elements on top, Calrod elements on bottom. (*Courtesy Lyle Industries, Inc.*)

Fig. 13-26. Microprocessor control provides precise and accurate temperature readings automatically, adjusting accordingly to bring sheet to the setpoint temperature. (*Courtesy Brown Machine, Div. John Brown*)

With the microprocessor, the ultimate in repeatability from cycle to cycle can be achieved, as well as on-screen readouts of oven functions.

VACUUM/AIR PRESSURE

During the thermoforming process, in order to form plastic sheet into a desired shape and force the material to follow the contours of the mold, one or a combination of these forces must be used: vacuum, pressure, or matched molds. The method chosen depends on the product size, quantity of products to be made, and cycle speeds desired.

Basic *vacuum forming*, the oldest method of thermoforming, relies on the sheet's self-sealing ability and evacuation of trapped air by applying vacuum. Natural atmospheric pressure fills the mold cavity, forcing the heated sheet into the evacuated space. A vacuum pump is necessary in this process.

Pressure forming is faster than vacuum forming and provides a more clearly defined detail on finished parts than straight vacuum forming. Furthermore, some materials require the faster forming method (e.g., oriented polystyrene/OPS). High air pressure is used, thus requiring molds capable of withstanding the pressures applied. Also, a sealable pressure chamber is needed, as well as equipment capable of withstanding the pressure forces. An ample supply of pressurized air with uniform pressure levels is required. The compressed air system must supply *dry* air to prevent machinery and mold damage.

Matched metal forming requires sufficient driving force of the platens, proper cavity air evacuation, and a realization of the depth-of-draw limitations. This process resembles compression forming.

Combination Vacuum/Air/Mold Forming. Today matched molds only have to be similar to each other. The female mold is precise with respect to the finished part. The male mold need only be similar to the female in order to assist or push softened plastic sheet into the female cavity—preforming it. These preforming molds are called plug assists. After the sheet is forced into the female mold, vacuum and/or air pressure is applied to finalize details of the formed part in the mold. This method allows adjustments to easily be made in air pressure, vacuum, mold closing speeds, and so on, to form a better part.

TOOLING

Tooling for the thermoforming process must combine well-designed molds with excellent temperature control and cooling efficiency to produce economical, good-quality parts.

The design of the form tooling is of utmost importance. A decision must be made about using male or female molds, depending on product detail, critical tolerances, undercuts, or product output requirements. Molds can be constructed of wood, epoxy, or polyester for prototype or short runs—or can be the preferred aluminum molds incorporating accurate temperature and vacuum and/or air-pressure control.

The tool design engineer must realize that wall thickness in the finished article will be reduced from the original sheet thickness in reverse proportion to the increase in area of the formed piece over the area of the original sheet. Proper air and/or vacuum passages are critical to the quality of finished products. Fast, complete, and continuous vacuum is a key factor in successful thermoforming. Another key is mold temperature control through air tubes or chambers.

Fig. 13-28. Multicavity female tooling with syntactic foam male plug assists. (*Courtesy Future Mold, Inc.*)

Unique tool design may incorporate engraving, hinges, and moving cores to produce logos, texture, ribs, and bosses. Different mold inserts may be installed in a mold base to accommodate a variety of products (see Fig. 13-27).

Plug assists, discussed above, usually are made of metal and heated to a temperature slightly below that of the hot plastic, so as not to cool the stock before it can reach its final shape. Instead of metal, today many processors use plugs constructed of syntactic foam (see Fig. 13-28). Also, wood or thermoset plastics such as phenolic or epoxy are used. These materials are poor conductors of heat and thus do not withdraw much heat from the sheet stock.

The formed part must be cooled in the mold so that it holds its new shape. The cooling time

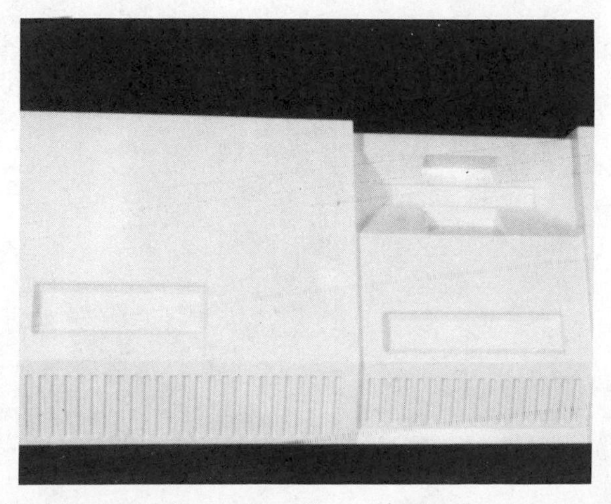

Fig. 13-27. Tooling configuration can produce various effects. Computer housing shown here incorporates textured effect, ribs, cutaways. (*Courtesy CAM/Central Automated Machinery, Inc.*)

will affect the total cycle time, but critical cooling temperature variances must be controlled to avoid unacceptable "chill marks" in the finished product.

Trim tooling is discussed in the following section.

TRIMMING

Unlike injection molding, 90% of thermoformed parts undergo some type of trimming operation. Because a sheet of material must be clamped on its edges to allow stretching of the sheet into a shape, edge trim must be removed. In the forming of multiple parts (e.g., cups or plates) a space is allowed between molds for clamping, leaving a skeleton-like web after the parts are trimmed or punched out. The reduction of edge trim or space trim could greatly affect the overall manufacturing costs, as an average trim scrap factor in thermoforming is 10 to 20%.

High-speed roll-fed thermoformers usually are run in-line with a high-speed trim press (see Fig. 13-29). Synchronized to the output of the thermoformer, this equipment can be microprocessor-controlled, and can incorporate parts ejection for deep draw parts, packing tables, stackers, and counters, or utilize downstream equipment such as packaging equipment and/or lip rollers. Scrap granulators are often placed under the press (see Fig. 13-30).

Trim tooling incorporates punches and dies. The web of formed parts is threaded into the trim press, either manually or automatically. From that point on, the press trims and ejects parts as the machine forms and ejects webs of parts.

Trim-in-place thermoforming machines are becoming more popular, where trimming takes place in the form station. This is especially desired in forming exotic, multilayer, barrier, or polypropylene materials. (See below.)

Steel-rule die trim stations are incorporated into many thermoforming machines to provide forming and trimming in one unit. Although not so fast and accurate as an off-line trim press, this is desirable for many nonfood items such as horticultural trays, box inserts, and blister packaging (see Fig. 13-31).

Fig. 13-29. New trim press developments provide increasing speed and trim accuracy through preloaded linear bearings and electric motor-driven fine tuning. (*Courtesy Lyle Industries, Inc.*)

Fig. 13-30. High-speed trim press shown incorporates four-row parts table and under-the-press granulator. (*Courtesy Brown Machine, Div. John Brown*)

Fig. 13-31. Typical steel-rule die trimmed parts where trim tolerances are not very critical and high-speed production is not a major factor. Usually the forming and trimming are accomplished in one unit. (*Courtesy Lyle Industries, Inc.*)

Trimming of cut-sheet parts usually is done off-line. New methods include robotics, lasers, and high-pressure water-jet systems. Other methods are routing and drilling, sawing, hot-wire, and deburring. New equipment developments are providing automated parts handling to an automated trimming station.

The method most often used in trimming (and punching) large, high-production, relatively flat articles, such as refrigerator door liners, requires handling by the operator, who places the untrimmed piece into a press, activates it through the cycle, and then removes the article and the trim scrap. This will be described more fully under equipment and tools.

Tools for trimming are described below:

- *Shear Dies.* These are the obvious tools for trimming large production runs of large articles if they are trimmed on a single level. These dies are built as if they were intended to trim metal articles, are mounted on die shoes, and are operated in metal-working presses. Some plastics require quite close tolerances between die members. The power required is about one-third of that required to shear mild steel of equal thickness. (See Fig. 13-32a.)
- *Steel-Rule Dies.* These are made of strip steel about $\frac{3}{32}$ inch thick and 1 inch wide, with one sharpened edge. The strips are formed to the shape of the trim line, and held to that shape by birch die stock. They are practical in small to medium runs, and for most thermoplastics in thin to medium thicknesses. Some of the more brittle plastics can be cut by this process only when warm, before they have cooled after forming, or by post-heating the part or the die. (See Fig. 13-32b.)
- *Walker Dies.* These dies are known also as envelope dies or high dies. They are a heavy-duty version of the steel-rule dies, in that they are forged to about $\frac{5}{16}$ inch thickness, and they are available up to 4 inches high, and thus may provide clearance for projections in the article to be trimmed.
- *Planetary Dies.* These dies provide for side motion and they progressively shear vertical flanges on the respective sides of an article. They require special machines. (See Fig. 13-32c.)

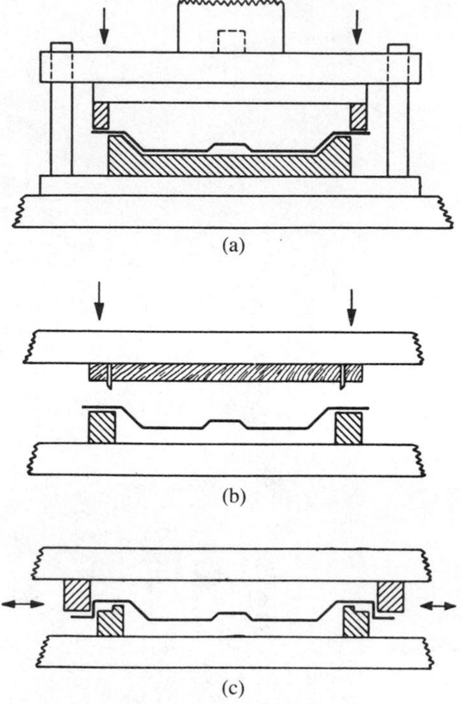

Fig. 13-32. Methods of trimming.

- *Machinery for Trimming.* It is obvious that tools similar or identical to those used in metal work should be used on the same type of equipment regardless of material. Therefore, trimming of plastics often is done on punch presses, press brakes, and other toggle-action machines. Hydraulic presses are entirely satisfactory, and frequently are moved into the forming area so that the operator who runs the forming machine is able to trim articles jut formed, within the duration of the next forming cycle.

The special contour dies such as the steel-rule and envelope types are used on clicker and dinker machines such as those used in the leather industry, on continuous or clutch-type toggle presses, and on hydraulic presses.

Saws are used for many operations where a vertical flange is to be trimmed. A radial-arm type is readily adjustable, and direct drive yields appropriate blade speed. If articles are registered from the table area, clearance beneath the saw blade can be provided. Bandsaws in both vertical and horizontal positions are efficient in trimming hat-shaped parts. Wood-shaper machines also are used, but are not so adaptable.

Problems of Shrinkage

Mold shrinkage occurs not only when the article is removed from the mold, but during a considerable period after removal. Therefore it is important, especially with large items that go into precise assemblies, that all trimming be done on schedule. If it is most practicable to trim as the article is removed from the mold, a shrinkage factor must be built into the trim die. If it is desired to "build a bank" of parts, a different factor must be experimentally determined. A "no-shrink" die for trimming large articles probably is practicable only after an overnight cooling period.

Trimming in the Mold

This innovation makes it possible to trim formed articles at the forming station. One method of trimming-in-place incorporates a knife edge around the periphery of the forming die and a movable heated ring directly over the die. After the object is formed in the conventional manner, but before it is cool, the heated ring drops down and presses against the knife edge in the mold and pinches the part from the web. The process may also work in reverse by placing a heated knife edge on the upper platen and bringing it in contact with the sheet against a flat mold surface. The problem with this method is that plastic may build up on the trimmer and reduce its effectiveness.

Another method uses shearing dies, which either punch a section of sheet that is then carried into the mold and formed, or trim the finished part before it is removed from the mold. These techniques are complex and require expensive tooling.

CONTROL

Manual-Control Equipment

Although complex, automated equipment is used for production forming, the need for economical prototype and short-run production of plastic parts has created markets for manual-control thermoforming machines. These machines also are often used to augment production equipment by enabling research and development of molds and techniques for use in high-speed production equipment.

With these machines, techniques for proper forming are developed to test tooling and various types of materials considered for use in the final production item. The relative cost of such machines enables the manufacturer to complete testing without interruption of normal production on the assembly line.

This type of equipment also is adaptable for educational purposes. There are presently hundreds of schools teaching plastics with R&D-type thermoformers.

Automation

Manufacturers of thermoforming machines have utilized technological advances in automation to produce machines for efficient performance, where each phase of the thermoforming process is accurately controlled. Automation can mean a single temperature-controlling device that maintains the sheet temperature at a given setting, but it also can mean a sophisticated system that controls all functions of extruding, thermoforming, and finishing.

Most thermoformers reach a compromise between these extremes to create a balance of the initial cost, man-hours, the need for technical skills, material costs, and maintenance, along with production output and production expense. These factors determine the equipment needed to enter a competitive area of mass production.

Some automated thermoforming equipment is designed to automatically control all thermoforming functions in a timed sequence. Thus, after the operator initiates the cycle, the machine functions on preset programming. (See Fig. 13-33.)

This programming generally includes:

1. Actuation of a clamping frame to hold the sheet during forming.

Fig. 13-33. Automated load/unload system places cut sheet onto a lift table. Upon completion of the forming process, the system automatically picks up the formed part and transports it to an awaiting conveyor for downstream trimming and finishing operations. (*Courtesy CAM/Central Automated Machinery, Inc.*)

2. Movement of the clamped plastic sheet to the heater section.
3. Application of controlled far infrared radiant heat in a timed cycle.
4. Movement of the heated sheet to the forming station.
5. The forming sequence. This portion of the forming cycle includes several sequenced and timed operations, depending on the thermoforming technique utilized. Automated controls include:
   (a) Platen movement to position. Top and bottom platens may move independently or in unison or by sequenced stages.
   (b) Air and/or vacuum pressure applied from either or both platens independently, in unison, or in sequence.
   (c) Cooling of the formed part through the mold and/or by convection.
   (d) Breaking the seal between the mold and the formed part.
   (e) Removal of the mold from the formed part.
6. Movement of the formed part by cut-sheet machines to the unloading–load station.
7. In roll-fed equipment, movement of the formed web into a trimming press intermittently in synchronization with the thermoformer.

Control Systems

Various systems are available to provide automated control:

Microprocessor Control. In today's thermoforming world, microprocessors are a desired means of control. Repeatability of operations from cycle to cycle and accurate control of all functions are essential. (See Fig. 13-34.) Preprogrammed setups can decrease cycle times, scrap, and labor. Processing conditions

Fig. 13-34. Thermoformer with microprocessor control provides repeatability of operations from cycle to cycle. Productivity is increased by preprogrammed setups and automated adjustment of processing functions during operation. (*Courtesy Lyle Industries, Inc.*)

are continually readjusted during operation to keep unacceptable product from being formed.

Although the initial cost of microprocessor control is high, the productivity savings can defray that initial investment over a short period of time.

PLC (Programmable Logic Controller). The PLC system is an up-and-coming means of computer control. Providing the same capabilities of repeatability, programmed operation, and increased productivity as a microprocessor system, the PLC has several advantages. Because a PLC uses ladder logic, less technically trained people are required, and troubleshooting is simplified. A microprocessor system uses computer language.

A disadvantage of the PLC is its high initial cost. Its capabilities for high-speed production requirements are less than that of a microprocessor. However, with the advances being made in the industry, we may soon see networking, with a PLC as a master control and microprocessors performing certain functions of the process—thus, combining the best of both worlds.

Electro-Mechanical Control. The oldest method of automatic control in thermoforming equipment is the electro-mechanical system. Multiple relays, individual timers, and limit switches initiate successive steps in the sequence. This method is used by a small percentage of thermoformers performing simple, nondemanding operations.

The sequence operation must be selected by an arrangement of toggle switches or multiplexers. Limit switches and operating cams are adjusted to provide interlocking of various functions such as platen advance and speed, vacuum bleed, and air blow. Safeguard features such as fully retracting platens and material clamp frame positions can be included. This system can use timers to monitor the duration of each stage of the thermoforming cycle. With these adjustments, the machine is now ready to operate in the production of the desired article; however, should a change in forming technique be required, the entire procedure of switching and adjusting must be repeated.

The end result is that this method of control is somewhat cumbersome and slow, requiring a proficient, experienced operator to set up the equipment. Its advantages include a relatively low initial cost.

THERMOFORMING OF COEXTRUDED STRUCTURES AND OTHER SPECIAL MATERIALS*

Thermoforming of many new materials such as coextruded multilayer barrier structures or CPET (crystallized polyethylene terephthalate) requires unique machinery features to ensure proper forming characteristics.

Important considerations for the proper heating of these materials to avoid wrinkling and deformation of formed product include: (1) zoning, (2) the preheat station, (3) precise oven temperature control, (4) oven length, (5) sheet sag control, (6) sheet tempering prior to forming, and (7) sheet transport through the oven, into the form station.

For the new materials, oven temperature control and oven length are the major considerations. In solid-phase forming, special ovens are required to keep the sheet flat. In melt phase forming, sheet support or special platen travel is required to handle sag. With that in mind, ideally a customer would desire to have an oven that is completely versatile in its zoning and its ability to profile across the sheet and with the sheet in the sheet direction.

This can be achieved in more than one way. An oven can be fitted with Calrod, ceramic, or quartz panel heaters, or any combination of these. Preferably, the oven will have zoning capability across the sheet and in the sheet direction—in particular, an arrangement that does not limit one to small increments, as with ceramic heaters. A bottom oven combination, of hot air convertible for sheet support and a standard heated oven, is an ideal all-purpose arrangement. (See Figs. 13-35 and 13-36.)

The length and the zoning of the oven must be accurate and adjustable, using the best control available to ensure proper material orientation.

*By G. E. Schwartz, President, Lyle Industries, Inc., Beaverton, MI.

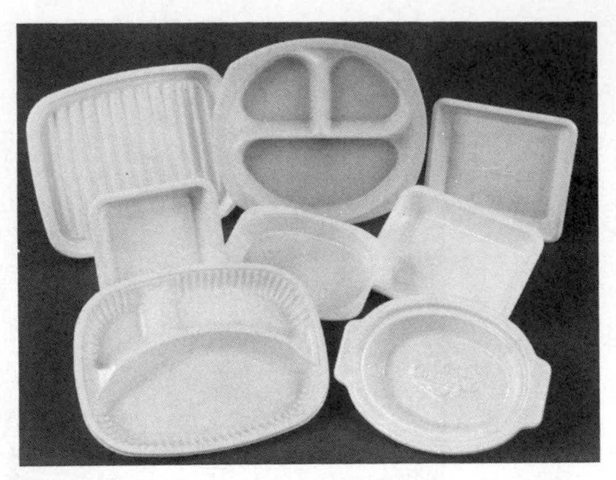

Fig. 13-35. Thermoformed food trays and containers represent only a portion of the broad range of microwavable CPET products requiring unique machine characteristics. (*Courtesy Lyle Industries, Inc.*)

Sheet conveyance control is another important consideration. Electronic index and sheet transport rails are designed for the demanding job of conveying tough materials, up to 125 mil thick, absorbing extraordinary or high stresses when the material is shrinking and pulling, and minimizing chain rail sag and deflection.

Special form station requirements must be taken into consideration for coining and trim-in-place.

Solid-phase forming or close-to-solid-phase forming requires clamping stations that will offer the higher tonnages required (wide bed equipment) and prevent platen alignment deflection as the coining rings are compressed. A standard toggle machine does not do this well; there can be too much clearance built up in all the links and bushings of the toggles. This ultimately leads to short trim-in-place tool life. The ideal forming station will offer extreme rigidity with multiple sequencing functions that will allow for precise coining, trim, and plug speed control as independent adjustments.

COOLING REQUIREMENTS OF THE THERMOFORMING PROCESS

Heat transfer is a fundamental element of plastics processing. In a variety of plastics pro-

*By Mark C. Stencel, AEC, Inc., Wood Dale, IL.

Fig. 13-36. Thermoformed food trays and containers represent only a portion of the broad range of microwavable CPET products requiring unique machine characteristics. (*Courtesy Lyle Industries, Inc.*)

Table 13-3. Thermoforming process parameters.

Material	*Typical starting temperature (°F)*	*Typical finish temperature (°F)*	*Specific heat (Btu/lb × °F)*
HDPE/LDPE	310	120	.83
PP	330	120	.72
PS	300	110	.41
PVC	295	110	.24

cesses, heat is transferred from plastic material (typically by a recirculating water system) to induce its conversion from a liquid to a solid state. The thermoforming process has unique characteristics, specifically less extreme material temperatures than extrusion or injection molding, which one must understand to successfully apply or design a cooling system for its requirements.

The heat transfer requirements for solidification of a plastic material are most easily defined by considering four critical parameters: the start temperature (°F) for the material, its the finish temperature (°F), the rate at which the material is to be processed (lb/hr), and an inherent physical characteristic of the material, specific heat (Btu/lb × °F). The equation used to express the relationship of these parameters in the definition of a cooling load is:

$$(\text{Start temperature} - \text{Finish temperature}) \times \text{Rate} \times \text{Specific heat} \times \text{Safety factor} = \text{Load (Btu/hr)}$$

A typical value used for the safety factor is 1.2. Other typical parameter values utilized for the thermoforming process are as shown in Table 13-3.

For example, to consider the thermoforming of 300 lb/hr of polypropylene, one could calculate the cooling load as follows:

$$(330 - 120)\ ^\circ\text{F} \times 300\ \text{lb/hr} \times .72\ \text{Btu/lb} \times {}^\circ\text{F} \times 1.2 = 54{,}432\ \text{Btu/hr}$$

A chilled water ton is expressed as 12,000 Btu/hr; so the application of a 5-ton chiller should be considered.

It is important that vendors of cooling equipment be supplied with data on the highest processing rates anticipated from thermoforming equipment. Although cooling equipment design typically allows for low-load conditions, physical limitations restrict the effective operation of cooling equipment upon overloading.

SUBSTRATE THERMOFORMING

A special application for thermoforming utilizes a preformed substrate to which softened plastic material is bonded. This process is ideally suited to vinyl-covered dash panels, consoles, door panels, window surrounds, and furniture.

A dual-wheel rotary thermoformer is used whereby one wheel carries the heat-softened decorative material, and the other wheel carries the substrate—meeting at the form station where permanent bonding occurs.

The main rotary wheel usually consists of a load station, a preheat station, a final heat station, and a form station. The secondary wheel consists of a load/unload station, a heat station, and a form station.

Material cut to the proper length is placed into the main load station, and carried through the preheat and final heat stations and into the form station.

Meanwhile, a substrate (usually porous particle board or wood—or injection-molded plastic or metal in which vacuum holes have been incorporated) is loaded onto the secondary wheel. Then it is rotated into the heat station, where an adhesive that has been preapplied to the substrate is activated by heating. Next the wheel rotates the substrate into the form station simultaneously with the heat-softened vinyl or decorative covering material. Here vacuum and air pressure are applied to form a permanent bonding. The clamp frames open, and the covered part is carried on the secondary wheel back to the unload station.

As with any thermoforming equipment, many options are available, such as automatic loading/unloading, sheet cutting, a multiple sheet turret for material selection, multiple molds, a tilting load/unload table, and downstream equipment such as conveyors, trimming systems, and so on. A new thermoforming system (continuous shuttle/rotary system) incorporates roll-fed material that is being fed into a shuttle thermoformer, heated, and simultaneously bonded onto a substrate that has been loaded onto a secondary rotary wheel. As in the dual-wheel rotary, the formed part then is carried on the secondary wheel back to the unload station, while the shuttle sheet car returns to accept new sheet. (See Figs. 13-37 and 13-38.)

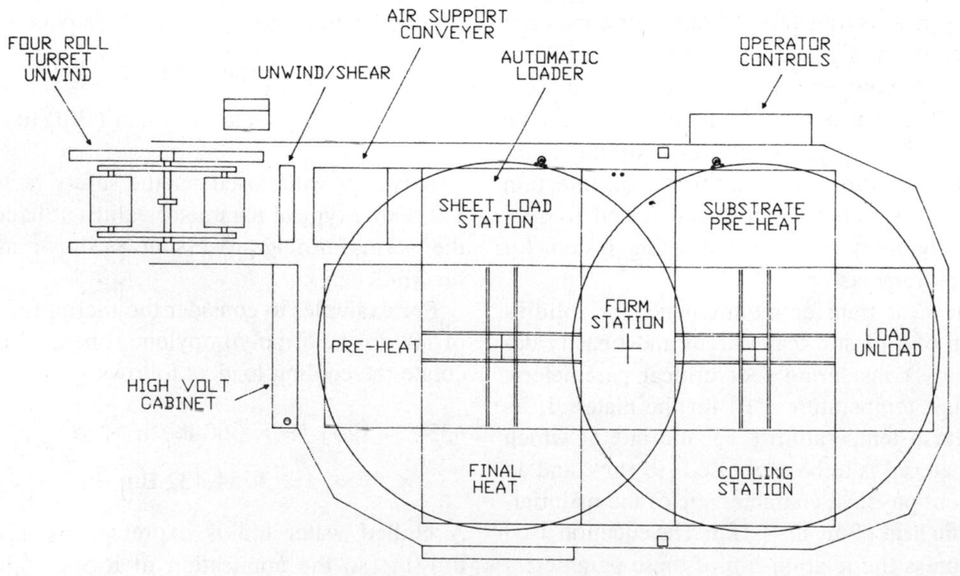

Fig. 13-37. Dual-wheel configuration for thermoforming decorative material onto a preformed substrate. (*Courtesy CAM/Central Automated Machinery, Inc.*)

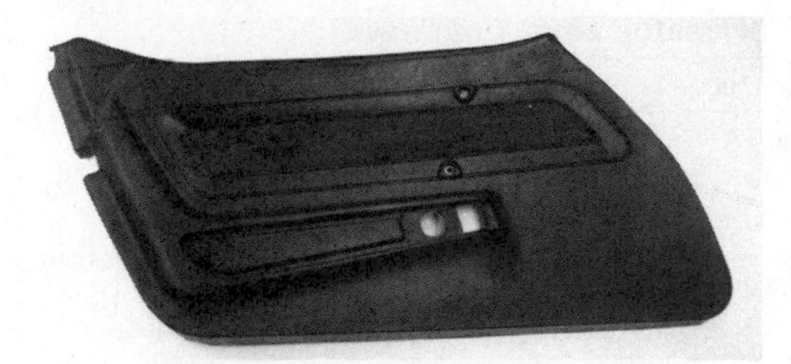

Fig. 13-38. Car door panel produced on a dual-wheel thermoformer. Softened vinyl is formed and permanently bonded to a substrate in one operation. (*Courtesy CAM/Central Automated Machinery, Inc.*)

TROUBLESHOOTING*

Checklist for Troubleshooting

Start troubleshooting by stopping the machine. Be sure that the sheet is not left in the oven (or if the oven retracts, be sure it is clear.) Look for any obvious problem. Review these potential causes before consulting the troubleshooting guide:

1. Pilot error
2. Mechanical problems
3. Sheet or film problems
4. Heating
5. Vacuum and/or compressed air
6. Controls and electrical systems

By William K. McConnell, Jr.; McConnell Co., Inc, Fort Worth, TX.

During troubleshooting and machine adjustments, change only one thing at a time.

Thermoforming Troubleshooting Guide

When the above investigation does not uncover the problem, start down the troubleshooting guide for an answer. The problems covered in the guide include:

1. Blister or bubbles.
2. Incomplete forming, poor detail.
3. Sheet scorched.
4. Blushing or change in color intensity.
5. Whitening of sheet.
6. Webbing, bridging or wrinkling.
7. Nipples on mold side of formed part.
8. Too much sag.
9. Sag variation between sheet blanks.
10. Chill marks or "mark-off" lines.
11. Bad surface markings.
12. Shiny streaks on part.
13. Excessive shrinkage or distortion of part after removing from mold.
14. Part warpage.
15. Poor wall thickness distribution and excessive thinning in some areas.
16. Nonuniform prestretch bubble.
17. Shrink marks on part, especially in corner areas (inside radius of molds).
18. Too thin corners in deep draws.
19. Part sticking to mold.
20. Sheet sticking to plug assist.
21. Tearing of part when forming.
22. Cracking in corners during service.

Troubleshooting Guide for Thermoforming

Problem	*Probable Cause*	*Suggested Course of Action*
1. Blisters or bubbles	A. Heating too rapidly	1. Lower heater temperature 2. Use slower heating 3. Increase distance between heater(s) and sheet
	B. Excess moisture	1. Predry 2. Preheat 3. Heat from both sides 4. Lower heater(s) temperatures 5. Do not remove material from moisture-proof wrap until ready to use 6. Obtain dry material from supplier

Troubleshooting Guide for Thermoforming (*Continued*)

Problem	*Probable Cause*	*Suggested Course of Action*
	C. Uneven heating	1. Screen for uniform heat by attaching baffles, masks, or screen 2. Check for heaters or screens out 3. Adjust individual heater temperatures for uniformity
	D. Wrong sheet type or formulation	1. Obtain correct formulation
2. Incomplete forming, poor detail	A. Sheet too cold	1. Heat sheet longer 2. Raise temperature of heaters 3. Use more heaters 4. If problem occurs repeatedly in same area, check for lack of uniformity of heat
	B. Clamping frame not hot before inserting sheet	1. Preheat clamping frame before inserting sheet
	C. Insufficient vacuum and/or compressed air	1. Check vacuum holes for clogging 2. Increase number of vacuum holes 3. Increase size of vacuum holes 4. Check for vacuum or air leak 5. Remove any 90° angles in vacuum system
	D. Vacuum not drawn fast enough	1. Use vacuum slots instead of holes where possible 2. Add vacuum surge and/or pump capacity 3. Enlarge vacuum line and valves avoiding sharp bends at tee and elbow connections 4. Check for vacuum leaks 5. Check vacuum system for minimum 25″ of Hg pressure
	E. Additional pressure needed	1. Use 20–50 psi air pressure on part opposite mold surface if mold will withstand this pressure 2. Use frame assist 3. Use plug, silicone slab rubber, or other pressure assist
3. Sheet scorched	A. Outer surface of sheet too hot	1. Shorten heat cycle 2. Use slower, soaking heat (lower temperature) 3. Move heater bands further from sheet
4. Blushing or change in color intensity	A. Insufficient heating	1. Lengthen heating cycle 2. Raise temperature of heaters
	B. Excess heating	1. Reduce heater temperature 2. Shorten heater cycle 3. If in same spot on sheet, check heaters

Troubleshooting Guide for Thermoforming (*Continued*)

Problem	*Probable Cause*	*Suggested Course of Action*
	C. Mold is too cold or hot	1. Heat mold or lower temperature
	D. Assist is too cold	1. Warm assist or use syntactic foam or felt covered plug
	E. Sheet is being stretched too far	1. Use heavier gauge sheet or more elastic, deep draw formulation 2. Change mold design 3. Change forming technique
	F. Sheet cools before it is completely formed	1. Move mold into sheet faster 2. Increase rate of vacuum withdrawal 3. Be sure molds and plugs are hot
	G. Poor mold design	1. Reduce depth of draw 2. Increase draft (taper) of mold 3. Enlarge radii
	H. Sheet material not suitable for job	1. Try different sheet formulation or a different plastic material
	I. Uncontrolled use of regrind	1. Control percentage and quality of regrind
5. Whitening of sheet	A. Cold sheet stretching beyond its temperature yield point	1. Increase heat of sheet; increase speed of drape and vacuum
	B. Sheet material dry colored	1. If above action won't correct, check with sheet supplier for availability of other types of coloring. Some colors do not lend themselves to dry or concentrate coloring 2. A hot air gun can be used to diminish or eliminate whitened surfaces on formed part
6. Webbing, bridging, or wrinkling	A. Sheet is too hot causing too much material in forming area	1. Shorten heating cycle 2. Increase heater distance 3. Lower heater temperature
	B. Melt strength of resin is too low (sheet sag too great)	1. Change to lower melt index resin 2. Ask sheet supplier for more orientation in sheet 3. Use minimum sheet temperature possible 4. Profile temperature of sheet
	C. Too much or too little sheet orientation	1. Have sheet supplier reduce or increase orientation
	D. Insufficient vacuum	1. Check vacuum system 2. Add more vacuum holes or slots
	E. Extrusion direction of sheet parallel to space between molds	1. Move sheet 90° in relation to space between molds
	F. Draw ratio too great in area of mold or poor mold design or layout	1. Redesign mold 2. Use plug or ring mechanical assist 3. Use female mold instead of male 4. Add take-up blocks to pull out wrinkles

Troubleshooting Guide for Thermoforming (*Continued*)

Problem	*Probable Cause*	*Suggested Course of Action*
		5. Increase draft and radii where possible 6. If more than one article being formed, move them farther apart 7. Speed up assist and/or mold travel 8. Redesign grid, plug, or ring assists 9. Use recessed pocket in web area
7. Nipples on mold side of formed part	A. Sheet too hot	1. Reduce heating cycle 2. Reduce heater temperature 3. Reduce temperature of sheet surface that contacts mold
	B. Vacuum holes too large	1. Plug holes and redrill with smaller bit 2. Use slot vacuum
8. Too much sag	A. Sheet is too hot	1. Reduce heating cycle 2. Reduce heater temperature
	B. Melt index is too high	1. Use lower melt index resin or different resin 2. Have sheet supplier put more orientation in sheet
	C. Sheet area is too large	1. Profile heat the sheet; use screening or other means of shading or giving preferential heat to sheet, thus reducing relative temperature of center of sheet
9. Sag variation between sheet blanks	A. Variation in sheet temperature	1. Check for air drafts through oven using solid screens around heater section to eliminate
	B. Wide sheet gauge variation	1. Replace sheet with proper gauge tolerance
	C. Sheet made from different resins; not a homogeneous mixture	1. Control regrind percentage and quality 2. Avoid resin mix-ups
10. Chill marks or ''mark-off'' lines	A. Plug assist temperature too low	1. Increase plug assist temperature 2. Use syntactic foam plug assist 3. Cover plug with cotton flannel or felt
	B. Mold temperature too low—stretching stops when sheet meets cold mold (or plug)	1. Increase mold temperature, not exceeding ''set temperature'' for particular resin 2. Relieve molds in critical areas
	C. Inadequate mold temperature control	1. Increase number of water cooling tubes or channels 2. Check for plugged water flow
	D. Sheet too hot	1. Reduce heat 2. Heat more slowly 3. Lower surface temperature of sheet

Troubleshooting Guide for Thermoforming (*Continued*)

Problem	*Probable Cause*	*Suggested Course of Action*
		4. Slightly chill surface of hot sheet contacting mold with forced air before forming
11. Bad surface markings	A. Pock marks due to air entrapment over smooth mold surface	1. Grit blast mold surface
	B. Poor vacuum	1. Add vacuum holes 2. If pock marks are in isolated area, add vacuum holes to this area or check for plugged vacuum holes or vacuum leak 3. Check entire vacuum system
	C. Mark-off due to accumulation of plasticizer on mold when using sheet with plasticizers	1. Use temperature-controlled mold 2. Have mold as far away from sheet as possible during heating cycle 3. If too long, shorten heating cycle 4. Wipe mold
	D. Mold is too hot	1. Reduce mold temperature
	E. Mold is too cold	1. Increase mold temperature
	F. Improper mold composition	1. Avoid phenolic or other "heat sink" glossy molds with clear transparent sheet 2. Use aluminum molds where possible
	G. Mold surface is too rough	1. Smooth surface 2. Change mold material 3. Sand blast mold surface with #30 shot grit
	H. Dirt on sheet	1. Clean sheet 2. Use ionized air blow
	I. Dirt on mold	1. Clean mold
	J. Dust in atmosphere	1. Clean thermoforming area; isolate area if necessary and supply filtered air 2. Use ionized air
	K. Contaminated sheet materials	1. If regrind is used, be sure to *keep* clean, and different materials stored separately 2. Check supplier of sheet 3. Use coex sheet with virgin
	L. Scratched sheet	1. Separate sheets with paper in storage 2. Polish sheet 3. Replace sheet
12. Shiny streaks on part	A. Sheet overheated in this area	1. Lower heater temperature in scorched area 2. Shield heater with screen wire to reduce overheating 3. Slow heating cycle 4. Increase heater to sheet distance
	B. Bad sheet	1. Check with sheet supplier

Troubleshooting Guide for Thermoforming (*Continued*)

Problem	*Probable Cause*	*Suggested Course of Action*
13. Excessive shrinkage or distortion of part after removing from mold	A. Removed part from mold too soon	1. Increase cooling cycle 2. Use cooling fixtures 3. Use fan or vapor spray mist to cool part faster on mold
	B. Too much sheet orientation or nonuniform orientation	1. Replace sheet
14. Part warpage	A. Uneven part cooling	1. Add more water channels or tubing to mold 2. Check for plugged water flow 3. Cool part at same rate on both sides
	B. Poor wall distribution	1. Improve prestretching or plugging techniques 2. Use plug assist 3. Check for nonuniformity of sheet heating 4. Check sheet gauge
	C. Poor mold design	1. Add moat to mold at trim line 2. Add vacuum holes 3. Check for plugged vacuum holes
	D. Poor part design	1. Break up large flat surfaces with ribs where practical or make concave or convex
	E. Mold temperature too low	1. Raise mold temperature to just below ''set-temperature'' of sheet material
	F. Too much or nonuniform orientation in sheet	1. Check supplier
15. Poor wall thickness distribution and excessive thinning or holes in some areas when sheet stretched	A. Improper sheet sag	1. Use different forming technique such as mounting mold on top platen 2. Use vacuum snap-back technique 3. Use billow vacuum snap back 4. Use billow-up plug assist or vacuum snap-back into female mold 5. Use different melt index resin 6. Try more orientation in sheet
	B. Variations in sheet gauge	1. Consult supplier regarding his commercial tolerances, and improve quality of sheet
	C. Bad sheet	1. Check supplier for calendering of extruded sheet causing thin spots 2. Lack of complete, homogenous sheet
	D. Hot or cold spots in sheet	1. Improve heating technique to achieve uniform heat distribution; screen or shade as necessary 2. Check to see if all heating elements are functioning

Troubleshooting Guide for Thermoforming (*Continued*)

Problem	*Probable Cause*	*Suggested Course of Action*
	E. Stray drafts and air currents around machine	1. Enclose heating and forming areas
	F. Too much sag	1. Use screening or other temperature control of center areas of heater banks
		2. Use lower melt index resin
		3. Use more orientation in sheet
	G. Mold is too cold	1. Provide uniform heating of mold to bring to proper temperature
		2. Check temperature control system for scale or other stoppage
	H. Sheet slipping out of frame	1. Adjust clamping frame to provide uniform pressure
		2. Check for variation in sheet gauge
		3. Heat frames to proper temperature before inserting sheet
		4. Check for nonuniformity of heat giving cold areas around clamp frame
16. Nonuniform prestretch bubble	A. Uneven sheet gauge	1. Consult sheet supplier
		2. Heat sheet slowly in a ''soak'' type heat
	B. Uneven heating of sheet	1. Check heater section for heaters out
		2. Check heater section for missing screens
		3. Screen heater section as necessary
	C. Stray air drafts	1. Enclose or otherwise shield or screen machine
		2. Check clamping frame air cylinders for leaks
	D. Nonuniform air blow	1. Baffle air inlet in prestretch box
17. Shrink marks on part, especially in corner areas (inside radius of molds)	A. Inadequate vacuum	1. Check for vacuum leaks
		2. Add vacuum surge and/or pump capacity
		3. Check for plugged vacuum holes
		4. Add vacuum holes
	B. Mold surface is too smooth	1. Grit blast mold surface with #30 grit
	C. Part shrinking away	1. May be impossible to eliminate on thick sheet with vacuum only; use 20–30 psi air pressure on part opposite mold surface if mold will withstand this pressure
		2. Add moat to mold just outside trim line

Troubleshooting Guide for Thermoforming (*Continued*)

Problem	*Probable Cause*	*Suggested Course of Action*
18. Too thin corners in deep drawers	A. Improper forming technique	1. Check other techniques such as billow-up plug assist, etc.
	B. Sheet too thin	1. Use heavier gauge
	C. Variation in sheet temperature	1. Profile sheet heating; adjust heating as needed by adding screens to portion of sheet going into corners or with panel heat lower temperature 2. Crosshatch sheet with markings prior to forming so movement of material can be accurately checked
	D. Variation in mold temperature	1. Adjust temperature control system for uniformity
	E. Improper material selection or poor sheet	1. Consult sheet supplier or raw material supplier to be sure proper material is correctly extruded
19. Part sticking to mold	A. Mold or sheet temperature too high	1. Increase cooling cycle 2. Slightly lower mold temperature, not much less than recommended by resin manufacturer 3. Lower surface temperature on sheet side that contacts mold
	B. Not enough draft in mold	1. Increase taper 2. Use female mold 3. Remove part from mold as early as possible; if above ''set temperature'', use cooling jigs
	C. Mold undercuts	1. Use stripping frame 2. Increase air-eject air pressure 3. Remove part from mold as early as possible; if above ''set temperature'', use cooling jigs 4. Change mold design for undercut to break away
	D. Wooden mold	1. Grease with Vaseline 2. Use Teflon spray or zinc stearate 3. Lower surface temperature on sheet side that contacts mold
	E. Rough mold surface	1. Polish corners or all of mold 2. Use mold release 3. Use Teflon spray or zinc stearate
	F. Different melt index resin used in color concentrate	1. Have supplier use same grade resin in color concentrate
20. Sheet sticking to plug assist	A. Improper metal plug assist temperature	1. Reduce plug temperature 2. Use mold release 3. Teflon-coat plug 4. Cover plug with felt cloth or cotton flannel 5. Use syntactic foam plug

Troubleshooting Guide for Thermoforming (*Continued*)

Problem	*Probable Cause*	*Suggested Course of Action*
	B. Wooden plug assist	1. Cover plug with felt cloth or cotton flannel 2. Grease with Vaseline 3. Use mold release compounds 4. Use Teflon spray or zinc stearate 5. Laminate wood plug surface with syntactic foam
21. Tearing of sheet when forming	A. Mold design	1. Increase radius of corner
	B. Sheet is too hot	1. Decrease heating time or temperature 2. Check for uniform heat 3. Preheat sheet
	C. Sheet is too cold (usually thinner gauges)	1. Increase heating time or temperature 2. Check for uniform heat 3. Preheat sheet
	D. Closing speed between mold and sheet	1. Reduce rate of closure
	E. Bad sheet	1. Check with supplier
22. Cracking in corners during service	A. Stress concentration	1. Increase fillets 2. In transparencies check with polarized light 3. Increase temperature of sheet 4. Be sure part is completely formed before some sections are too cool for proper forming, thus setting up undue stresses in these ares 5. Change to a stress-crack-resistance resin
	B. "Under-designed"	1. Reevaluate design
	C. "Cold"-formed	1. Vacuum *too* slow-speed up 2. Be sure "core" of sheet is at forming temperature and then rapidly drawn to mold with excellent vacuum and/or compressed air
	D. Wrong resin	1. Change to proper material
	E. Bad sheet	1. Check with supplier

14 Rotational Molding

OVERVIEW

In rotational molding, the product is formed from liquid or powdered thermoplastic resin inside a closed mold or cavity while the mold is rotating biaxially in a heating chamber. To obtain this mold rotation in two planes perpendicular to each other, the spindle is turned on a primary axis, while the molds are rotated on a secondary axis (Fig. 14-1).

Rotational molding (also popularly known as rotomolding) is best suited for large, hollow products requiring stress-free strength, complicated curves, a good finish, a variety of colors, a comparatively short (or very long) production run, and uniform wall thickness. It has been used for products such as fuel tanks, furniture, tilt trucks, industrial containers, storage tanks, portable outhouses, modular bathrooms, telephone booths, boat hulls, garbage cans, light globes, ice buckets, appliance housings, and toys (Fig. 14-2). The technique is applicable to most thermoplastics but is most widely used with polyethylene.

Rotational molding offers a number of advantages:

1. Virtually unlimited design possibilities (parts as small as a golf ball to a 22,500-gallon agricultural tank).
2. Relatively low machinery cost.
3. Low tooling costs.
4. Economical prototyping.
5. Strong outside corners in virtually stress-free parts.
6. Part finish from matte to high gloss.
7. Simultaneous processing of multiple colors.
8. Simultaneous processing of different parts.
9. Quick mold changes.
10. Possibility of molding in metal inserts.
11. Molded-in multicolor graphics.
12. Multilayer molding for chemical resistance or strength.
13. Double-walled parts molding for additional rigidity.
14. Possibility of minor undercuts.
15. Virtual 100% usage of material (no scrap).

How It Works

There are essentially four basic steps in rotational molding: loading, molding or curing, cooling, and unloading.

In the loading stage, either liquid or powdered plastic is charged into a hollow mold. The mold halves then are clamped shut and moved into an oven where the loaded mold spins biaxially. Rotation speeds should be infinitely variable at the heating station, ranging up to 40 rpm on the minor axes and 12 rpm on the major axes. A 4 : 1 rotation ratio generally is used for symmetrically shaped objects, but a wide variety of ratios are necessary for molding unusual configurations.

In the oven, the heat penetrates the mold, causing the plastic, if it is in powder form, to

Reviewed and updated by Harry Covington, President, Ferry Industries, Kent, OH.

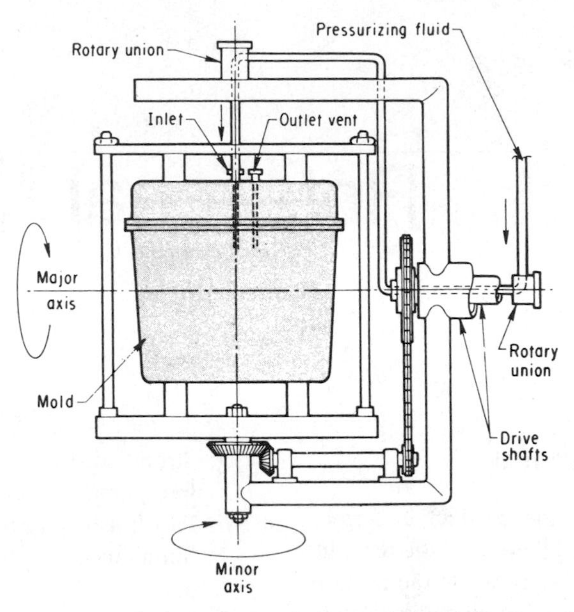

Fig. 14-1. In rotational molding, the product is formed inside a closed mold while the mold is rotated about two axes and heat is applied. The spindle is turned on a primary axis, while the molds are rotated on a secondary axis. (*Courtesy McNeil Femco, Div. of McNeil Corp.*)

become tacky and stick to the mold surface, or if it is in liquid form, to start to gel. On most units, the heating is done either by air (as in a gas-fired hot-air oven) or by a liquid of high specific heat, such as molten salt; where jacketed molds are used (see below), heating is done with a hot liquid medium, such as oil.

Because the molds continue to rotate while

Fig. 14-2. Rotationally molded polyethylene hobby horse. (*Courtesy Quantum Chemical Corp.*)

the heating is going on, the plastics gradually become distributed evenly on the mold cavity walls through gravitational force (centrifugal force is not a factor). As the cycle continues, the polymer melts completely, forming a homogeneous layer of molten plastic.

When the parts have been formed, the molds move to a cooling chamber where cooling is accomplished by a cold water spray and/or forced cold air and/or a cool liquid circulating inside the mold. The mold continues to rotate during the cooling cycle so as to ensure that the part does not sag away from the mold surface, causing distortion.

Finally, the molds are opened and the parts removed. This can be done manually or by using forced air or mechanical means to eject the part.

Cycle times typically range from 7 to 15 minutes, but can be as short as 5 minutes or as long as 30 minutes for very large parts. The wall thickness of the parts affects cycle time, but not in a direct ratio. Normally, on a plastic such as polyethylene, the cycle times increase by 30 seconds for each 25 mils of added thickness up to $\frac{1}{4}$ inch. Beyond $\frac{1}{4}$ inch the heat-insulating effect of the walls increases cycle time disproportionately for any further increase in thickness, and cycle times usually have to be determined experimentally.

MACHINERY

Machines for rotational molding generally are characterized by the weight in pounds of the maximum load supported by the arms, including the mold and the weight of the charge; and the spherical diameter, in inches, of the rotation that is possible by the mold extremities in the chambers.

Modern-day machines feature up to 5000-pound capacities, with some carousel-type machines sweeping out a 180-inch-diameter spherical swing. Large parts now being molded include 22,500-gallon agricultural tanks, 500-gallon industrial containers, and 200-pound refuse bins. For producing small parts, an arm may hold as many as 96 cavities.

The following paragraphs describe the major types of machinery in use today.

Batch-Type Machines

This type is the least expensive of the rotational molding machines because it is the least sophisticated and requires the most manual labor. In a typical batch operation, the charged mold is rolled into the oven for rotation and heating. At the completion of the cycle, the mold is removed and a newly charged one inserted in its place. The completed mold then is transferred manually on rollers to a cooling station for cure before removal of the parts.

Carousel-Type Machines

The most common rotational molding machine in use is the carousel unit, which is essentially a three-station rotary indexing type with a central turret and three cantilevered mold arms (see Fig. 14-3).

In operation, individual arms are involved in different phases simultaneously, so that no arms are idle at any time. The arms or mold spindles extend from a rotating hub in the center of the unit. Thus, while one arm is in the loading/unloading station, another is rotating within the oven, and the third is rotating within the cooling station. All operations are automated, and at the end of each cycle the turret is rotated 120°, thereby moving each mold arm to its next station.

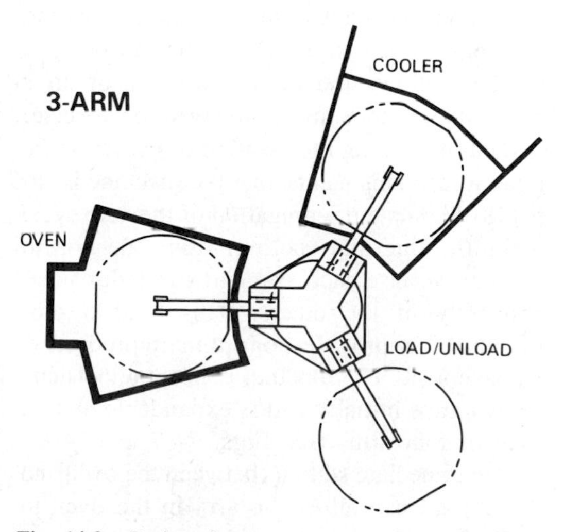

Fig. 14-3. Basics of carousel-type machine with three stations. (*Courtesy Ferry Industries, Inc.*)

Fig. 14-4. Four-arm carousel machine. (*Courtesy Ferry Industries*)

Newer carousel machines being offered today use four arms (see Fig. 14-4). The additional arm can be used in a second oven, cooler, or load station, depending on which is the most time-consuming part of the overall cycle (see Fig. 14-5). The four-arm carousel machines increase production by allowing the index from station to station to occur more frequently than could be managed on a three-arm machine.

Independent-Arm-Type Machines

The independent-arm-type machine offers even more flexibility. Featuring five stations—one oven, one intermediate station, one cooler, and two load/unload stations (see Fig. 14-6)—this machine allows individual indexing of arms from station to station, forward or reverse, without disturbing the position or cycles of the other arms. This means that the machine is capable of complete segregation of the cure cycle times (heating and cooling) from one arm to another because each arm/cart can index independently of the others. This built-in programmability provides for optimum production management. The modular concept allows one or two arms initially and is expandable up to a total of four arms (see Figs. 14-7 and 14-8). The intermediate station (between the oven and the cooler) can allow the arm in the oven to index from the oven even if the operator is not ready for the arm in the load station to enter the oven. This feature could prevent an overcure condition of the molds in the oven. The intermediate station also can be utilized as an additional oven or cooler as the process dictates. One load and one unload station provide for additional productivity in the case of a four-arm machine and provide extra work space for three-arm machines.

Straight-Line Machines

Used primarily for molding large parts, this type is a shuttle carriage machine that is generally a straight-line operation with the oven on one side, the loading/unloading station in the middle, and the cooling station on the opposite end. The carriage is guided on parallel tracks that ensure positive placement in the stations. Variations of this particular design, involving the placement and relationship of the oven, cooling stations, and loading/unloading stations, also are available for handling particular jobs.

Jacketed-Mold Machines

These units offer precise temperature control up to a 300°C heat-transfer medium temperature. This type of machine finds its biggest market in molding heat-sensitive polymers because of its accurate temperature control.

Key to the machines are double-walled jack-

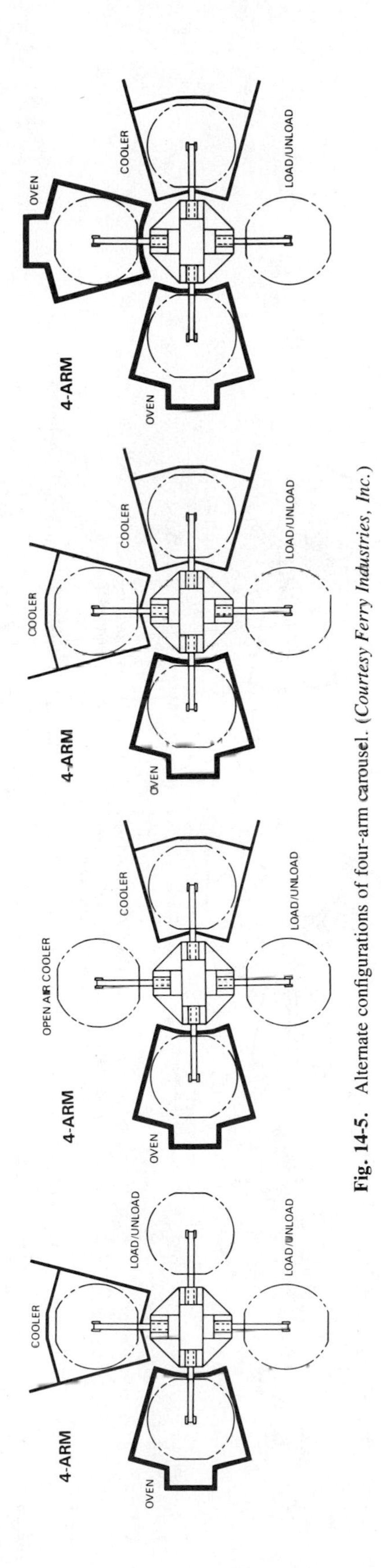

Fig. 14-5. Alternate configurations of four-arm carousel. *(Courtesy Ferry Industries, Inc.)*

eted molds that are charged with a hot liquid, usually oil, to attain temperature control. When the heating period is over, the oil is drained, and a coolant is introduced.

The molds rotate biaxially through all production operations. Significant savings are effected because the heating medium retains its heat, and the heat transfer is more rapid, resulting in faster thermal cycling at lower temperatures.

Thus, although the initial cost of machinery and molds is higher for hot oil machines, the more efficient thermal cycle and potentially lower operating cost could be an advantage. Because the temperature of the medium can be accurately controlled, this type of machine has been used to polymerize caprolactam directly in the mold.

MOLDS

Molds for convection units are generally inexpensive; however, this is entirely dependent upon the quality level of the end product required, the type of plastic being molded, and the operating temperature to be used in the process. Additionally, except for very large parts, multiple cavities are used (Fig. 14-9).

Cast aluminum molds are probably the most widely used, and are the most practical for small to medium parts requiring a number of cavities. Wall thicknesses vary from $\frac{1}{4}$ to $\frac{3}{16}$ inch for use in hot-air machines, and up to $\frac{1}{2}$ inch for molten salt machines. The initial cavity cost may be relatively high because a model and/or a pattern is required. However, subsequent cavities are moderate in price, and a fair reproduction of the mold surface on the finished part may be expected. The process used for casting aluminum molds is somewhat specialized; an experienced rotational casting moldmaker is required.

Electroformed nickel molds are best used where precise detail is required on the finished part, or where no parting line can be tolerated. This type of mold is not so durable as cast aluminum or sheet metal; however, it is widely used by molders of such objects as automotive headrests, armrests, and so forth, and by plastisol molders.

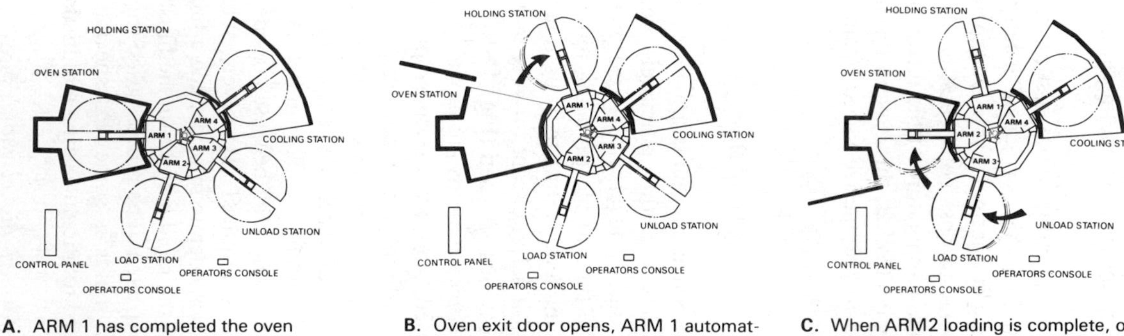

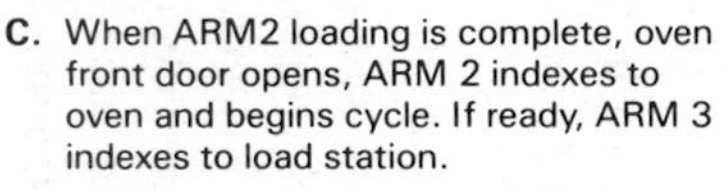

A. ARM 1 has completed the oven cycle, but ARM 2 loading operation is not complete.

B. Oven exit door opens, ARM 1 automatically indexes to the holding station continuing its rotation. **Part does not over cure.** Exit door closes. Cooling begins in holding station.

C. When ARM2 loading is complete, oven front door opens, ARM 2 indexes to oven and begins cycle. If ready, ARM 3 indexes to load station.

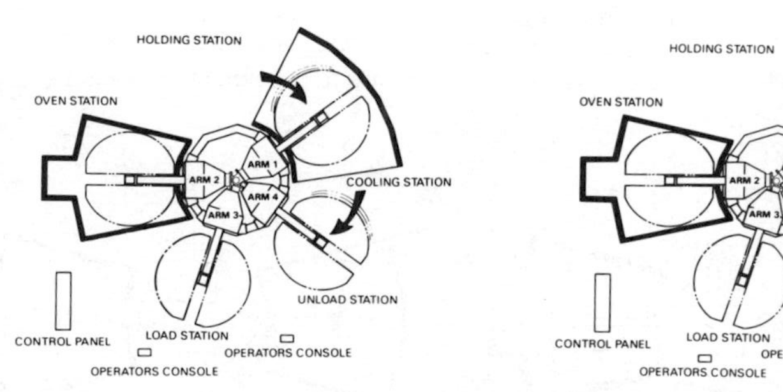

D. If cooling is completed, ARM 4 indexes to unload station. ARM 1 will then go into cooling station for cool cycle.

E. Machine is ready to repeat automatic sequence.

Fig. 14-6. Typical cycle: independent arm machine with four arms, five stations. (*Courtesy Ferry Industries, Inc.*)

Fig. 14-7. Large independent-arm machine. (*Courtesy Ferry Industries*)

Fig. 14-8. Center post and arm assemblies. (*Courtesy Ferry Industries*)

Sheet metal is appropriate for extremely large parts or single cavities requiring inexpensive tools, where a simple sheet metal mold is generally adequate. Prototype molds often are fabricated by this method for reasons of cost, though eventual production molds usually are made of cast aluminum.

Open areas may be molded by simply insulating the mold in the area where plastic is not desired. Some flash may result, but it can be easily trimmed. Inserts can be molded in place by locating the inserts inside cavities while loading the mold.

When it is essential to maintain atmospheric pressure inside the mold during casting, a tube may be inserted through the mold. In effect, this vents the mold and prevents a vacuum on the mold interior as the part is cooled. This vacuum would cause flat parts to warp or cause blow holes at the parting line of the mold.

Fig. 14-9. Multicavity rotational mold. (*Courtesy Kelch Corp.*)

Molds may be attached directly to the extremity of the major axis of the machine arm, or may be mounted on a ''spider'' or holding platform built onto the extended arm.

MATERIALS

Most thermoplastics can be used in rotational molding. The most popular materials, however, continue to be the polyethylenes, the first powdered plastics to be used in this field. Low-density, linear low-density, and medium-density polyethylene are being rotationally molded today (Fig. 14-10).

Other materials currently being rotomolded are: polyvinyl chloride (PVC) liquid or powder, nylon, polycarbonate, polyesters, ABS, acetal, acrylic, cellulosics, epoxy, fluorocarbons, phenolic, polybutylene, polypropylene, polystyrene, polyurethane, SAN, and silicone.

The physical properties of the molded parts usually differ considerably from those of injection-molded parts because of the long thermal cycles.

In composite rotational molding, two resins with different melting points are combined in the mold. One resin melts first, coating the mold and forming the outer layer of the product; then the second melts and fuses to the first. The following products are typical of those made by this process: a lamp globe with a tough outer skin of butyrate, combined with a polystyrene interior to produce a diffused light, and an ice bucket with a polystyrene core and a polyethylene skin.

Fig. 14-10. Rotationally molded high-density polyethylene sewer manhold. (*Courtesy Quantum Chemical Corp.*)

Also available for rotational molding is a thermally crosslinkable high-density polyethylene that offers outstanding environmental

Fig. 14-11. Rotationally molded recycling container made from crosslinked high-density polyethylene (*Courtesy Phillips 66.*)

stress cracking resistance, impact strength, and overall toughness (Fig. 14-11).

For other special applications, it is possible to add fillers to low- and medium-density polyethylene to rotomold a product with increased stiffness. For example, chopped strand fiberglass, about $\frac{1}{8}$-inch in length, can be added to higher melt index, low-density polyethylene resins. Satisfactory rotational moldings can be produced at about 8% glass in complex molds and up to 15% glass in simple molds. These reinforced articles approach the rigidity of high-density polyethylene.

Most resins used in rotomolding are in the powdered form, but PVC and polyurethane are available both as liquids and as powders. Liquid PVC (plastisol) cycles in a shorter time, but the properties of both are about equal. (See Chapter 16.)

PART DESIGN

Dimensional tolerances of $\pm 5\%$ are the present general limits of rotomolding, in both lineal dimension and wall thickness. Tolerances are a function of shrinkage in the mold. Low-density PE shrinks less than high-density PE; shrinkage rates for other plastics vary with no particular relationship to their density, melt index, or any other characteristic.

The final size of the part is affected by both the rate and the amount of shrinkage during cooling and the part configuration. The degree of shrinking also depends on the adhesion of the plastic to the mold during cooling and how this condition changes as the part cools. Machines that can apply internal air pressure to hold the part against the mold help minimize shrinking. Different shrinkage rates and wall thicknesses sometimes occur within a given part. (Flat sections may vary as much as $\pm 5\%$.) Like warping, this problem can be minimized by adding reinforcing ribs or frames behind the flat sections.

Wall thickness can be decreased or increased to the limits of rotomolding (0.030 to 0.5 in.) by simply adjusting the amount of the charge and the cycle time. Heat-insulating plugs can be used to reduce wall thickness or to eliminate the wall entirely in a given area. Wastebaskets, for example, are made by using an insulating disc at the open end to prevent plastic formation. Wall thickness can be maintained to a lesser degree by controlling the ratio of rotation so that the area that requires a thicker wall is at the bottom more often than the rest of the part. This technique might be used in making a tapered hopper, for example, with an opening at both ends.

Surface detail can be controlled only on the outside surface because the molds are female, and the inside of the part is in contact only with the air inside the mold. The cycle time and the mold temperature determine the inside finish. Too little time or heat leaves part of the powder unfused, causing a rough texture on both surfaces. Too much time or heat causes degradation of the plastic, discoloration, or scorching.

Generally, rotomolded parts have a good outer surface appearance without sink marks. Special finishes, particularly matte or any type of grain, are easily obtained. Uniform, high-gloss surfaces are more difficult. Surface gloss is simply a reflection of the glossiness of the mold, and practically any plastic will produce a glossy surface if the mold is smooth and polished. However, a glossy finish on the mold increases its price, but it is still much less than that of a comparable injection mold.

Injection molding is usually superior to rotomolding where surface detail includes sharp edges. The high pressure of injection molding forces the plastic into these mold cavities to produce sharper edges, whereas edges tend to become rounded off in rotomolding. Blow-molded parts, because of flash, trim, and thin

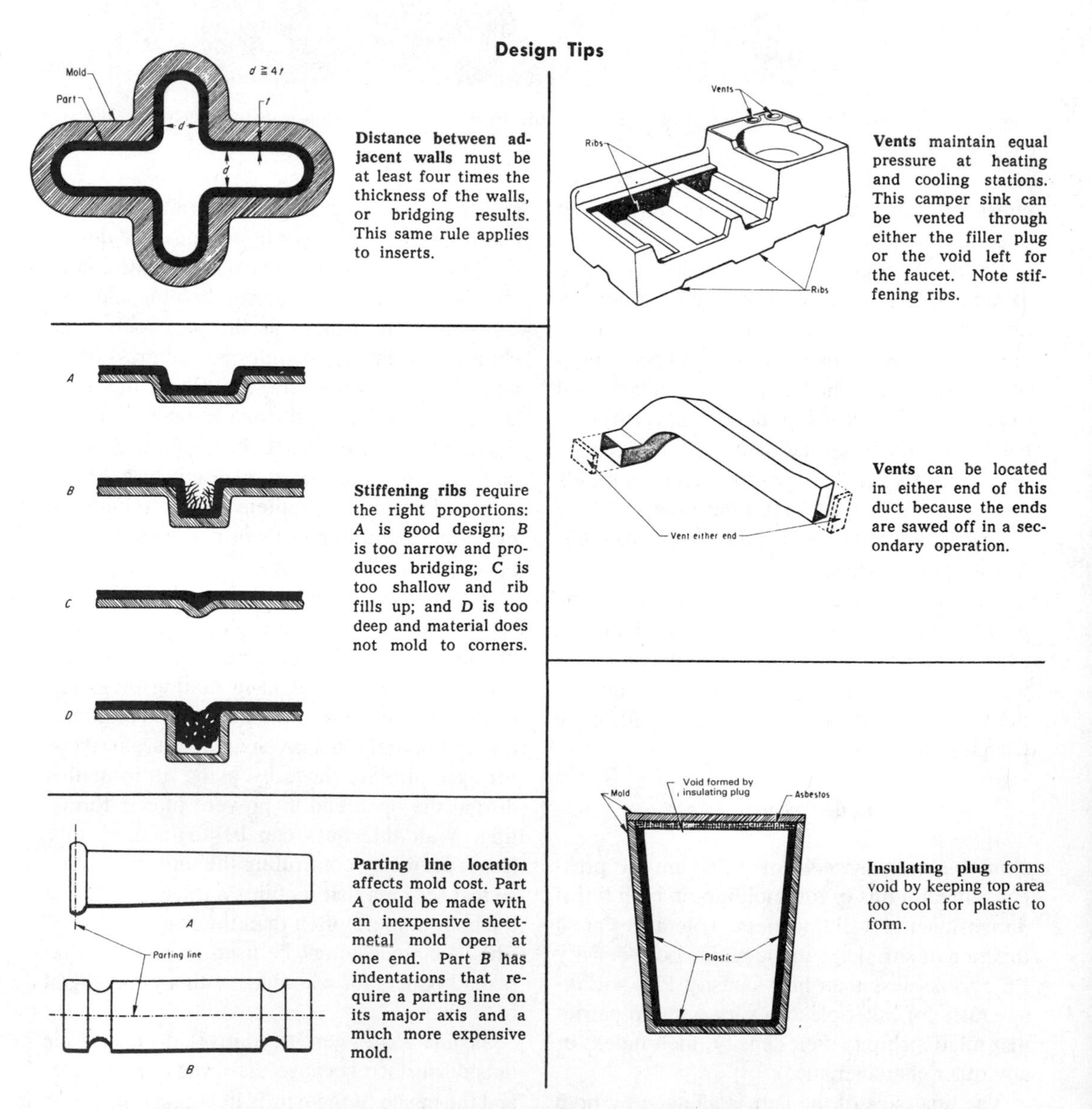

Fig. 14-12. Design tips for rotationally molded products. (*Courtesy McNeil Fenco, Div. of McNeil Corp.*)

spots at the corners, usually are inferior in appearance to rotomolded parts. Easy-flowing, high-melt-index plastics generally produce the best inside surface. Low-density polyethylenes give a better inside surface than the high-density versions.

Inserts are as easily added in rotomolding as in injection molding. They may be of any size, but they must be positioned so that they will be completely surrounded by plastic and anchored firmly. The insert also should be spaced away from any wall by at least four times the thickness of the wall to avoid bridging. Practically all inserts are metal, but higher-melting plastics also can be used. Nylon is the most common plastic insert material.

In some cases, it may be more economical to leave a void in the part and make the insert a secondary operation. One way to create a void is to use an insulating plug to keep the void area cool during the heating cycle. Another approach is to use a fluorocarbon plug to prevent the plastic from adhering to a specific area. In either case, there may be a small amount of flash to be trimmed.

Strength of rotomolded parts generally is high. Because only atmospheric pressure is used in the process, rotomolded parts are virtually stress-free. Material tends to build up in the corners where more strength is needed, as opposed to blow-molded products, which tend to thin out at the corners. The naturally uniform wall thickness of rotomolded hollow containers provides structural strength that is generally superior to that obtained in blow-molded products. (See also Fig. 14-12.)

15 Calendering

The calendering process, used in the production of plastic film and sheet, is a continuous, synchronized method for converting raw materials, fusing them and passing them through nips of a series of cooperating rolls, into a sheet of specified thickness and width. This web may be polished or embossed, as required, and may be of either a rigid or a flexible character, with the ''hand'' controlled by the formulation and the process.

Calendering is best known in the sheet formation of vinyl and ABS, and alloys of both. Vinyl may be formulated to be soft/flexible, for low/high-temperature performance, oil/U.V.-resistant, clear or opaque. Typically, the calender process gauge range is .002 to .045, and it has the advantage of higher throughput rates, compared to extrusion. Unlike extrusion or injection molding, a calender forms the web as a continuous extrusion between rolls and is not confined.

A typical calender plant starts with the inventorying of raw materials in either silos for powders or tanks for liquids. The only major ingredient not so handled is titanium dioxide, which has a tendency to cake if stored in bulk.

Minor ingredients such as expensive stabilizers and pigments are manually or automatically handled. Then, based on the compounder's recipe, various proportions of the material are delivered to the mixer/blender.

Reviewed and updated by John M. Coburn, Occidental Chemical Corp., Berwyn, PA.

Blending

Blending allows the distribution of all ingredients and, in the case of flexible formulations, permits the absorption of plasticizers into the resin to provide a free-flowing dry blend.

The blending of the recipe is critical, and two methods are employed to produce a *uniform*, colored, stabilized blend, of the desired flexibility, in powdered form.

1. High speed blending provides short cycles (5–10 minutes) and an intimate dispersion, suitable for both rigid and flexible compounding. Tandem vessels separate heating and cooling functions, with the latter sized adequately for blending two or more batches.
2. Ribbon blending has a lower capital cost but longer cycles (60 minutes) and usually accomplishes both heating and cooling in the same unit.

Selection of the unit is dependent upon cost and specific applications, such as product/color changes.

The process temperature and time become an important part of the process. The residence time at elevated temperatures must be uniform and controlled throughout the calendering process.

The feed to the blenders may be either manual or automatic, and in some cases, even computerized to weigh in sequence automatically. Vessels are jacketed to accelerate heating or cooling, and a second unit may be necessary to permit product and color changes.

Reground scrap is introduced in the ribbon blender or cooler, typically volumetrically.

After blending, the material is fed to a fluxing unit, where the powder is fused to a hot paste by mechanical/thermal energy in either a continuous or batch-type intensive mixer. Continuous mixers may be planetary design, single reciprocating screw, twin screw, or dual rotor mixers and will, if fed uniformly, deliver material at a uniform rate and stock temperature for direct feed to the calender. (See Fig. 15-1.)

Batch type mixers are composed of a pair of rotors encased in a chamber, where the work is done between the rotor tips and the body or between the rotors. (See Fig. 15-2) Cycle times vary from 60 seconds for rigids to four times that for highly plasticized formulations. Batch sizes are established by the mixer capacity, based on the chamber volume.

Scrap, color changes, and short runs are often better handled in the batch method; in continuous mixers, adequate size reduction of scrap is important to avoid problems caused by stratification and bridging.

Scrap and Cold Trim

Scrap and cold trim handling from the product line poses one of the more difficult problems. Scrap and trim represent from 10 to 40% of the product mix, depending on the trim width, rejects, and changes. A shorter flux rate and lower energy are required for remelting the scrap versus virgin blend. The potential for decomposition contamination, color drift, and gauge and flexibility variation exists whenever scrap is handled; yet for optimum uniformity, it is necessary to prepare the compound so that the blend of both new and old is fed *uniformly*

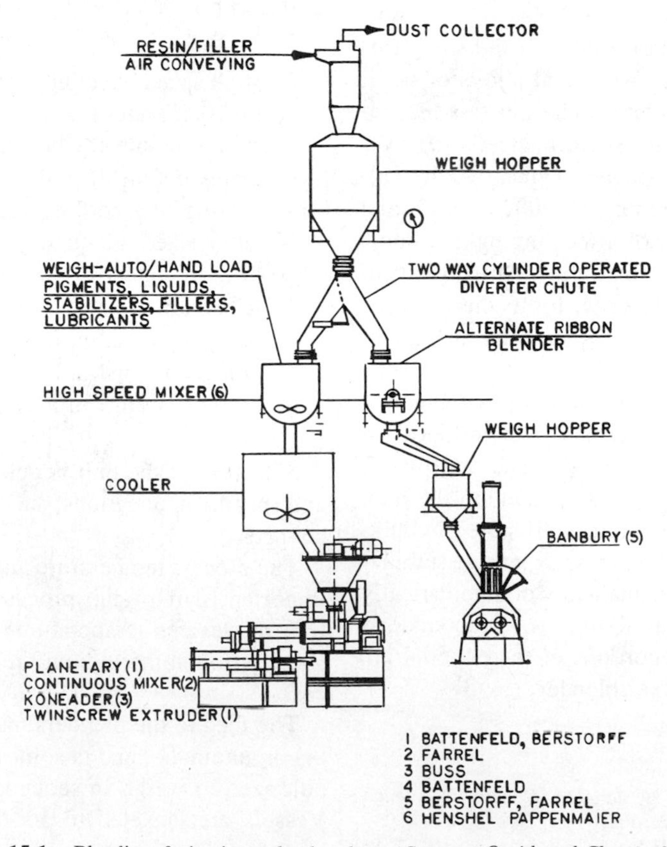

Fig. 15-1. Blending–fusion in a calender plant. (*Courtesy Occidental Chemical Corp.*)

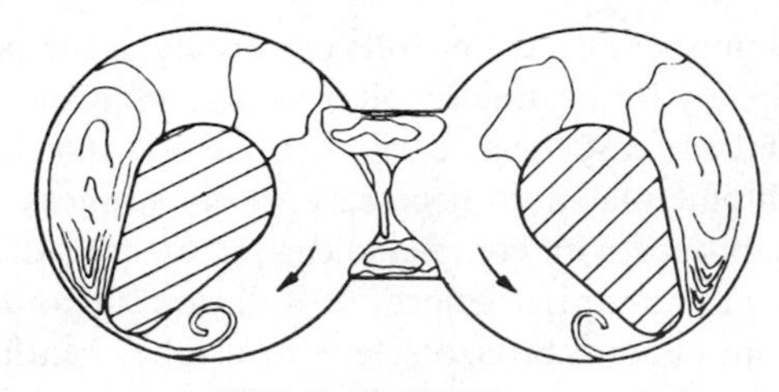

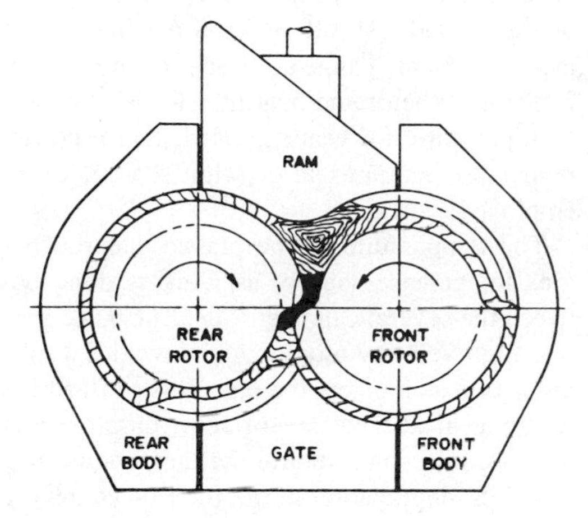

Fig. 15-2. Batch-type mixers

to the ribbon blender, mixer-cooler, or fluxing unit. Careful control of the scrap percentage in the total mix is essential to a quality product, and a reservoir or storage of scrap is necessary.

Mills and Strainers

The discharge from an intensive mixer may be fed to a mill, directly to a strainer, or directly to the calender for heat-sensitive materials such as rigid PVC. The mill provides a reservoir of material for the calender and can accept edge trim (hot or cold).

The mill has bored or drilled rolls, is oil- or steam-heated, with a single or double drive, and generally is a manual operation although progress is being made to automate the milling function and strip width control to feed the calender.

The size of the mill is determined by throughput rates, and if frequent order changes are required, a second or third unit will be necessary.

The mill then feeds the strainer, a variable speed, short L/D extruder. The molten material is cleaned by passing it through a wire mesh screen pack. To ensure that all metallic contamination has been removed, the plastic is conveyed through a metal detection unit as a rope or flat strip before entering the calender.

The strainer size should be determined by the throughput rate and the temperature buildup that can be tolerated. Frictional heat generation is to be minimized, and generally the strainer temperature rise should be limited to 15°F.

Calender

The material is fed to the nip of the calender, where, for the first time, it assumes the form of a sheet. The sheet is progressively pulled through two, three, or four subsequent banks in order to resurface each of the two sides. Plastic calenders generally are made in four basic configurations:

1. The "L" calender, wherein the offset roll is on the bottom, and the take-off is from the top roll (Fig. 15-3).
2. The inverted "L" calender, wherein the offset roll is on the top, and the take-off is from the bottom roll (Fig. 15-4).
3. The "F" calender, wherein the offset roll is on the top, and the take-off is from the middle roll, on the offset roll side (Fig. 15-5).
4. The "Z" calender, either flat or inclined. This calender has two offset rolls, and the take-off may be made from either the top

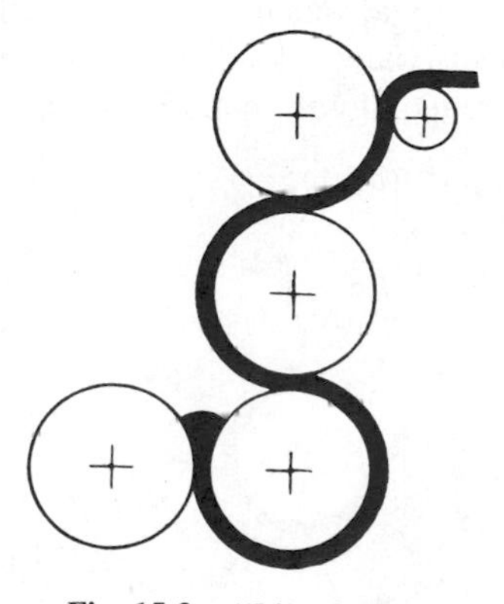

Fig. 15-3. "L" calender.

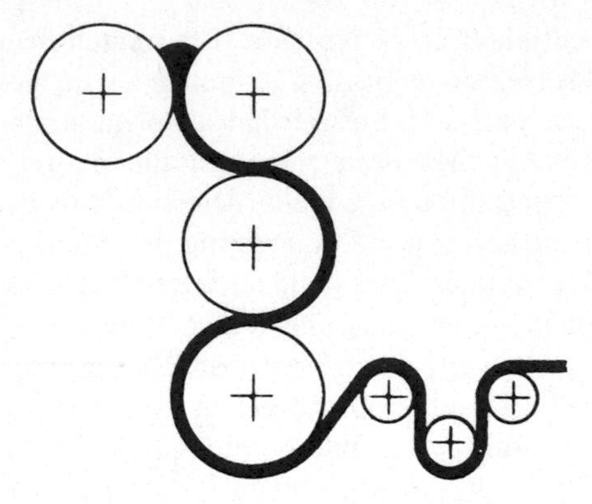

Fig. 15-4. Inverted "L" calender.

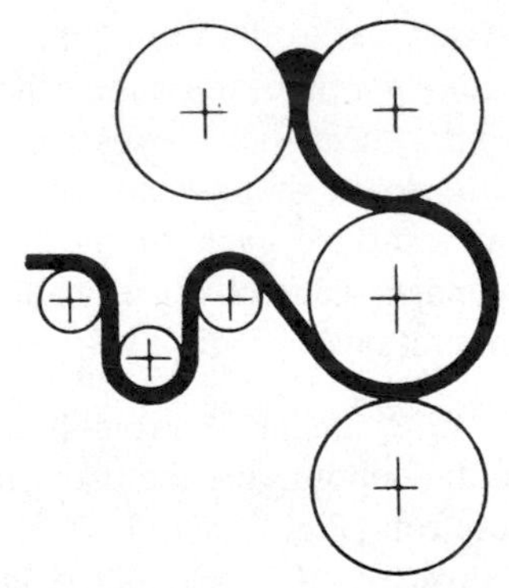

Fig. 15-5. "F" calender.

or the bottom offset roll, or from the back side of the stack roll (Fig. 15-6).

Typically, "L" type calenders (Fig. 15-3) are employed for rigid vinyls (for short feed transport) and inverted "L" calenders (Fig. 15-4) for flexible vinyls (to minimize volatiles/condensation). Rolls are individually driven, and design is advancing toward smaller-diameter models (new calenders) to enable smaller banks, reducing air entrapment and surface blemishes. Calender rolls are drilled at the periphery for heating by oil or water, are made of chilled cast iron or forged steel, and are chrome-plated if necessary. Roll stiffness is improved with compound casting or steel forging to minimize deflection. Rolls are supported with either a bronze sleeve or a roller bearing with T.I.R. targets .0003 inch runout. Roll bending on the number 2 roll, a crossing axis on the second-last roll, and roll bending on the last roll provide gauge accuracy in the range of 2 to 5%. When rapid heat transfer is required, high-pressure hot water is used, and for even higher temperatures (e.g., rigid PVC), oil is employed.

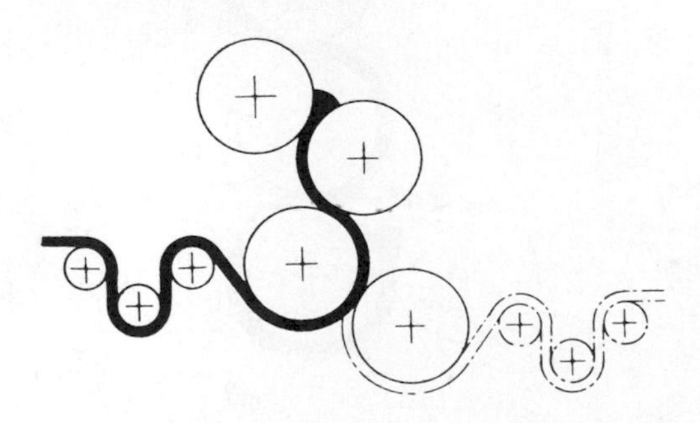

Fig. 15-6. "Z" calender.

The temperature of the plastic material begins to increase sharply as work is done between the several calender rolls. The bank goes into high-velocity rotation and develops a rolling action as it enters the nip. This causes high frictional heats that are of short duration and thus are tolerable. Should the temperature rise too high, degradation in the form of yellowing or burning begins to take place.

As the web moves through the calender nips, it widens out. The width of the sheet on the calender rolls is established by the final product trimmed width (Fig. 15-7). The sheet on the number two roll should be at least the width of the trimmed product in order to be uniform across the web.

A rule of thumb for processing is that material will be attracted to the hotter, faster,

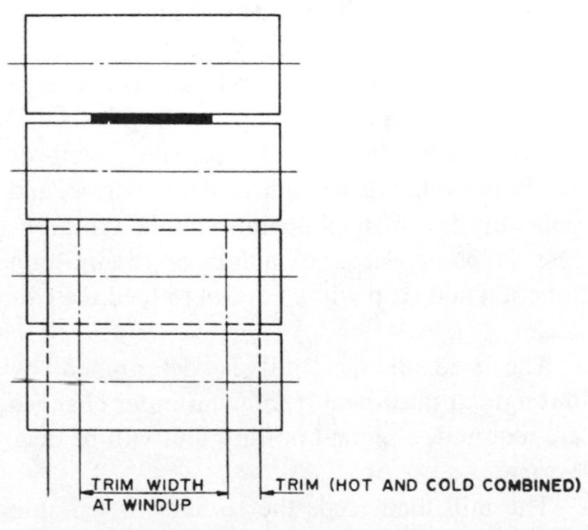

Fig. 15-7. Sheet width including trim.

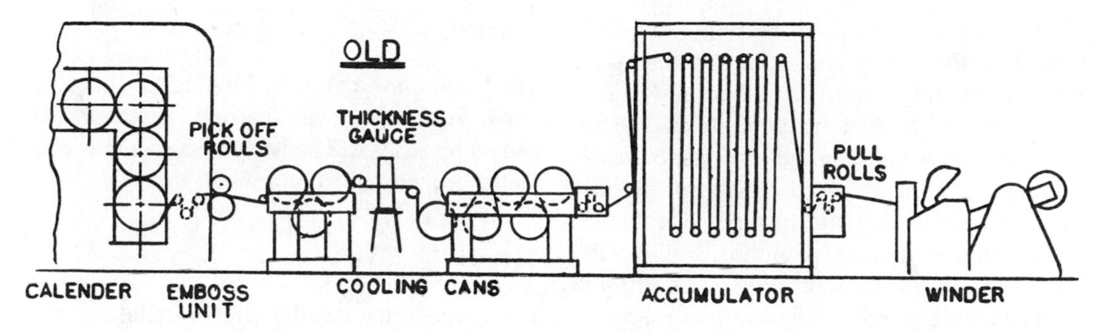

Fig. 15-8. Typical setup for inverted "L" calender and train. (*Courtesy Occidental Chemical Corp.*)

rougher roll. For rigid materials, the heat balance between the roll surface and the product occurs at much lower speeds than for plasticized material.

Because plastic webs release better from cold rolls than from hot rolls, it frequently is desirable or necessary to cool the surface of the roll from which the web is "taken off." This is difficult to accomplish when taking off the number three roll.

Substrates can be laminated in line and applied by nipping the fabric or the film at the last roll.

Take-off Stripper Section

Take-offs have been developed in Europe in recent years that improve on the older style, unheated, two- and three-roll units (Fig. 15-8). These are temperature-controlled, multi-drive, smaller-diameter systems that provide annealing, improved dimensional stability for thin-gauge films, and high film stretch. The increased stretch reduces calender surface speed, improves stock temperature control, and reduces air entrapment.

Embosser

After leaving the stripping section, the web passes directly to cooling or is fed into an embossing unit, which can be arranged either vertically or horizontally. The embosser consists of an engraved steel roll pressing against a rubber roll, designed for a quick roll change (15 minutes) to minimize process interruption. The hardness and thickness of the rubber covering, together with the pressure applied by the steel roll, determine the depth of indentations or embossing in the plastic sheet. In order to keep the rubber cover cool, cooling water is circulated inside the steel arbor and outside the rubber covering, by means of cold roll contact or a water bath. The water on the rubber surface must be squeegeed off before contacting the plastic to prevent watermarking of the film. The embossing roll is temperature-controlled, and this chilling effect sets the film surface.

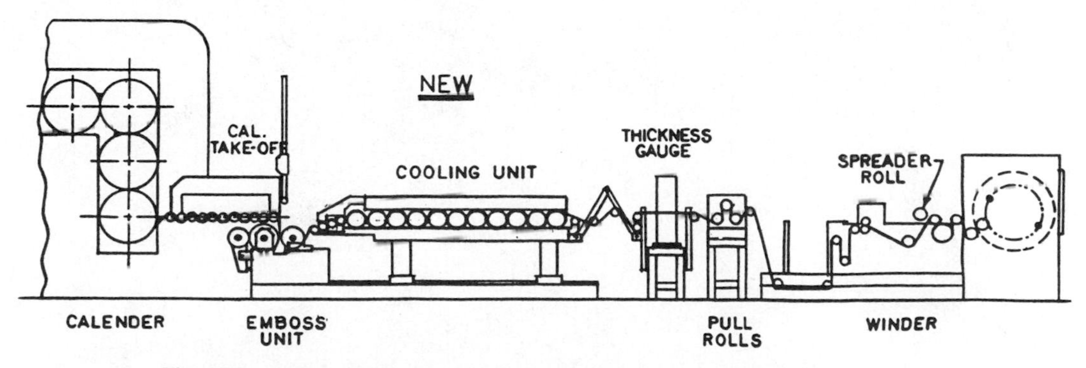

Fig. 15-9. Modern multi-drive calender train. (*Courtesy Occidental Chemical Corp.*)

From the embossing roll section, the web is annealed (Fig. 15-9) and is carried through a series of cooling drums and a beta ray gauge. The basis weight then can be correlated to the thickness, depending on the formulation used, and can, in turn, be used to control product thickness (final gap) and material distribution (roll bending/cross axis) automatically. SPC (statistical process control) and electronic maintenance data collection are options.

Controlled and complete cooling is essential; if the film is wound up too hot, it will have a tendency to cool on standing, resulting in wrinkles, reduction in emboss pattern depth, ''snap-back'' and tight rolls, leading to printing/laminating difficulties.

Winder

The winder has a very demanding function—to wind thick/thin, wide/narrow widths; fast/should be at least 2 inches per side, and is collected or automatically ground.

Trends

For a modern calender line installation, the capital costs without buildings are in excess of $10 million, and technological efforts are being directed toward increasing the quality of throughput and automation. Much of the progress in engineering and instrumentation refinement is taking place in Germany and Italy.

16

Vinyl Dispersions

OVERVIEW

Vinyl dispersions are fluid suspensions of fine-particle-size polyvinyl chloride resins in liquid plasticizer/diluent systems. The viscosity of the plastisol may range from a pourable liquid to a paste, depending on the formulation. When heated to about 150 to 170°C (302–338°F), the dispersion turns into a homogeneous hot melt. When the hot melt is cooled to around room temperature (25°C, 77°F), it becomes a tough vinyl product.

Most vinyl applications that use dry blend or pelletized vinyl compounds require heavy melt processing equipment such as calenders, extruders, injection molders, blow molders, and so on. With vinyl dispersions the processor can use convenient liquid handling techniques such as casting, spraying, strand coating, rotomolding, dipping, and so forth.

The term "plastisol" is used to describe a vinyl dispersion that contains little or no volatile thinners or diluents (Fig. 16-1). Plastisols range in viscosity from water-thin liquid to a heavy paste with a consistency of mayonnaise. The ingredients generally include PVC resins, plasticizers, stabilizers, fillers, and pigments. When no volatile thinners are included, the plastisol is considered as high solids (98–99%). There is always some evaporation of plasticizers during the curing cycle.

It is convenient in some instances to extend the liquid phase of a dispersion with volatile dispersants or thinners, which are removed during fusion. The term "organosol" applies to these dispersions (Fig. 16-1).

Approximately 10% of the polyvinyl chloride resin used in the United States is dispersion-grade resin. The major markets for PVC dispersion resins are floor coverings, carpet tiles, coated industrial fabrics, medical gloves, decorative films, coil coating, and automotive sealants.

Historical

The development of flexible vinyl compounds in fluid dispersion form was a goal of the early researchers, in 1930 through 1940. In Waldo L. Semon's patent disclosure[1] of the plasticization of poly(vinyl chloride) in 1933, he mentions application of highly plasticized combinations by the spread coating of hot melts. In 1939 and 1940, patents were issued in Great Britain[2] and the United States[3] that fully describe vinyl dispersions as we know them today.

The discovery of the plastisol method for processing polyvinyl chloride was one of the citations included in the 1968 presentation of the Morley Medal to Dr. Semon of The BFGoodrich Company by the Cleveland Section of the American Chemical Society.

The earliest commercial resins were a product of the Union Carbide Chemicals Co. fellowship at Mellon Institute. These emulsion-polymerized resins could be dispersed by grinding in an organic medium to make stable dispersions, provided that the suspending phase contained enough polar liquids to have some solvating action on the resin but not enough to

By Walter A. Edwards, Ashok C. Shah and Bela Mikofalvy, The BFGoodrich Chemical Co., Cleveland, OH.

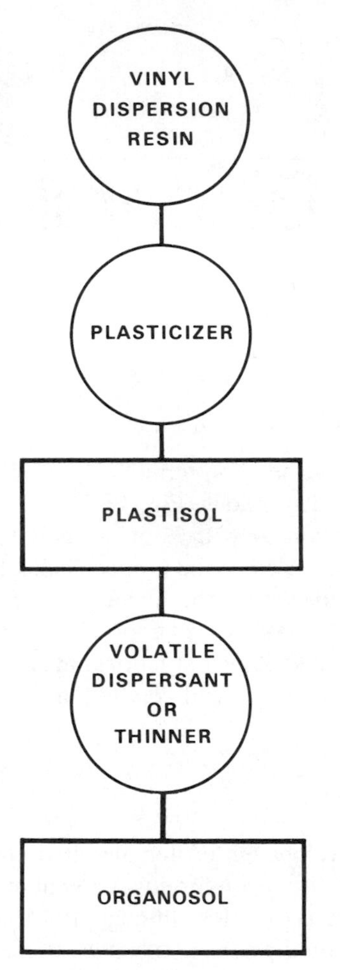

Fig. 16-1. Types of vinyl dispersions. (*All photographs and illustrations in this section, unless otherwise indicated, are courtesy of The BFGoodrich Co.*)

dissolve it. This technique was introduced to the cloth-coating industry in 1943, and the resin was introduced commercially in 1944. The significance of this development was recognized by the plastics industry with the award of the John Wesley Hyatt Medal in 1950 to G. M. Powell of Union Carbide.

Notable progress in the field of vinyl dispersions occurred in 1947 with commercial introduction, by The BFGoodrich Company, of a plastisol-type resin that simplified the preparation of dispersions. This resin could be mixed with relatively low levels of plasticizer by a simple stir-in technique, and volatile thinners no longer were required.

Through the 1950s and the 1960s improvements were made to these dispersion-grade PVC resins via the incorporation of copolymers, which lowered the processing temperatures of the plastisol. Higher-gloss products were developed as well as improved viscosity control and product reproducibility. Improvements in adhesion to difficult substrates such as metals and synthetics also were made.

During the 1970s the total PVC industry faced and overcame the occupational health and safety problems associated with the realization that vinyl chloride monomer was linked as a causal agent to one or more forms of cancer. This decade was dedicated to modifying manufacturing processes in order to contain vinyl chloride in the workplace and remove it from the PVC polymer.

Efforts in the 1980s continued to focus on safety and environmental issues in the form of waste disposal, fire, and smoke. Product improvements were made in areas such as heat stability of the polymer and low viscosity plastisols that do not become more viscous with age. The industry also underwent a significant consolidation, restructuring, and change of ownership of manufacturing companies.

Resins

Resins for vinyl dispersions are different from general-purpose suspension resins designed for calendering, extrusion, or injection molding. The latter make up about 90% of the resins consumed in the vinyl industry. The general-purpose suspension resins are produced via suspension or mass polymerization processes, and the average-particle-size diameter of the most common type of suspension resins falls between 50 and 250 microns. These resins are very porous and very coarse and will not readily stir-in to form paste below 100 phr plasticizer. Dispersion resins, on the other hand, are very small and nonporous. General-purpose suspension resins have very little or no utility in plastisol applications, as these resins do not form melts without sufficient mastication and high cure temperatures of around 380°F.

Vinyl dispersion resins are polyvinyl chloride homopolymers and copolymers produced via emulsion or micro-suspension polymerization processes. During polymerization, the distribution of particles formed is in the range of 0.2 to 15 microns. However, secondary particles (grains) in the range of 5 to 60 microns are formed during stripping, dewatering and drying, and grinding operations. Dispersion-grade resins are talc-like in quality, with finer particles clinging together and causing poor powder flow.

Special grades of suspension resins, called extender or blending resins, are significantly finer in particle size than general-purpose suspension resins which vary from 5 to 150 microns, the average particle size of the special grades being from 15 to 90 microns. The extender resins are less porous, form melts without mastication, and, when used in conjunction with dispersion resins, control paste viscosity, gloss, clarity, gelation, and other plastisol properties.

Over the last two decades there has been a significant reduction of vinyl dispersion and blending resin producers. In 1988 the list of resin producers was as follows:

Company	*Trade Name*
Borden Chemical	Borden VC
BFGoodrich	Geon
Formosa Plastics Corp. USA	Formolon
Goodyear	Pliovic
Georgia Gulf	EH
Occidental Chemical Co.	Oxy

Rheology

Various plastisol applications require a wide range of paste viscosities. Thin film applications where organosols are used require a viscosity as low as 200 to 500 cps. On the other extreme, applications for automotive sealants and primary adhesives for carpet tiles may require viscosities as high as 300,000 to 500,000 cps. Thus, control of the rheological properties of vinyl dispersions is of key importance.

Rheology is the science devoted to the study of flow and deformation. Expressed mathematically, viscosity (η), the resistance to liquid flow, can be defined as the ratio of shear stress (τ) to shear rate ($\dot{y}$) in laminar flow.

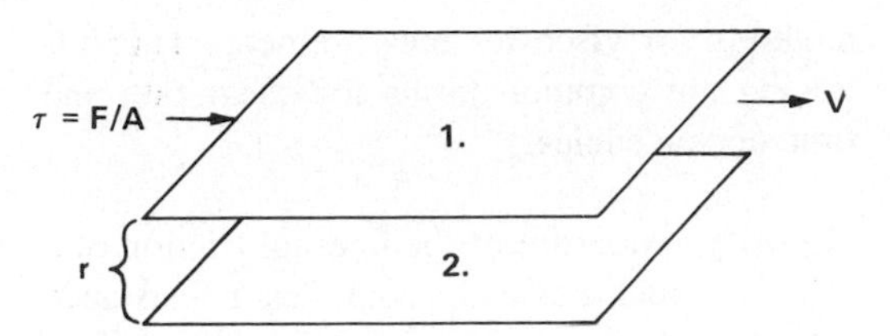

Fig. 16-2. Schematic illustrating laminar flow.

$$\eta = \frac{\tau}{\dot{y}} = \frac{\text{Shear stress}}{\text{Shear rate}}$$

Shear stress is the tangential force per unit area applied to a liquid layer, and shear rate is the ratio of the resulting velocity of the layer (No. 1 in Fig. 16-2) to the distance from a reference layer (No. 2). Shear rate is more precisely defined as the rate of change of velocity with distance in the system of the laminae; thus dv/dr.

Most simple liquid systems exhibit Newtonian flow, where viscosity is independent of shear rate. Most plastisols (solid dispersions) exhibit a more complex flow behavior (as shown in Fig. 16-3) such as shear thinning, shear thickening, or a combination of both, where the plastisol is shear-thickening at low shear rate and then shear-thinning at high shear rate.

The particles in vinyl dispersions interact to form temporary structures, and the dispersion viscosity reacts to the time necessary to break down and rebuild these structures. Time dependency, as well as shear-rate dependency, must be considered in selecting instruments and test

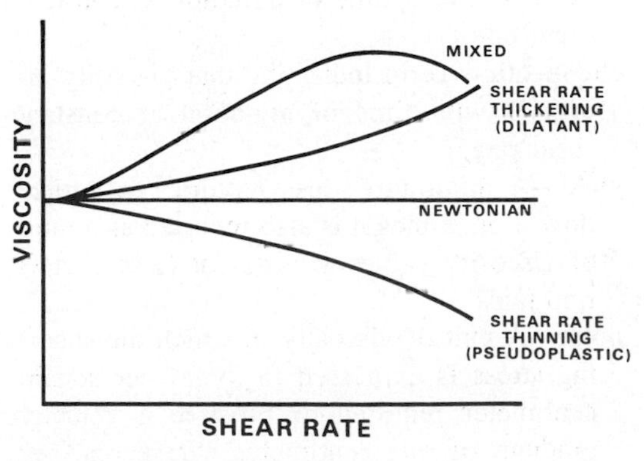

Fig. 16-3. Types of flow, illustrating shear rate dependence.

methods for viscosity measurement. The following are common terms for shear rate and time dependencies:

viscosity—Measure of the internal friction of a fluid, which basically connotes a resistance to flow. It also can be defined as the ratio of shear stress to shear rate.

apparent viscosity—Reporting of the viscosity of a fluid at various values of the spindle rpm instead of converting the spindle rpm to the specific shear rate.

shear stress—In laminar flow, the ratio of applied force to the area sheared. In order to overcome internal forces and start flow, a shear stress must be applied. Highly viscous liquids will require more force to move them than will less viscous liquids.

shear rate—The speed at which the surface bounding the fluid is moving. In a rotational viscometer the surface of the spindle exposed to fluid and the spindle rpm are used to calculate the shear rate. During coating, the speed of the line divided by the clearance will determine shear rate.

Newtonian fluid—A fluid in which the viscosity is independent of the shear rate (constant). The fluid's response to shear rate is proportional to the increase in shear stress.

dilatant (shear rate thickening)—Term indicating that viscosity increases with increasing shear rate.

pseudoplastic (shear thinning)—Term indicating that viscosity decreases with increasing shear rate.

thixotropic—Term indicating that viscosity decreases with time of agitation at constant shear rate.

rheopectic—Term indicating that viscosity increases with time of agitation at constant shear rate.

yield—A minimum stress required to initiate flow. Sometimes it is also reported as a ratio of viscosity values at different (viscometer) rpm rates.

poise—A unit of viscosity in which the shearing stress is expressed in dynes per square centimeter required to produce a velocity gradient of one centimeter per second per centimeter.

centipoise (CPS)—One hundredth of a poise.

One must always be careful to avoid formulating very dilatant plastisols, as they can damage mixing equipment and pumps during plastisol preparation and transfer. Such a plastisol will result in poor control of coating thickness, entrap air, and, in general, negatively affect most plastisol operations and product performance.

Because of shear-rate and time dependence of dispersion viscosity, the compounder should consider the shear rate and the expected plastisol age when developing and testing formulations.

The Brookfield Rotational Viscometer[4] and the Severs Extrusion Rheometer[5] are the two most common viscometers used for generating the shear rate–viscosity relationship of plastisols in the laboratory. Cone and plate or concentric cylinder viscometers[6,7] also are available for studying more complicated flow parameters.

Although the laboratory viscometers may not accurately determine and simulate the viscosities encountered during actual production situations, the exercise of using them will be very helpful and valuable in troubleshooting plant problems. By making comparison studies in the appropriate shear-rate range, one could select and check incoming raw materials and predict production performance quite accurately. Test methods, the rheological effects of compounding ingredients, and specific application requirements are discussed later in this chapter.

Fusion

The process in which a liquid dispersion is converted into a homogeneous solid upon the application of heat is called fusion. This conversion is a true fusion (melting) of the crystallite structures in the polymer particles, followed by solution of the molten polymer in the plasticizing vehicle. Overall, the mechanical properties (degree of fusion) development of the vinyl compound is a time–temperature phenomenon. The fusion temperature of plasticized vinyl is a complex function of the molecular weight, and amount of comonomer, plasticizer level, and polymer–plasticizer interaction parameter.[8] Because dispersions are fused with heat alone, without mechanical working, the particle size, particle size distribution, and interference from nonvinyl constituents at particle surfaces also affect the fusion temperature.

The crystallites in PVC melt over a fairly wide temperature range; so some strength is developed at temperatures below the ultimate fusion point. This strength cannot be increased by increasing the time of heating (Fig. 16-4).

As the plastisol is heated to the fusion point, it changes in several important ways (Fig. 16-5). In the early stages of heating, the viscosity is affected by two competing phenomena. Plasticizer viscosity drops with increasing temperature up to about 38°C (100°F). This causes a decrease in viscosity of the dispersion (A). As the temperature continues to rise, plasticizer begins to permeate the particle, the particles swell (B), and viscosity increases at an accelerating rate. Further heating and more plasticizer absorption lead to contact between swollen particles and a weak bond (C). If the material were cooled to room temperature at this point, it would have a weak cheesy structure and would be opaque. Additional heat would produce a clear matrix (D) and a strong product when cooled. Transparency, however, is not as good a criterion of complete fusion as is strength development.

In practice, the fusion temperature of any dispersion system usually is determined empirically from a curve of physical properties (usually the ultimate tensile strength and ultimate elongation at break) of cast film heated at various temperatures (Fig. 16-4). The "true" fusion temperature is considered to be the temperature that produces the best properties.

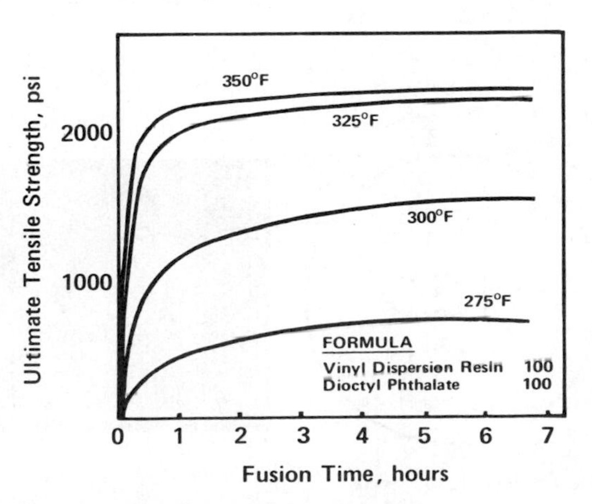

Fig. 16-4. Effect of time on plastisol fusion temperature–tensile strength relationship.

Formulating Vinyl Plastisols

In developing any vinyl compound, the vinyl formulator first must determine the end-product performance requirements, such as hardness, mechanical properties (tensile, elongation, modulus), low-temperature brittleness and flexibility, optical properties (gloss and clarity), flammability, smoke resistance, toxicity, and so on. In dispersions, the rheological contributions of the compounding ingredients are of major concern because of the rheological limitation of the intended fabrication process, as shown in Table 16-1.

Much information is available on the general properties of vinyl compounding ingredients.[9–11] Although a detailed discussion is beyond the scope of this chapter, a brief discussion of the effects of major plastisol ingredients such as resins, plasticizers, fillers, stabilizers, pigments, thickeners, flame retardants, and smoke suppressants on rheology, gelation, fusion, and end-product performance will be presented.

Dispersion and Blending Resins

There are probably more than 75 commercial vinyl dispersion resins available today, most of them homopolymers of polyvinyl chloride. They differ in particle size, molecular weight, and type of emulsifier, as well as the level of emulsifier remaining on the product.

The average particle size typically is in the range of 1 to 5 microns for most dispersion resins. The particle size distribution can be monomodal, bimodal, or multimodal. The particle size has a major impact on plastisol rheology, and its effect has been discussed in detail by Collins, Hoffman, and Soni.[12] If the particle size distribution is very narrow, the paste viscosities at both low and high shear will be very high and very dilatant. As the particle size distribution is broadened, the paste flow will be more shear-thinning (pseudoplastic) and in general lower than the paste viscosity at higher shear. Therefore, it is not uncommon to blend

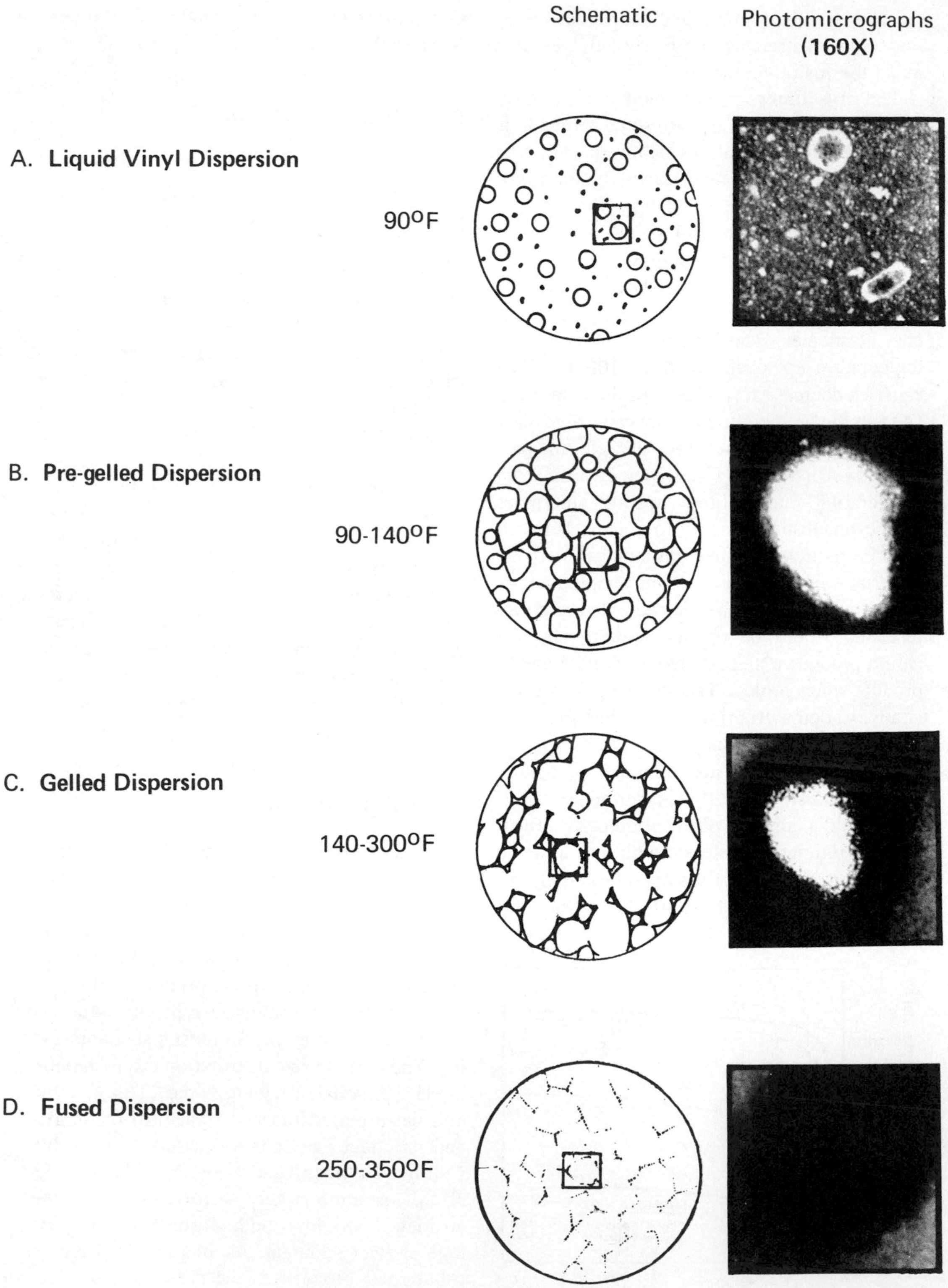

Schematic | Photomicrographs (160X)

A. **Liquid Vinyl Dispersion**

90°F

B. **Pre-gelled Dispersion**

90-140°F

C. **Gelled Dispersion**

140-300°F

D. **Fused Dispersion**

250-350°F

Fig. 16-5. Sequential effects of heating plastisols.

Table 16-1. Comparative rheological requirements for plastisol and organosol processing for various applications.

	Low shear		High shear
	Viscosity[a]	*Ratio*[b]	*Viscosity*[c]
Molding:			
Dipping–Hot	Low–medium	Medium	NC[d]
Cold	Medium–high	High	NC[d]
Slush molding	Low–medium	Medium	NC[d]
Rotocasting	Low	Low	NC[d]
Cavity molding	Low–medium	Medium	Low
Coating:			
Direct roll	Low	Low	Low
Reverse roll	Low	Low	Low
Fabric coating (knife)			
Open weave	Medium–high	High	Low–medium
Tight weave	Medium	Medium	Low–medium
Laminating adhesive	Low	Low–medium	Low
Spraying:			
Vertical	Low–medium	High	Low
Horizontal	Low–medium	Low	Low
Sealants	High	High	Low

[a]Low shear viscosity: Brookfield viscosity, 20 rpm,
Low = <100 poise (<10 Pa·s)
Medium = 100–300 poise (10–30 Pa·s)
High = 300–500 poise (30–50 Pa·s)
Ultrahigh = >500 poise (>50 Pa·s)
[b]Viscosity ratio: Brookfield viscosity at 2 rpm/20 rpm,
Low = 1–2
Medium = 2–3
High = >3
[c]High Shear Viscosity: Severs viscosity [80–100 psig (5.6–7.0 kgf/cm^2), orifice length 5 cm, diameter 0.17 cm]
Low = <50 poise (5 Pa·s),
Medium = 50–200 poise (5–20 Pa·s)
High = >200 poise (20 Pa·s)
[d]Not critical for this application.

more than one plastisol resin to control paste viscosities.

Special-purpose, fine-particle-size suspension or mass resins, also known as extender or blending resins, are very valuable for blending with dispersion resins to control paste viscosities and other plastisol properties. Blending resins afford lower paste viscosities, reduce dilatancy, facilitate air removal, and improve viscosity stability on aging. Because of the larger particle size, blending resin particles will be segregated and will settle faster than the dispersion resin particles on aging. The settling is accelerated when the paste viscosity is lower than 2000 cps. When blending resins are used, one must consider a way to stir the plastisol before use after it has set for a while.

Addition of blending resin reduces clarity and gloss and also retards gelation and reduces mechanical properties. In a highly plasticized formulation, blending resins provide a drier feel to the finished product. Use of a copolymer blending resin will help to accelerate gelation and fusion.

The inherent viscosity (ASTM D1243-66) is a common measure of molecular weight. Dispersion resins range in inherent viscosity (I.V.) from about 0.7 to 1.45. As molecular weight increases, many properties are affected. Increasing the molecular weight leads to increased fused strength, but slower gelation. It increases the viscosity of the vinyl melt, an effect that, in turn, has implications for reduced flowout on moldings and coatings and requires compound adjustment to give smooth expansion of chemically blown foam. Although changes in molecular weight affect the fused product strength, they have little influence on the fusion temperature per se. The fusion temperature is much more radically affected by the inclusion of comonomers such as vinyl acetate, which break up the crystalline structure in the polymer.

Copolymers are available with about 5 to 10% vinyl ester content. Increasing the vinyl ester content lowers the fusion point, but at the expense of the ultimately available strength (Fig. 16-6). Other effects of increasing the vinyl ester content are higher initial viscosity, more rapid viscosity aging, softer fused products, more rapid gelation, and poorer heat stability at any given processing temperature. The fact that the processing temperature drops with increasing comonomer content should be considered in making heat stability comparisons.

The residual emulsifier and/or surfactant left on the dispersion resins varies from as low as 0.5 phr to as high as 4.0 phr. As a rule, a resin with high residual emulsifier will afford higher paste viscosity, poor viscosity aging, and reduced gloss and clarity, and will not facilitate air release. The higher emulsifier may contribute to poor heat stability and to windshield fogging in autos, and it may affect taste, or acceptability, in food uses. The higher emulsifier also results in poor compatibility with most thermal stabilizers and accelerates plate-out.

Functionally active resins are available for

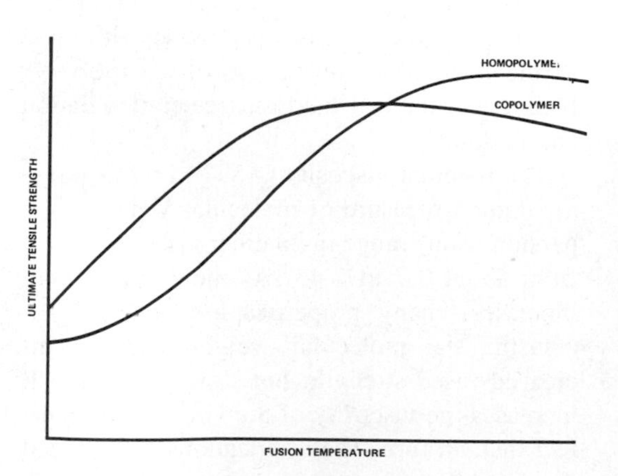

Fig. 16-6. Tensile strength development during fusion; homopolymer vs. copolymer dispersion resin.

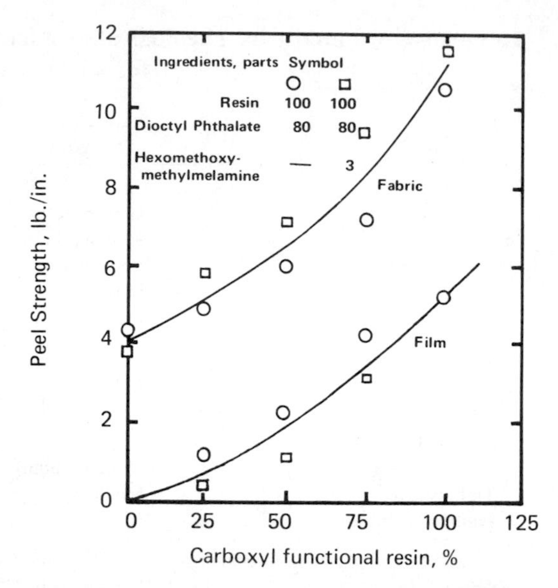

Fig. 16-7. Vinyl-to-nylon adhesion performance of simple laminating adhesives with various ratios of carboxyl-active and low fusion copolymer resins in a blend.

special properties requirements. For example, a report by Ward[13] describes the properties attributable to carboxyl groups on the vinyl chain. The carboxyl functionality increases adhesion to a number of metal and fabric substrates. (The effect on vinyl-to-nylon adhesion is shown in Fig. 16-7.) The increased demand for single-ply roofing and the need for high-performance industrial coated fabrics (awnings, geostructures, and tarpaulins) have been met with the use of carboxyl-functioning vinyl dispersion resins.

For most organosol applications, the conventional dispersion resin is adequate. However, when superior adhesion to metal is required, a small amount (10–20 phr) of solution resin is substituted for the dispersion resin. Solution resins are vinyl copolymer resins produced by the solvent process. These resins may have a carboxyl or a hydroxyl or other functionalities that, when crosslinked with epoxy, melamine, or adhesion promoters, provide superior adhesion to various metals.

Plasticizers

In dispersions, plasticizers provide fluidity and affect the finished properties. The plasticizer type and its level have an important influence on viscosity, viscosity aging, gelation and fusion characteristics, and end-product properties. The reader is directed to comprehensive industrial and reference works[14,15] for overall descriptions of the plasticizers available, as well as to the suppliers' literature. Most plasticizers utilized in vinyl dispersions are esters. They may be esters of long-chain alcohols with aromatics such as phthalic anhydride, with straight-chain dibasic acids such as sebacic, azeleic, or adipic, or with phosphoric acids; or they may be polyesters such as those based on propylene glycol and sebacic acid.[16,17]

As a general rule, the initial viscosity of a vinyl dispersion immediately after mixing directly correlates with the plasticizer viscosity, as shown in Fig. 16-8. On aging, the viscosity generally will increase somewhat, although the higher the proportion of the plasticizer component of a dispersion is, the lower the dispersion viscosity.

A theory that works quite well in predicting the effects of plasticizer type on viscosity aging involves the percent by weight of the plasticizer molecule present as an aromatic ring structure(s). Higher aromaticity means a higher relative ability of the plasticizer to interact with the resin. In Fig. 16-9 the plastisol viscosity versus temperature is traced for a variety of phthalate plasticizers as well as one based on sebacic acid.[18] As the alcohol component becomes a larger weight fraction of the ester or

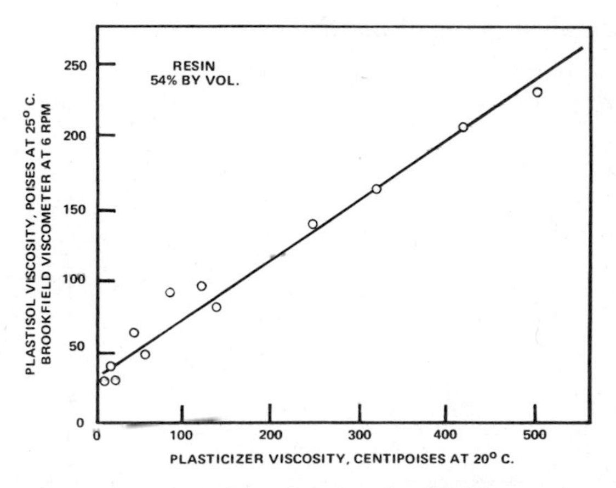

Fig. 16-8. Effect of the viscosity of a plasticizer on the initial viscosity of a plastisol.

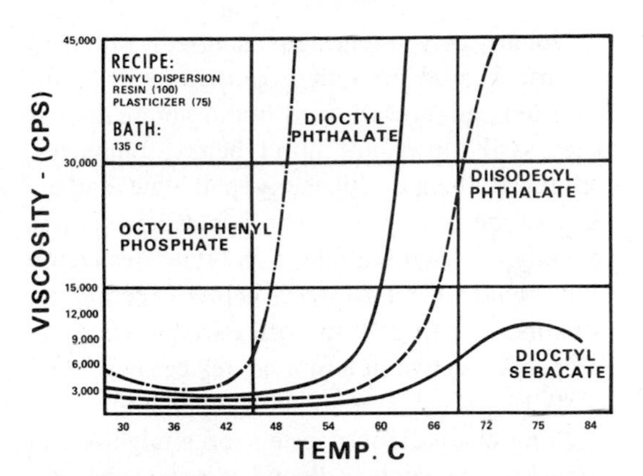

Fig. 16-9. Plastisol viscosity and gelation vs. temperature, using different plasticizer systems.

as the aromatic component disappears, the temperature at which a rapid viscosity rise takes place moves to a higher range. This type of measurement is an SPI Vinyl Division Standard Test Procedure (SPI-VD-T15).

Phthalate esters are used more often than any other plasticizer in the formulation of vinyl dispersions, the one most frequently used being di-2-ethylhexyl phthalate (DOP). DOP combines most of the desirable properties of a plasticizer, such as minimal interaction with the resin at room temperature, good fusion properties, low volatility, acceptable low-temperature flexibility in the product, and low cost.

Shorter-chain-alcohol phthalate esters, such as dihexyl and dibutyl phthalates, are more volatile but fuse faster in dispersions because of their higher percent aromaticity. Longer-chain-alcohol phthalates also are used, primarily for their greater permanence in the product because of lower volatility. Linear-alcohol phthalates, based on 7 to 11 carbon alcohols, have enjoyed success in recent years because they provide lower dispersion viscosities and somewhat better low-temperature properties than DOP at similar cost.

Phosphate esters, such as tricresyl phosphate (TCP) and octyldiphenyl phosphate, are highly solvating and find special use in vinyl dispersions. These esters, except for trioctyl phosphate (TOF), furnish rather high initial and aged viscosities and lower fusion temperature when compared with DOP. Phosphates contribute a degree of flame resistance through increasing char formation during combustion.

Other highly solvating plasticizers, designed to furnish good strength properties when the fusion temperature allowed by an application is necessarily low, are butyl benzyl phthalate, dihexyl phthalate, diisohexyl phthalate, and esters of benzoic acid. These plasticizers frequently are used with the copolymer dispersion and blending resins (see below, section on "Fabric Coating") to attain usable physical properties at fusion temperatures below 138°C (280°F).

Long-chain-alcohol esters of straight-chain dibasic acids, such as dioctyl adipate, sebacate, and dioctyl azelate, frequently are used in blends for their low initial and aged viscosity and excellent low-temperature flexibility in fused compounds. They are more volatile than DOP and have higher gelation and fusion temperatures.

Specialized monomeric esters that are gaining some favor in formulating are the trimellitates, including trioctyl trimellitate (TOTM), triisooctyl trimellitate (TIOTM), triisononyl trimellitate, and similar esters of linear alcohols. Their contribution is primarily very low volatility and resistance to extraction by solvents and oils.

Polymeric plasticizers, such as the polyester of propylene glycol and sebacic acid, find utility in vinyl dispersions where they contribute to very low volatility and excellent resistance to extraction from the fused product by solvents. These plasticizers, however, generally are very viscous and usually are blended with lower-viscosity plasticizers.

Other specialized materials also find some utility as vinyl dispersion plasticizers. Liquid nitrile rubber, though of high viscosity, contributes oil resistance and can be vulcanized. Low-volatility acrylate monomers such as triethylene glycol dimethacrylate contribute fluidity to the liquid dispersion and cure during fusion to a hard polymer.

Another class of plasticizers is differentiated from the foregoing by their limited compatibility, color, or odor. Used mainly for cost reduction in blends, they are hydrocarbons of both highly aromatic and highly aliphatic types. The aromatics provide moderate to high viscosity and fairly good compatibility with relatively poor heat and light stability. The aliphatics provide low viscosity but may exude from aged fused products. An example of this group is chlorinated hydrocarbons, which contribute some flame resistance and sometimes are used for this reason.

Epoxidized soya and linseed oils are used in small amounts because of their heat and light stabilization, as well as for their plasticization action.

Thus, the selection of a plasticizer or, more frequently, a blend is crucial to the overall performance of a vinyl dispersion. The primary goal remains a practical balance between liquid and fused properties. Table 16-2 presents an overall comparative performance rating of var-

Table 16-2. Overall performance rating of plastisols.

	General-purpose phthalates	*High molecular weight phthalates*	*Epoxy soya*	*TOTM**	*Polyesters*
Exudation	5	3	2	4	1
Humidity resistance	5	3	2	4	1
Acid environment	5	4	2	3	1
Dry blend rate	5	3	2	4	1
Fusion rate	5	2	1	4	3
Plastisol Viscosity	5	3	2	4	1
Volatility	1	2	4	5	3
Light stability	5	3	1	4	2
Migration resistance	1	2	4	3	5
Extraction (kerosene)	2	1	4	3	5
Fungal resistance	4	3	2	5	1
Total	43	29	26	43	24

Rating: 1 = fair to poor; 5 = excellent.
*Trioctyl Trimellitate

ious plasticizers, which can be helpful to a plastisol formulator as a guide to selecting the right plasticizer system for a particular application.

Stabilization

Polyvinyl chloride is inherently a very heat-sensitive polymer. If the polymer is processed without any stabilizers, it will start to degrade readily above 250 to 300°F and give rise to yellowing, followed by further discoloration, evolution of hydrochloric acid, and crosslinking; and eventually complete degradation will take place.

Polyvinyl chloride resins must be protected against more than just the heat of processing, as most vinyl products are exposed to varying degrees of heat during their normal service life. Vinyl products also need protection from long-term exposure to a harsh environment (weathering). Plastic products experience moisture in the form of rain, snow, and humidity; exposure to chemical pollutants; oxidation by sunlight; and biological attack.

An article by Jennings and Fletcher[19] describes polyvinyl chloride stabilizers. A primary stabilizer is described as one that, when employed as the sole stabilizer, imparts an acceptable degree of heat stability for commercial application. The primary stabilizers are leads, mixed metals, and organotin stabilizers. A secondary stabilizer is one that cannot be employed as the sole stabilizer, but can extend, complement, and synergistically improve the heat stability of PVC when used with a primary stabilizer. The secondary stabilizers are epoxidized oils and metallic soaps.

Stabilizers most commonly used for plastisols are barium, cadmium, and zinc salts of long-chain fatty acids such as stearic, oleic, benzoic, and lauric acids. Although lead-based materials such as dibasic lead phosphite, tribasic lead sulfate, and dibasic lead stearate are excellent thermal stabilizers, their use has been limited to wire and cable applications for reasons of toxicity. Organotins such as dibutyl tin dilaurates, alkyl tin mercaptides, and dioctyl tins are used when their high stabilizing effectiveness warrants their high cost.

Most stabilizers used for plastisols are liquids. If solid stabilizers are used, they must be predispersed in a plasticizer. The primary function of a stabilizer is to improve processibility, process heat stability, and long-term property retention. However, stabilizers can have such side effects as poor early color, a detrimental effect on clarity due to poor compatibility of the stabilizer with residual resin emulsifier, plate-out, volatility, and stain resistance.

A good plastisol stabilizer should provide good early color hold, good long-term stability at both high and moderate temperatures, and low viscosity, and should facilitate air release and provide little or no plate-out. Antioxidants and UV stabilizers may be added to minimize degradation on long-term exposure to UV light and weathering. Stabilizer manufacturers should be contacted about one's specific needs.

Thinners

Two types of volatile organic materials, called dispersants and diluents, are used extensively to control the rheology of organosols.

Diluents are poor or non-solvents and may be aliphatic, aromatic, or a mixture of both. They do not contribute to resin solvation and in most cases must be evaporated at temperatures below the complete fusion of the plastisol or organosol to prevent blisters, bubbles, or mud cracking. Varsol No. 2, Shell Sol 140, VM&P naphtha, and mineral spirits are aliphatic diluents, whereas xylene and toluene are aromatic diluents.

Dispersants are much stronger, volatile solvents. Unlike diluents, dispersants help to solvate the resin particle and are very helpful in developing good fusion properties. Because of their high solvency, dispersants tend to increase viscosity rapidly and thus are never used alone in organosols. Most organosols will contain a low-level plasticizer and a balance of diluents–dispersants, blended to obtain viscosity control and good fusion properties. Typical dispersants for organosol preparation are diisobutyl ketone, methyl isobutyl ketone, methyl ethyl ketone, isophorone, cyclohexahone, diacetone alcohol, butyl carbitol, and so on.

Frequently, when a plastisol is formulated to

provide the desired overall performance, the process calls for a further reduction in paste viscosity. This may be accomplished by adding small amounts of diluents (0–5 phr), which also helps to improve the plastisol's shelf stability.

Fillers

The most widely used fillers in plastisols are calcium carbonate and aluminum trihydrate. Other fillers include calcium sulfate, barium sulfate, clay, mica, and precipitated silicas.

Most fillers come in various particle sizes, shapes, and particle size distributions. Fillers with a finer particle size distribution with high surface areas and more irregular shapes will afford higher paste viscosities. As the filler loading is increased, the plastisol viscosity is increased along with thixotropy, and the plastisol flow becomes more dilatant. Aluminum trihydrate is used as a flame retardant and as a smoke suppressant.

Fillers in general are added to impart weight in carpet tile formulations and to improve properties such as insulation resistance, UV resistance, scuff resistance, heat deformation, and gloss reduction. However, fillers also have negative effects on tensiles, elongation, cut-through resistance, moisture resistance, and electrical insulation; they may also introduce unwanted color into the product because of metal impurities present in the filler.

Pigment

Pigments are incorporated into the vinyl plastisol, primarily to impart a given color to the final product. Color is very important to the end-user, and as such becomes a major consideration for the plastisol formulator, whose concerns must go far beyond simply providing the desired shade and intensity of color to the customer. The formulator also must provide the right degree of opacity, a color that will not change, a product that will not degrade no matter what the end-use environment, a safe product, a high-value-versus-price product, and a product that can be safely and efficiently processed in the manufacturer's plant. Materials that can be used to impart color to a plastisol-based item can be divided into several classifications. The first division includes those materials that are soluble in the PVC polymer and enter into the polymer matrix just as the plasticizer does during the fusion process; these can be broadly defined as dyes. Those materials that are insoluble in the PVC and the plasticizer, and thus maintain their particle structure in the fused part, generally are referred to as pigments. Pigments can be either organic in nature or inorganic but must be insoluble in the plastisol system. Included in the inorganic family of pigments are lead, cadmium, chrome, and titanium salts and oxides, titanium dioxide being by far the most used of any colorant material. Carbon black is perhaps the most common organic pigment, with azo- and phthalocyanines being other well-known families of compounds. Some special-interest products include those that produce metallic, pearlescent, and fluorescent effects in the vinyl.[20]

Of great importance to the final customer is the life of the product—how long the polymer system will maintain its integrity, and how long the product's color will remain unchanged. Of special concern here will be the effects of heat during processing, whether the pigment itself changes on exposure to heat, and whether the PVC changes. Products intended for outside use must be resistant to sunlight and moisture.

The formulator must be aware of the pigment's effects on the polymer or other ingredients in the system, whether positive or negative. For example, titanium dioxide is more often included in the formulation because of its ability to reflect UV light, which would rapidly degrade the PVC, rather than for the color its produces. On the other hand, pigments containing iron could cause the PVC to degrade. If the pigment were susceptible to attack by chemicals in its end-use or processing environment, such as alkali, or by contaminants in the air, then degradation of either the pigment, the polymer, or both could occur.

Migration of the pigment or dye to the surface of the vinyl causes numerous problems. If the migration occurs during the processing of the plastisol, the pigment can plate out or deposit on the surface of the mold or other processing equipment. Slow migration of the dry pigment to the surface, over time, is referred to as chalking or crocking, or by various other designations. The migration may result from solution of the pigment in the polymer or in another ingredient such as the plasticizer or the stabilizer carrier. The combination of ingredients present, the processing or end-use temperatures, and contact with other materials all can influence the migration of the pigment or dye.

The particle size, size distribution, and chemical nature of various pigments can cause processing problems. Because of the importance of achieving a uniform distribution of small pigment particles, getting that distribution can be one of the first problems encountered. The primary particles of most pigments are in the range of 1 micron, but it may be necessary to mill the pigment with a plasticizer in order to break down the agglomerated particles. Many pigment suppliers offer predispersed pigments in a paste form for the convenience of plastisol processors. Often processors who make numerous color changes will make large batches of nonpigmented plastisol and add a predispersed pigment paste to a small batch of the plastisol with the minimal mixing needed to disperse the pigment. Also, changes could occur in the rheology of the plastisol due to the incorporation of certain pigments, which would present processing difficulties. Extremely small particle sizes or chemical interactions of the pigment could result in significant increases in plastisol viscosity.[21]

Health and safety concerns always must be considered in dealing with pigments or any other ingredient in the formulation. Examples of pigments currently under scrutiny are those containing lead and cadmium. Good housekeeping and tight process controls always must be practiced, no matter what chemical is being handled.

Specialty Ingredients

Polymers, plasticizers, fillers, pigments, solvents, and diluents are considered main plastisol components. If the plastisol properties are to be modified, these ingredients should be the first ones altered; but at times changing one of these ingredients could have severe adverse effects on other properties. There also are many proprietary and generic chemical materials that are helpful in modifying the viscosity and the processing of plastisols and organosols when used at low concentration (1–2 phr), without significantly sacrificing other properties. These chemicals are classified as viscosity depressants and thickeners.

The chemical agents used to lower viscosity are called viscosity depressants. They are liquid surface-active agents that act by reducing the interparticle structure in the plastisol dispersion. They also are very effective when the plastisol is highly filled, by aiding the wetting of the filler surface. Most viscosity depressants are non-ionic or anionic type surface-active agents. A wide range of chemical compounds, such as ethoxylated alcohols, fatty acid esters, aliphatic alcohols, and lecithin-based derivities, have been found helpful in reducing plastisol viscosity. The advantages of viscosity reduction via modifiers must be weighed against such disadvantages as a possible loss in film clarity, detrimental effect on heat stability, reduction in electrical properties, blooming, and exudation. Generally, low levels (< 2 phr) will have minimal side effects. There are many plastisol applications, such as automotive sealants, dipping of a plastisol coating on a nonwoven open-weave substrate, or coating a thick film of plastisol on various substrates without penetration, where a high plastisol viscosity and a high degree of thixotropic properties are desired. If the above properties are not achieved by judicious selection of the resin and the plasticizers, small levels of thickening agents may be incorporated in a plastisol. Fumed and precipitated silicas, organo-metallic complexes (Ircogels), bentonite clay, special clays, and ultrafine calcium carbonates have found utility as thickeners in plastisol applications.

For fine-particle-size calcium carbonate, the viscosity increase is directly dependent on the oil-absorbing characteristics and surface areas. Fumed and precipitated silicas are very helpful in developing nondrip coating with high viscosity plastisols at very low levels (< 5 phr) of incorporation. The utility of Ircogel thickeners in various plastisol applications has been reported by the Lubrizol Company.[22] Ircogels do not impact the gross thickening achieved by silicas or ultrafine calcium carbonate filler. However, they provide sag-resistant plastisols with high degree of thixotropy and can be formulated in clear plastisols.

Liquid nitrile[23] rubbers have been used in plastisol applications such as roofing, glove dipping, conveyor belts, and so on, to reduce plasticizer migration and improve low-temperature impact, cut-through resistance, and abrasion resistance.

Plastisol Preparation

During the preparation of a plastisol, the primary goal of mixing is to break up the loose agglomerates present in the solid ingredients and to disperse them homogeneously throughout the liquid medium. To achieve a good dispersion, the solids must be subjected to shearing action in contact with the liquid medium. It is important that the temperature of the plastisol during mixing be maintained below 90°F; otherwise, it can result in too rapid solvation of the resin particles, poor viscosity control, and premature fusion of the resin particles.

Various types of mixers are available for plastisol and organosol preparation. Typically they are classified according to the intensity of mixing:

- Low to medium intensity
- High intensity
- Combination mixers

The Littleford mixer, universal mixer, conical screw mixer, paste and dough mixer, and double planetary mixer (Figs. 16-10 through 16-14) are typical examples of low- to medium-intensity mixers. These mixers are not as efficient as high-intensity mixers and are not recommended for plastisols with viscosities lower than 100 to 150 poises. The low-intensity mixers have an advantage in mixing plastisols with high viscosity (200–10,000 poises). Because of the low shearing action, very little

Fig. 16-10. Mixer equipped with plow-type blade and high-speed choppers that cause the dispersing, mixing action. (*Courtesy Littleford Brothers, Inc.*)

Fig. 16-11. Universal mixer in discharge position. (*Courtesy Baker-Perkins, Inc.*)

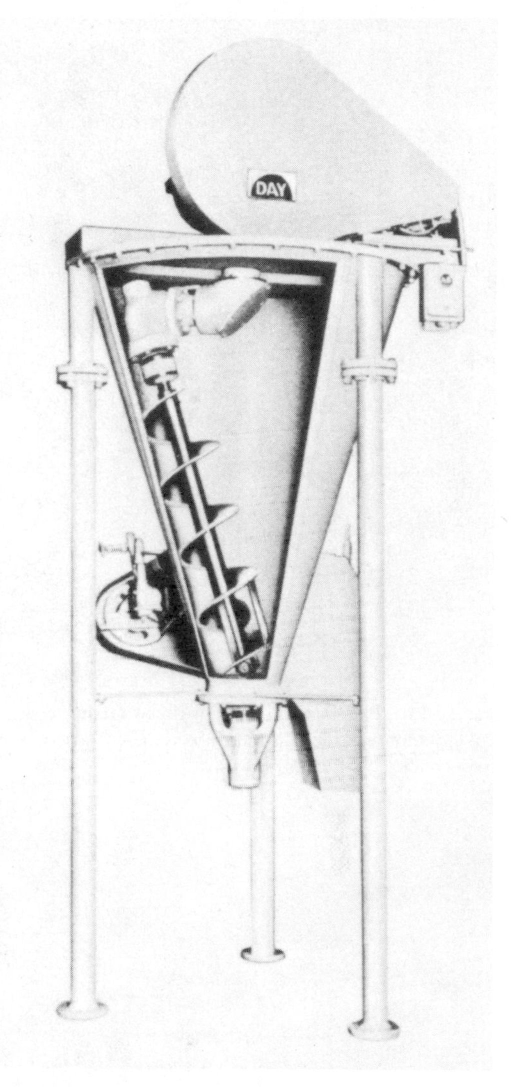

Fig. 16-12. Schematic of the conical screw mixer. (*Courtesy J. H. Day Co.*)

frictional heat is developed, and the plastisol has little tendency to trap air. Most of these mixers can be equipped with a cooling jacket and pulling vacuum if deaeration is required.

Figure 16-15 shows a high-intensity mixer. High-intensity mixers are very efficient, with mixing cycles that are much shorter than those of low- to medium-intensity mixers. These mixers are well suited to mixing plastisols and organosols with extremely low viscosities (< 150 poises). The heat buildup and aeration tendency are very high, so that a cooling jacket and deaeration capability are essential.

Figures 16-16 and 16-17 show two examples of combination mixers, the most versatile of all the plastisol mixers. These mixers combine the best attributes of low- and high-intensity mixing. They have two or more shafts that can be operated independently. In this way, during the initial mixing stage only the low-intensity blade is used to wet all the ingredients, and then high-intensity mixing is carried out for a short time to achieve the desired dispersion. A combination mixer will allow one to mix plastisols with viscosities ranging from 10 to 30,000 poises. Cooling and deaeration capabilities are recommended for combination mixers.

For most organosol preparations high-speed dispersers or combination mixers are sufficient. However, some inorganic pigments are very difficult to grind, and they may be predispersed on a three-roll mill. The plastisol or organosol can be further ground, if necessary, by passing it through a horizontal or a vertical sandmill (Fig. 16-18).

The plastisol or organosol generally will contain air unless the mixing is performed in a vacuum chamber. Air entrapment can be minimized by proper selection of plastisol ingredients, especially the right dispersion resin.

The use of a blending resin affords lower viscosity and facilitates deaeration. High-solvating plasticizers tend to increase viscosity and hinder air release, whereas viscosity depres-

Fig. 16-13. Planetary mixer with timed mixing control provides thorough blending and mixing. (*Courtesy Hobart Corporation, Troy, Ohio*)

Fig. 16-14. Planetary mixer for low-speed, high-shear mixing. (*Courtesy Charles Ross and Son Co.*)

Fig. 16-15. Typical high-speed, high-shear mixer used for organosols. (*Courtesy Morehouse Industries, Inc.*)

Fig. 16-16. Dual shaft mixer. (*Courtesy Morehouse Industries, Inc.*)

Fig. 16-18. Vertical sandmill. (*Courtesy Morehouse Industries, Inc.*)

Fig. 16-17. Dual shaft mixer, detail. (*Courtesy Charles Ross & Son Co.*)

Fig. 16-19. Typical vacuum deaerator system for batch deaeration. (*Courtesy Teknika, Inc.*)

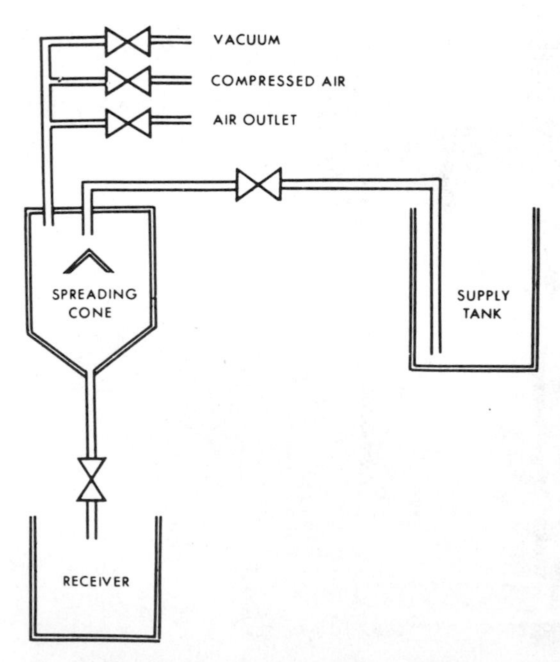

Fig. 16-20. Schematic of a deaeration system for continuous or batch-type operation. (*All photographs and illustrations in this section, unless otherwise indicated, are courtesy of The BFGoodrich Co.*)

sants and wetting agents reduce surface tension and facilitate air release. If a vacuum cannot be pulled in the mixing chamber, vacuum is applied to the plastisol in a separate vessel. Two equipment approaches are shown in Figs. 16-19 and 16-20.

Most plastisols and organosols prepared in the plant should be filtered through a 60 to 100 mesh metal screen or filter bags or socks to remove unwetted solid ingredients, dried-up solid particles, metal fragments, and other undesirable contaminants.

Property Determination

Many tests have been devised by raw material suppliers, users, and industry trade groups for characterizing vinyl dispersions. Tests are available for the properties of the liquid dispersions, for the transitional properties as the dispersions are heated, and for the physical properties of the finished, fused stock. Many tests are run to ensure the quality of the raw materials used in the dispersions and the uniformity of mixing the dispersions themselves. Table 16-3 lists some standard dispersion test procedures published by the Society of the Plastics Industry, which along with routinely updated ASTM procedures (also listed) offer methods of standardization to the industry.

The complex flow behavior, the sensitivity to the mixing technique, and the importance of the temperature history during mixing and storage all complicate the problem of measuring dispersion flow properties. If the viscosity tests listed in the table are to be used for the characterization of raw materials, it is important that the standardized mixing and storage methods included in these tests be adhered to. Because the viscosity of plastisols and organosols varies, depending on the shear rate, instruments are needed that will give an indication of the total behavior of the dispersion. There are several versatile but expensive wide-range cone and plate or capillary-type instruments that will measure viscosity across several decades of shear rate. In practice, the Brookfield viscometer for low-shear-rate measurement (SPI-VD-T1)-1959 and the Severs Extrusion Rheometer for high-shear measurement (SPI-VD-T2)-1959

Table 16-3. Standard dispersion tests.

	SPI Vinyl Dispersion Division tests
SPI-VD-T1 (1959)	Procedure for measuring the viscosity of a vinyl dispersion at low shear rates by the Brookfield viscometer.
SPI-VD-T2 (1959)	Procedure for measuring the viscosity of a vinyl dispersion at high shear rates by the Severs Extrusion Rheometer.
SPI-VD-T3 (1959)	Procedure for measuring the weight per gallon and specific gravity of a vinyl dispersion.
SPI-VD-T4 (1959)	Procedure for measuring Shore hardness of a vinyl dispersion.
SPI-VD-T5 (1959)	Tentative procedure for determining the heat-resistance of fused plastisol.
SPI-VD-T6 (1959)	Tentative procedure for the determination of the deaeration of vinyl dispersions.
SPI-VD-T7 (1959)	Tensile properties of thin plastic sheets and films.
SPI-VD-T8 (1959)	Volatile loss from plastic materials.
SPI-VD-T9 (1959)	Tear resistance of plastic film and sheeting.
SPI-VD-T10 (1962)	Procedure for the determination of the degree of dispersion of vinyl dispersions using the "precision vinyl dispersion gauge."
SPI-VD-T11 (1962)	Flammability of plastic foams and sheeting.
SPI-VD-T12 (1962)	Tentative procedure for measuring water extraction of plasticized sheeting-activated carbon technique.
SPI-VD-T13 (1962)	Tentative procedure for measuring loss by extraction of plasticized vinyl film, when immersed in mineral oil at 50.0°C.
SPI-VD-T14 (1962)	Statement on low-temperature flexibility methods.
SPI-VD-T15 (1962)	Method for the determination of the gelation characteristics of a plastisol using a Brookfield viscometer.
SPI-VD-T16 (1962)	Tentative migration test.
SPI-VD-T17 (1964)	Tentative procedure for the preparation of fused samples of vinyl dispersions for physical and chemical testing.
SPI-VD-T18 (1965)	Hot bench gelation temperature test.
SPI-VD-T19 (1968)	Procedure to determine a plastisol's apparent viscosity utilizing a torque rheometer (Brabender Plasticorder).
SPI-VD-T20 (1972)	Procedure for determination of air release.
SPI-VD-T21 (1972)	Procedure for determining nonvolatile content of vinyl dispersions–thin film technique.
SPI-VD-T22 (1972)	Procedure for measuring the force necessary to compress a fused plastisol 10% on the RPC test equipment.
SPI-VD-T23 (1975)	Procedure to determine a plastisol's or organosol's apparent gelation and fusion (hot melt) viscosity characteristics utilizing a torque rheometer with a programmed (increasing) mixer temperature and oil-heated mixer head.
	SPI Cellular Vinyl Division tests
SPI-CV-1 (1971-1)	Same as SPI-VD-T1 (1959)
SPI-CV-2 (1971-1)	Same as SPI-VD-T2 (1959)
SPI-CV-3 (1971-1)	Same as SPI-VD-T10 (1962)
SPI-CV-4 (1971-1)	Same as SPI-VD-T15 (1962)
SPI-CV-5 (1971-1)	Same as SPI-VD-T18 (1965)
SPI-CV-6 (1972-1)	Proposed test method for determining settling in plastisols.
SPI-CV-7 (1971-r)	Proposed test method for plastisol frothability on an Oakes Foamer.
SPI-CV-8 (1971-I)	Interim procedure for measuring froth density of mechanically frothed vinyl plastisol foams.
SPI-CV-9 (1971-I)	Interim procedure for measuring compression set of vinyl foam (Constant Deflection Method).
SPI-CV-10 (1971-I)	Interim procedure for measuring compression resistance of vinyl foam.
SPI-CV-11 (1971-I)	Interim procedure for determining density of vinyl foam.
SPI-CV-12 (1971-I)	Same as SPI-VD-T3 (1959).
SPI-CV-13 (1971-I)	Interim test method for resistance to exudation in vinyl foam.
SPI-CV-14 (1971-I)	Proposed test method for determining fusion of a vinyl foam.
SPI-CV-15 (1971-I)	Proposed procedure for measuring low-temperature flexibility of vinyl foams.
SPI-CV-16 (1971-I)	Interim test method for determining ash content of a vinyl foam.
SPI-CV-18 (1971-I)	Interim test method for measuring flammability of a vinyl foam by the Methanamine Pill Test.

Table 16-3. (*Continued*)

	SPI Vinyl Dispersion Division tests
SPI-CV-19 (1972)	Intermim test method for migration of ingredients in supported and unsupported vinyl foams.
	ASTM tests
D 1823-66	Test for apparent viscosity of plastisols and organosols at high shear rates by Castor-Severs Rheometer.
D 1824-66	Test for apparent viscosity of plastisols and organosols at low shear rate by Brookfield Viscometer.
D-1243-66	Test for dilute solution viscosity of vinyl chloride polymers.

represent a reasonable compromise in shear-rate range, cost, and ease of operation.

In the Brookfield viscometer, the principle involved is to rotate a suspended disc or cylinder at a predetermined rate in the liquid and measure the torque required. Depending on the model, the rotational speed can vary from as little as 0.5 to as much as 100 rpm. In any case, the shear rate is low, relative to high-speed strand coating or spread coating, and other instrumentation must be found for characterization at high shear. A single-point determination (at one speed) with the Brookfield gives an apparent viscosity that is useful in comparisons of low-shear viscosity. Methods have been devised for determining the relative thixotropy and the yield value of dispersion systems. These methods require changes in the spindle rotational speed and in the time that the spindle is allowed to rotate before torque is read.

The Severs Extrusion Rheometer is a simple, effective instrument for measuring high-shear characteristics of dispersions. The instrument consists of a precision capillary attached to a reservoir in which the dispersion can be subjected to pressures up to 100 psi. Fluids are assumed to pass through the capillary in laminar flow and thus obey Poiselle's law. The shear stress in this instrument can be controlled by varying the applied pressure, and the shear rate is directly proportional to the volume of the effluent liquid. As several capillary orifices are provided, ranging from a nominal $\frac{1}{16}$ inch to $\frac{1}{8}$ inch in diameter, a wide range of efflux and thus shear rates can be attained.

Viscosity measurement at any shear stress gives a single apparent viscosity reading. If two materials of different viscosity are to be compared at the same shear rate, it is necessary to make a series of measurements at various shear stress values and interpolate.

In many applications, it is important that a plastisol be readily able to release air trapped during pouring or pumping. Once air bubbles have coalesced and risen, for example, to the top of a dip tank, the tendency for them to break is a function of viscosity in the bubble film and surface tension of the plastisol. As these properties are quite difficult to measure, a number of more empirical-type methods have been developed to give an indication of the plastisol's ability to release air. One of these methods is an SPI test cited in Table 16-3 (SPI-VD-T20-1972). This method measures the time required for a plastisol film to break when it forms across various-sized holes in a steel cylinder. Another test simply stirs air into the plastisol and measures the time required for the bubbles all to rise to the surface and break.

SPI-VD-T20 (1972) provides a method for measuring the degree to which the agglomerates of PVC resin have been broken down and dispersed in the plastisol or organosol. A sample of the liquid is placed in the deep end of a channel that has been machined in a steel block. The channel depth tapers from 0 to 250 microns such that when the liquid is spread across the length of the channel with a drawing bar, the large particles will protrude above the surface at depths approximately equal to their diameter. Thus the size and number of agglomerates are determined by viewing the surface of the liquid in the channels. Similar tests are used where the maximum channel is other than 250 microns—for example, Hegman and North fineness tests, where the depth is 100 microns.

The rate of gelation of a plastisol depends, as does the fusion rate, on such things as particle size, resin composition, emulsifier coverage, and plasticizer solvating effectiveness. Gelation and fusion are not directly correlatable, however, and means of testing gelation independent of fusion are needed.

Three basically different methods of testing are used, all based on the principle that the plastisol undergoes an extremely rapid rate of change of viscosity with temperature just prior to the point where it solidifies to a dry but very weak gel. The methods differ mainly in that they shear the plastisol at varying rates during heatup. In the hot bench gelation test (SPI-VD-T18-1965), a 10-mil layer of plastisol is applied to a thermal gradient plate and allowed to heat for a specific period of time (usually 50 seconds). A strip of foil lightly applied along the gradient will adhere to the still-wet plastisol. When the foil is removed, the temperature at the line of demarcation between the wet and the dry material can be measured. In this test, the plastisol is in a completely static, unsheared state during the heatup.

The method for determination of gel rate using the Brookfield viscometer (SPI-VD-T15-1962) involves continuous measurement of plastisol viscosity while the material heats up in an oil bath. This apparatus (Fig. 16-21) provides a complete curve of viscosity versus temperature and is especially useful in characterizing materials that are in motion during fusion, but it is incapable of measuring consistency at, and just below, the point where solidification occurs. The highest-shear method of measuring gelation (and fusion) is the Brabender Plasticorder (SPI-VD-T19-1968 and SPI-VD-T23-1975). This instrument shear-heats plastisol in a mixing head and records torque versus temperature. It can cover the torques involved in mixing liquid plastisol through mastication of gelled material to mixing of molten stock. The shear rate is high but indeterminate; so tests with this instrument must be considered to be characterization tests.

In theory, fusion in vinyl material is defined as the minimum temperature at which the most perfect crystallite structures melt, and the matrix becomes amorphous. Practically, it is the

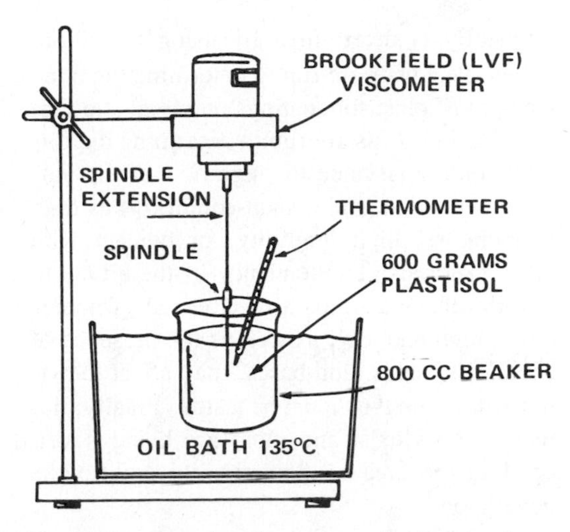

Fig. 16-21. Typical equipment setup for viscosity–time–temperature testing.

temperature required to give optimum physical properties. With experimental compounds the fusion temperature is easily determined by casting plastisol on a convenient surface, casting paper, or metal sheet, and heating successive samples at increasing temperatures. The measurement of physical properties (tensile strength, tear strength, etc.) yields a curve of strength versus fusion temperature, and the temperature at the maximum point can be easily determined. Tests of tensile strength require regular thick film or sheet samples. When a test is needed to determine the degree of fusion of a finished molded part, the processor often turns to one of the solvent-resistance tests. When a vinyl piece is underfused, and then stressed while in contact with a mild solvent such as acetone or ethyl acetate, it will break apart rather than simply swell, as would happen with well-fused material. The exact effect is a function of the solvent, and the common practice is to compare parts off the production line with parts that have been reheated in an oven to assure that they are fused.

Other physical tests performed on vinyl products include gloss and clarity, which simply are, respectively, the amount of light reflected from a surface and the amount of light transmitted through a film. The density of fused products is measured, and is especially important for foamed items. The hardness of the part frequently is determined by using the Shore scales. Peel tests are run to determine the bond strength of plastisol coatings on metal and fabrics. Various tests are run to determine the formulation's resistance to migration of the various ingredients under such conditions as high temperature, high humidity, or contact with other materials. The tendency of the article to absorb other materials also is tested. Temperature, high and low, as well as the presence of chemicals, acids, and bases, may affect the vinyl article and thus must be tested. Finally, numerous tests exist and more are being developed in the area of flammability and smoke generation.

SPREAD COATING PROCESSES AND APPLICATIONS

Plastisols and organosols can be applied by spread coating, molding, and other specialized processes. Spread coating processes include knife, roll, and curtain coatings and saturation. Molding is done by dipping, rotocasting, slushing, and cavity or in-place molding. Specialized processes include strand and spray coating and extrusion. All these techniques take advantage of the ease of convertibility from a liquid to a thermoplastic solid and the ability to formulate without volatiles. This section discusses spread coating applications, with molding the subject of the following section and specialized processes covered in a final section.

Spread Coating

Of the application methods, spread coating is the most important in terms of compound consumption. This process is widely used to make products as diverse as roll goods flooring, apparel fabric, and automotive padding. The relative processing ease makes plastisols quite competitive with calendered vinyl for short-to-medium runs of materials, for thin films, which can be made strain-free, and for chemically blown foams. Plastisols can be 100% solids with no solvent removal needed. This means that thick coatings can be applied in one pass, and that the cost of solvents and of the heat for the solvent vaporization is eliminated. On the other hand, plastisols depend on the plasticizer vehicle for their fluidity. Proper formulation and careful selection of materials are most important for successful coating.

Knife Coating

Fabric, paper, and metal substrates are coated by a process that, in its simplest form, consists of let-off and take-up equipment, a coating head, a fusion zone, an optional embosser, and a cooling zone (Fig. 16-22). There are at least 20 generic types of applicators used in applying coatings to webs.[24] Knife coaters, as in Fig. 16-22, and reverse roll coaters are the most important spreading mechanisms for applying dispersions. These coaters are more capable of handling the relatively viscous and somewhat viscoelastic dispersions than are, for example, wire-wound rod coaters, direct-roll coaters, and air knives.

Knife-coating heads are, in principle, the simplest applicators for applying dispersion coatings to flat webs. Figures 16-23, 16-24, and 16-25 illustrate the three most common types, which are differentiated by the method used for supporting the web. All depend on the clearance between a coating blade and the web for control of the coating thickness.

Knife-over-roll coaters (Figs. 16-23 and 16-26) apply from 3 to 125 mils of dispersion compound at the fixed gap between the knife and the backing roll. Elastomeric backing rolls usually are used in coaters intended for the heavier coating range, whereas more precise control of thin coatings is afforded by accurately ground, steel backing rolls. Coating speeds on knife machines range up to 100 ft/min. with rates of 60 ft/min. common for coatings around 10 mils. Because there is a fixed gap between the blade and the roll, any variations in the substrate thickness show up as variations in the coating gauge, and the result is a uniform-thickness composite that may vary in coating thickness. The knife angle has little effect on thickness, but the angle may be varied to change the quality of the coating.

Floating knife coaters (Fig. 16-24) carry the web, unbacked, between support channels or rolls. The coating weight is determined by the

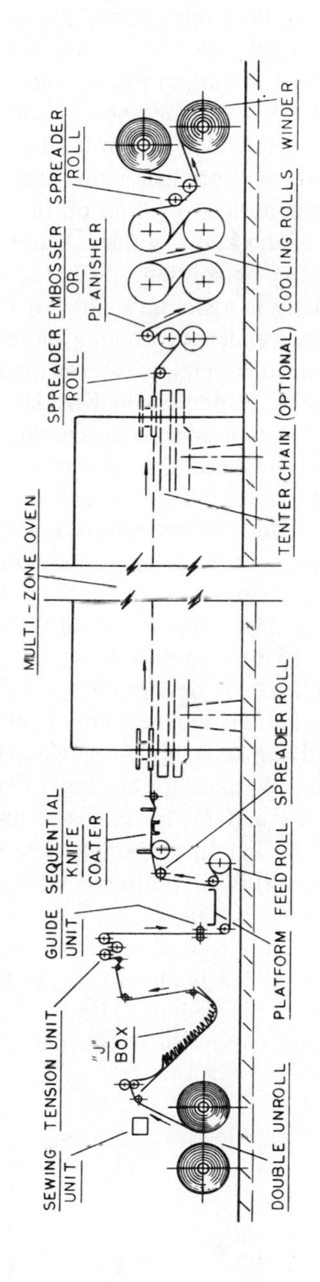

Fig. 16-22. A knife coating line. (*Courtesy Machinery Div., Midland Ross Corp.*)

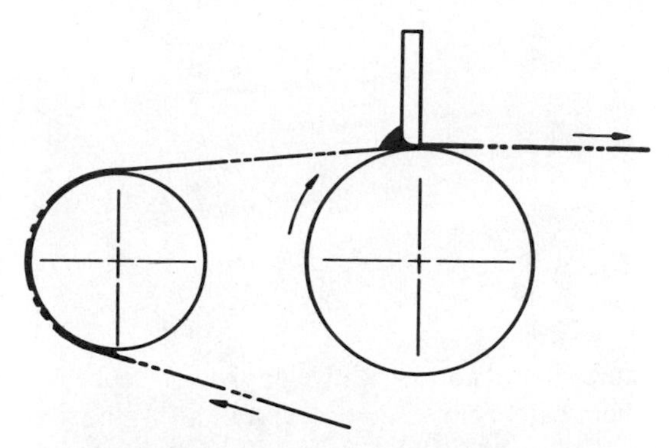

Fig. 16-23. Knife-over-roll coating. (*Courtesy Charles Ross and Son Co.*)

knife configuration, knife angle, and web tension (i.e., the pressure of the knife against the web). By controlling these variables, it is possible to apply uniform coatings to nonuniform substrates. Thick coatings are difficult to apply with floating knife coaters. Optimum results are obtained in the range of about 1 to 3 mils. In the coater shown in Fig. 16-22, a knife-over-roll and a floating knife coater are used in tandem. The knife-over-roll applies a heavy fill coat and levels the fabric, while the floating knife applies a light ''color coat.''

The knife-over-blanket coater (Fig. 16-25) represents a compromise between the knife-over-roll and the floating knife. In this coater, the web is supported on a driven rubber belt

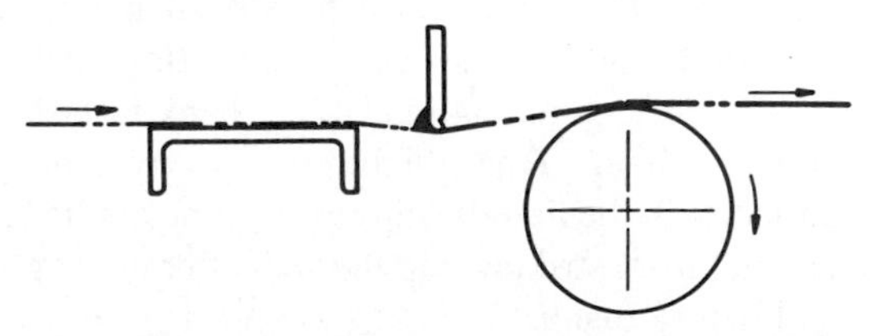

Fig. 16-24. Floating knife coating. (*Courtesy Charles Ross and Son Co.*)

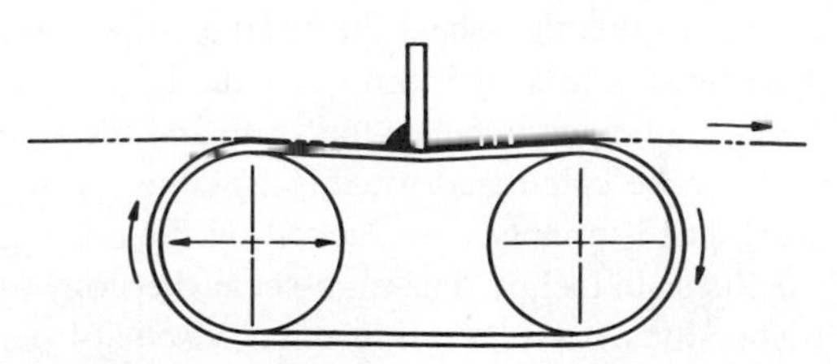

Fig. 16-25. Knife-on-blanket coating. (*Courtesy Charles Ross and Son Co.*)

Fig. 16-26. Sixty-inch knife/roll coater, web exit side. (*Courtesy Egan Machinery*)

rather than by its own tension, so that weak webs can be processed without tearing.

In formulating for knife coating, it is necessary to consider both the high and the low shear rheology of the compound.

Compounds with high yield value may be needed to prevent striking into the weave of open fabric before and immediately after the web has passed under the knife. True yield value can prevent the flow of material from the puddle behind the knife into the interstices of the web. Often, transverse backing dams are set behind the knife to retain the coating bank in a small area and minimize the bank-to-web contact time. True thixotropy (recoverable structure in the liquid) helps to keep the applied coating from striking into the web after coating and before fusion.

In many knife-coating processes, high rates of shear are induced into the dispersion as it passes under the knife. Thus, careful consideration of the high-shear rheology is important. The shear rate at the knife is directly proportional to the gap between the knife and the web and can be calculated roughly by dividing the web speed in inches per second by the coating thickness in inches. The high-shear rheology of dispersions has two parameters: viscosity per se at the coating knife and the extent to which the dispersion is shear-rate-thickening or shear-rate-thinning as it passes through the shear-rate range just prior to attaining the maximum at the coating gap. It is difficult to differentiate between the effects of these two parameters; so the usual goal of the compounder is to minimize both shear thickening and high-shear viscosity for optimum coatability. Compounds that are too shear-thickening or too viscous at high shear may actually act more like solids than liquids as they pass under the knife. They may actually shatter and produce skipped spots or droplets that deposit at random on the downstream side of the knife (termed ''spitting''). Unlike skipping or spitting, the streaking of coatings usually is a simple matter of foreign material lodged under the coating knife. This can be prevented by taking care in mixing to ensure complete dispersion and by screening the material before it goes to the coating head.

The proper choice of coating knife configuration can help prevent skipping and spitting in the high-shear range as well as striking through in the low-shear range. There are roughly five different knife shapes used in applying dispersions, and Fig. 16-27 shows four of them. The upstream or coating bank side of all of the knives in the figure is on the left.

The radius knife is the most commonly used knife for applying heavier coating weights and for smoothing coatings. Radius knives vary in blade thickness and in the radius of the upstream side. An extreme example of a radius knife (not shown) is the bull-nose knife with a 180° rounded surface in contact with the web. The bull-nose knife is used for smoothing very heavy coatings. The hook or ''j'' knife is a modification of the radius knife intended to prevent spitting from shear-thickening com-

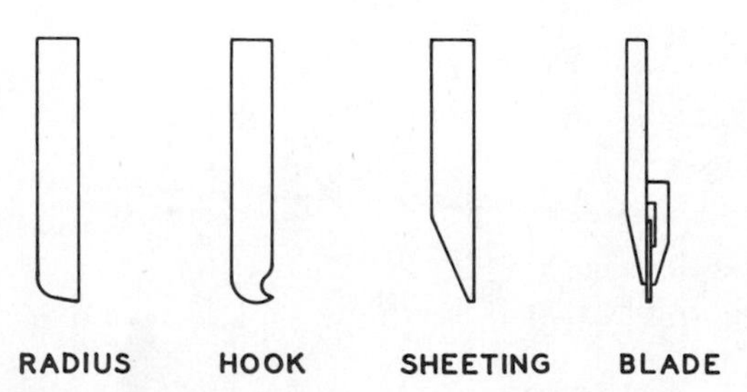

Fig. 16-27. Coating knife contours. (*Courtesy Charles Ross and Son Co.*)

pounds. In practice, it provides a pocket to collect globules of material that separate from the matrix, and it presents a very sharp trailing edge as the coating leaves the knife. The sharp trailing edge minimizes the forces that otherwise would tend to cause the departing coating to stick to the knife instead of adhering uniformly to the web.

For thin coatings and for coatings intended to fill surfaces without penetrating into the web, the sheeting or the blade knife is used. These knives scrape material off the web instead of pushing it into the web. The blade knife is most commonly used in Spanishing, a decorating technique that is described later. The knife angle and the gap usually are micrometrically adjustable. The knife can be rotated around an axis at the coating edge to vary the quality of the coating being deposited. This knife angle usually ranges from 85 to 90° relative to the upstream side of the web.

Roll Coating

As has been indicated, the coating weight is inherently difficult to control on knife coaters. Problems also can occur with strike-through and streaking because the coating is carried in excess on the web and is doctored off under a stationary knife. These problems can be minimized and faster coating rates can be obtained by applying the dispersion compound via a roll coater.

The simplest roll coating devices are the direct or forward roll coaters such as that illustrated in Fig. 16-28. In direct roll coating, liquid is picked up by or supplied to a roller that travels in the same direction as the web. The liquid film splits as it comes into contact with the web, and the resulting deposit must level before drying or fusion can take place.

Vinyl dispersions usually are high enough in viscosity and viscoelastic enough to make this splitting and leveling very difficult to control. As a result, the roll coating of dispersions usually is carried out by a reverse roll coater in which the complete deposit of coating on an applicator roll is wiped off on a web running in opposition to the roll.

Figure 16-29 illustrates a typical three-roll nip-fed reverse roll coater. Dispersion compound is metered through a precise nip between a slowly rotating metering roll and an applicator roll running in the opposite direction. The coating on the applicator roll then is wiped off onto a web traveling in the opposite direction. The applicator roll is run faster than the web travels, so that the entire amount of coating is wiped off, and no splits or kiss marks develop. The metering roll usually is rotated so that it constantly presents a fresh surface, and foreign matter cannot clog the nip and streak the coating. The amount of coating deposited by a reverse roll coater is a function of the amount carried by the applicator roll and the relative speed of the applicator and the web. The shear rate is controllable by varying these two parameters, and liquids of widely different viscosities can be adequately coated. It has been reported that viscosities ranging from water-thin to 100,000 cps can be handled by this type of ma-

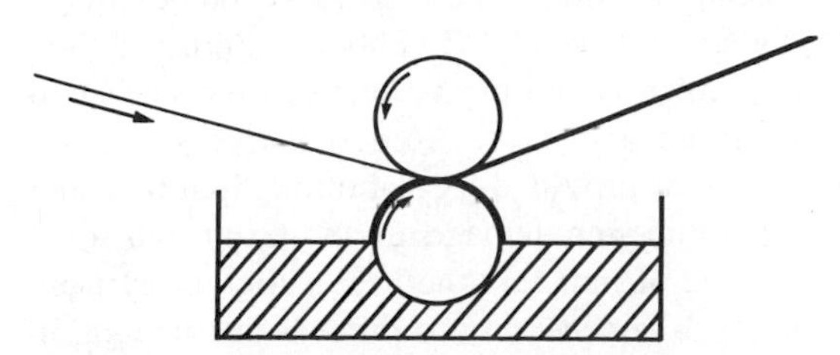

Fig. 16-28. Direct roll coater (useful only with very low viscosity dispersions). (*Courtesy The BFGoodrich Co.*)

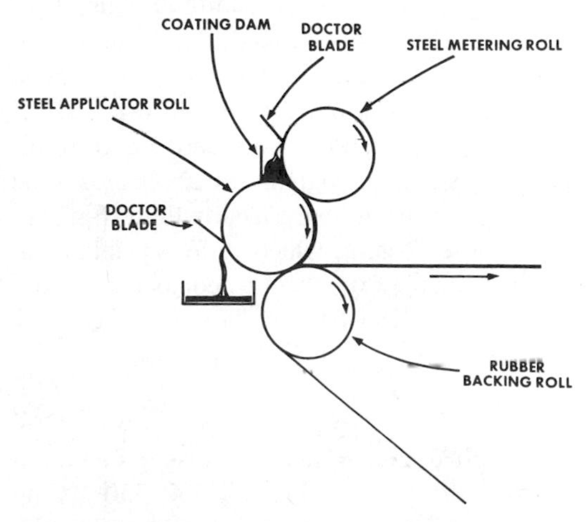

Fig. 16-29. Three-roll nip fed reverse roll coater. (*Courtesy The BFGoodrich Chemical Co.*)

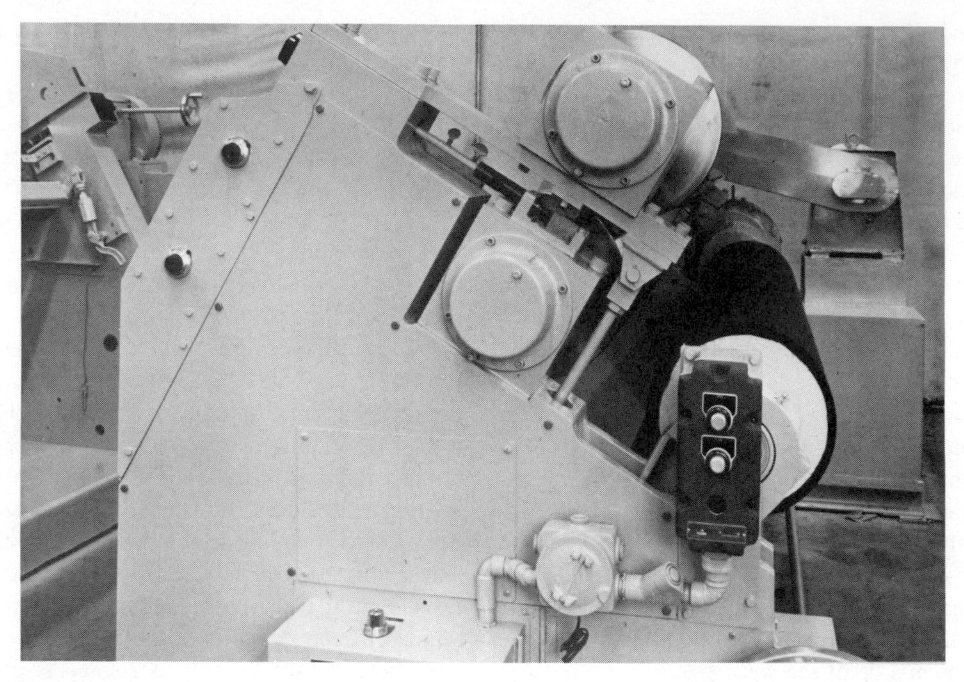

Fig. 16-30. Three-roll nip fed reverse roll coater. Detail shows precision adjustment for the metering roll. (*Courtesy Egan Machinery*)

chine and that, with highly fluid systems, coatings can be applied at rates up to 100 ft/min. Reverse roll coaters vary considerably in their design and cost, depending on the precision with which the coating must be applied.

Figure 16-30 shows a coater designed to deposit thin coatings precisely. In this coater, fine adjustments of nip spacing are made by flexing the beams that carry the metering roll bearings.

There are several other configurations of reverse roll coaters, designed, for example, to eliminate abrasion of the metering roll against a metal web, to provide a bottom reservoir of recirculatable coating compound, or to eliminate splashing of high-speed applicator rolls running in pans of low viscosity fluid (Fig. 16-31). These coaters, which differ in detail but not in principle from the nip-fed coater, are described fully elsewhere.[24–26]

Fusion

The applied coating must be fused. Temperatures of about 148 to 177°C (300–350°F) are required. Time at temperature is not a consideration, as fusion is a physical rather than a chemical phenomenon. However, time to reach temperature throughout the film is critical. The most commonly used oven type for web coating is the straight-pass tunnel oven (Fig. 16-32). These ovens vary in length from 15 to 250 feet, depending on the coating rate and fusion requirements. Frequently they are divided into two or three stages so that temperatures can be varied to drive off thinners prior to fusion. Because of the temperature requirements, steam heating is impractical, and oil-fired, gas-fired, electrical heating, or hot oil heat exchangers are most commonly used.

Although dispersions rarely need to reach more than 190°C (375°F) to fuse, the plenum temperatures in forced hot-air ovens used in spread coating lines usually range much higher—Up to 427°C (800°F). Some method of aiming the hot blast directly at the surface to be fused and away from sensitive substrates usually is provided. A fast throughput time and minimum web damage are the result, but some method of indexing the heat source away from the web is necessary to prevent overheating in case of line stoppage.

Sufficient exhaust must be provided for the

Fig. 16-31. Four-roll contracoater. (*Courtesy Black Clawson Co., Dilts Div.*)

Fig. 16-32. Plastisol coating line; tunnel type fusion oven at right. (*Courtesy Dawson Engineering*)

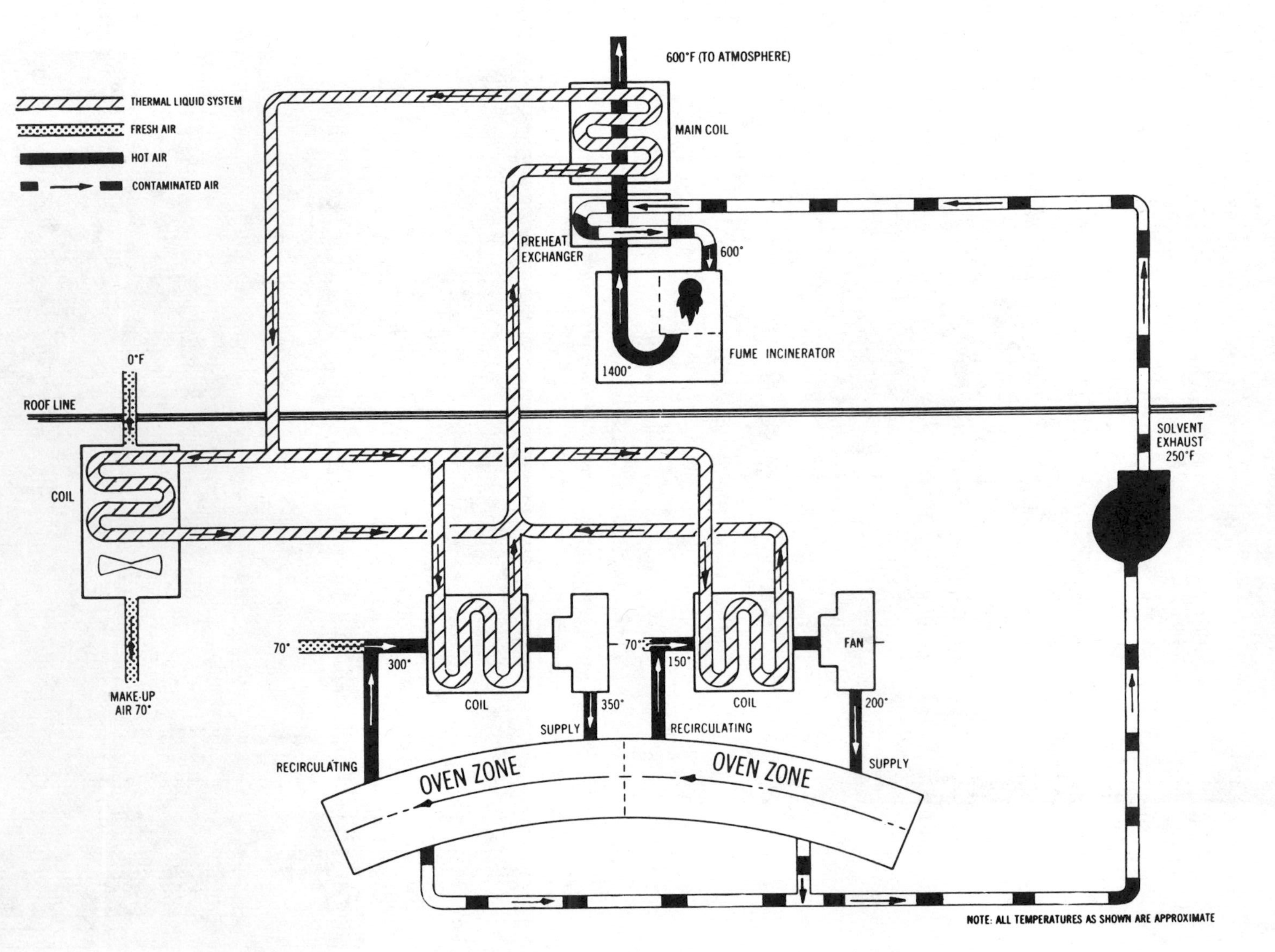

Fig. 16-33. Schematic of fume incinerator with heat reclaim package. (*Courtesy Charles Ross and Son Co.*)

removal of flammable organosol vapors. In the case of both plastisols and organosols, volatile plasticizer and thinner vapors either must be condensed and removed from the exiting gas, or they must be incinerated. Figure 16-33 is a schematic diagram of an installation for vapor incineration, with the heat of vapor combustion reclaimed and used to preheat the makeup and recirculating oven air.

Other fusion devices used in plastisol coating applications include heat platens, heat rolls, and radiant resistance rods or quartz lamp heating. Radiant heat units are especially useful in modifying existing equipment for greater heat input. These units usually operate at high temperatures and must be installed so that they can index away from the web in case of line stoppage.

Web Handling and Finishing

As a minimum, web handling equipment for plastisol coatings must include let-off and take-up equipment and accumulators for splicing rolls of material without stopping the line. "J" boxes of expandable dancer rolls (loopers) commonly perform this function. Cooling rolls, utilizing refrigerated water, bring the coating down to a temperature at which it will not block when rolled up.

Plastisol coating lines frequently provide for one or more finishing operations on the coated web. Embossing can be done in-line with an engraved steel roll applying pressure to the web as it passes over a rubber backup roll immediately at the oven exit. The fused plastisol must be chilled to set the emboss and allow it to release cleanly from the embossing roll. Alternatively, coated unembossed fabric can be reheated by passing it over heated platens and embossed in a separate operation (Fig. 16-34).

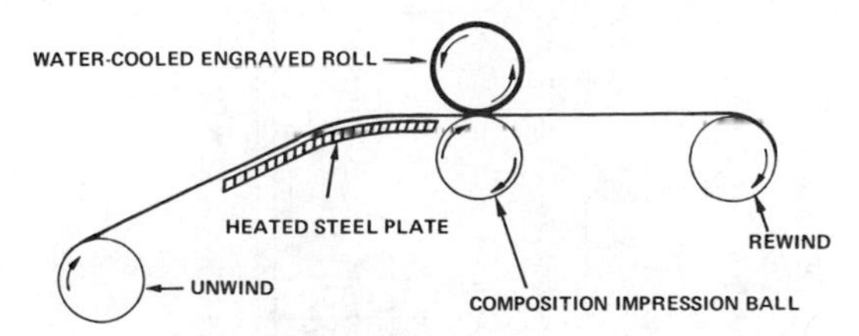

Fig. 16-34. Embosser for separate decoration of coated webs. (*Courtesy The BFGoodrich Co.*)

Separate finishing operations can provide economy by widening processing versatility and not tying the line speed to a slow operation.

Spanishing—application of a contrasting color in the emboss valley by means of a tight knife pass—and top coating usually are performed with solvent-based materials. Top coating protects exposed surfaces from abrasion, dirt, and stain pickup. Typically, top coatings are formulated with high molecular weight homopolymers to minimize plasticizer migration from the substrate. Acrylic solution polymers and cellulose acetate butyrate blended with the vinyl add to the blocking resistance and the dryness of the coating. Other typical finishing operations are printing and flocking. In flocking, copolymer or carboxy functional plastisol resin dispersions are used as the adhesive. Complete discussions of coating applicators and their auxiliary equipment appear in the literature.[24,25,27]

Fabric Coating

Complete lines can vary widely in complexity and cost, depending on the operation being performed and the substrate being coated. Perhaps the simplest operation is the coating of fabric for such products as shade cloth, wall covering, and fabrics for shoes and handbags. Base fabrics range from the lightweight cotton used in rainwear to the heavy duck used in awnings. A specially designed striping box is used in awning coating. Multicolored strips are laid down side by side simultaneously from separate slots in the applicator.

Vinyl dispersions are inherently adhesive to cotton and rayon; synthetics such as nylon and polyester fabrics require the use of adhesive primers. Nitrile latexes modified with borated casein and nitrile/phenolic solvent systems and carboxyl-modified vinyl dispersion resin systems have been found to be useful primers.[13] Carboxyl resin primer systems increase the adhesion of subsequent top coatings by as much as three or four times that obtained with vinyl chloride/vinyl acetate primers.

Most base fabrics are strong enough to be carried through the fusion oven with no support other than the oven rolls. Weak fabrics and nonheat-stable base materials are carried on tenters. One of the most useful fabrics for vinyl coating is knit cotton. Its stretch characteristics and hand make it ideal for upholstery, outerwear, and sundry applications such as handbags and boots. The open mesh and dimensional instability of knit backing make it necessary to use a radically different coating process for plastisol application. This process, commonly called transfer or cast coating, builds the coating in reverse (see Fig. 16-35). A casting substrate of thermoset-resin-treated paper, silicone-treated fabric, or metal is coated with the plastisol in multiple coats, starting with the eventual wear surface. Each coating is gelled before application of the next, and the final composite is heated just enough to leave a viscous, partially gelled surface. The base fabric is laid into this coating with light pressure, and the final assembly is fused. After cooling, the finished-coated knit fabric can be stripped from the carrier and rolled.

The type of carrier used is determined by the economics of the process and the finished product desired. Thermoset-resin-treated casting paper is the least expensive base, but it has a limited use life. This material cannot be expected to withstand more than about six cycles. The primary hazard is mechanical abuse. Casting papers are supplied in many finishes: high-gloss patent leather, matte, and a number of emboss patterns. For long runs of transfer coating, stainless steel belts, resin-treated fiberglass, or resin-treated cotton have the longest life expectancy. If emboss finishes are desired, they must be made separately with these casting surfaces. The high cost of stainless steel belting and the time consumed in chilling coated material to permit removal from the substrate are limiting factors.

Much of the coated knit fabric made for upholstery and apparel applications contains an intermediate layer of vinyl foam, about 20 mils thick, topped with a 6- to 8-mil wear surface. This construction gives a very supple leathery hand, which is generally quite superior in aesthetic properties to unexpanded coatings. Figure 16-35 illustrates a typical transfer (cast) coating line for foam fabric. This line has three coating stations. The wear surface is applied at Station No. 1, and this coating is then gelled in

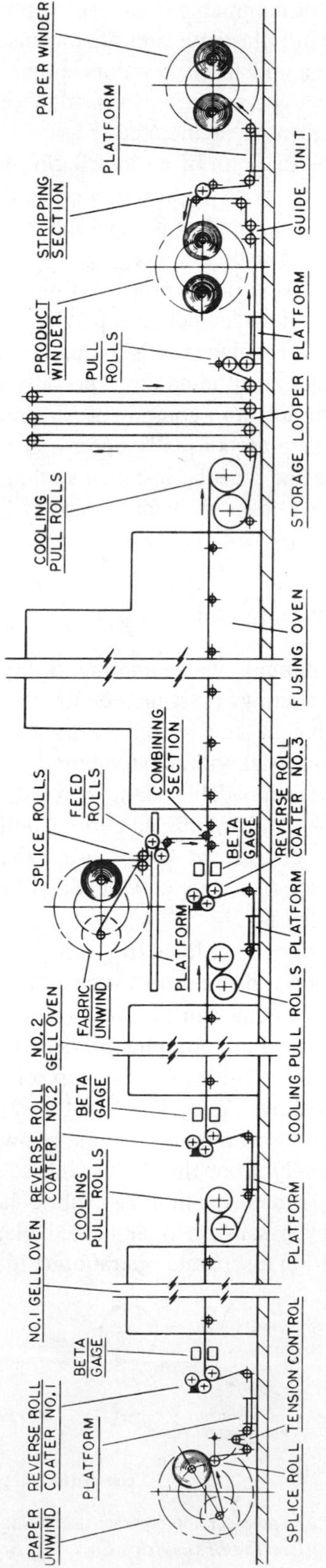

Fig. 16-35. Cast coating line for coating by transfer from paper carrier. *(Courtesy Charles Ross and Son Co.)*

Oven No. 1. After cooling, the foam innerlayer goes on at Station No. 2; it is gelled but not expanded on passing through Gell Oven No. 2. Because a third coating Station (No. 3) is available, the extent to which the foamable coat is gelled is not critical. A base coating of wet plastisol serves as the adhesive for the vinyl-to-knit back.

The latter comes off the roll just above and to the left of the fusing oven. The coated fabric—with its fused surface and expanded, fused inner layer—is separated from the casting paper in the stripping section, and product and paper are wound up. Formulating dispersions for cast coated foam/fabric requires coordinating the melt flow characteristics of the foam to the temperature at which the lowing agent decomposes to produce gas for expansion.

Another use of vinyl foam is as resilient backing for commercial carpet. Here a dispersion precoat of about $1\frac{1}{2}$ lb/sq yd is knife-applied to the back of carpet that is tufted through a polypropylene nonwoven fabric. A liquid froth of plastisol then is applied to this wet, or partially gelled, precoat at a rate of about 2 to $2\frac{1}{2}$ lb/sq yd, by using a stationary roller as the applicator. Foam coatings of this type must be fused at low temperatures, on the order of 135°C (275°F), to prevent heat shrinkage and melting of the polypropylene. Hence, materials for this application are formulated with copolymer resins and plasticizers of as high a solvating effectiveness as possible, compatible with the low viscosity necessary for foaming in mechanical foamers.

Screen printing and dot printing of plastisols onto fabrics produce a number of very interesting products. Soles for children's sleepwear and abrasion-resistant surfaces for mail sacks and cotton picking bags have been made by using coaters. A rubber squeegee knife forces the compound through perforations in a cylinder rotating in the web direction. The plastisol is formulated with yield-building agents, usually specialty fine-particle dispersion resins, so that the dots deposited will not slump before fusion.

Shapes other than simple round perforations have been used to produce decorative coatings for upholstery. Discontinuous coatings of this sort provide maximum abrasion protection for minimum coating weight and, in addition, pro-

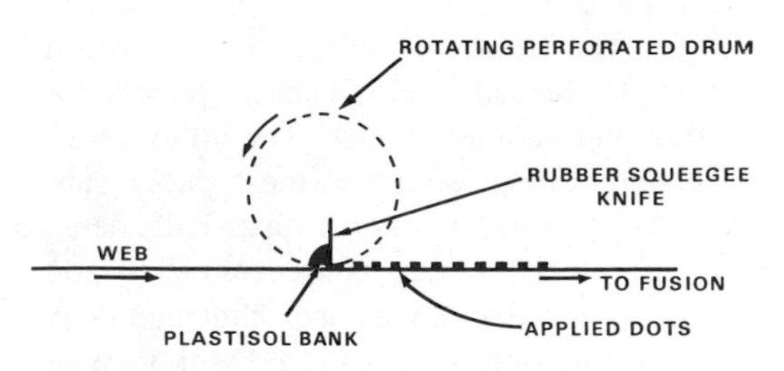

Fig. 16-36. Plastisol dot printer. (*Courtesy The BFGoodrich Co.*)

vide breathability and moisture vapor permeability, an added comfort factor.

Rotary wire screens replace the perforated cylinders of Fig. 16-36 when more intricate designs are needed. The screen can be blocked to produce a decorative pattern. Specialty floor and wall coverings have been made by screen-printing plastisol flock adhesive onto continuous vinyl-coated webs and then applying nylon or cotton flock via electrostatic or beater bar methods.

Some very soft fabric coatings are applied by a hot-melt coating technique. On fusion, vinyl dispersions pass from the fluidity of a two-phase dispersion through a weak cheesy gel to a homogeneous liquid hot melt. The viscosity of this melt depends on a number of factors, but notably the level of the plasticizer. Hence, with a high plasticizer content and the resulting very low dispersion viscosity, it is convenient to preheat and flux the dispersion and flow this melt from a hopper-type coater onto the substrate. Because the material must be held for a time at the melt temperature, excellent heat stabilization is a prime requisite in this type of application.

Coating Paper and Nonwoven Materials

Paper and nonwoven fabrics comprise another important family of casting substrates. In general, the paper adds dimensional stability and ease of application inexpensively. Products made from coated paper include packaging board, shelf lining papers, shoe liners, auto door panel finishes, and masking tape. For tapes, the plastisol coating acts as a release surface for the adhesive mass. The formulating requirements depend on the substrate, processing characteristics, and finished properties of the coating. The substrates vary, from box board to kraft to special latex saturated papers for flooring and automotive use. The adhesion of dispersion resin coatings to most paper substrates is good. In the case of super-calendered kraft, however, it may be necessary to add modified rosin derivatives and other tackifiers to ensure adhesion. Papers treated with latex or tapes utilizing rubber adhesive masses that are plasticizer-sensitive require special formulation of the coating. Polyester polymeric plasticizers are used here to prevent migration into and softening of the continuous nonvinyl polymer.

Nonwoven fabrics, scrims, and felts are used in a number of instances as the base for vinyl dispersion coatings. One application, coating carpet tufted through nonwoven polypropylene, has been mentioned. Another utilizes dots printed onto nonwoven polyester linings for apparel—as the melt adhesive system for attachment to the garment. Hot-iron pressing melts the dots of gelled plastisol adhesive, which then flow into the outer garment fabric to make a permanent bond. Latex systems are used for this same function when their somewhat higher cost is offset by the need for dry-cleaning resistance.

By far the most important coating operation involved with felts or nonwovens—in fact, the most important application of dispersion systems—is the manufacture of roll goods flooring. This application probably consumes from 20 to 25% of the total dispersion resin produced in the United States. Roll floorings have passed through about three evolutionary stages since vinyl supplanted oleoresinous wear surfaces in the early 1950s.[28] Such products are backed with either asphalt or latex-saturated cellulosic or mineral fiber felt. One of the earliest versions consisted of printed paper, coated with 4 mils of vinyl organosol and laminated to asphalted felt. In later constructions, asphalted felt was directly coated with barrier coats, fill coats, and print coats of latexes; after printing, a vinyl organosol or plastisol wear layer was applied and fused. Later, one of the most popular types of flooring consisted of an asbestos felt coated with about 20 mils of vinyl foam, which was printed and then top-coated with about 7 mils of hard, clear plastisol. Today paper felt or fiberglass mat are most commonly used as backings although all-vinyl flooring also is produced, where no fiber backing is used. A process of chemical embossing produces an embossed valley in perfect register with the printed pattern. One such technique[28] utilizes a chemical inhibitor that retards the gas evolution of the azodicarbonamide in the foamable layer during the final fusion stage.

The specific processes used in manufacturing resilient flooring are highly proprietary. Figure 16-37 is an idealized conception of a production line. Two reverse roll coating stations are provided to apply the foamable resilient inner layer and the wear coating. The decoration is printing onto the gelled, cooled foamable layer at two printing/drying stations using presses similar to the 154-inch three-roll press shown in Fig. 16-38 to make seamless flooring up to 12 feet wide.

The formulation of plastisol systems for the wear and resilient layers of this type of flooring requires a highly sophisticated technology. There is a high degree of interdependence between the foam compound, the embossing agent, and the wear layer compound. The requirements in the wear layer are stringent. The exact gloss needed must come from the compound itself, as post-planishing would destroy the emboss. Crystal clarity is needed for maximum visual impact of the printed pattern beneath the wear layer, and this clarity must be maintained despite contact with aqueous solutions of typical floor cleaning detergents. Maximum wear life requires that the top coating be high in hardness and abrasion-resistance, and that it resist indentation, marking, and staining from contact with shoe heels, furniture legs, and common household chemicals.

Thin-film coatings of polyurethane are used as surface layers to provide extra scuff resistance and gloss retention on some flooring products.

Film Casting

Many spread-coated products entail no permanently attached substrate. Decals and decorative roll goods films are manufactured by casting a vinyl dispersion on previously printed release paper, fusing, and then casting a solvent-based adhesive mass on the back of the

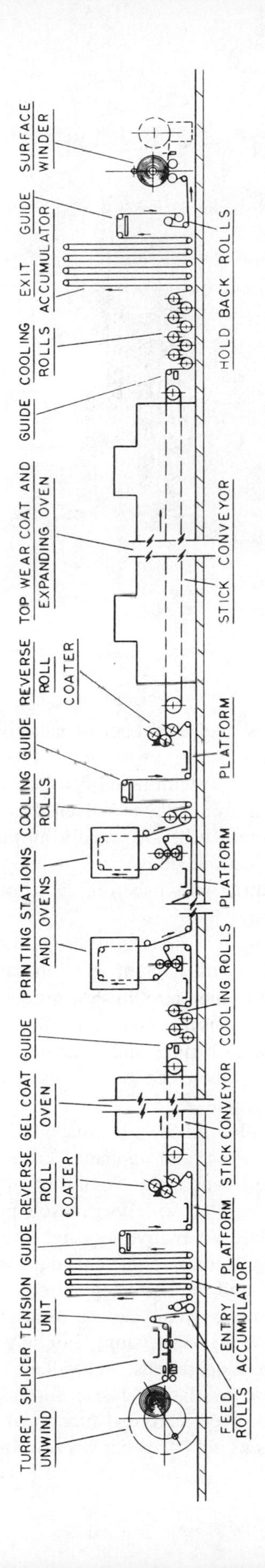

Fig. 16-37. Resilient vinyl floor covering line. *(Courtesy Charles Ross and Son Co.)*

dispersion. The release paper is designed so that the printing pattern transfers from the paper to the dispersion. When unrolled, the adhesive mass is temporarily attached to (originally) the bottom surface of the casting paper, and the product is right side up. Such contact adhesive films are convenient materials for household use and for automotive decorations (with more permanent adhesives) such as wood grain panels for station wagons.

Because dispersion systems can be cast on a substrate, fused, and cooled thoroughly before stripping, the likelihood of inducing strains by distorting warm or hot film is reduced to nearly zero. Thus, such film can withstand exposure under high ambient temperatures without experiencing shrinkage due to ''plastic memory.'' The production of strain-free, thin calendered, or extruded film is very difficult, and the use of more expensive dispersion systems is warranted.

The production of ''walk-off'' mats and runners is a related cast coating process. These mats contain carpet squares or continuous strips backed with and bordered by homogenous vinyl sheet about $\frac{3}{16}$ inch thick. The carpet serves as an excellent dirt pickup medium at entrances to buildings, and the heavy vinyl back prevents slipping and bunching up. In this process, plastisol is cast onto a continuous, treated fiber glass belt, the carpet is laid in, leaving the desired borders uncovered, and the composite is fused by a combination of platen heat from beneath and the surrounding oven heat. The fused, cooled matting is stripped and cut to shape. Because a moderate-size walk-off mat can contain as much as 6 pounds of vinyl compound, compounding requires a means of minimizing cost without jeopardizing floor surfaces from plasticizer migration or ruining the appearance of exposed borders. The heat needed for fusion usually is minimal, and fast-fusing copolymer resin systems are often used. (See Fig. 16-39.)

Spread Coating Metal Substrates

Except for the carboxyl-functional polymers, simple vinyl dispersion systems do not adhere to metal substrates. Proprietary dispersion systems have been developed with self-adhesive

Fig. 16-38. Three-roll press for printing wide webs of resilient flooring. (*Courtesy Egan Machinery*)

Fig. 16-39. Carpet tiles made with cast coated plastisol backing. (*Courtesy Collins and Aikman*)

characteristics, and a number of adhesive adjunct plasticizers are available.[29] The most common expedient with metal is to use one of the commercially available solvent-based adhesive primers. These are usually nitrile/phenolic/epoxy combinations and require flash baking or partial curing before application of the dispersion.

Coil coating is a process whereby metal is painted in flat form, coiled, and subsequently uncoiled and formed into finished articles. Despite the limitations that prepainting imposes on the subsequent shaping and fastening operations, it often is more economical than painting the finished article.[30]

Exterior siding finishes, downspouts, siding for mobile homes, and appliance covers represent markets where the flexibility of vinyl coatings is essential. Vinyl organosols provide a moderate cost alternative to vinyl or acrylic solutions. Some coil lines have tandem stations for priming and finishing operations,[31] and standard primer systems can be used in conjunction with vinyl dispersions. For single-station, one-pass operations, self-adherent formulations are available. These formulations combine dissolved carboxyl-functional solution vinyl resins with dispersion resin. Solvents

Fig. 16-40. Metals coil-coated with vinyl dispersions offer durable prepainted material for making appliance housings, metal furniture, and other formed metal products. (*Courtesy The BFGoodrich Co.*)

are selected to minimize swelling of the disperse phase. High gloss, one-mil, opaque, weatherable films result from coatings fused under fast coil-line cycles (e.g., 1 min. at 260°C [500°F]). (See Fig. 16-40.)

Miscellaneous Coating Applications

Vinyl dispersions are well suited for use as adhesives for vinyl calendered film to synthetic and natural fabric and for fabric-to-fabric laminations, especially when formulated with fast-solvating plasticizers. Thus, they will bite into and adhere to plasticized vinyl film at temperatures low enough to preclude damage to the film. Laminators using dispersion adhesives usually have a configuration such as that shown in Fig. 16-41. Here, one web of the two being laminated is coated with adhesive by a knife-over-roll coater. The two webs are brought together while the adhesive is still wet, and the combination is passed around a steam- or electrically heated drum with the least-heat-sensitive web in contact with the drum. Electric resistance rod units frequently are positioned above the drum for additional heat, if needed.

The technique of dip saturation (Fig. 16-42) is used to obtain thorough saturation in products such as conveyor belting. Passing the belt around stacked rolls, as in the illustration, opens up alternate surfaces and ensures good penetration into the interstices. Squeeze rolls positioned as shown drive out the air and yield a homogenous vinyl mass between the fibers. The excess compound is doctored off both surfaces of the belt before it enters the vertical fusion oven. Mine belting is wet-woven, then primed and top-coated in two such coating stations (Fig. 16-43).

In curtain-coating, liquid dispersion flows from a slit in a hopper-shaped container onto parts passing through on a conveyor beneath. The material bypassing the parts enters a catch basin beneath the conveyor and is recirculated. Curtain-coating is infrequently used with plastisols, as they generally are too viscous to be

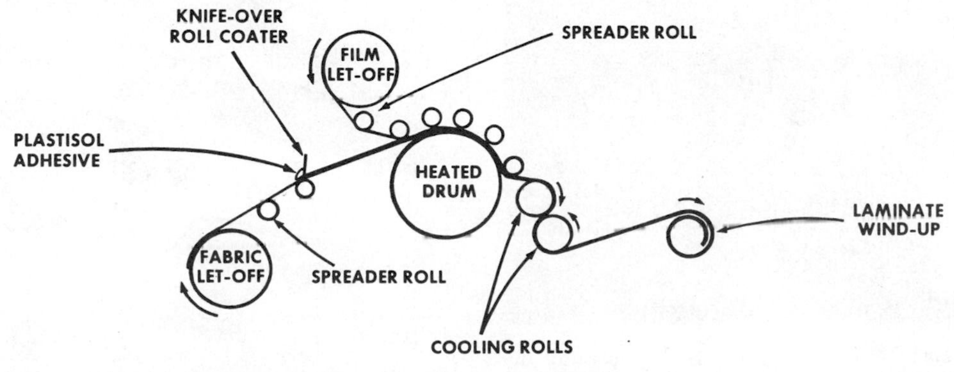

Fig. 16-41. Laminator for joining film to film or film to fabric with vinyl dispersion adhesives. (*Courtesy The BFGoodrich Co.*)

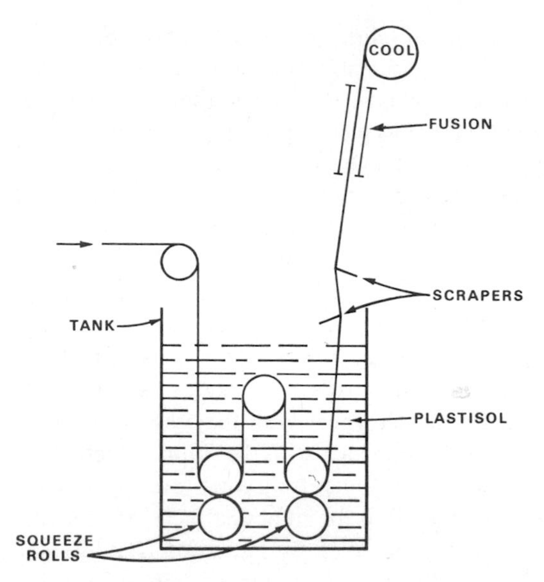

Fig. 16-42. Dip saturators are used for impregnation of heavy webs. (*Courtesy The BFGoodrich Co.*)

Fig. 16-43. Conveyor belting made by plastisol dip saturation techniques offers excellent flame and abrasion resistance in mines. (*Courtesy The BFGoodrich Co.*)

handled conveniently by this method. The process is, however, a good way to coat irregularly shaped parts. Specialty packaging applications, such as strippable foam coatings on automotive fenders and bumpers, have dominated.

MOLDING

Dip Coating and Dip Molding

Dip-coated or -molded products are readily produced by rapid, economical deposition of light or heavy layers of decorative, chemically resistant, tough plastisol coatings. The thickness of the coating can range from 1 mil to over 125 mils, or $\frac{1}{4}$ inch.

Basically, this process consists of dipping a mold into a plastisol and then fusing the coating that remains on the mold. There are many variations of the process: hot or cold dipping, or both, may be used; more than one dip may be used for heavier coatings or special effects; the coating may be stripped from the mold and the mold reused; or the coating may become a functional part of a finished product.

Generally, dip molding refers to the process of dipping, fusing, and stripping a mold of its plastisol coating to produce hollow products. Examples of plastisol dip-molding products are shown in Figs. 16-44 and 16-45. Figure 16-44 shows several designs of handlebar grips, and Fig. 16-45 shows automotive seat belt retractor boot covers. When the plastisol coating

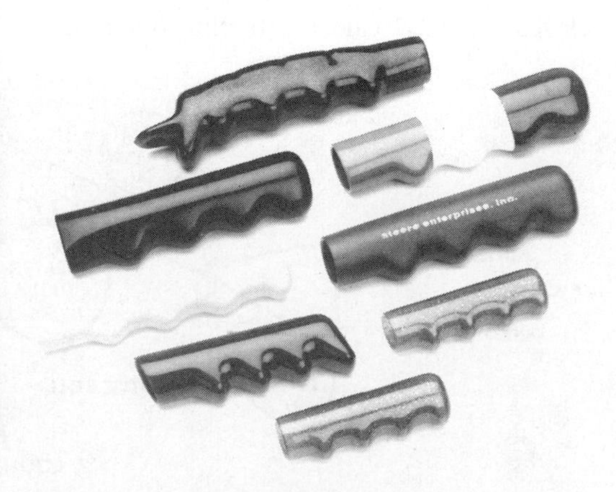

Fig. 16-44. Finger lug grips, dip molded by Steere Enterprises, Inc. (*Courtesy Steere Enterprises Inc.*)

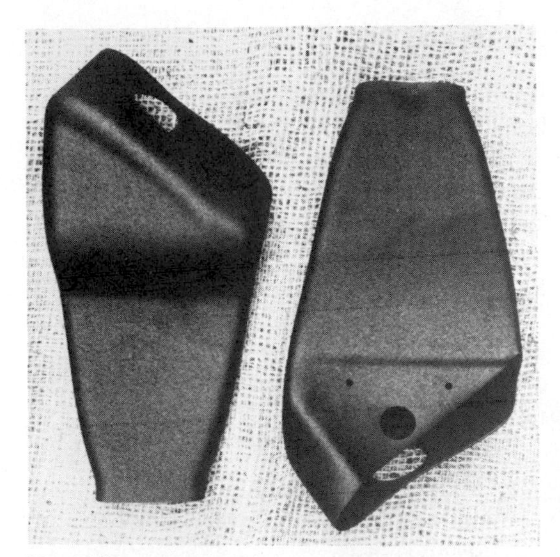

Fig. 16-45. Hot-dip-molded automotive seat belt retractor boot covers. (*Courtesy Steere Enterprises Inc.*)

is to become a functional part of the mold, the process is called dip coating, and the mold may or may not have an adhesive primer. Examples of dip coatings are shown in Fig. 16-46, where several different designs of glass bottles have plastisol protective coatings.

If adhesion to the substrate is important, and physical encirclement is not sufficient (e.g., in wire coatings), an adhesive primer will be required. Primers can be applied by brushing, dipping, spraying, or flow coating. Where the primer is to be applied selectively, such as that required on tool handles, the dipping method is most commonly used.

Fig. 16-46. Hot dip coated bottles. Coating offers protection as well as aesthetic appeal. (*Courtesy Chemical Products Corp.*)

Dip molding and coating may be done with a batch process or by a highly automated continuous process. The batch process is most frequently used where only a few similar parts, or a large number of dissimilar parts, are to be coated. The continuous dipping process is used for rapid production with minimum labor. Several molds mounted on a conveyor go through priming, preheating, coating, fusing, and cooling operations, all automatically. Some systems even strip the molds automatically with compressed air; in this case, the only labor required is for monitoring purposes. Figure 16-47 illustrates a highly automated, continuous dip-coating line for producing medical examination gloves.

Plastisol drips may occur at the lower edge or points of a mold after is has been withdrawn

Fig. 16-47. Automated continuous dip molding line for producing plastisol household and medical examination gloves. (*Courtesy McNeil Femco, Div. of McNeil Corp.*)

from the plastisol or during the gelation of the plastisol coating. For many applications, this is unimportant; for others, it is undesirable and should be avoided. Dripping can be controlled by proper compounding and by various processing techniques, including: (1) controlling the withdrawal rate of the mold from the wet plastisol; (2) inverting the mold just before the coating is gelled; and (3) jarring the mold to knock off the drips.

The molds used in dip molding usually are solid and are made from cast or machined aluminum, machined brass, steel, or ceramic.

Some examples of commercial dipping applications include plating racks, dish racks, electrical parts, spark plug covers, tool handles, traffic safety cones, work gloves, and many household and surgical examination gloves. Protective coatings are applied to many types of glass bottles, automotive bumpers, and hubcaps, and to protect the threads of carefully machined nuts and bolts.

Hot Dipping Process

Hot dipping is the most widely used technique because it takes advantage of the inherent heat gelling characteristics of a plastisol, which cause it to form a gelled coating on a hot mold. This technique is useful in building multiple coats and for coating over cured adhesives. Figure 16-48 shows glove molds coming out of the preheat oven and being dipped into a plastisol tank. The coating weight, or thickness of the coating, will depend upon the following factors:

1. Mass and shape (ratio of area to volume), not only of the whole object, but also of sections of it.
2. Temperature of the object at the instant of dipping.
3. Heat capacity and conductivity of the mold.
4. Dwell time of the mold in the plastisol.

Fig. 16-48. Automatic dip molding line for gloves. Preheated glove forms are shown going from the oven to the plastisol dip tank. (*Courtesy McNeil Femco, Div. of McNeil Corp.*)

5. Gelation characteristics of the plastisol.
6. Temperature of the plastisol at the time of dipping.
7. Withdrawal rate of the object from the plastisol.

The uniformity of the coating is controlled by balancing the withdrawal rate and the dwell time. The dwell time should be adjusted so that the plastisol is almost completely gelled on the surface at the conclusion of withdrawal. Flow-off and dripping are controlled by the withdrawal rate, which must be reasonalby slow and uniform. All other factors being equal, pickup is controlled primarily by the preheat temperature of the mold, which usually is in the range of 49°C (120°F) to 260°C (500°F). Slow-speed agitators and cooling coils are recommended for plastisol hot dip tanks to control the plastisol temperature. A significant rise in plastisol temperature, resulting from repeated hot dips, will increase plastisol viscosity and change the dipping requirements considerably.

Finally, the flow properties of the plastisol, or rather their change with change in temperature, may have a significant effect on the appearance and thickness of the fused coating. If the partially gelled film of plastisol undergoes a marked reduction in viscosity during fusion, sagging will result, with local thinning of the coating or an irregular surface appearance.

Cold Dipping Process

When a cold (or room temperature) mold is dipped into a plastisol, removed from the plastisol, and then fused, the process is referred to as cold dipping. The thickness of the coating depends upon the low-shear-rate viscosity and the yield value of the plastisol, in contrast to the process of gelling the plastisol with a preheated mold, as is done in hot dipping. Cold dipping is used when: (1) there is low heat capacity or conductivity in the mold to be dipped, and preheating is impractical; (2) the plastisol is extremely sensitive to heat, and very high viscosities would result from dipping hot molds; or (3) the mold is irregular and/or has fine surface detail that would not be reproduced in hot dipping because of premature plastisol gelation.

The withdrawal rates of the mold from the plastisol have a significant effect on the plastisol pickup and usually are slower than the withdrawal rates used in hot dipping. A careful balance must be made of the viscosity at low shear rates and the yield value of the plastisol so that it flows well enough to form a uniform coating and yet does not drip. Higher-viscosity plastisols require less yield value, but the important factor is the ratio of the viscosity to the yield value. To be effective, this ratio must not increase from the time when the dipping is completed until the coating has gelled. The desired viscosity range of the plastisol for a particular dipping application is obtained primarily by proper selection of the dispersion and blending resins and the plasticizer levels and type. Small amounts of volatile thinners sometimes are added to a plastisol to lower the viscosity. High yield values are obtained with gelling additives such as colloidal silicas, high-oil-absorbing fillers, and metallic–organic complexes.

Rotational Molding

Rotational molding is a unique process, compared to injection molding and blow molding; it allows the producer greater flexibility in end-product design, especially for hollow parts, double-wall constructions, and large sizes where conventional tooling would make end-product cost prohibitive. Products can be rotational-molded, ranging from small balls up to 17 ft × 17 ft × 8 ft modules, storage tanks, or shipping containers. Custom-designed machinery and molds are required for large objects.

The molding process involves four individual stages: loading the molds with raw materials, rotating and fusing the part, cooling the part, and unloading. In loading the molds, a predetermined amount of raw material is placed within the molds, and then the molds are closed and sealed. In the casting and fusing stage, the mold is rotated in two planes, perpendicular to one another, while subjected to heat (Fig. 16-49). As heat penetrates the mold, the raw material is gelled and builds up in an even dis-

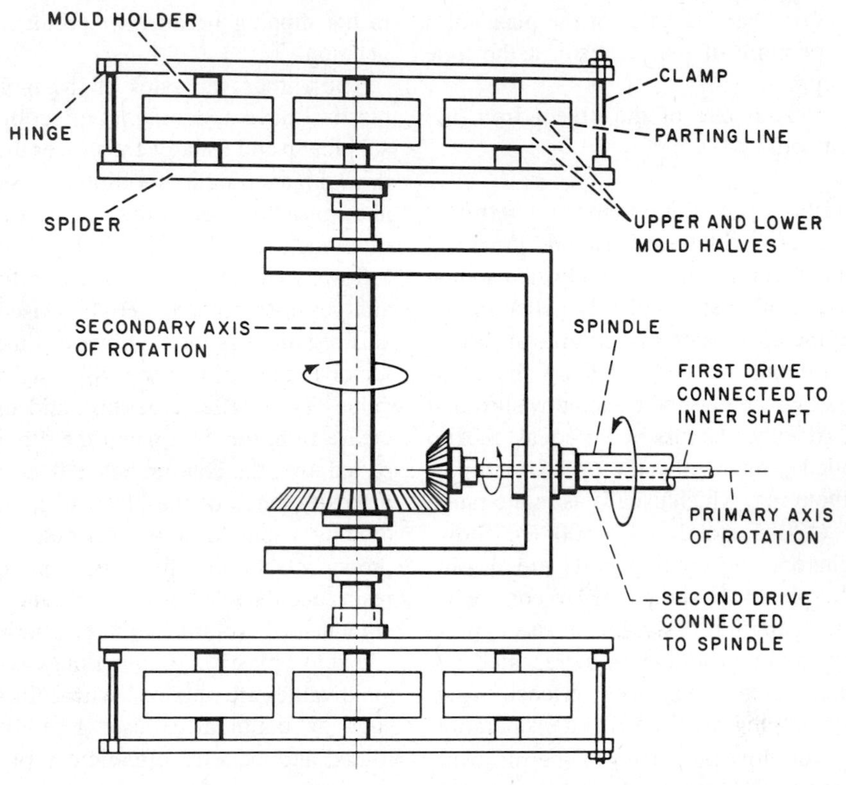

Fig. 16-49. Schematic presentation of a typical mechanical system used to obtain biaxial mold rotation in two perpendicular planes. (*Courtesy Quantum Chemical Corporation*)

tribution on the interior wall surface of the molds. After the heating cycle and while the mold is still rotating, it is exposed to cool air and water. Rotation is stopped, and the mold parts are removed.

Typical products rotationally molded from plastisols include boat bumpers, mannequins, dolls, squeeze toys, play balls, basketballs, footballs, and automotive armrests and headrests. Soft, flexible products with much detail, as in Fig. 16-50, or rigid functional products, such as the drum table in Fig. 16-51, are being molded from plastisols.

Molds. Molds used in rotational molding may be constructed from one or more of several metals. Cast aluminum often is used for rapid heat transfer and the lowest cost. Machined aluminum generally will give parts that are free of surface porosity or voids. Electroformed nickel is best for fine surface detail reproduction. Sheet metal is used for large, simple shapes that do not have undercuts. To obtain good surface and contour reproduction, powdered metal sprays can be applied to a model.

In general, regardless of the material used, the mold should meet the following requirements:

1. It should have good thermal conductivity.
2. The wall thickness should be the minimum necessary for strength and dimensional stability, and should be uniform.
3. Provision for venting (Fig. 16-52) should be made to allow for pressure equalization and also to aid in producing stress-free parts.
4. Provision must be made for clamping mold sections together firmly to prevent leakage.
5. The mounting must allow for free circulation of the heating medium so that all surfaces of the mold are heated equally to obtain uniform distribution of the plastisol within the mold. Uniform distribu-

Fig. 16-50. Rotationally molded toy banks used for promotional items. (*Courtesy Royalty Industries, Inc.*)

tion—regardless of the irregularities in configuration of the product—is perhaps the most important requirement in rotational molding.

Molds can be attached one at a time to the molding machine (common for large parts) or mounted several at a time on a "spider," which is then attached to the machine. Figure 16-53 illustrates open and closed sets of molds mounted on spiders.

The loading of the molds with plastisol is easily accomplished with many of the commercial metering, mixing, and dispensing units. Most systems are designed so that the operator can fill multiple-cavity molds easily and quickly without excess material being carried over or dripping onto the edges of the molds. Plastisol spills or drips on the mold exterior or parting lines will result in mold maintenance problems and poor moldings. Because there is very little flash or waste in molding plastisols, precise metering is required to get part-to-part uniformity.

Heating systems. The heat sources most used in rotational molding are forced hot air and sprayed molten salt. Other sources are open flame, infrared, and recirculating hot oil.

Recirculating hot air may range from 177 to 482°C (350–900°F). The higher the oven temperature within this range, the shorter the cycle. Plastisol fusion cycles may range from 4 to 30 minutes. Most cycles are 5 to 8 minutes at 288 to 371°C (550–700°F). It is important to accurately direct the air flow and control the velocity and uniformity of the temperature throughout the oven. Maximum heat transfer with a minimum of localized hot spots in the oven can be obtained with baffles and high-velocity air blowers. Excess heat capacity is desirable for the oven so that the required operating temperature can be attained rapidly after insertion of the cool, charged molds. Gas-fired, oil-fired, or electrically heated hot air ovens are available.

Heating by means of sprayed molten salt is used in the rotational molding of plastisols, where the molten salt is a melted standard mix-

Fig. 16-51. Rigid functional drum table rotocast from acrylic-modified plastisol. (*Courtesy Whittaker Corp., Advanced Structures Div.*)

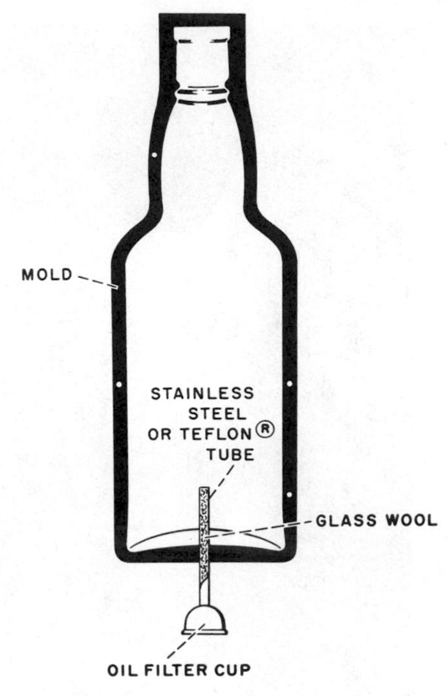

Fig. 16-52. System of venting a rotational mold for pressure equalization during heating and cooling. (*Courtesy Quantum Chemical Corporation*)

ture of low-melting-point heat transfer salts suitable for temperatures within the usual range of operations, from 177 to 288°C (350–500°F). The molten liquid then is pumped through a series of spray nozzles onto the rotating mold assembly, as shown in Fig. 16-54. In this heating system, heat is transferred by conduction rather than convection, as used with hot air. The major advantage of hot-liquid spraying is that heat transfer is very rapid, and plastisol fusion may be achieved faster with a lower-temperature heat source. This method is especially suitable for molding pieces with heavy walls and for the production of complex shapes. It is essential that molds fit tightly and form a good seal to keep the molten liquid from getting in and ruining the part. Molds usually are rinsed with water to remove the salts. Molten-salt equipment must have an efficient means of recovering and reusing the heating medium because that can mean the difference between profit and loss.

Special-purpose or prototype rotational molding equipment may use infrared or open-flame heating. Radiation heaters may be electrical or gas-fired. Open-flame heaters may be gas or oil-fed. Fusion of the part is accomplished by applying a direct flame to the exterior of the mold.

Recirculating hot oil with jacketed molds is used to some extent in rotational molding. This will be discussed under machinery types.

After completion of the fusion cycle, the molds may be subjected to a combination of cooling methods, including forced cold air, cold water spray, or cool liquid circulating through jacketed molds. It is important that the molds continue to rotate between fusion and cooling, and during cooling to control shrinkage and to prevent hot melt sag from occurring.

Machinery Setups. Rotational molding machines utilize different approaches to rotating, heating, and cooling of the molds. The mold-

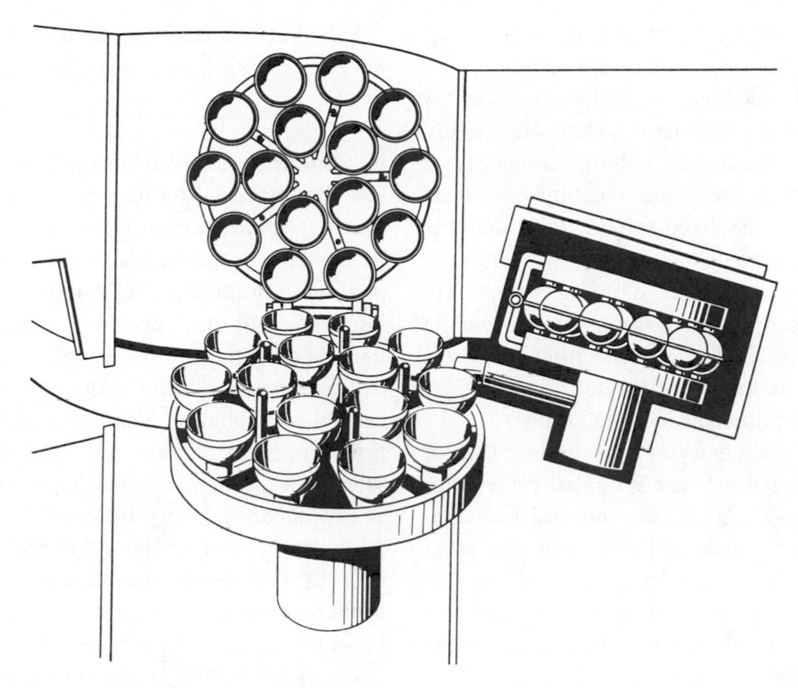

Fig. 16-53. Left: An open set of 15 molds mounted on a spider ready to be loaded. Right: Closed set of molds, containing molding material, entering fusion oven. (*Courtesy Quantum Chemical Corporation*)

Fig. 16-54. Mold carrier in an oven chamber with spray nozzles mounted above for delivery of molten salt. (*Courtesy E. B. Blue Co.*)

ing facility can be set up to be a batch-type continuous, shuttle, or jacketed-mold system.

Batch-type machines normally are used for small to medium-size parts. They are largely manual-type machines involving considerable operating labor. Two batch machines are used commercially: the fixed-spindle type and the pivoting-arm type.

Continuous rotational molding setups are most commonly of the carousel type, using multiple arms. This type of machine may have three, four, or five arms located in a circular pattern and equidistant from each other.

The arms, each carrying a group of molds or a single, large mold, are mounted on a common hub, and the molds are indexed from station to station. Each arm may function independently with respect to speed and ratio of rotation. All portions of the cycle are automatic with the exception of the loading and unloading of finished parts.

Shuttle-type rotational molding systems may be relatively simple, or quite complex. Shuttle machines contain features from both batch and continuous machines. This type of equipment generally is used for very large objects. Support of the mold may be on one or both ends. Two or more arms may be utilized in the same machine. The process is semiautomatic, with the molds shuttling on a continuous track, moving from station to station horizontally.

The jacketed mold type of rotational molding is expensive and sophisticated with respect to both the molds and the machinery. This is a single-station operation where the mold is cored, or has channels in the mold walls that will allow hot and cold liquids to circulate. After the mold is loaded and closed, it is rotated while a hot liquid, such as oil, is circulated through the mold walls and fuses the plastisol. Upon completion of fusion, a chilled liquid is circulated through the walls to cool the part. The mold does not change stations, and there is no oven or cooling chamber; thus, space requirements for this system are minimized. An insulated heating tank and cooling oil storage are required, along with a series of values and pumps. Fast cycles for long production runs of symmetrical-shaped objects are the primary advantages of this system.

For more information about rotational molding equipment, the reader is referred to Chapter 14.

Formulating Considerations. Plastisols allow the rotational molder to explore new markets with relative ease (Fig. 16-55). They can be custom-formulated, not only to meet end-product requirements but also to meet processing needs that may result from equipment limitations or difficult mold shapes.

The most important influences on the successful rotational molding of plastisols are the flow properties at room and elevated temperatures. The plastisols should have low viscosity at low shear and very little or no yield value. Also, appropriate gelation characteristics are necessary to obtain an even distribution of material without bridging in close tolerances. The optimum viscosity appears to be in the range of 1000 to 4000 cps.[32] This range provides the plastisol with the greatest opportunity to wet the mold prior to gelation. Plastisols lower than 1000 cps have a tendency to leak from the mold, while those greater than 4000 cps allow less and less control of the cycle during gelation.

Fig. 16-55. Large inflated vinyl balls with a variety of handle designs are rotationally molded from plastisols. (*Courtesy Sun Products Corp. Div. of Tally Industries*)

The gelation cycle is very critical in plastisol rotational molding. Gelation usually is completed within 2 minutes after the mold is placed in the oven; so distribution of the plastisol must take place within that time period.

In general, the larger the product to be rotocast from plastisol, the faster the gelation characteristics should be. Whenever a rotocasting trial is being made on a new product, complete fusion is essential in order to obtain optimum physical properties. Increasing the molding cycle time or temperature to the point of slight degradation and then backing off generally will give sufficient fusion for production development trials.

Slush Molding

Slush molding is an excellent method of producing hollow objects. A wide variety of products can be manufactured by this process, including rainboots, shoes, hollow toys and dolls, and automotive products such as the protective skin coatings on arm rests, head rests, and crash pads. The basic process of slush molding involves filling a hollow mold with plastisol, exposing the mold to heat, gelling an inner layer or wall of plastisol in the mold, inverting the mold to pour out the excess liquid plastisol, and heating the mold to fuse the plastisol. The mold then is cooled and the finished part removed. Slush molding can be a simple hand operation for limited production or an elaborate conveyorized system for long runs. This process can be a one-pour method, where finished or semifinished products can be made by one slushing, or a multiple-pour method, where two or more slushings are used.

In the one-pour method, molds are placed on a conveyor belt or system that has positions for as many molds as desired. Molds are filled to the top and carried through an oven where temperatures in the 93 to 316°C (200–600°F) range gel the plastisol next to the mold wall. The thickness of the gelled plastisol at a given oven temperature is determined by several factors: the thickness of the metal wall of the mold; the length of time the mold is in the oven; and the gelling characteristics of the plastisol. Next the mold is turned upside down, and the remaining liquid plastisol is dumped out, while the gelled plastisol next to the wall remains in place. The mold is conveyed to a second oven, where fusion of the plastisol is completed. Passage through a cooling chamber cools the mold sufficiently to permit removal of the finished piece by blowing it out with compressed air or collapsing it with vacuum. It is advisable to rotate the molds in the second oven to maintain an even temperature on all sides of the mold to evenly fuse the plastisol.

The multiple-pour method of slush molding is used when the mold has fine surface detail, a heavy wall thickness is desired, or different types of plastisol are used in the same product. For example, boots often are produced by the multiple-pour method. A cold mold is filed with plastisol and then emptied. The mold, with a thin coating of plastisol, is then heated, refilled while hot with the same plastisol or another type, and emptied again. The mold with the multiple coating of plastisol is heated to fuse the plastisol, and then it is cooled and the product removed. It is possible to reproduce fine detail of the mold surface with this method because of the cold pour and thorough wetting of the mold prior to gelation of the plastisol. The second of third pour may consist of a cellular plastisol that, upon fusion, will provide a soft lining and insulation properties, which are often necessary in cold weather boots.

Molds. Molds used in slush molding are produced from sprayed molten metal, spun aluminum, machined aluminum, or ceramics, or by electroforming. An electroformed mold is relatively inexpensive, easily reproduced and very durable. Some molds of this type have lasted for many years.

Regardless of the type of mold, the article molded from plastisol will faithfully reproduce the surface of the mold, whether matte or glossy. Molds occasionally have, or develop, surface porosity, in which case the finish will be affected and usually will have low gloss. In cases where a highly glossy surface is desired, or where transparency is necessary, molds can be coated on the interior with epoxies, phenolics, and even porcelain.

Formulating Considerations. Selection of a plastisol for a slush molding job will depend on three factors:

1. The molder's technique and equipment.
2. Variations in shape, size, and wall thickness of the article.
3. Physical properties required in the finished article, such as color, flexibility at various temperatures, grease resistance, electrical characteristics, and so on.

The plastisols used in slush molding are very often formulated similarly to those used in rotational molding. Brookfield viscosities may range from 1000 to over 10,000 cps, but usually a low Brookfield viscosity (under 5000 cps) is preferred. Good viscosity stability is desired because of the constant heating and cooling that the plastisol undergoes in slush molding.

The physical qualities of slush-molding plastisols allow exceptionally accurate work. Because the material is not fused under pressure, it exhibits very little shrinkage when molded. The color selection is almost unlimited. Plastisol articles may be readily cemented and heat-sealed. When properly formulated, they resist acids, alkalies, and a wide range of solvents. They also resist aging and sunlight, and they can be made to have excellent electrical insulating properties and flame resistance.

Cavity, In-Place, and Low-Pressure Molding

Plastisols are used in the manufacture of many different products that are solid moldings. Solid molded parts usually are made by cavity, in-place, and/or low-pressure molding.

The fusion time for solid molded parts depends upon the mold thickness and heat conductivity, the heat available, and the cross-sectional thickness of the article to be molded. Generally, for every $\frac{1}{4}$ inch of cross section, 15 minutes at 177 to 191°C (350–375°F) is sufficient to complete fusion. Plastisols to be molded in thick sections ($\frac{1}{2}$ inch or more) may require special heat stabilization to protect the vinyl from degradation during the long fusion cycle.

Cavity Molding. Cavity molding is the simplest process for the manufacture of solid molded parts. It consists of pouring the plastisol into a mold, heating the mold to fuse the plastisol, cooling, and stripping the molded part from the mold. Toy kits for making replicas of insects and for figurines use this process. Other products made by cavity molding include fishing lures and sink and disposal stoppers.

Plastisols used in cavity molding may be very fluid and pourable, or they can be formulated to have a stiff consistency like that of modeling clay. Wetting the mold surface with a thin film of plasticizer before pouring in the plastisol will aid in reproducing surface detail.

In-Place Molding. The excellent moldability, low cost, and excellent physical and chemical properties of plastisols have led to several large-volume applications for them in gasketing. These applications utilize in-place molding, a process of molding directly in, or onto, a finished article.

Largest-volume applications for in-place molding are jar lid and bottle cap liners, as shown in Fig. 16-56. Automatic cap production lines operate at very high speeds, and

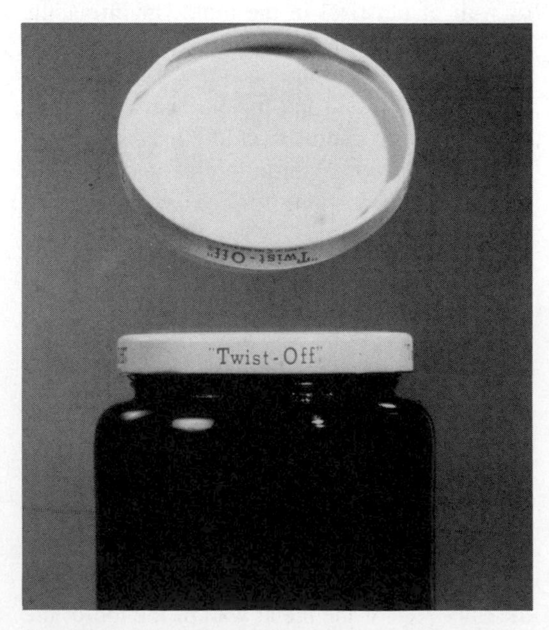

Fig. 16-56. Jar lid and bottle cap liners are molded in place from plastisols. (*Courtesy White Cap Div. of Continental Can Co.*)

thousands of plastisol gaskets are applied to jar lids and bottle crown caps per hour. Plastisols must be formulated to have low viscosity at the very high shear rates encountered in high-speed production lines.

The automotive air filters shown in Fig. 16-57 constitute another large-volume use for in-place molded plastisols. The filter paper and support elements are placed in a ring mold that has been filled with plastisol, and the mold is passed through an oven to fuse the plastisol. Once out of the oven, the filter is turned over, and the other end is placed in the ring mold with plastisol; and this, in turn, is fused in the oven. The fused plastisol provides an airtight gasket for assembly in the air cleaner of an automobile, and it also bonds the ends of the filter paper into a sealed assembly.

Sewer pipe gaskets have been made from in-place molded plastisols for many years. Special molds are placed over the ends of clay or ceramic sewer pipe, and plastisol is poured into the molds and fused around the pipe ends. The plastisol is formulated to a specific hardness range to permit the pipe ends to be joined together easily to form a tight seal. Solvents are used to lubricate the pipe ends during assembly and to cement the ends together upon drying. Sewer pipe can be laid in place very rapidly with this gasket system. Another advantage of the plastisol in this application is its ability to compensate for lack of roundness of the pipe.

Fig. 16-57. The end caps for most automotive air filters are molded in place from plastisols. (*Courtesy Chemical Products Corp.*)

Low-Pressure Molding. Plastisols can be injected into closed molds with low-pressure positive-displacement pumps. The molds then are placed in ovens (or they can be put into a heated press) to fuse the plastisol. The mold detail is accurately reproduced, and products up to 2 inches thick may be produced.

The molds must be well matched to form a tight fit, and they can be clamped or bolted together. To eliminate air pockets in filling closed molds, small bleeder holes must be correctly positioned in the mold so that all the air is ejected, and the cavity completely fills with plastisol. After filling, certain of these air bleeder holes and the inlet may have to be closed off so that the plastisol will not escape. This is particularly important because during heating of the plastisol the viscosity will decrease, and the plastisol may become quite fluid prior to gelation. Well-deaerated plastisols should be used in low-pressure molding to ensure void-free products.

Products made from the low-pressure molding of plastisols include shoe soles, printing plates, and encapsulated electronic parts.

SPECIALIZED PROCESSES

Strand Coating

Plastisols and organosols have been used for many years as protective coatings on various filaments, wires, and woven cords. Fibrous glass yarns are strand-coated at speeds of 500 to 600 ft/min. with the speed varying according to the thickness of the coating desired and the type of fusing heat used (radiant or convection). This coated fibrous glass is then woven into various types of screening.

There generally are three methods of strand coating: the set-die method, the floating-die method, and a method using no die. In the set-die method, the die is rigidly set and centered around the strand to be coated. The strand must

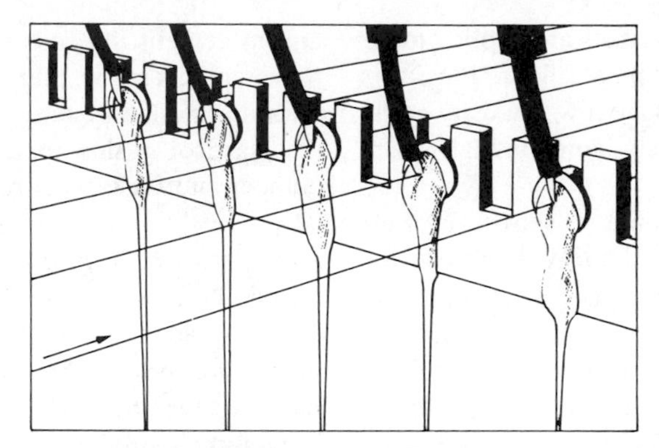

Fig. 16-58. Plastisol protective coatings on filaments, wires, and woven cords are applied at very high speeds via strand coating. (*Courtesy The BFGoodrich Co.*)

be held sufficiently taut so that all ends or imperfections are straightened out before it enters the die. Plastisol is poured or pumped over the face of the die, the strand passes through it, the die wipes the excess plastisol off (Fig. 16-58), and the strand then passes through a fusion tunnel. The excess plastisol that was wiped off is recirculated. Although this method will work for a wide range of viscosities, it is particularly suitable for viscous organosols or plastisols and thus will provide a heavier coating per pass.

In the floating-die method, the die is loosely held in place and allows a certain freedom of movement in one plane. The theory is that the die will center itself around any slight bend or imperfection. If the viscosity of the plastisol or the organosol is too high, the material will pile up under the die and restrict the die's freedom of movement. The plastisol or the organosol must be low enough in viscosity to flow freely off the die, with only slight accumulation.

With the method using no die, the strand is passed vertically through the plastisol or the organosol, the viscosity of which is adjusted to give a uniform pickup of desired thickness with no runbacks. A runback occurs if the wire or line is passing too fast through the coating material, or if the viscosity of this material is too high. In either case, there is too great an initial pickup of material, which tends to flow back down the strand and cause an uneven coating. This method generally is suitable for dispersions of relatively low viscosity. The maximum pickup without runbacks is ordinarily small, and several passes may be necessary to build thicker coatings.

The fusion of plastisol or organosol strand coatings is dependent upon the speed of travel, length of the oven or the tower, the fusion temperatures available, and the thickness of the coating.

In formulating for strand coating, both the requirements of the end use and the processing conditions must be considered. A formulation for coating fibrous glass for screening will be quite different from a formulation for coating electrical wire. In coating rope, thread, or woven cords, it must be determined whether the coating should penetrate the strand or merely coat the surface. These examples show the importance of selecting the proper ingredients for the vinyl dispersion to meet the requirements of the job.

Spray Coating

The spraying of vinyl dispersion has grown with the expanding use of plastisols in specialty applications and with the increased interest in new protective and decorative coatings on metal furniture, appliances, and building products. Spraying is the only sure way of coating many intricate or irregular objects. Sprayed vinyl dispersions can lay down a coating from a few mils up to 60 mils.

Organosols are widely used for spray coating because of their easier flow, but the extra body of plastisols permits heavy coats on vertical surfaces without sags in the coating. Standard spray guns can be used. By employing special spray heads, striking decorative veil and spatter patterns can be applied.

The flow properties of vinyl dispersions for spraying are very important for their successful application. Both low- and high-shear conditions are encountered. Standard air spray gun orifices used in plastisol work range from about 0.07 to 0.1 inch in diameter. In passing through these orifices at pressures ranging from about 30 to 100 psi, the materials undergo a high degree of shear. Then, upon being deposited at the work surface, the low-shear flow properties of the plastisol will dictate just how smooth a film will result and how much can be laid up without sags or runs. Balancing low-shear properties in the production of thick coatings is an exacting procedure. Too little yield for the coating thickness will cause the film to sag, curtain, and run off. Too much yield will inhibit leveling and result in an extreme orange peel surface.

Extrusion

Plastisols may be extruded as a means of preparing partially fused material for subsequent molding or as a means of producing finished goods. Plastisol extrusion is an ideal way to produce vinyl formulations of low durometer reading. These compounds are very difficult to produce by the conventional two-roll mill because of adhesion to the rolls and the low viscosity of the hot melt.

Almost any extruder will extrude plastisols, but the importance considerations are the quantity of production and the uniformity of output. The screw in an extrusion machine works as an inefficient high-pressure pump with plastisols, with the friction between the screw and the cylinder wall creating the forward driving force. Plastisols offer little resistance to the action of the screw forcing them forward; so it is possible to increase the screw speed two to three times its normal usage rate. If the extruder is being used for one step in the compounding of a plastisol, the material should be extruded incompletely fused so that it may be readily group up or cubed. Final fusion then will take place in an injection or a compression molding machine.

The method of feeding the plastisol to the extruder will depend upon its viscosity. It is desirable that the machine have a large opening or throat for high-viscosity materials so that they may flow into the hopper at a rate that will maintain a reservoir of material and avoid the entrapment of air or surging of the material. If the material is of low viscosity, it is desirable to suspend the drum of plastisol over the machine and pipe it to the hopper, with suitable gaskets to prevent leaking. It is possible to attach a cover to the drum and inject a safe amount of air to increase the flow of the plastisol in the extruder. The screw should have shallow flights, and, if possible, a heated screw should be used.

The advantage of plastisols over conventional molding compounds in extrusion lies in the production of low-durometer formulations. The extrusion of such formulations gives results more uniform than can be obtained from a mill. It is advantageous to use plastisols in extruding delicate shapes because their fluidity permits the use of low pressures.

REFERENCES

1. W. L. Semon, U.S. Patent 1,929,453 (Oct. 10, 1963), assigned to The BFGoodrich Company.
2. G. W. Johnson, British Patent 500,298 (Feb. 7, 1939), I. G. Farbenindustrie.
3. W. L. Semon, U.S. Patent 2,188,396 (Jan. 30, 1940), assigned to BFGoodrich Company.
4. "More Solutions to Sticky Problems," Brookfield Engineering Laboratories, Inc., Stoushton, MA.
5. A. C. Werner, "Spread Coating," *Encyclopedia of PVC*, Marcel Dekker, Inc., New York, 3(26), 1415–1475 (1977).
6. Technical Bulletin, "Rotovisco RV-20," Hanke Buchler Instruments, Inc. Saddle Brook, NJ.
7. Technical Bulletin, "Carri-Med, CS Rheometer," Mitech Corp., Twinsburg, OH.
8. L. E. Nielsen, *Mechanical Properties of Polymers*, pp. 30–35, Van Nostrand Reinhold Co., New York, 1962.
9. J. A. Sarvetnick, *Plastisol and Organosols*, Van Nostrand Reinhold Co., New York, 1972.
10. *Modern Plastics Encyclopedia*, 1972–1973.
11. Plastics World, *Directory of the Plastics Industry*, Vol. 30, No. 11 (Aug. 1972).

12. E. A. Collins, D. J. Hoffman, and P. L. Soni, "Rheology of PVC Plastisols," *Rubber Chemical Technology*, 52(3), 676 (1979).
13. D. W. Ward, "Reactive Functionality in Dispersion Resins," *SPE Journal*, 28, 44–50 (May 1972).
14. *Modern Plastics Encyclopedia*, 1988–1989.
15. J. K. Sears and J. R. Darby, *The Technology of Plasticizers*, John Wiley and Sons, New York, 1982.
16. P. D. Ritchie, *Plasticizers, Stabilizers, and Fillers*, The Plastics Institute, Iliffe Books, Ltd., London, 1972.
17. L. P. Whittington, A Guide to the Literature and Patents Concerning Polyvinyl Chloride Technology, Society of Plastics Engineers, Stamford, CT, 1963.
18. W. D. Todd, D. Esarove, and W. Smith, *Modern Plastics*, 34(1), 159 (1956).
19. T. C. Jennings and Charles W. Fletcher, "Actions and Characteristics of Stabilizers," *Encyclopedia of PVC*, 2nd ed., Marcel Dekker, Inc., New York, 1988.
20. W. V. Titow, "Colourants," *PVC Technology*, Fourth Edition, Chapter 11, pp. 401–419, Elsevier Applied Science Publishers, London and New York, 1984.
21. T. B. Reeve, "Pigment Colors for Vinyl Coatings," Pigments Department, E. I. du Pont de Nemours & Co., Wilmington, DE.
22. Technical Bulletin, "Ircogel," Lubrizol Diversified Products Group, Wickliffe, OH.
23. BFGoodrich Technical Bulletin No. HM-15, "Modifying Plastics with Hycar and Hydrin Elastomers," Cleveland, OH.
24. D. G. Higgins, "Coating Methods Survey," *Encyclopedia of Polymer Science*, Vol. 3, p. 766, John Wiley and Sons, New York, 1965.
25. R. J. Jacobs, "Fundamentals to Consider in Selecting Coating Methods," reprint from *Paper Film and Foil Converter*.
26. George L. Booth, "Take Your Choice of Coating Methods," *Modern Plastics Magazine* (Sept. and Oct. 1958).
27. George L. Booth, *Coating Equipment and Processes*, Lockwood Publishing Co., New York, 1970.
28. R. P. Conger, *SPE Journal*, 24(3), 43 (Mar. 1968).
29. Eastman Chemicals Publications L163, L164, and L165 on "Adhesion Promoting Plasticizers," Eastman Chemical Products Co., Kingsport, TN.
30. See, for example, G. H. Poll, "Painting Aluminum at 200 fpm," *Products Finishing*, pp. 2–9 (July 1969).
31. P. F. Bruins, *Basic Principles of Rotational Molding*, p. 276, Gordon and Breach, Science Publishers, New York, 1971.
32. Ibid., p. 278.

17 Powder Coatings

Introduction

Powder coatings have been developing as an environmentally acceptable finishing method since the mid-1950s. These coatings are essentially dry paints. They are coatings that are formulated with the same resins, pigments, and additives as paints, but without solvents. Solvents are used in paints simply as a vehicle to transport the resins and pigments to the piece that will be finished and then evaporated from the film. By eliminating the solvents, we dramatically reduce organic emission to the atmosphere and virtually eliminate effluent discharge from the factory.

There are two generic classes of powder coatings available in the marketplace: thermoplastic and thermosetting. Thermoplastic coatings are those that melt with the application of heat to form a continuous film. They do not change chemically, but simply harden as they cool. Thermosetting coatings, on the other hand, melt *and chemically react* with the application of heat. Thermosetting coatings make up the majority of powder coatings in everyday use.

The research and development of powder coatings has been accelerated by such potential advantages as minimized air pollution and water contamination, increased performance from the coating, and improved economics to the user in application and handling.

Powder coating is done in a closed loop system. All of the material that is purchased can be used for coating. There is no need to handle unusable materials, such as sludge from the paint booth. Powder coating also features fast film build when compared to liquid systems. This allows a reduced number of application units and simplifies the automation process. Consequently, the floor space for finishing is reduced. Because powder coating does not use solvents, there is no need for paint mixing and viscosity adjustments. Rejects due to color drift, cracking, sagging, or blistering also are eliminated. These factors reduce labor costs. Specialized and lengthy training in the application technique in the factory is unnecessary. Also because powder coatings lack solvents, the risk of paint-line fires is diminished. This difference can lead to lower insurance premiums.

However, it should be kept in mind that a powder coating is basically a chemical coating, and thus has many of the problems of solution paints. If not properly formulated, the powder coatings may sag at high film thickness, show poor performance when undercured, show film imperfections such as craters and pinholes, have poor hiding at low film thicknesses, and so on.

The following sections discuss the history, composition, application, and manufacture of powder coatings.

MANUFACTURING METHODS

Powder coatings are processed by one of three different methods: dry blend, melt mix, or solution.

Reviewed and updated by E. J. Duda, Ferro Corporation, Cleveland, OH.

Dry Blend

The dry blend method of manufacture is used primarily for thick film, thermosetting powders, and selected thermoplastics, such as polyvinyl chloride powders. A dry blended powder is a relatively uniform blend of all the raw materials in the formulation. Dry blending is a batch process method that utilizes high-intensity mixing or ball milling.

In the first case, all of the raw materials are added to a high-intensity mixer (see Fig. 17-1). The mixer cycles for several minutes at 1800 rpm. The material then is sieved through a 60 to 100 mesh screen and packaged.

A high grade of vinyl powder can be produced by dry blending (or agglomerate mixing, as it is also called). The process consists of loading a high-intensity mixer with the vinyl resin, pigments, and other solid additives. The mixer is run at approximately 1800 rpm for a predetermined cycle with or without heat added to the mixing jacket. External heat sometimes is added to shorten the cycle. The plasticizer and other liquid additives are added at a set rate. The batch temperature is normally around 100°C for optimum plasticizer absorption. At the end of the cycle, the batch is dropped into a cooling mixer.

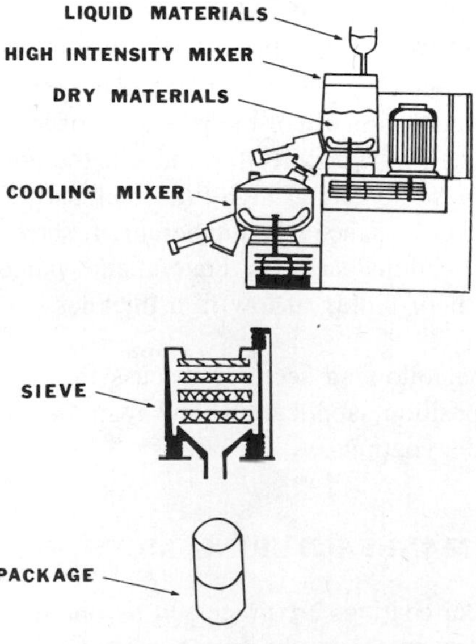

Fig. 17-1. Dry blend process.

The cooling mixer is a low shear unit, about 100 rpm, such as a ribbon blender. When the batch is at 35°C or less, the drying agent is added under low shear. On completion of the cycle, the batch is sieved through at least 40 mesh and packaged.

The ball mill also is a batch process. The formula ingredients are added to a ball mill, which tumbles for 10 to 20 hours. The cycle can be shortened by using a vibratory ball mill. Because of the temperatures developed with this technique, the hardener sometimes is added during the last 2 hours of processing to prevent partial curing. After the process, the batch is sieved and packaged.

Ball-milled powder is used in fluid bed and electrostatic application systems. The quality of the product is better than that from the high-intensity mixer but more expensive on a volume-per-unit-time basis. The quality of the product is inferior to that of material processed by melt mixing.

The first powders formulated for the pipe and electrical insulation industries were processed by the dry blend technique. The present decorative and functional coating market demands a quality of thermosetting powders that can be produced only by melt mix or solution techniques.

Melt Mix

The melt mix process (see Fig. 17-2) is the most popular for manufacturing thermoset powder coatings. It consists of the following steps:

- Premix
- Melt Mix
- Cool
- Flake or kibble
- Pulverize
- Sieve and package

In the melt mix process, all of the formula ingredients are added to the premixer, which may be a cone blender, ball mill, high-intensity mixer, ribbon blender, or similar piece of

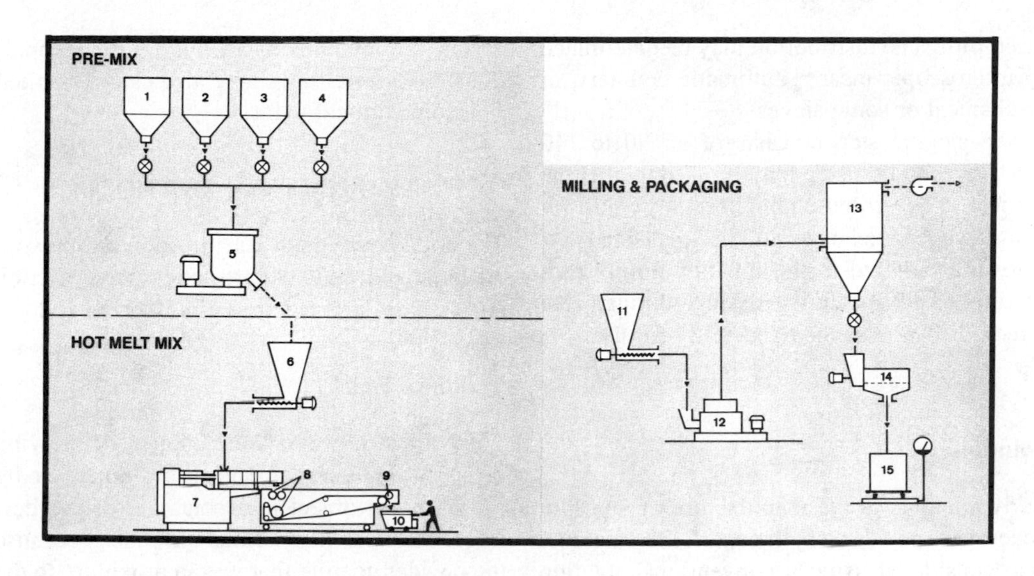

Fig. 17-2. The melt mix process.

equipment. The particle size of the mix should be between 20 and 60 mesh for a good mix and feed to the melt mixer (extruder).

After a satisfactory blend is achieved, the batch is transferred to the melt mixer. At this point, the process changes from a batch process to a continuous operation.

The melt mixer may be a kneader, a sigma blade mixer, a co-rotating or counter-rotating extruder, and so on. The mixed batch is metered into the feed section of the compounder at room temperature.

The following temperatures of the extruder depend on the rheological properties of the powder—the glass transition temperature, the melt temperature or melt viscosity, and the gel time. For example, for an epoxy with a WPE (weight per epoxide) of 1000 and a T_m of 105 to 120°C, the die head should be set at 100 to 110°C, the barrel next to the die at 80 to 100°C, the center section at 70 to 90°C, the feed section at 30°C, and the screw at 70 to 90°C. These temperatures represent a steady-state position; higher temperatures may be used at the start of the run.

From the die head, the molten mix is fed to a chill belt, cooled rolls, chilled air, or an equivalent system to drop the temperature of the material from 100 to 40°C for crushing. A coarse crusher prepares the material for subsequent pulverization into a fine powder.

The most common pulverizers are hammer, air classification, and fluid energy mills. The hammer mill generates large quantities of heat during the grinding process. Consequently, cryogenic media are commonly employed to keep the pulverized coating below its softening point. The air classification mill relies on large volumes of air to classify the particle size distribution of the powder and keep the pulverized coating cool. The disadvantage of this system is that a large-volume dust collector is needed, making the system bulky and difficult to clean. However, incorporation of high-efficiency cyclone collectors has reduced the time needed for cleaning. The fluid energy mills require large volumes of air at high pressures. The units usually have a low output and are bulky and difficult to clean from batch to batch.

The particle size distribution is a function of the type of pulverizer, the operation of the pulverizer, and the rheological properties of the powder. It is necessary to control the distribution because slight changes affect package stability, handling characteristics, electrostatic charge acceptance, and the appearance of the fused film. The distribution may be determined by microscopic means, automatic counters, or mechanical or sonic sieves.

The sieving step consists of an 80 to 140 mesh screen. This step may be used to alter the particle size distribution but normally is just for scalping oversized material. The oversided particles are returned to the mill for further processing. The material passing through the screen then is packaged and ready for use.

Solution

Early attempts in the manufacture of thin film decorative powder coatings (25–50 microns) also considered using a conventional solution coating technique. The process consists of manufacturing the coatings just like a conventional solvent-based coating. After the batch is standardized for quality, color, and so on, the solvent is removed by either spray drying, a devolatizer, or flocculation in water.

The spray drying approach is dependent on the percentage of solvent, the type of solvent, and the viscosity of the paint solution. The solution is reduced to 10 to 20% solids with methylene chloride or acetone. Spray drying produces a narrow particle size distribution that can be adjusted by varying the above parameters. For example, the higher the solids, the coarser the particle size distribution. The spray dry technique also produces a spherical particle, whereas the other techniques yield jagged, irregular-shaped particles. The spherical particles may be advantageous for handling and application of the powder. However, this technique generally is considered to be the most costly per pound output.

Devolatizers are of many designs, such as thin film evaporators. They are affected by the same parameters as the spray dryer. In this process, however, a film or flake normally is produced. The flake then is pulverized in the same manner as mentioned in the melt mix technique.

By the proper selection of solvents, the paint can be dispersed in water to form a powder slurry. The water can be removed by spray drying or devolatizing. Under certain conditions, however, this slurry also can be applied as a conventional paint.

APPLICATION METHODS

The four basic methods for applying powder are fluidized bed, electrostatic spray, friction static spray, and electrostatic fluidized bed.

Fluidized Bed

The earliest powders were applied by blowing or sprinkling powder onto a hot object or by rolling a hot object in a container of powder. These methods had serious drawbacks because the powder coating that was in proximity to the heated object would begin to fuse, sinter, or even react. The invention of the fluidized bed was the birth of serious application methods for powder coatings.

In the fluid bed process, powder is placed in the bed (see Fig. 17-3), which is basically a container with a false bottom. The bottom is a porous membrane, which does not allow the powder to fall through but does allow, and evenly distributes, an upflow of dried air.

The air rises through the powder contained in the bed, resulting in a fluid action of air and powder. This action is necessary to give an even distribution of powder and to allow an object to enter the bed.

The object to be coated is preheated to a tem-

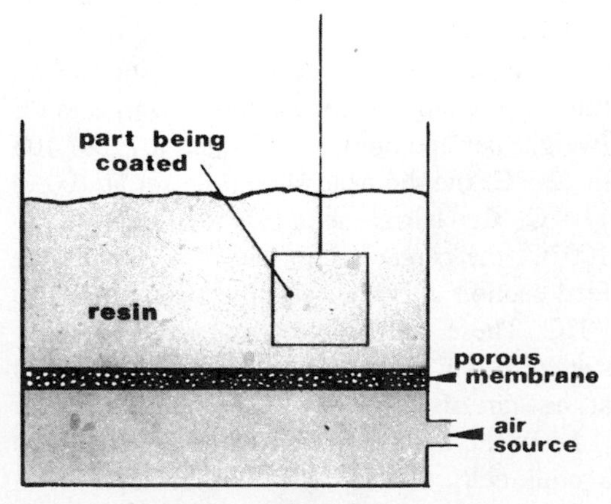

Fig. 17-3. Fluidized bed coating.

perature above the melting point of the powder coating. The preheated object then is immersed in the fluidized powder, where the powder particles melt and fuse together to form a film on the object. A post-heat sometimes is required to give the film more flow or an improved appearance or complete cure.

The final film thickness is determined by the following parameters: preheat temperature, the object's mass and its ability to retain heat, immersion time, the object's movement in the bed, and the velocity of fluidizing air. The thickness is in excess of 4 mils and generally is 10 to 15 mils.

The particle size of the powder also is important. A particle size that is too fine (e.g., around 30 microns) will be difficult to fluidize in the bed and will cause dusting over the bed.

The basic advantages of the bed techniques are the simplicity of the equipment and the technique, the low cost, and the ease of applying heavy film thicknesses. There also are disadvantages: it is not possible to apply thin films (less than 5 mils); the object to be coated is limited by its ability to retain heat from the preheat oven to the point of application; the type of object also is limited by the size of the bed; and color changes involve considerable cleaning unless multiple beds are used.

Electrostatic Spraying

In the 1960s, electrostatic charging principles were applied to the application of powder coatings. Most powders are insulators with relatively high volume resistivity values. Therefore, they accept a charge (positive or negative polarity) and are attracted to a grounded or oppositely charged object.

An electrostatic spray system (Fig. 17-4) consists of a powder reservoir, the powder feed mechanism, the gun design, a powder generator, the application booth, and powder recycling equipment. The powder reservoir usually is a fluidized bed or mechanically agitated hopper. From the reservoir, the powder is fed to the gun by means of a Venturi air pump. Arriving at the gun, the powder/air mixture is charged. The corona charge occurs internally, at the tip of the gun, or at an electrode close to the powder exit. The charge from the power pack varies from 60 to 120 kV and from 100 to 400 microamps. The pattern of the powder cloud may be controlled either by a specially shaped deflector or by swirling air nozzles at the spray gun tip.

The thickness of the applied powder is dependent on the object's speed through the cloud, the cloud pattern, the powder feed rate, the amount of charge applied to the powder particles, air patterns in the spray booth, the spray gun arrangement, and the resistivity of the powder.

The primary advantage of electrostatic powder spray is the ability to apply very uniform, low-thickness coating to a cold part. The system is readily automated, even if there are various-size parts. The initial transfer efficiency will vary from 50 to 85%, but the overall efficiency of a closed loop recovery system will approach 100% utilization.

The recycling system may be integrated or removed from the spray chamber. A nonintegrated system relies on a form of cyclone or

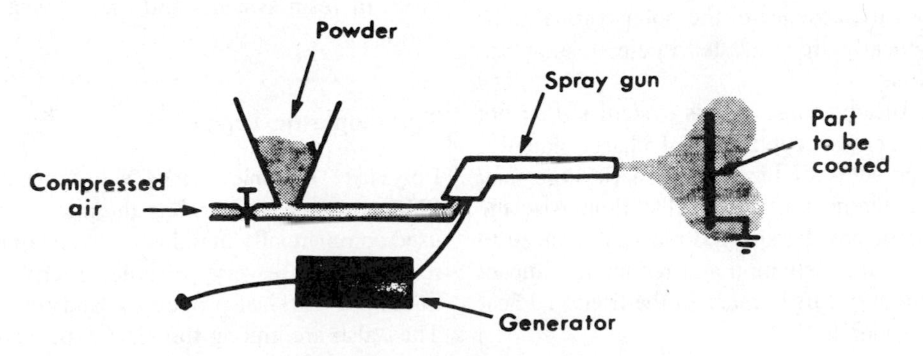

Fig. 17-4. Electrostatic spraying.

bag house to collect and remove the powder from the air cloud. The material then is transferred either manually or pneumatically to the powder feed hopper. The integrated system has the collection built into the spray chamber. This removes the need to transport the recovered material for reuse.

In all cases, however, the reclaimed material should be blended with fresh or virgin powder to maintain the appropriate particle size distribution.

A disadvantage of this method is that a quick color change is difficult to achieve. The reclamation equipment must be either cleaned or removed and replaced to accommodate a new color. Another disadvantage of electrostatic spray is the field effect that prevents powder deposition in small openings and tight angles (known as the "Faraday cage effect"). Also, because of the resistivity of the powder, thick films (in excess of 8 mils) can be difficult to deposit.

Friction Static Spraying

A method of powder application that was developed in Scandinavia during the 1970s is known as friction or tribo charging. Tribo charging relies on the powder particles colliding with the surfaces of the powder spray apparatus as well as each other; these collisions result in a random static charge being deposited on the powder particles. A device in the spray apparatus strips either the positive or the negative charge from the powder, allowing uniformly charged particles to exit the spray unit. The resulting static field allows the powder to enter "Faraday cage" areas more easily because of the absence of the polar corona fields that typically are generated in electrostatic applications.

The disadvantage of this system is that not all powders accept frictional charge equally. Even powders of the same generic type may show differences in charge retention. Also the volume of powder sprayed must be reduced to ensure charge retention and reduce the impact fusion that occurs because of the frictional heat that is generated.

Tribo charging application systems are best employed for parts that are difficult to coat with conventional or electrostatic methods.

Electrostatic Fluidized Bed

This process is a combination of the fluidized bed and electrostatic spray methods. A charged current of air passes through the permeable membrane, causing the powder particles to be repelled by each other and to form a cloud above the bed. The applied voltage determines the density of the powder cloud. As a grounded object passes over the bed, the charged powder is attracted to it.

The same electrostatic application principles are pertinent here. Thus, objects with deep recesses are difficult to coat. Also, long vertical or dimensional surfaces are not coated uniformly because of the inability of the unit to "throw" powders long distances.

Generally, because of the absence of liquid media, powder coatings are easier to apply than liquid-borne systems on a conductive substrate. Improvements in the design and performance of application equipment are continuing. Film thickness may be controlled to ± 0.2 mil, and powder delivery may be confined to ± 5 grams per minute. Future applications for powder will be high-speed coil and blank coating, webbing, and fencing. (See Fig. 17-5.)

POWDER COATING TYPES

As mentioned previously, there are two generic types of powder coating available in the marketplace, thermoplastic and thermosetting. This section gives a brief description of the various types of resin systems and their typical applications.

Thermoplastic Types

Polyvinyl Chloride (PVC). Thermoplastic PVC powders were among the first resin types used commercially in industry. Members of this resin family are very versatile in terms of formulation, physical properties, and processing. They also are among the least expensive resin systems used for coating.

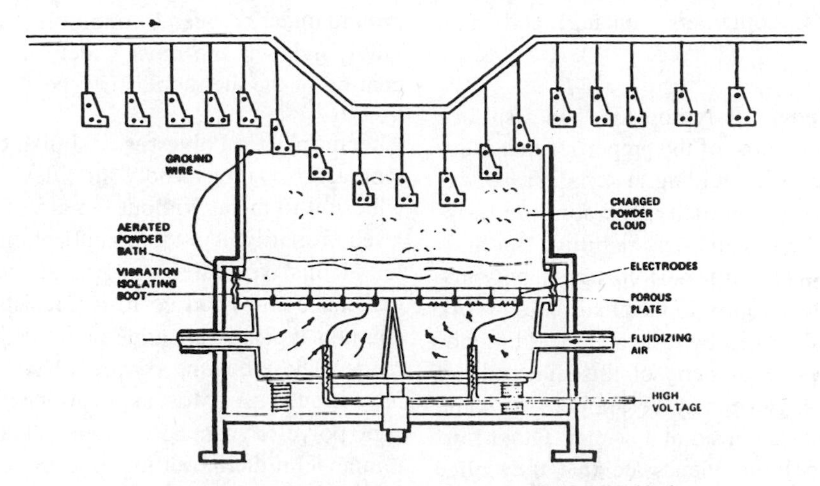

Fig. 17-5. A typical automated electrostatic fluidized bed.

Vinyl powders, used primarily in fluid bed processes, exhibit excellent exterior durability, high impact resistance, and good chemical resistance. As an encapsulation coating, PVC is most commonly used in conjunction with a primer because vinyl does not readily adhere to metal surfaces. When used over a properly formulated primer, vinyl powders exhibit outstanding resistance to corrosion, detergents, and humidity. They also have high dielectric strength and high abrasion resistance.

Typical applications for vinyls are coatings on dishwasher shelving, electrical grade insulators, and outdoor fencing.

The vinyl resin may be suspension- or bulk-polymerized; the main considerations are molecular weight, inherent viscosity, plasticizer absorption, and particle size. These formulations typically are made by using the dry blend method of manufacture. During blending, plasticizers, stabilizers, antioxidants, colorants, and inerts are blended with the base vinyl. After the initial high-intensity blending, the lumpy powder is cooled and coated with a light dusting of vinyl resin, after which the mixture is sifted. Vinyl systems can be formulated to be very flexible or very rigid, depending on the amount and type of plasticizer and stabilizer. The dusting resin is selected on the basis of its effect on film fusion during the heating cycle and the flowing properties of the bulk powder.

The major drawback of vinyl powders is that the normal applied film thickness usually is in excess of 5 mils, with 10 to 15 mils being common. Another problem is that they tend to become brittle with age because of exudation of the plasticizers.

Vinyl powders are used in the appliance industry because of their resistance to hot water, detergents, and heat. Their high dielectric strength and good corrosion resistance make them acceptable coatings for insulators, pipe, outdoor furniture, battery clamps, and so forth. They also are used for items with limited food contact.

Polyethylene. Polyethylene is another low-cost thermoplastic material. Polyethylene powders are relatively easy to formulate because of the wide range of densities and melt indexes available with these resins. Being the first powder coating offered to industry, polyethylene also offers excellent chemical resistance and high dielectric strength. Because of their smooth, slippery surface upon fusing, these coatings are easy to clean and often used for hospital and laboratory surfaces. Their major limitations are poor adhesion and poor abrasion resistance. Like vinyls, polyethylenes are used at relatively high film thicknesses.

Used primarily as a powder in rotational molding, polyethylene also has begun to emerge as a coating for the exterior of natural gas and oil distribution pipe. Its good corrosion resistance also makes it a suitable coating for marine parts, containers, fencing, and lawn furniture.

Polypropylene. Polypropylene, as a surface coating, offers most of the properties that it exhibits as a plastic molding material. It has excellent chemical, humidity, and abrasion resistance, as well as good heat stability. But as is true of most thermoplastic coatings, polypropylene adheres poorly to metal surfaces. Work has been undertaken by resin suppliers to modify the adhesive property of this material; so now polypropylene may be applied, in some cases, without the need of a primer. Other limitations of polypropylenes are that they offer poor weathering and impact performance.

Because polypropylene is essentially inert, it is commonly used for food contact items and hospital implements.

Cellulose Acetate Butyrate (CAB). Unlike most thermoplastic coatings, CAB can be applied at relatively low film thicknesses (3 mils). It does, however, require high fusion temperatures, normally 400 to 500°F. Coating films of CAB offer exceptionally high gloss and clarity. They also are very weather-resistant.

Once very popular as a coating material, CAB has fallen from favor as a base resin system and now is used primarily as an additive for other resins.

Nylon. Nylon coatings are tough, inert coatings that are very popular for use on parts that experience high abrasion. Powder coatings made of this resin system (usually nylon 11) offer excellent flow, an extremely low coefficient of friction, and good chemical resistance. This thermoplastic must be used with a primer to develop adhesion to a metal substrate.

Applied at film thicknesses of 5 to 10 mils, nylon powder commonly is used for coating kitchen implements, automotive engine components, machinery, and hospital equipment. Because of its relatively high cost, however, its use has been limited to specialized applications.

The fusion temperature needed for nylon is 400 to 500°F. It commonly is manufactured by using the melt mix technique, but it must be ground under cryogenic conditions. Along with vinyl, nylon is the most widely used powder coating of the thermoplastic type.

Thermoplastic Polyester. Unlike other thermoplastic coatings, polyester develops good adhesion to metal without the aid of a primer. Used primarily in outdoor applications, polyester of this type offers fairly good weathering resistance and good corrosion resistance. The coatings typically are applied in the 5 to 15 mil range. These coating powders have proved to have limited use. Most applications that can accept polyester can be accomplished with the thinner-film thermosetting grade.

As can be seen, thermoplastic powders generally are used for thicker-film, specialty applications. They do not typically compete for the same market applications as liquid-borne coatings. Thermoplastic powders can and do fill special coating requirements; but, generally, they make up less than 5% of the powder coatings in use today.

Thermosetting Types

Epoxy. Generally considered the workhorse of the industry, epoxy powder coatings now are the most widely used. Developed during the 1950s, epoxy coatings offer excellent chemical resistance, extraordinary hardness and abrasion resistance, and unsurpassed corrosion resistance. Hardness of 6H pencil is common while retaining reverse impact resistance in excess of 160 inch-lb. Epoxies also exhibit high dielectric resistance. Because of their high corrosion resistance, they are the coating of choice for such applications as automotive underbody components, natural gas and oil distribution pipelines, reinforcing steel bars used in highway and bridge construction, and appliance interiors and shelving. Newer multifunctional epoxies have found use in extreme-environment exposure, such as "down hole" oil drilling pipes. Also many epoxy formulations can be used in aggressive food contact applications, such as concentrated-citrus shipping containers.

Most epoxy powder coatings are based on Type 3 and Type 4 resins with a WPE of 1000 or greater. The higher the WPE, the greater the flexibility and package stability. Lower WPE resins demonstrate better flow and chemical resistance, but are less stable and more brittle.

Epoxy powders generally are made by using the melt mix process, but ball milling also is used. The formulating parameters are as wide as the resin types available, being very flexible as to the type, amount and variety of additives, colorants, inerts, and catalysts used.

The applied film thickness of epoxy powders can range from 0.8 mil to 50 mils on preheated metal. Formulated specular gloss can range from near flat (0%) to high gloss (95+%). The coating may be textured or ultrasmooth. This resin type also demonstrates outstanding charging capabilities for electrostatic or frictional charging.

The main problem of this panacea-like coating is that it has terrible weathering performance, degrading quickly in the presence of sunlight. Film chalking and color fade are quite prominent after a short period of time. However, for some strictly functional applications, its corrosion resistance far outweighs its unsightly appearance. Epoxies also tend to yellow in the presence of heat, especially in the presence of a natural-gas-fired heat source.

Epoxy Polyester Hybrids. This powder coating system originally was developed in an attempt to improve the weathering performance of the pure epoxy system. Specialty carboxylic (COOH)-terminated polyester resins were formulated to react with the epoxide group in the epoxy, to diffuse the photoreactive epoxide ring. These efforts were unsuccessful, but what did result was a new class of coating material. Hybrids offer some of the advantages of both epoxy and polyester chemistries.

This group of powder coatings should be considered part of the epoxy family, even though today many hybrid formulations contain much more polyester than epoxy resin. They are segregated here because their resultant properties differ substantially from either resin group used alone.

Hybrids demonstrate good corrosion resistance, impact, and flexibility, but they deliver a softer film than the epoxies. Although they may show the same performance in corrosion protection, they are more vulnerable to chemical attack. A major enhancement of hybrid systems is that they offer enhanced yellowing resistance.

Next to epoxies, hybrids have become the most commonly used powder coating material worldwide. The polyester resins are designed to react stoichiometrically with epoxy on a 40:60 to 90:10 basis, with cost versus performance being the major consideration. This puts them in the catchall category of "general purpose" coating materials.

Hybrids are manufactured by using standard melt mix and pulverizing techniques; so they are commonplace in the industry.

Hybrids are most easily compared to liquid alkyd paints, but like alkyds they have poor weathering and pervious-film integrity. They may be categorized as relatively low-cost, functional, decorative coatings.

Their main disadvantages are poor weathering, poor abrasion resistance, relatively poor flow and leveling, and moderate chemical resistance. However, they do maintain good heat resistance, good corrosion protection, and flexibility.

Polyester/Polyurethane. Thermosetting polyester powder coatings are further subdivided into polyester and polyurethane powders, as polyol ester may be reacted with a variety of curatives. The most common powder coating reactions involve polyester reacting with acids, epoxides, and isocyanates. Each reacted thermoset coating possesses unique properties.

The polyurethane powders are those that involve reaction of polyol ester with an isocyanate. Although not popular in Europe, this type is the largest-selling polyester powder in North America. Polyurethane powders offer exceptionally smooth films, good chemical and weathering resistance, good flexibility and impact resistance, and can be produced in a wide gloss range (0–100% reflected gloss).

They are limited in the film thickness range in which they may be applied (1–3 mils), primarily because of the chemical reaction that takes place. Polyurethane powders are similar to their liquid counterparts, the main difference being that the isocyanate is chemically blocked in the powder so that it will not react at room temperature. The blocking agent volatilizes at temperatures between 320°F and 350°F. Film thicknesses in excess of 3 mils may trap the blocking agent in the film, causing a pinholing effect. Also, because of the deblocking temperature, polyurethane powders will not react at temperatures less than 325°F. New developments in blocking agent technology may change these limitations in the near future.

Polyurethane powder coatings have found wide approval in the coating of automotive trim components, steel wheels, lighting fixtures, patio furniture, and appliance panels. Some appliance manufacturers are using polyurethane powders as coatings on precut blanks that are subsequently post-formed to a zero T-bend radius! Generally, any application that calls for very smooth films at low film thickness may use this coating type.

Usually prepared by using the melt mix or solution technique, polyurethane resin systems have characteristics that vary as to the melt and glass transition point, acid of hydroxyl value, and molecular weight. Each resin chemistry demonstrates unique gloss, flow, and reaction rate.

Glycidyl Polyester. Glycidyl reacted polyester powder is an improved weathering version of the polyester/epoxy hybrids previously mentioned. In this case, a low molecular weight glycidyl (or epoxide functional) curing agent is reacted with an acid-terminated polyester. Because of the low WPE, very little curative is needed to fully react with the polyester backbone (93 : 7 or 95 : 5). Due to the polyester content, these systems demonstrate equal or even better weathering resistance than polyurethanes.

Along with the good weathering, glycidyl polyesters impart high impact strength, good corrosion protection, and a wider cure response range and are easily applied over a variety of substrates. Because this coating is formed by using a condensation reaction, the applied film thickness can vary between 1 and 20 mils, making it useful as a decorative or a functional material.

Glycidyl polyesters tend to be softer than polyurethanes and cannot currently be formulated as low-gloss coatings (less than 30%). The flow and leveling of glycidyl polyester also are not as good as for polyurethanes. But as this new technology develops, these aspects of their properties are improving.

Glycidyl polyester is now the coating of choice for architectural aluminum profiles in Europe. Because of its good adhesion on various substrates, it also is widely used on galvanized steel. Automotive uses include base coat/clear coat systems for aluminum wheels. Its high dielectric strength makes it a good coating for transformers as well.

Manufactured by using the melt mix technique, glycidyl polyester powder coatings represent the latest technology development in polyester powders.

Acrylics. Acrylic powder coatings are being supplied as acrylic urethanes or as acid or epoxy reacted coatings. These powders are similar to their liquid counterparts; that is, they have excellent gloss and clarity, produce very hard films, and have excellent resistance to corrosive and chemical attack. They also provide smooth films at low film weights and apply well with electrostatic spray equipment.

Also like the liquid systems, however, acrylic powders tend to be brittle and do not have good impact strength when compared to other powder coatings.

Acrylic powders are gaining acceptance as the coating of choice for automotive, appliance, and sanitary ware applications. The fact that early formulations of acrylic were incompatible with other powder types slowed their growth in the marketplace, as their users had to dedicate a production facility to one powder type. New formulations are proving to be completely compatible with other powders, speeding their acceptance.

Manufactured by using melt mix or solution techniques, acrylic powders may cross the final hurdle as a powder coating, for use as an automotive top coat. Acrylic performance varies widely, depending on monomer and curative selection. Their cure response can be designed for reactions between 300 and 400°F. (See Table 17-1.)

Table 17-1. Powder coatings quick reference chart.

Property	*Std. Epoxy*	*Mod. Epoxy*	*Hybrid*	*Urethane (1)*	*Urethane (2)*	*Polyester*	*Acrylic*
Weatherability	P	P	F	P	VG	E	VG
Chemical Resistance	E	F	P	E	E	P	E
Detergent Resistance	F	P	G	E	E	G	E
Impact Resistance	E	E	E	VG	VG	E	G
Flexibility	E	E	E	VG	G	E	F
Hardness	2H–4H	H–2H	HB–2H	2H–6H	2H–6H	H–2H	2H–6H
Heat Stability	P	F	G	P	F	E	G
Gas Oven Stability	P	P	VG	P	F	E	F
Adhesion to: CRS	E	E	E	E	E	E	E
Aluminum	G	G	E	G	G	E	G
Galvanized	F	F	E	F	F	E	F
Copper	F	F	G	P	P	G	P
Film Thickness	1–4 mils	1–25 mils	1–15 mils	1–4 mils	1–4 mils	1–15 mils	2–3 mils
Gloss Range	0–100	70–100	50–100	5–100	5–100	50–100	10–100

Urethane (1) = TDI Cure Urethane (2) = IPDI Cure

THE APPLICATION LINE

The application of powder coatings is similar to that of liquid paints. One notable difference is that regardless of powder type, heating of the substrate is required. Consequently, powder coating is not practical for temperature-sensitive materials.

The first step in their use is substrate preparation. The substrate must be cleaned, chemically treated, or mechanically abraded in order to give maximum performance of the finished coating. The substrate then must be dried before coating. Most powders are not affected by moisture during the curing process; so dry-off ovens may be combined with the curing oven to reduce floor space and capital investment.

The next step, depending on the powder used and the film thickness required, is substrate temperature modification. Thick films normally require preheating of the substrate, whereas precisely controlled thin-film application calls for a relatively cool temperature and strict humidity control.

The material then is applied by using one of the methods previously described and then cooled or cured, depending on the powder type. Each of these factors contributes to the final performance of the coating. Thus care should be taken to closely specify each step of the application process. A typical application line is shown in Fig. 17-6.

ECONOMICS

Many items contribute to the economic advantage of powder coating, but one of the most important factors is that *the user can utilize all of the material that is purchased*. Besides material utilization, other factors that contribute to the advantage of powder coating are: energy savings, labor cost reduction, higher operation efficiency, reduced material handling, and lower insurance costs. This section will explain each of these advantages in detail.

The storage of finishing materials in the user's factory demonstrates a powder coating's advantage with respect to reduced floor space and hazardous material handling. The space needed for storage of 10,000 pounds of dry powder is much less than that needed for 20,000 pounds of liquid material that contains 10,000 pounds of dry solids. There is no need for construction of special rooms to store volatile material. This factor also affects shipping costs. A full truckload of liquid paint (i.e., 30,000 to 40,000 pounds) represents, on average, 15,000 to 20,000 pounds of dry solids coating. Freight is paid, however, on the full 40,000 pounds of liquid. Not only must this liquid medium be purchased and shipped; it must be disposed of after application.

The environmental costs (both literally and figuratively) are difficult to predetermine, but should be considered. The handling of paint

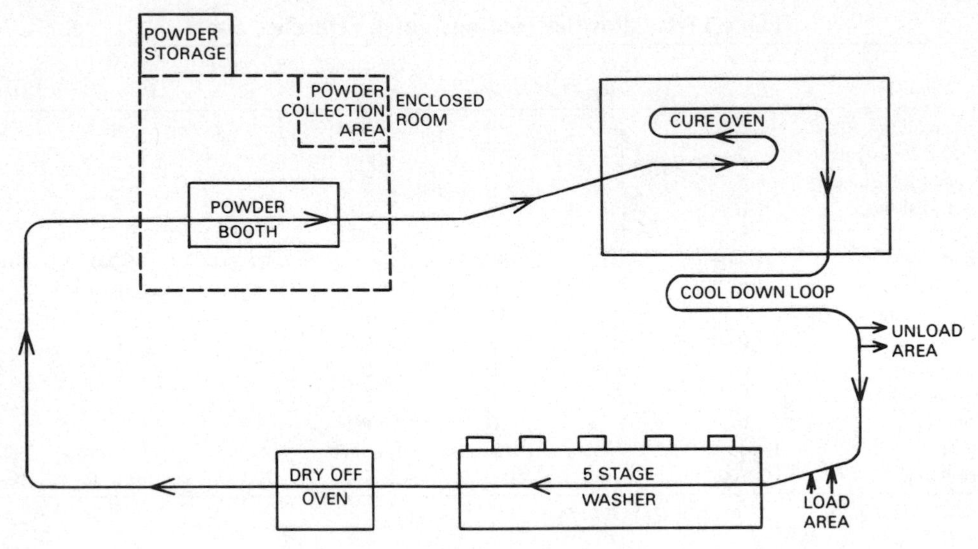

Fig. 17-6. A typical application line.

volatiles and the effluent discharge certainly represents an ever increasing cost to the user. Even coatings that are considered environmentally compliant, in terms of air pollution, still present user costs for handling of the paint sludge and the effluent discharge.

Another economic incentive of powder coating is that it features fast film build and ease of automation. Applications that call for film builds of 2 to 5 mils can be achieved in one application rather than multiple operations. Using the fluidized bed method, single-step film builds of 15 to 20 mils are common. Sagging, dripping, and solvent popping are no longer a concern with powder coating.

Although curing temperatures needed for powder coating are higher than those necessary for liquid systems, overall energy requirements are reduced. This is so because there is no longer a need to heat makeup air in the spray booth, and solvent or water flash-off zones are unnecessary. Exhaust air in the curing oven also is reduced. The normal air turnover of 15 to 20 times per hour for liquid-borne coatings can be reduced to 4 to 5 times per hour with powder because there is a greatly limited generation of volatile material.

Because powders come to the user "ready to use," there is no need for a paint mix room or a pumping station. Normal viscosity adjustments that are required for liquid systems during summer and winter months are not required for powders. This difference means lower space requirements and possibly lower equipment costs with powders.

Being in a ready-to-use state, powder coatings also eliminate the service of a skilled operator, making labor costs less intensive. Often, there is no demand for a manual coating application. Also, spray booth and ancillary equipment cleaning are greatly diminished, and there is no paint sludge to be disposed of or arrestor filters to be replaced. Savings also are realized by eliminating solvent recovery or afterburning equipment.

Last, but not least, is the economic consideration of factory safety. Because powder coatings do not contain solvents, the risk of fire is greatly diminished. Therefore, insurance costs are reduced. Any spillage outside of the application area may be cleaned by simple vacuum. There also is a reduced health hazard because factory operators are not breathing solvent-contaminated air.

A comparison of applied costs between liquid coating and powder coating is useful. The following calculations compare the cost per dry pound:

- *For liquid systems*:

Cost per dry pound

$$= \frac{\$\text{ Cost per gal.} + \$\text{ Cost of thinner}}{(\text{Weight per gal.}) \times (\%\text{ N.V. by weight})}$$

- *For powder systems:* cost per dry pound is the quoted price.

$$\text{II. } \frac{\text{Specific gravity} \times \text{Film thickness} \times \$/\text{lb} \times 100}{\text{Efficiency} \times 192.3} = ¢/\text{sq ft}$$

The actual application costs also can be compared:

- *Liquid*:

100% *Effective* efficiency

$$= \frac{\$\text{ Cost per dry lb. of paint}}{\text{Efficiency of liquid spray}}$$

- *Powder*:

100% *Effective* efficiency

$$= \frac{\$\text{ Cost per lb. of powder paint}}{\text{Efficiency of electrostatic spray}}$$

These formulas compare the actual application costs of liquid and powder paint according to the efficiency of the application.

As already indicated, there are a number of ways to save with powder coatings, such as less energy usage, less space required, no solvents, and higher efficiency of material usage. It is imperative here that the unit material and applications costs be determined. The following data detail unit cost calculations and the comparisons between powder coating and liquid system materials. These material unit costs along with the savings and applied unit costs are normally quite sufficient for the adoption of this technology.

Unit Cost Calculations:

(a) The coverage of one pound of powder can be determined by the following equation:

$$\text{I. } \frac{\text{Efficiency} \times 192.3}{\text{Specific gravity}} = \text{Coverage sq ft/mil}$$

(b) The cost per square foot can be found by using the following equation:

To determine your unit or part cost, merely multiply the cost per square foot by the area in feet of your part:

$$¢/\text{sq ft} \times \text{sq ft/unit} = ¢/\text{unit}$$

Example—Epoxy Powder Coating:

Film thickness: 1.4 mils
Specific gravity: 1.2
Cost: $2.10 per pound
Efficiency: 98%

Using equation (I):

$$\frac{.98 \times 192.3}{1.2} = 157 \text{ sq ft/mil}$$

or:

$$\frac{157 \text{ sq ft/lb}}{1.4 \text{ mil}} = 112 \text{ sq ft at } 1.4 \text{ mil}$$

Using equation (II):

$$\frac{1.2 \times 1.4 \times 2.1 \times 100}{.98 \times 192.3} = 1.8¢/\text{sq ft}$$

(c) The coverage of one gallon of paint can be determined by:

$$\text{III. Efficiency} \times \%\text{ Volume solids} \times 1604 = \text{Coverage sq ft/mil}$$

(d) The cost per square foot can be found by:

$$\text{IV.}\quad \frac{\$/\text{gallon} \times \text{Film thickness} \times 100}{\text{Efficiency} \times \%\ \text{Volume solids} \times 1604} = ¢/\text{sq ft}$$

Example—Liquid Paint:

Film thickness: 1.4 mils
Volume solids: 35%
Cost: $7.85 per gallon
Efficiency: 65%

Using equation (III):

$$.65 \times .35 \times 1604 = 365 \text{ sq ft/mil}$$

or:

$$\frac{365 \text{ sq ft/mil}}{1.4 \text{ mil}} = 261 \text{ sq ft at } 1.4 \text{ mil}$$

Using equation (IV):

$$\frac{7.85 \times 1.4 \times 100}{.65 \times .35 \times 1604} = 3.0¢/\text{sq ft}$$

As can be seen, the concept of powder coatings with plastics offers a number of economic and functional advantages. It is anticipated that it will continue to play an even more important role in the years to come.

18
Reinforced Plastics and Composites

A reinforced plastic, or composite, is a polymer resin matrix combined with a reinforcing agent or agents. The reinforcement improves strength and stiffness properties, compared to the neat resin. Almost every type of thermoplastic or thermosetting resin can be reinforced. ''Hybrid'' composites combine more than one type of reinforcement, such as carbon and glass fibers.

New reinforcing materials, resin matrices, and processing techniques are constantly being discovered or commercialized. At the same time, enhancements are continually engineered into existing materials. In addition, most resin types can be formulated or processed with a variety of fillers, additives, extenders, and chemicals. By selecting ingredients and setting process conditions, a designer or an engineer specifies a custom ''material system.'' The composite is engineered to meet specific application, performance, and/or processing needs.

Probably no other branch of plastics is more complex and diverse than the field of reinforced plastics and composites. The scope of this handbook does not permit an encyclopedic approach to the subject. The coverage herein is considered comprehensive but surely not all-inconclusive. Those desiring more details on specific subjects are encouraged to contact major suppliers and molders or The Composites Institute of The Society of the Plastics Industry.

By Richard D. Kaverman, Marketing Communications Manager, Owens-Corning Fiberglas Corporation, Toledo, OH.

OVERVIEW

A Brief History

Plastics were established for decades before researchers decided to process plastics with reinforcements. The synergy of fibers in polymers was a natural—as natural as cellulose in wood, straw in mud building blocks, and steel rebars in concrete.

Most industrial historians mention radomes and fuel tanks for World War II aircraft when asked to name the first major reinforced plastic applications. But the reinforced plastics industry did not start to blossom until postwar consumerism began to gain momentum. The first major application was the ''fiber glass'' boat, which completely transformed the pleasure boat industry. In 1953, Chevrolet introduced the Corvette with a ''fiber glass'' body (see Fig. 18-1).

The Corvette's success brought widespread attention to reinforced plastics, which continued to grow rapidly through the 1960s. Major uses included electrical insulating panels, bathtubs and shower stalls, piping and storage tanks, and fishing rods. Although low in volume, reinforced plastics applications for aircraft and aerospace were demonstrating the high-performance end of this family of materials. A distinct ''advanced composites'' branch evolved within the reinforced plastics industry.

Meanwhile, science and industry combined to develop ways to produce reinforced plastic

Fig. 18-1. Chevrolet Corvette convertible. (*Except where noted, photos and line drawings courtesy of Owens-Corning Fiberglas Corporation*)

components at higher volumes. The primary target of this attention was the automotive market. A major move toward increased automotive use was the development of sheet molding compound (SMC). The fiber glass and thermoset polyester sheet is compression-molded in matched metal dies. This and other processes are covered in more detail later in this chapter.

In the early 1970s, a small fender extender on a compact car was changed from sheet steel to SMC. Soon the material was specified for numerous front end and grille opening panels. Doughlike bulk molding compounds (BMCs) were being compression- and injection-molded into shrouds and covers for underhood parts. Simultaneously, usage of injection-molded reinforced thermoplastic compounds was starting to grow. By the mid-1970s, the transportation market passed the marine market as the number one consumer of reinforced plastics and composites.

In the second half of the 1980s, the reinforced plastics industry continued to post successive record years. This decade also saw a revolution in the research and development of new processes and materials. Labs also were introducing quantum improvements and advances in existing processes and materials. For example, during this period, the industry achieved a 60-second mold cycle for compression-molded SMC; and gains were made in the field of ''long fiber'' thermoplastic compounds for higher-strength composites.

In addition, structural automotive components were produced by the reaction injection and resin transfer molding processes. Thermoplastics, reactive resins, and phenolics were introduced to the pultrusion process. High-temperature thermoplastics were reinforced with high-modulus fibers to make aircraft components that met demanding performance requirements. An aircraft fuselage was produced by filament winding. Molders began to produce body panels for the General Motors APV minivan—the largest, mass-produced application of reinforced plastics in history.

Terminology

''Reinforced plastics'' and ''composites'' are just two of the many terms used to describe these materials. A glossary of designations includes the following:

advanced composites—Reinforced plastics with properties higher than ''conventional'' performance; usually have a reinforcing fiber with a higher modulus than E-glass fibers and a resin matrix with higher properties than polyesters.

composites—Used interchangeably with "reinforced plastics"; because the dictionary definition of composites is broader in meaning, the reinforced plastics industry often uses "composites" and "composite" with a modifier, as in "advanced composites" or "fiber glass composite."

fiber glass—Technically, glass in fibrous form; however, the term is often used to describe fiber glass-reinforced plastics and applications, as in "fiber glass" boat or "fiber glass" door. Fiberglas, a registered trademark of Owens-Corning Fiberglas Corporation, is often erroneously used in place of the generic term.

FRP—Used specifically for fiber glass-reinforced plastics.

GFRP—Glass fiber-reinforced plastic.

GRP—Widely used in Europe to mean glass-reinforced plastics or, more specifically, glass-reinforced polyester.

laminate—Originally reserved for material systems with layers, such as lay-up, but often used to generically describe reinforced plastics, as in pounds of "laminate" per part.

reinforced plastics—Polymer-based matrices that incorporate fibrous reinforcements, flakes, and other mechanical (as opposed to chemical) additives to enhance properties.

RP—Reinforced plastics.

RTP—Sometimes used to distinguish "reinforced thermoplastics" from reinforced thermosetting plastics.

Thermosetting Plastics

Reinforced plastics traditionally have been associated with thermosetting resins. Thermoset resins "cure" into a hardened, irreversible state, in which the molecular structure of the polymer is cross-linked. Cure can be accelerated, enhanced, or controlled by heat, radiation, or light. Many reinforced plastics molding processes use the application of pressure to improve surface finish and meet requirements for close dimensional tolerances.

Unsaturated polyesters are by far the most widely reinforced resin matrix. Thermoset polyesters provide a good combination for performance, handleability, and reasonable costs. The three major categories of thermoset polyesters are orthophthalics, isophthalics, and vinyl esters—listed in ascending order of cost and resistance to moisture, weathering, and chemicals. This family of materials also is extremely versatile. Because of their widespread and long-term use, thermoset polyesters often are relegated to merely a "conventional" status. However, a deeper understanding of polymerization has been accompanied by a growing number of ways to precisely manipulate, and replicate, chemical reactions. As a result, resin manufacturers are producing thermoset polyester systems that are engineered to address specific process and end-performance needs.

Polyester cure, or molecular cross-linking, starts when the resin and a monomer, such as styrene, are mixed with a catalyst. The catalyst for thermoset polyesters usually is an organic peroxide. Accelerators can be used to speed curing; inhibitors slow the cure rate for better control of the molding process and to give the resin a longer shelf life. Thixotropics and viscosity suppressants are used to control the flow of liquid resin to suit the molding process.

Epoxies bridge the cost and performance gap between polyesters and high-performance polymers. Because of an inherently adhesive nature, epoxies generally are more difficult to handle than polyesters. Epoxies perform well at elevated temperatures and in corrosive environments. This resin family exhibits low shrinkage when cured. Epoxy resins co-react with a hardener to achieve cross-linking, or cure. The major hardeners for reinforced epoxies are amines and anhydrides.

Phenolics typically are used for applications calling for heat resistance, good dielectrics, fire retardancy, and low-smoke characteristics. A phenolic resin cures through the catalyzed reaction of phenol and formaldehyde. Phenolics are available in either liquid or powder form. Elastomers, fluoropolymers, and graphite fibers are used in phenolic compounds for improved impact strength and lubricity.

Polyurethanes are predominant among the reactive thermosetting polymers. These resins cure when isocyanate and polyol components are brought together in an exothermic reaction.

Other reactive resins include polydicyclopentadiene and thermoset acrylics, based on methacrylates. These resins are used in the reaction injection molding processes (see below).

Other thermoset resins that are reinforced include terephthalic, alkyd, and diallyl phthalate polyesters, silicones, melamines, bismaleimides, and thermoset polyimides.

Thermoplastics

Thermoplastic polymers have a linear macromolecular structure that will repeatedly soften when they are heated to their melt temperature and harden, or cure, when they are cooled. Reinforced thermoplastics were virtually nonexistent in the early 1950s; but by the early 1990s, reinforced thermoplastics represented about one-fourth of the total reinforced plastics and composites market.

It would be impractical to list all the thermoplastic resins available for reinforcing. Within each thermoplastic family, there are often scores of different varieties. For example, a nylon (or polyamide) can be semicrystalline or amorphous, depending on how it is manufactured. There also are several commercially available constructions of polyethylenes, polypropylenes, styrenic resins, thermoplastic polyesters, fluoropolymers, liquid crystal polymers, and thermoplastic elastomers, to name a few. Other thermoplastic families are based on polyphenylenes, ketones, and sulfones. Also widely used are acetals, acrylics, polycarbonates, polyvinyl chlorides, and polyarylates. In addition, new thermoplastic materials are continually being developed through copolymerization, blending, and alloying. Some blends and alloys are made to differ simply by varying the percentages of the different resins used. Most thermoplastic polymers and alloys can be effectively reinforced to enhance performance and/or lower costs. (There is more on combining resins and reinforcements in the discussions of compounds, prepregs, and molding processes later in this chapter.)

Fillers and Additives

Fillers and additives are used in both thermosets and thermoplastics to change a resin system's performance, process, or cost parameters. A filler can be a simple resin extender—low-cost bulk that does not have a debilitating effect on the properties needed for the application. Some phenolic compounds, for instance, include wood ''flour'' ground from hardwoods and crushed nut shells. Other fillers provide a measured improvement in the performance of the compound and/or finished part. For example, calcium carbonate improves the surface gloss of rigid PVCs, and talc improves the heat deflection temperature of polypropylenes. Both these fillers see widespread use because of their low cost and their ability to be used at high filler loadings.

Adding alumina trihydrate is one approach to achieving flame-retardant characteristics. Another class of additives inhibits degradation from ultraviolet light. Conductive additives are used in reinforced plastics applications that require shielding from electromagnetic or radio frequency interference. Colorants provide integral color that will not chip or fade.

Metallic stearates are mixed into molding compounds to help finished parts release from the mold surface. Thermoplastic additives are used in thermoset formulations to control the shrinkage that occurs during cure. Very low shrinkage permits intricate molding to close tolerances and results in exceptionally smooth, Class A, surface finishes.

Reinforcements

The mechanical strength of a reinforced plastic component is largely dependent on the amount, arrangement, and type(s) of reinforcing fiber(s) in the resin matrix. Typically, the higher the fiber content is, the greater the strength. The arrangement has to do with the way that fibers are positioned. The three general arrangements for reinforcing fibers are:

1. *Unidirectional*—the greatest strength is in the direction of the fibers.
2. *Bidirectional*—two arrangements of fibers are at an angle to each other; strength is in each direction of fiber orientation.
3. *Multidirectional or isotropic*—strength can be uniformly distributed with chopped fibers arranged in all directions

or with various orientations of continuous fibers in a fabric.

The most common reinforcement type for plastics is glass in fibrous form. Numerous compositions and fiber orientations are available to meet specific cost, process, and performance needs (see Fig. 18-2). New technology continually enhances existing fiber glass forms and introduces new glass fibers to meet the changing needs of industry. Glass fiber reinforcements are made by drawing molten glass from a furnace through small, tubelike bushings. As a drawn filament cools, it is gathered together with other filaments to form a strand that is pulled by a winder. A surface treatment, or sizing, is applied to the fibers for two major reasons: first, like starch on textiles, the sizing protects fiber integrity during further handling steps; second, the sizing contains a chemical that helps the fiber bond to the resin matrix for which it is intended. Strands are processed into shippable fiber glass reinforcement types, described below.

Continuous roving is bundled, untwisted strands that are supplied as cylindrical packages, called doffs or balls. For some molding processes, continuous roving is chopped into specified lengths to provide isotropic strength properties within the resin matrix. For other processes, the fiber is used continuously within the matrix for exceptional strength in the fiber's longitudinal direction. ''Bulked'' continuous roving has inherent loops in both axial and transverse directions. The result is multiaxial-reinforced with uniaxial input.

Reinforcing *roll goods* include mats made with randomly dispersed chopped fibers or continuous fiber strands, laid down in a swirl pattern. In either case, the fibers are held together with resin-compatible binder. Woven roving is a heavy, drapable, fabriclike product that is woven from continuous rovings. Mats and woven rovings are sold in various densities, or weights per unit square. Reinforcement fabrics are woven from textured glass fiber yarns in a variety of weaves and weights per unit square. Nonwoven fabrics are made by knitting layers of fiber glass yarn together in unidirectional patterns. Combination products are made by stitching, needling, or bonding two or more different types of roll goods together. A typical combination is Bi-Ply(R) chopped strand mat/woven roving product. The benefit is the ability to lay up two different layers of reinforcing materials in a single step.

Fig. 18-2. Fiber glass roll goods, continuous rovings and yarns, and bulk goods.

Chopped strands are available in various lengths for compounding with resins and additives. The fiber length varies by compound but is usually in the $\frac{1}{8}$-inch (3.2 mm) to 2-inch (50 mm) range. Chopped strands are available with several different surface treatments, each designed for a specific thermosetting or thermoplastic resin.

Continuous filament yarns are used in woven and nonwoven reinforcing fabrics. These yarns are made by twisting and/or plying a number of fiber glass strands together. A proprietary yarn is engineered for the braiding of three-dimensional reinforcement structures. This special yarn does not require the user to burn off the sizing for handleability before adding a second sizing for resin compatibility. A single, dual-purpose sizing protects the fiber during braiding and still provides compatibility with polyester and epoxy resins.

As the name suggests, *milled fibers* are made by hammermilling fiber glass. The fibers are supplied in length categories by running them through different screen sizes. Milled fibers do not provide as good stiffness and strength properties as chopped strands, which are longer. Instead, the function of milled fibers is to control heat distortion of the resin and to improve the surface finish of molded parts.

Glass flakes are used in resinous coatings to increase resistance to permeability from moisture, vapors, and solvents. Glass flakes also have been used in urethanes to enhance the surface finish of reaction injection molded parts.

Conventional fibers are called E-glass. *S-2 Glass® fibers* provide higher tensile strength and modulus than E-glass. S-2 Glass reinforcements are lower in cost than other high-modulus fibers such as aramid and carbon.

Aramid fibers combine low density with high tensile properties. The fibers are made by extruding a solution of a liquid crystal thermoplastic polymer through a spinneret. Major forms available for reinforcing plastics are continuous filament yarns, rovings, and chopped fibers.

Carbon fibers are fine filaments made up mostly of elemental carbon. Fiber types range from amorphous carbon to crystalline graphite. Depending on its raw material input, a carbon fiber can have a stiffness or Young's modulus less than that of a glass fiber or three times that of steel. The most widely used carbon fibers for reinforcement have a modulus in the 30 to 40 million psi (207 to 268 GPa) range. Carbon reinforcements are available as short-length fibers, textiles, twisted and nontwisted yarns, continuous filament, and tows.

Other fiber reinforcements are made of high molecular weight polyethylene, nylon, thermoplastic polyester, and ceramic.

When two or more reinforcing fibers, such as glass and carbon, are used in a common resin matrix, the resulting composite is called a *hybrid*. These reinforcing arrangements are used to combine the higher performance of one fiber system with the lower cost of another.

Reinforcing fillers include mica platelets, fibrous and finely divided wollastonite, and hollow and solid glass microspheres.

Preforms, Compounds, and Prepregs

For some processes, reinforcing fibers are specially prepared prior to molding. Preforms are preshaped reinforcements that are combined with resin during molding. Compounds and prepregs are fiber–resin combinations that are prepared prior to molding.

A *preform* is an arrangement of fibers configured to replicate the shape of the finished part. The fibers are held together with a resinous binder. Preforms are placed in the mold cavity between male and female dies used in compression, resin transfer, or reaction injection molding. In the *directed fiber* approach to making preforms, fibers are chopped onto a rotating screen that is configured to simulate the shape of the finished part. The chopped fibers enter a sprayed stream of resinous binder that is directed toward the screen. The fibers are held in place as the preform is removed from the screen.

Preforms also can be made in a *plenum chamber*, where fiber roving is chopped over the preforming screen as air is drawn from behind the screen to hold the fibers in place. The fibers are oversprayed with the resinous binder.

Fig. 18-3. Braiding for skis.

Both directed fiber and plenum chamber preforms are placed in an oven where the binder cures, and the preform shape takes a set.

A more automated approach to making preforms is to use *thermoformable continuous strand mat*, which incorporates a binder system that allows the mat to take a shape when heated. For load-bearing applications, this mat ensures consistent, precise placement of reinforcement, especially for resin transfer and reaction injection molding.

Another way to make preforms is to *braid* the fibers into a replicate shape. Braids provide high, three-dimensional strength properties. Braided reinforcing structures can incorporate more than one fiber type, such as glass and aramid. (See Fig. 18.3)

Sheet molding compound (SMC) is an integrated, ready-to-mold composition of glass fibers, resin, and filler. SMC is made by metering a resin paste onto a thin plastic carrier film. The resin typically is thermoset polyester. SMCs also are made with vinyl ester, epoxy, phenolic, and polyimide resins. The compound is made by chopping continuous fibers onto the resin paste as it is conveyed on the film, and the glass and resin mixture is covered with an additional layer of resin on a second carrier film. Compaction rollers knead the fibers into the resin for uniform fiber distribution and wetting. (See Fig. 18-4.) The compound sandwiched between the carrier films is gathered into rolls or layered into containers and stored until it matures. Upon maturation, SMC is tack-free and has a leather-like feel; the carrier sheets are removed, and the sheet is prepared into a charge of predetermined weight and shape. The charge is placed on the bottom of two mold halves in a compression press. (For more on compression molding, see below, in the "Processes" section.)

SMC typically has 20 to 25% glass fibers by weight. Typical fiber length is 1 inch (25 mm), and fibers are deposited in a random pattern that provides isotropic orientation. However, strength properties can be enhanced by increasing the fiber content, increasing the fiber length, and/or introducing directionally oriented fibers.

Bulk molding compound (BMC) is a mixture of chopped glass fibers and resins with necessary fillers and additives. The premixed material has a doughlike consistency and is supplied in bulk form or as an extruded rope for ease of

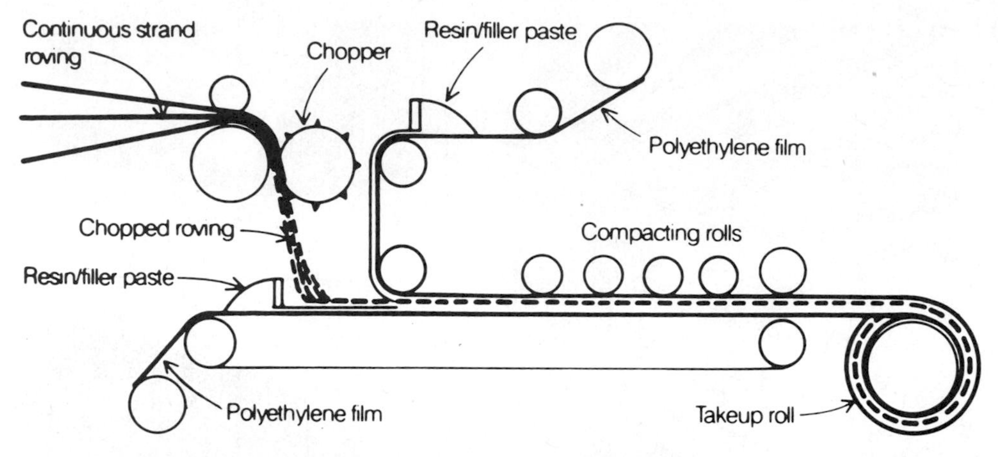

Fig. 18-4. Typical SMC machine. (*Reprinted with permission of Owens-Corning Fiberglas Corporation*)

handling. BMC is compression-, transfer-, or injection-molded.

Thermoset prepregs are made by combining liquid resin with woven and/or continuous fiber reinforcement. Through controlled chemical reaction, the resin is brought to its B-stage, which is a tack-free condition that meets particular handling and processing conditions. Prepregs are cured with heat and often are refrigerated to prevent premature cure. A prepreg also can be a three-dimensional braided structure of preimpregnated fibers. Prepreg tapes permit automated and precise placement of reinforcement in the finished part.

Several different techniques are used to produce *thermoplastic prepregs*. Because no chemical catalyst is needed to initiate thermoplastic cure, B-staging is not needed. Heat and pressure are applied to consolidate the reinforcing fiber and thermoplastic resin. Fibers can be preimpregnated or encapsulated in a resin solution, resin melt, or resin in powder form. Other thermoplastic prepregs are fabrics of commingled reinforcement and resin yarns. Thermoplastic prepregs also can be reinforced thermoplastic shapes that are re-formed under heat and pressure in a post-fabrication step. For example, filament winding and pultruding fibers with thermoplastics create composites that can be re-formed. Another thermoplastic prepreg, stampable sheet, is made by extruding thermoplastic polymer with reinforcement, generally continuous strand mat.

Pelletized compounds are made by extruding resin compound, usually thermoplastics, with chopped strands. The extrudate is cut to specific lengths to form pellets, and the pelletized compound is metered into injection molding machines or extruders and processed into an end-use component or profile. The pellet ensures good control of fiber glass content throughout the finished part. Compounds also can contain a variety of fillers and additives, depending on the cost/performance needs of the finished part. Each extruded pellet contains a near-equal amount of additive for good distribution throughout the molded part. Examples of additives are colorants, lubricants, flame retardants, and impact modifiers.

Long fiber compounds are used in applications needing additional strength and stiffness properties. These compounds can be blended with conventionally reinforced and unreinforced compounds. One technique for making long fiber compounds is to pull continuous reinforcement through the resin as it is being extruded. The resin-encapsulated strand then is cooled and chopped to length, typically more than one inch. Long fiber compounds also are made by pultruding continuous fiber through a thermoplastic resin bath. The cooled pultrudate then is chopped into a compound with specific-length fibers.

PROCESSES

Many molding and fabrication processes are available to those who want to benefit from reinforced plastics and composites. Each process has characteristics, advantages, and limitations relating to: (1) part size, complexity, and volume; (2) cost, performance, and appearance specifications; (3) cost, durability, and nature of the tooling or forming mold; and (4) compatibility with various resins, reinforcements, and additives. This section lists processes in six general molding categories: open, compression, injection, reactive and resin transfer, continuous, and other processes.

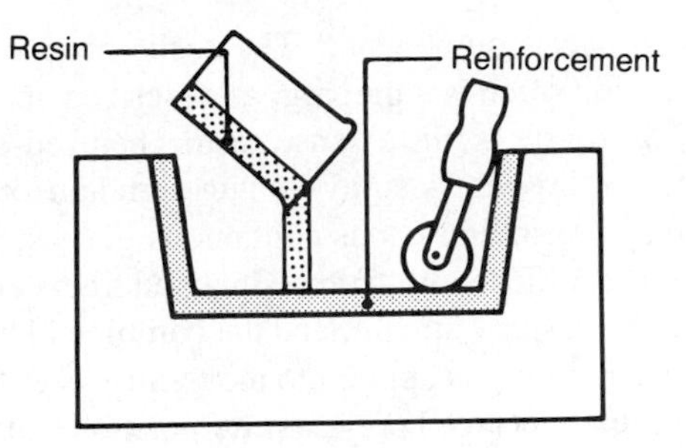

Fig. 18-5. Hand lay-up.

Open (or Contact) Molding

Lay-up is the most basic and one of the longest-used reinforced plastics molding process. The mold typically is a female cavity, and the mold surface is treated with a release agent, usually wax, that will facilitate removal of the finished part. A liquid gel coat of polyester, polyurethane, or other resin usually is applied to the mold surface; when the part is molded and removed, this gel coat will form the exterior surface of the part. In some cases, a vacuum-formed or thermoformed thermoplastic sheet can be used instead of a gel coat; then liquid resin, normally thermoset polyester, is combined with layers of reinforcement in roll form—mat, woven roving, fabric, or combination product. In practice, this is much like pouring concrete over steel reinforcing bars. To prevent voids in the finished part, air entrapped in the resinated reinforcement is worked out by using rollers or squeegees. Layers of reinforcement and resin are added to build the laminate to desired thickness. Usually, laid-up parts cure at ambient temperatures, but heat, catalysts, and accelerators can help speed resin cure.

Lay-up is suited to making large structures requiring high strengths. Traditionally, this process is called hand lay-up because layers of fiber glass mat or woven roving are manually placed in the mold cavity; but lay-up can be automated by running fiber glass woven roving through an impregnation machine, and then robotically laying the wet reinforcement in place. This automated technique is used to fabricate large yachts, commercial fishing vessels, and naval minehunters. However, most lay-up is accomplished manually. (See Fig. 18-5.) In fact, the labor-intensive nature of the process is a major limitation, especially for annual volumes above 1000 parts. Another disadvantage is that the process produces only one finished surface. The ''non-appearance'' side can be improved by pressing a film against the laminate before the resin cures, applying a special surfacing veil, or coating the surface in a post-mold operation. Advantages of the process include the ability to fabricate on-site, low-cost molds, and minimal start-up costs. Also, there is virtually no limit to the size of a laid-up part because, properly applied, succeeding layers of resin and reinforcement build into an integral laminate.

Bagging and autoclaving techniques can be used to improve a laid-up laminate. In vacuum bagging, a mold release film is placed over the reinforcement/resin composition. The film edges are sealed, and vacuum is drawn from within. Atmospheric pressure works the film toward the laminate to create a smoother surface, eliminate voids, and force out entrapped air and excess resin. In pressure bag molding, a deflated bag is placed on top of the film. An airtight seal is placed over the bag, and the bag is inflated. The resultant pressure against the laminate provides the same surface and void removal benefits as vacuum bagging. An advanced modification of bagging is to place the laminate and mold inside an autoclave. Pressures inside the airtight autoclave build to about 80 psi (0.55 MPa) as the mold release film is drawn to the laminate. The result is enhanced densification. The autoclave often can be heated to accelerate cure and increase productivity. Bag and autoclave assisted lay-up is used extensively to fabricate high-strength aircraft and aerospace components. The resin–fiber composition often is a prepreg, as described above.

Spray-up is, in a sense, a mechanized version of lay-up. A spray-up gun simultaneously sprays resin and chops continuous glass strand into specific length fibers. Chopped fibers enter the resin spray stream, and the combined fiber–resin material is applied to the mold. As in lay-up, the reinforced polymer composition usually backs up a gel coat or thermoplastic skin. Also like lay-up, the fiber–resin composition is rolled out to effect complete wetting of fibers and remove voids and bubbles before the resin cures. Spray-up and lay-up often are used together to combine the automated aspects of one process with the higher strength reinforcement of the other. Part volumes can be higher with spray-up, but, because of increased equipment needs, spray-up requires a higher initial investment than lay-up. Unless the spraying action is accomplished by computer-controlled robotics, process control of spray-up relies very much on operator skill. (See Fig. 18-6.)

Open molding is used extensively to produce boats, bathtubs, shower stalls and surrounds, and recreational vehicle exteriors. Because of low tooling costs, lay-up and spray-up are well suited to small runs of highly individualized applications. Open-molded laminates can have stiffeners and cores integrally molded in. Stiffener and core materials may be cardboard, balsa, resin-compatible foams, reinforced or unreinforced plastics, metal, or one of many honeycomb materials. One type of cored laminate is a sandwich construction—outer skins of reinforced plastics over a central core material. Sandwich laminates combine exceptionally high strength and stiffness with light weight. Honeycomb core/sandwich technology was used extensively to lighten the *Voyager* aircraft, the first to circle the globe nonstop on a single tank of fuel.

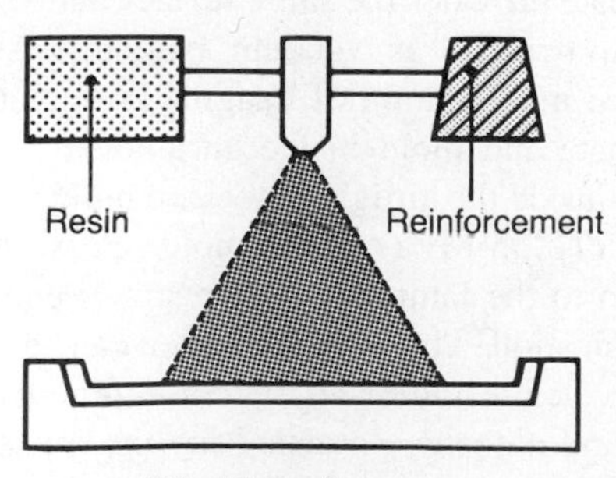

Fig. 18-6. Spray-up.

Compression Molding

Compression molding is accomplished with matched male and female dies, usually machined from tooling grade metals. During molding, the top mold half closes on the resin and reinforcement charge in the bottom mold half. As the mold halves come together, the charge fills the part cavity between them; the part forms and cures as the mold halves are brought together under heat and pressure. Thus the part cures while the press is closed. Cure cycles vary by material system, process temperature and pressure, and part thickness and complexity. The mold charge can be a sheet molding compound, bulk molding compound, resin-impregnated preform, or prepreg. (See Fig. 18-7.)

The most advanced form of reinforced plastics compression molding is *sheet molding compound* (SMC). The automotive industry makes widespread use of SMC, especially for car and truck body panels. In-mold coatings are applied to produce the smooth Class A surfaces specified by automotive engineers. SMC panels are bonded to each other and to metals. SMC molding cycles can be 60 seconds or less; bettering the one-minute threshold ensures that parts are compatible with the just-in-time delivery schedules established by automotive assembly plants.

Another variation of compression processing, *wet molding*, entails applying a liquid resin

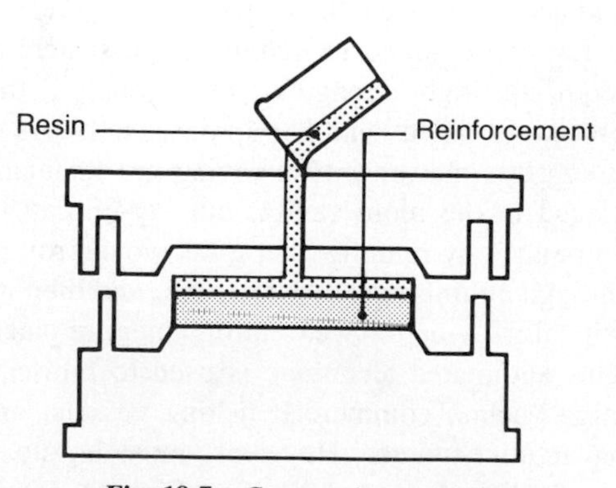

Fig. 18-7. Compression molding.

system to reinforcing materials or a preform. Resin is applied to the reinforcement just before or after it is placed in the bottom mold half. *Cold molding* is wet compression molding accomplished with ambient cure resins. The only heat comes from the exothermic reaction of resin cure, and pressures are low. The molds themselves often are reinforced plastics.

Transfer molding is related to compression molding. In this case, the reinforced thermoset compound first is heated in a chamber or pot separate from the matched dies. Then, a plunger or ram transfers the heated material into the closed mold. The part cures under compressive pressure. This technique facilitates the compression molding of small, intricate parts in multicavity molds.

The *injection-compression* molding combination starts by injecting the molding compound into a slightly open set of matched metal dies; compression of the mold halves is completed after the compound fills the mold cavity. The compound can be based on either thermoset or thermoplastic resins.

For *stamping*, or flow forming of reinforced thermoplastics, the charge usually is in sheet form. The sheet is preheated into a soft state, then placed between mold halves. Press closure occurs at rates faster than those of other compression molding variations; hence the term "stamping." However, unlike steel stamping, the press is held shut until the reinforced thermoplastic laminate has cooled. Although the strike time for thermoplastic sheet is longer, a single strike can accomplish complex configurations that require several progressive strikes with sheet metal.

Injection Molding

When pelletized compound is used, injection molding of reinforced *thermoplastics* is very much like injection molding of unreinforced thermoplastics. The compound is fed from a hopper into a heated cylinder where the resin softens into a flowable polymer melt. During a mold cycle, a rotating screw-ram injects the melt into the cavity formed when mold halves are brought together. For injection-molded thermoplastics, molds are kept relatively cool

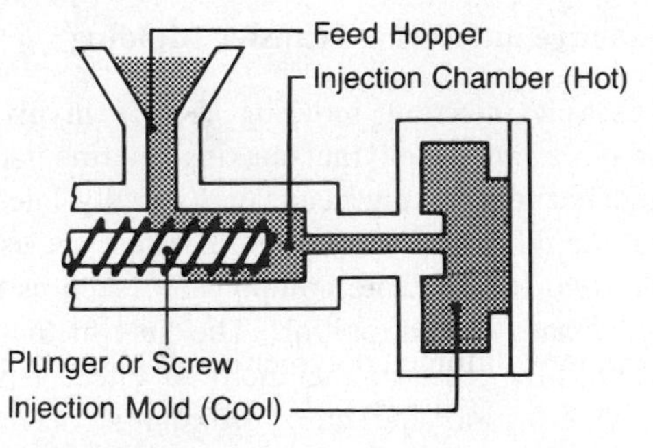

Fig. 18-8. Injection molding thermoplastics.

so the melt can freeze, or solidify, to form the finished part. (See Fig. 18-8.)

For injection-molded *thermosets*, the injection cylinder is cool instead of heated, and the molds are heated instead of cool. A plunger or low-shear screw-ram injects the reinforced compound into the mold cavity. The heated mold activates the thermoset cure chemistry. (See Fig. 18-9.)

Most injection molding is accomplished with molding compounds. There also are methods for *direct feeding* of fibers and other components just prior to molding. Examples include blending ingredients in a hopper, metering chopped fibers into the resin melt, or having continuous strand chopped by knobby protrusions on the screw that is mixing the resin melt.

Several technologies are used to vary the injection molding process. For example, inert gases are introduced during molding to either foam the polymer before cure or create hollow sections in the part. Structural foam plastics also are made by using blowing agents in the molding compound. These technologies are applicable to both reinforced and unreinforced plastics that are injection-molded.

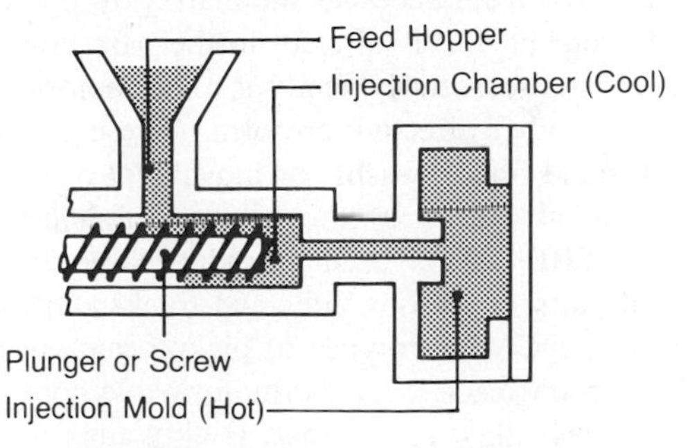

Fig. 18-9. Injection molding thermosets.

Reactive and Resin Transfer Molding

Reaction injection molding (RIM) involves bringing liquid, polymer-making intermediates together in a mixing head, then rapidly injecting the mix into a closed mold. There are usually two intermediate components, such as an isocyanate and a polyol. The intermediates chemically react in the mold to effect rapid polymerization and cure. Mixing is accomplished at relatively high pressures (up to 3500 psi or 24 MPa), but in-mold pressures are as low as 50 psi (345 kPa). Consequently, reactive molding requires lower-cost tooling and presses than those needed for conventional injection molding. However, reaction injection molding calls for closer process controls. Because the timing of the reaction must occur at the moment when the cavity is filled, reactive molding parameters must be consistent for every shot. A key advantage of the process is the low viscosity of the intermediates. Their liquid nature allows for the molding of complex parts.

Two different versions of reactive molding incorporate reinforcements. One of these processes, *reinforced reaction injection molding (RRIM)*, incorporates the reinforcing agent in the material system that is injected from the mixing head. The reinforcement—usually milled glass fiber or glass flakes—is added to one or both resin intermediates before resin components are mixed. It increases flexural modulus, improves thermal resistance and, for some applications, enhances the surface finish. (See Fig. 18-10.)

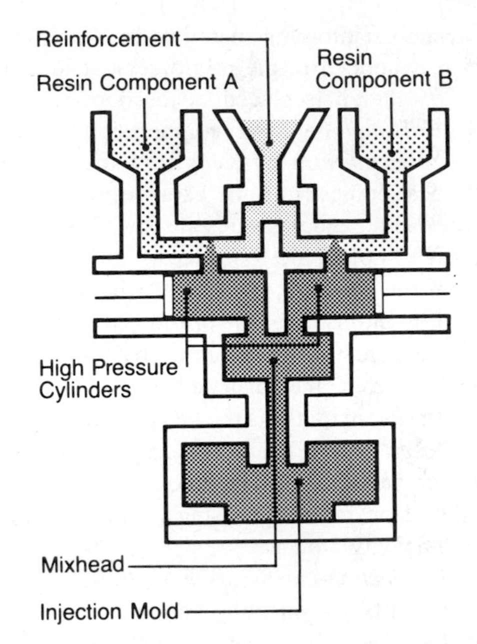

Fig. 18-10. Reinforced reaction injection molding.

In *structural reaction injection molding (SRIM)*, the reinforcing medium is placed in the mold before the mold halves are closed, and the mixed resin becomes the matrix for the reinforcement that is already in the part cavity. The reinforcement typically is a thermoformed mat or fiber-directed preform. These forms withstand fiber "wash" or movement of reinforcement caused by the pressure of injected resin. SRIM can be used to produce large structural parts for automobiles and trucks; for example, SRIM prototypes of pickup truck beds have been made with thermoformable continuous fiber glass strand mat. Highly automated equipment produces mat preforms at a rate fast enough to keep pace with the rapid reaction molding process.

Resin transfer molding (RTM) is similar to SRIM in that resin is introduced into a closed mold that already contains the reinforcing agent. In both processes, no press is needed, however, the resin chemistry for RTM is quite different from that of SRIM. Catalyzed resin is pumped into the mold to form the part; and the resin can be formulated to cure without heat, or heating coils can be incorporated into the mold to speed cure of an elevated-temperature resin formulation. The advantages of resin transfer molding are the relatively low investments needed for equipment and tooling and the ability to mold large, highly reinforced structures. RTM parts can incorporate cores, inserts, and encapsulates. (See Fig. 18-11.)

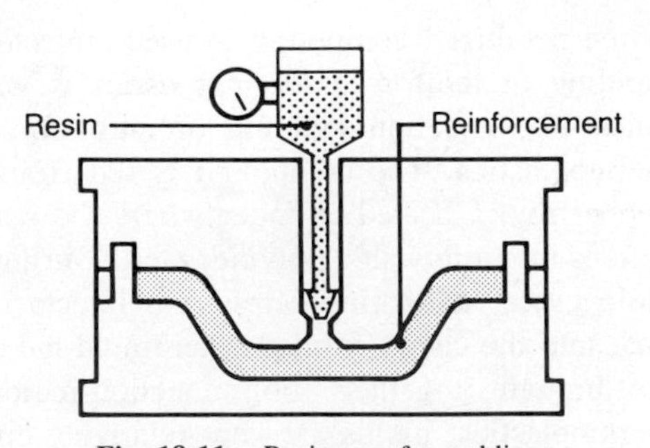

Fig. 18-11. Resin transfer molding.

Continuous Processes

Pultrusion uses one or several different forms of continuous reinforcement, such as rovings, mats, woven rovings, fabrics, and/or cloths, with the type and the amount of reinforcement depending on the desired properties. Surfacing veils are used to give the molded part a smooth, resin-rich finish. Typically, the reinforcement or reinforcement combination is drawn through a resin bath where the fibers are impregnated. After excess resin is stripped off, the fibers are sent through a forming station where fibers are oriented for proper entry into a heated steel die, inside which the fiber–resin laminate takes on its final cross-sectional shape. Heat completes the cure. Downstream from the die, pulling devices move the rigid profile forward to an area where profiles are cut to length. This same pulling action is used to continuously draw the fibers through the resin bath, forming station, and die. Pultrusion is used to produce a wide variety of profiles with a continuous cross section. It is possible to pultrude fiber and resin encapsulate over wood, foam, metal, and other cores. (See Fig. 18-12.)

Reinforcing fibers for pultrusion are predominantly glass. Graphite tow and aramid fibers also are used. Resins are usually thermoset, especially polyester and vinyl ester. Epoxy is used in pultrusion but is difficult to process. Advances in resin chemistry and adjustments in the process are expanding the possibilities of pultrusion beyond traditional resins. For example, acid-free phenolics have been developed to resist sticking. Fibers enter the die in a dry state, are preformed, and are pulled through a special chamber. There, fiber wet-out is accomplished by injecting the phenolic resin into the chamber. Other variations of the pultrusion process also eliminate the traditional resin bath impregnation. For instance, either reactive or thermoplastic resins can be polymerized in the die as the fibers are being pulled through. Thermoplastic pultrusion also is accomplished by pulling multiples of preheated thermoplastic prepregs through a heated die. Profiles pultruded with thermoplastic resins can be heated and re-formed. The result combines the high cross-sectional strength of pultrusion with the design flexibility of compression molding.

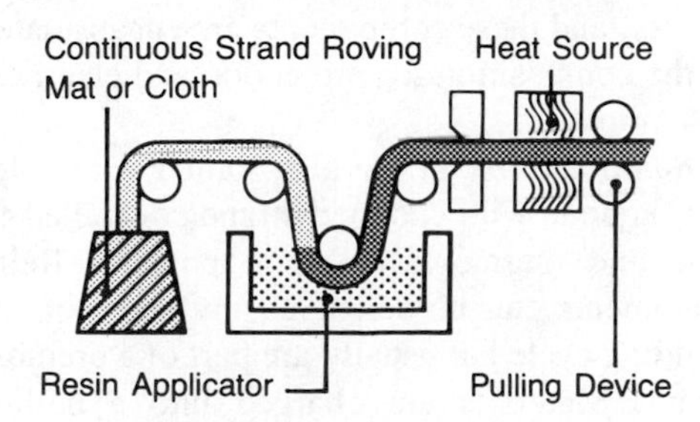

Fig. 18-12. Pultrusion.

Parts typically made by the pultrusion process include electrical safety ladder rails and rungs, grating profiles, nonconductive cable supports, and structural beams for corrosive environments. The process is also used to make automotive driveshafts, siding for buses and buildings, and structural framing for orbiting space satellites.

Pulforming is an offshoot of pultrusion. Resin-wet reinforcement is pulled through female mold cavities. The molds rotate on a turntable. The female cavities are enclosed by co-rotating male mold halves placed over them, or the female molds are covered by a rotating contact plate or belt. Each part cures as the closed mold rotates toward a cutoff station. Applications made this way include automotive leaf springs and hammer handles.

Filament winding is another continuous process. Continuous reinforcement is drawn through a resin-impregnation bath and wound wet under tension on a rotating mandrel. The angle of the fibers on the mandrel varies by application, and fiber lay-down patterns are controlled through a fiber delivery device. Fiber patterns typically criss-cross and overlap to produce good multidirectional strength properties. Possible fiber patterns include helical, circumferential, polar, and nonlinear. When the part is wound to the desired thickness, it is removed from the mandrel. Ambient and oven cure resins are used. (See Fig. 18-13.)

The traditional filament winding mandrel is cylindrical; and typical applications are storage tanks, piping, and chimney liners. However, spheres, "eggs," and nonsymmetrical shapes are possible as long as the shape can be formed by a continuous fiber path. The wind pattern can be computer-controlled over six different

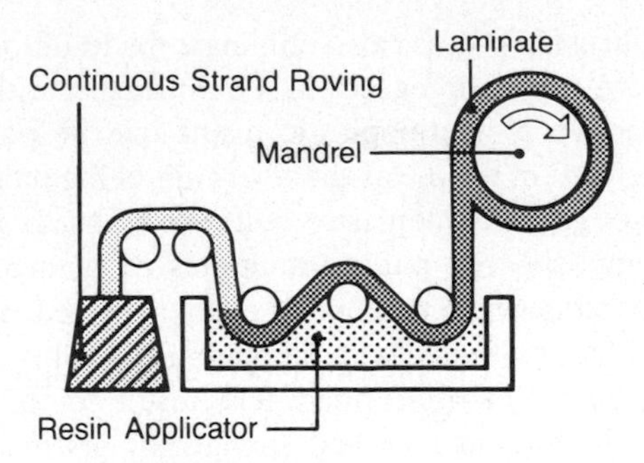

Fig. 18-13. Filament winding.

axes. In the *pin winding* variation, continuous fiber is wound around end pins, with winding accomplished inside a cavity of the desired shape. This technique is used to make helicopter blade spars.

Most filament winding is accomplished with E-glass fiber reinforcements. When achieving higher properties is cost-justified, S-2 Glass, graphite, or aramid fibers are wound. A cost-effective approach to added strength is to wind a hybrid composite of mostly E-glass fibers in combination with higher-modulus reinforcements. Chopped fibers may be deposited along with continuous fibers in a combination of filament winding and spray-up. Resins for this process are predominantly thermoset polyesters, vinyl esters, and epoxies. Some filament winding has been achieved with thermoplastics. In one approach, the thermoplastic is polymerized at the mandrel. Prepreg tapes made with either thermoset or thermoplastic resins can be applied in *dry filament winding*. Shapes wound with thermoplastic resin matrices can be heated and post-formed into complex shapes.

The *continuous lamination* process uses an in-line conveyor. Continuous fiber roving, usually glass, is chopped onto a bottom carrier film with a doctored layer of wet resin, usually thermoset polyester. The resin formulation usually incorporates many additives, such as UV stabilizers, colorants, flame retardants, and fillers. The film, with fibers and resin, is conveyed downstream, where a pressure roller works the fibers into the resin before a top film is applied. The film/composite/film sandwich is pulled through compression rollers to remove air and establish thickness. Finally, the laminate enters a forming/curing area. Most sheet is either made flat by special shoes or corrugated by shaping rollers. After cure, outer carrier films are stripped off, and the laminate is trimmed and cut to length. The light transmittance of reinforced plastic panels made this way is design-engineered for the application, and can range from 95% to complete opaqueness. Typical applications are skylights, greenhouse panels, building sidewalls, and truck-trailer liners.

Other Molding Processes

The *inflatable pressure bag* technique results in hollow reinforced plastics shapes, such as seamless tanks. The bag is placed inside a preformed fiber glass shape, and the bag-in-glass assembly then is placed inside an unheated, hollow mold. The mold is either open-ended or a two-part clamshell. After the fibers are resin-impregnated, the bag is inflated to push the reinforced composite against the mold walls. After resin cure, the bag is deflated and removed from the interior of the part.

Cylindrical piping, tubing, tanks, and other containers can be formed by using the *centrifugal casting* process. Chopped strand mat is placed inside a hollow, cylindrical mold, and resin is sprayed on the mat as the mandrel rotates slowly. One variation is to chop fibers inside the mandrel instead of using mat. When desired resin and glass contents are reached, the rotational speed is increased. Centrifugal force effects resin wet-out and fiber compaction, and heat accelerates the cure.

In *encapsulation*, chopped strands or milled fibers are mixed with catalyzed polyesters, epoxies, or silicones. The composition is poured into open cavities containing electrical or electronic components, such as terminal blocks, and these components are encapsulated in the composition for protection and electrical insulation.

Rotational molding, also called rotomolding, is done with chopped strands or milled fibers and thermoplastic resin powder. Reinforcements can be added at any stage of the molding cycle but usually are part of a premold blend. Materials are charged into a hollow heated mold that is rotated in one or two planes. Heat-fused resin encapsulates the fibers, and the composite takes shape against the inner mold walls. When ready, the mold and/or the part is chilled to "freeze" the part for removal. A water mist often is sprayed internally.

Other *reinforced thermoplastics composites* are made by extrusion, extrusion-blow molding, and thermoforming. These processes are covered in more detail in other chapters of this book. Originally developed for unreinforced thermoplastics, the processes are also run with fiber-reinforced compounds. Fibers bring about specific property enhancements, such as higher thermal resistance or impact strength. Extrusion and extrusion-blow molding materials typically are pelletized compounds. Premixed blends and direct feeding also are possible. Thermoforming is accomplished on extruded sheetstock.

FEATURES AND BENEFITS

Reinforced plastics and composites present a matrix of performance and cost options. This family of materials should be considered when application or market needs include one or more of the attributes discussed in the following paragraphs.

With so many molding and fabrication techniques available, the *design flexibility* obtainable with reinforced plastics exceeds that of virtually all other material systems. Parts can be as large and structural as a military minehunter vessel or as small and intricate as an electronic connector. Also, reinforced plastics and composites offer an almost limitless range of resins and reinforcement types, forms, and orientations.

Reinforced plastics provide an excellent combination of high strength with light weight. (See Table 18-1.) Because of *excellent strength-to-weight ratios*, tensile strengths by weight typically exceed those of many grades of iron and steel. In addition, flexural and impact strengths are equal to or better than those of many metals. Being lighter than alternatives, composites lower energy consumption needs for transportation uses, such as automotive body panels; and lighter-weight piping or structurals cost less to transport and are easier to install. Light weight also can reduce load-bearing stresses for better performance and lower costs.

Reinforced plastics offer very good *dimensional stability*. It often is this property that dictates the use of reinforced plastics when unreinforced thermoplastics meet other design requirements. The inclusion of reinforcing fibers elevates the resin's ability to maintain critical tolerances over a wide range of temperatures and physical stresses. Fibers also improve the material's thermal coefficient of expansion. In addition, a reinforced plastic enclosure or car panel will not take a "set" when impacted by a force that normally would dent sheet metal.

Through *parts consolidation*, a single molded reinforced plastics component replaces several parts that would require numerous assembly steps. This ability to combine parts ideally suits reinforced plastics to quality philosophies that emphasize just-in-time deliveries. Using reinforced plastics, fabricators deliver a "black box" modular subassembly that is easily integrated into the production line.

Reinforced plastics are well known for their *corrosion resistance*. The resistance to acids, alkalis, hydrocarbons, and solvents varies by resin matrix; thus the resin is selected to meet an application's need to resist specific levels of corrosion, pressure, and temperature. (See Table 18-2.) Even the most basic composite using an orthophthalic polyester resin will not rust or corrode. However, for more demanding environments, resin options include isophthalic polyesters, vinyl esters, and epoxies. Composites engineered for corrosion resistance typically use a surfacing veil or mat. These surface treatments create a resin-rich surface that resists corrosive media.

Reinforced plastics are *excellent dielectrics*. Like corrosion resistance, electrical resistance can be application-specific. Glass in any form, including fibrous, is a very low conductor of electricity. In fact, the designation of conventional glass fibers as E-glass is derived from the fibers' early use in electrical applications. Generally, thermoset resins offer better dielectric properties than thermoplastics. Reinforced polyesters are widely used in applications de-

Table 18-1. Properties comparisons of fiber glass composites and other materials.
(Source: Owens-Corning Fiberglas Corporation)

Material	Property	Glass Fiber By Weight	Specific Gravity	Density	Tensile Strength	Tensile Modulus	Elongation	Flexural Strength	Flexural Modulus	Compressive Strength
	Units	%		lbs/in³	10³ psi	10⁶ psi	%	10³ psi	10⁶ psi	10³ psi
	ASTM Test Method		D792		D638	D638	D638	D790	D790	D695
Glass Fiber Reinforced Thermosets	Polyester SMC (Compression)	30.0	1.85	.066	12.00	1.70	<1.0	26.00	1.60	24.00
	Polyester SMC (Compression)	20.0	1.78	.064	5.30	1.70	.4	16.00	1.40	23.00
	Polyester SMC (Compression)	50.0	2.00	.072	23.00	2.27	1.7	45.00	2.00	32.00
	Polyester BMC (Compression)	22.0	1.82	.065	6.00	1.75	.5	12.80	1.58	20.00
	Polyester BMC (Injection)	22.0	1.82	.065	4.86	1.53	.5	12.65	1.44	—
	Epoxy (Filament Wound)	80.0	2.08	.061	80.00	4.00	1.6	100.00	5.00	45.00
	Polyester (Pultruded)	55.0	1.69	.060	30.00	2.50	—	30.00	1.60	30.00
	Polyurethane, Milled Fibers (RRIM)	13.0	1.07	.038	2.80	—	140.0	—	.037-.053	—
	Polyurethane, Flaked Glass (RRIM)	23.0	1.17	.042	4.41	—	38.9	—	.15	—
	Polyester (Spray-up/Lay-up)	30.0	1.37	.049	12.50	1.00	1.3	28.00	.75	22.00
	Polyester, Woven Roving (Lay-up)	50.0	1.64	.059	37.00	2.25	1.6	46.00	2.25	27.00
Glass Fiber Reinforced Thermoplastics	Acetal	25.0	1.61	.058	18.50	1.25	3.0	28.00	1.10	17.00
	Nylon 6	30.0	1.37	.049	24.00	1.05	3.0	29.00	1.11	24.00
	Nylon 6/6	30.0	1.48	.053	23.00	1.20	1.9	35.00	.80	26.50
	Polycarbonate	10.0	1.26	.045	12.00	.75	9.0	16.00	.60	14.00
	Polypropylene	20.0	1.04	.037	6.50	.54	3.0	8.30	.52	25.00
	Polyphenylene Sulfide	40.0	1.64	.059	22.00	2.05	3.0	37.00	1.90	21.00
	Acrylonitrile Butadiene Styrene (ABS)	20.0	1.22	.044	11.00	.90	2.0	15.50	.87	14.00
	Polyphenylene Oxide (PPO)	20.0	1.21	.043	14.50	.92	5.0	18.50	.75	17.60
	Polystyrene Acrylonitrile (SAN)	20.0	1.22	.044	14.50	1.25	1.8	19.00	1.10	17.50
	Polyester (PBT)	30.0	1.52	.054	19.00	1.20	4.0	28.00	1.17	18.00
	Polyester (PET)	30.0	1.56	.056	21.00	1.30	6.6	32.00	1.25	25.00
Unreinforced Thermoplastics	Acetal	—	1.41	.051	8.80	.41	40.0	13.00	.38	16.00
	Nylon 6	—	1.12	.040	11.80	.38	30.0	15.70	.39	13.00
	Nylon 6/6	—	1.13	.041	11.50	.40	60.0	17.00	.42	15.00
	Polycarbonate	—	1.20	.043	9.50	.34	110.0	13.50	.34	12.50
	Polypropylene	—	.89	.032	5.00	.10	200.0	5.00	.13-.20	3.50
	Polyphenylene Sulfide	—	1.30	.045	9.50	.48	1.0	14.00	.55	16.00
	Acrylonitrile Butadiene Styrene (ABS)	—	1.03	.037	6.00	.30	5.0	11.00	.35-.40	10.00
	Polyphenylene Oxide (PPO)	—	1.10	.039	7.80	.38	50.0	12.80	.33-.40	12.00
	Polystyrene Acrylonitrile (SAN)	—	1.05	.038	9.50	.40	.5	14.00	.55	14.00
	Polyester (PBT)	—	1.31	.047	8.20	.28	50.0	12.00	.33-.40	8.60
	Polyester (PET)	—	1.34	.048	8.50	.40	50.0	14.00	.35-.45	11.00
Metals	ASTM A-606 HSLA Steel (Cold Rolled)	—	7.75	.280	65.00	30.00	22.0	—	—	65.00
	SAE 1008 Low Carbon Steel (Cold Rolled)	—	7.86	.280	48.00	30.00	37.0	—	—	48.00
	AISI 304 Stainless Steel	—	8.03	.290	80.00	28.00	40.0	—	—	80.00
	TA 2036 Aluminum (Wrought)	—	2.74	.099	49.00	10.20	23.0	—	—	49.00
	ASTM B85 Aluminum (Die Cast)	—	2.82	.102	48.00	10.30	2.5	—	—	48.00
	ASTM AZ91B Magnesium (Die Cast)	—	1.83	.066	33.00	65.00	3.0	—	—	33.00
	ASTM AG40A Zinc (Die Cast)	—	6.59	.238	41.00	10.90	10.0	—	—	41.00

These figures are presented only in an effort to suggest general ranges of properties for different materials. This is standard test data and does not reflect exact properties of any given finished part.

Impact Strength IZOD	Hardness	Flammability	Specific Heat	Thermal Coefficient of Expansion	Heat Deflection (DTUL)	Thermal Conductivity	Dielectric Strength	Volume Resistivity	Relative Permittivity	Arc Resistance	Water Absorption	Mold Shrinkage
ft-lbs/in notched@ 73°F	Rockwell (except where noted)		BTU/lb/F°	in/in/°F×10^{-6}	F°@264 psi	BTU/hr/ sq ft/F°/ft	V/mil	Ohm-cm	60 Hz	seconds	Percent 24 hours	in/in
D256	D785	UL-94		D696	D648	C177	D149	D257	D150	D495	D570	D955
16.00	Barcol 68	5V	.30	—	400+	—	500	5.7×10^{14}	4.40	188	.25	—
8.20	Barcol 68	5V	.30	—	400+	—	—	—	4.40	188	.10	.002
19.40	Barcol 68	5V	.30	9.4	400+	—	—	—	4.40	188	.50	—
4.26	Barcol 68	5V	.30	6.6	500	4.84	375	27×10^{14}	4.20	190	.20	.001
2.89	Barcol 68	VO	.30	6.6	500	4.84	375	—	4.20	190	.20	.004
45.00	M98	VO	.23	2.0	400+	1.92	300	$>10^{12}$	—	—	.50	.008
25.00	Barcol 50	VO	.28	5.0	—	4.00	200	10^{13}	4.40	80	.75	—
—	S.D.* 65-75	VO	—	78.0	85	—	—	—	—	—	—	—
2.10	—	VO	—	53.1	—	—	—	—	—	—	—	—
13.00-15.00	Barcol 50	VO	.31	12.0	400+	1.50	250	—	—	—	1.30	—
33.00	Barcol 50	VO	—	4.0	400+	—	350	10^{14}	4.20	—	.50	—
1.80	M79	HB	—	4.7	322	—	580	10^{14}	4.12	142	.29	.004
2.20	R121	HB	—	1.5	412	5.8-11.4	500	10^{13}	3.90	120	1.30	.004
2.20	M95	HB	.30	1.8	490	1.50	400	10^{15}	3.80	120	.50	.002
2.00	M80	V-1	.29	1.8	285	4.60	500	10^{16}	3.10	125	.14	.005
1.10	R103	HB	—	2.4	270	8.40	440	10^{15}	2.70	120	.05	.003
1.50	R123	V-0/5V	.25	1.1	510	2.00	380	4×10^{15}	3.00	125	.01	.002
1.20	R107	HB	—	2.1	210	1.40	465	10^{15}	3.20	80	.30	.002
1.80	R107	HB	(.2-.4)	2.0	290	3.80	500	10^{16}	3.20	70	.24	.003
1.10	R122	HB	—	2.1	215	2.80	490	10^{15}	3.50	70	.06	.002
1.80	R118	HB	.11	1.4	415	7.00	375	3.2×10^{16}	3.80	135	.06	.003
1.80	R120	HB	—	1.7	420	6.50	520	10^{16}	3.60	90	.05	.003
1.00	M78-M80	HB	.35	4.7	230	1.56	500	10^{15}	3.70	129	.22	.020
.60	R119	HB	.40	4.8	155-185	1.20	305	4.5×10^{13}	4.00	—	1.3-1.9	.005
.80	R120, M83	V-2	.30	4.5	167	1.70	385	1×10^{15}	4.00	120	1.0-1.3	.008
16.00	M70	V-2	.30	3.7	270	1.35	380	$>10^{16}$	2.96	120	.15	.005-.007
1-20	R50-96	HB	.45	3.8	115-140	1.21	600	$>10^{17}$	2.20	125	.03	.020
<.50	R123	V-0	—	—	275	1.67	380	10^{16}	—	—	<.02	.007
3-6	R107-115	HB	—	3.2	200-220	.96	350-500	2.7×10^{16}	2.80	—	.20-.45	.004-.009
5.00	R115	V-1	.2-.4	68.0	212	.92	400	$>10^{16}$	2.65	75	.07	.005-.007
.30-.45	M80-85	HB	.33	36.0	220	.70	515	4.4×10^{16}	3.50	65	.20-.35	—
.80	M68-78	HB	—	4.5	122-185	1.02-1.67	420-550	4×10^{16}	3.10	184	.08-.09	.016-.020
.25-.65	M94-101	HB	.34	—	100-106	.87	—	$>10^{16}$	—	—	.1-.2	.02-.025
—	B80	—	.11	6.8	—	25.00	—	—	—	—	—	—
—	B34-52	—	.10	6.7	—	35.00	—	—	—	—	—	—
—	B88	—	.12	9.6	—	9.40	—	—	—	—	—	—
—	R80	—	.21	13.9	—	92.00	—	—	—	—	—	—
—	Brinell 85	—	—	11.6	—	53.00	—	—	—	—	—	—
—	Brinell 85	—	.25	14.0	—	41.80	—	—	—	—	—	—
—	Brinell 82	—	.10	15.2	—	65.30	—	—	—	—	—	—

*Shore D

Table 18-2. Chemical resistance notations by materials family.

Thermosets

Epoxies are highly resistant to water, alkalis and organic solvents. Although standard epoxies have only a fair resistance to acids and oxidizing agents, they can be formulated for better resistance to these agents. They should be tested before use.

Polyester thermosets are available in many formulations. They can be formulated for good to excellent resistance to acids, weak alkalis, and organic solvents. However, they are not recommended for use with strong alkalis.

Polyurethane is normally used for its structural properties when molded. It is highly resistant to most organic solvents. It is not recommended for use with strong acids and alkalis, steam, fuels, and ketones.

Thermoplastics

ABS is highly resistant to weak acids and alkalis, and provides good resistance to most organic solvents. It is attacked by sulfuric and nitric acids, and is soluble in esters, ketones and ethylene dichloride.

Acetal is highly resistant to strong alkalis. Most organic solvents do not seriously alter its properties (test before use). It is not recommended for use in strong acids.

Nylons, as a group, are inert to most organic solvents. They resist alkalis and salt solutions and they are attacked by strong mineral acids and oxidizing agents.

Polyphenylene Oxide provides excellent resistance to aqueous media. It is softened by aromatic hydrocarbons.

Polycarbonates resist weak acids and alkalis, oil and grease. They are attacked by strong acids, alkalis, organic solvents, and fuels (test before use).

Thermoplastic Polyesters are resistant to most organic solvents, weak acids and alkalis. They are not recommended with strong acids and alkalis or for prolonged use in hot water.

Polypropylene provides good resistance to acids, alkalis and organic solvents even at higher temperatures. It is, however, soluble in chlorinated hydrocarbons.

Polyphenylene Sulfide provides excellent resistance to organic solvents (below 375°F). It is unaffected by strong alkalis or aqueous organic salt solutions.

Polystyrene provides good resistance to alkalis, most organic solvents, weak acids, and household chemicals. It is not recommended for use with strong acids, ketones, esters, and some chlorinated hydrocarbons (test before use).

Metals

Cold Rolled Steels are rusted by water, oxygen, and salt solutions. Not recommended for use with acids. They do, however, have good resistance to alkalis.

Stainless steels have poor acid resistivity (especially hydrochloride and sulfuric) as well as poor chloride solution resistivity. They do offer good resistance to alkalis and organic solvents.

Aluminum has poor acid resistance (especially hydrochloride and sulfuric) and also poor chloride salt resistance (must be chemically treated for appearance when exposed to weather).

Magnesium poorly resists acids except hydrofluoric, however, it has good resistance to alkalis. It corrodes in the presence of salt, salt spray or industrial atmospheres.

Zinc offers good atmospheric resistance but is not recommended for use with strong acids, bases, and steam.

signed to provide protection from electric shocks.

Integral finishes and surfaces are possible with reinforced plastics and composites. Colorants can be added to the resin formulation to produce a part that needs no painting; moreover, the color will not chip or scratch. Surfaces can incorporate textures, ridges, grids, and similar treatments replicated from the surface of the mold; and they can be mirrorlike to meet automotive body panel requirements for a Class A finish.

A key economic benefit of turning to reinforced plastics and composites is *moderate tooling costs*. Amortization of the tooling investment does not require production volumes in the hundreds of thousands. For lower pressures and volumes, the molds themselves often incorporate reinforced plastics. For high-volume molding, such as fiber glass/polyester automotive body panels, tooling costs are substantially lower than for the tooling needed for sheet metal. In addition, reproducing complex shapes in sheet metal requires several progressive stamping tools.

By offering *shorter lead times*, reinforced plastics are helping marketers to become more competitive. Short lead times mean that the product is becoming profitable sooner. Also, the compressed time frame from idea to finished part can make an important contribution toward simultaneous engineering. This strategy computer-integrates all essential product development programs to occur during the same, compressed time sequence. Decisions related to engineering, materials analysis, design of tooling, fixtures, and process control are made concurrently instead of successively. For one molder-assembler, this technique took a heavy-duty truck hood and fender assembly from concept to finished part in 12 months. The standard development cycle for this process of 24 to 36 months can prove costly to competitors in global markets.

Fig. 18-14. Chevrolet Lumina APV. (*Courtesy of Chevrolet Public Relations*)

MARKETS AND APPLICATIONS

The *automotive market* has been, is, and is projected to be the largest consumer of reinforced plastics and composites. The most visible applications are exterior body panels. The Chevrolet Corvette is an example of the continuous use of reinforced plastics for more than 35 years. The Pontiac Fiero demonstrated the ability of the reinforced plastics industry to meet the demand for high-volume production of automotive body panels. However, the GM minivan represents the highest-volume use of reinforced plastics by a single application. (See Fig. 18-14.)

When introduced for the 1990 model year, all APV body panels except the front fascia and small front fenders were fiber glass/thermoset polyester SMC. Each vehicle uses about 325 pounds (147.5 kg) of SMC. If the molded SMC panels were stamped in sheet metal, tooling costs would be about four times higher. The panels are assembled to a steel spaceframe using technology proved by the Fiero. The versatility of SMC allows molded components to be bonded together into subassemblies that are delivered to the final assembly site on a just-in-time basis.

In addition to their use on cars with plastics-intensive bodies, reinforced plastics body panels are assembled on vehicles alongside steel body panels. For example, fiber glass composite doors and liftgates offer light weight and corrosion resistance. Outer panels often are bonded to fiber glass composite inner panels that combine several parts into a single unit with molded-in assembly guides. Other reinforced plastic exterior panels on cars include grille opening panels, rear end/license plate panels, and spoilers.

A huge opportunity for reinforced plastics is emerging in structural automotive applications. Examples are composite drive shafts, wheels, and leaf springs. Developmental work with structural cross members may be the precursor to major chassis components molded of composites. (See Fig. 18-15.) In addition, each of the Big Three U.S. automakers has a program for making pickup truck beds of reinforced plastics. Processes under evaluation for this application are compression molding, structural RIM, and resin transfer molding.

Another area where reinforced plastics use is growing is underhood. Initially used in trucks, vinyl ester SMC oil pans, crankcases, and rocker covers are under study for automobiles.

Fig. 18-15. Thermoformable preform for developmental cross member support.

An air resonator for turbocharged Ford engines is blow-molded of a nylon 6/6 compound that is glass-reinforced. A modular windshield wiper system bracket is injection-molded of 30% glass fiber–reinforced thermoplastic polyester. In Europe, Ford's diesel-powered Escort has an air intake manifold molded of fiber glass/thermoset polyester. Inner chambers are formed with a melt-fusible and recoverable metallic core. There is even a reinforced phenolic engine under development.

Other automotive applications of reinforced plastics include headlamp housings, spare tire covers, and bumper beams. In addition to cars and light trucks, reinforced plastics are widely used for *other ground transportation vehicles*. Examples include hood/fender assemblies and engine components for heavy duty trucks, body panels and engine components for recreational vehicles, and pultruded body panels and air ducts for buses.

The *marine market* has been a large user of reinforced plastics for years. The use of fiber glass–reinforced polyester in pleasure boats is a mature application. (See Fig. 18-16.) However, some high-performance boats are made with enhanced-performance materials, such as S-2 Glass® fibers and vinyl esters. In addition to pleasure boats and sailboards, fiber glass–reinforced plastics are specified for the hulls and decks of commercial fishing boats and military minehunters.

Fig. 18-16. Pleasure boat. (*Courtesy of Wellcraft*)

Fig. 18-17. Underground gasoline storage tank.

Corrosion-resistant materials are a class of reinforced plastics often referred to as a market. However, these materials are engineered to resist corrosive environments in a variety of different markets. Major user-industries that benefit from special reinforced plastics are chemical processing, oil and gas, pulp and paper, waste water, air pollution control, and electroplating. Probably the best-known corrosion-resistant-material application is the underground gasoline storage tank. (See Fig. 18-17.) EPA concerns over fuel leakage into drinking water supplies has directed attention to fiber glass/thermoset polyester tanks. Other corrosion-resistant applications include grating, handrails, work platforms, walkways, structural beams, and even nuts and bolts.

Major *construction market* applications are paneling and skylights, bathtubs and shower stalls, and doors and windows. Reinforced plastics also are used in industrial and commercial buildings that need specific characteristics. For instance, radar installations or computer test sites benefit from good dielectric properties, and light weight is preferred for architectural assemblies that are crane-lifted. As noted above, there also are many corrosion-resistant applications in the construction market.

For reinforced plastics and composites, the *aircraft/aerospace/military market* is low in volume but high in value. Both commercial and military aircraft make extensive use of reinforced plastics materials that are referred to as "advanced composites." This designation is given to composites that consist of fibers with a higher modulus than conventional E-glass fibers and a resin matrix with higher properties than polyesters. (See Fig. 18-18.) One of the more impressive applications of advanced composites for aircraft is the fuselage for Beech Aircraft's Starship business turboprop. The pressurized fuselage is graphite fiber–reinforced epoxy prepreg over a honeycomb core. More than 70% of the aircraft's structural weight is advanced composites.

Advanced composites also are used for helicopter rotor blades, aircraft wing and tail components, and engine ducts. Also, because of their radar transparency, advanced composites play a key role in radar-evading "stealth" technologies. The National Aeronautics and Space Administration studied the performance of polymer-based composites in aircraft and helicopters for 13 years. NASA's report concludes, "In general, the composite aircraft components have performed better than com-

Fig. 18-18. S-2 Glass(R) fiber-reinforced composite intensive fighting vehicle.

parable metallic components.'' In another military use, properly engineered composites work well as protective armor against ballistics attack. For example, the U.S. Navy uses shipboard armor made of S-2 Glass® fiber-reinforced thermoset polyester. Reinforced plastics also are used for temporary runway pads for military aircraft and portable bridges for troops and artillery.

Electrical/electronic market applications vary widely. Composite ladder rails and rungs, stand-off poles, and conduit for power cables provide protection against electrical shock. One of the more unusual electrical uses is an all-composites building in which artificial lightning is generated to test the effects of lightning on aircraft and aerospace vehicles. Glass fabric-impregnated epoxy laminates are used to insulate printed circuitry. (See Fig. 18-19.) Lighting poles, electrical connections, and antennae also are molded of reinforced plastics.

Fig. 18-19. Printed circuit board with reinforcing fabric.

New assembly techniques create opportunities for increased use of reinforced plastics in the *appliance market*. The base of a large appliance, such as a refrigerator, now must act as an assembly pallet with molded-in snap-fit and alignment features. (See Fig. 18-20.) The refrigerator motor and compressor components are robotically assembled from the base up; so the base not only must be molded to close tolerances but must be strong and rigid. Reinforced plastics are well suited to these design parameters. Reinforced plastics also are an alternative to porcelainized metals for dishwashers, washers, dryers, ranges, and refrigerators. Other appliance applications are panels for toasters and toaster ovens, storage trays for refrigerators, housings for microwavable waffle irons, and skirts for steam irons.

The *business equipment market* chooses reinforced composites for parts consolidation, close molding tolerances, and heat resistance.

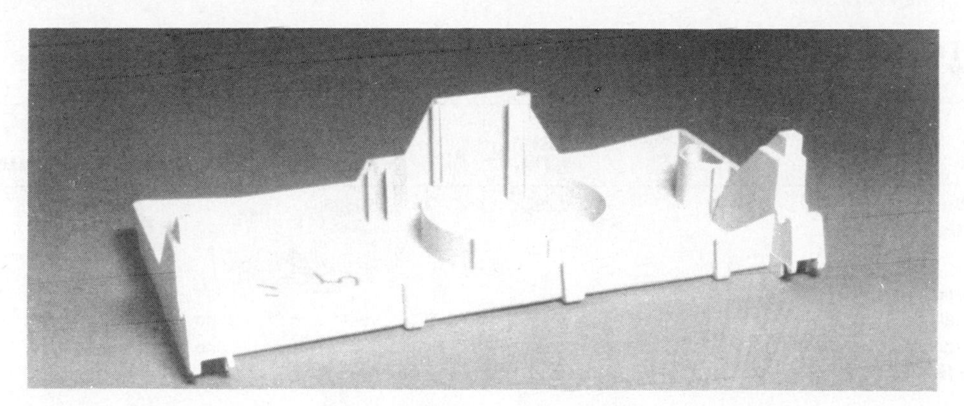

Fig. 18-20. Refrigerator base.

For example, in fiber glass/thermoset polyester, a computer printer mainframe is injection-molded as a unit. Previously, the part was an assembly of 15 machined, cast aluminum and sheet metal parts. The heat shield on a photocopier is glass-reinforced, RIM polyacrylamate. Another equipment example is a reinforced phenolic computer planar board with molded-in slots for adding computer cards.

Most composite applications in the *consumer goods market* are related to sports and recreation. Composite bicycle frames are lightweight and stiff; the increased stiffness reduces side motions that waste pedal power. Carbon fiber–reinforced arrow shafts result in higher arrow speeds and more accuracy, compared to aluminum shafts. Also, graphite fiber–reinforced epoxy baseball and softball bats can be fine-tuned to sound like wood or metal when the bat meets the ball. Reinforced plastic skis made with braided preforms have exceptional torsional stiffness. Other consumer applications for composites include dual-ovenable cookware, planters, and furniture.

In addition to their uses for the markets mentioned above, reinforced plastics are made into stadium and institutional seating, power saw tables, road repair plates, and decking for platform tennis. Medical applications include artificial joints and bones, casts, and equipment housings. Materials handling and industrial machinery also are potential markets for the growth of reinforced plastics because of their light weight, durability, and vibration dampening.

DESIGN REVIEW

The basic methodology for designing in reinforced composites is the same as it is for other materials. The initial steps in the development of a new product or component are more concerned with defining objectives than selecting a material and manufacturing process. (See Table 18-3.) However, as the design engineer or the industrial designer comes closer to making material and process decisions, the design benefits of reinforced composites become more obvious. It should be noted that configurations proven in metal, wood, and reinforced concrete may not be efficient in composites. Consequently, it is best to begin the design process with reinforced plastics "from the ground up," instead of as a replacement for another material.

Materials selection for composite products should be based on a clear definition of projected cost targets and all functional and performance requirements. Considerations include:

1. Structural strength and stiffness and the degree to which they are to be retained throughout the product's life cycle.
2. Nonstructural considerations, such as color, texture, and resistance to corrosion, electricity, moisture, and temperature extremes.
3. Process compatibility based on size and shape, as well as production volume and rate.

Table 18-3. Design considerations by process. (*Source: Owens-Corning Fiberglas Corporation*)

	SRIM RTM	Injection Molding	Pultrusion	RRIM
Minimum Inside Radius (in.)	0.25	0.0625	0.06	0.125
Holes:				
—Parallel[1]	yes[2]	yes[3]	no	yes
—Perpendicular[1]	yes[2]	yes	no	yes
Trim in Mold	no	no	no	yes
Built-in cores	possible	yes with retracting core	no	yes
Undercuts	possible with split mold	possible with slides	yes	yes with slides
Wall Thickness:				
—Minimum (in.)	0.08	0.04	0.08	0.08
—Maximum (in.)	1.0	0.5-1.0	0.5	0.5
—Variations (±in.)[4]	0.01	0.005	0.01	0.005
Minimum Draft:				
—to 6 in. depth	1°	1°	0-2°	1-3°
—over 6 in. depth	2°+1°/6 in.	3°+1°/6 in.	0-2°	3°
Corrugated Sections	yes	yes	yes	yes
Metal Inserts	yes	yes	no	yes
Bosses	difficult	yes	no	yes
Ribs	difficult	yes	yes	yes
Molded-in Labels	yes	no	no	no
Raised Numbers and Letters	yes	yes	no	yes
Surface Finish:				
—Number of finished surfaces	all	2	2	2
—Quality of surface	good to very good	excellent	fair to good	excellent

[1]Relative to ram action and mold closure. [2]Large areas possible; small holes post-drilled or routed. [3]Use core pulls or slides.

Hand Lay-up Spray-up	Compression: Sheet Molding Compound	Filament Winding	Compression: Bulk Molding Compound	Preform Molding
0.25	0.06	0.125	0.06	0.125
yes	yes	yes[2]	yes	not recommended
yes	undesirable	yes	undesirable	undesirable
no	yes	yes	yes	yes
possible	possible, however not recommended	possible with collapsible cores	possible, however not recommended	no
possible with split mold	possible with slides	no	possible with split mold	no
0.06	0.08	0.04	0.06	0.05
1.5-up	0.25	2.0	1.0	,25
0.02	0.005	0.01	0.005	0.005
0-2°	1-3°	3°	1-3°	1-3°
0-2°	3°+1°/6 in.	3°	3°+1°/6 in.	3°+1°/6 in.
yes	yes	difficult	yes	yes
yes	yes	yes	yes	not recommended
yes	yes	no	yes	not recommended
no	yes	no	yes	not recommended
yes	yes	yes	yes	yes
yes	yes	no	yes	yes
1 good to very good	2 very good to excellent	2 good	2 good	2 very good

[4]*Controlled by mold specs in closed molds, by operator skill in open molds.*

4. Assembly considerations, such as tolerances for robotics or thermal movement relative to fastening or bonding.

The size and configuration of components greatly affect the choice of composite materials. For example, for strength, stiffness, and impact resistance, especially over a wide temperature range, the resin selected is often a thermoset. For large parts, process tooling costs often favor reinforced thermosets over thermoplastics, whereas thermoplastics often are used when production rates and volumes are higher. The selection of the fiber reinforcement form (chopped fibers, woven roving, mat, etc.) must be compatible with the component configuration. For instance, the drape characteristics of mat or woven roving must be able to conform to the required shape.

Tradeoffs frequently are necessary to arrive at a final materials selection that achieves the best combination of performance and economics. Valid cost comparisons for composite components versus components made with alternative materials should include the timing of the investment and the value of the performance differences. Once the candidate materials have been determined, consideration should be given to alternative manufacturing processes.

All processes have inherent size and shape limitations. For instance, pultrusion is limited to profiles such as tubes, sheet, and shapes that form linear elements. In the production of high volumes of larger parts, the compression molding process is a faster method of producing products than hand lay-up or spray-up. Injection molding is limited by the high cost of the equipment and tooling for large components. Likewise, high-pressure molding is limited by the size and the capacity of available presses and often by the cost of the molds.

The remainder of this section provides general guidelines for use in designing for specific molding processes. The presentation points out how best to achieve various design features with reinforced plastics. This coverage of basic principles and recommendations is not intended to be a complete text, however; it is a guide for designers and engineers considering reinforced plastics for their next application.

Designing for Closed Mold Processes

The following characteristics pertain mostly to compression-, injection-, and transfer-molded composites.

Draft is a slight angle relative to the direction of the opening and closing of the mold. (See Fig. 18-21.) It is necessary to design the part so that all side surfaces, both interior and exterior, have draft. This enables the part to be removed from the mold without sticking or rubbing, which can degrade surfaces. A minimum draft angle of 1° on all surfaces parallel to mold movement is recommended for the first 3 inches (75 mm) of depth. Recommendations for parts with a deeper draw are: 2° for 3 to 4 inches (75 to 100 mm) of depth, 3° for 4 to 6 inches (100 to 150 mm), and one additional degree for every 2 inches (50 mm) thereafter. Zero draft can be achieved by the use of slides in the mold.

Radius is used in moldmaking to define the curvature established between two intersecting surfaces. (See Fig. 18-22.) More generous radii result in better material flow and less likelihood of "resin-rich" areas with low tolerance for abuse. A minimum radius of $\frac{1}{16}$ inch (1.6 mm) is recommended for both exterior and interior plane intersections. Radii should be designed so as to maintain a relatively uniform part thickness.

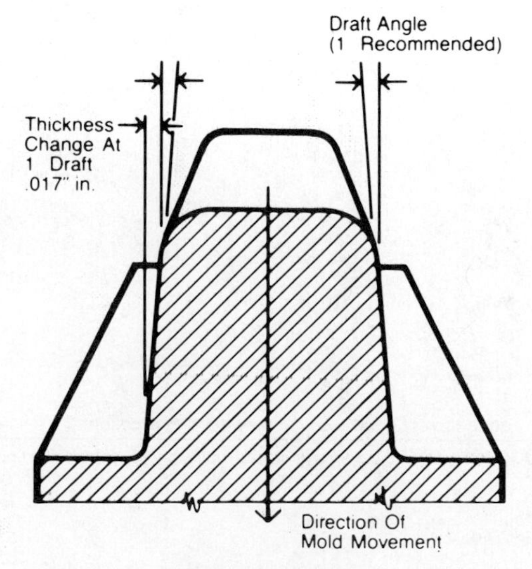

Fig. 18-21. Draft.

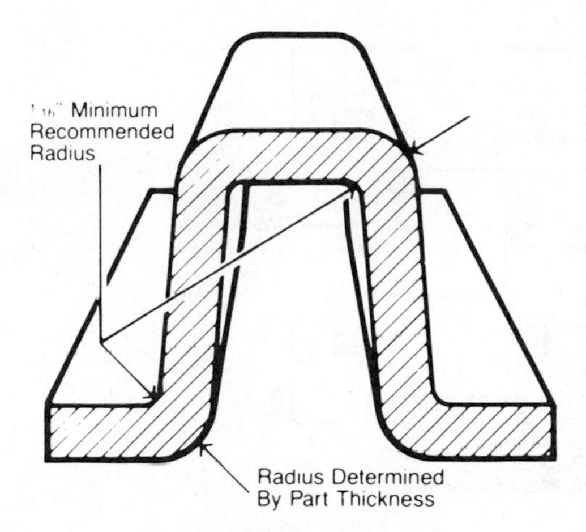

Fig. 18-22. Radius.

Edge stiffening is a design treatment applied to unsupported edges to prevent warping or bowing. Edge turning is preferable to edge thickening.

Ribs are linear projections from the adjacent plane surface of a part. Ribs are used to reduce the bulk and mass of a part while allowing it to meet strength and rigidity requirements. Ribbed designs also resist warpage or bowing on large plane surfaces. Ribs should be designed to maintain the nominal thickness established for the part. Ribs should have at least 1° of draft per side (2° total). A half degree per side (1° total) can work in some instances.

Bosses are projections from a plane surface of a part that provide for attachment and support of related components. Bosses may be solid or hollow or have molded-in inserts. Like ribs, bosses should be designed to maintain the nominal thickness established for the part, within the limitations determined by their function. If large-diameter bosses are required, they should be hollow or cored to maintain nominal thickness. Bosses should have at least 1° of draft all around; a half degree of draft is allowable in some instances.

Molded-in threads are difficult to mold in reinforced thermosets and require sophisticated molds and molding procedures. However, for molding threads in reinforced thermoplastics, there are some basic guidelines to assist the designer. External and internal threads may be molded in either plane of mold movement. However, internally threaded holes should be larger than $\frac{1}{4}$ inch (6.4 mm). If threads are required in the direction of mold movement, an insert is used and removed from the part after the part is removed from the mold. Rotating side cores are used to mold female threads in the plane of the parting line or male threads that must be free of parting lines. If parting lines do not affect the performance of male threads, these threads can be molded without side cores. (See Fig. 18-23.)

Undercuts, side holes, and similar details are located with respect to the direction of mold opening and closing. Achieving these features requires movable mold components to allow extraction of the part from the mold. Holes and undercuts in exterior surface perpendicular to the direction of mold movement require movable cores or slides in the mold. Holes and other details parallel to the direction of the mold movement can be produced with stationary mold elements. Interior holes are difficult and expensive to mold because it is difficult to design core pulling mechanisms in the interior of the mold.

Fastening, Welding, Bonding

Press or snap-fit fastening is possible with fiber glass composites, especially thermoplastics. They are difficult to design, however, because of the close dimensional tolerances required for reliable fit. Metal screws often are used for fasteners. Self-tapping screws are adequate for parts that rarely need to be disassembled, and should be used with thick-section bosses surrounding the hole. High–low thread cutting screws or shoulder bolts with washers are best for connecting two thermoset composite parts.

Thread-forming screws or bolts are used in thermoplastics. These screws and bolts displace thermoplastic material, and the displacement creates forces that lock the screw in place.

For repeated assembly and disassembly, *fitted metal inserts* should be molded into the part. Speed nuts and clips may be applied to threaded or unthreaded studs or post inserts to secure parts to each other.

Rivets are a low-cost, easy-to-install fastening method. With metal washers on both sides of the hole for stress relief, they provide very solid connections. However, they are not so

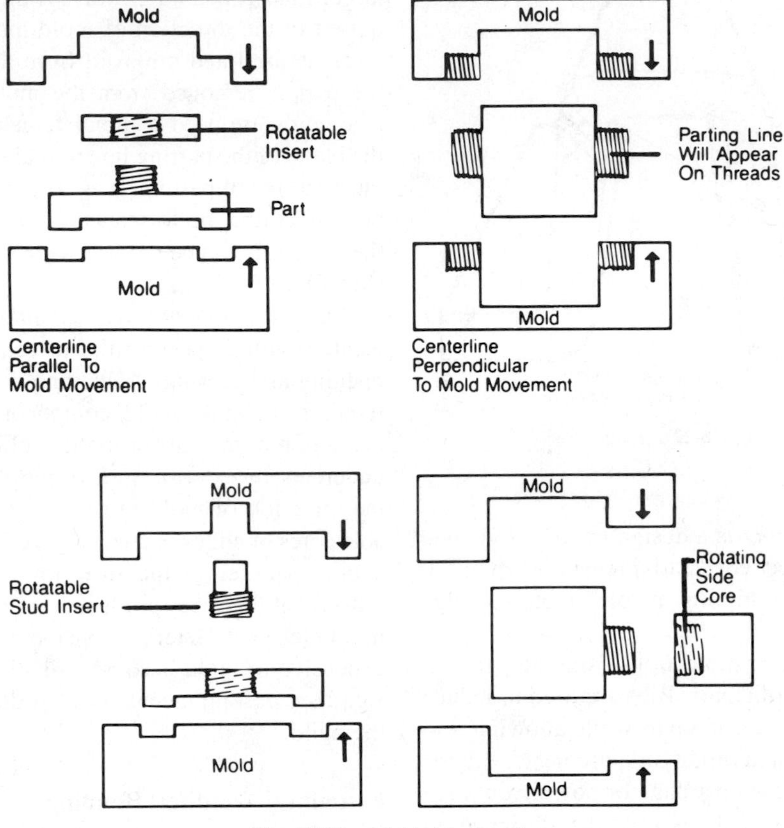

Fig. 18-23. Molded-in threads.

precise as threaded connections and no so reliable in supporting loads in tensions.

Fiber glass-reinforced thermosets and thermoplastics can be fastened with adhesives. *Adhesive bonding* often is used for joining fiber glass-reinforced thermosets. Suggested arrangements for adhesion include laps joints, butt joints, end caps, and tongue-and-groove designs.

Many thermoplastics can be *solvent-bonded*. The surfaces first are softened with an appropriate solvent and then the parts are joined together and become chemically bonded as the solvent evaporates. Several techniques join thermoplastic components together by taking advantage of the material's hot-melt characteristic. These techniques include ultrasonic welding, hot wire welding, spin welding, hot plate welding, hot gas welding, and electromagnetic bonding.

Designing for Hand Lay-up

Hand lay-up facilitates fabrication of large sections in one piece, yet, pieces must not be so large that fabricators cannot reach all points. Designers should keep body-movement limits in mind. Any point should be easily reached with resin and reinforcement and for roll-out. Ignoring this factor inevitably runs up job and tool costs. With parts having a large surface, it is possible to pivot the mold for easier accessibility. In large molds such as boat hulls, laminating frequently is accomplished from scaffolding. Design should be such that the part can be easily removed from the mold without damage. With molded-in grapple points, the part can be hoisted to facilitate separation of the part from the mold. Flanges can be molded onto the edge for wedging the part away from the mold. (See Fig. 18-24.) Ports designed into the mold

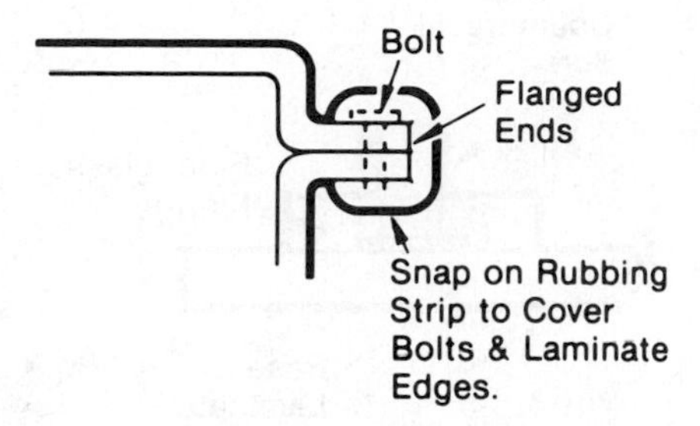

Fig. 18-24. Flanged edges.

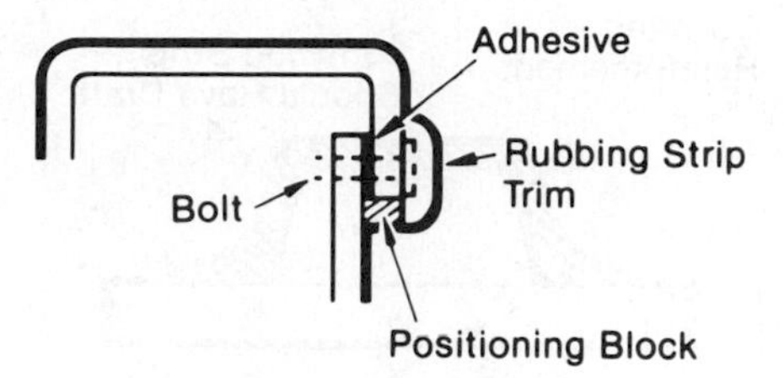

Fig. 18-25. Lapped sections.

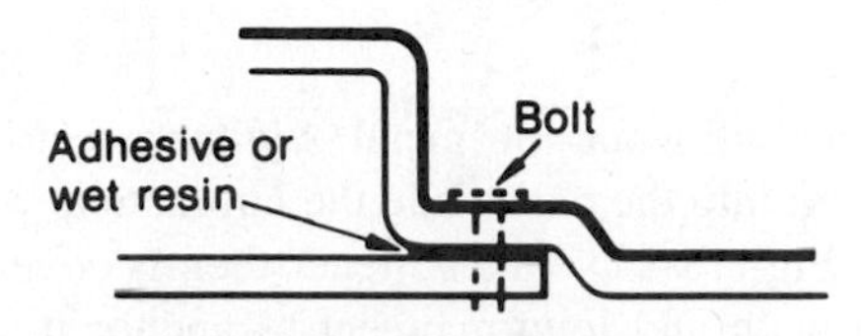

Fig. 18-26. Integral stiffening.

permit the use of air or water injection to initiate part release.

Laid-up parts may be *joined mechanically* to metal, wood, or other composite parts. Bolts and self-tapping threaded fasteners are most common. When threaded fasteners are used, they should be perpendicular to reinforcement plies. Edge fastening never should be used. When *threaded fasteners* are used, their distance from the edges and sides of the part should be at least $2\frac{1}{2}$ times the diameter of the fastener. Back-up plates or blocks are recommended. Then, spacing should exceed three times the fastener diameter. For *adhesive bonding*, flanges should be incorporated into the design of parts to be joined. The flanges should be large enough to provide a good base for the adhesive.

When premolded sheets are joined in a *simple butt joint*, edges should be angled to give more bonding area. Then, laminate layers are built up in thickness on both sides until the desired strength is achieved. A simple corner can be formed with premolded sheets of reinforcement and resin. For strength, a wood or metal angle can be incorporated. *Flanged edges* bolted together require trim to hide the joint. Bolts, self-tapping screws, or rivets can be used to hold hand lay-up parts together.

Lapped sections are held in place mechanically with adhesive or wet resin between sections as a seal. Trim hides the edges. The fastening devices can be bolts, screws, or rivets. (See Fig. 18-25.)

Large areas often require *stiffening* against deflection. The simplest—but not always most economical—way to get stiffness is to increase laminate thickness. Local stiffeners also may be added to the laminate during hand lay-up. Such stiffeners always should be designed into the part where localized loading occurs. (See Fig. 18-26.)

Built-up stiffeners incorporated in laminate may be cardboard, balsa, various foams, plastic, or sheet metal. They should be designed to spread loads over a wide area. Integral stiffeners are most easily incorporated as flanges at a joint to provide dual function. Pads or gussets are desirable where stiffeners intersect a flat, flexible area.

Designing for Spray-up

The spray-up process initially was developed for speed over hand lay-up. Properly applied, spray-up can impart other advantages as well. Spray-up molds are similar to those used for hand lay-up. Experience has shown that spray-up parts vary about ±20 mils (0.5 mm) in thickness for a common spray-up laminate with a nominal wall thickness of 125 mils (3.18 mm). This variation should be expected and considered in design.

Vertical planes may be incorporated in the design of parts. However, resins having non-drainage properties must be employed. If the designed thickness is in excess of 125 mils (3.18 mm), the part should be built up in 125-mil (3.18 mm) layers. Tapered wall sections and built-up radii in designs will serve to distribute stresses efficiently.

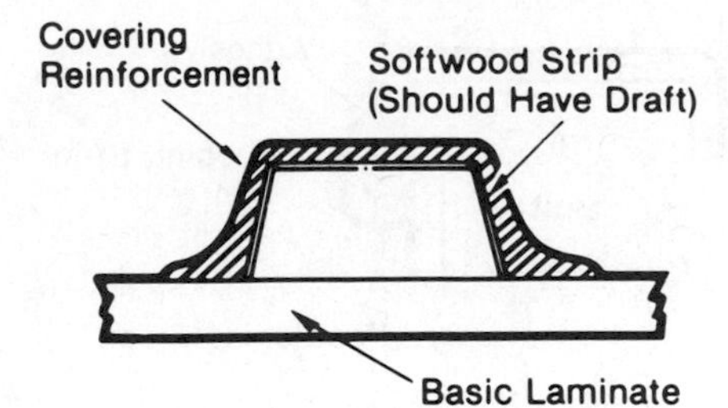

Fig. 18-27. Encased stiffener.

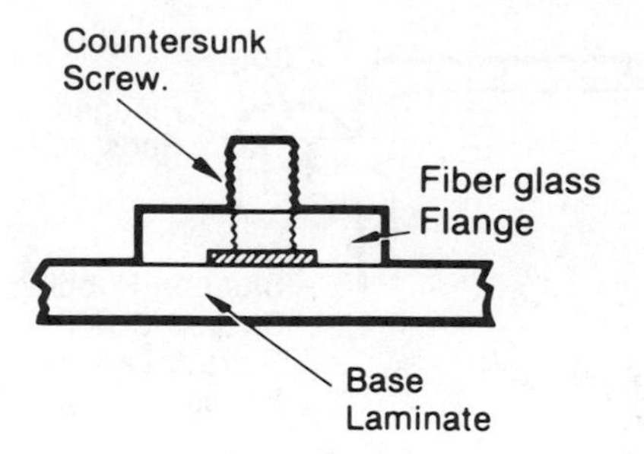

Fig. 18-28. Bolting flange.

Encased wood or metal *stiffeners* can be pressed into the part while the laminate is wet. (See Fig. 18-27.) The stiffener then is covered with resin and reinforcement to anchor it and to protect it from deterioration. Separately fabricated composite stiffeners can be pressed into the laminate while it is still wet. With proper surface preparation, fiber glass composite adheres well to itself using either additional spray-up or adhesives. In addition, lightweight, low-cost forms, such as cardboard or balsa wood, can be pressed into the laminate during spray-up to provide shape and stiffness to the finished composite.

A *bolting flange* can be made by cutting a section of a $\frac{1}{2}$-inch (12.7 mm)-thick laminate. (See Fig. 18-28.) This laminate can be tapped for studs or countersunk for cap screws. The assembly then is placed into a wet layer of spray-up material, and the flange can be strengthened by adding layers. To insert a *threaded coupling*, a collar of expanded metal mesh or coarse metal screen is tack-welded to a threaded pipe coupling. (See Fig. 18-29.) The assembly is pressed into a layer of wet spray-up, and cover layers are added for increased torque resistance. Large beads on the tack welds will act to "key" the laminate to prevent twisting.

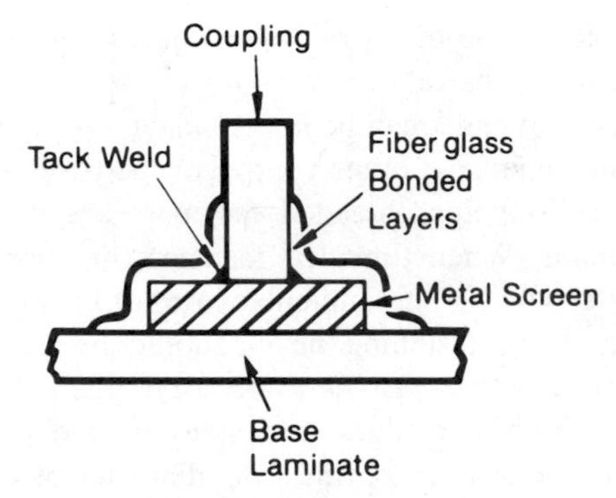

Fig. 18-29. Threaded coupling.

CONCLUSION

Reinforced plastics and composites are proven engineering materials uniquely capable of meeting a wide range of cost and performance requirements. Reinforced plastics and composites offer a combination of properties not found in other materials. The versatility of this family of materials is evident in its tens of thousands of different applications—from intricate electronic connectors to oceangoing vessels. With so many variables in reinforcing materials, resin matrices, and processes, it is possible to design-engineer a composite to be market- and application-specific.

19

Cellular Plastics

OVERVIEW

The term "cellular plastics" encompasses a range of materials with widely varying properties and fields of application. Virtually any polymer, thermoplastic or thermoset, can be made into a cellular or foamed form with the resulting products having densities ranging from 60 pounds per cubic foot (pcf) all the way down to 0.1 pcf.

Basically, cellular plastics can have either of two structural configurations: (a) the closed-cell type, in which each individual cell, more or less spherical in shape, is completely enclosed by a wall of plastic, or (b) the open-cell type, in which the individual cells are intercommunicating. The foams can be rigid, semirigid, or flexible. In general, the properties of the respective plastics are present in the foamed products, except, of course, those that are changed by conversion to the cellular structure.

In this chapter, the more common foamed plastics now in use are discussed. Readers are cautioned, however, that there may be little in common between many of these materials, other than some degree of cellular structure. Cellular plastics can be produced in the form of slabs, blocks, boards, sheets, molded shapes, and sprayed coatings. Some can also be "foamed-in-place" in an open cavity. The type of process used (casting, extrusion, injection molding) will affect the properties of the end-product.

As originally conceived, most foamed plastics were completely cellular in structure; today, it is possible to arrange the cells so that a product may have an essentially solid skin surface and a cellular core.

The ways in which the cellular structure is produced in plastics vary widely. The following are among those discussed in this chapter:

1. Air is whipped into a suspension or solution of the plastic, which is then hardened by heat or catalytic action or both.
2. A gas is dissolved in the mix and expands when pressure is reduced.
3. A component of the mix is volatilized by heat.
4. Water produced in an exothermic chemical reaction is volatilized within the mass by the heat of reaction.
5. Carbon dioxide gas is produced within the mass by chemical reaction.
6. A gas, such as nitrogen, is liberated within the mass by thermal decomposition of a chemical blowing agent.
7. Tiny beads of resin or even glass (e.g., micro-balloons) are incorporated in a plastic mix.

Finally, readers should be cautioned about the terminology used to identify cellular plastics. Unfortunately, it is loose and generally relates to the starting material (e.g., polyethylene) from which the foam is made. However, an expression such as "polyethylene foam" can have many meanings. This term can refer to special low-density polyethylene foams and special high-density polyethylene foams that are quite different in character. It could include cross-linked polyethylene foams, which differ even more, or it can refer to polyethylene foam films made by extruding low-density polyethylene with a nitrogen blowing agent. Or, to complicate matters still further, it also could encompass low-density polyethylene structural foams or high-density polyethylene structural foams that bear no resemblance or relationship to any of the other foams that have been mentioned thus far. In this chapter, an attempt will be made to maintain these distinctions and to point up the differences between the various cellular plastics whenever possible.

History

In a sense, the first foamed plastic was what resulted from the reaction of phenol and formaldehyde before Dr. Leo H. Baekeland, seeking to develop a nonporous resin, found means of preventing the development of voids. Thus the development of foamed plastics has involved learning how to do, under controlled conditions, the very thing that Baekeland so profitably avoided doing.

The development of foamed phenolics dates from about 1945 and the use of phenolic ''balloons'' in the special ''syntactic'' type of foam (see below) from about 1953.

Epoxy foams were first developed, in 1949, to meet the need for a suitable material of light weight for the encapsulation of electronic components. Further study led, in 1953, to the development of epoxy foam of a density of 2 pcf for the Bureau of Ships and for the aircraft industry.

At about that time there was also developed a special type of cellular plastic (''syntactic''), comprising tiny hollow spheres of phenolic resin embedded in epoxy resin. Later, similar hollow spheres of urea resin became available, and resins other than epoxy came to be used for foams of this type. Also, there was developed a further type in which tiny solid beads of polystyrene, containing dissolved gas, were mixed into epoxy resin and were expanded by the exothermic heat released during the cure of the epoxy.

Urea-formaldehyde foams were first produced in Germany, prior to World War II, in the form of slabs used as thermal insulation. During the war, such foam, manufactured in the United States, was used to float ammunition, food, and other supplies to shore during amphibious operations.

The first published reports on foamed vinyls appeared during World War II in Germany. The toxicity of the blowing agents thus disclosed delayed commercial development of chemically expanded vinyls in the United States, but subsequently nontoxic blowing agents became available. A process of mechanical blowing, by absorption and expansion of an inert gas, was developed in the United States prior to 1950. The development of foamed vinyls was greatly assisted by the availability of vinyl plastisols, which can be fused without application of pressure. Concurrently, pressure-molding techniques were developed for production of closed-cell vinyl foams. Extrusion methods more recently were developed for production of expanded shapes.

The technology of the urethanes was well developed in Germany by the mid-1930s, and in World War II urethane foams based upon alkyd-type polyesters came into wide use there—rigid foams as structural reinforcement in aircraft and ships; flexible foams in cushioning uses, replacing rubber foams for which latex was not obtainable.

Information brought from Germany by survey teams after the war, the licensing of German-owned patents, and service contracts to the aircraft industry stimulated research in the United States that resulted in the development of practical techniques, improvements in production equipment, and formulations based upon starting materials other than alkyd-type polyesters.

The extrusion of foamed polystyrene was based upon developments in Sweden in the mid-1930s. Commercial production in the United States was begun in 1944.

Expandable polystyrene first was developed in Germany about 1952. Subsequently, a similar product was developed in the United States, and introduced to the market in early 1954.

Development work on foams of silicone resins was started in 1950, to meet the need for a material of low weight that would retain good mechanical strength after long exposure to temperatures of 400 to 700°F. The first commercial product, which appeared in 1952, required compounding by the user, and was superseded in 1953 by premixed powders needing merely to be melted, whereupon they foam and are cured under the influence of heat. Because the temperature required for these powders is high (sometimes too high to be tolerated by other materials in an assembly), and the minimum density of the foam is high (10 pcf), attention was given to a newer series of resins, which can be expanded and cured at room temperature, and yield rigid foams having a density of 4 to 5 pcf.

Cellular polyethylene, a closed-cell foam, was developed in 1944, specifically as a low-loss insulation for wire and cable. Extruded polyethylene foam then became available in the form of slabs and blocks.

Cellular cellulose acetate was developed, at the time of World War II, primarily to meet the requirements of the U.S. Air Force for reinforcement in a monocoque construction, to supersede the balsa wood used in the British ''Mosquito'' bomber.

In the mid- and late 1960s, plastic foams made some important strides in terms of technology. A number of commercial techniques for producing structural foam plastics (e.g., the Engelit process, the Union Carbide process, the USM process, etc.) were introduced to industry. Advances also were made in the direct injection extrusion of polystyrene foam as opposed to the more conventional bead molding. In the polyolefins area, there were such developments as chemically- and radiation-cross-linked polyethylene foams, low density closed-cell polypropylene foams (also chemically cross-linked) for automotive cushioning, and polyethylene foam film (for bath mats, table mats, embossed wall coverings, etc.) extruded from conventional low-density polyethylene and a nitrogen blowing agent. The 1960s also saw the introduction of a rigid ABS foam, ionomer foams, and acrylic foams.

More recently, other materials have found their way onto the list of foamed plastics (i.e., styrene maleic anhydride copolymer, polyphenylene ether/polystyrene blends, polyesters, polyimides, and others). Phenolic foams have been gaining interest for insulation uses where flammability is an issue.

Processing has become more sophisticated, especially in the area of producing high-quality surfaces on foam parts. Sandwich molding and coextrusion are being used to manufacture ''foam core'' thermoplastic products with low density, rigidity, and smooth surfaces for a variety of applications.

On the negative side, the use of chlorofluorocarbon blowing agents to produce low-density polyurethane and polystyrene foams has come under fire because they have been linked to depletion of the stratospheric ozone layer. The industry currently is involved in massive programs to find environmentally acceptable alternatives.

CHEMICAL BLOWING AGENTS*

Knowledge of the chemistry of blowing agents and an understanding of their decomposition products can be helpful in selecting blowing agents for particular applications and can be valuable in approaching problems associated with their use.

The development of organic blowing agents has been a major contribution to the field of cellular rubber and plastics. These organic materials show important advantages over inorganic gas-forming agents such as sodium bicarbonate, which has long been used in the manufacture of open-cell expanded rubber. In contrast to ''soda'' the commercially available organic blowing agents are capable of producing a fine, closed-cell structure in rubber and plastics and their use has grown rapidly, particularly in cellular plastics.

Generally, commercial organic blowing agents are organic nitrogen compounds that are stable at normal storage and mixing temperatures but undergo controllable gas evolution at reasonably well-defined decomposition temperatures. Two comprehensive reviews of organic substances that have been suggested or used as blowing agents have been given by

*Based on the work of Dr. Byron A. Hunter, Senior Research Associate (retired), Uniroyal Chemical Co., Inc., as amended by Brendan J. Geelan, Specialty Chemicals Technical Service, Uniroyal Chemical Co., Inc.

Reed.[1] A more recent summary has also appeared.[2]*

Choice of Blowing Gas

Nitrogen is an inert, odorless, nontoxic gas. For this and other good reasons, nitrogen-producing organic substances have been preferred as blowing agents. Van Amerongen[3] measured the permeability of natural rubber to several gases and gave the following relative values (based on hydrogen as 100):

$$\begin{aligned} CO_2 &= 260 \\ H_2 &= 100 \\ O_2 &= 46 \\ N_2 &= 17 \end{aligned}$$

The organic blowing agents that have been offered vary widely in their properties. Importantly, they vary in the temperature at which they produce gas and in the nature of their decomposition products. Some produce odor, whereas others may yield colored or toxic substances on decomposition. The agents vary considerably in their response to other materials present in the expandable polymer that may function as activators or retarders. These factors are important in the selection of a blowing agent for a particular use.

$$(1)\quad 2(CH_3)_2C{=}O + H_2NNH_2 + 2\ NaCN + H_2SO_4 \rightarrow$$

$$\underset{\mid\atop CN}{(CH_3)_2C}-\overset{H}{N}-\overset{H}{N}-\underset{\mid\atop C}{C}(CH_3)_2 + Na_2SO_4 + 2H_2O \qquad N$$

$$(2)\quad (CH_3)_2\underset{\mid\atop CN}{C}-\overset{H}{N}-\overset{H}{N}-\underset{\mid\atop CN}{C}(CH_3)_2 + Cl_2 \rightarrow (CH_3)_2\underset{\mid\atop CN}{C}-N{=}N-\underset{\mid\atop CN}{C}(CH_3)_2 + 2\ HCl$$

The present discussion is concerned with the chemistry of those blowing agents that have attained commercial importance as expanding agents for rubber and plastics in the United States. Reference also will be made to a number of products that have not achieved commercial status. The discussion will emphasize the chemistry and decomposition mechanisms of chemical blowing agents insofar as this information has been made known.

Azobis(isobutyronitrile)

An interesting blowing agent came out of Germany during World War II.[4] This compound, azobis(isobutyronitrile):

$$(CH_3)_2\underset{\mid\atop CN}{C}-N{=}N-\underset{\mid\atop CN}{C}(CH_3)_2$$

was introduced for the production of sponge rubber articles. Later the material was recommended for expanding polyvinyl chloride. The material is completely nondiscoloring and nonstaining and yields white PVC foam of fine uniform cell structure. The compound is manufactured from hydrazine, acetone, sodium cyanide and acid, using chlorine to oxidize the intermediate hydrazo bis(isobutyronitrile):

The decomposition of azobis(isobutyronitrile) proceeds as follows:

$$(3)\quad (CH_3)_2-\underset{\mid\atop CN}{C}-N{=}N-\underset{\mid\atop CN}{C}(CH_3)_2 \xrightarrow{\Delta} N_2 + \underset{\text{(toxic)}}{(CH_3)_2\underset{\mid\atop CN}{C}-\underset{\mid\atop CN}{C}(CH_3)_2}$$

The decomposition occurs rapidly at temperatures above 100°C and more slowly at lower

*References follow, at end of section.

temperatures. Complete decomposition yields 137 ml of gas per gram (measured at standard conditions). The decomposition residue, tetramethyl succinonitrile, is a toxic substance, and precautionary measures must be taken to eliminate this hazard.[4c] There has been reluctance in the United States to use the chemical as a blowing agent, and the principal U.S. manufacturer does not recommend the material for this use. It has been stated in Europe[1] that azobis(isobutyronitrile) can be successfully employed to expand PVC without accident if adequate ventilation is provided.

The decomposition of azobis(isobutyronitrile) yields free radicals and this substance has been employed quite widely as a polymerization initiator.

Dinitroso pentamethylene tetramine

$$
\begin{array}{ccccc}
 & CH_2 & - N - & CH_2 & \\
 & | & | & | & \\
ON - & N & CH_2 & N & - NO \\
 & | & | & | & \\
 & CH_2 & - N - & CH_2 &
\end{array}
$$

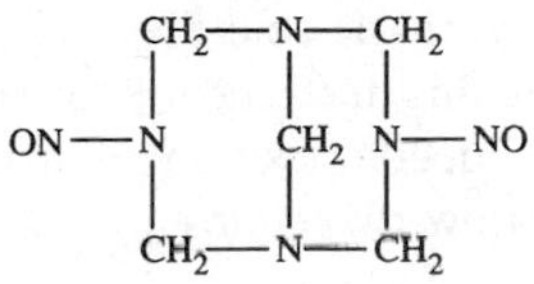

This compound is prepared by the nitrosation of hexamethylene tetramine (reaction product of formaldehyde and ammonia):

(4)

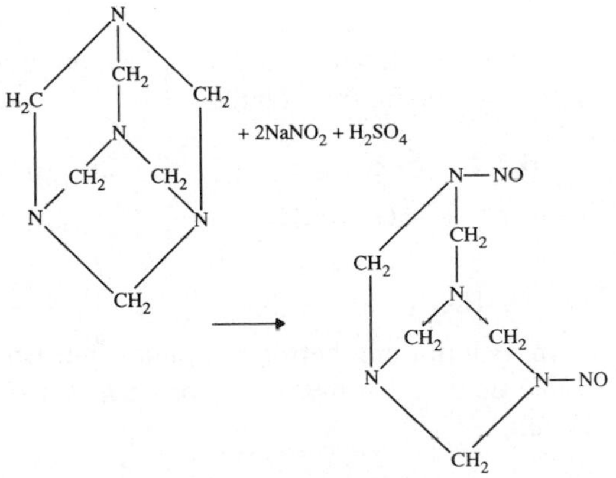

$$\text{hexamethylene tetramine } (N_4(CH_2)_6) + 2NaNO_2 + H_2SO_4 \longrightarrow \text{dinitroso pentamethylene tetramine } ((CH_2)_5N_4(NO)_2)$$

The chemical, when heated alone or in the presence of inert diluents, decomposes near 195°C (383°F); but when used in rubber or plastics in the presence of certain activators, it produces gas within a temperature range of 130 to 190°C (266–374°F).[5] The quantity of gas produced from the undiluted material is close to two moles per mole of blowing agent (near 265 ml/g measured at standard conditions).[2]

The decomposition products of dinitroso pentamethylene tetramine have not been fully elucidated. It has been reported[1] that nitrogen and nitrous oxide* are formed as well as amines and water. The presence of formaldehyde can also be detected in the decomposition of the dry material. The amine residue gives rise to a characteristic "fishy odor" in the expanded product. The odor can be at least partially suppressed by the addition of urea, melamine, and certain amino compounds.

The temperature of decomposition of dinitroso pentamethylene tetramine can be substantially lowered and the decomposition accelerated by acidic substances such as salicylic acid and phthalic anhydride, which are commonly used as activators. Urea (BIK)† and hydroxy compounds (as ethylene glycol) have also been employed as activators.

Dinitroso pentamethylene tetramine is used widely as a blowing agent in the rubber industry but is of limited use in plastics because of the high decomposition exotherm and the unpleasant odor of the residue.[2] The chemical is nondiscoloring and nonstaining.

Azodicarbonamide

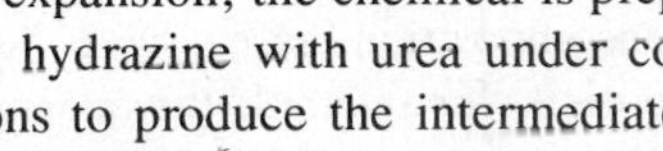

$$H_2N-\underset{\underset{O}{\|}}{C}-N=N-\underset{\underset{O}{\|}}{C}-NH_2$$

First suggested as a potential blowing agent for plastics in Germany during World War II,[4] azodicarbonamide was introduced in the United States in the early 1950s. Important today in plastics expansion, the chemical is prepared by reacting hydrazine with urea under controlled conditions to produce the intermediate hydrazodicarbonamide,[6] which is oxidized to azodi-

*One manufacturer has found no evidence of nitrous oxide but reports that ammonia is present.

†A surface-treated urea supplied by Uniroyal Chemical.

carbonamide:

$$(5)\quad H_2NNH_2 + 2H_2N{-}\underset{\underset{O}{\|}}{C}{-}NH_2 \underset{H+}{\overset{\Delta}{\rightarrow}} H_2N{-}\underset{\underset{O}{\|}}{C}{-}\overset{H}{N}{-}\overset{H}{N}{-}\underset{\underset{O}{\|}}{C}{-}NH_2 + 2NH_3$$

$$(6)\quad H_2N{-}\underset{\underset{O}{\|}}{C}{-}\overset{H}{N}{-}\overset{H}{N}{-}\underset{\underset{O}{\|}}{C}{-}NH_2 \overset{(O)}{\rightarrow} H_2N{-}\underset{\underset{O}{\|}}{C}{-}N{=}N{-}\underset{\underset{O}{\|}}{C}{-}NH_2 + H_2O$$

Effective oxidants are dichromate, nitrates, nitrogen dioxide, chlorine, and so on.

Azodicarbonamide (azobisformamide) is a yellow crystalline solid which decomposes to produce a high yield of gas (220–240 cc/g, STP). Unlike most other blowing agents, azodicarbonamide does not support combustion and is self-extinguishing. The white decomposition residue is odorless, nontoxic, nondiscoloring, and nonstaining.

The dry decomposition temperature is high (195–216°C, depending on the mode of preparation). This range can be lowered by a variety of activators (e.g., lead, cadmium, zinc and barium salts, etc.), thereby making azodicarbonamide a successful agent in the expansion of rubber as well as plastics.

$$(7)\quad H_2N{-}\underset{\underset{O}{\|}}{C}{-}N{=}N{-}\underset{\underset{O}{\|}}{C}{-}NH_2 \rightarrow NH_3 + CO + N_2 + HNCO$$

$$(8)\quad 2\,H_2N{-}\underset{\underset{O}{\|}}{C}{-}N{=}N{-}\underset{\underset{O}{\|}}{C}{-}NH_2 \rightarrow H_2N{-}\underset{\underset{O}{\|}}{C}{-}\overset{H}{N}{-}\overset{H}{N}{-}\underset{\underset{O}{\|}}{C}{-}NH_2 + N_2 + 2HNCO$$

$$(9)\quad H_2N{-}\underset{\underset{O}{\|}}{C}{-}\overset{H}{N}{-}\overset{H}{N}{-}\underset{\underset{O}{\|}}{C}{-}NH_2 \rightarrow \text{urazole (HN–NH, O=C C=O, N–H ring)} + NH_3$$

The decomposition gases have been analyzed. In diethylene glycol, the decomposition produces a gaseous mixture of nitrogen (62%), carbon monoxide (35%), and ammonia and carbon dioxide (total 3%). Reed[1] has reported a similar analysis of the gases from dry decomposition (190°C) of azodicarbonamide and has also analyzed the solid decomposition products. He found gases (noted above) to be 32% by weight; the weights of the solid products as percentages of azodicarbonamide were: solid residue, 41% (urazole, 39%, biurea, 2%), and sublimate, 27% (cyanuric acid, 26%, cyamelide, 1%).

However, Reed found the solid products of decomposition at 190°C under liquid paraffin were markedly different in proportion (% by weight): gaseous products, 34%; solid residue, 61% (urazole, 27%, biurea 34%); and sublimate, 5% (cyranuric acid 5%, cyamelide 0%).

To explain this difference, Reed suggests that the primary decomposition of azodicarbonamide can follow two courses:

Urazole can result from the elimination of a mole of ammonia from biurea:

An equilibrium between cyanic and isocyanic acid and trimerization of each can be written:

$$(10)\quad HO{-}C{\equiv}N \rightleftharpoons HNCO$$

Cyanuric acid (ring: OH–C, N, N, HO–C, C–OH, N) ← HO–C≡N; HNCO → Cyamelide (ring: HN=C, O, C=NH, O, C=NH, O)

Cyanuric acid

Cyamelide

Moisture can affect the blowing characteristics of azodicarbonamide. It is known that hydrolysis of the chemical can occur at high temperatures in the presence of either acid or base to produce biurea, nitrogen, and carbon dioxide.[4]

Benzene Sulfonyl Hydrazide

Benzene sulfonyl hydrazide is the simplest aromatic compound in the class of sulfonyl hydrazides. It is prepared by treating benzene sulfonyl chloride with hydrazine in the presence of a base (such as ammonia):

$$(11)\quad C_6H_5SO_2Cl + H_2NNH_2 + NH_4OH \longrightarrow C_6H_5SO_2NHNH_2 + NH_4Cl + H_2O$$

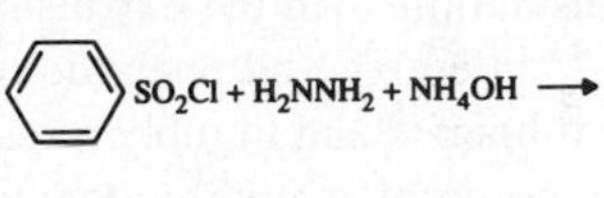

The product is a white crystalline solid that melts and begins to decompose near 105°C. It is capable of producing a white unicellular foam when incorporated into a PVC plastisol.[7] Unfortunately the odor is strong and unpleasant, reminiscent of thiophenol.

Interestingly, experiments to detect thiophenol in decomposition vapors have been negative. Instead, the odor may actually be due to diphenyl disulfide. Decomposition involves an internal oxidation–reduction of the sulfonyl hydrazide group. A possible mechanism of the gas forming reactions has been suggested:[8a]

$$(12)\quad C_6H_5SO_2N(H)NH_2 \longrightarrow H_2 + \left[C_6H_5SOH\right] + H_2O$$

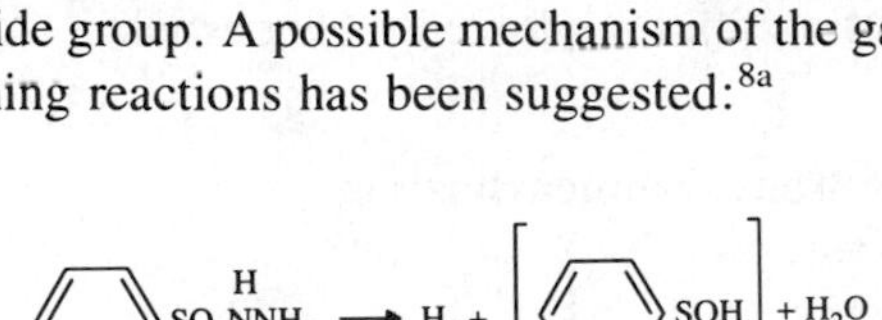

$$(13)\quad 4\left[C_6H_5SOH\right]\ (\mathrm{I}) \longrightarrow C_6H_5S{-}SC_6H_5\ (\mathrm{II}) + C_6H_5S{-}S(=O)_2{-}C_6H_5\ (\mathrm{III}) + 2H_2O$$

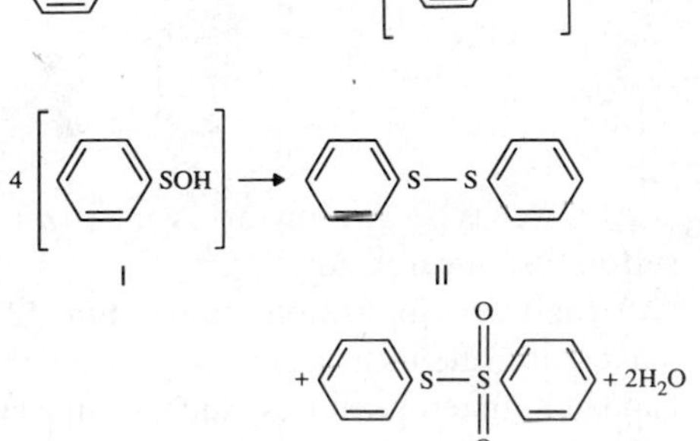

The hypothetical intermediate benzene sulfenic acid (I) apparently is incapable of existence and immediately disproportionates to diphenyl disulfide (II) and phenyl benzene thiosulfonate (III) as shown in Equation (13).

Diphenyl disulfide was identified in the decomposition products by early investigators.[9] A thiosulfonic ester was identified in a recent study in the similar decomposition of *p*-toluene sulfonyl hydrazide[10] (discussed next).

The strong odor where it is a blowing agent in PVC vanishes when the chemical is used to expand rubber. A likely explanation is that in rubber expansion the sulfur-containing residues (or their reactive intermediates) react with the rubber, leaving nonodorous combinations. Supporting this is the observation that certain difunctional sulfonyl hydrazides can function as curing agents in rubber compositions.[11]

p-Toluene Sulfonyl Hydrazide

The compound *p*-toluene sulfonyl hydrazide is similar to benzene sulfonyl hydrazide except that the melting point and decomposition temperatures are higher (120°C and up) for this compound. Their decomposition mechanisms and odor problems are similar. Deavin and Rees[10] isolated ditolyl disulfide and *p*-tolyl *p*-toluene thiosulfonate as main decomposition products. Other decomposition products are *p*-toluene sulfinic acid, hydrazine, and a small quantity of the *p*-toluene sulfonyl hydrazide salt of *p*-toluene sulfinic acid (see Fig. 19-1). The presence of *p*-toluene sulfinic acid and the salt of *p*-toluene sulfonic acid can be explained by the reaction shown in Fig. 19-1 and by the disproportionation reactions of sulfenic and sulfinic acids.[12]

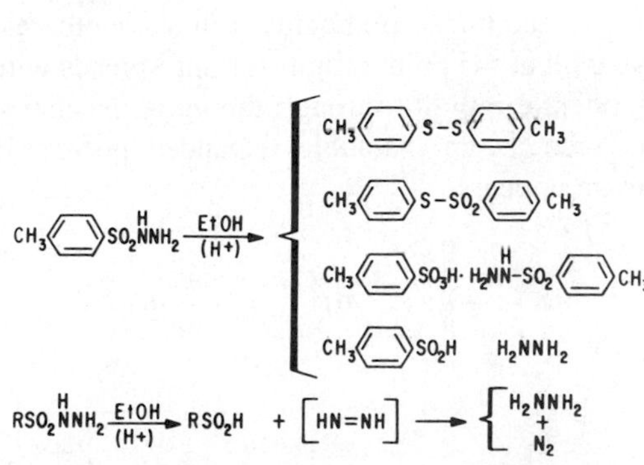

$$CH_3C_6H_4SO_2N(H)NH_2 \xrightarrow[(H^+)]{EtOH} \begin{cases} CH_3C_6H_4S{-}SC_6H_4CH_3 \\ CH_3C_6H_4S{-}SO_2C_6H_4CH_3 \\ CH_3C_6H_4SO_3H\cdot H_2N\overset{H}{N}{-}SO_2C_6H_4CH_3 \\ CH_3C_6H_4SO_2H \quad H_2NNH_2 \end{cases}$$

$$RSO_2\overset{H}{N}NH_2 \xrightarrow[(H^+)]{EtOH} RSO_2H + \left[HN{=}NH\right] \longrightarrow \begin{cases} H_2NNH_2 \\ + \\ N_2 \end{cases}$$

Fig. 19-1. Decomposition of *p*-toluene sulfonyl hydrazide in boiling ethanol (acid present).

p,p-Oxybis(Benzene Sulfonyl Hydrazide)

For all practical purposes this blowing agent eliminates odor in applications with cellular plastics. It is prepared by chlorosulfonation of diphenyl ether with chlorosulfonic acid and subsequent reaction with hydrazine in the presence of a base:[14]

$$(14)\quad C_6H_5\text{–O–}C_6H_5 + 4Cl \xrightarrow{SO_3H} ClSO_2\text{–}C_6H_4\text{–O–}C_6H_4\text{–}SO_2Cl + 2H_2SO_4 + 2HCl$$

$$(15)\quad ClSO_2\text{–}C_6H_4\text{–O–}C_6H_4\text{–}SO_2Cl \xrightarrow[2OH^-]{2N_2H_4} H_2N\text{–}NH\text{–}SO_2\text{–}C_6H_4\text{–O–}C_6H_4\text{–}SO_2\text{–}NH\text{–}NH_2 + 2Cl^- + 2H_2O$$

The product[8c] is a white crystalline solid that melts with decomposition at 164°C, and lower, in solution or in the presence of rubber or plastics. The rate is slow below 120°C (248°F). The gas yield is 125 cc nitrogen gas per gram at STP, which corresponds to the theoretical amount of gas.

Decomposition of *p,p′*-oxybis(benzene sulfonyl hydrazide) can be postulated by analogy with Equations (12) and (13) for benzene sulfonyl hydrazide.[8c]

The polymeric residue (IV) (see Equation 17) exhibits practically no odor. That the residue is indeed polymeric can be shown by carefully igniting the blowing agent with a heated glass rod. Decomposition begins with a smooth gassing off at the point of ignition and spreads with moderate rapidity through the mass. No flame is seen as an insoluble expanded polymeric foam appears.

$$(16)\quad H_2N\text{–}NH\text{–}SO_2\text{–}C_6H_4\text{–O–}C_6H_4\text{–}SO_2\text{–}NH\text{–}NH_2 \xrightarrow{\Delta} [HOS\text{–}C_6H_4\text{–O–}C_6H_4\text{–}SOH] + 2N_2 - 2H_2O$$

$$(17)\quad n\,[HOS\text{–}C_6H_4\text{–O–}C_6H_4\text{–}SOH] \rightarrow [\text{–S–}C_6H_4\text{–O–}C_6H_4\text{–S–} + \text{–S–}C_6H_4\text{–O–}C_6H_4\text{–}SO_2\text{–}]_{n/2} + nH_2O$$

IV

Upon cooling, the residue readily crumbles to an insoluble polymeric ash. Decomposition within a polymer is more controllable, and the residue remains colorless.

This agent is widely used in rubber and plastics: in the extrusion of cellular polyethylene for wire insulation,[15] in the expansion of PVC plastisols,[13] in epoxy and phenolic resins,[16] in expanded rubbers[13] and in rubber–resin blends. It presents no problems with odor, toxicity, or discoloration. It also has the unusual ability of simultaneously expanding and crosslinking rubbery polymers in the absence of any other conventional curatives.[11]

This agent will decompose to produce gas in an alkaline latex system at steam temperature. Under these conditions the blowing agent forms acidic decomposition products (probably sulfinic or sulfonic acids), that coagulate or gel the latex in the foamed state and provides a unique method of producing an expanded rubber product.[17] Impregnation of paper or nonwoven fabric with latex so treated can produce an interesting fiber-reinforced rubber article.

Sulfonyl Semicarbazides

$$CH_3\text{–}C_6H_4\text{–}SO_2\text{–}NH\text{–}NH\text{–}C(=O)\text{–}NH_2$$

V

A newer type of blowing agent is *p*-toluene sulfonyl semicarbazide (V).[18–20]

A high decomposition temperature (235°C) makes this chemical useful in plastics that expand at high temperatures, such as high-density polyethylene, polypropylene, rigid polyvinyl chloride, ABS polymers, polycarbonates, and nylon, among others.[19,20] A similar chemical is *p,p′*-oxybis(benzene sulfonyl semicarbazide) (VI), which decomposes at 215°C:

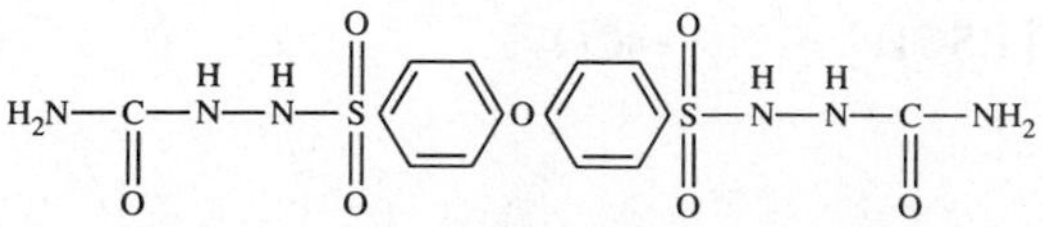

$$H_2N-\underset{\underset{O}{\|}}{C}-\overset{H}{N}-\overset{H}{N}-\overset{\overset{O}{\|}}{\underset{\underset{O}{\|}}{S}}-C_6H_4-O-C_6H_4-\overset{\overset{O}{\|}}{\underset{\underset{O}{\|}}{S}}-\overset{H}{N}-\overset{H}{N}-\underset{\underset{O}{\|}}{C}-NH_2$$

VI

Both can be used for rubber expansion because activators lower their decomposition temperatures. Both have relatively high gas yields (around 143–145 cc/g, measured at STP). This corresponds to about 1.5 moles of gas per mole in the case of *p*-toluene sulfonyl semicarbazide. On the other hand, *p,p'*-oxybis(benzene sulfonyl semicarbazide, being a difunctional molecule, gives nearly 3.0 moles of gas per mole (1.5 moles gas per sulfonyl semicarbazide group). This strikingly high gas yield is due to a considerable volume of carbon dioxide as well as nitrogen in the decomposition gases.

Sulfonyl semicarbazides can be made by two methods:[18,20]

(1) The reaction of a sulfonyl chloride with semicarbazide in the presence of an acid sequestering agent:

$$RSO_2Cl + H_2N-\overset{H}{N}-\underset{\underset{O}{\|}}{C}-NH_2 \rightarrow RSO_2\overset{H}{N}-\overset{H}{N}-\underset{\underset{O}{\|}}{C}-NH_2 + HCl \quad (18)$$

(2) The reaction of a sulfonyl hydrazide with a source of cyanic acid:

$$RSO_2\overset{H}{N}NH_2 \xrightarrow{(HOCN)} RSO_2\overset{H}{N}-\overset{H}{N}-\underset{\underset{O}{\|}}{C}-NH_2 \quad (19)$$

The decomposition products from sulfonyl semicarbazides vary substantially from those formed from the parent sulfonyl hydrazides. The presence of considerable carbon dioxide in the decomposition gases has already been mentioned.

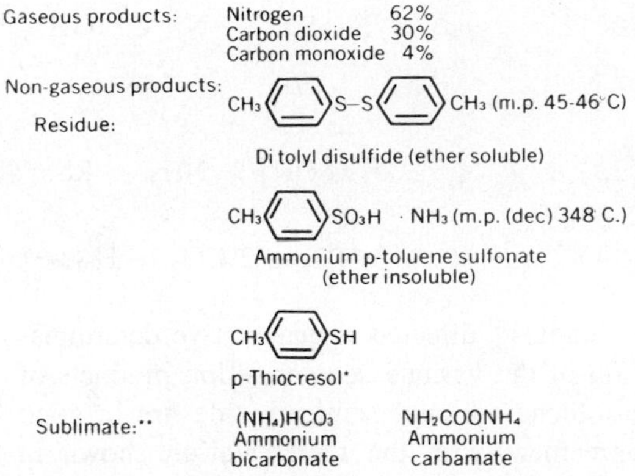

Gaseous products:	Nitrogen	62%
	Carbon dioxide	30%
	Carbon monoxide	4%

Non-gaseous products:
Residue:

$CH_3-C_6H_4-S-S-C_6H_4-CH_3$ (m.p. 45-46°C)

Di tolyl disulfide (ether soluble)

$CH_3-C_6H_4-SO_3H \cdot NH_3$ (m.p. (dec) 348 C.)

Ammonium p-toluene sulfonate (ether insoluble)

$CH_3-C_6H_4-SH$

p-Thiocresol*

Sublimate:** $(NH_4)HCO_3$ Ammonium bicarbonate; NH_2COONH_4 Ammonium carbamate

Fig. 19-2. Products of dry decomposition of *p*-toluene sulfonyl semicarbazide at 235°C.

The nongaseous products also are quite different, indicating a distinct difference in the decomposition mechanisms between the two classes of blowing agents. The greater stability (higher decomposition temperature) of the sulfonyl semicarbazides is significant in this connection.

The products of dry decomposition of *p*-toluene sulfonyl semicarbazide at 235°C are shown in Fig. 19-2.

Decomposition Products of *p*-Toluene Sulfonyl Semicarbazide[20]

It would appear that the carbon dioxide and the ammonium bicarbonate (or carbamate) must result from hydrolysis of the carbonamide group. The necessary water could result from an oxidation–reduction of the sulfonyl hydrazide structure in the molecule. The mechanism postulated is indicated in Equations (20) through (24). Brackets indicate speculative transitory intermediates.

$$R-\overset{\overset{O}{\|}}{\underset{\underset{O}{\|}}{S}}-\overset{H}{N}-\overset{H}{N}-\underset{\underset{O}{\|}}{C}-NH_2 \rightarrow R-\underset{\underset{O}{\|}}{S}-N=N-\underset{\underset{O}{\|}}{C}-NH_2 + H_2O \quad (20)$$

$$R-\underset{\underset{O}{\|}}{S}-N=N-\underset{\underset{O}{\|}}{C}-NH_2 + H_2O \rightarrow R-\underset{\underset{O}{\|}}{S}-N=N-\underset{\underset{O}{\|}}{C}-OH + NH_3 \quad (21)$$

$$\text{(22)} \qquad R{-}\underset{\underset{O}{\|}}{S}{-}N{=}N{-}\underset{\underset{O}{\|}}{C}{-}OH \rightarrow [RSOH] + N_2 + CO_2$$

$$\text{(23)} \qquad 5[RSOH] + NH_3 \rightarrow RS{-}SR + RSO_3NH_4 + 2\,H_2O$$

$$\text{(24)} \qquad 3\,NH_3 + 2\,CO_2 + H_2O \rightarrow (NH_4)HCO_3 + NH_2COONH_4$$

Studies[20] directed at quantitative determination of the various decomposition products of *p*-toluene sulfonyl semicarbazide are in good agreement with the representation shown in Equation (25):

(25)

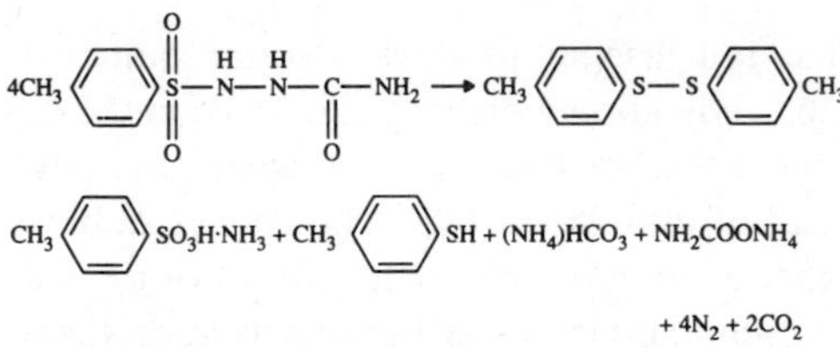

$$4CH_3C_6H_4{-}SO_2{-}NH{-}NH{-}\underset{\underset{O}{\|}}{C}{-}NH_2 \rightarrow CH_3C_6H_4{-}S{-}S{-}C_6H_4CH_3 + CH_3C_6H_4SO_3H{\cdot}NH_3 + CH_3C_6H_4SH + (NH_4)HCO_3 + NH_2COONH_4 + 4N_2 + 2CO_2$$

The formation of six moles of gas from four moles of the blowing agent agrees with the observed evolution of 1.5 moles of gas per mole. The ratio of nitrogen to carbon dioxide (2:1) is in close agreement with the analysis (62% nitrogen, 30% carbon dioxide). The proportions of ditolyl disulfide (VII) and ammonium *p*-toluene sulfonate (VIII) isolated from the decomposition residue agree reasonably well with the values required by Equation (25). *p*-Thiocresol was not isolated in the study, but titration of the products of decomposition of *p*-toluene sulfonyl semicarbazide in dioctyl phthalate with standard iodine solution gave values in good agreement with that expected for the oxidation of one mole of *p*-thiocresol as indicated in Equation (25). The sublimate, believed to consist of ammonium bicarbonate, believed to consist of ammonium bicarbonate and ammonium carbamate, was not subjected to rigorous analysis.

Equations (20) through (25) are useful in representing the chemistry and stoichiometry of the decomposition of *p*-toluene sulfonyl semicarbazide but probably do not reflect the actual mechanism of the decomposition. The rapid decomposition at 235°C suggests that a concerted mechanism is involved. Spacial representation of the sulfonyl semicarbazide molecule as shown in Fig. 19-3 indicates opportunities for strong hydrogen bonding, which may account for the relatively high heat stability. At the high decomposition temperature the excitation of the molecule is sufficient to produce electron shifts, resulting in breaking and making of bonds in a concerted manner to give the products indicated. Figure 19-4 indicates a similar proposed mechanism, resulting in the formation of some carbon monoxide.

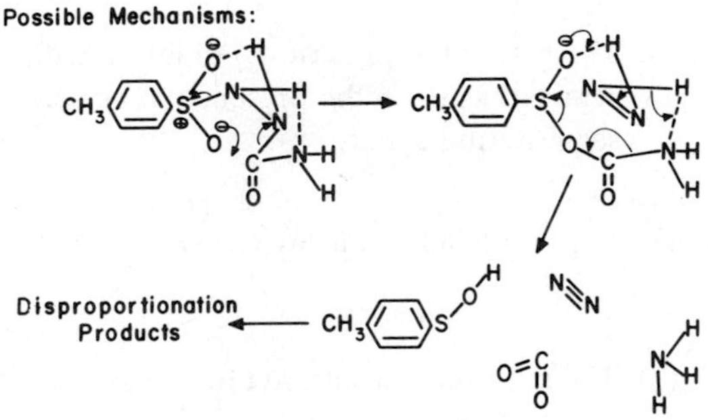

Fig. 19-3. Spatial representation of the sulfonyl semicarbazide molecules, indicating opportunities for strong hydrogen bonding.

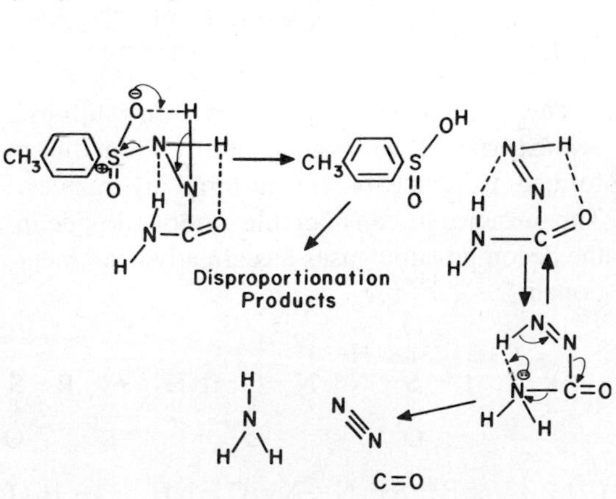

Fig. 19-4. A proposed mechanism similar to Fig. 19-3, resulting in the formation of some carbon monoxide.

5-Phenyl Tetrazole

With the increased use of engineering plastics in recent years, the need for chemical blowing agents having higher decomposition temperatures to match the processing temperatures of these materials has also gown. One of the newer entries into this arena is 5-phenyl tetrazole (5PT), which is a white crystalline powder with a melting point of 212°C. Upon heating to 245°C, 5PT will decompose to yield approximately 200 cc of gas (measured at STP) per gram of blowing agent. This gas is predominantly nitrogen.[21]

Hydrocarbyl tetrazoles (the class of materials to which 5PT belongs) may be synthesized via the reaction of an aromatic amine with a metal azide in the presence of a Friedel-Crafts catalyst.[22] Hence, 5PT can be obtained from the reaction of benzonitrile and sodium azide:

(26) 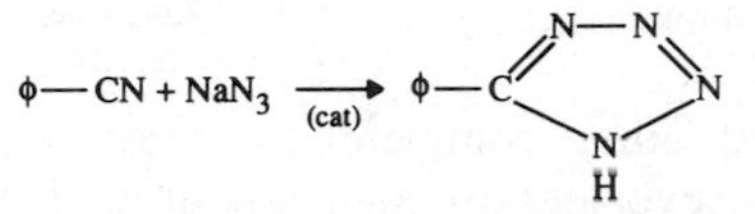 $\phi\text{—CN} + NaN_3 \xrightarrow[\text{(cat)}]{} \phi\text{—C}$ (ring: N—N, N, N–H)

5-phenyl tetrazole

An alternative route to 5PT has been demonstrated via the cyclization of benzamidrazone with nitrous acid:[23]

(27) 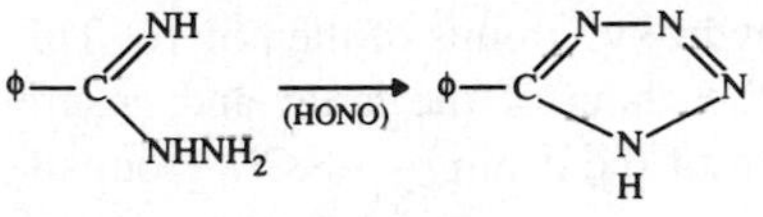 $\phi\text{—C}(=NH)NHNH_2 \xrightarrow[\text{(HONO)}]{} \phi\text{—C}$ (ring: N—N, N, N–H)

The most commonly reported solid decomposition product of 5PT is 3,5 diphenyl-1,2,4 triazole.[21,23] Other solid decomposition products claimed are aminodiphenyltriazole and triphenyl-S-triazine.[23] Structures for these materials are shown below:

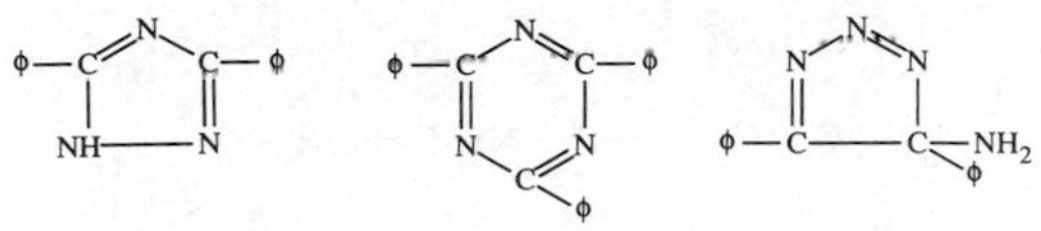

3,5 diphenyl-1,2,4 triazole triphenyl-S-triazine aminodiphenyltriazole

The decomposition mechanism of 5PT has not been extensively studied, although the nature of the decomposition products permits some speculation. Clearly, more than one reaction pathway is allowed. Also, the decomposition must involve some interaction between pairs of 5PT molecules and/or decomposition intermediates. As is generally the case with this sort of mechanism, the gas yield and the residue composition can be affected by the conditions under which decomposition occurs (e.g., heating rate and decomposition media).

Salts of 5-Phenyl Tetrazole

Some of the metal salts of 5PT have found use as chemical blowing agents where even higher decomposition temperatures than the parent compound are required.[24]

The calcium salt of 5PT is a white crystalline powder with a reported decomposition temperature of 350°C.[25] The barium salt is similar in appearance and has a decomposition temperature of 375°C.[26]

These salts may be obtained by first converting the 5PT to a soluble salt by, for instance, treatment with sodium hydroxide. The soluble salt thus obtained then can be reacted with the appropriate metal halide (such as calcium chloride), whereupon the insoluble salt is separated via precipitation.

References

1. (a) R. A. Reed, *Plastics Progress*, pp. 51–80, Iliffe and Sons, Ltd., London, 1955; (b) R. A. Reed, *Brit. Plastics*, 33(10), 469 (1969).
2. H. R. Lasman, *Modern Plastics Encyclopedia*, pp. 394–402, New York, 1966.
3. G. L. van Amerongen, *Rubber Chem. & Tech.*, 20, 503 (1947).
4. (a) B.I.O.S. Final Report 1150(PB79428), pp. 22–23. (b) F. Lober, *Angew, Chem.*, 64, 65–76 (1952). (c) Stevens and Emblem, *Industrial Chemist*, 27, 391–394 (1951).
5. U.S. Patent 2,491,709 (ICI).
6. T. H. Newby and J. M. Allen, U.S. Patent 2,692,281 (Uniroyal).
7. F. Lober, F. Bogemann, and R. Wegler, U.S. Patent 2,626,933; British Patent 691,142 (Farbenfabriken Bayer).
8. (a) B. A. Hunter and D. L. Schoene, *Ind. Eng. Chem.*, 44, 119 (1952). (b) B. A. Hunter and R. S. Stander, British Patent 686,814 and British Patent 693,954 (Uniroyal). (c) D. L. Schoene, U.S. Patent 2,552,065 (Uniroyal).
9. T. Curtius and F. Lorenzen, *J. Prakt. Chem.* (ii), 58, 160 (1898).

10. A. Deavin and C. W. Rees, *J. Chem. Soc.*, 4970 (1961).
11. U.S. Patents 2,849,028; 2,873,259 (Armstrong Cork Co.)
12. N. Kharasch, S. J. Potempa, and H. L. Wehrmeister, *Chem. Revs.*, 39, 269–332 (especially 276) (1946).
13. Bulletin, "Celogen OT," Uniroyal Chemical Co., Middlebury, CT.
14. (a) N. K. Sundholm, U.S. Patent 2,640,853 (Uniroyal).
    (b) G. H. Stemple, U.S. Patent 2,830,086 (General Tire Co.).
15. U.S. Patent 3,068,532 (Union Carbide Corp.).
16. Bulletin SC 53-52-R (12/54), "Foamed Epon Resins," Shell Chemical Co.
17. J. T. Fairclough, U.S. Patent 2,858,282 (Uniroyal Inc.).
18. B. A. Hunter, U.S. Patents 3,152,176 and 3,235,519 (Uniroyal Inc.).
19. Product Bulletin, "Celogen[R] RA," Uniroyal Chemical Co., Middlebury, CT.
20. B. A. Hunter, F. B. Root and G. J. Morrissey, *J. Cellular Plastics*, 3(6), 268 (1967).
21. Product Bulletin, "Expandex[R] 5-PT," Uniroyal Chemical Co., Middlebury, CT.
22. U.S. Patent 3,442,829 (Borg-Warner Corp.).
23. *Rodd's Chemistry of Carbon Compounds*, 2nd ed., Vol. IV, Part D, M. F. Ansell, ed., Elsevier.
24. U.S. Patent 3,873,477 (Stepan Chemical Co.).
25. Product Bulletin, "Expandex[R] 150," Uniroyal Chemical Co., Middlebury CT.
26. Product Bulletin, "Expandex[R] 175," Uniroyal Chemical Co., Middlebury, CT.

POLYURETHANE FOAMS*

The family of polyurethane foams is one of the most versatile members of the cellular plastics group and is growing more versatile each year. Depending on the starting ingredients, it is possible to produce a range of products from extremely soft flexible foams to tough rigid foams to integral-skinned foams.

Chemistry

Polyurethane foams are prepared by reacting hydroxyl-terminated compounds called polyols (usually of the polyester or polyether family) with a polyisocyanate.

*Reviewed and revised by Mobay Corporation.
Tables and figures in this section courtesy Mobay Corporation unless otherwise noted.

The structural formula of a typical polyester polyol is as follows:

$$HO-(CH_2)_2-O-(CH_2)_2-O-\overset{\overset{\displaystyle O}{\|}}{C}-(CH_2)_2 \cdots (CH_2)_2-OH$$

The structural formula of a typical polyisocyanate used in the production of urethane foams—in this case, a mixture, 80/20 of the 2,4- and 2,6-isomers of toluene diisocyanate (TDI)—is as follows:

CH$_3$ / NCO / NCO

2,4-isomer

CH$_3$ / OCN / NCO

2,6-isomer

The other commercially most significant polyisocyanates are members of the MDI family based on diphenylmethanediisocyanate and its higher molecular weight analogues.

In the production of urethane foam, two reactions occur. The first is a reaction of the polyisocyanate, which is present in excess, with the hydroxyl groups of the polyol. This lengthens the chain of the latter and ensures termination of the chain by -NCO groups:

$$\underset{\text{polyisocyanate}}{O=C=N-R-N=C=O} + \underset{\text{polyol}}{H^*O-R \cdots R-OH^*} +$$

$$\underset{\text{polyisocyanate}}{O=C=N-R-N=C=O} + \underset{\text{polyol}}{H^*O-R \cdots R-OH^*}$$

$$\downarrow$$

$$O=C=N-R-\underset{\underset{\displaystyle H^*}{|}}{N}-\overset{\overset{\displaystyle O}{\|}}{C}-OR \cdots$$

$$R-O-\overset{\overset{\displaystyle O}{\|}}{C}-\underset{\underset{\displaystyle H^*}{|}}{N}-R-\underset{\underset{\displaystyle H^*}{|}}{N}-\overset{\overset{\displaystyle O}{\|}}{C}-OR \cdots R-OH^*$$

polyurethane

The second reaction, which generates the blowing gas and produces the expanded foam structure, may be chemical or physical in nature. In the chemical reaction, the polymeric structure of the foam is formed by the reaction of the polyisocyanate with the polyol. Simultaneously, the polyisocyanates react with added water contained in the reaction mixture to form an intermediate product, carbamic acid, which decomposes to give a primary amine and carbon dioxide gas, which functions as the blowing agent to expand the polymer matrix.

The physical reaction involves the volatilization of a blowing agent (i.e., an inert, low-boiling chemical) to provide expansion. Typically chlorofluorocarbons, such as CFC-11 or CFC-12 or environmentally safer compounds (e.g., HCFCs), are used as physical blowing agents in combination with water or alone. The blowing agent is added to the formulation and the foam structure is formed by volatilization of the low-boiling chlorofluorocarbon under the exothermic heat produced by the polyol–isocyanate reaction. Of special interest in some applications (e.g., insulation) is the fact that the gas formed by vaporization of the chlorofluorocarbon is entrapped in the cells, imparting a very low thermal conductivity (*K*-factor) to the foam. Other low-boiling solvents, such as methylene chloride, can also be used in certain flexible urethane foams for density and hardness control.

Polyols and Polyisocyanates

It is possible to vary the two major ingredients of the urethanes—polyols and polyisocyanates—to create a wide range of products.

Polyisocyanates in use today include: toluene diisocyanate, known as TDI, which has found wide use in flexible foams; diphenylmethanediisocyanate, known as MDI; various polymeric MDI types—often described as "PMDI"—which have become important in one-shot processing, cold-cure foams, and so on; and different blends, such as TDI/PMDI blends.

Polyols, the other major ingredient of a polyurethane foam, are active hydrogen-containing compounds, primarily variations of polyesters and polyethers. It is possible to prepare many different types of foams by simply changing the molecular weight of the polyol, as it is the molecular backbone that supplies the reactive sites for crosslinking, which in turn is the principal factor in determining whether a given foam will be flexible, semirigid, or rigid. For example, using a single isocyanate, it is possible to prepare soft, flexible foams by using polyols with a hydroxyl equivalent weight of about 1000 to 2000, or very dense foams suitable for rigid furniture parts by using polyols with functionalities of more than two and a hydroxyl equivalent weight of about 100 to 200.

As a rule of thumb, high molecular weight, low functionality polyols produce molecules with a low amount of cross-linking and, consequently, a flexible foam. Conversely, low molecular weight polyols of high functionality produce a structure with a high degree of cross-linking, a rigid foam. Of course, it is possible to vary the formulation to produce any degree of flexibility or rigidity between the two extremes.

Special polyols have been introduced recently that contain solid organic fillers finely dispersed in the polyol. The PHD polyols contain polyurea as filler, whereas the polymer polyols are dispersions of polymerized olefinic monomers. Both types enhance processing and foam properties in flexible foams.

Modifiers and Additives

In addition to the basic polyol, polyisocyanate, and blowing agents, urethane formulations may include certain other ingredients:

Catalysts, primarily organotin compounds and tertiary amines, are added to control and accelerate the rate of reaction, so that maximum rise is synchronized with gelation to regulate cell size and prevent collapse of the foam.

Surfactants control surface tension and are used to develop a fine, uniform cell structure, or, in some cases, to produce large-celled foams. They are usually silicone copolymers.

Modifiers can be added as inexpensive fillers that reduce the cost of the foam or improve a specific physical characteristic such as processing or strength.

Modifiers such as halogenated or phosphorous compounds also can be added to the polyurethane foam formulation, especially to modify combustibility properties of the resulting foams.

Dyes and pigments can be added to color the polyurethane. Also, work is continuing on the concept of using reinforcing materials such as fibrous glass to improve the physical characteristics of the various foams.

One-Shots and Prepolymers

The reactions by which urethane foams are produced can be carried out in a single stage or in a multi-stage sequence. The two principal methods are one-shot and prepolymer.

In the one-shot method, all of the ingredients (polyol, polyisocyanate, blowing agent, catalyst, etc.) are mixed simultaneously, and the resulting mixture is allowed to foam.

In the prepolymer method, a portion of the polyol is prereacted with an excess of polyisocyanate to yield a prepolymer. The prepolymer subsequently is mixed with additional polyol, catalyst, and other additives to effect foaming. In the quasi-prepolymer method, the polyol is reacted with polyisocyanate to form one component, and the additional polyol plus other additives are mixed together to form the second component. The two components can then be mixed to effect foaming. They usually are mixed in equal quantities.

Rigid Polyurethane and Polyisocyanurate Foam

The high crosslink density and closed-cell content of rigid polyurethane (PUR) and polyisocyanurate (PIR) foams differentiate them from flexible polyurethane foams. PUR foams are produced by the reaction of low molecular weight polyols with polymeric isocyanates in the presence of catalysts, surfactants, and blowing agents. PIR foams are produced by the trimerization of polymeric isocyanates in the presence of polyols, catalysts, surfactants, and blowing agents. The major difference between PUR and PIR foams is the higher inherent thermal stability of PIR foams due to their isocyanurate ring structure. Because their other properties are so similar, both polyurethane and polyisocyanurate rigid foams will be referred to generically as PUR in this discussion.

The normal density range for PUR thermal insulation is 1.5 to 3.5 pcf. However, in packaging foam applications the density can go as low as 0.4 pcf, whereas structural applications such as certain furniture parts involve densities upwards of 20 pcf. Many of these higher-density foams belong to the category of structural RIM foams (see below).

PUR foams have an outstanding resistance to heat transfer (*R* value*).

The stabilized *R* value is 7.1 to 8.3 for foams with nonpermeable facings, and it is 5.6 to 6.2 for foams without facings or with facings having little resistance to permeation by atmospheric gases. PUR foams have several other valuable properties: excellent compressive strength, good dimensional stability, low water vapor transmission, and a high strength-to-weight ratio.

Typical PUR foam properties are shown in Table 19-1.

The construction industry, particularly commercial roofing and residential sheathing, is the largest market area for PUR foam, consuming over 50% of the total production (Fig. 19-5). PUR boardstock laminates are produced with a variety of flexible and/or rigid facers; aluminum foil, foil–paper composites, felt, glass fiber mat, perlite, gypsum, and waferboard are commonly used. Furthermore, the adhesive character of foam-in-place PUR foam and its rigidity make it an excellent core material for structural metal building panels, extruded aluminum windows, and metal entry doors. Pipe and tank insulation, as well as cold storage warehouses, also are included in the construction market area.

The next largest market area is appliance insulation, which consumes over 15% of production. PUR foam is used in domestic and commercial refrigerators and freezers, partly

*For a 1-inch-thick specimen, $R = h \cdot ft^2 \cdot °F/Btu$ at 75°F mean test temperature.

Table 19-1. Typical PUR foam properties.

Density (pcf)	*Compressive strength (psi)*	*Compressive modulus (psi)*	*Shear strength (psi)*	*Shear modulus (psi)*
1.5–2.0	20–60	400–2000	20–50	250–550
2.1–3.0	35–95	800–3500	30–70	350–800
3.1–4.5	50–185	1500–6000	45–125	500–1300
4.6–7.0	100–350	3800–12,000	65–180	850–2000
7.1–10.0	200–600	5000–20,000	125–275	1300–3000

Fig. 19-5. Polyurethane foam being sprayed over old felt and asphalt roof.

because of its thermal resistance and partly because it simplified the manufacturing process for these appliances. These same factors apply to water heaters as well as picnic and water coolers. PUR foam also imparts structural strength to these parts, which permits significant reduction in sheet metal gauge or in the thickness of plastic liners. The transportation industry has greatly benefited from the unique properties of PUR foam. Refrigerated trucks and railcars carry more payload because less space and weight are consumed by PUR thermal insulation compared to competitive materials. Furthermore, composite panels of PUR foam as a core between aluminum or reinforced polyester skins are becoming a factor in the manufacture of shipping containers and semitrailers.

The marine industry also utilizes the thermal and structural properties of PUR foam. Its light weight and insulating efficiency are advantageous in fishing boats as well as liquefied natural gas (LNG) tankers. Its buoyancy and the rigidity it imparts to "sandwich" construction allow it to be used in vessels as diverse as river barges and pleasure craft.

PUR foam complements flexible foam in packaging a wide variety of fragile items from electronic typewriters and ceramic vases to engine crankshafts. It exhibits the low resilience, slow recovery, and shock absorbency necessary for protective packaging as well as ease of handling.

Processing Rigid Polyurethane Foams

Three basic techniques are used to produce PUR foams that meet the requirements of their many applications: pour, spray, and froth. These processes are discussed briefly in this section. A more extensive review of the equipment used in processing rigid and flexible foams can be found in a later section on "Polyurethane Processing Equipment" (see below).

In the pour technique, a liquid urethane chemical mixture is metered into a cavity. The reactive liquid flows to the bottom of the cavity and starts to expand, filling the void with a strong seamless bond. The cavity can be anything from the space between two walls of a refrigerator or a refrigerated semitrailer to the space between the inner and outer hulls of a river barge. The cavity also can be the space between flexible or rigid facers. Here, the liquid mixture is metered onto the lower facer. Before the foam expansion is complete, a second facer or skin is applied, and the "sandwich" is usually run through a double belt conveyor to control the thickness of the boardstock laminate or metal building panel. The pour technique can be done in a factory or at the job site. The spray technique is well suited for application in the field. Specially formulated systems are pumped through heated hoses and mixed in two-component spray guns. The reacting foam mixture is sprayed directly onto the surface to be insulated, forming a monolithic seal against heat transfer.

The froth technique is a variation of the pour technique in that the foam mixture is dispensed in a partially expanded state (like shaving cream) rather than as a flowing liquid. It is based on the use of a mixture of blowing agents that produce a two-step expansion. The first blowing agent is a low-boiling agent that expands almost immediately to blow the foam mixture into a froth; the second blowing agent has a higher boiling point, which requires the exothermic heat that is produced by the PUR and PIR reactions. Thus, there is a slight delay between the dispensing of the foam as a froth and the final blow. This is especially useful when the cavity cannot restrain a great deal of pressure.

Flexible Polyurethane Foams

Flexible foams differ from the aforementioned rigid foams in that they can be elastically deformed and return to their original contours when the force is removed. They are usually in the 1 to 6 pcf density range.

In flexible foam formulations, high molecular weight polyethers are used, generally triols based on trimethylol propane or glycerine and propylene and ethylene oxide. Specialty foams also can be produced from linear or slightly branched polyesters. High resilience flexible foams often incorporate active, organic-filler-containing polyethers, referred to as PHD or polymer polyols. The former contain finely suspended aromatic, substituted polyureas, whereas the latter incorporate suspensions of styrene/acrylonitrile polymers.

Flexible foams generally are made from toluene diisocyanate, but in some instances blends with PMDI or with modified TDIs are employed to produce special properties.

Tables 19-2 and 19-3 show typical foam properties for commercial products and indicate their wide ranges of density and load bear-

Table 19-2. Typical properties, standard flexible slabstock.

	Special polyester	*Soft polyether*	*Low-density polyether*	*High-density polyether*	*High-load-bearing polyether*
Density, pcf	0.7	1.0	0.9	1.8	2.3
IFD, 25% *R*, lb/50 in.^{2}	44	9	29	36	85
IFD, 65% *R*, lb/50 in.^{2}	80	17	55	73	186
SAG factor	1.8	1.9	1.9	2.0	2.2
Tensile, psi	32	8	13	8	22
Tear, pli	4.6	1.4	1.8	1.5	1.5
Elongation, %	310	190	180	150	90
90% Comp. set. %	—	6	—	3	5

Table 19-3. Typical properties, high-resilience flexible slabstock.

	Soft	*Medium*	*Firm*	*Ultrafirm*
Density, pcf	1.7	2.5	3.0	5.2
IFD, 25% *R*, lb/50 in.^{2}	10	32	70	120
IFD, 65% *R*, lb/50 in.^{2}	27	77	166	300
SAG factor	2.6	2.4	2.4	2.5
Recovery, %	75	84	81	85
Tensile, psi	14	17	27	30
Tear, pli	1.9	1.5	1.4	1.2
Elongation, %	227	150	125	110
75% Comp. set, %	9	5	5	3
Ball rebound	56	70	54	45

ing, expressed as Indentation Force Deflection (IFD) at 25% and 65% compression.

Processing Flexible Polyurethanes

The two major production techniques in the flexible foam area are the slabstock and the molded foam processes.

Slabstock Foam Process. The wide variety of slabstock foams are manufactured by using either the traverse or the trough machine method. These two production techniques are similar, differing primarily in the way the foam is poured. In the traverse method, the individual ingredients are fed into a mixing head, mixed, and deposited evenly onto an angled, moving conveyor from the traversing mixhead. Using the trough method, the ingredients are fed into a mixhead, mixed, and pumped into a trough that evenly deposits the partially reacted material onto a fall plate down to a flat moving conveyor. The resultant continuous-length buns can be 8 feet wide and 45 to 50 inches high, depending on the throughput of the equipment.

After the foaming step is completed, the foam is cut into various lengths and stored to cure before fabrication. Foam fabricating involves reducing the larger foam block, up to 200 feet long, into many different smaller shapes and sizes, which is accomplished by a variety of cutting and shaping machines. Horizontal, vertical, and curvilinear cuttings are possible. Large sections of foam can also be peeled to a variety of thicknesses and rerolled. The fabrication process is geared to the end-use, market application. Flexible slabstock foam is used in the furniture, transportation, bedding, carpet underlay, textile, packaging, and novelty markets. In all these applications it is primarily the outstanding comfort it provides that has made flexible polyurethane the most popular cushioning material.

Flexible Foam Molding Process. The molding process is advantageous for the production of parts with such complex shapes that fabrication from slabstock would not be economically or technically feasible. In many cases this process is chosen because of a need to mold inserts in place for the fixture of covers, or for easier assembly of seat constructions.

Earlier, most molding techniques for flexible foams involved metering the liquid ingredients into a mold cavity (with or without inserts such as springs, frames, etc.) and then using large amounts of heat for cure. In a typical application, molds would be filled and closed, then heated rapidly to 300 to 400°F to allow demolding in 10 to 14 minutes without deformation. This process, which came to be known as "hot-molding," is still used, but it is being largely replaced by the "cold-cure" technique.

Through the use of more reactive polyol components, cold-cure foams do not require such severe curing conditions as "hot-foam." Optimum mold temperatures are in the range of 120 to 160°F with demold times of 2 to 5 minutes possible, allowing for much faster production rates and lower energy requirements. Typical properties of a cold-cure foam are shown in Table 19-4.

Recent developments in flexible molded foam include: dual-hardness seating foams (shown in Fig. 19-6), in which the seating and wing area can be produced at different hardness

Table 19-4. Typical properties of molded flexible foam.

	English	*SI*
Density	1.8–2.7 pcf	29–43 kg/m^2
ILD (50%)	11–78 lb/50 in.^{2}	50–350 N/323 cm^2
Tensile strength	20–30 psi	138–207 kPa
Elongation	125–150%	125–150%
Tear strength	1.2–1.7 pli	210–298 N/m
Compression set (50% 2A)	10–15%	10–15%

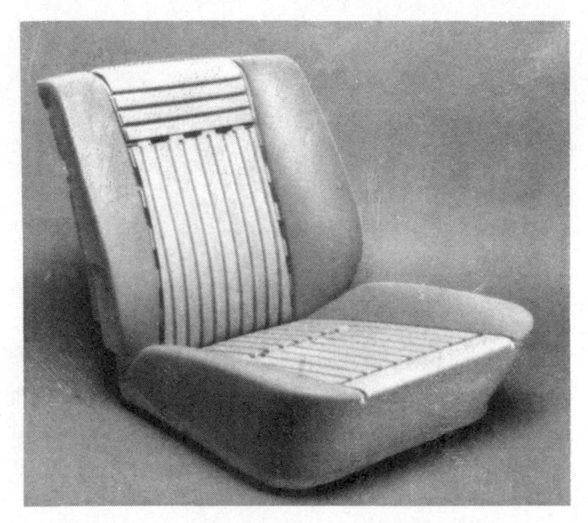

Fig. 19-6. Auto seating from dual-hardness foam for maximum comfort and support.

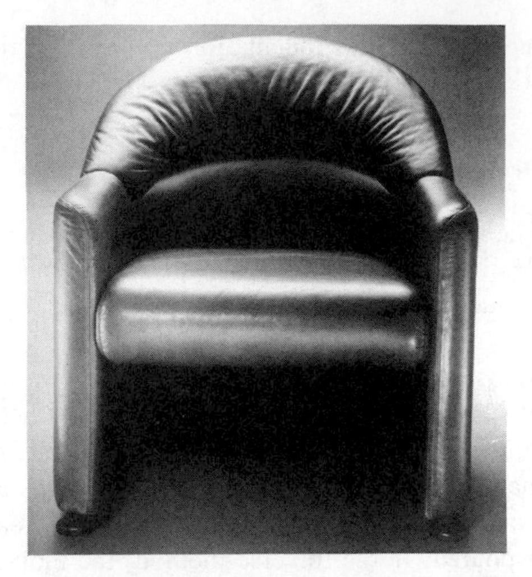

Fig. 19-7. Club chair uses cushioning of high-resilience polyurethane foam.

values for improved lateral support; and foam-in-cover technology, where the foam is molded directly onto the seat cover, reducing trim costs.

The primary markets for molded foams are the furniture and automotive industries. For furniture applications, a large variety of molded flexible foam parts are being produced—seat cushions, seat backs, and even full foam sofas and chairs (Fig. 19-7). The automotive industry uses molded seat cushions, seat backs, armrests, headrests, and knee bolsters (examples are shown in Fig. 19-8).

Semirigid Foams

These foams are characterized by low resilience, in that they recover very slowly from compression, and by high-energy-absorbing

Fig. 19-8. Auto interior with instrument panel, seating, and headliner from polyurethane foam.

characteristics. As such, they have found a prime outlet in the automotive industry in applications such as safety padding, arm rests, sun visors, horn buttons, and so on. These foams also are cold-curing in nature and usually involve special polymeric isocyanates. They usually are higher in density than regular flexible foam (up to 12 pcf) and generally are applied behind vinyl or ABS skins. The liquid ingredients are simply poured into a mold into which the vinyl or ABS skins and metal inserts for attachments have been laid. The foam fills the cavity, bonding to the skin and inserts.

It is also possible to produce an "integral skin" semirigid (or semiflexible) foam, in which the foam comes out of the mold with a continuous skin that can replace the separate vinyl or ABS surface (see below).

REACTION INJECTION MOLDING

Introduction

Reaction injection molding, or RIM as it is commonly called, was conceived in the laboratories of Bayer AG, Leverkusen, West Germany, about 1964. This early work centered around the formation of a high-density skin and a low-density core in a single operation to give the high strength-to-weight ratio of a laminate. Initially, work was centered on rigid urethane foam; later programs carried this same technology over into elastomeric structures.

The RIM process produces a sandwichlike structure consisting of a solid, nonporous skin and a lower-density microcellular core. The core and the surface layers consist of the same material and are formed in a single operation.

An integral-skin material will have a density gradient as shown in Fig. 19-9. The shape of the gradient, that is, the amplitude, will vary with the type of system and the molding technique. The important consideration is the thickness and quality of the skin, as this is the prime determinant of both mechanical properties and final surface finish. The properties of these foams can be tailored readily to a wide variety of end-uses by variation of the chemistry of the new components or processing conditions.

In RIM, as shown in Fig. 19-10, one starts

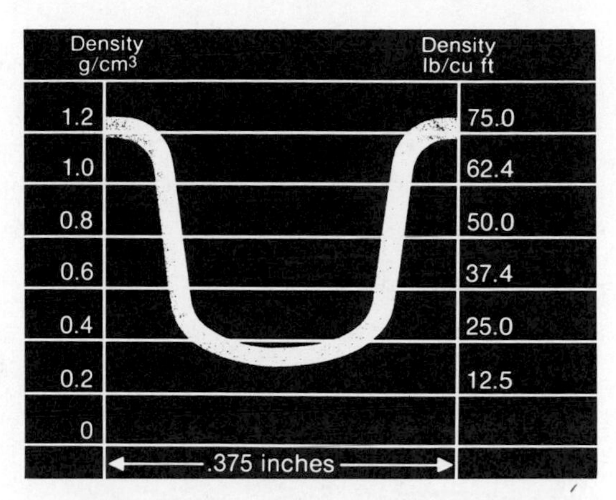

Fig. 19-9. Density gradient in RIM polyurethane structural foam.

with two low viscosity liquid monomers, an isocyanate component and a polyol component. These components are kept separate and continually recirculating at low pressure. The shot cycle of a RIM machine involves the following sequence:

1. Low-pressure recirculation (normal mode).
2. High-pressure recirculation.
3. Shot.
4. Return to low-pressure recirculation.

The period of high-pressure recirculation, usually 5 to 10 seconds, ensures that the material is at the proper temperature and uniformity for injection into the mold. The two liquid components are metered in the correct proportions into a mixing chamber, where they are initially mixed and injected at atmospheric pressure into a mold. Afterwards a mechanical piston cleans out the mixing chamber so the mixhead is ready for the next shot. The part weight is determined by a shot timer.

The chemical reaction between the two components to produce the polyurethane begins in the mold. This reaction is exothermic, and the heat evolved during the reaction vaporizes the blowing agent, a low-boiling solvent contained in the polyol component, causing the liquid mass to foam. The reaction mixture becomes progressively more viscous and passes through a gelation point, and, at the completion of the

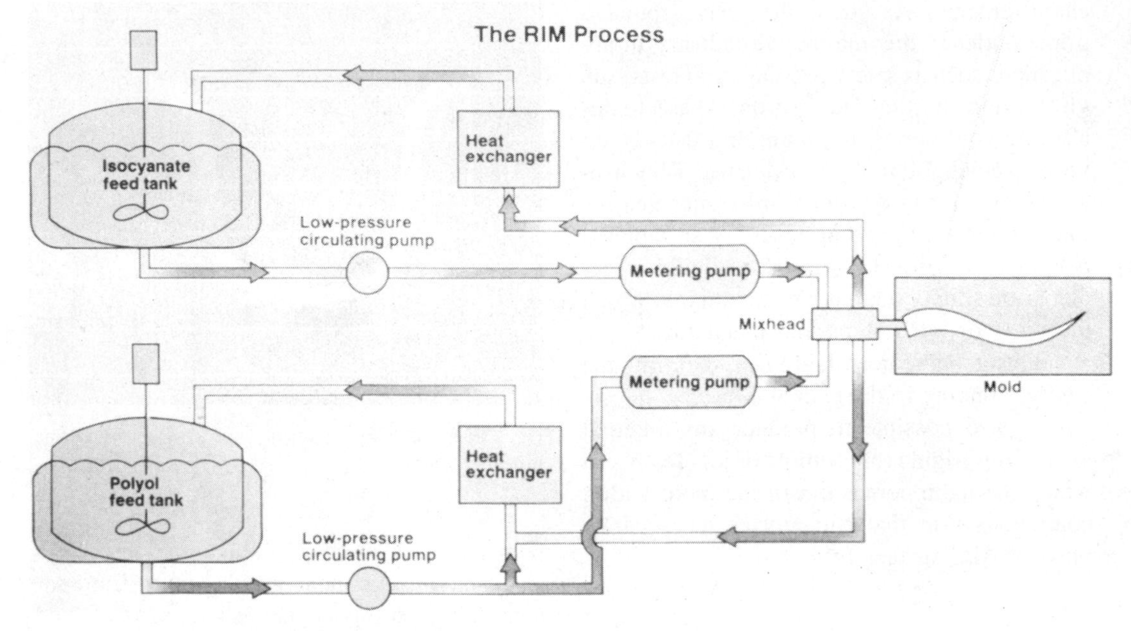

Fig. 19-10. The reaction injection molding (RIM) process.

reaction, it is transformed into an extensively crosslinked thermoset polymeric material.

The skin is formed as a result of the blowing agent's condensing on the relatively colder mold surface and the increasing pressure of the expanding foam. Accordingly, skin formation and thickness will be affected by factors that influence the condensation of the blowing agent at the mold interface. The major factors having an effect are:

1. *Shot weight*: Increased shot weight produces thicker skins.
2. *Formulation*: A high blowing agent concentration increases skin thickness.
3. *Mold temperature*: The usual temperature range is 40 to 80°C. Lower temperatures lead to thick skins.
4. *Mold thermal conductivity*: A highly heat-conductive mold is necessary to control skin thickness.

The smooth, hard skins provide high strength-to-weight-ratios; their strength is what separates polyurethane RIM materials from the well-known decorative urethane foam parts. The typical part thickness of RIM structural foams is 0.250 inch, and the skin thickness is 0.0625 inch.

The Process

RIM differs significantly from thermoplastic foam processes. Thermoplastic foam molding usually requires one injection machine for every sizable mold in production; conversely, the RIM system allows one metering machine to service a number of presses, giving the processor the flexibility to economically produce large- or small-quantity runs with the same machine. (See Fig. 19-11.) The production of large quantities of the same part is accomplished by utilizing duplicate molds in the system, whereas smaller quantities can be produced by the installation of a single mold.

The polyurethane structural foam (RIM) process develops approximately 30 to 70 psi internal mold pressure. Because of these low pressures, cast aluminum, aluminum, and steel can be used in production. Thermoplastic structural foams generally develop much higher internal mold pressures, which require a more expensive tool for production and usually cost anywhere from 30 to 40% more than the molds for RIM.

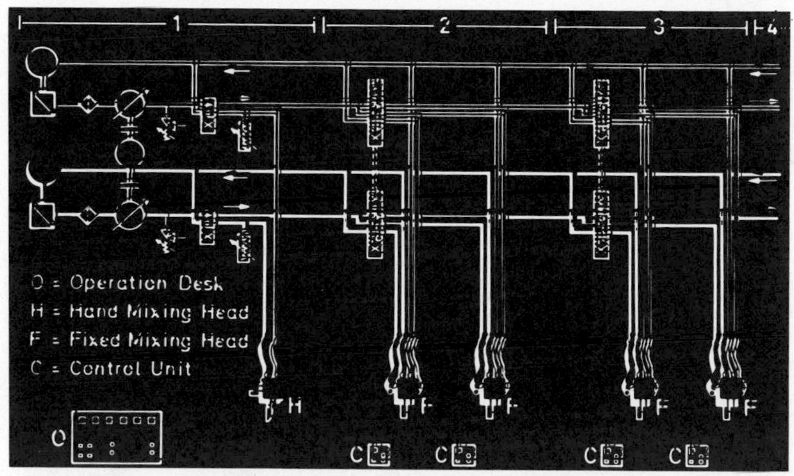

Fig. 19-11. Multiple mixhead arrangement.

Polyurethane structural foam offers several other savings, among them (1) the ability to attain a UL-Subject 94 V-O* specification at a lower cost per cubic foot and (2) the production of smooth, swirl-free surfaces that greatly reduce finishing costs.

Another important advantage in the use of RIM is its lower energy consumption. In recent years, energy supplies and energy consumption have become a major concern of government and industry. In the future, it is not unreasonable to assume that products and processes will be scrutinized according to their negative effect on energy and raw material depletion. Those processes that use the least energy will be viewed most favorably.

The RIM process, compared with thermoplastic processing, uses less than half as much energy to manufacture a molded part.

Materials

As mentioned earlier, there are two basic types of RIM materials, the well-known elastomeric type and the structural foam type. Table 19-5 compares some of their typical properties. As can be seen, the materials are quite different. One might expect their chemistry also to be different. The overall reaction is that of isocyanate with urethane; but to see how it is achieved, one must explore the world of RIM chemistry.

Table 19-5. Property comparison of RIM materials.

	Structural foam	*Elastomer (unfilled)*
Specific gravity	0.3–0.8	0.8–1.0
Typical part thickness	0.125–1.5	0.100–0.250
Flex modulus (unreinforced)	10–250,000	5000–120,000
Typical demold times	1–4 min.	20 sec–1 min.

A typical elastomer system would consist of the following materials:

- Long-chain polyether polyol: 4000 to 6000 M.W. diol or triol.
- Chain extender: ethylene glycol, butanediol, or an amine.
- Amine or tin catalysts.
- Modified MDI isocyanate.

This type of chemistry results in a material structure composed of unique domains of hard and soft segments, and results in a tough resilient material. Physical properties of a typical elastomeric system can be seen in Table 19-6.

Of the physical properties, the flex modulus, impact strength, and heat sag are the most important. In fact, the flex modulus frequently is

*Flammability results are based on small-scale laboratory tests and do not reflect the hazard presented by these or other materials under actual fire conditions.

Table 19-6. Physical properties of typical elastomeric system.

Flow direction	*II*	*L*	*Unfilled*
% Glass flakes	20		
Specific gravity	1.15	1.15	1.01
Flexural modulus (psi)			
RT	200,000	175,000	75,000
−30°C	400,000	323,000	140,000
+65°C	144,000	107,000	45,000
Modulus ratio (−30°C/+65°C)	2.9	3.0	3.1
Tensile properties			
Ultimate stress (psi)	4700	4700	3500
% Elongation	20	20	110
Heat sag inches			
4″ OH @ 250°F, 1 hr	0.10	0.10	0.08
6″ OH @ 250°F, 1 hr	0.22	0.22	0.50
Notched Izod (ft/lb-in.)	2.0	2.0	4.0
CLTE × 10^6/°F	31	35	80
Dart impact (2.7 joules)			
@ RT	Pass		Pass
@ −30°C	Surface cracks only		Pass

*Post-cured @ 121°C for one hour.

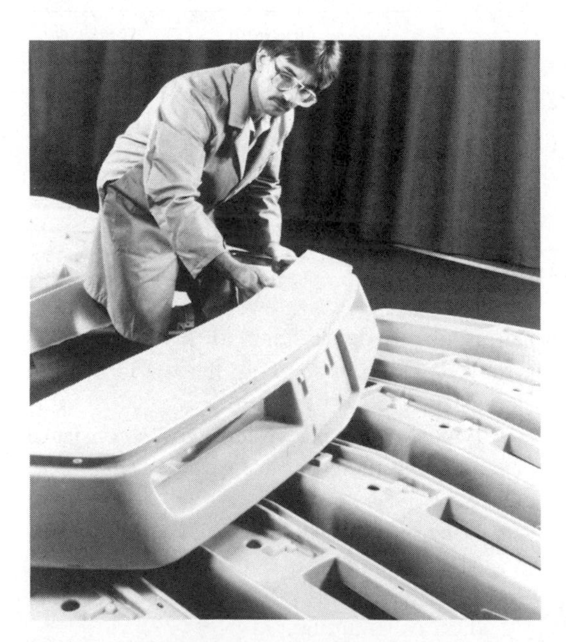

Fig. 19-12. Molded polyurea auto fascia with Class A surface.

used to identify a material. The use of RIM materials as automotive fascias, with the ability to absorb a 6 mph impact at −30°C with no damage, is evidence of the materials' toughness (Fig. 19-12). In addition, automotive paint bake cycles of 250°F for 1 hour necessitate good thermal resistance to sag.

As more automotive body parts are produced by the RIM process (i.e., fenders, decks, and door panels), the coefficient of linear thermal expansion (CLTE) will play an important role in ensuring good fit and proper expansion. To achieve a reduction in CLTE, the addition of fillers (i.e. glass, mica, etc.) has been used; the effect of these materials has been to increase stiffness and decrease CLTE.

The RRIM process also introduced an entirely new machine technology concept, which will be discussed later.

A structural-foam-type RIM material typically consists of:

- Polyether polyol blend
- Surfactants
- Amine or tin catalysts
- Flame retardants
- Blowing agent
- Polymeric isocyanate

This type of formulation results in an amorphous type of structure with no defined domain. Physical properties of such a system are shown in Table 19-7.

For a RIM structural foam, three primary properties are the impact strength, heat distortion, and flammability rating. RIM structural foams have long been looked on as the brittle cousins of RIM elastomers; however, that be-

Table 19-7. Physical properties of typical RIM structural foam.

Property	*Units*	*Value*	*ASTM Method*
Thickness	inch	0.25	—
Specific gravity		0.85	D-792
Flexural modulus	psi	24,000 ± 10,000	D-790
Flexural strength	psi	9,500 ± 500	D-790
Tensile strength	psi	5,800 ± 300	D-638
Elongation at break	%	10	D-638
Compressive strength	psi	6,000 ± 100	D-695
Heat deflection temperature			
under load @ 66 psi	°F	212 ± 5	—
@ 264 psi	°F	181 ± 5	D-648
Charpy impact (unnotched)	ft-lb/in.^{2}	25 ± 5	—
Gardner impact	in.-lb	72 ± 5	—

lief is changing fast because of the new materials available today (Table 19-8).

The impact strength of the polyurethane structural foam is superior to that of the foamed thermoplastics—polystyrene, ABS, and modified polyphenylene oxide. Another important property of RIM structural foams is their ability to be combustion-modified. Many applications require varying degrees of flame retardency. Typical ratings are the UL-94 V-O* and 5-V* ratings, which are common for the business machine industry.

RIM structural foams can now utilize the same designs as the thermoplastic structural foams. In addition, faster cycle tlmes can be obtained with RIM structural foams; the newest systems enable part demolding after 1 to 2 minutes.

Equipment

The heart of the RIM process lies in the metering and the mixing. In the RIM process, the mixing occurs by impingement of two streams in the mixhead; then the reactants are injected at atmospheric pressure into a closed mold to which the head is attached. The mold cavity is filled to from 30 to 95% with a low viscosity reaction mixture. The RIM process requires:

- Precise metering
- Intimate mixing
- Temperature control
- Laminar flow

*Flammability results are based on small-scale laboratory tests and do not reflect the hazard presented by these or other materials under actual fire conditions.

For precise delivery of large amounts of chemicals, there are several types of pumping systems, which range from piston types to axial, radial, or the currently used hydromatic pumps. It should be kept in mind that these pumps are intended only for use with unfilled systems. The pumps are produced in a variety of sizes with capacities ranging from 5 to 300 lb/min. Maximum machine throughput is achieved at a 1:1 component ratio. (See Fig. 19-13.)

Practically instantaneous, highly efficient mixing is an absolute necessity for the trouble-free production of RIM parts. Insufficient mixing and/or a lead/lag problem inevitably results in molded parts with some surface defects, which may range from light and dark streaks to an ''alligator skin'' finish.

The first high-pressure mixheads were of a hand-held design, featuring injection nozzles that were spring-loaded and actuated by pressure increases on the component acting against the spring. They were not satisfactory for the RIM process, as it was impossible to properly synchronize the pressure buildup of the liquid to prevent lead/lag problems. Thus, a new mixhead technology had to be developed. The machine manufacturers have developed various mixheads to solve this problem, all of which have the following characteristics:

- Recycling of both components through the head for precise temperature control.
- A mechanical means of very rapid and simultaneous changing of the polyol and isocyanate components from a recycle to

Table 19-8. RIM structural foam properties, as compared to foamed thermoplastics.

Material	*RIM structural foam**	*Polystyrene*	*ABS*	*Modified PPO*
Specific gravity	0.85	0.85	0.89	0.85
Thickness (in.)	0.25	0.25	0.25	0.25
Flexural modulus (psi)	240,000	275,000	240,000	261,000
Flexural strength (psi)	9,500	5,700	6,000	6,800
Tensile strength (psi)	5,800	2,300	3,000	3,400
Elongation (%)	10	20	—	16
Gardner impact** (in-lb)	74	28	10	—
Falling ball (ft-lb)	—	—	10	—
Dart impact (ft-lb)	—	—	177*	205
Heat Distortion Temperature				
@ 66 psi (°F)	212	183*	177*	205
@ 264 psi (°F)	180	168*	162*	180

**Gardner Impact test results from Mobay
*Unannealed
All of the data for the thermoplastic structural foams listed, except for Gardner impact test results, were taken directly from Dow, and General Electric data sheets.

a pour mode, thus eliminating lead/lag problems.

- Self-cleaning.
- Direct attachment to the mold in such a manner that all of the polymer can be demolded with the part, providing for the next shot without any preparation.

Figure 19-14 shows a mixhead design that meets these requirements. This mixhead was designed to switch from the recycle mode to the pour mode by simultaneously opening the pour ports into the mix chamber. A cleanout piston mechanically cleans out the chamber after the shot. As is required, a demoldable sprue is produced, which is flush with the mixhead and flush with the parting line of the mold cavity.

In processing fillers, some modifications in equipment are necessary in order to handle those fillers that are very abrasive. These modifications usually involve hardening the nozzles and mixing chambers and using metering cylinders rather than pumps. In this case, a metering pump meters oil flow to the cylinders, which then force the components to the mixhead with the required output and pressure.

As has been pointed out, the liquid reaction mixture fills only part of the mold. When the mixture expands to fill the total cavity, it must replace the air within that cavity. In order to make good parts, the mold must be oriented so that the air can be channeled to the highest point in the mold cavity, where it can be removed through vents placed in the parting line. Vents are necessary because the foam fills the cavity under its own power and does not have enough force to move air out of deep pockets. In many parts, it is not immediately apparent what the best mold position will be. To more rapidly find the correct position, mold clamping and positioning devices were developed that could rotate the mold on two axes (Fig. 19-15).

The presses with two-dimensional rotating capability are primarily used for prototyping and special applications. In production, normally the proper mold position would first be determined on this type of press, and then the mold would be mounted at the optimum angle in a simpler, one-dimensional press for production.

Fig. 19-13. High-pressure piston dispensing equipment for filled and unfilled RIM polyurethane and polyurea systems.

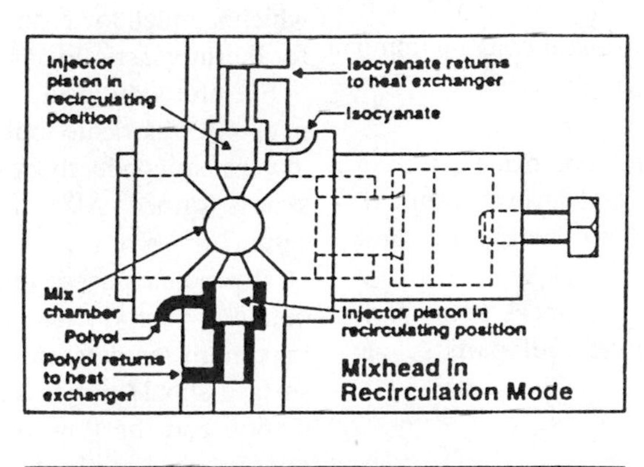

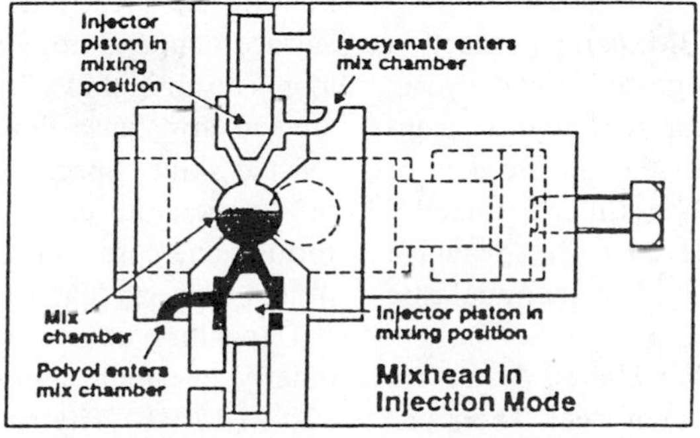

Fig. 19-14a,b. Mixhead design for RIM.

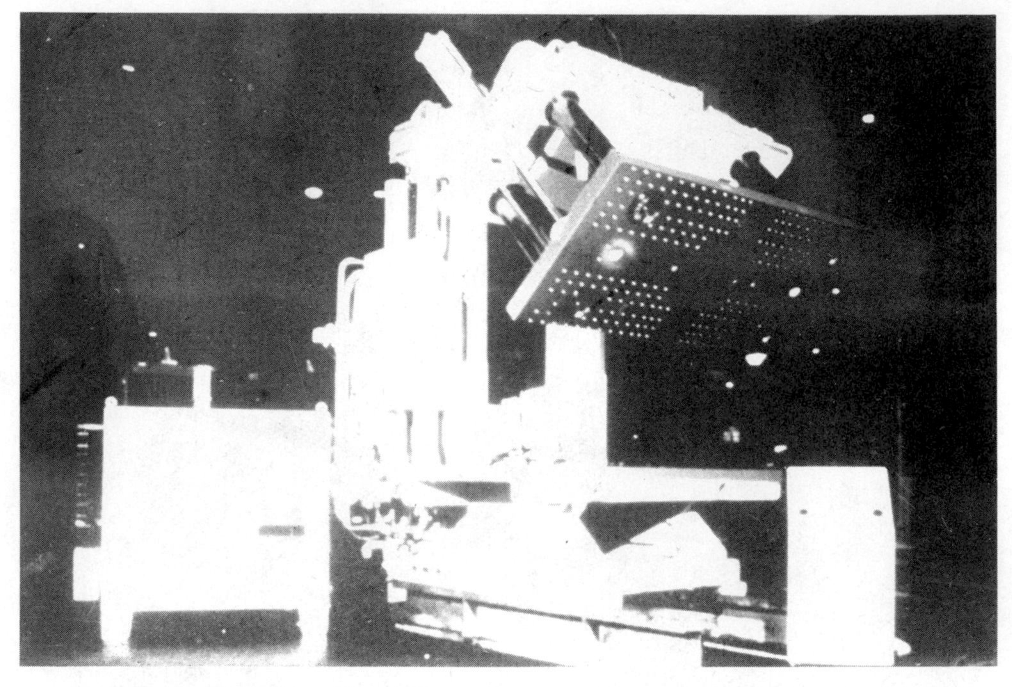

Fig. 19-15. RIM press with tilting clamp.

The Mold

The choice of material for mold construction is determined by the following criteria:

- The number of parts to be made.
- Dimensions, shape, and surface texture of the parts (multicavity mold; movable cores).
- Surface quality and tolerances of the part.
- Mold life (total number of parts; cycle times).
- Mold cost.

The molds used in RIM parts production have been continuously upgraded over the years to the point where they can no longer be considered inexpensive. Even though the cost of RIM molds has increased, it still can generally be said that such molds have a price advantage compared to the molds used in thermoplastic structural foam molding.

Most molds used in the United States for RIM are made of aluminum or steel. Their surface porosity should be at a minimum. The molds should be built to withstand 100 psi, which is much lower than the pressure required for thermoplastic foams.

For prototyping, it is possible to make parts in epoxy or kirksite tooling, but because of the lower heat conductivity of epoxy resin, a poor skin is formed. Also, the mold life is very uncertain.

The molds must be temperature-controlled to obtain a reproducible production cycle. The maximum temperature variations on the mold surface should not exceed $\pm 2°C$, as skin formation and the flow behavior of the reaction mixture are greatly dependent upon the mold surface temperature. Excessively high temperatures result in thin skins, and temperatures that are too low cause thick, brittle, insufficiently reacted skins. Special care should be given to the temperature control of mold sections that form inside corners because high temperatures in this area can lead to thin skin formations.

Depending upon the system and the part geometry, the mold temperature can range from 120 to 170°F. The heat of reaction (approximately 200 Btu/lb) must be conducted away through the mold surface. If the surface is a poor conductor, heat builds up at the surface of the part and prevents proper skin formation. Therefore, only materials that conduct heat well should be used for RIM molds, and uniform temperature control of the mold must be maintained at all times. The best medium for controlling the mold temperature has proved to be tempered water.

Knockouts are essential and should be numerous, especially when lower-density parts are made, and the pins should be placed above or next to skin sections in the knockout direction whenever possible. Pneumatic demolding aids can also be used on large surfaces.

Mold Gating

Early in the RIM process development, the biggest problem was to eliminate entrapped air from the foam, and much effort has been expended in isolating and correcting the problem areas. The most important development was in the area of sprue designs, and was directed toward elimination of turbulence in the liquid stream as it enters the mold cavity. The less air beaten into the reaction mixture during the initial injection, the easier it is to obtain good parts.

The sprue cross sections should be designed so that the mixture does not lose contact with the sprue wall, causing eddies and turbulence to occur. The sprue area in the parting line should seal extremely well, as a bad seal can have an aspirator effect, sucking air into the reaction mixture. Inside the cavity, the material cannot tumble over protrusions or sharp edges. To prevent this condition, it is always good practice to feed the material from the lowest point in the cavity under the liquid level. Fan sprues and rod sprues are used. A barrier normally is placed between the entrance channel and the sprue to redirect the stream and improve the flow distribution.

Part Design

In comparison with other plastic processes, polyurethane RIM materials give the easiest production of large parts, with the least expensive tooling. The processor is limited only by the capacity of the foam machine and the size of the clamping units. If the size of the clamping units for extremely large parts is a limitation, then self-contained molds can be used. Parts weighing over 200 pounds have been produced. No other molding system can claim the production of larger parts more economically than the RIM process.

Large variations in wall thickness are possible without any problem of sink marks. The minimum wall thickness can range from 0.100 to 1.5 inches, depending on the material and the application. The maximum wall thickness depends on the system, and lies between 1 and 2 inches. The upper practical limit is mostly determined by the necessary cure time. Wall thicknesses of 1 inch and above generally require demold times in excess of 5 minutes. Localized thick sections can be incorporated without influencing the demolding time, but should be avoided when possible. In ribbing, for example, the best results are obtained by using ribs with the same thickness as the bulk of the wall. For ease of demolding, generous draft should be given to the ribs, and all inside corners should have a radius of 1/8 inch.

Bosses can be molded in, the only problem being the danger of air pockets during foaming. The bosses should be connected with the walls or should be shallow with sufficient draft to give the air an easy escape route. In the design of a part, it is always good practice to make sure that a two-part mold can be used.

Sectional parts can be assembled in several ways. For a one-time operation, wood screws can be used, screwed directly into the foam part. The screw's holding power is dependent on its size and the diameter of the pilot hole. Self-tapping screws are preferred. The holding power can be improved by molding the pilot hole, which gives extra skin material for the thread to bite into.

Permanent connections can be made by gluing normally. A two-component adhesive is required, and the bonding surface should be mitered so that it is at least three times as wide as the wall thickness. The surfaces should be roughened mechanically. The adhesive bond then will be as strong as the foam part.

Fig. 19-16. Door panels and front fenders of the Pontiac Fiero were molded from polyurea RIM materials.

Applications

RIM urethane has been used in a wide variety of applications. In the automotive area, over 200 million pounds of PUR were used worldwide in 1989 to produce front and rear bumper parts. In 1990 the first mass-produced plastic bodied van was introduced, the General Motors APV, with several RIM vertical body panels. (See Fig. 19-16.)

In the beverage industry, beer kegs are being produced by the hundreds of thousands, using polyurethane RIM materials to encapsulate a stainless steel liner.

In the appliance market, RIM polyurethane is used to produce large evaporative coolers that must withstand temperature extremes in the desert regions of the United States.

In the electronic market, RIM systems compete with thermoplastic structural foams. (See Fig. 19-17.)

The future of RIM is challenging, but with

Fig. 19-17. Structural RIM polyurethane foams compete with thermoplastics in applications such as this computer housing.

new development in the areas of internal mold release, EMI shielding, and new systems for recycling, RIM materials will remain viable candidates for new applications.

POLYURETHANE PROCESSING EQUIPMENT

Certain considerations should be addressed before the purchase of any polyurethane processing equipment:

- The most efficient method of manufacture in terms of:

  1. Product dollar output per dollar invested. A good example of this is the difference between molded items and slab products. For an equal investment cost, the product dollar output of a molding line may be one-tenth that of a slab process. This must be balanced against a greater product loss from the slab process.
  2. Degree of mechanization or instrumentation required. There are two main factors: manual labor beyond a certain basic amount directly reduces productivity per unit of time, and it is extremely difficult to reproduce product quality without a basic level of mechanization and instrumentation.

- The versatility desired in the original equipment installation. This refers to the capability of the equipment to be adapted to changing production requirements or to take advantage of new or different technology.
- The cost. This section lists most of the types of equipment that are currently available to the industry, together with very generalized comments on their type, capability and cost. Most equipment manufacturers today offer a full line of standard equipment. Generally, however, the bulk of equipment sold falls into the semicustom or custom category; other variations are possible.

This section is divided into five main subsections based primarily on the type of application for which the equipment is used, plus a final subsection on accessory items.

The types of equipment discussed include:

- Equipment for spray applications.
- Equipment for pour-in-place applications.
- Equipment for molding applications.
- Equipment for the production of slab foam.
- Equipment for cast elastomer applications.

Equipment for Spray Applications

Spray machines are required to apply mixed foam chemicals in a largely atomized condition onto a vertical, horizontal, or other type of surface. The equipment available today includes machines designed primarily for field use, in-plant or factory use, laboratory use, or combinations of all of these. These units generally are designed for two components and are available with recirculating-type systems when desirable and where applicable (Fig. 19-18). The type of metering systems available cover the entire range from pneumatic and hydraulic piston pumps through vane pumps and gear pumps, both internal and external. Mixing types include both external and internal systems, with high- and low-pressure machines being used.

The production capacity of the usual spray unit usually lies between 2 and 10 pounds of chemicals per minute. Some of the machines are capable of covering this entire output range with any formulation. Others are fixed at some point within this range and may be limited as to the ratio and viscosity of the formulations with which they can be used.

Accessories often required for successful application include temperature control devices for the chemical streams, instrumentation, automatic level control of the chemicals, solvent flush systems, air dryers and filters, additional hose length, and so forth.

It should be noted that some of the spray machines currently available, particularly those with positive internal mixing, are capable of

Fig. 19-18. Typical two-component, gear-pump-type spray machine. (*Courtesy Martin Sweets Co.*)

being used for spray applications, pour-in-place applications, and froth applications merely by substituting tips and turning the atomizing air off.

Also, although the greatest interest in the spray field lies with rigid and semirigid foams, the required equipment, techniques, and technology are available to spray flexible foams. Recently, there also has been a great deal of development work done on elastomeric materials in spray applications. Both high-pressure and low-pressure machines are used for spraying elastomers, but high-pressure is the more desirable type because of the surface quality obtained.

Equipment for Pour-In-Place Applications

Pour-in-place applications are those in which the liquid is poured into an existing cavity and allowed to bond itself intricately to the walls of the final product. The cavity may be brought to a fixed in-plant installation to be filled, or the equipment may be taken to the site of the cavity.

Equipment is available for field use, in-plant or factory use, laboratory use, or combinations of these uses. Those machines used for the production of rigid foams are primarily two-component designs. Full recirculating-type equipment generally has the advantage here, in terms of reproducibility and reliability, because the product rarely can be removed for inspection without destroying it; thus, the urethane mix must be applied correctly each time.

Flow rates of equipment in this category can vary from as little as a few grams per minute up to several hundred pounds per minute, with many installations that are regarded as multiples having several machines pouring simultaneously. The bulk of the work, however, would be accomplished in the 20 to 100 lb/min. range. It should be further noted that there has been a steady increase in the average flow rate of rigid processing machines as technological advances have been made in the industry.

Metering pumps generally are piston-type pumps, either in-line, axial, or radial. The piston pumps are used for their reliability and high degree of accuracy. The pumps typically are variable-volume, and in some instances are connected to variable-speed motors to further increase their flexibility in ratio and output.

The mixheads used for the pour-in-place application typically are high-pressure impingement-type, self-cleaning heads. Earlier heads used solvent, flush air, or a combination of both for cleaning the mixing chamber; but with current environmental laws and technological advancement in self-cleaning mixheads, they are the preferred choice.

The production capacity of this type of machine ranges from that of the very smallest available, less than a quarter of a pound per minute total chemical flow rate, up to that of some of the largest rigid machines, more than 500 lb/min. The majority of the applications seem to require units in the 20 to 100 lb/min. range, with capacity determined by both the size of the cavity and the cream time of the formulation used, as this determines the time available for pouring the quantity desired. Again, it should be noted that the average flow rate of rigid foam machinery is on the increase.

Accessories often required for successful applications include temperature control devices, instrumentation, automatic level control of chemicals, bulk storage tank farms, filters, additional components, air bleed systems, special seals, special pumps, and so forth.

Equipment for Molding Applications

Molding applications are those in which the liquid mixture is poured into a cavity in an object from which the foamed article is later removed. The equipment is very similar in design and capacity to the pour-in-place type of equipment, differing primarily in being somewhat more sophisticated with a greater degree of instrumentation and control than the pour-in-place type has.

Molding is used here as a generalized designation to include flexible, rigid, and RIM applications. Different-type mixheads are normally used for the flexible and rigid applications—heads that have an additional "mixing area" after the initial mixing chamber. The method for achieving this additional "mixing area" differs, depending on the manufacturer, but the mixheads are all high-pressure impingement types.

RIM applications require heads of a relatively simple design because there is usually some type of aftermixer in the mold, and the systems are sometimes rather easy to mix. Recent advancements in polyurethane chemistry for RIM systems dictate that the heads operate as quickly as possible; so the simpler the head, the better.

The metering equipment used for molding applications includes both the conventional pumping units and new lance-cylinder-type machines. The need for fillers in some of the chemical systems required something other than pumps, which would be subject to excessive abrasion and wear. The lance cylinders deliver a desired rate and amount of material to the mixhead without major concern about wear. The mixheads used for filled systems also have to be specially made, with abrasion-resistant materials in the high-wear areas.

Equipment for the Production of Slab Foam

Foam-slab-producing equipment is designed to produce a continuous length of foam, generally from 4 to 7 feet wide and from 2 to 3 feet high. As usually configured, it consists of the chemical handing, metering, and mixing equipment, the spreading equipment for distributing the chemicals on the conveyor, paper-handling equipment for lining the conveyor continuously with paper and removing it from the foam, ventilating hoods, various types of curing equipment, and various types of cutting and shaping equipment. (See Figs. 19-19 and 19-20.)

The most common process for producing slab foam today is the Maxfoam procedure. A typical installation consists of a mixhead that pours the reacting material into a trough. The froth moves up and cascades out of the trough onto an inclined fall plate or weir. The foam action

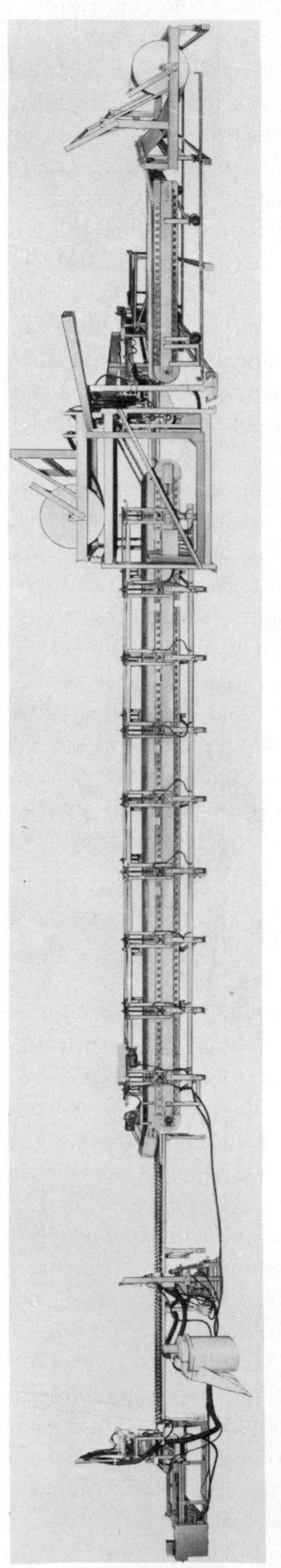

Fig. 19-19. Full-length view of conveyor line (minus foam machine) for the continuous production of urethane foam core insulation boards. Any roll material, such as paper, foil, plastic, and so on, can be used as the skin of the board.

progresses in a downward direction along the fall plate, allowing the top surface to remain level as the bun moves forward from the mixing head.

The advent of flat topping equipment, of which this is an example, brought an exciting new dimension to flexible slab polyurethane production, notably eliminating the conventional rounded top. This permits the foam producer to reduce scrap losses dramatically.

The various segments required for foam production can be bought separately or as a complete package. If purchased separately, however, they must be knowledgeably integrated with each other. Most reliable equipment manufacturers are able to offer turnkey plants, equipped with a guarantee to turn out a product of previously agreed-upon quality. This type of equipment is generally much more complex than the previous classifications and includes what is generally called "boardstock machines" of various types, as well as the open and closed-top flexible and rigid slab complexes.

The flow rate of these machines varies from as little as 50 lb/min. up to and somewhat above 1000 lb/min., with the actual flow rate being determined more from the size of the product desired than from any economic considerations. The capacity, of course, is directly related to the height and width of the foam block that is to be produced.

Most of these installations receive their raw materials by tank car or truck; some utilize shipping containers directly for daily storage, with in-plant storage of sufficient size to permit switching from one tank car to the next without interruption of flow and without mixing two tank car deliveries of material.

In the larger installations, preblending or mixing has been minimized through the use of multiple-component metering units able to handle each chemical separately.

Equipment for Cast Elastomer Applications

Elastomer casting can be defined as a technique whereby a liquid resin is poured or otherwise dispensed into an open mold where it

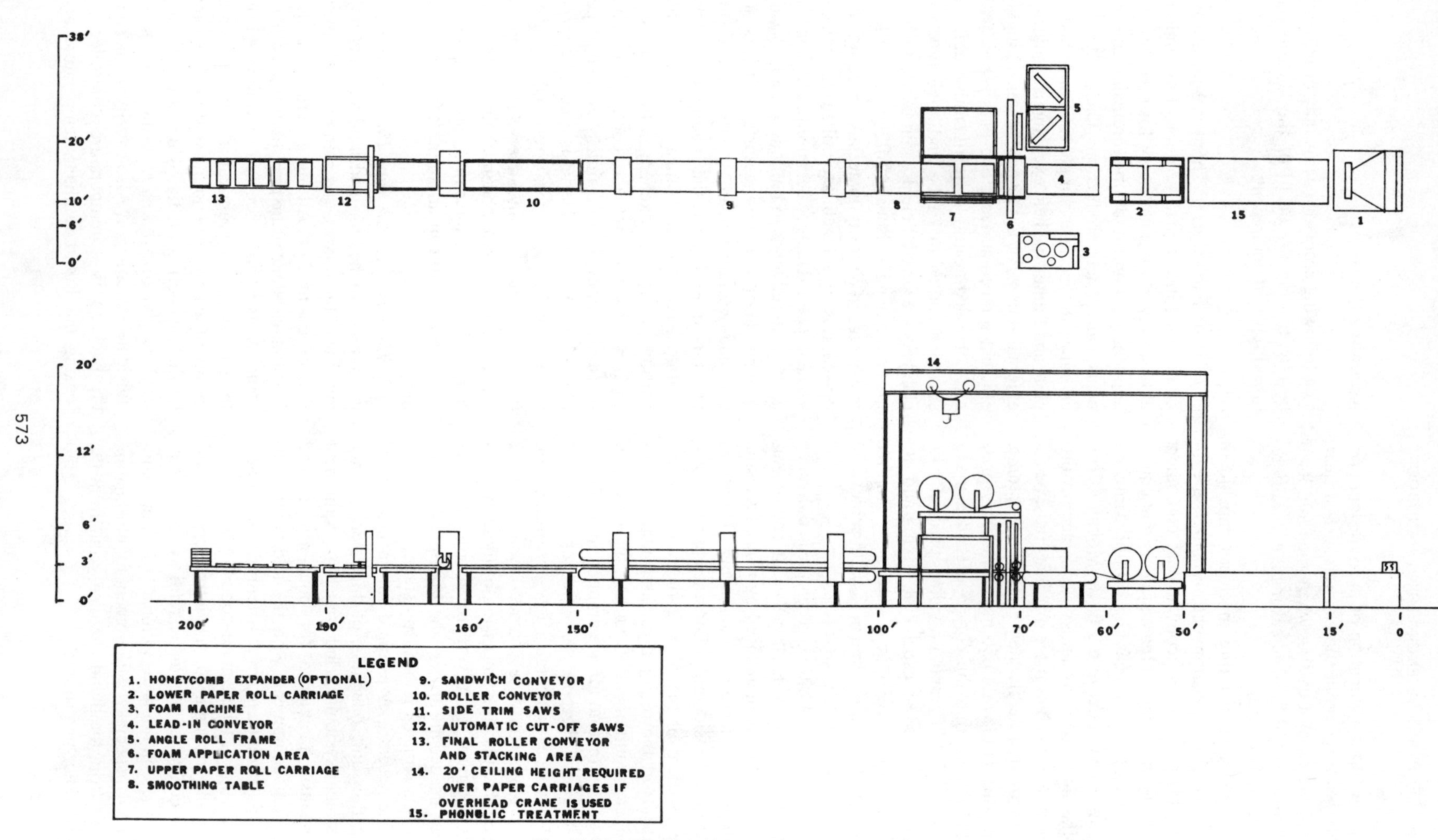

Fig. 19-20. Top and side view of typical panel production line.

cures without application of pressure, but possibly with the application or removal of heat. A special case of casting where the mold itself becomes part of finished casting is generally referred to as *potting*.

Mixing and metering machines capable of processing low-performance elastomers are almost identical to the conventional equipment for handling urethane foam systems, but equipment capable of processing high-performance elastomer systems (e.g., Bayer/Mobay's Vulkollan and Baytec Systems) is considerably different, inasmuch as these machines must be capable of thoroughly degassing the elastomer components to within 50 mm of mercury pressure, maintaining the components at a maximum of 285°F, and providing precise ratio control, as well as capable of mixing and depositing the system in a bubble- or air-free manner. There are different ways to achieve all these capabilities.

Degassing generally is accomplished in much the same manner on all makes of equipment. Ratio control is maintained by using precision pumps and electronic monitoring. However, there are considerably different approaches to the problem of maintaining the various components at the required elevated temperatures and the requirement of pouring the elastomer system so that no air is entrapped in the liquid itself during the pour. Some recent technological advances concerning low-performance or room-temperature elastomer systems have made the machine processing of these systems less complex. However, the elastomer equipment needed to process the high-performance systems remains rather sophisticated.

The accessories required for successful application and operation of high-performance elastomer systems include much the equipment previously mentioned for pour-in-place and molding equipment, with the addition of degassing equipment, ovens, temperature-conditioned tanks, temperature-conditioned material lines and mixing head, specialized and temperature-controlled automatic refill systems, temperature-conditioned pumps (if not submerged), specialized equipment for eliminating all air from the liquid chemicals and from the liquid pour, and so forth.

Accessory Items

The following accessory items may or may not be required for any or all of the above-mentioned types of equipment:

1. Heat Exchangers. These should always be included in competitive quotations, and are almost universally mandatory for good-quality temperature control of the urethane components and efficient production. These exchangers range in type from simple tube and shell or plate and frame type heat exchangers, requiring external sources of hot or cold water, to highly efficient scraped-surface, one-pass types, with self-contained heating or refrigeration units. It should be noted that tube and shell type exchangers are very efficient when utilized with a material having a viscosity of 2500 cps or less. When higher-viscosity materials are being processed, it is advantageous to utilize the scraped-surface type exchanger because the higher-viscosity material tends to build up a thick coating of material on the cooling surface; if it is not continually scraped clean as in the scraped-surface type of exchanger, the coating reduces the heat transfer efficiency drastically. Also, the portion of the heat exchanger that comes in contact with the urethane component being processed should be of stainless steel construction. If an alternate material is used in the construction of the exchanger, care should be taken that it will not have any adverse effects on the liquid urethane component being processed (e.g., a brass exchanger in an amine metering system may result in an amine complex problem).

2. Heating and Cooling Units. There are many types of heating and cooling units provided with the mixing and metering equipment now on the market. All are generally claimed to maintain temperature control of the urethane components being processed to within ±1°F. In general, this equipment consists of some means of heating or cooling and pumping the heating or cooling media, which are generally ethylene glycol and water, to the above-mentioned heat exchanger. Also usually included in the package is a temperature controller for sensing the liquid urethane temperature and controlling the water, whether hot or cold, required in the exchanger. These units will range from a totally manual unit, where the hot and cold water must be manually valved into the tube and sheet heat exchanger, to totally automatic units, which will sense the need for a change of temperature of the water in the tube and shell and make the change automatically. Even in the totally automatic units, one must be careful about how this operation is accomplished, as some units use manually adjustable flow control valves or solenoid valves to valve hot or cold water into the tube and shell heat exchanger, creating a bucking condition between the hot and cold water conditioning units. In contrast, there are totally automatic units with single dial control that utilize a common heating/cooling medium (water–antifreeze mix) throughout the heating/cooling cycle, with a null point controller to prevent sine-wave or over/undershoot controlling, which is predominant in units using flow control valves or solenoid valves with separate hot and cold water supplies.

3. Instrumentation. Product reproducibility is difficult, if not impossible, without adequate production records showing how previous products were made. Desirable information includes temperature, pressure, flow rates, process speed rates, and so forth, for all components. The instrumentation needed for this can consist of anything from a simple dial thermometer (to monitor temperature) and a pressure gauge at the mixing head (to monitor material pump pressure) to a complete solid-state computerized metering and monitoring system. Such a system could be operated by one person, and it would be fully automated by controlling multiple streams to 1/2% of the total flow rate of each individual component. Systems such as this generally feature a digital readout in pounds per minute for each material stream and a master total in pounds per minute dispensed, an audible and visual alarm system to warn the operator of any irregularities in the system, a digital readout in material pressure on each component being pumped, and a digital readout of the material pump rpm of each component.

4. Other Accessories. Accessories that are sometimes necessary or desirable are liquid level controls and/or indicators for the chemical tanks, air dryers and filters, additional or special-type hoses, air bleed systems, special pumps and seals, "know-how" books, and so forth.

The following accessory items generally are associated with urethane slab stock equipment:

1. Traverse or Spreader Units. These units are used to spread the mixed chemicals on the casting surface, whatever it might be. Most usual types involve a reciprocating carriage on a frame above the casting surface, which carries the mixing head back and forth. Those available range from simple and reliable pneumatic motor-driven types to elaborate mechanical, electrical, or hydraulically driven types. The pneumatic motor type is popular because of its capability of immediately discharging the energy absorbed in the reversal of the mass during the traversing cycle. However, the hydraulic system, when installed as a two-speed system, is also popular because of its capability of controlling the lay-down pattern of the liquid mix to a very fine degree.

2. Paper Handling Equipment. This equipment generally is used to line the conveyor on slab production units. These machines generally take the paper from one or more rolls and shape it wrinkle-free in the form of a wide "U," on which the mixed chemicals are spread. Multiple paper roll systems are available that enable the operator to cover both sides, top and bottom, of the conveyor with individual continuous sheets of paper.

3. Curing Equipment. These systems are usually of a conventional nature, and their use is so diverse as to make generalizations impractical. Almost any heat source, from infrared lamps to a sophisticated steam, gas, or electrical oven, can be used.

4. Cutting Equipment. This equipment generally is associated with foam slab and rigid boardstock production. Types available are widely diversified, and the product range is virtually unlimited. Usually, such equipment includes: horizontal incline types (both automatic and manually adjusting in nature), vertical incline types (conveyorized or manual push type), convoluters, scooping cutters, peelers, borers, shoulder pad cutters, hot wire cutters, and so forth.

THERMOPLASTIC STRUCTURAL FOAMS*

By definition, *structural foam* is a plastic product having integral skins, a cellular core, and a high-enough strength-to-weight ratio to be utilized in a load-bearing application. This definition allows for many variations of the general ''structural foam'' concept. It includes both thermoplastic and thermoset polymers and covers a wide range of density.

The thermoplastics that can be produced as structural foams already run a wide gamut—ABS, acetals, acrylics, styrenes, polyethylenes, nylon, PVC, polycarbonate, modified polyphenylene oxide, polypropylene, polysulfone, thermoplastic polyesters, and various glass-reinforced nylons, polyethylenes, polypropylenes, and other thermoplastics. Among the thermosets, urethane structural foams are most in use (covered in detail in the previous section on RIM). This section will deal exclusively with thermoplastic structural foams.

There are many systems that produce such a structure, and a growing number of equipment manufacturers are producing machines to supply such products. Variety is found not only in the pressures used to inject the plastic into the mold, but in the number of nozzles, the blowing agents, and the presses themselves.

Why would a design engineer choose a foamed product at all? Why not use the tried-and-true method of injection molding a solid, conventional product? The true value of foam lies in the increase obtained in the strength-to-weight ratio and in the reduction of sink marks in areas opposite thick sections. Other values include a significant reduction in tooling costs due to the lower pressures encountered in the cavity during most types of processing.

The structural integrity of a properly formed part also is greatly improved because of a lowering of stresses as well as a general increase in stiffness due to increased wall thickness. Higher-pressure foams with thinner walls retain some of these properties in addition to improved surface conditions.

The main disadvantage of most structural foams remains the need to finish the product, in order to achieve an acceptable appearance for such applications as business machines.

History

In the late 1960s, Union Carbide commercialized a process for producing a molded foam product from thermoplastics. Calling the product ''structural foam,'' Union Carbide succeeded in licensing some 30 molders to utilize the process and the product. Such products as furniture components, automotive fan shrouds, pallets, and road barriers were common. Other work was being done by other companies. A patent filed in 1964 by a Frenchman, T. P. Engle, was purchased by Phillips Petroleum. This process used a chemical blowing agent and was designed to produce small items such as paintbrush handles.

Allied Chemical announced its egression process. United Shoe Machinery was showing an expanding mold technique that gave a fine cell, a smooth skin, and good density control. AMF people in Connecticut worked on a process and published several papers. Dow's TAF process was developed. ICI pushed a sandwich molding technique using a nonfoaming skin shot into a mold, followed by a foamable melt that provided a cellular core.

Borg-Warner developed a foundry system. In this unique approach, unheated resin containing a chemical blowing agent was introduced into a mold, which then was sealed and run through an oven to heat the resin and activate the blowing agent. Once the heat cycle was complete, the mold was moved through a cooling shower and finally to a demolding area, where it was pulled apart and the part removed; and then the mold was readied for the next cycle. The part size attainable with this technique was unlimited.

Structural foam today has basically two markets. Housings for computers, business ma-

*Portions of this section adapted from the brochure ''Structural Foam,'' issued by the SPI's Structural Plastics Division.

chines, and medical instruments are number one in terms of sales dollars; whereas in pounds of product the material handling industry consumes the greatest amount, supplying such needs as pallets, large bins, waste receptacles, and so on.

New technology is bringing about many changes, and the industry must be prepared to grow with them. The old ways of molding a cellular product may not be the best. Application areas once thought to be exclusively the right place for low-pressure thermoplastic foams may be challenged by newer materials and processes.

Properties of Structural Foams

Basically, structural foam molding is intended to produce tough, rigid, lightweight thermoplastic products with a solid skin and a cellular core. The solid skin gives the structural foam its strength and rigidity; the cellular core helps structural foam to resist impact.

The importance of structural foam can best be understood by reviewing the general practices followed in designing a solid molded part that is found to have insufficient stiffness. This inadequacy generally is overcome by increasing the thickness of the part or adding ribs. Either solution adds material and cost to the product, and the second approach sometimes significantly detracts from its appearance. However, the thickness of the part can be increased without a corresponding increase in weight if a foam structure is employed. This is an excellent method of increasing rigidity, as stiffness is a function of thickness to the third power. The integral skins also improve the stiffness of the foam cross section. A linear decrease in the modulus of elasticity is assumed.

The physical properties of materials molded of structural foam are difficult to define. The part configuration, process used, equipment, and molder expertise are all variables that influence the end result. Therefore, a careful analysis should be made of the physical and functional needs of the application. Material suppliers should review the project and make recommendations. After a careful study of materials and processes, a selection can be made. Table 19-9 provides a list of general properties for several commonly used structural foam materials.

The physical properties of all materials can be altered to some extent by the addition of various fillers, such as glass fibers, talc, or other inert materials. The expected effect of each of these additives will vary, depending upon how they are compounded. Therefore, it is recommended that physical needs be reviewed with several material supplies in order to match properties with cost.

Combustibility

Fire is an ever-present environmental danger. In use, all organic or polymeric materials will burn under sufficient fire stress; so fire safety is a matter of the relative behavior of systems in real applications subjected to a range of real-fire circumstances. Foam plastics thus have certain combustibility characteristics that should be considered in determining their suitability for particular uses.

Much of the combustibility data reported on structural foam materials has been the result of the large use of structural foam materials in office and business machine enclosures. Listed in Table 19-10 are UL-94 values for many structural foam materials. Because the thickness influences the ease with which a material will respond to a small ignition source, all materials are reported at $\frac{1}{4}$-inch thickness.

Making Structural Foams

There are many different systems, and an even more diverse range of machines available, for producing structural foams. Low-pressure structural foam systems involve the use of nitrogen gas or chemical blowing agents as expansion devices. In a low-pressure system (so-called because the molds are under very low pressure), the molds are only partially filled (i.e., a ''short shot'') with the melt; the melt then is expanded by nitrogen gas or by the gases released by the decomposing agents to fill the mold. In the processes involving nitrogen gas, the resin, in pellet form, is fed into an extruder where it is plasticated and mixed with the gas

Table 19-9. Physical properties of thermoplastic structural foam.

(at .250 Wall With 20% Density Reduction)

Property	*Unit*	*Method of Testing*	*High Density Polyethylene*	*ABS*	*Modified Polyphenylene Oxide*	*Polycarbonate*	*Thermoplastic Polyester*	*Polypropylene*	*High Impact Polystyrene*	*High Impact Polystyrene w/FR*
Specific gravity	lbs./ft.^{3}	ASTM-D-792	.60	.86	.85	.90	1.2	.67	.70	.85
Deflection temperature under load	°F@66 psi	ASTM-D-792	129.6	187	205	280	405	167	189	194
	°F@264 psi		93.5	172	180	260	340	112	176	187
Coefficient of thermal expansion	in./in./°F × 10^{-5}	ASTM-D-696	12	4.9	3.8	2	4.5	5.2	9	4.5
Tensile strength	psi	ASTM-D-638	1,310	3,900	3,400	6,100	9,910	1,900	1,800	2,300
Tensile modulus	psi	ASTM-D-638		2,500,000	235,000	300,000	1,028,000	79,000	141,160	245,000
Flexural modulus	psi	ASTM-D-790	120,000	2,800,000	261,000	357,000	1,000,000	80,400	200,321	275,000
Compressive strength (10% deformation)	psi	ASTM-D-695	1,840	4,400	5,200	5,200	11,300	2,800	3,447	
Combustibility rating		UL Standard 94°		V-0	V-0/5V	V-0/5V	V-0	HB	HB	V-0

*This rating is not intended to reflect hazards presented by this or any other material under actual fire conditions.
Material properties given above are typical and vary from supplier to supplier. It is recommended that an end user contact his supplier and/or molder to obtain specific properties for use in a given application.

Table 19-10. UL Standard 94 ratings for structural foam products at $\frac{1}{4}$ inch.

Material	*Rating**
ABS	HB
FR-ABS	V-0
Acetal	HB
Modified PPO	V-0, 5V
Nylon	HB
PBT	HB
FR-PBT	V-0
Polycarbonate	V-0, 5V
Polypropylene	HB
FR-Polypropylene	V-0
Polystyrene	HB
FR-Polystyrene	V-2 to V-0, 5V
FR-Polyurethane	V-0

*This numerical flame spread rating is not intended to reflect hazards presented by this or any other material under actual fire conditions.

before injection into the mold. When chemical blowing agents are used, they are either already incorporated into the molded pellet or dry-mixed into the resin by the processor or supplied in the form of an agent/resin concentrate.

The machinery involved in low-pressure foam molding systems also varies widely. In the nitrogen process, special equipment is required. With chemical blowing agents, it is possible to use conventional injection molding machines with minor modifications or machines modified especially for structural foam molding with such equipment as oversized injection pumps, large platen areas, multiple-mold stations, and so on.

The most widely used low-pressure system in the United States is the multi-nozzle method. This type of machinery features horizontal, vertical, or even multiple clamp units. Characteristics include the use of a physical blowing agent, usually nitrogen gas, and an interchangeable multiple nozzle manifold that allows several molds to be run in concert.

A much less frequently used process in structural foam molding is the high-pressure molding system. As distinguished from the low-pressure system, it involves injecting the polymer melt and the blowing agents under higher pressures into the mold cavity to completely fill the mold; the mold then expands or mold inserts are withdrawn to accommodate the foaming action. Because tooling is the critical element in this system, conventional machines can be adapted; special machines with accumulators also are available.

Other newer processes offer unique differences from conventional foam. One common advantage of all processes in this category is that they improve the surface finish.

Coinjection, also known as sandwich molding, requires specialized equipment. The clamp unit is identical to those found on most molding machines. The injection unit requires two, two-stage injectors, allowing the injection of two dissimilar plastics. One injection unit will shoot solid plastic, and the other a plastic containing a blowing agent. The solid plastic injection stage begins a split second before the foamed plastic, so that the foamed plastic moves to the inside of the solid material. The resultant product will display solid skins with a foamed cellular core. It also is possible to use a less expensive plastic in the core than is used in the skin.

Counterpressure foam molding has been known to European molders for many years. The process is based on delaying the blowing agent expansion until the mold has been filled and a smooth surface formed. Prior to melt injection, the mold is pressurized just above the expansion pressure of the blowing agent. Once the plastic/blowing agent mixture is in the mold, the pressure is released, and the foaming action takes place. The result is a structural foam part with a smooth surface and excellent physical properties.

Low-pressure structural foam currently leads the industry as compared to high-pressure and newer processes.

There is some blurring of the advantages and disadvantages in comparing one particular low-pressure system to another. For example, the newest multi-nozzle machines feature fast injection speed and improved melt control.

No one system is always best; so the end-user's requirements and expectations must be carefully considered in order to select the best system for a particular application. In most cases, all systems are capable of manufacturing the product; the question in one of part cost, mold expense, and surface quality.

Process Details

Low-Pressure/Nitrogen. In the nitrogen process, the equipment consists of an extruder, a nozzle, a hydraulic accumulator, and a mold. Resin in pellet form is fed into the extruder, where it is melted and mixed with nitrogen gas that is injected through the extruder barrel. This material is pumped by the extruder into the accumulator and held there under pressure. When the accumulator has enough material to fill the mold, the nozzle is opened, and the plastic and gas mixture is forced out of the accumulator and into the mold. The nozzle then is closed, a step that causes the accumulator to refill while the part is cooling in the mold. At the end of the cycle, the part is cool and ready for ejection. When the mold is empty and reclosed, the accumulator is full, and the valve is opened, so that the mold is filled again. (See Fig. 19-21.)

Accumulator pressures are generally between 2000 and 3000 psi, with the mold pressure about 300 psi. Only enough plastic to fill about one-half of the mold is delivered by the accumulator; but because it contains gas, it expands and fills the mold with foam. As the foam flows through the mold, the surface cells collapse, and solid skins are formed. These skins are beneficial because the maximum tensile and compressive stresses occur on the surface when a member is subject to bending.

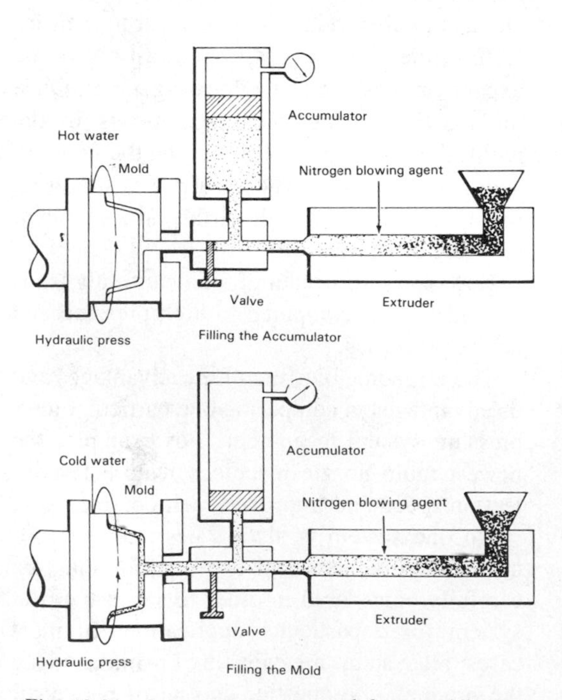

Fig. 19-21. Low-pressure structural foam molding system, using nitrogen gas as the blowing agent.

Because the mold never is packed with solid material, the pressures are low, and lower-cost molds can be used (see subsection on "Tooling," below).

Low-Pressure/Chemical Blowing Agent. In this low-pressure system, resin is combined with a chemical blowing agent by one of several techniques. The processor, for example, can simply tumble resin pellets with dry blowing agents. A more sophisticated approach uses continuous on-stream mixers for proportioning exact amounts of the two materials. Blowing agents are also available in concentrate form or already compounded into the resin.

The percentage of blowing agent used depends on the type used and on other factors (part wall thickness, desired density, molding process, etc.), but it generally ranges from 0.2 to 5.0%. No matter how the blowing agent is added to the resin, however, the molding process proceeds in the same way. A metered short shot is injected into the mold as rapidly as possible, partially filling the cavity. During processing, the blowing agent decomposes with heat and releases gas through the melt, expanding the plastic to fill the mold completely.

Standard injection molding machines can be modified to mold structural foams with CBAs. Hydraulic booster circuits or other means must be supplied to increase injection speed, and shutoff nozzles have to be provided to prevent material from drooling during the plastication part of the cycle. Because of the different requirements for structural foam versus conventional injection molding, specialized equipment has been developed. Generally, these machines provide higher-capacity injection ends and larger platen/lower tonnage clamp units. Rotary machines for structural foam molding (see Fig. 19-22) have gained favor for some applications because of the long cooling cycles for thick wall sections.

Foam Characteristics. Low spots or sink marks are not found in structural foam parts.

Fig. 19-22. Rotary structural foam molding machine. (*Courtesy Wilmington Plastics Machinery*)

Because the gas exerts an internal pressure, the parts tend to swell or puff where there is a thick section adjacent to a thin section. This can be controlled by proper cooling. These parts generally have perfectly flat surfaces. Sink marks at T-shaped intersections have been eliminated.

The integral solid skins on the surface of structural foam parts are formed in the molding cycle, requiring no separate operations. As the foam flows through the mold, the cells that rub the mold surface rupture and densify, forming a solid skin. Pressure exerted on this skin by the cellular core also helps to solidify it. The melt temperature, mold temperature, amount of gas in the melt, and speed of melt injection all contribute to the formation of the skins. These conditions can be varied to produce the desired skin thicknesses. Parts with very thin skins have low stiffness, whereas parts with thick skins are stiffer but become quite heavy. The preferred skin thickness on a $\frac{1}{4}$-inch part ranges from 15 to 50 thousandths of an inch.

Structural foam parts can be molded to have the same weight as injection-molded parts, but they will be three to four times as stiff or rigid. However, they may also be molded significantly lighter than injection-molded parts and still be one and one-half to three times as rigid. Thus they have great economic advantage over injection-molded parts. Structural foam has been used successfully in some commercial items that previously required wood or metal for high rigidity. Injection-molded plastic could not compete in these areas.

Limitations. Low-pressure foam parts generally have a swirl surface pattern, a texture that is preferred for some applications. For furniture the part can be molded with a wood-grained surface and painted, as is done with injection-molded parts. Because of the inherent ruggedness of this material, it has many industry uses where a glossy, smooth surface appearance is not necessary. Some pigments hide the swirl pattern more than others, and molding conditions can be adjusted to vary the appearance. By increasing the temperature of the mold, the surface can be greatly improved, but the cooling cycle then is lengthened. It also is possible to hide the swirl by special finishing techniques.

High-Pressure Systems. A high-pressure foam process is characterized by injecting a "full" shot of foamable melt into a closed mold at near-normal injection molding pressures. Foaming takes place either by allowing the escape of excess material back into a runner system or by moving a plate or plates to open the mold cavity itself.

The density of a molded part produced by this process can be accurately and consistently controlled by establishing the ratio of the starting volume to the end-product volume. Because the process completely fills the mold with plastic before beginning the foaming action, a solid skin is formed that is free from surface gas splay marks. The solid skin promotes fine reproduction of whatever detail is required—wood grain, miter joints, tooled leather, and so on. Even the farthest corner of the mold is filled by a known level of injection pressure, ensuring excellent surface definition on all portions of the part.

Foam core densities can vary from 0.25 to solid, an advantage for functional component design calling for acoustical, impact, and strength properties. Solid sections can be selectively designed into a part to meet functional requirements or to provide maximum shear or impact properties.

A major advantage of this process is the uniformity, consistency, and size of core foam cells. Structural integrity is significantly improved by eliminating voids in the foam core, and design requirements are eased for fastener placement, as foam exhibits strong holding power for staples, screws, nails, or other fasteners.

In addition, the advantage of using multicavity molds and short cycle times broadens the economic impact of structural foam molding. It is now feasible to foam-mold very large parts as well as smaller, high-production parts.

How the Process Works. The process consists of plasticizing a polymer foaming agent blended in a specially modified reciprocating screw injection molding machine, and then injecting it into a specially designed expandable mold.

Unlike other processes, the molding machine is dual-purpose. For conventional molding, the simple action of turning a switch in the main control panel selects the proper hydraulic and electrical operating sequence. The modifications and improved technology needed for foam molding also considerably enhance solid-part molding.

The mold expansion cycle is the point at which the machine noticeably departs from conventional practice (see Fig. 19-23). The platen motion begins at the completion of the mold fill, and expands the cavity volume at a controlled rate.

Parts produced by the process exhibit excellent flexural strength and stiffness-to-weight ratio. The mass of the solid skin, separated by a low-density foam core, simulates an I-beam and provides the most efficient structure for strength per unit of weight.

Gas Counterpressure. The gas counterpressure molding process was developed to improve the surface quality of structural foam parts. It involves the pressuring the mold up to 500 psi with an inert gas (nitrogen), which prevents the blowing agent from breaking through the surface of the melt during injection. A solid skin forms as the core expands, and the surface of the part is not marred by the swirls typical of low-pressure processes. Another advantage of gas counterpressure is that it produces foams with a uniform cell structure and, thus, improved mechanical properties.

The process and equipment for counterpressure molding are a bit tricky because the molds have to be sealed to contain the pressurized gas, and controlled venting is required to allow expansion of the foam. Conventional machinery can be modified fairly easily for the process by adding a tank of nitrogen gas with regulator, pressure gauge, inlet, outlet, and relief valves. The schematic in Fig. 19-24 shows one way to

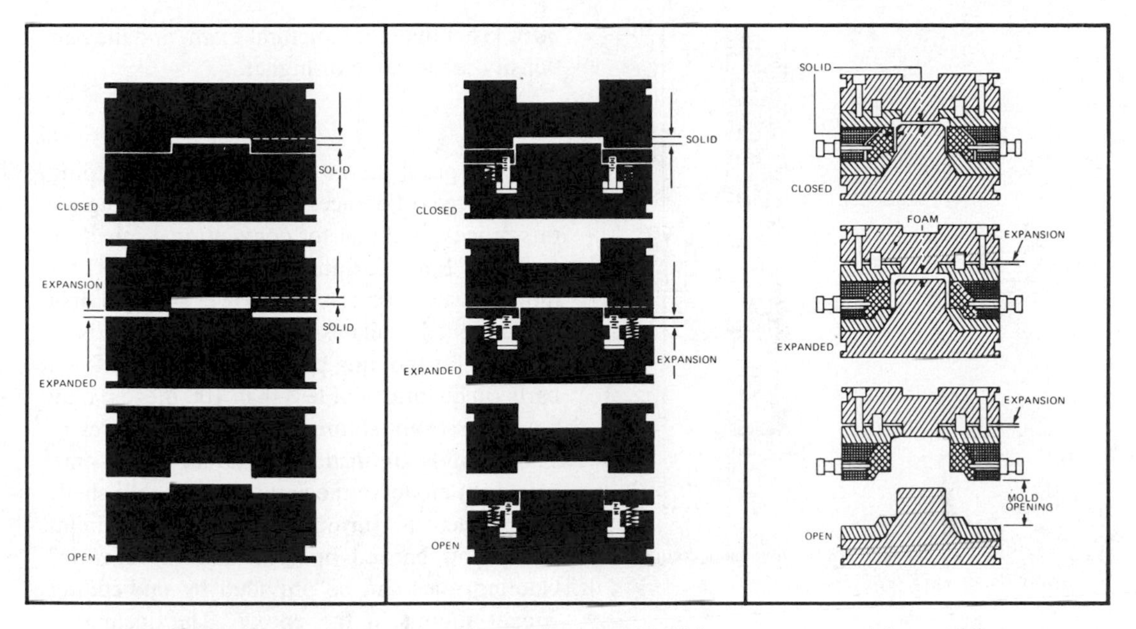

Fig. 19-23. Three techniques for mold expansion as used in high-pressure structural foam molding.

accomplish gas counterpressure. The auxiliary equipment can be either attached to the molding machine or mounted on a mobile cart.

Sealing of the mold itself usually is accomplished by placing an O-ring in a specially machined groove. A gas inlet and outlet to the cavities is provided inside the O-ring. Poppet relief or flow control valves can be positioned in hard-to-fill areas of the mold to ''steer'' material during injection.

The amount of counterpressure needed to overcome the gas pressure generated by the blowing agent will increase the clamping requirements by that amount. Because the entire area inside the O-ring must be included in the projected molding area, careful calculation of the tonnage needed should be made during the design process.

Coinjection. The coinjection or sandwich molding process for structural foam molding, developed by ICI in England, produces the skin and the core from two different injection units. The sequential formation of a solid skin and a cellular core is accomplished by using a special two-channel nozzle patented and commercialized by Battenfeld in the mid-1970s. The principle is shown in Fig. 19-25.

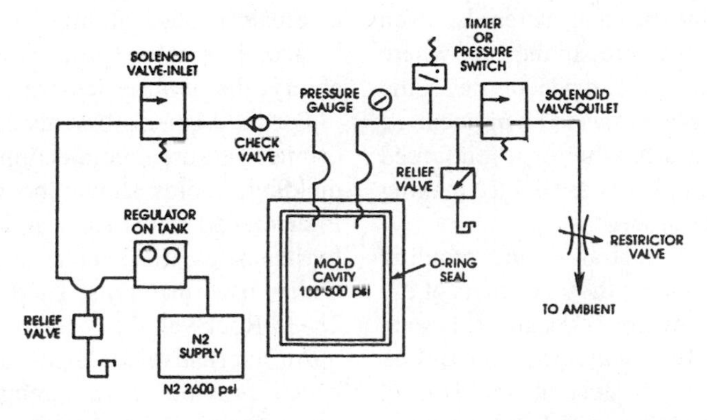

Fig. 19-24. Schematic for gas counterpressure process. (*Courtesy GE Plastics*)

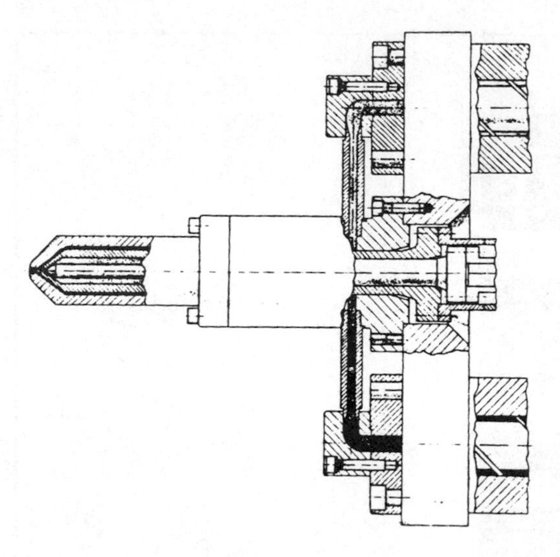

Fig. 19-25. Two-channel nozzle for foam coinjection. (*Courtesy Battenfeld Corp.*)

Because the skin is molded from unfoamed resin, the surface of the part is free of the swirl marks typical of low-pressure structural foam. The process will reproduce mold details as well as conventional injection molding, eliminating the need for secondary finishing operations. If the part is to be painted, there is no need to age it as there is with structural foam molding. Painting can be done immediately after molding without the danger of blistering caused by outgassing. With low-pressure structural foam parts, it is recommended that one wait for a day or two.

In coinjection molding, each material is plasticated in its own extruder; this allows the inner and outer portions of the structure to be made of totally different resins. Costs can be reduced by using inexpensive materials, even recycled resins, for the core. In the case where the skin and the core are made of the same resin, only the outer layer has to be pigmented. The option of using a flexible or a reinforced material for the skin also is available, making this a most versatile process.

Compared with structural foam, molding pressures are about double those for parts of the same thickness. Cavity pressures are still much lower than with conventional injection molding of heavy-walled parts. In general, molds used for structural foam can be used for coinjection with little or no change. Cycle times are 10 to 30% faster than for structural foam, and the part density is the same or higher.

Tooling

The low pressures of structural foam molding create some differences between the tooling for this process and that for conventional injection molding. For one thing, the materials used for mold construction can be of lower strength than the hardened steels used for injection molds.

Tools for molding prototype structural foam parts (in quantities of less than 100 pieces) may be made from aluminum-filled epoxy resin. These molds are made by spraying metal onto a wooden model of the part to form a thin shell, which then is surrounded by an aluminum frame and backed up with the filled resin. Cooling lines can be provided by embedding copper tubing in the epoxy. The insulating properties of this type of mold result in longer cycles than those obtained with metal production tooling.

The least expensive production molds for structural foam parts are made from cast aluminum or kirksite, and provide moderate strength and excellent cooling rates. The cooling lines can be cast directly into the mold. Cast molds are not recommended for long runs, however, or for applications where tolerances are critical.

The majority of structural foam molds for the low-pressure process are made from machined aluminum. Using aluminum plates and machined mold components (cavities and cores) provides accuracy and replaceability. For low cost and weldability (for future modifications or repairs), 6061 aluminum is the material of choice. For extra hardness and easy machinability, the more expensive 7075 alloy is best.

For the higher pressures encountered in gas counterpressure, coinjection, and thin-wall molding, molds should be made from steel. Prehardened steels such as P-20 are adequate for plates, cavities, and cores, and are the most widely used materials. P-20 has a hardness of 28–35 Rockwell C.

An alternative for high-volume production, which provides high strength and excellent conductivity, is beryllium copper, which casts well and reproduces detail accurately. However, the material is expensive, and the machining costs are high.

The cooling capability in structural foam molds is an important factor because of the insulating properties of the cellular material. Unless generous cooling is provided, cycles in steel molds will be substantially slower than in aluminum tools. Water lines should be placed at close intervals, and baffles, bubblers, or heat pipes should be used for optimum cooling.

Designing with Structural Foam

Designing with foam is similar to designing for any molded product. Some general rules include:

1. Keep undercuts to a minimum unless you are prepared to pay for slides, cams, and cylinders.
2. Use as generous a draft as possible on vertical walls while still keeping wall thicknesses as uniform as possible.
3. Keep material flow to a minimum.
4. Molded foam parts are still notch-sensitive, so use generous radii whenever possible.
5. Coring thick sections is advantageous in two ways. First, there is a reduction of material; second, this same coring also cuts the cycle time and reduces molding cost.

Material Selection. Choosing the right material for an application is a major responsibility of the design engineer. An evaluation of the environment in which a part must perform is the first step in selecting a material. After this analysis is complete, the physical properties of different materials must be examined in order to choose the material with the best combination of properties.

Combining of Part Functions. One of the key advantages of any molded plastic part is that it is possible to combine functions of a set or an assembly of parts into one molded part. This is true of structural foam and should be one of the first aspects a designer should review once a project is started. Questions to be asked include:

1. Which parts, which are now assembled with the part in question, can be combined?
2. What are all the possible ways that a part can be designed to do the job specified?
3. What does study of the function of the existing part reveal about what the other parts with which it comes in contact do? Can these parts be combined?

Tolerances. When close tolerances are needed, most structural foam processes will provide good reproducibility, assuming that the mold is right in the first place. Tolerance specifications are extremely important, as they can directly affect both part cost and ability to perform properly. The practice of blanket tolerances can result in unnecessary cost, both in the part and in the tool.

The final dimensions of a part are affected by thermal expansion and contraction, processing conditions, mold dimensions and design, part configuration, and material and selection.

The addition of fillers such as glass or talc will reduce shrinkage, and this technique often can be used to bring tolerances into specification.

Ribs. Ribs can be used to increase the rigidity and the load-bearing capability of a structural foam part without increasing the wall thickness or the part cycle time. Because structural foam parts are much less subject to sink marks, thicker ribs can be incorporated than in injection molding. A general guide for rib thickness as a function of the surrounding wall is shown in Table 19-11.

Draft angles on the rib should be about 0.5 to 1.5°, and fillet radii at the base should be 0.06 to 0.10 inch *R*.

Wall Thickness. The structural foam process permits molded parts with sections thicker than can be realized in injection molding and

Table 19-11. Guide to rib thickness.

Wall Thickness	*Rib Base Thickness as % of Surrounding Wall*
.157″-.175″	75%
.176″-.215″	85%
.216″-.300″	100%
.300″	120%

without sink marks and warpage. Traditionally, structural foam parts were designed with 0.250 inch wall thicknesses. Now with modified engineering resins, parts can be designed with wall sections as low as 0.125 inch. The design criteria of a structural foam part must be considered before the optimum wall thickness is selected.

Draft Angles. In injection molding, sufficient draft angles are necessary; but because of the lower pressures involved in foam molding, smaller draft angles can be tolerated. Generally, an angle of 0.5 to 3° will provide sufficient draft to release the part.

Textured surfaces on sidewalls generally require an additional 1° draft per 0.001 inch depth of texture. For best results, one should consult the engraver before specifying the draft requirements.

Fillets and Radii. Sharp corners create points of stress concentration and restrict material flow. They are often a cause of part failure. One should use as large a radius as possible on inside and outside corners to minimize this stress concentration and to aid in mold filling. In most parts, the minimum inside radius should be 0.060 inch. If the section is under load or subject to impact, a minimum radius of 0.125 inch should be used.

A radius equal to 0.6 times the wall thickness will provide a desirable fillet for most situations.

Bosses. Bosses can be easily incorporated into structural foam parts to accept fasteners and support components. In many applications, the addition of molded-in bosses, mounting pads, standoffs, and retainers can replace costly brackets and miscellaneous small metal part assemblies. In general, boss diameters should be two times that of the cored hole. This recommendation will vary, depending on the resin used and the boss wall thickness.

Bosses should be cored to prevent the formation of a thick section in the part. Generous fillet radii should be used at the base of the boss to avoid stress concentration and resist torque loading.

Transition Sections. Transition sections from thick to thin walls are more easily achieved without sink marks in structural foam than in injection molding. Still, uniform wall thickness should be maintained to minimize restrictions to material flow.

The transition from thick to thin walls should be tapered for proper processing. In molding parts with wall sections of varying thicknesses, it often is desirable to gate the part in the thin section and allow the material to flow into the thicker area.

Hinges. Properly designed, integral structural foam hinges offer a fastening technique while eliminating costly bracketry and assembly time. Hinges can be designed to be either hidden or visible, depending on their location.

Snap-fits. The superior rigidity and strength of structural foam permits increased utilization of snap-fits for assembly and for mounting heavy components in bases. A quick and extremely economical assembly method, snap-fitting eliminates the need for added screws, brackets, and fasteners, significantly reducing labor.

Mechanical Properties

Tensile Strength. The apparent tensile strength of any foam material is less than that of the same material in a solid configuration. The strength is reduced considerably because of the density reduction and stress concentrations caused by each individual cell. Therefore, tensile stresses should not exceed the proportional limit of the material.

Compressive Strength. The compressive strength of a foam material is higher than its tensile strength. In bending, a compressive failure—although extremely rare—normally involves buckling of the skin and collapse of the cellular core.

Stiffness-to-Weight. The high stiffness-to-weight ratio of structural foam is the primary advantage of this material has over metal and standard (nonfoamed) injection-molded plastics. An equivalent weight of 0.250 inch foam can have over seven times the rigidity of steel and thirteen times the rigidity of zinc. Compared to an equivalent weight of solid plastic, 0.250 inch foam can have twice the rigidity.

Flexural Properties. Structural foam's distribution of material at any cross section makes for excellent flexural properties. Solid resin is situated at the outermost regions of the cross section, where maximum flexural stresses occur. Because of this distribution and excellent resin properties, a high flexural modulus is maintained at elevated temperatures. Again, the wall thickness of the material is an important factor.

Coefficient of Expansion. Contraction and expansion of plastic parts should be considered in any design situation where plastic is secured to metal. Plastic materials will expand much more over a given temperature gradient than a metal part. If they are tied together by fasteners, this expansion will bow the system out of shape. Slotted fastening holes normally will take care of this problem.

Special design problems are encountered with the molding of larger and larger parts. Shrinkage over a 4- to 6-foot part can be excessive, and quick ejection from the mold is critical to keep the part from tearing itself apart.

Impact. There are many tests available to measure the impact strength of a material (Charpy, Izod, tensile impact, Gardner, falling ball, etc.). Falling dart impact tests generally provide data that are especially practical in terms of actual part performance.

The impact strength of a structural foam part usually increases with increasing wall thickness. Temperature also has a large effect on the impact strength of a material. As temperatures decrease, there will be a decrease in impact performance.

Load-bearing Environments. In designing for a load-bearing environment, standard engineering formulas can be used to determine the stresses, strains, and deflections seen in the structural foam part. Some error is introduced into these analyses due to the nonhomogeneous nature of the foam cross section (solid skin and inner cellular core). The effect of this error will be small as long as material supplier recommendations concerning maximum working stresses and moduli are followed. With thermoplastic foams, material characteristics vary with time, temperature, rate of loading, part density, and wall thickness. In general, as temperature increases, physical properties such as tensile strength, flexural strength, and flexural modulus will decrease. As part density increases, physical properties generally will increase.

The time–temperature behavior of a material or the rate of creep must be examined in designing a part for long-time loading. Creep, which is the deformation that occurs under load after an immediate elastic deformation, can be compensated for by working with the material's apparent modulus, which is used to predict performance at given points in time after the initial application of a load. The apparent modulus either is provided by the material supplier or can be obtained from creep curves of a material. The apparent modulus can be determined from the following equation:

$$E_A = \frac{S}{e_1 + e_2}$$

where:

E_A = Apparent modulus
S = Induced stress (psi)
e_1 = Initial strain (in./in.)
e_2 = Strain due to creep of material at given stress and time (in./in.)
$e_1 + e_2$ = Total strain

Extreme Temperature Environments. As mentioned above, the physical properties of a material will vary with temperature. One must also consider whether the material can withstand either momentary or long-term exposure to extreme temperatures without significant drops in performance. Several properties of materials should be examined when one is selecting a material to operate in extreme temperature environments: (1) distortion temperature under load; (2) UL continuous-use temperature rating; (3) cold temperature impact strength; (4) coefficient of thermal expansion.

The DTUL (distortion temperature under load) of a material can be used as a guide for momentary exposure to high temperatures; the UL CUTR (UL continuous-use temperature rating) is more applicable to longer-term exposures.

Chemical Environments. Normally, foamed parts will exhibit lower molded-in stresses and better chemical resistance than parts made in solid resins. However, if parts are heavily stressed in assembly or heavily loaded in the end-use application, their resistance to chemicals may decrease.

It is important to note that environmental and chemical resistance are complex phenomena; the many variables in part design, molding conditions, operational stresses, temperature levels, and cycles are often difficult to predict. There is no dependable substitute for careful testing of prototypes or production parts in typical operating environments.

Finishing

In order to properly identify the benefits of structural foam, it is necessary to take a systems approach. The surface quality of structural foam varies with the material grade, mold and part design, processing conditions, and equipment. With most low-pressure systems, a swirl pattern is evident on the surface. Finishing, the addition of a good-quality coating to the part, offers not only uniformity of color, but environmental, chemical, and abrasion resistance as well.

There are many ways for a structural foam part to be finished. However, the majority of applications are finished with durable paint systems developed specifically for structural foam.

Part Preparation. There are several steps involved in painting a structural foam part. First, sprue or gate removal is accomplished. Patching, when necessary, is done by using techniques similar to that employed on wood products. Properly molded parts can be finished with little surface preparation; often, sanding is recommended to promote paint adhesion, reduce the severity of surface imperfections, or remove contamination. Before structural foam is painted, it often is necessary to let the internal gas pressures equalize to ambient air pressure. This process, known as outgassing, normally takes 24 to 96 hours.

The paint system in combination with the condition of the substrate and application techniques will determine the number of coats. Textured surfaces can be obtained by using either a sprayed texture over a color coat, or a color coat over a molded-in texture. The paint system employed usually depends upon end-user specification. Commonly used systems include urethanes, epoxies, and acrylics. Higher-solids and waterborne paints are becoming more prevalent because of environmental regulations and costs.

Alternative Surface Finishes. Vapor polishing is a finishing technique that offers potential cost reductions for conventional low-pressure structural foam parts. The procedure involves immersing the part in solvent vapors, which dissolve the exposed surface, resulting in a ''high gloss'' appearance. This technique, used with a textured tool in combination with custom-colored resins, may provide an alternative to painting.

Custom-colored resins, with or without texturing, often provide acceptable surfaces for structural foam parts. This is especially true in industrial and material handling applications.

EMI Shielding. Electromagnetic interference (EMI) is undesirable interference from radiation, including such sources as electronic devices, electric motors, or other sources of electromagnetic energy. Most normal interference problems are generated in the radio frequency (RF) range from 10 kHz to 100 GHz.

Plastic enclosures, being extremely good insulators, do not provide any resistance to the flow of EMI radiation, and will not shield electronic assemblies from its effects. Similarly, they will not absorb or reflect radiation that the devices themselves emit.

An additional problem resulting from the use of plastics is static electricity. The metal cabinets formerly used provided effective protection against static buildup because of their natural conductivity; plastics will not do so.

To meet certain requirements, structural foam enclosures must be shielded from the effects of EMI in applications where susceptible devices are used. Effective shielding of most devices is achieved with one or more of the following common shielding methods:

- *Conductive coatings:* Nickel, silver, or copper particles are suspended in an organic vehicle. They are applied by conventional spray techniques.

- *Conductive composites:* Conductive fillers are molded directly into the plastic part.
- *Electroless plating:* Autocatalytic deposition of a metallic coating is carried out, with the metal applied to a sensitized plastic part.
- *Ion plating:* Deposition of metal coating on the part is obtained by use of an ion beam in a vacuum chamber.
- *Vacuum metallizing:* Deposition of metal coating on the part is achieved by evaporating a metal (usually aluminum) in a vacuum chamber.
- *Wire spray:* Deposition of metal coating on the part is obtained by melting wire (usually zinc) by either flame or an electric arc and applying it to the part by spraying.

A wide range of cost and shielding effectiveness values has been obtained with the above-described techniques. It is suggested that the user study carefully the applicability of each method to his or her specific substrate prior to choosing one.

Foam Extrusion

The European plastics industry has for some time extruded high-quality structural foam profiles used in such products as window frames, cladding, panels, balcony boards, and door frames. Because of their hard-skinned surfaces, profiles offer the qualities of medium-density foams (i.e, lower weight, thermal and sound insulation, rigidity, volume reduction, resistance to thermal expansion, etc.), in combination with many of the usual advantages of solid, unexpanded plastics (i.e., mechanical strength, moisture resistance, outdoor weathering, intricate shapes with good dimensional tolerance, easy assembly, wide range of colors and finishes.)

Two techniques for the extrusion of foam are free-foaming and controlled foaming. In free-foaming, thermoplastic material containing blowing agents is extruded through a conventional die. After the die, the profile is expanded almost to its final dimension, and then it is calibrated through a cold shaper (sometimes called a cold die) to achieve its final dimensions.

The controlled-foam (Celuka) extrusion process is rather different from free-foaming. Briefly, the process consists of extruding a special thermoplastic formulation through a die having about the same outer dimensions as those desired in the finished profile (Fig. 19-26). In addition, there is a mandrel in the center of the die so that the extruded profile comes out in the shape of a formed "tube." This extrudate immediately goes into a shaper, which also has the same dimensions as the die. The surface of the extrusion is quickly cooled to form a solid, unexpanded skin; at the same time, expansion of the thermoplastic takes place internally, thereby forming a uniform cellular core. Upon leaving the shaper, the profile goes through the normal downstream equipment for cooling takeup. Standard extrusion lines can be used. Only specially designed dies and shapers

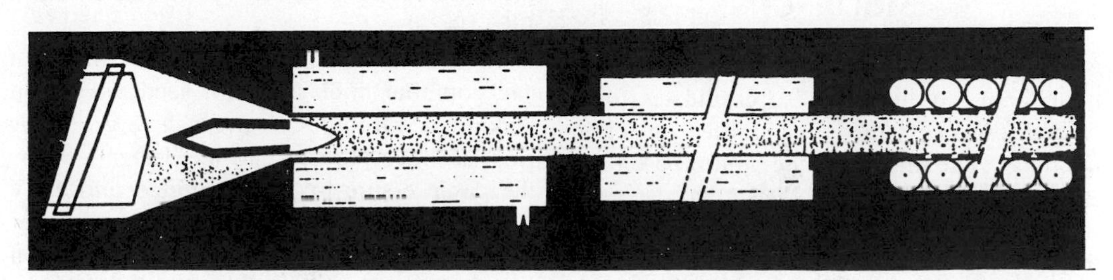

Fig. 19-26. The Celuka process is a controlled-foam extrusion process covered by patents around the world, including U.S. Patents No. 3,764,642 (October 9, 1973) and No. 3,879,505 (April 22, 1975). The Celuka process is a registered trademark of Produits Chimiques Ugine Kuhlmann, 25 boulevard de l''Amiral Bruix, Paris, France.

Table 19-12. Controlled-foam extrusion process: typical extrusion and weight rates, PVC.

PROFILE SIZE	Small	Medium	Large	Large	Large
SHAPE	Simple	Simple	Complex	Complex	Simple
EXAMPLE	Multiple Outlet Mouldings	Boards Siding	Door and Window Frames	Hollow Panels	Boards
CROSS SECTION AREA					
sq. mm.	160–580	1290–2580	1290–6450	3225–7740	10,000–25,000
sq. in.	.25–.9	2.4	2.10	5–12	15–39
LINEAR RATE					
m./min.	7–3	1.8–1.2	1.2–0.9	1.0–0.8	1 – 6
ft./min.	23–10	6–4	4–3	3.5–2.5	3–2
DENSITY					
gr./cc.	0.4	0.4	0.55	0.5	.05
lb./cu. ft.	25	25	34	31	31
WEIGHT RATE					
Kg./hr.	20–52	57–75	52–195	103–177	180–450
lb./hr.	44–115	125–166	114–429	228–390	400–1,000

and a slight downstream alteration are necessary to make the process work.

Almost any thermoplastic can be handled by this process, although most of the current work has concentrated on PVC (e.g., Table 19-12), polystyrene, and the polyolefins. Products made by this technique include: profiles to replace prefinished wood moldings, hollow wall-panel systems, door frames, window frames, siding, and picket fence posts. It also is possible to use this process with a crosshead on the extruder. This enables an expanded thermoplastic to be coated over a metal core, such as copper wire or tubing.

GAS-ASSISTED INJECTION MOLDING*

Many of the types of products suited for structural foam molding are also candidates for a related process called gas-assisted injection molding. With this technique, hollowed-out injection moldings are produced by the controlled injection of an inert gas (nitrogen) into the polymer melt. The gas does not mix with the plastic but instead forms continuous channels through the hotter, less viscous, sections of the stream. The gas maintains pressure throughout the molding cycle. During the cooling phase, the gas ensures positive contact between the polymer and the surface of the mold. There are no gas bubbles escaping to cause swirl marks on the surface of the molded part in this process, which thus avoids a major shortcoming of low-pressure structural foam.

Variations on the gas injection process have been patented. Detroit Plastic Molding (DPM) holds exclusive U.S. rights to inject gas into the machine nozzle with a sprue break via the Friedrich patent (U.S. Patent No. 4,101,617), as shown in Fig. 19-27a. Cinpres, Ltd., UK, holds patents for injection of gas directly into the mold cavity or runner system, as shown in Fig. 19-27b.

A hollow gas-injection-molded component can combine thick and thin sections without sinking or warping. Gas channels (e.g., hollow ribs) are incorporated into the design to allow the gas pressure to be transmitted uniformly. The hollow channels formed by the gas act effectively as box sections, which further increase the stiffness/weight ratio of the molded part. Wall sections can be as thin as 0.007 inch. Sink marks, usually a problem in molded parts with varied cross sections, normally are elimi-

*Adapted from ICI Advanced Materials Technical Newsletter, "Hollow Gas Injection Molding," 1989.

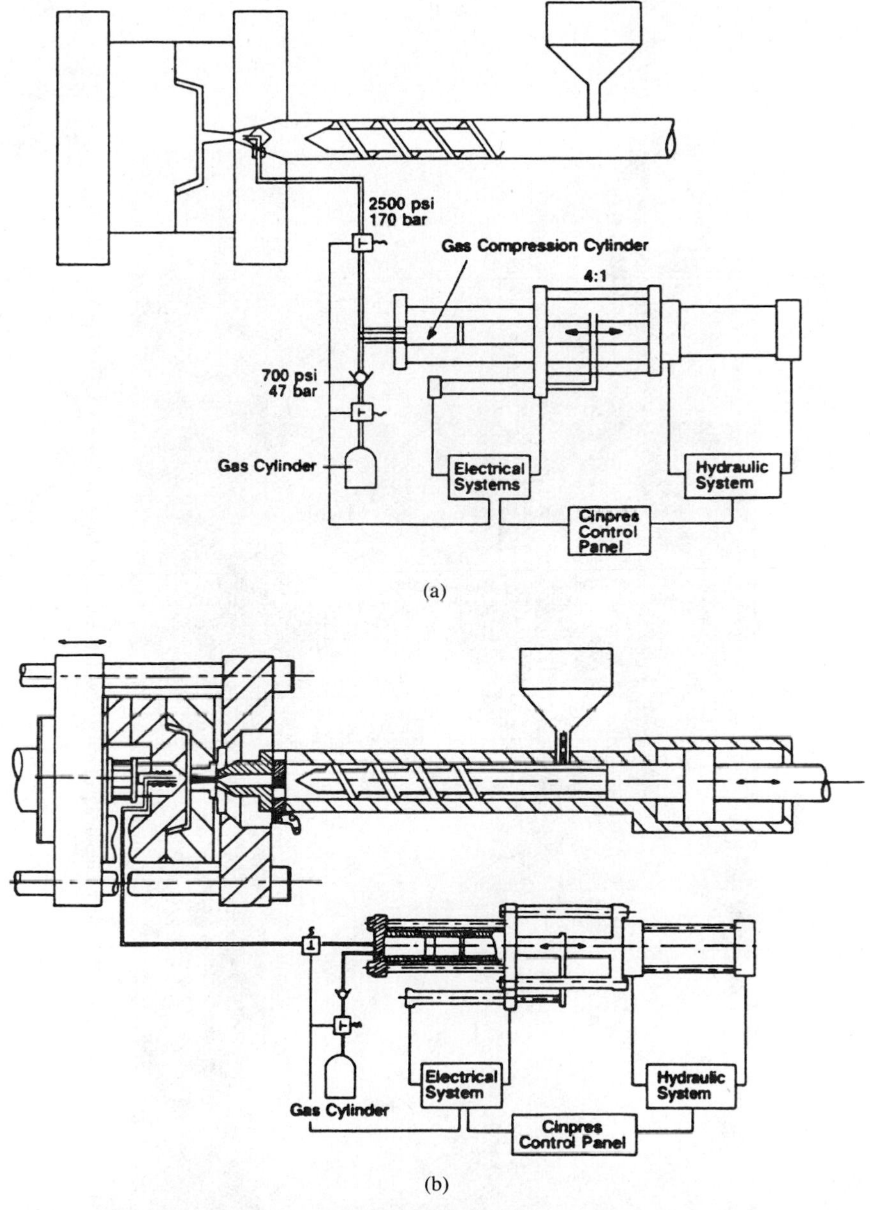

Fig. 19-27. Gas-assisted injection molding processes: (a) into the nozzle and (b) into the mold. (*Courtesy ICI Advanced Materials*)

nated by the internal gas pressure. (See Fig. 19-28.)

There is little or no laminar flow in the polymer during the hollow gas injection molding cooling cycle. This means that molded-in stress and post-molding warpage can be significantly reduced. The pressure within a hollow gas injection molding can be as low as 15% of that needed for an equivalent injection-molded component. As a result, required press clamp forces are much lower. The investment in molding machinery can be significantly reduced, or the effective clamping force of an injection molding machine can be uprated by a factor of 6 to 7.

The hollow gas injection molding cycle usu-

Fig. 19-28. Hollow center in gas-assisted injection molded part reduces weight and cycle time. (*Courtesy Sajar Plastics*)

ally is slightly shorter than that of conventional injection molding and often half that of structural foam molding.

The hollow gas injection molding process has been employed successfully with most grades of thermoplastic polymers. Neat and reinforced amorphous and crystalline polymers have been successfully hollow-gas-injection-molded, including long-fiber-reinforced grades.

General Design Considerations

1. Highlight the thick sections to be cored out by the gas by marking a model or a drawing.
2. Determine the polymer gate location.
3. Determine the location of gas nozzle(s), taking into account the filling pattern of the part.
4. Determine the number of gas nozzle(s) required, based on part geometry, material, and gate location.
5. Because the gas will follow the path of least resistance, the tool design and nozzle location should be such that:
   a. Sections do not fill prematurely.
   b. The gas is not forced to flow against the polymer flow.
   c. Air is not entrapped between wall sections of varying thicknesses.
   d. Overflow wells are provided to uniformly core out end sections.

Application Areas

Applications have included the coring of ribs in rigid panels such as crates, tables, frames, base plates, mounting plates, and covers. The process also has been used for hollowing out entire parts such as tennis racquet handles, wheels, pedals, support brackets, bent tubes, and toilet seats.

POLYSTYRENE FOAMS*

Polystyrene, which is widely used in injection molding and extrusion, is also one of the most versatile raw materials for the manufacture of plastic foams. Polystyrene foams generally are categorized as either extruded (board, billet, or sheet) or expanded bead (loose-fill or molded).

A number of factors determine the suitability of a polymer for producing a commercially acceptable expanded plastic foam, including cost, availability, processibility, and resultant physical properties. Polystyrene is a moderately priced, widely available plastic that is readily processed at relatively low temperature. Expanded polystyrene foam exhibits excellent physical properties for both protective packaging and thermal insulation applications, including low thermal conductivity, good resistance to water, and a high strength-to-weight ratio. These factors combine to make polystyrene foam a material of choice for a wide variety of applications, including:

- Thermal insulation for residential walls, ceilings, and foundations.
- Thermal insulation for commercial roofing, walls, foundations, and floors.
- Molded, fabricated, or loose-fill protective packaging.
- Buoyancy for floating marinas, docks, and rafts.
- Geothermal insulation below roadways, airport runways, and concrete floors.
- Insulating consumer items such as foam cups and picnic coolers.
- Foamed sheet applications, including food service trays, egg cartons, and insulated shipping containers.

Extruded Polystyrene Foam

Extruded polystyrene foam has been commercially manufactured for over 40 years, and is produced in both boardstock and billet form. Extruded polystyrene is made by continuously extruding molten polystyrene containing a blowing agent under elevated temperature and pressure into ambient or vacuum conditions, allowing the mass to expand into a lightweight, closed-cell foam. Insulating boards produced in this fashion have a natural high-density "skin" on uncut surfaces, enhancing such physical properties as water resistance and me-

*By Mark C. Braemer, Senior Development Engineer, The Dow Chemical Company, Granville, OH.

chanical strength. Following edge treatment and trim, boardstock leaves the extruder line ready for use. Common fabrication tools include bandsaws, hot-wire cutters, planers, and routers. Boardstock may be fabricated with a planed surface for special applications, or left with a natural high-density skin as manufactured. Edge treatments for thermal insulation applications include tongue-and-groove, shiplap, and square edges.

Because of the nature of the manufacturing processes, extruded polystyrene foam is available in a wide range of board and billet sizes. Boardstock generally is manufactured up to 4 inches thick, 4 feet wide, and in lengths of 8 feet or more. Billets may vary in size but are typically available up to 12 inches by 24 inches by 9 feet long.

Molded Bead Polystyrene Foam

Molded expanded polystyrene (MEPS) foam is not manufactured by an extrusion process, but rather by expansion and subsequent molding of expandable resin particles or "beads." Unexpanded beads consist of approximately 90% polystyrene, 5 to 8% blowing agent (usually normal pentane or isopentane), and minor amounts of additives such as processing aids and flame retardants. When exposed to sufficient heat, the beads expand from an original density of 40 pcf to a density of 1 pcf or less. Special bead shapes or strands to be used for loose-fill packaging are not processed beyond expansion.

Molding MEPS typically involves three distinct processing steps:

1. Expansion of the raw beads to low density by steam or another heating medium.
2. Aging of the expanded beads or "prepuff" to allow stabilization and drying.
3. Fusion of the aged prepuff in a heated mold to form a shaped part or billet.

During processing, the small (i.e., less than 1 mm diameter) transparent beads grow significantly larger, turning translucent and pure white in color.

Bead Expansion

During this first step of processing, MEPS beads are expanded to the approximate desired part density by operator control of the expander temperature, bead throughput rate, and agitator speed. The expansion operation may be either a continuous or a batch process. As molding does not significantly change the product's bulk density, final part density is controlled and fixed during expansion.

The continuous steam expander remains the most widely used commercial method of bead expansion. In the continuous steam expander, a venturi or screw auger carries the beads at a controlled rate to the bottom of the vessel. As the beads expand, the rotating agitator prevents them from fusing together as the lighter-density, expanded product is forced to the top of the vessel and out to the pneumatic takeaway system, to storage. A typical small-capacity continuous expander is shown in Fig. 19-29. Commercial equipment ranges in capacity from roughly 200 to over 2000 lb/hr, depending on expander size and product density.

Fig. 19-29. Continuous expandable polystyrene pre-expander. Unfoamed beads enter, along with steam, at the bottom of the expander, are agitated by the rotating blade, and exit through the chute at the top of the unit. (*Photos and illustrations courtesy Dow Chemical USA*)

Loose-fill packaging shapes and strands usually are expanded on a wire mesh conveyor belt passing through a steam chamber or in an inclined rotating drum expander. To achieve the very low target densities for loose fill, two or three passes through the expander are necessary.

Batch expansion techniques are used for producing smaller quantities of carefully-controlled-density product. As with continuous equipment, bead density is controlled by the bead residence time and the expander temperature. Batch expanders typically consist of a steam-jacketed expansion chamber with an agitator to avoid lumping. Vacuum often is used in conjunction with indirect heating to assist in achieving desired densities at reasonable output rates (50 to 150 lb/hr).

A typical batch expander is shown in Fig. 19-30.

Molding Shaped Parts

The molding of MEPS is the critical step in determining the mechanical integrity and aesthetic appearance of the molded part or billet. Sufficient heat must be applied to the expanded beads in the closed mold to cause them to soften, re-expand, and fuse together. Overheating during molding may result in product melting and/or collapse.

The preferred heating medium for molding is steam, directly diffused around the beads within the mold. Other techniques involve indirect, conductive heating through the mold wall with steam or hot air as the heat source.

The typical steam chest or jacketed mold, shown in Fig. 19-31, is double-walled, with a perforated inner wall to allow steam entry into the mold cavity. As shown in Fig. 19-32, the mold, utilities, and controls are mounted on a frame to form a press. Depending upon the size of the mold, one or multiple parts may be molded per molding cycle. The molding cycle generally consists of the following steps:

1. Close mold, and fill it with expanded beads.
2. Apply vacuum to exhaust air from the mold.
3. Steam with venting to heat and soften the beads.
4. Steam under pressure to fuse the beads.
5. Cool the mold, and release the part.

The steam chest method of MEPS molding is highly versatile, as molds are interchangeable, and a single press may be used for any

Fig. 19-30. Expander for free-flowing loose-fill packaging material of expandable polystyrene. Unfoamed particles enter at the left and exit at the chute on right.

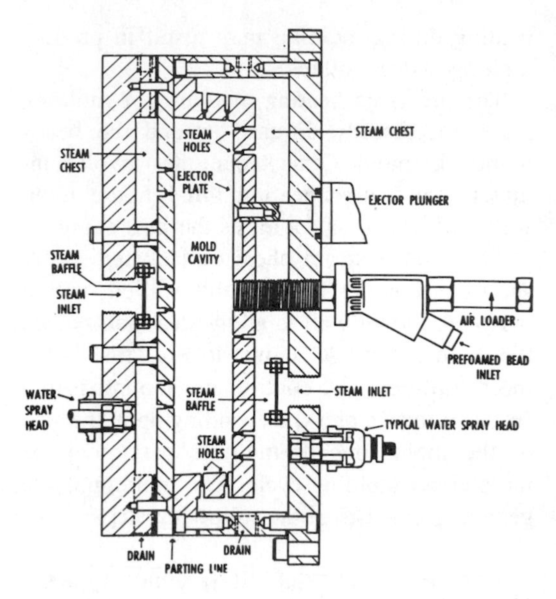

Fig. 19-31. Cross section of typical steam chest mold. Steam is injected into the mold cavity through the steam holes in the mold plate.

number of molding jobs. Molded parts from $\frac{1}{8}$ inch to over 24 inches thick, ranging from 1 to 5 pcf density are produced by this method. Recent advances in steam chest molding technology include vacuum-assisted molding, transfer molding, and feedback (vs. clock timer) control system.

Block Molding

Large billets of MEPS foam are manufactured by a modified steam chest technique in a block mold. After gravity filling of the mold with expanded beads, steam is directly injected into the mold through the lid, bottom, and faces. As with shape molding, a vacuum step may be included to assist in air evacuation prior to steaming. Fusion of the beads throughout the entire block is critical to ensure a good-quality product. Block cooling is accomplished by cooling the outside of the mold, often in conjunction with vacuum-assisted evaporative cooling.

Fig. 19-32. Typical steam chest molding machine for producing small packaging or novelty parts of expandable polystyrene.

MEPS block molds typically measure 2 feet by 4 feet by 16 feet, although sizes vary. Block-molded bead polystyrene billets usually are hot-wire-cut into sheets for use as thermal insulation (beadboard). Other uses for billets are flotation, landfill, and fabrication into smaller shapes.

MEPS foam cups are molded in special conductive heating molds, from very small expanded beads at high density (4 to 5 pcf).

Other types of MEPS molding techniques include continuous board production, continuous block production, and steam probe molding.

Expanded Polystyrene Sheet

Foamed polystyrene sheet and film are manufactured by a tubular extrusion process utilizing expandable polystyrene feedstock, or by the extrusion of standard polystyrene with injection of blowing agents directly into the extruder barrel. In both techniques the extrudate passes through an annular tubing die and is expanded into a tubular "bubble." The foamed polystyrene tube subsequently is slit on both sides to form two foamed sheets, which, after rolling up, may be further processed by vacuum or thermoforming.

Tandem extrusion equipment is usually employed for manufacturing polystyrene foam sheet. In the first extruder, the resin is melted, and the blowing agent is injected under pressure with high precision. Most of the mixing takes place in the first extruder. The second extruder generally has deep flights and turns slowly to pump the material into the die. Another major function of the second extruder is to cool the melt to the point where it is strong enough that rupturing of the cellular structure will not occur when the pressure is reduced at the die lips. It is possible to accomplish the same thing with a single, long L/D extruder, but the tandem setup does it more efficiently.

Until recently, polystyrene foams were blown with CFC-12 or other fully halogenated chlorofluorocarbons. Because such chemicals have been implicated in the destruction of the protective ozone layer in the stratosphere, blowing is now accomplished with HFC-22, a hydrochlorofluorocarbon.

Polystyrene Foam Properties

Extruded polystyrene and expanded bead polystyrene differ not only in manufacturing technique but also in physical properties. Key property data for four ASTM types of thermal insulation boardstock may be seen in Table 19-13. See ASTM C578-87a for a complete listing of material types.

The choice of which type of polystyrene foam to use depends upon the application, as well as such factors as desired strength, water resistance, and economics.

Applications for Extruded Polystyrene Foam

The major application for extruded polystyrene foam is thermal insulation for the construction industry. Key applications include residential sheathing, below-grade foundation insulation, commercial walls, and commercial roof insulation. The high moisture resistance of extruded polystyrene foam allows its use above roofing membranes, insulating the roof and protecting the membrane as well.

A well-established market for extruded polystyrene foam is thermal insulation for low-temperature buildings used as commercial refrigerators and freezers. Walls, floors, and roofs may be insulated with the material.

In the agricultural area, extruded polystyrene is used to insulate poultry, livestock, and produce buildings, protecting the contents from wide temperature variations and reducing energy costs.

Extruded polystyrene foam also is used as the core material for structural sandwich panels for building construction, as well as for recreational vehicles and mobile homes.

Other below-grade applications include the insulation of highways and airport runways in cold climates, protecting them from damage due to frost heave. The water and freeze-thaw resistance of extruded polystyrene foam makes

Table 19-13. Physical property requirements of polystyrene foam thermal insulation (per ASTM C578-87a).

		Extruded PS		*Expanded Bead PS*	
Property	*ASTM method*	*Type IV*	*Type VI*	*Type XI*	*Type I*
Density (pcf) (min)	C303	1.60	1.80	0.70	0.90
Thermal resistance @75°F @ 1" thick (min)	C518	5.00	5.00	3.10	3.60
Compressive strength @10% (psi) (min)	D1621	25.0	40.0	5.0%	10.0
Flexural strength (psi) (min)	C203	50.0	60.0	10.0	25.0
Water vapor permeance (perms) (max)	E96	1.1	1.1	5.0	5.0
Water absorption (% volume) (max)	C272	0.3	0.3	4.0	4.0
Dimensional stability (% change) (max)	D2126	2.0	2.0	2.0	2.0
Oxygen index (% vol) (min)	D2863	24.0	24.0	24.0	24.0

it an excellent and popular candidate for this type of severe exposure.

White and colored extruded polystyrene foams are used in the floral and craft industries, as well as for display boards and hobbyists.

Fabricated extruded polystyrene foam also is used as thermal insulation for low-temperature (i.e., below 165°F) pipelines, above or below ground.

Applications for Polystyrene Sheet

Extruded polystyrene foam sheet in the thickness range of $\frac{1}{16}$ to $\frac{1}{4}$ inch has found significant use in meat and produce trays, egg cartons, and food service plates and containers. The foam sheet is a good thermal insulator, is nonporous and sanitary, and has excellent cushioning characteristics.

Extruded polystyrene billets, boards, and sheet are readily recyclable after use, and may be recycled into nonfood applications such as thermal insulation or nonfoamed polystyrene products.

Applications for Bead Polystyrene Foam

Molded bead polystyrene foam (beadboard) is often used in thermal insulation applications where reasonably high *R*-value at low cost is desired. MEPS finds major application in commercial roofing, below single-ply membranes, and on the exterior walls of commercial buildings.

Other applications include flotation (usually completely encapsulated), insulating building panels (Fig. 19-33), and protective packaging (Fig. 19-34).

The moldability of MEPS makes it an excellent choice for use in making toys, displays and novelties.

One of the largest markets for MEPS is the insulated hot and cold drink cup market. Bead foam cups are shown in Fig. 19-35.

Molded drip trays for refrigerators, freezer evaporators, and shields often are manufactured from MEPS.

Applications for Loose-Fill Polystyrene Foam

Expanded polystyrene foam strands, "peanuts," stars, and saddles are extensively used as a protective loose-fill cushioning medium for fragile items. Their low density, resiliency, cushioning characteristics, and recyclability have made them an ideal choice as a packaging material.

Fig. 19-33. Polystyrene foam building panels provide insulation for home construction. (*Courtesy Dow Chemical USA*)

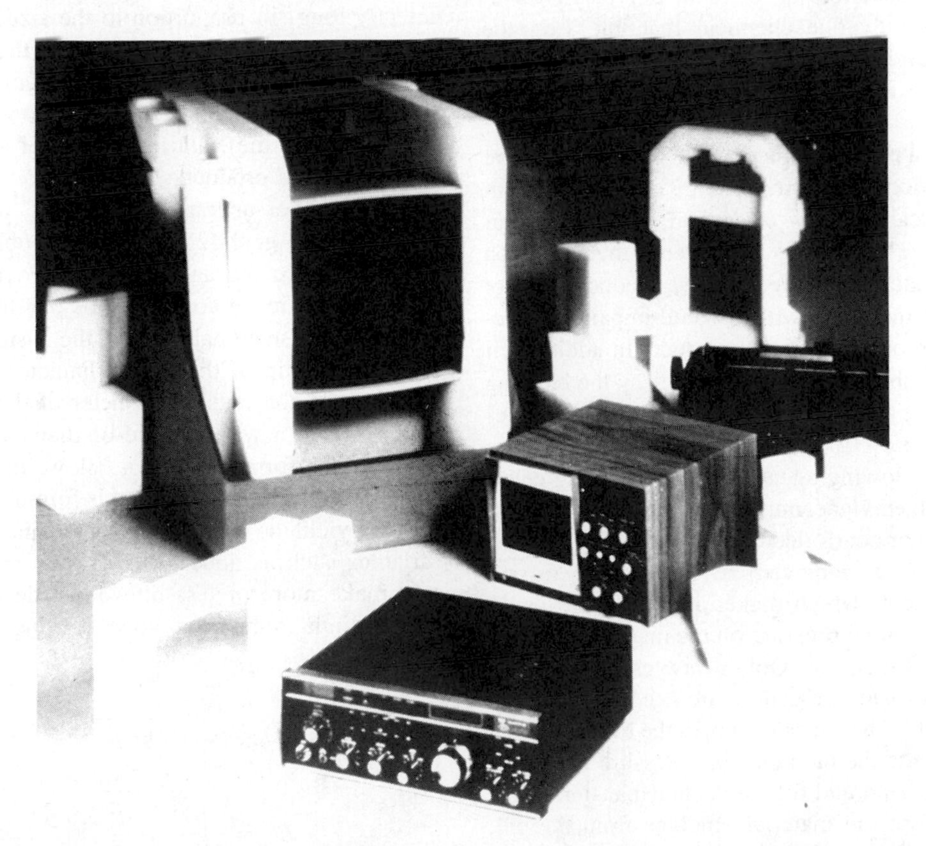

Fig. 19-34. Molded polystyrene foam cushioning protects sensitive electronics items. (*Courtesy Dow Chemical USA*)

Fig. 19-35. Hot and cold drink cups are major outlet for molded beads.

CELLULAR POLYETHYLENE

Production

The production of cellular polyethylene involves only one chemical reaction, i.e., the thermal decomposition of a blowing agent at a specific temperature, which liberates an inert gas.

For a product for electrical service, i.e., wire insulation, the correct choice of blowing agent is critical, in view of several unusual requirements. The blowing agent itself, the gas which it liberates and the residual by-product must not absorb moisture, which would impair the electrical properties of the product. In addition, it is desirable that the residue left by the blowing agent be nonpolar, to avoid losses at high frequencies.

The blowing agent used in producing cellular polyethylene must be made to liberate its gas under controlled conditions. This can most readily be achieved in standard extrusion equipment, which makes it possible to maintain continual pressure on the material up to the point of foaming. Only a few changes in normal extrusion procedure are necessary.

Heat in the extruder causes the blowing agent to liberate the inert gas, but pressure within the barrel, head and die prevents expansion of the ga before the material emerges from the die. When the extruder is performing properly, the finished cellular product has a smooth surface, and the cells are of a uniform size. Temperatures higher than necessary result in nonuniformity of size of cells, roughness of surface, and difficulties during cooling of the product. To establish the optimum operating temperatures for a given machine, the temperatures of barrel and head should be set first at 150°C, and the specific gravity of the product measured. Then the temperature of the barrel is raised, by small increments, until the usual specific gravity of 0.42 is achieved.

For best results, the ratio of length of screw to diameter should be at least 16 : 1. Screws with high compression ratio have been found somewhat more effective than those having a low ratio.

To prevent premature expansion of the gas, sufficient pressure must be maintained in the barrel and head. A head temperature lower than the compound temperature also tends to prevent premature expansion. Pressure in the head can be maintained by making the die land sufficiently long, in proportion to the size of the annular opening between the die and the wire, or core. Premature expansion is recognized by "die plating" (a deposit of the compound on the face of the die), and roughness of surface of the extruded product.

Under proper operating conditions, the expansion (to sp gr 0.42) more than doubles the volume. Hence the annular area between the die and the wire, or core, must be one-half the desired cross-sectional area of the insulation. The relationship of the inside diameter of the die (D_1) and the outside diameter of the insulation (D) on a wire or core of diameter d is given by the formulas shown below. Practical experience has indicated that this formula often does not yield the optimum diameter since other variables such as adhesion and back pressure may make more or less drawdown desirable. The formula remains a good starting point, however.

$$D = \sqrt{2D_1^2 - d^2}$$

$$D_1 = \sqrt{\frac{D^2}{2} + \frac{d^2}{2}}$$

The temperature of the screw should be as low as needed to maintain uniform head pressure and temperature. Some cooling is always recommended for best output rates (at 150°C maximum) and product uniformity.

The conductor should be preheated to 100 to 150°C before it enters the die head, to promote a smooth surface on the insulation. Lower temperatures may reduce the foaming and conductor adhesion. Higher temperatures result in over-foaming at the conductor with possible voiding and flattening of the insulation.

Several seconds of air cooling before first water contact is recommended in order to allow full expansion of the polymer. Hot or cold water can be used after this.

Cellular polyethylene is normally made to have a specific gravity of 0.42. For uses requiring higher density, the increase is best achieved by blending granules of cellular stock with granules of straight polyethylene. This lowers the concentration of the blowing agent. Extruding at lower temperature to limit the degree of expansion cannot be controlled to yield constant results. Extrusion must always be conducted under conditions which will ensure essentially complete liberation of the gas available from the blowing agent.

Properties

Since cellular polyethylene comprises about equal volumes of polyethylene and gas, its properties differ from those of ordinary polyethylene. Comparisons of several properties are given in Table 19-14.

Cellular polyethylene offers the advantage of a much lower dielectric constant (hence lower electrical losses). The composition of polyethylene (dielectric constant 2.3) and an inert gas (dielectric constant 1.0) has a dielectric constant of 1.5. In terms of electrical insulation, this lower dielectric constant permits a reduction in space between inner and outer conductors without changing the characteristic impedance. Consequently the attenuation may be reduced by increasing the size of the inner core without increasing the over-all diameter, or the weight may be reduced by decreasing the over-all diameter without decreasing the size of the inner conductor.

The lower density of the cellular material presents the advantages of lower cost and weight, per unit of volume.

Cellular polyethylene is of closed-cell type. Hence its permeability to moisture, while sev-

Table 19-14. Typical properties of cellular polyethylene insulation and comparison with solid polyethylene.

Property	Solid Polyethylene	Cellular Polyethylene
Tensile strength, psi at 23°C	2180	620
Elongation, %*	600	300
Dielectric strength, v/mil, ASTM D-149-55T short-time at 0.125 in.	550	150
Specific gravity, 23/23°C	0.92	0.42
Mandrel bend at −55°C and 2X†	no failure	no failure
Dissipation factor		
at 1 kc	0.00020	0.00025
10 kc	0.00020	0.00025
50 kc	0.00020	0.00025
Dielectric constant		
at 1kc	2.28	1.51
10 kc	2.28	1.51
50 kc	2.28	1.51

* #14 AWG wire + 93 mils.

† #14 AWG wire + 32 mils; coiled at twice the outside diameter of the insulation.

eral times as large as that of solid polyethylene, is still desirably low.

Applications

The ease of handling of cellular polyethylene in modern wire-coating equipment, and the economies which it provides in size and weight of insulated conductors, indicate its utility in many electrical applications. Major applications are in coaxial cables (CATV, military, and other) and in twin leads.

LOW DENSITY POLYETHYLENE FOAMS

Low density polyethylene foams have experienced high commercial interest because of their unique properties. They possess all the advantages of the base polymer, such as excellent resistance to most chemicals, both organic and inorganic. The low density polyethylene foams are generally closed cell and are classed as "semi-rigid" and "flexible" depending upon their densities and shapes, and can be made to feel as "soft" as low density flexible urethane foam and as "hard" as rigid polystyrene foam.

There are basically two types of low density polyethylene foam: extruded and cross-linked. The extruded foam is produced in a continuous process by first blending molten polyethylene polymer with a foaming or blowing agent (usually a halogenated hydrocarbon gas) under high pressure, conveying this mixture in a temperature controlled screw extruder through a die opening to a continuous conveyer exposed to atmospheric pressure (Fig. 19-36). When the hot viscous liquid-gas solution is exposed to atmospheric pressure the gas expands to form individual cells. Simultaneously the mass is cooled to solidify the molten polyethylene, thus trapping the blowing agent in the intersticial cells. The degree of expansion, the cell size, and cell orientation can be controlled by varying flow rate, heating and cooling temperatures, gas-liquid ratio, and pressure drop through the die opening.

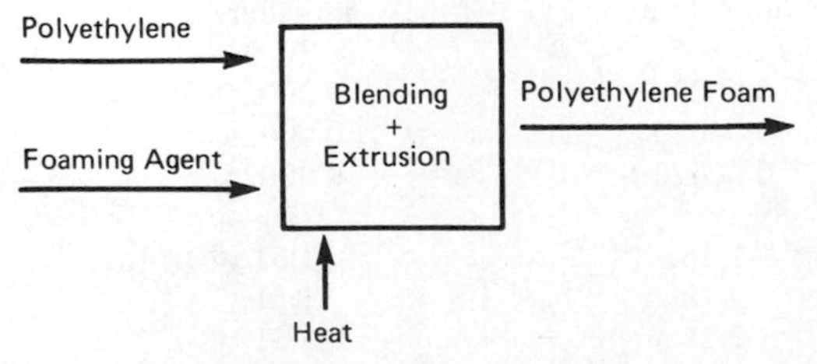

Fig. 19-36. Schematic of conventional extruded polyethylene foam.

In this process, cross sections of up to around 50 in.^{2} are attainable with tolerances acceptable to most end uses. Presently low density polyethylene foam sheet having densities from 2.0 lb/ft^3 and sizes up to $\frac{1}{2}$ in. thick by 72 in. wide can be produced.

"Planks" of densities from 2 to 15 lb/ft^3 and cross sections of 1 × 14 in. to 4 × 12 in. are also attainable in this process. "Rounds" from $\frac{1}{4}$ to 8 in. diameter, ovals, and other nonstandard cross sections are also products which have been successfully produced by the extruded process.

The other basic type of low density polyethylene foam is called "cross-linked" and can be produced by batch and continuous processes and cross-linking accomplished by chemical or irradiation methods.

The chemical cross-linked polyethylene foam is produced in a batch process (Fig. 19-37) and because of production economics is limited to producing "plank" products. Presently, plank sizes up to 48 × 48 × 3 in. are manufactured by this process. Solid polyethylene is blended with a chemical cross-linking agent, usually a solid which decomposes at a certain temperature T_{c1} and with a chemical foaming agent, usually a solid which decomposes at a temperature T_{f1} which is a higher temperature than T_{c1}. The blend is then subjected to a temperature T_{c2} higher than the decomposition temperature (T_{c1}) of the cross-linking agent to initiate cross-linking of the polymer. It is important that this temperature is maintained below the premature foaming. After the cross-linking has proceeded to desired levels the temperature is raised to T_{f2} (higher than T_{f1}) to initiate and maintain foaming. Generally these two unit operations are carried out in the same vessel (the mold) and under pressure to help control the rates of decomposition of the cross-linking and foaming agents. Presently only 2 lb/ft^3 density productions are made in this process.

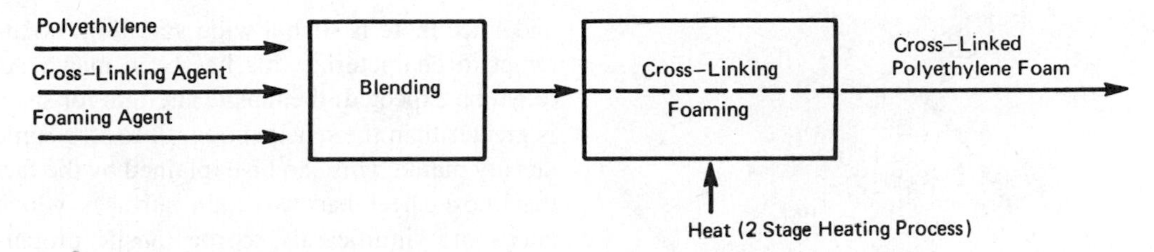

Fig. 19-37. Schematic of chemically cross-linked polyethylene foam process.

The radiation cross-linking process (Fig. 19-38) allows the continuous production of cross-linking polyethylene foam but is limited to producing relatively thin cross sections (up to $\frac{3}{8}$ in.) or sheet products. The maximum width possible with this process is determined by practical considerations in the densified sheet extrusion and in the radiation operations. Generally, widths up to 48 in. are practical.

The chemical, mechanical, and thermal properties of the extruded and the cross-linked low density polyethylene foams are very similar. The largest difference between the two is that the cell size of the cross-linked foam is generally smaller and more uniform than the extruded products. Also the gauge tolerance or thickness dimension is more tight and uniform in cross-linked sheet as compared to extruded sheet. In addition, the cross-linked products possess a softer "feel" than the extruded products.

The properties of cross-linked plank (chemically cross-linked) generally are not as uniform throughout the plank and from plank to plank, whereas the extruded plank possesses more uniformity within a given plank, and from lot to lot.

The mechanical properties (compressive, tensile, dynamic cushioning, etc.) of sheet products are usually much different than those of the plank.

In the interest of generalizing typical properties and characteristics of low density polyethylene foams, the following sections have been written by categorizing sheet versus plank. Where no significant differences exist, the category is "polyethylene foam."

To obtain exact values of properties for specification or design use, the reader is encouraged to contact suppliers of low density polyethylene foams.

Compression Characteristics

Compression Deflection. As mentioned previously, polyethylene foam exhibits properties similar to flexibles and semi-rigid foams as evidenced in compression deflection characteristics. A typical stress-strain relationship for plank is shown in Fig. 19-39. The compression or compressive strengths of polyethylene sheet are a bit lower than the respective density plank products and as one might expect the compressive strength increases as density increases both in the plank and sheet products.

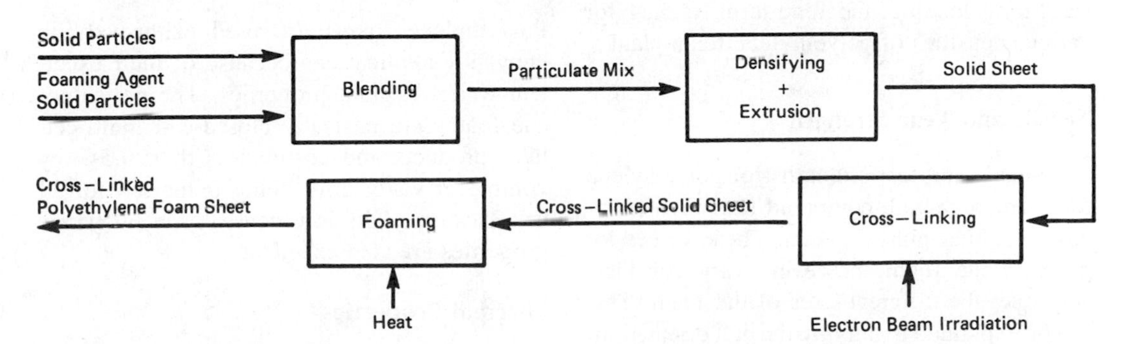

Fig. 19-38. Schematic of radiation crosslinked polyethylene foam process.

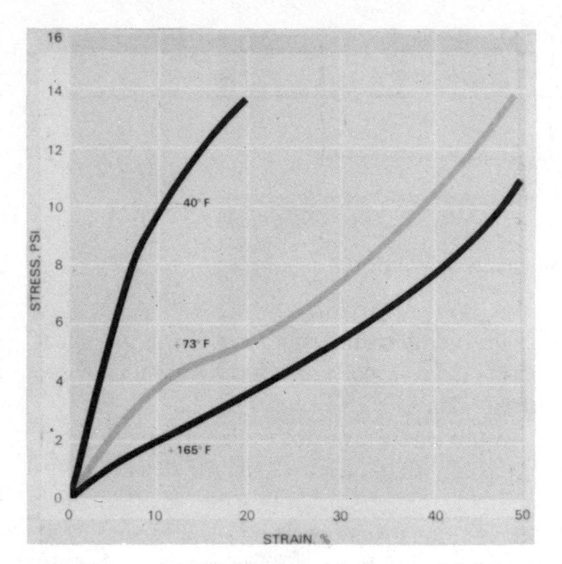

Fig. 19-39. Compressive stress-strain relationship for 2.2 pcf polyethylene foam plank at −40°F, 73°F, and 165°F.

Compressive Creep. When any material is loaded continuously over a period of time, it tends to creep or lose a portion of its original thickness. Generally, lower density products exhibit this more than high density products. The creep characteristics are important because they reflect long-term load carrying ability and affect the cushioning ability of the material. In Fig. 19-40 the compressive creep characteristics of various density polyethylene foam planks are presented, along with some data at elevated temperatures.

Compression Set Recovery. Compression set is the amount of thickness a material fails to recover after compressive creep has been experienced and the load is removed. Figures 20-50 and 20-51 show this property for both short-term loading and long-term loading for various densities of polyethylene foam plank.

Tensile and Tear Strength

The tensile and tear strengths for polyethylene foam are very high compared to similar properties of other plastic foams. These values for polyethylene foam, however, vary considerably over the different axes of the foam. This can be explained mainly by the cell orientation. Also, tensile elongation varies considerably from polyethylene foam to polyethylene foam and since there is such a wide variation, no attempt to characterize this has been made. As might be expected, the tensile strength for sheet is greater than the tensile strength for the same density plank. This can be explained by the fact that most sheet has two skin surfaces which contribute significantly to the tensile properties. This is particularly true of extruded polyethylene foam.

Dynamic Cushioning

One of the largest uses of polyethylene foams is in the packaging area to protect product from damage during handling, shipping, and storage. Polyethylene foams are excellent energy absorbers and because of the wide range of dynamic cushioning properties afforded with different densities, a wide spectrum of packaging demands can be met with polyethylene foams.

Dynamic cushioning curves, which are the plot of peak deceleration values versus static stress, are generated by dropping a series of known weights onto a foam sample from a specified height and measuring the shock transferred to the foam. Each point in the curve represents the loading a product of known weight will apply to the cushion and how much shock the cushion will transmit to the product. Changes in cushioning ability at extremely high or low temperatures have to be considered if the packaging will be subjected to those conditions.

Polyethylene foams provide optimum cushioning characteristics for objects exerting static stresses up to about 20 psi.

Water Properties

Polyethylene foams are used extensively in buoyancy applications because of their excellent water-resistant properties. The polyethylene foams are basically closed-cell multi-cellular products and absorb less than 0.5% by volume of water after being immersed for 24 hr. Because of its low density, the buoyancy properties are also excellent.

Thermal Properties

Thermal Conductivity. The measure of heat transmission through a foam material is deter-

Compressive creep of 2.2 pcf foam plank.

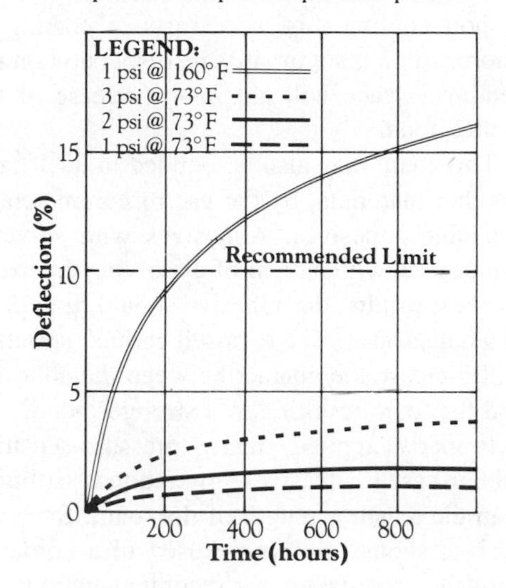

Compressive creep of 6.6 pcf foam plank.

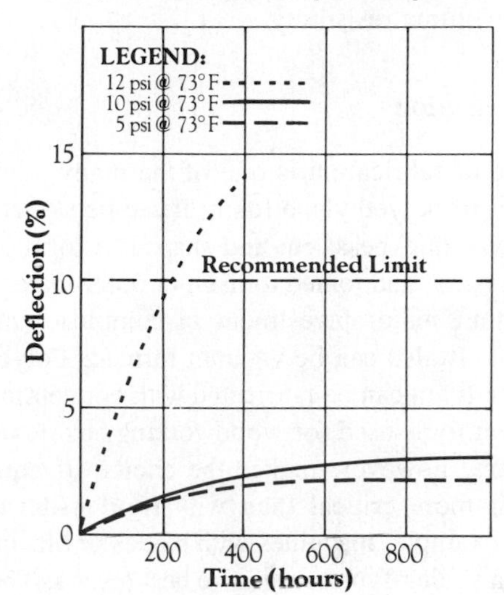

Compressive creep of 9.5 pcf foam plank.

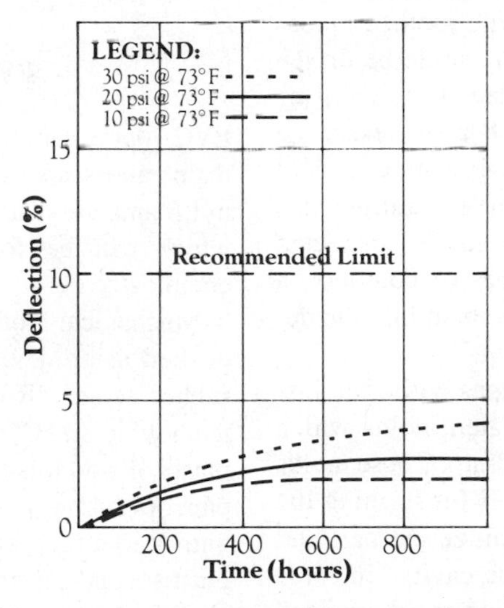

Fig. 19-40. Compressive creep characteristics of polyethylene foams (*Courtesy of Dow Chemical USA.*)

mined by many factors; the base material, the cell size, degree of closed cell structure, and others. This property, called *thermal conductivity*, varies somewhat between the different polyethylene foams; however, within the limits of a respectable insulation at a range of .36 to .57 Btu-in./hr. ft^2 °F.

Thermal Stability. The thermal stability of polyethylene foams is dependent upon the load imposed on the foam, the exposure time, and the temperature involved. Therefore, no specific maximum use temperature can be given for polyethylene foams without qualifying these requirements.

Electrical Properties

The excellent dielectric characteristics of polyethylene are retained when this plastic is expanded to make foam. Expanded polyethylene foam is a candidate for many electrical material uses requiring good properties of dielectric strength, dielectric constant, dissipation factor and volume resistivity.

Fabrication

Ease of fabrication is one of the many advantages of polyethylene foam. It can be skived to precise thickness, cut and shaped to form custom parts, and joined to itself or other material, without major investment in complex equipment. It also can be vacuum formed. Polyethylene foam can be fabricated with conventional power tools used for woodworking. Its flexible nature, however, makes the choice of equipment more critical than with rigid materials. For example, machines with blades or bits having a slicing-type action give best results. These include band saws with scalloped edge blades, slab splitters, and router bits with spiral cutting edges. Circular saws and dados, straight edge router bits, and band saws with toothed blades can be used successfully, but should be driven with as high a peripheral speed as is practical.

The use of electrically heated resistance wires is another method of fabrication. A single wire can suffice, but to make multiple cuts simultaneously, several wires can be connected in parallel. For special shapes or contours, a heavier resistance wire can be bent into the desired form.

Longer runs of special shapes will often justify the use of contoured heated molds which form cavities by melting the foam. These molds are heated from 250 to 350°F for forming the part, and then cooled to stabilize the part and facilitate its removal from the cavity. In some cases, a hot mold and a chilled mold are used to achieve a smooth part with minimum cycle time. Hot molds must be covered with a release agent to prevent adhesion of the polyethylene to the hot metal surface.

Bonding

Expanded polyethylene will adhere to itself by the use of heat alone. Hot air or a plate heated to approximately 350°F can be used to simultaneously heat the surfaces of two sections of foam to be joined. When softening begins, the pieces are quickly joined with moderate pressure and an excellent bond is formed, with only a short cooling time necessary. A coating of fluorocarbon resin or silicone dispersion on the heating surface will aid in the release of the melted foam.

This foam may also be bonded to itself, and to other materials, by the use of commercially available adhesives. Adhesives with solvents can be used without fear of dissolving the foam. For best results, the adhesive should be applied to a cut surface. The exposed cellular structure will increase the contact between the adhesive and the foam resulting in a stronger bond.

Properly applied, many pressure-sensitive rubber-based adhesives give bonds stronger than the tensile strength of the foam. Such adhesives should always be used in accordance with the manufacturer's recommendations for open time, temperature range, and water resistance.

PVC FOAMS

PVC foams can be made from both dispersion resins and suspension resins. Most flexible vinyl foams are made from dispersion plastisols, which can be foamed either physically or chemically.

Mechanical frothing of plastisols is accomplished using devices originally used to make rubber foams. Rotating sets of intermeshing teeth whip air, CO_2, or nitrogen into the liquid plastisol which is then cast onto a web (release paper or fabric, plastic film or sheet, or directly onto carpeting) and cured. The amounts of plastisol and gas are metered to control the density of the foam. Mechanically frothed plastisol has an open, very fine cell structure. Typical end products are air filters, gaskets, and carpet backing.

Plastisols foamed with chemical blowing agents can have either open-cell or closed-cell structures, depending on the particular resin and blowing agent used and on processing conditions. Chemically blown plastisols are used in the manufacture of resilient vinyl flooring, crown cap liners, foam weather-stripping and artificial leather.

Chemically blown plastisols are also used in molding applications. In one process, the plastisol and blowing agent (azodicarbonamide) are held in a mold under pressure while the temperature is raised to decompose the blowing agent and fuse the plastisol. Under the pressure conditions, the gas dissolves in the melt in the form of small bubbles. The part is cooled and removed from the mold; subsequent heating results in expansion of the part to its final density. The process is used to make closed-cell products like floor mats, life preservers and other marine products, and vibration damping equipment.

In another molding process, plastisol and blowing agent are injected into a mold and expanded with heat. Venting is required to remove air displaced by the expanding foam. This type of process is used to mold slipper soles directly onto a fabric upper.

Expanded unplasticized PVC suspension resins generally are produced via extrusion techniques. The Celuka process for making PVC structural foam profiles is described in detail elsewhere in this chapter. Extruded rigid PVC foams are used as wood substitutes in molding and trim products for the building industry. Foam core structures created using coextrusion techniques are being used in the manufacture of PVC pipe and other profiles.

IONOMER FOAMS

Conventional processes for extrusion or foam injection molding of ionomers are used to produce a tough, closed-cell foamed structure. The low melting point, high melt strength of the resin, and compatibility with nucleating agents or fillers combine to provide systems that have utility in athletic products, footwear, and construction.

A low melting/freezing point and melt hot tack can provide product advantages. These same resin properties, however, have to be recognized in fabrication as limitations in certain processing equipment. In extrusion, it is necessary to consider air/water circulation for the inside cooling drum to prevent hot foam from sticking as it exits the die. In injection molding, the lower freezing point will result in a longer cycle time on single station equipment. On the other hand, these same melt characteristics also help to produce a tougher skin in the low density foamed sheet and a better surface finishing the higher density injection molded products. The higher tensile strength and low melt point characteristics of an ionomer give stronger heat seal seams for fabricated sections used in packaging applications.

Products ranging in density from 3 lb/ft^3 are tougher and more solvent-resistant than equal density foams made from polyethylene or polystyrene. The ionomer foamed sheeting may be vacuum formed, laminated, stitched, glued, and modified for flame-retardant requirements.

In foam injection molded parts, the smooth skinned, resilient foam produces a tough, energy-absorbing structure.

Ionomer foams are competitive with polystyrene, polyethylene, and some urethane structures. Where the combination of light weight, toughness, and/or direct paint adhesion are prerequisites, the foamed ionomer systems are more than competitive by reducing the overall cost through lower part weight and less involved finishing.

EPOXY FOAMS

Epoxy formulations can be foamed into open- or closed-cell structures using low boiling liquids, chlorofluorocarbons, or substituted sulfonylyhydrazides as blowing agents. The process involves proper balancing of the formulation and control of the exothermic reaction to yield the desired uniformity of structure. Curing of the resin should ideally take place at the moment of maximum expansion; in this connection, control of temperature is important.

The presence of solvent-diluents (which modify the viscosity of the mass, and may function as auxiliary blowing agents) and of other additives influences the thermal behavior of the mixture, as also do the initial temperature and the size and shape of the batch.

Epoxies can be cured without formation of byproducts and hence with little shrinkage, by use of a large variety of curing agents and hardeners. They are also compatible with many modifiers, creating a wide spectrum of formulations and the opportunity to achieve a desired combination of properties. Cured epoxies are dimensionally stable and filled resins can be used in continuous service at up to 250°F.

Epoxy foams adhere strongly to metals, glass, ceramics, plastics, and concrete. They are resistant to alkalies, most acids, solvents, and moisture.

Foamed epoxies have excellent electrical properties and are used to encapsulate electrical and electronic equipment. They are also used for waterproofing and insulation. Sprayable formulations are used to apply protective linings to tanks and tank cars.

PHENOLIC FOAMS

Phenolic foams generally are made from resol, or thermosetting, phenol formaldehyde resins. While enough water is produced in the exothermic reaction to act as a blowing agent, foaming normally is aided by adding a low-boiling liquid such as a fluorocarbon or n-alkane. Because of their friable and hygroscopic nature, the commercial use of phenolic foams has been limited. One of the largest applications, for example, has been in manufacturing sponges to support floral displays. Recently, there has been a resurgence of the use of phenolics in foams because of increasing demand for high-temperature, fire-resistant polymers in the transportation and other industries. Phenolic foams are characterized by low flammability and low smoke generation.

Use of surfactants to control cell formation and structure has made it possible to produce phenolic foams in which a large percentage of the cells are closed. For additional impact strength, they can be modified with vinyl or elastomeric additives. The stronger materials are suitable for making building board and other insulating products. Phenolic foams also can be sprayed "in situ" and, filled with perlite, made into polymer concrete.

SILICONE FOAMS

Room temperature vulcanizing (RTV) silicone formulations are based upon the chemical reaction between two components, curing agent and base compound, which generate hydrogen gas when mixed together. The gas acts as a foam blowing agent and the reaction also initiates crosslinking to cure the resin.

Elastomeric foams with densities down to 12 pcf are now being produced commercially and development work indicates that it will be possible to reach densities as low as 5 pcf. Such foams would be very competitive with organic foams such as the polyurethanes.

Compared with organic foams, silicones have the advantage that they do not burn, nor do they contain halogens which create toxic smoke in fire conditions. Silicone foams have received UL94 V0 ratings at thicknesses of $\frac{1}{4}$ inch. In a fire, they form a self-extinguishing char on the surface. They are resistant to heat, cold, chemicals, moisture, UV, and ozone, and have excellent electrical properties.

Silicone foams are being used increasingly in automotive applications, particularly under the hood, where their heat resistance can be utilized. The foams also are used as gasketing and insulation materials in building and electronics products. Other applications take advantage of their good adhesion to a variety of substrates, e.g., aluminum foil, fabrics, carpeting, etc.

20
Radiation Processing

Many of the other chapters of this handbook contain references to the use of radiation techniques for curing plastics, crosslinking plastics, and so on. Because of the importance of these techniques in plastics processing, this chapter presents a brief introduction to the subject, covering both high-energy ionizing radiation (e.g., via radioisotopes or particle accelerators) and the so-called non-ionizing systems such as ultraviolet, infrared, induction, dielectric, and microwave systems.

IONIZING RADIATION*

In the mid-1940s, interest in applied radiation chemistry was stimulated by the availability of high activity radioisotopes. High costs, however, prevented any significant commercial application in the plastics industry. Later, as prices started to come down, a number of plastics processing techniques making use of radiation evolved. The uses included: curing thermosets, crosslinking thermoplastics, making graft copolymers, producing wood/plastic and concrete/plastic composites, and curing plastics coatings. Radiation also is used to sterilize medical disposables; because it involves a cold process, it is especially applicable to plastics disposables, as it will not distort them.

In these applications, two types of radiation sources are used: either a radioisotope (e.g., cobalt as Co^{60}) giving gamma radiation or a particle accelerator (e.g., direct electron beam). The intensity of the latter is a function of the design.

Gamma radiation has the advantage, when compared with accelerators, of being directly available with no need for constant supervision; also, it does not involve shutdowns, has greater penetration, and is currently suitable for high-capacity production where relatively long dwell times are allowable. Disadvantages are the high cost of installation, an unwieldy design resulting from thick shielding requirements, and slow transmission of energy.

Electron radiation produced by machine acceleration has the advantage, when compared with radioisotopes, of its capacity for being readily switched on or off. Also, shielding requirements for it are less than with gamma radiation in relation to radiation output. (Accelerator beams are directional; gamma radiations are spherical in geometry.) Electron radiation provides high dose rates and is well suited for on-line applications with high-speed processing. Units are available with hundreds of kilowatts of power at up to 5 million volts.

There are four types of electron accelerator available *industrially*:

1. Dynamitron.
2. Modified Cock-Croft-Walton.
3. Linear accelerators.
4. Single gap.

Crosslinking Thermoplastics for Wire and Cable

An interesting phenomenon occurs when thermoplastics are subjected to high energy radia-

*Portions of this chapter adapted from PLASTEC Report R41, "Applications of Ionizing Radiations in Plastics and Polymer Technology," published by Plastics Technical Evaluation Center, Picatinny Arsenal, Dover, NJ.

tion—the formation of free radicals that then combine to crosslink the molecular chains. What this means in practice is that it is possible to change the characteristics of the base material (e.g., polyethylene) to improve its temperature resistance, boost tensile strength, improve chemical resistance and weatherability, and lower dielectric losses at high temperatures.

This has made radiation especially useful as applied to wire and cable coatings or to polyethylene film used as an insulating wrap. Polyethylene is the major material so far adapted to crosslinked wire coatings, but vinyl is being increasingly used.

In the form of insulating film, crosslinked polyethylene is especially advantageous in terms of high-temperature resistance, resistance to stress cracking, and superior electricals (as compared to nontreated polyethylene). Irradiated polyethylene film tape also shows some interesting and useful shrink characteristics. When the tape is wound upon a conductor, coil, or other object, under moderate tension, and then heated above 110°C (preferably to 135–150°C), the inherent orientation in the tape causes it to shrink in the lengthwise direction, exerting pressure and causing the layers underneath to bond together into a substantially uniform sheath of excellent electrical and physical properties.

It also is possible to produce irradiated polyethylene tubing. In this form, the elastic memory of irradiated polyethylene makes the tubing of value as an insulative protective sheath for soldered terminations, cable markers, capacitor covers, line splices, and insulated connectors.

Figure 20-1 shows several applications for radiation processing.

Crosslinking Thermoplastics for Packaging Film

By the same token, polyethylene film also can be irradiated by electrons to produce a heat-shrinkable packaging wrap. Irradiation cross-

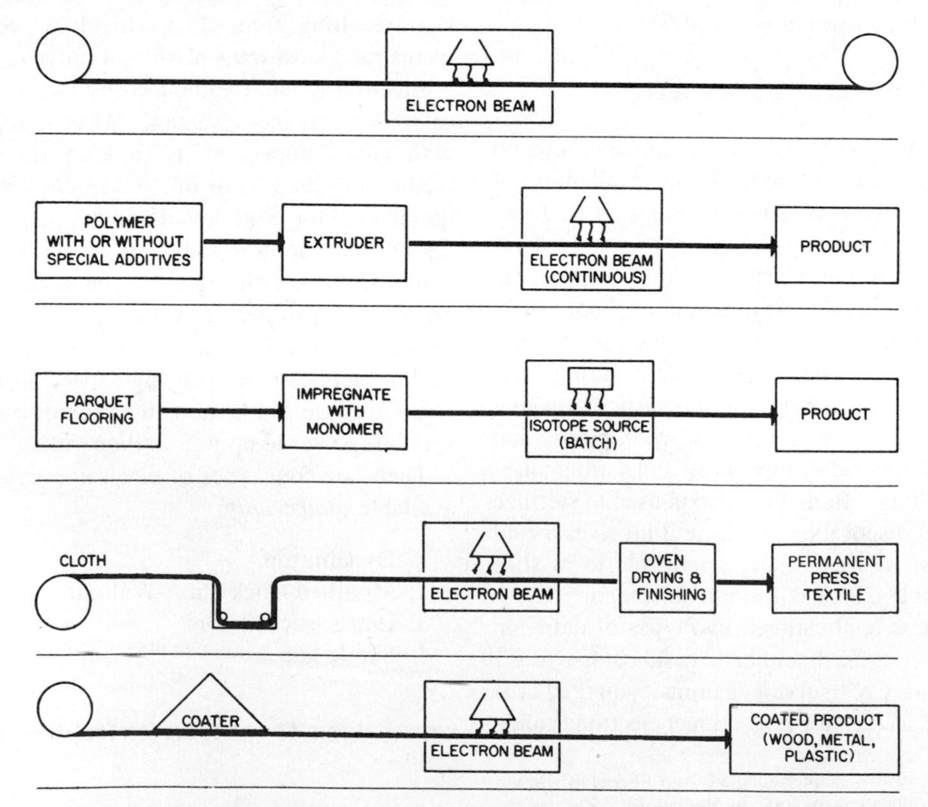

Fig. 20-1. Five applications for radiation processing (top to bottom): crosslinking film; crosslinking wire insulation; making wood–plastic composites by irradiation; finishing textiles (grafting); and curing of coatings (crosslinking). (*Courtesy Plastics Technology Magazine*)

links the polyethylene to the point where the film is stretchable but does not become fluid at its original melting temperature of 105 to 110°C. This means that the film can be placed around the object to be wrapped and heated to 100°C so that it will shrink tightly around the object.

Other types of applications using the elastic memory of the irradiated plastics also have been made commercial. One such use is a series of fasteners, rivet devices, or other mechanical joining system. The fasteners are produced by irradiating polyethylene rods (by either electron or gamma irradiation) and stretching them under heat. The cooled rods, now longer and thinner, are cut to size and inserted into the appropriate openings (i.e., through the two parts to be joined). Upon the application of heat, the inserted rivets expand back to their original shape (i.e., thicker), filling the opening and providing an effective lock.

Crosslinked Foamed Resins

A number of thermoplastic foams crosslinked by radiation are available. Crosslinked styrene foam, for example, offers superior mechanical and chemical properties. On a typical production system for producing such foams, a mixture of polystyrene, a functional monomer, and a blowing agent is irradiated with 4 to 6 Mrads of electron radiation, at temperatures of nondissociation of the foaming agent, in order to crosslink the resin. Then, by further heating, the foaming agent is dissociated, to give the foamed product. Alternatively, the heating for dissociating of the foaming agent is done while the molding is irradiated.

A more popular base for crosslinked foam structures, however, is polyethylene. These crosslinked foams offer higher thermal stability and mechanical strength, better insulation characteristics, and improved energy-absorbing characteristics (as compared to conventional polyethylene foams). Most can be thermoformed, embossed, printed, laminated, or punched by using conventional equipment.

There are a number of techniques for producing crosslinked thermoplastic foams. One technique proceeds as follows: A mixture is made that incorporates: one or more of the monomers selected from the group consisting of acrylamide, methacrylamide, acrylic acid, and methacrylic acid; a foaming agent in solid or liquid state at room temperature; a crosslinking agent having more than two polymerizable double bonds in the molecule; and any monomer, if necessary, that is copolymerizable with the monomers mentioned above. The mixture is polymerized to a thermoplastic body containing the foaming agent homogeneously, and the plastic body then is heated to a temperature above the softening point of the plastic in order to give, simultaneously, the foaming and the crosslinking reaction between the polymer molecules. Radio-induced polymerization or catalytic polymerization followed by irradiation reportedly produces the best results. (See also Chapter 19.)

Graft Copolymers

The use of radiation processing to manufacture graft copolymers is finding increasing use in the production of improved plastic materials and tailored products. These copolymers differ from other copolymers in that a main polymer backbone actually is attached to another homopolymer by grafting on the side chain.

Because of this, the properties of the homopolymers are retained to a great extent, but they can be improved by the addition of the second homopolymer. One example would be grafting a small amount of hydrophilic material to a hydrophobic material—not to make it hydrophilic, but to give it a high tolerance for water (e.g., cellulose acetate grafted to styrene, which has been used for desalinization of water to remove up to 99% of salt and other impurities in one pass). Another potential is combining materials for optimum UV resistance and gas permeability.

Grafting can be effected with powders or polymers in solution; it also can be applied to materials in fiber and film form. Films reportedly are particularly easy to graft by two techniques: (1) pre-irradiation, where the film is irradiated, then passed into a liquid or vapor monomer environment; or (2) mutual irradiation, where the film is irradiated directly in the presence of the monomer.

Radiation-Induced Curing of Coatings

Electron beams are beginning to find wide use in curing coatings on both metal and plastics. With irradiation, the need for elevated temperature cure is eliminated, production rates are increased, and the resulting coating has less porosity.

One of the more important new commercial techniques is the so-called electrocure process, originally developed for curing painted plastics in the automotive industry. In operation, the coatings, generally an acrylic-based material, are cured via electron beams, at a rate reportedly 750 times faster than conventional techniques. With this method, the manufacturers claim superior resistance to chipping and peeling, no need for antipollution equipment (because the uncured resin contains no conventional solvents), and a simplified production line. The electron beam curing system also allows paints to adhere to polypropylene, which is generally difficult with most curing systems. The system further permits greater use of large plastic body parts, which ordinarily would distort under the heats involved in conventional thermal curing methods.

Irradiated Wood–Plastic Composites

Wood–plastic combinations are made by first removing the air and moisture from the wood by means of vacuum and heating techniques. Subsequently, the wood is impregnated with a liquid monomer. The impregnated wood is irradiated with gamma radiation, generally from a Co^{60} source, resulting in polymerization of the monomer and creating a unique type of composite material. Much of the initial work has been done with methyl methacrylate and vinyl acetate monomers, but styrene, ethyl acrylate, butyl acrylate, acrylonitrile, and 2-ethylhexyl acrylate also have been considered.

The major advantage of the combination is that although it retains the grain and appearance of wood, it can be two to three times harder than untreated wood. Other advantages include: higher mar, abrasion, and scratch resistance; improved resistance to warping and swelling; the fact that no finishes are required; and increased compression/static bending/shear strengths.

Irradiated Concrete–Plastic Composites

Similarly, it is possible to impregnate concrete with plastic and polymerize the plastic by radiation to produce an unusual combination material. In application, preformed concrete is dried, evacuated, and monomer-soaked or coated. The monomer is polymerized in situ, either by Co^{60} gamma radiation or by thermal-catalytic initiation (or a combination of both methods). For premix concrete, part of the water is replaced with monomer, or the monomer is added to the fresh concrete mix. Monomers tested include vinyl acetate, acrylonitrile, methyl methacrylate, styrene, styrene-acrylonitrile, polyester-styrene, and epoxy-styrene.

The use of polymer loading improved the compressive strength (four times better than untreated concrete), tensile strength, modulus of elasticity, and modulus of rupture. Freezing and thawing resistance also improved, while water permeability and water absorption were significantly decreased. Corrosion resistance was further improved.

Applications of high potential for concrete-polymer systems include: pipe (irrigation water, sewage, municipal/industrial water), housing (beams, wall/floor panels, load-bearing columns), structures resistant to chemical attack (desalination, corrosive wastes), underwater use, and prestressed pressure vessels. Colors can be incorporated in the material to give it aesthetic and architectural advantages.

"NON-IONIZING" RADIATION*

These techniques go by such names as ultraviolet, infrared, induction (magnetic loss), dielectric, and microwave radiant energies. They generally are not considered radiation systems

*Portions of this chapter adapted from PLASTEC Report R43, "Plastics Fabrication by Ultraviolet, Infrared, Induction, Dielectric, and Microwave Radiation Methods," published by Plastics Technical Evaluation Center, Picatinny Arsenal, Dover, NJ.

in the way that the ionizing radiation methods described above are regarded. And, unlike ionizing radiation, they have been in use by the plastics industry for many years. They are, however, playing an increasingly important role in the plastics industry and are included here simply to put the subject in perspective.

The approximate order of wavelengths and frequencies of the various radiations discussed in this section are as follows:

Radiation	*Wavelength (millimicrons)*	*Frequency, cycle/sec (Hertz)*
Ultraviolet	10^1 to 10^2	$(>10^{16})$ to $(>10^{15})$
Infrared	10^3 to 10^5	$(>10^{14})$ to $(>10^{11})$
Induction	10^{11} to 10^{12}	$(>10^6)$ to $(>10^5)$
Dielectric	10^9 to 10^{10}	$(<10^8)$ to $(>10^5)$
Microwave	10^7 to 10^8	$(<10^{10})$ to $(<10^9)$

Each is covered briefly below, and most of these techniques are discussed in further detail elsewhere in this handbook.

Ultraviolet Radiation

Ultraviolet radiation has found application in the plastics industry in the curing of polyesters and other polymers. Usually, the light sources used in such reactions are mercury lamps. More recently, glass-polyester prepregs have been made available that can cure simply by direct exposure to sunlight.

Work also has been done in systems for rigidizing expandable resin-impregnated structures in space. It is claimed that space ultraviolet (also in conjunction with natural infrared radiation) would be suitable for crosslinking diallyl phthalate polyester resin containing benzoin (as a photosensitizer) and benzoyl peroxide. In operation, a typical structure (e.g., a communications satellite) could consist of a sandwich of flexible polyester film for the skins and a lightweight fabric (impregnated with an ultraviolet-activated polyester resin) as the core. Once in space, the polyester would cure on exposure to UV, stiffening and rigidizing the structure.

Ultraviolet energy also has been used in curing coatings (based on polyester and other polymers) and in curing orthopedic casts based on plastic-impregnated bandages.

Infrared Radiation

Typical plastics fabrication or finishing techniques that use infrared energy include: thermoforming, film extrusion, orientation, embossing, coating, laminating, ink drying (on printed plastics), and fusing.

One of the major outlets for infrared as in the heating of plastic sheets for thermoforming. (See Chapter 13.)

It also has found use in curing filament-wound tubular structures (using glass roving impregnated with such resins as epoxy, polyester, and nylon-phenolic) and in the manufacture of the ''split or slit'' polypropylene yarns. In this latter application, polypropylene film is slit into strips that subsequently are heated for stretching and orientation into yarns and fibers. Take-up machines twist and wind the fibers onto bobbins. A radiant heater tunnel is used in many installations for heating such fibers or yarns.

Paint drying and baking, however, is still the largest single use for infrared energy systems.

Inductive (Magnetic) Energy

The prime use for inductive energy in the plastics industry is in various induction bonding techniques. These are discussed in Chapter 24.

Dielectric Energy

Dielectric (or radio-frequency) energy has found a number of uses in the plastics industry, including the following:

1. Radio-frequency preheating of molding materials (e.g., phenolic) prior to compression or transfer molding. (See Chapter 9.)
2. Radio-frequency curing of reinforced plastics pipe (e.g., glass/epoxy-phenolic systems).
3. Dielectric preheating of cast epoxy resins prior to pouring the resin into the mold

(to decrease gel time and increase production rates).

4. Dielectric heat sealing of thermoplastics film and sheeting. (See Chapter 24.)
5. Radio-frequency curing of cast furniture and other decorative parts based on filled polyester and high-density urethane foam. Electronic heating is used in this instance to shorten the cure.
6. Dielectric heating and drying of inks and coatings on web substrates.
7. Radio-frequency energy for molding expandable polystyrene beads. (See Chapter 19.)
8. Flow molding. A die-cut plastic blank (either solid or expanded vinyl or vinyl-coated substrate) is placed in a mold (usually silicone rubber), and power is applied via a high-frequency RF generator to melt the plastic so that it flows into the mold to the desired shape and with the desired texture.

Microwave Energy

Microwaves can be used to create heat in, and thus accelerate the curing of, such materials as polyesters, epoxies, urethanes, and so on. Among the possibilities that have been investigated are the microwave curing of cast epoxies and polyesters and of urethane elastomers.

Work also has progressed in using microwaves for cast rigid urethane foam furniture parts. Microwaves have proved especially applicable in this instance because most materials used in mold construction for rigid foams are generally poor conductors of heat. Because heating with microwaves is independent of thermal conductivity, the temperatures at which the urethane foam is poured into the molds can be carefully controlled with this type of system.

Other potential uses for microwaves are in the molding of reinforced plastics (it may be possible to improve the resin penetration and viscosity by maintaining fine temperature control through the use of a microwave applicator), in the faster and more accurate curing of molded flexible urethane foam parts (such as automotive seat backs), and in the bonding of untreated polyolefin surfaces. In the latter area, microwave radiation units are used as an energy source to melt a filled polyolefin layer between two untreated polyolefin surfaces. This melted layer provides a structural bond between the surfaces.

21
Material and Parts Handling

With increased demands for higher production rates and material consumption, the volume of material moving through a typical processing plant has grown tremendously. Trying to keep up with this demand by hand-loading machines is often well beyond both physical and practical limits, thus creating a need for automation.

It is also an economic necessity that the processor be able to maximize product quality during every cycle of every machine, so that the amount of scrap being generated is minimized. One condition that plays a role in quality control is the proper preconditioning of moisture-sensitive materials before they enter the processing machine.

These two elements—automation in material flow and preconditioning—offer many secondary benefits (in addition to saving material, labor, and time) for the processor. Prime among these benefits for those processors who have storage facilities (i.e., outdoor silos) and in-plant coloring capabilities is the economic advantage inherent in purchasing uncolored materials in bulk quantities at lower prices, in saving storage space, and in reducing inventory problems.

Preheating and drying materials automatically at the processing machine can be more effective than overdrying, which involves hand loading and unloading of trays and may expose material to ambient moisture on its trip from the oven to the machine. Automatically conveying regrinds back into the processing operation as they are generated offers secondary benefits by eliminating the need for hand-emptying grinder bins and for extra storage space, as well as regrind inventory problems.

The choice of materials handling equipment depends on (1) the type of material—pellets, powders, and so on; (2) the amount of material needed to keep up with processing; (3) the vertical and horizontal distances over which material is moved; and (4) the special functions required of the equipment to meet processing requirements—meter in colorants, proportion regrinds with virgin, and so forth.

By Charles E. Morgan, Director, Corp. Communications, The Conair Group, Franklin, PA.

Vacuum Loaders

Vacuum conveyors are by far the most popular units for moving plastic pellets and powders from shipping containers to processing. Small integral motor/pump units, which convey 250 to 1700 lb/hr of pellets or regrinds require no floor space and operate on standard 115 volt. They are fully automatic and load on demand. (See Fig. 21-1.)

Larger, integral motor hopper loaders are available for loading at rates to 2000 lb/hr. These units are recommended for carrying pellets over short distances, up to 50 feet.

When considerable regrind is generated, virgin–regrind ratio loaders will automatically keep regrind used up, while the processor maintains control over the amount of the regrind mixed with virgin. Such loaders have two feed tubes, one for the virgin material and one that is mounted in the grinder or the regrind source.

Fig. 21-1. Integral motor loader. (*Courtesy AEC, Inc.*)

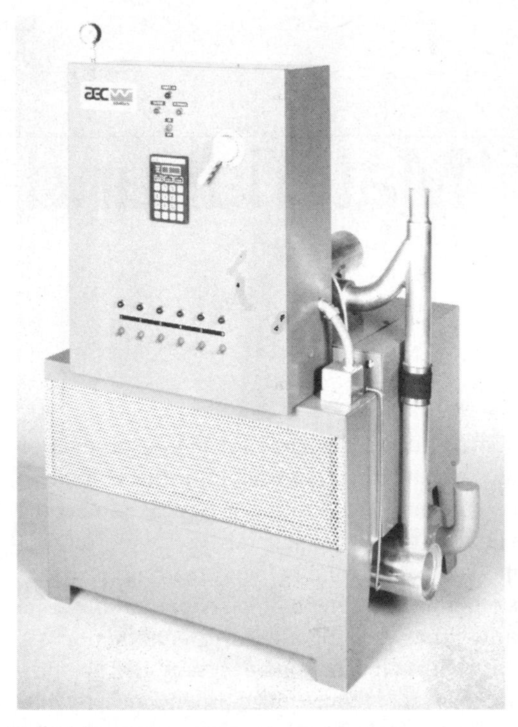

Fig. 21-2. Positive-displacement vacuum pump. (*Courtesy AEC, Inc.*)

Positive-Displacement Vacuum Loaders. Processors who require high-capacity loaders, or must convey materials over long distances, or must move free-flowing powders (such as PVC), will find that positive-displacement vacuum loaders may best satisfy their needs. Vacuum is supplied by a positive-displacement pump driven by a three-phase motor. (See Fig. 21-2.) Such pumps handle up to 15,000 lb/hr and distances up to 600 feet. These larger units frequently are used for unloading bulk railcars to silo storage.

The positive-displacement vacuum unit also can power a multiple machine loading installation, with the cost of the pump being spread over a number of processing machines.

A common vacuum line joins all of the stations with the central vacuum pump. Individual material feed lines are used at each loading station, and where a common material is fed to more than one station, a common material feeding system may be used. For these systems, material valves are used at each station to block the movement of material to all but the loader to be fed.

As the same pump will be required to provide the vacuum for a number of loading stations, a control system is required that works with all of the loaders and the single pump. A level switch that typically switches on the loader when the receiving vessel beneath the loader requires material is used to trigger the pump and create vacuum. Cables connect the pump with all of the loading stations, and the controls that determine the amount of loading time required for each loader may be provided in two different formats, centralized or local.

Centralized controls are the most common type used in the industry today. The system consists of a set of timers that provide the vacuum on/vacuum off times required for the pump starter. The system starts with a signal from any loader that indicates the need for material. Once that particular station is "satisfied," the signal from the central control scans the condition of the level switch of each loader in the system. A valve on the common vacuum line works in sequence with each loader to direct full vacuum power to each station and shut off the vacuum to all of the other loaders.

The more sophisticated versions of central control panels are equipped with separate timers for each loading station so that individual loading conditions may be programmed for each loader and material. Central controls are

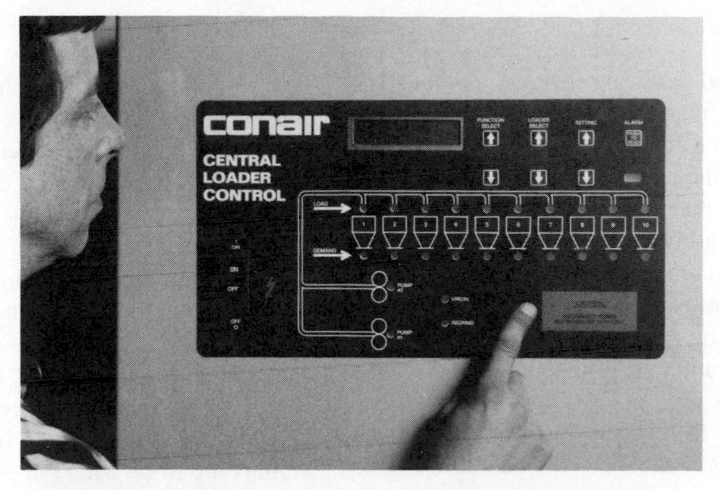

Fig. 21-3. Central loader control panel. (*Courtesy The Conair Group*)

also available with connections for computer integrated manufacturing (CIM) so that all programming and cumulative data may be interchanged with a computer or other form of central control system.

Centralized loader control systems are provided in sizes limited to the number of loaders that may be controlled and are wired to the systems they operate with cables that reach from the central control to each of the loading stations. Popular models provide 5 to 30 stations of control. (See Fig. 21-3.)

Local controls, on the other hand, provide control of the loading station from the station itself, as the control is part of the loader. A single length of cable reaches from the pump to loader number one to loader number two, and so on, in a series configuration, instead of from each loader to the single control as in a centralized system. Usually, a local system provides a greater degree of flexibility, as the number of controls possible in a system is dependent on only the capacity of the pump, not the given number of controls provided with the initial purchase. Stations may be added on freely by simply extending the control cable to the new station and extending the vacuum header. (See Fig. 21-4.)

Fig. 21-4. Vacuum loader utilizing local controls. (*Courtesy The Conair Group*)

Coloring Loaders

Machine-mounted coloring loaders combine virgin materials, regrinds, and one or two colorants or additives (such as slip agents, inhibitors, etc.); mix the materials by tumbling; and drop the mixture into the process hopper. Some coloring loaders allow the use of dry powdered colorants, color concentrates, or liquid colorants through the same unit without major equipment alterations. (See Fig. 21-5.)

The coloring units are self-loading and are mounted directly over processing machines so there is no need to manually handle virgin or regrind materials and risk contamination and waste. When powdered dry pigments are used, they are placed in a canister in a separate color room, and the filled canister then is mounted on the coloring loader.

Color concentrates usually are conveyed automatically into a color storage bin on the coloring loader. The volumetric colorant augers meter in the desired percentage of color for each fixed-volume batch of natural virgin or virgin ratio loaded with regrinds. Of course, the regrinds are already colored, so they are not considered in the color calibration.

Freeze-dried colorants, which are conglomerations of powdered dry colors in a clean and easily handled particle form, can be successfully metered through standard coloring equipment.

Consideration of the amount and quality of regrinds and thorough incorporation of colorants with both virgin and regrind materials are important to successful on-the-machine coloring, regardless of the colorant type. It is important, too, that an accurate, constant volume of the basic resin be delivered into the system if colorants are to be added in preset amounts. The color calibration depends on it.

Toward this end, coloring loaders are equipped with fill sensors within their volumetric vacuum loading hoppers. These sensors are positioned and adjusted so they provide a "full-load" signal to the vacuum loading process. This assures that a repeatable, dependable quantity of virgin and/or regrind materials is loaded each time. An insufficient load stops the automatic coloring process and alerts the operator with a signal.

Fig. 21-5. Automatic coloring loader. (*Courtesy The Conair Group*)

Disposable Machine Hoppers. Although the coloring loaders minimize labor and time costs during color changes, the machine hopper always presented a major down-time factor because it must be wiped clean before the next color was used. To get at the machine hopper, it was usually necessary to move the loader out of the way, which slowed the process further.

A properly sized loading unit will meet material usage demands of any processing machine; so all that is needed is a funnel to direct the material into the machine throat as it falls from the loader. Large supplies of material above the machines, whether colored or not, can be a disadvantage when changes are frequent.

A clear plastic cone held in a steel frame is available, on which either a regular hopper loader or a coloring loader is mounted. (See Fig. 21-6.) The cone is low-cost, so it can either be thrown away or be saved for the color the next time it is run. It usually can be changed

Fig. 21-6. Disposable machine hopper in use wlth dual coloring units. (*Courtesy The Conair Group*)

from floor level, eliminating the need to climb to clean large machine hoppers.

Metering accuracy and the ability to handle a wide range of colorants and material types should be provided in any coloring unit design. It is usually advisable to pretest materials through units before deciding if they will meet processing requirements.

Where both the loading and/or the color metering process need to be more flexible, or the cost for a coloring loader is prohibitive, an at-the-throat metering system may be installed. (See Fig. 21-7.) These simple devices are mounted directly to the throat of the processing machine with a bolt-on adaptor that, in turn, supports the machine hopper. The metering device provides variable-speed feeding of any additive to the machine throat, usually into the flow of a virgin material provided by the standard machine hopper, located above the material device adaptor.

Feeding is accomplished with a volumetric auger, turned by a dc motor. A control system is provided that works with the signal provided from the processing machine. The signal may simply start the device at a given time, as in the case of an injection molding operation, or may provide a voltage that relates to the speed of the process, as in extrusion. For injection models, a timer is included that allows the metering cycle to be regulated in time with the molding cycle.

A variety of augers usually are available to closely match the required additives' flow characteristics and throughput. Some models allow the installation of more than one feeder on each adaptor housing so that a number of materials may be fed at once. Some sophisticated models also are equipped with sampling ports for calibration and a control that allows timed sampling cycles to be triggered for calibration ease. The additive hopper may be equipped with special flow aids to assure movement of the additive into the feeding auger. For free-flowing, high-volume additives, an automatic loader may be provided on the additive hopper.

On-the-Machine Blenders

With the demand for more sophisticated blends increasing, many new systems have been introduced in recent years that will allow custom

Fig. 21-7. At-the-throat metering feeder. (*Courtesy The Conair Group*)

formulation to take place right on the processing machine. (See Fig. 21-8.) These systems usually consist of a number of supply hoppers configured around a blending chamber mounted to the throat of the machine. The supply hoppers are positioned above metering devices that feed all ingredients, at very specific individual rates, to the chamber, allowing them to be agitated and to gravity-flow into the process. The metering method may be the popular volumetric auger feeding, as described above, or feeding via rotating, cavitated plates, or belt feeding. Microprocessor controls are commonplace in this sophisticated blending market, and the best controls offer built-in conveniences for the blender operator. Some control features are:

- Direct readouts of individual metering rates.
- Master blending rate control.
- Overall metering rate changes that do not interrupt the individual material ratios.
- Cumulative material usage reports.
- On-line calibration.

The most sophisticated blender on the market today features gravimetric metering con-

Fig. 21-8. On-the-machine volumetric blender. (*Courtesy The Conair Group*)

trol. In these systems, the weight of the material is used as a basis for measurement, instead of the volume. These systems provide the most precise blends possible because they are not affected by minor material flow irregularities. (See Fig. 21-9.)

Gravimetric, or loss-of-weight, blenders operate with load cell transducers that report the exact weight of the supply hopper to the control system instantly. The control, in turn, meters material to the blend according to the prescribed weight ratio. The most advanced gravimetric systems, like those of the volumetric variety, are equipped with sophisticated controls that simplify programming, calibration, and operation for the operator.

An important plus for the gravimetric blender is ease in generating material usage reports. Because the material is gauged by its weight, total throughputs may be derived easily from the loss-of-weight control system.

Most blending systems are equipped with automatic loading systems for the supply hoppers. These loaders generally operate as conventional loaders do, with the exception of those designed to work with a gravimetric system. In these loss-of-weight systems, the loader-filling procedure is extremely critical because the control system is monitoring the weight of the supply hopper at all times. The loader must be mechanically isolated from the supply hopper, and the dumping sequence of the loader must be connected to the control system of the blender, to prevent unscheduled changes in supply hopper weight as a result of the material dumping sequence.

Fig. 21-9. Gravimetric blender. (*Courtesy The Conair Group*)

Bulk Vacuum Flow Systems

Since the first bulk railcar shipments of plastics were sent to a wire coating company in 1958, thousands of conveying installations have been erected to satisfy bulk material buying, storage, and automatic in-plant distribution.

Properly designed, a bulk conveying system enables processors to realize the lowest price for raw materials and to add savings from efficient space utilization and elimination of hand-trucking material and hand loading, as well as to minimize material waste through spillage, bag breakage, and contamination. Systems range from only one, truck-filled silo to a "tank farm" of many silos sized to hold railcar shipments. (See Fig. 21-10.)

Because nearly all free-flowing plastic materials can be shipped in bulk quantities, most processors using 100,000 lb/month of one type can justify such a program. Even materials usually shipped in smaller containers, such as moisture-sensitive acrylics, can be moved in bulk truck or trainload quantities. They can be delivered dry, kept dry in storage silos by adding a dehumidifying dryer to maintain dry air in the silo void, and brought to the machine in a closed system where final drying and preheating can be accomplished.

Although the bulk trucks mentioned above are self-unloading with their own pressure-type blowers, railcar shipments usually are unloaded by positive-displacement vacuum pumps. Similar equipment then distributes the stored materials within the plant as needed,

Fig. 21-10. Storage silos, equipped to be loaded from trucks. (*Courtesy The Conair Group*)

whether directly to processing or into batch weighing and blending stations.

However, a bulk system can require rates that exceed the practical limits of vacuum-type conveyors. Pressure systems have capacity advantages because of their higher power supply and continuous, rather than cyclic, operation, but they also cost considerably more because of their higher horsepower, inclusion of cyclone separators (or similar air/dust separation), and rotary feeder valves.

A major deterrent to the processor's taking advantage of bulk conveying is the use of a material that is not free-flowing. Especially problematic are the powdered materials, which may include fillers, the recent "granular" resins, or additives that make them cake when compressed in storage and nearly impossible to get moving through a conveying system.

Powdered PVC resins, for example, can fall into the hard-to-handle category. Although many dry blends flow and are conveyed readily from the manufacturer, they then may require in-plant blending where their characteristics are changed when lubricants and additives are put into the blend. Such materials will require special equipment to allow their smooth handling into and through the processing plant.

Plastics Material Drying

Some plastics (e.g., nylon, polycarbonate, etc.) are hygroscopic in nature and sensitive to moisture, which means they need controlled preheating and thorough drying prior to processing to ensure the surface and strength qualities for which they were selected.

Although nonhygroscopic materials may not require dehumidified drying, they will carry surface moisture that should be removed before processing by use of hot air dryers. This preheating also removes one variable from processing, as the materials are maintained at a constant temperature year-round. This can mean improved cycling and increased production.

Dehumidifying Dryers. Dehumidifying dryers are available in two distinct configurations: the twin tower and the rotating cartridge or Somos style. (See Figs. 21-11 and 21-12.)

The twin tower variety consists of two massive containers of desiccant material that provide drying of the circulated air, one at a time. As one tower of desiccant is on-stream, removing moisture from the closed loop of air that flows through the plastic material, the other

Fig. 21-11. Dehumidifying dryer with drying hopper. (*Courtesy The Conair Group*)

Fig. 21-12. Interior of "Somos" dehumidifying dryer with rotating cartridges. (*Courtesy The Conair Group*)

tower is heated with a separate flow of very hot air that purges the trapped moisture from the desiccant. Air flow is controlled between the two towers by a diverter valve system that operates by compressed air.

Air pressure is usually very high in these systems because the towers of desiccant represent a high resistance to the air flow, which must force the dry air through the plastic material as well.

The drying temperature for the plastic material in the twin tower variety is regulated by a temperature control system that is set according to the plastic material's specifications. The desiccant drying, or regeneration, temperature is an independent, locked temperature setting designed to rid the desiccant of absorbed moisture with intense heat. As a result, the temperature of the regenerated tower is considerably higher than the on-stream, or processing, temperature.

When the timed cycle indicates the need to place the regenerated tower back on-stream, the temperature must be reduced. This is usually done by adding a third segment to the timed cycle, which stops the production of regeneration heat during the last portion of the regeneration cycle. During this unheated phase of regeneration, called "cooling," the valve system switches to allow the entrance of ambient air to be blown through the regenerated tower so its temperature may be reduced prior to its being placed back in the process. Some drying systems simply stop the regeneration function and allow the desiccant to cool statically.

In contrast to the twin tower design, the Somos variety of dryer is equipped with three or more cartridges of desiccant material instead of two towers. The number and the size of these cartridges depend on the rated capacity of the dryer. The cartridges actually rotate within the dryer on a timed cycle between: (1) drying the plastic in the drying hopper (process), (2) being purged of moisture by a separate flow of air (regeneration), and (3) lowering the desiccant temperature for proper moisture absorption (cooling).

Indexed into position between Teflon-coated bed plates, the cartridges hold significantly less desiccant material than does the twin tower configurations, as the cartridges are of a hollow core design that exposes a broader cross section of desiccant to the air flow. This decreased quantity of desiccant requires less air flow for operation, as the tanks rotate frequently. This equates to a greater exposure of process air to the desiccant.

In addition, the Somos style of dryer is equipped with three distinct phases of operation instead of the twin tower's two. The cooling cycle contains its own stop on the bed plate rotation sequence, and the cool-down air is drawn off the process air cycle instead of using ambient air. This allows for more efficient use of the desiccant's moisture-absorbing capability.

Many recent advances have been made in the control systems of dehumidifying dryers, which have permitted massive reductions in energy consumption of these heat-generating material preconditioners. Options that sense the actual moisture levels within the process air (and desiccant) have allowed several of the cycles within the dryer to be modified and used conservatively, as needed. For example, the regeneration cycle usually generates 425°F heat for the drying of the off-stream cartridge or tower. With an energy-saving option employing a moisture-sensing device located within the air stream, it can be determined whether the air is already dry and does not require further operation of the regeneration cycle. If so, the regeneration heaters are automatically shut off, along with their blower, saving energy. (See Fig. 21-13.)

Similar systems are designed for the process

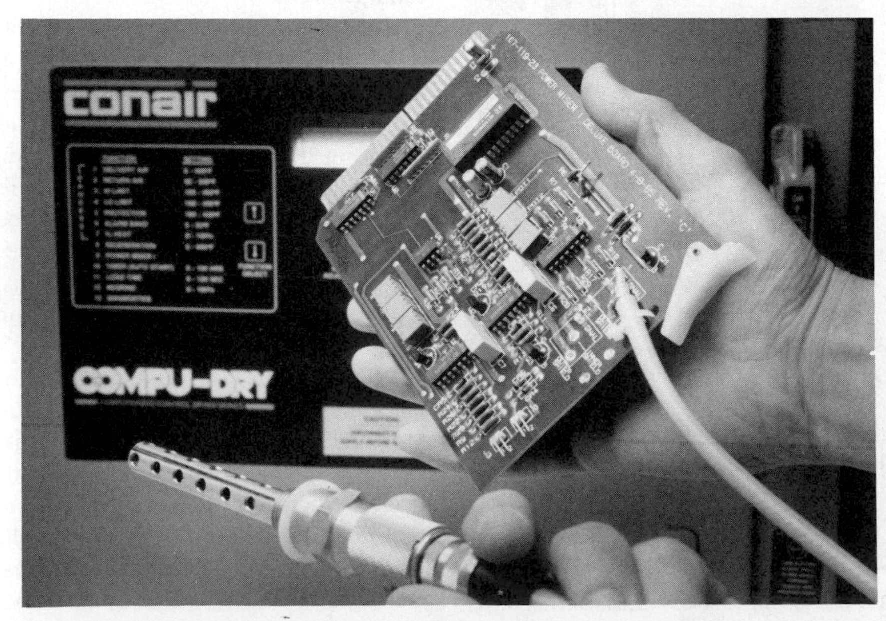

Fig. 21-13. Energy saver microprocessor board and moisture sensor. (*Courtesy The Conair Group*)

heating and blowing cycle, and the result is considerable potential savings in electrical dollars.

Alternate Energy Sources for Dryers. Recent advances have introduced two new possible means of supplying the energy required for the dehumidifying process: natural gas and RF (radio frequency). Both are very new to the world of plastics processing, and they will continue to be developed as viable energy sources for the future.

Natural Gas. The use of natural gas in dehumidifying dryers was predicated on the intention of reducing the cost of energy consumption required for the drying of plastic resin. As explained above, the drying process requires a great deal of electrical heat, and natural gas offers an economical alternative, as the price of natural gas is considerably less than that of electricity.

The principle is relatively simple. The heaters of a standard Somos-style dryer are replaced (along with their associated electrical contactors) with clean, efficient, flameless gas burners. The air-aspirated burners respond to the same microprocessor signals as the conventional dryers, but use cost-saving natural gas for heat instead of electricity. (Electricity still is used for the blowers, controllers, and so on.) Energy costs are projected to be down to 30% of comparable electrical units, which is especially beneficial for higher-throughput models. (See Fig. 21-14.)

Radio Frequency Energy. RF energy has been the object of considerable research in an attempt to find an alternative to traditional moisture purging through the application of external heat. The RF principle, similar to that of microwave technology, uses the natural polar orientation of the moisture within the plastic material to generate heat with an applied RF field.

The future looks bright for RF drying, particularly in the fields of powder drying and recrystallizing PET, currently very difficult with conventional air systems. Salable units may be expected on the market in the early 1990s.

Fig. 21-14. Natural gas dryer. (*Courtesy The Conair Group*)

Low-Throughput, Automated Dehumidifying Dryers. With the need to automate even the smallest of applications, dryer manufacturers have met the challenge with very small, yet totally automatic dryers. (See Fig. 21-15.) Machine-mounted and drying at throughputs down to only 12 pounds an hour, many of these smaller models operate in the same way as the

Fig. 21-15. Low throughput automatic dehumidifying dryer. (*Courtesy The Conair Group*)

large, twin tower configuration. Many, however, minimize the need for duplicate components, and save on machine-mounted space, by using a push–pull drying system for automatic operation.

The push–pull system utilizes only one blower for air flow, along with an air flow valve assembly. The valve and its associated timer allow the single blower to push dehumidified air through the integral drying hopper to dry the plastic in one portion of the cycle, and then to redirect the air, either by blower reversal or valve positioning, to perform regeneration and cooling of the desiccant. The valve assembly usually is electrically operated and works in conjunction with a series of flow valves that circulate the blower's air through the proper phases according to the direction of air flow. These systems usually are very simple and economical to operate, yet offer a degree of automation and control sophistication comparable to the conventional, stand-alone, larger two-piece, hopper/dryer units.

Injection Molding Parts Handling

The move to automate a majority of the previously manual jobs in the typical molding plant has included the distribution and treatment of raw materials, as discussed above. Recently, however, with the need to find cost-cutting procedures that can directly affect a molder's bottom line, robotics are being employed, in order to accomplish the following:

- To speed up cycle times.
- To eliminate the human element from the molding process.
- To reduce damage to fragile parts.
- To perform secondary operations.
- To position molded parts for downstream operations.
- To maintain cavity separation.

From basic sprue and runner separators to complicated multi-task, programmable robots, designed to perform virtually any repeated industrial function, robotics in the world of injection molding is a wide-open field.

Any level of sophistication is available today. The trend of recent years, however, has been to utilize specific robotic packages designed for the industry in which they are to be used. For injection molding, the choices are three:

1. Limited-motion pickers.
2. Mechanically linked robotic arms.
3. Highly accurate, sophisticated, pick and place robots.

Pickers. Simple pickers come in a variety of sizes to cover virtually the entire range of injection presses available today from 15 to 500 tons. (See Fig. 21-16.) Their main purpose is simply to retrieve sprue and runner systems from the face of the mold to allow for parts separation. In these instances, the part itself usually falls to a conveyor below the press, and the scrap is dropped into an open-top granula-

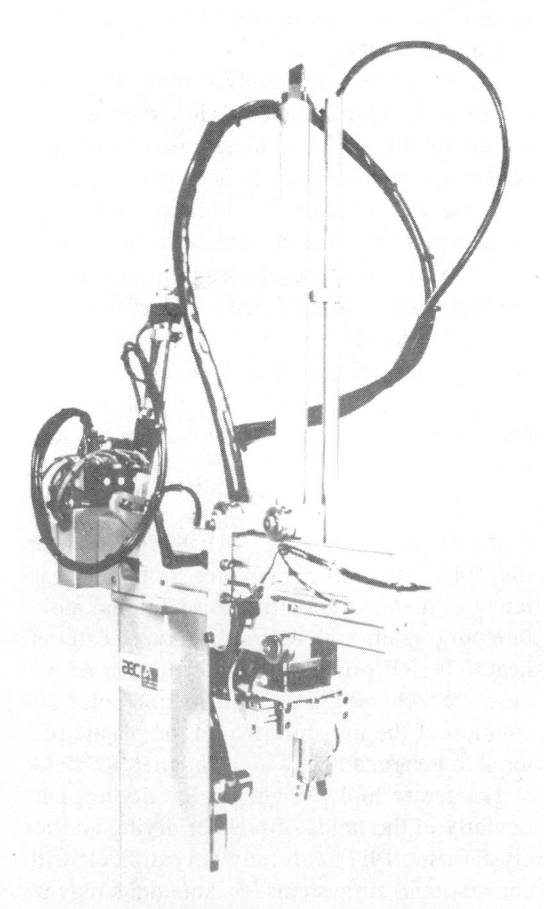

Fig. 21-16. Sprue/part picker. (*Courtesy AEC, Inc.*)

tor. The reprocessed granulate then can be vacuum-conveyed back to the press hopper loader through the use of a ratio loader or proportioning blender, accomplishing closed-loop scrap recycling. An alternative arrangement is to program the picker to grasp the part and allow the scrap components to fall. The scrap then may be taken by conveyor directly into a granulator for recycling, and the part dropped onto a separate conveyor or into a workstation.

Pickers are provided in a number of formats, yet operate very similarly to each other. A central base mounts the picker to the stationary platen of the press. Pneumatically operated cylinders on the picker are equipped with limit switches that are set to provide an extension of the main arm into the open mold as determined by a timing function, electrically coordinated with the press to ensure safety. A mechanical gripping tool, fitted to the end of the robot arm, also operates with compressed air to grasp the sprue at the predetermined, programmed portion of the press's cycle. The arm retracts with the part/sprue, reaching a fully up limit switch and allowing the mold to close for another cycle; the arm then performs a swift motion to move the part/sprue out of the platen area, to the side of the press. This side motion is usually a pivoting of the picker's main arm that allows the arm to extend once again, this time to the side, to drop the part/sprue as required.

Pickers may be equipped with simple vacuum plate tooling or special grippers to remove parts. Part pickers generally incorporate a wrist motion to rotate the parts parallel to the floor for placement on an elevator table, conveyor, or slide for transfer to a work table or storage bin. Tubes or dividers may be employed to ensure cavity separation for quality control purposes.

More sophisticated pickers are equipped with many conveniences that provide a great deal more versatility in setup, operation, applications, serviceability, and length of service. Adjustable pick points are available on some models, allowing field matching to specific press and part requirements while eliminating the need for customized mounting bases. Microprocessor controls are standard on these units, allowing the selection of motion sequences, pick points, and timing sequences to meet almost any application requirement.

Mechanical Robotic Arms. Robotic arm-style robots are driven by the motion of the press itself. Usually mounted on the moving platen, the arm is driven by a linkage to the stationary platen; so as the moving platen moves back and forth, the arm moves in and out of the mold area. The form of the linkage varies by manufacturer, but the most popular type is a spherical cam shaft system. (See Fig. 21-17.)

Spherical cam robots have been used extensively for injection molding because of their smooth operation, durability, and speed. The arm, which is attached to the cam rod, rotates to the center of the mold as it is pulled through a helix track by the opening and closing of the press platens. The twisting motion of the cam rod provides smooth travel for the robotic arm, allowing it to move quickly and precisely into place when the mold is open. Its low profile makes it ideal for low-overhead applications. In addition to traditional sprue and runner removal applications, spherical cam robots have been fitted with customized end-of-arm tooling and options allowing them to be specified for the highly precise compact disc market. Because its major motions are mechanical, precision is assured, and the use of pneumatic cylinders is minimized, eliminating contamination from exhaust air and allowing a clean operation.

Pick and Place Robots. Providing an evolutionary step between automated parts removal and downstream robotic systems are the flexible pick and place robots. With the new capabilities of microprocessor controls, pick and place units are performing more and more sophisticated functions as an adjunct to their initial parts removal role.

These special-purpose robots generally are more rugged than simple pickers and may be equipped with more sophisticated "hands" (end-of-arm tooling), for not only the removal of parts, but the removal and sprues and runners, placement of the parts, placement of the

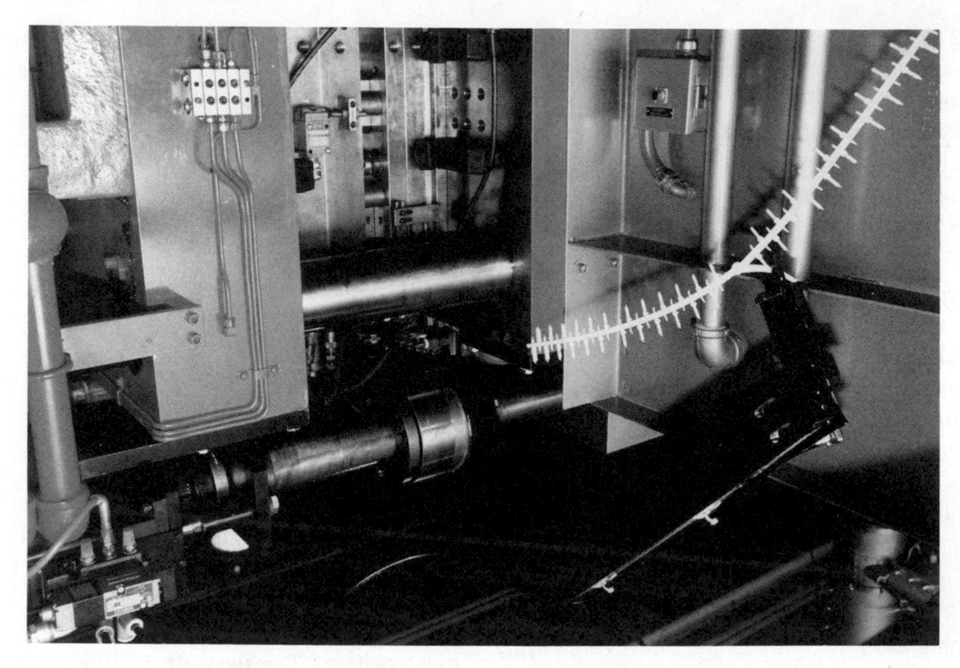

Fig. 21-17. Mechanical robot arm. (*Courtesy The Conair Group*)

scrap, and even the loading of inserts into the mold prior to molding. (See Fig. 21-18.)

Truss and beam construction is standard among the entire range of these robots, consisting of a central beam positioned above and parallel to the stationary platen. The beam provides a central mounting point for all of the robot's hardware, as well as the railed path the robot will follow in its part-removal function. A moving, pneumatic assembly glides along the beam from the center point of the mold to a position outside the mold area where secondary operations will take place. Depending upon the size and reach of the robot, the beam may be totally supported by the stationary platen or be provided with outboard support, connected to the floor frame of the injection press.

Because of the larger size of these beam-type robots and the job they are required to perform, the motions usually are much more precise than those of typical pick and place-style robot systems. The central carriage that holds all of the pneumatic actuators is firmly supported on the track system and allows precision placement within a tenth of an inch. Payloads are dramatically larger, too, permitting removal of parts, gates, sprues, runners, the end-of-arm tooling that is required to separately grip all of them, integrated cutters to separate them, and possibly an adjacent configuration that could grip inserts that will be loaded in between removal cycles. Larger models will permit total loads of up to 100 pounds or more.

Control systems for sophisticated pick and place robots are selected from a vast array of off-the-shelf industrial control systems, as well as a selection of manufacturers' custom-designed systems. In most cases, even the custom-designed systems are configured around conventional industrial controls, such as Allen Bradley, General Electric, Gould, Mitsubishi, and so on.

Of primary concern in the field of robot control are the interface requirements with the press, as well as alternative secondary systems. Safety lockouts must be provided that will absolutely prevent any possible interference with the mold or motion of the platens. In most cases, this safety interlock system is quite simple and positive. Sophisticated alarms and alternative robot functions based on certain conditional parameters require the robot control to

Fig. 21-18. Beam style pick and place robot. (*Courtesy The Conair Group*)

work closely with the press's control system to provide ample signals, parallel to the contingent operations.

As mentioned above, the sophistication achieved by these more versatile part removers allows the introduction of other functions that may become a part of the control system governing the robot. For instance, the insert loading function that may be incorporated into the robotic hand may also require the hand-loading mechanism to be controlled from the robot controller. In addition, downstream functions such as indexing conveyors, labelers, painters, packers, and welders may all be controlled by the robot controller system. Extra memory and easy programming are all worthy criteria now being considered by informed parts-removal-robot specifiers.

Extruded Parts Handling

The wide range of profiles produced by the extrusion process is as diverse as the range of molded products. Profile extrusion, however, has the advantage of possessing generally the same basic shape—a part with a long dimension that is produced continuously. As a result, the involvement of downstream equipment of the part-handling variety plays a much more significant role for the profile extrusion process than for molding processes.

Unlike injection molding, which relies almost solely on the operation of the press itself to produce the basic dimensional part, the profile extrusion process depends a great deal on the performance of downstream equipment. Not only is part length determined by downstream functions, but all critical size specifications are so calculated, including the ultimate processing speed, as determined by extrusion pullers that draw the profile from the die.

It is for this reason that most profile extrusion operations are established as complete systems, designed to produce a single product or a related group of products. The extruder and its related downstream auxiliaries all must work together to produce the desired results, and they generally are specified at the same time for that job.

Viewing the downstream profile extrusion operation as a series of continuous part handlers allows us to segment the equipment available into the following groups:

1. Sizers
2. Cutters
3. Pullers
4. Coilers

As mentioned above, each one of these devices would be specified to work directly with the extruder and die to produce the desired results.

Cooling Tanks and Sizers. An essential element of the retention of the product dimension in the extrusion process is careful product handling in the cooling mode. Cooling tanks provide water immersion for the product as it comes out of the extruder die, and, when equipped, also provide the required final sizing for the product as it cools. (See Fig. 21-19.)

Cooling tanks typically are long water baths fitted with a die at the extruder end that is very similar to the die used for product forming. The die encourages product size maintenance and keeps water within the tank. The water is kept to an appropriate temperature with temperature control equipment that removes the heat from the product at the proper rate.

When sizing is required, the tank is fitted with a special system supplying differential pressures to the inside and the outside of the part. The inside, positive pressure usually is applied at the extruder die, and vacuum pressure is applied within the tank through a vacuum pump system. These systems are known as differential pressure calibrators, and the result is a carefully controlled product spec that retains its shape throughout the cooling process.

Cutters. Essential to the continuous extrusion of plastic parts is the size selection process, which requires a cutting tool to determine, on a continuous basis, the length of the final product.

Cutters may be supplied in the form of blades, circular or reciprocating saws, or, in the case of very large pipe, runaround saws that actually cut through the extrusion with a saw that travels around the circumference of the piece. The preferred method depends on the product to be cut, considering the size, material, geometric shape, and required frequency of cuts or final product length. (See Fig. 21-20.)

The cutter must have an effective means of repeatedly measuring the extrudate and performing a clean cut while the part is moving because the extrusion process is continuous.

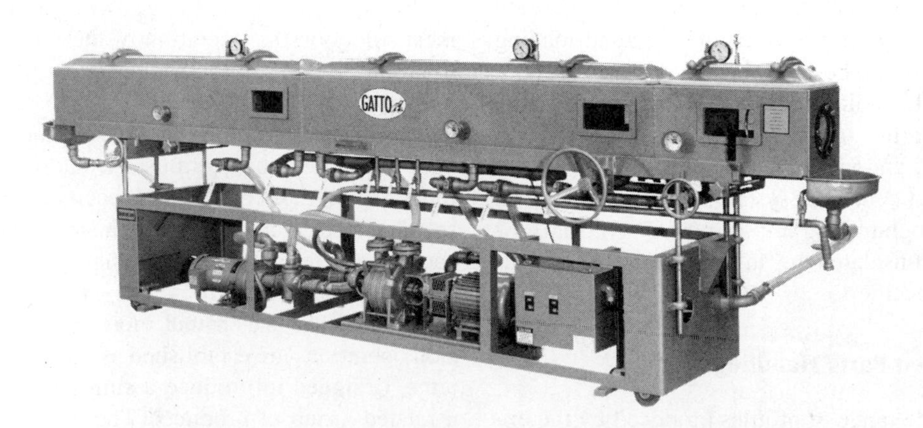

Fig. 21-19. Vacuum sizing tank. (*Courtesy The Conair Group*)

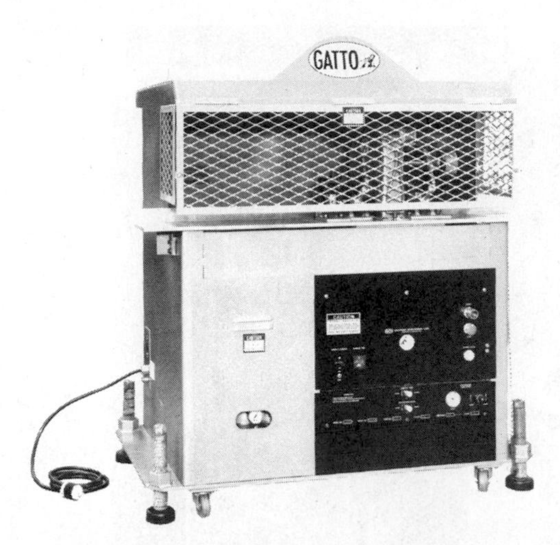

Fig. 21-20. High speed extrusion saw. (*Courtesy The Conair Group*)

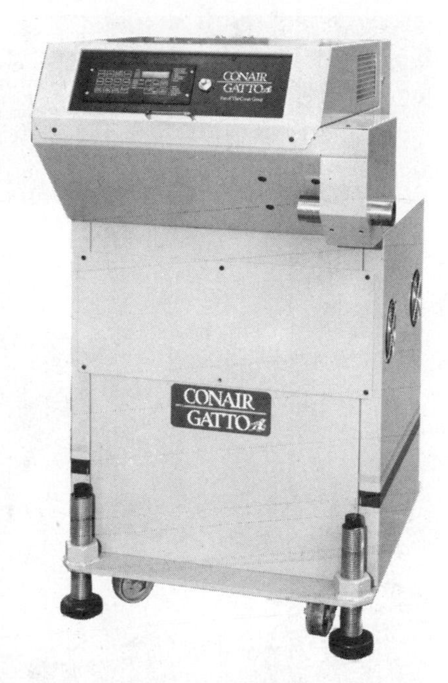

Fig. 21-21. On demand, microprocessor controlled cutter. (*Courtesy The Conair Group*)

The advancements in controls in recent years have allowed the programming of these parameters to be easier and more flexible than ever. Changes may be made to the part length "on-the-fly" so that there are no wasted programming time and no scrapped cuts. An example of an application where this is helpful is a job that uses the same product in different lengths. The two lengths may be programmed in and then implemented at the touch of a button. (See Fig. 21-21.)

Once again, the importance of considering the entire extrusion line is clear when one is selecting a cutter. Its speed and operating range must be adaptable to the job required by the entire line.

Pullers. To control the movement of the extrudate as it emerges from the extruder die in a soft state, a puller is positioned after product cooling to draw the profile along at a safe, regulated speed. (See Fig. 21-22.)

Pullers consist of a pair or a series of belts or rollers that gently grasp the part and provide driven movement in pace with the speed of the extrusion line. As with the cutters, the pace is slaved to the speed of the line and must increase or decrease as production rates are modified. Smooth movement is extremely important because the puller has a key role in determining the final product size in concert with the cooling/sizing equipment. With too fast a puller speed, the part will be drawn below dimensional spec; with too slow a speed,

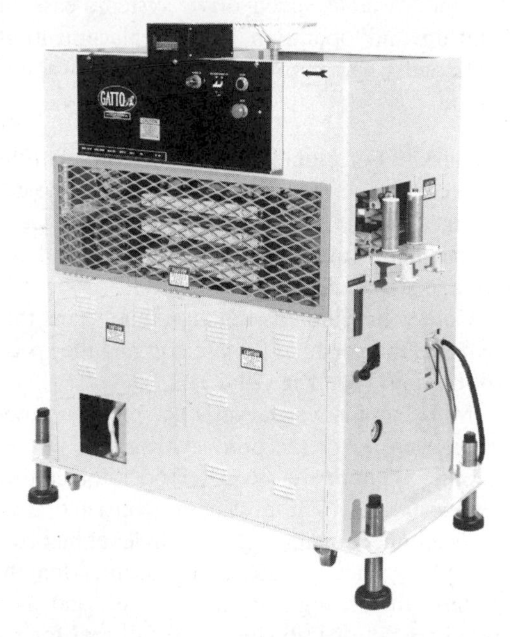

Fig. 21-22. Extrusion puller. (*Courtesy The Conair Group*)

Fig. 21-23. Automatic coiler for extruded tubing. (*Courtesy The Conair Group*)

the part will broaden because of insufficient take-away speed.

Key elements in the design of a quality puller are the variable speed drive system, ease of start-up and operation, easy replacement of belts and wear items, and flexibility of application.

Coilers. For efficient take-away of small-diameter tube, coilers are the preferred choice. Unlike profile extrusions that are usually cut to a specified length, tubing may be coiled for storage, movement, and so on.

Coilers provide this function in a format that matches the needs of the product and the speed of the line. (See Fig. 21-23.)

More sophisticated versions provide automatic transfer of the product from one coil to another without the need for operator cutting and restarting. These systems consist of two reels mounted on an automatic indexer that cuts the tubing once it reaches the specified length, restarts the tubing onto a new reel, and then provides the operator with the full reel for removal and replacement with an empty one.

Granulators and Shredders

Scrap is a fact of life in plastics processing plants. Runners and sprues, reject molded parts, and trim from film extrusion and thermoforming operations, represent significant amounts of material that must be dealt with by the manufacturer. Whether the scrap is to be disposed of or reused, it is usually put through a size reduction process for easier handling. In the case of thermoplastics, scrap is easily reused by reducing its size to the point where it can be fed into processing equipment in the same manner as virgin resin.

Most size reduction is accomplished using granulators, which have rotating knives that cut the plastic into small pieces. Granulators can either be placed next to or under each machine, or in a central location to handle scrap from several machines. Material can be fed into the granulator either manually or by being conveyed directly from the machine (see Fig. 21-24). Depending on the job to be done, units may range from 1 hp to as large as 500 hp. The cutting action takes place between intermesh-

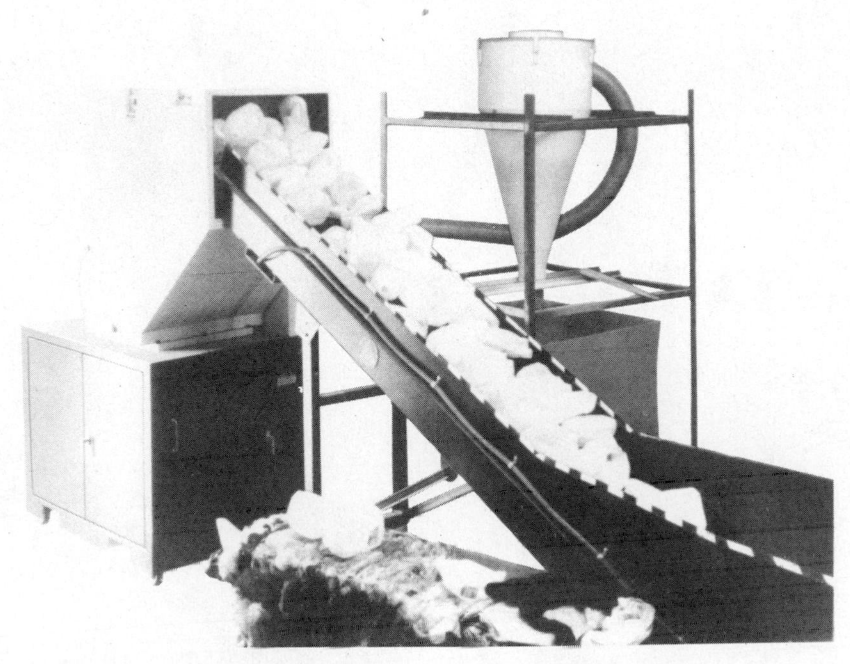

Fig. 21-24. Scrap granulator with conveyor field. (*Courtesy The Sterling Group*)

Fig. 21-25. Shredder over granulator. (*Courtesy AEC, Inc.*)

ing rotating and stationary blades at about 600 rpm. When the pieces get small enough, they fall through a screen at the bottom of the chamber.

For large parts like blown containers, pipe and profiles, and rolls of film, two-stage size reduction may be the best procedure. In these cases, a shredder can be used to chop the material into pieces that then go into the granulator (see Fig. 21-25). Shredders operate at lower speeds and with less power consumption and they reduce shock and load requirements on the granulator. Large machines used for grinding chunks and other bit parts are often equipped with flywheels to reduce the stresses of shock loading.

Size reduction equipment is also used extensively in recycling operations to return post-industrial and post-consumer plastic scrap into remanufacturing operations.

22 Compounding

Up to this point, we have referred to plastics in their basic resin form. Obviously, in actual practice, many resins involve the addition of various chemicals and additives to make them tougher (impact modifiers), rigid (reinforcing fibers), more flexible (plasticizers), more resistant to ultraviolet (stabilizers), colorful (colorants), and so on. Most of these additives are added to the resin by various compounding techniques, either at the resin supplier level or by specialists in compounding or, as has been the more recent case, by plastics processors themselves. This chapter presents a brief review of compounding.

COMPOUNDING OVERVIEW

Few resins are useful in their natural form; so they are mixed with other materials to improve and enhance their properties and thereby make them more useful for a variety of applications. The process by which ingredients are intimately melt-mixed together into as nearly a homogeneous mass as is possible is known as *compounding*. Because of the nature of both the resin and the other ingredients, compounding requires a wide range of mixes—dry powders, slurries, pastes, doughy consistencies, for example, and a corresponding range of mixing operations.

The task of mixing becomes one of changing the original distribution of two or more nonrandom or segregated masses, so that an acceptable probability distribution of one mass throughout the other(s) is achieved. Thus the problem becomes one of deforming or redistributing masses in order to achieve this desired probability distribution, in the absence of diffusion or other random molecular motions. A problem arises if the ultimate particles are not independent of one another, but instead exert interparticulate forces leading to particle agglomerations. Then, external force must be exerted on such agglomerates to allow them to mix. Dealing with these forces, or stresses, is at the heart of dispersion processes.

In mixing and dispersing processes involving thermoplastic melts, the thermoplastic material is regarded as essentially a fluid subjected only to laminar flow, capable of being deformed. Thus, the problem of mixing in thermoplastics is that of subjecting such materials to laminar shear deformations in such a manner that an initially nonrandom distribution of ingredients approaches some arbitrary scale of randomness. Mixing usually is complicated by the fact that ingredients do exhibit interparticulate forces, so that stresses accompanying the deformation must be considered, as well as the deformation process itself.

Mixing Theory

The general theory of mixing usually considers a nonrandom or segregated mass of two components, and their deformation by a laminar or shearing deformation process. The object of shearing is to mix the mass in such a way that samples taken from the mass exhibit minimal variations, ultimately tending to zero. The shearing process is generalized; that is, it is not

Reviewed and updated by Dr. Paul Fenelon, Comalloy International Corp., Brentwood, TN.

limited to any particular kind of shearing action or mixing device, but applies to all such devices.

Three basic principles are given by the general theory of mixing:

1. The interfacial area between different components must be greatly increased in the mixing process because it results in a decrease in average striation thickness.
2. Elements of the interface must be distributed uniformly throughout the mass being mixed. Uniformity can be defined by the intensity of segregation, or by the average deviation of particles from a mean concentration.
3. The mixture components must be distributed so that for any unit of volume, the ratio of components within the unit is the same as that of the whole system.

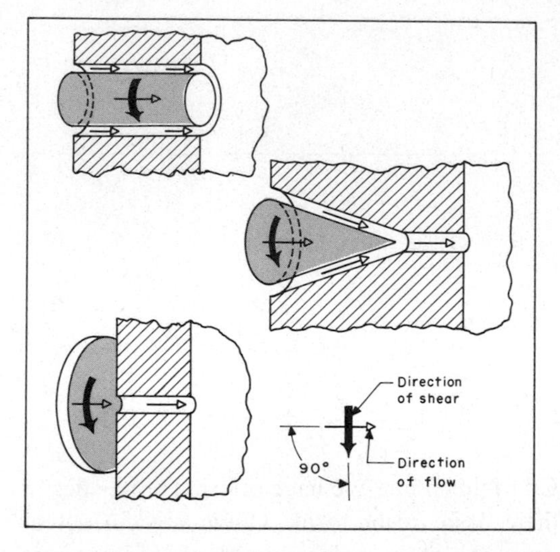

Fig. 22-1. Three ideal mixing devices. Shear is applied at right angles to the direction of flow, the shear rate is varied by varying the speed of the rotating element, and the particles follow a spiral path to the exit. The shear rate can be regulated by controlling the gap size or the rotational speed. The gap size is varied by axially displacing the rotating element.

The first and second principles obviously require, for any particular mixing process or device, that the dynamics of fluid or particulate flow within the device ideally obey the three requirements of efficient mixing, namely, that the shear be at right angles to the direction of flow, that the shear rate be controlled by varying the gap between the stationary and rotating elements, or by varying rotational speed, or both, and that the particles each experience the same shear history (Fig. 22-1). They also suggest that the optimum initial orientation of the interface between components should be normal to the flow lines or streamlines within the device, and that all the streamlines of flow should lead to a region of maximum shear. The third principle implies that the scale of sampling must be large in comparison to the ultimate particle size (Fig. 22-2).

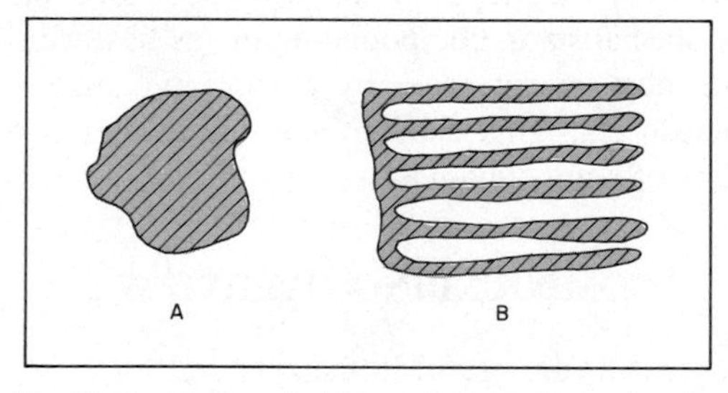

Fig. 22-2. The interfacial area in (A) is increased to that of (B) to pass through each volume element of the mixture. The material in each volume element is in the same ratio as in the mixture.

Limitations of Mixing Theory

Being generalized, the mixing theory suffers from limitations that must be recognized when it is applied to actual plastic materials and present mixing devices:

1. It is limited to purely viscous (Newtonian) materials, whereas actual plastic materials are viscoelastic, plastic, or otherwise non-Newtonian. However, the qualitative concepts that hold for simple viscous bodies are of great value in furthering an understanding of the mixing of real materials.
2. It is assumed that there are no van der Waals or other forces between particles.
3. The initial orientation of components is assumed to be known although this is almost never true in practice.
4. The mixing process is assumed to be isothermal, yet the heat generated in mixing most plastic materials is sufficiently high

that the energy dissipation during mixing is considerable.

The major gap between theory and practice lies in our inability to fully describe the rheological behavior of complex plastic melts and the dispersion of powdered materials, particularly because the flow within actual mixing devices is very complex. Still, the application of qualitative and semiquantitative concepts has thrown considerable light on the process of mixing.

Mixing Evaluation

A mixture is described in terms of the statistical deviation of a suitable number of samples from a mean, the sample sizes being dependent upon some length, volume, or area characteristics of the mixture or its properties. For example, if color is imparted by adding pigment, and homogeneity is measured by visual impression, the characteristic length is the resolving power of the eye, say, 0.001 inch. A completely mixed compound exhibits pigment streaks no greater than 0.001 inch. On the other hand, the color value for any given series of samples would appear uniform to a spectrophotometer that integrates over a 1-inch-diameter circle even if the streaks were 0.1 inch thick. Likewise, the intensity of color difference between streaks would affect the resolving power of the eye—or the spectrophotometer—and, hence, the characteristic length.

The means for measuring the degree of mixing are varied. In commercial practice, inspection for color homogeneity or color comparison for specks, streaks, or spots of unmixed filler or resin is visual. Frequently, changes in physical properties, such as tensile strength, modulus, or density, are used to evaluate the degree of mixing. Because these properties may be affected during polymer degradation and by thermal effects on the mixture, property changes do not strictly evaluate the mixing as such.

Some of the more fundamental measures evaluate changes in rheological properties, chemical reactions, and electrical conductivity. These evaluations most nearly approach the ideal measurement of the basic criteria of mixing.

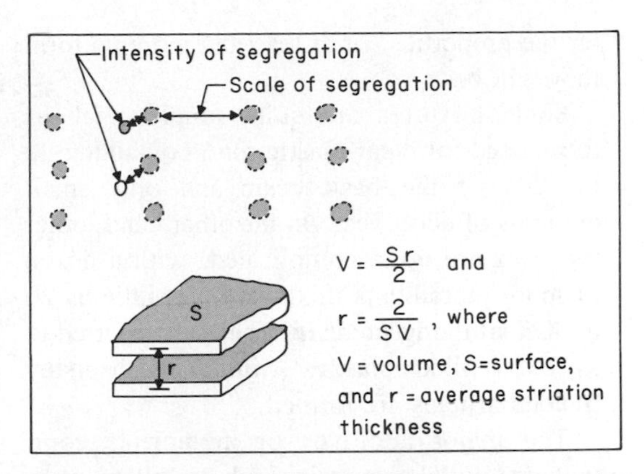

Fig. 22-3. The average striation thickness (r) is computed from the product of the interfacial surface area between the components and the total volume of the mixture. The average striation thickness, in effect, measures the scale of segregation, which is the average distance between clumps of the same component. The intensity of segregation is the average deviation of particles from the mean concentration.

The only direct measure of a basic criterion of mixing is striation thickness, though this property is measurable at present only under certain conditions (Fig. 22-3).

Because of the many variables of mixing and the scantiness of criteria for measuring mixing effectiveness, the current theory of mixing of heavy viscous materials is less than satisfactory. Often practical and extensive tests are the only means of determining the suitability of a particular mixing machine for a specified plastics mixture. Most compounding probably can be classified as an art.

The factors that contribute most to the ultimate properties of plastic compounds are many, depending on the type of basic resin, type of compound, types and amounts of other ingredients, and desirable degree of homogeneity. The basic resin may vary considerably chemically, that is, as to molecular weight and configuration and particle size. The compound may range from an adhesive or coating solution to a molding powder. In addition, there are a large number of secondary plastics materials that may be incorporated with the plastics resin in the mixture, such as plasticizers or solvents, solids such as fillers or pigments, and others such as stabilizers, dyestuffs, and lubricants. Finally, the uniformity of the mixture contributes greatly to the properties of a plastic; for usually the more homogeneous the mixture is, the better the properties, or at least the more uniform they will be.

Some mixtures are quite simple, such as those used for clear plastic film, containing 90 to 95% of the basic resin and only small amounts of additives. On the other hand, other mixtures are quite complicated, with a dozen or more ingredients mixed with as little as 20 to 30% of the basic resin. These are referred to as the "filled plastics" from which many molded articles are formed.

The major additives or ingredients compounded with basic resins include: fillers, plasticizers, colorants, heat stabilizers, antioxidants, ultraviolet light absorbers, antistatic agents, flame retardants, blowing agents, and lubricants.

Fillers

Fillers are classified as inorganic, organic, mineral, natural, and synthetic. They are more commonly used with the thermosetting resins such as the phenolics, ureas, and melamines although they fill some thermoplastics as well. Large amounts of fillers are commonly referred to as extenders, because they decrease the cost of a plastic. Because the properties of an extended plastic often suffer, the use of fillers is limited to less critical applications. Fillers normally endow plastics with specific mechanical, physical, and electrical properties such as strength characteristics, hardness, density, and dielectric strength. Fillers also tend to increase the resistance of a plastic to one or more of the various service conditions. A filler usually comprises between 10 and 50% of the weight of the mix. Some of the common fillers and their functions are listed in Table 22-1.

One of the most widely used fillers, especially for the thermosetting plastics, is wood flour, which consists of a finely ground powder of one or more of the hardwoods or sometimes of nut shells. Wood flours are readily available, cheap, lightweight, strong (because of their fibrous nature), and easy to compound, being easily wet by the resin; but they also exhibit low thermal resistance, low dimensional stability (due to high moisture content), and poor electrical characteristics.

Mineral fillers are varied, comprising a large number of natural minerals as well as refined minerals and inorganic pigments. Virtually every type of rock has been used in powder form. Many are fairly inexpensive and easily available, but more expensive minerals also are used, such as the refined metal oxides, sulfates, and other inorganic pigments, to impart both color and hardness to the plastic. The purer forms of silica, such as mica and quartz, also provide good heat and electrical insulation. However, inorganic and mineral fillers used in substantial amounts result in increased brittleness. The considerations for choosing a filler are:

- Cost, availability, and uniformity.
- Compatibility or wettability with the resin.
- Moisture absorption.
- Physical properties.
- Thermal stability to mold temperatures.
- Resistance to chemicals.
- Abrasiveness.
- Effect on plastic flow characteristics.

Plasticizers

Plastics may need to be plasticized to enhance flexibility, resiliency, and melt flow. Without the addition of a plasticizer, it would not be possible to make plastics sheeting, tubing, film, and other flexible forms of plastics. In theory plasticizers enable the molecular chains of polymers to move freely with respect to one another, with a minimum of entanglement or internal friction. A plasticizer thus acts as an internal lubricant, overcoming attractive forces between the chains and separating them to prevent intermeshing. The higher the temperature is, the greater the penetration of plasticizer between chains and the greater the melt flow or moldability.

A plasticizer can be anything incorporated into plastic but not chemically linked to it. Any solvent that would dissolve PVC, for example, could be called a plasticizer. But in reality, a plasticizer must meet very exacting requirements. Also, the use of more than one plasticizer in a single formulation is not uncommon in today's applications.

Table 22-1. Common fillers and their uses.

Filler/reinforcement	*Uses*
Alumina trihydrate	Extender; flame retardant; smoke suppressant
Barium sulfate	Used as a filler and white pigment; increases specific gravity, frictional resistance, chemical resistance
Boron fibers	High tensile strength and compressive load-bearing capacity; expensive
Calcium carbonate	Most widely used extender/pigment or filler for plastics
Calcium sulfate	Extender; also enhances physical properties, increases impact, tensile, compressive strength
Carbon black	Filler; used as pigment, antistat agent, or to aid in crosslinking; conductive
Carbon/graphite fibers	Reinforcement; high modulus and strength; low density; low coefficient of expansion; low coefficient of friction; conductive
Ceramic fibers	Reinforcement; very high temperature resistance; expensive
Feldspar and nepheline syenite	Specialty filler; easily wet and dispersed; enables transparency and translucency; weather and chemical resistance
Glass reinforcement (fiber, cloth, etc.)	Largest volume reinforcement; high strength, dimensional stability, heat resistance, chemical resistance
Kaolin	Second-largest volume extender/pigment, with largest use in wire and cable, SMC and BMC, vinyl flooring
Metal fillers, filaments	Used to impart conductivity (thermal and electrical) or magnetic properties or to reduce friction; expensive
Mica	Flake reinforcement; improves dielectric, thermal, mechanical properties; low in cost
Microspheres, hollow	Reduces weight of plastic systems; improves stiffness, impact resistance
Microspheres, solid	Improves flow properties, stress distribution
Organic fillers	Extenders/fillers, like wood flour, nutshell, corncobs, rice, peanut hulls
Polymeric fibers	Reinforcement; lightweight
Silica	Fillers/extenders/reinforcements; functions in thickening liquid systems and making them thixotropic, in helping avoid plate-out with PVC, in acting as flatting agents
Talc	Extenders/reinforcements/fillers; higher stiffness, tensile strength and resistance to creep
Wollastonite	High loadings possible; can improve strength, lower moisture absorption, elevate heat and dimensional stability, improve electricals

Table courtesy of *Plastic Compounding Redbook*, an Edgell Plastics Publication.

Although nearly 80% of all plasticizers are used in PVC, they also are often used in cellulosics, nylon, ABS, and polystyrene, among other resins.

Most plasticizers are liquids although a few are solids that melt at compounding temperatures. All must exhibit good compatibility with the resin they modify. They usually are colorless and odorless, and have low vapor pressure and good thermal stability. Unfortunately, they decrease the strength characteristics, heat resistance, dimensional stability, and solvent resistance of resins.

The most popular general-purpose plasticizers are the phthalates, although epoxies, phosphates, adipate diesters, sebacates, and polyesters are also in common use. With some 500 different plasticizers to choose and combine from, an unlimited number of formulations is possible.

The choice of a plasticizer is exacting because it affects the physical properties of the end product as well as processibility. A good starting point is to decide on the properties required in the end product. The choice of plasticizer often comes down to finding one that satisfies the end-product property requirements, is compatible with the resin, and is the least expensive.

As a general rule the higher the ratio of plasticizer to resin, the greater the flexibility achieved. Plasticizers might compose up to 50 to 60% of a final PVC product.

Epoxy plasticizers, for example, add heat and light stabilization. The phosphates improve flame resistance, but may impair heat and light stability. The polyesters and trimellitates are used where durability is vital, but where low-temperature properties are less important. Aliphatic diesters, on the other hand, impart good low-temperature properties.

Other considerations include: federal approval for food packaging, specific gravity, and compatibility with other additives in the formulation.

Secondary plasticizers can be used in smaller quantities to impart properties not achieved with primary plasticizers.

Colorants

The use of colorants makes it possible to produce a great variety of materials in colors varying from pastels to deep hues, as well as the varicolors and marblelike shades. Broadly speaking, there are two types of colorants used in plastics, namely, dyes and pigments, both organic and inorganic, the essential difference between them being that of solubility. Dyes are fairly soluble in plastics, whereas pigments, being insoluble, are dispersed throughout the mass. The choice of either depends on resin compatibility or the need for solubility. Of almost equal importance is color stability—that the dyestuff or the pigment is stable at molding temperatures and on exposure to the light, moisture, and air expected in the end-use. Colorants also are chosen for strength, electrical properties, specific gravity, clarity, and resistance to migration (bleeding). The darker phenolic resins need bright pigments to hide them, whereas the lighter-colored ureas and melamines call for pastel shades to take advantage of their attractive translucency and glasslike transparency. Typical examples are the acrylics, which require soluble dyes that will not dim their clarity.

With some exceptions, colorants are supplied as dry powders of various specific gravities and bulking values. Soluble dyes are the easiest to use. Powders are simply tumbled or rolled in a drum with preweighed quantities of resin powder and distributed evenly throughout the mix. As the resin liquefies in an extruder or an injection molding machine, the dye dissolves in the resin.

All dyes are transparent. Most have relatively poor light fastness and limited heat stability, but give bright shades in transmitted light and thus are widely used where transparency is desired. Dyed plastics often hold their color much better against fading than pigmented plastics, a fact attributed to the greater depth of color imparted with the dyes. The surface color actually may fade, but the more deeply placed undamaged dye will show through and obscure the faded dye.

Dyes have much lower thermal stability than pigments, particularly inorganic pigments. Thus they are much more sensitive to molding temperatures and use conditions, and they may easily become discolored under the higher heats. With some of the newer engineering plastics requiring higher molding temperatures and longer mold cycles for the larger parts, the problem of the thermal stability of organic dyes and pigments becomes most significant. With molding temperatures of 400 to 600°F not uncommon, choosing a colorant is limited to a few organic pigments and a great number of inorganic pigments that can withstand high heat without yellowing or browning to some extent.

Dyes also are subject to color migration or bleeding, a condition that not only lessens the beauty of a color but requires selection of government-certified colors for toys, cosmetic containers, nursing bottles, plastics film for food packaging, and similar consumer items. In the same way, color migration may be a serious problem in costume jewelry, buttons, or tableware, for example, where the dye can stain the skin, a dress, or table linen, or cause an allergic reaction.

Pigmented plastics are opaque and light-sensitive, fading on the surface and obscuring the unchanged pigment beneath. Coloring with pigments is more difficult, particularly with organic pigment. Incorporating pigment into the various resins often requires special technology and equipment.

A pigment first of all must meet given end-use requirements, such as lightfastness, transparency or opacity, brilliance, or color, down to the specific shade. It also must disperse well. The dispersion of a pigment is a process by which pigment particles are "wetted" down by the resin in the liquid or molten stage. How well a pigment disperses depends on the temperature at which the two materials are mixed, the particle size of the pigment, and the molecular weight of the polymer. The mixing time and equipment are very important.

Improper dispersion will affect the color shade, as color strength is lost. Often the undispersed particles will detrimentally affect the physical properties of the plastic. It is possible to compensate for loss of color strength by using more pigment, but this substantially increases cost.

Moreover, a large number of undispersed pigment agglomerates can block extruder screens and eventually cause them to burst under the pressure. In these circumstances, the surfaces of extruded material may lose smoothness, and in films these particles also will cause pinholes. Dispersion is the key to successful coloring.

Pigments can be only broadly classified in terms of dispersibility. Large-particle-size materials such as titanium whites are easiest to handle. Generally, inorganic pigments have fairly large particle sizes, but only a few are as easy to work with as titanium pigments.

By contrast, organic pigments are usually the most difficult to disperse. Many can be dry-blended with resins and extruded or injection-molded on conventional equipment, but the preparation prior to the dry blending often requires skill. The blending sequence as well as the expert use of dispersing aids can be important.

Some organic pigments are so difficult to disperse that only the most efficient equipment can be effective. Organic pigments are very light and fluffy and carry electrostatic charges, all of which makes dry blending very difficult.

Smaller-particle-size pigments carry greater coloring strength over the entire powder, but their specific gravity creates problems in metering or weighing. Too, these colorants also require proper handling facilities and procedures to ensure color uniformity and prevent contamination of other plant equipment and material.

Color concentrates, furnished with the natural plastics molding compounds, enable the molder to color plastics by mixing some of the color with the molding powder in a blender, or by adding it directly in the hopper of an extruder press. The use of color concentrates improves color control and greatly facilitates handling. Color concentrates are particularly useful in wire coverings, where a large variety in color is desirable, especially for identification in coaxial cables.

It is crucial that colorants be used economically. The key questions are: Should preblended or predispersed pigments be used, or should dry powder be purchased directly from pigment manufacturers and blended and dispersed in-plant? The choice is obvious when large volumes of a single pigment or a single color are involved, particularly if the pigment is inorganic; but there are, unfortunately, only a few areas where this is the case. However, the resin supplier is often willing to supply precolored resin at only a nominal extra cost if the volume is large.

Most color manufacturers supply a variety of colors, but only rarely will a single pigment produce a desired shade; so blending of several pigments is necessary. An experienced color matcher must be on hand, and accurate weighing of colorants must be the rule to ensure batch-to-batch uniformity of a shade.

For frequent short runs, one cannot ignore clean-up, the increased frequency of risks, and maintaining an inventory of many different colors when calculating costs. Even infrequent mistakes can be very costly. Mistakes can result from switching production from one machine to another, variations in equipment size, inaccuracy of temperature-measuring instruments, and differences in speed, dwell time, and the size of a part. These mistakes can produce different shades of color even for the same resin and pigment formulation.

Only large operations can afford a colorant specialist, fairly large capital investment in special or modified equipment, and a practical colorant inventory. Smaller or even medium-size operations very often find it more economical to consult specialists in color blending, color compounding, or masterbatching. Color compounders maintain close contact with suppliers and fairly extensive information files on all pertinent materials. The larger compounders also maintain sophisticated, electronic color matching systems and well-equipped testing laboratories.

Heat Stabilizers

Stabilizers are used to prevent the degradation of resins during processing, when melts are subjected to high temperatures, or to extend the life of end products of which they become a part. Polyvinyl chloride is particularly vulnerable to degradation during processing, and thus a prime consumer of heat stabilizers. Other resins requiring stabilization are chlorinated polyethylene and blends of ABS and polyvinyl chloride.

In polyvinyl chloride, stabilizers are chosen for their ability to prevent changes in color during processing. Organotin, barium, and other systems are chosen for their specific effects on end products. Generally, liquid stabilizers are used for flexible polyvinyl chloride. Barium/zinc liquids are the workhorses in the vinyl industry, particularly in the area of packaging and handling. In recent years, the trend has been toward the bulk handling of plastic additives, both liquids and solids.

The first widespread move toward bulk handling involved use of the tote bin, a packaging system regarded as an important first step in the bulk handling of materials ranging anywhere from straight stabilizers to multicomponent blends containing a stabilizer, epoxy plasticizer, and UV absorbers. Additive mixtures are particularly suitable for operations involving large quantities and constant ratios, as, for example, in vinyl flooring compositions and plastisol foam constructions.

Use of the tote bin is just a small step away from the handling of these materials in bulk. Liquid barium cadmium/zinc stabilizers, as is well recognized, are antioxidants. Many of them are also subject to hydrolysis. When blended with epoxy plasticizer, they can be handled by the same kind of storage tanks, piping, and other equipment used for straight epoxy plasticizer. However, for the bulk handling of straight liquid stabilizers, special precautions are important, such as the use of stainless steel tanks, pipes, pumps, and other handling devices, and the provision of an inert atmosphere, because liquid barium/cadmium/zinc stabilizers, as antioxidants, essentially are prone to degradation through the action of oxygen and moisture.

Antioxidants

Antioxidants are used to protect materials from deterioration through oxidation brought on by heat, light, or chemically induced mechanisms. Deterioration is evidenced by embrittlement, melt flow instability, loss of tensile properties, and discoloration.

The three main preventive mechanisms to control deterioration of polymers operate by:

- Absorbing or screening ultraviolet light.
- Deactivating metal ions.
- Decomposing hydro-peroxides to nonradical products.

The mechanisms can very often be compounded when two or more antioxidants are used in one system. For example, because many chain-terminating antioxidants are also metal decomposers and some peroxide decomposers are radical terminators, the same result can be achieved in more than one way.

Although antioxidants are not ultraviolet deactivators, the decomposition by-products of antioxidants are effective hydrogen conjugators, that is, capable of hydrogen bonding—a reaction that permits degrading the electromagnetic energy of incoming light and redistributing it as thermal dissipation in the polymer without the formation of free radicals. In this aspect, it compares with the mechanism of ultraviolet absorbers such as the benzotriazoles and benzophenones.

To counteract the oxidizing effect of metal ions, through the formation of unstable complexes with hydroperoxides, followed by an electron transfer and the emission of free radicals, antioxidants remove the ability of metal ions to complex with free radicals, thereby effectively nullifying their detrimental reaction.

Materials subjected to heat, as they are during processing, undergo chain scission and crosslinking and the formation of hydroperoxides and free radicals, the rates of each being dependent on the polymer and processing conditions. Antioxidants deactivate these sites by decomposing the hydroperoxide or terminating the free radical by hydrogen abstraction and radical conjugation. The free radicals remaining within the antioxidant do not react toward other hydroperoxides or free radicals to continue the chain reaction.

Peroxide-decomposing antioxidants are normally used as secondary additives. They differ from primary phenols or amines in that they undergo decomposition when reacted with hydroperoxide instead of containing the hydroperoxide. Secondary antioxidants do not form free radicals when they induce peroxide decomposition; so the products of decomposition differ from those produced by free-radical trapping.

Synergism, or the cooperative effect of two antioxidants that may react in the same or alternative methods, often produces more effective results than one system at the same concentration. One theory has it that the synergist regenerates the primary antioxidant; the other holds that the secondary antioxidant is destroyed. Both theories are accepted.

The total effect of synergism in polymeric materials, particularly in polyethylene and polypropylene, can be to improve antioxidant efficiencies as much as 200%.

Ultraviolet Light Absorbers

Virtually every plastic degrades in sunlight in a number of ways, the most common being discoloration and the loss of physical properties. Particularly susceptible to this type of degradation are the polyolefins, polystyrene, PVC, ABS, polyesters, and polyurethanes. The job of stabilizing color and lengthening the life of a product falls to ultraviolet light absorbers.

Black, in any form, whether carbon black, black paint, or black dye, is the most effective UV absorber, but it cannot be universally used. Hence a variety of chemicals are used instead.

Benzophenones are used in clear polyolefin systems; but they are in fact good general-purpose UV absorbers and can also be used in pigmented systems, although they normally defer to a nickel complex of an alkylated hindered phenol. The nickel complex functions as a UV absorber, an antioxidant, and, in polypropylene, a dye acceptor. The benzotriazoles are used almost exclusively for polystyrene.

Both the benzotriazoles and the benzophenones are used as UV absorbers in the polyesters. The formulations of both are varied.

The choice of UV absorber depends on the particular application, the basic resin, the ultimate effect on color, and the required durability of the end product, as UV absorbers are used up in time. Only a small number have federal approval for applications involving food and drugs. Concentrations in any formulation are on the order of 0.25 to 1%.

Hindered amine light stabilizers (HALS) are a relatively new type of UV protector. They operate by a free-radical trapping process, which also contributes to heat and radiation stabilization. HALS are effective in protecting both pigments and polymers from UV degradation. They are suitable for use in polyolefins, styrenics, PVC, polyurethanes, and polycarbonate.

Antistatic Agents

Antistatic agents, sometimes called destaticizers, are used to reduce the buildup of electrostatic charges on the surface of plastics materials by increasing surface conductivity.

Plastics particularly susceptible to the accumulation of electrostatic charge are polyethylene, polypropylene, polystyrene, nylon, polyesters, urethanes, cellulosics, acrylics, and acrylonitriles.

The most common antistatic agents for plastics are amines, quaternary ammonium compounds, phosphate esters, and polyethylene glycol esters.

Antistatic agents are selected on the basis of application, durability, necessary concentration, FDA approval, where applicable, and effectiveness at low humidity.

Flame Retardants

Flame retardants are used to affect combustion in plastics. There are many flame retardants, whose choice depends primarily on the resin to which they are being added. For example, the flame retardant might be added to keep temperatures below a given combustion level, or to smother a reaction between the material and oxygen or other combustion-aiding gases, or, finally, to work on combustion through various types of vaporization.

Flame retardants thus work on four basic principles: either they insulate, create an endothermic cooling reaction, coat the product thereby excluding oxygen, or actually influence combustion through reaction with materials that have different physical properties.

Flame retardants can be inorganics such as alumina trihydrate (ATH), antimony oxide, or zinc borate, or organics such as phosphate esters and halogenated compounds of various types.

ATH is one of the most common flame retardants in use today. Used at high loading levels, it is effective in thermosets (polyesters, phenolics, epoxies, etc.) and some thermoplastics, functioning as a heat absorber and by creating steam. It also stifles combustion by forming a char layer on the burning material. Magnesium hydroxide and magnesium carbonate function in a similar manner, but at higher temperatures, and can be used on their own or in combination with ATH.

Halogenated flame retardants, chlorine and bromine compounds, are commonly used in a variety of thermosets and many thermoplastics, from polyethylene, polypropylene, and PVC, through ABS, and into engineering materials. Toxicity concerns may eventually result in other agents being substituted for halogenated compounds where possible.

Antimony oxide often is used in combination with halogenated flame retardants. Synergism between the materials improves their performance and allows the amount of halogen to be reduced. Zinc borate may be substituted for part of the antimony oxide in a formulation to reduce costs.

Phosphorus-based flame retardants (phosphate esters, et al.) are used in flexible PVC, polyurethanes, polyesters, and epoxies. Recently, they have been generating increased interest as potential replacements for halogenated compounds in certain thermoplastics.

Blowing Agents

A blowing or foaming agent is used alone or in combination with other substances to produce a cellular structure in a plastic mass. The term covers a wide variety of products and techniques, but compounders are limited to chemical agents, which decompose or react under the influence of heat to form a gas. Chemical blowing agents range from simple salts such as ammonium or sodium bicarbonate to complex nitrogen-releasing agents.

The basic requirements for an ideal chemical blowing agent are:

- The gas-release temperature must be within a narrow range.
- The release rate must be controllable but rapid.
- The gas must not be corrosive.
- The compound must be stable in storage.
- The residue should be colorless, nonstaining, and free of unpleasant odors.
- The compound and residue must be nontoxic.
- The residue must be compatible with the plastic to be foamed and have no effect on its properties.

Nitrogen-releasing compounds dominate the field of chemical blowing agents. Some of the products commercially available are the azo compounds, such as azodicarbonamide (azobisformamide), the N-nitroso compounds, and sulfonyl hydrazides.

Azodicarbonamide has been for years the principal choice of the chemical blowing agent family. The temperature range in which azodicarbonamide liberates coincides with the temperature at which a number of polymers have a melt viscosity suitable for foaming; so it has been used extensively in vinyl plastisols and calendered vinyl, as well as during the early stages of extrusion and injection molding of structural foams based on polyethylene, polypropylene, vinyl, styrene, and ABS.

Other blowing agents and blowing agent systems are being developed, to meet the challenge of foaming engineering polymer systems or to solve such problems as plate-out screw and barrel corrosion, corrosion of the mold, and uneven cell structure.

With the continuing emphasis on new engineering thermoplastics for high-temperature applications, the requirements for blowing agents with high decomposition temperature are increasing. With a decomposition temperature between 500 and 550°F, and a gas yield of approximately 175 ml/g at 536°F, the THT type of blowing agent is being used commercially to expand nylons, thermoplastic polyesters, and polycarbonates.

The blowing agent or system used in foamable glass-fiber-reinforced thermoplastics is generally a proprietary mixture.

For a detailed discussion of chemical blowing agents, the reader is referred to Chapter 19.

Lubricants

Lubricants are used to enhance resin processibility and the appearance of end products. Effective lubricants are compatible with the resins in which they are used, do not adversely affect the properties of end products, are easily combined, and have FDA approval, where applicable.

Lubricants fall into five categories: metallic stearates, fatty acid amides and esters, fatty acids, hydrocarbon waxes, and low molecular weight polyethylenes.

Clear grades of polystyrene require a clear-melt grade of zinc stearate, although bis-stearamides are generally adequate. ABS systems require metallic stearates in combination with glycerol monostrearate; styrene acrylonitrile requires only fatty acid amides. Polyvinyl chloride needs large concentrations of calcium stearate and hydrocarbon waxes, one serving as an internal lubricant and the other as an external lubricant. Low molecular weight polyethylenes are reported to be among the most efficient external lubricants available. Most systems incorporate polymeric processing aids for faster fusion and higher gloss.

In polyolefins, lubricants tie up catalyst residues, generally calcium stearate. Stearates and ethylene bis-stearamide waxes are sometimes used in the processing of fine powdered polyolefins.

For phenolics and melamines, the choices are ethylene bis-stearamide, calcium, and zinc stearate, the last particularly for molding grades. Zinc stearate also is widely used as a lubricant or release agent for reinforced polyester compounds, although calcium stearate is used in certain applications.

Underlubrication will cause degradation and frequently higher melt viscosities, but overlubrication can cause too much slippage and lower outputs. An imbalance of lubricant and stabilizer can cause plate-out or the migration of pigment from a melt system.

METHODS OF COMPOUNDING

Strictly speaking, compounding involves the fusion of different materials into a homogeneous mass, uniform in composition and structure. The actions in this type of mixing consist of smearing, folding, stretching, wiping, compressing, and shearing. However, component materials may be dry-blended either in preparation for compounding or for direct product fabrication.

In compounding, the order of addition of ingredients is most important, in order to maintain as liquid a mix as possible at the beginning, and then progressing stepwise to a stiff plastic mixture. The completeness of the mixing is limited by the toughest and most viscous ingredient, which supplies a supporting structure and folds about the other ingredients as the mixing proceeds. Thus the more mobile materials are locked up inside a tough, viscous, doughy envelope.

In other instances, extensive exposure of the mixing surfaces produces a drying or hardening effect that causes striations. The use of heat, either internally generated by shear or externally applied, is needed to obtain compound uniformity. Because it is always necessary to minimize the heat history during compounding, it is often necessary to supplement heat transfer through machine elements with a cooling system. This is particularly true for thermosetting compounds, to avoid overpolymerization of the resin through excessive heating.

With this in mind, we next consider the major types of plastics compounding equipment in current use.

Intensive Dry Mixers

Intensive dry mixers are used for dry-blending powdered resins, such as PVC, with plasticizers and other additives. A typical mixer consists essentially of a high-speed propeller-like impeller located at the bottom of a container (Fig. 22-4). Heat generated during the blending cycle is continuously removed to stabilize the dry blend and improve flow characteristics. In one design, the blend is cooled by impeller action. Heat is transferred from the constantly moving fluidized bed of material as it makes continuous and intimate contact with a cooling jacket wall. Other designs make use of cooling plate coils.

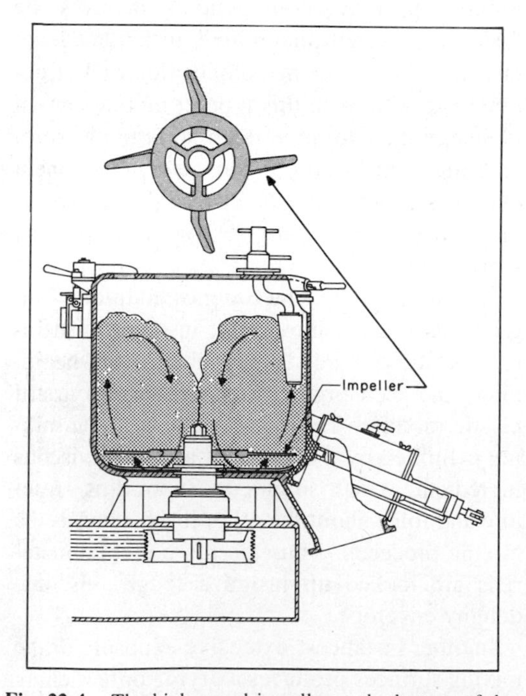

Fig. 22-4. The high-speed impeller at the bottom of the container housing is used to dry-blend powdered resin, such as PVC, with plasticizers and other additives. Heat and volatiles are removed during blending.

Impellers operate at the relatively high speed of 80 to 160 rpm. Moisture volatilized during the blending cycle is sucked into a vortex formed in the center of the mixer and vented from there. High-speed chopper blades operating at about 3600 rpm can be used to break up agglomerates and lumps and thereby assure the rapid incorporation of fluids.

Large-capacity systems generally combine high-intensity mixing units with low-intensity cooling units. The cooling units usually exceed the mixing units in capacity to maximize operating efficiency. Mixing cycles are short, averaging 5 to 10 minutes.

Internal Intensive Batch Mixer

Internal batch mixers have been widely used in the production of vinyl film and sheeting since the introduction of these products. In fact, every large manufacturer of vinyl film and sheet uses internal mixers for dispersion and fluxing. High-speed operation, which is used extensively, results in overall cycles of 2 to 4 minutes. Internal mixers also are used for processing plastics such as vinyls, polyolefins, ABS, and styrene, along with thermosets such as melamines and ureas because they can hold materials at a constant temperature.

Essentially, internal mixers consist of cylindrical chambers of shells within which materials to be mixed are deformed by rotating blades or rotors (Fig. 22-5). In most cases a shell actually consists of two adjacent cylindrical shells with a rotor describing an arc concentric with each shell section. Because the rotor blade tips clear the cylindrical segment of the shell by a small amount, the blade motion causes the mixture to be sheared between the blade tip and the shell. The blades also interact between themselves to cause folding or "shuffling" of the mass. The shuffling is further accentuated by the helical arrangement of the blade along the axis of the rotor, thereby imparting motion to the mass in the third, or axial, direction. Frequently, the blade is divided into two helices of opposite direction of pitch

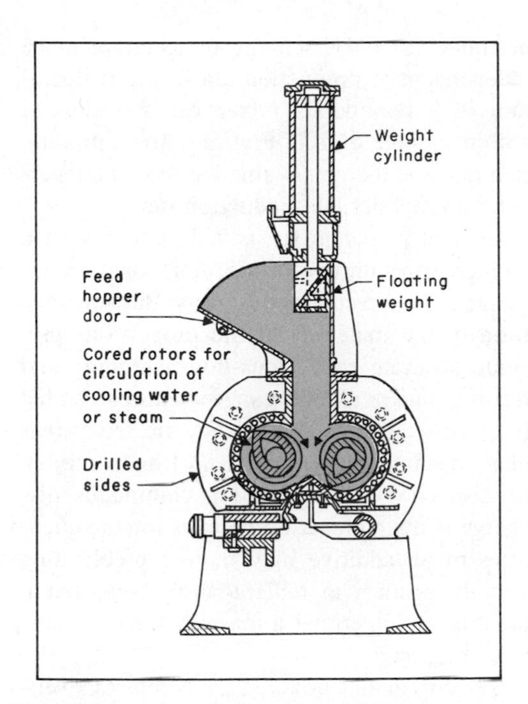

Fig. 22-5. Rotors within cylindrical chambers deform material in the tight clearances between the rotor blade tips and the chamber walls. Rotated at fairly high speeds, the blades provide kneading, shearing, smearing, pulling, and compressing action. Cored rotors provide channels for cooling water or steam. The batch is held confined by a ram in the feeding neck.

in order to further the shuffling of components within the mixture. In some cases, the rotor blades are Z-shaped arms instead of elliptical. Still further variations are haben and fishtail blades. The shell and rotor are cored or otherwise provided with means for heating or cooling for the purpose of controlling the batch temperature.

The principle of internal batch mixing was first introduced in 1916 with the development of the Banbury mixer. The mixer consists of a completely enclosed mixing chamber in which two spiral-shaped rotors operate, a hopper to receive the material for mixing, and a door for discharging the batch. Revolving in opposite directions and at slightly different speeds, the two rotors keep the stock in constant circulation. The ridge between the two cylindrical chamber sections helps force mixing; and the acute convergence of the rotors and the chamber walls imparts shearing. This combination of intensive working produces a highly homogeneous mix. The batch is confined within the sphere of mixing action by an air-operated ram in the feeding neck. The ram also greatly facilitates the flow between the blade tip and shell, where shearing is most extensive.

Cooling water or steam is circulated through the cored rotors and through a multiplicity of drilled holes in the mixing chamber liner to provide the optimum machine temperature.

Internal mixers range in size from laboratory models with a capacity of about 2 pounds of mixture to models of approximately 100 to 150 pounds capacity. Comparable power ranges are from 10 to 100 hp.

The thermoplastic mixture discharged from the internal mixer is usually in the form of large shapeless lumps; it frequently is convenient to roll these lumps into sheets with a two-roll mill for eventual grinding or dicing, or to feed them directly into an extruder-pelletizer. Thus a mill or extruder-pelletizer is placed beneath the internal mixer. However, they may also perform additional mixing and dispersing.

More recently, mixers have been built with the capability of dewatering and devolatizing of a variety of plastics. Slurries containing more than 70% moisture can be dewatered in 3 to 4 minutes. Most of the liquid is drained or pumped out of the machine; the remainder is vaporized by heat produced by mechanical action of the stock and contact with the mixer's heated working elements. Steam escapes through an exhaust system.

Color and formula changes can be made easily in internal mixers because mixing chambers are self-cleaning. Thus a variety of stocks of different color or formulas can be handled in succession. Many internal mixers in operation today are used for repeated changes of color in the same machine—from white to green, to gray, to red, to brown, to black, all in the course of one day. The same adaptability applies to formula changes.

Continuous Mixer

A continuous mixer is one whose rotors and mixing action are similar to those of the Banbury. Raw material is automatically fed from feed hoppers into the first section of the rotor,

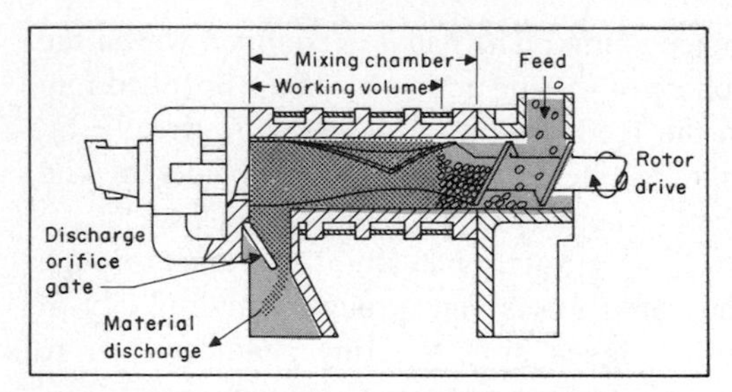

Fig. 22-6. The working volume is that portion of the mixer volume between the point where fluxing occurs and the discharge end. The working volume of the material in the mixer body is affected by the production rate and other operating conditions. Raw materials coming into the mixing chamber through the feed screws are converted to a homogeneous mass somewhere between the end of the last feed screw flight and the rotor apex (change in direction of pitch). Intensive shear occurs between the rotor and the chamber wall. As in the Banbury batch mixer, material is interchanged between the two bores of the mixing section.

which acts as a screw conveyor, propelling the material to the mixing section where it undergoes intensive shear between the rotors and the chamber wall, kneading between the rotors, and a rolling action (Fig. 22-6). Interchange of material between the two bores of the mixing section is an inherent feature of rotor design. The amount and quality of mixing can be controlled accurately by adjustment of the speed, feed rates, and orifice opening. Because the feed screw is constantly starved and the mixing action is rotary, there is little thrust or extruder action involved. The stock discharge temperature is controlled by the rotor speed and the discharge orifice; production rates are controlled primarily by feed rates.

The continuous mixer can maintain optimum dispersion or mixing over a wide range of production rates. On most materials, a production rate range of 5 to 1 or even 10 to 1 is possible. For example, a mixer designed to produce over 10,000 lb/hr PVC sheeting compound has proved capable of handling the same material, with equivalent quality, at rates of 1000 or 2000 lb/hr. Such operation is desirable in a calendering line where gauge changes in the sheet require different production rates.

In practically all mixing, the discharge temperature is critical, because it is either a limiting factor, a measure of the amount of work done, or a goal. To maintain versatility, it is desirable for the discharge temperature to be independent of production rates. The temperature of a continuous mixer can be adjusted within a range of 100°F at any given production rate and then maintained at the set temperature throughout the production run.

Generally, the mixer is fed—either with a preblend or with two, three, four, or more ingredients—on a continuous basis. But when the form of the materials or the proportions preclude accurate continuous-flow weighing and feeding, one or all of the ingredients can be fed in batch weighings. When they are fed at regular time intervals, there is no loss in mixing or dispersion quality, and a continuous discharge is maintained. Sometimes minute quantities of an additive may require preblending with the primary ingredient before being fed to the mixer. Sometimes a masterbatch may have to be prepared.

The continuous mixer is not completely self-purging, but one stock can be followed by another, without cleaning, with contamination of only a small amount of material. Still, before following with a different compound, it is generally desirable to clean the orifice unit, which is removable as a complete assembly.

Two-Roll Mills and Pelletizers

As stated earlier, after being processed in a mixer, material may be dropped directly onto a mill for sheeting and dicing or into an extruder or pelletizer. Arrangements vary. Where a mixer is located directly above the sheeting mill, the discharged stock drops into the bite of the rolls and can be taken off in a continuous strip. In other installations, where the sheeting mill is at a distance from the mixer, conveyors of various types are used to carry the mixed batch to the mill. High-production mixers require mechanical means of handling the large quantities of materials passing through them, both before and after mixing.

The two-roll mill consists of two opposite-rotating parallel rollers placed close to one another with the roll axes lying in a horizontal plane, so that a relatively small space or nip between the cylindrical surfaces exists. Material reaching the nip is deformed by friction

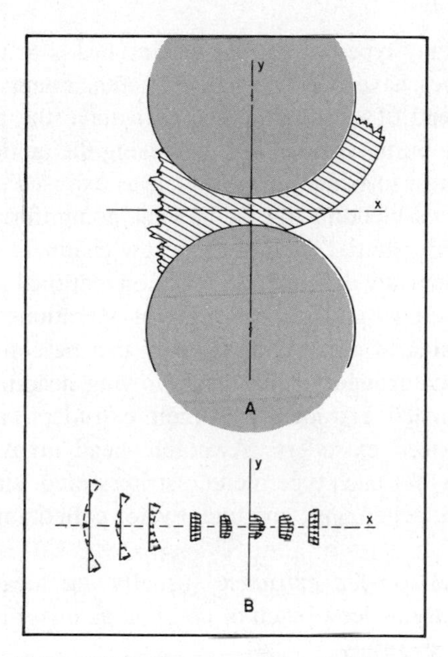

Fig. 22-7. (A) Rolls act on thin, flowing wedge of material simultaneously compressed and forced to flow through the roll nip. Rolls rotate in opposite directions. Material leaving the nip adheres to either roll if rolls rotate at equal speeds. (B) Velocity profiles of flows are symmetrical about the *X*-axis within the nip area for equal roll speeds.

forces between itself and the rollers and made to flow through the nip in the direction of roll motion (Fig. 22-7). Liquid plasticizers or finely divided solid ingredients also are placed in the nip and are incorporated into the resin through shearing action. Strip cutters are used to cut ribbons of stock from the roll for feeding to a cooling tank and dicer.

Usually, by adjusting the roll temperature the compound can be made to adhere to one of the rolls as a relatively thin sheet. The rolls are heated or cooled by a heating or cooling medium introduced into their hollow cores. They usually are rotated at different speeds to facilitate the formation of a sheet or band on one of them. During mixing, the band frequently is cut and manually pulled loose from the roller. The gap between the rollers generally may be adjusted manually by means of hand-driven or motor-driven screws.

Roll mills vary greatly in size from very small laboratory machines with rollers about 1 inch in diameter and driven by fractional-horsepower motors to very large mills with rollers nearly 3 feet in diameter and 7 or 8 feet in length and driven by motors of over 100 hp.

Special-purpose roll mills are available with three, four, or five rolls, in which the material is caused to pass from one nip to the next, in succession. Likewise, rolls may be arranged in pairs within a single frame, each independently adjustable in clearance. Material is led through the various pairs in tandem in a cascade arrangement for continuous mixing.

Underwater Pelletizer

The underwater pelletizer adapts to a variety of materials, such as polyethylene, linear polyethylene, polystyrene, polypropylene, color masterbatch, and plastic blends. The pellets can be easily handled, accurately weighed, and conveniently stored. In operation, the mixer discharges directly into the pelletizer, where the material, after being worked by the extruder screw, is forced through a straining screen and then pelletized in the head. Water circulated through the head separates the pellets and conveys them to the dryer. For light materials, the pellets are carried upward by flowing water to a dewatering system. For heavier materials, spray nozzles are used to flush the pellets to the bottom of the chamber and out through a side discharge port.

Single-Screw Extruders

Although developed to form thermoplastics, the single-screw extruder also functions as a mixing device because it subjects materials to laminar-flow deformation. Thus, although the extruder is not primarily used as a mixer, it is frequently used to add ingredients to a resin during melt extrusion and thereby to take advantage of their inherent mixing action. It is quite common for colorants to be added to extruded products in this way. Both the colorant and the resin are metered into a feed hopper, and a suitably colored extrudate frequently is obtained.

Reclaimed scrap stocks of different colors often are mixed satisfactorily in the extruder, provided that the materials are suitably metered in the feed hopper. Again, a masterbatch or premixed concentrate of colored thermoplastics may be fed to the extruder simultaneously with the resin to be colored. Here the extruder becomes an auxiliary mixer.

The amount of mixing a given volume of resin receives may be expressed in terms of the total amount of shear to which the resin is exposed. It is the product of shear rate and the extruder residence time, and is a measure of the relative displacement of one particle with respect to its neighbor.

The total amount of shear in an operation can be increased in one of several ways: by diminishing the channel depth of the screw, by letting the screw helix angle approach 0 or 90°, or by increasing the amount of pressure flow and leakage flow through increased die restriction (Fig. 22-8).

The total amount of shear to which a particular element of liquid resin is exposed depends upon its initial position in the screw channel, as the flow path, and thus both residence time and shear rate, vary according to the initial position of the material in the cross section of the screw channel. Material near the center of the channel receives less mixing than material near the screw surface or the barrel wall. This is because of the shorter residence time of the material near the center of the channel. It is also important to note that no motion of material backward along the screw axis ever takes place. Hence, single-screw extruders provide no "turnover" and little bulk mixing in the axial direction.

Many types of special mixing heads or torpedoes have been designed as attachments to the end of extruder screws. All have the primary purpose of further increasing the amount of shear to which the extrudate is exposed and of crossblending the somewhat nonuniformly mixed material leaving the screw channel.

There are different kinds of conventional single-screw extruders for different operations. In the area of compounding, they can be conveniently grouped under the following headings: mixer-fed extruders, hot-melt extruders, and cold-feed extruders. Available head arrangements for each type include strip, strand, slug, cone, pelletizing, and underwater pelletizing.

- *Mixer-fed extruders* usually are located directly under a batch or continuous mixer and fed by gravity.

Operation is automatic; a variable speed drive is used to synchronize the extruder speed with the production rate of the mixer in order to keep the screw always full of stock and the discharge uniform and unbroken.

Extruder cylinders generally are jacketed for steam or liquid heating, but they may be electrically heated, as required. Extruder size is dependent on the production rate of the continuous mixer or the size and mixing cycle duration of the batch mixer. Drive requirements are determined to a great extent by the compound to be handled.

- *Hot-melt extruders* take the resin from materials suppliers' reactors or receivers and

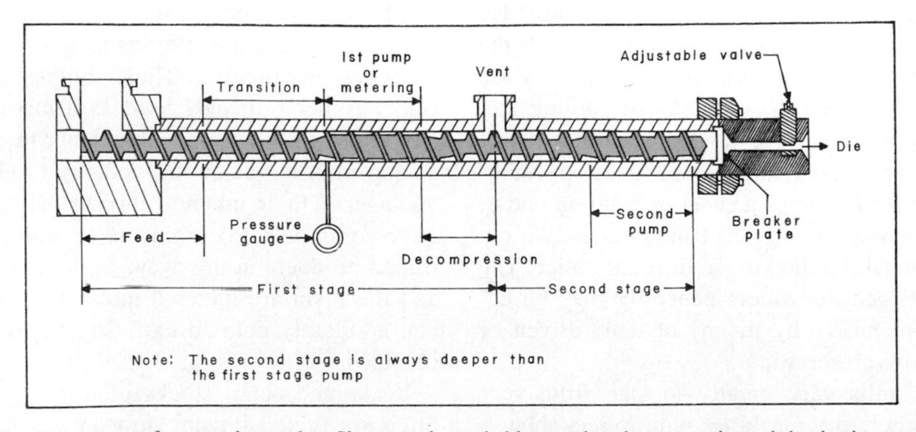

Fig. 22-8. Two stages of a vented extruder. Shear can be varied by varying the screw channel depth, the screw helix angle, or the amount of pressure flow and leakage flow through die restriction. The second-stage pump is always deeper than the first-stage pump.

deliver it in flat-strip or pelletized form. A great many hot-melt extruders have been built in recent years, most of them for handling polymers within the range of 0.2 to 50 melt index.

Extruder size is determined by the hourly production rate desired. Horsepower requirements, throughput rate, and screw design are dependent on material characteristics.

• *Cold-feed extruders* are available for working high-density and low-density polyethylene, plasticized PVC, and other compounds. The machines are generally cube- or pellet-fed and deliver extrudate to pelletizing heads. Most are of a completely jacketed construction, designed for use with pressurized water circulating systems. They are started up with circulating hot water, and then, as material flow is established, the water temperature is gradually reduced to obtain the best operating conditions for the resin being handled.

Compounder-Extruder

One modification of the single screw is a single-step ''enforced order'' variable-intensity compounder and extruder operating over a stepless gradient from extensive to intensive mixing, with a running cutoff on the intensive mixing end at any desired shear rate between 30 and 3000 reciprocal seconds.

The unusual design incorporates a rotor, or single rotating screw, inside a series of stators, or stationary screws, which replace the smooth barrel of a conventional extruder. The stator and rotor grooves are opposite-handed and alternatively increase or decrease in depth. Both rotor and stator grooves are shaped to help the material vortexing.

The cross-sectional diagram of the rotor and stators shows six extensive mixing stages, a vent section for removing volatiles and entrapped air, and a wide-landed, conical, high-intensity, variable-shear section (Fig. 22-9). The material starts to vortex in the grooves of the rotor as a solid as soon as it reaches the barrel. Because it is immediately transferred to the stators where it makes contact with the heat transfer walls of the stator grooves, it starts to melt and immediately is mixed with the bulk of the solid material, melting further resin.

This transfer of material from rotor to stator, and vice versa, can be broken down into different actions (namely, vortexing), in which the vortex of material in the giver screw diminishes and increases, because in the taker screw the outside layer of the vortex of the giver screw *unpeels* and then becomes the outside of the vortex in the taker screw. In this way, all of the material comes in contact with the heat transfer walls of both rotor and stator, once per stage. (A stage is defined as one transfer from rotor to stator, and back again.)

Besides the vortexing in the grooves, the material in the rotor moves forward, more or less

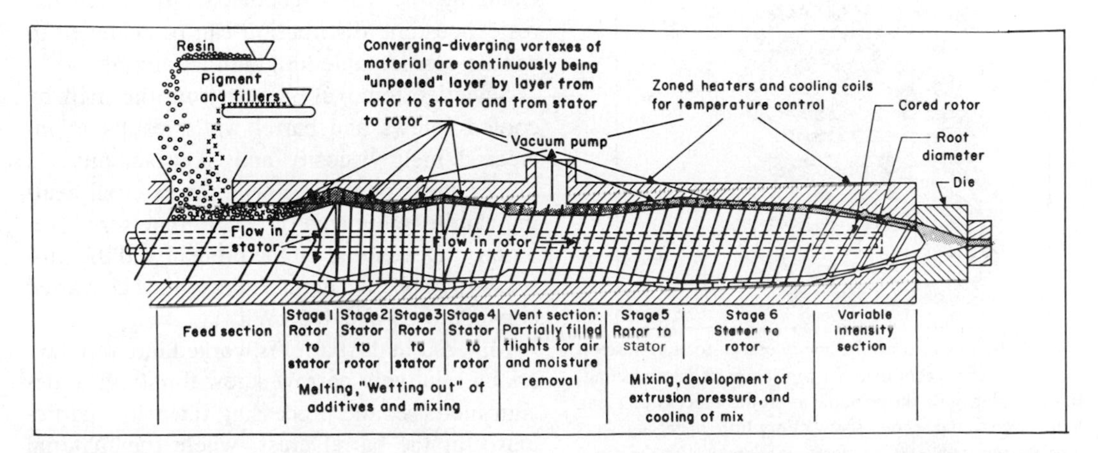

Fig. 22-9. Material taken up at the inlet by the rotor helix is peeled off by the stator helix and alternately by the rotor helix, once each stage. The vortexing action of the rotor and the stator transfers material to and from rotor and stator heat transfer surfaces. Pumping is achieved by enforced-order interaction of material in opposite-handed screws.

in a straight line or gentle helix. But as material is transferred into the stator, it must also follow the exact helical path of the stator. Thus there is a *change in direction or flow* with every pass from rotor to stator and back.

It is this change of direction of flow that is responsible for the pumping or extruding capability of the machine, regardless of the nature of the material. If the material tends to turn with the rotor, it is nevertheless pumped forward by grooves of the stator. If it does not turn with the rotor, it is propelled forward like a nut on a bolt.

Twin-Screw Extruder

In twin-screw extruders, two screws are arranged side by side. One design incorporates co-rotating screws that are intermeshing and self-wiping. Because the screws rotate in the same direction, material moves helically along the inside barrel wall in a figure-8 path from the feed section to the discharge point (Fig. 22-10).

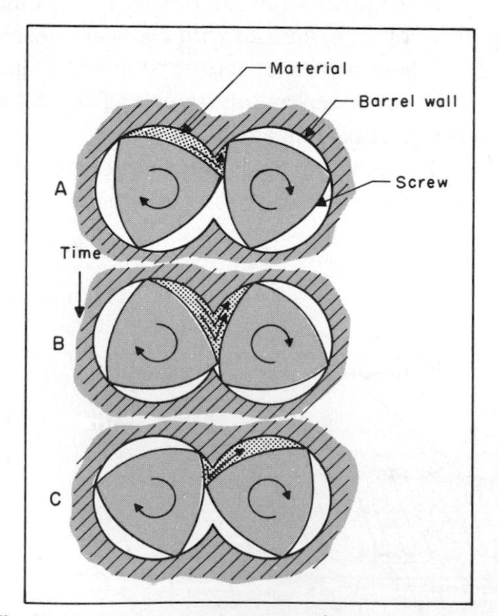

Fig. 22-10. Twin-screw geometry of a co-rotating system. (A) Screw transports material to a point of intermeshing. (B) A majority of the material is in the process of being transferred to the second screw. Very little material passes between the screws. (C) Transfer is complete. The material follows a figure-8 path around the screws as it moves down the length of the barrel.

The geometry of the screw components is such that the root of one screw is constantly wiped by the flight tip of the second screw, with a uniformly small clearance between them at every point. Thus dead spots are eliminated, the residence time of each melt particle is uniform, and purging times are shortened. Because of the positive conveyance of material, all types of material—molten polymers, pastes, flakes, powders of low bulk density, and fibrous materials—are efficiently conveyed, irrespective of friction coefficients. With positive conveying, operation with partially filled screw flights is possible while degassing is going on in some zones, or additional components are introduced into others.

Screw configurations are used to vary conveying efficiency, throughput rate, the degree of filling, and pressure buildup. Screws with reversed flights are used to generate localized high pressure. Hydraulically operated dynamic valves between barrel sections can be used to allow pressure variations during operation. Staggered stepped screws provide intensive transversal mixing by shear forces of varying intensity. Various mixing and kneading effects are obtained by suitable screw design.

The residence time is determined by the barrel length, screw lead, screw speed, and throughput rate. The residence time may vary between 20 seconds and 10 minutes, depending on the process and operating conditions. Residence time distribution may be influenced by changing the screw geometry, by which the residence time distribution can be widened to handle considerable longitudinal mixing.

Selective removal of heat from the melt by cooled screws and barrel walls results in increased melt viscosity and, consequently, in higher shear rates. The need for external heating can be minimized by appropriate screw geometry. Depending on the product and the process, the specific input energy may be varied between 0.05 and 1.2 kWh/kg.

Material in the barrel is worked into thin layers by relatively narrow screw flights. The design increases the processing intensity, particularly at the barrel crest, where the material transfers from one screw to the other, and the layers ar continuously mixed and inverted.

Volatiles are removed through vent ports at different locations along the barrel length. Fluids or melts of high or low viscosity or free-flowing solids may be introduced into the barrel at various points by means of suitable metering devices.

Twin-screw extruders with counter-rotating screws also are available and, despite similarities to the co-rotating screw types, there are some significant differences in the way they handle the melt.

With counter-rotating screws, screw fights carry material by friction in such a direction that all of it is forced toward the point where the two screws meet, there forming a bank of material similar to that of a two-roll mill, although some material slips through the gap between the two screws. As with the two-roll mill, the theory is to feed material through the nip from the bank on top of the nip gradually and statistically, in order that each particle be processed equally over a period of time.

Counter-rotating twin screws can put the material passing through the nip under an extremely high degree of shear, in the manner of a two-roll mill. Varying the clearance between the screws varies the portion of material fed between the screws against the portion of material accumulated in the bank and simply moving down the barrel. Material passing through the nip can be subjected to great shear forces by narrowing the clearance, and the portion of material in the bank also is greater.

Statistically, however, only a portion of the material is subjected to the high-shear condition of the nip, and only a portion to the low-shear condition of the bank. Counter-rotating twin screws are not totally self-cleaning.

In co-rotating twin screws, one screw transports the material around up to the point of intermeshing, where, because of the existence of two opposing and equal velocity gradients, a great majority of the material is transferred from one screw to the other along the entire barrel length in the figure-8 path mentioned earlier. Because the figure-8 path is relatively long, the chances of controlling the melt temperature are much better with this design.

The amount of shear energy developed at the point of defection can be regulated within very wide limits by choosing the depths of the screw flights.

Self-cleaning screws, aside from the obvious advantage of facilitating color change, provide control over residence time distribution. Control of residence time also is of great importance for heat-sensitive resins and pigments, or for operating at higher processing temperatures. A short and uniform residence time is essential to minimize heat and shear history, and, thereby, to maximum quality.

Compounding Lines

By now it is apparent that a compounding line involves the careful selection, arrangement, and location of the foregoing equipment, if it is to achieve a maximum effectiveness (Fig. 22-11). Ideally, a compounding line will deal with the questions of shear, temperature sensitivity of resins and additives, and the volume of the operation.

Compounding operations vary from 500 to 10,000 lb/hr in throughput capacity, whereas the capacities of in-line compounding systems rarely exceed 1000 lb/hr. Generally, compounders preblend the material in drum tumbling systems, ribbon mixers, paddle mixers, or double-cone twin-shell blenders, but they may also use devices set up on their extruders. The throughput rate of such systems is generally only 50 to 500 lb/hr.

Most PVC compounds can be processed from a dry blend, that is, one already containing plasticizer, and fed into cable coating extruders, provided they are large, with throughput rates in excess of 250 lb/hr. When coating extruders are small, handling all kinds of thin wire, separate compounding systems are preferred.

Single-screw compounding extruders are adequate in most applications, particularly when the heat history is less critical. With advances in screw design and in devices that localize and control the introduction of shear, they will do the job probably 70% of the time.

Twin-screw compounding extruders are one answer to the need for very good control of shear and melt temperature, and the removal of relatively large quantities of volatiles. In appli-

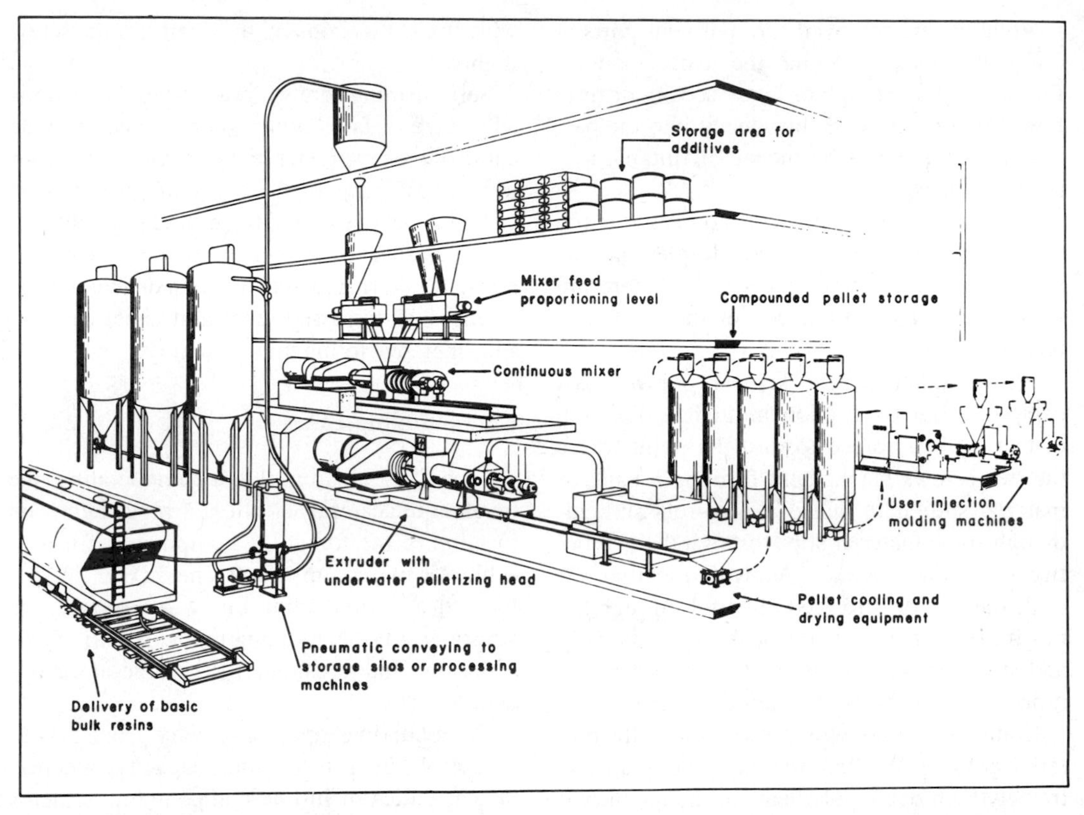

Fig. 22-11. Typical in-plant compounding system for an injection molder. Bulk resins are delivered in tank cars and pneumatically conveyed to silos for storage or directly to machines for processing. Resins and additives are proportioned and fed to the continuous mixer. The output of the mixer goes to the extruder and underwater pelletizer and then to pellet storage silos for subsequent use in an injection molding machine or a fabricating extruder.

cations involving the use of heat-sensitive material, as in fluxing PVC, twin-screw compounding extruders may be used to perform different compounding operations, that is, provide different levels of shear, in different sections of the same machine. Twin-screw compounding machines are generally smaller than single-screw extruders, requiring less floor space. For example, the overall length of a 2000 lb/hr twin-screw compounding extruder is around 12 feet, whereas the length of a comparable single-screw extruder would probably exceed 20 feet.

Feeding and preblending equipment have achieved high levels of performance; but, commercially, volumetric or gravimetric feeders, while able to satisfy any number of requirements, usually are subject to short-term inaccuracy, being unable to maintain minimal acceptable percent variations for the common residence times of 20 to 30 seconds. Short-term feed rate fluctuations can be overcome, however, through continuous preblending, by which volumetric or gravimetric feeders feed their output into an intermediate device, from which a continuous preblended steam of material is fed to a compounding extruder. Preblenders also perform intimate mixing at ambient temperatures, thereby minimizing the heat history.

The majority of pelletizing operations still involve the practice of extruding strands of melt into a water bath for cold cutting. For throughput rates of 1500 lb/hr or more, underwater pelletizers tend to be the rule. Underwater cutters must contend, however, with a delicate heat balance between the heated die plate and the body of water into which the melt is extruded, especially during startup and shutdown and with higher-melting-temperature polymers, where the temperature difference between the die plate and water exceeds 200°F.

In-line compounding is an alternative to separate compounding systems, the idea being to obtain the benefit of separate compounding and extrusion while avoiding the operational and inventory complications of a separate compounding operation. Typical examples of inline compounding operations are found in PVC pipe extrusion, foamed polystyrene sheet extrusion, and the production of siding material from certain plastic alloys. It generally is desirable to combine preblending for multiple inline compounding lines to a single unit in order to make the best use possible of the rather expensive metering and blending equipment. Here the problem is one of designing a distribution and hold system to inventory preblended materials.

A second alternative to separate compounding is on-the-machine compounding, where the plasticating section of an extruder or injection molding machine performs the compounding. On-the-machine blenders are used to meter in correct proportions of ingredients, but are limited in their capability of introducing a controlled amount of shear into a polymer mixture.

Another approach to on-the-machine compounding is that of mounting a continuous mixer on top of a large injection molding machine. Such a step has been used successfully in compounding a thermoplastic and an inorganic filler for the large-scale production of high-quality parts.

A compounding plant can be set up on the ground floor of a conventionally built warehouse, provided that the ceilings are at least 16 to 18 feet high. The compounding machine can be fed by gravity from a preblending and mixing room located on a mezzanine. Because of its general height, the building can be used to store incoming raw materials in silos and outgoing material in bags and containers. Only one operator is needed, who can usually supervise two lines. Personnel are needed more in formula preparation and product bagging. The use of continuously fed systems that are electrically monitored can eliminate operator mistakes. The same holds true downstream. Automated metering and blending, continuous systems, and automated packaging are economical in the long run.

An overhead rail parallel to the axis of the compounding machine will facilitate screw and die changes. The rail can be equipped with multiple die arrangements to stay abreast of color changes.

Water-cooled machines generally operate better in closed-circuit cooling systems, as the use of untreated plant water in the cooling channels eventually will lead to formation of hard-to-remove scales and possibly to plugging.

Control cabinets should be away from the main extruder, protected from the heat generated and mechanical vibrations. They also should be purged and kept under positive pressure. To conserve space, they should be designed with front access and placed against walls.

The amount of space required for a compounding operation depends largely on how feedstocks arrive—whether by bulk shipment or in bags or other containers—and how much of the product mix has to be kept in inventory. A typical product mix of a commercial compounding unit producing 2000 lb/hr of compound will require a floor space of 10,000 sq ft. The figure includes laboratory space, but not offices and off-site facilities.

23
Finishing and Machining Plastics

Many plastic products will require some degree of machining and finishing after they have been processed and before they are put in the marketplace.

A basic purpose of many finishing and machining operations is to remove such imperfections as scratches, dents, dull spots, marks of misalignment, slight ridges where the mold sections joined, or distinct flash. In transfer or injection moldings, rough spots may be left by breaking off the gates. Extruded pieces may exhibit longitudinal scratches from the edge of the extrusion die. Thermoformed pieces may show surface defects that were present in the original sheet. Cast plastic formed in a rigid mold will have a slight mark where the halves of the mold are joined; even where flexible molds are used, a gate mark may be left where the fluid plastic entered the mold.

Many finishing operations on molded articles can, of course, be avoided by careful design of the product or the mold (e.g., placing flash lines and gates in accessible positions); but where finishing or machining is necessary, a wide range of techniques are available for use with plastics, including tumbling, filing, sanding, buffing, tapping, turning, sawing, piercing, trimming, routing, grinding, and so on.

Obviously, repair work and defect removal are not the only reasons for finishing and machining. In some applications, it may be necessary to saw, drill, tap, and so on, simply to prepare the plastic product for assembly or mounting. In other instances, machining and finishing may be used to meet very narrow tolerances by molding a critical dimension oversize, then machining it down to the desired tolerances.

Machining also can be useful in the early stages of design engineering for the creation of prototypes or in low-volume production. Prototypes can provide material and performance data to aid in material selection, or in successful final part production if the material has already been decided upon. Short production runs can be entirely machined, practically and economically, instead of committing relatively large sums of money for complex molds or dies. High-speed automatic screw machines, for example, are increasingly being used for low-volume runs.

Another aspect of the finishing operation is the decorating of the plastic part, which will be discussed in Chapter 26. This chapter is concerned with the various aspects of finishing and machining other than decorating.

FINISHING

In the general nomenclature, finishing operations sometimes are distinguished from machining operations in that they include such techniques as filing, grinding, sanding, buffing, and polishing, as used to finish-off, smooth-out, or polish-up the plastic part. Of

Reviewed and updated by Donald S. Swavely, Polymer Corporation, Reading, PA.

course, such a distinction is rather vague in that some of the so-called finishing operations, such as grinding, also can be used to fashion a part (i.e., machine it to a different shape), just as machining operations, such as cutting, can be used to trim off flash (i.e., finish the product). For purposes of this discussion, however, we shall follow this arbitrary distinction, assuming that the reader will be able to apply any of the techniques covered for his or her own particular needs.

Filing

Files are used for finishing molded articles, and for beveling, smoothing, burring, and fitting the edges and corners of sheets of plastic. For removal of flash, tumbling is to be preferred if feasible, but the shape, size, or contours of the article may require filing to remove both heavy and thin flash as well as gates of a heavy section, or burrs left by machining operations such as drilling and tapping. Filing by hand is more costly than using special machine setups, but may be more economical on short runs.

Selection of the proper file is of great importance, as the shape, cut, size, and pattern of the file will determine the ease and speed of removal of stock, as well as the appearance of the filed surface.

The type of file selected must be adapted to the plastic—its hardness, brittleness, flexibility, and heat-resistance. The size and shape of the file are determined by the size, shape, and contour of the article to be filed. For removal of flash, files should have very sharp, thin-topped teeth that will hold their edge, well-rounded gullets to minimize the tendency to clog, and the proper rake for clearing the chips.

Machine Filing. Removal of flash from circular or cylindrical articles in large production runs sometimes is accomplished by semiautomatic machines. A circular revolving table carries from six to ten work-holding stations past a series of fixed stations equipped to remove flash, to polish, and to buff. The work-holding stations revolve on their individual axes as they pass the workstations. If the articles are perfectly round, the workstations may be equipped with tool-holders to accommodate files, and so on. If the articles are slightly out of round, or vary in size, as may be the case with pieces produced in multicavity molds, then motor-driven sanding belts, spring-loaded files, or buffing wheels are used. The efficiency of these machines is limited by the quality of the molding. Excessively heavy flash or even a slight mismatch will seriously reduce production or make hand finishing necessary.

The edges of round articles can be finished in a simple button machine, consisting of a drive chuck that rotates at high speed and can be moved axially by means of a pedal. The work is put against a stationary center, the drive chuck is brought forward to engage and rotate the work, and then the operator applies a file, usually about 70° from the flash plane, to remove the fin. Frequently as many as three different flash edges may be smoothed in one handling. When the pedal is released, the work falls into a tray, and the operator inserts another piece.

An excellent way to remove material rapidly is by using a rotary file or burr. Standard medium-cut, high-speed steel burrs operated at 800 to 1000 surface feet per minute are quite effective. Ground burrs provide better chip clearance than hand-cut rotary files and are, therefore, preferred. Carbide burrs offer no advantage over high-speed steel burrs. Because of their tooth geometry, carbide burrs remove less material than the high-speed steel burrs at the same speed, and when operated at higher speeds, tend to cause excessive frictional heat buildup.

Filing Thermoplastics. With thermoplastics, filing has become less popular than it was in the early days, as flash is more effectively removed by the three-square scrapers described below.

For the soft thermoplastics, fine files should be avoided, because they become clogged. Coarse, single-cut, shear-tooth files should be used, with teeth cut on a 45° angle, and in flat or half-round shape. The combination of coarse teeth and a long angle promotes self-cleaning. Shear-tooth files are used with long sweeping light strokes, to avoid running off of the work.

Clogging is not encountered with the harder thermoplastics, and fine files may be used (although the shear-type, coarser files still are generally recommended). It is also well to note that some plastics with unusual toughness and abrasion resistance, such as nylon, are not easily filed.

For filing edges of sheet stock, milled-tooth files are recommended. The files should be held at approximately a 20° angle with the edge.

In general, when filing thermoplastics, it would be wise to check with suppliers on handling the various types, as some distinctions are evident. For ABS, for example, files are usually of the coarse texture type. For polycarbonate, a single-hatched file is better than a cross-hatched file because the latter has a tendency to clog under heavy pressure. A No. 2 pillar file has no tendency to clog when used with light pressure. For acetal, best results are obtained by using the coarsest file consistent with the size of the surface and the finish required. Milled curved-tooth files with coarse, single-cut, shear-type teeth are particularly suitable.

A more up-to-date method of removing flash and parting lines and breaking sharp corners is to use a high-speed router equipped with a solid carbide or carbide-tipped cutter. For flat or square stock, it is recommended that the router be mounted to a table with guides and that the work be moved instead of the router.

Scraping Thermoplastics. Scraping involves smoothing a plastic surface by the use of a sharp-bladed tool, which is held so that the blade rides over the surface and has no tendency to dig under it.

A standard scraper is a three-edged tool of triangular cross section, the sides making up a scraping edge (at an angle of 60°). The blade is tapered so that the scraping edge runs from a fairly long straight section at the handle end to a curved section near the point. This variation in shape allows the careful scraping of a great variety of straight, concave, and convex surfaces.

Another type of scraper contains a small triangular sharp-edged insert mounted on the end of a handle. The insert or scraper is secured by retracting a collet, inserting the scraper, and releasing the collet. A variety of inserts are available, depending on the type of edge or material to be deburred or deflashed. The scraper is drawn by hand along either a straight or a contoured work edge, thereby removing a light but continuous curl of material from the entire edge. The result is a smooth, $\frac{1}{64}$ to $\frac{1}{32}$ inch, chamfered edge.

As with the other tools, the edges of the scraper must be free from burrs, dents, or scratches, which might mar the surface of the plastic.

Filing Thermosetting Plastics. Articles molded from thermosetting plastics always require some finishing operation for removal of flash from parting lines (or the removal of gates).

Flash should be filed off in such a way as to break it toward a solid portion rather than away from the main body, in order to prevent chipping. The file is pushed with a firm stroke to break off the flash close to the body, and then filing is continued to smooth the surface.

Selection is made by trial and error; so it is well to have a variety of files on hand.

Mill files in bastard and western cuts are used extensively to remove the flash from flat or convex surfaces of molded articles and on the corners and edges of sheets, to remove the burrs from sawing, or to bevel the corners. The western cut is slightly the coarser.

Milled-tooth files are recommended for large areas or for beveling the corners and edges of large sheets. They are relatively coarse files with curved teeth, and are available in both flexible and rigid types.

Various shapes of Swiss-pattern files are used, coarse enough to effect rapid removal of material, and fine enough to leave a good finish. These small files of various shapes, in cuts from No. 00 to No. 4, are used for small, intricate moldings, in which the surfaces to be filed are hard to reach. Round and half-round files are used for cleaning out holes or slots with rounded surfaces; knife and warding files are used to reach down between flutes and into narrow slots and grooves.

Files are designed to cut in one direction only; pressure should be applied in that direction, and relieved on the return stroke. Experience has shown that on some fine, intricate articles a steady pressure on both strokes may result in a better finish, but it is damaging to the file. As much as possible of the filing surface, on both length and width, should be used, so that the file will wear uniformly.

The life of the file is greatly shortened by improper selection, use, and care. Because many resins and fillers cause rapid wear of cutting tools, it is imperative that the files be given proper care to retain their sharpness. Proper care must be given in storage also. Files should never be thrown into a drawer with other tools, or stacked on top of each other, as such treatment ruins the cutting edge of the teeth. They should be stored standing with the tangs in a row of holes, or hanging on racks by their handles, and in a dry place so that they will not rust. Files should be kept clean of filings or chips, which collect between the teeth during use, by tapping the end after every few strokes. A file card or brush should be used to remove the chips before storage. Oil or grease is removed by applying chalk and then brushing. When a file becomes dull, but is otherwise in good condition, it should be sharpened. Files can be resharpened as often as four times, even though they do not then do as good a job as when new.

Tumbling

Thermoplastics. Tumbling is used to round corners, to remove stumps of gates, and to apply finish to surfaces. It is the cheapest way of doing these things, for the equipment is not expensive, and the only labor involved is in the loading and unloading of barrels. It is applicable chiefly to small objects that do not have projections that are easily broken off. Tumbling does not produce as high a finish as that obtainable by ashing and polishing, but for many articles a very high polish is not necessary and is not worth the higher labor cost involved.

The articles to be tumbled are placed in barrels, 20 to 30 inches in diameter, which may be divided into two or more compartments to permit the handling of two or more colors or shapes simultaneously. The barrels are mounted on a horizontal axis and are usually run at a speed of 15 to 30 rpm.

Abrasives and hardwood pegs are put into the barrels with the articles to be tumbled. The pegs serve to rub the abrasive against the articles during the tumbling. Sawdust and pumice are used in the first stage because of their rapid abrasive action; finer abrasives and polish are used later to give better finishes.

Tumbling of thermoplastics can be done wet, as well as dry. For wet polishing of a thermoplastic such as acetal, for example, a system might consist of aluminum oxide chips with a high-sudsing, burnishing compound (i.e., containing soap, detergent, alkaline cleaning agents, etc).

Automatic vibratory finishing machines are also available.

Thermosetting Plastics. Tumbling is used on all kinds of thermosetting materials to remove flash. It is done in barrels of various types, with several different materials as filing agents.

A cylindrical ("cement-mixer") barrel, running at speeds of 15 to 25 rpm, is used for light articles that have little or very thin flash. The articles are allowed to roll by themselves, from 2 to 5 minutes as necessary.

An octagonal barrel with alternate closed and open sides, running horizontally, is used for heavier articles and where more positive action is needed. The open sections are covered by screens of suitable coarseness to let the fragments of flash fall out. Lignum vitae balls from $\frac{1}{2}$ to 1 inch in diameter are used to give a rolling action to the articles and to prevent chipping. It has been found that a mixture of two parts of balls to one part of moldings, by volume, generally gives good results. Hardwood blocks or scrap molded parts also are used on some jobs. The speed of this barrel should be variable from 5 to about 30 rpm. The time required varies with the work, some jobs running as long as 2 to 3 hours.

Tumbling may be used to reduce the size of molded articles. This is done in a closed barrel running at a fairly high speed, and employs a cutting agent. Rubber impregnated with an abrasive is cut into various shapes and sizes and tumbled with the articles to give a satiny finish without a harsh cutting action. For fast removal of stock, strips of abrasive cloth or abrasive paper, mixed with the articles, are used. The barrel may be lined with abrasive cloth. For very light cuts, especially before polishing, pumice and small pegs are used. Octagonal barrels running at 20 to 35 rpm have proved best for this work.

With articles having grooves or projections, it is more difficult to achieve a uniform complete polishing, and the polishing agent tends to accumulate in grooves or other pockets on the surface. Such articles are best polished by being tumbled in barrels with string mops that have been impregnated and coated with specially compounded wax by being tumbled in clean barrels with balls of the wax. The tumbling together of articles and mop is done in a closed octagonal barrel having no screens, and no metal on the inside, which would scratch the surface of the articles. The barrel, on a horizontal axis, is rotated at speeds from 15 rpm (for large articles) up to 30 rpm (for small articles).

Besides plain tumbling, most thermoset deflashing is accomplished in machines using the impellor wheel design. Using nonabrasive media such as vegetable grain (walnut shell, corn cob, apricot pit, etc.) or plastic pellets (such as polycarbonate), these machines effectively handle large loads in a matter of minutes. In essence, these units are based on the use of a conveyor that carries the articles, in tumbling motion, through the stream of pellets projected at high velocity by a rotating bladed wheel (Fig. 23-1). A modified version of this type of equipment utilizes a rotary table or platform arrangement to pass parts under the impellor wheel without tumbling (Fig. 23-2). Production rates are slower with this type of equipment, but damage to delicate or fragile parts is avoided by eliminating the tumbling operation.

Deflashing also can be carried out by using pressure blast type systems with individual blasting guns. A number of configurations can be used for handling parts, from simple hand machines through batch-type tumbling systems to the use of conveyors, rotary tables, and so on.

Cryogenic deflashers use liquid nitrogen to freeze flash on molded parts, making it more brittle and, thus, easier to break off. The parts are tumbled or vibrated, either alone or with a deflashing medium.

Grinding and Sanding

These operations, if conducted dry, require an exhaust system to dispose of dust.

Thermoplastics. Standard sanding machines of belt and disc types, run wet or dry, are used for form sanding, or for long production runs.

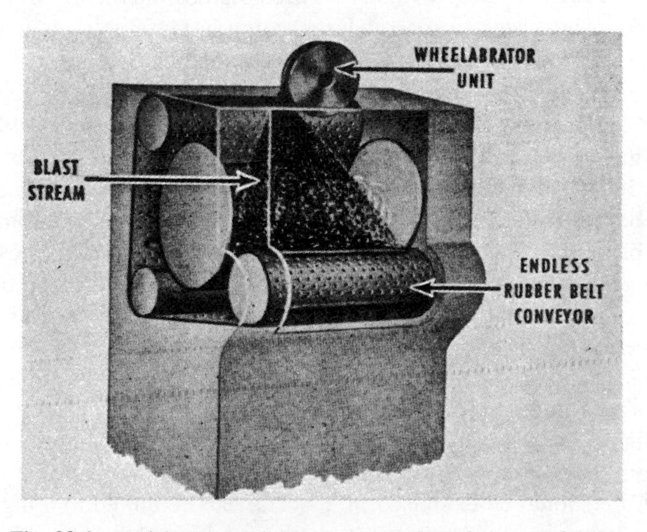

Fig. 23-1. Articles are carried by conveyor through stream of pellets.

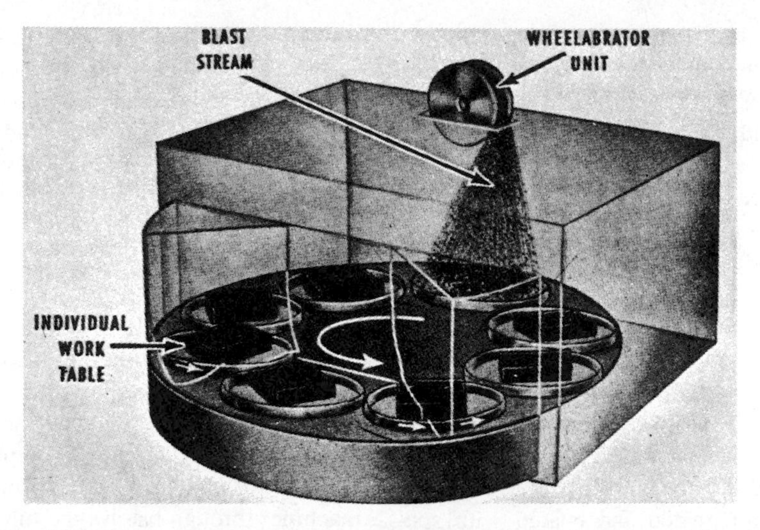

Fig. 23-2. Parts on a rotary table pass under an impellor wheel without tumbling.

A variable-speed control provides the proper speed in accordance with the amount of material to be removed.

Belts carrying coarse abrasive, and run dry, may be used for fast, rough cutting, if the speed is kept down to avoid excessive heating. In most cases, however, particularly in fine sanding, belts are run wet, as a precaution against overheating.

On *methyl methacrylate* resin (acrylic), the finest sandpaper that will remove the scratch or other defects (no coarser than grade 320) is used first. The paper, which should be of a waterproof type, is wrapped around a soft block of felt or rubber, and the area is rubbed lightly with a circular motion, using water, or soap and water, as a lubricant. An area having a diameter two or three times the length of the defect should be followed by similar treatment with progressively finer grades of sandpaper (grade 360A, 400A, and 500A, or 600A), each of which removes the deeper scratches left by the preceding one. The plastic must be washed after each sanding operation. Where a large amount of polishing is to be done, ashing compounds may be used in place of sandpaper.

When *polystyrene* is sanded, the heat developed should be kept to a minimum to prevent gumming and loading of the abrasive belt or disc. For this reason, wet sanding is the best type of operation. Care should be exercised in filtering the coolant to prevent abrasive grit contamination of the surface of the molded article.

A vertical belt sander is most frequently used, with belts having grits to suit the operation. For rough work, 80 grit is suggested. For medium and fine work, 220 to 280 grit and 400 to 600 grit belts are best.

ABS can be centerless, surface, and spindle ground or sanded. Many abrasives can be used to obtain the desired stock removal or surface finish. Coolants and lubricants used to reduce frictional heat include water, soap, detergents, or soluble oil solutions. A variety of industrial abrasives are available for grinding parts to desired flatness or other preferred shapes. Wet sanding is preferable, as it dissipates heat and retards loading of the abrasive belt. Silicon carbide belts or pads of No. 82 to 130 grit are ideal. Light pressure should be used to avoid overheating. Surface speeds of grinding should be determined by starting at speeds of 4500 surface feet per minute (sfpm) and adjusting the surface speed, feed rate, pressure, grit size, and coolant to obtain the desired finish or stock removal rates. Avoid excessive heat generation, or melting may occur.

Acetals can be wet-sanded by using conventional belt and disc sanding equipment. Moderate feeds should be used in sanding to prevent overheating. After sanding to a smooth finish, acetal can be buffed to a high surface luster.

In working with *polycarbonate*, sandpaper or

silicon carbide (emery) abrasive-coated belts may be used to remove sprue projections or toolmarks or for flat-surfacing sheet stock. Coarse abrasives will produce scratches that are difficult to remove, whereas too-fine grades will cause the belts to fill. For a scratch-free, highly polished surface, a grit size not over No. 180 is suggested. Hence, it may be desirable to use a coarse-grit sandpaper first and follow with a finer one. Belt speeds of approximately 3000 sfpm are adequate, but somewhat faster speeds (4000 sfpm) can be used so long as the pressure applied is not excessive.

Hand sanding must be done with light pressure in order to avoid clogging, especially with extremely fine grades. For example, 100 grit carborundum paper works well with light pressure, but clogs with heavy pressure. A 240 grit paper exhibits slight clogging even with very light pressure, but continues to produce a fine satin finish. Crocus cloth also shows a tendency to clog, but the use of kerosene lubricant prevents clogging and produces a fine satin finish.

Thermosetting Plastic. Belt sanding is commonly used for the removal of heavy flash and sprue projections, and for flat surfacing and beveling. The belts should carry a silicon carbide abrasive bonded by waterproof synthetic resin. Popular grit sizes are 50, 120, 180, 220, and 400—the coarse for heavy flash and sprues, the fine for lighter operations. Speeds recommended are from 2000 to 5000 linear fpm, but 4000 is most commonly used.

With a good exhaust system to remove dust and do some cooling by pulling air past the edge being sanded, thermosetting plastics can be successfully sanded at medium speeds on dry belts, thus avoiding the washing and drying required after wet sanding. However, wet sanding offers the advantages of freedom from dust and overheating, longer life of belts, freedom from clogging of the belt, and the finer surface produced because of the lubrication.

It is seldom necessary to resort to sanding the surfaces of well-machined pieces, but sanding can be done with any good abrasive cloth or abrasive paper that can be used in belt form or in rotating discs. The finer the paper and the slower the cutting, the less pronounced are the marks left by the abrasive. Facets and other surfaces can be cut in this way, with care to avoid overheating and discoloring the work. The surface speed for sanding cast phenolics is about the same as, or slightly less than, that for sanding wood.

Ashing, Buffing, and Polishing

The finishing department requires polishing lathes, buffing wheels, and suitable compositions for the ashing, polishing, cutdown buffing, and luster buffing that may be required.

Lathes for ashing, buffing, and polishing are available in types ranging from low-powered bench models, which are essentially converted bench grinders, to 50-hp floor models,

For finishing plastics a popular machine is a 2- or 3-hp floor-type lathe, with motor in the base and V-belt drive. This is preferred over the motor-on-spindle lathe because its speed can be changed by merely changing the diameters of the pulleys. Where speeds must be changed frequently, variable-speed lathes are available, in 3-, 5-, and $7\frac{1}{2}$-hp sizes, which permit changes of speed between 1500 and 3000 rpm, generally without stopping the machine. With these machines, it is possible to operate at the most efficient peripheral speed, regardless of the diameter of the buff.

For production buffing of articles of simple contours, automatic buffing machines are available, which must, however, be engineered for the particular job.

All dry buffing and polishing operations require an efficient exhaust system. Suitable sheet metal hoods should enclose as much of the wheel as is practicable, and be connected to the exhaust piping. The exhaust should pass through a dust-collector rather than to open air. Because many of the dusts are combustible, care should be taken so that, if steel inserts are ground, incandescent metal particles are not drawn into the exhaust system.

Buffing wheels for finishing plastics are generally made up of sections of muslin discs, either with or without sewing, depending on the flexibility required. The cloth should be a high-count sheeting, such as 84 × 92, for the faster cutting, and a lower count, such as 64 × 68, for buffing and polishing. For waxing, canton flannel and 48-48 muslin are both very popular.

Where the contours of the article are regular, and fast cutting is desired, the buffing wheel is composed of sewed buffing sections with stitching spaced at $\frac{1}{4}$ or $\frac{3}{8}$ inch. Wider spacings, and narrower, down to $\frac{1}{8}$ inch, are available. The wider spacings give the softer wheels. The next softer medium is the pocketed or folded buff, which presents pockets of cloth to the work and which makes for greater cooling of the surface, a faster cut than that given by a loose full-disc buff, and greater flexibility than that of conventional sewed buffs. Loose buffs made of full discs of muslin, while not cutting as fast as these, have a greater flexibility and give a smoother, more even, intermediate finish than do the harder wheels, and are generally to be preferred, especially for articles with curved or irregular contours.

For special cases, extremely soft wheels are needed, such as the packed buff or the string brush. The packed buff is made up to large discs of cloth alternated with smaller discs in the proportion of 1 : 1, 2 : 1, or 1 : 2, depending on the degree of hardness needed. The string brush is like a bristle brush, with cotton string substituted for the bristles. The wheel offers the maximum in flexibility, and in conjunction with greaseless compound is recommended for smoothing the edges of intricate articles.

Thermoplastics. In the ashing and polishing of thermoplastics, overheating must be carefully avoided because it may soften and distort the surface into ripples. Hence, it is necessary to avoid excessively hard buffing wheels, excessive speed of wheels, and excessive pressure of the work against the wheel.

Ashing is frequently required for the removal of "cold spots," teardrops, deep scratches, parting lines, and so on, from irregular surfaces that cannot be smoothed by wet sanding. A satin-smooth but dull surface is usually produced, requiring further finishing.

For ashing, wet pumice, grade No. 00 to No. 1, is used, on a loose muslin buff running at about 4000 linear fpm. The buff is often packed to increase its flexibility. The buff must be well hooded because the wet pumice does not adhere to it and tends to be thrown off. After wet ashing, the articles must be washed and dried before being polished.

As wet ashing is essentially a messy operation, attempts are continually being made to get more cut in a buffing operation than is usual in cut-down buffing, so as to obviate the need for wet ashing. Certain commercial compounds are available that perform a fast cutting job, which can be called semi-ashing, on cellulose acetate, cellulose acetate butyrate, and acrylics.

For the buffing of thermoplastics, fine silica powders in special grease binders, differing considerably from mixtures that have been developed for finishing metals, generally are most successful. Compositions must be formulated to give sufficient lubrication to prevent excessive heating. Pocketed and ventilated buffs are in aid in the prevention of overheating, although soft packed buffs are preferred for final luster buffing. Speeds generally run from 3000 to 4000 surface fpm.

Thermoplastics sometimes can be polished with greaseless compounds such as are used with thermosetting resins, but only with very soft wheels and with special caution against overheating.

ABS. In buffing ABS, a two-step method is the general practice. First, the part is cut down or polished, using commercially available wheels made of unbleached cotton discs. The second step involves wiping or coloring to increase the luster, using the same kind of wheels. Both the polishing and the wiping wheel should be about 6 inches thick and run at 1200 to 3000 rpm. A final softer wheel, kept free of polishing compound, is used for finishing off or wiping.

For ashing, a high-speed wheel (400 sfpm), built up of muslin discs that have been sewn together, is employed. Water and pumice are used for the cutting action and heat production.

In polishing ABS, an exceptionally high-gloss finish may be obtained by the use of muslin discs, felt, or sheepskin, with various polishing compounds. Detergents and soap solutions are used to reduce frictional heat and static electricity buildup.

Acetal. The ashing operation removes deep

scratches, parting lines, and harsh surface imperfections. In many instances, the ashing operation in buffing can be bypassed if the surface imperfections on the part are not severe. Ashing is done on a wheel made up of alternating 12- and 6-inch-diameter muslin discs wetted with a slurry of pumice and water. The wheel speed during the operation should be kept at approximately 1000 rpm. The acetal part should be held lightly against the wheel and kept in constant motion to prevent burning and irregular ashing.

The polishing and wiping operations are carried out on the same type of buffing wheel as is used for ashing. Instead of using a wet wheel, however, one-half of the wheel should be operated dry and the other half-lubricated with a polishing compound, such as jeweler's rouge. The acetal part should first be held against the lubricated half for polishing, and then wiped clean of polishing compound with the dry half. Wheel speeds of 1000 to 2000 rpm are suggested.

Styrene. Ashing wheels are built up, alternating spaces and discs from muslin. A suitable mixture of #00 pumice mud is used as an abrasive and may be applied to the wheel.

In buffing, two operations are used. The first consists of polishing the piece, or "cutting it down." The second involves wiping or "coloring" to increase the luster. Wheels for the first operation usually consist of unbleached cotton discs laid up alternately with two layers of 5-inch and two layers of 12-inch discs. The wiping wheel is composed of two layers of 12-inch and four layers of 5-inch unbleached cotton discs laid up alternately. Both wheels should be approximately 6 inches thick and run at about 1200 to 3000 rpm.

After polishing the piece with a suitable compound, such as a rough or greasy tripoli, a clean wheel that is made up softer than the polishing wheel is used for finishing off or wiping. This wheel should be kept clean, or it will tend to accumulate compound. Line or chalk "coloring" compound can be applied to the wiping wheel to remove any grease and bring out a lustrous surface. The pieces need only be wiped on this wheel.

In polishing styrene, care should be exercised to prevent a temperature rise to over 175°F, or else crazing and gumming will occur. As a precaution, avoid polishing under too great a pressure, on too stiff a wheel, or with insufficient cooling.

Thermosetting Plastics. For thermosetting plastics, a polishing operation with greaseless compositions on muslin buffing wheels is recommended for the removal of surface defects, light or residual flash, and marks from machining operations, and for the smoothing of irregularities left by the belt sander. The cutting face is formed and maintained by periodic transfer of greaseless compound in bar form to the face of the revolving wheel. A wheel coated in such a manner presents a faster-cutting face similar to the surface of emery cloth, and because of its resilience has the ability to smooth irregularly shaped parts without distorting or gouging. Although conditions and materials differ somewhat with the individual job, it is common procedure to use a No. 220 greaseless compound on a full-disc loose muslin buff at 5000 linear fpm.

Cut-down buffing is a procedure that converts a dull sanded surface into a smooth semigloss, preliminary to final luster buffing. In cases where a very high luster is not needed, this becomes the final operation. For cut-down buffing of thermosetting plastics, compositions or "waxes," are used, composed essentially of a fast-cutting buffing powder in a grease binder. A dry, fast-cutting bar with no free grease is well adapted to this operation, and produces a minimum of buffing dirt. Sewed, pocketed, or full-disc loose buffs are used, depending on the imperfections to be removed. Speeds range from 4000 to 6000 linear fpm.

Luster-buffing compositions or "waxes" are composed essentially of the finest abrasive buffing powders, such as levigated alumina, in a grease binder. The powders are finer and contain less grease than those used for cut-down buffing. Loose muslin buffs at 4000 to 5000 linear fpm generally are used.

Pigments can be added to white luster-buffing bars to match the shade of the plastic being buffed. These pigments have the distinct advantage of coloring spots of the filler, such as wood flour, that may be exposed, and of not being noticeable if not thoroughly removed from the molded article. In cut-down bars, black pigments often are incorporated for the same reason, but light or bright-colored pigments are not effective, because of the dark color of the buffing powders used in fast-cutting bars.

If compositions with excess grease have been used, the soft residual film of grease can be removed by wiping with a clean, dry, soft buff to expose the lustrous surface.

MACHINING PLASTICS

Each type of plastic has unique properties, and therefore can be assumed to have different machining characteristics—far different from those of metallic materials familiar to many manufacturing engineers. The principal considerations are described in the introduction that follows.

Introduction*

Lower Modulus of Elasticity (Softness). Thermoplastics are relatively resilient when compared to metals; therefore, the forces involved in holding and cutting must be adjusted accordingly, and the material properly supported to prevent distortion. Even within each family of plastics this characteristic will vary to some extent. For example, TFE-fluoroplastic and polyethylene have relatively low resistance to deformation, whereas polystyrene is harder and usually more brittle than either of the other two.

Plastic Memory. Elastic recovery occurs in plastic materials both during and after machining, and provision must be made in the tool geometry for sufficient clearance to provide relief. This is because the expansion of compressed material, due to elastic recovery, causes increased friction between the recovered cut surface and the relief surface of the tool. In addition to generating heat, this abrasion causes tool wear. Elastic recovery that occurs after machining also explains why, if proper precautions are not used, drilled or tapped holes in plastics often are tapered or become smaller than the diameter of the drills that were used to make them and also why turned diameters often become larger than the dimensionals that were measured just after cutting.

Low Thermal Conductivity. Heat conductivity of plastics is substantially less than that of metal. Essentially all of the heat generated by cutting friction between the plastics and metal cutting tool will be absorbed by the cutting tool. The small amount of heat conducted into the plastics cannot be transferred to the core of the shape, and the temperature of the surface layer will rise significantly. This heat must be kept minimal or be removed by a coolant to ensure a good job.

Coefficient of Thermal Expansion. The coefficients of thermal expansion of plastics are greater, roughly by 10 times, than those of metals. Expansion of the plastics caused by the heat generated during machining increases friction, and consequently, the amount of heat produced. Here again, adequate cutting-tool clearances are necessary to avoid rubbing.

Softening Point. The softening, deformation, and degradation temperatures of plastics are relatively low. Gumming, discoloration, poor tolerance control, and poor finish are apt to occur if frictional heat is generated and allowed to build up. Thermoplastics having relatively high melting or softening points, such as nylon or TFE-fluoroplastic, have less tendency to become gummed, melted, or crazed in machining than do plastics with lower melting points. Heat buildup becomes more critical in plastics with lower melting points.

General Property Considerations. The modulus of elasticity for metals is 10 to 60 times greater than that of every plastic. Therefore, tool forces cause much greater deflection during the cutting operation for plastics. Tool

*This Introduction contributed by the Machinability Data Center, operated by Metcut Research Assoc., Inc., 3980 Rosslyn Dr., Cincinnati, OH 45209 for the U.S. Dept. of Defense, Defense Supply Agency. It originally appeared in the 1972–1973 *Modern Plastics Encyclopedia* issue, published by McGraw-Hill, Inc., New York, NY and is reprinted through the courtesy of the publishers. Tables 23-1 through 23-13 in this chapter are reprinted from *Machining Data Handbook*, second edition, 1972, published by the Machinability Data Center.

forces increase considerably as tools become dull; therefore, it is essential that one use sharp tools. Another factor having considerable influence on tool force is the rake angle. Both the cutting force and the thrust force are higher for negative or zero rake angles than for the recommended positive rake angles.

The high coefficients of expansion of the plastics will cause problems in several ways. Dimensional control will be a problem because small variations in temperature will cause considerable dimensional change. The temperature rise during cutting, caused by heat generation from the shear zone and friction, will cause the workpiece to expand and rub, thus causing more heat.

Consideration of the properties of the work material is important in specifying the best speeds, feeds, depth of cuts, tool materials, tool geometries, and cutting fluids. The machining data given in Tables 23-1 through 23-13 represent starting recommendations provided by the *Machining Data Handbook*. It should be recognized that some of the materials may be cut at higher cutting speeds without resulting in loss of reasonable tool life. But higher speeds usually result in thermal problems with plastics.

Five general guidelines for tool geometry when machining plastics are:

1. To reduce frictional drag and temperature, it is desirable to have honed or polished surfaces on the tool where it comes into contact with work.
2. The geometries of tools should be such that they generate continuous-type chips. In general, large rake angles will serve this purpose because of the force directions resulting from these rake angles. Care must be exercised so that rake angles will not be so large that brittle fracture of workpieces result and chips become discontinuous.
3. Dr. A. Kabayashi has data* to indicate that there exists a critical rake angle in single point turning of plastics. He has found that the critical rake angle is dependent on depth of cut, cutting speed, and type of plastics material.
4. Drill geometry for plastics should differ from that for metals in two major respects: (a) wide polished flutes combined with low helix angles should be used to help eliminate packing of chips that causes overheatings, and (b) the normal 118° point angle is generally modified to 70 to 120°. Table 23-1 gives drill-point geometry recommendations for several plastics materials.
5. Cutting-tool geometry for multiple tooth cutters can be arrived at by using the principles discussed above.

Drilling and Reaming

Tables 23-1 and 23-2 provide starting recommendations for drilling many thermosetting and thermoplastics materials. Drilling is a very severe operation because of restricted chip flow, inherently poor rake angles, and variable cut-

**Machining of Plastics*, published by McGraw-Hill, Inc., New York.

Table 23-1. Drill geometry.*

Material	Helix Angle	Point Angle	Clearance	Rake
Polyethylene	10°-20°	70°-90°	9°-15°	0°
Rigid polyvinyl chloride	27°	120°	9°-15°	–
Acrylic (polymethyl methacrylate)	27°	120°	12°-20°	–
Polystyrene	40°-50°	60°-90°	12°-15°	0°
Polyamide resin	17°	70°-90°	9°-15°	–
Polycarbonate	27°	80°	9°-15°	–
Acetal resin	10°-20°	60°-90°	10°-15°	–

*Reprinted from *Machining of Plastics* by Dr. A. Kabayashi, published by McGraw-Hill, Inc.

Table 23-2. Drilling.*

Material	Hardness	Condition	Speed fpm	Feed, in./revolution Nominal hole diameter, in.								HSS Tool Material
				1/16	1/8	1/4	1/2	3/4	1	1 1/2	2	
Thermoplastics												
Polyethylene	$31R_R$	Extruded,	150-200	0.002	0.003	0.005	0.010	0.015	0.020	0.025	0.030	M10
Polypropylene	to	molded										M7
TFE-fluorocarbon	$116R_R$	or cast										M1
Butyrate												
High impact styrene	$83R_R$	Extruded,	150-200	0.002	0.004	0.005	0.006	0.006	0.008	0.008	0.010	M10
Acrylonitrile-butadiene-styrene	to	molded										M7
	$107R_R$	or cast										M1
Modified acrylic												
Nylon	$79R_M$	Molded	150-200	0.002	0.003	0.005	0.008	0.010	0.012	0.015	0.015	M10
Acetals	to											M7
Polycarbonate	$100R_M$											M1
Acrylics	$80R_M$	Extruded,	150-200	0.001	0.002	0.004	0.008	0.010	0.012	0.015	0.015	M10
	to	molded										M7
	$103R_M$	or cast										M1
Polystyrenes	$70R_M$	Molded	150-200	0.001	0.002	0.003	0.004	0.005	0.006	0.007	0.008	M10
	to	or										M7
	$95R_M$	extruded										M1
Thermosets												
Paper or cotton base	$50R_M$	Cast,	200-400	0.002	0.003	0.005	0.006	0.010	0.012	0.015	0.015	M10
	to	molded										M7
	$125R_M$	or filled										M1
Fiber glass, graphitized, and asbestos	$50R_M$	Cast,	200-250	0.002	0.003	0.005	0.008	0.010	0.012	0.015	0.015	M10
base	to	molded										M7
	$125R_M$	or filled										M1

*Reprinted from *Machining Data Handbook*, 2nd ed. (1972), published by Machinability Data Center, Metcut Research Associates Inc.

ting speed across the cutting edge. Wide polished flutes will help provide chip clearance, and the wide flutes also provide easy entrance for the cutting fluid. High clearance angles will help to prevent the drill flank from rubbing in the bottom of the hold. All tool surfaces in contact with the workpiece should be honed or polished to reduce frictional heat. Table 23-1 gives the helix angle and drill point geometry.

Drilling of Thermoplastics. In view of the variety of materials and operations involved, the following discussion is of a general nature. It is followed by some special instructions for specific thermoplastics.

Standard horizontal or vertical drill presses, single, gang, or multiple, can be used for drilling thermoplastics. Also, drills are commercially available that are especially designed for plastics.

Drills having one or two wide and highly polished or chrome-plated flutes, narrow lands, and large helix angles are the most desirable, as they expel chips with minimum friction, and hence with minimum overheating and gumming.

Points should have an included angle of 60 to 90° and a lip clearance of 12 to 18°. A substantial clearance on the cutting edges makes for a smoother finish. Drill points must be sharpened frequently and carefully, with care to avoid loss of the desired point angles. Carbide-tipped points will hold cutting edges longest, may be used at high speeds, and in some applications do not have to be cooled by liquid.

The use of liquids as lubricants or cooling agents should be avoided if practicable, as it necessitates a subsequent cleaning of the articles. In the drilling of most holes, an air blast will suffice to help clear chips and to prevent overheating, which would cause clogging.

The speeds used in drilling holes in thermoplastics will depend upon the type of material and size and depth of the holes. In general, speeds will be decreased with an increase in the size of the hole, and increased with increasing hardness of the material.

Drilling equipment must be in good condition for accurate holes to be obtained. Loose spindle bearings or bent or poorly sharpened drills will give inaccurate results with any material. The speed should be the greatest that will not cause burning or gumming, and the feed should be slow and uniform, to produce a smooth hole of uniform diameter. Chips should be removed by frequently withdrawing the drill from the work to clear the hole. In deep holes, the application of cutting oil or other cooling agent will prevent sticking of the chips. Pressure should be relaxed near the termination of through holes, to prevent breakthrough. In deep holes, an intermittent relaxing of the drill pressure will reduce clogging and runoff.

Drilling can be expedited by specially designed drill jigs. It is also possible to automate machining operations via self-feed drills that are ganged up around a product and can simultaneously drill $\frac{5}{52}$- to $\frac{1}{4}$-inch-diameter holes at the rate of four to five products per minute. The operator loads a shell into the machine, pushes the start button, and the units advance, drill, and retract.

CNC-programmed drilling machines automate the drilling operation and provide extremely fast movement of the table with the fixtured part attached. Although the actual drilling speeds and feeds cannot be altered greatly because of material properties, the time of movement between holes and other operations may be as high as 780 in./min., compared to standard manual/power feed machines at 80 in./min. The result is more uniform work and higher production because of machine capabilities and the elimination of operator fatigue.

Cellulose Acetate and Acetate Butyrate. Standard twist drills developed for wood or metal are frequently used on acetate or butyrate, but the drills especially designed for plastics, mentioned above, are preferred.

Drilling with the conventional drills of the type used for metal requires much slower speeds and feeds to give a clean hole and to keep the material from gumming, and more frequent backing out of the drill to clear chips. The quality of work can best be controlled by air-operated feeds and mechanical feeding devices.

Reaming of holes drilled in acetate or butyrate is not recommended. Where accurate dimensions are required in thin sections, good results are obtained by drilling to within about 0.001 inch of size and then running a hardened polished rod through the hole to smooth it.

Polystyrene. The most important factor is the efficient removal of chips by the drill during operation. Chips often tend to pack in the flutes and fuse together because of the frictional heat developed and its effect upon styrene, which is thermoplastic. To minimize this fusing, highly polished flutes with a slow helix are used. Generous side relief also will help reduce friction, but the cutting speed and feed are the prime factors in drilling a clear, true hole.

The suggested drilling speed for styrene ranges from 75 to 150 ft/min. for high-speed steel drills.

The most satisfactory type of drills used have been high-speed steel with a thinned web and drill point angle of 90° for small holes, increasing the angle to 118° for large holes. The major manufacturers of twist drills supply special drills for use with plastic parts.

For accurate work and to minimize breakage of drills, guide bushings should be used (see Fig. 23-3). The design and heat treatment of these bushings are the same as used for drilling ferrous and nonferrous metals. Ample chip clearance is necessary if a through hole is drilled. Some provision in the jig is, of course, essential to ensure proper location of the work.

The use of coolants is suggested. These coolants have been successfully applied by passing the drill through a felt wick wet with the coolant, a technique that minimizes cleaning of the part.

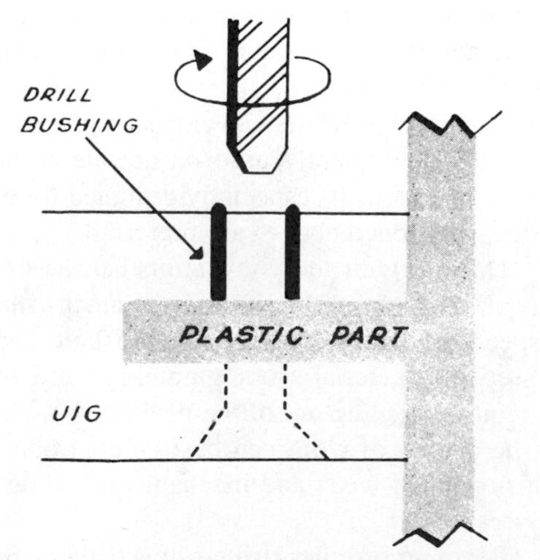

Fig. 23-3. When drilling polystyrene, guide bushings should be used to ensure accuracy and to minimize breakage of drills. (*Courtesy Monsanto*)

Where a blind hole is drilled, a faster helix often facilitates chip removal.

Acrylics. Because of the transparency of acrylics, it usually is desirable that the inside of a drilled hole have a high finish. Hence the drilling of acrylics requires extra care, and may call for special drills.

Though standard metal-type twist drills are satisfactory for the average drilling job in acrylics, they have a tendency to grab in large or deep holes. Better results are obtained with a sharp drill having a flute angle of 17 to 18°, an included lip angle of 70°, and a lip clearance angle of 4 to 8°. The lands of the drill should be highly polished and about one-fourth the width of the heel. Special drills for acrylics are commercially available, and usually have a slow spiral with highly polished flutes. Outstanding results can be achieved with a jet drill, the point of which is cooled with a lubricant fed through a hole in the drill. Holes of large diameter can be cut with either hollow-end mills or fly cutters. It is important that tools for drilling acrylics be kept free of nicks and burrs. All types of tools can be used on standard vertical or horizontal presses.

Rates of feed can be determined only by experience. The proper feed will result in smooth, continuous spiral chips or ribbons. The feed should be slowed as the depth of the cut increases.

In drilling deep holes in acrylics, a lubricant is needed to prevent clogging and possible burning or scarring of the wall of the hole. An air blast to cool the drill will be found beneficial in all cases.

Nylon. Nylon can be drilled satisfactorily with conventional twist drills, but more rapidly with special drills, designed for plastics, and having deep flutes, highly polished to facilitate removal of chips. In some, the flute leads are much longer than in conventional drills.

Polycarbonate. Standard high-speed twist drills perform satisfactorily (Fig. 23-4). The drill life in polycarbonate is five to six times greater than in low carbon steel. For the even

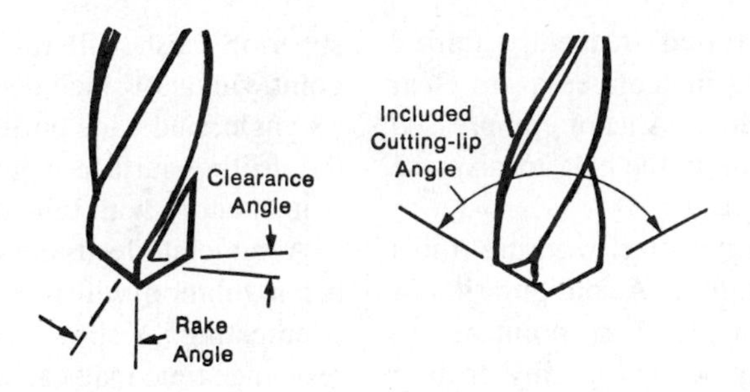

THESE ANGLES ARE:

Included angle	90° with coolant 60° without coolant
Clearance angle	15°
Rake angle	Depends on drill spiral. May be ground-to-zero or five negative for thin sheet drilling.

Fig. 23-4. Recommended drill design for polycarbonate. (*Courtesy GE Plastics*)

longer life and sharper cutting edges required for high-speed work, carbide-tipped drills are recommended. An added advantage of carbide-tipped drills is the absence of gumming, even without air or liquid coolants. There is no tendency for the drill to break out of the bottom of the piece or chip the edge of the hole, even when the drill is forced. Holes can be enlarged with larger drills without hogging or chipping.

Turning speeds for drilling will vary with the surface finish desired and the degree of induced surface strain acceptable. The best hole is obtained with surface speeds of 200 to 300 in./min. for drills less than $\frac{1}{4}$ inch in diameter and speeds of 350 to 450 in./min. for drill from $\frac{1}{4}$ to $\frac{1}{2}$ inch in diameter, when machined dry. A cooling medium should be used with speeds of 500 to 700 in./min. for drills under $\frac{1}{4}$ inch in diameter, and 1500 to 1600 in./min. for drills $\frac{1}{4}$ or $\frac{1}{2}$ inch in diameter. A feed rate of 0.001 to 0.0015 inch per revolution will give the best results.

No matter what type of drill is used, cutting edges must be kept sharp. Dull drills give poor surface finish and undersized holes. Burrs tend to heat up the work and induce machined-in stresses.

The use of a coolant is suggested to lower induced strain. Commonly used cooling media are: air, an air–water spray mist, or very light machine oil. One should avoid using standard cutting oils as they are not compatible with polycarbonate resin.

Annealing of drilled parts lowers strains and ensures optimum part performance.

Acetal. Standard twist drills and special "plastic" drills are suitable.

Standard twist drills are normally supplied with drill point angles of 118° and lip-clearance angles of about 12°. These drills work better on acetal if the included drill point angle is reduced to about 90°, although the 118° angle can be used. The lip clearance angle should be maintained within a range of 10 to 15°.

"Plastic" drills are furnished with included point angles of 60 and 90°, and with lip-clearance angles of 10 to 15°. The cutting edges of these drills usually are flattened slightly to provide zero or negative rake. They have extra-wide, highly polished flutes and a low helix angle to ease the removal of chips. For best performance with acetal, the cutting edges of the drill should be sharpened in the usual manner (not flattened) so that a normal positive rake is obtained. A 60° included angle is of some advantage when drilling through thin sections (approximately $\frac{1}{16}$ inch thick), but the 90° point is better for thicker work.

During drilling, the work should be firmly supported and securely held. For deep holes, the drill should be raised frequently during drilling—about every $\frac{1}{4}$ inch of depth—to clear the drill and hole of chips. A jet of compressed air should be directed into the hole to disperse chips and cool the drill.

Polysulfone. Normal steel-working tools work well with polysulfone. A configuration of 12 to 15° clearance angle, 118° point angle, and 5° rake angle may be used for any drilling operation.

Small holes can be enlarged readily without chipping. When drilling completely through a piece of polysulfone, there is a tendency for the drill to break out of the bottom of the piece or chip the edge of the hole. This can be eliminated by backing up the piece and reducing the rate of feed.

PPO Resins. Standard high-speed steel twist drills with a rake angle of 5° and an included angle of 118° perform well. It is important to use sharp drills to maintain the quality of machined parts. Recommended drilling speed are shown in Table 23-3.

ABS. ABS may be drilled effectively because of its hardness and rigidity. The most important factor is the removal of chips by the drill. Chips will tend to pack in the flutes and, as a result of frictional heat, fuse together. To minimize fusion, highly polished flutes with a slow helix are recommended, with generous side relief, and the use of coolants. Cutting speeds and feed rates also influence the drilling of clear, true holes.

A standard drill press is adequate for ABS; however, the drill bit should be ground to scrape rather than cut. Customarily high-speed, double-fluted twist drills are satisfactory, but superior finish will result from bits having a point with a 90° included point angle, 300° helix angle, and wide polished flutes. Backing up the drilling surface is generally desirable, and is mandatory with thin stock.

Table 23-3. Recommended drilling speeds for unfilled and glass-reinforced PPO resins.

	SPEED (IN./MIN.)	
Diameter (in.)	Unfilled PPO	Glass-Reinforced PPO
1/8	150	700
1/4	300	750
3/8	450	750
1/2	500	750
5/8	650	875
3/4	775	800

A moderate feed rate of 0.001 to 0.005 inch per revolution will tend to avoid burring and overheating. A slow feed rate and consequent residence time may cause heat buildup and tool drift. A drill surface speed of 60 to 180 sfpm is recommended. Holes with a depth of up to five times diameter may be cooled with a jet of forced air; for deeper holes, cooling with water is recommended.

Single-fluted drills, with the required frictional characteristics for piloting and chip removal capacity, allow relatively good heat dissipation with surface speeds of 300 to 500 sfpm for small diameters, depth-to-diameter ratios of 25:1, and plunging feed rates.

Burnishing double-fluted, small twist drills with 118° point angles and 0.003 to 0.005 inch off center, allow 65 to 100 sfpm with feed rates of 0.010 to 0.055 inch per revolution.

Hole saws with a skip tooth blade can be mounted in the drill press to cut larger holes.

Reaming Thermoplastics. Even though drilled holes are easy to make, they occasionally lack the precision or finish required for optimum performance and appearance. The solution is reaming, which like drilling, is a simple process requiring only conventional tools.

Fluted reamers provide both accuracy and good finish. Properly sharpened, they assure shearing of material from sidewalls rather than trouble-making dislocation. Standard high-speed steel units, straight or fluted, are usable without alteration of the cutting edge, chamfer, or rake angles. The normal chamfer angle is 45°; the normal rake angle is 5°.

The helically fluted reamer is preferred over the straight-flute type because it provides a smoother cut, finer finish, and can be used in both through and blind holes. The straight flute reamer is limited to through holes.

Although reaming can be done dry, the use of coolants will produce better finishes. Water is the preferred coolant, but light machine oils can be used. Standard cutting oils should be avoided.

Recommended conditions for reaming thermoplastics and thermosets are shown in Table 23-4.

Drilling of Thermosetting Plastics. The following table gives a general basis for selecting speeds of drills for conventionally molded thermosetting plastics:

Drill Size	*rpm*
No. 1 through No. 10	1700
No. 11 through No. 27	2500
No. 28 through No. 41	3000
No. 42 and up	5000
$\frac{1}{16}$ inch	5000
$\frac{1}{8}$ inch	3000
$\frac{3}{16}$ inch	2500
$\frac{1}{4}$ inch	1700
$\frac{5}{16}$ inch	1300
$\frac{3}{8}$ inch	1000
$\frac{7}{16}$ inch	600
$\frac{1}{2}$ inch	600
A and B	1700
C through O	1300
P through Z	1000

To drill holes in molded articles or to remove flash or fins in molded holes, it is best to use standard high-speed steel drills with deep flutes. Nitrided high-speed drills do not require frequent sharpening, and will last a long time. Drill points should be ground to an included angle of 70 to 90°, and have a lip clearance or relief made by grinding the back away to $\frac{1}{16}$ inch wide, which reduces friction between the drill and the work and gives clearance for the chips. Backing off the cutting lip (rake angle) prevents the grabbing that occurs with a drill with a normal point, and will sometimes prevent chipping of the hole when the drill breaks through the under side of the work.

Most drilling is done without a lubricant, but a blast of air at the drill point will keep the drill and work cool, prolong the life of the point, and help clear away chips. Drill speeds should be from 100 to 150 fpm, or faster if proved by trial. Drills should be about 0.002 to 0.003 inch oversize. For drilling thin sections, the point of the drill can be ground with a sharper included angle to stop chipping around the hole.

Some manufacturers will make to order special drills for use in long-run or automatic drilling operations. These drills are made on slow-twist blanks and tipped with tungsten carbide. This is about the most economical type of drill for phenolics if the production will warrant the cost, and the best for very deep holes.

In drilling deep holes, good results are obtained, however, with steel drills having specially polished flutes and 0.0001 to 0.0002 inch of chromium plate.

For drilling through holes in canvas-filled materials, the drill may be specially ground. The end is ground like that of a wood drill. The outer edges of the drill are cut like circle-cutting tools, while the center acts as a pilot. Thus, at the breakthrough the cutting is done through a thin section supported on both sides by heavier areas, and the final chip is a disc with a hole.

Reducing the friction between the drill and the material by grinding the drill off-center (which results in a slightly larger hole) will often prolong the life of the drill.

Cast Phenolics. For small holes, drill speeds of about 2800 to 12,000 rpm are commonly used. Drills of $\frac{1}{4}$ inch or more diameter should have large flutes, for efficient removal of material, and the cutting edges should be ground with a negative rake. For the drilling of small holes for self-tapping screws, the hole is usually one drill size smaller than the screw.

Rapid production is obtained by a multiple drill assembly. Multiple drill heads are likewise effective. Where neither is available, drilling with a jig can be made both fast and accurate.

Tapping and Threading

General tapping recommendations for thermoplastics and thermosets are given in Table 23-5. Finish ground and polish flute taps are recommended because less frictional heat will be generated. It is recommended that oversized taps be used because of elastic recovery of plastics materials. Oversized taps are designated:

H1: Basic—basic + 0.0005-in.
H2: Basic + 0.0005 basic + 0.0010-in.

Table 23-4. Reaming.*

				High-Speed Steel Tool								Carbide Tool					
				Feed Inches Per Revolution†								Feed Inches Per Revolution†					
				Reamer Diameter Inches								Reamer Diameter Inches					
Material	Hardness BHN	Condition	Speed (fpm)	⅛	¼	½	1	1.5	2	Tool Mtl	Speed fpm	⅛	¼	½	1	1.5	2	Tool Mtl
Thermoplastics	$31R_R$	Extruded,	250							M2	500							
Polyethylene	to	molded,	to							M7	to							
Polypropylene	$116R_R$	or cast	300	.006	.008	.010	.010	.012	.015	M1	600	.006	.008	.010	.010	.012	.015	C-2
Fluorocarbons																		
Butyrates																		
Nylon	$79R_M$		250							M2	350							
Acetals	to	Molded	to	.004	.006	.008	.010	.012	.015	M7	to	.004	.006	.008	.010	.012	.015	C-2
Polycarbonates	$100R_M$		300							M1	450							
Acrylics	$80R_M$	Extruded,	200							M2	300							
	to	molded,	to	.006	.008	.010	.010	.012	.015	M7	to	.006	.008	.010	.010	.012	.015	C-2
	$103R_M$	or cast	300							M1	400							
Thermosets	$50R_M$	Cast,	200							M2	250							
Paper and cotton base	to	molded	to	.003	.003	.004	.005	.005	.005	M7	to	.003	.003	.004	.005	.005	.005	C-2
reinforced	$125R_M$	or filled	250							M1	300							
Fiber glass and	$50R_M$	Cast,	100							M2	150							
graphitized base	to	molded,	to	.002	.002	.002	.003	.005	.005	M7	to	.002	.002	.002	.003	.005	.005	C-2
reinforced	$125R_M$	or filled	150							M1	200							

*Reprinted from *Machining Data Handbook*.
†Based on a 6-flute reamer.

Table 23-5. Tapping.*

Material	Hardness	Condition	Speed (fpm)	HSS Tool Material
Thermoplastics	$31R_R$ to $125R_M$	Extruded, molded, or cast	50	M10, M7, M1
Thermosets	$50R_M$ to $125R_M$	Extruded, molded, or filled	50	M10, M7 M1
Reinforced plastics Silica fiber-reinforced phenolic resin (Refrasil)	55† to 75	Molded	25	M10, M7, M1

*Reprinted from *Machining Data Handbook.*
†Barcol hardness.

H3: Basic + 0.0010 basic + 0.0015-in.
H4: Basic + 0.0015 basic + 0.0020-in.
H5: Basic + 0.0020 basic + 0.0025-in.
H6: Basic + 0.0025 basic + 0.0030-in.

The amount of oversize depends on elastic recovery properties of the material and sizes of holes. The number of flutes determines the chip space and the chip load per tooth; therefore, some compromise must be made. In general, the two-flute taps are preferred for holes that measure up to $\frac{1}{8}$ inch.

Thermoplastics. Unless special accurate tapping machines with lead screws are available, it is unwise to attempt Class 2 or 3 fits. In any case, a higher percentage of rejects and higher costs may be expected, especially with nylon.

United States Standard (American Coarse Thread Series), Whitworth Standard (British Standard Series), and Acme are generally satisfactory. Sharp V-threads are to be avoided because the apex is easily broken. Coarse-pitch threads are preferred for their strength.

Bottom taps should be avoided whenever possible. If a bottom tap must be used, it should be used in a second operation done by hand, and only when a Class 2 or 3 fit is required. For maximum strength and dimensional stability, all tapped parts should be annealed to relieve the stresses set up by the tapping.

Before tapping, it is recommended that the hole be drilled to such size as to permit not more than 75% of a full thread, to minimize difficulty in clearing the tap.

To obtain effective clearance of chips with a minimum of friction, large, highly polished flutes are recommended. Taps should nitrided or chrome-plated. All new taps should be stoned to remove burrs.

Taps for all thermoplastics should have maximum back clearance. In most cases, the pitch diameter should be 0.002 inch oversize. For tapping nylon, 0.005 inch oversize is recommended, unless a tight fit is desired.

Designs and speeds for several thermoplastics are as follows:

	Number of Flutes	*Cutting Speed (fpm)*
Cellulose acetate	2 or 3	50 to 100
Methyl methacrylate	4[a]	35 to 75
Nylon	3 or 4[b]	75 to 125
Polystyrene	3 or 4[c]	25 to 35

[a] Grind back rake angle to about 2° positive.
[b] Or use No. 2 flute spiral.
[c] Grind to zero rake.

The use of air or a lubricant is not essential in tapping, but it facilitates clearing the chips and permits faster tapping. The tap should be backed out before enough chips are formed to block the cooling agent.

In threading or tapping thermoplastics to fit a metal bolt or nut, allowance should be made for the difference in thermal expansion between the two materials. A slight increase over normal metal clearances usually is ample; but if variations in service temperature are to be extreme, dimensional changes will be too great to be accommodated in this way, and threading should be avoided.

Instead of being tapped, thermoplastics may be threaded on conventional lathes or screw machines. On automatic and semiautomatic machines, with self-opening dies, chasers should be ground to zero rake, highly polished, and chromium-plated. For nylon, a conventional rake, as for mild steel, is recommended. In most cases, two passes should be made, and the work flooded with water containing a high percentage of mild soap or other cooling agent that will not attack the plastic. If a single-point tool is used in a lathe, the point should have a 2° side rake and a zero back rake (for nylon, a zero side rake and a 2° back rake). The best possible results will be given by diamond-pointed tools.

Polystyrene. It is generally advisable to tap threads only in impact (modified) type styrene. As a general rule, National Coarse (NC) threads are preferred for the following reasons: greater strength, as more plastic area results due to the smaller minor diameter for the internal threads; and a slower helix, which results from the smaller number of threads per inch, making chip removal easier.

Spindle speeds approximately one-half of drilling speed are suggested for tappings, with the use of a coolant.

A three-flute tap is self-centering and offers easier chip removal than a four-flute type. Holes to be tapped should be slightly larger diameter than for metal in order to leave about 75% of a full thread. This will prevent the top of the thread form breaking or peeling off.

Polycarbonate. Standard steel working taps are recommended. Taps that produce threads with slightly rounded root diameters are preferred, and the use of a light machine oil during tapping is recommended to overcome resistance and reduce tap wear.

Self-tapping screws of the Parker-Kalon B-F National Screw—Type 25 thread cutting may be used with polycarbonate resin where environmental conditions permit. To ensure the best performance, it is important that the diameter of the hole be in proper relation to the diameter of the screw to be used. For example, the correct diameter of the hole for a No. 8 × $\frac{1}{2}$ inch screw is 0.147 inch, and the ideal penetration is 0.75 inch, or about four full threads on the screw. The wall thickness from the screw to an edge should be at least equal to the diameter of the screw.

The torque resistance of threads tapped in polycarbonate resin is high. A $\frac{1}{2}$-inch-diameter bolt of 13 threads per inch, penetrating a block of resin to a depth of $\frac{9}{16}$ inch, requires 48 ft lb of torque to strip the threads.

ABS. Standard metal-working tools are used. Taps should have a slight negative rake, and should produce threads with the root diameters slightly rounded in order to avoid any possibility of notch-effect weakening of the part. Taps and dies for copper and brass may be used effectively. Lubricants should always be used, and turning speeds should be very slow.

Thermosetting Plastics. Phenolics may be tapped with standard taps. The most durable are commercial ground taps with rather short chamfer and with 0.0001 to 0.0002 inch of chromium plate. If it is required to hold Class 2 or Class 3 threads, the taps should be oversize by 0.002 to 0.003 inch. Holes should be chamfered to the maximum diameter of the thread. Here again is an instance where a planned mold design may help. Frequently a hole may be spotted or molded to a shallow depth, to be drilled and tapped to final depth later. In such cases, if the hole is to be tapped, the chamfer may be molded in at the same time as the hole, to eliminate one operation.

For long production runs, a high-speed nitrided and chromium-plated tap, having three flutes rather than four, is recommended. Solid carbide taps will pay for themselves if used in a machine equipped with torque-control. A negative rake of about 5° on the front face of the land and ample clearance are necessary to ensure accurate cutting and to prevent binding and chipping during backing out. For tapping mineral-filled material, which tends to dull the tap very rapidly, sometimes a carbide tap can be used, provided that the work is clamped tightly under a torque-controlled tapping spindle so as to prevent breakage of the tap.

Flutes of taps can be opened by grinding, to make room for clearance of chips. Most tapping is done dry, but oils can be applied as lubricants. Paraffin wax sometimes is applied to the point of the tap to help to prevent heating, but air blasts on the tap, operated by the stroke of the tapping head, will help to clear the chips and cool the tap and the work. This minimizes overheating, prolongs the life of the tap, and promotes greater production per tap.

Peripheral speeds for tapping molded phenolics are from 50 to 80 fpm for taps up to $\frac{1}{4}$ inch diameter. Taps larger than this are impractical in phenolics.

A blind hole should not be machine-tapped unless there is plenty of clearance at the bottom for the tap.

Cast Phenolics. Cast phenolics may be easily tapped on vertical or horizontal tapping machines with standard taps. To provide strength, fairly coarse threads should be used. Tapped holes should be checked with plug gauges, as the abrasive action of the material causes wear. Standard machine screws can be used for assembly.

Cutting of threads of large diameter or coarse pitch, such as on bottle caps or jar covers machined from solid material, is done on a thread-milling machine, with a small milling cutter of the proper shape.

Turning and Milling

Tables 23-6 and 23-7 give the starting recommendations for turning many thermosets and thermoplastics. Table 23-6 covers turning, single-point and box tools; Table 23-7 covers turning, cutoff and form tools. The definitions are as follows: (1) *single-point turning*: using a tool with one cutting edge; (2) *box tool turning*: turning the end of a workpiece with one or more cutters mounted in a boxlike frame, primarily for finish cuts; (3) *turning cutoff*: severing the workpiece with a special lathe tool; and (4) *form turning*: using a tool with a special shape.

Chatter, with resulting problems in tool life, finish, and tolerance, can be encountered in turning operations due to workpiece flexibility. The low modulus of elasticity of plastics makes it desirable to support the work to prevent deflection of stock away from the cutting tool due to cutting forces. Close chucking on short parts and follow rests on long parts are beneficial. Box tools are designed to support long turning operations, and should be used where possible. Water-soluble cutting fluids should be used, unless they react adversely with the work material, to reduce the surface temperature generated at the shear zone and the tool–chip interface. To reduce frictional drag and temperature, tools should be honed or polished at work contact points.

Face Milling. Table 23-8 gives starting recommendations for face milling plastic materials. The fixtures used should provide adequate support for workpieces. Milling cutters usually have multiple cutting edges. In Table 23-8, feed is given in in./per tooth, which must be converted to in./min of table travel. Too high a table feed will cause a rough surface; too low a feed travel will generate excessive heat that can cause melting, surface cracks due to high temperature, loss of dimensions, and poor surface finish. Use of mist-type water-soluble cutting fluids is recommended unless workpiece-cutting fluid compatibility problems are known to exist. To reduce frictional drag and temperature, tools should be honed or polished where they contact the workpiece.

Face milling also can be accomplished by using a flycutter. The flycutter is held in a collet or chuck in the milling machine and contains a special lathe cutting tool mounted at an angle of approximately 15° to the work surface. The flycutter rotates across the work and cuts over the entire workpiece as it advances. This results in a smooth surface over the entire work area, as opposed to the several passes required by the face of a smaller end mill.

Tables 23-9 and 23-10 give starting recommendations for end milling—slotting and end

Table 23-6. Turning, single-point and box tools.*

Material	Hardness	Condition	Depth of Cut (in.)	High-Speed Steel Tool Speed (fpm)	High-Speed Steel Tool Feed (ipr)	High-Speed Steel Tool Tool Material	Carbide Tool Speed (fpm)	Carbide Tool Feed (ipr)	Carbide Tool Tool Material
Thermoplastics									
Polyethylene	$31R_R$	Extruded,	0.150	250-350	0.010	M2, T5	400-450	0.010	C-2
Polypropylene	to	molded							
TFE-fluorocarbon	$116R_R$	or cast	0.025	300-400	0.002	M2, T5	450-500	0.002	C-2
Butyrates									
High impact sty-	$83R_R$	Extruded,	0.150	250-350	0.015	M2 T5	400-450	0.015	C-2
rene; acrylonitrile-	to	molded							
butadiene-styrene;	$107R_R$	or cast	0.025	300-400	0.005	M2 T5	450-500	0.005	C-2
modified acrylic									
Nylon	$79R_M$	Molded	0.150	300-400	0.010	M2, T5	500-600	0.015	C-2
Acetals	to								
Polycarbonate	$100R_M$		0.025	400-500	0.002	M2 T5	600-700	0.005	C-2
Acrylics	$80R_M$	Extruded,	0.150	250-300	0.008	M2, T5	450-500	0.010	C-2
	to	molded							
	$103R_M$	or cast	0.025	300-400	0.005	M2, T5	500-600	0.005	C-2
Polystyrenes, low	$70R_M$	Molded	0.150	75-100	0.005	M2, T5	200-300	0.010	C-2
and medium impact	to	or							
	$95R_M$	extruded	0.025	150-200	0.001	M2 T5	350-400	0.002	C-2
Thermosets	$50R_M$	Cast,	0.150	500-1000	0.012	M2 T5	750-2000	0.012	C-2
Paper and cotton	to	molded							
base	$125R_M$	or filled	0.025	1000-2000	0.005	M2, T5	1000-3000	0.005	C-2
Fiber glass and	$50R_M$	Cast,	0.150	400-500	0.012	M2, T5	500-1000	0.012	C-2
graphite base	to	molded							
	$125R_M$	or filled	0.025	500-1000	0.005	M2 T5	750-1500	0.005	C-2
Asbestos base	$50R_M$	Molded	0.150	650-750	0.012	M2, T5	700-1000	0.010	C-2
	to								
	$125R_M$		0.025	750-1000	0.005	M2, T5	750-2000	0.005	C-2
Silica fiber-	55†	Molded	0.050	–	–	–	200	0.010	C-2
reinforced phenolic	to								
resin (Refrasil)	75		–	–	–	–	–	–	–

*Reprinted from *Machining Data Handbook*.
†Barcol hardness.

Table 23-7. Turning, cutoff and form tools.*

Material	Hardness BHN	Condition	Speed (fpm)	Feed—Inches Per Revolution: Cutoff Tool Width—Inches .062	.125	.250	Form Tool Width—Inches .500	.750	1.00	1.50	2.00	Tool Material
Thermoplastics† Polyethylene Polypropylene Fluorocarbons Butyrates	$31R_R$ to $116R_R$	Extruded, molded or cast	250 to 350	.003	.003	.003	.002	.002	.001	.001	.001	M2, M5 HSS or C-2 Carbide
High-impact styrene Acrylonitrile-butadiene-styrene Modified acrylic	$83R_R$ to $107R_R$	Extruded, molded, or cast	250 to 350	.003	.003	.003	.002	.002	.001	.001	.001	M2, M5 HSS or C-2 Carbide
Nylon Acetals Polycarbonates	$79R_M$ to $100R_M$	Molded	300 to 400	.005	.005	.005	.005	.004	.003	.002	.002	M2, M5 HSS or C-2 Carbide
Acrylics	$80R_M$ to $103R_M$	Extruded, molded, or cast	250 to 350	.003	.003	.003	.002	.002	.001	.001	.001	M2, M5 HSS or C-2 Carbide
Polystyrenes medium and low impact	$70R_M$ to $95R_M$	Molded or extruded	150 to 200	.002	.002	.002	.001	.001	.0005	.0005	.0005	M2, M5 HSS or C-2 Carbide
Thermosets Paper and cotton base	$50R_M$ to $125R_M$	Cast, molded, or filled	400 to 500	.003	.003	.003	.002	.002	.002	.002	.002	M2, M5 HSS or C-2 Carbide
Fiber glass and graphitized base	$50R_M$ to $125R_M$	Cast, molded, or filled	400 to 500	.003	.003	.003	.002	.002	.002	.002	.002	M2, M5 HSS or C-2 Carbide
Asbestos base	$50R_M$ to $125R_M$	Molded	400 to 500	.003	.003	.003	.002	.002	.002	.002	.002	M2, M5 HSS or C-2 Carbide
Silica fiber-reinforced phenolic resin (Refrasil)	55‡ to 75	Molded	100 to 200	.003	.003	.003	.002	.002	.002	.002	.002	C-2 Carbide

*Reprinted from *Machining Data Handbook*.
†The width of the form tool should not be greater than the minimum diameter of the part unless the part is supported to prevent any deflection.
‡Barcol hardness.

Table 23-8. Face milling.*

				High-Speed Steel Tool			Carbide Tool		
Material	Hardness	Condition	Depth of Cut,† (in.)	Speed (fpm)	Feed (in./tooth)	Tool Material	Speed (fpm)	Feed, (in./tooth)	Tool Material
Thermoplastics	$31R_R$ to $125R_R$	Extruded, molded, or cast	0.150 0.060	500-750 750-1000	0.016 0.004	M2, M7 M2, M7	1300-1500 1500-2000	0.020 0.005	C-2 C-2
Thermosets	$50R_M$ to $125R_M$	Extruded, molded, or cast	0.150 0.060	200-300 400-500	0.015 0.005	M2, M7 M2, M7	1300-1500 1500-2000	0.015 0.005	C-2 C-2
Reinforced plastics	55‡ to		–	–	–	–	–	–	–
Silica fiber-reinforced phenolic resin (Refrasil)	75	Molded	0.060	–	–	–	1300	0.009	C-2

*Reprinted from *Machining Data Handbook*.
†Depth of cut measured parallel to axis of cutter.
‡Barcol hardness.

milling—peripheral. End milling generally is accomplished with a tool having cutting edges on its cylindrical surfaces as well as on its end. In end milling—peripheral, the peripheral cutting edges on the cylindrical surface are used; whereas in end milling—slotting, both end and peripheral cutting edges remove metal. Face milling, in contrast, applies to milling a surface perpendicular to the axis of the cutter.

Specifically, end mills are used on thermo-

Table 23-9. End milling—slotting.*

					Feed—inches per Tooth				
					Width of Slot—Inches				
Material	Hardness BHN	Condition	Depth* of cut (in.)	Speed (fpm)	1/4	1/2	3/4	1 to 2	HSS Tool Material except as noted
Thermoplastics	$31R_R$	Extruded,	.250	270 to 450	.002	.003	.005	.008	M2, M7
	to	molded,							
	$125R_R$	or cast	.050	300 to 500	.001	.002	.004	.006	M2, M7
Thermosets	$50R_M$	Cast,	.250	125 to 180	.002	.003	.005	.008	M2, M7
	to	molded,							
	$125R_M$	or filled	.050	140 to 200	.001	.002	.004	.006	M2, M7
Silica fiber-reinforced phenolic resin (Refrasil)	55		.250	760	–	–	.004	.005	C-2 Carbide
	to	Molded							
	75†		.050	850	–	–	.003	.004	C-2 Carbide

*Reprinted from *Machining Data Handbook*.
†Barcol hardness.

Table 23-10. End milling—peripheral.*

Material	Hardness BHN	Condition	Depth of Cut† (in.)	High Speed Steel Tool						Carbide Tool					
				Speed (fpm)	Feed—Inches Per Tooth, Cutter Diameter Inches				Tool Mtl.	Speed (fpm)	Feed—Inches Per Tooth, Cutter Diameter Inches				Tool Mtl.
					1/4	1/2	3/4	1 to 2			1/4	1/2	3/4	1 to 2	
Thermoplastics	$31R_R$	Extruded,	.050	500 to 750	.004	.006	.010	.015	M2 M7	1300 to 1500	.004	.006	.010	.020	C-2
	to	molded,													
	$125R_R$	or Cast	.015	750 to 1000	.003	.005	.008	.010	M2 M7	1500 to 2000	.003	.005	.008	.010	C-2
Thermosets	$50R_M$	Cast,	.050	200 to 300	.005	.006	.010	.015	M2 M7	1300 to 1500	.005	.006	.010	.015	C-2
	to	molded,													
	$125R_M$	or filled	.015	400 to 500	.004	.005	.008	.010	M2 M7	1500 to 2000	.004	.005	.008	.010	C-2
Silica fiber-	55‡			–	–	–	–	–	–	–	–	–	–	–	–
reinforced phenolic	to	Molded													
resin (Refrasil)	75		.015	–	–	–	–	–	–	1300	.004	.005	.008	.010	C-2

*Reprinted from *Machining Data Handbook*.
†The depth of cut is measured perpendicular to the axis of the cutter. The width of cut parallel to the axis can be equal to the cutter diameter, up to 1 inch maximum.
‡Barcol hardness.

plastics to finish off the surface when a part has been degated from the sprue. The end mill may be placed in a drill press and used as in a drilling operation. A jig should be used for safe practice and quality work.

In addition, a side mill or a reamer may be used to remove a heavy parting line from a molded thermoplastic piece. The removal of unsightly seams due to clamping and cementing may be done in the same manner.

The best results are obtained in milling when using two-flute end mills for greater chip clearance and less work contact, to avoid burning and discoloration of material.

Finally, milling may be used to shape or fabricate a plastic rod or bar.

Thermoplastics. For turning and milling of thermoplastics, four cutting materials fill most needs: high-speed steel, high-speed steel chrome-plated, tungsten carbide, and diamond. They may be rated as shown in the tabulation, where the range is arbitrary from 1 (best) to 10 (worst).

Tools of standard carbon steel or high-speed steel may be used for short runs if their cutting edges are kept very sharp and their faces highly polished. However, carbide-tipped or diamond tools are almost essential for long runs because they hold a keener edge for a longer time. Such tools make possible a highly polished surface finish. To minimize surface friction at the maximum cutting speed, the side and front clearances of the tools should be somewhat greater than those of standard turning tools. Both steel and diamond end-mills should have the same cutting angles and rakes as carbide tools.

The cutting edges of all tools should be kept honed very keen. Standard carbide-tipped tools are very satisfactory except that clearance angles should be ground to from 7 to 12°. The top surface should be lapped to a bright finish. A soft iron wheel impregnated with a fine grit of powdered diamond often is used. Diamond tools are usually designed like the standard carbide tools described above, except that for turning nylon a sharp-pointed tool with a 20° positive rake is used.

A circular tool is economical in quantity production wherever it can be used.

With most thermoplastics, surface speeds may run as high as 600 fpm with feeds of from 0.002 to 0.005 inch. The speed and feed must be determined largely by the finish desired and the kind of tool used. It is difficult to make fine finish cuts on nylon except with a carbide- or diamond-tipped tool.

Good results are obtained without lubricating the cutting and turning tools if feed pressure is relaxed often enough to avoid overheating. On turret lathes, good lubrication is obtained, if necessary, by flowing a mixture of equal parts of soluble oil and water, or a soap solution, over the tools.

Milling operations must be held to very light cuts, not over 0.010 inch at normal feeds. End milling with a centercutting tool may be performed on an ordinary drill press, as long as the work is rigidly supported and the feed is steady and slow. Feeds may be increased with the aid of a suitable cooling agent. It is entirely practicable to use hand feed, but for the best finishes, power feed is advisable.

Vibration of the machine will cause a poor finish.

Tool Design. In turning *polycarbonate* parts, a very sharp tool of high-speed steel or carbide is best suited for clean smooth cuts. Suggested tool angles for this particular use are as follows: back rake—5 to 10°; side rake—15°; side clearance—15°; and front clearance—5 to 10°. Polycarbonate parts can be cut without coolant at turning speeds of 1500 to 2500 in./min. A coolant, preferably water, should be used with higher speeds to minimize strains and reduce surface galling.

Cutting depths range from 1 to 100 mils. When the finish or dimensions are critical, the depth should not exceed 15 mils. The cross-feed rate, which is dependent on the cutting depth, varies from 1 to 3 mils per revolution.

The finest surfaces are obtained by using a round-tip cutter, a high turning speed, a shallow cut, and a low cross-feed rate. Radii of 15 to 30 mils are suggested for round-tip cutters.

In turning *polystyrene* parts, tools should be ground to produce a straight nontwisting ribbon cut. The tool should have a slight negative rake (about 2°), a clearance angle of from 10 to 15°,

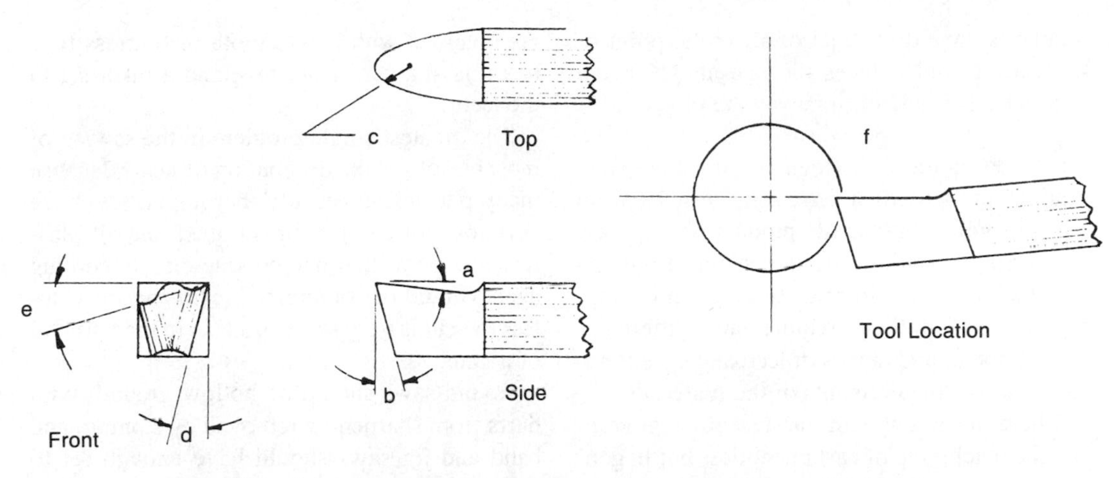

Fig. 23-5. General-purpose tool recommendations for ABS. (*Courtesy GE Plastics*)

and a side rake of about 10°. By setting the cutting edge at 1 to 2° above the center of the work and not directly at the center, the tendency of the work to climb is minimized. In all turning and most especially the facing of large diameters, coolants are essential.

Pieces should be run at speeds at which there is no vibration. This can only be determined by trial and error. Speeds up to 1300 rpm, if the piece is held rigidly, are practicable for light cuts. Either a liquid or an air blast coolant will do.

For inside turning or boring, if the diameter is large, comparable speeds may be used; but in feeding coolant to small hole operations, surface speeds may have to be decreased. To avoid crazing of the surface, the surface heat should never be raised above 175°F. Accordingly, the tools must be kept sharp and cool.

In turning *ABS*, high-speed steel tools are satisfactory, but for longer service life, carbide-tipped tools are recommended. A general-purpose tool design is shown in Fig. 23-5, with recommended angles as follows: (a) positive rake angle—0 to 5° (less frictional heat buildup occurs using a positive rake because work is cut rather than scraped); (b) front or end clearance angle—10 to 15° (this angle must be great enough to prevent the tool heel from contacting the work); (c) nose or end radius—$\frac{1}{16}$ to $\frac{3}{16}$ inch (in general, the greater the nose radius, the better the surface finish at a given speed or feed rate; cutting may be reduced, however, and heat buildup increased as the radius increases); (d) side clearance angle—10 to 15° (this dimension is not critical, except that a clearance must be allowed to minimize frictional heat buildup); (e) side rake angle—0 to 15° (to aid in removal of curls); and (f) tool angle in relation to work—approximately 2° above centerline of work (this will minimize tendency of work to climb or chatter).

Thermosetting Plastics. Molded phenolics and ureas should not be machined unless it is impossible to form the desired shape by molding. Because machining of molded articles destroys their lustrous surface, machining is restricted to articles made for utility rather than appearance. Only with a few special materials can the machined surface be satisfactorily polished.

Fillers cause difficulties in machining in the order (1) wood flour (least), (2) fabric, (3) cotton, (4) mineral, and (5) glass (most).

One of the greatest problems in machining phenolics is their inability to dissipate heat. Another is their tendency to chip.

The procedures for machining phenolics are similar to those for machining brass. The cutting action is more a scraping than a peeling. The speed in feet per minute is high and compares with the speeds for brass. The principal difference is in the abrasive action; phenolics dull the regular steel tool very quickly.

Most turning and boring of the phenolics on

a lathe is done dry. A jet of air at the point of the cutting tool reduces the rate at which the tool is dulled, and clears away the chips.

Cast Phenolics. Regular machine-shop equipment is used in producing articles from cast phenolics. In quantity production on screw machines, operations similar to those applied to metal, such as drilling, turning, threading, tapping, and milling, require only differently ground tools, and ranges of feed and speeds adjusted to the requirements of the materials.

There are a few hard-and-fast rules governing the machining of cast phenolics, but in general high speeds and light cuts are preferable. Nearly all the work is done dry. Nonalkaline cooling agents may be employed, but rarely are required. In producing articles on an automatic screw machine, where the taper of cast phenolic rod may hamper the feeding of it, the rod may first be centerless-ground accurately to uniform diameter.

Turning tools are sharpened very much as for brass. There should be plenty of clearance, 10 to 20°, and a slightly negative or zero rake is desirable. The tool should be set 1 to 2°, and a slightly negative or zero rake is desirable. The tool should be set 1 to 2° above the center of the material. Cutting edges should be in condition to produce long ribbon shavings or chips. Honed tools produce the smoothest cuts. Diamond, carbide, or high-speed steel may be used to get the longest runs without resharpening or resetting.

For maximum efficiency and best working conditions, dust and shavings should be removed by a blower.

Spindle speeds in turning operations range from 450 to 6000 rpm, depending on the specific work being done and the diameter of the material, and should be regulated to give a surface speed of about 600 fpm.

Sawing

Tables 23-11 and 23-12 give general recommendations for the power band sawing of plastics. Band sawing is recommended because it is more versatile for making irregular or curve contours as well as straight cuts. Best results are obtained with a skip tooth or buttress-type tooth having zero front rake and a raker set to the teeth.

The greatest single problem in the sawing of most plastics is the dissipation of heat. Because many plastics, especially thermoplastics, have very low softening temperatures, and all plastics are poor thermal conductors, a cooling agent (liquid or compressed air) is needed, unless the cut is very short, such as in the removal of a gate.

Round saws should be hollow-ground, with burrs from sharpening removed by stoning, and band and jig saws should have enough set to give adequate clearance to the back of the blade. This set should be greater than is usual for cutting steel. It is always best to relieve the feed pressure near the end of a cut, to avoid chipping.

The proper rate of feed is important, and, because most sawing operations are hand-fed, it can be learned only through experience. Attempts to force the feed will result in heating of the blade, gumming of the material, loading of the saw teeth, and an excessively rough cut. Before the next cut is made, the saw must be cleaned. Chromium plating of the blade reduces friction and tends to give better cuts.

Above all, the saw, whether band or circular, must be kept sharp. Frequent sharpening of circular saws pays for itself. Dull bandsaws should be replaced.

Circular saws are usually from $\frac{1}{32}$ to $\frac{1}{8}$ inch thick. The width of bandsaws is usually $\frac{3}{16}$ to 1 inch.

Both thermoplastic and thermosetting resins can be sawed also by use of cutoff machines having abrasive wheels. This equipment is used for cutting rods, tubes, and so forth, into lengths, for slicing profiled bars of cast phenolics, and for removing large gates. Narrow-faced wheels are used (0.02 to 0.125 inch), usually containing a silicon carbide abrasive in the range of No. 36 to No. 50. With an appropriate wheel properly used, clean cuts can be made. If necessary, water is used to prevent overheating.

Thermoplastics. As already indicated, the sawing of thermoplastics will require a wide

Table 23-11. Power band sawing, vertical, high-speed steel blades.*

Material	Hardness BHN	Condition	Material Thickness (in.)	Tooth Form†	Pitch (teeth per in.)	Band Speed (fpm)
Thermoplastics						
Polyethylene	$31R_R$	Extruded,	$<\frac{1}{2}$	P	10-14	4000‡
Polypropylene	to	molded,	$\frac{1}{2}$-1	P	6	3000‡
Fluorocarbons	$116R_R$	or cast	1-3	B	3	2500‡
Butyrates			>3	B	3	1500‡
High-impact styrene	$83R_R$	Extruded	$<\frac{1}{2}$	P	10-14	2300‡
Acrylonitrile-butadiene-	to	molded,	$\frac{1}{2}$-1	P	8	2000‡
styrene	$107R_R$	or cast	1-3	P	6	1500‡
Modified acrylic			>3	B	3	1000‡
Nylon	$79R_M$	Molded	$<\frac{1}{2}$	P	10-14	3000‡
Acetals	to		$\frac{1}{2}$-1	P	6	2500‡
Polycarbonates	$100R_M$		1-3	B	3	2000‡
			>3	B	3	1500‡
			$<\frac{1}{2}$	P	10-14	4000‡
Acrylics	$80R_M$	Extruded,	$\frac{1}{2}$-1	P	6	3000‡
	to	molded,	1-3	B	3	2500‡
	$103R_M$	or cast	>3	B	3	1500‡
			$<\frac{1}{2}$	P	8-10	2300‡
Polystyrenes,	$70R_M$	Molded	$\frac{1}{2}$-1	P	6	2000‡
medium and low impact	to	or	1-3	B	3	1500‡
	$95R_M$	extruded	>3	B	3	1000‡
			$<\frac{1}{2}$	P	8-10	2300‡
Fluorocarbons	$31R_R$	Extruded,	$\frac{1}{2}$-1	P	6	2000‡
	to	molded,	1-3	B	3	1500‡
	$116R_R$	or cast	>3	B	3	1000‡
			$<\frac{1}{2}$	P	10-14	4000‡
Cellulose acetate	$49R_R$	Molded	$\frac{1}{2}$-1	P	6	3000‡
	to		1-3	B	3	2500‡
	$112R_R$		>3	B	3	1500‡
			$<\frac{1}{2}$	P	10-14	3000‡
Nylon	$79R_M$	Molded	$\frac{1}{2}$-1	P	6	2500‡
	to		1-3	B	3	2000‡
	$100R_M$		>3	B	3	1500‡
			$<\frac{1}{2}$	P	10-14	5500‡
Thermosetting Plastics	$114R_M$	Unfilled,	$\frac{1}{2}$-1	P	8	5000‡
Melamine-	to	molded	1-3	B	3	4000‡
formaldehyde	$119R_M$		>3	B	3	3000‡
	$50R_M$	Cast, molded,	$<\frac{1}{2}$	P	10-14	5500‡
	to	laminated-	$\frac{1}{2}$-1	P	6	5000‡
	$93R_M$	asbestos or	1-3	B	3	4000‡
		fabric filler	>3	B	3	3000‡
	$100R_M$	Cast or	$<\frac{1}{2}$	P	10-14	5500‡
	to	molded,	$\frac{1}{2}$-1	P	6	5000‡
	$105R_R$	no filler	1-3	B	3	4000‡
			>3	B	3	3000‡
Phenolics	$108R_M$	Cast, molded	$<\frac{1}{2}$	P	10-18	75‡
	to	or laminated	$\frac{1}{2}$-1	P	8	50‡
	$115R_M$	with glass	1-3	P	6	50‡
		or mineral	>3	–	–	–
		filler				
			$<\frac{1}{2}$	P	10-14	5000‡
			$\frac{1}{2}$-1	P	6	4000‡
Urea-formaldehyde	$94R_E$	Molded	1-3	B	3	3500‡
			>3	B	3	2200‡

*Reprinted from *Machining Data Handbook*.
†P—Precision; C—Claw; B—Buttress.
‡Recommended speed for high carbon blades.

Table 23-12. Power band sawing, high carbon saw blads.*

Material	Hardness	Condition	Pitch, teeth/in. Minimum thickness of material, in.			Speed (fpm)
			1/4 and under	1/4-1 1/2	1 1/2 and over	
Thermoplastics						
Cellulose acetate	$49R_R$ to $112R_R$	Molded	10	6-10	3-6	3000
Acrylics	$80R_M$ to $103R_M$	Extruded, molded, or cast	14	6-10	3-6	3000
Nylon	$79R_M$ to $100R_M$	Molded	14	4-10	3-4	2500
Polystyrenes	$70R_M$ to $95R_M$	Molded or extruded	10	4-6	3-4	2000
Thermosets						
Melamine-formaldehyde	$114R_M$ to $119R_M$	Unfilled molded	14	6-10	3-4	4500
Phenolics	$50R_M$ to $93R_M$	Cast, molded laminated asbestos or fabric filler	14	6-10	3-6	4500
	$100R_M$ to $105R_M$	Cast or molded, no filler	14	6-10	3-6	4500
	$108R_M$ to $115R_M$	Cast, molded or laminated with glass fiber or graphitized cloth	14-18	8-10	6-8	2000
Urea-formaldehyde	$94R_E$	Molded	14	6-10	3-6	4500

*Reprinted from *Machining Data Handbook*, Research Associates Inc.

range of conditions, depending upon the type of resin involved. To illustrate, the following are recommendations for sawing styrene and for sawing polycarbonate, one of the so-called engineering plastics with greater toughness and heat resistance.

Polystyrene. Of the three types of sawing operations possible on styrene, namely circular jig or band sawing, the most preferable technique is that of band sawing. This method lends itself most readily to sawing styrene since the heat build-up can be most easily dissipated.

- *Band sawing:* The use of coolant (usually water) usually is advisable here. Nevertheless, it has been found that for styrene sections from $\frac{1}{8}$ inch up, it is possible to increase the cutting efficiency and reduce the extent of lubrication by using a skip-tooth band, four to six teeth per inch, instead of the usual metal cutting band. No coolant is required for $\frac{1}{8}$-inch thicknesses of styrene. Metal cutting band saws can be employed for cuts in sections of thickness smaller than $\frac{1}{8}$ inch. Ample coolant should be used for the latter; otherwise, the band will be seized by gumming and cause chipping at the cut. The operating speeds for band sawing styrene should be from 465 to 550 fpm.

In sawing thick sections of styrene (4–5 inches) it is desirable to use a wide band slot to permit clearing of waste and also to remove the slot disc. A helpful device in sawing thick sections is to support the work on two wooden cleats which straddle the blade so that the specimen is held about $\frac{3}{4}$ inch from the base of the saw. For thick pieces, operating speeds of 1100 rpm and a 36-inch band saw are recommended.

- *Circular sawing:* It is advantageous to use a hollow ground saw for this purpose, although this is not always necessary if all the other precautions are taken. Blade thicknesses may vary from 0.040 inch for sections up to $\frac{1}{8}$ inch, increasing to $\frac{3}{32}$ inch for heavier pieces. For general work a $\frac{1}{16}$-inch blade is a good compromise. Saw diameters may be from 4 to 10 inches. Smaller saws derive their advantages from the fact that they have lower peripheral speeds at any given spindle speed. Consequently, they have less tendency to heat themselves and the work. The thickness of the work determines, for the most part, the number of teeth per inch. Another consideration is the diameter of the saw. For a 4-inch blade, a good compromise is from 8 to 10 teeth per inch; for thicker work sections and larger-diameter blades, the number of teeth may increase to 15 per inch.

The use of coolant is almost always recommended for this operation. The saw should be flooded with coolant directly above and at the point of operation. Customary operating speeds are from 1800–2000 rpm. Nevertheless, if an uninterrupted flow of coolant is fed to the saw, speeds as high as 3450 rpm can be reached without damaging the material or the tool. Only when small cuts (of 2 inches or less) are taken can coolant be dispensed with in circular saw operations using normal-type blades. In so doing, the feed must be unforced. A helpful device in circular sawing is to have the clearance between the blade and the saw table as narrow as possible to prevent the material from pulling into the gap.

• *Jig sawing:* This operation is not a recommended one for styrene. The jig saw blade, because of its reciprocal action, cannot handle the liberal quantity of coolant required and consequently tends to splash it up toward the operator on the upward stroke of the blade. It is generally preferable to rout holes out rather than resort to jig sawing. For sharp-cornered holes, band filing without coolant is recommended, provided that the feed is light. In those instances where it is absolutely necessary to use a jig saw, standard blades may be used at a speed of about 875 cycles per minute.

Polycarbonate. Parts molded of polycarbonate can be cut with common band and circular saws as well as with hand or power hack saws. Special blades rarely are required. Blade speeds and cutting rates are not so critical as with other thermoplastics because of the high heat resistance of polycarbonate.

Special attention to blades and cutting rates is needed only when sawing is the final or only machining operation, or when very thick sections are involved. For material 1 inch or more thick, blade and feed speeds should be carefully chosen, just as they would be for soft metals. Band saw blades with 10 to 18 teeth per inch are satisfactory. When the blades have a set, the preferred range is 20 to 30 mils. Tooth spacing for circular saw blades ranges from large (for cutting thick sections) to very small (for cutting thin sections). The best procedure is to use the smallest spacing that gives satisfactory performance.

The recommended clearance angle for circular saw teeth is between 20 and 30°. The rake angle should be 15°.

High-speed steel or carbide-tipped circular saws with alternate beveled and straight milled teeth are preferred. The use of designs that eliminate the possibility of the blade body rubbing the stock during sawing improves the results, particularly when working with thick sections.

The table gives recommended sawing conditions for polycarbonate.

Thermosetting Plastics. The sawing of these plastics is confined to band saws (solid or inserted-segment), carbide circular saws, and abrasive cutoff discs.

Nearly all of these plastics are extremely abrasive. Some of them, notably the glass-filled alkyds, with dull a circular saw of high-speed steel in a cut of a few inches in $\frac{1}{4}$-inch material. In production runs of this material, it is probably best to use an abrasive cutoff disc, 0.040 to $\frac{1}{16}$ inch thick, 6 to 20 inches in diameter, run at from 3500 to 6000 rpm.

Phenolics, ureas, and melamines usually are cut by band saws, which have the advantage of dissipating the heat.

An air jet should be used at the point of contact, and a suction hose to remove the dust.

Piercing, Trimming, and Routing

Thermoplastics. Small holes may be punched through thin sheets of thermoplastics on standard hand-operated arbor presses, but an ordinary punch press or shearing machine generally is used. Thin sheets may be processed cold, but thicker sheets should be heated. For acrylics, about 185°F will be required for the heavier sections, but the other materials should not be heated above about 120°F, in order to prevent damaging the surface and finish.

It is common practice to remove ring gates from injection-molded articles by shearing, even to the extent of making multiple-cavity punching dies for multiple-cavity moldings. It is important that the punch and die fit closely, to avoid producing ragged edges.

Blanking or shaving dies frequently are used for removing parting lines and flash lines. The punch and die sometimes are heated to leave the best possible finish. Routing and shaping are done on standard woodworking equipment. For fine cuts at high speed, single- or multiple-bladed fly cutters should be used.

Routing is a grooving or milling operation. It is done by means of small-diameter cutters having three or more teeth of the desired profile and mounted on vertical spindles that project above a metal tabletop. These cutters revolve at high speeds (e.g., 12,600 rpm) so as to produce smooth cuts.

The work is held in a special fixture equipment with a metal master guideplate that is kept in contact with a metal collar on the table, directly below the cutter. The fixture is moved by hand.

For bringing rough-sawed articles of flat stock to size and shape, edge-molders are used, of the type used for finishing articles of wood. A cutter of the desired profile is mounted on a verticle spindle, which revolves at 7000 to 10,000 rpm, depending on the size of the cutter and the amount of stock to be removed. The work is held in a fixture equipped with a metal master guideplate, which is kept in contact with a collar on the spindle below the cutter. The fixture is moved by hand. Fixtures of semiautomatic type have been developed to accomplish this same result; the operator loads and unloads the fixtures as they advance to and from the cutters. The cut is made rapidly to prevent burning the material.

Thermosetting Plastics. If the quantities are sufficiently large to warrant the cost of piercing tools, they provide the best method for removing flash from holes. A simple pad die with stripper plate may be made to pierce all holes and trim the outside shape of an article, at the rate of about 700 per hour. Small articles with several holes may be pierced in a dial press at the rate of about 1200 per hour. The piercing punches for removal of flash should be round-nosed and should have two or three helical grooves around the working end. These grooves should pass by the flash line in the hole, to rake the flash out of the hole. If the flash is on the surface of the article, it should be punched into the hole rather than out of it in order to prevent chipping.

Although thin sheets ($\frac{1}{4}$ to $\frac{3}{16}$ inch) of cast phenolic may be blanked with a steel-rule die, this is not recommended because of the difficulty in obtaining accurate cuts. Blanks are better and more economically obtained by casting a bar of the desired profile and slicing or cutting it with standard slicing equipment or abrasive cutoff equipment.

Blanking and Die-Cutting

Many thermoplastics sheets are thermoformed into end products such as luggage, housings, toys, trays, and so on. Most of these products are easily trimmed or punched with hardened steel dies.

Other types of trimming operations may include the use of steel rules, punch dies, clicker dies, and hi-dies, depending on the gauge of the material. Punch presses, kick presses, and clicker or dinking machines are used in actuating the different die-cutting methods.

Dies should be kept as sharp as possible. If they are not kept sharp, ragged or burred edges will result, possibly allowing cracks to propagate when the stock is under stress. Dies must also be kept free of grease and oils, some of which may create stress crazing or cracking in various plastics.

Many light-gauge plastics sheets may be trimmed with dies made of steel rules formed to the desired shape and placed in a wooden or metal frame. Steel rule generally is used to trim shallow drawn parts, but it can be built up economically into dies for cutting parts with draws more than 2 inches deep. Backup surfaces in many cases (e.g., for the styrene sheets) may be hard rubber or wood blocks oriented to use the end grains.

Heavier-gauge sheet may require stronger dies. Tool steel up to $\frac{1}{4}$ inch thick can be bent or forged to shape to make clicker dies. The hi-die is also suitable for heavy-gauge material and long runs. Parts with draws as deep as 7 inches may be trimmed without building up these dies.

Counterboring and Spotfacing

Counterboring involves the removal of material to enlarge a hole for part of its depth with a rotary, pilot-guided, end cutting tool having two or more cutting lips and usually having straight or helical flutes for the passage of chips and the admission of a cutting fluid.

In spotfacing, one uses a rotary, hole-piloted, end facing tool to produce a flat surface normal to the axis of rotation of the tool on or slightly below the workpiece surface.

Recommendations for counterboring or spotfacing plastics are indicated in Table 23-13.

Laser Machining

A technique that has gained importance in the machining of plastics is the use of lasers, particularly the carbon dioxide laser. The carbon dioxide laser emits radiation at a wavelength of 10.6 microns, which is in the infrared region and is strongly absorbed by most plastics. In essence, the laser concentrates the invisible infrared beam through focusing lenses to create a point in space with tremendous power density. If a workpiece is held stationary under the laser, it will drill a hole; if the piece is moved, it slits the material. The induced heat is so intense and the action takes place so quickly, that little heating of adjacent areas takes place.

Laser machining has been used for cutting and drilling such plastics as acrylic, polyethylene film, polystyrene foam, acetal, and nylon. Most plastics can be machined with the technique (epoxies and phenolics do not perform well and vinyl also has problems).

Jigs, Fixtures, and Automatic Feeding Devices

The primary purpose of jigs and fixtures is to facilitate machining, in order to reduce manufacturing costs and to make practicable various operations that cannot be performed by hand. With their aid it is possible to produce more pieces per hour, to use less skilled labor, to improve the accuracy of dimensions, and to reduce the fitting necessary in assembling.

It generally is agreed that a fixture is a device that holds the article while the cutting tool is performing the work, whereas a jig is a device that not only holds the object but also incorporates special arrangements for guiding the tool to the proper position. It would follow, therefore, that jigs are used principally for drilling, boring, and so on, whereas fixtures are used in milling and grinding. Regardless of the nomenclature, the most important feature of any jig or fixture is its ability to perform the work for which it was made.

Simplicity of design makes for low cost, long life, and inexpensive maintenance. With this in mind, the following rules may be used as a guide to the proper design of jigs and fixtures:

1. First, it should be shown on the basis of cost studies, that the cost of designing and making a proposed tool will be less than the gross savings that its use will yield.
2. All clamping devices should be quick-acting, easily accessible, and so placed as to give maximum resistance to the direction of the force of the cutting tool.
3. The clamping pressure must not crack or distort the article of plastic.
4. Locating points should be visible to the operator when positioning the article, in order to minimize loading time.
5. All bearings should be of the sealed type, for protection against the highly abrasive plastic chips and dust.

Table 23-13. Counterboring and spotfacing.*

Material	Hardness BHN	Condition	Speed (fpm)	Feed – Inches Per Revolution: Nominal Hole Diameter – Inches								Tool Material
				¼	½	¾	1	1.5	2	2.5	3	
THERMOPLASTICS												
Polyethylene Polypropylene Fluorocarbons Butyrates	$31R_R$ to $116R_R$	Extruded, Molded, or cast	175	.001	.0015	.002	.002	.0025	.003	.0035	.0045	M2, M3 HSS
			400	.002	.002	.003	.003	.003	.004	.004	.005	C-2 Carbide
High-impact styrene Acrylonitrile-butadiene-styrene Modified acrylic	$83R_R$ to $107R_R$	Extruded, molded, or cast	175	.001	.0015	.002	.002	.0025	.003	.0035	.0045	M2, M3 HSS
			400	.002	.002	.003	.003	.003	.004	.004	.005	C-2 Carbide
Nylon Acetals Polycarbonates	$79R_M$ to $100R_M$	Molded	200	.002	.003	.004	.006	.008	.009	.010	.012	M2, M3 HSS
			450	.002	.004	.005	.007	.009	.010	.011	.012	C-2 Carbide
Acrylics	$80R_M$ to $103R_M$	Extruded, molded, or cast	200	.002	.003	.004	.006	.008	.009	.010	.012	M2, M3 HSS
			450	.003	.004	.005	.007	.009	.010	.011	.012	C-2 Carbide
Polystyrenes, medium and low impact	$70R_M$ to $95R_M$	Molded or extruded	150	.001	.0015	.002	.002	.0025	.003	.0035	.004	M2, M3 HSS
			300	.002	.002	.003	.003	.003	.004	.004	.005	C-2 Carbide

THERMOSETS												
Paper and cotton base reinforced	$50R_M$ to $125R_M$	Cast, molded, or filled	300	.002	.003	.004	.006	.008	.009	.010	.012	T15, M33 M41 Thru M47 HSS
			500	.003	.004	.005	.007	.009	.010	.011	.012	C-2 Carbide
Silica fiber-reinforced phenolic resin (Refrasil)	55† to 75†	Molded	25	.003	.004	.005	.007	.010	.012	.014	.016	T15, M33, M41 Thru M47 HSS
			150	.004	.005	.006	.008	.011	.013	.015	.017	C-2 Carbide
Fiber glass, graphitized, and asbestos base	$50R_M$ to $125R_M$	Cast, molded, or filled	200	.002	.003	.004	.006	.008	.009	.010	.012	T15, M33, M41 Thru M47 HSS
			450	.002	.004	.005	.007	.009	.010	.011	.012	C-2 Carbide

*Reprinted from *Machinery Data Handbook*.
†Barcol hardness.

6. Proper consideration should be given to cooling the cutting tool.
7. The fixture must be foolproof, so that the operator cannot position the article in any way but the correct one.
8. Lubrication must be provided for moving parts.
9. The tool should be made as light as possible by using light materials or coring out unnecessary metal.
10. Holes should be provided for the escape of chips or dirt.

When one is designing jigs or fixtures, it is well to keep in mind the possibilities of converting to automatic operation if production runs will be large enough to make this worth the cost. By coupling electrical or air-powered devices and automatic feeders to the fixture, it is possible to eliminate the need for an operator except to load the hopper and remove the finished pieces. Because one operator can service a number of automatic fixtures, this may reduce the manufacturing cost per piece to a very small fraction of the original cost.

A discussion of automation as it applies to machining, assembly, and decoration can be found in the section on ''Automation'' in Chapter 24.

24

Joining and Assembling Plastics

This chapter is concerned with basic techniques for joining plastics to themselves, to other plastics, and to other materials.

The techniques vary considerably in terms of the plastic involved and the application for which it is intended. Adhesives, for example, are widely used in plastics assembly; but they, in turn, are subdivided into solvent or dope cements, which are suitable for most thermoplastics (not thermosets), and monomeric or polymerizable cements, which can be used for most thermoplastics and thermosets. It should also be noted that there are some plastics with outstanding chemical resistance, such as the polyolefins, that preclude the use of many cements and generally require some form of surface treatment prior to adhesion.

To accommodate these distinctions, this chapter is subdivided into sections on adhesive bonding of plastics, solvent cementing of thermoplastics, adhesives, cementing of specific plastics, cementing of thermosetting plastics, welding of thermoplastics, ultrasonic assembly, mechanical joints, and automation.

A good starting point is Table 24-1, which summarizes available bonding and joining techniques in terms of their basic advantages and limitations.

ADHESIVE BONDING OF PLASTICS

Adhesive bonding has found wide use in the assembly of plastics by virtue of low cost and adaptability to high-speed production. In addition, adhesives provide a more uniform distribution of stresses over the assembled areas and a high strength/weight ratio.

Three different types of cements are commonly used in bonding plastics:

(1) Solvent cements and (2) dope cements are used for most thermoplastics and function by attacking the surfaces of the adherends so that they soften and, on evaporation of the solvent, will join together. The dope cements, or bodied cements, differ from the straight solvents in that they also contain, in solution, a quantity of the same plastic that is being bonded. In drying, these cements leave a film of plastic that contributes to the bond between the surfaces to be joined.

(3) Monomeric or polymerizable cements consist of a reactive monomer, identical with or compatible with the plastic to be bonded, together with a suitable system of catalyst and promoter. The mixture will polymerize either at room temperature or at a temperature below the softening point of the thermoplastic. In order to hasten setting and to reduce shrinkage, a quantity of the solid plastic may be dissolved in the monomer. In addition to chemical bonding (specific adhesion), there may or may not be a degree of solvent attack by the monomer on the plastic. Adhesives of this type may be of an entirely different chemical type from the plastics being bonded; for example, many thermoplastics may be bonded by means of a liquid mixture of resin and hardener based upon epoxy resins, where the chemical reactivity and hy-

Table 24-1. Bonding and joining techniques for plastics materials.*

Technique	*Description*	*Advantages*	*Limitations*	*Processing Considerations*
SOLVENT CEMENT AND DOPES	Solvent softens the surface of an amorphous thermoplastic; mating takes place when the solvent has completely evaporated. Bodied cement with small percentage of parent material can give more workable cement, fill in voids in bond area. Cannot be used for polyolefins and acetal homopolymers.	Strength, up to 100% of parent materials, easily and economically obtained with minimum equipment requirements.	Long evaporation times required; solvent may be hazardous; may cause crazing in some resins.	Equipment ranges from hypodermic needle or just a wiping medium to tanks for dip and soak. Clamping devices are necessary, and air dryer is usually required. Solvent recovery apparatus may be necessary or required. Processing speeds are relatively slow because of drying times. Equipment costs are low to medium.
THERMAL BONDING				
1. Ultrasonics	High-frequency sound vibrations transmitted by a metal horn generate friction at the bond area of a thermoplastic part, melting plastics just enough to permit a bond. Materials most readily weldable are acetal, ABS, acrylic, nylon, PC, polyimide, PS, SAN.	Strong bonds for most thermoplastics; fast, often less than 1 sec.	Size and shape limited. Limited applications to PVCs, polyolefins.	Converter to change 20 kHz electrical into 20 kHz mechanical energy is required long with stand and horn to transmit energy to part. Rotary tables and high-speed feeder can be incorporated.
2. Hot plate and hot tool welding	Mating surfaces are heated against a hot surface, allowed to soften sufficiently to produce a good bond, then clamped together while bond sets. Applicable to rigid thermoplastics.	Can be very fast, e.g., 4–10 sec in some cases; strong bonds.	Stresses may occur in bond area.	Uses simple soldering guns and hot irons, relatively simple hot plates attached to heating elements up to semiautomatic hot plate equipment. Clamps needed in all cases.
3. Hot gas welding	Welding rod of the same material being joined (largest application is vinyl) is softened by hot air or nitrogen as it is fed through a gun that is softening part surface simultaneously. Rod fills in joint area and cools to effect a bond.	Strong bonds, especially for large structural shapes.	Relatively slow; not an "appearance" weld.	Requires a hand gun, special welding tips, an air source and welding rod. Regular hand gun speeds run 6 in./min.; high-speed hand-held tool boosts this to 48–60 in./min.

4. Spin welding	Parts to be bonded are spun at high speed, developing friction at the bond area; when spinning stops, parts cool in fixture under pressure to set bond. Applicable to most rigid thermoplastics.	Very fast (as low as 1–2 sec); strong bonds.	Bond area must be circular.	Basic apparatus is a spinning device, but sophisticated feeding and handling devices are generally incorporated to take advantage of high-speed operation.
5. Dielectrics	High-frequency voltage applied to film or sheet causes material to melt at bonding surfaces. Material cools rapidly to effect a bond. Most widely used with vinyls.	Fast seal with minimum heat applied.	Only for film and sheet.	Requires RF generator, dies, and press. Operation can range from hand-fed to semiautomatic with speeds depending on thickness and type of product being handled. 3–25 kW units are most common.
6. Induction	A metal insert or screen is placed between the parts to be welded, and energized with an electromagnetic field. As the insert heats up, the parts around it melt, and, when cooled, form a bond. For most thermoplastics.	Provides rapid heating of solid sections to reduce chance of degradation.	Because metal is embedded in plastic, stress may be caused at bond.	High-frequency generator, heating coil and inserts (generally 0.02–0.04 in. thick). Hooked up to automated devices, speeds are high. (1–5 kW used.) Work coils, water cooling for electronics, automatic timers, multiple position stations may also be required.
ADHESIVES[a]				
1. Liquids—solvent, water base, anaerobics	Solvent- and water-based liquid adhesives, available in a wide number of bases—e.g., polyester, vinyl—in one- or two-part form, fill bonding needs ranging from high-speed lamination to one-of-a-kind joining of dissimilar plastics parts. Solvents provide more bite, but cost much more than similar-base water-type adhesive. Anaerobics are a group of adhesives that cure in the absence of air, with a minimum amount of pressure required to effect the initial bond. Adhesives are used for almost every type of plastic.	Easy to apply; adhesives available to fit most applications.	Shelf and pot life often limited. Solvents may cause pollution problems; water-base not as strong; anaerobics toxic.	Application techniques range from simply brushing on to spraying and roller coating-lamination for very high production. Adhesive application techniques, often similar to decorating equipment. Anaerobics are generally applied a drop at a time from a special bottle or dispenser.

Table 24-1. (*Continued*)

Technique	*Description*	*Advantages*	*Limitations*	*Processing Considerations*
2. Mastics	Highly viscous single- or two-component materials, which cure to a very hard or flexible joint depending on adhesive type.	Does not run when applied.	Shelf and pot life often limited.	Often applied via a trowel, knife, or gun-type dispenser; one-component systems can be applied directly from a tube. Various types of roller coaters are also used.
3. Hot melts	100% solids adhesives that become flowable when heat is applied. Often used to bond continuous flat surfaces.	Fast application; clean operation.	Virtually no structural hot melts for plastics.	Hot melts are applied at high speeds via heating the adhesive, then extruding (actually squirting) it onto a substrate, roller coating, using a special dispenser or roll to apply dots or simply dipping.
4. Film	Available in several forms including hot melts, these are sheets of solid adhesive. Mostly used to bond film or sheet to a substrate.	Clean, efficient.	High cost.	Film adhesive is reactivated by a heat source; production costs are in the medium–high range, depending on heat source used.
5. Pressure-sensitive	Tacky adhesives used in a variety of commercial applications (e.g., cellophane too). Often used with polyolefins.	Flexible.	Bonds not very strong.	Generally applied by spray with bonding effected by light pressure.
MECHANICAL FASTENERS (Staples, screws, molded-in inserts, snap fits and variety of proprietary fasteners.)	Typical mechanical fasteners are listed on the left. Devices are made of metal or plastic. Type selected will depend on how strong the end product must be, appearance factors. Often used to join dissimilar plastics or plastics to nonplastics.	Adaptable to many materials; low-to-medium costs; can be used for parts that must be disassembled.	Some have limited pull-out strength; molded-in inserts may result in stresses.	Nails and staples are applied by simply hammering or stapling. Other fasteners may be inserted by drill press, ultrasonics, air or electric gun, hand tool. Special molding—i.e., molded-in-hole—may be required.

[a]Because of the thousands of formulations available within the various adhesive categories, it is not practical to attempt an analysis here of which adhesives will satisfy a particular application's need. However, typical adhesives in each class are: Liquids: (1) solvent—polyester, vinyl, phenolics, acrylics, rubbers, epoxies, polyamide; (2) water—acrylics, rubber, casein; (3) anaerobics—cyanacrylate. Mastics—rubbers, epoxies. Hot melts—polyamides, PE, PS, PVA. Film—epoxies, polyamide, phenolics. Pressure-sensitive—rubbers.

*Reprinted from the August 1970 issue of *Plastics Technology*, published by Bill Communications, "Bonding and Joining Plastics," by J. O'Rinda Trauernicht.

drogen bonding available from the epoxy adhesive contribute to excellent specific adhesion to many materials.

Techniques

Regardless of the composition of the cement employed, the following general rules should be observed in cementing plastic materials:

1. The surfaces to be cemented must be clean; a slight film of oil, mold-release agent, water, or polishing compound will cause poor bonding.
2. The surfaces must be smooth, and aligned as nearly perfectly as practicable.
3. Where solvent or dope is used, it must be sufficiently active to soften the surfaces to such a depth that when pressure is applied, a slight flow occurs at every point in the softened area.
4. The solvent or cement must be of such composition that it will dry completely without blushing.
5. Light pressure must be applied to the cemented joint until it has hardened to the extent that there is no movement when it is released. (If the required clamping process actually deforms the parts, the stress due to spring-back of the flexible parts must be relieved before the joint hardens thoroughly, or else the joint may subsequently fail. This may be accomplished by releasing the clamps while the adhesive is just slightly ''wet.'')
6. Subsequent finishing operations must be postponed until the cement has hardened.
7. Care must be taken that the vapor from solvent cements is not confined, in order to prevent the surface of the molded piece from becoming etched.
8. Adequate ventilation and attention to fire hazards also are required for the protection of personnel.

In the application of adhesives, it is very important that the surfaces of the joint be clean and well matched. Poor contact of mating surfaces can cause many troubles. The problem of getting proper contact is aggravated by warpage, shrinkage, flash, marks from ejector pins, and nonflat surfaces.

Care must be taken to prevent application of adhesive to surfaces other than those to be joined, in order to avoid disfigurement of the surface. The adhesives should be applied evenly over the entire joint surface in sufficient quantity to preclude voids. The assembly should be made as soon as the surfaces have become tacky, which usually means within a few seconds after application. Enough pressure should be applied to ensure good contact until the initial bond strength has been achieved. Stronger bonds result when the adhesive is applied to both pieces of an assembly.

Joints

The design of the joint plays a large part in the effectiveness of the cement, the appearance of the joint, and the ease and cost of assembly. Butt, lap, and tongue-and-groove joints are in most frequent use, but angled or scarf joints and V-joints are used in some assemblies (see Fig. 24-1). If the strength of the joint is to be equivalent to that of the adjacent wall, the area

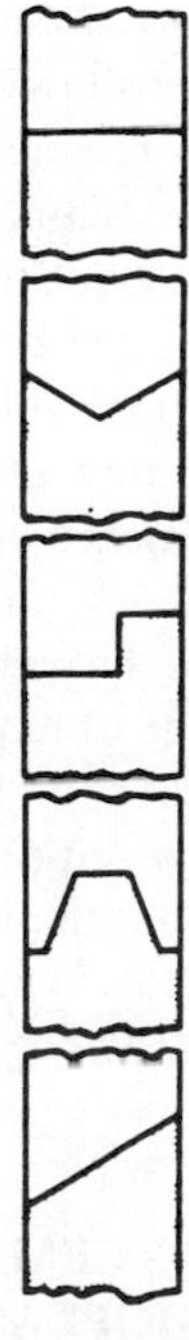

Fig. 24-1. Joint designs. From top to bottom: butt, ''V,'' lap, tongue-and-groove, scarfed. (*Courtesy GE Plastics*)

of the joint should be increased by at least 50% over that of the original contiguous edge. A joint that combines both shear and tensile strength is most effective.

Lap Joint. The lap joint has some advantages over the others, in regard to both appearance and, to some extent, strength. For best appearance, the solvent should be applied by a felt pad to the half of the joint that fits on the inside of the article. Then no exudation of adhesive will appear at the outer parting line.

When two pieces of annular cross section are to be cemented with a lap joint, it is advantageous to make the outer lip thin in proportion to the inner (e.g., in a ratio of 1 : 4), as the greater periphery thus gained adds to the area, and hence the strength, of the joint. Lap joints are generally used for adhesive joining, whereas butt joints are preferred for solvent cementing.

Butt Joint. A butt joint gives less contact area than most lap joints. The butt joint often has the disadvantage of unsightly bond lines, but it is probably the easiest type to provide for in molding. The butt joint is not a self-locating joint, and locating pins or fixtures often are required to prevent slipping during clamping.

Tongue-and-Groove Joint. A tongue-and-groove joint is self-positioning, but unless it is very shallow, it will require application of adhesive by other than felt-pad methods. If the adhesive is applied by a flow gun into the groove, the amount must be controlled so that after assembly the fluid will come to the edge of the joint, but not flow out to mar the outer appearance of the finished article.

Other Joints. The V-joint and the scarfed joint are variations of the preceding joints. They have the disadvantage of requiring a somewhat more complex method of application of cement. The V-joint is self-positioning, and the scarfed bond is self-positioning on certain types of articles. It is difficult to mold the component parts to the close tolerance required by these joints.

SOLVENT CEMENTING THERMOPLASTICS

Structural bonds of up to 100% of the strength of the parent material are possible with solvent bonding, a technique suitable for the amorphous plastics such as polystyrene and acrylic. It cannot be used for the polyolefins. Cellulosics, vinyls, polycarbonate, polysulfone, and polyphenylene ether copolymers are all good candidates for solvent cementing.

Solvent cementing basically involves softening the bonding area with a solvent or a solvent containing small quantities of the parent material, referred to as *dope* or *bodied cement*, generally containing less than 15% resin. A bodied cement serves several purposes: it can slow the solvent evaporation rate, it can provide a more workable solution, and it can fill in spaces where mating surfaces are not perfectly matched. (This last use should not be considered a final solution to mismatched parts that match up as closely as possible.)

A resin dissolves best in a solvent whose solubility most closely approaches its own. (See Table 24-2 for a list of solvents for various plastics.) Solvent bonding of dissimilar mate-

Table 24-2. Solvents for cementing of plastics.

ABS	Methyl ethyl ketone, methyl isobutyl ketone, tetrahydrofuran, methylene chloride
Acrylics	Methylene chloride, ethylene dichloride, trichloroethylene
Cellulosics	Methyl ethyl ketone, acetone
Nylons	Aqueous phenol, solutions of resorcinol in alcohol, solutions of calcium chloride in alcohol
PPO/PPE	Trichloroethylene, ethylene dichloride, chloroform, methylene chloride
PVC	Cyclohexane, tetrahydrofuran, dichlorobenzene; methyl ethyl ketone
Polycarbonate	Methylene chloride, ethylene dichloride
Polystyrene	Methylene chloride, ethylene ketone, ethylene dichloride, trichloroethylene, toluene, xylene.
Polysulfone	Methylene chloride

These are solvents recommended by the various resin suppliers. A key to the selection of solvents is how fast they evaporate: a fast-evaporating product may not last long enough for some assemblies; too-slow evaporation could hold up production.

rials is possible where the materials can be bonded with the same solvents.

The suitability of solvents or solution-dispersed adhesives may depend upon: the mechanical structure of the plastic; the importance of optical clarity in the bond; the presence or absence of molding stresses, which could result in crazing when released by solvent; the composition of the solvent; and the presence of highly soluble compounding ingredients in the plastic.

Built-in molding stresses must be relieved, or else solvent crazing is likely to occur. The avoidance of solvents of a low boiling point or diluting the solvent through the addition of polymer is generally desirable, and at the very least delays the appearance of crazing cracks. To minimize strains and resulting stresses that may be caused by flexing or by change of temperature, the layer of adhesive should always be as thin as possible, and no more rigid than the adherends.

Application

The solvent usually is applied by one of the devices or techniques discussed in the following paragraphs.

Dip Method. In the dip method, one of the two parts to be joined is dipped into the cement just enough so that the solvent will act on the desired area for the maximum time. After dipping, the parts should be assembled immediately and held under light pressure for a short time. Bonding takes place by the formation of a "cushion" of solvent-swollen plastic surface. To avoid squeezing out this cushioned material from the dipped plastic surface before the solvent can act on the mating dry surface, about 15 to 30 seconds should be allowed before pressure is increased.

Specially designed dipping fixtures, clamps, and conveyors should be considered for each job. The area of solvent attack on the piece can be controlled by laying a roll of glass rods in the cementing tray, on which the piece to be cemented can be rested. The level of the cement should be kept equal to the diameter of the rods. Skillful operators can assemble pieces by this method without spilling or wasting cement. The areas of contact of cement and solvent can be controlled also by lining the bottom of the cement tray with a felt pad. Pieces to be cemented then are placed on the felt pad, which is kept thoroughly wet with the liquid solvent cement.

Capillary Method. For some applications where the surfaces to be cemented fit very closely, it is possible to introduce the cement by brush, eyedropper, or hypodermic needle into the edges of the joint. The cement is allowed to spread to the rest of the joint by capillary action. In other cases it is possible to insert fine wires into the joint when the parts are assembled in the jig. The cement then is introduced into the joint. After the cement has reached all parts of the joint, the wires are removed. This procedure is useful also in removing air bubbles and in filling voids in joints made in other ways.

Soak techniques are available in which pieces are immersed in the solvent, softened, removed, and quickly brought together. Areas adjoining the joint area should be masked to prevent them from being etched. After soak, parts should be assembled rapidly.

No matter which technique is selected, the surfaces to be joined should be clean (use of a cleaning solvent, e.g., acetone, often is recommended), not below 65°F, and fit together as flat surfaces. If the parts do not fit well, the time for a part to soften after application of cement must be increased to aid in obtaining a satisfactory fit.

Once the parts are mated, they must be fastened together by spring clamps, C clamps, or toggle clamps so that uniform pressure is exerted on bonded areas. Pressure should be low—100 to 200 psi often is recommended—so that the part will not be flexed or stressed. The pressure should be sufficient to squeeze all air bubbles from the joint and to assure good contact of the mating surfaces.

When the clamps are removed, an elevated-temperature cure is often called for; for example, acrylic to 120°F and heating ABS to 130 to 150°F, PPO over 200°F, and polycarbonate up to 175°F.

Precautions. Cementing trays should be made of materials that are inert to the cementing solvents. Care should be taken to avoid excessive contact of the hands or skin with cement solvents, as most of them will remove natural oils from the skin and may cause chapping or even mild dermatitis after too-frequent exposure. Where glacial acetic acid is used, contact should be carefully avoided because of its corrosive nature.

Working areas must be well ventilated. Breathing of concentrated vapors of ethylene dichloride, methylene chloride, glacial acetic acid, and so on, may cause toxic effects. Cementing tanks, trays, and storage containers should be equipped with tight covers to reduce evaporation when they are not in use.

Because solvent vapors as well as solvents themselves can cause crazing, assemblies within enclosed spaces should be adequately ventilated. An air line can be inserted to sweep the vapors out, or vacuum can be used, along with a tube to bleed fresh air to the side opposite the vacuum takeoff. Where it is impossible to remove solvent vapors from the joint itself, after the bond has been made, the joint should be designed so that solvent vapors can be excluded from the adjacent area.

ADHESIVES*

Adhesive bonding is a process in which the adhesive acts as an agent to hold two substrates together (as opposed to solvent welding, where the parent materials actually become an integral part of the bond). Because of the smooth surfaces of plastics, adhesion must be chemical rather than mechanical.

Table 24-3 lists various adhesives and their typical applications in plastics. The list is not complete, but it does give a general idea of what types of adhesives are used where. One should remember, of course, that thousands and thousands of variations of standard adhesives are available off the shelf.

Among the most important criteria in selecting an adhesive are:

1. Substrates involved, including additives used in the plastic.
2. Joint configuration and size.
3. Stresses the joint will be subjected to in use.
4. Environmental elements the joint can be expected to encounter.
5. Economics (performance/cost ratio).

*Adopted from "Bonding and joining plastics," by J. O'Rinda Trauernicht, *Plastics Technology*, August 1970. Reprinted with permission from the publishers.

Application

Just about anything from a finger to high-speed spray and laminating equipment is used to apply adhesives.

Before bonding, most plastics require surface preparation. Common surface treatment could include: (1) chemical cleaning with a solvent such as acetone; (2) abrasive cleaning by sandblasting, vapor honing, sandpaper, or wire brush; (3) degreasing to remove residual contaminants from surfaces to be joined. In cases where a mold release agent has been used, one should be certain that no traces remain.

Some key application techniques are listed below.

Tube, Knife, Spatula, Brush, Roller, Squeeze Bottle. These hand-type techniques are used for the liquids and the mastics. Equipment is simple, and costs are low. Besides the actual application device, equipment should consist of mixing vessels (for multicomponent materials), clamps where required, and dryers where elevated cure is necessary. Depending on production requirements, labor costs are generally high. However, for some adhesives (e.g., some epoxies and silicones), this is the only practical method of application. Some of the less viscous adhesives can be applied via a self-feeding brush that speeds hand application somewhat. One epoxy supplier offers a two-part adhesive in a two-component pouch. The seal between the components is broken, and the components are mixed and then dispensed right from the pouch.

Hot Press Bonding. Use of a hot press is common for reactivating sheet-type adhesives. It is also used to reactivate dual-component adhesives that work by applying one component of the adhesive to a substrate and the second to the second substrate, and then allowing the adhesives to dry. This is a widely used lamination technique. When the substrates are brought together under the heat and pressure of the press, reactivation and bonding take place. A hot tool (e.g., an iron) can be used for some field and low-volume production.

Meter/Mix/Dispense. This type of equipment is widely used for accurate handling of multicomponent adhesives, especially those with long cure times or requiring elevated cures. Some of the newer fast-cure adhesives must be applied by one of the hand techniques to ensure that they do not set in the dispensing equipment.

Spray Application. Spraying is fast and provides a means of reaching inaccessible areas quickly. It is often used for high-speed production application of liquids, and for application of rubber-type contact adhesives. Spraying is largely confined to one-component types. It is identical to the process used for spray painting, except that a simplified nozzle, called a siphon nozzle, is used and the process is slower than painting. During spraying, it is essential that checks be made to assure that proper coverage is reached.

Roll Coaters. Essentially, roll coating consists of a glue reservoir, a roll that automatically picks up the glue from the reservoir, and a second roll that picks up the glue from the first roll and transfers it to a moving substrate. A doctor blade can be used to control the amount of glue on the roll. Other variations include a single-roll setup, a system in which the substrate is dipped into the glue and the roll used to remove the excess adhesive. An engraved roll can be used to deposit a noncontinuous pattern of glue, generally in an even pattern of dots.

The actual lamination, using either water- or solvent-based adhesives, can be done in one of two ways: wet combining or thermoplastic mounting. In wet combining, an adhesive is applied to either substrate, and the materials are joined immediately and stacked for drying. In mounting, an adhesive is applied to one substrate (the board in panel lamination), allowed to dry, and then reactivated to a tacky state and mated. Thermoplastic mounting is said to be faster and more versatile than wet combining, although it is more expensive.

Curtain and Screen Coating. Both curtain (applying a molten sheet of adhesive to a substrate passing underneath the flow) and screen (forcing adhesive through a specially prepared screen onto the substrate below) techniques can be adapted for adhesive processing. Curtain coating is commonly used for packaging, and screen processing for high-quality printing. Both, when used with adhesive instead of molten plastic or printing inks, provide a closely controlled deposition process.

Hot Melt Application. Over a half-dozen techniques can be used to apply hot melts, with some suppliers requiring that their equipment be used and others recommending a wide range of commercial equipment. The job of a hot-melt applicator is to melt the material (most melt from 300 to 500°F), and to apply it at a controlled rate to the substrate.

Two basic types of application equipment are melt-reservoir and progressive feed. In melt-reservoir, the adhesive is put into the melting pot and heated to a predetermined temperature; in progressive feed, a rope of hot melt is fed through the heating system, keeping the amount of adhesive molten to a minimum at any one time.

Some application methods are: (1) extrusion—actually squirting the molten resin onto the substrate via a pistol-type or automated, air-operated dispenser; (2) hot-drop application with either manual or automatic dispensing, in which a wide variety of droplet patterns can be applied; (3) hot-melt coating using a roller.

Adhesive Forms

The form that adhesives take (i.e., liquids, mastics, hot melts, etc.) can have a bearing on how and where they are used, as explained in the following paragraphs.

Liquids. Liquid adhesives lend themselves very well to high-speed processing techniques such as spray and automatic roller coating. They may be solvent- or water-based; the anaerobics (cyanoacrylate-based bonding agents that cure in the absence of air) also fall into this class, as do a number of epoxies.

An extensive area of activity in liquids is the increased use of wood-grain printed vinyl sheet bonded to particle board or other inexpensive solid substrate. Both solvent-based and water-based adhesives are used. The solvent-type adhesive gives a better bite (14–20 psi peel

Table 24-3. Typical adhesives for bonding plastics.*

	ABS	Acetal	Acrylic	Cellulosics	Fluoro-carbons	Nylon	PPO	PVC	Poly-carbonate
Metals	23	3,23	2,3	2,3	22,23	2,23	2,4,23	2,3,15, 23,36, 43	23
Paper		3,23	42	42	22,23	3,41	6,23	42	36
Wood	23	23	2,3,42	3	23	2,3	2,4,23	3,23, 36,42	23,36
Rubber	23	3	1-4	1-4	23	2	2,4,23	3,4, 15	4,36
Ceramics		23	2,3	3	23	3,23	4,23	3,4	23,36
ABS	23,43	2,4, 11,23				21,23		23	4,23
Acetal	2,4, 11,23	3,23	23		23	23		23	
Acrylic	2,4, 11,23	2,4,11, 23	S		23	2,3, 15,22		3,4, 11,23	
Cellulosics				3,4 14,36	23	2,3, 15,22		23	
Fluorocarbons		23			22,23			23	
Nylon	21,23	21,23	2,3, 15,22	2,3, 15,22		2,22, 23,36	21,23		
PPO							4,23, 43		
PVC	23	23	3,4, 11		23			3,4,11, 36,42	
Polycarbonate	23							3,4,11, 36,42	S
Polyethylene	23				23				
Polypropylene									
Polystyrene					23				
Polysulfone									
Alkyds							4,23, 43	23	
Epoxy							4,23, 43	23	
Melamines and Ureas							4,23, 43		
Phenolics	23,43	43	4,21, 23	4,21, 23	23		4,23, 43	4,15	
Polyesters			23	23			4,23, 43	23	
Polyurethanes			4,23		23		4		

Table 24-3. (*Continued*)

Poly-ethylene	Poly-propylene	Poly-styrene	Poly-sulfone	Alkyds	Epoxy	Melamines & Ureas	Phenolics	Polyesters	Poly-urethanes
2,31, 41	1	31	4,23		4	3,43	2	4	3,4
41	1,41	4,31, 36			3	41,42	42	41	4,36
2,41	1,41	31,36			23,31	2,3	2,42	2	36
2,41	1,41	5			3	2,3,43	2,3	1-4	4,36
2,41	1,41	41,42			23,31	2	2	2	3
23									
							23		
							4,21, 23	23	4,23
							23		
23		23							23
				21,23	21,23	21,23	21,23, 43	21,23	
				4,23, 43	4,23, 43	4,23, 43	4,23, 43	4,23, 43	4
			23				4,15	23	
23,31 41			4						
	23,31 41								
		4,5,13, 23,31, 36					4,23, 43		
4			4,23						
					2,3,23, 31,36				
						2,3,23, 31,36			
		4,23				3,21, 23,24	2-4,23, 31,43		
								3,23, 31,36	
									3,4, 23,36

(*Key and footnotes overleaf*)

Key and footnotes for Table 24-3.

Elastomeric		
1. Natural Rubber	2. Neoprene	3. Nitrile
4. Urethane	5. Styrene-Butadiene	6. Silicones
Thermoplastic		
11. Polyvinyl acetate	12. Polyvinyl alcohol	13. Acrylic
14. Cellulose nitrate	15. Polyaminde	
Thermosetting		
21. Phenol Formaldehyde (Phenolic)	22. Rescorcinol, Phenol-Rescorcinol	23. Epoxy
24. Urea-Formaldehyde		
Resin		
31. Phenolic-Polyvinyl Butyral	32. Phenolic-Polyvinyl Formal	33. Phenolic-Nylon
36. Polyester	37. Acrylic	
Other		
41. Rubber Latices	42. Resin Emulsions	43. Cyanoacrylate
S. Solvent only recommended		

NOTE: This information contains a compilation of suggestions and guidelines offered by various adhesive and materials and manufacturers and plastics molders. It is intended only to show typical adhesives used in various applications. Lack of a suggested adhesive for two materials does not mean that these materials cannot be bonded, only that suppliers do not commonly indicate which adhesive to use. Before making a final selection, consult both materials supplier and materials manufacturer. Key for Table and part of the data were supplied by USM Corp., Chemical Div.

*Reprinted from "Bonding and joining plastics," by J. O'Rinda Trauernicht, *Plastics Technology*, August 1970.

strength compared to 4–6 psi for the water-based), but also will be more expensive. Application is by direct roll coaters, with speeds of 50 to 80 ft/min being typical.

The anaerobics, which can give some very high bond strengths and are usable with all materials except PE and fluorocarbons, are dispensed by the drop. A thin application of the anaerobics is said to give better bond strength than a heavy application.

Mastics. The more viscous, mastic-type cements include some, although not all, of the epoxies, urethanes, and silicones. Mastics normally can be applied at low-to-medium production speeds, but in some cases can be heated for higher production speeds.

Epoxies adhere well to both thermosets and thermoplastics, but are not recommended for most polyolefin bonding.

Developments in epoxies include the introduction of more fast-cure types (pot life of 5 minutes or less), and lower-temperature cure requirements for some of the elevated-cure materials.

Urethane adhesives have made inroads into flexible packaging, the shoe industry, and vinyl bonding. Polyester-based polyurethanes have surpassed polyether systems because of their higher cohesive and adhesive properties.

Silicones are especially recommended where both bonding and sealing are desired.

Hot Melts. Hot melts are 100% solids adhesives that are heated to produce a workable material. They bond simply by cooling and solidifying in place. Thermoplastic hot melts can be based on polyolefins (polyethylenes and EVAs), polyamides, or polyesters. For high-temperature applications, there are thermoset hot melts based on epoxies and/or polyurethanes. Hot melts generally are supplied in granule, pellet, chunk, or rope form, depending on the type of dispensing equipment used.

Major advantages of hot melts over solvent-based adhesives are their speed of curing and the elimination of potentially dangerous volatiles.

Hot melts are employed widely in packaging applications, shoe manufacturing, furniture assembly, and other commercial uses. They can be applied on high-speed lines on either a continuous or intermittent basis.

Pressure-Sensitives. Contact-bond adhesives—usually rubber-base—provide a low-strength, permanently tacky bond. They have a number of consumer applications (e.g., cellophane tape); but they also are used in industrial applications: (1) where a permanent bond is not desirable, and (2) where a strong bond may not be necessary. The adhesive itself is applied rapidly by spray. Assembly is a mere matter of pressing the parts together.

Film Adhesives. Film adhesives overlap several areas, but exist as a separate group because of their processing similarity. All of the materials in this group require an outside means such as heat, water, or solvent to reactivate them to a tacky state. Among the film types are some hot melts, epoxies, phenolics, elastomers, and polyamides. Production rates for film adhesives are generally slow but very precise.

Film adhesives can be die-cut into complicated shapes to ensure precision bonding of unusual shapes.

Applications for this type of adhesive include bonding plastic bezels onto automobiles, attaching trim to both interiors and exteriors, and attaching nameplates on luggage.

Pretreatment

Although most materials bond without trouble once the proper adhesive has been selected, a few, notably polyolefins, fluorocarbons, and acetals, require special treatment prior to bonding.

Untreated polyolefins adhere to very few substrates; this is why polyethylene is such a popular material for packaging adhesives. Treating methods include electronic or corona-discharge, flame treating (especially recommended for large, irregular-shaped articles), acid treating by dipping the articles in a solution of potassium dichromate and sulfuric acid, and solvent treating. A rubber-based adhesive that remains permanently tacky is generally recommended.

Fluorocarbons must be cleaned with a solvent such as acetone and then treated with a special etching solution.

Acetals can be prepared for bonding by several techniques. One method involves immersing articles in a special solution composed mainly of perchloroethylene, drying at 250°F, rinsing, and then air drying.

CEMENTING OF SPECIFIC PLASTICS

The same basic handling techniques apply to almost all thermoplastic materials. In the following discussion, each thermoplastic will be treated separately, with mention of the specific cements most suitable for each. However, the first subsections that follow (on the cementing of acrylic) include discussions of bonding jigs and bonding techniques that are equally applicable, in most cases, to all other thermoplastics. To avoid duplication, this detail has been largely omitted from the subsections dealing with other thermoplastics. These sections on acrylics should be read in connection with the cementing of any of the other thermoplastics.

Cast Acrylic Sheeting

Articles of considerable size and complexity can be fabricated from methyl methacrylate plastics by cementing sections together. The technique described in this subsection applies to cast sheeting. The cementing of articles made from methyl methacrylate molding powders of extruded rod, tubing, or other shapes is more specialized, and reference should be made to the instructions applicable to these materials in the following subsection.

With care and practice, the transparency of acrylic resin can be retained in cemented joints; the joint will be clear and sound as a result of the complete union of the two surfaces brought into contact. In order to accomplish this, the surfaces to be cemented are thoroughly softened by means of a solvent, to the extent that a soft layer or cushion is produced on each; then the uniting and hardening will effect a homogeneous bond. This principle underlies all solvent-cementing of acrylic resins.

Usually one of the two surfaces to be joined is soaked in the cement until a soft layer has been formed upon it. This soft surface then is pressed against the surface to be attached and, in contact with it, softens it also, by means of the excess of cement contained in the soaked area. (See Fig. 24-2).

For some purposes it may be desirable to dissolve clean shavings of methyl methacrylate resin in the cement, in order to raise its viscosity so that it may be handled like glue. However, even with a thickened cement it still is necessary to establish a sound bond.

Most of the conventional types of joint construction can be employed, such as overlap,

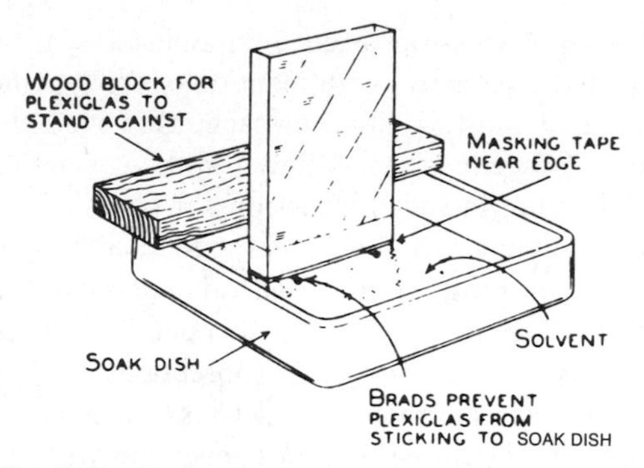

Fig. 24-2. Cementing acrylic. Edges are soaked until swollen into a cushion. Then the edges are joined, and pressure is applied. (*Courtesy Cadillac Plastic and Chemical Co.*)

butt, rabbet, miter, scarf, and so on, depending on the service requirements of the article. The area of the joint must be large enough to develop sufficient strength, and must be so designed as to give even distribution of stresses.

An accurate fit of mating pieces is essential in cementing acrylic resins. This need is primarily due to their rigidity, and the inability of a solvent cement to compensate for discrepancies in fit. In butt joints, it is necessary that both edges be true and square before cementing. Flat surfaces are more easily cemented than curved ones; so, wherever possible, it is desirable to rout or sand curved surfaces to form flat ones. When two curved surfaces are to be cemented, both must have the same radius of curvature. It is not good practice to force either piece in order to bring the surfaces into complete contact. The only exception to this is the case of very thin sheeting, such as 0.060 or 0.080 inch, and even here the deviation must not be great.

The surfaces to be cemented should be smooth and clean, but need not be polished. A smooth machined surface is most satisfactory for cementing.

When ribs are to be cemented to curved pieces, such as panels for airplane enclosures, each rib should be machined from flat sheet stock, heated in an oven to shaping temperature, and then, without cement, and with the aid of a jig, shaped against the surface to which it is later to be cemented, and allowed to remain in contact until cool. Then the curved rib is soaked in cement, and cemented to the curved panel with the aid of the same jig.

In soaking the plastic in the cement, the softening action should be confined to the immediate region of the joint, by masking the rest of the surface with a tough paper or cellophane tape coated with a pressure-sensitive adhesive. The tape should be applied firmly, and with special care at the edges. The tape used should be impervious to the cement, and should not be applied long before cementing, as its adhesive may loosen if it is allowed to stand too long.

The role of the cushion formed by soaking the surface in the cement is solely to enable the two surfaces to be brought into complete conformity, so as to exclude air bubbles and to compensate for lack of perfect fit between the two surfaces. The thickness of the cushion must be adequate to provide complete contact but preferably no greater than this, as an excess reduces the strength of the joint and prolongs unnecessarily the setting of the bond. With solvent cements and unplasticized sheets, a soaking time of 15 minutes is usually enough to form an adequate cushion. Inadequacy of the cushion is one cause of bubbles or uncemented areas in the finished joint.

The soaked surface, after removal from the cement, should be brought rapidly, while still wet, into contact with the surface to which it is to be joined because soaked surfaces that have become dry do not wet the mating surface adequately. The softened surface, however, should not be dripping because superfluous cement will run off and cause smears. The tray of cement, and the work, should be so situated that the transfer from cement to jig can be made conveniently and rapidly. A soaked surface that has become dry can be wetted again by brushing additional solvent cement on it.

Too-early application of pressure on the joint will result in squeezing out some of the solvent needed to soften the dry surface. After the two surfaces have been brought into contact, only very slight pressure should be applied at first. Then, after about 15 to 30 seconds, the cementing jig can be tightened, to apply pressure.

Small pieces often can be weighted with bags of shot, tied with cord, or taped to supply the required pressure. However, assembly is best

Fig. 24-3. Typical clamping method for cementing a rib on acrylic sheet. (*Courtesy Cadillac Plastic and Chemical Co.*)

accomplished in well-designed cementing jigs (Figs. 24-3, 24-4, and 24-5) capable of holding the pieces firmly together until the joint is hard, without forcing either of them out of shape. Excessive pressure should be avoided because it would be likely to cause stress crazing. The pressure should be adequate to squeeze out bubbles or pockets of air from the joint, evenly applied over the entire joint, and maintained throughout the period of setting. Those requirements are met through the use of spring clips or clamps, either alone, in the case of simple assemblies, or in conjunction with cementing jigs of wood or metal in more complicated cases. For most joints, a pressure in the neighborhood of 10 psi is suitable, provided that it does not force either of the parts appreciably out of shape.

After assembly of the work in the jig, any excess cement that has been extruded from the joint should be removed by scraping it onto the

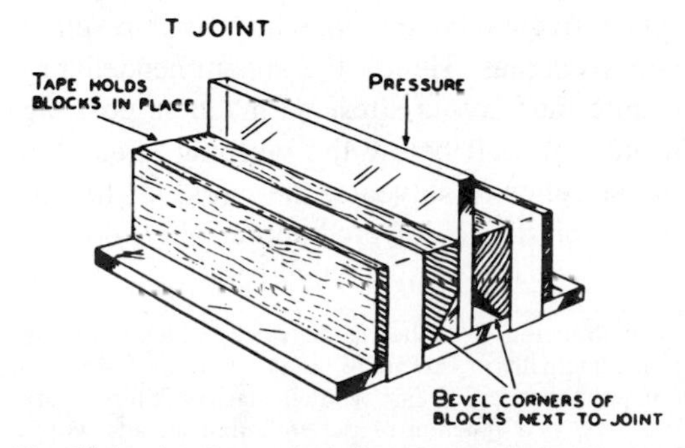

Fig. 24-4. Another typical clamping method. (*Courtesy Cadillac Plastic and Chemical Co.*)

masking paper, which is subsequently to be removed. Prompt cleaning up of the excess of soft cement will save time in finishing after the bond has set.

The assembly should be allowed to remain in the jig for from 1 to 2 hours, and an additional 4 hours should elapse before the work is subjected to heat treatment or to finishing operations. These times are approximate only, depending on the complexity of the job.

Methylene chloride, a solvent that produces joints of medium strength, requires a soaking time of 3 to 10 minutes and sets very quickly. Ethylene dichloride is a trifle slower and is less apt to produce cloudy joints.

The most universally applicable type of solvent cement is the polymerizable type, comprising a mixture of solvent and catalyzed monomer. These mixtures are mobile liquids, volatile, rapid in action, and capable of yielding strong sound bonds. They should be used with adequate ventilation. An example is a 40–60 mixture of catalyzed methyl methacrylate monomer and methylene chloride.

Additional cement formulations, and their adaptability to specific types of acrylic resin, are detailed in the wealth of literature available directly from the resin manufacturers.

Heat treatment or annealing of joints made with solvent cements is highly desirable because it greatly increases the strength of the joint, but it frequently is not necessary because joints are adequately strong without it. Solvent joints never become completely dry; that is, they are never entirely freed of solvent through evaporation and diffusion. If the cemented piece is placed in an oven, the penetration of the diffusing solvent progresses farther away from the bond, with some resulting increase in the strength of the cemented joint. The heat treatment of joints made with solvent cement should be done carefully, so as not to warp the assembly or to approach the boiling point of the cement and thus produce bubbles.

The optimum and maximum heat-treating (annealing) temperatures depend upon the specific plastic. The recommendations of the resin manufacturer should be consulted.

Hardening of the joint, with or without heat treatment, should be completed before finish-

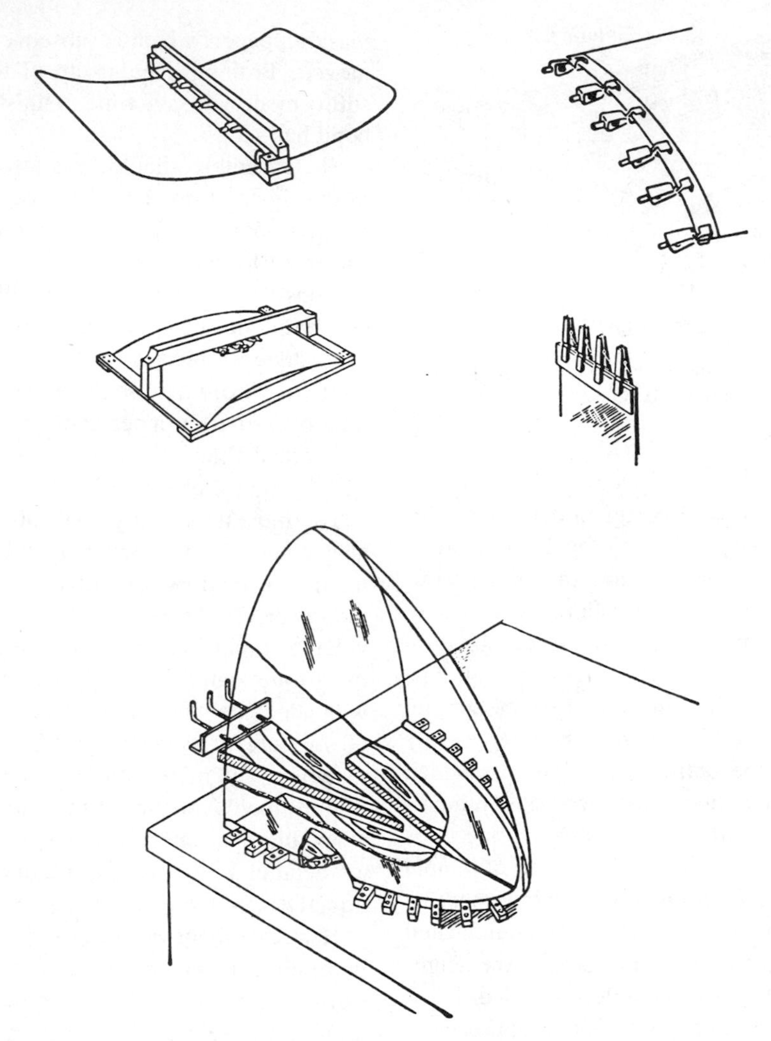

Fig. 24-5. Clamping setups for various contoured acrylic parts. (*Courtesy Cadillac Plastic and Chemical Co.*)

ing operations such as sanding, machining, or polishing are carried out. Otherwise, the softened material in the joint will subsequently recede into the joint.

Molded Acrylic Pieces

With molded acrylics, cemented joints are generally not as satisfactory as with cast sheeting. Fabricators frequently experience considerable difficulty with these joints, and many precautions must be taken to obtain satisfactory results. It is preferable to design molded articles to employ mechanical fasteners or closures rather than cemented joints, but this is not always feasible, and recourse sometimes must be had to solvent cement.

Cemented joints in molded methyl methacrylate frequently are unsightly as a result of stress crazing. Hence the recommended procedure must avoid stress as much as possible, to keep it well below the working stress.* In the presence of solvents, the working stress of thermoplastics is lowered appreciably. Stresses

**Working stress* is defined as that stress which can be applied, internally or externally, to an article for a period of time, on the order of years, without causing failure. Working stress is a function of the environment, and will be lowered by exposure to most solvents and chemicals or by thermal strain.

built in during fabrication, combined with the effect of exposure to solvent during the bonding operation, may exceed the working stress and cause cracking or crazing. Mechanically built-in stresses may be greatly reduced by annealing.

If crazing appears, the cracks may be of such a size as to be easily visible to the eye, or so minute as to appear merely as surface clouding. In either case, crazing spoils the appearance of the article, promotes mechanical failure, and in some cases degrades electrical properties.

Crazing may occur either locally or generally. It may be caused by excessive local mechanical stresses (internally or externally applied), or it may be the result of localized loss of volatile constituents. The crazed cross section must withstand tensile loads and elongations beyond its ability, and it finally fails.

When articles must be cemented, stress avoidance should be considered in the design of the dies because the conditions of molding and the design and location of the gates determine to a large extent the degree of stress that will be present in the molded articles. The gate should be located away from the area to be cemented because frozen stresses usually are greatest near the gate. In designing two pieces to be cemented, the exact dimensions should be very carefully considered, so that the pieces will fit together easily and accurately without forcing, as stresses developed by forcing during assembly will promote solvent crazing.

When a cover is to be cemented into a recess or a groove, right-angled corners should, if possible, be replaced by bevels. This promotes the escape of air, when the surfaces are brought together, and reduces the excess cement that must be wiped away.

A cover that must be cemented should be made circular, if possible, to permit it to be twisted in place, immediately after cementing, to remove air bubbles.

The finishing operations that will be required should be anticipated when the original design is made. The design of the gate should be such as to minimize chipping during its removal because surface fractures occasion solvent crazing. For the same reason, degating of the molded article should be very carefully done.

The conditions of molding are important. To minimize stress, injection-molded articles should be run with the mold as hot as possible and with as short a cycle as will permit ready removal of the shot from the mold.

The selection of a cement is of less importance than the foregoing precautions to minimize strain. A minimum amount of cement should be used. The cement generally is applied to well-fitting molded pieces by a brush, but in some cases the use of a hypodermic needle is advantageous. Solvent-type cements may be employed.

Cellulosic Plastics

Sheeting and molded articles of cellulose acetate can be cemented readily to pieces of the same plastic with a bond practically as strong as the material itself. The usual precautions needed for the best results in cementing plastics must be observed.

The cements used with cellulose acetate plastics are of two types: (1) solvent-type, consisting only of a solvent or a mixture of solvents; and (2) dope-type, consisting of a solution of cellulose acetate plastic in a solvent or mixture of solvents.

The solvent-type cement generally is employed when the surfaces to be cemented are in a single plane and simple in nature, and can be readily held to a perfect fit. The dope-type cement is used when the surfaces are irregular or not easily accessible.

Acetone and mixtures of acetone and methyl ''Cellosolve''* are commonly used as solvent cements for cellulose acetate. Acetone is a strong solvent for the plastic, but evaporates rapidly. The addition of methyl ''Cellosolve'' retards the evaporation, prevents blushing, and permits more time for handling the parts after application of the cement.

A cement of the dope type for cellulose acetate plastic, by virtue of its containing plastic in solution, upon drying leaves a film of plastic that forms the bond between the surfaces to be joined. These cements are generally used when

*''Cellosolve,'' in this chapter, denotes ethylene glycol monoethylether (2-ethoxyethanol).

an imperfect fit of the parts requires filling. Dope cements, satisfactory for many purposes, can be made by dissolving cellulose acetate plastic in solvents. A typical formula is:

	Parts by weight
Cellulose acetate plastic	130
Acetone	400
Methyl "Cellosolve"	150
Methyl "Cellosolve" acetate	50

A general formula, suitable for use with a wide variety of ingredients, would be the following:

	Parts by weight
Cellulose acetate (low viscosity)	8–12
Low- and medium-boiling solvents and diluents (under 100°C)	45–75
High-boiling solvents (over 100°C)	20–50

After being cemented, the pieces being united should be held under light pressure for 1 to 10 minutes, depending on the nature of the bond and the type of cement used. The assembly should be allowed to stand at least 24 hours before subsequent operations are performed, such as sanding, polishing, testing, and packing.

Cellulose acetate butyrate and propionate. Plastics are cemented in accordance with the technique described for cellulose acetate.

In the case of dope cements, the plastic to be dissolved in solvents is cellulose propionate.

The strongest bonds between pieces of ethyl cellulose plastic are made by solvents or by solvents bodied with ethyl cellulose plastic.

Nylon

The recommended adhesives for bonding nylon to nylon are generally solvents, such as aqueous phenol, solutions of resorcinol in alcohol, and solutions of calcium chloride in alcohol, sometimes bodied by the inclusion of nylon in small percentages.

Aqueous phenol containing 10 to 15% water is the most generally used cement for bonding nylon to itself. The cement is prepared by melting phenol (available from chemical supply houses) and stirring in the necessary water. The bond achieved by use of this cement is water-resistant and flexible, and has strength approaching that of nylon.

The surfaces to be cemented should be cleaned thoroughly with a conventional degreasing agent such as acetone or trichlorethylene. The cement is applied with a brush or cotton swab to the mating surfaces, which should be wetted completely using one or more coats as needed.

If the parts fit well, they can be assembled immediately. However, if the fit is poor or loose at the interface, a waiting period of 2 to 3 minutes to soften the surface will help in obtaining a satisfactory fit. A few experiments at this point will determine the time required to form an adequately softened surface for the formulation of the best joint.

The parts are fastened together by means of spring clamps, C clamps, or toggle clamps in such a way that uniform pressure is exerted on the bonded areas.

If the phenol is exposed to moist air, the cemented joint may turn white as it cures. A precure after the parts are fastened, for 1 hour at 100°F or overnight in a dry atmosphere at ambient temperature, will prevent the joint from whitening.

The assembly next is immersed in boiling water in order to remove the excess phenol and to ensure a strong bond. The immersion time is dependent on the area of the bonded section. Five minutes of immersion is used for $\frac{1}{8}$-inch-wide bonds; for 1-inch-wide surfaces, boiling as long as 1 hour may be necessary. As a general guide, the assembly should be boiled until the odor of phenol has disappeared from the joint.

In cases where it is objectionable to immerse the sample in boiling water, the sample may be cured by prolonged air exposure at ambient temperature, or an oven temperature of 150°F. Adequate bond strength for handling will be reached in 48 hours, and maximum strength in 3 or 4 days. The curing time is dependent on the quantity of phenol used.

Calcium chloride–ethanol bodied with nylon

is recommended for nylon-to-nylon joints where there is potential contact with foods, or where phenol or resorcinol would be otherwise objectionable. Such solvents produce bonds that are equivalent in strength to those obtained with phenol or resorcinol, except when used with high molecular weight nylons. This adhesive also has the added advantages of forming a nonbrittle bond, of being less corrosive than phenol or resorcinol, and of being easy to apply.

Bonding Nylon to Metal and Other Materials. Various commercial adhesives, especially those based on phenol-formaldehyde and epoxy resins, sometimes are used for bonding nylon to other materials. (Although these commercial adhesives also can be used for bonding nylon to nylon, they usually are considered inferior to solvent-type adhesives in that application because they result in a brittle joint.)

Epoxy adhesives, for example, have been used to produce satisfactory joints between nylon and metal, wood, glass, and leather.

These adhesives are two-part systems, which are prepared before each use by mixing a curing agent with the adhesive resin according to the manufacturer's description. Thorough mixing is essential to obtain optimum performance.

Surfaces to be bonded must be thoroughly cleaned with a suitable degreasing agent. For best results, the nylon should be roughened with an emery cloth, and the metal should be sandblasted.

The catalyzed epoxy adhesives are best applied with a wooden paddle or spatula. They will cure at room temperature in 24 to 48 hours. To accelerate the cure, they may be heated in a circulating air oven for 1 hour at 150°F.

Polycarbonate

Solvent cementing of parts molded of polycarbonate may be effected by the use of a variety of solvents or light solutions of resin in solvents.

The selection of the solvent system best suited for a particular application depends upon the area to be bonded, the speed with which parts can be assembled, and the practical limitation of necessary drying conditions.

The following solvents are recommended for solvent-cementing of parts molded of polycarbonate:

1. Methylene chloride, which has a low boiling point of 40.1°C and an extremely fast evaporation rate. This solvent is recommended for most temperate climate zones.
2. A 1 to 5% solution of polycarbonate in methylene chloride, which can be used in extreme cases to obtain smooth, completely filled joints where perfectly mated bonding areas are impossible to obtain. It also has the advantage of a decreased evaporation rate. Higher concentrations of this solution are not recommended because of the great difficulty in obtaining completely bubble-free joints.
3. A mixture of methylene chloride and ethylene dichloride, with a maximum of 40% ethylene dichloride, which may be used where it is difficult to join both halves fast enough to prevent complete evaporation of methylene chloride.

The evaporation rate of methylene chloride to ethylene dichloride is 6.7 to 1. This is important because methylene chloride, after bonding, will dry completely in a much shorter time than ethylene dichloride.

For solvent cementing parts molded of polycarbonate, one should use the minimum amount of solvent necessary for good adhesion. This is the opposite of the procedure recommended for other thermoplastics, where one or both halves are soaked in solvent for a considerable period. Solvent should be applied to only one of the bonding surfaces; the other half remains dry and ready in the clamping fixture.

Parts must fit precisely without pressure or any contact-inhibiting irregularities. Locating pins, tongue-and-groove shapes, or flanges may be used to align bonding areas and to effect rapid part matching after the application of solvent. These devices should be kept shallow to avoid trapping solvent in the mating surface.

As soon as the two parts have been put together, pressure should be applied immediately. A pressure between 200 and 600 psi is suggested for best results. The holding time in the pressure fixture should be approximately 1 to 5 minutes, depending on the size of the bonding area.

Although the cemented parts may be handled without damage after the holding time in the pressure fixture, sufficient bond strength for room-temperature use normally is developed by drying the cemented parts for 24 to 48 hours at room temperature. If the solvent-cemented part is to be used at high temperatures, a longer solvent-removal program is necessary.

Adhesive Bonding of Polycarbonate. Parts molded of polycarbonate can be bonded to other plastics, glass, aluminum, brass, steel, wood, and other materials, using a wide variety of adhesives. Generally, the best results are obtained with solventless materials such as epoxies and urethanes. However, each application has unique requirements for flexibility, temperature resistance, ease of application, and appearance, requiring careful adhesive selection. Room-temperature-curing products such as RTV silicone elastomers work well, as do a number of air-curing and baking-type products.

Parts should be cleaned vigorously before adhesive bonding. All oil, grease, paint, mold releases, rust, and oxides must be removed by washing with compatible solvents, such as petroleum ether, isopropyl alcohol, heptane, VM and P naphtha, or a light solution of detergents, such as Joy. Bond strength may be improved by sanding, sandblasting, or vapor honing the bonding surfaces.

Polyethylene

The good solvent resistance of polyethylene and other polyolefins precludes the use of solvent-type cements, and maximum bonds can be made only when the surfaces of the parts are prepared to give anchor points for selected adhesives.

One technique for surface preparation is to dip polyethylene in a chromic acid bath (made up of concentrated sulfuric acid, 150 parts by weight; water 12 parts; and potassium dichromate 7.5 parts) for about 30 seconds at 160°F. The parts are rinsed with cold water after this treatment.

Still another effective surface treatment for producing cementable surfaces on polyethylene is electrical discharge. The discharge and arc resulting from the electrical breakdown of air are an effective source of energy. Several types of commercial electrical discharge treating equipment are available. The open oxidizing flame method also is used extensively to give cementable surfaces of molded and extruded polyethylene.

Several commercial rubber-base adhesives produced moderate adhesion with polyethylene that had been surface-treated. Cements were applied with a cotton swab, and the coated areas were air-dried for 3 minutes before assembly. After assembly, the samples were conditioned overnight at room temperature.

Polystyrene

Complex assemblies of polystyrene, usually molded in sections, may be joined by means of solvents and adhesives. Polystyrene plastics are soluble in a wide variety of organic solvents, and solvent action on the surface is the usual means by which parts are held together.

Care must be exercised in the fabrication (especially by injection molding) of polystyrene pieces that are subsequently to be joined by the use of solvents, to keep internal stress as low as possible.

External loads should not be applied until the residual solvent in the joint has evaporated completely. This may require drying at room for a week, or as long as a month, before the article is placed in service, the drying time depending upon the amount of solvent used and the opportunity for escape.

Typical solvents, listed in Table 24-4, may be divided into three groups according to their relative volatilities. Those with a low boiling point, and thus high volatility, are "fast-drying." They are low in cost, are readily available, and dry rapidly; but they are unsatisfactory for transparent articles of polystyrene because they cause rapid crazing. Solvents of the second type are classified as "medium-drying," and have higher boiling temperatures.

Table 24-4. Solvent cements for polystyrene.

Solvent	*Boiling Point (°C)*	*Tensile Strength of Joint (psi)*
Fast-drying		
Methylene chloride	39.8	1800
Carbon tetrachloride	76.5	1350
Ethyl acetate	76.7	1500
Benzol	80.1	—
Methyl ethyl ketone	79.6	1600
Ethylene dichloride	83.5	1800
Trichloroethylene	87.1	1800
Medium-drying		
Toluol	110.6	1700
Perchlorethylene	121.2	1700
Ethyl benzene	136.2	1650
Xylols	138.4–144.4	1450
p-Diethyl benzol	183.7	1400
Slow-drying		
Amylbenzol	202.1	1300
2-Ethylnaphthalene	251	1300

High-boiling or "slow-drying" solvents often require excessive time for the development of sufficient bond strength to permit handling; but these solvents will not cause crazing to appear quickly, and thus may be useful where a "clean" joint is essential. It is desirable to add up to 65% of a fast- or medium-drying solvent to a slow-drying solvent, to speed up the development of initial tack without greatly reducing the time before crazing appears.

A bodied, or more viscous, solvent may be required for certain joint designs and for producing airtight or watertight seals. These adhesives can be easily made by dissolving polystyrene in a solvent. Usually 5 to 15% of polystyrene by weight is adequate. Grinding the plastic to a powder will aid in making a solution. The choice of solvent depends upon the properties desired, such as setting time, flammability, safety hazard, and so on.

Solvents may be applied by dipping, a felt pad, a syringe, and so on. Bodied solvents are best applied by brush, flow-gun, doctor-knife, and so forth. The assembly should be made as soon as the surfaces have become tacky, usually within a few seconds after application. Enough pressure should be applied to ensure good contact until the initial bond strength has been achieved. Stronger bonds result when solvent is applied to both surfaces to be joined. Experience will show how long the joint must remain clamped in order to develop enough strength to be handled. Considerable additional time will be required before the bond can be expected to withstand service loads.

The impact grades of polystyrene are not so readily soluble as general-purpose polystyrene; hence good bonds are more difficult to make with them. Most of the impact polystyrenes can be successfully joined by using the medium-drying solvents and those solvents with the highest boiling points in the fast-drying range.

Numerous commercial adhesives, mostly of the solvent-dispersed, lacquer, or rubber-based type, are available for polystyrene plastics.

Polystyrene plastics can be joined also by conventional heat-sealing techniques such as friction spinning and welding.

Polystyrene-to-Wood. Solvent-based contact cements provide the strongest bond between polystyrene and wood. These adhesives all have a neoprene (polychloroprene) base and a ketonic–aromatic solvent system. The significant difference among solvent-based contact cements is in the solids content. The resistance of the joint to cleavage-type stresses increases with an increase in the solids content of the adhesive.

Solvent-based contact cements should exhibit excellent stability during exposure to the temperatures of a finish drying oven. However, the use of this type of adhesive can present the manufacturer with a potential problem, as excessive amounts of the adhesive can induce crazing of polystyrene because it contains polystyrene-active solvents (ketones and aromatics). This crazing could propagate into cracks and thereby produce premature failure of a polystyrene component during end-use.

Water-based contact cements produce joint strengths between polystyrene and wood only slightly inferior to that obtained with solvent-based contact cement having a similar solids contents. Despite the slightly lower bond strength values, a comparable amount of wood failure was noted during the testing of joints with water-based contact cements.

A definite disadvantage of these adhesives is that they do not wet the surface of the polystyrene, and the water must be totally absorbed by the wood. (Polystyrene samples joined with a water-based contact cement were still wet after 24 hours of drying at room temperature.) Consequently, the bond strengths of joints made with water-based contact cement shortly after assembly (3 hours) are considerably less than that exhibited by joints made with solvent-based contact cements of similar solids contents. In assembly situations where the speed of adhesive set is relatively unimportant and the degree of solvent crazing is critical, water-based contact cements with high solids contents (40%), should produce a satisfactory bond between polystyrene and wood.

PPO Resins

Modified PPO (Noryl, General Electric) may be solvent-cemented to itself or a dissimilar plastic using a number of commercially available solvents, solvent mixtures, and solvent solutions containing 1 to 7% PPO resin.

The choice of the cementing technique best suited to each application is dependent upon the evaporation rate of the solvent (or assembly time), the type of joint to be cemented, and the area to be bonded. One should follow these recommended cementing procedures to obtain maximum bond strength:

1. Remove all traces of grease, dust, and other foreign matter, using a clean cloth dampened with isopropyl alcohol.
2. Lightly abrade the surface with fine sandpaper or treat the surface with chromic acid (E-20 etchant, Marbon Chemical Company). For best results, immerse the area to be bonded in an 80°C acid bath for 30 to 60 seconds, depending upon part design.
3. Wipe the surface a second time with a cloth dampened with isopropyl alcohol.
4. Apply solvent to bond surfaces, and quickly assemble the two parts. The rapid connection of bonded surfaces will prevent excessive solvent evaporation.
5. Apply pressure as soon as the parts have been assembled. The amount of pressure required will generally depend upon part geometry; however, moderate pressure (100–600 psi) usually is sufficient. Clamping pressure should be sufficient to ensure good interface contact, but not be so high that the materials are deformed, or that solvent is extruded from the joint.

Maintain pressure for 30 to 60 seconds, depending upon the part design, bonding area, and part geometry. Bonded parts may be handled safely after the original hold time, although maximum bond strength is not usually attained for 48 hours at room temperature or 2 hours at 100°C. The cure time necessary to achieve maximum bond strength will vary with the shape of the part.

Apply solvent in an area of good ventilation, and avoid direct solvent contact. Isopropyl alcohol may be used to clean clothing and equipment.

Adhesive Bonding of PPO. Parts molded of PPO resins may be bonded to one another as well as to dissimilar materials, using a wide range of commercially available adhesives. Because adhesive bonding involves the application of a chemically different substance between two molded parts, the end-use environment of the assembled unit is of major importance in selecting an adhesive. Operating temperatures, environments, bond appearance, and shape and flexural and tensile properties must all be taken into consideration. Epoxy adhesives, because of their versatility, are generally recommended.

The following recommendations should be followed in selecting an adhesive:

1. The cure temperature of the adhesive must not exceed the heat deflection temperature of the resin.
2. Adhesives containing solvents or catalysts that are incompatible with the resins should be avoided.
3. Adhesives should be tested for compatibility and bond strength under expected operating conditions.

With the exception of the holding pressure and the cure cycle, the bonding procedures used for solvents can be used with adhesives. One should be sure that part surfaces are free of dirt, grease, dust, oil, or mold release agents. The surface of the part should be sanded or chromic acid–etched before bonding for maximum strength. To ensure against misalignment during the cure cycle, only ''finger-tight'' pressure should be applied. One should follow recommended cure times and temperatures outlined by the adhesive manufacturer.

Polyvinyl Chloride and Vinyl Chloride–Acetate Copolymers

Greater diversity of composition, and, correspondingly, of the ability to be cemented by solvents, exists among the vinyl chloride–acetate copolymer resins than in most of the other plastics that are cemented. This is due to the relative insolubility of the polyvinyl chloride component of the copolymer.

As the percentage of vinyl acetate is increased in the copolymer resins, the effect of solvents is markedly increased. For this reason, the cements that depend upon solvent action for strength will be the more effective with copolymers containing the greater percentage of vinyl acetate. In general, cementing by use of solvent is less satisfactory with the vinyl copolymers than with the more soluble plastics, such as those of the cellulose esters.

Copolymers resins are available also in plasticized forms, and these, particularly in the more highly plasticized formulations, are more rapidly cemented than are the unplasticized resins.

In the case of cements depending on their tackiness alone for adhesion, little difference between the resins will be encountered. In heat-sealing operations, also, there will be little difference in weldability, provided that the optimum temperature for the particular composition is used.

Cements for copolymer resins are usually of two types: (1) solvent cements (also called laminating thinners) and (2) dope-type cements.

Cementing with Solvent. Where smooth, rigid surfaces are to be joined, the solvent adhesives can be readily used. The adhesive is applied to the edges of the two pieces, which are held closely together, and flows between them by capillary action. Initial bonding takes place rapidly, usually within a few seconds, but the full strength of the joint is not reached until the solvent has completely evaporated. The ultimate strength of properly prepared bonds is practically as high as that of the original plastic.

The vinyl chloride–acetate copolymer resins are most rapidly dissolved by the ketone solvents, such as acetone, methyl ethyl ketone, and methyl isobutyl ketone. The cyclic ketones, such as cyclohexanone, form solutions of the highest solids content, but they evaporate slowly and ordinarily are used only for copolymers of high molecular weight and straight polyvinyl chloride polymers. Propylene oxide also is a very useful solvent in hastening solution of copolymer resins, especially those of high molecular weight, and of straight polyvinyl chloride. Mixtures of ketones and aromatic hydrocarbons have a more rapid softening action than do the ketones alone. Mixtures of solvents and nonsolvents are preferable to solvents alone, and additions of aliphatic hydrocarbons and alcohols are sometimes advantageous. Two percent glacial acetic acid, added to the solvent cement, improves wetting and speeds the capillary flow of the cement between the two surfaces being joined. It should not be used, however, in formulations containing propylene oxide.

Embrittlement by the solvent is one of the most troublesome factors in the bonding of rigid sheets by the use of laminating thinners. This is probably caused by molecular orientation and release of stress when the solvent is applied. Two procedures that minimize this embrittlement have been used in solvent bonding, with considerable success. In the first, mixtures based on the less powerful solvents are used; these apparently do not penetrate the sheet at a rate rapid enough to bring about embrittlement. In the second method, resin solutions are used for the bonding; their viscosity apparently minimizes penetration by the solvent.

The formulations of a number of typical laminating thinners are shown in Table 24-5. The

Table 24-5. Laminating thinners.

	A	B	C	D
	(parts by weight)			
Dioctyl phthalate	5	2.5	2.5	
Methyl acetate (82%)	58			
Ethyl acetate (85%)	10			
Butyl acetate	10			
Methyl ethyl ketone		63		40
Dioxane		20		
Isophorone		2.5	2.5	
Methylene chloride			50	
Ethylene dichloride			43	
Cyclohexanone				40
Propylene oxide				20
Petroleum solvent (p.p. 94.4-121.7°C)	15			
Acetic acid	2	2	2	
Methanol		10		
	100	100	100	100

solvents are arranged in order of increasing solvent power and decreasing rate of setting of the joint.

Propylene oxide penetrates vinyl chloride–acetate copolymer resins very rapidly and, in amounts up to about 20 to 25%, improves the "bite" into the resin. Under conditions where propylene oxide evaporates too rapidly, acetone may be substituted. Solutions containing propylene oxide should be stored with care and well stoppered to prevent evaporation, since its boiling point is only 34°C.

The chlorinated hydrocarbons also are excellent solvent for the vinyl chloride–acetate copolymer resins, and are suitable for use in cements.

Cyclohexanone and isophorone are extremely high-boiling solvents which impart slow drying and prolonged tackiness. By themselves they are very slow penetrants, and hence they are usually used in combination with low-boiling solvents.

Since many of the solvents discussed above present possible toxicity hazards, they should always be used with adequate ventilation.

Dope-Type Cements. These are prepared by the addition of small amounts of resin to solvent cements to thicken them. Dope-type cements are used where the surfaces to be cemented are in contact only over a very small area or are mated so inaccurately that a thin solvent cement will not fill the gap.

Small percentages of vinyl chloride–acetate resin, in the form of shavings, turnings, or chips, may be dissolved in the previously mentioned solvent cements to impart viscosity. In conjunction with the vinyl resin there should be included 0.2 to 0.5% of propylene oxide to help stabilize the solution against discoloration by light and heat when stored. Adhesion may be increased by the addition of about 5% of a plasticizer, such as tricresyl phosphate.

In general, the dope-type adhesives set slowly with vinyl chloride–acetate copolymer resins, and have little strength immediately after application. The strength of the bond develops as the solvent evaporates.

The vinyl chloride–acetate resin cements do not possess a high degree of tackiness when wet. However, this can be greatly increased by incorporation of certain other compatible resins, to yield cements of improved adhesion. Data covering resins suitable for this purpose may be obtained from suppliers.

TFE-fluoroplastic

As there are no solvents for TFE-fluoroplastic (Teflon, Du Pont), techniques other than conventional solvent cementing must be used. The surfaces of parts fabricated of TFE are first modified with strong etching solutions, making them receptive to conventional self-curing adhesive systems. Actually, self-curing adhesives of the epoxy and phenolic types are the most suitable means of bonding modified TFE to itself and to other materials.

CEMENTING OF THERMOSETTING PLASTICS

The cementing of thermosetting materials to themselves, or to other materials, poses problems that are not inherent in the cementing of thermoplastics. The insolubility of thermosetting materials makes it impossible to apply the solvent techniques used with thermoplastics, and the surface smoothness of molded thermosetting plastics adds to the difficulty of cementing them.

The surfaces to be joined must mate perfectly, unless a gap-filling cement can be used. The smooth surface must be roughened; if machining is not required for mating, then the surfaces should be sanded. This process removes the gloss and the mold-release agent. However, where the specific adhesive strength to the plastic is high, sanding occasionally reduces the net bond strength by providing nuclei for cohesive failure of the plastic.

Cementing of Dissimilar Materials

The great strength of reinforced and thermosetting plastics, and their usefulness under a wide range of environments and conditions, have led to many applications, frequently in combination with metals and other rigid materials.

The new synthetic resin and elastomer adhesives are chemically related to the new plastic materials, particularly those used in "reinforced" applications, which also are being adapted for structural (load-bearing) applications. Adhesive bonding is, for various reasons, the logical method of fastening for the production of structures composed of reinforced or rigid plastics, as explained in the following paragraphs.

In applications where exposed metal must not be used because of corrosion, metallic fasteners obviously must not be used with the plastic; nor can a plastic be used to cover or protect a metal in such cases. The plastic article may not be able to accept the piercing and the concentration of stress involved with a metallic fastener.

Many plastics, and particularly the crosslinked and thermoset reinforced plastics, cannot be bonded to themselves, or to other materials, by thermal welding or heat sealing. However, adhesives are available for joining dissimilar materials, and for bonding plastics to very thin metal structures. Their use provides means of achieving smooth contours, free from projecting fasteners, and also of ensuring uniform distribution of stress over the area bonded. The adhesive contributes sealing action, insulation, and damping of vibration. In many cases, the use of an adhesive rather than metallic fasteners reduces the cost of production, reduces the weight of the assembly, or provides longer life in service.

Adhesive Formulations. The general chemical similarity of adhesives to plastics not only contributes to their good specific adhesion to plastics, but also assures that properly formulated adhesives closely match the plastics with regard to modulus, flexibility, toughness, resistance to thermal degradation, resistance to solvents and chemicals, and cohesive strength.

In cemented assemblies of reinforced plastics and metals, where structural strength is generally desired, the adhesive must be more rigid than those used for bonding plastic to plastic; that is, it must have a modulus, cohesive strength, and coefficient of thermal expansion between those of the plastic and the metal. In many cases, such adhesives are stronger than the plastic itself.

As most reinforced plastics are quite resistant to solvents and heat, heat-curing solvent-dispersed adhesives may be used; obviously, these could not be considered for bonding many thermoplastics. Such adhesives consist of reactive or thermosetting resins (e.g., phenolics, epoxies, urea-formaldehydes, alkyds, silicones, and combinations of these), together with compatible film-formers such as elastomers or vinyl-aldehyde condensation resins. Isocyanates are frequently added as modifiers to improve specific adhesion to surfaces that are difficult to bond. These adhesives may be applied not only in solvent-dispersed form, but also in the form of film, either unsupported, or supported on fabric, glass mat, and so on.

One type of solvent-dispersed formulation includes a monomer (such as vinyl or acrylic) as the solvent; a dissolved polymer of the monomer as the film-former, to increase viscosity and decrease shrinkage; and a catalyst, packaged separately and added just before use, to polymerize the monomer. Crosslinking monomers in small proportion, such as divinylbenzene and triallylcyanurate, improve the bond strength of adhesives of this type, as well as their resistance to heat and solvents. It will be recognized that these adhesives are related to the monomer cements previously discussed.

A great many outstanding formulations are based on epoxy resins. The use of polymeric hardeners such as polyamides and polyamines, phenolics, isocyanates, alkyds, combinations of amines with polysulfide elastomers, and so on, and the ''alloying'' of the epoxy with compatible polymeric film-formers, such as polyvinyl acetate and certain elastomers, has led to the development of a broad spectrum of adhesive formulations with a wide range of available properties. In general, these adhesives are characterized by toughness, high cohesiveness, and specific adhesion to a great many materials, including most plastics. In addition, many of these adhesives contain no solvent.

By the selection of the proper hardener (and several chemical families of various degrees of reactivity are available: monomeric polymeric, or mixed), the adhesive can be formulated to cure at room temperature, ''low'' temperature, or elevated temperature. Generally speaking, the phenolic–epoxy and the elastomer–phenolic–epoxy blends are characterized by outstanding resistance to high temperature.

WELDING OF THERMOPLASTICS

Welding by heat, or thermal welding, provides an advantageous means of joining most thermoplastics. All such techniques use heat to soften (or melt) the bonding areas of the mating surfaces to fusion temperature. After joining, the area is cooled, causing the plastic to harden (resolidify) and forming the joint or weld.

Several techniques are available for welding thermoplastics, including: (1) dielectric heat sealing; (2) thermal heat sealing (non-high-frequency); (3) hot gas welding; (4) hot tool welding; (5) hot plate welding; (6) induction heating; (7) spin welding; and (8) ultrasonic assembly (this last discussed below in a separate section).

In determining the techniques best suited for a job, several factors must be considered: the strength required; the shape and size of the parts to be joined; the plastics involved; the equipment to be used (i.e., whether to use available machines or invest in new facilities); and whether or not the joint must be concealed. By careful part design, it should be possible to weld most thermoplastics with strong joints.

With each type of weld, however, it is important to establish time, temperature, and pressure combinations that melt the polymers at the mating surfaces and hold them firmly in contact while cooling. Excessive heating in any type of welding can result in melt flow, and in some instances, even resin degradation. Insufficient heating, on the other hand, can produce weak joints.

High-frequency or Dielectric Heat Sealing

In its simplest explanation, dielectric sealing involves the use of electricity or electronic waves from an ac generator passing through a plastic sheet or film. These waves cause a high level of molecular ''friction'' within the plastic, so that (given a plastic with sufficient dielectric loss) heat will be generated within the plastic and the plastic will melt at the point of contact (i.e., where the energy field is concentrated). By applying pressure at this point and allowing sufficient cooling time, a permanent bond is formed.

Dielectric sealing can handle many of the thermoplastics: ABC, acetate, polyester film, PVC, polyurethane, acrylics, to mention several; it is even possible to dielectrically weld some of the tougher engineering plastics, such as acetal film, but in many such instances, where the plastic has a relatively low loss factor, fairly high frequencies must be used, necessitating more expensive equipment. The biggest use for this technique is in sealing vinyl film and sheeting into such products as rainwear, inflatables, shower curtains, swimming pool liners, blister packages, medical bags, auto interior panels, and loose-leaf binders.

The most usual line for a high-frequency sealing operation consists of an RF generator (with an oscillator tube), a sealing press to apply pressure to the bonding area, and heat sealing dies set up in the press to determine the shape of the sealing area (commonly fabricated from brass strips bent into the desired pattern). Operating ranges from 1 to 100 kW are available, but the most commonly used range is from 3 to 25 kW; the higher the frequency, the faster a material with a given dielectric constant, can be fused or bonded.

Materials handling systems for feeding the sealing machines also vary widely. The most simple, of course, involves hand-feeding the part to be sealed into the press. More complex setups mount the sealing equipment on trolleys that move on a track alongside the film to be sealed or use conveyors or moving the tables to automatically feed the film into the press.

Another design for increasing production the rotary table machine with automated indexing work areas. A typical unit of this type might incorporate a rotary table with four workstations: loading, sealing, cooling, and unloading. Such machines can seal as many as 400 to 500 units an hour.

Over the years, the dielectric sealing technique has become an extremely versatile process. Although originally developed for sealing plastic to itself (i.e., PVC to PVC), the technique is now available for sealing different materials together. In the automotive industry, for example, door panels based on vinyl overlays welded to hardboard sheets with a urethane foam core in between, are being dielectrically sealed together. In many of these applications, vinyl chloride latex adhesives are involved. These adhesives are bonded directly to a solid skin or impregnated into a porous material such as urethane foam. Once dried, the treated parts can be bonded dielectrically.

Finally, dielectric sealing has also come into use as a method of decorating plastic products. For example, by sealing two sheets of vinyl together in a predetermined pattern, it is possible to achieve a textured quilted effect. Another method is known as applique bonding. Basically, this is achieved by applying the "tear seal" system. In a tear seal, the sealing die is forced farther into the plastic than is normally the case, causing it to thin out (as the melted plastic is displaced by the die) to the point where excess material can easily be torn off or separated by hand; the shape of the die, of course, determines which areas will be torn off. Thus, if a separate film is laid on top of the base film and the tear seal technique is applied (i.e., cutting through the top film and not the base film), it is possible to create an applique pattern by tearing off the excess material from around the sealing area and leaving a pattern defined by the sealing die.

Thermal Sealing of Film and Sheet

Non-high-frequency thermal sealing of film and sheet also is available, and it does not require the high-frequency energy that goes into dielectric sealing. This method usually involves a press with at least one moving platen bar. Bars are heated by low-voltage, heavy electric current. Heat is applied to the outer surface of the product and must travel through the material to the interfaces to make the weld. A variation of this process, similar to high-energy induction bonding, involves putting a metal wire or other insert between the two sheets to be sealed, to make the heating step more efficient.

Some of the most common types of heat-sealing-equipment are:

- *Jaw-type bar sealers:* Sealing takes place between an electrically heated bar and a stationary base; TFE or other high-temperature-resistant coverings should be used to cover the hot bar to prevent molten material from sticking to it.
- *Rotary sealers:* For continuous operation, rotaries operate by passing the film over a revolving drum that acts as a base for the heated sealing bar to press against. In an alternative method, the material is carried on belts between thin heated rollers or wheels revolving in opposite directions.
- *Band rotary sealers:* In this process, the film moves between metal bands, traveling first between heated jaws and then between pressure rollers, and finally between cooling jaws.
- *Impulse-type sealers:* As mentioned above, impulse sealing involves the use of a metal wire. Heat is supplied on an intermittent basis so that the heat exposure time is reduced.
- *Hot knife or side weld sealers:* A seal is made by pressing a continuously heated knife-edge bar across the film.
- *Multi-point sealing:* Heat is transmitted to the product at closely spaced points so that only a small portion of the surface of the film is actually sealed. The result is a peelable seal that will not open under normal conditions.

As with most sealing operations, temperature and pressure are the important factors. Low- and medium-density PE melt at about 230 to 240°F, and linear PE melts about 25°F higher than this.

Hot Gas Welding

Most thermoplastics can be joined via hot gas welding, but PVC is still the major material being assembled by this technique, into such products as tank liners and ducting. Other candidates for hot gas welding are polyethylene, polypropylene, acrylics, some ABS blends, polystyrene, and polycarbonate. The process is not good for filled materials and is used only to join materials thicker than $\frac{1}{16}$ inch. Parts thicker than $\frac{3}{8}$ inch are not generally gas welded although they have been. Applications are usually in large, structural parts (such as signs, piping, fans, ducts, and fume scrubbers), and equipment manufacturers note that the weld is not an appearance joint but a functional bond.

Basically, heated gas (generally air, but nitrogen is used for PE and acetals, which oxidize in air, and is recommended for polypropylene that will be subjected to UV exposure) is used to soften both the mating surfaces and the welding rod. This rod is made of the same material as the parts being joined, and as the welding tool moves down the joint, the softened rod is deposited into the mechanically beveled joint area. On cooling, the mating surfaces and the rod filling the joint between are fused into an integral weld.

The following is a more detailed description of the hot gas welding process.*

Joints. The types of welded joints used in hot gas thermoplastic welding (Fig. 24-6) are similar to those in metal welding. The same material preparations are required in plastic welding as in metal welding, except that beveling of thermoplastics edges is *essential.* Flux is not required; however, some materials weld more satisfactorily in an inert gas atmosphere.

In thermoplastics, the materials do not melt and flow; they simply soften, and the welder must apply pressure to the welding rod to force the softened surface into the joint to create the required permanent bond.

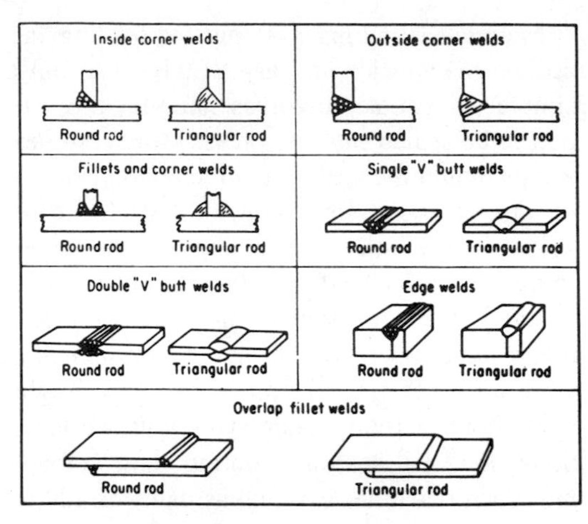

Fig. 24-6. Joint designs produced by hot gas welding. (*Courtesy Kamweld Products Co., Inc.*)

Triangular welding rods can be used in "V" welds or fillet welds up to $\frac{3}{8}$ inch wide, which can be filled with one pass. Thus the weld can be completed in less time and with better appearance, especially inside tank corners. When triangular rod is used, the chances of porosity (or leaks) occurring between one round rod and another one are eliminated.

Triangular rod can be hand-welded in the same way as hand welding of round rod; however, it can be applied more evenly with a properly designed high-speed tool. A speed weld also produces a much neater appearance than a hand weld. It must be understood that the use of heavy triangular rod, especially in horizontal runs, can be more tiring to the welder; but, at the same time, the weld can be accomplished in one pass rather than in a series of passes, as with round rod. In some cases, a small-diameter round rod is welded in the bottom of the "V" to join the pieces together and prepare the joint for the larger triangular rod.

Welding guns contain a heating unit (electric or gas) through which is passed a stream of compressed air or inert gas that reaches the nozzle or tip at a temperature of between 400 and 900°F. For production work, a gun of the type whose air hose also encloses its electrical cable is the more efficient and flexible in operation. The wattage of the heater determines the effective range of available heat; the air

*Adapted from an article by S. J. Kaminsky, Kamweld Products Co., Norwood, MA 02062, appearing in the 1971–1972 issue of *Modern Plastics Encyclopedia*, reprinted with permission of the publishers, McGraw-Hill, Inc., New York, NY.

pressure determines the actual amount of heat attained at the tip.

Tack welding is used to hold pieces in place for final welding. It can be effective in reducing or eliminating the need for clamps, jigs, or additional personnel during assembly. Tack welds are produced simply by running a heated tacker tip along the joint at intervals or in a continuous line.

In hand welding, the welder uses both hands, one to hold the welding gun and the other the welding rod. Using a round tip on the gun, the welder directs a hot air stream with a fanning motion onto both sides of the joint to be welded, as well as onto the welding rod. After the weld is started, the welding rod is bent back slightly and enough pressure is exerted uniformly to fuse the heated surfaces.

Only experience and careful observation can assure good joints with maximum strength and minimum degradation of the material through overheating. Underheating makes a "cold weld" with poor tensile strength. The aim is always to produce adequate pressure at the joint to permit complete fusion, yet not so much as to cause the welding rod to "catch up" with the welding tip. This will cause the rod to stretch, resulting in a poor weld.

High-Speed Hand Welding. Speed of making welds can be greatly increased through the use of welding gun tips that hold the welding rod in the correct position and simplify applying optimum pressure while making the weld.

Flat strip may be used instead of round rod in a tip for welding tank linings and similar applications. Only a single pass is made, the wide strip providing adequate surface to give strength in the finished joint.

Hot Plate Welding*

Hot plate welding can readily join plastic parts that have complex contours: a hermetic seal is easily obtained if a proper joint design and material are selected. The hot plate welding process works extremely well with a wide variety of plastic materials and is often the welding process chosen for welding softer plastics such as polypropylene, polyethylene, and the thermopolastic polyimides, where solvent-based adhesives and ultrasonic welding are ineffective. It also does an excellent welding operation on large parts such as automotive vent ducts, fuel tanks, liquid reservoirs, and washing machine tubs, agitators and spray arms.

Each welder is designed to accommodate a specific range of product requirements. Depending on the requirements, machines are available for hot plate welding on either horizontal or vertical planes. In most cases, machines can handle a range of varying part designs. A change in part design may mean a change in holding fixtures, but not necessarily a change in machines. The hot plate welding process consists of six steps as shown in Fig. 24-7.

For accurate mating and alignment, the hot plate welding process relies on holding fixtures to support the parts to be joined. Collets, gripping fingers, mechanical devices or vacuum cups may be used for folding or gripping the parts.

To plasticize the part edges, the fixtures press the parts against a heating platen. As the platen melts the part's mating surface, plastic material is displaced. This initial melting produces a smooth mating edge by removing surface imperfections, warps, and small sinks. Melted material continues to be displaced until the "melt stops" on the platen contact the "tooling stops" on the holding fixture. Once the melt stops and tooling stops are in contact, material ceases to be displaced and the parts are held fixed against the platen until each part's edge is plasticized to a predetermined depth. Melt depth is regulated by the contact time of the plastic part against the platen, usually one to six seconds, and by the placement of the melt stops.

The temperature at which the platen is set depends on the type of plastic being welded. Each thermoplastic has a characteristic time/temperature curve for melting, and a bond or weld can be produced at any temperature on the curve. Typically, the highest possible temperature (and, thus, the shortest time) is selected to maximize production speed. The normal

*Courtesy Forward Technology Industries, Inc., Minneapolis, MN.

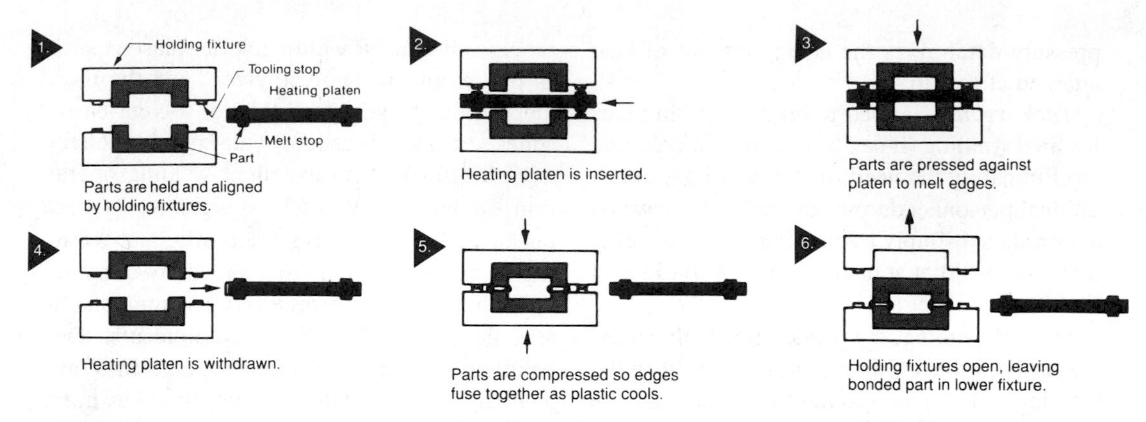

Fig. 24-7. Six steps for hot plate welding thermoplastics. (*Courtesy Forward Technology Industries, Inc.*)

temperature range is 300 to 700°F. Aluminum platens are used for temperatures to 500°F and are usually coated with PTFE-impregnated fiberglass cloth to prevent parts from sticking. Bronze and steel platens are used for temperatures over 500°F.

After the part edges are plasticized, the holding fixtures open and the heating platen is withdrawn. The fixtures then close, forcing the two parts together until the tooling stops on the fixtures come into contact. The parts are held together under pressure, allowing the melted plastic material to cool and weld together. Cooling time is normally from three to six seconds. When cooling is complete, the gripping mechanism on one of the holding fixtures releases the part, the fixtures open, and the finished part is manually or automatically removed.

Total cycle time, from start to finish, generally is twenty seconds or less, well within the cycle range of injection-molding systems often used in conjunction with hot plate welders.

Hot plate welding joint design is limited only by the designer's imagination. Figure 24-8 illustrates a few of the more commonly used joints.

In the hot plate welding process, a small amount of material is displaced from each of the mating parts. This displaced material creates a weld bead. From the designer's point of view, displacement must be taken into consideration for two reasons: 1) In intricate assemblies where dimensional tolerance is critical, material needs to be added at the mating surface. 2) In parts requiring external aesthetics, the weld bead must be hidden or removed. If there is insufficient design flexibility to hide the bead, it can be easily removed in the welding process.

The hot plate welding process produces a joint which, in many cases, is consistently equal to or stronger than any other area of the part. As a result, the weld area can be exposed to the same strains and stresses as any other area of the part.

Induction Welding

Induction welding is exceptionally fast and versatile. Heat is induced by a high-frequency electrodynamic field in a metallic insert placed in the interface of the areas to be joined, and this heat brings the surrounding material to fu-

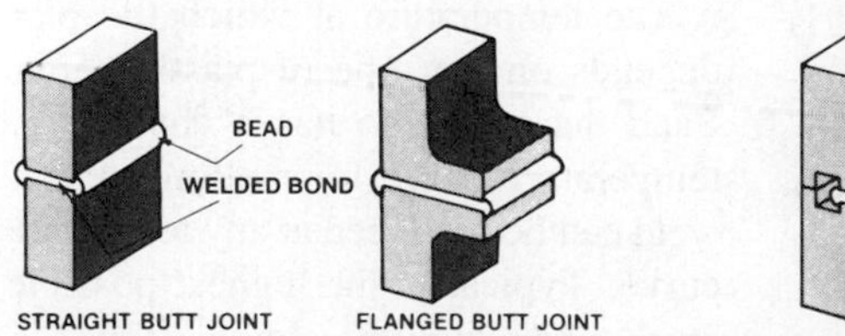

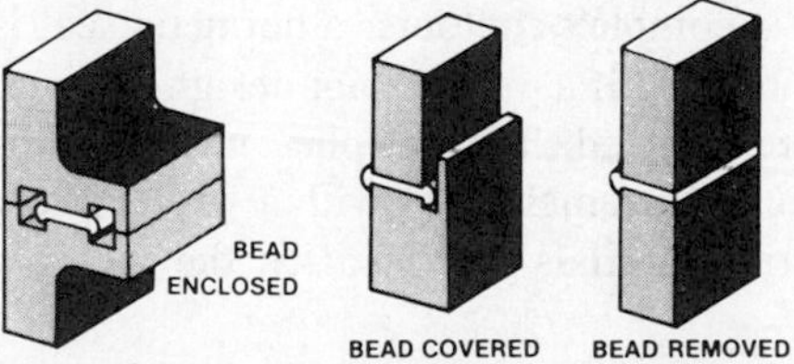

Fig. 24-8. Possible joint designs for hot plate welding. (*Courtesy Forward Technology Industries, Inc.*)

sion temperature. Pressure is maintained as the electromagnetic field is turned off, and the joint solidifies. The weld strength depends on the size and geometry of the metal insert.

Induction welding is one of the fastest methods of joining plastics. Some applications require as little as 1 second of welding time; typical applications require from 3 to 10 seconds. The rate of production generally is limited only by the speed with which the parts may be assembled in and removed from the welding jig.

Successful welds can be obtained by using stamped foil inserts, standard metallic shapes, such as wire screen, or various other configurations of conductive metal. The shape of the insert is not limited to the normal pattern of a closed resistance circuit; thus it may be made in a shape that provides a decorative effect in the finished article (e.g., stars or letters). Wire can be used, preferably of diameter from 0.010 to 0.030 inch. Printed or metallized inserts may be feasible, but have not yet been fully evaluated.

The higher the pressure applied on the joint during welding, the higher the permissible temperature. The feasible pressure appears to be limited only by the strength and stiffness of the materials being welded. In most applications a pressure of 100 psi of joint is required, but greater pressure may be needed with some plastics, to minimize development of bubbles from thermal decomposition at the joint. The joint should be designed so that pressure is distributed uniformly throughout, and over the metal insert. The insert should be located as closely as possible to the generator coil. It should be centered accurately within the coil, to avoid its being pulled out of position by the attraction of the coil. A tongue-and-groove joint or a similar configuration is desirable, to locate and hold the metal insert. In any case, the insert must be located in the interface so that no portion of it is exposed to air, as such exposure could cause rapid heating and subsequent disintegration of the insert.

Strengths of induction welds are limited by the relative smallness of the area in which the actual weld is formed. Welding occurs only in the area immediately adjacent to the metallic insert. In most cases this is a band, $\frac{1}{16}$ to $\frac{1}{8}$ inch wide, along the periphery of the insert. Very little if any strength is obtained by metal-to-plastic bonding over the area of the insert.

Weld strengths can be determined by multiplying the unit weld strength by the area over which actual welding takes place. With wire screen as the insert between polyethylene slabs, weld strengths of better than 50% of that of the parent material can be consistently produced. Acrylics have been processed equally well, and from all indications welds of significant strength can be obtained in almost all thermoplastics.

Another adaptation of induction bonding involves replacing the metal insert with a metal-filled plastic. In essence, magnetic materials, in the form of micron-size particles, are dispersed within a thermoplastic matrix that is compatible with the material to be joined (e.g., a polypropylene matrix when polypropylene parts are being joined). The matrix can be supplied in the form of a liquid, or it can be molded into a shape (e.g., a gasket) that will conform to the shape of the weld area.

The metal-filled plastic bonding agent is simply applied to the parts to be joined and then subjected to the high-frequency magnetic field of an induction heating coil. The magnetic flux reaches through the work and generates heat in the bonding. Heating is the result of internal energy losses within the material that cause the temperature to rise. It is, of course, the metal particles (with magnetic properties) within the plastic that generate these losses through hysteresis and eddy current. (See Figs. 24-9 and 24-10).

As a consequence, the thermoplastic matrix in which the particles are confined melts and is subsequently fused to the two abutting surfaces, forming a structural bond on cooling.

The process is available for polypropylene, polyethylene, nylon, ABS, polystyrene, and PVC. Its advantages include: adaptability to bonding most of the thermoplastics; speed of bonding; the ability to make structural bonds in either butt, peel, or shear; adaptability to hermetic seals; flexibility in bonding area in terms of size (can be a small spot weld or a long, continuous weld) and shape (irregular as well as symmetrical); and the fact that the physical

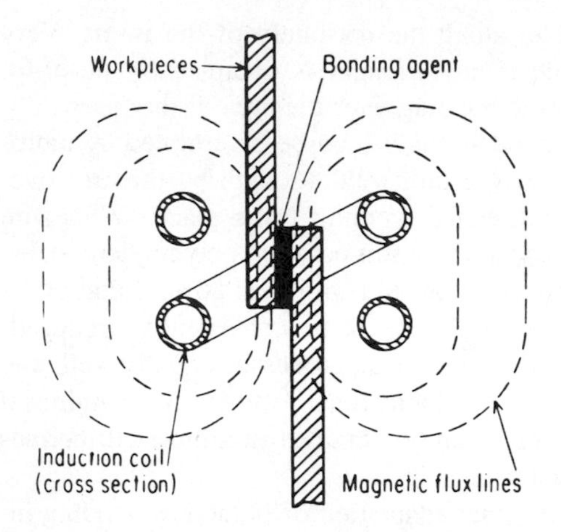

Fig. 24-9. Electrical induction heat is produced by the high-frequency magnetic field of an induction heating coil. The magnetic flux (dashed lines) reaches through the work and produces heat in the bonding agent. (*Courtesy McGraw-Hill*)

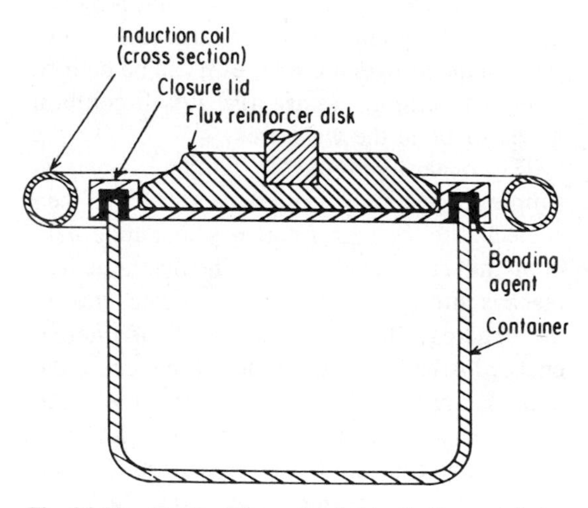

Fig. 24-10. Plastics lids are bonded to plastics containers using the induction bonding method. (*Courtesy McGraw-Hill*)

and chemical properties of the bond area are similar to those of the material being joined. It is also claimed that by using small particles of metal as opposed to larger pieces of metal or wire, stresses are not molded into the bond area.

Spin or Friction Welding

In this process, the two surfaces of thermoplastic are rubbed together until enough friction heat develops to fuse and weld them. The strength of the welded joint approaches that of the parent material.

Typical welding cycles are 1 to 2 seconds. High-speed feeding devices (e.g., vibratory feeders) can be hooked into the operation for continuous production.

Nearly all the rigid thermoplastics can be spin-welded, although there are some problems with the softer rigids, such as low-density polyethylene or ethylene vinyl acetate.

The advantages of the process are: (1) the strength and good appearance of the joint; (2) the exclusion of oxidizing effect, in that the heated surfaces are directly in contact with each other throughout the process; and (3) the economy resulting from adaptability to standard shop equipment, such as drill presses and lathes.

Its disadvantages are: (1) limitation to a circular or annular area (but sometimes a circular area of weld can be made part of the design of a noncircular article); and (2) the squeezing out (flashing) of soft material beyond the area of the weld before the weld is completed (but this can frequently be made unobjectionable by designing so that the flashing is directed to an internal area of the article).

Another drawback has generally been considered to be the technique's limitation to fairly small parts. However, work has been done in spin-welding parts up to 20 inches in diameter.

Spin welding involves the rotating of one of the pieces against the other, which is held stationary. Rubbing contact is maintained at a speed and a pressure sufficient to generate frictional heat and melt the adjacent surfaces. The frictional heat is sufficient to effect almost immediate melting of the two surfaces without substantially affecting the temperature of the material immediately beneath the surfaces. When sufficient melt is obtained, the spinning is stopped, and pressure is increased to squeeze out all bubbles and to distribute the melt uniformly between the two surfaces. Pressure is maintained until the weld solidifies. In many instances, a single pressure setting has been found adequate. For thermoplastics having sharp melting points, braking sometimes is required, in order to stop the rotation of the parts quickly to avoid tearing of the partially solidified weld.

Spin Welding Equipment.* Prototype or low-production-volume spin welding can be performed on a standard drill press with a driving tool and chuck. High-volume applications require either a drill press equipped with a spin-welding tool, air cylinder, valves, and timer or specially designed spin-welding equipment.

The basic equipment requirements for a spin welder are diagrammed in Fig. 24-11. Parts to be welded are shown held in a chuck. A spinning tool is attached to an electric motor by a drive shaft and V-belt. An air cylinder advances the tool and applies force to the parts during welding. A timer and an air valve are used to control the air cylinder. A motor speed control, motor brake, and clutch may be required.

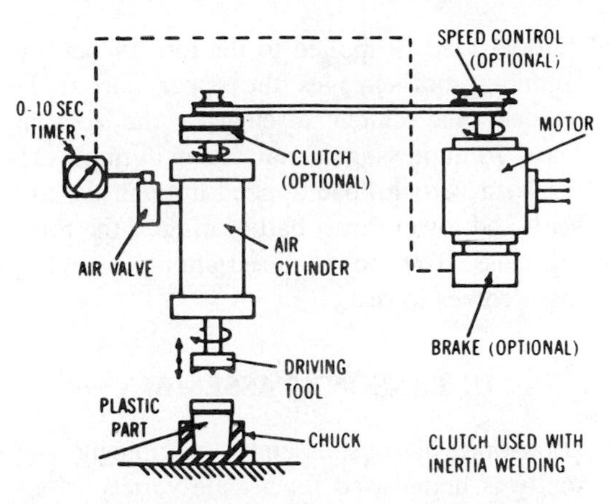

Fig. 24-11. Basic equipment requirements for spin welding. (*Courtesy E. I. du Pont de Nemours & Co., Inc.*)

Variables of Spin Welding. Basic variables of spin-welding are rotational speed, joint pressure, and spin time.

Rotational speed depends on the joint diameter.

Joint pressure is a convenient means for establishing the thrust required to generate heat between spinning parts. During welding, joint pressure forces bubbles, contamination, and excess material from the weld. The actual thrust will depend on the joint design.

Spin time is the duration of relative motion between the rotating surfaces before melting takes place. Once melting has occurred, relative motion must be stopped to allow the weld to solidify under pressure because failure to arrest motion will cause tearing of the weld and result in low weld strength.

Spin Welding Methods.* There are two basic methods for spin welding, which differ only in the type of driving tool used: pivot and inertia methods. With both methods, rotational speed and joint pressure are adjusted similarly to control the welding process. Spin times are controlled differently, however. With the pivot method, the spin time is controlled by a timer, which activates withdrawal of the pivot tool. In the inertia method, the spinning tool is disengaged from the motor after the start of the weld cycle. The kinetic energy of the freely spinning inertia tool is converted into heat energy during welding. Thus, the spin time may be varied markedly by adjusting the mass of the inertia tool or more finely by varying the speed of rotation. Pivot tools are generally used for joints having a projected area of $\frac{1}{2}$ in.^{2} (3 cm^2) or less. Inertia tools are better for joints of large area because they offer fine control of input energy.

Essential parts of the pivot tool are the driving element, such as teeth or rubber facing, and the pivot pin. The pivot tool turns at constant speed throughout the weld cycle. A load is applied to parts by the spring-loaded pivot pin before the driving element engages the parts. At the end of the cycle, the tool retracts, and the driving element disengages while joint pressure is maintained by the pivot pin. The relative motion between parts stops almost instantly, and the weld solidifies.

When part design does not permit the use of a pivot pin, a thrust bearing may be used. A rubber facing may be used on the driving elements if no tooth indentations can be tolerated on the welded part.

Essential elements of the inertia tool are the driving element and the rotating mass or flywheel. The clutch for disconnecting the inertia tool from the motor may be part of the main welder drive system, or it may be built into the inertia tool. With this inertia tool design, the rotating mass is supported by needle and ball bearings around a central shaft. A conical clutch connects the shaft and mass when no

*Adapted from a Du Pont technical data brochure on machining Delrin acetal.

vertical load is applied to the tool. When the spinning tool engages the upper part to be welded, the clutch disengages the rotating mass. Joint pressure is maintained during welding by a vertical load applied through the tool shaft and lower thrust ball bearing to the rotating mass. The tool is raised after the rotating mass comes to rest.

ULTRASONIC ASSEMBLY*

Ultrasonic plastics assembly equipment currently is being used for a wide variety of applications in many different industries.

Typical ultrasonic processes are welding injection molded plastics, inserting metal into plastics, staking over plastic studs or bosses, spot-welding sheet material, continuous film or synthetic fabric sealing and cutting, degating parts from a runner system, swaging, and forming.

The basic principle of ultrasonic assembly is that when two pieces of thermoplastic are vibrated together fast enough with an appropriate amplitude, heat will be developed, and a flow of plastic will occur at the joint interface. The method can provide strong (95–100% strength), hermetically sealed components at a high production rate. It is clean, fast (20 to 60 parts/min.), and relatively inexpensive as compared with other assembly methods. Materials handling equipment can easily be interfaced with an ultrasonic system. (Figure 24-12 shows a microprocessor-controlled ultrasonic welder.)

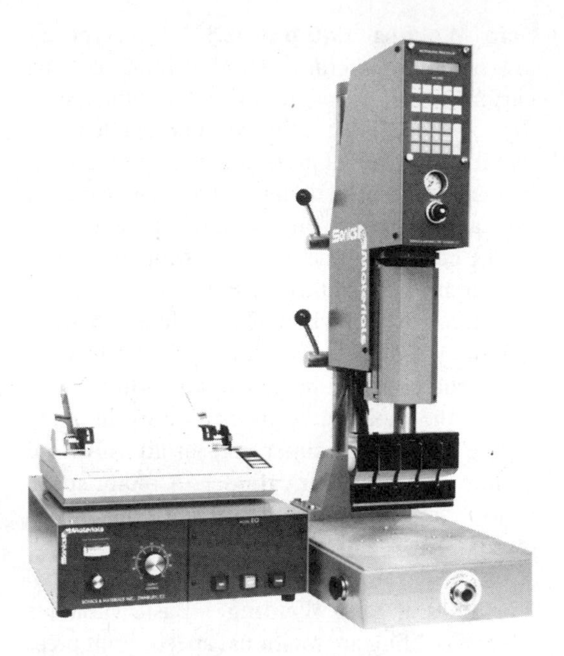

Fig. 24-12. Microprocessor-controlled ultrasonic welder with 2000 watt, 20 kHz power supply and a printer for recording various parameters. (*Courtesy Sonics and Materials, Inc.*)

Equipment

An ultrasonic assembly system can be broken down into seven major components (Fig. 24-13).

1. Power supply or ultrasonic generator
2. Converter or transducer
3. Booster horn
4. Horn
5. Stand or pneumatic press
6. Welding controls
7. Fixture

Power Supply. The power supply transforms 50/60 Hz electrical power from a wall outlet to ultrasonic power at a given frequency. The most common frequency used is 20 kHz. Other frequencies that are commercially available are 15, 30, 36, 40, and 70 kHz.

With lower frequencies such as 15 kHz, one can use very large horns and, therefore, weld large parts. Another important advantage of using 15 kHz is that there is significantly less attenuation through a thermoplastic, permitting the welding of many softer plastics and welding at a greater distance from the horn contact area. Higher frequencies allow the welding of small delicate parts with no disturbance of the adjoining plastic material.

A majority of welding applications are accomplished by using a power supply with an output of 1000 watts. Power supplies are available with power ranging from 50 to 3000 watts.

*By Robert S. Soloff, President, Sonics and Materials, Inc., Danbury, CT.

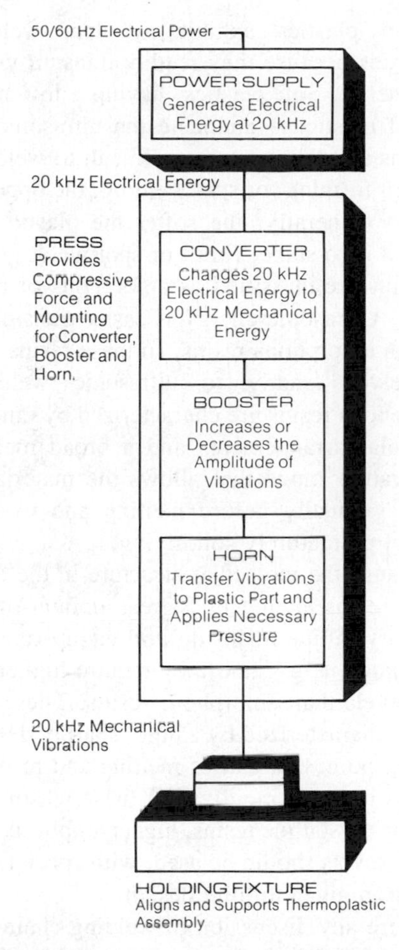

Fig. 24-13 Ultrasonic assembly system. (*Courtesy Sonics and Materials, Inc.*)

Converter. The electrical power output from the power supply is introduced into a stack of piezoelectric ceramics (PZT), causing the PZT material to expand and contract at a high frequency. This phenomenon converts the 20 kHz of electrical energy into 20 kHz of mechanical energy or vibrations. The distance that the mechanical vibrations move back and forth is called the amplitude. Typical amplitudes, at the converter output face at 20 kHz, are between 0.0006 inch and 0.0008 inch peak to peak.

Booster Horn. The success of an ultrasonic plastics assembly application depends on having the proper vibrational amplitude at the horn face. As it is impractical always to design the correct amplitude into the horn, booster horns are necessary to either increase or decrease the amplitude to produce the proper amount of vibrational heat at the weld area of the plastic parts. The booster horn, which is a half-wavelength resonant element, is made of either aluminum or titanium and is mounted between the converter and the horn. Boosters usually are color-coded or marked to indicate their ratio.

Horn. The horn also is a tuned half-wavelength resonant element, which transfers the mechanical vibratory energy from the converter and booster horn to the workpiece. The horn can further modify the booster horn's amplitude output by the nature of its design. The horn's surface also must be machined to conform to the shape of the plastic part. Horns are made of materials that have a high strength-to-weight ratio and low losses at ultrasonic frequencies. Titanium has the best acoustical properties of the high-strength metals, whereas aluminum is the best of the lower-strength metals. Heat-treated steel horns are used when wear-resistant surfaces are required, but they are subject to higher ultrasonic losses. A variety of horn shapes are available, depending on the size, shape, and weldability of the plastic part being welded.

Press (Stand). The press consists of a base, column, actuator, and control panel, all housed in a frame and interconnected by cables. The base contains dual non-tie-down palm buttons (the operator must maintain both hands on the palm buttons during the weld cycle) and an emergency stop button, both for operator safety. The actuator housing contains the converter, booster horn, and horn, and is pneumatically raised and lowered to apply ultrasonics at a predetermined pressure and time to the workpiece. Each application requires different settings for an optimum weld. The weld pressure, trigger pressure, stroke speed, and stroke distance, among other functions, are all adjusted in the stand. In certain configurations, some of the controls are located in the power supply.

Timer or Microprocessor Controller. The variety of controls and the repeatability of each of them determine the effectiveness of the ultrasonic system. The timer, which can be located in the power supply or in the stand, controls the weld and hold times. Typical weld times are usually less than 2 seconds, with most being under 1 second. The hold time is usually less than 0.5 second.

An alternative to the timer and a more effective means of controlling the various functions in the weld cycle is to use a small computer or microprocessor controller. Microprocessor controls were introduced for use with ultrasonic assembly in 1985. The advantage of the microprocessor is its ability to weld to an energy value rather than just to a preset time. Most microprocessors have the ability to set limits on preset parameters and sound an alarm or activate a reject control switch in order to sort out rejects.

Job storage, force triggering, rejects and part counts, internal calibration, lockout codes, self-diagnosis, and printer and computer interfaces are other features usually provided in the welder's microprocessor.

Welding to a preset distance, either from a fixed reference point or measuring the collapse of the plastic part during welding, is accomplished by using a linear encoder mounted on the welding press. When the preset distance is reached, a signal is sent to the controls to terminate the weld.

Fixtures. The fixture, mounted on the base of the stand, must hold the plastic parts in proper alignment during the weld cycle. Fixtures usually are made from chrome-plated aluminum and steel or poured epoxy to minimize tool wear and marking of the plastic parts.

Materials

The characteristics of a plastic determine its weldability. Thermoplastic materials generally are well suited to ultrasonic welding because they melt within a specific range, whereas thermosetting materials are not, because they degrade when heat is applied. The characteristics of thermoplastic materials, such as melt temperature, modulus of elasticity, and structure, determine the energy requirements of the weld.

Rigid plastics exhibit excellent welding properties because they readily transmit vibratory energy. Soft plastics, having a low modulus of elasticity, attenuate the ultrasonic vibrations and thus are more difficult to weld. In staking, forming, or spot welding, the opposite is true. Generally, the softer the plastic, the easier it is to stake, form, or spot-weld.

Resins are classified as amorphous or crystalline. Ultrasonic energy is easily transmitted through amorphous resins, so these resins lend themselves readily to ultrasonic welding. Amorphous resins are characterized by random molecular arrangements and a broad melting temperature range that allows the material to soften gradually before melting and to flow without prematurely solidifying.

Because the molecular structure in the crystalline resins attenuates a great amount of energy, crystalline resins do not readily transmit ultrasonic energy, and they require higher energy levels than amorphous resins. These resins are characterized by a high, sharply defined melting point that causes melting and resolidification to occur rapidly. For these reasons, in welding crystalline resins, higher amplitude and energy levels should be used, with special consideration given to joint design.

Before any discussion of welding characteristics, the difference between near-field and far-field welding must be understood. Near-field welding refers to welding a joint located $\frac{1}{4}$ inch (6 mm) or less from the area of horn contact. The greater the distance from the point of horn contact. The greater the distance from the point of horn contact to the joint, the more difficult it will be for the vibration to travel through the material and for the welding process to take place.

The differential, if any, in the melt temperature of the materials being welded should not exceed 30° (17°C), and the materials' molecular structures should be compatible (i.e., blends, alloys, copolymers, and terpolymers).

The moisture content, mold release agents, lubricants, plasticizers, fillers, reinforcing agents, regrinds, pigments, flame retardants, and resin grade all can influence weldability. The moisture content of parts molded from resins that are hygroscopic (moisture-absorbent) can be problematic. Nylon (and to a lesser degree polycarbonate and polysulfone) present most of the problem, and parts molded in these resins should be stored in sealed polyethylene bags with an appropriate desiccant immediately after molding. If moist parts are welded, the escaping vapors may cause voids and fissures in the molten material, resulting in a weld of poor integrity.

Mold release agents such as zinc stearate, aluminum stearate, fluorocarbons, and silicones are not compatible with ultrasonic welding. If it is necessary to use a mold release agent, the paintable/printable grades that permit painting and silk screening should be considered. Other release agents should be removed with either TF Freon for crystalline resins or a 50/50 solution of water and liquid detergent.

Lubricants, whether waxes, stearates, or fatty esters, reduce intermolecular friction within the polymer and inhibit the ultrasonic assembly process. However, because they generally are dispersed internally, their effect usually is negligible.

Plasticizers, which usually impart flexibility and softness to a resin, can interfere with a resin's ability to transmit vibratory energy. FDA-approved plasticizers do not present as much of a problem as metallic plasticizers, but experimentation is recommended.

Although fillers and reinforcing agents such as glass and talc can increase the ultrasonic weldability of thermoplastic considerably, they should be used judiciously. When the additive content exceeds 10%, premature horn wear may result, and specially treated steel or carbide-faced titanium horns might be required. When the filler content approaches 35%, there may be insufficient resin at the surface to obtain hermetic seals. When the filler content exceeds 50%, insufficient plastic is present at the interface to form a positive bond. Reinforcement composed of long glass fibers always is more problematic than reinforcement composed of short glass fibers.

Ultrasonic assembly is one of the few methods that permit regrinding of parts, as no foreign substance is introduced into the resin. Ultrasonically assembling parts have been manufactured from regrind parts presents no problem, provided that the percentage of regrind is not excessive and the plastic has not been degraded. Regrind limitations suggested by the resin suppliers should be observed.

Although most pigments do not interfere with the ultrasonic assembly process, some oil-based colorants can adversely influence weldability. Non-oil-based pigments should be used.

Flame retardants greatly affect the weldability of thermoplastics, and the effects of the various additives should be investigated experimentally prior to resin selection. The grade of resin can have a significant influence on weldability. There is a great difference between injection/extrusion grades and cast grades; their molecular weight, melt temperature, and modulus of elasticity are quite different. Injection/extrusion grades should be used only with injection/extrusion grades and cast grades used only with cast grades.

Table 24-6 lists the welding characteristics of thermoplastics.

Figure 24-14 shows typical ultrasonic assembly processes.

Joint Design

Perhaps the most critical facet of part design for ultrasonic welding is joint design (the configuration of the two mating surfaces). It should be considered when the parts to be welded are still in the design stage, and incorporated into the molded parts. There are a variety of joint designs, each with specific features and advantages. Their selection is determined by such factors as type of plastic, part geometry, weld requirements, machining and molding capabilities, and cosmetic appearance.

Butt Joint with Energy Director. The butt joint with energy director is the most common joint design used in ultrasonic welding, and the easiest to mold into a part. The main feature of this joint is a small 90° or 60° triangular-shaped ridge molded into one of the mating surfaces. This energy director limits initial contact to a very small area, and focuses the ultrasonic energy at the apex of the triangle. During the welding cycle, the concentrated ultrasonic en-

Table 24-6. Ultrasonic welding characteristics of thermoplastics.

Material	*Spot Welding*	*Staking Swaging*	*Inserting*	*Field of Welding*	
				Near	*Far*
AMORPHOUS:					
ABS	E	E	E	E	G
ABS/Polycarbonate	G	G	G	G	F
ABS/PVC	G	G	F	G	F
Acrylic	G	F	G	G	F
Acrylic Multipolymer-XT Polymer	G	G	G	G	F
Acrylic/PVC	G	G	F	G	F
Acrylic—Impact Modified	F	F	P	F	P
Butadiene—Styrene (BDS)	G	G	G	G	F
Cellulosics—CA, CAB, CAP	P	G	E	P	—
Modified Phenylene Oxide	E	E	E	E	G
Polyarylate	F	F	G	G	F
Polycarbonate	G	F	G	G	F
Polyetherimide	G	G	E	E	G
Polystryene, G.P.	F	F	G	E	E
Polystyrene, Impact Modified	F	F	G	G	P
PVC—Rigid	F	G	E	P	P
PVC—Flexible	P	—	—	P	—
SAN—NAS—ASA	F	F	G	E	E
Styrene-Maleic-Anhydride	E	E	E	E	G
Sulfone Polymers	F	F	G	G	F
CRYSTALLINE:					
Acetal Copolymer	F	F	G	G	F
Acetal Homopolymer	F	F	G	G	F
Fluoropolymers	—	—	—	P	—
Nylon	F	F	G	G	F
PC-PET	G	G	E	E	G
Polyester—PBT	F	F	G	G	F
Polyester—PET	F	F	G	G	F
Polyetheretherketone	G	G	E	E	G
Polyethylene (LDPE, HDPE)	G	F	G	P	P
Polyethylene (UHMW)	—	—	—	—	—
Polymethylpentene	G	F	E	F	P
Polyphenylene Sulfide	F	P	G	G	F
Polypropylene	E	E	G	F–P	P

E-Excellent
G-Good
F-Fair
P-Poor
—Not suitable for ultrasonic assembly

ergy causes the ridge to melt and the plastic to flow throughout the joint area, bonding the parts together.

For easy-to-weld resins—amorphous polymers such as ABS, SAN, acrylic and polystyrene—the size of the energy director is dependent on the area to be joined. Practical considerations suggest a minimum height between .008 and .025 inch (.2 and .6 mm).

Crystalline polymers, such as nylon, thermoplastic polyesters, acetal, polyethlene, polypropylene, and polyphenylene sulfide, as well as high-melt-temperature amorphous resins, such as polycarbonate and polysulfones, are more difficult to weld. For these resins, energy directors with a minimum height between .015 and .020 inch (.4 and .5 mm) with a 60° included angle generally are recommended.

The 90° included angle energy director height should be at least 10% of the joint width, and the width of the energy director should be at least 20% of the joint widths. Figure 24-15 shows a butt joint with a 90° included angle energy director. With thick-walled joints, two

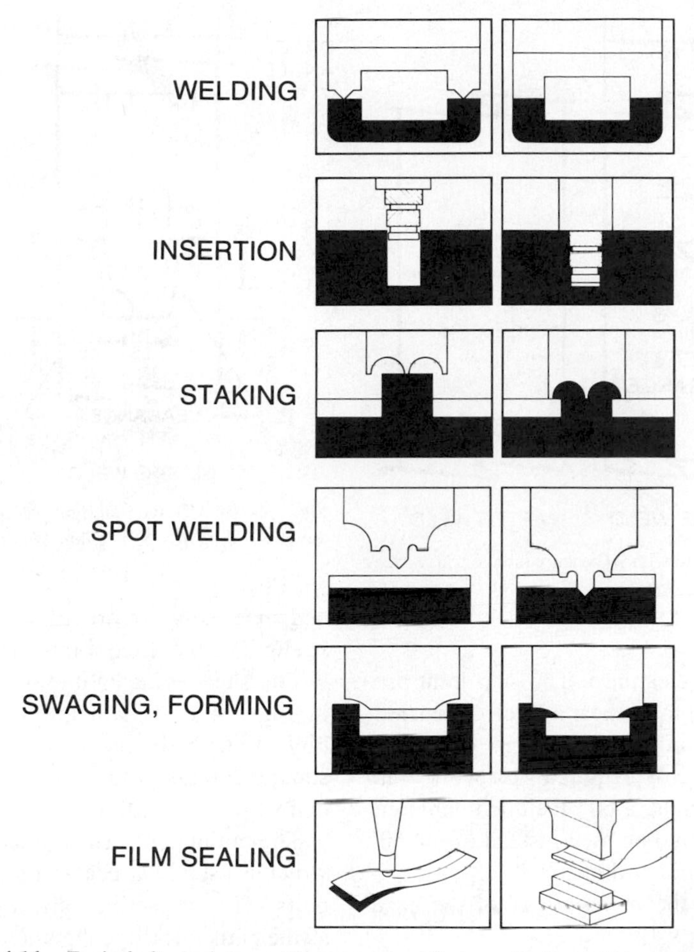

Fig. 24-14. Typical ultrasonic assembly processes. (*Courtesy Sonics and Materials, Inc.*)

or more energy directors should be used, and the sum of their heights should equal 10% of the joint width.

To achieve hermetic seals when welding polycarbonate components, it is recommended that a 60° included angle energy director be designed into the part. The energy director width should be 25 to 30% of the wall thickness.

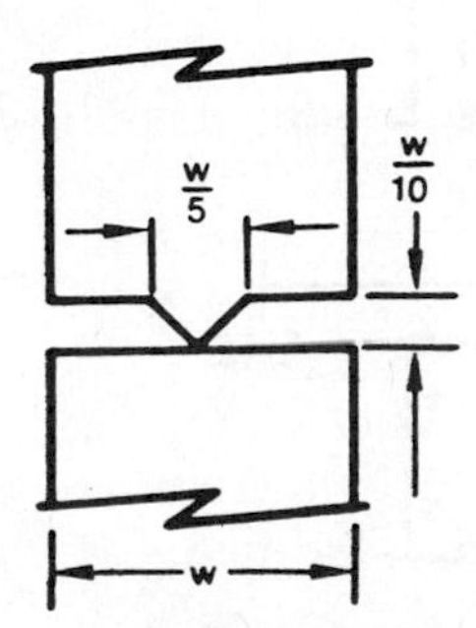

Fig. 24-15. Joint design. Butt joint with energy director. (*Courtesy Sonics and Materials, Inc.*)

With assemblies whose components are made of identical thermoplastics, the energy director can be designed into either half of the assembly. However, in designing energy directors into assemblies consisting of a part made of copolymers or terpolymers, such as ABS, and another part made of a homopolymer, such as acrylic, for maximum strength, the energy director should always be incorporated into the half of the homopolymer assembly.

The step joint with energy director is illustrated in Fig. 24-16. This joint molds readily, and provides a strong, well-aligned joint with a minimum of effort. This joint is usually stronger than a butt joint because material flows

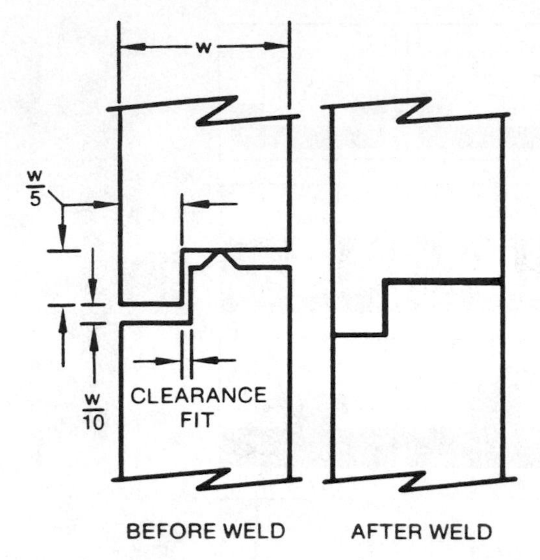

Fig. 24-16. Step joint with 90° energy director. (*Courtesy Sonics and Materials, Inc.*)

into the vertical clearance. The step joint provides good strength in shear as well as tension, and often is recommended where good cosmetic appearance is required. In work with crystalline materials, a 60° included angle energy director should be used instead of the 90° included angle energy director.

Figure 24-17 shows variations of the basic step joint design.

The tongue-and-groove joint with energy director is illustrated in Fig. 24-18. This joint is used primarily for scan welding, self-location of parts, and prevention of flash both internally

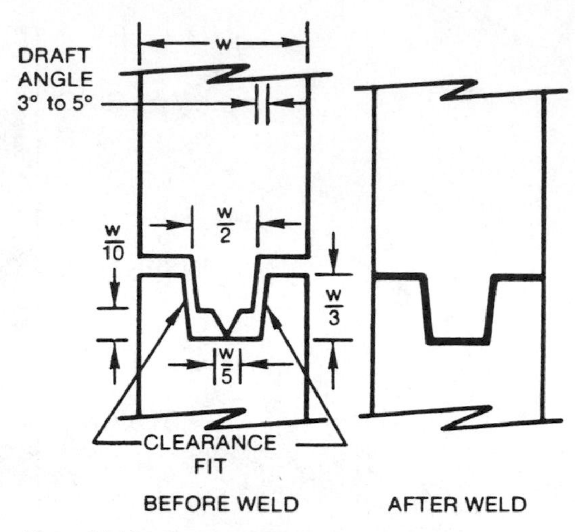

Fig. 24-18. Tongue-and-groove joint with energy director. (*Courtesy Sonics and Materials, Inc.*)

and externally. It provides the greatest bond strength of the three joints discussed so far.

The shear joint or interference joint shown in Fig. 24-19 is generally recommended for high-strength hermetic seals on parts with square corners or rectangular designs, especially with crystalline resins.

The initial contact is limited to a small area, which is usually a recess or step in either of the parts. The contacting surfaces melt first; then as the parts telescope together, they continue to melt along the vertical walls. The smearing action of the two melt surfaces eliminate leaks and voids, making this the best joint for strong hermetic seals.

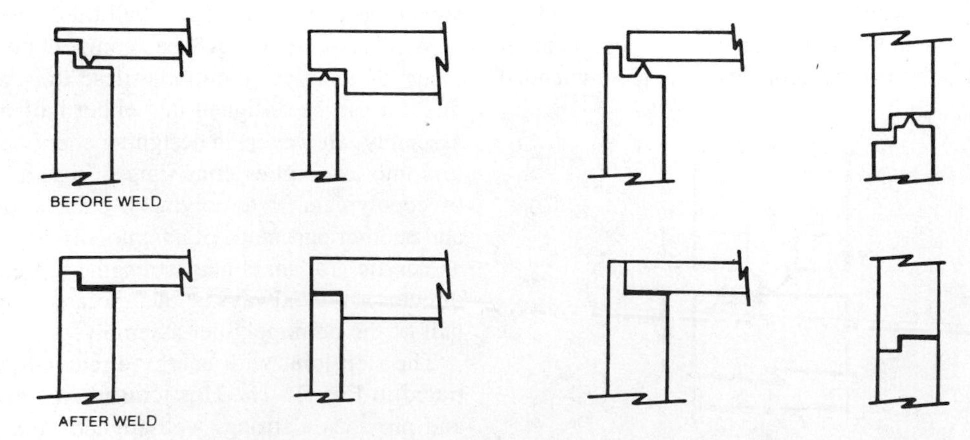

Fig. 24-17. Step joint variations. (*Courtesy Sonics and Materials, Inc.*)

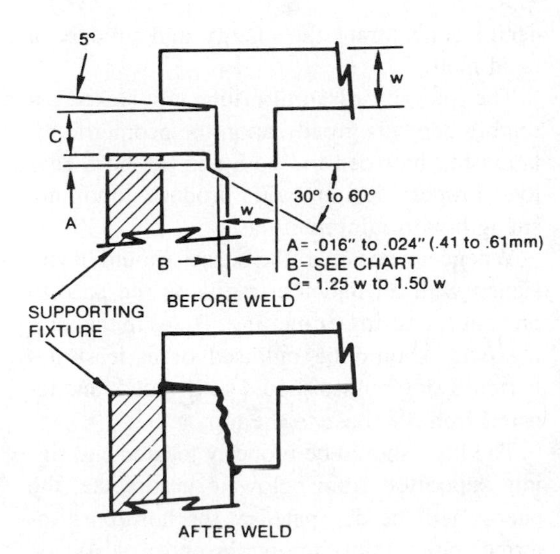

Fig. 24-19. Shear joint. (*Courtesy Sonics and Materials, Inc.*)

Several important aspects of the shear joint should be considered: (1) the top part should be as shallow as possible, (2) the outer walls should be well supported by a holding fixture, (3) the design should allow for a clearance fit, and (4) a lead-in (A) should be incorporated. The chart shows part and interference joint dimensions.

Maximum Part Dimension	*Interference (B)*
Less than 0.75″ (19 mm)	0.008″ to 0.012″ (0.2 to 0.3 mm)
0.75″ to 1.50″ (19 to 38 mm)	0.012″ to 0.016 (0.3 to 0.4 mm)
Greater than 1.50″ (38 mm)	0.016″ to 0.020″ (0.4 to 0.5 mm)

The shear joint requires weld times in the range of three to four times that of other joint designs because larger amounts of resin are being melted. In addition, a certain amount of flash will be visible on the surface after welding.

When flash cannot be tolerated for aesthetic or functional reasons, a well similar to the ones shown in Fig. 24-20 should be incorporated.

Modified joints, such as those shown in Fig. 24-21, should be considered for large parts of parts where the top piece is deep and flexible.

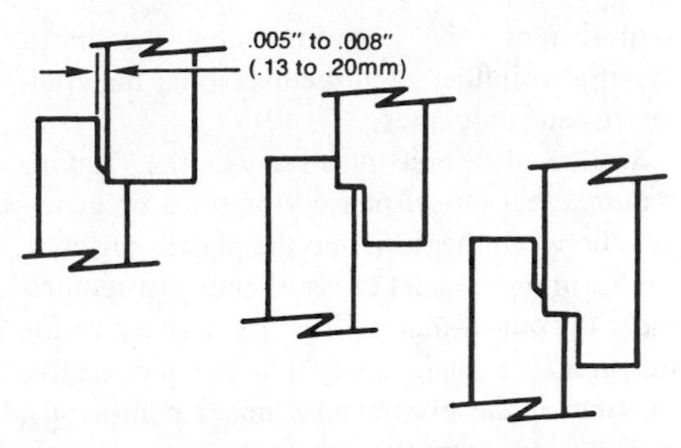

Fig. 24-20. Shear joints with flash wells. (*Courtesy Sonics and Materials, Inc.*)

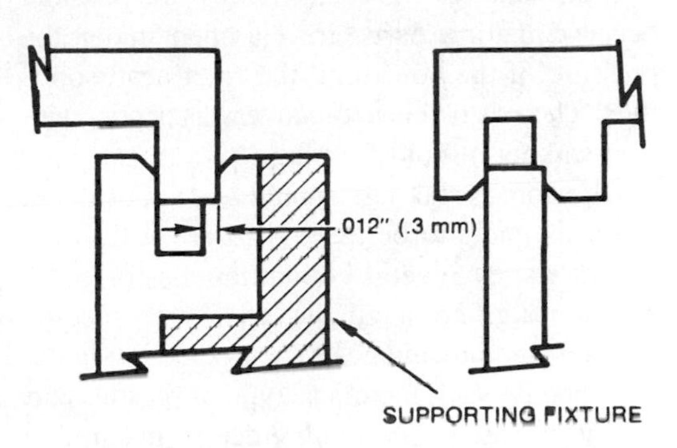

Fig. 24-21. Shear joint variations with large parts. (*Courtesy Sonics and Materials, Inc.*)

Ultrasonic Staking

In ultrasonic staking, also referred to as ultrasonic heading or riveting, the controlled flow of the molten plastic is used to capture or retain another component, usually of a different material, in place.

Ultrasonic staking provides an alternative to welding when the two parts consist of dissimilar materials that cannot be welded, or when simple mechanical retention of one part relative to another is adequate (i.e., as distinct from molecular bonding).

The most usual application involves the attachment of metal to plastic. A hole in the metal part receives a pre-molded plastic boss. The horn tip, vibrating at high frequency, contacts the boss and, through friction, creates localized heat. As the boss melts because of frictional heat, light pressure from the horn forms a head to a shape determined by the horn tip configuration. When the vibrations stop, the plastic material solidifies, and the dissimilar materials are fastened together.

Unlike ultrasonic plastics welding, staking requires that out-of-phase vibrations be generated between the horn and the plastic surfaces. Light, initial contact pressure thus is a requirement for out-of-phase vibratory activity within the limited contact area. It is the progressive melting of the plastic boss under continuous, but light, pressure that forms the head. In staking, low pressure rather than high pressure usually is recommended.

With staking, tight assemblies are possible because mating parts are clamped under the pressure of the horn until the rivet head solidifies. There is no elastic recovery as occurs with heat staking or cold forming.

Ultrasonic staking should be considered when the parts to be assembled are still in the design stage. Several configurations for boss/cavity design are available, each with specific features and advantages. Their selection is determined by such factors as type of plastic, part geometry, assembly requirements, machining and molding capabilities, and cosmetic appearance. The principle of staking is the same for each: the area of initial contact between the horn and the boss is kept to a minimum, in order to concentrate the energy and produce a rapid melt.

The integrity of an ultrasonically staked assembly depends greatly upon the geometric relationship between the boss and the horn cavity. Proper design will produce optimum strength with minimum flash.

Whenever possible, the bosses should be designed with an undercut radius at the base to prevent fracturing or melting. Holes in the mating parts should be radiused or at least deburred. Long bosses should be avoided, and tapered from the base to the top.

The boss should be properly located and rigidly supported from below to ensure that the energy will be dissipated at the horn/boss interface rather than exciting the entire plastic assembly and fixture.

Best staking results are obtained when the ultrasonic vibrations are started before the horn contacts the boss. This prevents cold forming and allows for the gradual reforming of the boss. The pretriggering of the ultrasonic vibrations normally is accomplished by using a pretrigger switch.

To obtain repeatable results when staking, the distance that the horn travels should be consistent and limited by the positive stop adjustment. (See Figs. 24-22 through 24-26.)

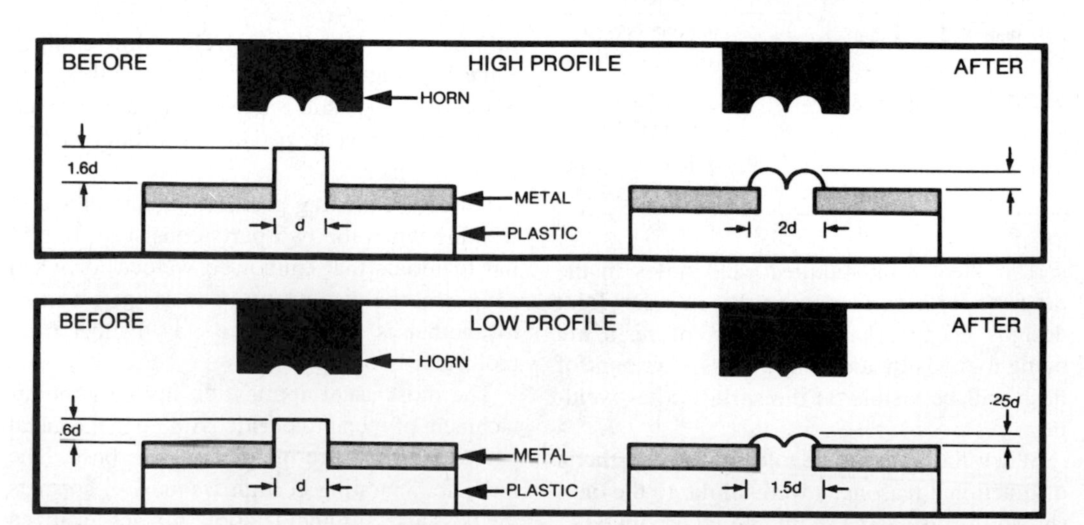

Fig. 24-22. Standard flared stake: The standard flared stake satisfies the requirements of most applications. This stake is recommended for bosses with an O.D. of $\frac{1}{16}$ inch (1.6 mm) or larger, and is ideally suited for low-density, nonabrasive amorphous plastics. (*Courtesy Sonics and Materials, Inc.*)

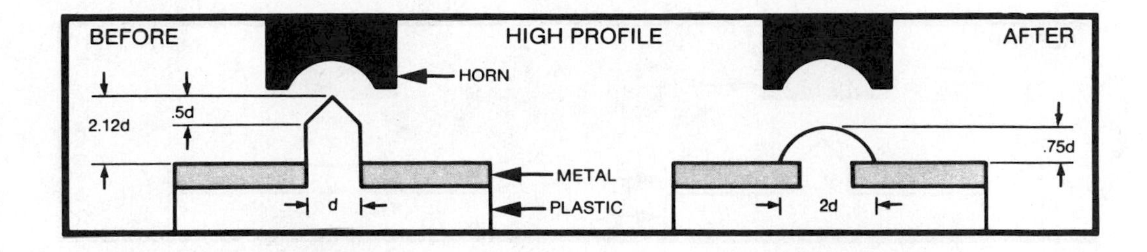

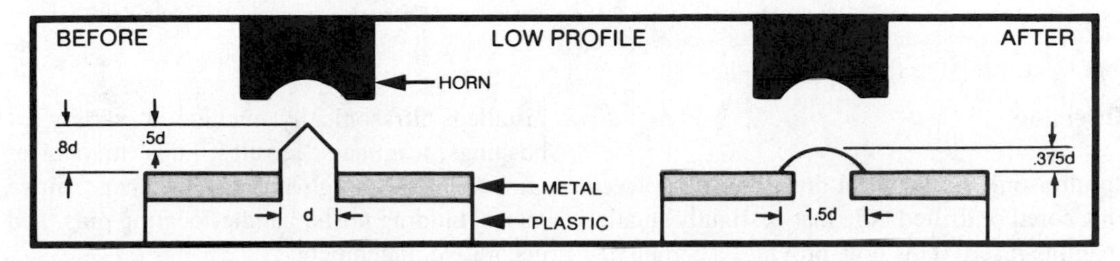

Fig. 24-23. Spherical stake: The spherical stake is preferred for bosses with an O.D. less than $\frac{1}{16}$ inch, and is recommended for rigid crystalline plastics with sharp, highly defined melting temperatures, for plastics with abrasive fillers, and for materials that degrade easily. (*Courtesy Sonics and Materials, Inc.*)

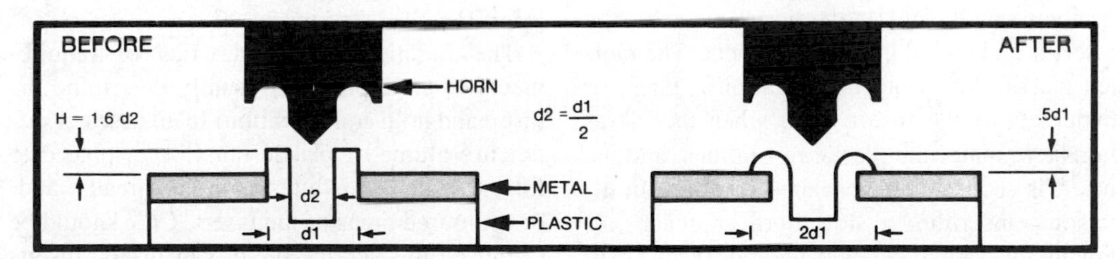

Fig. 24-24. Hollow stake: Bosses with an O.D. in excess of $\frac{5}{32}$ inch (4 mm) should be made hollow. Staking a hollow boss produces a large, strong head with no need to melt a large amount of material. Also, the hollow stake avoids a sink mark on the opposite side of the component, and enables the parts to be reassembled with self-tapping screws should repair and disassembly be necessary. (*Courtesy Sonics and Materials, Inc.*)

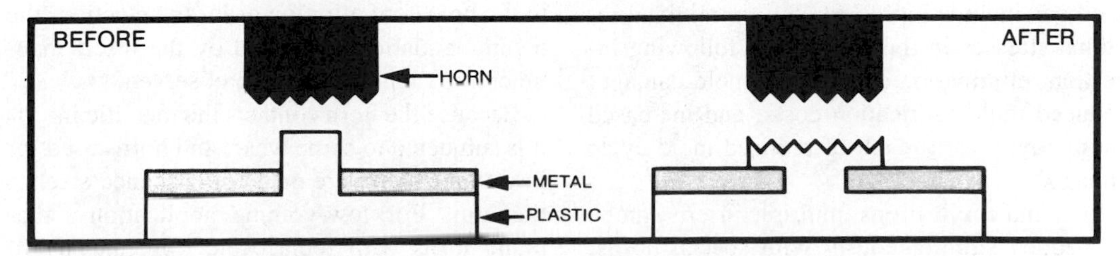

Fig. 24-25. Knurled stake: The knurled stake is used in applications where appearance and strength are not critical. Because alignment is not an important consideration, the knurled stake is ideally suited for high-volume production, and is often recommended for use with a hand-held ultrasonic spot welder. Knurled tips are available in a variety of fine, medium, and coarse configurations. (*Courtesy Sonics and Materials, Inc.*)

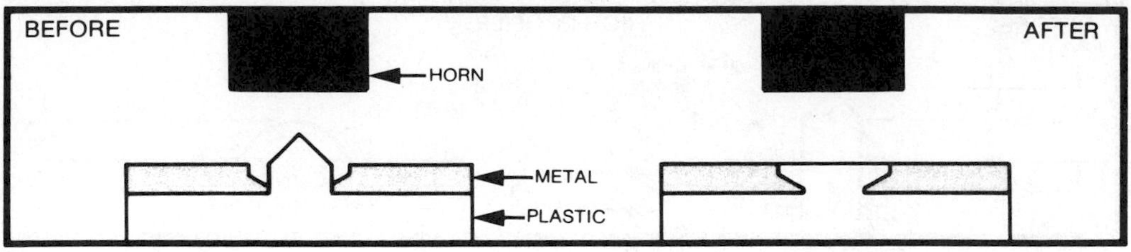

Fig. 24-26. Flush stake: The flush stake is used for applications requiring a flush surface. The flush stake requires that the retained piece have sufficient thickness for a chamfer or counterbore. (*Courtesy Sonics and Materials, Inc.*)

Inserting

In ultrasonic insertion, a metal insert is placed in a cored or drilled hole that is slightly smaller than the insert. This hole provides a certain degree of interference and also serves to guide the insert into place. The vibrating ultrasonic horn contacts the insert, and the ultrasonic vibrations travel through the insert to the interface of the metal and plastic. Heat, generated by the insert vibrating against the plastic, causes the plastic to melt, and as the horn advances, the insert is embedded in the component. The molten plastic flows into the serrations, flutes, or undercuts of the insert, and, when the vibrations terminate, the plastic resolidifies, and the insert is securely encapsulated in place. In ultrasonic insertion, a slow horn approach, allowing the horn to develop a homogeneous melt phase, is preferable to "pressing" the insert.

Ultrasonic insertion provides the high-performance-strength values of a molded-in insert while retaining all of the advantages of postmolded installation. Inserts can be ultrasonically installed in most thermoplastics. Some of the advantages of ultrasonic inserting over other methods include rapid installation, minimal residual stresses in the component following insertion, elimination of potential mold damage, reduced mold fabrication costs, and increased productivity as a result of reduced mold cycle times.

In some applications, multiple inserts can be embedded simultaneously with special horns, increasing productivity and further reducing assembly and manufacturing costs.

Ultrasonic insertion is not restricted to standard-type threaded inserts. Inserts that can be installed ultrasonically include a variety of bushings, terminals, ferrules, hubs, pivots, retainers, feed-through fittings, fasteners, hinge plates, binding posts, handle-locating pins, and decorative attachments.

Typically, the plastic component is fixtured and the insert driven in place by the horn (Method 1, Fig. 24-27). However, in some cases, the part configuration might prohibit insert contact by the horn, and the horn is made to contact the plastic component instead of the insert (Method 2, Fig. 24-27).

The functional characteristics or requirements of an application usually determine the insert and hold configuration. In all cases, a sufficient volume of plastic must be displaced to fill the undercuts, flutes, knurls, threads, and/or contoured areas of the insert. Care should be exercised in selecting the proper insert. Inserts are designed for maximum pullout strengths, torque retention, or some combination of both. Inserts with horizontal protrusion, grooves, or indents usually are recommended for high-pullout-strength requirements, whereas inserts with vertical grooves, or knurls are usually recommended for high torque retention. In regard to the hole configuration or insert selection, the recommendations provided by the insert manufacturer should always be observed.

Because the horn contacts the metallic insert, it is subjected to some wear, and horns used for insertion usually are made of hardened steel or titanium. For low-volume applications, titanium horns with replaceable tips can be utilized.

Ideally the diameter of the horn should be twice the diameter of the insert.

To prevent a "jack-out" condition, the top

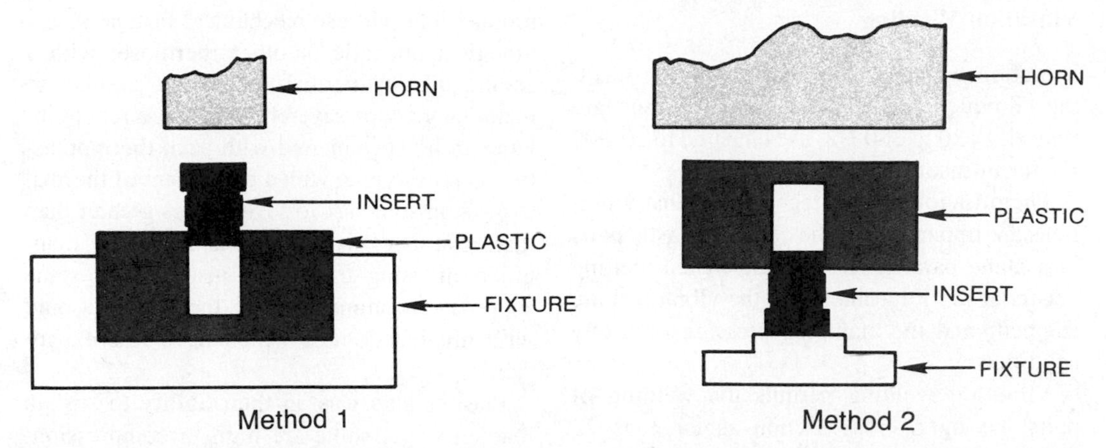

Fig. 24-27. Inserting. Typically, the insert is driven in place by the horn (Method 1). However, in some cases, the horn is made to contact the plastic component instead of the insert (Method 2). (*Courtesy Sonics and Materials, Inc.*)

of the seated insert should be flush or slightly above the surface of the part. Rigid fixturing should be placed directly under the insert.

In most instances, it is necessary to initiate ultrasonic vibrations prior to horn contact with the insert.

To maintain an accurate depth of insertion, the total distance the horn travels should be limited, either mechanically by a positive stop or electrically by a lower-limit switch, or both.

Spot Welding

Using an ultrasonic spot welder and standard replaceable tips, large thermoplastic parts and those with hard-to-reach joining surfaces can easily be welded together.

Vibrating ultrasonically, the pilot of the tip penetrates the top sheet and enters the bottom sheet to a depth of one-half the top sheet thickness. The molten plastic displaced is shaped by a radial cavity in the tip to form an annular formation around the weld. Simultaneously, the molten plastic displaced from the second sheet flows into the preheated area and forms a permanent molecular bond (Fig. 24-28).

Film and Fabric Welding and Cutting

Film and fabric sealing is done on a continuous-feed basis with the material being pulled under the horn. The horn usually has a radiused face, and the anvil can be a rotating wheel or stationary block. A stitch pattern can be engraved into the rotary anvil to produce ultrasonically ''sewn'' fabrics.

Applications include the sealing of polypropylene and polyester films and woven and nonwoven fabrics. Ultrasonics is the preferred way to join Mylar film. There also are other applications where ultrasonics is used to cut and seal the edges of woven and nonwoven materials.

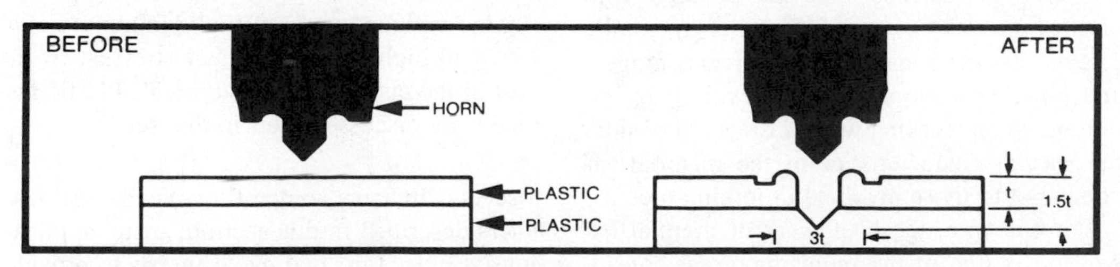

Fig. 24-28. Spot welding. (See text for description.) (*Courtesy Sonics and Materials, Inc.*)

Vibration Welding

Vibration welding is a form of friction welding. Frequencies used for this technique are typically 120 to 300 Hz, as compared to 20,000 Hz for ultrasonic welding.

The principle of this technique is that vibrations are imparted into the clamped plastic parts in a plane parallel to the joint. When melting occurs at the joint interface, the vibrations are stopped, and the clamping pressure is briefly solidified.

Vibration welding permits the welding of parts having a cross section as large as 22 inches by 16 inches. Crystalline resins such as nylon, acetal, polyethylene, polypropylene, and thermoplastic polyesters are excellent candidates, along with many filled resins, for this technique. All the plastics that can be welded ultrasonically can also be assembled with vibration welding.

Typical applications for this technique include automobile bumpers, fuel tanks, pump impellers, expansion tanks, and emission canisters.

MECHANICAL JOINTS

There are a variety of methods of providing mechanical fastenings and various types of joints in all kinds of plastics. This subsection covers: (1) fasteners, both in metal and in plastics; (2) such techniques as swaging, press-fitting, cold-heading, and so on, which represent mechanical means of fastening; and (3) unique plastic fastening concepts such as the integral or ''living'' hinge.

Mechanical Fasteners

The type of fasteners that can be used with plastics covers a wide range: clips, nails, screws, assorted inserts, bolts, rivets, hinges, and pins, to name a few. However, it is important when working with plastics to consider the special characteristics of the material, as compared to those involved in joining metals.

For example, the coefficient of thermal expansion is one of the more important considerations in selecting a fastener or designing a product that will use mechanical fasteners. Although a phenolic or other thermoset with a coefficient of thermal expansion similar to metal may not be severely affected, a real problem can be encountered with such thermoplastics as polystyrene with a coefficient of thermal expansion some six to seven times greater than that of steel. This problem is especially magnified in using molded-in inserts, and some suppliers recommend using these inserts only with filled materials, which have less expansion.

Plastics also vary in their ability to sustain loading (e.g., some are high in compression, others in tension), and plastics deflect more than metals under the same loading (so metal and plastic should not share loads in parallel).

The following discussion contains specific guidelines relating to the various types of mechanical fasteners. However, because of the many variables among different plastic materials, it is suggested that users contact materials manufacturers for precise assembly data. Also, for a complete catalog of mechanical fasteners, a useful reference is the *Fasteners Handbook*, by J. Soled (Van Nostrand Reinhold Co., 1957).

Among the advantages offered by mechanical fasteners are these: (1) ultimate bond strength is attained immediately; (2) a movable joint can be attained, if desired; (3) disassembly is possible, with some types of fasteners; and (4) capital equipment costs are low.

It should also be noted that mechanical fasteners are available in plastics as well as in metals. Fasteners made of plastic are particularly useful in problems of fastening involving (1) corrosion resistance, (2) color matching, (3) sealing, (4) protection of painted or porcelained finishes, and (5) electrical insulation. They should not normally be considered where service temperatures are high, joints are subjected to high tensile or shear stresses, or the cost of the fastener itself is critical. Plastic fasteners are discussed later in this section.

Inserts. In using some of the mechanical fasteners described in this section, many applications will use threaded metal inserts to provide threaded holes in the plastic parts. These in-

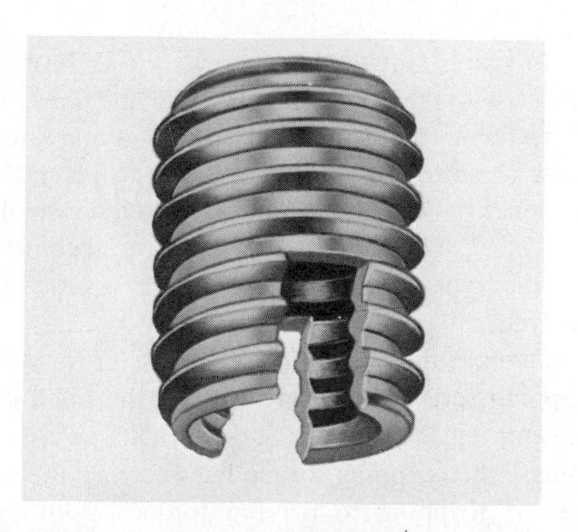

Fig. 24-29. Coarse external thread brass insert is ideal for both thermoplastics and thermoset materials. The wider thread spacing reduces the installation torque and provides stronger threads in weak or brittle plastics. Regular slotted case-hardened steel inserts are recommended if the plastic contains highly abrasive fillers. (*Courtesy Groov-Pin Corp.*)

serts are discussed in greater detail in Chapter 25. Basically, the major types of inserts include (1) molded-in inserts and (2) post-molding inserts.

Post-molding inserts may be subdivided into:

- *Self-tapping inserts:* There are two types of such inserts—thread-cutting and thread-forming. Thread-cutting inserts cut their own mating threads in the plastic material. They can be used in both thermoplastic and thermosetting materials. For most applications in plastics, the slotted type can be used. These inserts have coarse external threads that reduce installation torque and provide strong threads even in brittle plastics (Fig. 24-29).

  In very hard plastics, or those with abrasive fillers, a case-hardened steel insert is used. This is the same type that is used in many metal applications and is available with internal locking features.

  Thread-cutting inserts also are available with male threads. This self-tapping stud or male insert is useful where alignment or ease of assembly is a consideration.

  Installation of thread-cutting inserts consists of two steps: drilling or coring the holes and driving the insert. Because only two steps are needed and tapping is not required, thread-cutting inserts can be economically installed. Equipment available in most shops can be readily adapted for volume production.

  Thread-forming inserts create threads by mechanically displacing material as the insert is driven.
- *Press-fit inserts* (generally knurled).
- *Expansion inserts:* These inserts are placed in predrilled or premolded holes and expanded against the walls by insertion of a screw. Torsion resistance and pullout retention are obtained by embedding the knurls into the material.
- *Inserts installed ultrasonically:* Friction caused by ultrasonic vibrations melts a thin film of resin at the metal–plastic interface. Pressure from the ultrasonic tool directs the insert into the cored or machined hole.

Standard Machine Screws. For applications where assembly and disassembly are expected to be frequent, it is generally recommended that the part design incorporate metal threaded inserts (as described above) to accept a mating threaded screw.

Where the application involves infrequent assembly (less than six times), a hole in the plastic part with machined or molded-in threads to receive the mating threaded screws can be used. Most thermoplastics can be readily machined after molding with standard metal-working tools, and this procedure is generally recommended for achieving threads below $\frac{1}{4}$ inch in diameter.

Larger internal threads may be formed by a threaded core pin that is unscrewed either manually or automatically. External threads may be formed in the mold by splitting the thread along its axis or by unscrewing the part from the mold. When a parting line or the slightest flash on the threads cannot be tolerated, the second method (unscrewing the part from the mold) must be used. Mold designs with automatic unscrewing operations are feasible.

Coarse threads can be molded more easily than fine threads and thus are preferable. Threads finer than 28 pitch or closer than class 2 should not be specified. The roots and crests of all threads should be rounded with a 0.005 to 0.010 inch radius to reduce stress concentration and provide increased strength.

For inserting a screw-type fastener, a hand-held or automatic drill is preferred, although it can be a simple hand insertion. Many screw insertions make use of commonly available hand-held drills.

Standard taps can be used with phenolic parts. The predrilled or premolded hole should be chamfered to the maximum diameter of the thread. For threading or tapping a thermoplastic to receive a metal fastener, a slight increase over normal metal clearance is required for the difference in thermal expansion coefficients.

Self-Tapping Screws. These screws are available only in metal and are designed to produce their own mating threads when driven into a material. The unthreaded hole for a self-tapping screw can be either molded or drilled into the molding. The entering edge usually is suitably chamfered to prevent spalling.

Self-tapping screws provide a substantial cost savings by simplifying the molding operation and reducing assembly costs (molding or machining threads into the part is not necessary because the self-tapping screw forms its own). They are not recommended, however, when frequent disassembly is required. Reassembly is generally limited to about five or six times.

There are two types of such screws: thread-cutting and thread-forming. Thread-cutting screws tap or cut a mating thread as the screw is driven. They are slotted to provide a channel for disposal of chips; therefore, the depth of the hole should be slightly deeper than the screw (generally around $\frac{1}{32}$ inch) to provide a depository or reservoir for chips. Thread-cutting screws usually are suggested for use with plastics with higher flexural modulus (over 2 × 105 psi). They are used with thermosets and various high-performance engineering plastics, such as the acetals, nylon, ABS, and so on. The stress factor with thread-cutting screws is relatively low.

Thread-forming screws create threads by mechanically displacing material as the screw is driven. They are used with the more ductile thermoplastics.

In using self-tapping screws, it is best to check with the screw manufacturer and the resin supplier as to installation specifications (i.e., how close to the edge of the part to place the fastener, how wide and how deep a hole should be, etc.). These specifications vary according to the type of fasteners and the plastic involved.

Overall, the optimum hole diameter for self-tapping screws in plastic parts is the pitch diameter of the screw. Smaller hole sizes increase the stripping torque but decrease the ratio of stripping to driving torque. Both the stripping torque and the strip–drive ratio are important factors in fastener design. Larger holes reduce drive torque, strip torque, and vibration resistance.

In general, the number of screw threads engaged in the plastic is determined by the depth of the hole. The threads on the taper of a self-tapping screw provide little, if any, holding power. Therefore, the strength of a self-tapping screw is dependent upon the length of the cylinder of plastic that extends from the lowest full thread engaged in the plastic up to the top of the hole.

In designing for a self-tapping screw in thermoplastics, the taper length and the length tolerance of the screw must be subtracted from the depth of penetration available in the plastic. As a rule, the hole should be deep enough to provide a minimum of two or three full threads of engagement. The practical upper limit of hole depth is a hole deep enough to provide a stripping torque equal to the torsional strength of the screw. For fine-pitch threads, this is approximately eight to twelve threads. For coarse threads, three to eight full threads are required.

Bosses. The use of bosses may be desirable in many applications and they are quite commonly used in plastics. Bosses are protruding studs or pads used for the reinforcement of holes or for mounting and assembly. In general, when working with thermoplastics, the boss diameter is usually from two to three times the outside diameter of the screw (the average is $2\frac{1}{2}$ times). This is generally sufficient to take the possible hoop stresses developed due to screw insertion (see Fig. 24-30).

Troubleshooting Tips. The three most com-

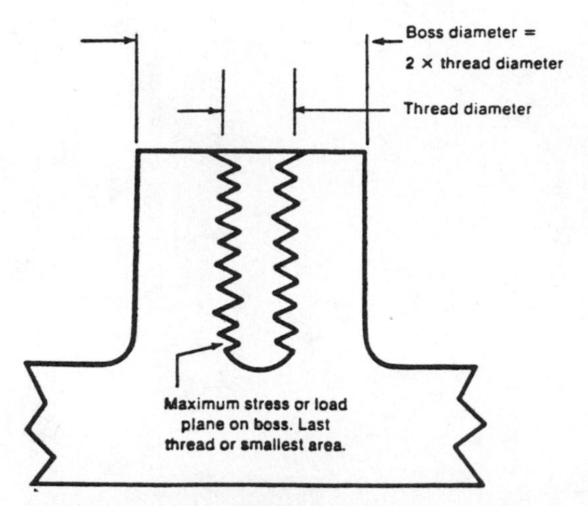

Fig. 24-30. In applications requiring the use of a metal screw and a threaded polycarbonate boss, the boss diameter must be equal to twice the thread diameter. This recommendation should generally be followed for screws up to $\frac{3}{8}$ inch. (*Courtesy GE Plastics*)

mon problems associated with the use of self-tapping screws in plastic parts are: (1) fracture of the part as the screw is inserted, (2) stripping of the plastic threads as the screw is inserted, and (3) loosening of the screw after a period of time.

- *Part fracture:* The basic cause of this problem is a high stress as the screw is inserted. The user should determine whether he or she is employing the right type of screw (e.g., thread-forming screws, as noted above, develop more stresses in the plastic than thread-cutting screws).

A second major cause of part fracture is the use of small-diameter holes. With a small hole, the tap of the screw takes a deeper bite into the plastic material, and this may again produce excessive stress in the boss.

- *Stripped threads in the plastic:* Often, the cause of this problem is an insufficient strip–drive ratio. The initial step in correcting this difficulty is to determine the hole size in the plastic. Larger-diameter holes yield a lower value of stripping torque but increase the strip–drive ratio.

If the hole depth in the plastic is too shallow, the screw may bottom out before the screw head seats. The hole should be deep enough to allow for the length tolerance of the screw.

If the screws are inserted with a power-driven screwdriver, the high driver speed may cause failure of the plastic thread. Slowing down the driver tends to increase the accuracy of torque setting and decrease the rate of loading of the threads in the plastic.

For some applications, a finer pitch screw may correct thread stripping, as the finer pitch is less efficient in converting applied torque to compressive loadings of the threads in the plastic. The hole size should be checked when the screw pitch is changed.

If all of the above have failed to correct the problem, redesign may be necessary. The shear area i the plastic can be increased with a larger-diameter screw, a longer screw with a deeper hole, or a larger-diameter head on the screw.

- *Screw loosening:* The common cause of this problem is insufficient thread engagement. The area of thread engagement may be increased by any or all of the following: smaller-diameter holes; longer screws and deeper holes; or screws with finer pitch. Also, a careful re-evaluation of the application environment should be made to determine that estimates of temperature and moisture conditions are valid.

Drive Screws. These screws are designed to be driven into place with an impact and are used for permanent fastening. A distinct advantage with this type of screw is alignment. Drive screws are ideal for conduit installation or other operations where installation space or alignment may be a problem. They also can be used to join thin-gauge materials (Fig. 24-31).

Bolts and Nuts. Bolts and nuts of standard configuration, made of metal or of plastic, can be used where mechanical requirements of the joined area permit.

Rivets. Rivets, made of metal or of plastic, can be used to hold parts together securely, or to hold two or more parts together and permit motion between them.

Rivets are noted for their low cost and simple mechanical installation that lends itself to automation. High clamp load or tension load is limited, and accuracy of location is not as good as with other mechanical fasteners. Rivets can

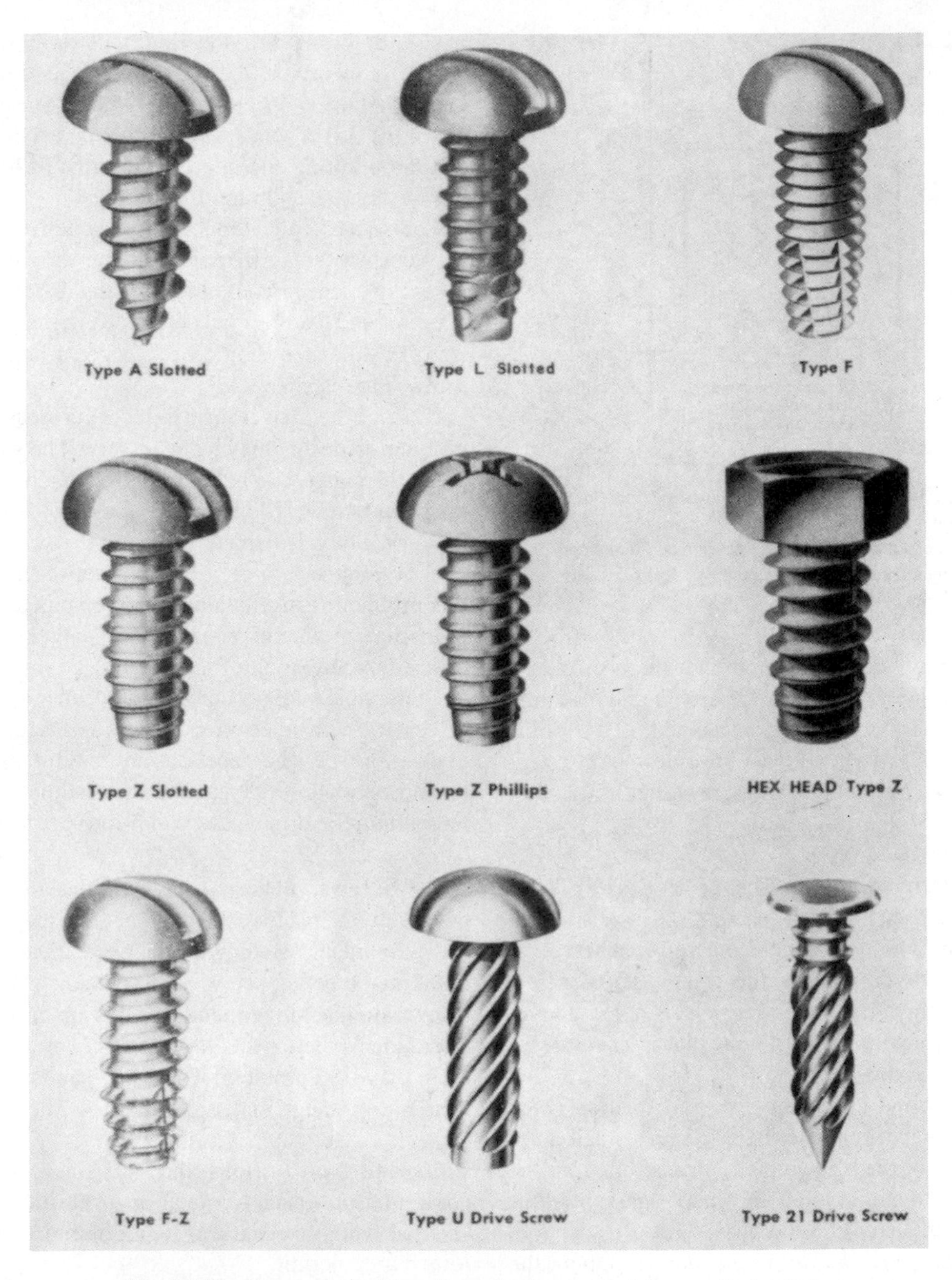

Fig. 24-31. Some typical self-tapping and drive screws. (*Courtesy Parker-Kalon*)

be installed to allow for dimensional change with temperature. Rivets are particularly applicable for use with plastics of high impact strength. To join very thin sections of plastics, metallic eyelets may be used instead of commercial rivets. Eyelets are particularly advantageous in providing a bushing for a rivet, or metal shaft that must rotate in service. The eyelet minimizes wear of the plastic.

Of the rivets used in joining plastics, the blind rivet is probably the most common. Blind rivets are available in both metal and plastics, and are designed for installation from one side only. Though there are a variety of proprietary designs for such rivets, essentially they consist of a hollow body and a solid pin. The setting of the rivet is accomplished by driving or pulling (depending on the specific rivet being considered) the solid pin through the hollow shank and thus flaring the shank on the blind side of the rivet, to effect a positive locking action (Fig. 24-32).

Chemically expanded rivets have hollow shanks in which a chemical charge is reacted to expand the hollow shank. Heat is applied to the rivet head with an approved soldering gun.

Chemically expanded metal rivets can be used for joining plastics as well as the light metals. In joining materials of relatively low elastic moduli, such as polyethylene and polytetrafluoroethylene resin, applications are limited to those where joints of low strength are permissible because the expansion of the exploding rivet shank distorts most soft plastics.

Spring Clips and Nuts. A wide variety of proprietary metallic spring clips and nuts provides inexpensive methods of rapidly fastening plastics. They range from simple spring-type fasteners, which are forced over a molded stud, to multi-perforated rings, tubular devices, and irregular shapes. These fasteners can be applied unattached, or attached to one of the parts to be joined. If the spring device is attached to

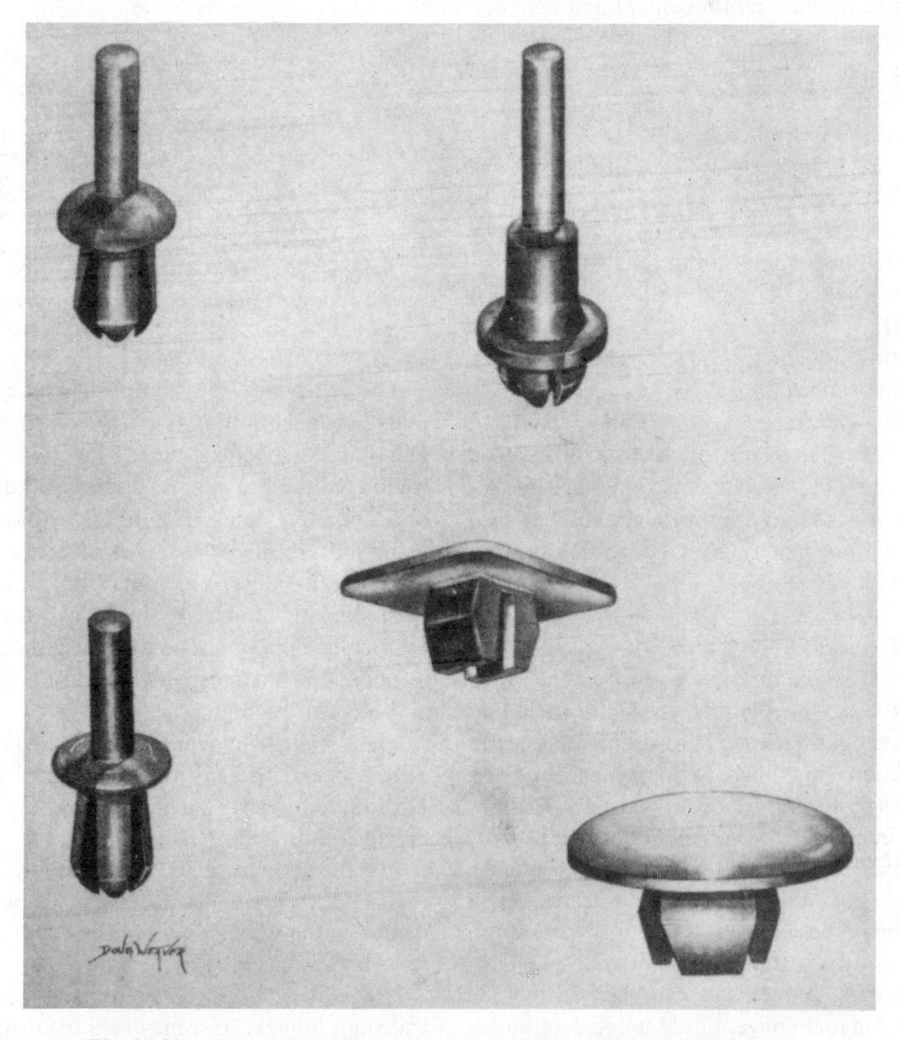

Fig. 24-32. Blind rivets. (*Courtesy Shakeproof, Div. of Illinois Tool Workers*)

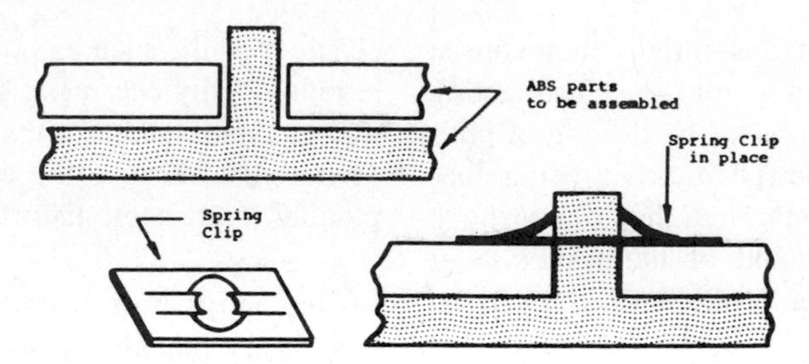

Fig. 24-33. Tinnerman spring clips to lock ABS parts together. (*Courtesy GE Plastics*)

one of the parts, rapid assembly can later be completed from one side only. (See Fig. 24-33.)

Speed nut stampings are used in combination with screws. The mating screw is usually of the coarse-thread sheet-metal variety. The speed nut acts to provide reinforcement and load distribution, and resists vibration. The speed clip provides push-on attachment over a molded boss or stud. The load capacity usually is low and reassembly limited.

Pins can be effectively used in plastic. The physical characteristics of plastics related to loading and residual stress propagation must be considered in design. Spring pins with their inherent resilience combine well with most plastic materials. Pins have a variety of usages such as locating and locking devices, bearing surfaces, and hinges.

A number of means of providing movable assemblies, such as hinges, latches, snap locks and bead-chain attachments, are largely derived from standard or proprietary devices used with common materials of construction other than plastics.

Other Devices.

Hinges. Several types of hinges are available for use with plastics products, including integral hinges such as those described later in this section (in which two plastics parts are joined together by a strip of plastic). Another type of integral hinge consists of a molded ball-and-socket type of hinge assembled through a snap-fit (Fig. 24-34). As an alternative, a socket may be molded for a drilled-through hole. In terms of more conventional mechanical fasteners, hinges that can be used include lug-and-pin and conventional hinge assemblies such as piano hinges.

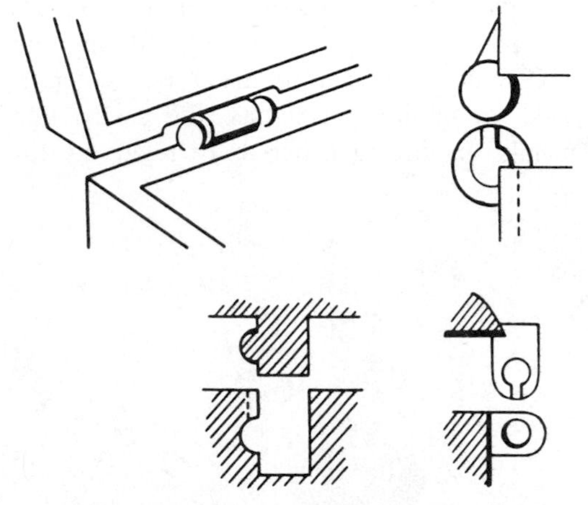

Fig. 24-34. Molded ball-and-socket type hinge assembled through a snap-fit. As an alternative, a socket may be a molded or drilled-through hole. (*Courtesy Monsanto*)

The lug-pin-rivet type consists of a molded male lug (generally found in the cover of a box) and a molded female slot. The cover and the bottom of the box are assembled and drilled as one complete unit. The hinge is formed by using a drive-pin or a hinge-pin.

To reduce cost, holes may be drilled from both sides of the box by using two drills, diametrically opposed. The drill unit should have a kinematic linkage, which controls the depth and feed of the hole.

The two drills may also be operated by small air cylinders instead of a mechanical linkage. Drilling fixtures with hardened bushings should always be used for best quality.

For the rivet type, the design may be such that no lugs and slots are required. A hole is drilled in the vertical wall of the cover and bottom, and is fastened by a rivet.

Hinge assemblies can be divided into hinges and snap-hinges. In some cases the usual hinge is a piano hinge screwed to the plastic cover with drive screws or swaged to molded plastic lugs. In designing the box, however, adequate wall thickness must be allowed around the molded or drilled hole. Otherwise, assemblies may crack when the drive screws are forced into the plastic.

In addition to the piano hinge, there are shorter-type hinges assembled by riveting; also certain specialty designs are available whereby molded undercuts hold the hinge assembly to the plastic component.

Nails. As plastics continue to show up in applications originally intended for wood (e.g., furniture, shoe heels, etc.), there has been an increasing use of conventional nails as an assembly device. Many styrene structural forms used in furniture, for example, have been specially formulated to accept and hold nails. For ABS plastics used in heels, conventional woodworking equipment (i.e., a nailing machine) has been successfully used to attach the heels to the shoe. It is recommended in this latter case, however, that nails be located at least five diameters from an edge so that strains may be readily distributed.

Before using nails, however, one preferably should check first with the resin supplier. The stresses involved and the nail-holding ability of the plastic must be taken into consideration.

Swaging or Peening

Swaging or peening is commonly used to fasten or connect molded pieces or metal parts to molded thermoplastics. It can also be used to provide an upset-type head (like a common nail) on a molded shaft, pin, or vee-slot.

Swaging is useful to assemble metal to plastic, or plastic to plastic, where motion between the parts is required. This operation is best performed when the plastic is heated to a softening temperature and formed under high pressure. If the temperature is high, less pressure is needed, but the quality of the swage is reduced.

The simplest form of tool used to swage is a soldering iron. For greatest production, automation can be employed by using a kick-press or air-operated cylinder.

In addition, an air-operated indexing table can be used to feed the work and position it under the heated tips. The use of air will provide the necessary pressure and minimize operator fatigue.

The most commonly used soldering iron tips are $\frac{1}{4}$ to $\frac{5}{16}$-inch-diameter copper. However, they are often soft and deform or bend under continued usage. Hardened brass, which has been nickel-plated and then flash-chrome-plated, is most satisfactory. The brass is easy to machine into a concave tip to control the flow. Chrome improves the life and mold release.

During swaging operations a thin film of polymer may stick to the tip. This can be removed by a wire brush or a solvent.

Standard-type mold release agents on the heated tips will minimize sticking.

Press or Shrink Fitting

These techniques are universally applicable to joining similar and dissimilar thermoplastics, and require no foreign elements such as cement or metal inserts in the finished joints. Properly applied, they produce serviceable joints with good strength at a minimum cost.

Plastics are press-fitted in the same manner as metals and other materials, but interfaces are generally increased to compensate for the relatively low elastic modulus of most plastics. For maximum joint strength, interferences should be made as large as possible without restricting assembly or stressing a piece beyond its yield point. Theoretical relationships of interference and stress level are based on geometry and the properties of the materials. Interferences can be calculated by standard stress-analysis procedures.

The relationships of maximum stress, caused by press-fitting a shaft or insert in a plastic hub, to diametral interference (ΔD) is expressed as:

$$\Delta D = \frac{S_d D_s}{L}\left[\frac{L + \mu_h}{E_h} + \frac{L - \mu_s}{E_s}\right]$$

and:

$$L = \frac{1 + \left(\frac{D_s}{D_h}\right)^2}{1 - \left(\frac{D_s}{D_h}\right)^2}$$

where:

D = Diametral interference, inches
S_d = Design stress, psi. Refer to AWS under "Design Parameters"
D_h = Outside diameter of hub, inches
D_s = Diameter of shaft, inches
E_h = Tensile modulus of elasticity of hub, psi
E = Modulus of elasticity of shaft, psi
μ_h = Poisson's ratio of hub material
μ_s = Poisson's ratio of shaft material
L = Geometry factor

Where large interferences are not required, shrink fitting may be suitable. Interferences for shrink fitting are determined by adding shrinkage of the hub to expansion of the shaft. In some applications it may be practicable to shrink-fit immediately after molding, while the article is still hot, to eliminate the necessity for reheating the hub.

Residual joint strengths in press- or shrink-fitted articles are affected by complex variables such as apparent modulus and coefficient of friction. For most thermoplastics, variation in apparent modulus becomes negligible after a year, so that joint strength becomes constant. Because the coefficient of friction is affected by variables such as lubrication, moisture, temperature, and stress level, the coefficient under each of these conditions must be known in order to calculate accurately the strength of the joints. Axial strength and torsional strength of joints then can be calculated by standard equations. When torsional strength is critical, a ribbed shaft should be used; when axial strength is critical, rings or threads should be used. When both torsional and axial strength are critical, a knurled shaft or combinations of rings and ribs provide a good balance of properties.

When pieces are to be press-fitted for maximum holding power immediately after molding, they should be free from internal stresses, which can be reduced by annealing. Also in designing such joints, environmental conditions should be carefully considered. Expansion by heat and moisture can be compensated by designing for expected growth at the worst conditions, by considering expected expansion as an addition to the interferences selected for the desired joint strength. In plastics, a press-fitted assembly experiences its highest level of stress immediately after fittings; subsequent environmental dimensional shrinkage usually is more than compensated by creep, and thus generally can be neglected.

Internal stresses in a press-fitted article may tend to promote crazing in some plastics such as acrylics and polystyrenes, and may reduce impact strength. Where a pressed article is expected to withstand impact it should be tested under actual conditions to determine the feasibility of this type of joint.

The force required to press-fit two parts may be approximated by using the following equation;

$$F = \pi f P D_s T$$

and:

$$P = \frac{S_d}{L}$$

where:

F = Assembly force, lb
f = Coefficient of friction
P = Joint pressure, psi
D_s = Diameter of shaft, inches
T = Length of press-fit surfaces, inches
S_d = Design stress, psi
L = Geometry factor

Snap Fitting

Snap fitting is a simple and rapid means of assembly, as all the elements for assembly (i.e., undercuts and corresponding lips on the mating part) are molded directly into the products so that they are ready to be joined together as they come out of the mold. Basically, snap fits involve an undercut on one part engaging a molded lip on the other to retain the assembly.

Joints of this type are strong, but are usually not pressure-right unless other features such as an O-ring are incorporated in the joint design. The critical factor in any snap fit is the amount of undercut that can be molded or machined into the part. Generally, undercuts can be snapped out of the mold by a standard ejection system; the critical depth of the undercut will depend upon the material chosen, the wall thickness, part configuration, and so on. Where undercuts are too deep to be ejected conventionally, collapsible cores sometimes can be used. It is also possible to machine a deep undercut into a part.

In terms of design, undercuts and mating lips on snap-fit parts may be fully cylindrical or consist of flexible cantilevered lugs. Cylindrical snap-fits are generally stronger but require greater assembly force than cantilevered-lug snap-fits. For complex parts, the use of cantilevered lugs may simplify molding the parts.

The two general types of snap fits are the snap-on and the snap-in. The snap-on is accomplished by molding an undercut on one part and a corresponding lip on the mating part (Fig. 24-35). The snap-in fittings generally incorporate molded-in prongs on one part that are pressed through holes in the mating part.

The force required to assemble parts by snap fitting depends upon part geometry and on the coefficient of friction between the materials. This force may be divided arbitrarily into two elements—the force required initially to expand the hub, and the force required to overcome friction during the press-fit stage. As the beveled edges slide past each other, the maximum force for the expansion occurs at the point of maximum hub expansion and is approximated by:

$$F_e = \frac{(1 + f) \tan (\alpha)\ S_d \pi D_s L_h}{W}$$

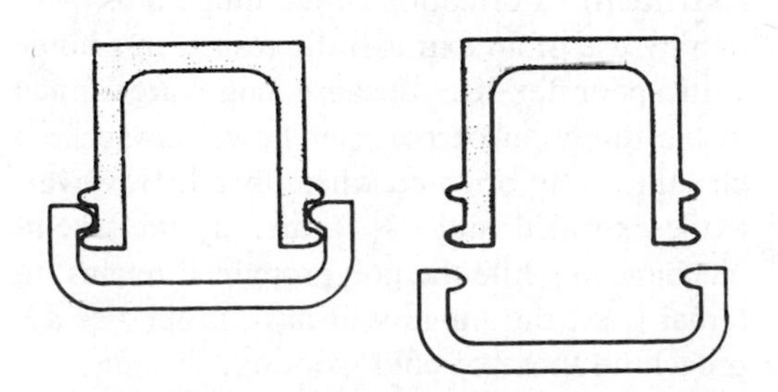

Fig. 24-35. Snap-on fit is accomplished by molding an undercut on one part and a corresponding lip in the mating part. (*Courtesy GE Plastics*)

where:

F_e = Expansion force, lb (kg)
f = Coefficient of friction
α = Angle of beveled surfaces
S_d = Stress due to interference, psi (kg/cm^2)
D_s = Shaft diameter, inches (cm)
L_h = Length of hub expanded, inches (cm)
W = Geometry factor

For blind hubs, the length of hub expanded, L_h, may be approximated by twice the shaft diameter. For short open hubs such as in pulleys, the length of hub expanded will equal the axial hub length. For other part geometries or where greater accuracy is required, tests on actual parts are recommended. The force required to overcome friction during the press-fit stage is approximated by:

$$F_f = \frac{f S D_s L_s}{W}$$

where:

F_f = Friction force, lb (kg)
f = Coefficient of friction
S_d = Stress due to interference, psi (kg/cm^2)
D_s = Shaft diameter, inches (cm)
L_s = Length of interference sliding surface, inches (cm)
W = Geometry factor

Generally, the friction force is less than the force for hub expansion for most assemblies.

Integral Hinges

Integral hinges offer a great potential for savings. Because they are molded into the item, they cost almost nothing to produce and eliminate much of the assembly cost with standard hinges.

Many approaches can be used. One basic style is the knuckle-and-pin design, in which the set of knuckles that make up the hinge pivot on an axial pin. The knuckles can be either hooks or eyes or a combination of the two. The hook-and-eye combination does not require a pin, but the hinge can disengage if the cover is swung back too far.

A popular hinge, widely used on small boxes, is the ball-grip design. One half of the hinge consists of two balls, which snap into depressions in a projection molded into the mating part.

Still another approach is the integral strap hinge. The top and base are molded simultaneously, connected by straps spaced along one edge. A typical strap is about 0.25 inch wide and 0.35 inch thick.

The living hinge, made possible by polypropylene, has become extremely popular. This design is much neater in appearance than the strap hinge and is much longer-lived. Properly designed, a living hinge will flex over 300,000 times without failure.

Techniques used to fabricate integral hinges include: molded-in, cold-worked, extruded, and coined, as described below.*

Molded-In. The integral hinge can be injection molded by conventional techniques, provided that certain factors are kept in mind.

The desirable molecular orientation is transverse to the hinge axis. This can be best achieved by fast flow through the hinge section using high melt temperatures. Because these requirements are consistent with good molding practices, optimum production rates can be maintained.

The main concern in integral-hinge molding is to avoid conditions that can lead to delamination of the hinge section. These include filling too slowly; too low a melt temperature; a nonuniform flow-front through the hinge section; material contamination such as pigment agglomerates; and excessively high mold temperatures near the hinge area.

The integral hinge also can be produced by post-mold flexing. The hinge section is molded, then subjected to stresses beyond the yield point immediately after molding by closing the hinge, which creates a "necking-down" effect. Stretching the oriented polymer molecules on the outer surface of the hinge radius provides the remarkable flex strength of the thinned-down hinge section.

Flexing of the molded hinge must be done while it is still hot, through an angle sufficient to stress its outer fibers. This post-molding step provides maximum and uniform orientation in the hinge area with a minimum of applied stresses. The thinness of the hinge area requires that pigments be well dispersed so that agglomerates will not provide focal points of weakness in the hinge structure.

Cold-Worked. Where parts are heavy or complex, it may be impractical to force the necessary quantity of resin through the hinge sections. In these cases, integral hinges can be obtained by cold-working the molded parts With this process, the molecules are properly oriented during forming; so the direction of polymer flow in molding is not critical.

A press, home-made toggle job, or hot-stamping machine can be used to perform the cold-working operation. The male forming die should be at about 270 to 280°F. Pressure is maintained for about 10 seconds although the time can be reduced if the part still retains residual molding heat or is preheated. The recommended preheating temperature is from 175 to 230°F.

Die backings may be either hard or flexible. With a hard backing such as steel, the softened polypropylene is die-formed into the desired hinge contour. Thinner hinges usually are made by using flexible backing such a stiff rubber. The deformation of this type of backing produces the hinge contour by stretching the softened plastic, and it generally results in thinner cross sections.

Extruded. Formation of the hinge cross section by use of an extruder die results in a hinge with a poor flex-life. Because hinges are formed in the direction of polymer flow, they cannot be sufficiently oriented when flexed. However, if the extruded hinge is formed by the takeoff mechanism while the polypropylene retains internal heat, the hinge will have properties approaching those of cold working.

Cold-Forming or Coining Hinges. It is possible to create hinges in some of the tougher

*Information supplied by Exxon Chemical Co.

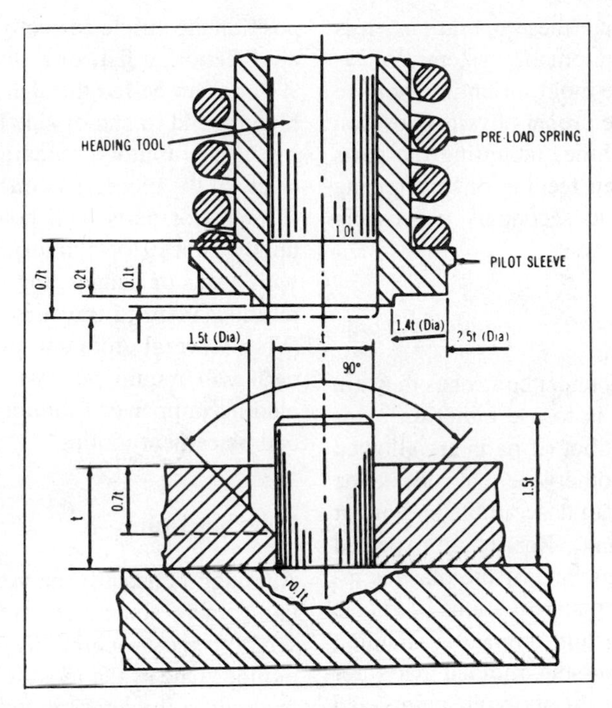

Fig. 24-36. Cold heading of acetal. (*Courtesy E. I. du Pont de Nemours & Co., Inc.*)

engineering thermoplastics by coining techniques. A molded or extruded part is placed in a fixture between two coining bars. Pressure is applied to the bars, and the part is compressed to the desired thickness, elongating the plastic. Coining is effective only when the material is elongated beyond the tensile yield point. The process usually is used for such materials as acetal and nylon, which cannot normally be molded in a sufficiently thin section for a strong, durable hinge.

Cold or Hot Heading

This technique is useful for joining thermoplastics to similar or dissimilar materials.

In cold heading or staking, a compression load is applied to the end of a plastic shaft (while holding and containing the shaft body). As the load exceeds the shaft's yield strength, the shaft (or stake) will cold-flow, forming a rivet-like head. The shaft, of course, can be molded right onto the part itself to facilitate assembly, or a separate shaft can be used to join two components together. In this case, double heading of the shaft (at each end) is required.

Cold staking is normally done at room temperature (73°F). However, if a head is formed at a higher temperature than is expected in use, it will not recover unless the temperature at which it was formed is exceeded.

Cold staking may be accomplished by using equipment ranging from a simple arbor press and hand vise to rivet-setting machines (Fig. 24-36).

Hot staking is similar to the cold staking method. It is accomplished by using a heated (400–500°F) flaring tool. This tool is frequently concave in shape, and will effectively form a rivet-like head.

AUTOMATION

As is evidenced by the coverage of secondary processing operations (machining, finishing, joining, decorating) in this handbook, automated procedures are of considerable importance. It is appropriate, then, to conclude this chapter with a few words on the subject.

In each of the chapters on secondary processing, we note the availability of automatic equipment to mark, decorate, seal, join, drill, and so on. To automate these operations, it is necessary to devise automatic systems for removing parts from the mold, orienting them so that they will be in the proper physical position to be fed to the machine, mounting the parts appropriately, and then feeding or transferring them to the automatic secondary processing unit.

Robots

One way to keep parts under control as they are taken from the mold is to use either an automatic extractor or a robot (if parts are allowed to fall free or are otherwise ejected, some means must be added to unscramble and orient them for proper feeding). Robots are expected to influence many segments of the plastics industry. Thus far, the main functions of robots have been in loading and unloading molding machines, in orienting the molded parts for secondary operations, in applying paints and coatings, and even in spraying up fibrous glass for reinforced plastic parts.

Using hydraulically controlled ''hands'' or grippers, a robot can perform most of the highly repetitive simple operations that humans can. Some of the more advanced types can handle a number of different articulations or movements. For a more complete discussion of robotic parts handling, the reader is referred to Chapter 21 of this handbook.

Feeding Parts

If parts are not assembled immediately after being removed from the molding machine by a robot, they will have to be fed into position by other mechanical means. If orientation is required, the parts must have a molded-in device that can be sensed by the feeder and used to position the article correctly. The device can be a projection, a flat, or a slot on the part.

Parts can be fed through tracks to an assembly unit end to end or side by side, using gravity or a vibratory feeder, depending on the shape of the piece. The tracks can be as simple as rods (for parts with holes in the center) or hollow tubes (for cylindrical items). For parts with heads or flanges, the track may have rails to allow them to travel in a hanging position. When several different parts are assembled, each will require its own feeding device, including hopper or vibratory feed bowl, track, and placement tooling.

Indexing Units

Although constantly moving conveyors or rotating table sometimes are used for parts assembly, most operations require intermittent motion. One of the most widely used assembly methods is the rotary indexing table, which revolves 360° in a number of steps, depending on how many stations are required. For example, a six-station table will rotate 60° at a time. The first and last stations are usually for loading and unloading, and the remainder are for the various assembly steps.

Rotary tables can be powered pneumatically or mechanically. Air-powered units are lower in cost, but their abrupt start-and-stop motion may not be suitable for all jobs. Mechanically driven tables use cams to regulate acceleration and deceleration for smooth operation.

Where more working room is required, linear indexing units can be used. The number of stations and the space allotted to each are limited only by the length of the belt. As with the rotaries, the drive system can range from simple (ratchet) to sophisticated (servo), depending on the need for accuracy in positioning.

25 Design Standards for Inserts

INTRODUCTION*

Although the subject of inserts is touched on in other chapters, this section deals specifically with design standards for inserts.

It would be well first to review the various mechanical fastening methods. The available fastening methods are as follows:

1. Cored or drilled hole with either thread-cutting or thread-forming self-tapping type screws.
2. Molded or tapped threads used in conjunction with regular machine screws.
3. Molded-in inserts of either the internal threaded bushing or external threaded stud varieties.
4. Post-molding inserts that are self-tapped, pressed, cemented, or expanded into the parent material. Also included in this group are those inserts that are installed into tapped or molded threads as well as those installed in thermoplastics by remelting using ultrasonics or some other heat-generating principle.

Thread-Cutting or Thread-Forming Self-Tapping Screws

The basic difference between the two types of screws is that the thread-cutting type cuts through the material, whereas the thread-forming screw displaces the material to form the thread. Thread-cutting screws generally are used with thermosetting plastics, whereas thread-forming screws generally are used with thermoplastic materials.

Tap holes and clearance hole sizes and lengths are important variables. Three factors must be analyzed in determining the proper hole sizes: ease of driving the screw, cracking of the boss (especially important is the long-term effect of stresses created), and adequate assembly strength. The optimum hole size must be thoroughly tested under all conditions, and must be controlled within close limits for the more friable materials. As is typical with interference fit systems, relatively small changes in size will have a drastic effect on the torque produced and stresses created in driving the screw into the material.

Advantages:

1. Least expensive.
2. Fastest to assemble.

Limitations (especially applicable to most thermosetting plastics):

1. Questionable holding power is obtained in comparison to standard machine screw used with a female embedded insert in the molded part.
2. Noticeable loss of holding power occurs after removal from the self-tapped hole and reassembly.
3. Technique may cause chipping and

*By Donald P. Viscio, Manager Advanced Engineering and Product Development, and Richard L. Davis, Manager Design Engineering, Heli-Coil Products Div. of Emhart Corp., Danbury, CT.

cracking of the molded part unless hole and screw dimensions are closely held.
4. Reentry of screw can easily damage threads by improper entry.

Molded or Tapped Threads with Machine Screws

The standard sizes and types of machine screws (either metal or plastic) can be used.

The internal threads can be prepared by either tapping or molding. Tapping using conventional methods can be achieved by having the proper tap drill size molded into the part or drilled. Factors such as complexity of the mold, number of holes, and difficulty of machining by drilling should be carefully evaluated. Many plastics are abrasive in nature, and both drilling and tapping may require special tooling. Oversize taps, chrome-plated taps, and even carbide taps have proved to be well worthwhile for large-volume production or when the material is extremely abrasive, such as a glass-filled plastic.

Molding threads into the part is considered only when the resultant thread class of fit, surface finish, and strength are critical, or if the threaded connection forms the essential function of the part. Generally, molded threads are limited to one to three per part, because of the increased complexity of the mold, and the extra operations required to unscrew the threaded core pins for each cycle. Another factor must be considered: if split mold sections are used, a parting line or flash develops, which may have to be tapped out. Molded internal threads no finer than 32 pitch, or longer than $\frac{1}{2}$ inch, should be used; otherwise, too much operator time will be required to disassemble the part from the mold.

Advantages:

1. Relatively inexpensive; screw cost is less than that for self-tapping style.
2. Holding power somewhat better than that of self-tapping drive screws.
3. Minimal stress in material due to thread forming.
4. Easy assembly of screw.

Limitations:

1. Separate drilling and tapping operations are costly, due to tap wear and labor involved.
2. Strength is not equal to that of molded-in or post-molding insert.
3. Threads may chip and are easily damaged by improper entry of the screw.
4. Molding-in threads increases mold cost significantly. Also, cycle time is increased, especially when more than one thread is molded.

Molded-in Inserts

Where there is repeated disassembly, or where strength of assembly is important, molded-in inserts can be specified (Fig. 25-1). Careful attention must be given to design with molded-in inserts for the following reasons:

1. There are unpredictable variations in molding.
2. Possibility exists of developing excessive strains on the part, due to differential cooling rates of the plastic and the insert.
3. Some plastic materials will crack around the insert after they have aged. Other materials will creep in aging, or, if two inserts are rigidly located together in a mating part, one or the other of the inserts will pull out as the plastic ages and shrinks.
4. Delicate or light inserts may be damaged or dislodged during the molding cycle.

Female inserts molded through the part frequently are not threaded when compression molding is used because the plastic will flow into the threads, and a retapping operation will be required. If injection or transfer methods are used, the female insert may be tapped prior to molding for most materials because the mold is closed on the insert before the compound is forced into the mold.

Inserts that are molded in and intended to provide a sealed, leakproof connection should incorporate grooves and gaskets or O-rings.

The location of the insert and the boss con-

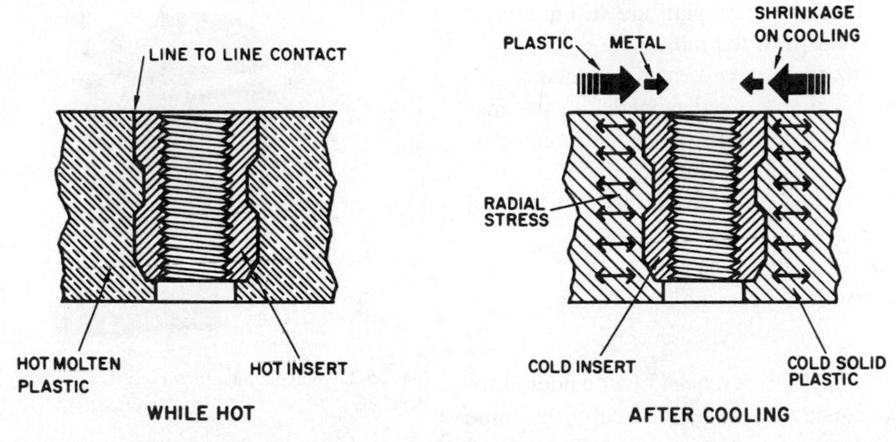

Fig. 25-1. Molded-in insert. (*Courtesy Heli-Coil Productions*)

figuration in the part are also important. The minimum boss diameter is equal to two times the metal insert outside diameter.

For those materials prone to cracking after cooling or aging, the following suggestions are offered:

1. Keep knurls to a minimum.
2. Knurls should not be too sharp.
3. Blind inserts should have a round head.
4. Remove all sharp corners.

For optimum location of inserts within the part itself, the following factors should be taken into account:

1. The floor under the insert should be at least one-sixth the diameter of the insert to prevent a sink mark.
2. Inserts should be longer or shorter than the boss itself.
3. The mold recess for male inserts should be at least 0.020 inch from the edge of the mold if cracking of the mold is to be avoided.
4. Inserts molded in opposite sides of a thermosetting plastic should be no closer than $\frac{1}{8}$ inch.
5. Avoid bringing a boss down to a narrow fin around the insert.

Advantages:

1. Very good holding power of screw and insert is possible.
2. Subsequent reassembly may be achieved without the loss of screw holding power.

Limitations:

1. It is difficult to charge the mold with inserts on an automatic cycle without a costly feeding, loading, and indexing mechanism.
2. If a part is produced with the insert missing or misplaced, the entire part must be rejected.
3. Inserts must be cleaned of all flash especially in threaded inserts. This can be costly.
4. The technique may create residual stresses in the plastic after cooling and aging. Also, creep and shrinkage affect insert retention.
5. Floating inserts can cause extensive mold damage.
6. Large inserts require preheating above the mold temperature to pre-expand them and improve flow and cure-in.
7. Removal of inserts is required prior to regrind of rejected or unused parts.

Post-molding Inserts

The inserts that fall into this category have gained acceptance in recent years. There are six basic types of inserts in this category:

A. Press-in inserts, which are usually assembled while the parts are still hot, after removal from the mold.

B. Inserts that are cemented in place.
C. Inserts that are self-tapped into the material. They can be either the thread-cutting or the thread-forming type.
D. Inserts that are installed into tapped holes.
E. Expansion-type inserts.
F. Inserts for ultrasonic installation.

Group A and B inserts can be considered together, because they have practically the same advantages and limitations.

Advantages:

1. They can be used with parts produced by automatic molding.
2. No cleaning of flash from inserts is needed.
3. Installation mechanics are simple, requiring no investment for special tools.

Limitations:

1. Insert holding power is not equal to that of molded-in inserts.
2. Insert retention reliability may be questionable, because of the reliance on close tolerances in the case of the press-in insert, and reliance on clean surfaces properly bonding to the cement under all conditions for the cemented insert.
3. Press-in bushings may crack the material if holes are slightly undersize, or if the bushing is slightly oversize.

Group C inserts are of either the solid bushing or carbon steel coiled wire types. They are driven into a plain hole that has been prepared by either molding or drilling to the proper diameter. In both cases, the insert taps its own thread, and, when assembled, the internal threads form the desired female thread. Self-tapping inserts have somewhat greater holding strength than expansion types. (See Fig. 25-2.)

Fig. 25-2. Self-tapping metal insert. (*Courtesy Heli-Coil Products*)

Advantages:

1. Strong assembly strength (resistance to pullout) because of shearing of a large area.
2. Good-quality internal threads achieved.

Limitations:

1. Insert cost is usually quite high in comparison to expansion types.
2. Installation is the slowest on a per piece basis for all types of fasteners.
3. High driving torque, or poor insert retention, can result from extremes in hole–insert tolerance variations.
4. Size range generally restricted to #8 through $\frac{1}{2}$ inch.

Group D type inserts, which are installed into threaded holes, offer high quality and precision assembly. The tapped holes are either molded or tapped to close tolerances (Class 3B limits may be achieved if necessary). Also, improved strength of assembly will result from this group, because of its increase in shear area over the self-tapping group mentioned above. Generally available with inserts in this group is a screw locking device to prevent loosening, or for adjustment screws. Insert retention principles vary with each manufacturer, but the basic principle is to ''lock-in'' the insert with a collar, keys, prongs or rings, nylon pellets, or pins, or by spring expansion of the stainless steel helically coiled wire type insert. (See Fig. 25-3.)

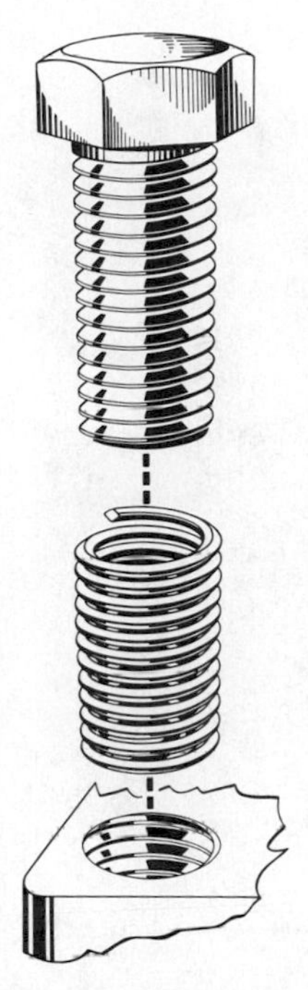

Fig. 25-3. Stainless steel, helically coiled wire type insert. (*Courtesy Heli-Coil Products*)

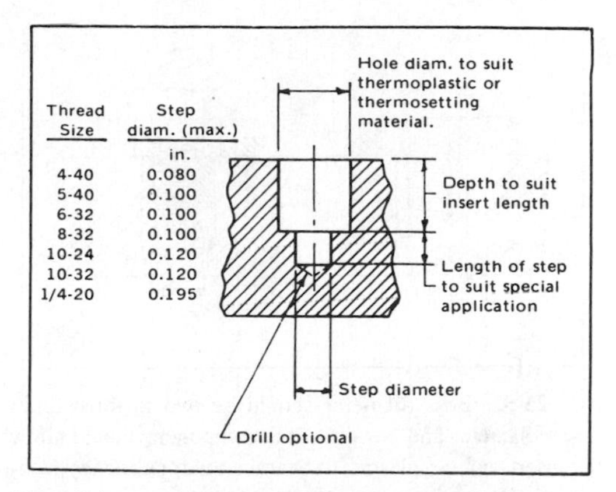

Fig. 25-4. Recommended design for stepped holes for use with post-molding inserts. Stepped holes are used to avoid sink marks opposite insert holes. (*Courtesy Heli-Coil Products*)

Advantages:

1. Highest quality of internal thread.
2. Strongest insert strength of all types.
3. Screw locking feature available.
4. Hard, wear-resistant internal threads.

Limitations:

1. Highest cost of all other types of inserts.
2. Installation slow, usually requiring expensive equipment if automatic installation is desired.

Group E inserts are relatively recent. Expansion-type inserts are installed into cored or drilled holes, and are secured by expanding a knurled portion of the insert. The expansion is achieved by either the entry of the screw or a spreader plate, the distinction being that the former relies on full penetration through the insert by the screw to expand the knurls, whereas the latter is retained by the integral expansion feature of the insert itself. Both types distribute the locking stresses evenly in the hole and provide ample resistance to torsional and tensile loads. They are easily installed, either manually or automatically.

The strength of the assembled expansion-type insert, although not as high as that for self-tapping inserts, is sufficient in most plastics to cause the screw to shear before pulling out the insert. The strength of the expansion types is approximately halfway between that of press-in and self-tapping inserts. Loading values for a given insert will vary with the properties of the material.

In work with post-molding inserts, the effect of hole size on torsional and tensile assembly strength varies greatly among plastics, and rather wide variations in hole dimensions result in less loss of assembly strength than might be expected. In short, the plastic used has far more effect on the assembly strength than has the hole-to-insert fit. (See Fig. 25-4.)

The boss diameters of the plastic should be at least two times the diameter of the insert, and, in many cases, 2.5 to three times the insert diameter is required. In designing clearance diameters of component parts, it is extremely important to see that the insert, not the plastic, carries the load. (See Fig. 25-5.)

Several examples of expansion-type inserts are shown in Figs. 25-6 through 25-9.

Advantages:

1. Lower insert installation cost, by maximum utilization of press time.

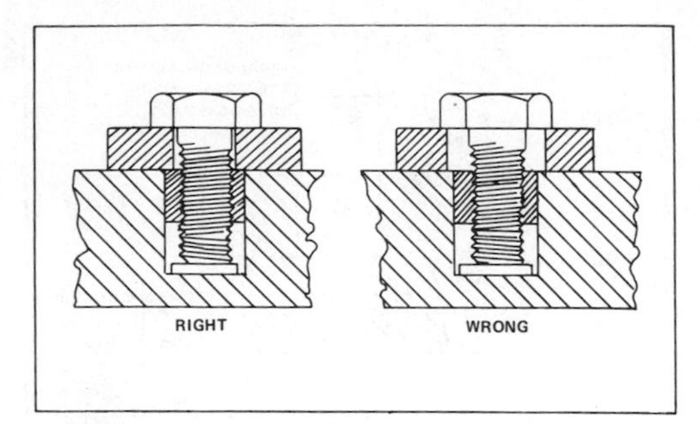

Fig. 25-5. Boss diameter should be two to three times insert diameter and design of the component should allow the insert, not the plastic, to bear the load. (*Courtesy Heli-Coil Products*)

2. Rework to remove flash in threads eliminated.
3. Elimination of floating inserts, damage to mold, breaking of core pins.
4. Adaptability to fast-cycling automatic installation.

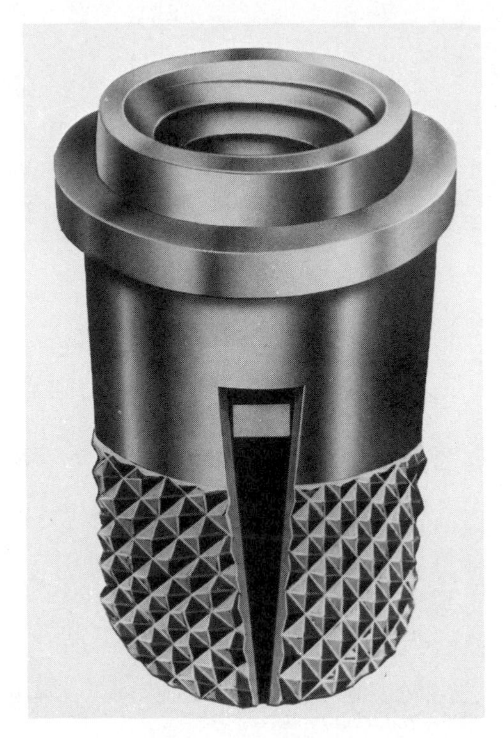

Fig. 25-7. Clinch insert features a pilot (which may be clinched over a terminal, eyelet fashion) and a flange (which offers a broad surface for electrical contact).

Fig. 25-6. Standard insert provides brass threads in plastics parts of all types after molding. (*Illustrations of insert types courtesy Heli-Coil Products*)

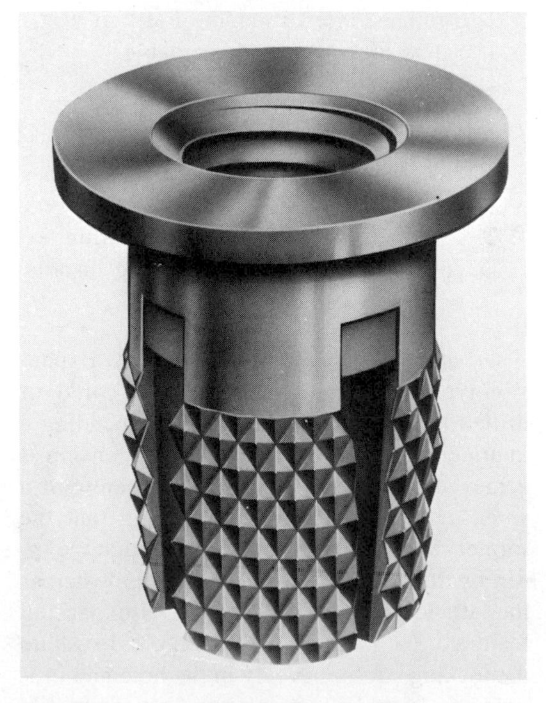

Fig. 25-8. Flange insert features a flange with large surface for effective electrical contact or for holding down mating parts.

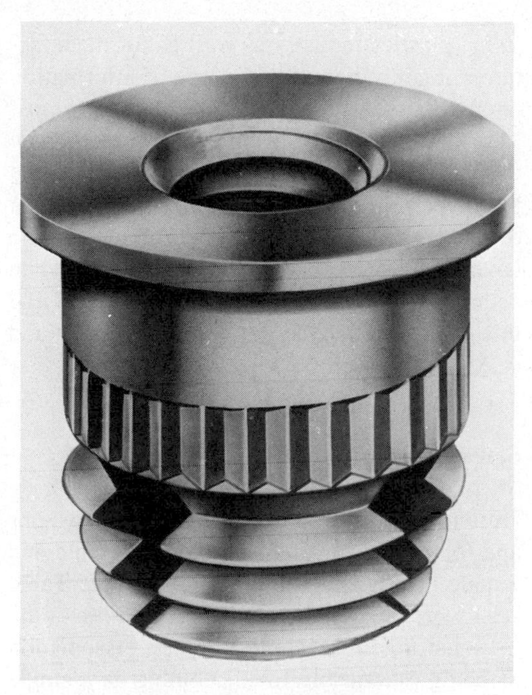

Fig. 25-9. In the wedge insert, the large wedge configurations of the insert body are forced deeply into the wall of the drilled hole. Design is especially applicable for use in soft plastics. (*Courtesy Heli-Coil Products*)

Fig. 25-10. Tapered metal insert engineered for ultrasonic installation. (*Courtesy Heli-Coil Products*)

Fig. 25-11. Ultrasonic insert with parallel sides for installation into straight holes. (*Courtesy Heli-Coil Products*)

Limitations:

1. Strength not as high as that of mold-in types but higher than press-in types.
2. Cost slightly higher than for mold-in types.

Group F: Recently many manufacturers have been turning to inserts for ultrasonic installation into thermoplastics. The theory behind this method is quite simple: high frequency vibration develops frictional heat at the insert/plastic interface. The heat thus created remelts the plastic in a narrow zone around the insert. As the plastic cools, it shrinks away from the warm metal outward toward the cooler plastic. After this shrinking, the diameter of the plastic hole is increased, creating a microscopic relief zone that in no way decreases the holding strength of the insert. This relief zone, however, does prevent the compressive stresses normally created by the mold-in method. Another advantage of the ultrasonic installation of specially engineered inserts is the avoidance of flash in the insert threads. The hole that is to receive the insert may be either tapered (which is easier to mold) or straight. A tapered hole, which is best for ease of insert installation, can be either cored or drilled with comparatively liberal tolerances.

The ultrasonic installation cycle time for inserts in the #2 through #10 insert sizes is usually less than one second in many of the common thermoplastics.

Metal inserts designed specifically for ultrasonic installation are shown in Figs. 25-10, 25-11, and 25-12.

Advantages:

1. Achieve a high strength stress-free assembly without the problems normally

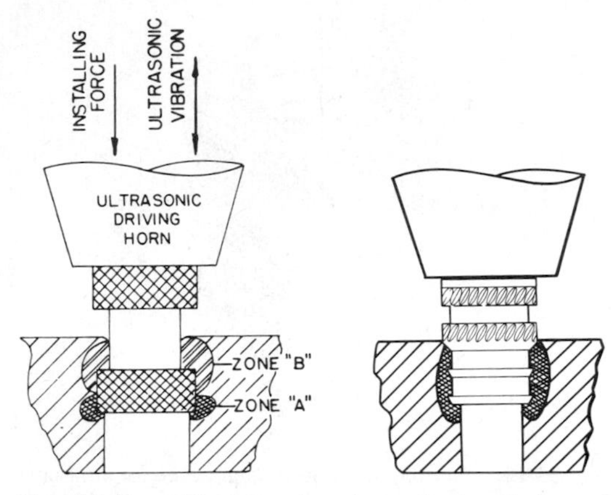

Fig. 25-12. Differences in ultrasonic installation of conventional insert (left) and inserts designed for ultrasonic use (right). (*Courtesy Heli-Coil Products*)

encountered with mold-in or the high cost of inserts installed into tapped threads.
2. Can be installed quickly and easily into cored or drilled, straight or tapered hole. Tolerances are quite liberal.
3. Can be used in small-diameter bosses or where wall thickness is limited. All other inserts require a much thicker wall to prevent bulging and cracking. This is especially significant in stress-sensitive polymers such as polycarbonates or acrylics.

Limitations:

1. Insert cost is slightly higher than for mold-in or expansion type.
2. Original cost of installation equipment is comparatively high, but it can be amortized in a relatively short run.

DESIGN STANDARDS

Inserts of many types are used—those made on screw machines and those made by cold forging, stamping, and drawing. The discussion in this chapter is based, rather loosely, on these types, with general instructions and precautions given under broad headings that are largely applicable to all the types. Inserts of various designs are shown in Fig. 25-13.

Maintaining proper accuracy in various dimensions of inserts has always been a problem in the plastics industry as well as for manufacturers, mainly because of a lack of information on design and standardization of dimensions. The technicians who were selected to prepare the engineering standard for inserts endeavored to draw on their own knowledge as well as that of the entire plastics industry. Engineers who have a reasonable knowledge of plastics and an acquaintance with inserts and their use will find this standard of value in the proper design and selection of inserts.

Screw-Machine Inserts

Dimensions and Tolerances. Dimensions and tolerances for the usual types of male and female inserts, shown in Fig. 25-14 and Table 25-1 were compiled with the cooperation of the National Screw Machine Products Association as being practicable for machining as a single operation on an automatic screw machine, and thus are most economical.

Note that the dimensions given for tapped inserts apply only to nonferrous metals where the depth of usable tapping is not more than $1\frac{1}{2}$ times the tap diameter. On A-2 (minor diameter) and C (length of tapped inserts) the maximum "regular" tolerance should be specified whenever possible. However, for closer tolerances, "precision" can be specified when necessary. To maintain the "precision" tolerance, reaming and other additional operations will be necessary, at additional cost. In certain cases, the thread on a stud can be rolled, reducing length B-1 (Fig. 25-14) but generally not increasing the cost.

If steel inserts are required, Fig. 25-14 and Table 25-1 cannot be used in design without several modifications that will increase the cost over that of inserts made of brass or, in special cases, of aluminum.

The minimum wall thickness of metal in the inserts depends entirely upon the desired accuracy of the inside dimensions of the insert. If too thin a wall of metal is used, the combination of stress caused by shrinkage of the plastic and by molding pressure may collapse the wall of the insert, so that the inside diameter will be out of the range of specified tolerances. Table

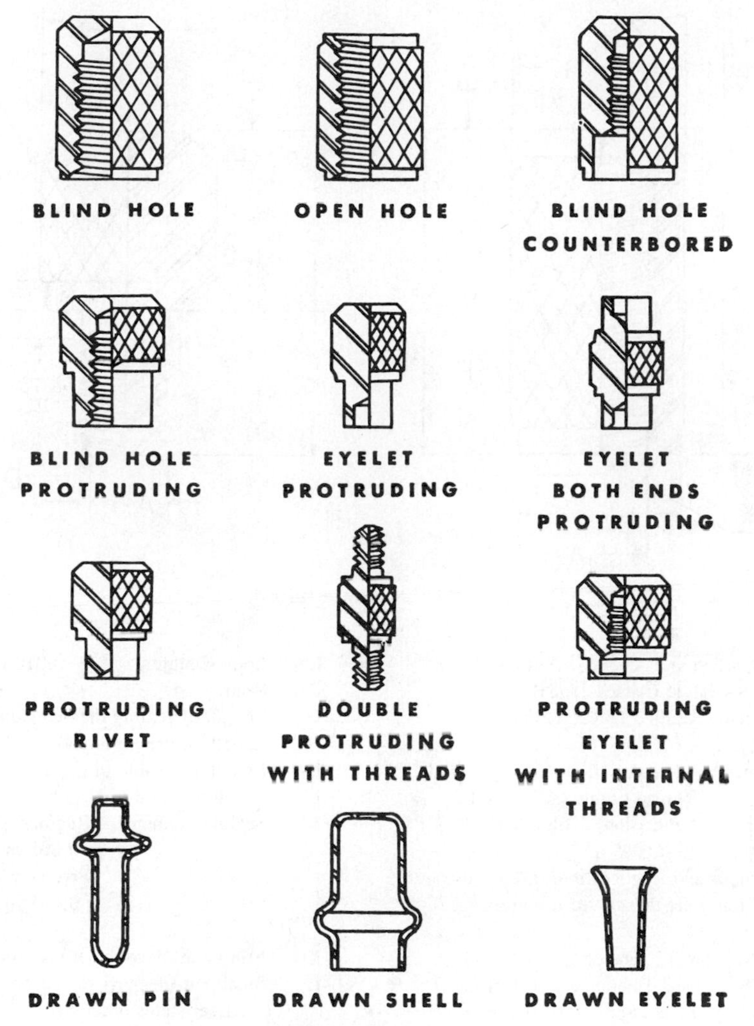

Fig. 25-13. Usual types of inserts.

25-1 shows the minimum recommended diameters of bar stock for various sizes of inserts.

Cold-Forged Inserts

In general, the volume or quantity needed to ensure economical production by cold-forging is about the same as that needed by other processes. When second operations are required, such as turning, drilling, tapping and others, larger quantities are needed.

In the discussion below, of specific problems, the utility of cold-forged inserts will become apparent. None of the cold-forged inserts shown has been machined, but they could be, by either automatic or single-purpose equipment. If machining, drilling, reaming, or tapping is involved, the tolerances are the same as those for screw-machine inserts given in Table 25-1.

There are no specific formulas controlling the individual relationships of the diameters and widths of collars to the shank or the kind and variety of shapes, such as ribbed, finned, pinchneck, hexagon, and so on, which may be combined with other symmetrical or unsymmetrical shapes in one piece. For each problem, a solution should be reached through the cooperation of the designer of the molded piece and manufacturer of the insert.

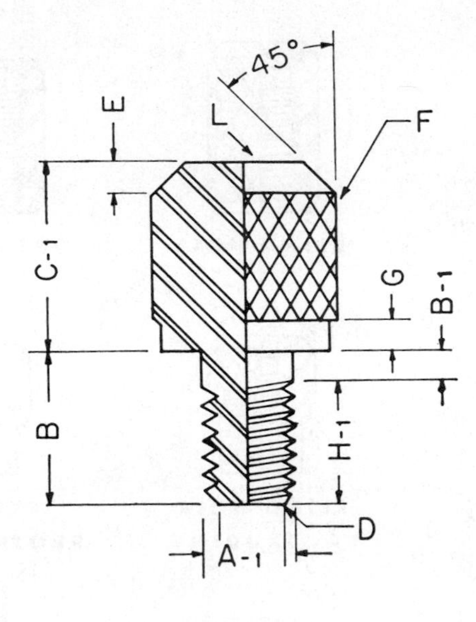

Fig. 25-14. Tolerance index.*

A Thread, "Unified and American National" (ASA B1.1) Classes 2A/2B
A-1 Major diameter, Unified Thread Limits (ASA B1.1)
A-2 Minor diameter, "Regular" Tolerance, Unified Thread Limits (ASA B1.1) "Precision" Tolerance ± 0.0005 in.
B Depth of minor and length of major diameter
B-1 Number of unusable threads from bottom and top
C Length, "Regular" Tolerance ± 0.010 in. "Precision" Tolerance ± 0.001 in.
C-1 Length of body, male insert, ± 0.010 in.
D* Thread chamfer, 45° ± 0.005 in.
E* Body chamfer, 45° ± 0.010 in.
F Knurl
G Length of sealing diameter, minimum 1/32 in.
H Length of usable thread, 1.5 × diameter
H-1 Length of usable thread H-1 + B-1 = B
I Amount to add to H to obtain C, H + I = C
J Sealing diameter, "Regular" Tolerance ± 0.003 in. "Precision" Tolerance ± 0.001 in.
K Minimum diameter of bar stock
L Small cut-off burrs are acceptable unless purchaser states otherwise

*See Table 25-1.

Materials. Almost any metal can be cold-worked, but cold-working grades of the following are preferred in the order named:

1. Aluminum and aluminum alloys
2. Brass
3. Copper and copper alloys
4. Carbon steels
5. Alloy steels
6. Stainless steels
7. Silver and other precious metals

Tolerances without Finishing Operations. The tolerances given for the inserts in the layouts of Figs. 25-15 and 25-16 are those ordered, although closer ones could be met if necessary. Tolerances for any element such as length or diameter vary with the material and with the sizes and proportions of the piece, which in turn determine the equipment or method of heading to be used.

In general, the following tolerances can be considered as commercial without finishing operations, although in some cases special care must be exercised to meet them:

- Length: ±0.010 inch (maximum)
- Fillets: sharp or rounded, as specified

Table 25-1. Dimensions and tolerances.*

Nonferrous inserts which have a usable thread length not more than 1½ times the tap diameter

A		K	J	Tap		A-1					
Coarse	Fine	Minimum	Maximum	Drill	A-2	Maximum	Minimum	B-1	I	D and E	Knurl
2-56		3/16	9/64	#50	0.0700	0.0860	0.0820	3	3/32	1/64	Fine
	2-64	3/16	9/64	#49	.0730	.0860	.0822	3	3/32	1/64	Fine
3-48		7/32	5/32	#45	.0820	.0990	.0946	3	7/64	1/64	Fine
	3-56	7/32	5/32	#45	.0820	.0990	.0950	3	3/32	1/64	Fine
4-40		7/32	11/64	#43	.0890	.1120	.1072	2½	7/64	1/64	Fine
	4-48	7/32	11/64	#42	.0935	.1120	.1076	2½	7/64	1/64	Fine
5-40		1/4	3/16	#37	.1040	.1250	.1202	2½	7/64	1/32	Med.
	5-44	1/4	3/16	#37	.1040	.1250	.1204	2½	7/64	1/32	Med.
6-32		1/4	13/64	#33	.1130	.1380	.1326	2½	5/32	1/32	Med.
	6-40	1/4	13/64	#32	.1160	.1380	.1332	2½	9/64	1/32	Med.
8-32		9/32	7/32	#29	.1360	.1640	.1586	2½	5/32	1/32	Med.
	8-36	9/32	7/32	#28	.1405	.1640	.1590	2½	9/64	1/32	Med.
10-24		5/16	1/4	#23	.1540	.1900	.1834	2½	3/16	1/32	Med.
	10-32	5/16	1/4	#20	.1610	.1900	.1846	2½	5/32	1/32	Med.
12-24		3/8	5/16	#16	.1770	.2160	.2094	2½	13/64	3/64	Med.
	12-28	3/8	5/16	#13	.1850	.2160	.2098	2½	11/64	3/64	Med.
1/4-20		13/32	11/32	#6	.2040	.2500	.2428	2	13/64	3/64	Coarse
	1/4-28	13/32	11/32	7/32	.2187	.2500	.2438	2	11/64	3/64	Coarse
5/16-18		15/32	13/32	G	.2610	.3125	.3043	2	7/32	3/64	Coarse
	5/16-24	15/32	13/32	I	.2720	.3125	.3059	2	13/64	3/64	Coarse
3/8-16		9/16	15/32	O	.3160	.3750	.3660	2	1/4	3/64	Coarse
	3/8-24	9/16	15/32	Q	.3320	.3750	.3684	2	7/32	3/64	Coarse
7/16-14		5/8	17/32	U	.3680	.4375	.4277	2	9/32	3/64	Coarse
	7/16-20	5/8	17/32	25/64	.3906	.4375	.4303	2	1/4	3/64	Coarse
1/2-13		11/16	19/32	27/64	.4218	.5000	.4896	2	5/16	1/16	Coarse
	1/2-20	11/16	19/32	29/64	.4531	.5000	.4928	2	17/64	1/16	Coarse
9/16-12		3/4	21/32	31/64	.4843	.5625	.5513	2	11/32	1/16	Coarse
	9/16-18	3/4	21/32	33/64	.5156	.5625	.5543	2	9/32	1/16	Coarse
5/8-11		13/16	23/32	35/64	.5469	.6250	.6132	2	3/8	1/16	Coarse
	5/8-18	13/16	23/32	37/64	.5781	.6250	.6168	2	5/16	1/16	Coarse

*See Fig. 25-14.

- Diameter: 0.002 inch (minimum)
- Squareness, shoulders or collars with shank: ±1° maximum

Tolerances with Finishing Operations. Whatever tolerance is needed can be met by adding finishing operations. For example, aircraft studs, bolts, and special inserts are commonly made today to tolerances as close as 0.0005 inch and even less.

Special Inserts

It would be an endless task to try to cover the entire field of special inserts, but some of the more important phases of their design are included in this section.

The design of special inserts for various applications requires as much engineering as other phases of preliminary work do, if not more. In many cases too little significance is attached to planning special inserts. The design engineer, the manufacturer of the insert, and the molder must cooperate to obtain simplicity of design, which will result in the production of satisfactory articles and promote economical production.

Typical applications of special inserts are commutators, wire-and-insert connections on telephone handsets, and radio resistors where carbon or other elements are molded inside the plastic. A radio condenser is a good example of a built-up laminated insert.

Selection of Metal for Inserts

The correct selection of metal for inserts is essential because of differences in the coefficient

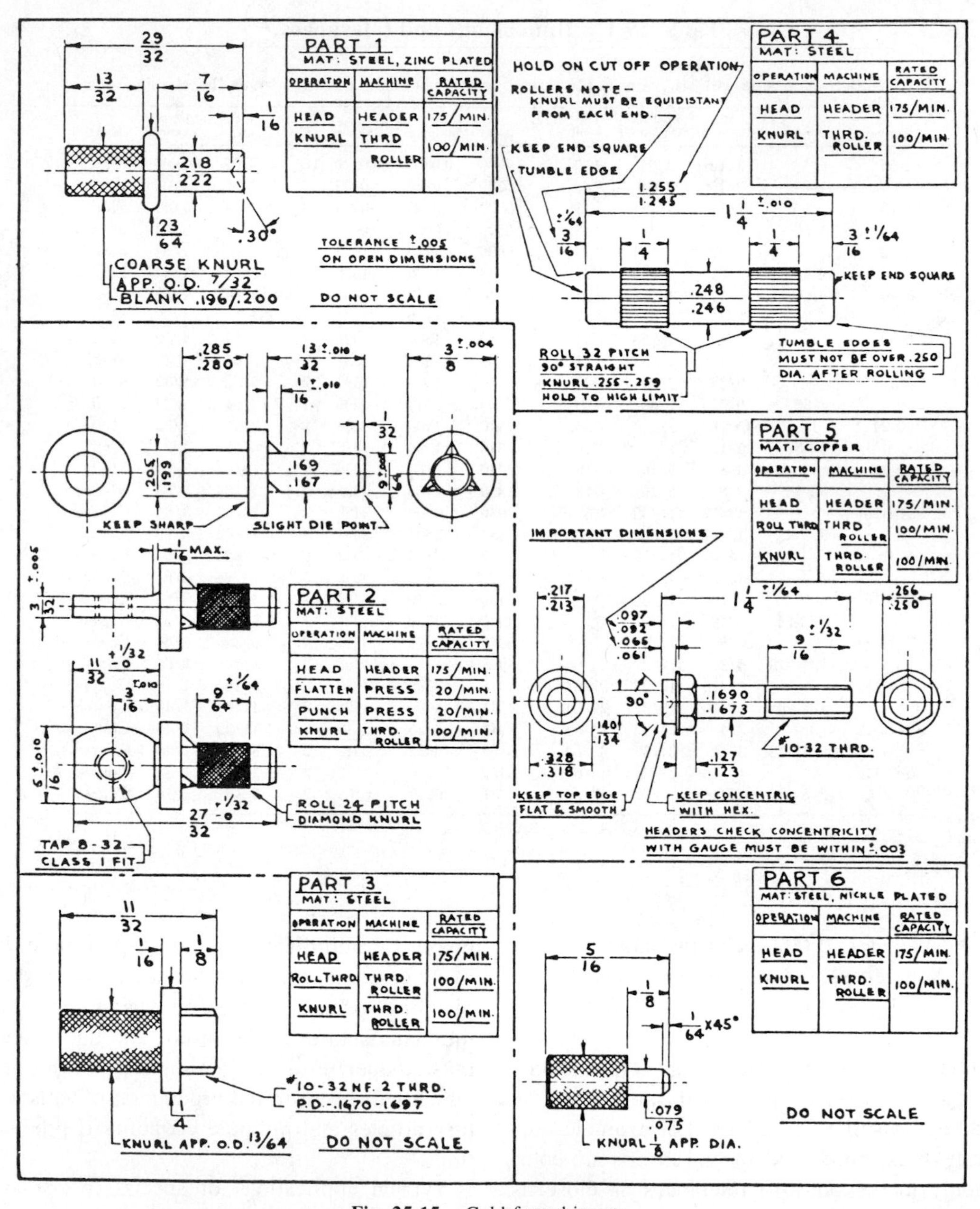

Fig. 25-15. Cold-forged inserts.

of expansion among the various metals and plastics (see Table 25-2). It is impossible to keep the plastics listing in Table 25-2 up to date because of the wide variety of new materials being produced. Therefore it is suggested that the technician refer to data sheets of the material suppliers.

Minimum Wall Thickness of Material around Inserts

The thickness of the wall of plastic required around inserts depends upon (1) whether the material is thermoplastic or thermosetting, (2) the type of material within each group, (3) the

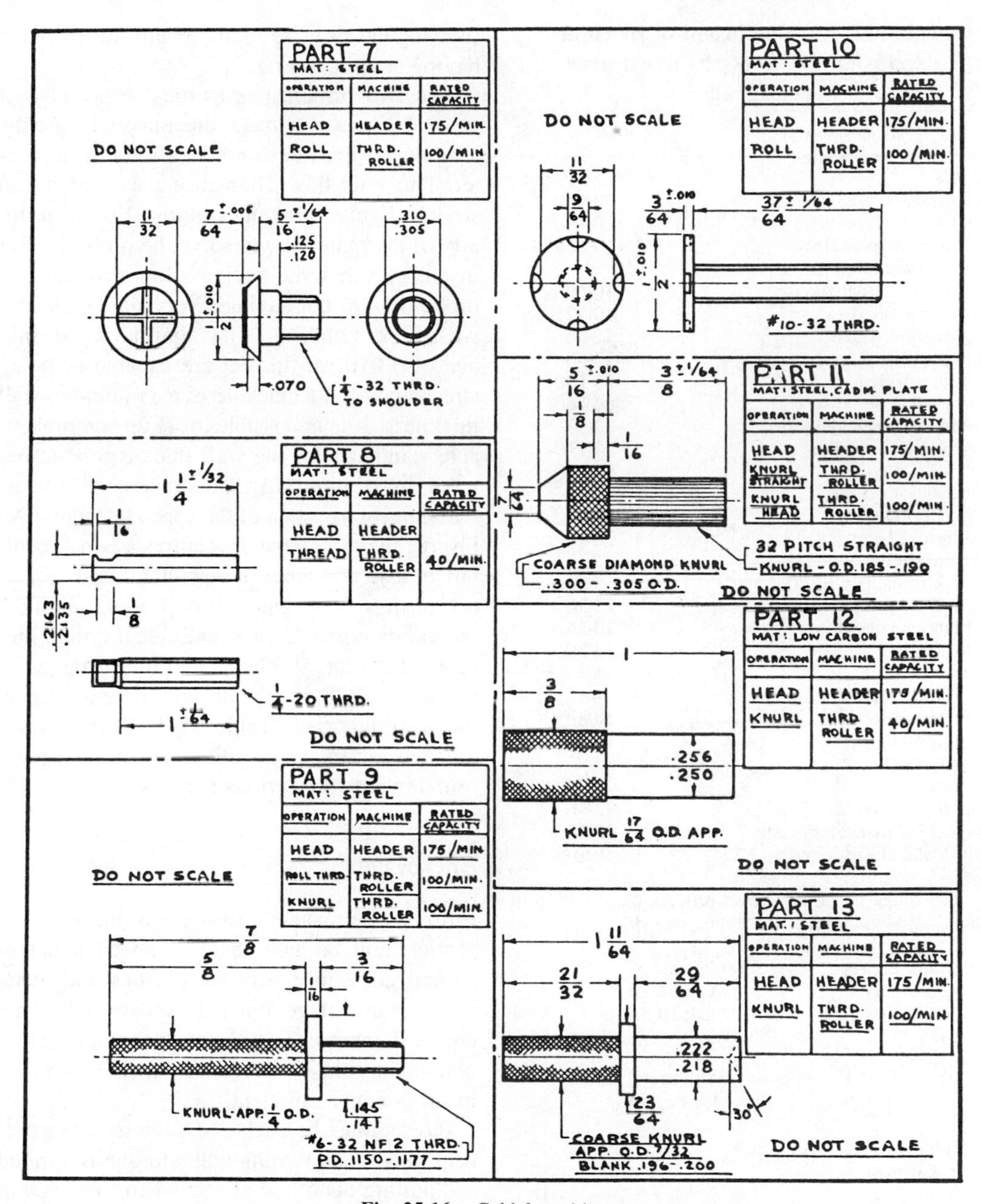

Fig. 25-16. Cold-forged inserts.

shrinkage of the material, (4) the modulus of elasticity and (5) the coefficient of expansion of the material, (6) the coefficient of expansion of the metal used in the inserts, (7) the temperature range over which the molded article will have to function, (8) the moisture-sensitivity of the plastic, (9) any loss of flexibility caused by aging and especially (10) the design of the insert, and (11) allowance for desired electrical properties.

Very often the molded article is designed first, and the necessary inserts then are fitted into the remaining space. If inserts are required, they should be considered first and then

Table 25-2. Coefficient of thermal expansion (30 to 60°C) per degree centigrade.

Typical Material	Coefficient x 10^6
alkyds	25-35
cellulose acetate	80-160
cellulose acetate butyrate	110-170
cellulose propionate	110-170
diallyl phthalate:	
synthetic-fiber-filled	50-60
mineral-filled	40-42
glass-fiber-filled	32-36
epoxy, filled	20-60
epoxy, unfilled	40-100
ethyl cellulose	100-160
melamine-formaldehyde	20-57
methyl methacrylate	54-110
nylon	90-108
phenolics:	
general-purpose	30-45
improved-impact	30-45
medium-impact–CFI-10	29
high-impact–CFI-20	22
medium-heat resistant	15-30
high-heat-resistant	20-35
low-loss	19-26
arc-resistant	49
polyester, colored	48
polyester, premixes	30-80
polyethylene	110-250
polystyrene	60-80
silicones	8-50
urea-formaldehyde	22-36
vinyl chloride-acetate resin	50-185
vinylidene chloride resin	190

The range of values shown reflects the variation in fillers and variations in resins.

Typical Material	Composition	Coefficient x 10^6
aluminum 2S	99.2% Al	23.5
brass, ordinary	67 Cu, 33 Zn	18.8
bronze, commercial	90 Cu, 10 Zn	17.4
copper	99.9+	16.7
C.R. steel		11.7
Monel	67 Ni, 30 Cu, 1.4 Fe, 1 Mn	14.
nickel		13.3
phosphor bronze		17.
silver, German		18.
silver, sterling	92.5 Ag, 7.5 Cu	18.
solder, half-and-half		24.
stainless steel	90-2 Fe, 8 Cr, 0.4 Mn, 0.12 C	11.
steel	99 Fe, 1 C	10.8
zinc	95 Zn, 5 Al	28.

the molded article designed around them. The shape and form of the insert govern the wall thickness of the plastic to a great degree, especially when the inserts are of irregular shape (rectangular, square, star, or any other shape having sharp corners).

The two principal properties, especially of phenolic, urea, and melamine materials, are the modulus of elasticity and the ability of the material to cold-flow after curing so that it can stretch slightly without cracking. No one property of the material will solve the problem. For instance, a material having a low shrinkage of 0.002 in./in. but having a very rigid character will crack. Other materials that have a shrinkage of 0.010 in./in. but are capable of being stretched will not crack despite a minimum wall thickness. It is impossible to set up comprehensive standards for the wall thickness of a material in relationship to diameters of inserts, particularly for some of the special designs. An insert $\frac{1}{4}$ inch in diameter requires a $\frac{1}{8}$-inch wall, but an insert 6 inches in diameter might require a $1\frac{1}{2}$-inch wall of material, depending upon the factors mentioned. Each individual article presents different problems and must be engineered according to the design of the insert and the material used. Table 25-3 shows recommended minimum wall thicknesses with plain round inserts for various plastics.

Anchorage

Firm and permanent anchorage of inserts is essential; and because there is no chemical or natural adherence between plastics and metal inserts, anchorage must be obtained by mechanical means. The slight anchorage that is obtained by the shrinkage of plastic around the insert is never sufficient.

Inserts must be anchored sufficiently to prévent them from turning when torque is applied and pulling out of the plastic when subjected to tension. However, internal stresses in the molded plastic must be kept to a minimum.

In the early days of plastics, it was customary to use hexagonal stock for inserts (Fig. 25-17). This approach is mechanically incorrect except in some special applications. Hexagonal stock provides torsional anchorage only, and grooves must be machined to obtain sufficient anchorage in tension. Combinations of sharp corners and grooves on hexagonal stock

Table 25-3. Minimum wall thickness of material (in.).

Diameter of Inserts (in.):	1/8	1/4	3/8	1/2	3/4	1	1-1/4	1-1/2	1-3/4	2
phenolics:										
general-purpose	3/32	5/32	3/16	7/32	5/16	11/32	3/8	13/32	7/16	15/32
medium-impact	5/64	9/64	5/32	13/64	9/32	5/16	11/32	3/8	13/32	7/16
high-impact (rag)	1/16	1/8	9/64	3/16	1/4	9/32	5/16	11/32	3/8	13/32
high-impact (sisal)	5/64	9/64	5/32	3/16	1/4	9/32	5/16	11/32	3/8	13/32
high-impact (glass)	1/16	3/32	1/8	1/8	3/16	3/16	1/4	1/4	5/16	5/16
high-heat-resistant, general-purpose type	1/8	3/16	7/32	1/4	11/32	3/8	13/32	3/16	15/32	1/2
high-heat-resistant, impact type	5/64	9/64	5/32	13/64	9/32	5/16	11/32	3/8	13/32	7/16
low-loss	5/32	7/32	1/4	9/32	3/8	13/32	7/16	15/32	1/2	17/32
special for large inserts	3/64	7/64	1/8	5/32	7/32	1/4	9/32	5/16	11/32	3/8
polyester, colors	3/32	5/32	3/16	7/32	5/16	11/32	3/8	13/32	7/16	15/32
polyester, sisal-filled	5/64	9/64	5/32	3/16	1/4	9/32	5/16	11/32	3/8	13/32
polyester, glass-filled	1/16	1/8	9/64	3/16	1/4	9/32	5/16	11/32	3/8	13/32
diallyl phthalate:										
(a) "Orlon"-filled	1/8	3/16	7/32	5/16	11/32	3/8	13/32	7/16	15/32	1/2
(b) mineral-filled	3/32	5/32	3/16	7/32	5/16	11/32	3/8	13/32	7/16	15/32
(c) glass-filled	5/64	9/64	5/32	3/16	1/4	9/32	5/16	11/32	3/8	13/32
cellulose acetate	1/8	1/4	3/8	1/2	3/4	1	1-1/4	1-1/2	1-3/4	2
cellulose acetate butyrate	1/8	1/4	3/8	1/2	3/4	1	1-1/4	1-1/2	1-3/4	2
ethyl cellulose	1/16	3/32	1/8	5/32	3/16	7/32	1/4	9/32	5/16	11/32
urea formaldehyde	3/32	5/32	3/16	7/32	5/16	11/32	3/8	13/32	7/16	15/32
*melamine formaldehyde (a)	3/32	5/32	3/16	7/32	5/16	11/32	3/8	13/32	7/16	15/32
(b)	1/8	3/16	7/32	5/16	11/32	3/8	13/32	7/16	15/32	1/2
vinylidene chloride resin	3/32	1/8	3/16	1/4	3/8	1/2	1/4	9/32	5/16	11/32
methyl methacrylate resin	3/32	1/8	3/16	3/16	7/32	1/4	5/8	3/4	7/8	1
polystyrene	3/16	3/8	9/16	3/4	1-1/8	1-1/2	1-7/8	2-1/4	2-5/8	3
polyethylene	1/16	3/32	1/8	5/32	3/16	7/32	1/4	9/32	5/16	11/32
nylon:										
"Zytel" 101 or equiv.	1/16	3/32	1/8	5/32	3/16	7/32	1/4	9/32	5/16	11/32
" 31 " "	3/32	1/8	5/32	7/32	1/4	5/16	11/32	13/32	7/16	15/32
" 63 " "	3/32	5/32	3/16	1/4	5/16	11/32	13/32	7/16	1/2	9/16
" 69 " "	1/8	7/32	9/32	11/32	7/16	1/2	19/32	21/32	23/32	13/16
" 105 " "	1/16	3/32	1/8	5/32	3/16	7/32	1/4	9/32	5/16	11/32
" 211 " "	3/32	1/8	5/32	7/32	1/4	5/16	11/32	13/32	7/16	15/32
" 42 " "	1/16	3/32	1/8	5/32	3/16	7/32	1/4	9/32	5/16	11/32
vinyl chloride-acetate resin	3/32	1/8	3/16	1/4	3/8	1/2	5/8	3/4	7/8	1

* Melamine formaldehyde (a) mineral-filled melamine ignition material
(b) cellulose-filled melamine, electrical grade

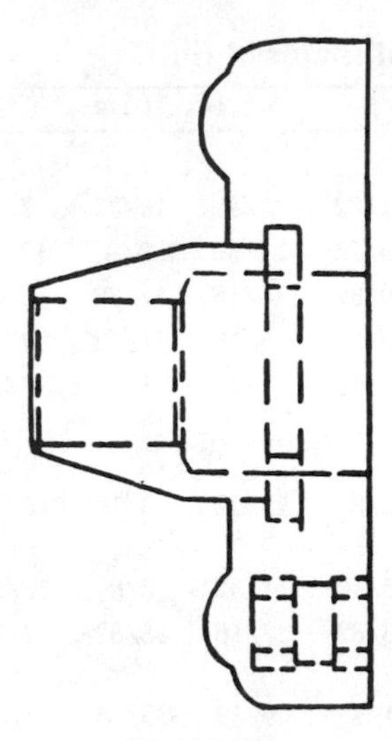

Fig. 25-17.

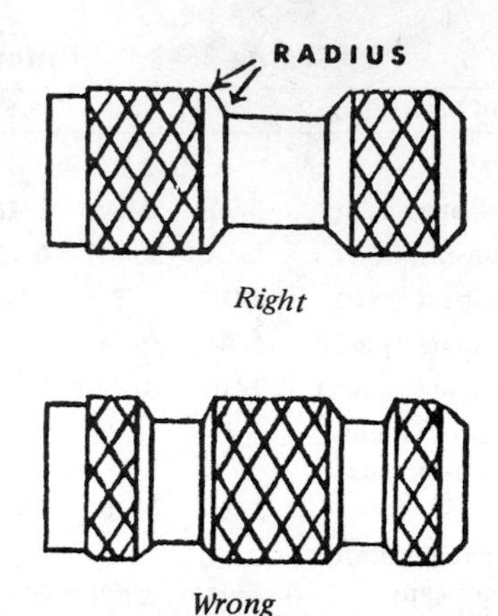

Fig. 25-19. (*Courtesy Heli-Coil Products*)

set up certain stresses in the plastic that often result in cracking. In practically all instances, round stock is recommended. Diamond knurling minimizes possible cracking around the insert. Diagonal knurling provides the best combination of torque and tensile strength.

Knurling of inserts is best accomplished in screw machines with end-knurling tools. The stock sizes given in Table 25-1 are ample to allow end knurling and to leave sufficient stock for a proper sealing diameter free of knurling at the open end of the insert. Cold-forged inserts are knurled on reciprocating or rotary rollers.

Grooves can be used in conjunction with diamond knurling (Fig. 25-18) and also with diagonal or straight knurling. Sharp corners must be avoided in machining the grooves. When grooves are used, one wide groove should be provided in the center of the insert rather than two grooves, one on each end. The center groove allows the material to shrink or creep toward the center and minimizes strain within the piece, and thus possible cracking. Correct and incorrect designs are illustrated in Fig. 25-19.

See below (in this section) for a discussion of the anchorage of special inserts.

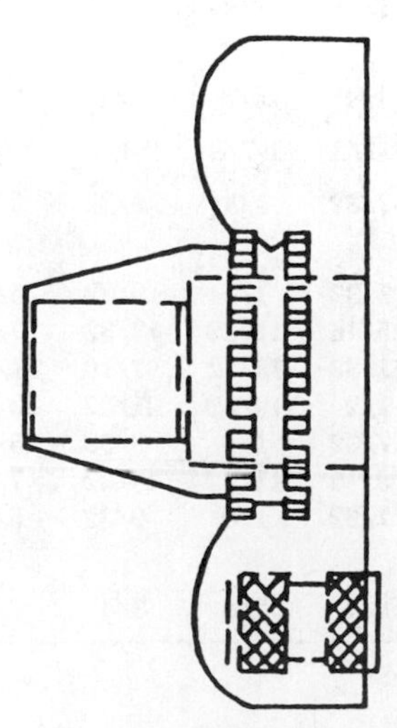

Fig. 25-18.

Insert Testing

There has been considerable comment on and evaluation of various testing procedures for inserts. They have not yet been standardized in the industry, but several companies have suggested the following terminology and procedures:

- *Tensile strength:* Axial force (in pounds or Newtons) required to pull the insert out of the material at least 0.020 inch.
- *Jack-out torque:* Rotational force (in inch-pounds or Newton-meters) applied to a mating screw that pulls the insert out of the material through a washer with ade-

quate clearance for the insert outside diameter.

- *Insert rotation torque:* Rotational force required to turn the insert in the parent material. It is a good comparative measure of overall strength of the assembly.
- *Clamping torque:* Rotational force applied to a mating screw when the insert is allowed to contact a nonrotating plate, and the insert diameter is larger than the clearance hole in the plate. This simulates the usual application condition. It is important to know what material the plate will be made of for accurate values.

Sketches of the various procedures are indicated in Fig. 25-20.

Problems in Molding with the Usual Types of Inserts

Most of the following data relate to the use of inserts in thermoset molding. Portions, however, are applicable to thermoplastics.

Floating of Inserts. Floating of inserts can be controlled or prevented by several methods:

1. The retaining pins may be tapered slightly, by starting the taper at the fillet and carrying it up to one-third of the length of the pin. If too much taper is allowed, making the insert too tight on the retaining pin, the insert may pull out of the material.
2. It has been found that a straight knurl on the retaining pin provides a sufficient holding surface.
3. Square retaining pins can be used.
4. Split pins are practical for blind-hole inserts.
5. Spring tension pins can be used, in which the retaining pin is slotted, and music wire is inserted into the slot. This method presents difficulties because the slightest flow of material into the slot prevents the spring from functioning properly.
6. An extended shoulder can be provided on the insert, shown in Fig. 25-14 as J (sealing), and this should allowed to enter into the mold proper. This is an ideal method of preventing the insert from floating although it is not permissible when inserts must be flush with the surface of the material.
7. On male inserts, a tapered hole can be provided for a drive fit if close accuracy of inserts is maintained. In such a case, a taper of 0.0005 to 0.001 inch for the depth of the hole is sufficient. If the insert is long enough, a small side hole can be drilled in the pin, and music wire inserted to provide spring action. This spring action prevents the insert from floating in practically all method of molding. The same method can be used for holding inserts in the top half of the mold.
8. When precise location of the insert is essential, removable threaded pins are provided in the mold. Inserts are screwed to these pins. However, this procedure increases the cost of production. Subsequent removal of the flash from the thread is avoided in most cases.
9. In some instances, location of a male or a female insert may be effected in the top half of the mold by use of a ball-and-socket arrangement. The necessary ball,

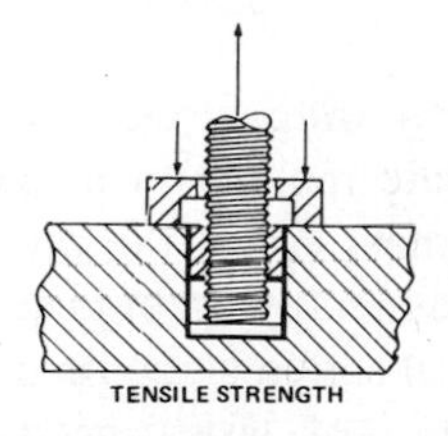

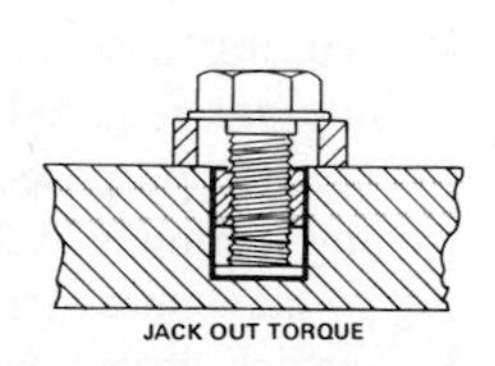

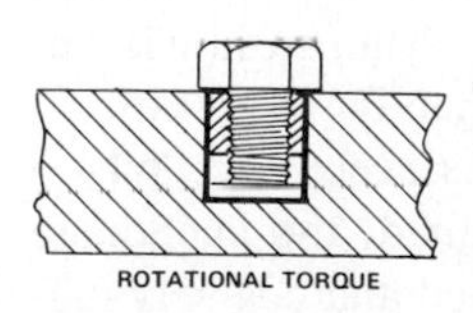

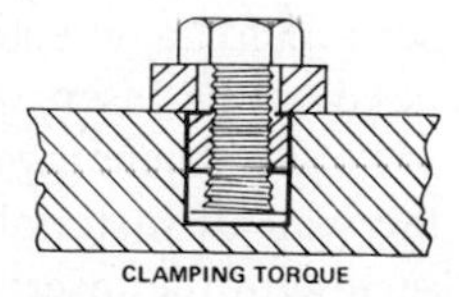

Fig. 25-20. Some suggested procedures for insert testing.

spring, and plug for this arrangement are commercially available. The insert normally is pushed into the hole, engaging the ball, and the side pressure exerted by the ball will hold the insert in place during molding.

10. Male inserts sometimes can be held in the upper half of the mold by use of a magnetized section in the mold. Of course, the size and the type of material used for the insert determine the practicability of this method.

Crushing of Inserts. In transfer molding, there are very few difficulties with crushing of inserts if close tolerances on the length are maintained. In compression molding, however, when the insert must show on both sides and is molded vertically or in line with the press motion, crushing of inserts can be prevented by the use of preforms with holes that allow the preform to slip over the insert. Sliding pins are provided in the force plug, operated by spring, air, or hydraulic action. These pins are in a down position when the mold is being closed, and they contact the surface of the inserts before the flow of material takes place. Because the pins are under constant pressure, no material can enter the insert. This method can be applied to blind- or open-hole inserts, and either top or bottom pins. Considerable pressure can be applied on the inserts. Actual tests on a brass insert $\frac{1}{2}$ inch long, 6 × 32 thread, with $\frac{1}{16}$ inch wall, show that the insert withstands six cycles of 500 pounds total pressure with a reduction of 0.0005 inch in length. When the inserts are not of the through type, solid preforms may be used. Preheating of the material is recommended.

Flow of Material into an Open-Hole Through-Type Insert. In transfer molding, there is very little difficulty with material flow if the length of the insert is maintained from 0.002 to 0.004 inch oversize. When the mold is closed, the insert is pinched in the mold, and it is impossible for the plastic to enter. In compression molding, however, it is impossible to prevent material from entering the hole unless pressure-type pins are used, as above, to prevent crushing of inserts. If pressure pins cannot be used, it is advisable, especially on larger inserts, to tap the inserts undersize before molding, and then to retap to proper size after molding. Extreme care should be taken in retapping to prevent stripping of threads, especially if considerable material has flowed into the thread. Small inserts are most economically molded with a drilled hole and tapped after molding.

Flow of Material into a Blind-Hole Insert. Difficulties with the flow of material into blind-hole inserts are not as numerous as with the open-hole types. In most cases such flow is caused by loose retaining pins that allow the insert to float with the flow of material, uneven machining on the face of the insert, or knurling on the entire outside diameter of the insert, leaving extended burrs on the face that do not permit the insert to rest flat on the surface of the mold or the surface of the retaining pin. In all cases, it is good practice to provide a slight recess in the mold, accommodating the outside diameter of the insert. When the J diameter (Fig. 25-14) of the retaining pin is the same as that of the insert, and sharp corners can be retained in the hole, a 0.005-inch depth is sufficient to prevent the plastic from flowing in. This method allows the insert to protrude above the surface of the molded article, and this is desirable, especially when electrical contacts are to be made.

Protruding Inserts. Protruding inserts frequently are required and are molded in place for specific purposes. In most cases, the protruding section is used for assembly or for bearing points where mechanical action is required. In special cases, especially of large inserts where the molded article is subjected to considerable torque in order to obtain a tight connection, it is advisable to allow a hexagonal section of the insert to protrude above the molded surface for a wrench grip. Thus strain is applied on the insert rather than on the plastic.

Perfect anchorage also is necessary. Where the wall of material is limited, the anchorage section of the insert is turned and coarsely diamond-knurled. A groove can be added to in-

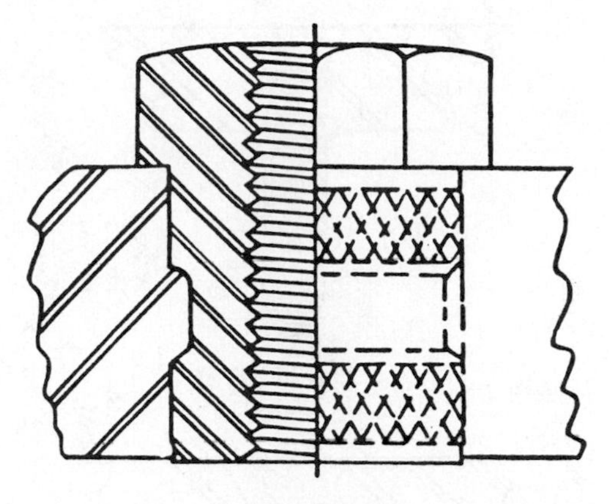

Fig. 25-21.

crease anchorage for tension. However, sharp corners must be avoided. In the event that the hexagonal shape is used for anchorage, sharp corners must be reduced by turning, and grooves provided for tension anchorage. Figure 25-21 shows a recommended design.

Anchorage of Special Inserts

Most of the following data relate to molded thermosets; in some cases, they are also applicable to molded thermoplastics.

Thin Tubular Inserts. These inserts are extremely difficult to anchor properly. If a tubular insert is molded partway up a molded article, it is possible to invert a bead that will act as a satisfactory anchorage. The bead can be used on outside or inside inserts as shown in Figs. 25-22 and 25-23, respectively. A perforated surface around the circumference also can be used where permissible. In molding an outside tubular insert, it often is necessary to coat the

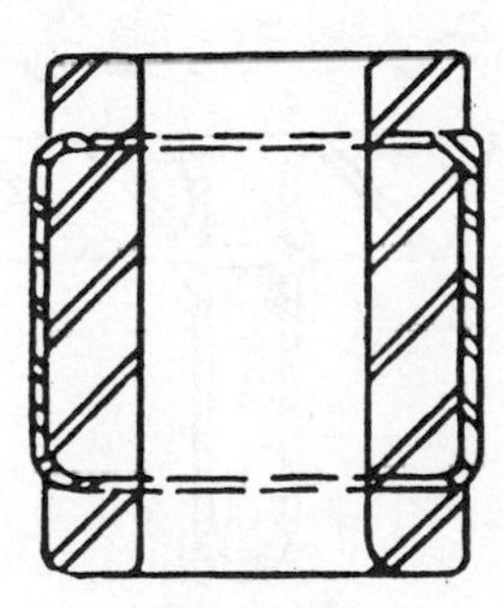

Fig. 25-22.

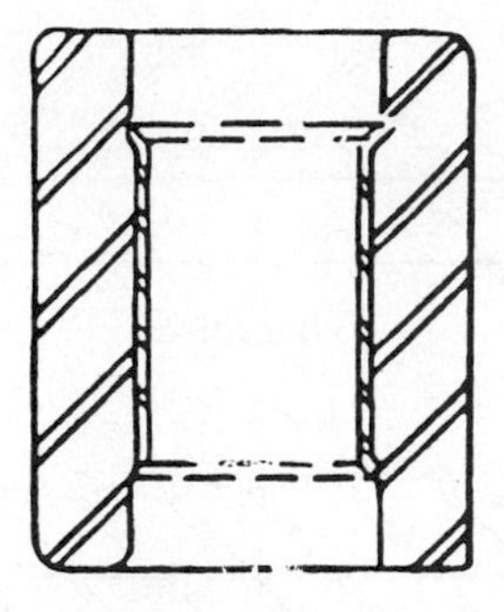

Fig. 25-23.

inside of the insert with neoprene or vinyl to improve the bonding.

Flat Plate-Type Inserts. These inserts can be anchored by means of countersunk holes wherever permissible. It is necessary to bevel all edges of the insert, or, if certain sections of the insert are not required for the functioning of the article, the section can be partially cut out and bent over to provide anchorage. This method is illustrated in Fig. 25-24. If metal inserts must be thick, bosses can be extruded and slightly flared to provide satisfactory anchorage. Anchorage may be obtained also by spot-welding lugs to the underside of the insert.

Drawn Shell-Type Inserts. Where a minimum wall thickness of plastic is specified, and

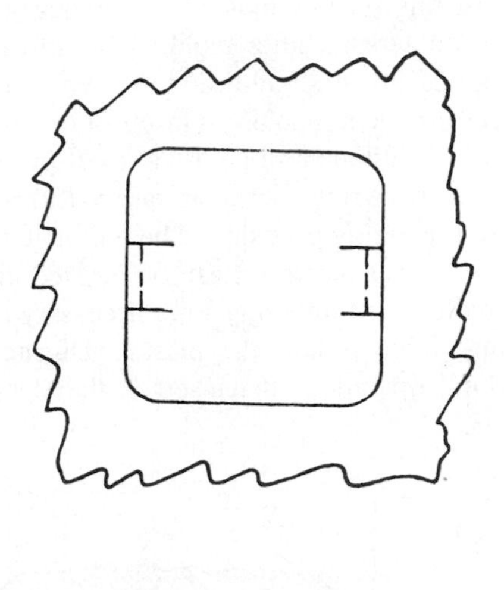

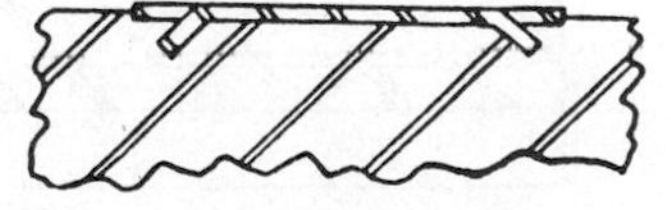

Fig. 25-24.

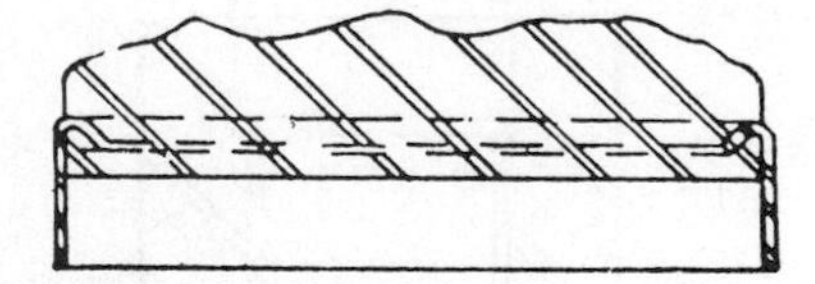

Fig. 25-25.

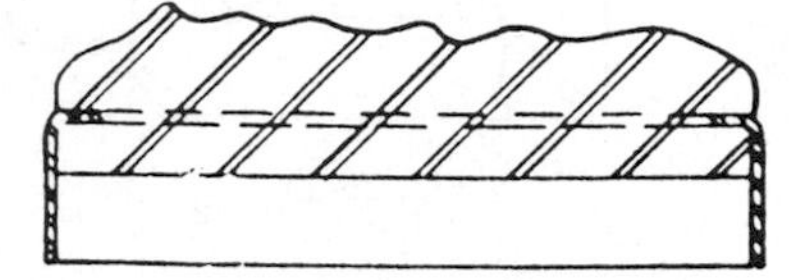

Fig. 25-26.

an insert of this type is used, extreme caution must be exercised to provide proper anchorage. Figure 25-25 shows an unsatisfactory anchorage that provides insufficient wall thickness of plastic to avoid cracking. Figure 25-26 shows an insert that is fairly well designed and could be used to good advantage. In an insert of this type, the plastic has a chance to slide over the insert. However, to provide the best possible anchorage, the insert should be flared in slightly, as shown in Fig. 25-27. With this design, the plastic actually has a chance to anchor the insert and to creep while shrinking.

Drawn Pin-Type Inserts. Very often an insert of this type is molded into a plastic and then countersunk after molding, as illustrated in Fig. 25-28. A slight bead provided as an undercut for anchorage on an insert of this type is entirely insufficient to hold the insert properly. Wherever possible, when an insert of this type is used, piercing pins should be provided in the mold, so that the insert can be pierced during the molding operation and the necessary countersink molded into the plastic. During this piercing operation, the insert is flared out to

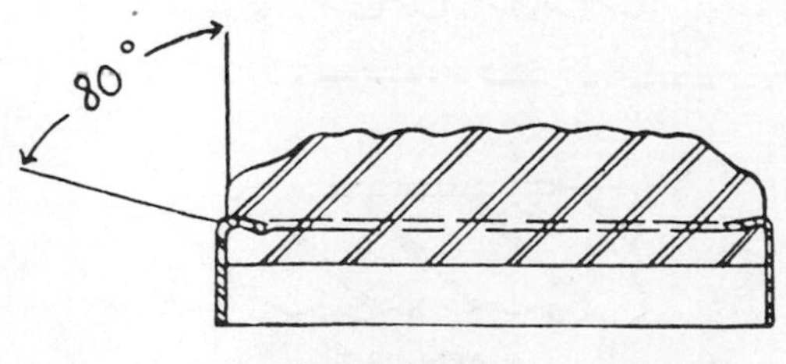

Fig. 25-27.

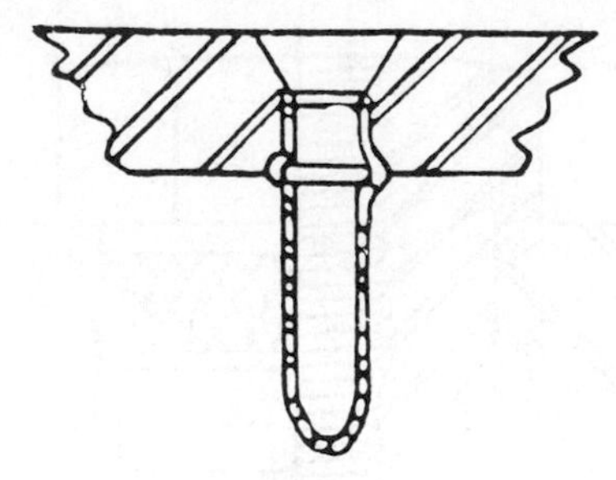

Fig. 25-28.

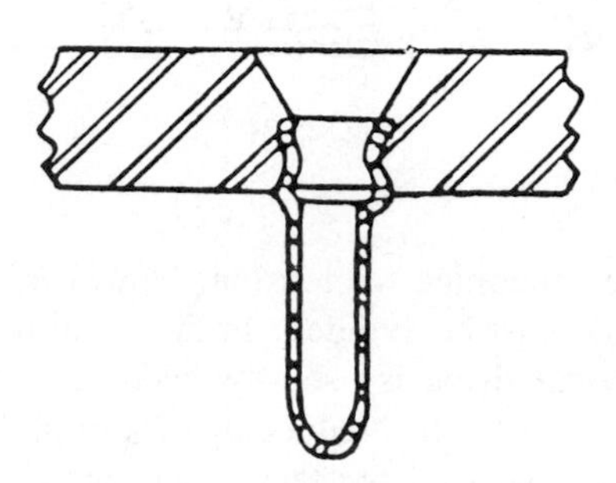

Fig. 25-29.

provide proper anchorage, as shown in Fig. 25-29.

Figure 25-30 shows a drawn-type pin with an open end. Partial anchorage is obtained by shearing and folding two segments during the molding operation. Floating pressure-type piercing pins in the mold are recommended to minimize the flow of plastic into the insert.

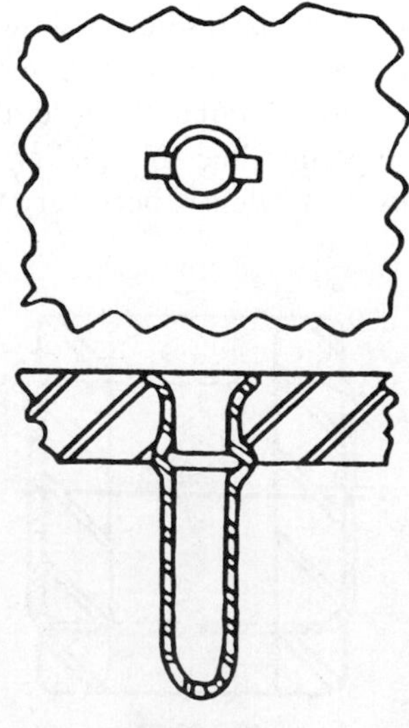

Fig. 25-30.

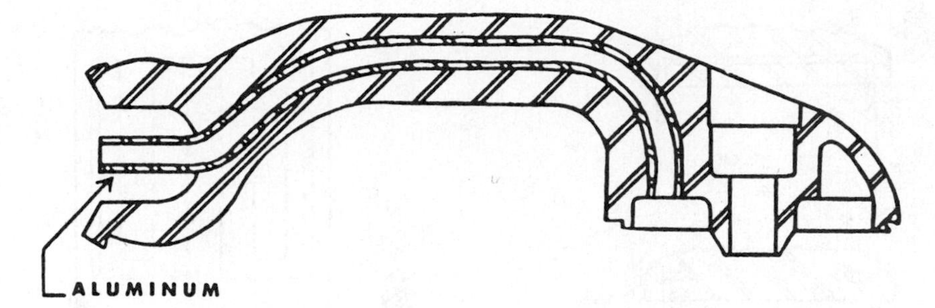

Fig. 25-31.

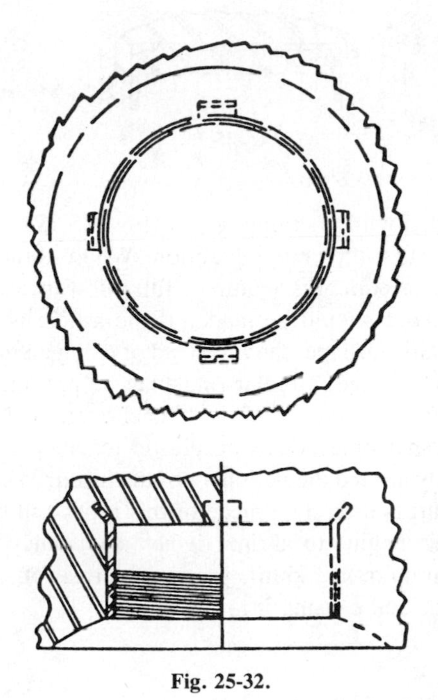

Fig. 25-32.

Drawn-Shell Threaded Inserts. As illustrated in Fig. 25-31, these inserts are often used in large molded articles where it is not necessary to have 75% of thread, or where insert space is limited. Because the shell is usually thin, approximately 50% of the depth of thread is obtained. The four flared lugs provide a satisfactory anchorage in every respect. It is impossible to provide sealing points on an insert of this type; hence flow of material into the thread must be expected. Tapping after molding is recommended for most satisfactory results.

Intricate Inserts. An intricate insert is shown in Fig. 25-32. Considerable difficulty with cracking of the plastic was encountered until aluminum inserts were selected. Actually there were two factors in favor of aluminum—its coefficient of expansion and its ability to yield or spring slightly when the plastic was shrinking.

Large-Surface Inserts. It is often necessary to mold one or more large-surface inserts on one side of the plastic, as illustrated in Fig. 25-33. Inserts of this type cause nonuniform shrinkage of plastic and considerable warpage. Even when shrinkage or cooling fixtures are used, it is certain that surface A will be convex and B concave after the piece is allowed to cool and age. If a flat surface is required, the surface must be machined. Best results will be obtained when the articles are allowed to age or, if possible, are baked in an oven for at least 72 hours, at suitable temperatures, before being machined.

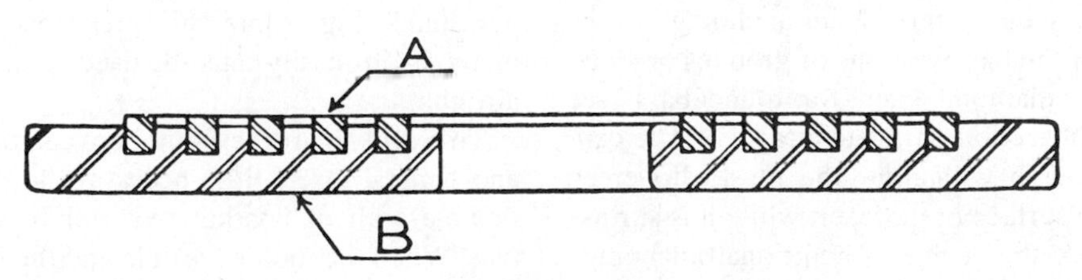

Fig. 25-33.

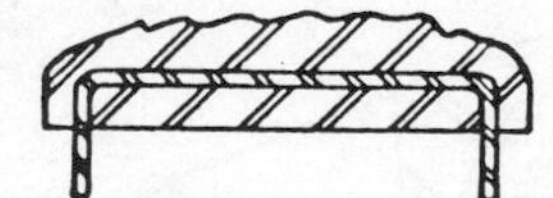

Fig. 25-34.

Fig. 25-35.

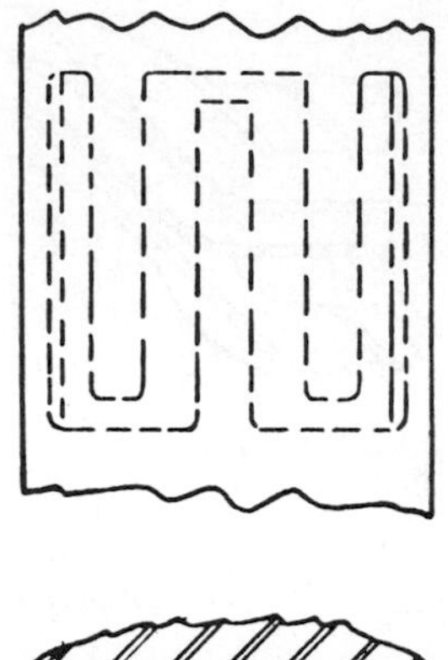

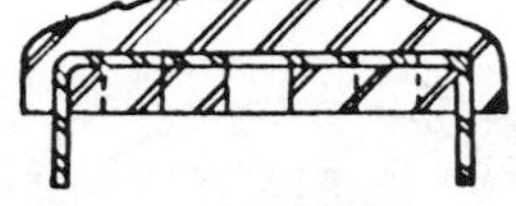

Fig. 25-36.

Large Inserts with a Minimum of Wall Thickness of Material. Where a minimum thickness of a thermosetting plastic is allowed around a large insert, a special noncracking type will generally have to be used. Extreme care must be taken in the design of the insert to avoid sharp corners or other features that might create local stresses.

Irregular-Shaped Inserts. These inserts cause the greatest difficulty. Figure 25-34 shows a U-shaped insert approximately $1\frac{1}{2}$ inches long, on which two rib projections are required. From the standpoint of economy in forming this insert and loading it into the mold, it can be made in one piece, but it will cause difficulty, with cracking of the plastic. It would be more economical in the long run to make two separate inserts, as shown in Fig. 25-35. If electrical contact is required, a wire can be fastened between the two inserts; or, if a more solid connection is desired, the insert can be made solid with cutout slots, as shown in Fig. 25-36. Provision must be made in one half of the mold to prevent these slots from being filled with plastic. When these slots are open, there will be a slight give in the insert when the plastic shrinks. This will consequently reduce or eliminate the possibility of cracking the plastic.

When a long bar-type insert is used, it is always advisable to provide an anchorage in the middle of the bar by means of grooves or slots, or coarse diamond knurl for round bars (see above, subsection on "Anchorage"). The center anchorage will allow the plastic to creep along the surface of the insert while it is shrinking toward the center. If additional anchorage is desired on round bars, the ends can be knurled with a straight knurl (Fig. 25-37) and still retain the creeping action. Where dimensional accuracy is required, full allowance for shrinkage should be made. If the article is of cylindrical shape, then, instead of using a knurl for anchorage, circular rings can be provided, which will give satisfactory anchorage and at the same time allow the plastic to creep uniformly around the periphery of the insert. When a knurl is used on a piece of this type, and the plastic begins to shrink, it has a tendency to climb up on the knurl, producing stress on the plastic and causing it to crack.

Leakproof Inserts

Because of the difference in the behavior between plastics and metals (e.g., difference in coefficient of thermal expansion), the characteristics of some plastics, and the problems of providing proper adequate anchorage, it is usually impossible to make an insert remain airtight within the plastic even under small pressures. If inserts are used in articles that must withstand high internal pressures, special methods ordinarily must be used to make them airtight.

To retain an airtight joint between the plastic and a metal insert, it is necessary to provide a flexible wall of another material between the two. When the molded article and the insert expand and contract, this flexible material, al-

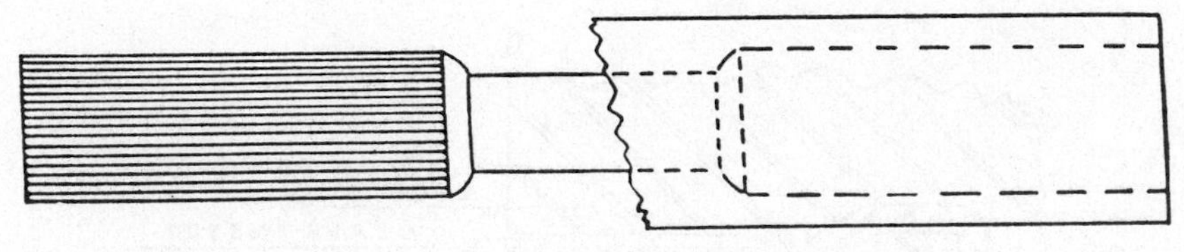

Fig. 25-37.

though it consists of only a very thin coating, will compensate for the difference in coefficient of thermal expansion between metal and plastic.

A few successful methods are recommended. The insert is knurled the same as for normal anchorage and is provided with at least two grooves, about $\frac{1}{32}$ inch wide and 0.020 inch deep. The head or anchorage part is dipped in neoprene, polyvinyl chloride-acetate, or other rubbery synthetic material, and then oven-dried before use. This process will supply sufficient coating on the insert to give it the necessary cushioning action.

It is possible also, especially on round inserts, to provide a groove in the anchorage head of the insert large enough that a neoprene washer can be used. Under normal molding conditions, the washer will produce satisfactory results. On some applications, a retaining groove is molded or machined between the insert and the plastic. The groove is filled with alkyd resin and allowed to dry at room temperature, or is oven-baked.

Special Inserts for Reinforcement

It is often necessary to mold inserts into plastics as reinforcements to provide greater strength, greater rigidity, greater safety (as in automobile steering wheels), or greater dimensional accuracy.

In molding a thermoplastic housing, for example, instead of molding a thick wall to obtain rigidity, a sheet-metal reinforcement can be molded on the inside of the housing. This will not only produce greater rigidity with a minimum of wall thickness, but it will also assist in maintaining better dimensional accuracy. Various materials can be used as reinforcement—molding board, laminated phenolics, perforated metal, or metal screens.

Nonmetallic Inserts

Inserts of various nonmetallic materials are used successfully. The use of wooden inserts in applications such as doorknobs or automobile gearshift knobs saves considerable material and shortens the molding cycle.

Glass inserts are being successfully molded into thermoplastics by injection and into thermosetting materials by transfer. Difficulties can be reduced during the initial engineering of the mold design. The worst problems are caused by the nonuniformity of contours and dimensions. Glass, being of a brittle nature, does not lend itself to the application of full clamping pressures during molding. It is necessary to provide a cushion, by means of springs or rubber, to compensate for the normal irregularities in dimensions of glass inserts. In some cases, paper is glued to the surface of glass to provide additional cushion, as well as to protect the surface from scratching during handling.

Locating the insert in the mold is difficult. Figure 25-38 illustrates the use of a sleeve-type ejector. The inside diameter of the sleeve is the same as the outside diameter of the glass insert. The insert is located by placing it inside the ejector sleeve when it is protruding in ejected position. Figure 25-39 illustrates a step-molded article. The diameter of one of the steps must be the same as the outside diameter of the insert. A sleeve-type ejector is used on this step, to locate the insert, as in the preceding case.

Preparation of Inserts Before Molding

Cleaning of Inserts. Considerable significance should be attached to the cleaning or washing of inserts prior to molding, especially screw-machine inserts. If inserts are improperly washed, even though they appear clean, there may be loose metal chips hanging on the threads, or fine metal dust in the knurls. The

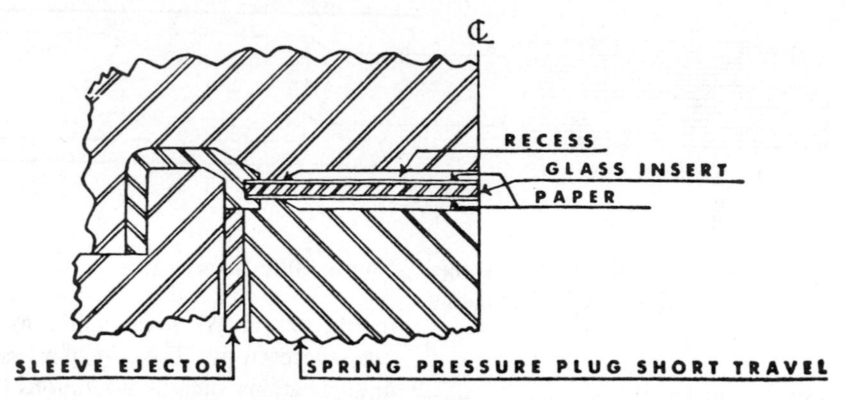

Fig. 25-38.

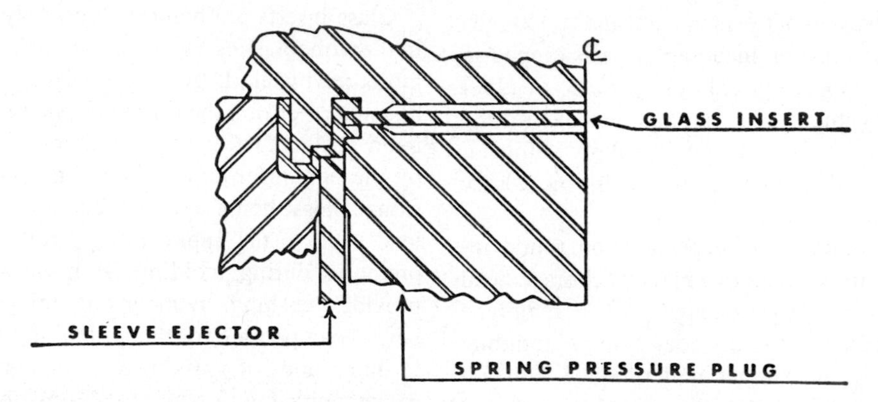

Fig. 25-39.

latter is often rolled into the surface by the process of knurling, and is not easily washed off, but it will be loosened by the flow of the plastic. The metal chips may flow up to the surface and impair the appearance of the molded article. The most serious difficulty, however, is in electrical applications, where a small particle or a slight amount of metal dust may cause a total breakdown electrically. Grease and oil also are detrimental to the appearance of molded articles and should be thoroughly washed off.

Processes of cleaning are divided into three types:

1. Mechanical, including hand polishing, tumbling, shot- or sandblasting, or washing with solvent or alkali.
2. Chemical, such as removal of iron rust and silver tarnish by an acid bath.
3. Those using electrolytic cleaners.

Oil and machining chips can best be removed by a well-stirred alkali bath followed by a rinse with hot water, except where the nature of the metal, such as aluminum, rules out the use of alkali in favor of degreasing with a solvent.

In many cases a reasonable amount of tarnish can do no harm, but where the function or the appearance of the piece demands chemically clean inserts, an acid dip is necessary. For brass and bronze, a mixture of nitric acid and sulfuric acids or nitric alone is commonly used.

Silver tarnish can be removed with nitric acid or a diluted solution of one of the cyanides. Trisodium phosphate has been found to be an efficient remover of iron rust.

Preheating of Inserts. Large inserts should be preheated (above the mold temperature if possible) prior to molding. This will permit maximum expansion and improve the flow and cure of the plastic. With thermoplastic materials, preheating of inserts will reduce the likelihood of weld marks, which often result in cracking of the plastic after molding.

Cleaning Flash from Inserts

Most of the difficulty with flash can be avoided in the design of the article and the insert by providing sealing points so that the flow of plastic is cut off or at least minimized. However, even with the best design there will be some material on the inserts, especially when the mold becomes worn, or close tolerance on inserts is not maintained. Several methods are recommended to minimize this, particularly lubricating the insert, prior to molding, with wax, soap, grease, or oil. For thermoplastics it is recommended that a mineral oil of viscosity SAE 100 or greater be used. Plating and polishing the inserts minimize the adherence of flash.

To remove the flash, cut it close to the molded article and peel it off. In most cases, a mold solution of caustic soda will loosen the flash so that it can be easily removed. This method, however, requires extreme caution because too lengthy contact or too strong a solution will harm the surface of the article, and may even loosen the insert in its anchorage.

Salvage of Inserts

When the inserts are of the through type, they can be knocked out by means of a foot press and fixture. When they are anchored partway in the material, a strong solution of caustic soda will loosen the inserts in thermosetting material so that they can be picked out. For thermoplastic material it is necessary to use suitable solvents for the plastic, or to soften the articles in an oven and pull out the inserts. Reclaimed inserts should be inspected before reuse.

Relieving Molding Stresses Around Inserts

Considerable stresses are set up in molded articles or irregular design, such as those having both thin and thick sections, and especially those with metal inserts. The best method to relieve stresses is to allow the article to cool slowly. The ideal condition would be to carry the articles on a conveyor through an oven that has various stages of temperatures, starting at 50°F below the molding temperature, then gradually decreasing until the article is cooled to room temperature. This method, however, requires special equipment.

The next best method requires two ovens, one at approximately 225°F and one at 150°F. The molded article remains in each oven successively until its temperature is reduced to oven temperature.

The final step is cooling to room temperature. In case of thermoplastic materials, molding stresses are relieved by annealing in an oven or a water bath at suitable recommended temperatures.

26
Decorating Plastics

INTRODUCTION

The surface of a plastic article, film, or sheet may be coated, marked, mechanically finished, or otherwise altered for appearance or performance reasons. A large variety of coatings and inks are used, as well as many different techniques for their application. The choice of the most suitable system can be quite difficult, because of the multitude of factors involved: appearance, durability, cost of the materials, and investment in the application equipment. The product geometry might eliminate some application methods; the type of plastic and its surface condition might require pretreatment and a choice of coatings or inks that adhere well to the required substrate. The decorating process might be performed in line with other processes, and the rate of decorating should not slow down the operation. A plastic part may be of irregular shape, as many injection-molded parts are, or it can be processed in the form of a film or sheet. The choice of decorating method might be determined by the shape of the surface.

Decorating methods may be subdivided into several basic classifications:

1. *Painting*, which involves application of a liquid coating onto the plastic surface. The coating is dried or otherwise solidified after it has been applied to the surface in the required manner and at the required thickness.
2. *Printing*, which is the application of a liquid coating over a relatively small surface area. It is identical to the painting process except that different application techniques are used.
3. *Vacuum coating techniques*, where a coating, usually metallic, is applied as a vapor or as small particles to the surface.
4. *Plating*, which involves deposition of metal by either electrolytic or electroless processes.
5. *Dyeing*, where dye molecules penetrate below the surface.
6. *Labels and decals*, which involve affixing of a predecorated material to the plastic surface.
7. *Mechanical surface changes*, such as polishing, grinding, and embossing.

By D. Satas, Satas & Associates, Warwick, RI.

SURFACE APPEARANCE MEASUREMENT

Surface appearance is mainly perceived visually, and it is subdivided into two main categories: color and light distributive properties. The color depends on selective light absorption and can be described by its tristimulus values or simply by comparison to a collection of samples, such as the well-known Munsell system, which classifies color in terms of hue, lightness, and saturation. Color principles have been described in numerous books.[1]

Light distributive properties can be described optically, but they are much more difficult to relate to human perception than color. The degree of surface gloss as perceived by the human eye is a complex phenomenon, and the gloss level, surface uniformity, sheen, and other associated properties are very important in achieving the desired appearance.[2]

COATABILITY AND COATING ADHESION

Good adhesion between the coating or printing ink and the plastic substrate is required in order to obtain a lasting product. Coating adhesion is often tested by the tape test, in which an adhesive tape is applied over the coating and peeled away at a sharp angle. Sometimes the coating area is cross-hatched; poorly adhered coatings are removed by the tape. This test is a useful and simple qualitative method to eliminate poorly adhered coatings. Several quantitative methods are also available. They involve adhering a probe to the coating surface and then measuring the force required to pull away the probe, or employing a probe to scratch the surface of the coating.[3]

In order to obtain a good coating adhesion between the coating and the substrate, the coating must wet the substrate. Wettability is determined by the difference between the surface tension of the liquid coating and the surface energy of the solid substrate. The coating viscosity must be low enough to allow the coating to flow. Usually the coating is applied by some mechanical means, and thus is spread over the surface by the forces acting during coating. If the surface energies are not favorable for wetting, the coating will recede from the substrate. The surface tension of the liquid coating must be lower than the critical surface tension of the solid substrate. These concepts have been discussed in detail in many publications.[4]

The critical surface tension of the solid surface can be estimated by employing liquids of varying surface tension and observing which liquid just barely wets the surface (ASTM Standard D2578-67). Favorable wetting conditions can be achieved by either decreasing the surface tension of the liquid coating or increasing the critical surface tension of the plastic substrate. The addition of surface-active agents to the liquid coating lowers its surface tension. Water-soluble surface-active agents are added to aqueous coatings and solvent-soluble agents to solventborne coatings. There are many surface treatment and cleaning processes aimed at increasing the surface tension of the plastic substrate or removing surface contamination. Surface contaminants will interfere with coating adhesion by preventing the coating from contacting the plastic substrate. Contaminants that are soluble in the coating may be tolerated.

The liquid coating applied over the substrate is not perfectly smooth but may have various ridges and surface irregularities caused by the mechanical forces to which the coating was subjected during its application. A well-formulated coating will level immediately after its application, before it solidifies. In these applications, surface tension is the force causing leveling, and viscosity is the property that resists the leveling. Low viscosity coatings level much more rapidly and easily than high viscosity ones. The details of leveling have been discussed in several papers and summarized in books.[4]

Various surface blemishes, bubbles, and other disturbances may be introduced during the drying or curing process and require a proper adjustment of conditions in order to avoid the formation of such surface blemishes.

SURFACE TREATMENT

A plastic surface might require cleaning if it has become contaminated during processing. The removal of release agents used in molds might be necessary. Some plastics, especially polyolefins, may contain low molecular weight materials that have a tendency to accumulate at the surfaces and thus interfere with coating adhesion. Some other incompatible additives may migrate to the surface and require removal.

Some plastics have a critical surface tension that is too low for paint wettability and adhesion, and such surfaces must be treated in order to raise their critical surface tension. Table 26-1 lists the critical surface tensions of some plastic materials.

Surface cleaning-treatment may be carried out by various means as listed below:

- Water washing

Table 26-1. Critical surface tension of various plastic surfaces.

Material	*Critical Surface Tension (dynes/cm at 20°C)*
Poly(1,1-dihydrofluorooctyl methacrylate)	10.6
Polyhexafluoropropylene	16.2–17.1
Poly(tetrafluoroethylene-*co*-hexafluoropropylene)	17.8–19
Polytetrafluoroethylene	18.5
Poly(tetrafluoroethylene-*co*-chlorotrifluoroethylene)	20–24
Polytrifluoroethylene	22
Silicone rubber	22
Polydimethyl siloxane	24
Polyvinylidene fluoride	25
Poly(tetrafluoroethylene-*co*-ethylene)	26–27
Butyl rubber	27
Polyvinyl fluoride	28
Polyvinyl butyral	28
Polybutyl acrylate	28
Polyvinyl methyl ether	29
Polyurethane	29
Polyisoprene	30–31
Polyethylene	31
Polychlorotrifluoroethylene	31
Polypropylene	31
Polystyrene	33–35
Polyethyl methacrylate	33
Polyvinyl chloride	33–38
Polymethyl methacrylate	33–34
Polyacrylamide	35–40
Cellulose	35.5–41.5
Rubber hydrochloride	36
Polyvinyl ethyl ether	36
Polyvinyl alcohol	37
Polyvinyl acetate	37
Polyisoprene, chlorinated	37
Chlorosulfonated polyethylene	37
Nylon 6	38–42
Chloroprene	38
Polyvinyl chloride, rigid	39
Cellulose acetate	39
Polyvinylidene chloride	40
Polymethyl acrylate	40
Polyoxyphenylene	41
Polysulfone	41
Nylon 6,6	42.5–46
Polyethylene terephthalate	43
Casein	43
Polyacrylonitrile	44
Cellulose, regenerated	44–45
Phenol-resorcinol adhesive	52
Polyamide-epichlorohydrin resin	52
Urea-formaldehyde resin	61

- Solvent cleaning and etching
- Mechanical abrasion
- Chemical etching
- Priming
- Flame treatment
- Corona discharge
- Plasma treatment
- Ultraviolet and other irradiation

Surface cleaning, flame, and corona discharge are the most commonly used methods of surface treatment.

Plastic surfaces are easily electrically charged, and such static charges attract dust and dirt. The elimination of such surface charges is required in order to minimize the contamination of plastic surfaces. Ionized air blowers are used to remove the surface charges and to clean accumulated dust from the surface. Charge development is minimized at high humidity conditions, and plastics cleaning and painting areas might be humidified by water spray.

Washing and Cleaning

Water washing may be used to remove surface contaminants prior to coating. Alkaline cleaners are used to remove general contaminants; an aqueous hydrocarbon solvent emulsion may be employed to remove water-insoluble contaminants, such as mold release agents.

Solvent cleaning is carried out by wiping, immersion, spraying, or vapor degreasing in either cold or hot solvents. Wiping is the least effective method, and might lead to the redistribution of contaminants rather than their removal. The immersion process may be enhanced by either mechanical or ultrasonic scrubbing.

Vapor degreasing is an effective process for cleaning plastic surfaces. The object is placed in the vapor zone, and vapor condenses on the cooler plastic surface and runs off with the dissolved impurities. The cleaning process ceases when the surface temperature reaches the solvent's boiling point. The advantage of vapor degreasing is that the solvent that reaches the plastic surface is always clean. Vapor degreasing may be used as the last step, in conjunction with other cleaning methods.

Solvents may swell or etch the plastic surface, thus improving its receptivity to the coating, but they also may cause crazing and other surface distortions.

Mechanical Abrasion

Mechanical abrasion, usually sandblasting, removes surface impurities in addition to increasing the surface area. Both actions help to improve adhesion.

Chemical Etching

Chemical etching is the treatment of plastic surfaces by exposure to solutions of reactive chemical compounds, usually oxidizers. The treatment may cause a chemical change of the plastic surface, and it also may introduce microroughness by removing some material. Surface wettability and bondability are thereby improved. Chemical etching is difficult to carry out and is used for difficult-to-adhere plastics (e.g., fluorocarbon polymers) and in processes where such treatment can be easily integrated (e.g., electroplating). Sulfuric and chromic acid treatments are used for many different plastics. Fluorocarbons may be effectively treated by sodium naphthalene solution. Sodium removes fluorine from the surface, replacing it with carboxyl or carbonyl groups.

Priming

Priming is a deposition of a thin coating of another polymer between the plastic surface and the paint, to improve the paint's adhesion.

Flame Treatment

Flame treatment is the most often used process to improve the bondability of molded plastic articles. It is especially useful in treating irregularly shaped parts (see Fig. 26-1), although flame treatment of polymeric films sometimes is also used. An oxidizing flame at a temperature of 1090 to 2760°C contacts the plastics surface for a period of less than a second, causing its oxidation among other surface changes.

Corona Discharge

This process is used most often for the continuous surface treatment of plastic films. The corona discharge system consists of a generator, a transformer, and a treater. The generator raises the frequency of the 60 Hz alternating current to a much higher level—25 to 30 kHz is a commonly used frequency. The transformer steps up the voltage to the 15,000 to 30,000 volts required to produce a corona discharge consisting of many small, uniformly distributed sparks.

The treater is a capacitor with the plastic material to be treated placed between the electrodes. The arrangement of the treater elements varies, depending on the equipment design. A common design employs a dielectric covered roll and a bare electrode (see Fig. 26-2). Another design features a quartz-covered electrode and a bare metallic roll. This arrangement allows the treatment of conductive (metallized) films.

The processes that take place during the corona discharge treatment are not entirely under-

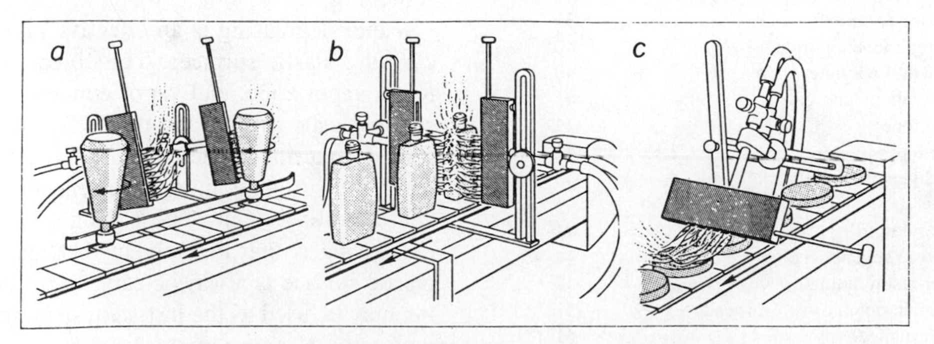

Fig. 26-1. Flame treatment of plastic bottles and lids. (*Courtesy Quantum Chemical Corp.*)

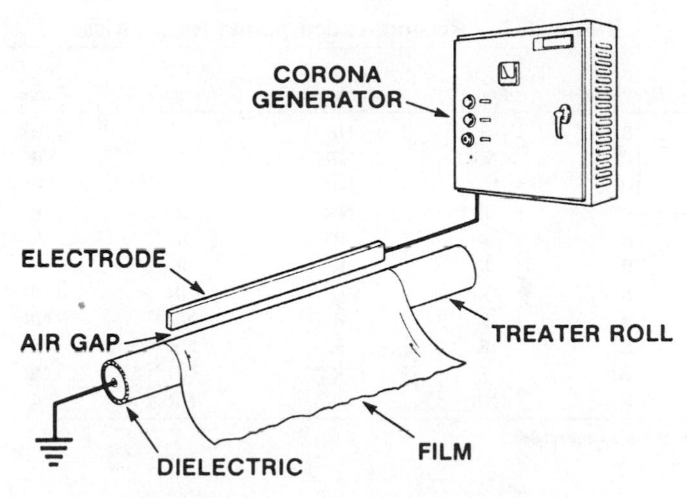

Fig. 26-2. Corona dicharge treating system.

stood. Electrons, protons, excited atoms, and ions in the corona discharge break carbon-to-carbon and carbon-to-hydrogen bonds on the plastic's surface, causing radical formation. The radicals react with oxygen and nitrogen in the air, introducing polar groups to the plastic surface. Other processes also take place during corona discharge treatment.

Plasma Treatment

Cold plasma technology is increasingly used for the surface treatment of plastic parts. Plasma treatment is quite effective, but it is a batch process carried out in an evacuated chamber. Continuous systems are expensive and have been made only for narrow webs. Plasma treatment causes ablation (microetching), surface cleaning, crosslinking, and surface activation by reaction of plasma gases with the plastic surface. Various activated gases may be used.

PAINTS AND COATINGS

Numerous paint systems suitable for plastics are available. Such coatings may be solventborne, waterborne, high solids, or 100% reactive solids, the last of which can be thermally or radiation-cured.

Air pollution laws favor the use of high solids, 100% solids, or waterborne paints.

High solids coatings are compounded by employing liquid polymers that are low molecular weight polymers with either random or end-group (telechelic) functionality. Such polymers are reactive and solidify on the application of heat. Only a small quantity of solvents is required to maintain sufficient fluidity during the coating operation. A large variety of high solids coating compositions have been developed employing liquid, reactive polymers.

The 100% solids coating are similar to high solids coatings, but they do not contain any solvents at all. Such coatings are sufficiently liquid for application and are usually solidified by radiation, although thermally cured coatings also are available. They are reactive multicomponent systems containing oligomers and monomers. Acrylic and modified acrylic systems dominate the composition of these coatings.

Waterborne paints are mainly aqueous emulsions of polymers such as acrylic, polyvinyl acetate, polyvinyl chloride, and their copolymers, which are dried by removal of water and can be postcured by the application of heat.

Solventborne coatings are solutions of polymers in appropriate solvents. Other additives also may be dissolved or dispersed in the coating. Drying takes place by solvent evaporation although some crosslinking-curing may also take place upon baking. If clear, such coatings often are called lacquers.

Coatings are colored by the addition of dis-

Table 26-2. Recommended paints for plastics.

Plastic	*Urethane*	*Epoxy*	*Polyester*	*Acrylic Lacquer*	*Acrylic Enamel*	*Waterborne*
ABS	R	R	NR	R	R	R
Acrylic	NR	NR	NR	R	R	R
PVC	NR	NR	NR	R	R	NR
Styrene	R	R	NR	R	R	R
PPO/PPE	R	R	R	R	R	R
Polycarbonate	R	R	R	R	R	R
Nylon	R	R	R	NR	NR	NR
Polypropylene	R	R	R	NR	NR	NR
Polyethylene	R	R	R	NR	NR	NR
Polyster	R	R	R	NR	NR	NR
RIM	R	NR	NR	NR	R	R

*R = recommended; NR = not recommended.

persed pigments. Usually a high degree of opacity is desired, but sometimes tinted transparent coatings are required. In such cases, dyes or transparent iron oxide pigments are used.

Coatings may be based on different polymers: polyurethanes, epoxies, polyesters, acrylics, modified acrylics, alkyds, and others. Table 26-2 lists some of the paint systems and shows their suitability for various plastics. The choice of the best suitable paint system depends on a number of factors. The resistance of the plastic to solvents may determine the type of solvent used in the solventborne paint. The heat distortion temperature and the heat resistance of the plastic will determine whether the bake-type paints can be used and the maximum baking temperature that can be tolerated.

Coated Products

Plastics are coated for many different applications. The automotive industry is an important consumer of molded plastics, and many of them are painted or metallized. Structural foam is used very widely for electronic equipment housing and similar applications. The surface of such rigid foams often exhibits swirl and other imperfections and requires extensive finishing, including priming and painting.

Transparent plastics are susceptible to surface damage by abrasion and often are coated with hard, either thermally or radiation-cured, coatings.

Coating Techniques

Paint and other coatings are applied to plastic surfaces by many different techniques. The choice of the most suitable technique is determined by: the geometry of the plastic part (irregularly shaped parts require different coating methods from those used for flat films or sheets); the type of coating used, especially its viscosity; the coating thickness required; and a few other parameters. The coating of a plastic surface is by far the most important technique used to change its appearance (e.g., its color and gloss level), as well as its physical properties (e.g., abrasion and scuff resistance, resistance to weathering) and many other functional properties.

Spraying. Spraying is the method most often used to coat plastic parts. Its main advantages are its suitability for coating irregularly shaped parts and the possibility of applying relatively thin coatings. Coating thickness control and coating utilization efficiency can be poor. Spraying may be carried out manually by small hand-held spray guns, or automatically in large installations.

Various coating types are applied by spraying: aqueous, solventborne, and hot melt. The main requirement is that the coating be of sufficiently low viscosity to be sprayable.

Spraying equipment may be subdivided into several categories:

- Air spraying heads
- Airless spraying heads
- Air-assisted airless
- Centrifugal spraying disks and cones
- Electrostatically assisted spraying heads

Air Atomization. A liquid jet will break up into small particles if introduced into a high-velocity airstream. The mechanism of atomization is a two-step process: the formation of ligaments by air friction, followed by the collapse of these ligaments with drop formation. Air atomization is the best method if a fine particle size is required; it yields droplets below 15 μm in diameter, and the smallest droplets may be in the range of 1 to 5 μm.

The construction of an air gun is shown in Fig. 26-3. Liquid is introduced into a high-velocity airstream that is discharged through an annular orifice around the opening for the liquid. A pneumatically operated needle is placed in the fluid outlet, and it closes and opens the fluid outlet and cleans it in the process. Air guns are equipped with a cap containing drilled air passages. The air delivered by the cap shapes the spray cone into a fan, which facilitates overlap and helps to provide uniform coverage. Air guns may be mounted on reciprocating or rotary carriages for automated large spray-coating stations.

Airless Spray. Forcing a liquid through a nozzle may break it into drops if a certain velocity is exceeded, and this principle forms the basis for airless spray guns. Many design variations of such airless nozzles are available. Airless spraying generally produces coarse particle sizes, and low-viscosity fluids are best suited to it. A finer particle size may be obtained by providing a low-velocity airstream to assist the atomization and to guide the spray.

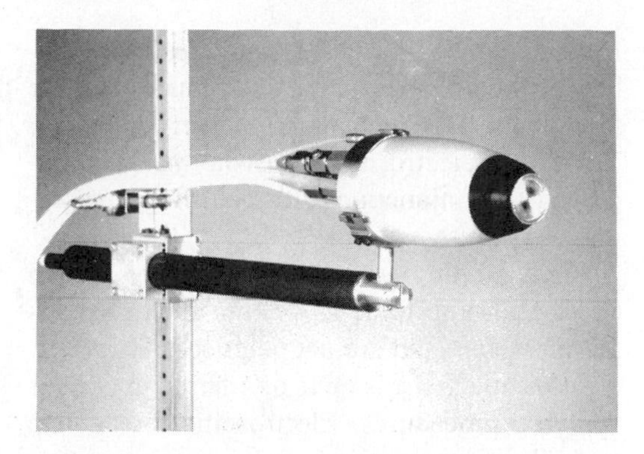

Fig. 26-4. Centrifugal atomizer. (*Courtesy Interrad Corp.*)

Centrifugal Spraying Heads. Rotational disk or bell atomizers cause the breakdown of a liquid stream coming off the edge of the disk or bell by centrifugal force. Bells may be 7.5 to 9 cm and the disks 15 to 50 cm in diameter. The atomizer usually is driven by an air turbine and rotates at a speed of 15,000 to 20,000 rpm. The rotational speed can be decreased if electrostatic assist is used. A typical atomizer is shown in Fig. 26-4. Centrifugal atomizers are the most efficient type in terms of the energy input to achieve the required atomization, whereas air atomizers are the least efficient.

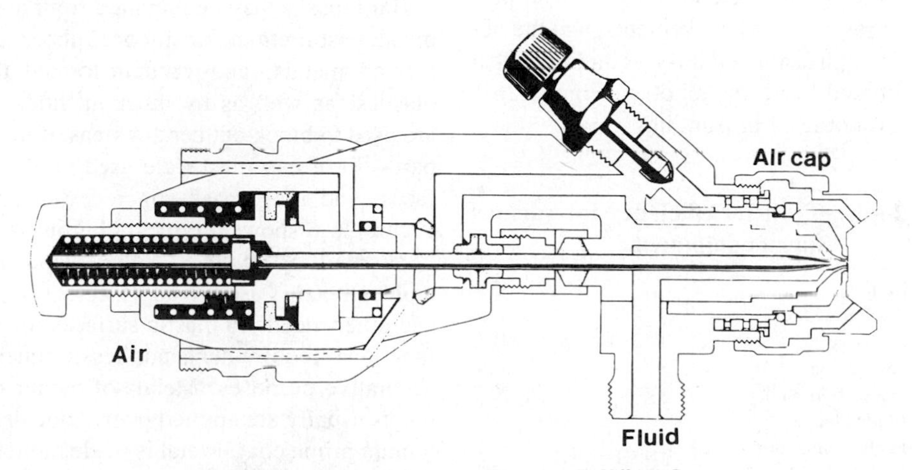

Fig. 26-3. Air atomizing gun. (*Courtesy DeVilbiss Co.*)

Electrically charging the liquid particles enhances their disintegration and reduces particle size, as well as increasing transfer efficiency. Therefore, electrostatic assist normally is employed in conjunction with centrifugal spraying.

Electrostatic Assist. Liquids can be broken by electrostatic force alone, but such methods are inefficient and are not practiced. However, electrostatic assist is quite often used in various spraying procedures. Electrostatic forces help to facilitate atomization, and, more important, the electrostatic assist improves transfer efficiency. The object to be coated usually is kept at ground potential, and the spray particles are charged. This procedure directs the particles to the object and helps to prevent them from going astray. The electrostatic assist method is almost always used with disk or cone sprayers, and is often used with air or airless atomizers.

Transfer Efficiency. This parameter indicates the percentage of the paint that is deposited on the target. The transfer efficiency is lower for small and irregular objects than it is for flat sheets. Also, it decreases with increasing atomization. Table 26-3 shows the transfer efficiencies for some spraying methods.[3]

Auxiliaries. A spraying operation requires various auxiliary equipment. Large-volume spraying requires many spray guns, which may be arranged on reciprocating or rotating carriages. Such spraying arrangements are especially suitable for coating films or sheets. Also, spraying was one of the first manufacturing areas where robots were introduced. The process requires repetitive movements, and the operation is unpleasant and dirty—conditions that are addressed well by robotry. Figure 26-5 shows a robot used in painting.

Table 26-3. Transfer efficiency for various spray equipment.

Air atomization	35%
Air atomization with electrostatic charge	55%
Airless atomization	50%
Airless atomization with electrostatic charge	70%
Disk and bell atomization with electrostatic charge	90%

Fig. 26-5. Painting robot. (*Courtesy DeVilbiss Co.*)

Spraying produces a considerable overspray, which is collected in the spray booth, often by a water curtain that cleans the air from particulate material. Solvent vapor, if present, can be removed by standard methods of incineration or adsorption in activated carbon beds.

Masking often is required in the painting of plastic parts in order to paint a multicolor design or to establish break lines. It is accomplished with simple paper and tape application or with more elaborate fixtures, such as fabricated hard masks. Tape masking requires special tapes resistant to solvents and to the baking cycle.

Hard masks may be prepared from machined metal, cast urethane or silicone rubber, electroformed metals, and vacuum-formed thermoplastics, as well as by other methods. Masks are used to block out certain areas of the plastic part—lip or cap masks are used on raised surfaces, and plug masks on recessed surfaces. Figure 26-6 shows a typical plug mask.

Thermal Spraying. Spray deposition of molten metal is used to apply zinc, tin, pewter, and other metals to plastic surfaces for various functional requirements and less frequently for decorative purposes. Metals of higher melting points usually are applied over a zinc or an aluminum prime coat. Metal is made molten in the gun by an electric arc or a gas flame, having

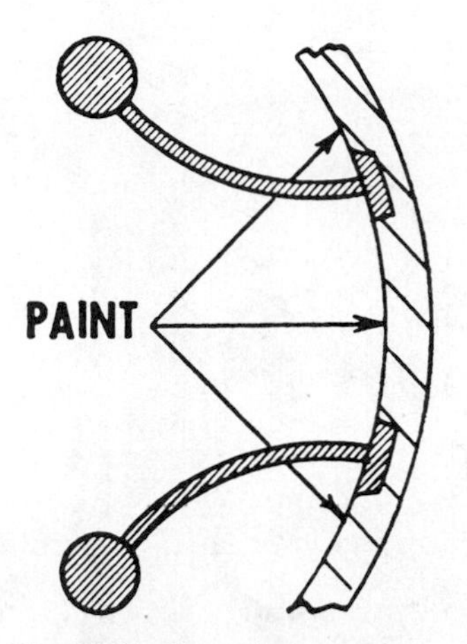

Fig. 26-6. Plug painting mask.

been fed as wire or powder into the gun. Molten metal is atomized by compressed air in very much the same way a regular paint gun operates. The substrate surface temperature does not increase above 50 to 70°C, and most plastics can be coated by this technique without heat damage.

Dip Coating

The part is immersed in a paint-filled vessel and then withdrawn, and excess paint is allowed to drain. Parts are air-dried and baked if necessary. The amount of material deposited depends on the paint viscosity and other properties; the coating appearance is influenced by the rate of withdrawal, which should be as slow as possible. Paint viscosity may range from 20 to 100 mPa–sec and the rate of withdrawal from 15 to 30 cm/min. A dip coating covers all surfaces, and a sharp demarcation line between the dipped and uncoated portion is obtained.

Various plastics can be dip-coated: ABS, polypropylene, acrylic, polyphenylene oxide, polycarbonate, polystyrene, and many other thermoplastics. The base and topcoats are applied by dipping on many plastic parts used in the automotive industry, such as headlamp bezels and grilles.

Dip coating is easily automated by moving the parts on an overhead conveyor from the dipping into drying or curing areas. The paint is continuously agitated and recirculated. Filtration is necessary to remove the particular contamination introduced from the surroundings and by plastic parts.

Flow Coating

In flow coating, the paint is poured or sprayed over the article and allowed to drain by gravity or by the application of centrifugal force. The process resembles dip coating except that the paint may be applied only to one side of the object. Completely enclosed flow coating units have been constructed (see Fig. 26-7) to minimize the exposure of coated parts to outside contamination.

Curtain Coating

Curtain coating consists of passing the object under a free-falling vertical paint curtain. Flat sheets are best coated by this method, although some surface irregularities can be tolerated. The main problem is that air may be entrapped in such surface irregularities, preventing paint–surface contact and thus adhesion.

A curtain coater consists of the following parts: head, circulating pump, overflow collector, and conveyor. The head consists of a hex with an adjustable-width (1–2 mm) slot on the bottom that forms the curtain. Two types of head construction are used: pressure type and weir type. In the former the pressure on the liquid is adjusted by an air cushion (see Fig. 26-8), and in the weir-type coater the pressure is adjusted by the liquid level in the head. A liquid curtain wider than the object falls down, and the excess coating is collected and recirculated. The curtain stretches as it falls. The object is moving on the conveyor at a speed slightly higher than that of the falling curtain, as some stretch of the curtain is desirable. The coating weight is affected by the following process variables: width of the gap, head pressure, coating head distance from the work, and conveyor speed.

Coating speeds of 60 to 130 m/min. normally are used, but such coaters have run as fast as 400 m/min. The curtain width varies between 30 and 400 cm.

Curtain coating is a useful method where a relatively high coating weight is applied over a

Fig. 26-7. Flow coater for automotive bezels. (*Courtesy Deco Tools, Inc.*)

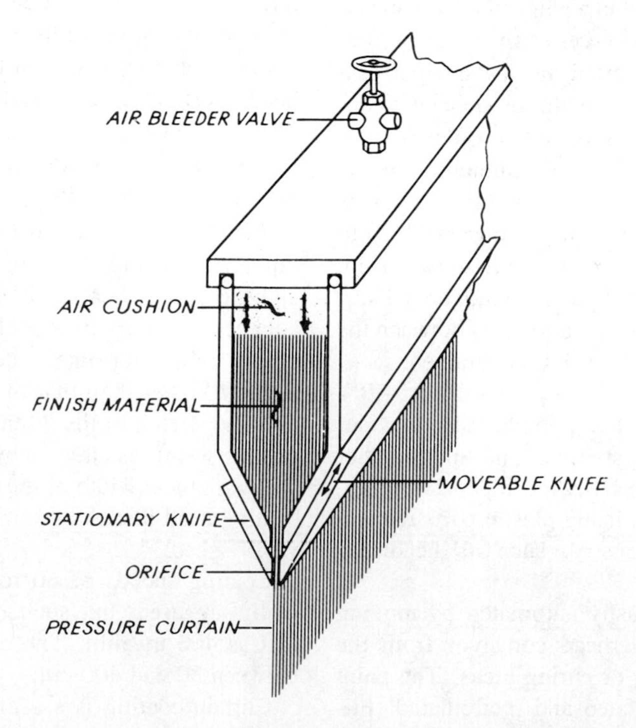

Fig. 26-8. Curtain coater pressure head. (*Courtesy Technical Products Div., George Koch Sons, Inc.*)

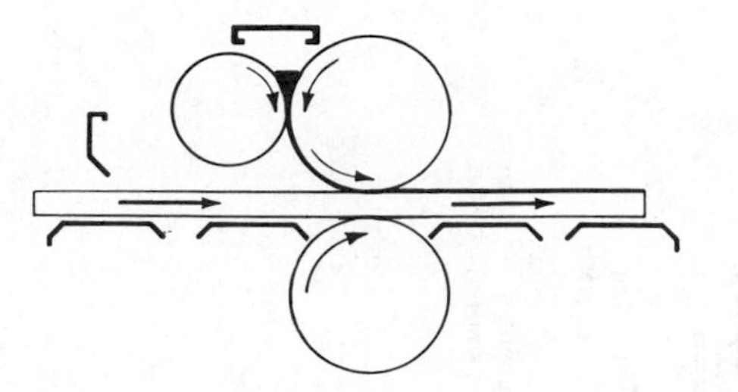

Fig. 26-9. Roll coater. (*Courtesy The Black Bros. Co.*)

flat surface. Also, separate sheets and irregular objects are handled well by curtain coating. Solventborne, aqueous, and hot melt coatings can be run.

Roll Coating

Roll, knife, and bar coating of continuous films is carried out on many different machines, whose descriptions are beyond the scope of this chapter. Such equipment has been reviewed elsewhere.[4] These machines are not suitable for the coating of separate sheets, and basically only one coater design (Fig. 26-9) is available for sheet coating. The coating reservoir lies between the smaller doctoring roll and the applicator roll, and the sheets are run between the applicator and the supporting roll. Such roll coating is suitable for low viscosity fluids and for the application of relatively lightweight coatings.

PRINTING

Printing is a variation of the coating process in which ink or paint is deposited in small selected areas over the plastic surface.

Screen Printing

Screen printing is widely used for plastics decorating for several reasons: the equipment is simple and inexpensive; good-quality multicolor printing is possible, including the coverage of large surface areas; and the coating is fairly heavy so that good coverage and good abrasion resistance are obtained. However, the process is a slow one, mainly suited for short runs.

The screen, which is the most important component of the screen printing equipment, consists of a stencil bonded to a fine-weave fabric and tensioned in a frame. Ink is pushed through the screen openings by a squeegee and deposited onto the plastic surface (Fig. 26-10). The squeegee is simply a shaped flexible blade in a wood or metal holder. Polyurethane elastomers are most often used to make squeegees.

Screen printers can be flat bed, rotary, or cylinder configurations (Fig. 26-11); and they can be simple, manually operated units or completely automated machines. An automated machine can print about 8000 items per shift. The automation may include automatic feed and takeoff, static eliminators, flame treaters, nest carriers to hold objects to be printed, a means for automatic registration, expandable mandrels for soft plastics or tubes, an in-line dryer, and other accessories.

Various ink types are used. Screen inks are specially formulated to behave as a solid and not drip when placed on the screen. The ink should flow easily through the screen when a force is applied by the squeegee, and it should level well after its deposition but should not ex-

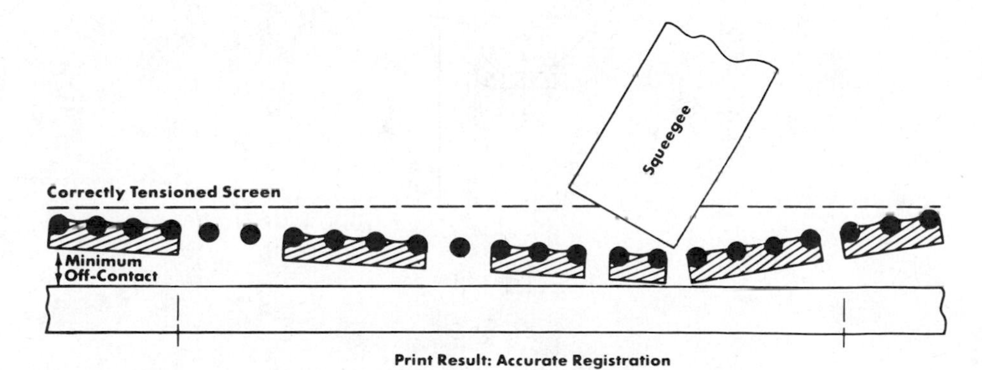

Fig. 26-10. Mechanics of the screen printing process. (*Courtesy Tetko, Inc.*)

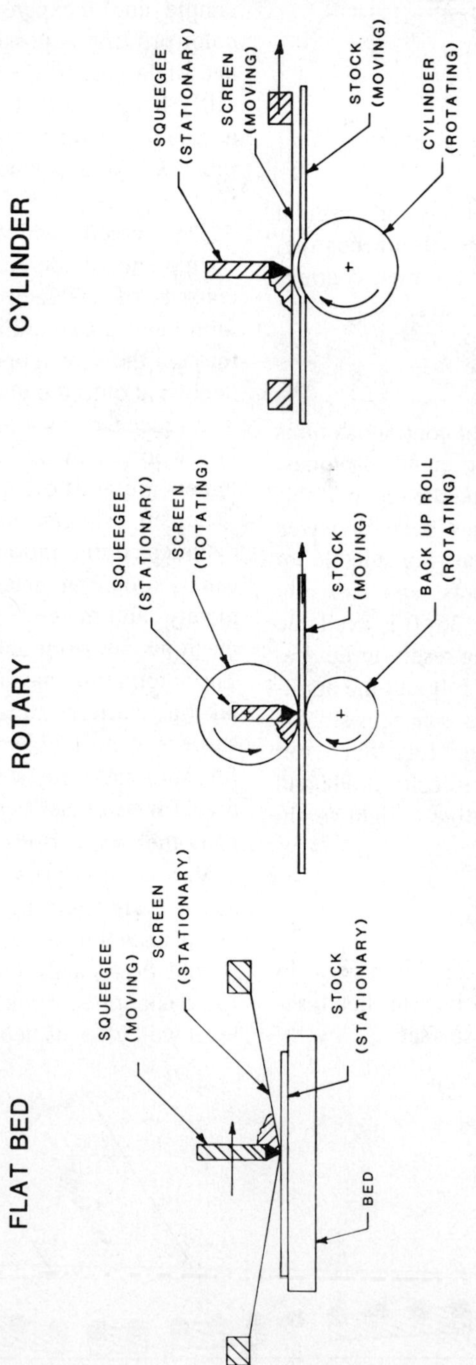

Fig. 26-11. Basic types of screen printers. (*Courtesy Sheldahl, Inc.*)

hibit excessive flow, in order to give a sharp image. It also should adhere to the substrate. Solventborne inks require drying, and the printing unit usually is followed by a dryer unless air drying is possible. One-part (alkyds) or two-part (epoxies and urethanes) reactive inks may be used, and they may harden without heat application. UV-radiation-curable inks also are used for screen printing, because of their rapid cure. Hot melt inks are applied molten and solidify immediately upon cooling. Aqueous screen inks and coatings also are available, but they are not yet widely used. In addition, many inks have been designed for special effects: texturing, abrasion resistance, selective gloss level, dead front, and so on.

Screen printing may be applied over flat or cylindrical surfaces, but it is not suitable for the printing of irregularly shaped surfaces.

Flexographic Printing

Flexography is a variation of the letterpress printing process in which ink is carried on the raised portions of an engraved plate (Fig. 26-12). The plates used in flexography are made from a resilient elastomeric material rather than from rigid metal as in letterpress printing. Flexography grew with the growing use of plastic films for packaging and developed from a rather crude printing method to a high-quality printing process that dominates in flexible packaging printing.

Printing plates are made from either rubber or photopolymers, with the rubber plates ranging in Shore A durometer from 20 to 90. Photopolymer plates are made by imaging with light through a photographic negative; the light causes crosslinking of the polymer, and the

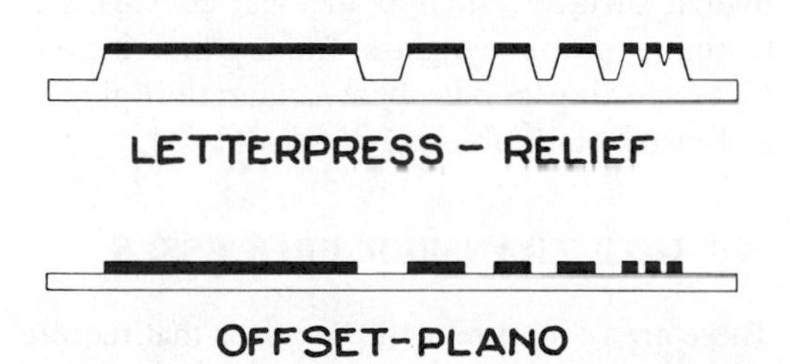

Fig. 26-12. Profile of letterpress and flexography printing plates.

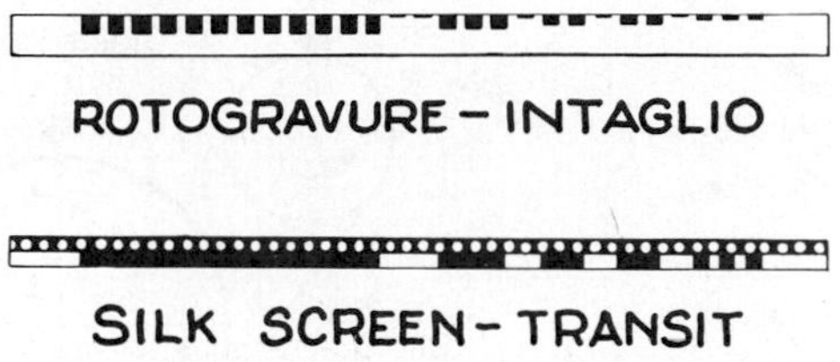

ROTOGRAVURE - INTAGLIO

SILK SCREEN - TRANSIT

Fig. 26-13. Profile of gravure printing plates.

nonirradiated areas are removed by solvent etching.

Solventborne inks are used in flexography.

Gravure Printing

Gravure printing is based on the intaglio process of transferring ink from recessed cells of an engraved cylinder (Fig. 26-13). Gravure printing is widely used for plastic films, competing with flexography for such applications. Gravure cylinders are more expensive than flexographic plates, however; so flexography is preferred for short runs. Also, gravure printing is not suitable for the printing of small plastic items such as bottles, but flexography can be used for such items as well as for continuous film printing.

A schematic diagram of offset gravure printing is shown in Fig. 26-14.

Pad Printing

Pad printing is basically an offset gravure printing method specially adapted for the printing of small and irregular objects. The ink is applied over an engraved plate (cliche); a squeegee removes the excess; then a soft silicone pad picks up the ink from the cliche and transfers it to the surface to be printed. The soft silicone pad allows ink deposition into recesses and over irregularly shaped surfaces. Traditional pad printing is a reciprocating process; recently, rotary pad printing has become increasingly common. Reciprocating units operate at about 30 to 40 cycles per minute, whereas rotary pad printing is much faster and can process 300 to 400 pieces per minute. The pad printing process is shown schematically in Fig. 26-15.

Multicolor wet-on-wet printing is possible,

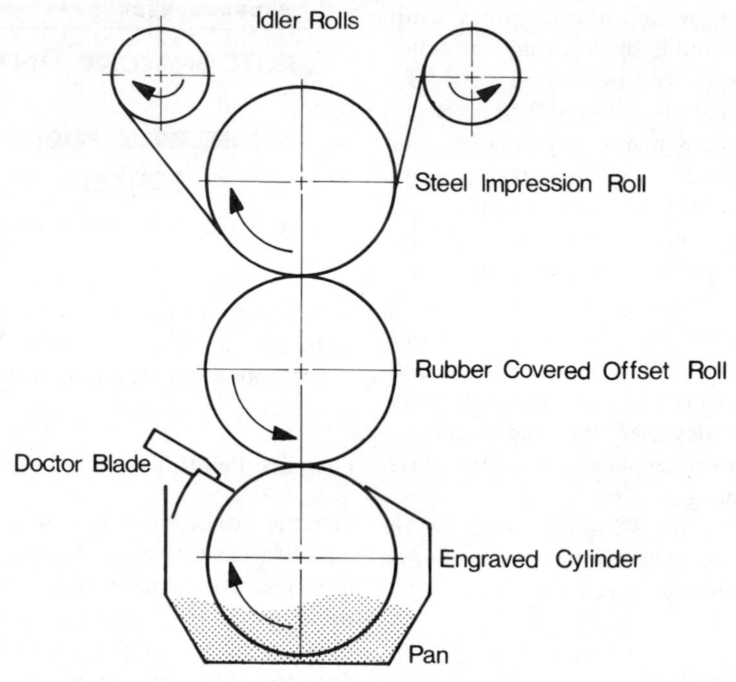

Fig. 26-14. Offset gravure printer.

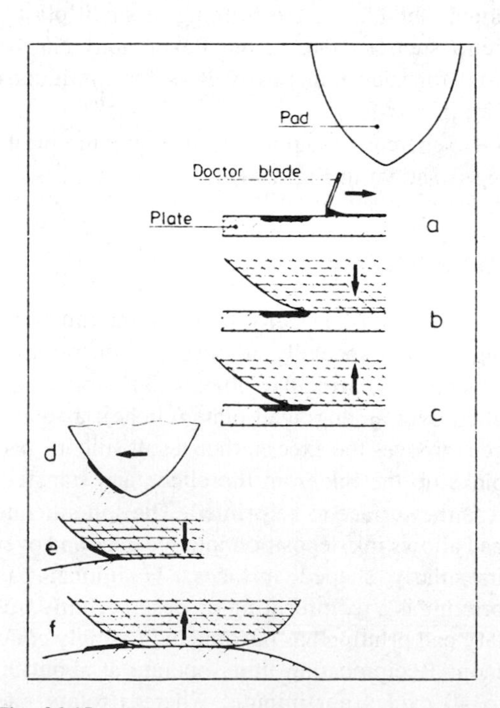

Fig. 26-15. Stages of pad printing: (a) doctoring, (b) initial contact with cliche, (c) ink removal from cliche, (d) pad moving to printing surface, (e) initial contact of printing surface, and (f) removal of the pad.

requiring a printing station for each color. The object usually travels on a rotating table.

Pad printing is most useful for printing small areas; it reproduces fine line engraving. It has been shown, however, that reasonably large (up to 25–40 cm^2) areas also can be printed by this method. The coating thickness is low; so the abrasion resistance is not high. Pad printing has been found useful in printing with diffusible dyes. Pad printing inks are similar to those used for screen printing. Radiation-curable inks also are used.

Pad printing equipment is relatively simple and inexpensive. The process has found applications mainly for printing small items with irregular surfaces, such as medical devices, automotive parts, eyeglass frames and lenses, toys, sporting goods, pens, cigarette lighters, and many other objects.

HOT TRANSFER PROCESSES

There are several related processes that require heat to transfer an image to the plastic surface.

Hot stamping employs a specially prepared foil consisting of: (a) a thin film carrier (usually polyester film); (b) a release coating, which is activated by the application of heat; (c) a decorative coating, often metallic, although it could be in any solid color; and (d) an adhesive coating, which is formulated to adhere to the plastic surface on the application of heat and pressure, and is quite specific, depending on the material to which the foil must adhere. If improved abrasion resistance is required, another coating may be employed, located between the decorative and adhesive coatings. In the case of metallic coatings, which consist of vacuum-deposited aluminum, colors may be introduced by adding a dye or a transparent pigment to the protective coating.

The foil is transferred to the plastic surface by bringing a hot metal or rubber die, which bears the required design or lettering, into contact with the foil and the surface to be decorated. The application of heat melts the release coating and activates the adhesive; also, it usually melts the surface of the plastic. After a short contact (less than 1 second), the die pressure is released and the film carrier separated from the plastic surface, leaving a decorative coating transferred in the area of the die. The plastic is slightly debossed during the hot stamping operation, which helps to protect the decoration from abrasion and scuffing. Similarly, a flat or roll rubber die can be used to transfer the foil to the raised areas of a previously embossed or molded plastic surface.

A similar process is employed to transfer preprinted decals. A multicolored design is transferred to the surface by the application of heat, using a flat or roll rubber die. Such decals have an advantage over roll leaf application: multicolored designs are applied in registration in a single pass, whereas roll leaf application would require a separate pass for each color. The decal is placed on the flat surface without an embossing effect.

Wood-grain roll leaf is applied in a similar way over large surfaces by employing a preprinted roll leaf and a hot rubber roller. Radio and TV cabinets often are decorated by this process.

Another related process is printing with sublimable dyes, a process adopted from textile printing. A pattern printed on release paper employing sublimable dyes is transferred to the plastic surface by heat, causing the dye to vaporize and diffuse into the plastic. Such decorations have good abrasion resistance because the dye penetrates the plastic's surface.

Two basic types of equipment are used for hot transfer decoration. Reciprocating presses are employed to apply the decoration to a flat surface, or to a cylindrical surface if the decoration does not extend beyond 25% of the cylinder's circumference. Rotary machines are used to decorate cylindrical objects if the decoration extends beyond 25% of the circumference, or if the decorated surface is larger than 50 in.^{2}. A cylindrical die helps to eliminate air entrapment underneath the foil.

LASER MARKING

Laser marking is a noncontact process that allows well-defined marks to be made on plastics and other substrates. A high-energy laser beam is directed to the plastic surface, causing it to melt, vaporize, or change color in the restricted area of laser beam application. Figure 26-16 shows cross sections of different types of laser marks on plastics. Figure 26-16a shows a laser mark caused by vaporization; Fig. 26-16b shows a mark where the material was vaporized and also thermally transformed; Fig. 26-16c shows a case where the material was thermally transformed without vaporization; and Figs. 26-16d and 26-16e show the effects of engraving in laminates where the removal of the plastic from the top surface exposes either the metal or a plastic surface of another color.

The volume of laser-decorated plastics is small, but the process is being used increasingly for the decoration of pens and similar items.

DYEING

Some plastic articles are colored by dyeing. Buttons, buckles, and other apparel accessories, sunglasses, and watch crystals often are

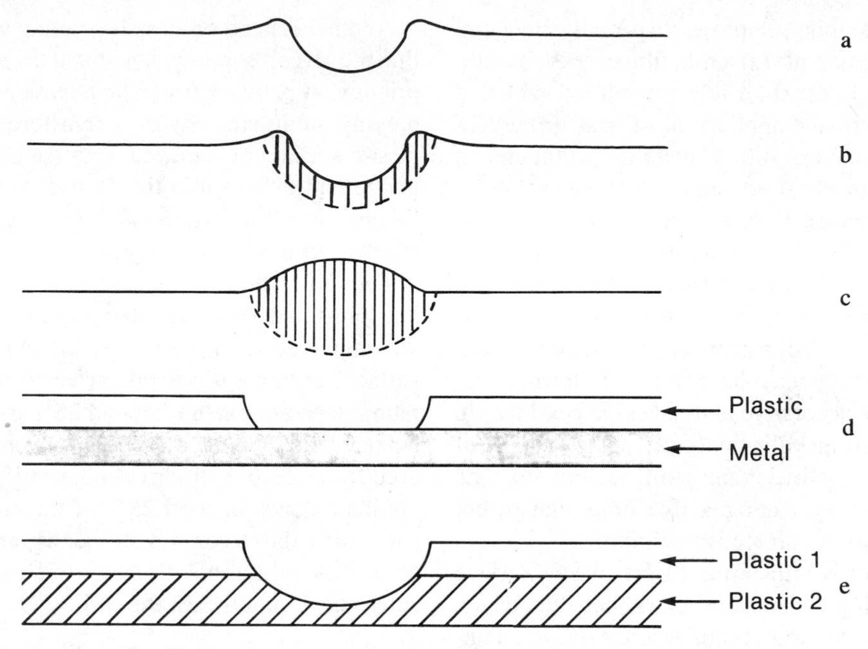

Fig. 26-16. Cross sections of different types of laser marks on plastics.

dyed. Dyeing is carried out by immersion in an aqueous or other solvent dye solution, and usually is a low-volume batch operation. Dye diffuses into the plastic; so such colors are more permanent than surface printing. Various plastics can be dyed—thermosetting polyesters, nylon, cellulose acetate, casein, and acrylates being among the most often dyed materials.

ELECTROLESS PLATING

In electroless plating, metallic coatings are formed as a result of a chemical reaction between the reducing agent present in the solution and metal ions. Such coatings can be deposited on electrically nonconductive materials (i.e., plastics). The deposition process is quite simple, involving immersion of the part in the solution after pretreatment of the surface.

Electroless plating is quite widely used for the coating of plastic surfaces, to produce conductive coatings for subsequent electroplating, coatings for printed circuit boards (PCBs), coatings for radio frequency interference shielding (RFI), and decorative coatings. A thin coating (0.25–1 μm) of electroless nickel or copper usually serves as a base coat for these applications.

ELECTROLYTIC PLATING

Electrolytic plating is the deposition of a metal on a conductor using an electric current. The plastic surface must be made conductive in order to be electrolytically plated. This is accomplished by applying a thin layer of electroless plating or by the use of additives such as carbon filler. The object to be electrolytically plated is immersed in a solution of metal salts connected to a cathodic direct current source, and an anodic conductor is immersed in the bath to complete the electrical circuit. Current flows from the cathode to the anode, and the electron flow reduces dissolved metal ions to pure metal on the cathodic surface. The electroplating bath is a complex solution of salts, including various additives for improved brightness, leveling, and other properties. The anode usually is made from the same metal, which dissolves during the electroplating process and replenishes the plating bath.

A commercial electroplating process in-

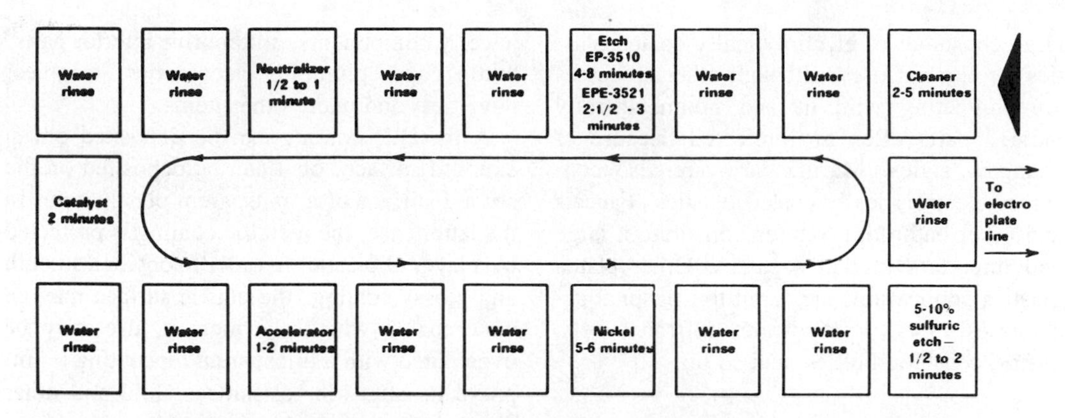

Fig. 26-17. Flow chart of electroless nickel plating process. (*Courtesy GE Plastics*)

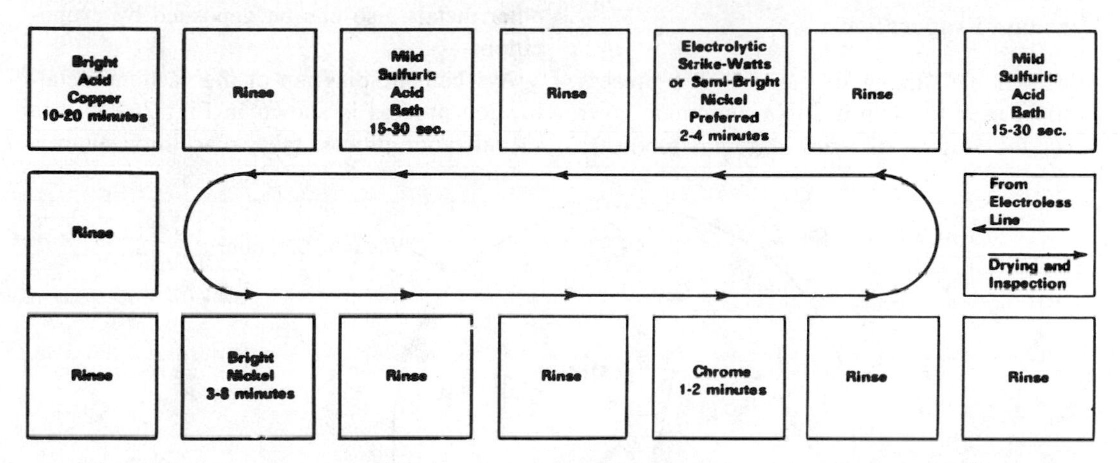

Fig. 26-18. Flow chart of a typical electrolytic plating line for decorative plating. (*Courtesy GE Plastics*)

volves many additional steps such as surface cleaning and etching, rinsing between plating steps, and so on. Such typical steps for electroless and electrolytic coating are shown in Figs. 26-17 and 26-18. Parts to be electroplated are moved on an overhead conveyor from one tank to another.

Electrolytic coatings not only give a metallic appearance to the plastic, but also increase its surface hardness, solvent resistance, and many other properties.

Many different plastics are electroplated commercially, ABS and foamed polyphenylene ether copolymers (PEC) being most often used. Table 26-4 lists commercially platable plastics. Nickel/chrome electrolytic coatings are applied most often, deposited over a thin layer of electroless copper. The automotive industry is a

Table 26-4. Commercially platable plastics.

ABS*	Epoxy/glass†	Polyacetal
PEC*	Polyimide†	Polypropylene
ABS/polycarbonate*	Nylon	Polysulfone
Polystryene	Polyester	Teflon®/glass†
	Phenolics†	Polycarbonate

*Highest volume for plating on plastics.
†Typically plated for printed circuits.

large consumer of electrolytically coated plastics for exterior uses, although other processes are competing with it, and nonmetallically painted parts often are preferred because of changing styles. Marine hardware also consumes electrolytically coated plastics. Faucets and other bathroom fixtures constitute a large and important market segment. Other plated plastics applications are furniture, hospital accessories, toys, kitchenware, mirror bezels, knobs, cosmetic bottles, and so on.

VACUUM METALLIZING

Vacuum Evaporation

Vacuum metallizing of plastics is a process widely used to deposit a thin aluminum layer over the surface of many different products: jewelry components, automotive interior parts, bottle caps, plumbing accessories, trophies, novelties, and many other items.

A metallic coating can be deposited on an exposed surface, or it can be deposited on the second surface of a transparent plastic item. In the latter case, the metallic coating is protected by a layer of plastic. In order to obtain a smooth and glossy coating, the plastic surface may be undercoated with a lacquer. It also may be overcoated with a transparent topcoating to improve its abrasion resistance. Gold and other colors are obtained by dyeing the topcoating. The metallic coating is usually aluminum, but other metals also can be deposited by evaporation.

A schematic diagram of the vacuum metallization process is shown in Fig. 26-19. The metallization process takes place in vacuum in

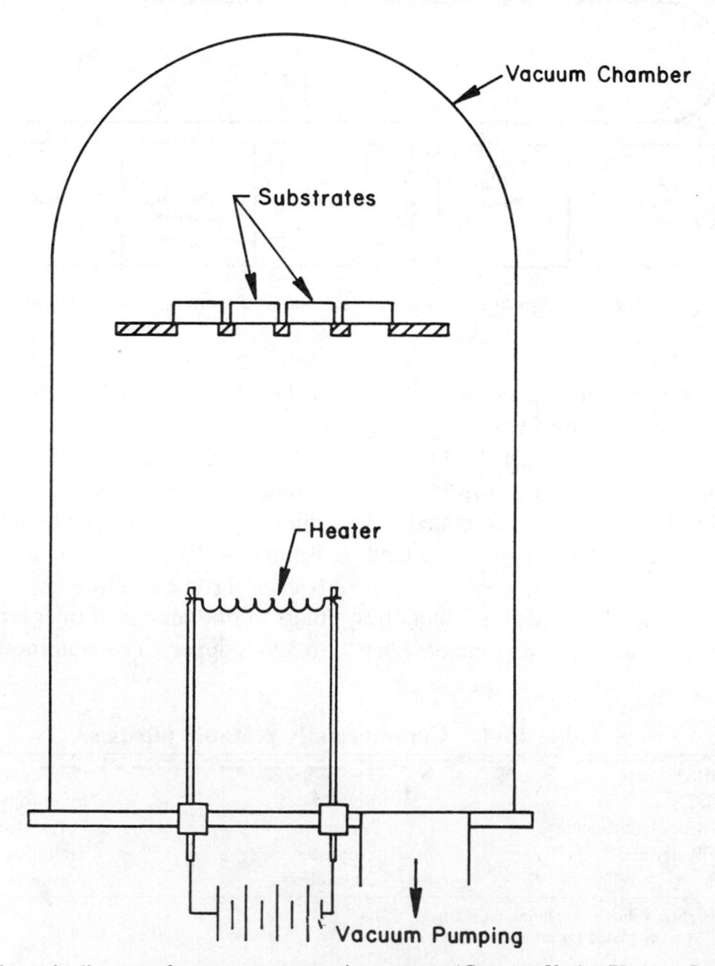

Fig. 26-19. Schematic diagram of a vacuum evaporation system. (*Courtesy Varian/Vacuum Products Division*)

order to allow the metal vapor to reach the plastic surface before the vapor is oxidized by the oxygen in air. The metal is heated to the point at which it will vaporize under vacuum conditions. The substrate is at a sufficiently low temperature to allow the condensation of metal vapor. Parts to be metallized are held in an appropriate fixture that might rotate in order to expose all surfaces to be metallized.

Several metal evaporation sources are used. Resistance-heated tungsten filaments are used most often for applying decorative coatings. These filaments are placed in the required positions to obtain uniform coverage, and aluminum chips or staples are placed on the filaments. Several power sources are used: resistance heating, induction evaporators, or electron beam guns.

Plastic films are vacuum-metallized for packaging uses as well as for hot stamped foils. The film is unwound and rewound inside the vacuum chamber, and as it passes over a cooled drum, the metallic vapor condenses on the film surface.

Sputter Deposition

Sputter deposition is a vacuum coating process in which atoms of the coating material are displaced by impact with a heavy inert gas plasma, such as argon. Any material, including metal alloys, can be sputtered because vaporization of the coating is carried out by a mechanical process on an atomic scale, rather than by heating. Metal alloys are deposited in the same ratio as in the original alloy; separation due to the metals' different vaporization rates does not take place.

Sputtered coatings have a better adhesion and are more resistant to abrasion than vapor deposition coatings. Sputtered chrome-coated plastics have been used in the automotive industry for exterior components where only electrolytically coated plastics had been used before. Complex multilayer sputtered coatings are used for window films; titanium nitride coatings, because of their goldlike appearance and high hardness, are used for jewelry.

There are many variations of the basic sputtering process, including reactive sputtering, where a gas is used to react with metal (titanium with nitrogen to yield titanium nitride, argon/oxygen with aluminum to yield aluminum oxide coatings). In the ion plating process, the substrate is changed because of bombardment with a flux of energetic particles.

LABELS AND DECALS

Labeling is a process of affixing a preprinted, precut flexible material to a surface; thus labeling competes with direct printing or decorating. Hot transfer decals are basically labels. Most labels are attached by an adhesive; pressure-sensitive adhesives dominate the label field, but heat-activated adhesives also are used. Plastic films often are used for label construction, and many plastic items, such as bottles, toys, and so on, are decorated by labeling.

A pressure-sensitive label or decal consists of a thin film that is printed, metallized, or otherwise decorated, and that has a pressure-sensitive adhesive coating on one side protected by a silicone-coated release liner. The release liner is removed before use, and the label is affixed either manually or automatically. Printed transparent film labels with a clear adhesive are used on cosmetic and pharmaceutical bottles, rather than direct screen printing. It is difficult to distinguish a labeled from a direct-printed bottle.

Water-Transfer Process

Preprinted designs such as marble, wood-grain, and other patterns can be applied to irregularly shaped plastic items by a water-transfer process. A water-soluble film, usually polyvinyl alcohol, is printed with the desired pattern. Then the film is floated on the surface of a warm water bath, where it swells and is contacted with the plastic surface that is to be decorated; the surface contacts the inked side of the film. The ink is designed to adhere to a properly prepared surface and is tackified by a solvent wipe. Complete transfer is accomplished by immersing the plastic item in the bath; the highly plasticized film conforms to the irregularly shaped article and covers all surfaces. The residual water-soluble film is removed by washing, and only the ink remains adhered to the surface.

IN-MOLD DECORATING

Plastics parts may be decorated by various in-mold processes, thus eliminating a separate decorating step. In-mold decorating may consist of coating the mold walls prior to molding or the incorporation of a preprinted insert or label.

Preprinted resin-impregnated papers have been used for a long time for decorating melamine dishes and for urea and phenolic plastics made by compression or transfer molding. In-mold decorating of thermoplastic materials is accomplished by insertion of a preprinted, die-cut foil of a compatible material, usually the same as the molding. The foil is placed against the mold wall, the mold is closed, and the plastic is injected. The foil can be placed manually, or the insertion can be mechanized. In-mold insertion has been used on various thermoplastic materials: polystyrene, polyethylene, polypropylene, acrylics, ABS, polycarbonate, and others.

In-mold foils are limited to a draw of about 15 mm. Deeper draws (40 mm) can be accommodated by employing thermoformed inserts. Wood-grain or brushed aluminum decorated inserts are most often used.

Plastic bottles often are in-mold labeled, eliminating a separate labeling process and also reducing the plastic's weight, as the label becomes an integral supporting part of the bottle. Several variations of label-inserting robot arms have been developed to facilitate the molding process.

In-mold finishing of sheet molding compound (SMC), thermoplastic structural foam, and reaction-injection-molded (RIM) parts simplifies the subsequent finishing operations.

FLOCKING

Flocking is the deposition of short fibers onto an adhesive-coated surface. Plastics often are flocked to achieve a soft textilelike feel and appearance. Flocked plastics are used for automobile interiors, toys, novelties, and other applications.

Flock may be applied by several processes. Mechanical flocking involves the feeding of a fiber dust onto an adhesive surface that is vibrated by beater bars (Fig. 26-20). Mechanical flocking may be assisted by electrostatic forces—by charging the particles and maintaining the substrate at ground potential, which improves the fiber alignment. Electrostatic-pneumatic flocking is especially useful for three-dimensional objects (Fig. 26-21). Flock is introduced into the airstream, conveyed along a

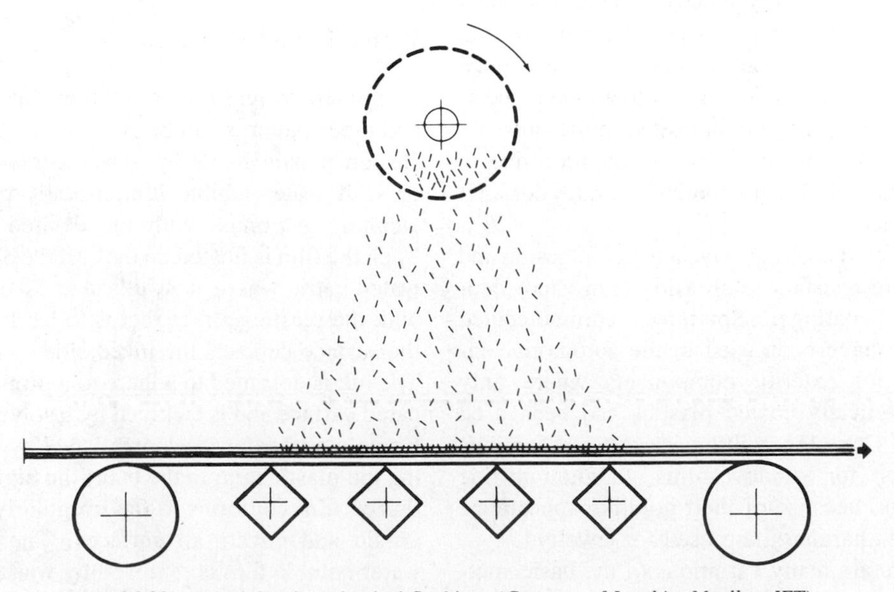

Fig. 26-20. Principle of mechanical flocking. (*Courtesy of Joachim Mueller, IFT*)

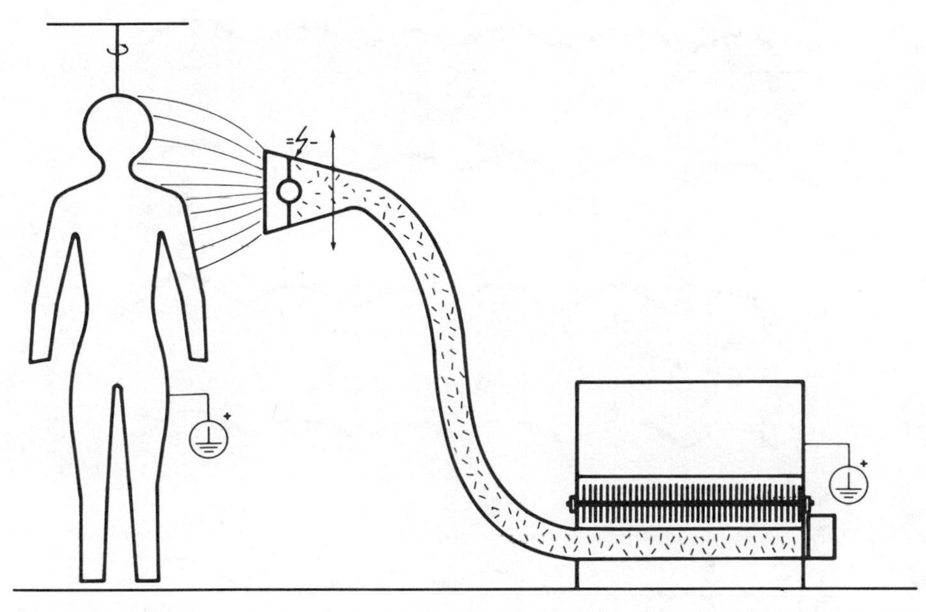

Fig. 26-21. Principle of flocking three-dimensional objects with electrostatic-pneumatic flocking unit. (*Courtesy of Joachim Mueller, IFT*)

flexible tube to the applicator nozzle, and charged by means of an electrode.

Flock may consist of different fiber types and dimensions. Rayon flock is available as ground or precision-cut (0.3–0.4 mm in length). Nylon flock is used most widely; ground and cut flock (0.3–12 mm) is available. Generally flock longer than 3 mm is not important. Cotton flock is marketed only as ground flock. Polyester flock is gaining in automotive applications.

EMBOSSING AND SURFACE TEXTURING

The surface texture markedly contributes to the visual appearance and tactile characteristics of a plastic surface. Surface texture affects the gloss level and also may be used in decorative patterns, including imitations of the texture of natural surfaces such as wood or leather.

Molded plastic articles accept the surface characteristics of the mold, and mold surfaces can be polished or textured to impart required surface finishes. Textured surfaces hide smudges, scratches, and flow and sink marks.

The embossing of plastic films and sheets may be carried out by casting the plastic over pre-embossed release-coated paper. This technique often has been used to manufacture polyurethane and vinyl film and sheet with a leather grain.

Plastic films and sheets also can be embossed by pressure embossing. This is achieved by passing the plastic between an engraved embossing roll and a supporting roll. The supporting roll may be a rubber-covered roll, a matched filled paper roll, or even a matched steel roll. The material that is to be embossed may be preheated and then chilled by a cold embossing roll. Vinyl and other thermoplastic films are embossed in this way. Some materials are best embossed by mechanical deformation, where the film or sheet is not preheated and passes through a heated nip. A flat press also may be used to emboss plastic sheeting or laminated materials.

Vacuum embossing may be used to obtain a deep pattern. A preheated sheet is placed against a roll or a plate with recessions. A vacuum is pulled through the porous roll, and the preheated sheet conforms to the pattern of the vacuum roll.

Dielectric embossing involves dielectrically heating a film in restricted spots and deforming it at those spots. Similarly, ultrasonic embossing is carried out by heating a plastic sheet in

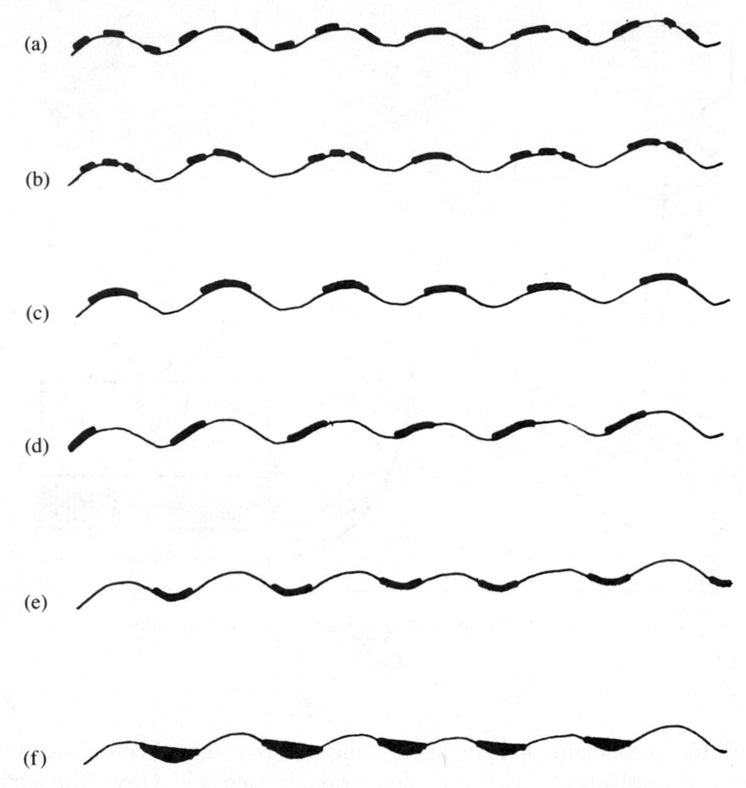

Fig. 26-22. Various printing-embossing effects; (a) embossing after printing, (b) printing after embossing, (c) tipping, (d) shading, (e) valley printing, (f) spanishing.

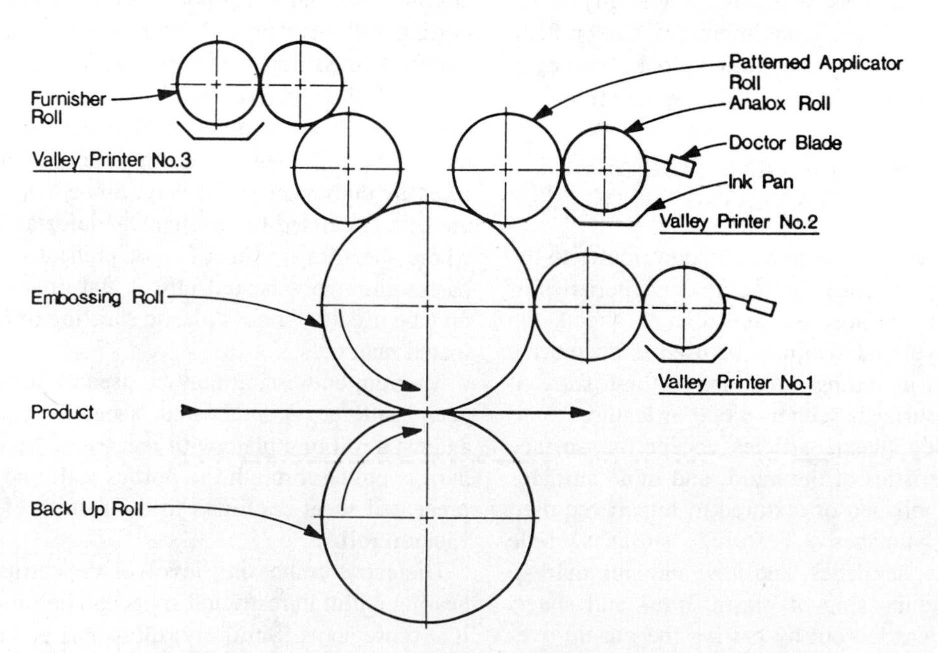

Fig. 26-23. A schematic diagram of valley embossing equipment. (*Courtesy Liberty Machine Co., Inc.*)

selected spots by friction against a vibrating ultrasonic horn. Chemical embossing involves foaming of a coating in selected areas.

Several processes combine embossing and printing to decorate the plastic surface. This is especially useful in obtaining realistic leather, textile, or wood-grain patterns. Various printing-embossing effects are shown schematically in Fig. 26-22. Printing over a pre-embossed surface allows the deposition of ink on the raised portions only. Valley printing and spanishing allow ink deposition in the depressed portions of the surface. Valley printing is carried out simultaneously with embossing as shown in Fig. 26-23. Multicolor printing is possible.

GRINDING AND POLISHING

Grinding is a process during which material is removed from a surface by the action of abrasive grains. The grinding of plastics is different from the grinding of materials. In the case of metals, the removal of shavings result in a smooth and well-defined furrow. During plastics grinding, the plastic material crumbles and flakes under the action of abrasive grains, and the furrows are cracked.

Polishing is a process in which no material is removed from the surface, but the surface is smoothed out by melting a thin surface layer and redistributing the material.

Plastics are ground by one of three methods: employing grinding media bonded to a belt, employing loosely bonded grinding grains in wax, or employing a drum. Wet or dry belt grinding may be used. Rough and strong moldings may be ground with a coarse belt of grit 180. Where only the edges have to be ground, a finer belt of grit 220 or 320 is used. Grinding with waxes employs cloth disks. This method is milder than that employing belts. Plastics should be ground under the mildest possible conditions.

Small plastic objects may be polished in a drum. The grinding/polishing process may consist of several steps: coarse grind, fine grind, polishing, and finishing. Wooden pegs serve as a carrier for the abrasive, and are covered with grinding powder in an oil. For polishing operations, a polishing cream or a special powder in oil is applied to the pegs. Plastic parts are either loosely placed into the drum or mounted on special rigs. Rotation of the drum causes friction between the plastic items and the abrasive-covered pegs.

REFERENCES

1. F. W. Billmeyer and M. Saltzman, *Principles of Color Technology*, 2nd ed., John Wiley and Sons, New York, 1981.
2. R. S. Hunter and R. W. Harold, *The Measurement of Appearance*, 2nd ed., John Wiley and Sons, New York, 1987.
3. D. Satas, ed., *Plastics Finishing and Decoration*, Van Nostrand Reinhold Co., New York, 1986.
4. D. Satas, ed. *Web Processing and Converting Technology and Equipment*, Van Nostrand Reinhold Co., New York, 1984.

27
Performance Testing of Plastics Products

Plastics engineers frequently are required to evaluate the physical and other characteristics of objects made from plastics. The purpose may be control of quality in production, acceptance testing against specifications, establishment of data for engineering and design, or other economically important activities.

Whatever the purpose, the question of the reliability of the evaluation cannot be avoided; and when the economic importance of the evaluation is great, the reliability of the evaluation must be in proportion to it. Neither the authority of the engineer nor the computations performed on his or her data can produce reliable predictions from unreliable data; there is no substitute for a valid testing method, which must be selected before evaluation begins.

Plastics, perhaps more than most other classes of materials, require specialized testing methods. The characteristics of the resins and compounds, and of the production processes used to manufacture the articles economically, interact to determine the final properties of fabricated articles. An engineer who tests an object made from plastic tests not only its material, but also the way the object was made. There are inherent variations in the articles, from lot to lot, and within each lot, and also from place to place within each individual article. This place-to-place variation establishes the rule that the characteristics of each article made from a plastic are governed not only by the material of which it is made, but also by its shape and dimensions, that is, its design. These variations are rooted in the nature of the production process itself.

The very word "plastic," implying the ability to be shaped, suggests the economic basis of the processes of fabrication. The manufacture of articles from plastics must be a repetitive process of maximum rapidity, involving rapid changes in temperature and in pressure, and hence necessarily abrupt transitions between fluid (or plastic) and solid conditions. The quality of the finished product is notably sensitive to any irregularity in temperature, pressure, or time in the production cycle. Hence, the uniformity of performance of a finished product, from lot to lot, and even within a lot, depends not only upon the uniformity of the plastic used but also very considerably upon the uniformity of the operation of manufacturing the article, that is, upon precision of control of all factors in the operation.

To establish the degree of uniformity that can be expected, or to determine the reproducibility of a method of test, it may be necessary to test all, or a large percentage, of the articles in a lot.

But, particularly because many performance tests are destructive, considerations of cost usually restrict testing, for control purposes, to a very small number of samples from a lot. In this connection, it must be emphasized that the results of testing only a few samples from a lot

---

Reviewed and updated by Mark Schlack, *Plastics World* magazine, Newton, MA.

can be reliable only when the lot as a whole has been manufactured under constant conditions—uninterrupted production, with effective instrument control over all variables.

Performance tests, realistically pertinent to the conditions and hazards of service of the article, may become a very potent tool for:

1. Choosing the kind, and specific grade, of plastic for the job, or an acceptable alternative material.
2. Establishing, at the start, the proper operating conditions for production (preferably at the time when samples are made for approval).
3. Proving the soundness of design of the product.
4. Checking the uniformity of the material, and of the article as made.
5. Checking the effects of subsequent changes in tooling or machinery or operating conditions (e.g., enlarging of gates; change to a larger machine; change in any factor in the cycle of manufacture).

There is a great deal of useful literature on test methods for measuring the physical and other characteristics of the various plastics as materials. Almost without exception, these methods utilize specimens of standard dimensions and shapes, prepared specifically for the purpose. The resulting data are valuable in identifying and determining the uniformity of the materials themselves, but usually the published standard data derived from such test specimens are not reliably and directly applicable to design calculations.

Hence it is difficult to predict from such standard test data the performance of a specific article, made from a specific plastic, when it is subjected to the stresses and exposures that it is expected to withstand in service.

Few commercial articles will resemble, even remotely, in shape or dimensions, the standard test specimens. And even when it is feasible to cut specimens of standard dimensions from the larger object that is to be evaluated, the results of tests on such specimens rarely resemble those obtained on standard molded specimens; nor can such results be relied upon to indicate precisely the probable performance of the whole object itself.

The plastics engineer, then—recognizing the difficulty of evaluating with reliability the probable performance of the article, on the basis of either the standard test data on the material or tests of standard-sized specimens cut from the article itself—must arrive at a method of test that is adaptable to the specific case at hand, and may have to design one. Prime consideration must be given to the relevance of the test to service conditions of the article. Ideally, the result of the performance test should correlate perfectly with the actual performance of the article in service.

The literature on methods of testing objects of nonstandard shapes and dimensions is not—nor, by definition, can it ever be—anything like the orderly and compact, yet comprehensive, set of guides covering the testing of plastics simply as materials. Not only are there as many or more properties to be measured, but there is an infinity of possible different shapes and, for each possible shape, a wide range of possible dimensions. Given the nature of the material used, and the process by which the article is made, each article is likely to have its own unique set of characteristics. Hence, in the testing of articles made from plastics there can be no complete codification of methods, apparatus, or acceptable levels of performance.

This chapter represents an attempt to describe tests and equipment, in several categories, that have been effectively used for testing specific articles, and for measuring certain important characteristics of plastics in the form of finished articles. Although the methods outlined may not meet the requirements of any given case, it is suggested that the general principles set forth herein be used in devising variations or new methods that will be suited to the particular article to be tested.

PERFORMANCE TESTING*

One test may measure a single property or several properties at once. In every case the test

*Based on "Standard Tests on Plastics, Bulletin GIC" issued by Hoechst Celanese Corp.

has been devised to be as accurate as possible. After many years of work by thousands of technical specialists in the plastics industry, the tests presently used are generally deemed suitable. Nevertheless, further improvement is constantly sought. The ASTM, the source of the most widely used plastics tests, regards testing as a dynamic science always receptive to further improvement. To make tests more accurate, more reproducible, and more meaningful, the plastics industry and interested university and independent groups are continually perfecting the existing tests and developing better ones for future use.

Of course, tests are not ends in themselves, but rather means of extracting knowledge about materials. The real test of a material comes with actual service. Once a plastic product is taken home and used by the consumer, it no longer matters whether its tensile strength is 6500 or 6600 psi. The product succeeds entirely, or it fails. To assure the success of toys, housewares, industrial products, and automotive components, the properties of likely materials are studied by design engineers who, through experience and judgment, balance material characteristics and service requirements against the amount of material needed in a part to give an adequate safety margin. It is in this area of product design and specifications that the tests themselves are tested.

In certain cases, the service requirements may be so complex that the suitability of a material for that service can be determined only in actual use. A plastic material considered for large-scale use, as in automobiles, is fabricated into finished parts for in-test service. A maker of plastic dishes may drop them on concrete, run them through dishwashers repeatedly, and otherwise put them through informal ''practical'' tests.

The tests included here may be found in the ASTM books. They are described as briefly as possible (far too briefly to serve as a laboratory guide) in order to give interested persons a general idea of what they are about. Fully detailed procedures for all ASTM tests on plastics are available from the American Society for Testing and Materials, 1916 Race Street, Philadelphia, PA.

Density by Density Gradient Technique (ASTM D1505)

Specimen: Any small piece may be a specimen, as long as its shape does not encourage entrapment of air bubbles, it does not contain voids, and it permits accurate determination of the center of volume. Conditioning is required only if specimens might change in density more than the limits of accuracy would allow.

Procedure: A density gradient column contains the densest liquid at the bottom and decreasingly dense liquid at higher levels. A group of such tubes can contain a range of densities, which may be as low as 0.80 g/cc and may go as high as 2.89 g/cc. Into such columns, standard glass floats of various known densities are inserted. When they reach equilibrium levels, the density gradient within the tube is established. Plastic specimens then may be placed into the columns, and, at equilibrium level, their density may be read from the centerline of their volume. The whole system is kept at a constant temperature of 23°C.

Significance: Density determinations by this method are very accurate and quick, and it is a widely used technique. It requires very careful preparation and handling.

Specific Gravity and Density (ASTM D792)

Specimen: The determination may be made on any shape with a volume of more than 1 cc. Conditioning of specimens is optional. Extrudate or molded specimens are preferred over molding pellets, as the latter may contain voids. Molded specimens are not tested for specific gravity until after post-molding shrinkage has been accomplished—usually 24 hours.

Procedure: In Method A, a piece of the article is held by a fine wire, weighed, and submerged in water. While it is in the water, it is weighed again. From the weight difference the density can be calculated.

In Method B, 5 grams of the pellets or powdered material is added to a measured volume in a pyconometer, and the specific gravity is calculated from the weight and volume change at 73.4°F. This method requires that all air be evacuated. Also, with such specimens as film this can be troublesome, and there is a risk of error from bubbles that may remain.

Significance: Specific gravity is a strong element in pricing and thus is of great importance. Beyond the price/volume relationship, however, specific gravity is used in production control, both in raw material production and in molding and extrusion. Polyethylenes, for instance, may have density variation, depending upon the degree of "packing" during molding or the rate of quench during extrusion.

Although "specific gravity" and "density" frequently are used interchangeably, there is a very slight difference in their meaning:

Specific gravity is the ratio of the weight of a given volume of material at 73.4°F (23°C) to that of an equal volume of water at the same temperature. It is properly expressed as "Specific Gravity, 23/23°C."

Density is the weight per unit volume of material at 23°C and is expressed as "D23C, g per cm^3."

The discrepancy enters from the fact that water at 23°C has a density slightly less than *one*. To convert specific gravity to density, the following factor can be used:

D23C, g per cm^3 = specific gravity, 23/23°C × 0.99756.

Flow Rate (Melt Index) by Extrusion Plastometer (ASTM D1238)

Specimen: Any form that can be introduced into the cylinder bore may be used (e.g., powder, granules, strips of film, etc.).

The conditioning that is required varies, and is specified for each material.

Procedure: The apparatus (Fig. 27-1) is preheated. Material is put into the cylinder, and the loaded piston is put into place. Temperature and pressure conditions also vary according to the material. After 5 minutes, the extrudate issuing from the orifice is cut off flush, and again one minute later. These cuts are discarded. Cuts for the test are taken at 1, 2, 3, or 6 minutes, depending on the material or its flow rate. The flow rate is calculated and given as g/10 min.

The flow rate sometimes is called the melt index, particularly for polyolefins. The melt index is defined by ASTM as the flow rate of polyethylene when a 190°C temperature and 2.16 kg load are used to perform the flow rate test.

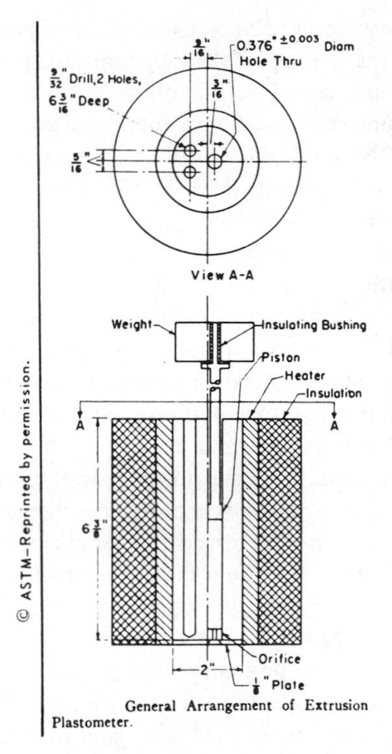

Fig. 27-1. General arrangement of extrusion plastometer. (*Copyright ASTM—reprinted by permission*)

Significance: The flow rate test is primarily useful to raw material manufacturers as a method of controlling material uniformity. Although the data from this test are not directly translatable into relative end-use processing characteristics, the flow rate is nonetheless strongly indicative of relative "flowability" of various kinds and grades of PE.

The property measured by this test is basically melt viscosity or "rate of shear." In general, the materials that are more resistant to flow are those with higher molecular weight.

Water Absorption (ASTM D570)

Specimen: For molding materials the specimens are discs 2 inches in diameter and $\frac{1}{8}$ inch thick. For sheet materials the specimens are bars 3 inches × 1 inch × thickness of the material. Rod or tube specimens also may be used, according to the dimensions specified by the standard.

The specimens are dried 24 hours in an oven at 50°C, cooled in a desiccator, and immediately weighed.

Procedure: Water absorption data may be obtained by immersion for 2 hours, 24 hours, or longer in water at 73.4°F. Data also can be obtained by immersion for 30 minutes or 1 hour in boiling water, or by immersion for 48 hours in water at 50°C. Upon removal, the specimens are wiped dry with a cloth and immediately weighed. The increase in weight is reported as percentage gained.

For materials that lose some soluble matter during immersion—such as cellulosics—the sample must be redried, reweighed, and reported as "percent soluble matter lost." The % gain in weight + % soluble matter lost = % water absorption.

Significance: The various plastics absorb varying amounts of water, and the presence of absorbed water may affect plastics in different ways.

Electrical properties change most noticeably with water absorption, and this is one of the reasons why polyethylene, which absorbs almost no water, is highly favored as a dielectric.

Materials that absorb relatively large amounts of water tend to change dimension in the process. When dimensional stability is required in products made of such materials, grades with less tendency to absorb water are chosen. Some hygroscopic polymers, such as nylon, show increased impact resistance after they absorb moisture.

The water absorption rate of acetal-type plastics is so low as to have a negligible effect on properties.

Environmental Stress Cracking (ASTM D1693)

Specimen: This test is limited to polyethylenes. Specimens measure $\frac{1}{2} \times 1\frac{1}{2}$ inch and are $\frac{1}{8}$ inch thick for polyethylenes up to 0.926 g/cc density; they are 0.070 to 0.080 inch thick for higher-density polyethylenes. After conditioning according to Procedure A, ASTM D618, the specimens are nicked according to the directions given.

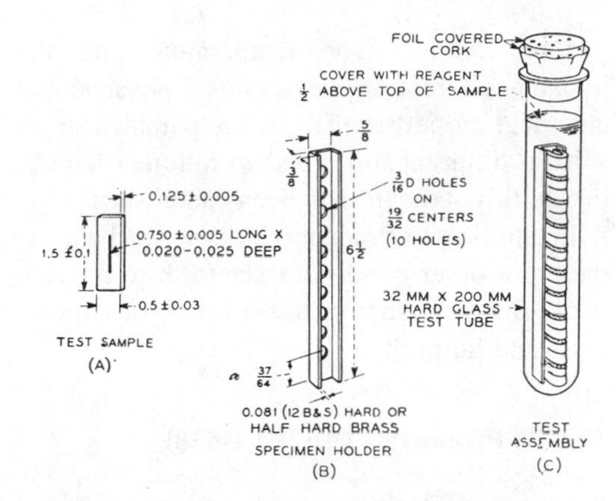

Fig. 27-2. Environmental stress cracking.

Procedure: The specimens are bent into the specimen holder and next inserted into a test tube, which then is filled with fresh reagent (Igepal). The tube is stoppered with an aluminum-covered cork (as shown in Fig. 27-2) and placed in a constant temperature bath at 50°C (100°C for high-density polyethylenes). The specimens are inspected periodically, and any visible crack is considered a failure. The duration of the test is reported along with the percentage of failures. Time to 50% failure sometimes is used as a reference point.

Significance: The cracking obtained in this test is indicative of what may be expected from a wide variety of other stress cracking agents. The information cannot be translated directly into end-use service prediction, but serves to rank various types and grades of PE in categories of resistance to ESC.

Conditioning Procedures (ASTM D618)

Procedure: Procedure A for conditioning test specimens calls for the following periods in a standard laboratory atmosphere (50 ± 2% RH, 73.4 ± 1.8°F):

Specimen thickness, inch	*Time, hr*
0.25 or under	40
Over 0.25	88

Adequate air circulation around all specimens must be provided.

Significance: The temperature and the moisture content of plastics affect physical and electrical properties. To get comparable test results at different times and in different laboratories, this standard has been established.

In addition to Procedure A described above, there are other conditions set forth to provide for testing at higher or lower levels of temperature and humidity.

Tensile Properties (ASTM D638)

Specimen: Specimens can be injection-molded or machined from compression-molded plaques. They are given standard conditioning.* Typically 0.13 to 0.28 inch thick, their size can vary; their shape is exemplified in Fig. 27-3.†

Procedure: Both ends of the specimen are firmly clamped in the jaws of a universal testing machine. The best results are obtained when an extensometer is attached to both ends of the narrow neck (Fig. 27-4). The jaws may move apart at rates of 0.2, 0.5, 2, or 20 inches a minute, pulling the sample from both ends. Stress is plotted against strain (elongation) to produce a stress–strain curve that describes the material's response to tensile loading.

Significance: Tensile properties are the most important single indication of strength in a material. The force necessary to pull the specimen apart is determined, along with how much the material stretches before breaking.

The elastic modulus ("modulus of elasticity" or "tensile modulus") is the ratio of stress to strain below the proportional limit of the material (the point at which the material begins to deform). It is the most useful tensile datum because parts should be designed to accommodate stresses well below the proportional limit. For very flexible materials, the secant modulus sometimes is used. The secant modulus is the slope of the stress–strain curve in the region between a no-strain condition and an arbitrary elongation, typically 5%.

*See ASTM D618—"Conditioning Plastics."

†For polyethylene, the Standard (D-1248-72) requires the use of specimens described in ASTM D638 as Type IV. This specimen is smaller and allows for the much greater elongation of PE.

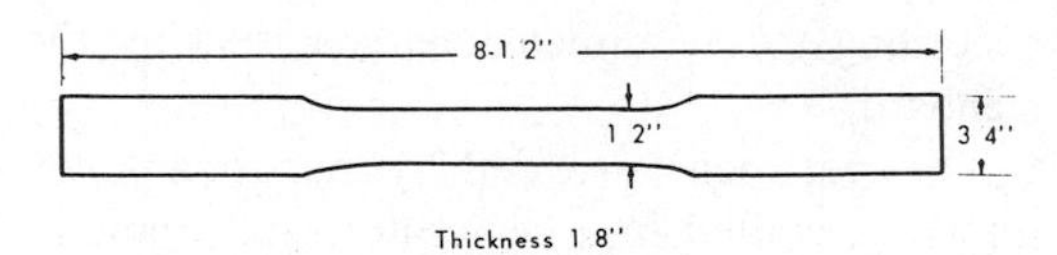

Fig. 27-3. Tensile test specimen.

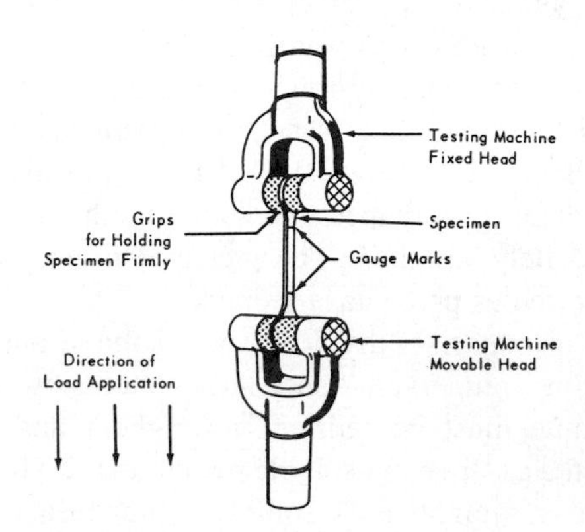

Fig. 27-4. Extensometer specimen holders.

For some applications where almost rubbery elasticity is desirable, a high ultimate elongation may be an asset. For rigid parts, on the other hand, there is little benefit in the fact that they can be stretched extremely long.

There is great benefit in moderate elongation, however, as this quality permits absorption of rapid impact and shock. A material of very high tensile strength and little elongation would tend to be brittle in service. Thus the total area under a stress–strain curve is indicative of overall toughness.

Flexural Properties of Plastics (ASTM D790)

Specimen: Usually specimens measure $\frac{1}{8} \times \frac{1}{2} \times 5$ inches, but sheet or plaques as thin as $\frac{1}{32}$ inch may be used. The span and width depend upon the thickness. Specimens are conditioned according to Procedure A, ASTM D618.

Procedure: There are two test *methods* and two test *procedures*. In Method I, the specimen is placed on two supports, spaced 4 inches apart (Fig. 27-5). A load is applied in the center at a specified rate, and the loading at failure (psi) is the flexural strength. For materials that do not

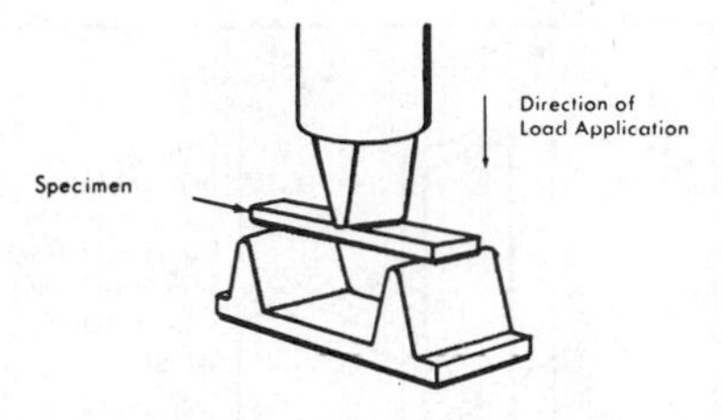

Fig. 27-5. Flexural properties of plastics, (Method I).

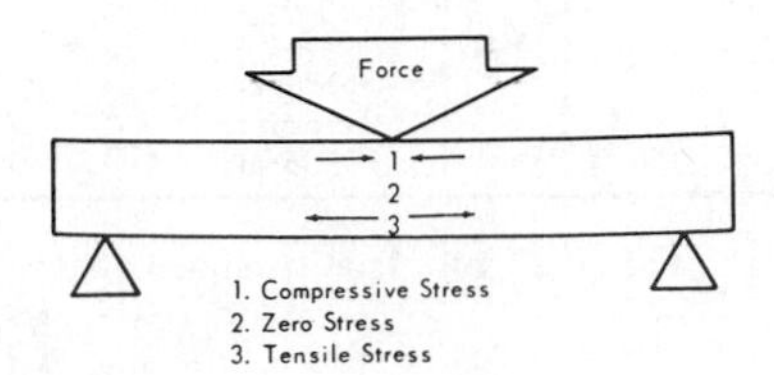

Fig. 27-6. Stresses in flexed sample.

break, the flexural property usually given is flexural stress at 5% strain. In Method II, a four-point support system is used. Unlike Method I, which places maximum stress at the contact point, Method II distributes the maximum stress between the loading noses.

Because of the wide range of flexibility of plastics, two procedures, consisting of different rates of applying the load, have been devised. Procedure A is used for semirigid and rigid plastics. Procedure B uses crosshead rates ten times greater than Procedure A, and is for materials that deflect large amounts during testing.

Significance: In bending, a beam is subject to both tensile and compressive stresses, as indicated in Fig. 27-6.

Because most thermoplastics do not break in this test even after being greatly deflected, the flexural strength cannot be calculated. Instead, stress at 5% strain is calculated—that is, the loading in psi necessary to stretch the outer surface 5%.

The flexural modulus, which is the ratio of stress to strain within the elastic limit of a material, also is commonly used as an indicator of the stiffness of a material.

Stiffness in Flexure (ASTM D747)

Specimen: The specimens must have rectangular cross section, but dimensions may vary with the kind of material.

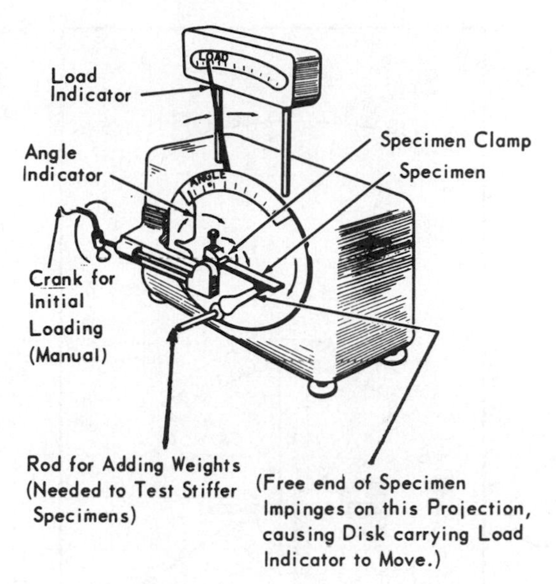

Fig. 27-7. Stiffness in flexure apparatus.

Specimens are conditioned according to Procedure A, ASTM D618.

Procedure: The specimen is clamped into the apparatus (see Fig. 27-7), and a 1% load is applied manually. The deflection scale is set at zero. The motor is engaged and the loading increased, with deflection and loading figures recorded at intervals. A curve is drawn of deflection versus load, and from this is calculated stiffness in flexure in pounds per square inch.

Significance: This test does not distinguish the plastic and elastic elements involved in the measurement; so a true elastic modulus is not calculable. Instead, an apparent value is obtained and called "stiffness in flexure." (See Fig. 27-8.) It is a measure of the relative stiffness of various plastics, and taken with other pertinent property data it is useful in material selection.

Izod Impact (ASTM D256)

Specimen: The specimen usually measures $\frac{1}{8} \times \frac{1}{2} \times 2$ inches. Specimens of other thicknesses can be used (up to $\frac{1}{2}$ inch), but $\frac{1}{8}$ inch frequently is used for molding materials because it is representative of the average part thickness.

A notch is cut on the narrow face of the spec-

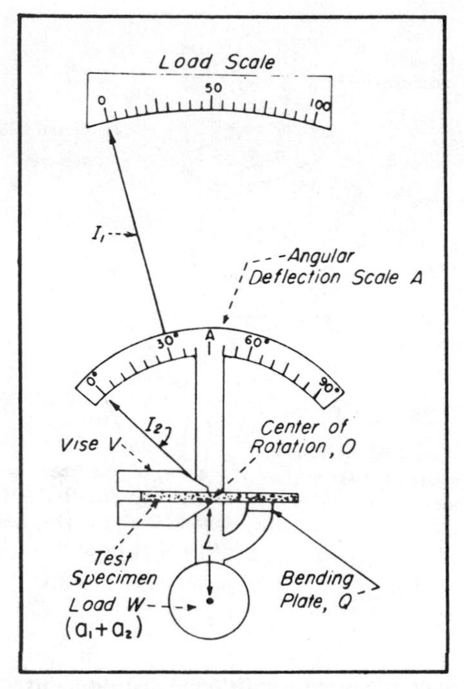

Fig. 27-8. Mechanical system of stiffness tester. (*Copyright ASTM—reprinted by permission*)

imen, which is conditioned according to Procedure A of ASTM D618.

Procedure: A sample is clamped in the base of a pendulum testing machine (see Fig. 27-9) so that it is cantilevered upward with the notch facing the direction of impact. The pendulum is released, and the force consumed in breaking the sample is calculated from the height the pendulum reaches on the follow-through.

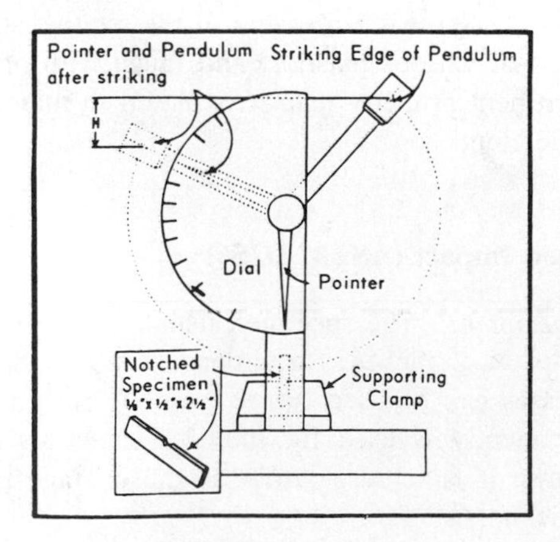

Fig. 27-9. Izod impact.

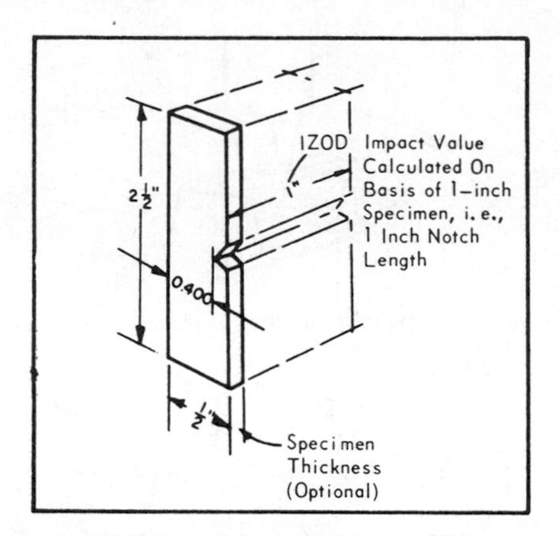

Fig. 27-10. Izod specimen.

Significance: The Izod impact test indicates the energy required to break notched specimens under standard conditions. It is calculated as ft-lb per inch of notch, usually on the basis of a one-inch specimen although the specimen used may be thinner in the lateral direction. (This is indicated in Fig. 27-10.)

The Izod value is most useful in comparing different batches of the same plastic. In comparing one plastic with another, however, the Izod impact test should not be considered a reliable indicator of overall toughness or impact strength. Some materials are notch-sensitive and derive greater concentrations of stress from the notching operation. The Izod impact test may indicate the need for avoiding sharp corners in parts made of such materials. For example, nylon and acetal-type plastics, which in molded parts are among the toughest materials, are notch-sensitive and register relatively low values on the notched Izod impact test.

Tensile Impact (ASTM D1822)

Specimen: Small standard specimens of "tensile bar" shape measuring $2\frac{1}{2}$ inches long are used.

Specimens are conditioned according to Procedure A, ASTM D618.

Procedure: The specimen is mounted between a pendulum head and a crosshead clamp on the pendulum of an impact tester (see Fig. 27-11). The pendulum then is released, and it swings past a fixed anvil that halts the crosshead clamp. The pendulum head continues for-

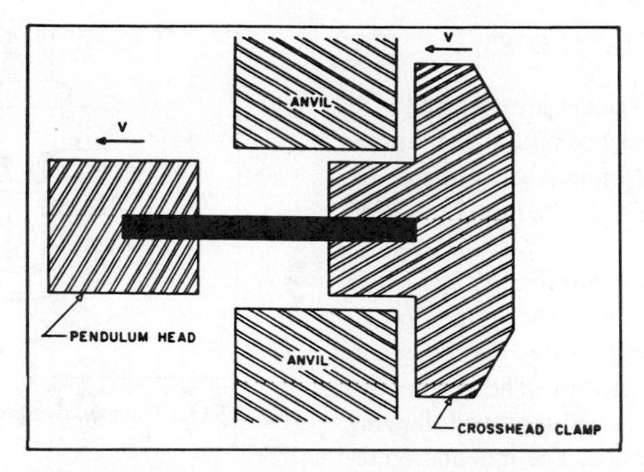

Fig. 27-11. Specimen-in-head tension impact machine (schematic). (*Copyright ASTM—reprinted by permission*)

ward, carrying the forward portion of the ruptured specimen.

The energy loss (tensile impact energy) is recorded, as well as whether the failure appeared to be a brittle or a ductile type.

Significance: This test was adopted by ASTM in 1961. Possible advantages over the notched Izod test are immediately apparent: the notch sensitivity factor is eliminated, and energy is not used in pushing aside the broken portion of the specimen.

Falling Dart Impact (ASTM DD1709, ASTM D2444, ASTM D3029)

Specimen: The specimen can be film (D1709), pipe (D2444), or sheet or a molded part (D3029).

Procedure: A weight, known as a tup or falling dart, is dropped from a tower onto a specimen. The weight can be a cylinder with a rounded nose or a steel ball. Usually the weight hits the specimen directly, but in some tests the impact is indirectly transferred by a metal impactor resting on the specimen.

The amount of weight and the drop height are varied until at least half of the specimens break. Because the weight of the dart and the height are known, the total energy absorbed to cause failure can be calculated.

Significance: Because the specimens are unnotched, the true impact behavior of the material is more correctly observed by this test than in the Izod test. But because the test produces only a single value—total energy to failure—brittleness and ductility are not evaluated. The value of the test, then, is mainly limited to comparing different batches of the same material.

Instrumented Impact

Specimen: Although there are standard specimens for instrumented testing, it can also be done on parts. In the latter mode, instrumented testing can more closely simulate actual service conditions than any other impact test.

Procedure: Instrumented tests can be done with either an Izod-type pendulum or a falling dart apparatus. The pendulum or dart must be equipped with a sensor, usually a load-cell. A computer acquires data from the sensor, which is used to calculate the deformation of and load on the plastic as it is being impacted.

Significance: Instrumented tests break down the impact event into an energy absorption and a failure phase. With that information, engineers obtain an estimate of the service conditions a material can endure, whether it will fail in the ductile or the brittle mode, and how the part will fail (ductile failure vs. catastrophic failure). Instrumented tests give engineers insight into the margin of safety a material provides between initial crack propagation and total failure.

For these reasons, instrumented impact tests have gained adherents for comparing different materials, particularly for engineering purposes. However, interpreting the data is itself a skill.

Rockwell Hardness (ASTM D785)

Specimen: Sheets or plaques are at least $\frac{1}{4}$ inch thick. This thickness may be built up of thinner pieces if necessary. Normally specimens are conditioned according to Procedure A, ASTM D618.

A steel ball under a minor load is applied to the surface of the specimen. (See Fig. 27-12.) This indents slightly and assures good contact. The gauge is then set at zero. The major load is applied for 15 seconds and removed, leaving the minor load still applied. The indentation remaining after 15 seconds is read directly off the dial. This value is preceded by a letter (R, L, M, E, or K) representing the Rockwell hardness scale used.

The size of the balls used and loadings vary, and values obtained with one set cannot be correlated with values from another set.

Significance: Rockwell hardness can differentiate the relative indentation hardness of different types of a given plastic. But elastic recovery is involved as well as hardness, so it is not valid to compare the hardness of various kinds of plastic entirely on the basis of this test.

Rockwell hardness is not an index of wear qualities or abrasion resistance. For instance, polystyrenes have high Rockwell hardness values but poor scratch resistance. The indentation hardness of thermoset materials at room temperature, however, is generally a marker of the degree of cure of the materials. Undercured materials typically are softer than the materials are when fully cured.

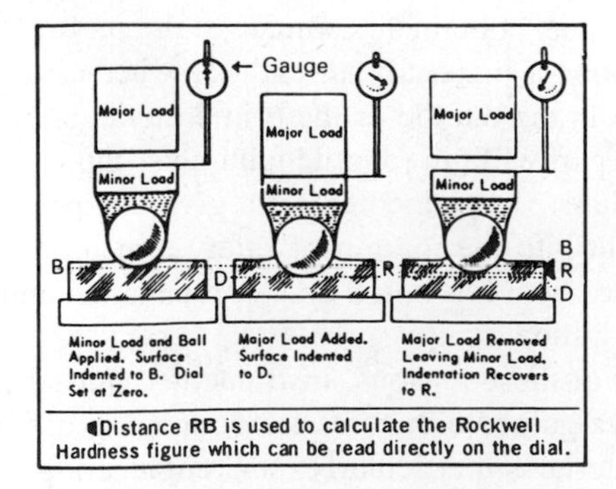

Fig. 27-12. Rockwell hardness.

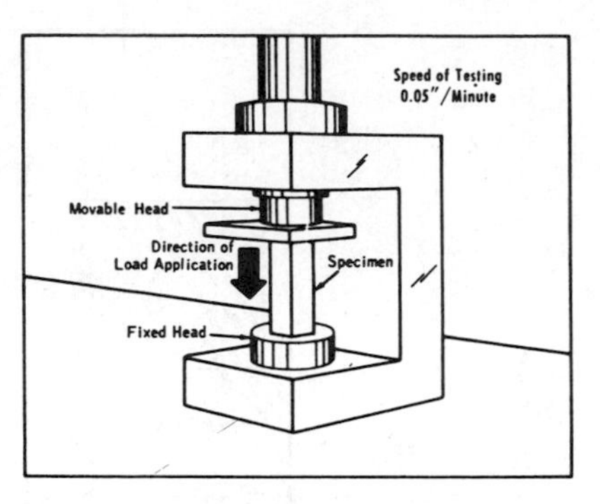

Fig. 27-13. Compressive properties of rigid plastics.

Compressive Properties of Rigid Plastics (ASTM D695)

Specimen: Prisms $\frac{1}{2} \times \frac{1}{2} \times 1$ inch or cylinders $\frac{1}{2}$ inch diameter $\times$ 1 inch.

Specimens usually are conditioned according to Procedure A, ASTM D618.

Procedure: The specimen is mounted in a compression tool between testing machine heads, which exert a constant rate of movement. An indicator registers loading. (See Fig. 27-13).

The compressive strength of a material is calculated as the pressure required to rupture the specimen or deform the specimen a given percentage of its height. It can be expressed as psi either at rupture or at a given percentage of deformation.

Significance: The compressive strength of plastics is of limited design value because plastic products (except foams) seldom fail from compressive loading alone. At the other end of the spectrum, very rigid plastics can shatter from compression. Knowing their compressive strength at failure is necessary in order to avoid catastrophic part failures. More generally, compressive strength figures may be useful in specifications for distinguishing between different grades of a material, and also for assessing, along with other property data, the overall strength of different kinds of materials.

For more flexible materials, an alternate method tests their recovery from compressive load. Only 100 psi pressure is used, and the specimen is removed immediately after compression. It remains at test temperature for one hour and at room temperature for an additional half-hour before deformation is measured.

Durometer Hardness (ASTM D2240)

Specimen: Specimens must be $\frac{1}{4}$ inch thick, although $\frac{1}{8}$-inch specimens may be used if it is shown that results are the same. Normally the specimens are conditioned according to Procedure A, ASTM D618.

Procedure: The durometer instrument has an indenter projecting below the base (face) of the pressure foot. Two types of indenters can be used: blunt-ended (Type A) for softer materials and pointed (Type D) for harder materials. When the indenter is pressed into the plastic specimen so that the base rests on the plastic surface, the amount of indentation registers directly on the dial indicator.

Significance: This test measures the indentation into the plastic of the indenter under load, on a scale of 0 to 100. There is no unit of measurement, and readings above 90 on the A scale may overlap with readings below 20 on the D scale. Readings taken immediately after application may vary from those taken after pressure has been held for a time, because of creep. This test is preferred for polyethylene because the Rockwell test loses meaning when excessive creep is encountered. It also is the standard test for elastomers. For other materials (acetate, acetal, etc.) the Rockwell hardness test still is the standard.

Shear Strength (ASTM D732)

Specimen: Sheets or molded discs from 0.005 to 0.500 inch thick are used. The conditioning method is Procedure A, ASTM D618.

Procedure: The 2-inch-square or 2-inch-diameter round specimen is mounted in a punch-type shear fixture, and the punch (1 inch *D*) is pushed down at a rate of 0.005 in./min. until the moving portion of the sample clears the stationary portion.

Shear strength is calculated as the force/area sheared.

Significance: Shear strength is particularly important in film and sheet products where failures from this type of load may often occur.

For the design of molded and thicker extruded products, it would seldom be a factor.

Deformation under Load (ASTM D621)

Specimen: A $\frac{1}{2}$-inch cube is used, either solid or composite. (The cube can be built up from thinner sheets.) Specimens are conditioned according to Procedure A, ASTM D618.

Procedure: The specimen is placed between the anvils of the testing machine and loaded. (See Fig. 27-14.) The gauge is read 10 seconds after loading, and again 24 hours later. The deflection is recorded in mils. The original height is calculated after the specimen is removed from the testing machine by adding the change in height to the height after testing. By dividing the change in height by the original height and multiplying by 100, the percent deformation is calculated. This test may be run at 73.4, 122, or 158°F and 1000, 500, and 250 psi load. However, the combination of temperature and pressure should not produce more than 25% deformation, as values greater than that are of questionable importance.

Significance: This test on rigid plastics indicates their ability to withstand continuous short-term compression without yielding and loosening when fastened, as in insulators or other assemblies by bolts, rivets, and so on. It does not indicate the creep resistance of a particular plastic for long periods of time.

It also is a measure of rigidity at service temperatures and can be used as identification for procurement.

For nonrigid materials, the alternate method is a measure of their recovery from deforming stress. As such, it also indicates how well a part with elastic properties will follow associated parts during use.

Vicat Softening Point (ASTM D1525)

Specimen: Flat specimens must be at least $\frac{1}{2}$ inch wide and $\frac{1}{8}$ inch thick. Two specimens may be stacked, if necessary, to get the thickness,

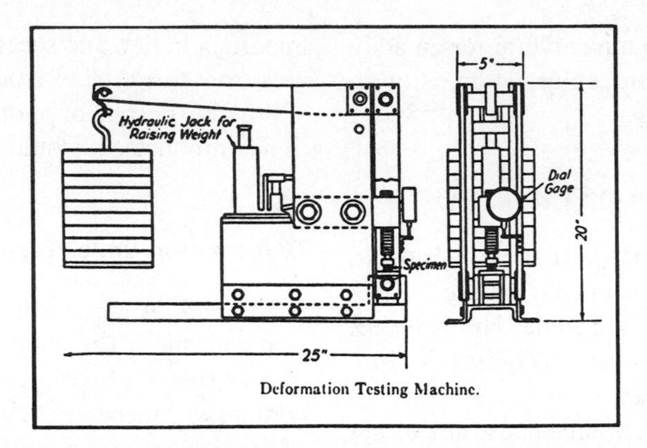

Fig. 27-14. Deformation testing machine. (*Copyright ASTM—reprinted by permission*)

and the specimens may be compression- or injection-molded.

Procedure: The apparatus for testing the Vicat softening point consists of a temperature-regulated oil bath with a flat-ended needle penetrator so mounted as to register the degree of penetration on a gauge. The needle point has a 1000-gram load pressing down on it. (See Fig. 27-15.)

A specimen is placed with the needle resting on it. The temperature of the bath (preheated to about 50°C lower than the anticipated Vicat softening point) is raised at a rate of 50°C/hour or 120°C/hour. The temperature at which the needle penetrates 1 mm is the Vicat softening point.

Significance: The Vicat softening temperature is a good way of comparing the heat-softening characteristics of polyethylenes and polystyrenes.

It also may be used with other thermoplastics, but is not recommended for ethyl cellulose, nonrigid polyvinyl chloride, polyvinylidene chloride, or materials with a wide Vicat softening range.

Deflection Temperature under Flexural Load* (ASTM D648)

Specimen: Specimens measure 5 × $\frac{1}{2}$ inch × any thickness from $\frac{1}{8}$ to $\frac{1}{2}$ inch. Specimens are conditioned according to Procedure A, ASTM D618.

*This property used to be called the "heat distortion temperature."

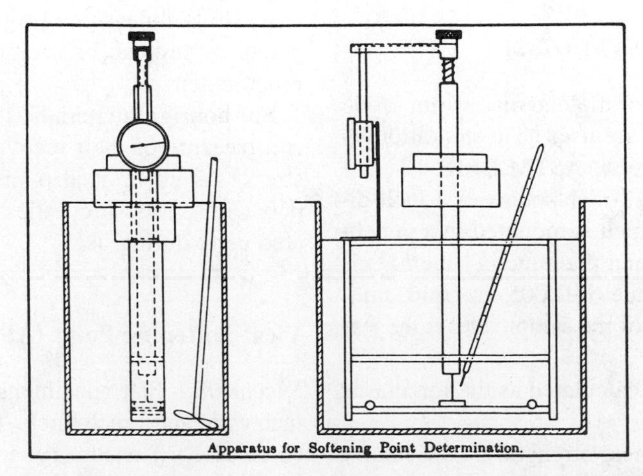

Fig. 27-15. Apparatus for softening point determination. (*Copyright ASTM—reprinted by permission*)

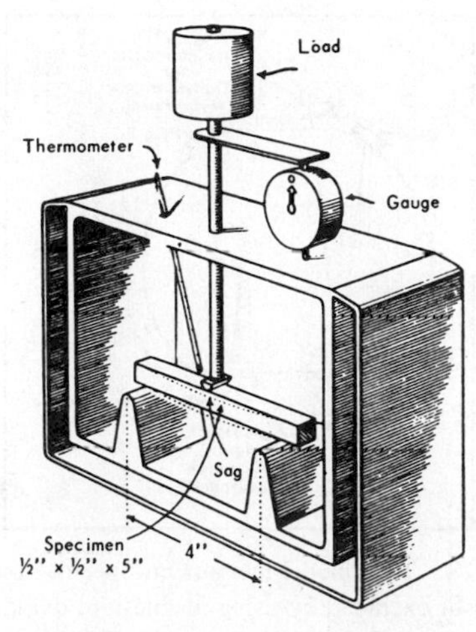

Fig. 27-16. Deflection temperature.

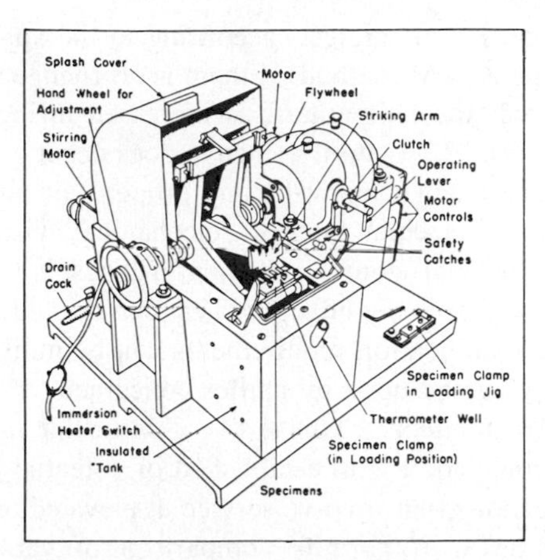

Fig. 27-17. Motor-driven brittleness temperature tester. (*Copyright ASTM—reprinted by permission*)

Procedure: The specimen is placed on supports 4 inches apart, and a load of 66 or 264 psi is placed on the center. (See Fig. 27-16.) The temperature in the chamber is raised at the rate of 2° ± 0.2°C per minute. The temperature at which the bar has deflected 0.010 inch is reported as "deflection temperature at 66 (or 264) psi fiber stress."

Significance: This test shows the temperature at which an arbitrary amount of deflection occurs under established loads. It is not intended to be a direct guide to high-temperature limits for specific applications. It may be useful in comparing the relative behavior of various materials in these test conditions, but it is primarily useful for control and development purposes.

Brittleness Temperature (ASTM D746)

Specimen: Pieces used are $\frac{1}{4}$ inch wide, 0.075 inch thick, and $1\frac{1}{4}$ inch long. Specimens also may be T-shaped, with the longer section being 1 inch long and 0.100 inch wide. They are conditioned according to Procedure A, ASTM D618.

Procedure: The conditioned specimens are cantilevered from the sample holder in the test apparatus (Fig. 27-17), which has been brought to a low temperature (that at which specimens would be expected to fail). When the specimens have been in the test medium for 3 minutes, a single impact is administered, and the samples are examined for failure. Failures are total breaks, partial breaks, or any visible cracks. The test is conducted at a range of temperatures producing varying percentages of breaks. From these data, the temperature at which 50% failure would occur is calculated or plotted and reported as the brittleness temperature of the material according to this test.

Significance: This test is of some use in judging the relative merits of various materials for low-temperature flexing or impact. However, it is specifically relevant only for materials and conditions specified in the test, and the values cannot be directly applied to other shapes and conditions.

The brittleness temperature does not put any lower limit on the service temperature for end-use products. The brittleness temperature sometimes is used in specifications.

Permanent Effect of Heat (ASTM D794)

Specimen: Any piece of plastic or molded plastic part may be used.

Procedure: The specimens are placed in an air circulating oven at a temperature (multiple of 25°C) that is thought or known to be near the temperature limit of the material. The material may be maintained at that temperature, or cycled between two temperatures. The exposure time is determined by the tester and can range from minutes to weeks. At the end of the exposure period, the specimen is tested for the properties of interest, according to the appropriate ASTM method. If there is no change observed, the temperature is increased in increments of 25°C until a change does occur.

The change might be any property or properties of special interest—mechanical, visual, dimensional, color, and so on. The test is written so that many effects of heat can be studied, and specification requirements can be individually agreed upon by parties concerned.

Significance: This test is of particular value in connection with established or potential applications that involve service at elevated temperatures. It permits comparison of various plastics and grades of one plastic in the form of test specimens, as well as molded parts in finished form.

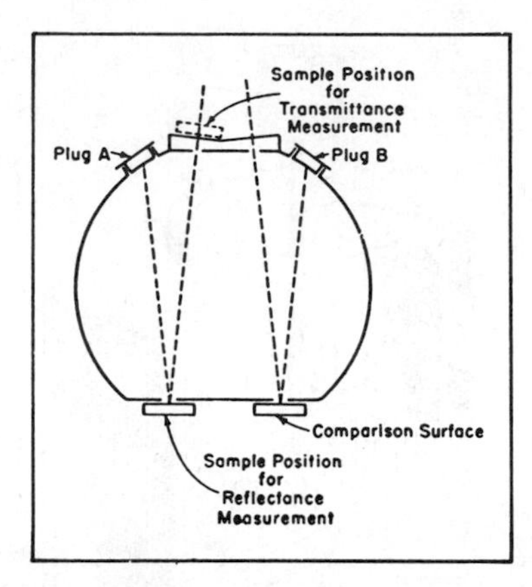

Fig. 27-18. Late model instrument. Specular component included or excluded by using magnesium oxide or black velvet plugs, respectively. (*Copyright ASTM—reprinted by permission*)

Haze and Luminous Transmittance of Transparent Plastics (ASTM D1003)

Specimen: Transparent film and sheet samples are used, or molded samples that have parallel plane surfaces. A disc, 2 inches in diameter, is recommended. No conditioning is required.

Procedure: Procedure A is followed when a hazemeter is used in the determinations.

Procedure B is followed when a recording spectrophotometer is used.

Significance: In this test, the haze of a specimen is defined as the percentage of transmitted light that, in passing through the specimen, deviates more than 2.5° from the incident beam by forward scattering.

Luminous transmittance is defined as the ratio of transmitted to incident light.

These qualities are considered in most applications for transparent plastics. They form a basis for directly comparing the transparency of various grades and types of plastic.

From the raw material manufacturing standpoint, they are important control tests in the various stages of production.

Luminous Reflectance, Transmittance and Color (ASTM E308)

Specimen: Opaque specimens should have at least one plane surface. Translucent and transparent specimens must have two surfaces that are plane and parallel. The piece must be at least 2 inches in diameter.

Procedure: The sample is mounted in the instrument, and along with it a comparison surface (white chalk). (See Fig. 27-18.) The samples are placed in the instrument, and light of different wavelength intervals is impinged against the surface. Reflected or transmitted light then is measured to obtain the property values listed below.

Significance: This test is the primary method of obtaining colorimetric data. Properties determined include the following:

1. Total luminous reflectance or the luminous directional reflectance.
2. Luminous transmittance.
3. The chromaticity coordinates x and y (color).

Tests for Electrical Resistance (ASTM D257)

Specimen: Specimens for these tests may be any practical form, such as flat plates, sheets, and tubes.

Procedure: These tests describe methods for determining the several properties defined below. Two electrodes are placed on or embed-

Table 27-1. Electrical resistance properties.

Property	Definition	Significance
Insulation Resistance	Ratio of direct voltage applied to the electrodes to the total current between them; dependent upon both volume and surface resistance of the specimen.	In materials used to insulate and support components of an electrical network, it is generally desirable to have insulation resistance as high as possible.
Volume Resistivity	Ratio of the potential gradient parallel to the current in the material to the current density.	Knowing the volume and surface resistivity of an insulating material makes it possible to design an insulator for a specific application.
Surface Resistivity	Ratio of the potential gradient parallel to the current along its surface to the current per unit width of the surface.	
Volume Resistance	Ratio of direct voltage applied to the electrodes to that portion of current between them that is distributed through the volume of the specimen.	High volume and surface resistance are desirable in order to limit the current leakage of the conductor which is being insulated.
Surface Resistance	Ratio of the direct voltage applied to the electrodes to that portion of the current between them which is in a thin layer of moisture or other semiconducting material that may be deposited on the surface.	

ded in the surface of a test specimen. The properties listed in Table 27-1 are calculated.

Significance: See Table 27-1.

Dielectric Constant and Dissipation Factor (ASTM D150)

Specimen: The specimen may be a sheet of any size convenient to test, but should be of uniform thickness. The test may be run at standard room temperatures and humidity, or in special sets of conditions as desired. In any case, the specimens should be preconditioned to the set of conditions used.

Procedure: Alternating current is applied through electrodes attached to opposite faces of the test specimen. Frequencies from less than 1 Hz to over 100 Mz can be used. The test is relatively insensitive to voltage levels, as long as the voltage used is less than that required to break down the plastic. The capacitance and dielectric loss are then measured by comparison or substitution methods in an electric bridge circuit. From these measurements and the dimensions of the specimen, the following quantities are computed: dielectric constant (permittivity), dissipation factor, loss index, power factor, phase angle, and loss angle.

Significance: This test characterizes a plastic's ac electrical characteristics. The permittivity, or dielectric constant, is important in determining the charge-storing qualities of a material. If a plastic is to be used as an insulator, low permittivity is desirable; capacitor materials require higher permittivity. Most electrical applications require plastics with low ac loss (indicated by the dissipation factor).

Arc Resistance (ASTM D495)

Specimen: Standard specimens are $\frac{1}{8}$ inch thick, but molded parts also may be used.

Procedure: Two electrodes are placed on the specimen, spaced $\frac{1}{4}$ inch apart. A 12,500-volt, 10 milliampere current is applied, which creates an arc between the electrodes. The arc is increased in duration and/or power every 60 seconds until failure occurs. At failure, the arc disappears into the material. The arc resistance is the number of seconds required to produce failure.

Significance: This test shows the ability of a material to resist the action of an arc of high voltage and low current, close to the surface of the insulation, measured by the time required to form a conducting path therein.

The arc resistance values are of relative value only, in distinguishing materials of nearly identical composition, such as for quality control, development, and identification.

Dielectric Strength (ASTM D149)

Specimen: Specimens are thin sheets, molded parts, or plates having parallel plane surfaces and of a size sufficient to prevent flashing over. The dielectric strength varies with thickness; so the specimen thickness must be reported.

Because temperature and humidity affect results, it is necessary to condition each type of material as directed in the specification for that material. The test for dielectric strength must be run in the conditioning chamber or immediately after removal of the specimen from the chamber.

Procedure: The specimen is placed between heavy cylindrical brass electrodes that carry electrical current during the test. There are two ways of running this test for dielectric strength:

1. *Short-time test:* The voltage is increased from zero to breakdown at a uniform rate—0.5 to 1.0 kV/sec. The precise rate of voltage rise is specified in governing material specifications.
2. *Step-by-step test:* The initial voltage applied is 50% of the breakdown voltage shown by the short-time test. It is increased at rates specified for each type of material, and the breakdown level is noted.

Breakdown by these tests means passage of sudden excessive current through the specimen, which can be verified by instruments and visible damage to the specimen.

Significance: This test is an indication of the electrical strength of a material as an insulator. The dielectric strength of an insulating material is the voltage gradient at which electric failure or breakdown occurs as a continuous arc (the electrical property analogous to tensile strength in mechanical properties). The dielectric strength of materials varies greatly with several conditions, such as humidity and geometry, and it is not possible to directly apply the standard test values to field use unless all conditions, including specimen dimensions, are the same. Because of this, the dielectric strength test results are of relative rather than absolute value as a specification guide.

The dielectric strength of polyethylenes is usually around 500 volts/mil. The value will drop sharply if holes, bubbles, or contaminants are present in the specimen being tested.

The dielectric strength varies inversely with the thickness of the specimen.

Outdoor Weathering (ASTM D1435)

Specimen: No specified size. Specimens for this test may consist of any standard molded test specimen or cut pieces of sheet or machined samples.

Procedure: Specimens are mounted outdoors on racks facing south. The angle of the racks may be equivalent to the geographic latitude of the site (for maximum ultraviolet exposure), 45°, 90°, or horizontal (appropriate for roof membranes, artificial turf, etc.). It is recommended that concurrent exposure be carried out in many varied climates to obtain the broadest, most representative total body of data. Sample specimens are kept indoors as controls and for comparison.

Reports of weathering describe all changes noted, areas of exposure, and period of time. Exposure times can be as short as one week or last for several years.

Significance: Outdoor testing is the most accurate method of obtaining a true picture of weather resistance. The only drawback of this test is the time required for several years of exposure.

A large number of specimens usually are required to allow periodic removal and to run representative laboratory tests after exposure.

Accelerated Weathering (ASTM G23) (Recommended Practice)

Specimen: Any shape; size up to 5 × 7 × 2 inches.

Procedure: Artificial weathering has been defined by ASTM as "The exposure of plastic to cyclic laboratory conditions involving changes in temperature, relative humidity, and ultraviolet (UV) radiant energy, with or without direct water spray, in an attempt to produce changes in the material similar to those observed after long-term continuous outdoor exposure."

Three types of light sources for artificial weatherings are in common use, as shown in Table 27-2. Selection of a light source involves many conditions and circumstances, such as what material is being tested, the proposed end-use, previous testing experience, the type of information desired, and so on.

Table 27-2. Accelerated weathering light sources.

Source	UV Energy Output, Approx. (X Sunlight)
Enclosed UV Carbon Arc	7.5
Open-flame Sunshine Carbon Arc	3
Water-cooled Xenon Arc	1*

* (ASTM G26).

Xenon arcs have been shown to have a spectral energy distribution, when properly filtered, that closely simulates that of sunlight. However, because the energy emitted by xenon lamps decays with time, and the parameters of temperature and water do not represent a specific known climatic condition, the results of laboratory exposure are not necessarily intended to correlate with data obtained by outdoor weathering or by using other devices.

Significance: Because weather varies from day to day, year to year, and place to place, no precise correlation exists between artificial laboratory weathering and natural outdoor weathering. However, standard laboratory test conditions produce results with acceptable reproducibility that are in *general* agreement with data obtained from outdoor exposures. Fairly rapid indications of weatherability thus are obtainable on samples of known materials that, through testing experience over a period of time have general correlations established. There is no artificial substitute for *precisely* predicting outdoor weatherability on materials with no previous weathering history.

Accelerated Exposure to Sunlight Using the Atlas Type 18FR Fade-Ometer® (ASTM G23)

The Atlas Type 18FR Fade-Ometer is used primarily to check and compare color stability. Besides determining the ability of various pigments needed to provide both standard and custom colors, the Fade-Ometer is helpful in preliminary studies of various stabilizers, dyes, and pigments compounded in plastics to prolong their useful life. It is designed to test materials used in articles subject to indoor exposure to sunlight.

The Fade-Ometer was extensively used in the development of UV-absorbing acetate film for store windows to protect merchandise displayed in direct sunlight.

Exposure in the Fade-Ometer cannot be directly related to exposure in direct sunlight,

Table 27-3. Pipe tests.

Test	Procedure	Significance
Sustained Pressure	Specimens are filled with water and put under pressure and temperatures indicated for each type of pipe and held for at least 1000 hours.	Service installations of plastic pipe may be expected to give many years of service. This test, by using relatively high pressures, indicates whether the materials will hold up in normal service. The test advises continuation for a year or longer for further reliability.
Incremental Pressure	Specimens are filled with water and brought to pressure level indicated. Pressure is raised in steps until maximum indicated pressure is attained. No more than 2 of 6 specimens can fail.	Most pipe in service bears pressure of varying amounts sometimes abruptly changed. This can place a greater strain on pipe than gradual changes. The incremental pressure tests show ability of pipe to withstand relatively sudden increases in pressure to levels above "normal" service ratings.
Environmental Stress Cracking	Lengths of pipe are put under pressure and coated with Igepal CO 630. After 3 hours, 4 of 6 specimens must retain full pressure.	Susceptibility to ESC is checked as a quality control measure, to assure that excessive levels of internal strain in the pipe will not contribute to failure in service contacts with such agents as detergents.
Density	Density of base resin is calculated by the density gradient technique. (Amount of carbon black is given as percentage.)	Many properties are density related. The base resin density in pipe determines classification and size requirements.

partially because other weather factors are always present outdoors.

Pipe Tests (Commercial Standard CS 255*)

Specimen: Sections of pipe are used in these tests. The lengths are specified for each type, grade and dimension of pipe.

*This Standard lists pipe wall thicknesses required for various materials and kinds of service, general appearance and finish requirements, and tolerances, as well as the tests is described here. (The pressure test apparatus is described in ASTM D1598-63T.) Passing all tests is a requirement for marketing pipe as meeting this commercial standard.

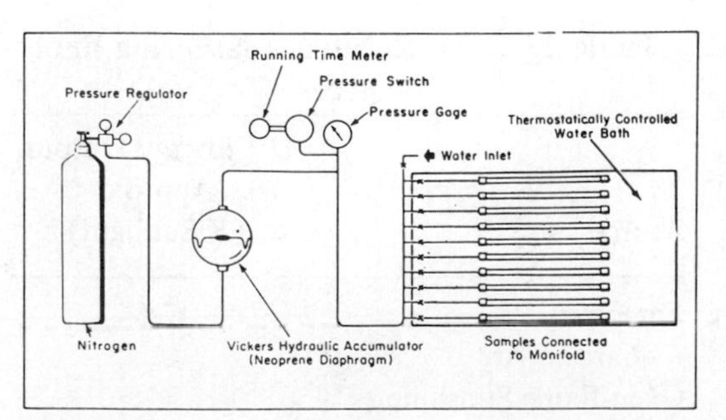

Fig. 27-19. Diagram of pipe stress rupture equipment. (*Copyright ASTM—reprinted by permission*)

Procedure and Significance: Table 27-3 summarizes the test procedures and their significance.

AUTOMATION OF TESTING

The advent of inexpensive computers in the 1980s changed the way performance testing is done, and more changes are on the way.

The first change that occurred was the use of microprocessors to control testing equipment. Universal testers, for example, were equipped with on-board computers that allowed the crosshead speeds and loads exerted on the specimen to be both closely controlled and readily changed.

As robotized sample loaders have been added to computer-controlled testing equipment, the drudgery of repeated testing has been greatly reduced. Highly repetitive tests now can be performed without human intervention.

The freedom to perform many test routines without human involvement opens the door to overcoming some of the limitations of performance tests. Most mechanical tests, such as tensile, flexural, and compressive strength tests, typically are performed at a single load or crosshead speed. The result is single-point data that may actually be misleading when comparing different materials. With equipment that can be programmed to follow a sequence of different test parameters, and with robots to change samples, laboratory personnel have the ability to perform multiple test regimens and gather a family of data that is more revealing than single-point data. Those data then can be used by engineers to gain more insight into the likely real-world performance of a material.

For quality control laboratories, computerization has imposed a new challenge: gathering and analyzing the data from different computerized instruments. Hardware and software systems have been developed to collate test results from multiple instruments and assemble them into a unified report that documents many aspects of a material's performance characteristics.

28

Standards for Molding Tolerances

This chapter contains guidelines for practical tolerances for the dimensions of articles molded from a variety of thermosetting and thermoplastic materials. The information is presented in the form of tables based on data compiled from a survey of plastics molders, parts assemblers, and materials suppliers to establish the values most consistent with good industry wide molding practices. The tables were developed by the Molders Division of the Society of the Plastics Industry, Inc. in cooperation with the Manufacturing Resource and Productivity Center of Ferris State University.

Ranges of acceptable specifications have been determined using the simplest of parts for reference. It must be stressed that these tables are not to be construed as offering hard-and-fast rules applicable to all conditions. Complicated and non-uniform designs must be considered individually. The guidelines can be best used as a basis for communication between engineers, designers, molders, and customers.

Materials covered in the pages following are:

It should be recognized that extreme accuracy of dimensions in molded articles is expensive to achieve. The closer the tolerances demanded, the greater will be the cost of the molds, and also the greater will be the operational costs of molding because of the extra care required to maintain uniform conditions. In some cases, a further expense may arise from the need to use cooling fixtures after parts are removed from the mold.

Dimensional tolerances in a molded article are the allowable variations, plus and minus, from a nominal or mean dimension. Fine tolerance represents the narrowest possible limits of variation obtainable under close supervision and control of production. Standard tolerance is that which can be held under average conditions of manufacture.

Using the Tables

The basic format of the tables is similar for each of the materials that are covered—whether for thermoplastics or thermosets.

Users will find that two separate sets of values are represented. The commercial values represent common production tolerances that can be achieved at the most economical level. The fine values represent closer tolerances that can be held, but at a greater cost. The selection will depend on the application under consideration and the economics that are involved.

By referring to the hypothetical molded article and its cross-section illustrated in the table, and by then using the applicable code number (e.g., A represents the diameter) in the first column of the table and the exact dimensions as indicated in the second column, readers can find the recommended tolerances either in the chart at the top of the table or in the two columns underneath. (Note that the typical article shown in cross-section in the tables may be of round or rectangular or other shapes. Thus, dimensions A and B may be diameters or lengths.)

For example, an ABS part with a diameter (A) of 2 inches would show tolerances of plus-or-minus 0.004 inches (fine value) and 0.007 inches (commercial value). If the dimensions go up to 6 inches, the tolerances would change to plus-or-minus 0.008 inches (fine) and 0.013 inches (commercial). If the dimension is greater than 6 inches, however, the tolerance for 6 inches is increased by the amount indicated on the lines directly below the chart in the table. Thus, if dimension A is 10 inches, the tolerance (commercial) is plus-or-minus 0.025 inches (i.e., 0.013 inches for a 6-inch dimension plus 0.003 for each of the additional four inches).

Other tolerances for the various dimensions shown in the typical cross-section are as indicated in the two columns running under the chart.

Special notes for users of the charts:

(1) The tolerances indicated in the tables for diameter and hole size do not include allowance for aging characteristics of the particular plastics material under consideration.
(2) Tolerances are based on $\frac{1}{8}$-inch wall sections.
(3) For depth, height, and bottom wall dimensions, parting line must be taken into consideration.
(4) In terms of side wall dimensions, flatness, and concentricity, part design should maintain a wall thickness as nearly constant as possible. Complete uniformity in this dimension is impossible to achieve.
(5) In determining hole size depth and draft allowance per side, care must be taken that the ratio of the depth of a cored hole to the diameter does not reach a point that will result in excessive pin damage.
(6) Values for fillets, ribs, and corners should be increased whenever compatible with desired design and good molding technique.
(7) Where surface finish and color stability are concerned, customer and molder must come to an understanding as to what is necessary for the particular job under consideration prior to the decision to begin molding.

Design Considerations

"Do I need, and can I afford, the tolerances I am specifying?" If the designer will also ask himself these two questions before specifying tolerances, excessive mold and processing costs can often be avoided with no penalty in part performance. Regardless of the economics, it may be unreasonable to specify close production tolerances on a part when it is designed to operate through a wide range of environmental conditions. Dimension changes due to temperature variations alone can be three to four times as great as the specified tolerances. Also, in many applications, close tolerances with plastics are not as critical as with metals because of the resiliency of plastics.

Many factors control the production of precision parts and it is beyond the scope of this handbook to completely review this complex subject. However, realizing that the proper specification of tolerance is important to the designer, the following suggestions are offered:

(1) A design for a plastic part should indicate the conditions under which the dimensions shown must be held. For example, a drawing should state that dimensions shown are to be as specified after molding, after annealing, after moisture conditioning, etc.

(2) Overall tolerances for a part should be shown in inches per inch, not in fixed values. As an example, a title block might read, "All decimal dimensions ±0.00X in. per in. (±0.00X mm/mm), unless otherwise specified."

(3) Only those tight tolerances required for specific dimensions should be labeled as such. Less important dimensions can be controlled by overall tolerances.

(4) Generous molding tolerances should be allowed in any areas which will be machined after molding. Production variables such as the number and size of cavities in a mold also affect tolerances. Where compromises in tolerances may be acceptable from a performance standpoint, discussion of such tolerances with the molder may result in economies.

For example, the use of a multi-cavity mold is usually an economical production method. But, as the number of cavities per mold increases, so must the tolerances on critical dimensions. An increase of 1–5% per cavity is about average. For example: dimensions of a part produced in a single-cavity mold may be held to ±0.002 in. per in. (±0.002 mm/mm). When the number of cavities is increased to 20, the closest tolerances obtainable may be ±0.004 in. per in. (±0.004 mm/mm), an increase of 5% per cavity, or a total increase of 100%.

Standards & Practices of Plastics Molders

Material
Alkyd
(Thermoset)

Note: The *Commercial* values shown below represent common production tolerances at the most economical level. The *Fine* values represent closer tolerances that can be held but at a greater cost. Any addition of fillers will compromise physical properties and alter dimensional stability. Please consult the manufacturer.

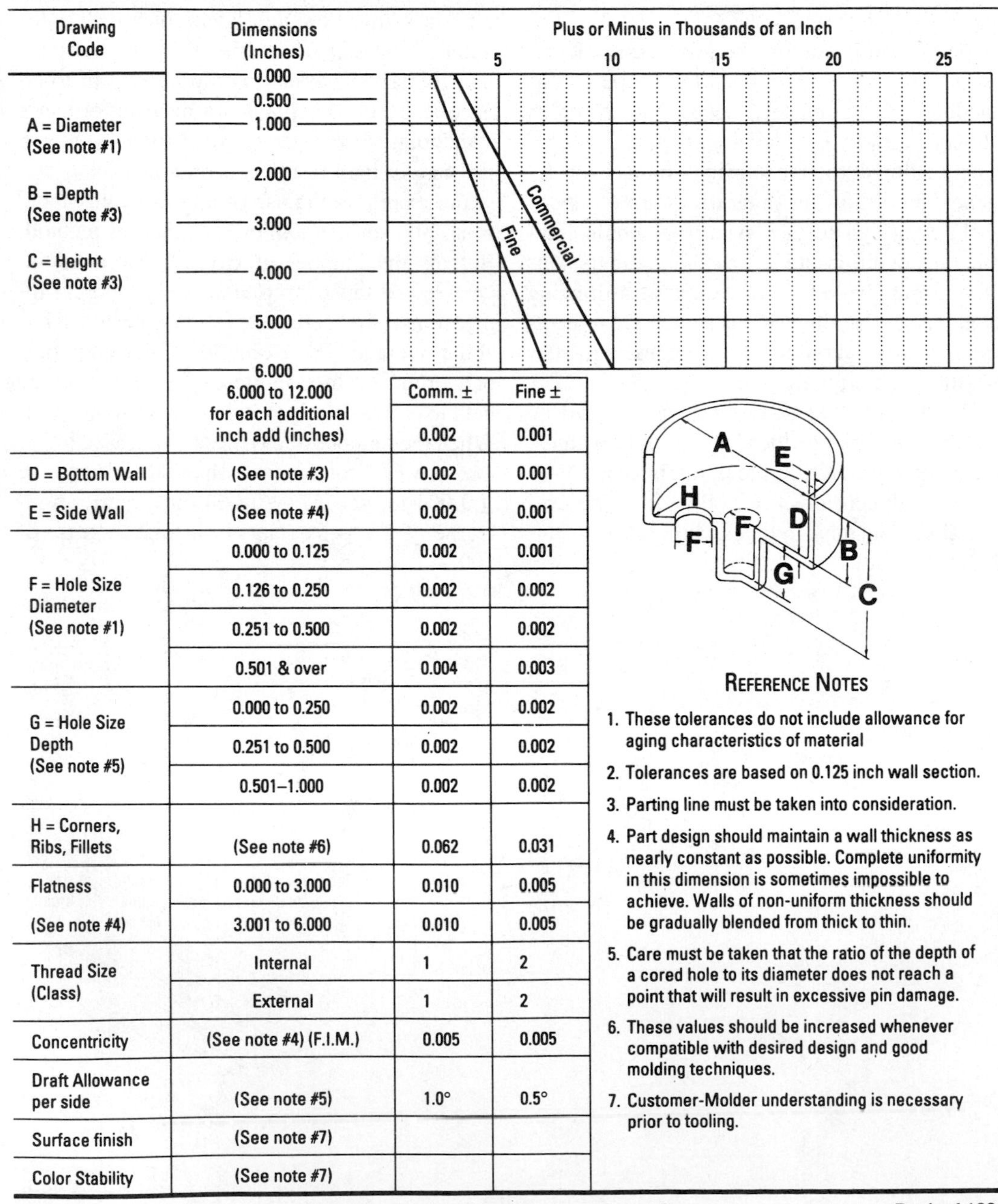

Drawing Code	Dimensions (Inches)	Comm. ±	Fine ±
	6.000 to 12.000 for each additional inch add (inches)	0.002	0.001
D = Bottom Wall	(See note #3)	0.002	0.001
E = Side Wall	(See note #4)	0.002	0.001
F = Hole Size Diameter (See note #1)	0.000 to 0.125	0.002	0.001
	0.126 to 0.250	0.002	0.002
	0.251 to 0.500	0.002	0.002
	0.501 & over	0.004	0.003
G = Hole Size Depth (See note #5)	0.000 to 0.250	0.002	0.002
	0.251 to 0.500	0.002	0.002
	0.501–1.000	0.002	0.002
H = Corners, Ribs, Fillets	(See note #6)	0.062	0.031
Flatness	0.000 to 3.000	0.010	0.005
(See note #4)	3.001 to 6.000	0.010	0.005
Thread Size (Class)	Internal	1	2
	External	1	2
Concentricity	(See note #4) (F.I.M.)	0.005	0.005
Draft Allowance per side	(See note #5)	1.0°	0.5°
Surface finish	(See note #7)		
Color Stability	(See note #7)		

Reference Notes

1. These tolerances do not include allowance for aging characteristics of material
2. Tolerances are based on 0.125 inch wall section.
3. Parting line must be taken into consideration.
4. Part design should maintain a wall thickness as nearly constant as possible. Complete uniformity in this dimension is sometimes impossible to achieve. Walls of non-uniform thickness should be gradually blended from thick to thin.
5. Care must be taken that the ratio of the depth of a cored hole to its diameter does not reach a point that will result in excessive pin damage.
6. These values should be increased whenever compatible with desired design and good molding techniques.
7. Customer-Molder understanding is necessary prior to tooling.

Standards & Practices of Plastics Molders

Material
Diallylphthalate
(DAP)

Note: **The *Commercial* values shown below represent common production tolerances at the most economical level. The *Fine* values represent closer tolerances that can be held but at a greater cost. Any addition of fillers will compromise physical properties and alter dimensional stability. Please consult the manufacturer.**

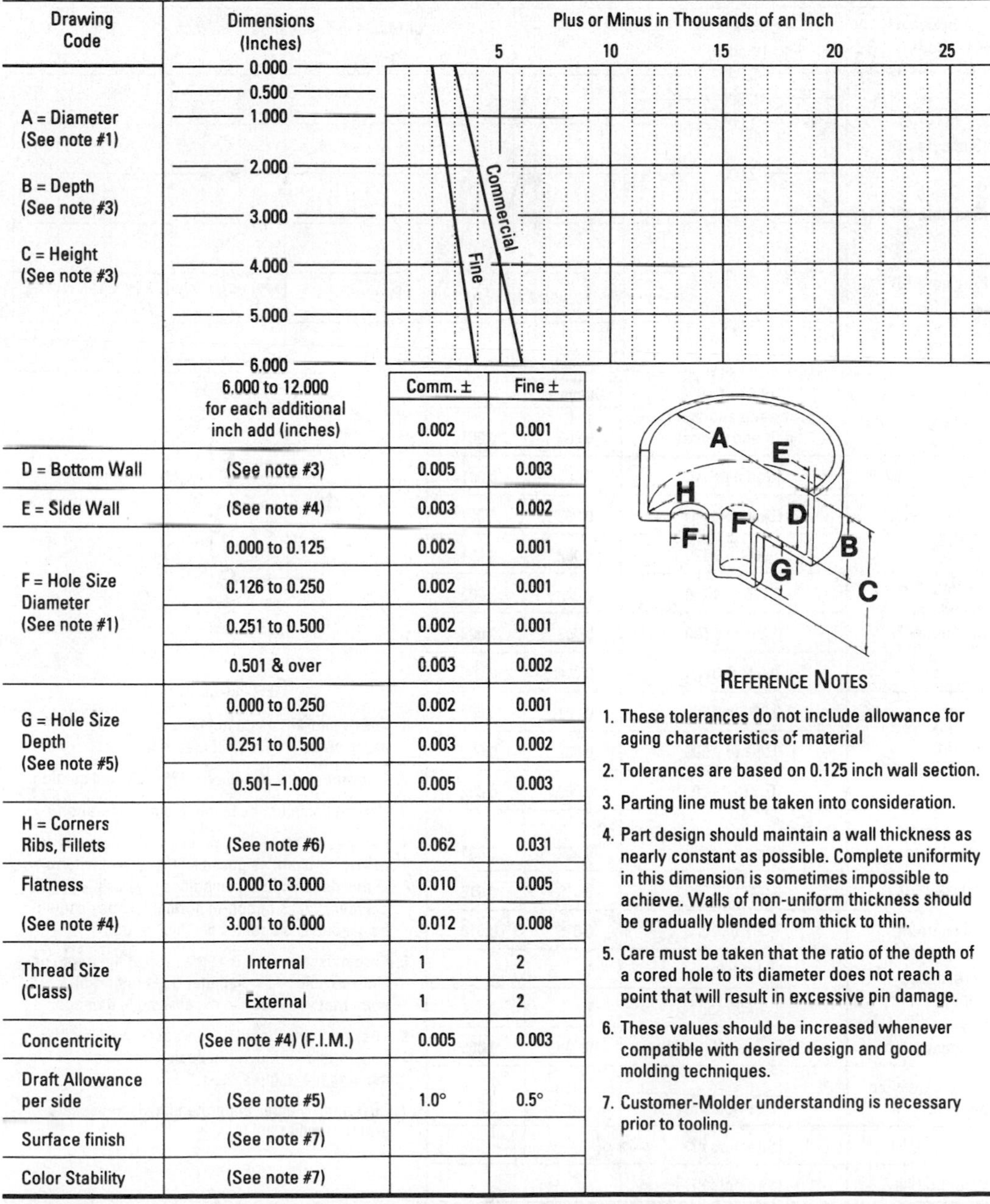

Drawing Code	Dimensions (Inches)	Plus or Minus in Thousands of an Inch	
	0.000	5 10 15 20 25	
	0.500		
A = Diameter (See note #1)	1.000		
B = Depth (See note #3)	2.000	Commercial	
C = Height (See note #3)	3.000	Fine	
	4.000		
	5.000		
	6.000		
		Comm. ±	**Fine ±**
	6.000 to 12.000 for each additional inch add (inches)	0.002	0.001
D = Bottom Wall	(See note #3)	0.005	0.003
E = Side Wall	(See note #4)	0.003	0.002
F = Hole Size Diameter (See note #1)	0.000 to 0.125	0.002	0.001
	0.126 to 0.250	0.002	0.001
	0.251 to 0.500	0.002	0.001
	0.501 & over	0.003	0.002
G = Hole Size Depth (See note #5)	0.000 to 0.250	0.002	0.001
	0.251 to 0.500	0.003	0.002
	0.501–1.000	0.005	0.003
H = Corners Ribs, Fillets	(See note #6)	0.062	0.031
Flatness (See note #4)	0.000 to 3.000	0.010	0.005
	3.001 to 6.000	0.012	0.008
Thread Size (Class)	Internal	1	2
	External	1	2
Concentricity	(See note #4) (F.I.M.)	0.005	0.003
Draft Allowance per side	(See note #5)	1.0°	0.5°
Surface finish	(See note #7)		
Color Stability	(See note #7)		

Reference Notes

1. These tolerances do not include allowance for aging characteristics of material
2. Tolerances are based on 0.125 inch wall section.
3. Parting line must be taken into consideration.
4. Part design should maintain a wall thickness as nearly constant as possible. Complete uniformity in this dimension is sometimes impossible to achieve. Walls of non-uniform thickness should be gradually blended from thick to thin.
5. Care must be taken that the ratio of the depth of a cored hole to its diameter does not reach a point that will result in excessive pin damage.
6. These values should be increased whenever compatible with desired design and good molding techniques.
7. Customer-Molder understanding is necessary prior to tooling.

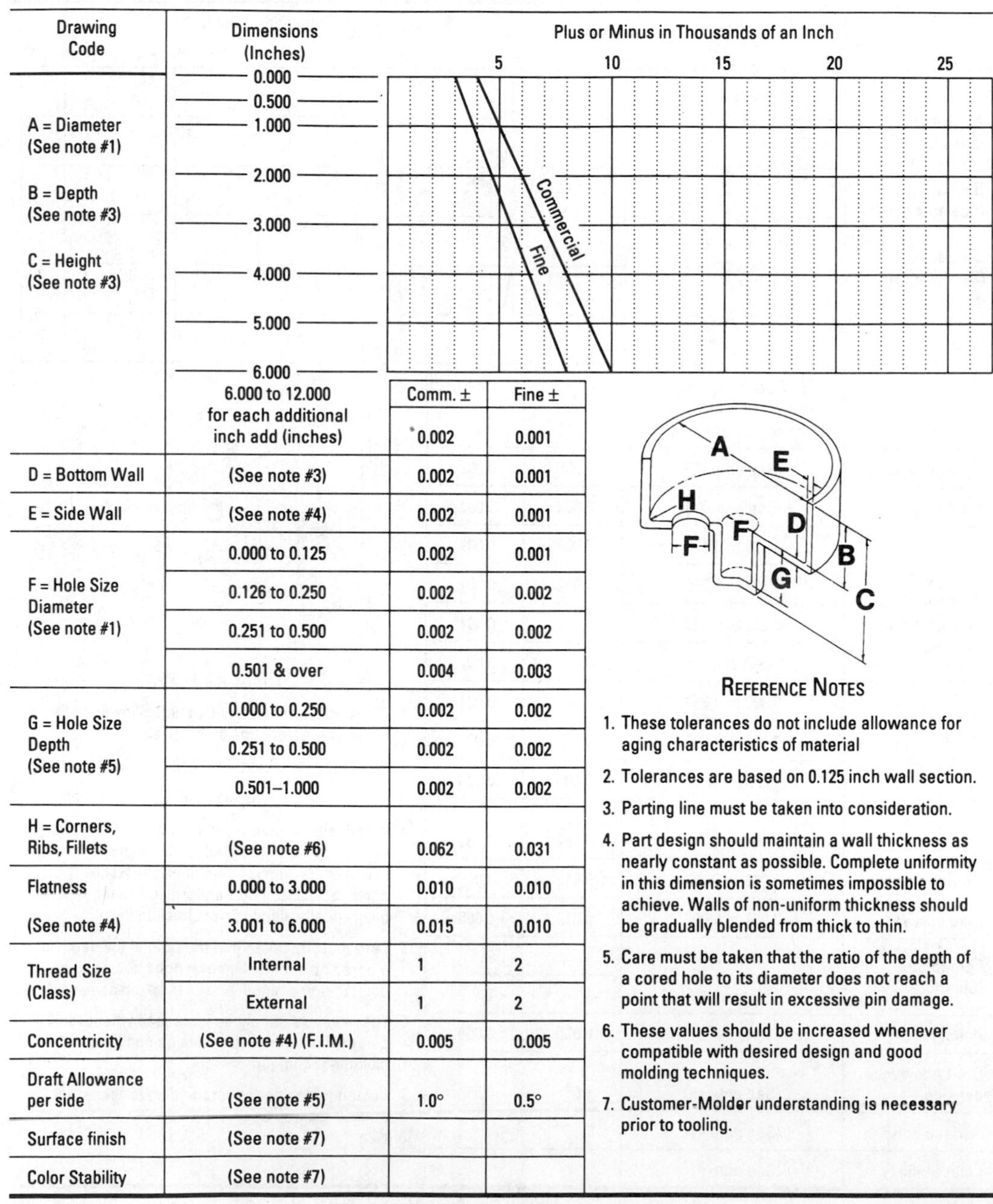

Standards & Practices of Plastics Molders

Material
Epoxy
(EP)

Note: The *Commercial* values shown below represent common production tolerances at the most economical level. The *Fine* values represent closer tolerances that can be held but at a greater cost. Any addition of fillers will compromise physical properties and alter dimensional stability. Please consult the manufacturer.

Drawing Code	Dimensions (Inches)	Comm. ±	Fine ±
	6.000 to 12.000 for each additional inch add (inches)	0.002	0.001
D = Bottom Wall	(See note #3)	0.002	0.001
E = Side Wall	(See note #4)	0.002	0.001
F = Hole Size Diameter (See note #1)	0.000 to 0.125	0.002	0.001
	0.126 to 0.250	0.002	0.002
	0.251 to 0.500	0.002	0.002
	0.501 & over	0.004	0.003
G = Hole Size Depth (See note #5)	0.000 to 0.250	0.002	0.002
	0.251 to 0.500	0.002	0.002
	0.501–1.000	0.002	0.002
H = Corners, Ribs, Fillets	(See note #6)	0.062	0.031
Flatness (See note #4)	0.000 to 3.000	0.010	0.010
	3.001 to 6.000	0.015	0.010
Thread Size (Class)	Internal	1	2
	External	1	2
Concentricity	(See note #4) (F.I.M.)	0.005	0.005
Draft Allowance per side	(See note #5)	1.0°	0.5°
Surface finish	(See note #7)		
Color Stability	(See note #7)		

Reference Notes

1. These tolerances do not include allowance for aging characteristics of material
2. Tolerances are based on 0.125 inch wall section.
3. Parting line must be taken into consideration.
4. Part design should maintain a wall thickness as nearly constant as possible. Complete uniformity in this dimension is sometimes impossible to achieve. Walls of non-uniform thickness should be gradually blended from thick to thin.
5. Care must be taken that the ratio of the depth of a cored hole to its diameter does not reach a point that will result in excessive pin damage.
6. These values should be increased whenever compatible with desired design and good molding techniques.
7. Customer-Molder understanding is necessary prior to tooling.

Standards & Practices of Plastics Molders

Material
Melamine - Urea
(MF-UF)

Note: The *Commercial* values shown below represent common production tolerances at the most economical level. The *Fine* values represent closer tolerances that can be held but at a greater cost. Any addition of fillers will compromise physical properties and alter dimensional stability. Please consult the manufacturer.

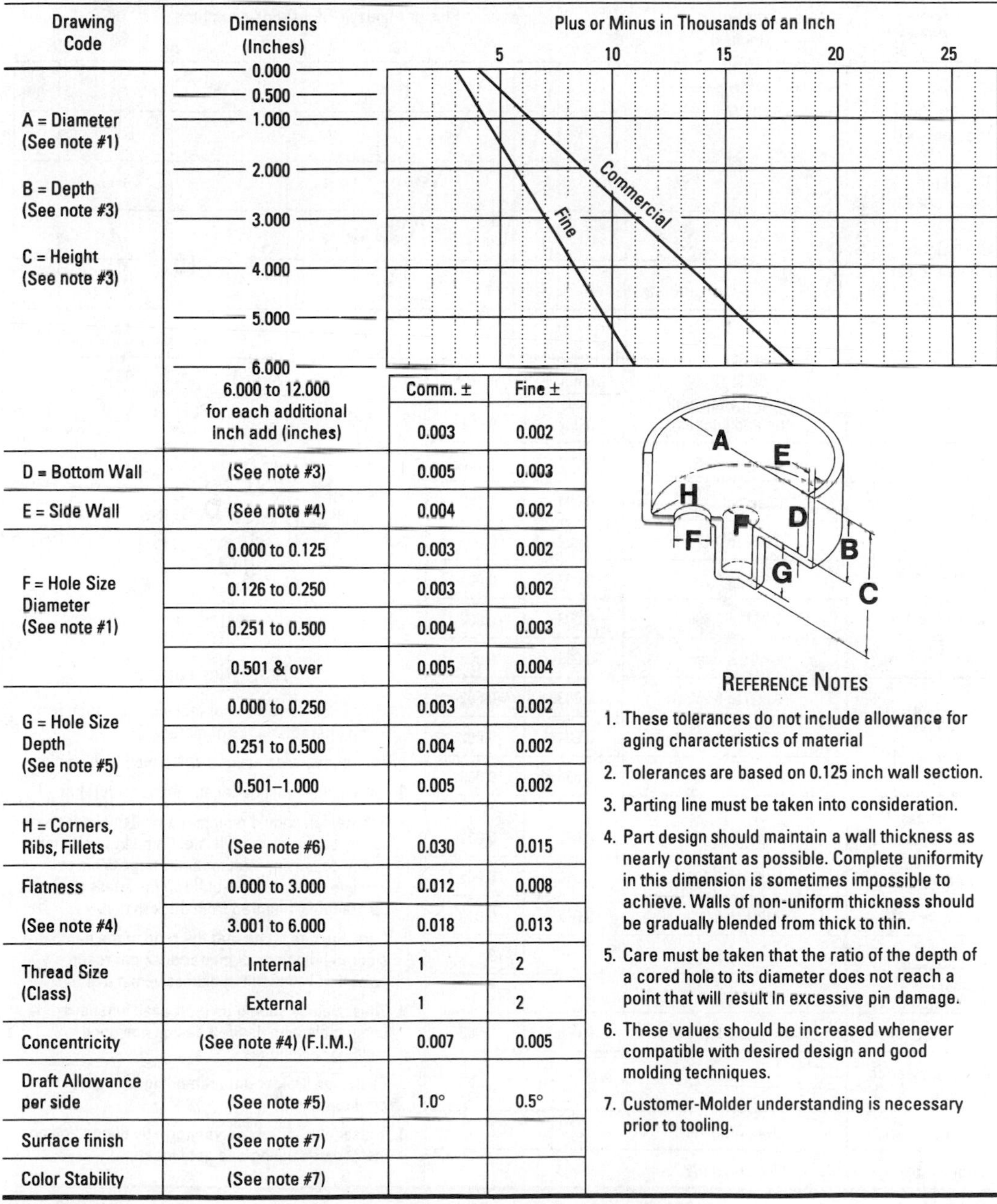

Drawing Code	Dimensions (Inches)	Comm. ±	Fine ±
A = Diameter (See note #1)	0.000 – 6.000 (see chart)		
B = Depth (See note #3)			
C = Height (See note #3)			
	6.000 to 12.000 for each additional inch add (inches)	0.003	0.002
D = Bottom Wall	(See note #3)	0.005	0.003
E = Side Wall	(See note #4)	0.004	0.002
F = Hole Size Diameter (See note #1)	0.000 to 0.125	0.003	0.002
	0.126 to 0.250	0.003	0.002
	0.251 to 0.500	0.004	0.003
	0.501 & over	0.005	0.004
G = Hole Size Depth (See note #5)	0.000 to 0.250	0.003	0.002
	0.251 to 0.500	0.004	0.002
	0.501–1.000	0.005	0.002
H = Corners, Ribs, Fillets	(See note #6)	0.030	0.015
Flatness	0.000 to 3.000	0.012	0.008
(See note #4)	3.001 to 6.000	0.018	0.013
Thread Size (Class)	Internal	1	2
	External	1	2
Concentricity	(See note #4) (F.I.M.)	0.007	0.005
Draft Allowance per side	(See note #5)	1.0°	0.5°
Surface finish	(See note #7)		
Color Stability	(See note #7)		

Reference Notes

1. These tolerances do not include allowance for aging characteristics of material
2. Tolerances are based on 0.125 inch wall section.
3. Parting line must be taken into consideration.
4. Part design should maintain a wall thickness as nearly constant as possible. Complete uniformity in this dimension is sometimes impossible to achieve. Walls of non-uniform thickness should be gradually blended from thick to thin.
5. Care must be taken that the ratio of the depth of a cored hole to its diameter does not reach a point that will result in excessive pin damage.
6. These values should be increased whenever compatible with desired design and good molding techniques.
7. Customer-Molder understanding is necessary prior to tooling.

Standards & Practices of Plastics Molders

Material
Phenol-Formaldehyde (PF) (Phenolic) General Purpose

Note: The *Commercial* values shown below represent common production tolerances at the most economical level. The *Fine* values represent closer tolerances that can be held but at a greater cost. Any addition of fillers will compromise physical properties and alter dimensional stability. Please consult the manufacturer.

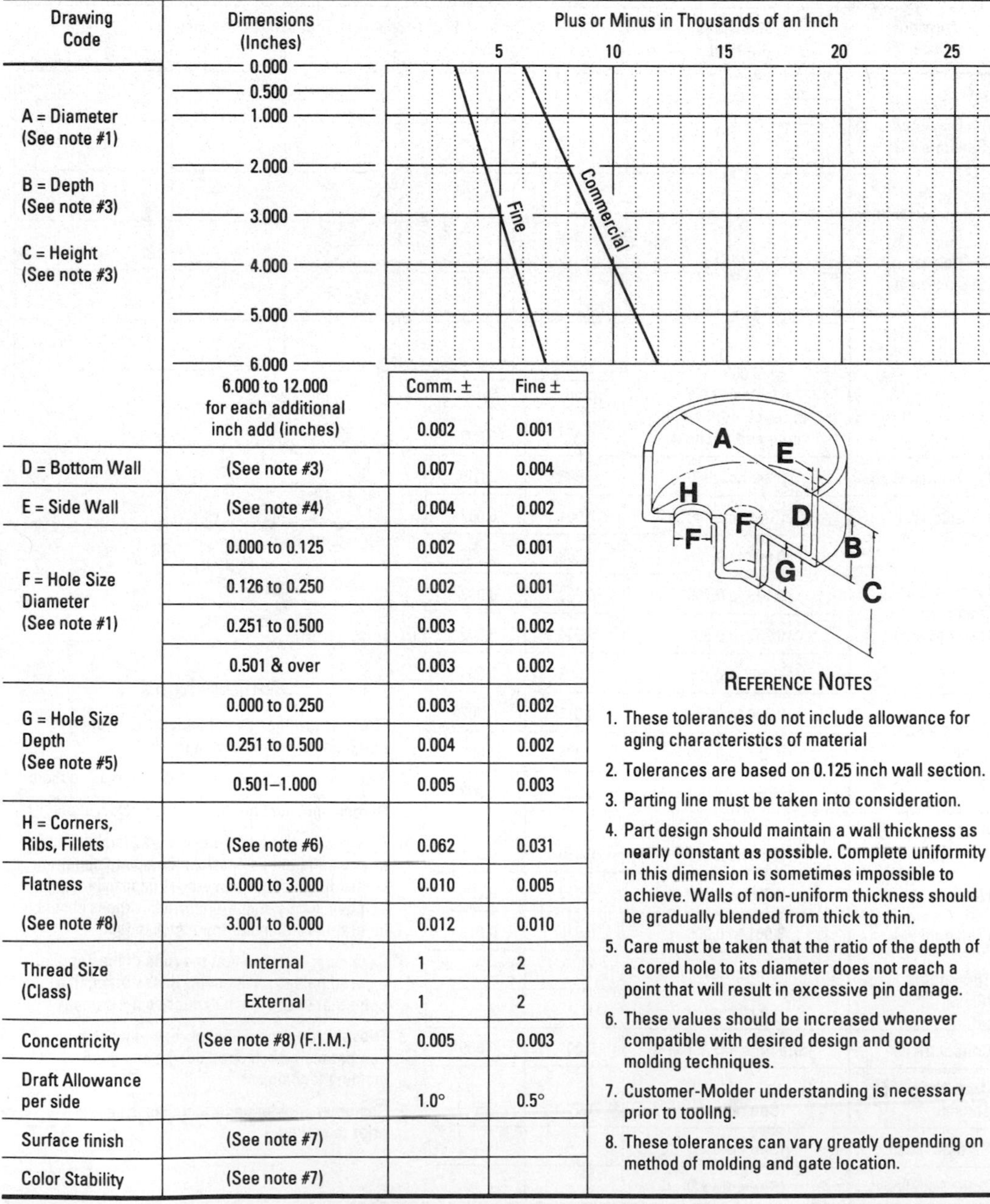

Drawing Code	Dimensions (Inches)	Comm. ±	Fine ±
A = Diameter (See note #1) B = Depth (See note #3) C = Height (See note #3)	6.000 to 12.000 for each additional inch add (inches)	0.002	0.001
D = Bottom Wall	(See note #3)	0.007	0.004
E = Side Wall	(See note #4)	0.004	0.002
F = Hole Size Diameter (See note #1)	0.000 to 0.125	0.002	0.001
	0.126 to 0.250	0.002	0.001
	0.251 to 0.500	0.003	0.002
	0.501 & over	0.003	0.002
G = Hole Size Depth (See note #5)	0.000 to 0.250	0.003	0.002
	0.251 to 0.500	0.004	0.002
	0.501–1.000	0.005	0.003
H = Corners, Ribs, Fillets	(See note #6)	0.062	0.031
Flatness (See note #8)	0.000 to 3.000	0.010	0.005
	3.001 to 6.000	0.012	0.010
Thread Size (Class)	Internal	1	2
	External	1	2
Concentricity	(See note #8) (F.I.M.)	0.005	0.003
Draft Allowance per side		1.0°	0.5°
Surface finish	(See note #7)		
Color Stability	(See note #7)		

Reference Notes

1. These tolerances do not include allowance for aging characteristics of material
2. Tolerances are based on 0.125 inch wall section.
3. Parting line must be taken into consideration.
4. Part design should maintain a wall thickness as nearly constant as possible. Complete uniformity in this dimension is sometimes impossible to achieve. Walls of non-uniform thickness should be gradually blended from thick to thin.
5. Care must be taken that the ratio of the depth of a cored hole to its diameter does not reach a point that will result in excessive pin damage.
6. These values should be increased whenever compatible with desired design and good molding techniques.
7. Customer-Molder understanding is necessary prior to tooling.
8. These tolerances can vary greatly depending on method of molding and gate location.

Standards & Practices of Plastics Molders

Material
Acrylonitrile Butadiene Styrene (ABS)

Note: The *Commercial* values shown below represent common production tolerances at the most economical level. The *Fine* values represent closer tolerances that can be held but at a greater cost. Any addition of fillers will compromise physical properties and alter dimensional stability. Please consult the manufacturer.

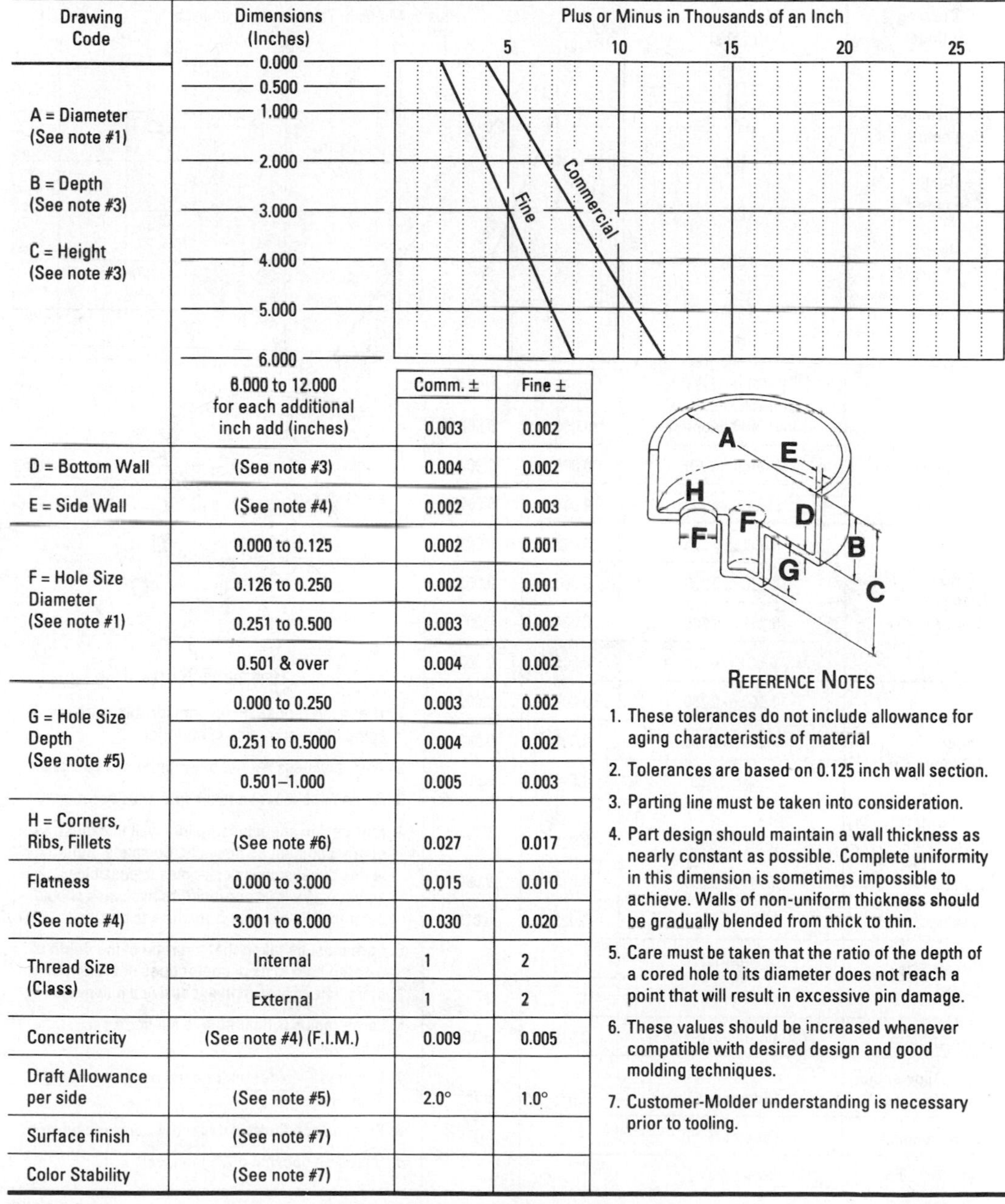

Drawing Code	Dimensions (Inches)	Comm. ±	Fine ±
A = Diameter (See note #1) B = Depth (See note #3) C = Height (See note #3)	0.000 0.500 1.000 2.000 3.000 4.000 5.000 6.000		
	6.000 to 12.000 for each additional inch add (inches)	0.003	0.002
D = Bottom Wall	(See note #3)	0.004	0.002
E = Side Wall	(See note #4)	0.002	0.003
F = Hole Size Diameter (See note #1)	0.000 to 0.125	0.002	0.001
	0.126 to 0.250	0.002	0.001
	0.251 to 0.500	0.003	0.002
	0.501 & over	0.004	0.002
G = Hole Size Depth (See note #5)	0.000 to 0.250	0.003	0.002
	0.251 to 0.5000	0.004	0.002
	0.501–1.000	0.005	0.003
H = Corners, Ribs, Fillets	(See note #6)	0.027	0.017
Flatness	0.000 to 3.000	0.015	0.010
(See note #4)	3.001 to 6.000	0.030	0.020
Thread Size (Class)	Internal	1	2
	External	1	2
Concentricity	(See note #4) (F.I.M.)	0.009	0.005
Draft Allowance per side	(See note #5)	2.0°	1.0°
Surface finish	(See note #7)		
Color Stability	(See note #7)		

Reference Notes

1. These tolerances do not include allowance for aging characteristics of material
2. Tolerances are based on 0.125 inch wall section.
3. Parting line must be taken into consideration.
4. Part design should maintain a wall thickness as nearly constant as possible. Complete uniformity in this dimension is sometimes impossible to achieve. Walls of non-uniform thickness should be gradually blended from thick to thin.
5. Care must be taken that the ratio of the depth of a cored hole to its diameter does not reach a point that will result in excessive pin damage.
6. These values should be increased whenever compatible with desired design and good molding techniques.
7. Customer-Molder understanding is necessary prior to tooling.

Standards & Practices of Plastics Molders	**Material** Acrylic

Note: The *Commercial* values shown below represent common production tolerances at the most economical level. The *Fine* values represent closer tolerances that can be held but at a greater cost. Any addition of fillers will compromise physical properties and alter dimensional stability. Please consult the manufacturer.

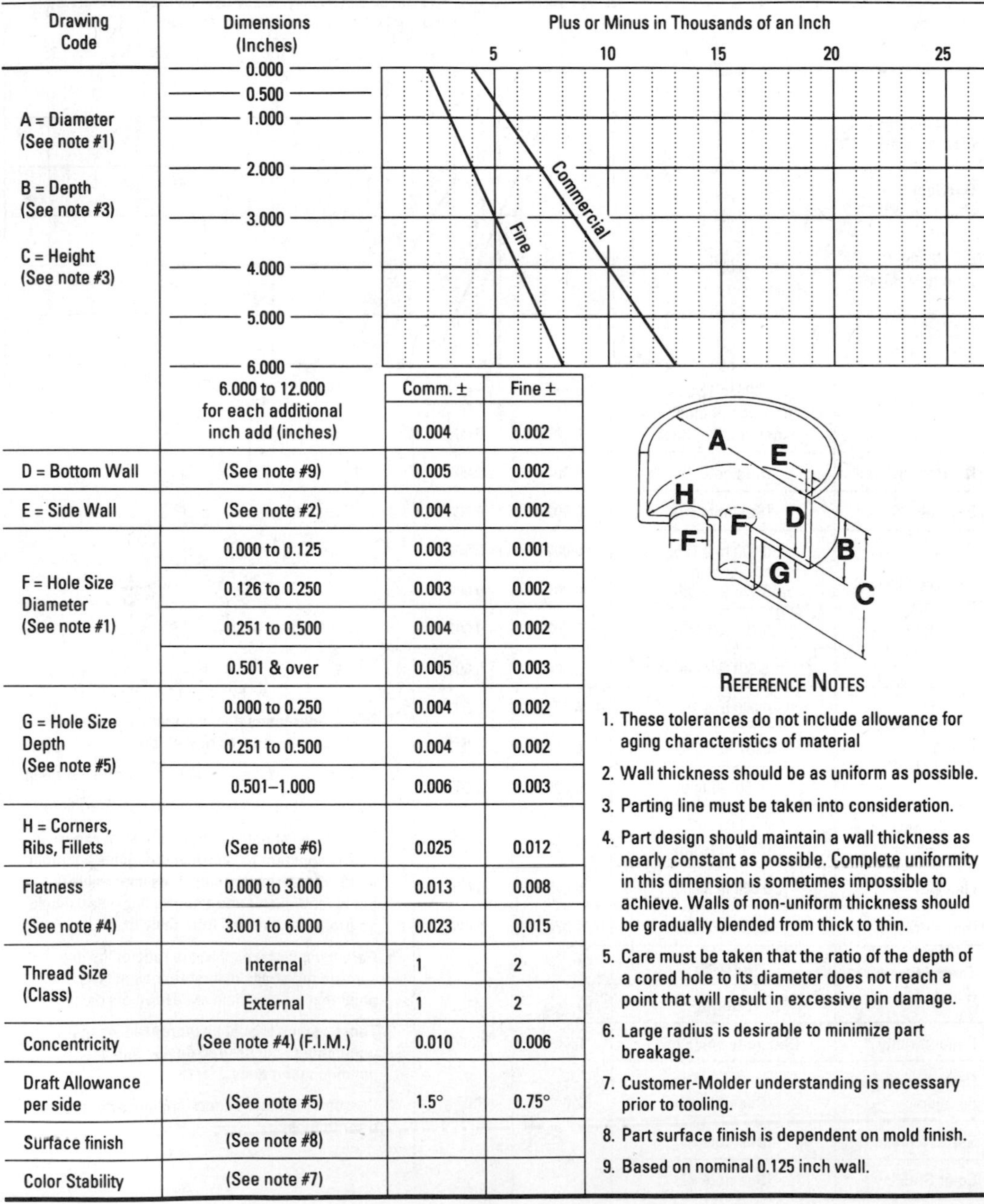

Drawing Code	Dimensions (Inches)	Comm. ±	Fine ±
A = Diameter (See note #1)	0.000		
B = Depth (See note #3)	0.500		
C = Height (See note #3)	1.000		
	2.000		
	3.000		
	4.000		
	5.000		
	6.000		
	6.000 to 12.000 for each additional inch add (inches)	0.004	0.002
D = Bottom Wall	(See note #9)	0.005	0.002
E = Side Wall	(See note #2)	0.004	0.002
F = Hole Size Diameter (See note #1)	0.000 to 0.125	0.003	0.001
	0.126 to 0.250	0.003	0.002
	0.251 to 0.500	0.004	0.002
	0.501 & over	0.005	0.003
G = Hole Size Depth (See note #5)	0.000 to 0.250	0.004	0.002
	0.251 to 0.500	0.004	0.002
	0.501–1.000	0.006	0.003
H = Corners, Ribs, Fillets	(See note #6)	0.025	0.012
Flatness (See note #4)	0.000 to 3.000	0.013	0.008
	3.001 to 6.000	0.023	0.015
Thread Size (Class)	Internal	1	2
	External	1	2
Concentricity	(See note #4) (F.I.M.)	0.010	0.006
Draft Allowance per side	(See note #5)	1.5°	0.75°
Surface finish	(See note #8)		
Color Stability	(See note #7)		

REFERENCE NOTES

1. These tolerances do not include allowance for aging characteristics of material
2. Wall thickness should be as uniform as possible.
3. Parting line must be taken into consideration.
4. Part design should maintain a wall thickness as nearly constant as possible. Complete uniformity in this dimension is sometimes impossible to achieve. Walls of non-uniform thickness should be gradually blended from thick to thin.
5. Care must be taken that the ratio of the depth of a cored hole to its diameter does not reach a point that will result in excessive pin damage.
6. Large radius is desirable to minimize part breakage.
7. Customer-Molder understanding is necessary prior to tooling.
8. Part surface finish is dependent on mold finish.
9. Based on nominal 0.125 inch wall.

Standards & Practices of Plastics Molders

Material

Cellulosics

Note: The *Commercial* values shown below represent common production tolerances at the most economical level. The *Fine* values represent closer tolerances that can be held but at a greater cost. Any addition of fillers will compromise physical properties and alter dimensional stability. Please consult the manufacturer.

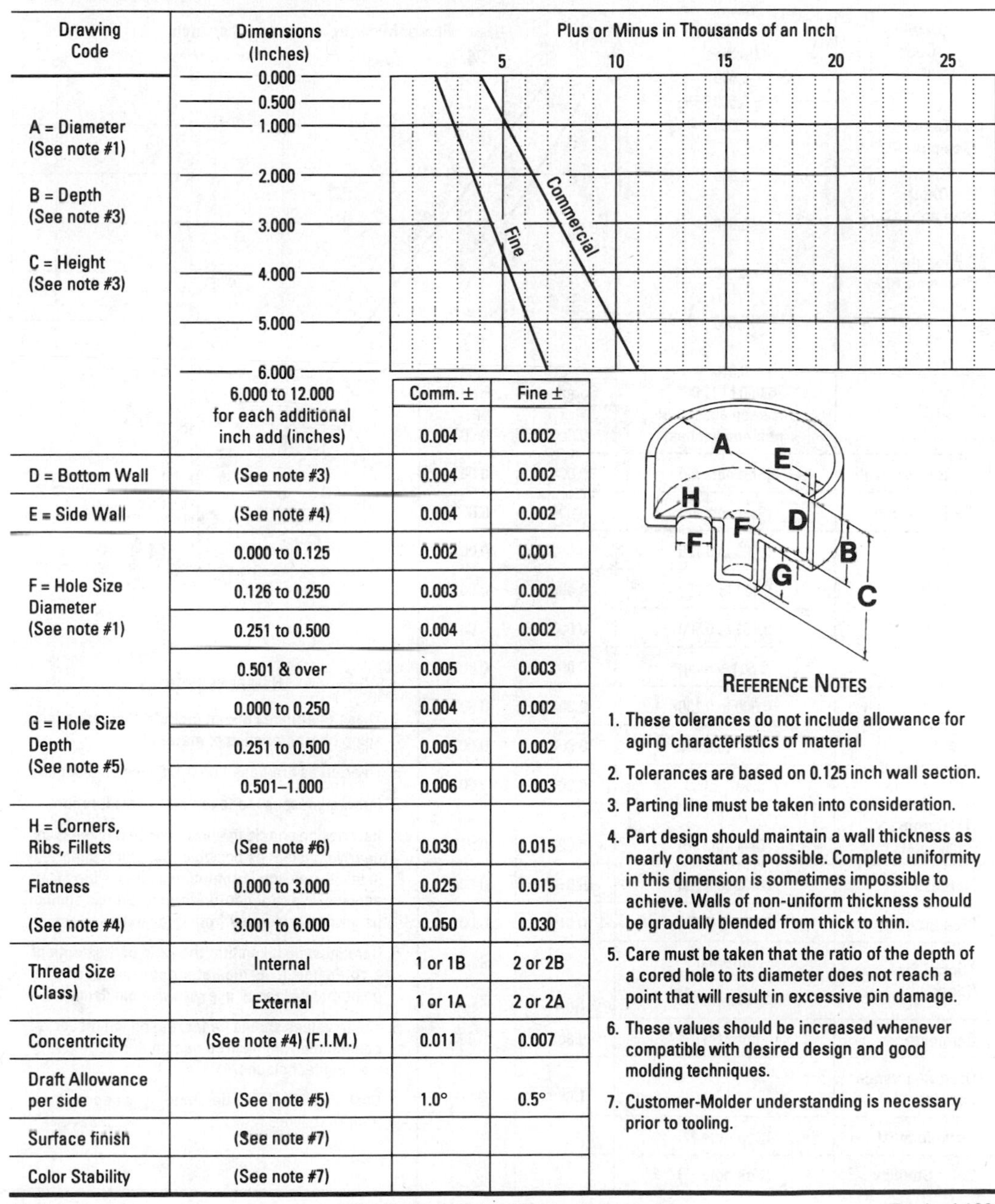

Drawing Code	Dimensions (Inches)	Comm. ±	Fine ±
A = Diameter (See note #1) B = Depth (See note #3) C = Height (See note #3)	0.000 – 6.000 (see chart)		
	6.000 to 12.000 for each additional inch add (inches)	0.004	0.002
D = Bottom Wall	(See note #3)	0.004	0.002
E = Side Wall	(See note #4)	0.004	0.002
F = Hole Size Diameter (See note #1)	0.000 to 0.125	0.002	0.001
	0.126 to 0.250	0.003	0.002
	0.251 to 0.500	0.004	0.002
	0.501 & over	0.005	0.003
G = Hole Size Depth (See note #5)	0.000 to 0.250	0.004	0.002
	0.251 to 0.500	0.005	0.002
	0.501–1.000	0.006	0.003
H = Corners, Ribs, Fillets	(See note #6)	0.030	0.015
Flatness (See note #4)	0.000 to 3.000	0.025	0.015
	3.001 to 6.000	0.050	0.030
Thread Size (Class)	Internal	1 or 1B	2 or 2B
	External	1 or 1A	2 or 2A
Concentricity	(See note #4) (F.I.M.)	0.011	0.007
Draft Allowance per side	(See note #5)	1.0°	0.5°
Surface finish	(See note #7)		
Color Stability	(See note #7)		

Reference Notes

1. These tolerances do not include allowance for aging characteristics of material
2. Tolerances are based on 0.125 inch wall section.
3. Parting line must be taken into consideration.
4. Part design should maintain a wall thickness as nearly constant as possible. Complete uniformity in this dimension is sometimes impossible to achieve. Walls of non-uniform thickness should be gradually blended from thick to thin.
5. Care must be taken that the ratio of the depth of a cored hole to its diameter does not reach a point that will result in excessive pin damage.
6. These values should be increased whenever compatible with desired design and good molding techniques.
7. Customer-Molder understanding is necessary prior to tooling.

Standards & Practices of Plastics Molders

Material
Polyamide (Nylon)
(PA)

Note: The *Commercial* values shown below represent common production tolerances at the most economical level. The *Fine* values represent closer tolerances that can be held but at a greater cost. Any addition of fillers will compromise physical properties and alter dimensional stability. Please consult the manufacturer.

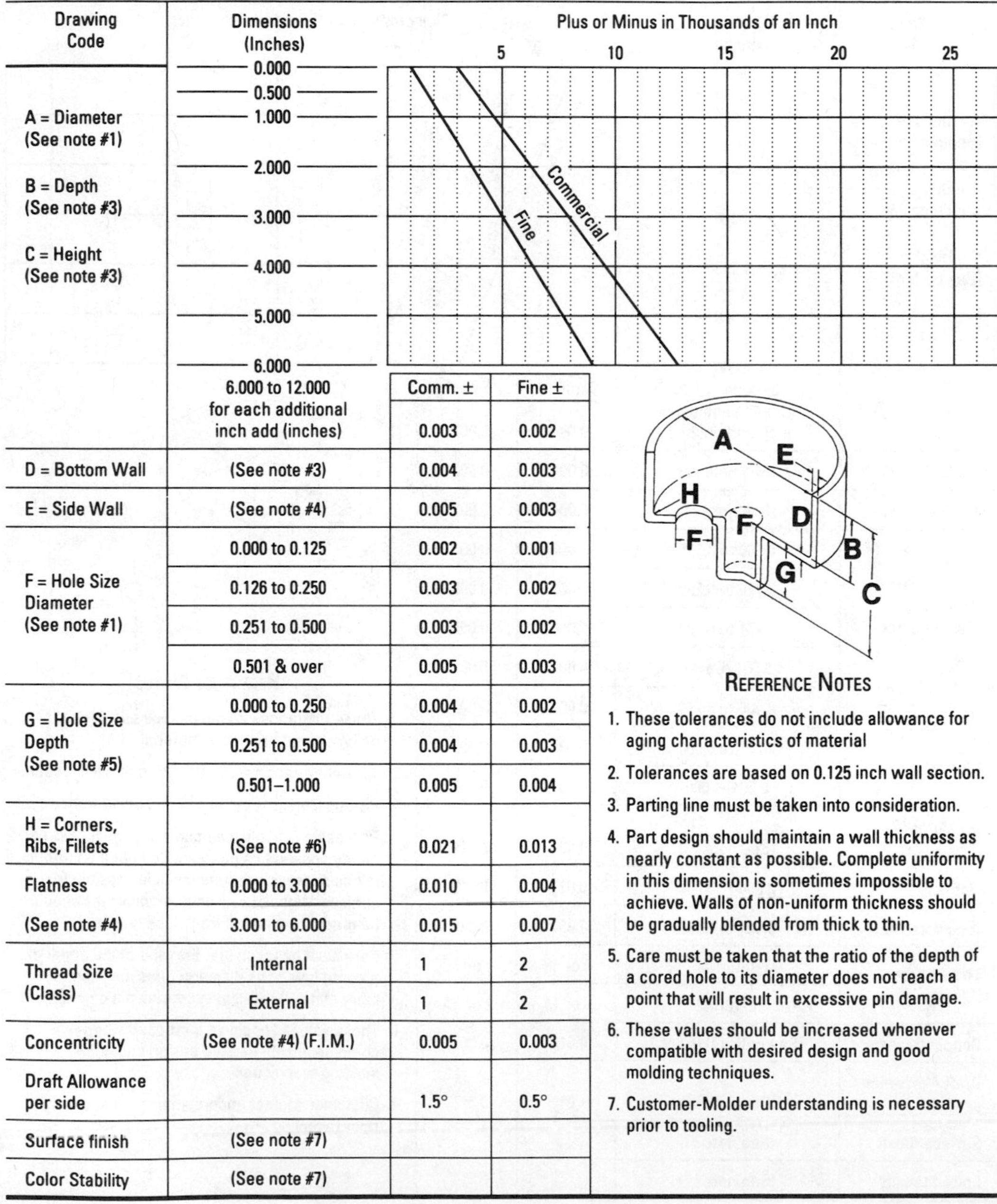

Drawing Code	Dimensions (Inches)	Comm. ±	Fine ±
A = Diameter (See note #1)	0.000		
	0.500		
	1.000		
B = Depth (See note #3)	2.000		
	3.000		
C = Height (See note #3)	4.000		
	5.000		
	6.000		
	6.000 to 12.000 for each additional inch add (inches)	0.003	0.002
D = Bottom Wall	(See note #3)	0.004	0.003
E = Side Wall	(See note #4)	0.005	0.003
F = Hole Size Diameter (See note #1)	0.000 to 0.125	0.002	0.001
	0.126 to 0.250	0.003	0.002
	0.251 to 0.500	0.003	0.002
	0.501 & over	0.005	0.003
G = Hole Size Depth (See note #5)	0.000 to 0.250	0.004	0.002
	0.251 to 0.500	0.004	0.003
	0.501–1.000	0.005	0.004
H = Corners, Ribs, Fillets	(See note #6)	0.021	0.013
Flatness (See note #4)	0.000 to 3.000	0.010	0.004
	3.001 to 6.000	0.015	0.007
Thread Size (Class)	Internal	1	2
	External	1	2
Concentricity	(See note #4) (F.I.M.)	0.005	0.003
Draft Allowance per side		1.5°	0.5°
Surface finish	(See note #7)		
Color Stability	(See note #7)		

Reference Notes

1. These tolerances do not include allowance for aging characteristics of material
2. Tolerances are based on 0.125 inch wall section.
3. Parting line must be taken into consideration.
4. Part design should maintain a wall thickness as nearly constant as possible. Complete uniformity in this dimension is sometimes impossible to achieve. Walls of non-uniform thickness should be gradually blended from thick to thin.
5. Care must be taken that the ratio of the depth of a cored hole to its diameter does not reach a point that will result in excessive pin damage.
6. These values should be increased whenever compatible with desired design and good molding techniques.
7. Customer-Molder understanding is necessary prior to tooling.

 Revised 1991

Standards & Practices of Plastics Molders

Material
Polycarbonate
(PC)

Note: The *Commercial* values shown below represent common production tolerances at the most economical level. The *Fine* values represent closer tolerances that can be held but at a greater cost. Any addition of fillers will compromise physical properties and alter dimensional stability. Please consult the manufacturer.

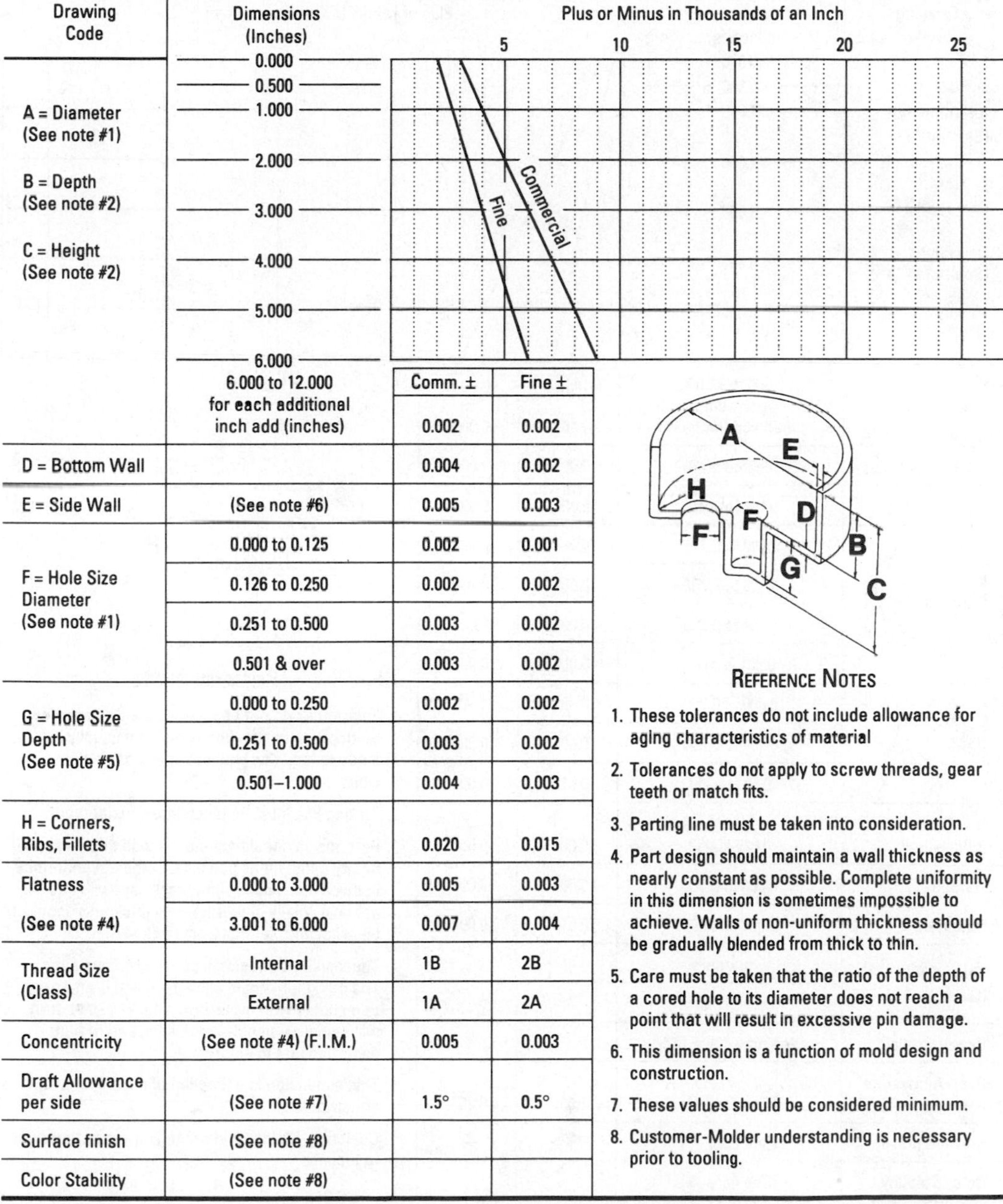

Drawing Code	Dimensions (Inches)	Comm. ±	Fine ±
A = Diameter (See note #1) B = Depth (See note #2) C = Height (See note #2)	6.000 to 12.000 for each additional inch add (inches)	0.002	0.002
D = Bottom Wall		0.004	0.002
E = Side Wall	(See note #6)	0.005	0.003
F = Hole Size Diameter (See note #1)	0.000 to 0.125	0.002	0.001
	0.126 to 0.250	0.002	0.002
	0.251 to 0.500	0.003	0.002
	0.501 & over	0.003	0.002
G = Hole Size Depth (See note #5)	0.000 to 0.250	0.002	0.002
	0.251 to 0.500	0.003	0.002
	0.501–1.000	0.004	0.003
H = Corners, Ribs, Fillets		0.020	0.015
Flatness	0.000 to 3.000	0.005	0.003
(See note #4)	3.001 to 6.000	0.007	0.004
Thread Size (Class)	Internal	1B	2B
	External	1A	2A
Concentricity	(See note #4) (F.I.M.)	0.005	0.003
Draft Allowance per side	(See note #7)	1.5°	0.5°
Surface finish	(See note #8)		
Color Stability	(See note #8)		

Reference Notes

1. These tolerances do not include allowance for aging characteristics of material
2. Tolerances do not apply to screw threads, gear teeth or match fits.
3. Parting line must be taken into consideration.
4. Part design should maintain a wall thickness as nearly constant as possible. Complete uniformity in this dimension is sometimes impossible to achieve. Walls of non-uniform thickness should be gradually blended from thick to thin.
5. Care must be taken that the ratio of the depth of a cored hole to its diameter does not reach a point that will result in excessive pin damage.
6. This dimension is a function of mold design and construction.
7. These values should be considered minimum.
8. Customer-Molder understanding is necessary prior to tooling.

Standards & Practices of Plastics Molders

Material
Polyetherimide (PEI)

Note: The *Commercial* values shown below represent common production tolerances at the most economical level. The *Fine* values represent closer tolerances that can be held but at a greater cost. Any addition of fillers will compromise physical properties and alter dimensional stability. Please consult the manufacturer.

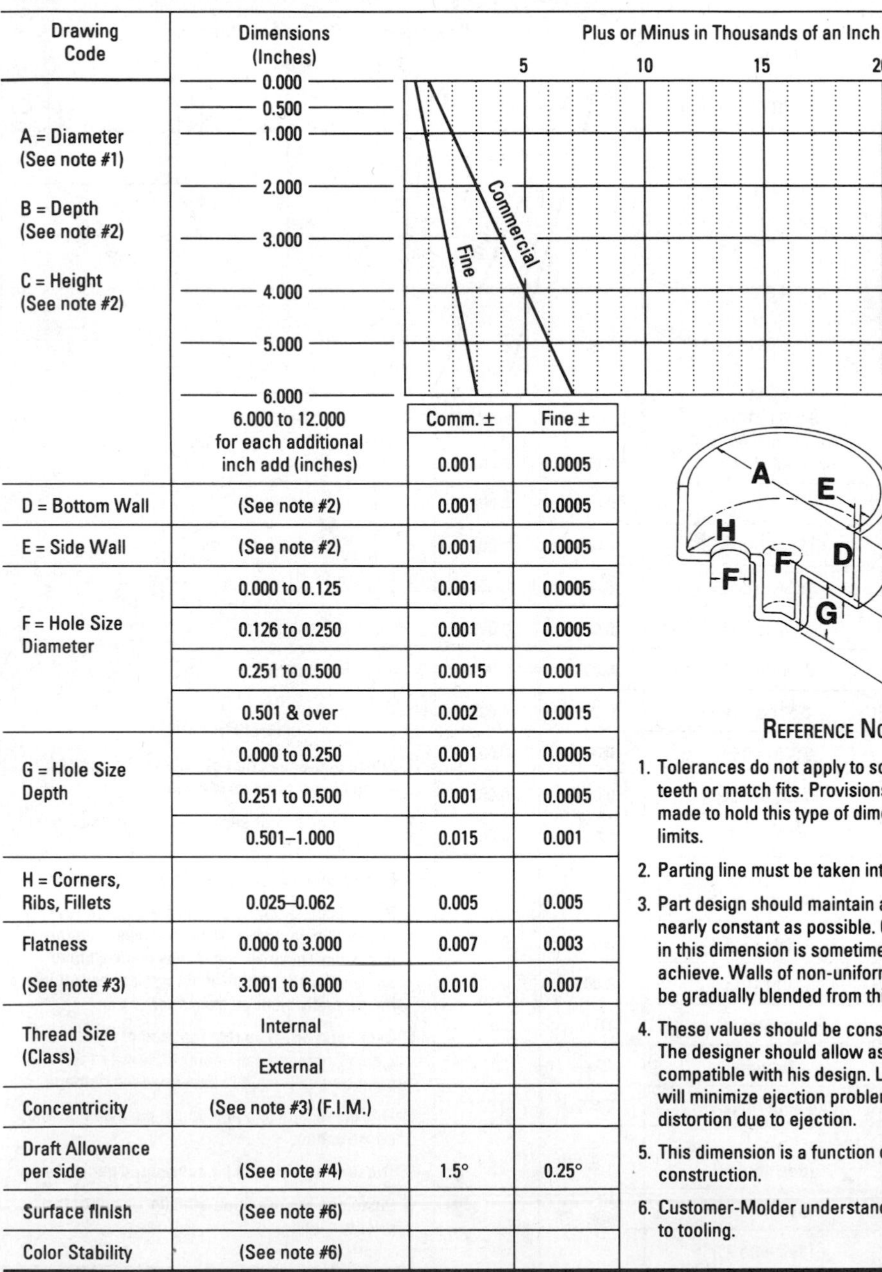

Drawing Code	Dimensions (Inches)	Comm. ±	Fine ±
A = Diameter (See note #1) B = Depth (See note #2) C = Height (See note #2)	6.000 to 12.000 for each additional inch add (inches)	0.001	0.0005
D = Bottom Wall	(See note #2)	0.001	0.0005
E = Side Wall	(See note #2)	0.001	0.0005
F = Hole Size Diameter	0.000 to 0.125	0.001	0.0005
	0.126 to 0.250	0.001	0.0005
	0.251 to 0.500	0.0015	0.001
	0.501 & over	0.002	0.0015
G = Hole Size Depth	0.000 to 0.250	0.001	0.0005
	0.251 to 0.500	0.001	0.0005
	0.501–1.000	0.015	0.001
H = Corners, Ribs, Fillets	0.025–0.062	0.005	0.005
Flatness	0.000 to 3.000	0.007	0.003
(See note #3)	3.001 to 6.000	0.010	0.007
Thread Size (Class)	Internal		
	External		
Concentricity	(See note #3) (F.I.M.)		
Draft Allowance per side	(See note #4)	1.5°	0.25°
Surface finish	(See note #6)		
Color Stability	(See note #6)		

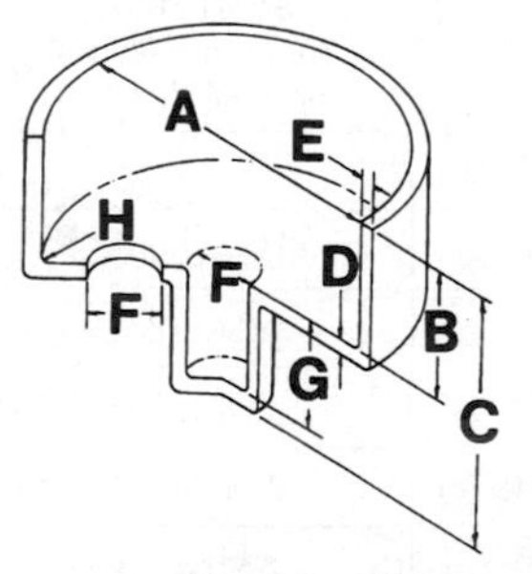

Reference Notes

1. Tolerances do not apply to screw threads, gear teeth or match fits. Provisions can usually be made to hold this type of dimension to close limits.
2. Parting line must be taken into consideration.
3. Part design should maintain a wall thickness as nearly constant as possible. Complete uniformity in this dimension is sometimes impossible to achieve. Walls of non-uniform thickness should be gradually blended from thick to thin.
4. These values should be considered minimum. The designer should allow as much draft as is compatible with his design. Liberal use of draft will minimize ejection problems, and reduce distortion due to ejection.
5. This dimension is a function of mold design and construction.
6. Customer-Molder understanding necessary prior to tooling.

Standards & Practices of Plastics Molders

Material
High Density Polyethylene (HDPE)

Note: The *Commercial* values shown below represent common production tolerances at the most economical level. The *Fine* values represent closer tolerances that can be held but at a greater cost. Any addition of fillers will compromise physical properties and alter dimensional stability. Please consult the manufacturer.

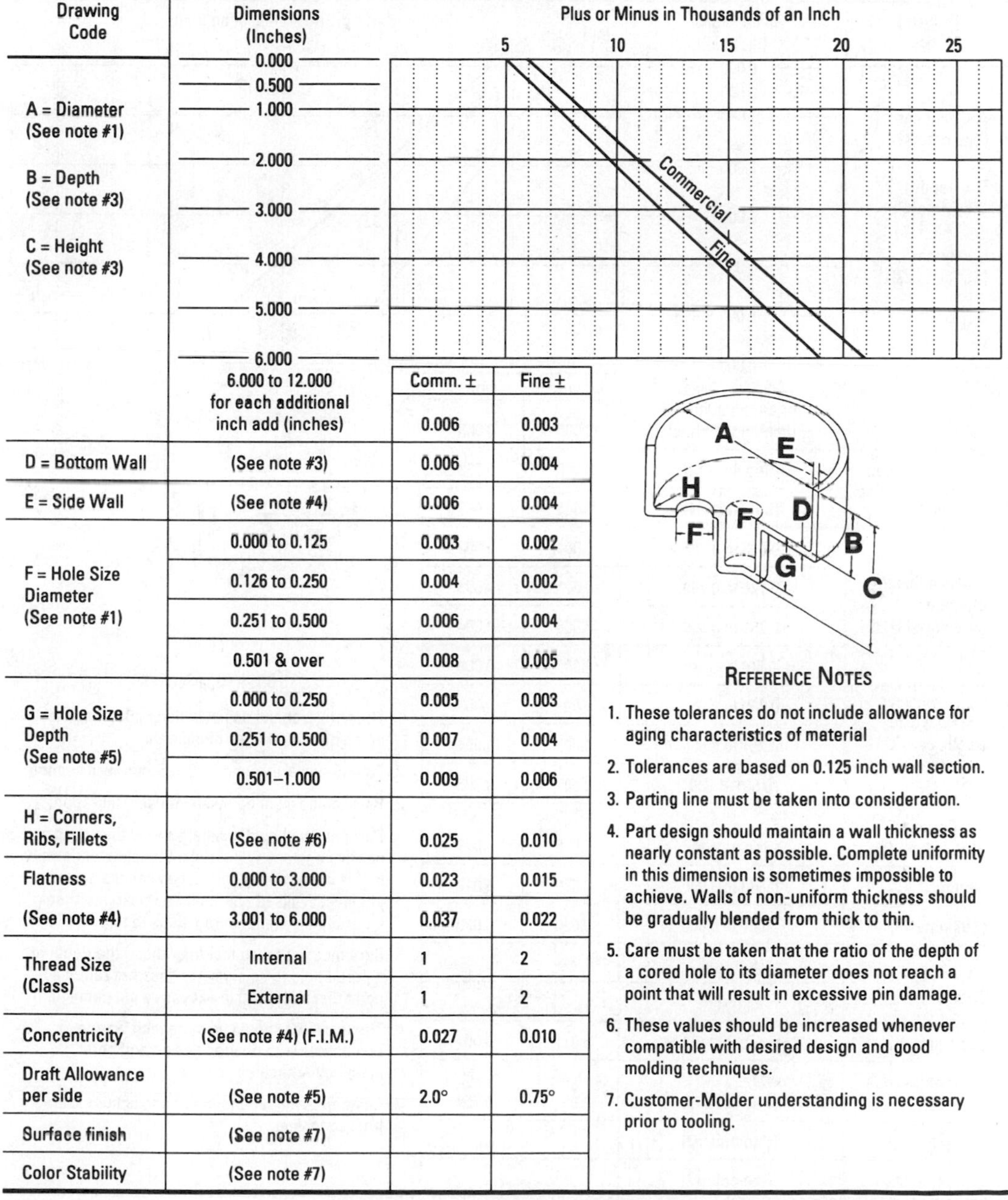

Drawing Code	Dimensions (Inches)	Comm. ±	Fine ±
A = Diameter (See note #1) B = Depth (See note #3) C = Height (See note #3)	0.000 0.500 1.000 2.000 3.000 4.000 5.000 6.000		
	6.000 to 12.000 for each additional inch add (inches)	0.006	0.003
D = Bottom Wall	(See note #3)	0.006	0.004
E = Side Wall	(See note #4)	0.006	0.004
F = Hole Size Diameter (See note #1)	0.000 to 0.125	0.003	0.002
	0.126 to 0.250	0.004	0.002
	0.251 to 0.500	0.006	0.004
	0.501 & over	0.008	0.005
G = Hole Size Depth (See note #5)	0.000 to 0.250	0.005	0.003
	0.251 to 0.500	0.007	0.004
	0.501–1.000	0.009	0.006
H = Corners, Ribs, Fillets	(See note #6)	0.025	0.010
Flatness	0.000 to 3.000	0.023	0.015
(See note #4)	3.001 to 6.000	0.037	0.022
Thread Size (Class)	Internal	1	2
	External	1	2
Concentricity	(See note #4) (F.I.M.)	0.027	0.010
Draft Allowance per side	(See note #5)	2.0°	0.75°
Surface finish	(See note #7)		
Color Stability	(See note #7)		

Reference Notes

1. These tolerances do not include allowance for aging characteristics of material
2. Tolerances are based on 0.125 inch wall section.
3. Parting line must be taken into consideration.
4. Part design should maintain a wall thickness as nearly constant as possible. Complete uniformity in this dimension is sometimes impossible to achieve. Walls of non-uniform thickness should be gradually blended from thick to thin.
5. Care must be taken that the ratio of the depth of a cored hole to its diameter does not reach a point that will result in excessive pin damage.
6. These values should be increased whenever compatible with desired design and good molding techniques.
7. Customer-Molder understanding is necessary prior to tooling.

Standards & Practices of Plastics Molders

Material
Low Density Polyethylene (LDPE)

Note: The *Commercial* values shown below represent common production tolerances at the most economical level. The *Fine* values represent closer tolerances that can be held but at a greater cost. Any addition of fillers will compromise physical properties and alter dimensional stability. Please consult the manufacturer.

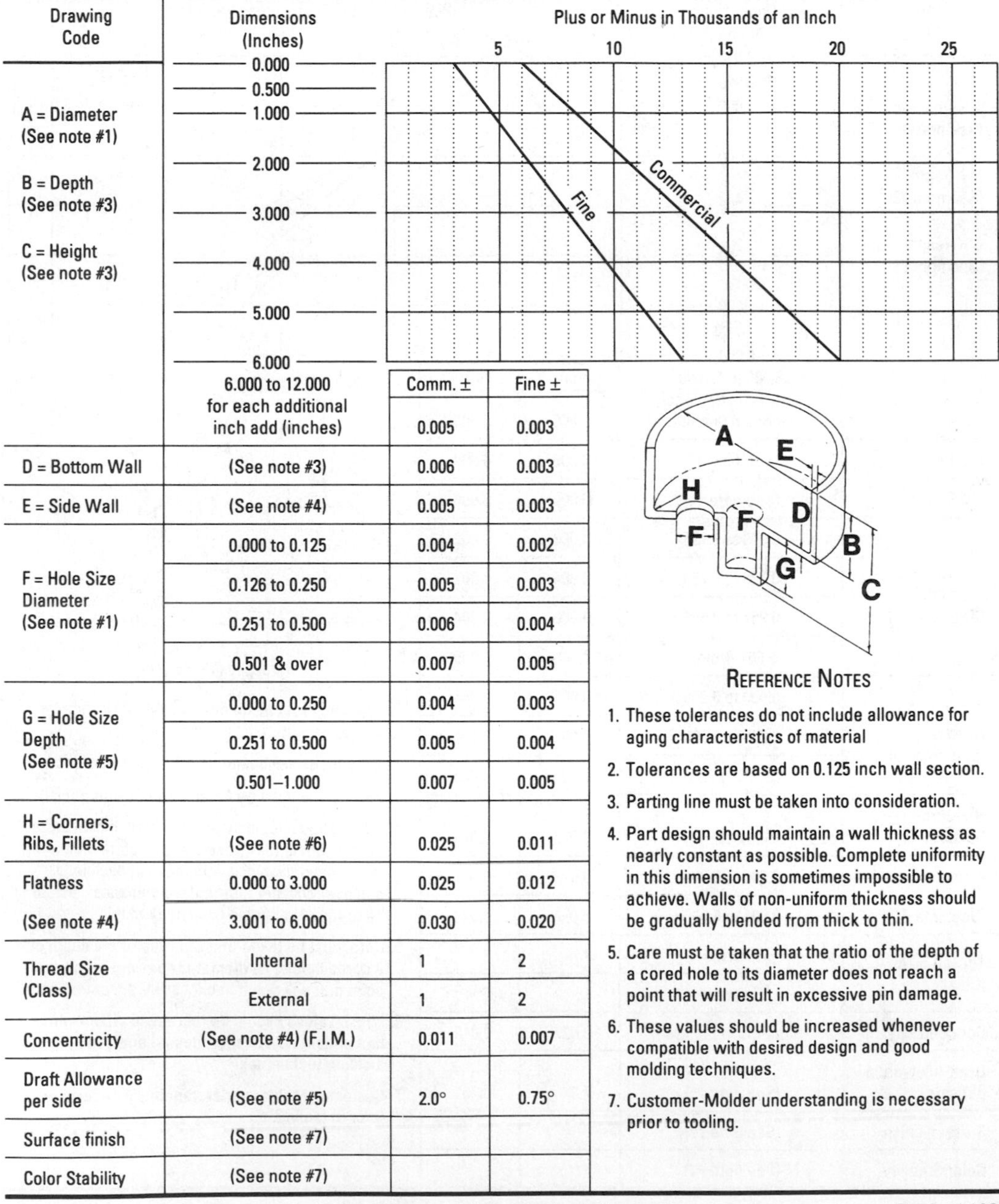

Drawing Code	Dimensions (Inches)	Comm. ±	Fine ±
A = Diameter (See note #1) B = Depth (See note #3) C = Height (See note #3)	6.000 to 12.000 for each additional inch add (inches)	0.005	0.003
D = Bottom Wall	(See note #3)	0.006	0.003
E = Side Wall	(See note #4)	0.005	0.003
F = Hole Size Diameter (See note #1)	0.000 to 0.125	0.004	0.002
	0.126 to 0.250	0.005	0.003
	0.251 to 0.500	0.006	0.004
	0.501 & over	0.007	0.005
G = Hole Size Depth (See note #5)	0.000 to 0.250	0.004	0.003
	0.251 to 0.500	0.005	0.004
	0.501–1.000	0.007	0.005
H = Corners, Ribs, Fillets	(See note #6)	0.025	0.011
Flatness (See note #4)	0.000 to 3.000	0.025	0.012
	3.001 to 6.000	0.030	0.020
Thread Size (Class)	Internal	1	2
	External	1	2
Concentricity	(See note #4) (F.I.M.)	0.011	0.007
Draft Allowance per side	(See note #5)	2.0°	0.75°
Surface finish	(See note #7)		
Color Stability	(See note #7)		

REFERENCE NOTES

1. These tolerances do not include allowance for aging characteristics of material
2. Tolerances are based on 0.125 inch wall section.
3. Parting line must be taken into consideration.
4. Part design should maintain a wall thickness as nearly constant as possible. Complete uniformity in this dimension is sometimes impossible to achieve. Walls of non-uniform thickness should be gradually blended from thick to thin.
5. Care must be taken that the ratio of the depth of a cored hole to its diameter does not reach a point that will result in excessive pin damage.
6. These values should be increased whenever compatible with desired design and good molding techniques.
7. Customer-Molder understanding is necessary prior to tooling.

Standards & Practices of Plastics Molders

Material
Polyethylene-Terephthalate (PETE)

Note: The *Commercial* values shown below represent common production tolerances at the most economical level. The *Fine* values represent closer tolerances that can be held but at a greater cost. Any addition of fillers will compromise physical properties and alter dimensional stability. Please consult the manufacturer.

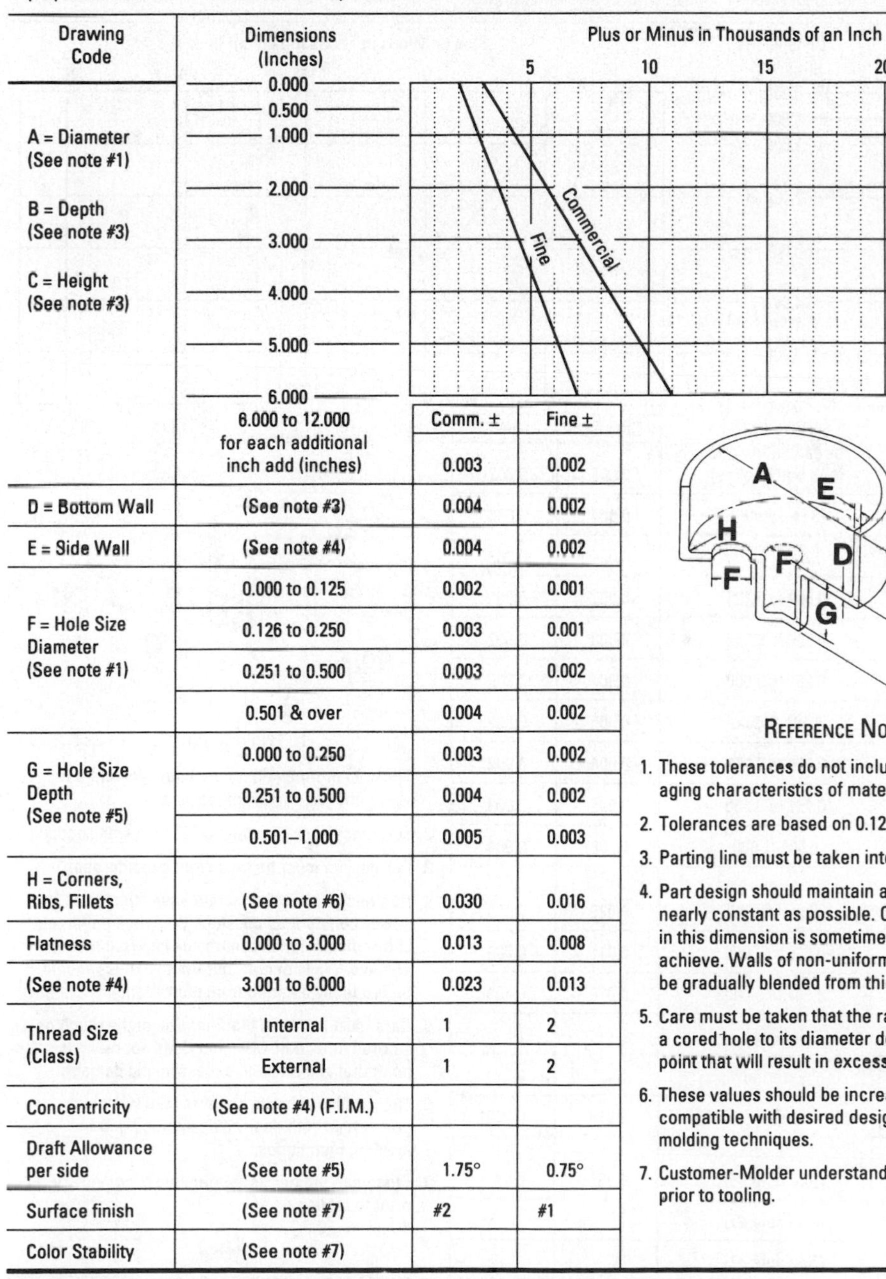

Drawing Code	Dimensions (Inches)	Comm. ±	Fine ±
A = Diameter (See note #1)	0.000		
	0.500		
	1.000		
B = Depth (See note #3)	2.000		
	3.000		
C = Height (See note #3)	4.000		
	5.000		
	6.000		
	6.000 to 12.000 for each additional inch add (inches)	0.003	0.002
D = Bottom Wall	(See note #3)	0.004	0.002
E = Side Wall	(See note #4)	0.004	0.002
F = Hole Size Diameter (See note #1)	0.000 to 0.125	0.002	0.001
	0.126 to 0.250	0.003	0.001
	0.251 to 0.500	0.003	0.002
	0.501 & over	0.004	0.002
G = Hole Size Depth (See note #5)	0.000 to 0.250	0.003	0.002
	0.251 to 0.500	0.004	0.002
	0.501–1.000	0.005	0.003
H = Corners, Ribs, Fillets	(See note #6)	0.030	0.016
Flatness	0.000 to 3.000	0.013	0.008
(See note #4)	3.001 to 6.000	0.023	0.013
Thread Size (Class)	Internal	1	2
	External	1	2
Concentricity	(See note #4) (F.I.M.)		
Draft Allowance per side	(See note #5)	1.75°	0.75°
Surface finish	(See note #7)	#2	#1
Color Stability	(See note #7)		

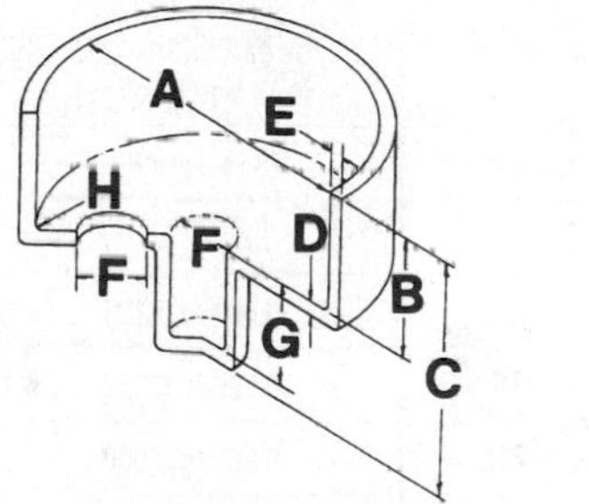

Reference Notes

1. These tolerances do not include allowance for aging characteristics of material
2. Tolerances are based on 0.125 inch wall section.
3. Parting line must be taken into consideration.
4. Part design should maintain a wall thickness as nearly constant as possible. Complete uniformity in this dimension is sometimes impossible to achieve. Walls of non-uniform thickness should be gradually blended from thick to thin.
5. Care must be taken that the ratio of the depth of a cored hole to its diameter does not reach a point that will result in excessive pin damage.
6. These values should be increased whenever compatible with desired design and good molding techniques.
7. Customer-Molder understanding is necessary prior to tooling.

Standards & Practices of Plastics Molders

Material
Polyoxymethylene
(Acetal) (POM)

Note: The *Commercial* values shown below represent common production tolerances at the most economical level. The *Fine* values represent closer tolerances that can be held but at a greater cost. Any addition of fillers will compromise physical properties and alter dimensional stability. Please consult the manufacturer.

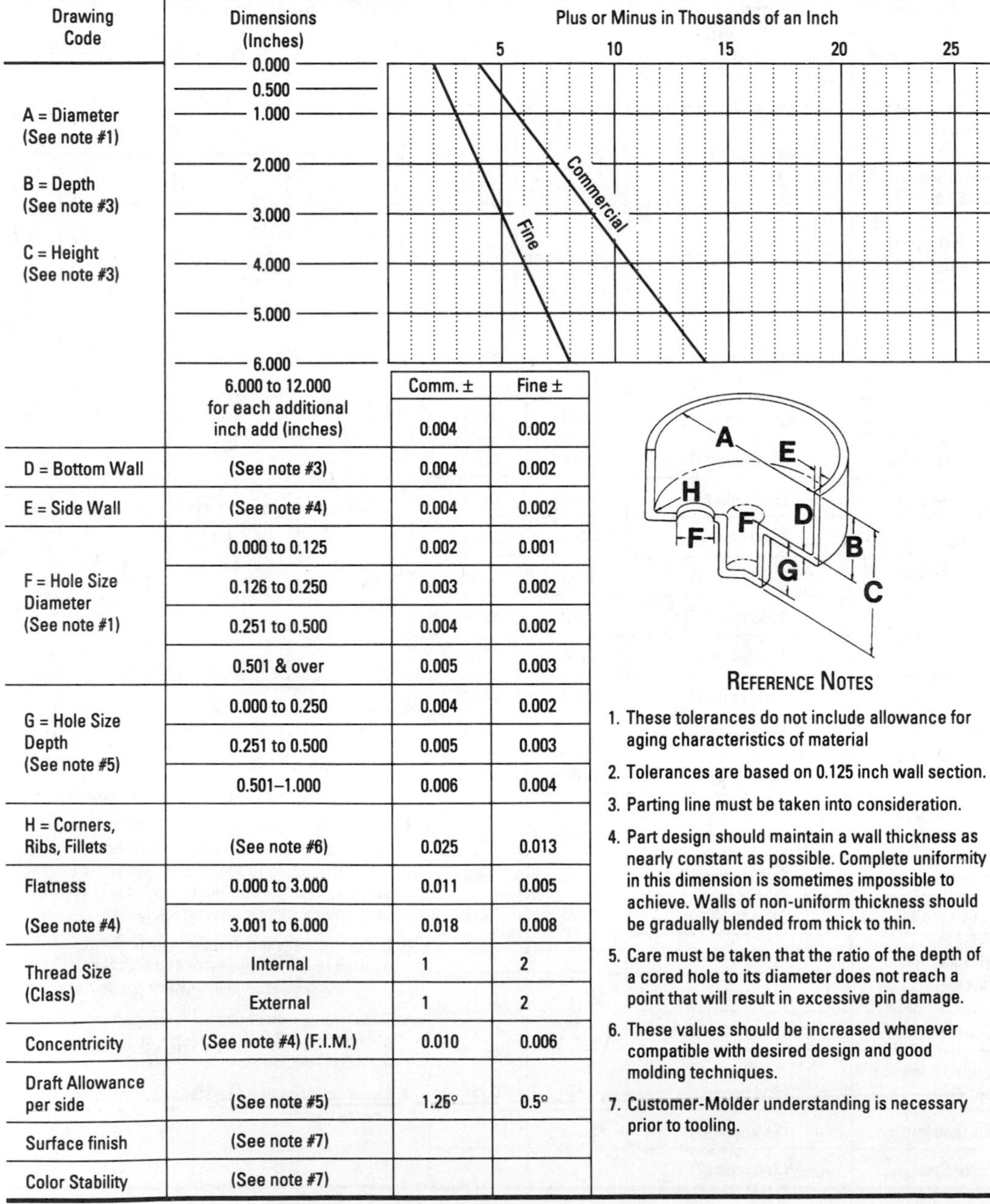

Drawing Code	Dimensions (Inches)	Comm. ±	Fine ±
A = Diameter (See note #1); B = Depth (See note #3); C = Height (See note #3)	0.000 / 0.500 / 1.000 / 2.000 / 3.000 / 4.000 / 5.000 / 6.000		
	6.000 to 12.000 for each additional inch add (inches)	0.004	0.002
D = Bottom Wall	(See note #3)	0.004	0.002
E = Side Wall	(See note #4)	0.004	0.002
F = Hole Size Diameter (See note #1)	0.000 to 0.125	0.002	0.001
	0.126 to 0.250	0.003	0.002
	0.251 to 0.500	0.004	0.002
	0.501 & over	0.005	0.003
G = Hole Size Depth (See note #5)	0.000 to 0.250	0.004	0.002
	0.251 to 0.500	0.005	0.003
	0.501–1.000	0.006	0.004
H = Corners, Ribs, Fillets	(See note #6)	0.025	0.013
Flatness (See note #4)	0.000 to 3.000	0.011	0.005
	3.001 to 6.000	0.018	0.008
Thread Size (Class)	Internal	1	2
	External	1	2
Concentricity	(See note #4) (F.I.M.)	0.010	0.006
Draft Allowance per side	(See note #5)	1.25°	0.5°
Surface finish	(See note #7)		
Color Stability	(See note #7)		

Reference Notes

1. These tolerances do not include allowance for aging characteristics of material
2. Tolerances are based on 0.125 inch wall section.
3. Parting line must be taken into consideration.
4. Part design should maintain a wall thickness as nearly constant as possible. Complete uniformity in this dimension is sometimes impossible to achieve. Walls of non-uniform thickness should be gradually blended from thick to thin.
5. Care must be taken that the ratio of the depth of a cored hole to its diameter does not reach a point that will result in excessive pin damage.
6. These values should be increased whenever compatible with desired design and good molding techniques.
7. Customer-Molder understanding is necessary prior to tooling.

Standards & Practices of Plastics Molders

Material
Polyphenylene Ether
(PPE)

Note: The *Commercial* values shown below represent common production tolerances at the most economical level. The *Fine* values represent closer tolerances that can be held but at a greater cost. Any addition of fillers will compromise physical properties and alter dimensional stability. Please consult the manufacturer.

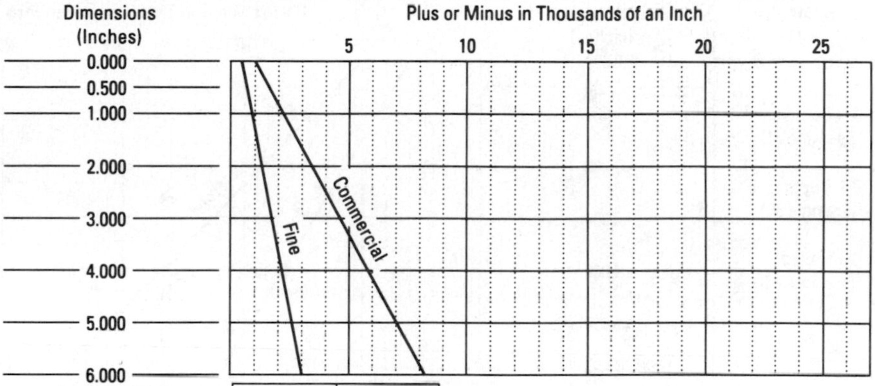

Drawing Code: A = Diameter (See note #1); B = Depth (See note #2); C = Height (See note #2)

Dimensions (Inches): 0.000, 0.500, 1.000, 2.000, 3.000, 4.000, 5.000, 6.000

Drawing Code	Dimensions (Inches)	Comm. ±	Fine ±
	6.000 to 12.000 for each additional inch add (inches)	0.001	0.0005
D = Bottom Wall	(See note #3)	0.001	0.0005
E = Side Wall	(See note #6)	0.001	0.0005
F = Hole Size Diameter	0.000 to 0.125	0.001	0.0005
	0.126 to 0.250	0.001	0.0005
	0.251 to 0.500	0.0015	0.001
	0.501 & over	0.002	0.0015
G = Hole Size Depth	0.000 to 0.250	0.001	0.0005
	0.251 to 0.500	0.001	0.0005
	0.501–1.000	0.0015	0.001
H = Corners, Ribs, Fillets		0.005	0.005
Flatness	0.000 to 3.000	0.010	0.007
	3.001 to 6.000	0.020	0.014
Thread Size (Class)	Internal		
	External		
Concentricity	(See note #4) (F.I.M.)	0.005	0.002
Draft Allowance per side	(See note #7)	2.0°	0.5°
Surface finish	(See note #5)		
Color Stability	(See note #5)		

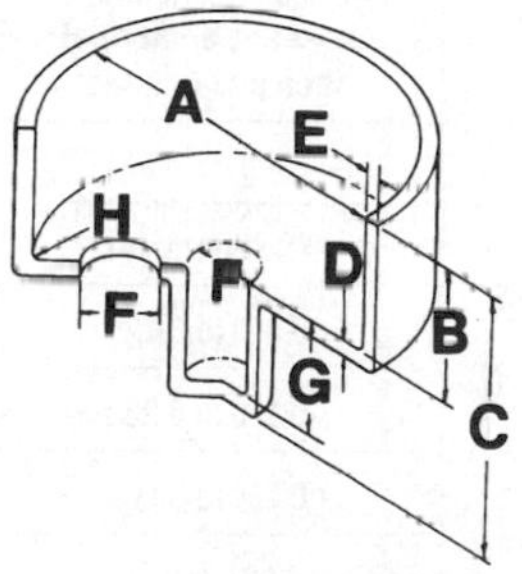

Reference Notes

1. These tolerances do not include allowance for aging characteristics of material
2. Tolerances do not apply to screw threads, gear teeth or match fits.
3. Parting line must be taken into consideration.
4. Part design should maintain a wall thickness as nearly constant as possible. Complete uniformity in this dimension is sometimes impossible to achieve. Walls of non-uniform thickness should be gradually blended from thick to thin.
5. Customer-Molder understanding is necessary prior to tooling.
6. This dimension is a function of mold design and construction.
7. These values should be considered minimum.

Standards & Practices of Plastics Molders

Material
Polypropylene
(PP)

Note: The *Commercial* values shown below represent common production tolerances at the most economical level. The *Fine* values represent closer tolerances that can be held but at a greater cost. Any addition of fillers will compromise physical properties and alter dimensional stability. Please consult the manufacturer.

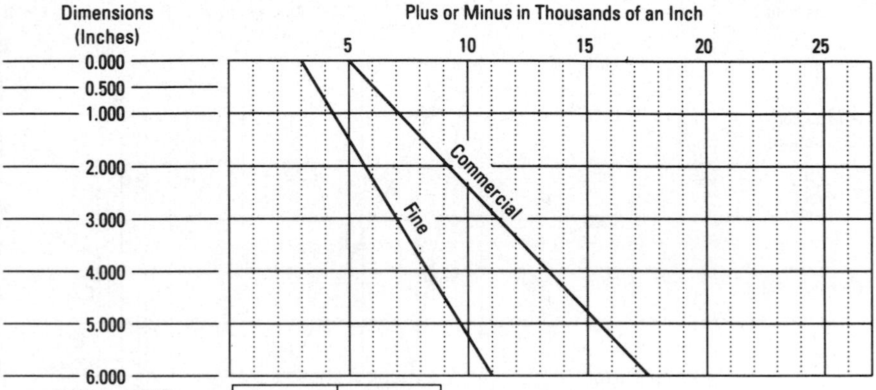

Drawing Code	Dimensions (Inches)	Comm. ±	Fine ±
A = Diameter (See note #1) B = Depth (See note #3) C = Height (See note #3)	0.000 to 6.000 (see chart)		
	6.000 to 12.000 for each additional inch add (inches)	0.005	0.003
D = Bottom Wall	(See note #3)	0.006	0.003
E = Side Wall	(See note #4)	0.006	0.003
F = Hole Size Diameter (See note #1)	0.000 to 0.125	0.003	0.002
	0.126 to 0.250	0.004	0.003
	0.251 to 0.500	0.005	0.004
	0.501 & over	0.008	0.006
G = Hole Size Depth (See note #5)	0.000 to 0.250	0.005	0.003
	0.251 to 0.500	0.006	0.004
	0.501–1.000	0.009	0.006
H = Corners, Ribs, Fillets	(See note #6)	0.029	0.016
Flatness	0.000 to 3.000	0.022	0.014
(See note #4)	3.001 to 6.000	0.036	0.021
Thread Size (Class)	Internal	1	2
	External	1	2
Concentricity	(See note #4) (F.I.M.)	0.015	0.012
Draft Allowance per side	(See note #5)	1.5°	0.5°
Surface finish	(See note #7)		
Color Stability	(See note #7)		

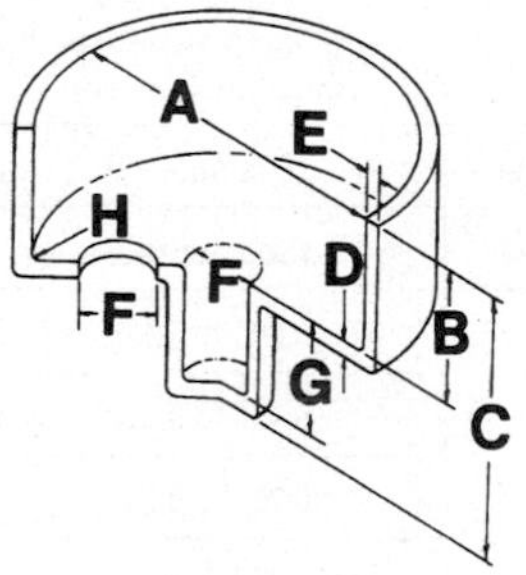

Reference Notes

1. These tolerances do not include allowance for aging characteristics of material
2. Tolerances are based on 0.125 inch wall section.
3. Parting line must be taken into consideration.
4. Part design should maintain a wall thickness as nearly constant as possible. Complete uniformity in this dimension is sometimes impossible to achieve. Walls of non-uniform thickness should be gradually blended from thick to thin.
5. Care must be taken that the ratio of the depth of a cored hole to its diameter does not reach a point that will result in excessive pin damage.
6. These values should be increased whenever compatible with desired design and good molding techniques.
7. Customer-Molder understanding is necessary prior to tooling.

Standards & Practices of Plastics Molders

Material
Polystyrene
(PS)

Note: The *Commercial* values shown below represent common production tolerances at the most economical level. The *Fine* values represent closer tolerances that can be held but at a greater cost. Any addition of fillers will compromise physical properties and alter dimensional stability. Please consult the manufacturer.

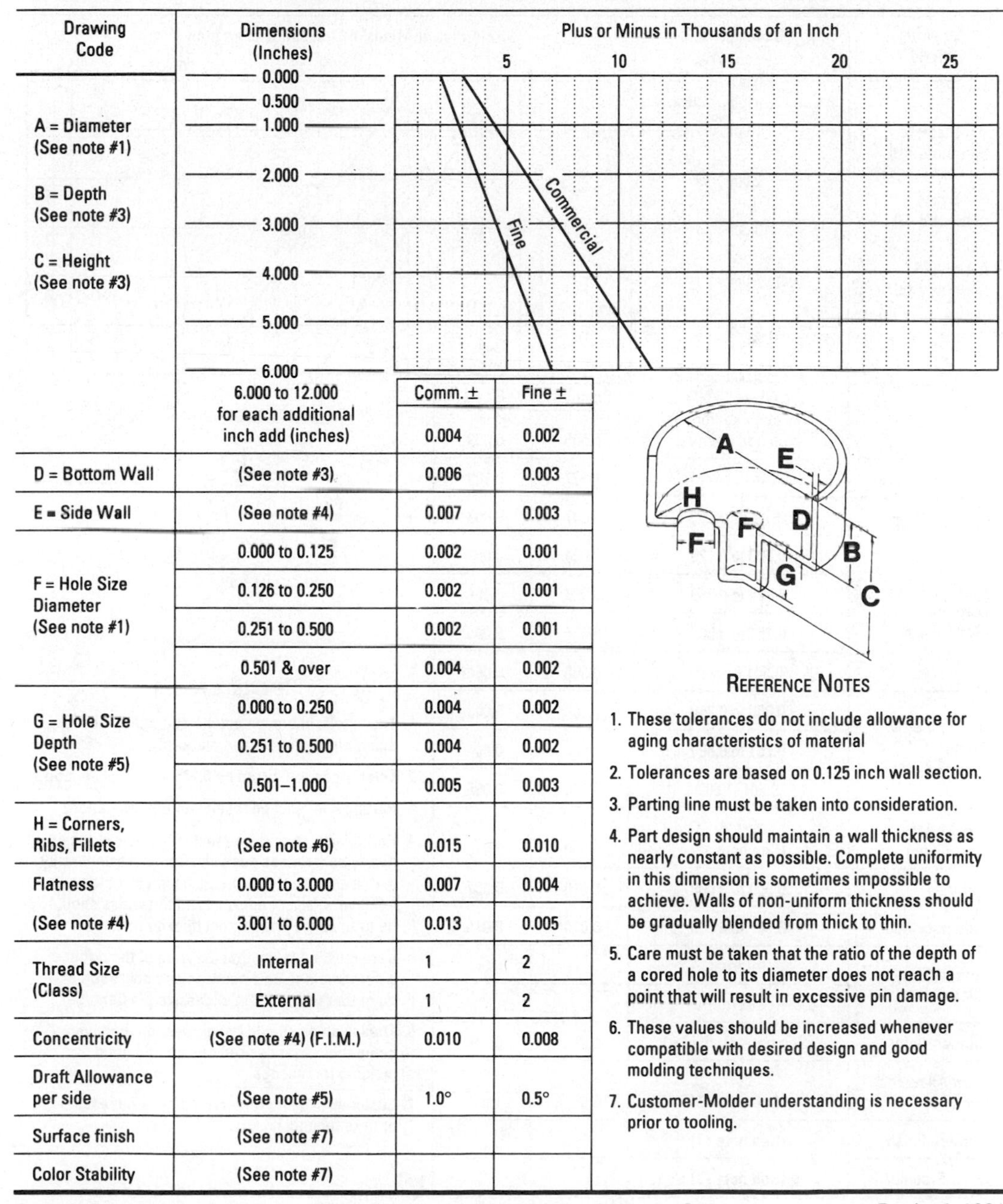

Drawing Code	Dimensions (Inches)	Comm. ±	Fine ±
A = Diameter (See note #1) B = Depth (See note #3) C = Height (See note #3)	0.000, 0.500, 1.000, 2.000, 3.000, 4.000, 5.000, 6.000 (see chart)		
	6.000 to 12.000 for each additional inch add (inches)	0.004	0.002
D = Bottom Wall	(See note #3)	0.006	0.003
E = Side Wall	(See note #4)	0.007	0.003
F = Hole Size Diameter (See note #1)	0.000 to 0.125	0.002	0.001
	0.126 to 0.250	0.002	0.001
	0.251 to 0.500	0.002	0.001
	0.501 & over	0.004	0.002
G = Hole Size Depth (See note #5)	0.000 to 0.250	0.004	0.002
	0.251 to 0.500	0.004	0.002
	0.501–1.000	0.005	0.003
H = Corners, Ribs, Fillets	(See note #6)	0.015	0.010
Flatness	0.000 to 3.000	0.007	0.004
(See note #4)	3.001 to 6.000	0.013	0.005
Thread Size (Class)	Internal	1	2
	External	1	2
Concentricity	(See note #4) (F.I.M.)	0.010	0.008
Draft Allowance per side	(See note #5)	1.0°	0.5°
Surface finish	(See note #7)		
Color Stability	(See note #7)		

Reference Notes

1. These tolerances do not include allowance for aging characteristics of material
2. Tolerances are based on 0.125 inch wall section.
3. Parting line must be taken into consideration.
4. Part design should maintain a wall thickness as nearly constant as possible. Complete uniformity in this dimension is sometimes impossible to achieve. Walls of non-uniform thickness should be gradually blended from thick to thin.
5. Care must be taken that the ratio of the depth of a cored hole to its diameter does not reach a point that will result in excessive pin damage.
6. These values should be increased whenever compatible with desired design and good molding techniques.
7. Customer-Molder understanding is necessary prior to tooling.

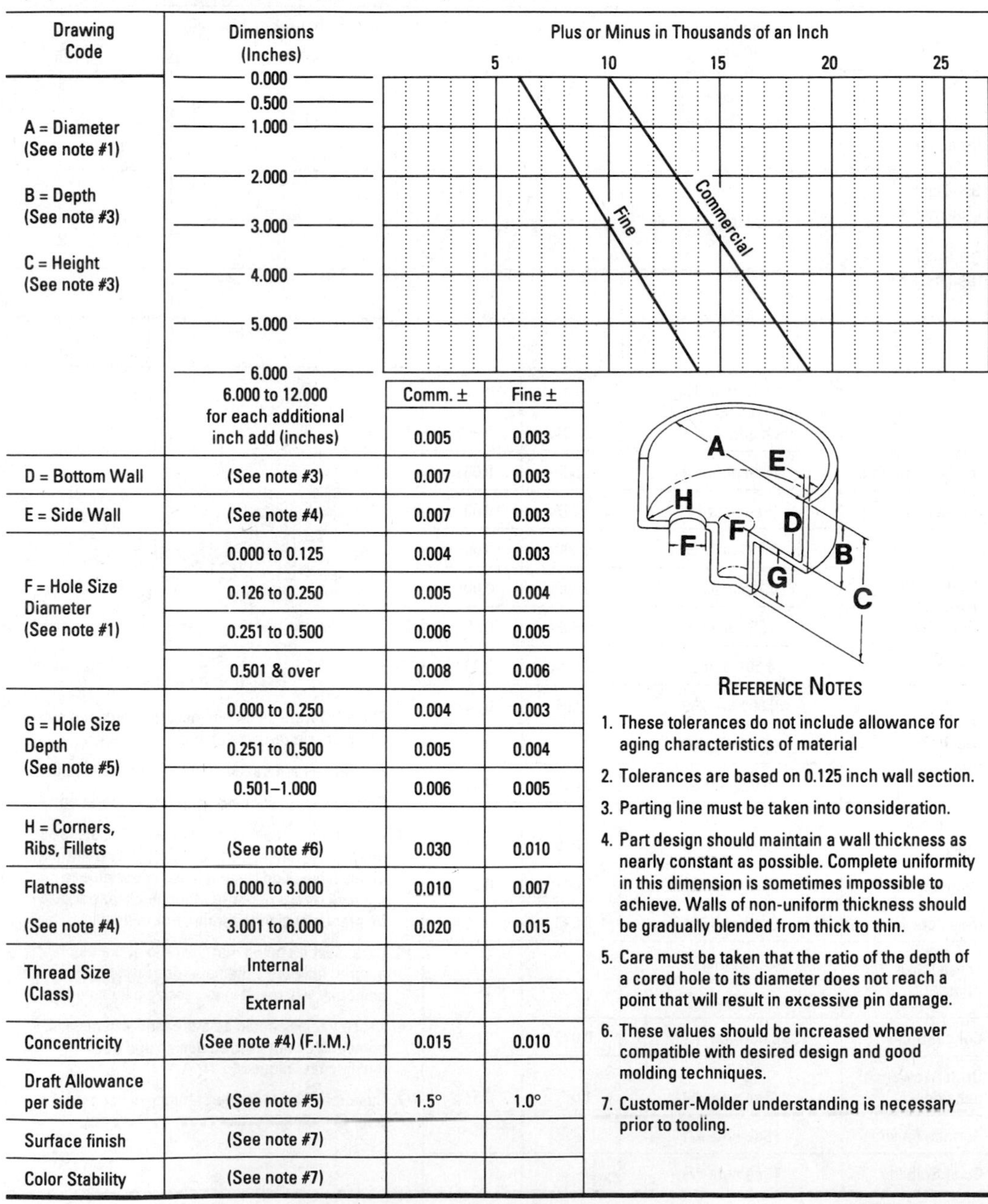

Standards & Practices of Plastics Molders

Material
Polyvinyl Chloride (PVC) (Vinyl) (Flexible)

Note: The *Commercial* values shown below represent common production tolerances at the most economical level. The *Fine* values represent closer tolerances that can be held but at a greater cost. Any addition of fillers will compromise physical properties and alter dimensional stability. Please consult the manufacturer.

Drawing Code	Dimensions (Inches)	Comm. ±	Fine ±
	6.000 to 12.000 for each additional inch add (inches)	0.005	0.003
D = Bottom Wall	(See note #3)	0.007	0.003
E = Side Wall	(See note #4)	0.007	0.003
F = Hole Size Diameter (See note #1)	0.000 to 0.125	0.004	0.003
	0.126 to 0.250	0.005	0.004
	0.251 to 0.500	0.006	0.005
	0.501 & over	0.008	0.006
G = Hole Size Depth (See note #5)	0.000 to 0.250	0.004	0.003
	0.251 to 0.500	0.005	0.004
	0.501–1.000	0.006	0.005
H = Corners, Ribs, Fillets	(See note #6)	0.030	0.010
Flatness (See note #4)	0.000 to 3.000	0.010	0.007
	3.001 to 6.000	0.020	0.015
Thread Size (Class)	Internal		
	External		
Concentricity	(See note #4) (F.I.M.)	0.015	0.010
Draft Allowance per side	(See note #5)	1.5°	1.0°
Surface finish	(See note #7)		
Color Stability	(See note #7)		

Reference Notes

1. These tolerances do not include allowance for aging characteristics of material
2. Tolerances are based on 0.125 inch wall section.
3. Parting line must be taken into consideration.
4. Part design should maintain a wall thickness as nearly constant as possible. Complete uniformity in this dimension is sometimes impossible to achieve. Walls of non-uniform thickness should be gradually blended from thick to thin.
5. Care must be taken that the ratio of the depth of a cored hole to its diameter does not reach a point that will result in excessive pin damage.
6. These values should be increased whenever compatible with desired design and good molding techniques.
7. Customer-Molder understanding is necessary prior to tooling.

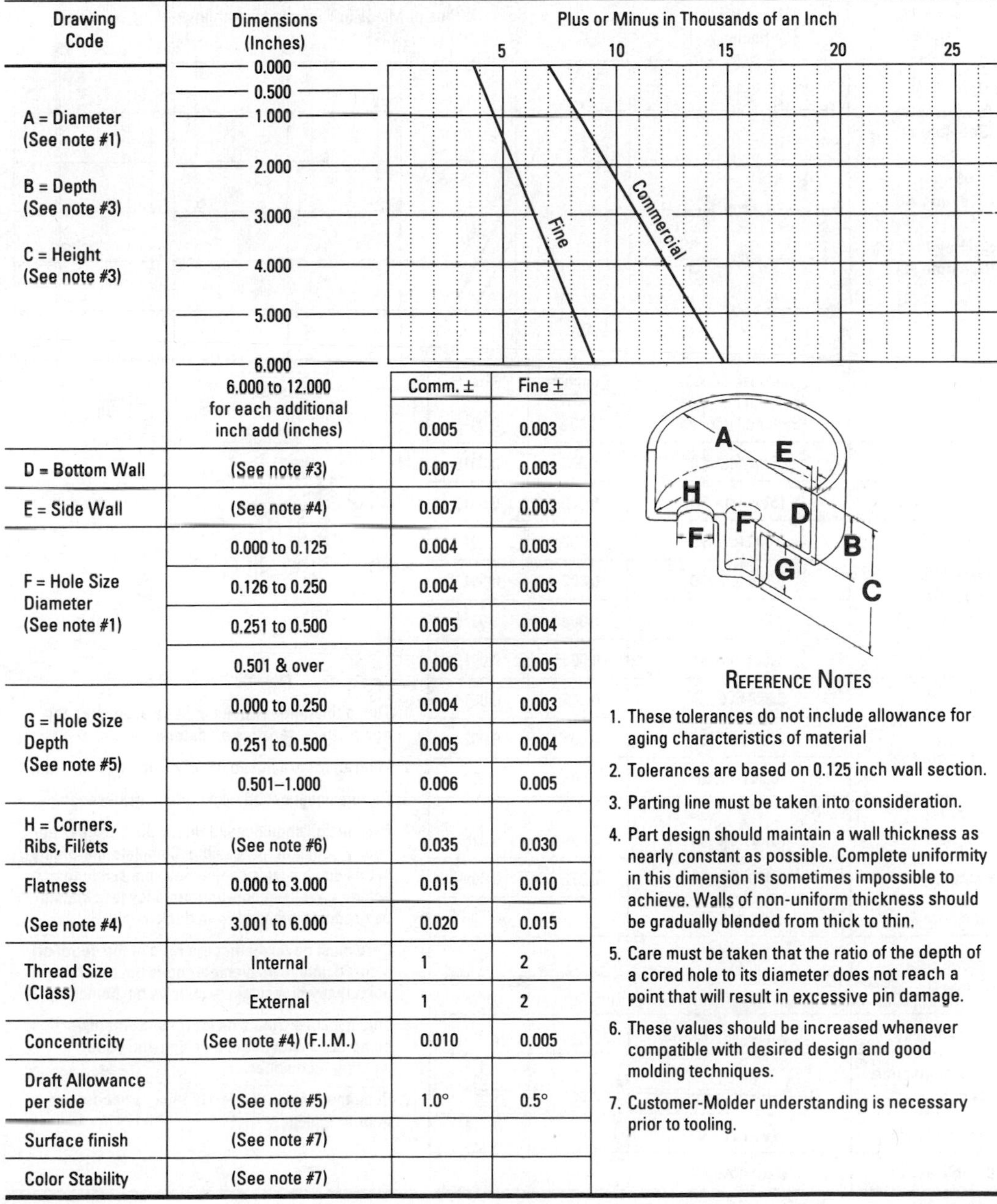

Standards & Practices of Plastics Molders

Material
Polyvinyl Chloride (PVC) (Vinyl) (Rigid)

Note: The *Commercial* values shown below represent common production tolerances at the most economical level. The *Fine* values represent closer tolerances that can be held but at a greater cost. Any addition of fillers will compromise physical properties and alter dimensional stability. Please consult the manufacturer.

Drawing Code	Dimensions (Inches)	Comm. ±	Fine ±
	6.000 to 12.000 for each additional inch add (inches)	0.005	0.003
D = Bottom Wall	(See note #3)	0.007	0.003
E = Side Wall	(See note #4)	0.007	0.003
F = Hole Size Diameter (See note #1)	0.000 to 0.125	0.004	0.003
	0.126 to 0.250	0.004	0.003
	0.251 to 0.500	0.005	0.004
	0.501 & over	0.006	0.005
G = Hole Size Depth (See note #5)	0.000 to 0.250	0.004	0.003
	0.251 to 0.500	0.005	0.004
	0.501–1.000	0.006	0.005
H = Corners, Ribs, Fillets	(See note #6)	0.035	0.030
Flatness	0.000 to 3.000	0.015	0.010
(See note #4)	3.001 to 6.000	0.020	0.015
Thread Size (Class)	Internal	1	2
	External	1	2
Concentricity	(See note #4) (F.I.M.)	0.010	0.005
Draft Allowance per side	(See note #5)	1.0°	0.5°
Surface finish	(See note #7)		
Color Stability	(See note #7)		

Reference Notes

1. These tolerances do not include allowance for aging characteristics of material
2. Tolerances are based on 0.125 inch wall section.
3. Parting line must be taken into consideration.
4. Part design should maintain a wall thickness as nearly constant as possible. Complete uniformity in this dimension is sometimes impossible to achieve. Walls of non-uniform thickness should be gradually blended from thick to thin.
5. Care must be taken that the ratio of the depth of a cored hole to its diameter does not reach a point that will result in excessive pin damage.
6. These values should be increased whenever compatible with desired design and good molding techniques.
7. Customer-Molder understanding is necessary prior to tooling.

Standards & Practices of Plastics Molders

Material
Styrene-Acrylonitrile (SAN)

Note: The *Commercial* values shown below represent common production tolerances at the most economical level. The *Fine* values represent closer tolerances that can be held but at a greater cost. Any addition of fillers will compromise physical properties and alter dimensional stability. Please consult the manufacturer.

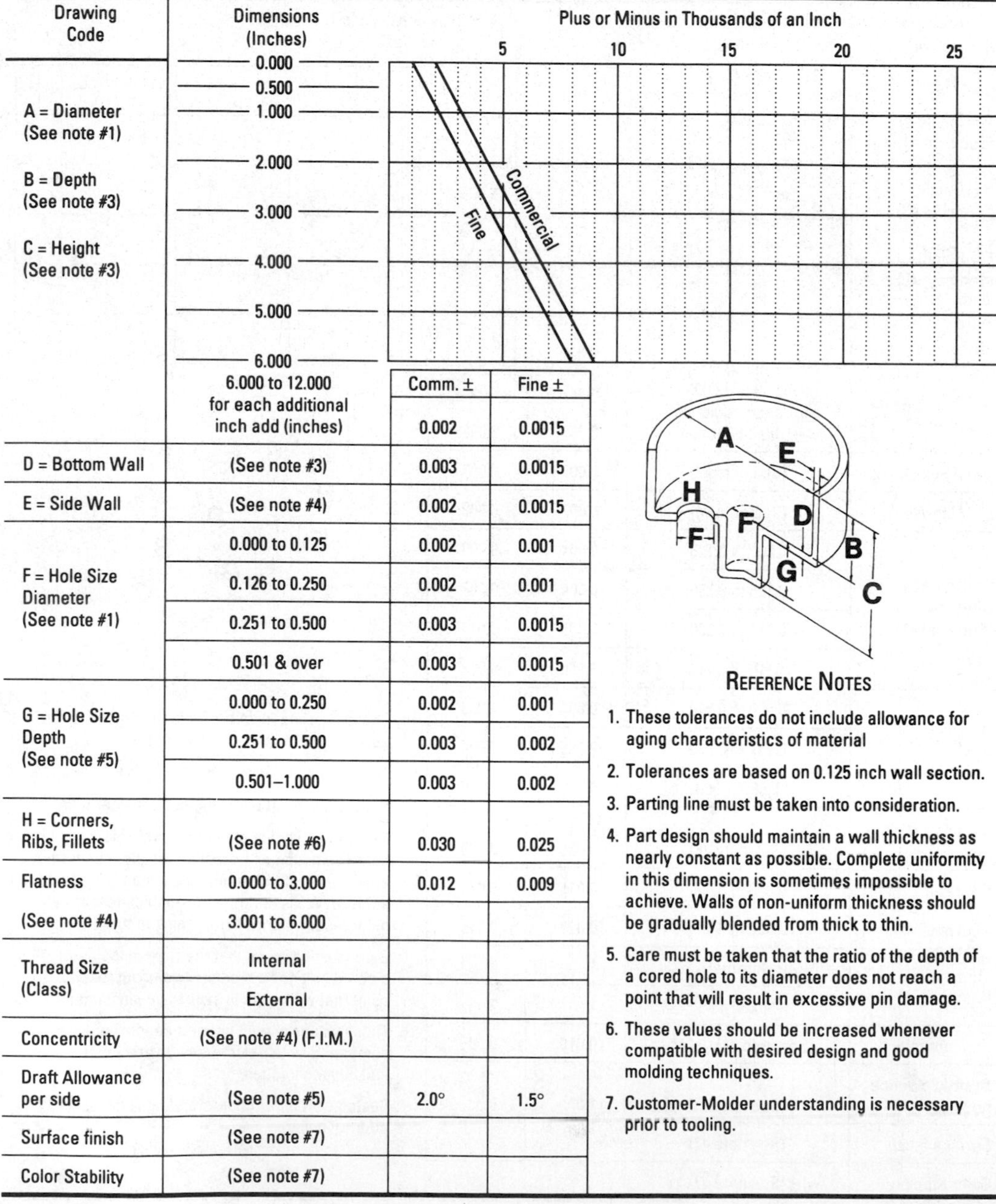

Drawing Code	Dimensions (Inches)	Comm. ±	Fine ±
A = Diameter (See note #1); B = Depth (See note #3); C = Height (See note #3)	0.000 to 6.000 (see chart)		
	6.000 to 12.000 for each additional inch add (inches)	0.002	0.0015
D = Bottom Wall	(See note #3)	0.003	0.0015
E = Side Wall	(See note #4)	0.002	0.0015
F = Hole Size Diameter (See note #1)	0.000 to 0.125	0.002	0.001
	0.126 to 0.250	0.002	0.001
	0.251 to 0.500	0.003	0.0015
	0.501 & over	0.003	0.0015
G = Hole Size Depth (See note #5)	0.000 to 0.250	0.002	0.001
	0.251 to 0.500	0.003	0.002
	0.501–1.000	0.003	0.002
H = Corners, Ribs, Fillets	(See note #6)	0.030	0.025
Flatness	0.000 to 3.000	0.012	0.009
(See note #4)	3.001 to 6.000		
Thread Size (Class)	Internal		
	External		
Concentricity	(See note #4) (F.I.M.)		
Draft Allowance per side	(See note #5)	2.0°	1.5°
Surface finish	(See note #7)		
Color Stability	(See note #7)		

Reference Notes

1. These tolerances do not include allowance for aging characteristics of material
2. Tolerances are based on 0.125 inch wall section.
3. Parting line must be taken into consideration.
4. Part design should maintain a wall thickness as nearly constant as possible. Complete uniformity in this dimension is sometimes impossible to achieve. Walls of non-uniform thickness should be gradually blended from thick to thin.
5. Care must be taken that the ratio of the depth of a cored hole to its diameter does not reach a point that will result in excessive pin damage.
6. These values should be increased whenever compatible with desired design and good molding techniques.
7. Customer-Molder understanding is necessary prior to tooling.

Index